AERODYNAMIC COMPONENTS OF AIRCRAFT AT HIGH SPEEDS

I. Thermodynamics and Physics of Matter. Editor: F. D. Rossini
II. Combustion Processes. Editors: B. Lewis, R. N. Pease, H. S. Taylor
III. Fundamentals of Gas Dynamics. Editor: H. W. Emmons
IV. Laminar Flows and Transition to Turbulence. Editor: C. C. Lin
V. Turbulent Flows and Heat Transfer. Editor: C. C. Lin
VI. General Theory of High Speed Aerodynamics. Editor: W. R. Sears
VII. Aerodynamic Components of Aircraft at High Speeds. Editors: A. F. Donovan, H. R. Lawrence
VIII. High Speed Problems of Aircraft and Experimental Methods. Editors: A. F. Donovan, H. R. Lawrence, F. Goddard, R. R. Gilruth
IX. Physical Measurements in Gas Dynamics and Combustion. Editors: R. W. Ladenburg, B. Lewis, R. N. Pease, H. S. Taylor
X. Aerodynamics of Turbines and Compressors. Editor: W. R. Hawthorne
XI. Design and Performance of Gas Turbine Power Plants. Editors: W. R. Hawthorne, W. T. Olson
XII. Jet Propulsion Engines. Editor: O. E. Lancaster

VOLUME VII

HIGH SPEED AERODYNAMICS

AND JET PROPULSION

AERODYNAMIC COMPONENTS OF AIRCRAFT AT HIGH SPEEDS

EDITORS

A. F. DONOVAN

H. R. LAWRENCE

PRINCETON, NEW JERSEY

PRINCETON UNIVERSITY PRESS

1957

London: Oxford University Press

L. C. Card 56-8843

Printed in the United States of America by
The Maple Press Company, Inc., York, Penna.

FOREWORD

On behalf of the Editorial Board, I would like to make an acknowledgement to those branches of our military establishment whose interest and whose financial support were instrumental in the initiation of this publication program. It is noteworthy that this assistance has included all three branches of our Services. The Department of the Air Force through the Air Research and Development Command, the Department of the Army through the Office of the Chief of Ordnance, and the Department of the Navy through the Bureau of Aeronautics, Bureau of Ships, Bureau of Ordnance, and the Office of Naval Research made significant contributions. In particular, the Power Branch of the Office of Naval Research has carried the burden of responsibilities of the contractual administration and processing of all manuscripts from a security standpoint. The administration, operation, and editorial functions of the program have been centered at Princeton University. In addition, the University has contributed financially to the support of the undertaking. It is appropriate that special appreciation be expressed to Princeton University for its important over-all role in this effort.

The Editorial Board is confident that the present series which this support has made possible will have far-reaching beneficial effects on the further development of the aeronautical sciences.

Theodore von Kármán

PREFACE

Rapid advances made during the past decade on problems associated with high speed flight have brought into ever sharper focus the need for a comprehensive and competent treatment of the fundamental aspects of the aerodynamic and propulsion problems of high speed flight, together with a survey of those aspects of the underlying basic sciences cognate to such problems. The need for a treatment of this type has been long felt in research institutions, universities, and private industry and its potential reflected importance in the advanced training of nascent aeronautical scientists has also been an important motivation in this undertaking.

The entire program is the cumulative work of over one hundred scientists and engineers, representing many different branches of engineering and fields of science both in this country and abroad.

The work consists of twelve volumes treating in sequence elements of the properties of gases, liquids, and solids; combustion processes and chemical kinetics; fundamentals of gas dynamics; viscous phenomena; turbulence; heat transfer; theoretical methods in high speed aerodynamics; applications to wings, bodies and complete aircraft; nonsteady aerodynamics; principles of physical measurements; experimental methods in high speed aerodynamics and combustion; aerodynamic problems of turbo machines; the combination of aerodynamic and combustion principles in combustor design; and finally, problems of complete power plants. The intent has been to emphasize the fundamental aspects of jet propulsion and high speed aerodynamics, to develop the theoretical tools for attack on these problems, and to seek to highlight the directions in which research may be potentially most fruitful.

Preliminary discussions, which ultimately led to the foundation of the present program, were held in 1947 and 1948 and, in large measure, by virtue of the enthusiasm, inspiration, and encouragement of Dr. Theodore von Kármán and later the invaluable assistance of Dr. Hugh L. Dryden and Dean Hugh Taylor as members of the Editorial Board, these discussions ultimately saw their fruition in the formal establishment of the Aeronautics Publication Program at Princeton University in the fall of 1949.

The contributing authors and, in particular, the volume editors, have sacrificed generously of their spare time under present-day emergency conditions where continuing demands on their energies have been great. The program is also indebted to the work of Dr. Martin Summerfield who guided the planning work as General Editor from 1949–1952. The cooperation and assistance of the personnel of Princeton University Press and of the staff of this office has been noteworthy. In particular,

Mr. H. S. Bailey, Jr., the Director of the Press, and Mr. R. S. Snedeker, who has supervised the project at the Press and drawn all the figures, have been of great help. Special mention is also due Mrs. E. W. Wetterau of this office who has handled the bulk of the detailed editorial work for the program.

Coleman duP. Donaldson
Joseph V. Charyk
General Editors

PREFACE TO VOLUME VII

It is the primary concern of this volume, along with the first part of Volume VIII, to present the aerodynamic knowledge that is of most direct interest to the designers of high speed aircraft. Size limits this volume primarily to aircraft components, i.e. wings (in both steady and nonsteady motion), bodies, interference effects, propellers, diffusers, and nozzles. Items dealing with the aircraft as a whole, such as performance, aeroelasticity, and stability and control, are found in Volume VIII.

The volume begins with a comprehensive treatment of the steady state aerodynamic theory of wings by R. T. Jones and Doris Cohen. The tables of flow functions in this section should be particularly useful to the practicing aerodynamicist. Clinton Brown next presents a section on the aerodynamics of bodies at high speed, which is noteworthy for including bodies of noncircular cross section—a type which is usually given too little attention. Interactions between wings, body, tail, and propeller are treated in a section by C. Ferrari, who was aided in translation by R. Cramer. This section is notable for bringing together much information which otherwise can only be found scattered widely through the literature. Propellers at high speeds, including supersonic propellers, are covered in a section by C. B. Smith. A section on diffusers and nozzles by J. C. Evvard completes the treatment of the external aerodynamics of high speed propulsion systems. The basis for the treatment of aeroelastic problems is presented in a section by I. E. Garrick on nonsteady wing characteristics. Finally, a section by C. W. Frick on the experimental aerodynamics of wings clearly brings out the accomplishments of aerodynamic theory and the limitations imposed by assumptions.

The volume editors wish to express their appreciation for the fine cooperation they have received, and for the assistance received from the General Editor and his staff. Our thanks are extended to all.

Allen F. Donovan
Herbert R. Lawrence
Volume Editors

CONTENTS

AERODYNAMIC COMPONENTS OF AIRCRAFT AT HIGH SPEEDS

SECTION A

AERODYNAMICS OF WINGS AT HIGH SPEEDS

ROBERT T. JONES
DORIS COHEN

CHAPTER 1. FUNDAMENTAL CONSIDERATIONS IN THE DEVELOPMENT OF WINGS FOR HIGH SPEEDS

A,1. Review of Wing Theory for Low Speeds.

Function of the Wing. The chief advantage of air transportation over other forms of travel is the great speed that can be achieved with a relatively moderate cost in terms of energy, or fuel expended per mile of flight. An airplane of efficient aerodynamic form may have a drag less than one twentieth of its weight. The energy required in steady flight is therefore less than one twentieth of the weight times the distance flown. By proper adjustment of the wing loading and altitude this value may be maintained up to speeds approaching the speed of sound. Beyond the speed of sound the efficiency of flight tends to be reduced—but it seems probable that advances in aerodynamics and in jet propulsion can overcome the loss, at least for moderately supersonic speeds.

It is easily seen that the efficiency of the airplane is the result of the favorable aerodynamic properties of the wing. If one contemplates travel by a rocket, which overcomes gravity and achieves its distance solely by virtue of the kinetic energy imparted at the beginning of the motion, then it is found that the energy requirement is much greater—of the order of the whole weight times the distance. The energy expenditure of a wingless rocket or projectile thus corresponds to a lift-drag ratio of about one, a figure which can be surpassed easily by almost any form of winged body.

Early Explanations of Wing Flows. The possibility of the favorable aerodynamic properties of airfoils was not apparent to early students of hydrodynamics and, in fact, prior to actual demonstrations of mechanical flight, solutions of the hydrodynamic equations were used as the basis of predictions that such flight would prove impossible. This situation is perhaps not surprising when it is recalled that the equations of hydrodynamics or gas dynamics do not provide a definite solution to the problem of the airflow around a wing, but, in fact, furnish a multiplicity of possible solutions. The identification of the proper physical solution in every case requires additional information not contained in the equations of motion nor in readily determined boundary conditions.

Helmholtz' solution. An early attempt to explain the action of wings on the basis of hydrodynamic theory is that of Helmholtz [*1*]. Helmholtz adopted as a representative shape an inclined flat plate. The simplest

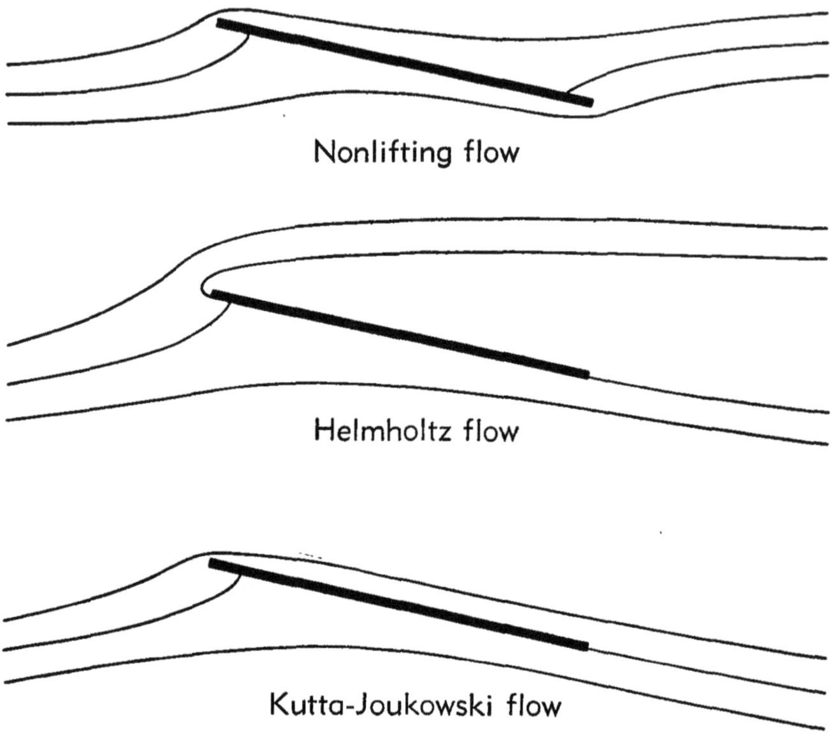

Fig. A,1a. Hydrodynamic solutions for the flow over a flat plate.

mathematical solution in this case depicts a flow without lift (Fig. A,1a, top) which involves infinite velocities around the edges of the plate. To avoid this physically unrealistic situation, Helmholtz devised a solution in which the streamlines leave the leading and trailing edges tangentially, forming external surfaces of discontinuity which enclose a body of "dead" air above the wing (Fig. A,1a, middle). Such a configuration is certainly not contrary to experience. However, when the aerodynamic forces associated with the Helmholtz flow are compared with those that can be achieved experimentally with properly designed wing shapes, the Helmholtz flow is seen to give relatively small lift and an excessive amount of drag. The lift force developed is

$$L = \frac{\pi}{4} \alpha \rho_\infty U^2 c \tag{1-1}$$

or, in the usual coefficient form,

$$C_L = \frac{\pi}{2} \alpha \tag{1-2}$$

for small angles of attack α. Since the resultant force is at right angles to the plate, a drag equal to the product of the lift and the angle of attack is developed.

Kutta-Joukowski flow. A more suitable starting point for the description of wing flows was found later by Kutta and Joukowski. Kutta [*2*] prescribed Helmholtz' condition of tangential flow at the trailing edge, but not at the leading edge (Fig. A,1a, bottom). In Joukowski's solution [*3*], the airfoil was rounded to permit flow around the leading edge without separation, and was terminated in a cusp so that Kutta's condition could be applied to determine a unique value of the lift for each angle of attack. The Kutta-Joukowski flow consists of two independent solutions of the equations of motion, one determined directly by the normal components of velocity of the solid surface and the other—the circulatory flow—involving only a parallel sliding motion around the surface. The entire lift depends on the secondary flow, which cannot be determined directly from the normal boundary conditions, but only indirectly from the condition of tangential velocity at the trailing edge. For thin flat sections at small angles the lift developed is, in coefficient form,

$$C_L = 2\pi\alpha \tag{1-3}$$

and in two-dimensional frictionless motion the drag is zero.

PHYSICAL CHOICE OF BOUNDARY CONDITIONS.

Effect of the boundary layer. The occurrence of the Kutta-Joukowski type of flow depends physically on the action of the boundary layer, which causes the flow to leave the trailing edge tangentially. Separation of the flow around the leading edge can be avoided for a small range of angles of attack because of the thinness of the boundary layer in this region. The essential difference between a leading edge and a trailing edge is evident in the case of a flow through a pipe. Fig. A,1b shows schematically two examples of streamlines provided by the perfect-fluid theory for such a flow. However, the boundary layer formed inside the pipe will prevent the turning of the streamlines at the exit, so that the actual flow will resemble that depicted in the bottom part of the figure, rather than in the top part.

Leading and trailing edge conditions. In the Helmholtz flow, the velocity is constant along the streamlines bounding the dead air region,

Thus the surfaces of discontinuity leaving the leading and trailing edges may be represented by vortex sheets of uniform strength. In the development of the circulatory flow, on the other hand, a single surface of discontinuity or vortex surface is formed along the trailing edge as the wing starts from rest. For each increment in the strength of the circulation around the wing there appears a corresponding increase of the total vortex strength of this wake. However, as the flow approaches the steady state, the rate of growth of the circulation, and therefore the strength of the shed vortices, diminishes. In the final steady state of the two-dimensional flow no additional vortices leave the trailing edge.

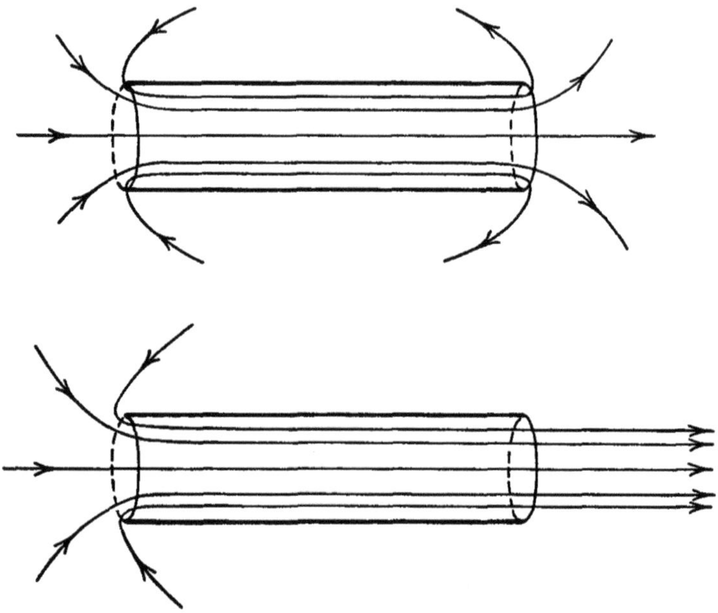

Fig. A,1b. Flows through a pipe.

In the three-dimensional flow around a wing of finite span (Fig. A,1c) the vortex surface remains attached to the wing, but in the final steady state the discontinuity appears only in the lateral velocity v, so that the elementary vortices are left behind the wing in a direction parallel to the direction of motion.

Side edges. The sharp distinction between the leading edge and the trailing edge in the Kutta-Joukowski flow cannot be carried over completely to actual wings of finite planform, especially if part of the outline of the wing lies parallel or nearly parallel to the stream direction. A satisfactory solution for the flow around such a side edge has not as yet been obtained. The calculations that have been made thus far are designed to satisfy the condition of zero lift at the side edge, but the solutions show infinite velocities *around* the edge (Fig. A,1d, top). A considera-

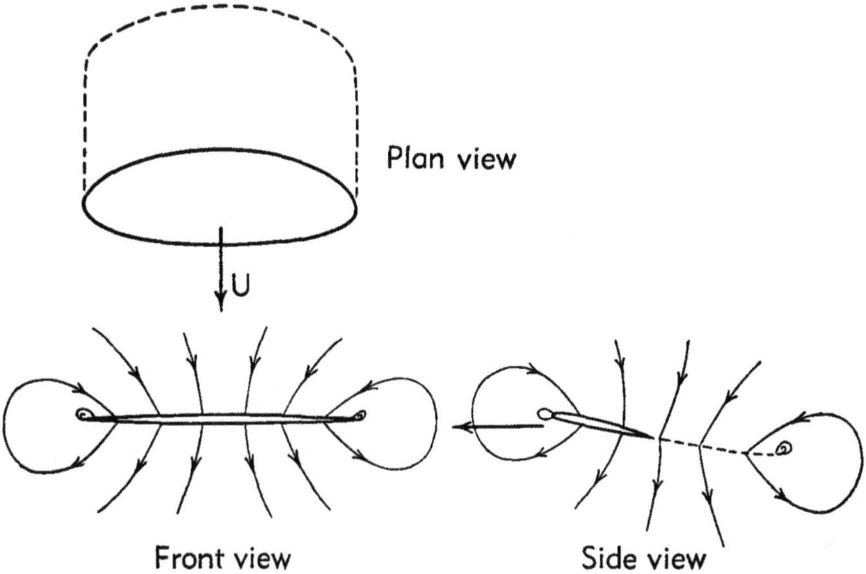

Fig. A,1c. Wake left by wing starting from rest.

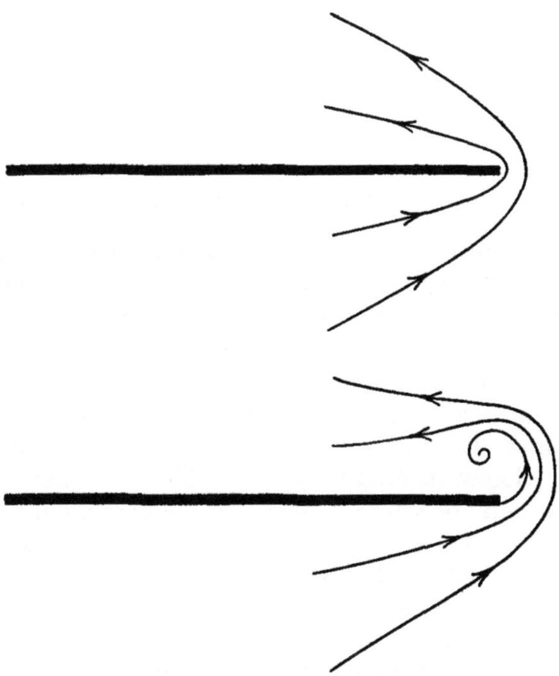

Fig. A,1d. Flows around wing tip.

tion of the stresses involved, on the other hand, leads to the conclusion that the condition of tangential flow should, strictly speaking, be applied at the side edges as well as at the trailing edge (Fig. A,1d, bottom).

Experimental demonstration of airfoil efficiency. According to the Kutta-Joukowski hypothesis the two-dimensional frictionless flow around an airfoil results in the production of a lifting force without accompanying drag. The extent to which this ideal, dragless type of flow can be realized in practice under suitable conditions is well illustrated by an experiment performed by Jacobs [4] at the Langley Field Laboratory of the NACA. The airfoil (Fig. A,1e) in this experiment was of 5-foot chord, 12 per cent thickness, and extended completely across the wind tunnel. At an airspeed of 50 miles per hour and an angle of attack of 7°, a lift force of 100 pounds was measured. The drag amounted to only slightly more than one third of a pound—corresponding to a ratio of lift to drag of nearly 300 to 1. For comparative purposes a round rod having approximately the same drag in pounds as the lifting airfoil is also shown.

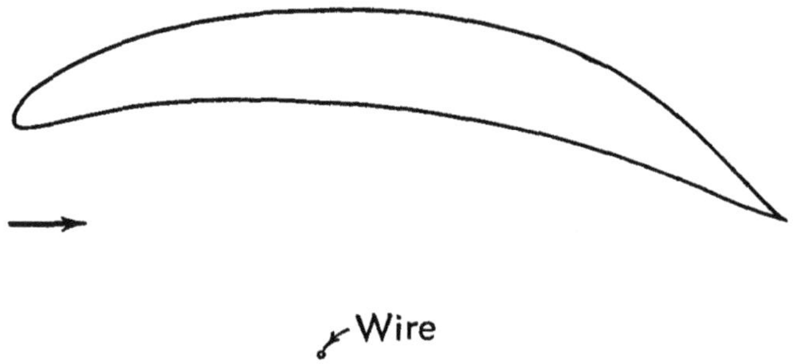

Fig. A,1e. Wing compared to circular wire of equal drag.

In modern wind tunnel experiments on more practical airfoils, a drag coefficient of 0.0045 and a lift-drag ratio of 100 to 1 is not unusual (see [5]). Such a drag coefficient, in the case of a 15 per cent thick airfoil, amounts to the force of the impact pressure acting on an area equal to only 3 per cent of the projected frontal area of the airfoil. Since this force is actually almost equal to the estimated surface friction, it is concluded that the type of pressure distribution obtained from the theory of perfect fluids can be very closely realized in practice. This of course is not true for bodies of a general or arbitrary shape, where because of flow separation the actual flow is frequently an exceedingly complex phenomenon, apparently not corresponding to any simple or intelligible solutions of the hydrodynamical equations.

Origin of Drag in Three-Dimensional Flow. In the experiments cited above, the conditions of two-dimensional flow were closely simulated by extending the wing completely across the wind tunnel. In free air the production of lift on a wing of finite span is accompanied by a pressure drag or "induced drag" which is associated with the continual extension of the vortex wake behind the wing discussed above. As a result of the induced drag, the lift-drag ratios of wings of practical proportions fall considerably below the values obtainable in two-dimensional flow.

Concept of induced drag. The flow around a lifting wing is characterized by an upwash in the stream ahead and a downwash behind. These upwash and downwash velocities may be represented by the action of a distribution of vortices in the airfoil; at a large distance the flow may be approximated by the velocity field of a single vortex superimposed on the velocity of the stream. It is easily seen that if the span of the wing extends to infinity the influence of the lifting vortex will be symmetrical ahead of and behind the center of lift. The absence of drag in two-dimensional lifting flow is related to this fore-and-aft symmetry of the downwash field.

If the span of the wing is limited, however, the fore-and-aft symmetry of the flow disappears. As explained by Lanchester [*6*], the vortices representing the circulatory flow around the wing must turn in the vicinity of the tips and extend backward, following the streamlines of the flow behind the wing. Turning the vortices from a lateral to a backward position diminishes the upwash in the field ahead of the wing and increases the downwash in the region behind. As a result of the downward inclination of the flow the lift of the section is reduced and inclined rearward, giving rise to a drag which is termed the "induced drag" [*7*].

Lifting line theory. It is seen that the reduction of upwash in the stream ahead of the wing is similar in effect to a reduction of the angle of attack. Prandtl [*8*] found that for wings of high aspect ratio a satisfactory representation of the flow and the drag could be obtained on this basis by applying a correction to the angle of attack in the two-dimensional formulas. The quantitative theory of this correction, known as the lifting line theory, is based on the assumption that the effect of the finite span is to give rise to a motion on a relatively large scale of dimensions, which can be superimposed on the smaller, locally two-dimensional wing section flows. For the purpose of computing this downwash the wing section flow is replaced temporarily by the flow around a single straight vortex at the center of pressure. Since a straight lifting vortex induces no downwash at its own axis, the downwash computed along the "lifting line" will arise solely from the vortices in the wake. At a great distance behind the wing, the wake forms a two-dimensional field of motion, and, since in the first order theory the form of the wake does not change with distance, it can be shown from symmetry that the value of the downwash at the wing is one half the final value in the two-dimensional field of the wake. Through the Helmholtz vortex theorem the distribution of vorticity in the wake is related to the gradient of the spanwise distribution of lift or circulation at the wing. Hence the induced downwash correction and the induced drag depend only on the spanwise distribution of the lift and are independent of the chordwise distribution of lift.

Although in its derivation the lifting line theory appears to involve drastic approximations, a comparison with more exact three-dimensional solutions, which will appear later in this section (Art. 7), indicates that in cases of straight narrow wings with smooth load distributions the theory is well founded. It may be shown from symmetry that the lifting line theory actually gives the correct value (in the sense of linearized lifting surface theory) of the downwash along the center of pressure of any distribution of lift that is symmetrical ahead of and behind a straight line at right angles to the direction of flight. However, for curved or yawed lifting lines and especially for wings of low aspect ratio, the lifting line theory is no longer adequate. As will be shown later in the present section the effect of compressibility at higher speeds is also such as to require a more complete treatment of the lifting surface in three dimensions.

A,2. Incompressible Flow in Two Dimensions. The Theory of Thin Airfoils.

Introduction. The necessity of minimizing the induced drag leads to the choice of a high aspect ratio for the proportions of practical wings designed for subsonic flight speeds. With such long narrow wings and especially at low speeds, the flow in the vicinity of the wing sections can be expected to resemble closely the flow around an infinitely long cylindrical wing of the same section profile. This fact has led to extensive use of the concept of two-dimensional flow in both theoretical and experimental studies of wing forms.

Mathematical formulation of two-dimensional flow fields. On the theoretical side, the assumption of two-dimensional flow simplifies the problem mathematically to the extent that the determination of the flow field and pressure distribution around any arbitrarily specified airfoil shape becomes a practical calculation. The mathematical simplification arises from the fact that the motion of an incompressible fluid in two dimensions can be represented by an analytic function of a single variable. Thus it can be shown by direct differentiation that the function

$$\varphi = F(\alpha x + \gamma z) \tag{2-1}$$

satisfies Laplace's equation

$$\frac{\partial^2 \varphi}{\partial x^2} + \frac{\partial^2 \varphi}{\partial z^2} = 0 \tag{2-2}$$

provided $\alpha^2 + \gamma^2 = 0$. The choice $\alpha = 1$, $\gamma = i$ leads to

$$\varphi = F(x + iz) \tag{2-3}$$

Every analytic function of such a complex variable can be separated into a pair of distinct functions, one real and one imaginary; thus

$$\varphi = \varphi_1(x, z) + i\varphi_2(x, z)$$

and both φ_1 and φ_2 will be solutions of Laplace's equation. Since φ_1 and φ_2 are orthogonal, they may represent the equipotential lines and the streamlines of a two-dimensional flow.[1]

Linearization of boundary conditions. In the case of thin airfoils with small camber or angle of attack, a still further simplification of the two-dimensional potential flow theory is possible. In incompressible flow this simplification is the basis of the well-known Munk theory of thin airfoils [*9*], while at supersonic speeds it forms the basis of the Ackeret theory [*10*]. At the present time, engineering calculations for wings at high speeds are based almost exclusively on the assumptions of the thin airfoil theory, especially since, as will appear later in the present section, this theory offers a practical approach to the problem of compressible potential flow in three dimensions.

In its mathematical derivation the thin airfoil theory is shown to correspond to the limiting case of infinitesimal thickness or angle of attack. In this limiting case the boundary condition of the airfoil can be replaced by a specification of horizontal or vertical components of the perturbation velocity along a straight "chord line" of the airfoil, without regard for the actual displacement of the airfoil boundary away from this chord line. If the shape of the airfoil is given by the ordinates $z_1(x)$, then the requirement of tangential flow along the surface reduces to

$$\left(\frac{w}{U}\right)_{z=0} = \frac{dz_1}{dx} \tag{2-4}$$

The pressure distribution throughout the field is obtained from the horizontal perturbation velocity u by the relation

$$\frac{\Delta p}{\frac{1}{2}\rho_\infty U^2} = -\frac{2u}{U} \tag{2-5}$$

which is an approximate form of Bernoulli's equation.

The upper and lower surfaces of the chord line correspond to the upper and lower surfaces of the airfoil, and, since the velocities on these two surfaces are different, there will arise a discontinuity in the velocity across this line. A discontinuity in the vertical component w results from the thickness distribution of the airfoil, while a discontinuity in the horizontal component u arises from the camber or angle of attack of the mean line, and is related to the distribution of lift.

[1] The corresponding solution of Laplace's equation for three dimensions is

$$\varphi = F(\alpha x + \beta y + \gamma z)$$

with $\alpha^2 + \beta^2 + \gamma^2 = 0$. In this case, however, it is not generally possible to separate the solution into three parts corresponding to three distinct units α, β, γ.

With a fixed surface independent of the solution at which to consider the boundary velocities, and with a linear equation for the motion of the fluid (Laplace's equation, in this case), the airfoil problem becomes a linear one. Airfoil ordinates may be superimposed in any linear combination and the resulting velocity distribution may be obtained by the addition of the velocity fields corresponding to the elements of the combination. This simplification makes it possible to divide airfoil problems into two classes, i.e. those involving only camber or angle of attack of the mean line (considered to be an infinitely thin airfoil), and those corresponding to distributions of thickness disposed symmetrically above and below the chord line. In this way the discontinuities in u and w can be treated separately. In the former case the horizontal perturbation velocity u will have equal and opposite values above and below the chord line, resulting in a lifting pressure given in coefficient form by

$$C_p = \frac{\Delta p_{\text{lower}} - \Delta p_{\text{upper}}}{\frac{1}{2}\rho_\infty U^2} = \frac{4u}{U} \tag{2-6}$$

where u is taken on the upper side. With a symmetrical distribution of thickness, the horizontal perturbation velocity u will be continuous, giving rise to equal values of the pressure above and below the chord plane and hence to no lift. The vertical velocity w is discontinuous and is related to the thickness by Eq. 2-4.

Complex velocity functions. As noted above, any analytic function of a complex variable can represent the streamlines and equipotential lines of a two-dimensional flow field. Differentiation for the velocity components shows that the conjugate of the complex velocity vector can be given directly by such a function; i.e.

$$\frac{u - iw}{U} = f(x + iz) \tag{2-7}$$

Thin airfoil flows can hence be obtained by constructing certain functions f which vanish at infinity and which show the desired distribution of velocities along the chord line. For convenience in writing such functions the chord line is taken along the real axis between ± 1.

As an example, suppose it is desired to obtain the solution for a flat surface. Since $w/U = dz/dx = \text{const}$, it is necessary to find a function whose imaginary part remains at a fixed value along the real axis between ± 1. One such function[2] is

$$\frac{u - iw}{U} = \frac{1}{\sqrt{1 - \xi^2}} \tag{2-8}$$

[2] It should be noted that the problem is not completely specified and that other flows about a flat surface are possible.

where $\xi = x + iz$. The real part is discontinuous; i.e.,

$$u(x + 0 \cdot i) = -u(x - 0 \cdot i)$$

between the points ± 1, while the imaginary part is continuous all along the real axis. Since the imaginary part is zero between ± 1, the solution represents a lifting flow over a flat plate at zero angle of attack and is, in fact, the well-known circulatory flow around the flat plate. The total circulation is given by

$$\Gamma = \int_{-1}^{+1} 2u(x + 0 \cdot i)dx = 2\pi U \tag{2-9}$$

in this case.

The following table lists a number of complex velocity functions for thin airfoils[3] and shows the corresponding distributions of horizontal and vertical perturbation velocities along the chord line.

Application to airfoils with finite thickness. The interpretation of thin airfoil flows in terms of airfoils of finite thickness can be facilitated by a

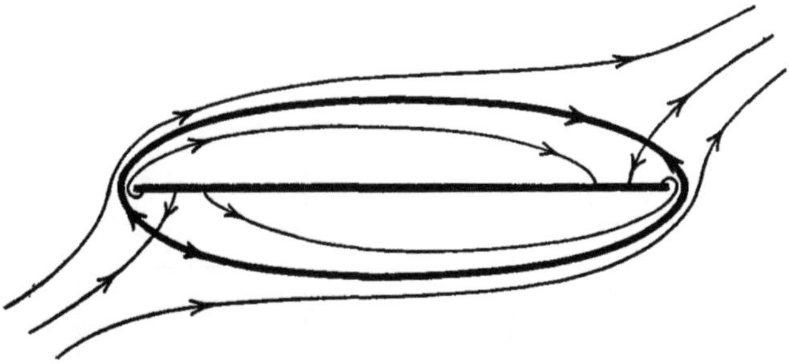

Fig. A,2a. Internal streamlines.

comparison of these flows with the known exact solutions for typical shapes. Most simple shapes such as the Joukowski airfoil and the ellipse can be produced by the action of a continuous distribution of sources and vortices along a mean line inside the body. The internal flows in such cases therefore show a discontinuity of velocity along this line. When comparison is made with the velocity distribution given in each case by thin airfoil theory, a surprisingly close similarity is found to exist. The velocity distributions given by the thin airfoil theory along the line $z = 0$ may thus be considered as the velocities of the internal flow along the mean line of an airfoil of finite thickness. (See Fig. A,2a.)

In the case of the elliptical section, it is found that the velocity function actually satisfies the boundary condition for an elliptical shape of slightly different proportions than those given by the approximate

[3] Most of the functions listed may be derived from formulas given in *NACA Rept. 833* by H. J. Allen, *Fundamentals of Fluid Dynamics for Aircraft Designers* by M. M. Munk, or *The Elements of Aerofoil and Airscrew Theory* by H. Glauert. (See Bibliography, Art. 16.)

Table A,2. **Two-dimensional flow functions.**

Velocity function $f(\xi) = \frac{u - iw}{U}$	Components u/U, w/U along chord line	Airfoil shape $z = \int \frac{w}{U}\,dx$
1. $\frac{i}{\sqrt{\xi^2 - 1}} = \frac{-1}{\sqrt{1 - \xi^2}}$	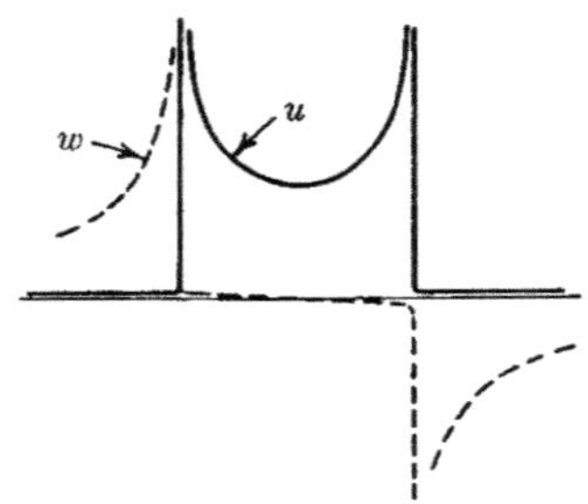	
2. $i\left(1 - \frac{\xi}{\sqrt{\xi^2 - 1}}\right)$	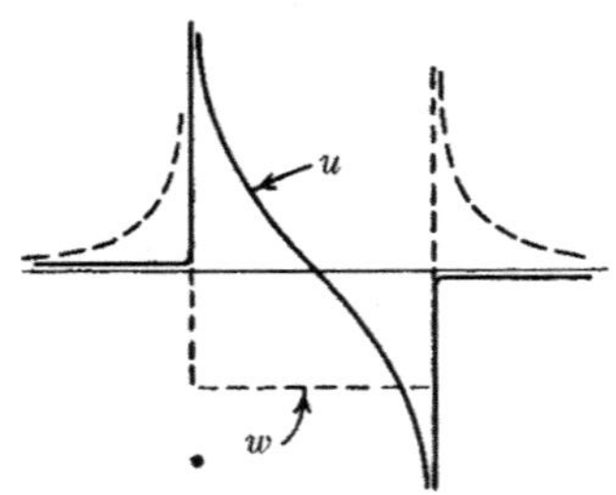	

3. $i - \sqrt{\frac{1-\xi}{1+\xi}}$	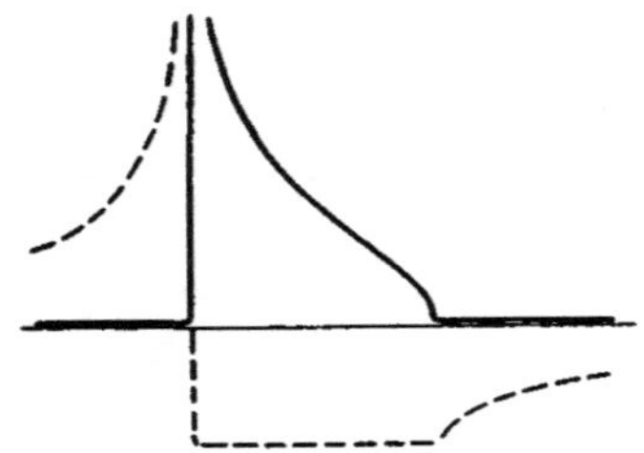	
4. $iA_n(\xi - \sqrt{\xi^2 - 1})^n$	$x = \cos\theta$ $\frac{u}{U} = A_n \sin n\theta$ $\frac{w}{U} = -A_n \cos n\theta$	
5. $i(\xi - \sqrt{\xi^2 - 1})$	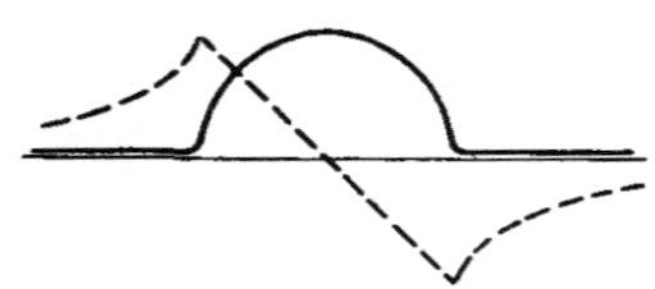	

Table A,2 (continued)

	Velocity function $f(\xi) = \frac{u - iw}{U}$	Components u/U, w/U along chord line	Airfoil shape $z = \int \frac{w}{U}\,dx$
6.	$-i(\xi - \sqrt{\xi^2 - 1})^2$	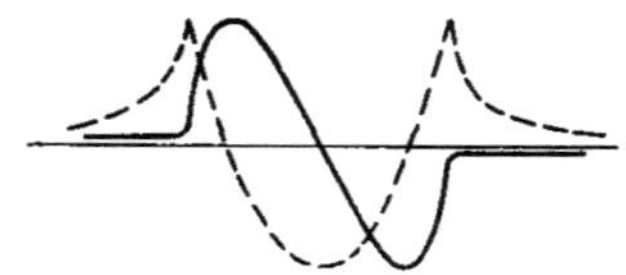	
7.	$\frac{2i}{\pi} A_n Q_n(\xi)$	$\frac{u}{U} = A_n P_n(x)$ $\frac{w}{U} = \frac{2}{\pi} A_n Q_n(x)$	
8.	$\frac{2i}{\pi} Q_0(\xi)$	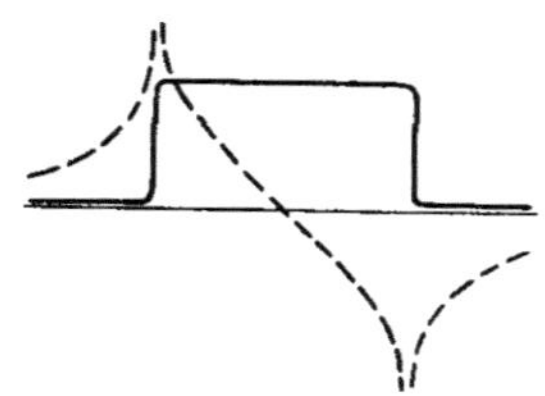	

9. $-\frac{2i}{\pi} Q_1(\xi)$	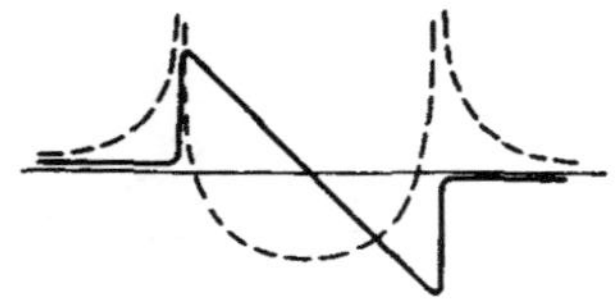	
10. $\frac{2}{\pi} \cosh^{-1} \frac{1}{\xi} - i$	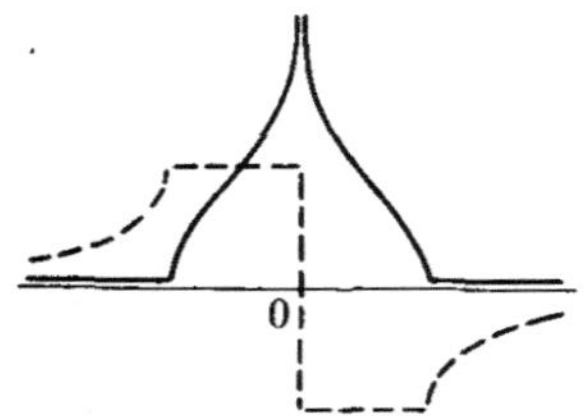	
11. $\frac{2}{\pi} \left(\cosh^{-1} \frac{1 - x_0 \xi}{\xi - x_0} - i \cos^{-1} x_0 \right)$	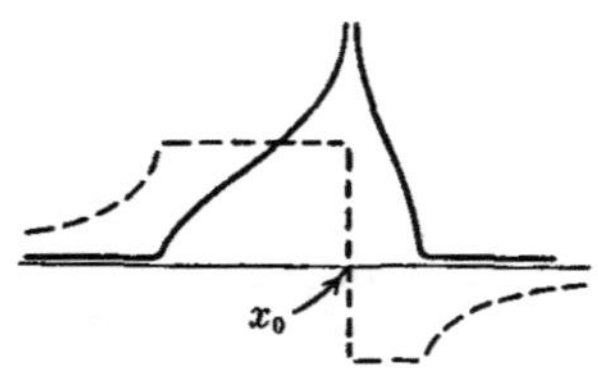	

Table A,2 (continued)

	Velocity function $f(\xi) = \dfrac{u - iw}{U}$	Components u/U, w/U along chord line	Airfoil shape $z = \int \dfrac{w}{U}\, dx$
12.	$\dfrac{1}{\sqrt{\xi^2 - 1}} = \dfrac{i}{\sqrt{1 - \xi^2}}$	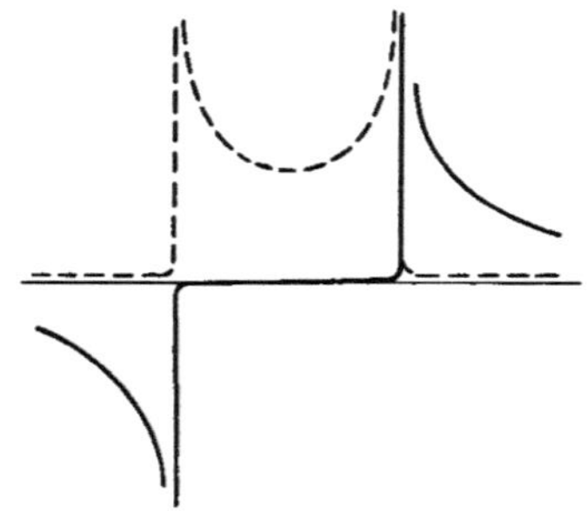	
13.	$1 - \dfrac{\xi}{\sqrt{\xi^2 - 1}}$	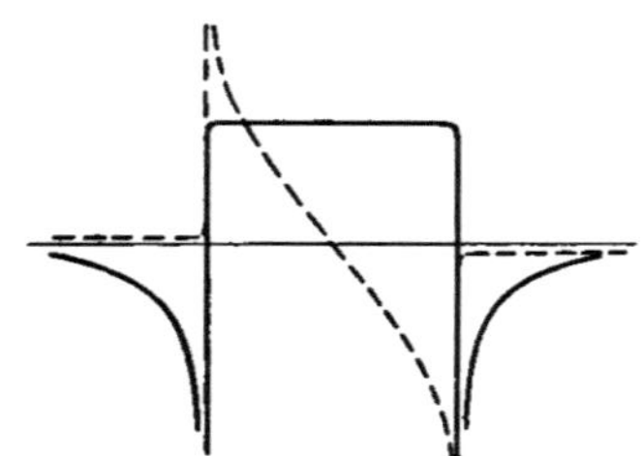	

14. $\dfrac{1-\xi}{\sqrt{\xi^2-1}} + 2(\sqrt{\xi^2-1} - \xi) + 1$	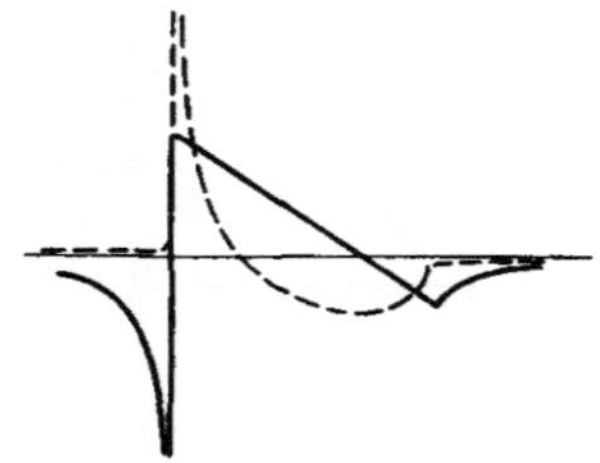	
15. $A_n(\xi - \sqrt{\xi^2-1})^n$	$x = \cos\theta$ $\dfrac{u}{U} = A_n \cos n\theta$ $\dfrac{w}{U} = A_n \sin n\theta$	
16. $\xi - \sqrt{\xi^2-1}$	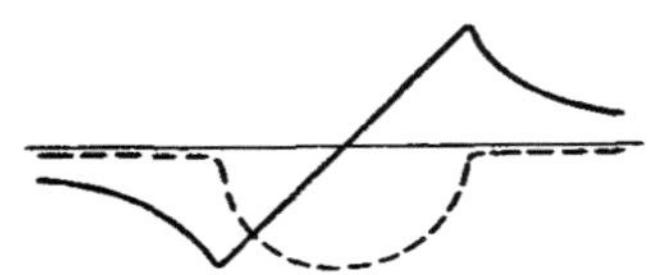	

Table A,2 (continued)

	Velocity function $f(\xi) = \dfrac{u - iw}{U}$	Components u/U, w/U along chord line	Airfoil shape $z = \int \dfrac{w}{U}\, dx$
17.	$-(\xi - \sqrt{\xi^2 - 1})^2$		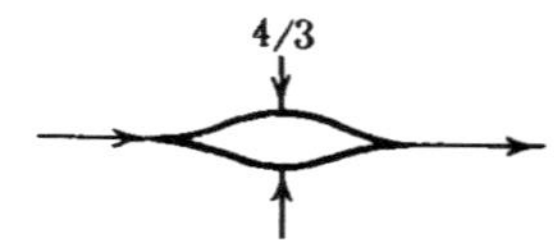
18.	$\dfrac{2}{\pi} A_n Q_n(\xi)$	$\dfrac{u}{U} = \dfrac{2}{\pi} A_n Q_n(x)$ $\dfrac{w}{U} = A_n P_n(x)$	
19.	$\dfrac{2}{\pi} Q_0(\xi)$	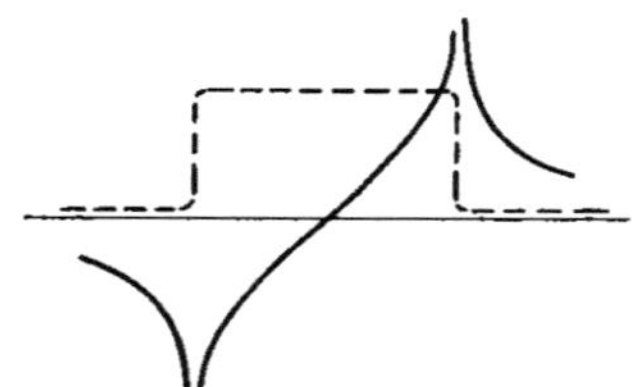	

20.	$-\frac{2}{\pi} Q_1(\xi)$	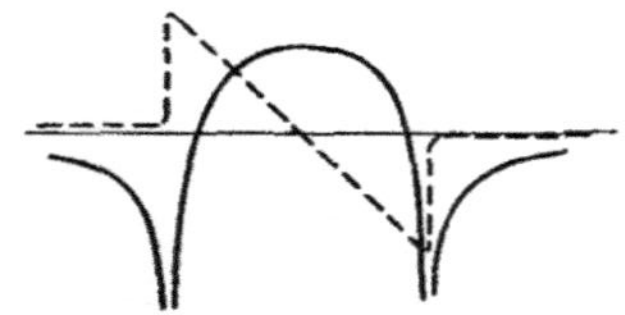	
21.	$\frac{i}{\pi} \int_{-1}^{+1} \frac{u(x_1 + 0i)}{\xi - x_1} dx_1$	$w(x) = \frac{1}{\pi} \int_{-1}^{+1} \frac{u(x + 0i)}{x_1 - x} dx_1$	
22.	$\frac{1}{\pi \sqrt{1 - \xi^2}} \int_{-1}^{+1} \frac{w(x_1) \sqrt{1 - x_1^2}}{\xi - x_1} dx_1$	$u(x + 0i)$ $= \frac{1}{\pi \sqrt{1 - x^2}} \int_{-1}^{+1} \frac{w(x_1) \sqrt{1 - x_1^2}}{x - x_1} dx_1$	 $(\Gamma = 0)$
23.	$\frac{1}{\pi} \int_{-1}^{+1} \frac{w(x_1 + 0i)}{\xi - x_1} dx_1$	$u(x) = \frac{1}{\pi} \int_{-1}^{+1} \frac{w(x_1 + 0i)}{x_1 - x} dx_1$	
24.	$\frac{i}{\pi \sqrt{1 - \xi^2}} \int_{-1}^{+1} \frac{u(x_1) \sqrt{1 - x_1^2}}{\xi - x_1} dx_1$	$w(x + 0i)$ $= \frac{1}{\pi \sqrt{1 - x^2}} \int_{-1}^{+1} \frac{u(x_1) \sqrt{1 - x_1^2}}{x - x_1} dx_1$	 $(z(1) = 0)$

boundary condition. The boundary of the actual ellipse extends beyond the ends of the chord line, so that the points ± 1 occur at the foci of the elliptical section. By substituting the actual coordinates of the elliptical boundary into formula 13 of Table A,2, the velocity distribution around the surface may be obtained. This same function can be made to satisfy the boundary condition for every confocal ellipse by adjusting the magnitude of the velocity discontinuity, or source strength, in relation to the stream velocity U.

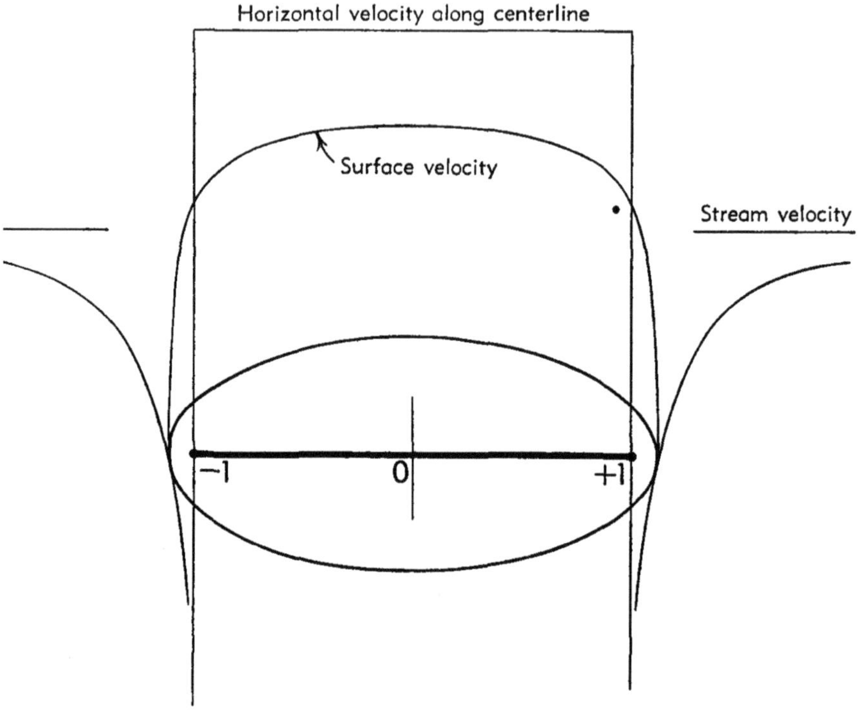

Fig. A,2b. Comparison of velocity distributions along the chord line and at the surface of an elliptical section.

Fig. A,2b illustrates the nature of the correspondence between the velocities along the axis and the surface velocities in the case of a thick section. The effect of proceeding from the chord line to the surface may be described as an attenuation with distance. For thin sections the attenuation is small, but is always more pronounced in regions of high local velocities or accelerations. Thus the infinite velocities at the ends of the chord line, ± 1, are reduced to finite values at the position of the nose. The fact that such infinite velocities arise only in the fictitious "internal" flow removes a difficulty in the physical interpretation of the theory and is especially important in the application of the theory to compressible fluids.

Total lift and drag. Leading edge forces. The total lift may be obtained, to the first order, by an integration of the pressure distribution along the chord line, and the total drag by integrating the product of the pressure and the local inclination of the airfoil surface. In the integration for the drag, however, care must be taken to evaluate the effects of singularities which appear—usually at the leading edge—in the thin airfoil flows. Such singularities can be included by integrating along a path $\xi = x + 0 \cdot i$ displaced a vanishingly small distance in the z direction and also slightly extended so as to pass above the point of infinite velocity at the leading edge. Since $\Delta p = -\rho_\infty uU$ and $dz/dx = w/U$, the integral for the drag may be written, considering both upper and lower surfaces,

$$D = 2 \int \Delta p \frac{dz}{dx} dx = -2 \int \rho_\infty uw dx \tag{2-10}$$

For the two-dimensional velocity field, the product occurring in the integral on the right may be denoted by

$$-2uw = \text{I.P.}\ (u - iw)^2 \tag{2-11}$$

In the case of the flat plate at an angle of attack α, the integral for the drag becomes (see formula 3, Table A,2)

$$D = \text{I.P.} \int \rho_\infty U^2\alpha^2 \left(\frac{1-\xi}{1+\xi} - 2i\sqrt{\frac{1-\xi}{1+\xi}} - 1\right) d\xi \tag{2-12}$$

The leading edge force is obtained by evaluating the contribution to this integral in passing just above the singular point at $\xi = -1$. In the limit, only the first term in the parentheses is effective, and its integral approaches the value $2 \ln (1 + \xi)$. In crossing over the point -1, the vector $(1 + \xi)$ rotates through $-180°$, so that the imaginary part of the logarithm changes by $-\pi$. Hence the contribution of the leading edge singularity to the drag is

$$D_{\text{l.e.}} = -2\pi\rho_\infty U^2\alpha^2 \tag{2-13}$$

Since

$$L = 2\pi\rho_\infty U^2\alpha \qquad (\text{chord} = 2)$$

the leading edge force is equal and opposite to the product of the total lift and the slope of the wing surface, and the total drag in this case is zero.

The foregoing development can be generalized as follows: Whenever the pressure in the vicinity of the leading edge ($x = -1$) can be written in the form

$$-\rho_\infty U \left[\frac{f(x)}{\sqrt{1+x}} + \text{nonsingular terms}\right]; \quad f(-1) \text{ finite}$$

or, from Eq. 2-7, the complex velocity function approaches, in the neighborhood of the leading edge,

$$u - iw = \frac{f(\xi)}{\sqrt{1+\xi}} + \text{nonsingular terms}$$

the leading edge force will be

$$D_{\text{l.e.}} = \text{I.P. } \rho_\infty \int \left[\frac{f^2(\xi)}{1+\xi} + \text{nonsingular terms}\right] d\xi$$

In the limit this gives

$$D_{\text{l.e.}} = -\pi\rho_\infty C^2 \tag{2-14}$$

in which C is the value, at the leading edge, of u times the square root of the distance to the leading edge.

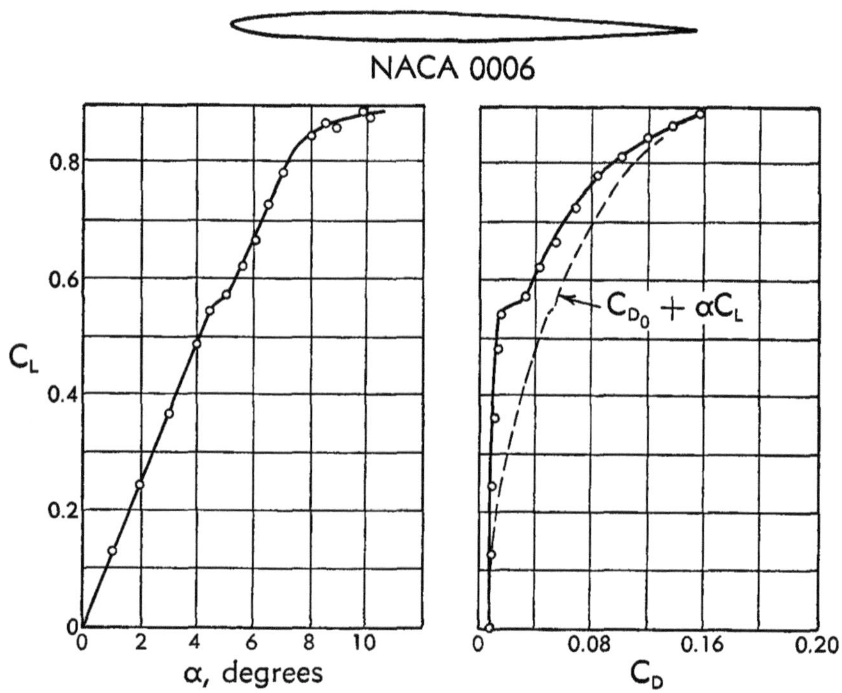

Fig. A,2c. Experimental lift and drag curves for thin section.

A similar limiting process must be considered in connection with the velocity functions associated with distributions of thickness with rounded leading edges, such as formula 14, Table A,2. In these cases, Eq. 2-14 will be found to give a positive component of drag.

With an airfoil of finite thickness and a continuously curved surface, the leading edge force cannot of course be clearly separated from the force developed on other parts of the airfoil surface. If the airfoil is made thinner, however, the suction developed in the upward flow around the leading edge becomes greater, while the forwardly projecting area over which this suction can act becomes less. The two factors change in such a way that their product remains essentially constant right down to the limit of zero thickness. In the limiting case of zero thickness the pressure stress in the vicinity of the nose tends toward infinity.

Such a limit cannot of course be approached in practice. It is found experimentally that if the radius of the nose is made too small, flow separation will occur at a relatively small angle of attack. Increases of angle of attack beyond this point cause a rapid increase in drag coefficient.

Physical realization of foregoing type of flow. Fig. A,2c, taken from experimental data given in [*11*], illustrates the rapid rise in drag encountered on an airfoil of 6 per cent maximum thickness when the flow ceases to follow the sharp curvature around the nose. It will be noted that the drag remains nearly constant up to a lift coefficient of 0.55 (angle of attack of 5°). This behavior illustrates the action of the leading edge suction force. A small increase of the angle of attack beyond this point, however, causes flow separation at the nose, and the resultant force on the airfoil falls back toward a perpendicular to the chord direction, as illustrated by the rapid increase of the drag.

In general, the range of angles of attack and thickness-chord ratios within which an unseparated flow, resembling the potential flow pattern, can be maintained constitutes the range of practical flight conditions. Outside this range the drag becomes excessive, while the control and flying qualities are usually impaired by unsteadiness of the flow and buffetting.

A,3. Effects of Compressibility at Subsonic Speeds.

Introduction. As long as the speed of flight is small compared with the velocity of sound and the pressures generated are small relative to the ambient atmospheric pressure, the influence of the compressibility of the air can safely be neglected. The development of power plants of increased capacity, however, continually makes possible flight at higher altitudes and higher speeds. It is found that, up to a certain limit, the higher speeds can be achieved without loss of efficiency or without increase of the amount of fuel expended per mile of flight. Beyond this limit, however, the favorable aerodynamic properties of conventional wing forms, and hence the high ratio of lift to drag of the airplane, deteriorate rapidly. Since the limitation is imposed by the effects of compressibility, the study of such effects has become a primary task of present-day aerodynamic research.

Physical Considerations. Before proceeding to a quantitative study of the effects of the compressibility of air on the behavior of an airfoil, it is desirable to review certain physical aspects of the flow of an incompressible fluid. Suppose the airfoil, as in Fig. A,3a, to be immersed in a uniform stream moving to the right. Since its thickness has the effect of diminishing the cross section of the oncoming stream, the flow around the midsection of the airfoil occurs with an excess of velocity and in this region the mass flux per unit area must be increased to prevent the accumulation of mass in the region ahead. Assuming a frictionless fluid, the motion of each element of volume must be produced by the action of pressure, or buoyancy, forces. Thus the flow is accelerated past the midsection of the airfoil by entering a region of reduced pressures. The areas of positive and negative pressure are symmetrically disposed, so as to conduct the flow around the airfoil without the occurrence of any pressure drag. It is also readily apparent from geometric considerations that the widening of the stream tubes near the nose and the contraction of width in the regions of increased velocity above and below the airfoil lead to a progressive reduction in curvature of the streamlines, and hence to an attenuation of the flow disturbance with distance from the airfoil.

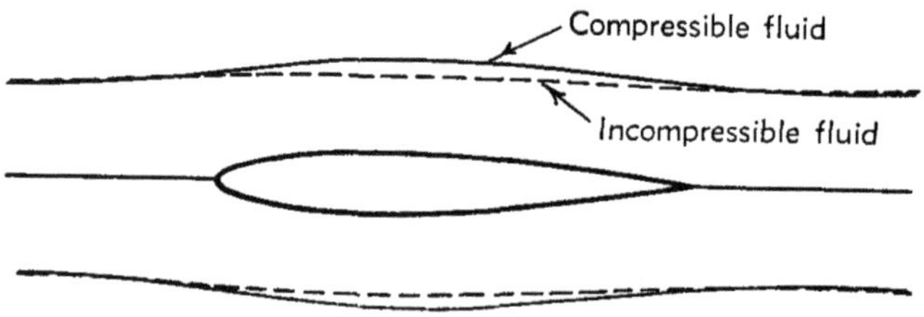

Fig. A,3a. Expansion of streamlines in compressible flow.

In the corresponding flow of an elastic fluid, the streamline pattern is modified by the expansion of the volume of the fluid on passing through the regions of low pressure above and below the airfoil and, to a lesser degree, by the compression occurring in regions of higher pressure. Since its volume is increased in the regions of high velocity and low pressure, the elastic fluid requires a still higher velocity in order to get past the midsection of the airfoil, as well as more space in order to maintain the given total mass flow. The expansion of the minimum cross section of the stream tubes forces the streamlines outward to conform more nearly to the curvature of the airfoil surface, with the result that the disturbance caused by the airfoil extends vertically to a greater distance.

ONE-DIMENSIONAL ANALYSIS. In spite of its fairly simple physical aspect, the problem of determining the configuration of the flow field around an airfoil so as to agree with the dynamical equations and with a realistic equation of state for air has turned out to be one of great mathematical difficulty. A quantitative insight into the phenomenon can be gained, however, through the so-called "one-dimensional" analysis [*12*]. This analysis is based on the selection of stream tubes of infinitesimal cross section, so that the change in any quantity across the width of the stream tube is negligible. The velocity, pressure, and density can then be considered as functions of the distance along the stream tube. Such one-dimensional relations must be satisfied throughout any steady flow, and hence form a necessary condition on the field as a whole—but they are not sufficient to determine the complete pattern because of the additional requirement of dynamical balance between adjacent stream tubes.

Assuming frictionless motion, we have for the dynamical equation

$$u\frac{du}{dx} = -\frac{1}{\rho}\frac{dp}{dx} \tag{3-1}$$

where the velocity u and the distance x are measured along the stream tube. If A is the cross-sectional area of the stream tube, the equation for continuity of mass flow is

$$\frac{d}{dx}(\rho A u) = 0 \tag{3-2}$$

By making use of the expression for the velocity of sound in an elastic medium, i.e.

$$a = \sqrt{\frac{\partial p}{\partial \rho}} \tag{3-3}$$

Eq. 3-1 and 3-2 can be combined into a single relation between the velocity of flow, the rate of change of cross section, and the local Mach number u/a, viz.

$$\left(1 - \frac{u^2}{a^2}\right)\frac{du}{dx} = -\frac{u}{a}\frac{dA}{dx} \tag{3-4}$$

Eq. 3-4 shows that the stream tubes become narrower as the velocity increases, so long as the ratio u/a remains less than one. Just at the speed of sound ($u = a$) the quantity dA/dx becomes zero, and at this speed a decrease of pressure in the direction of flow causes an expansion of volume just sufficient to counteract the increased flow velocity, so that the streamlines are no longer reduced in width by an increase of velocity. When the local velocity reaches sonic speed the rate of mass flow per unit of cross-sectional area ρu reaches an absolute maximum and can no longer be increased by the action of a pressure gradient. At this speed any change of velocity through the action of pressure forces tends to diminish the mass flowing through a given cross section.

At the exact speed of sound the idealization of a steady flow which vanishes at infinity breaks down, since mass would continually accumulate ahead of the obstacle. This accumulation makes itself apparent in the phenomenon of "choking" in the wind tunnel and in the extensive bow wave developed ahead of an airplane when flying near the speed of sound [*13*].

LINEARIZED COMPRESSIBLE FLOW THEORY.

Similarity of compressible and incompressible flows at subsonic speeds. Within the regime of continuous subsonic motion, it is found that the flow of a compressible fluid retains a basic similarity to an incompressible flow. In particular, the compressible flow can for many purposes be considered an irrotational potential motion. In addition to the absence of friction, the existence of potential motion in a compressible fluid depends on the existence of a unique relation between the pressure and the density. In a frictionless flow the fluid elements are accelerated entirely by the action of the pressure gradient. If the density is a function of the pressure alone, the pressure gradient will coincide in direction with the density gradient at all points. The force on each element will then be aligned with the center of gravity of the element and the pressure forces will introduce no rotation.

Prandtl's linerarized equation for steady flow. Prandtl has shown [*14*] that the assumptions of the thin airfoil theory lead to an essential simplification of the equation for the steady potential flow of a compressible fluid. Since successful flight at high speeds actually depends on the use of thin wings and slender bodies, the resulting linearized theory, while not suitable for the study of compressible flows in general, appears to provide a suitable basis for the practical calculations of aeronautical engineering.

It is instructive to derive Prandtl's approximate linearized equation of motion from the more exact one-dimensional flow equation (3-4). It will be evident that the rate of change of the cross-sectional area A of a stream tube can be related kinematically to small lateral and vertical velocities v and w superimposed on the main velocity u. By such a kinematic relation the quantity $(u/A)(\partial A/\partial x)$ is found to be equal to $\partial v/\partial y + \partial w/\partial z$, or, in other words, to the divergence of the velocity pattern observed by looking along the local direction of the streamlines. Hence Eq. 3-4 can be written

$$\left(1 - \frac{u^2}{a^2}\right)\frac{\partial u}{\partial x} + \frac{\partial v}{\partial y} + \frac{\partial w}{\partial z} = 0 \qquad (3\text{-}5)$$

Eq. 3-5 must be satisfied locally at every point in the flow field and, in order to apply it to the flow field as a whole, variations of the local Mach number u/a as well as the inclination of the axes x, y, z (which have been assumed to lie parallel to the local flow direction) would have to be considered. However, if the airfoil is thin and its angle of attack small, so that the streamlines are nearly horizontal, little error is made in the differential lengths if a single set of x, y, z axes aligned with the main stream velocity U is used for the entire flow field. Furthermore, if the resultant velocity throughout the field departs only slightly from the free stream velocity, the local Mach number u/a will depart only slightly from the free stream Mach number M_∞.

In addition to the one-dimensional equation for the flow along the streamlines, it is necessary to introduce the condition of dynamical balance between adjacent streamlines. Thus we must have

$$\rho_\infty U \frac{\partial w}{\partial x} = -\frac{\partial p}{\partial z} \tag{3-6}$$

in order for the pressure gradients to balance the centrifugal forces. As in the case of incompressible frictionless flow, it is found that to the first order $\Delta p = -\rho_\infty u U$, so that Eq. 3-6 becomes

$$\frac{\partial w}{\partial x} = \frac{\partial u}{\partial z} \tag{3-7}$$

The latter equation expresses the condition of irrotationality and leads to the existence of a velocity potential ϕ. Introducing $\partial u/\partial x = \phi_{xx} = \varphi_{xx}$, etc. into Eq. 3-5 results in

$$(1 - M_\infty^2)\varphi_{xx} + \varphi_{yy} + \varphi_{zz} = 0 \tag{3-8}$$

which is the equation given by Prandtl.

Relation to acoustic equation. Eq. 3-8 may be shown to be a special case of the equation for sound waves of small amplitude, i.e.

$$\varphi_{xx} + \varphi_{yy} + \varphi_{zz} = \frac{1}{a_\infty^2}\varphi_{tt} \tag{3-9}$$

From this point of view the moving airfoil is considered to give rise to small velocities and pressures in air that is otherwise at rest. If the airfoil has been moving at constant velocity for a sufficiently long time it may be assumed to have developed a fixed, or steady flow pattern, which travels with the airfoil at the velocity of flight U. By adopting axes fixed in the moving airfoil, so that $x' = x + Ut$, φ_{xx} may be replaced by $\varphi_{x'x'}$, and φ_{tt} by $U^2\varphi_{x'x'}$. Then, transposing the term on the right, replacing U^2/a_∞^2 by M_∞^2, and dropping the primes reduces Eq. 3-9 to Eq. 3-8. The steady flow around the thin airfoil may thus be considered as a type of sound disturbance that travels at a velocity U without change of form.

It is of interest to note the connection between the one-dimensional flow equation and the theory of acoustic waves. Such waves present an infinite variety of motions, but at the wave front, which travels at the acoustic velocity a, all lateral or vertical perturbation velocities disappear and the disturbance consists solely of longitudinal motions. Hence a sound wave is commonly referred to as a "longitudinal" wave. For that part of the wave which travels at the acoustic velocity a the streamlines are thus straight and parallel, as required by the one-dimensional equation.

Correspondence between compressible and incompressible flows. Eq. 3-8 can be reduced to Laplace's equation by a simple transformation of coordinates amounting to a change in the ratio of the horizontal to the vertical and lateral dimensions of the potential field. This may be seen by rewriting Eq. 3-8 in the form

$$\varphi_{xx} = -\frac{(\varphi_{yy} + \varphi_{zz})}{1 - M_\infty^2} \tag{3-10}$$

Comparison with Laplace's equation

$$\varphi_{xx} = -(\varphi_{yy} + \varphi_{zz}) \tag{3-11}$$

shows that in the compressible flow the second derivative of φ in the x direction is increased by the factor $1/(1 - M_\infty^2)$. This is just the effect

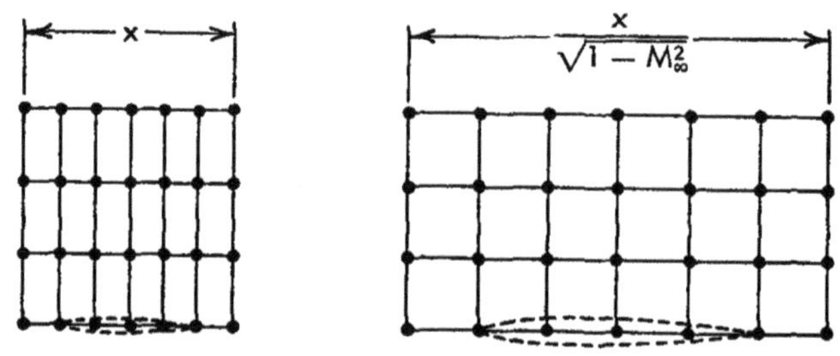

Fig. A,3b. Distribution of points having equal values of ϕ in a Prandtl transformation.

that would be produced by taking a distribution of φ and compressing it uniformly in the x-wise direction by the factor $\sqrt{1 - M_\infty^2}$ (see Fig. A,3b). Thus, if a solution of Laplace's equation for incompressible flow is known, a corresponding solution for the linearized compressible flow equation can immediately be produced.

If the potential distribution satisfying Laplace's equation and the corresponding foreshortened distribution (which satisfies the compressible flow equation) are compared at points having the same values of φ, it is seen that $u = \partial\varphi/\partial x$ is increased in the ratio $1/\sqrt{1 - M_\infty^2}$, while the lateral and vertical velocities, corresponding to the lateral and vertical gradients of φ, remain unchanged.

Although the construction of a flow field for a compressible medium from a known incompressible flow is relatively simple, care must be taken in the determination of the boundary conditions satisfied by the compressible flow field. Suppose now a solution of Laplace's equation satisfying the boundary condition for a certain thin airfoil is given. If the airfoil chord line and the surrounding potential field are foreshortened together, the values of $w = \partial\varphi/\partial z$ will remain unchanged at corresponding points along the chord line. Considering the same value of U in the two cases, it follows that the slopes of the streamlines emanating from the chord plane will be the same at equal fractional distances along the chord in the two cases. Within the usual approximation of the thin airfoil theory this slope is equal to the slope of the airfoil surface and the two airfoil sections are thus geometrically similar. It is clear that in three-dimensional flow a change in aspect ratio will be involved. In two-dimensional flow, however, the foreshortening of the chord corresponds merely to a change in the absolute size of the wing, so that the effect of compressibility on an airfoil section of given slope is simply to increase the horizontal perturbation velocities over the airfoil surface uniformly by the factor $1/\sqrt{1 - M_\infty^2}$. This factor was also found by Glauert [*15*] and is frequently referred to as the Prandtl-Glauert correction.

More Accurate Representations of Subsonic Compressible Flow. Although the Prandtl-Glauert method provides an exceedingly simple means for calculating the flow around an airfoil in two dimensions, it should be borne in mind that this method generally underestimates the effect of compressibility on the magnitude of the disturbance with airfoils of finite thickness. As the stream Mach number approaches 1, the quantity $1/\sqrt{1 - M_\infty^2}$ approaches infinity, with the result that the ratio of the horizontal perturbation velocity u to the airfoil thickness also approaches infinity. Obviously for any finite thickness a point will be reached at which the perturbation velocity u cannot be considered small relative to the stream velocity. Hence the Prandtl-Glauert formula begins to show increasing departures from reality at some Mach number below unity, the exact value of which depends on the thickness or angle of attack of the airfoil.

More accurate representations of the two-dimensional flow around airfoils of finite thickness at subsonic Mach numbers have been the subject of extensive investigation (see VI,E and F). In these investigations the assumption of isentropic potential motion is retained and solutions of the steady flow equation [*14*]

$$\left(1 - \frac{u^2}{a^2}\right)\frac{\partial u}{\partial x} + \left(1 - \frac{w^2}{a^2}\right)\frac{\partial w}{\partial z} - \frac{2uw}{a^2}\frac{\partial u}{\partial z} = 0 \tag{3-12}$$

together with the irrotationality relation

$$\frac{\partial u}{\partial z} = \frac{\partial w}{\partial x} \tag{3-13}$$

are sought. In these equations u and w refer, of course, to the components of the resultant velocity rather than to small perturbation velocities, and the velocity of sound a is a function of the local velocity at each point.

Exact solutions of Eq. 3-12 and 3-13 are supplied by the hodograph method of Molenbroeck [*16*] and Chaplygin [*17*]. Although the method cannot be carried out exactly for an airfoil of fixed shape, its application to the airfoil problem has been studied by von Kármán and Tsien [*18*], and others (see VI,F,5). The results of these investigations show that the compressible flow over an airfoil section for a wide range of Mach numbers can be obtained, at least approximately, by a simple correction applied to the velocities and pressures of the incompressible flow over the same shape. In terms of the pressure coefficient $C_p(=\Delta p/\frac{1}{2}\rho_\infty U^2)$, von Kármán and Tsien give the correction

$$C_p = \frac{C_{p_0}}{\sqrt{1 - M_\infty^2} + \frac{C_{p_0}}{2}(1 - \sqrt{1 - M_\infty^2})} \tag{3-14}$$

where M_∞ is the free stream Mach number and C_{p_0} is the pressure coefficient at $M_\infty = 0$. It appears that the correction is greater in regions of higher local velocity, a result in agreement with the one-dimensional flow equation (3-4). Experiments show that the Kármán-Tsien correction affords a noticeable improvement over the linearized theory for airfoils of finite thickness in two-dimensional flow, but the method likewise fails after a certain critical Mach number is exceeded.

A method [*19*] adaptable to sufficiently thin shapes involves the direct solution of Eq. 3-12 by an expansion of the velocity potential in powers of the thickness or angle of attack. This method amounts to a successive improvement of the linear approximation, which corresponds to the *first* power of the thickness-chord ratio t/c. The result obtained by Kaplan for the lift curve slope of elliptical sections at $\alpha = 0$ may be written

$$\frac{L_c}{L_i} = \mu + \frac{t/c}{1 + (t/c)}\left[\mu(\mu + 1) + \frac{\gamma + 1}{4}(\mu^2 - 1)^2\right] \tag{3-15}$$

where $\mu = 1/\sqrt{1 - M_\infty^2}$ and γ is the ratio of specific heats. A similar expression was found by Hantzsche and Wendt [*20*] for Joukowski sections.

If the assumptions of infinitesmal perturbation velocities and two-dimensional flow are retained and the stream velocity is allowed to approach sonic speed, it is found that the zone of disturbance expands vertically without limit and that the ratio of the longitudinal to the vertical perturbation velocities increases indefinitely. A simplification of the differential equation which makes use of this change in the geometric and kinematic proportions of the flow field has recently been studied by von Kármán [*21*], Guderley [*22*], and Oswatitsch [*23, 24*]. If the velocity field is supposed to be nearly horizontal and to consist of small increments above and below the critical velocity of sound, an especially simple nonlinear differential equation is obtained. This equation, known as

Tricomi's equation, seems to express the essential features of a transonic flow field about thin sections in two-dimensional flow and, at the present time, investigations of the practical validity of the method are under way (VI,A,7 and VI,F,11).

Limitations of Present Methods of Analysis. Thus far, theoretical investigations of flows at high subsonic velocities have been limited for the most part to cases of two-dimensional motion. As the critical Mach number is approached, it is found that the region of reduced density and increased velocity expands vertically at a rapid rate, so that the size of the disturbed region grows out of proportion to the size of the body and the relation of the pattern to the shape of the body tends to become somewhat indefinite. As noted above, the field begins to resemble that of a longitudinal sound wave which, once started, travels without the assistance of a moving solid. In part, the difficulty of the solution must be attributed to the assumed two-dimensional character of the motion, since this assumption effectively limits the expansion of the flow to the vertical dimension. It can be expected that in three dimensions the critical flow condition will be delayed to a higher Mach number and this is indeed found to be the case experimentally [*25*]. Studies of simplified flow equations for partially supersonic flows in three dimensions have recently been made by Spreiter [*26*] and Berndt [*27*].

When solutions obtained on the basis of steady potential flow are compared with observed flow patterns over sections of normal thickness, a fundamental discrepancy is encountered. Experiments in two-dimensional flow, or with straight wings of high aspect ratio, show that the continuous regular flow pattern around the airfoil changes rather abruptly to a discontinuous and frequently unsteady motion shortly after the appearance of a zone of supersonic velocity in the flow field. The supersonic zone expands rapidly and is followed by a vertically spreading shock wave. The fore-and-aft symmetry associated with the incompressible, or purely subsonic, type of flow is then lost, and a rapid increase of the drag coefficient appears. Solutions of the differential equations (3-12 and 3-13), however, continue to show reversible flow patterns and an absence of drag for flows containing extensive regions of supersonic velocity. It has not been possible to find experimentally the conditions under which this dragless type of flow might be realized in practice. (See VI,F,13.)

As far as can be determined from the differential equation (3-12), the condition of fore-and-aft symmetry would also exist in the flow over a symmetrical body in supersonic flight. Thus the solution would show Mach waves extending forward from the body as well as to the rear, and would be obviously incorrect from the physical standpoint. No such definite criterion for discarding the symmetrical solution in the subsonic case has as yet been established, although recent investigators [*28,29,30*]

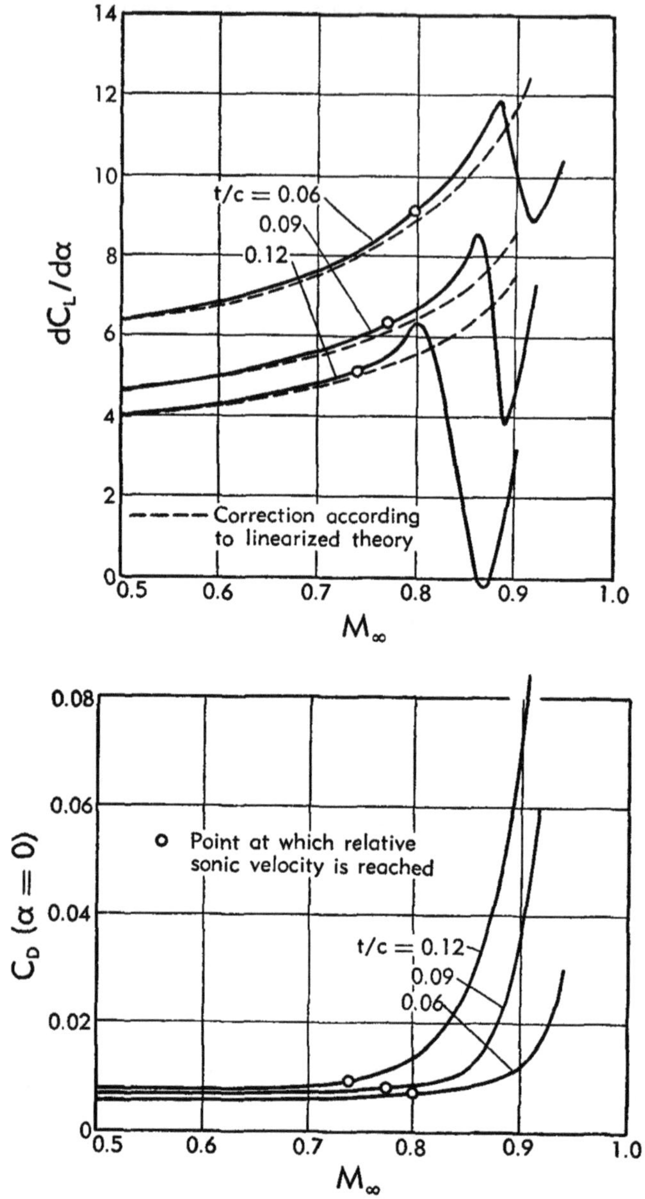

Fig. A,3c. Effect of airfoil thickness on lift curve slope and drag at subsonic Mach numbers [*31*].

have indicated that this type of configuration may be unstable if small disturbances in the flow or in the shape of the boundary are considered (VI,F,14 and 15).

INDICATION OF EXPERIMENT. Wind tunnel measurements on wings of conventional form and thickness show a large increase of the drag coefficient and erratic variations of the lift and center of pressure after the development of shock waves, and schlieren photographs show that the shock wave usually causes flow separation.

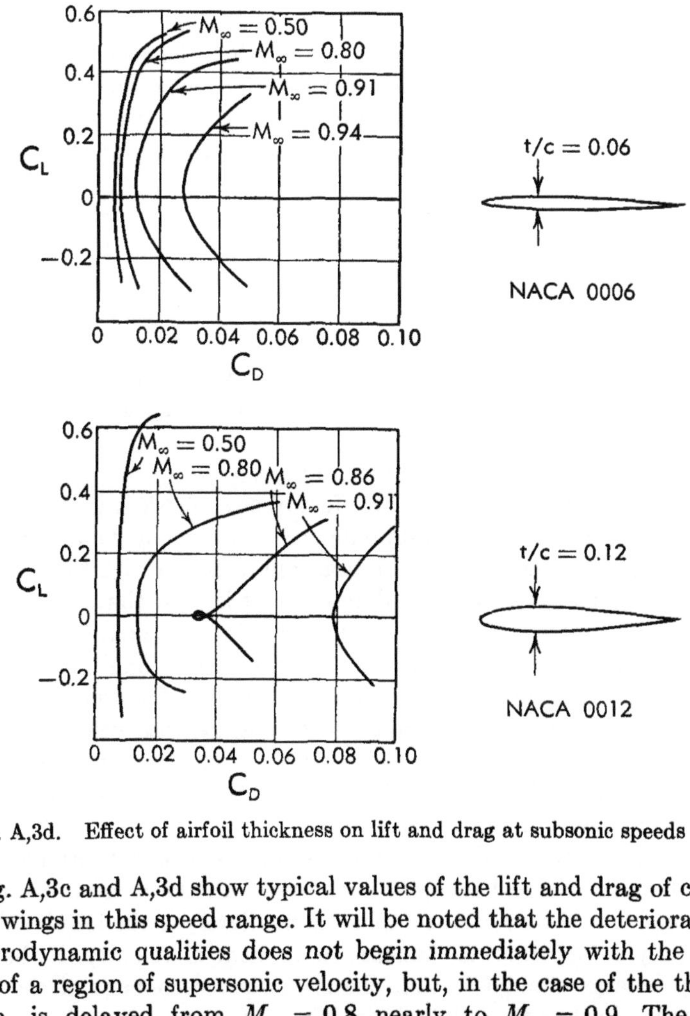

Fig. A,3d. Effect of airfoil thickness on lift and drag at subsonic speeds [*31*].

Fig. A,3c and A,3d show typical values of the lift and drag of conventional wings in this speed range. It will be noted that the deterioration of the aerodynamic qualities does not begin immediately with the occurrence of a region of supersonic velocity, but, in the case of the thinnest section, is delayed from $M_\infty = 0.8$ nearly to $M_\infty = 0.9$. The worst effects occur with thicker sections. In the present experiments the airfoil with 12 per cent thick section actually suffered a reversal of the lift at a Mach number of 0.86. Some improvement in the aerodynamic characteristics can be achieved by refinement of the section shape. Much greater effects result, however, from changes in the planform of the wing. These will be taken up in the next article.

A,4. Effect of Sweepback.

Introduction. For the range of speeds within which the Kutta-Joukowski type of flow can be maintained, the straight wing of high aspect ratio shows the greatest aerodynamic efficiency. However, the advantage of this design depends in a critical way on the absence of pressure drag in the flow over the sections, and when the speed exceeds that for which this type of flow is possible the straight wing no longer appears to be the most suitable form. In the present article it will be shown that better efficiencies can be maintained by placing the wings at an oblique angle to the flow.

The use of sweepback to improve the efficiency of wings at supersonic speeds was first considered by Busemann [*32*] in 1935. Later, Betz, in Germany, pointed out that the critical Mach number of wings at subsonic speeds could be increased by the use of sweep. More recently it has been shown [*33*] that continuous flow of the Kutta-Joukowski type may be maintained at supersonic speeds and that the wave drag associated with such speeds disappears in the case of an infinite wing yawed at an angle greater than the Mach angle.

Theory of Oblique Cylindrical Flow.

Principle of independence. The effect of sweepback is best visualized by considering the effect of sideslipping an infinitely long wing of constant cross section. Suppose the wing is initially in a stream with the oncoming velocity at right angles to the leading edge. It is clear that, if the flow is frictionless and all sections of the wing are alike, an axial or lengthwise motion may be introduced without causing any additional motion of the fluid, since the axial motion results only in a sliding of the surface parallel to itself. The combination of crosswise and lengthwise velocities results in an oblique relative motion. Conversely, in the oblique motion of a cylindrical wing the flow will be determined entirely by the crosswise, or normal, component of the velocity.

Since this principle of independence applies to any frictionless flow governed by the normal boundary condition, all the results of the two-dimensional flow theory can be applied immediately to the infinite oblique wing simply by subtracting the axial component of velocity. If the angle of sweep is large enough so that the crosswise component of the Mach number remains less than the critical Mach number of the wing sections, then the reversible subsonic type of flow over the sections can be preserved and the adverse effects of compressibility delayed to a higher speed. The maintenance of the essentially dragless subsonic type of flow at supersonic speeds requires that the wing be yawed at an angle greater than the Mach angle.

The elimination of drag at supersonic speeds is of course a feature of the two-dimensional flow. As later calculations will show, a drag does arise with wings of limited dimensions, chiefly as the result of departures from two-dimensional flow at the apex of the wing.

Further discussion of flow field. Since the effect of sweepback is of fundamental importance in the study of wings for high speeds, it will be worthwhile to consider it from several different points of view.

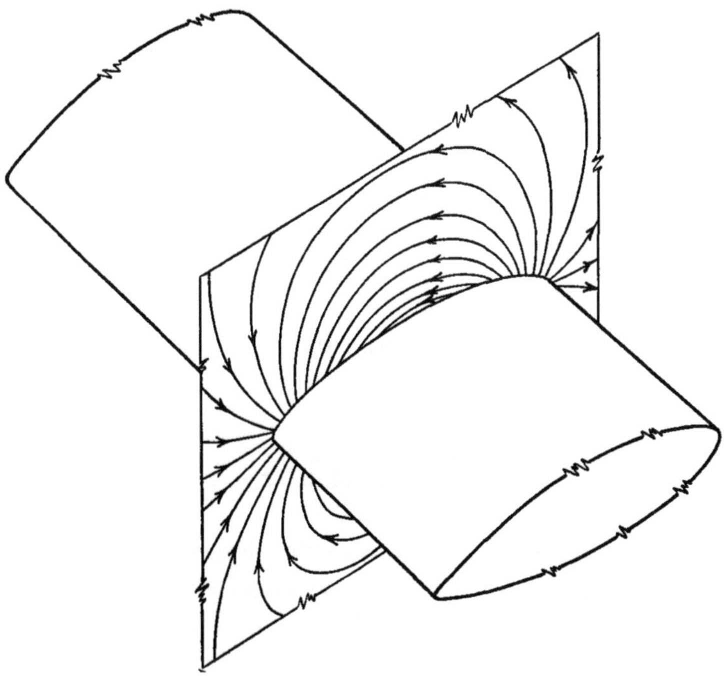

Fig. A,4a. Velocities produced by wing in oblique motion.

If observed from a point at rest relative to the air, the motion produced by oblique translation of the wing would be confined in direction to planes cutting the wing at right angles to the leading edge (see Fig. A,4a). If observed from a point fixed relative to the wing, however, the velocity of flight would appear added vectorially to the disturbance velocities produced by the wing, and since the latter are generally small compared to the flight velocity, particles in crossing near the wing would be observed to depart but little from planes parallel to the flight direction.

With the wing at rest in an oncoming stream, the flow is steady and the cross sections of the individual stream tubes must follow the law of variation given by the one-dimensional flow equation (3-4). According to this equation the change in cross section of the stream tubes with changes in pressure bears a fixed relation to the local Mach number, independent of the shape of the body. However, the streamline pattern and the pressures around the sections of the yawed wing must be identical with those for a smaller Mach number, and the actual variation in stream tube area therefore must differ from the variation that appears in the flow over the sections. As shown in Fig. A,4b this difference appears in the plan view of the streamlines. Suppose for example that the oncoming stream has a supersonic velocity but the wing is yawed sufficiently so that the component of velocity normal to its leading edge is subsonic. Then the flow over the sections will have the configuration appropriate to the subsonic velocity, and the streamlines in this view will become narrower in the

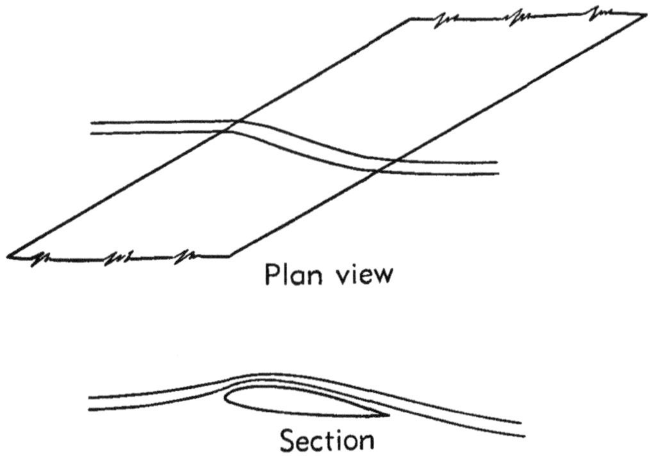

Fig. A,4b. Streamlines crossing oblique wing.

region of increased velocity and reduced pressure just above the upper surface of the wing. However, since the pressure gradients induced by the wing lie in a direction oblique to the main flow, the streamlines are curved laterally as they pass over the wing and this curvature has the effect of increasing the width of the stream tubes in the plan view in regions of excess velocity. Since the total velocity is supersonic, the resultant cross-sectional area must increase in these regions and hence the lateral widening observed in the plan view must predominate.[4]

The occurrence of shock waves and the consequent rapid rise in pressure drag on the oblique wing is to be expected when the crosswise component of the Mach number exceeds the critical Mach number of the

[4] It is of interest to note that the theory of oblique cylindrical flows provides a simple construction for the form of a channel required to produce the regular and continuous deceleration of a sonic or supersonic stream to a subsonic velocity. Starting with a subsonic two-dimensional channel flow, the addition of a constant velocity at right angles to the plane of the two-dimensional flow results in stream channels which show the required curvature and widening in the third dimension.

sections. It should be noted that the pressure force on an oblique wing of cylindrical form will act in the direction at right angles to its axis, and hence only a component of this force will be effective in the flight direction. Thus the pressure drag is reduced by yaw approximately in the ratio $\cos^3 \beta$, while the lift is reduced in the ratio $\cos^2 \beta$.

Relation to acoustic equation. The treatment of the infinite oblique wing by the thin airfoil theory is simplified by the solution of the acoustic

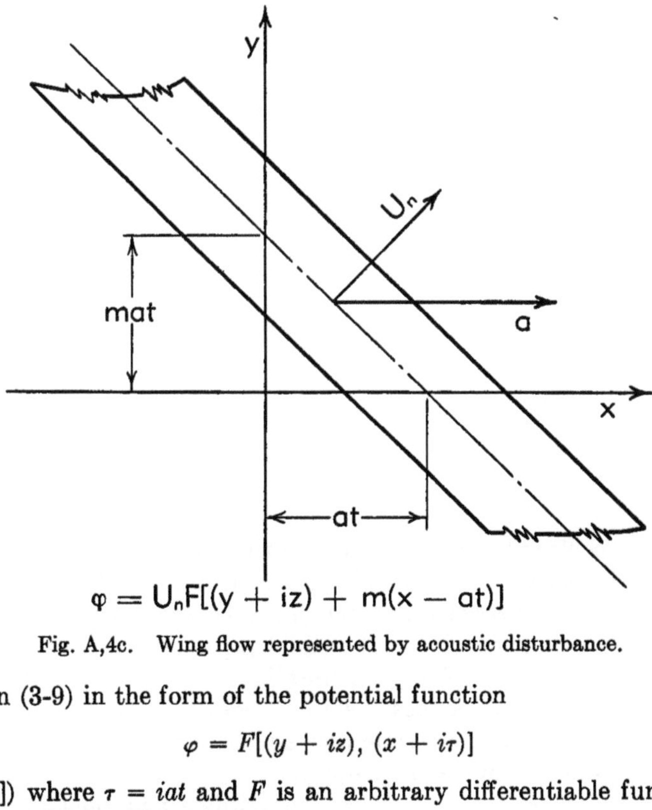

Fig. A,4c. Wing flow represented by acoustic disturbance.

equation (3-9) in the form of the potential function

$$\varphi = F[(y + iz), (x + i\tau)] \tag{4-1}$$

(see [34]) where $\tau = iat$ and F is an arbitrary differentiable function of the two variables. The solution

$$\varphi = U_n F[(y + iz) + m(x - at)] = U_n F(\xi) \tag{4-2}$$

represents a disturbance of cylindrical form traveling with the velocity a along x, but with the normal velocity U_n (see Fig. A,4c). Differentiation for the velocities in the y, z plane results in

$$v = \frac{\partial \varphi}{\partial y} = U_n F'(\xi) \tag{4-3}$$

and

$$w = \frac{\partial \varphi}{\partial z} = U_n i F'(\xi) \tag{4-4}$$

If only the real solution is considered

$$\frac{v - iw}{U_n} = F'(\xi) = f(\xi) \tag{4-5}$$

and the function f may be selected with the aid of examples given in Table A,2. As required by the one-dimensional flow equation, the divergence of the velocity field is zero in planes at right angles to the direction in which the apparent velocity is the speed of sound (Fig. A,4c). By making use of the acoustic equation for the pressure, we obtain

$$\frac{\Delta p}{\frac{1}{2}\rho_\infty U_n^2} = \frac{2(\partial\varphi/\partial t)}{U_n^2} = \frac{f(\xi)}{\sqrt{1 - (U_n^2/a^2)}} \tag{4-6}$$

The formulas are of course independent of the resultant velocity U, which is not shown on Fig. A,4c, since it may lie at any angle β to U_n. If we introduce $U_n = U \cos \beta$ and $U_n/a_\infty = M_\infty \cos \beta$, there results

$$\frac{\Delta p}{\frac{1}{2}\rho_\infty U^2} = \frac{\cos^2 \beta}{\sqrt{1 - (M_\infty \cos \beta)^2}} f(\xi) \tag{4-7}$$

Applying formula 3 of Table A,2 and integrating for the total lift, one obtains

$$C_L = \frac{2\pi\alpha \cos \beta}{\sqrt{1 - (M_\infty \cos \beta)^2}} \tag{4-8}$$

for the lift coefficient of a thin oblique airfoil in two-dimensional flow.

It will be noted that the derivations just given furnish a simple graphical relation between the compressible and incompressible flow fields around a thin airfoil whether in straight or oblique motion. The resultant perturbation velocity vectors lie in the plane of the normal velocity U_n. Projection of these velocity vectors on the y, z plane yields the corresponding incompressible flow pattern as given by the Prandtl-Glauert transformation (see Fig. A,3b).

Experimental verification. Experimental determinations of the pressure and force characteristics of oblique wings show substantial agreement with the foregoing theoretical concept when allowance is made for the fact that perfectly two-dimensional flows cannot be realized experimentally.

Fig. A,4d shows the results of drag measurements made by the NACA on straight and sweptback wings attached to a falling body. The wings were attached internally to electrical balances and measurements of the drag force acting on the wings were transmitted to the ground by radio during the free fall [*35*]. The bodies were released from an altitude of approximately 40,000 feet and accelerated to a Mach number of nearly 1.3 before impact. The drags of the forward and rear wings differed slightly and the curves plotted on the figure are average values from the two sets of data. The results show the large gain predicted by the theoretical considerations. The departure of the experimental curve for the sweptback wing from the two-dimensional theory is seen to be satisfactorily explained by three-dimensional flow considerations. More detailed tests to check the theory were made in the wind tunnel of the Institute for Gas Dynamics, Braunschweig, Germany [*36*]. In these experiments a model (Fig. A,4e) which completely spanned the tunnel and which could be set at angles of yaw up to 40° was used. Pressure measurements

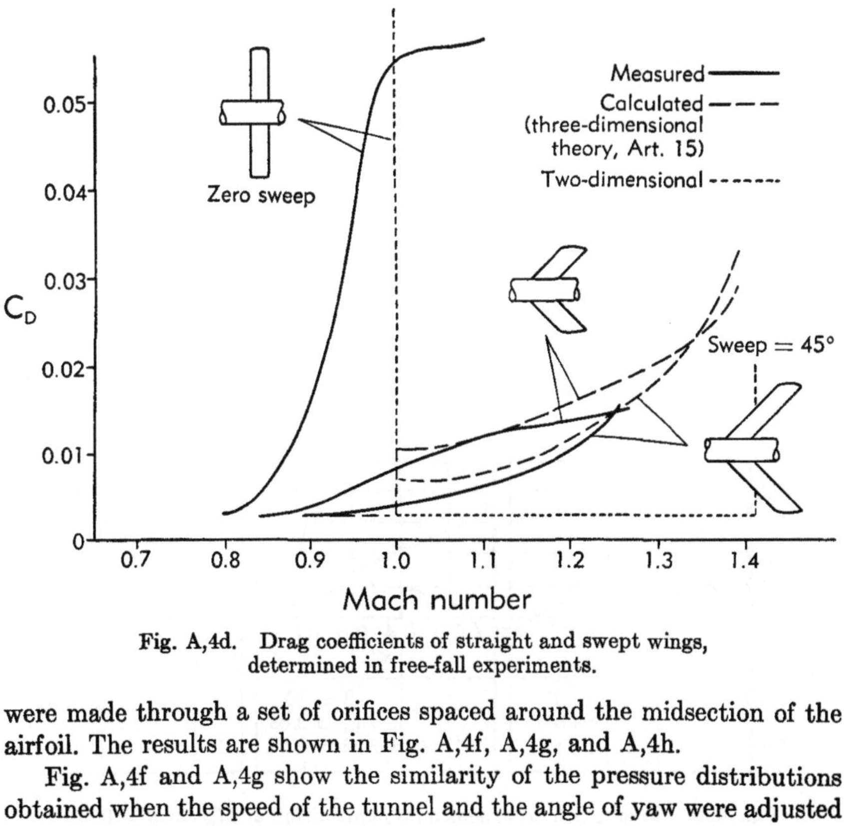

Fig. A,4d. Drag coefficients of straight and swept wings, determined in free-fall experiments.

were made through a set of orifices spaced around the midsection of the airfoil. The results are shown in Fig. A,4f, A,4g, and A,4h.

Fig. A,4f and A,4g show the similarity of the pressure distributions obtained when the speed of the tunnel and the angle of yaw were adjusted to maintain a constant value of the crosswise velocity component. The last two pressure distributions, taken at a crosswise velocity component corresponding to $M_\infty \cos \beta = 0.77$, indicate the existence of a shock wave on the upper surface at about 40 per cent of the chord from the leading edge.

In Fig. A,4h the variations of lift curve slope with Mach number for the three angles of yaw are compared with the variations given by linear theory. In these experiments the angle of attack α was measured as the angle of rotation of the airfoil about its axis, and the variation of the lift curve slope is therefore given by

$$\left(\frac{dC_L}{d\alpha}\right)_{M_\infty,\beta} = \left(\frac{dC_L}{d\alpha}\right)_{0,0} \frac{\cos^2 \beta}{\sqrt{1 - (M_\infty \cos \beta)^2}} \tag{4-9}$$

rather than by Eq. 4-8.

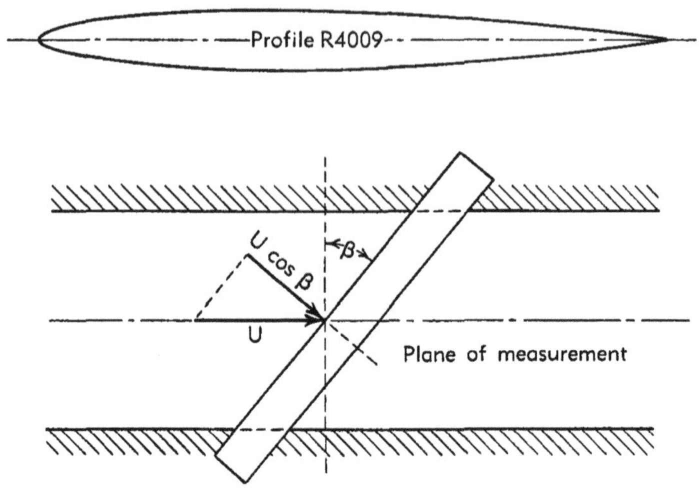

Fig. A,4e. Wind tunnel experiments with oblique airfoil [*36*].

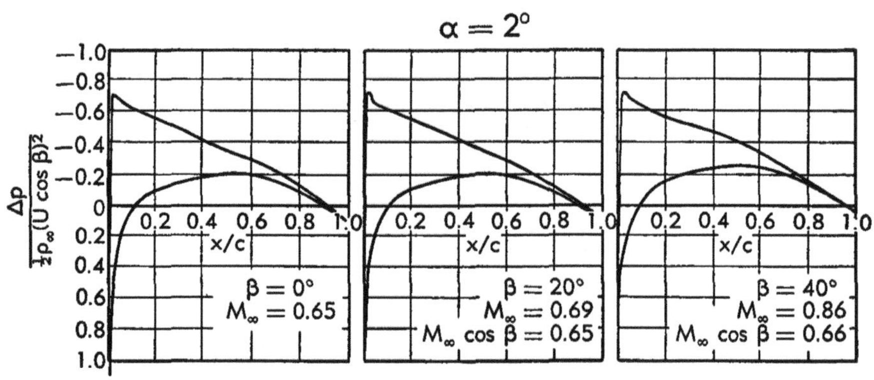

Fig. A,4f. Pressure distributions on oblique airfoil at $M_\infty \cos \beta = 0.65$.

Oblique Flows with Viscosity.

Infinite cylindrical wing. The foregoing discussion has been based on the assumption of frictionless flow. Consideration of the effects of viscosity in oblique flows shows, however, that the most important relations for the yawed wing remain unmodified. In particular it can be shown that so long as the linear law of viscous stress usually adopted in the Navier-Stokes equations [*1*, pp. 571–577] remains valid, the pattern of the cross flow over an infinite oblique cylinder remains independent of

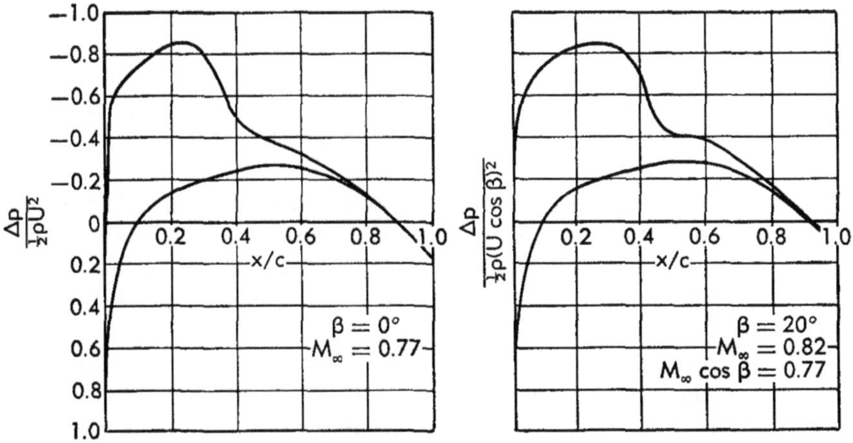

Fig. A,4g. Pressure distributions on oblique airfoil at $M_\infty \cos \beta = 0.77$.

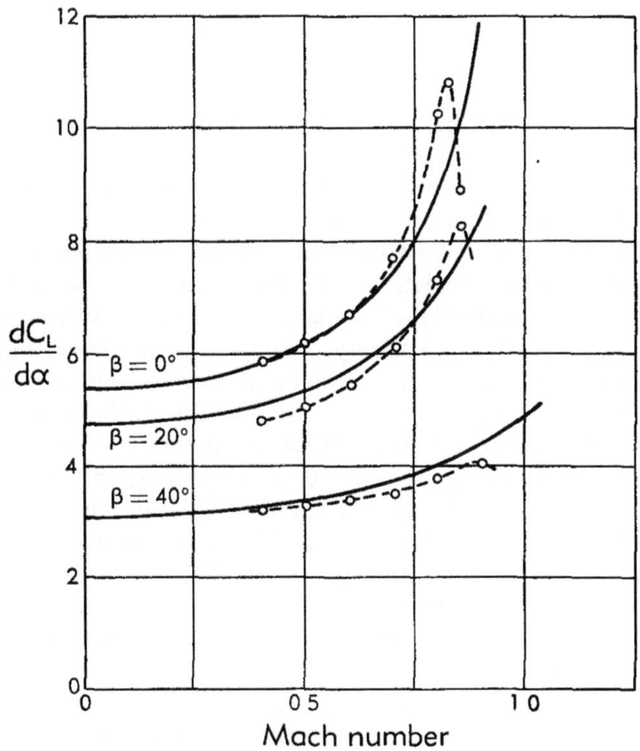

Fig. A,4h. Variation of lift curve slope with Mach number for oblique airfoils.

the axial motion of the cylinder [*37,38,39*]. This law of independence may be verified by writing the Navier-Stokes equations for a three-dimensional distribution of velocities and introducing the assumption that the form of the velocity field is independent of the axial position along the cylinder. The two equations for the cross flow then become independent of the third equation and are identical with the two-dimensional equations applicable to an unyawed cylinder traveling at a reduced velocity and reduced Reynolds number. The third equation remains dependent on the other two, however, and hence it is concluded that the cross flow is independent of the axial flow, but the axial flow is not independent of the cross flow.

The question of how far the principle of independence can be utilized depends on the range of validity of the linear stress relation. At very high supersonic speeds a shearing motion in one direction increases the temperature of the fluid and hence in effect modifies the coefficient of viscosity, so that the shearing stress in other directions is modified and a strict independence cannot hold. In the case of the turbulent boundary layer it seems that the nature of the internal stresses is not sufficiently well understood to establish the independence from first principles. The theory appears to be borne out, however, in the case of the flow over an oblique circular cylinder, which involves a large turbulent wake. Fig. A,4i shows the variation with yaw of the cross force developed by a circular wire. Since the test conditions entailed a range of crosswise Reynolds numbers in which the drag coefficient is known to be constant, the theoretical variation of the cross force coefficient in this case is almost exactly as $\cos^2 \beta$. The theoretical variation is shown in the figure by the solid line, while the experimental values from [*40*] are indicated by circles.

Finite aspect ratio. The theory of the action of viscosity in oblique flows leads to conclusions of considerable practical importance in the design of wings. First, it is to be noted that the thickness and velocity profile of the laminar boundary layer on an oblique wing are determined by the crosswise component of velocity alone. In the case of thin wings, experiment shows that the stalling angle, or angle of maximum lift, is determined by the phenomenon of laminar separation. If a straight wing and an oblique wing are now compared, it is seen that laminar separation can be expected at the same values of the lift coefficient and Reynolds number if these are based on the crosswise component of velocity. Thus, if a long narrow wing is placed at increasing angles of yaw β, the value of the lift at which flow separation takes place can be expected to diminish with $\cos^2 \beta$. With a sufficiently long wing the maximum lift can be expected to diminish in the same ratio. Such a wing thus behaves in nearly every respect as if it were flying at the reduced velocity $U \cos \beta$.

With sweptback wings of normal proportions it is observed experimentally that the relation between flow separation and maximum lift is not the same as it is with the straight wing. This difference can be understood if it is remembered that the similarity of flows has been demonstrated only in connection with the cross flows, or in other words with the flows observed by looking in the direction of the long axis of the wing. Separation occurs when a substantial volume of fluid adjacent to the upper surface of the wing is brought to rest or reversed in direction by the retarding action of the pressure field. Since the isobars are nearly parallel to the axis of the wing they can have little effect on the component of velocity parallel to the axis. Hence, at the point of flow separation on the

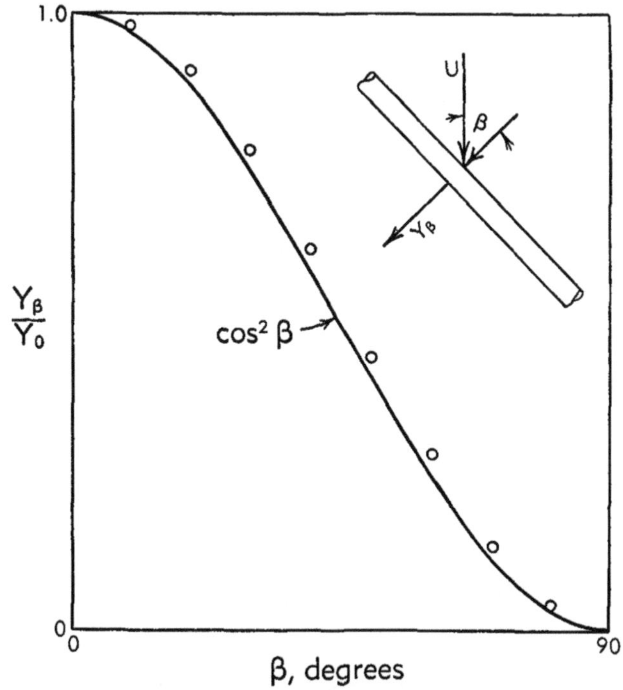

Fig. A,4i. Theoretical and experimental variations of the cross force on a circular wire.

oblique wing, the fluid merely loses the momentum carrying it toward the trailing edge—and flows in the direction parallel to the axis with the approximate velocity $U \sin \beta$ (see Fig. A,4j). It will be evident that once this endwise flow is developed on a wing of limited length the flow can no longer be approximated by that over an infinite wing. The separated air over the sections of the finite wing escapes at the downstream tip and hence does not remain to fill out an extended region or wake behind each section, such as appears behind the sections of a straight wing. The similarity of the flows can be expected to hold in practice for wings of normal proportions only so long as the boundary layer remains thin. After separation the similarity is lost.

As observed experimentally, the result is that flow separation on the straight wing is followed by a sharp drop in lift, whereas the lift of a wing with yaw or sweep of as much as 45° continues to rise with increasing angles of attack. With more extreme sweepback and especially with wings of low aspect ratio, the lift has been observed to increase up to angles of attack of 45°.

Fig. A,4k, taken from experiments made in the 40 by 80 ft wind tunnel of the Ames Aeronautical Laboratory, NACA, shows the results of

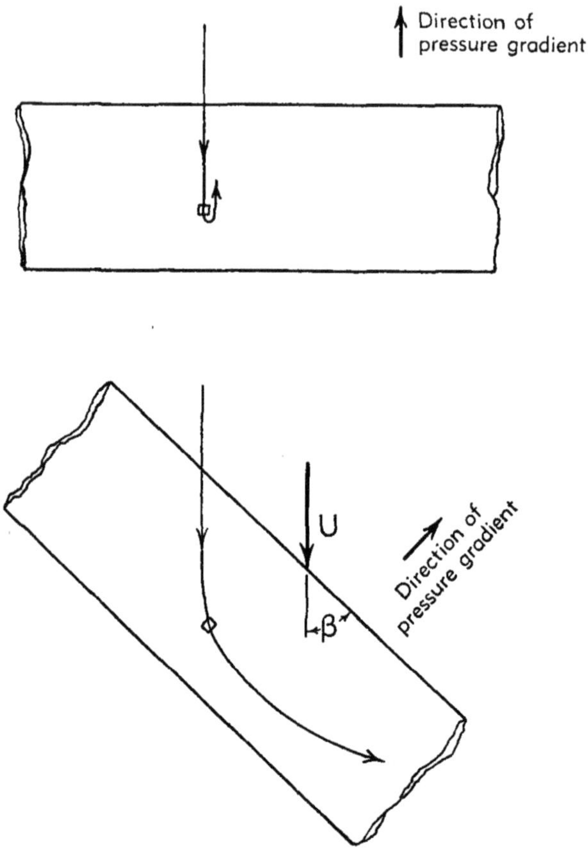

Fig. A,4j. Effect of yaw on the path of a particle in a separated boundary layer.

force measurements made on a model having approximately 60° of sweep. The tests were made at low speed, but the Reynolds number was of the order of 8 million. It will be noted that the drag curve follows the law of variation indicated by induced drag theory, viz.

$$C_D = C_{D_{\min}} + \frac{C_L^2}{\pi \mathcal{R}} \tag{4-10}$$

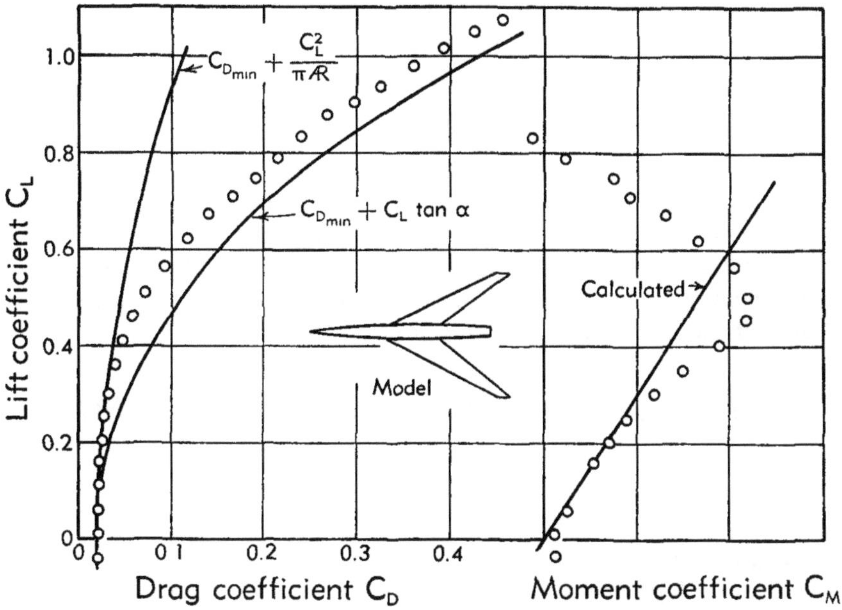

Fig. A,4k. Force measurements indicating flow separation on model with large angle of sweep.

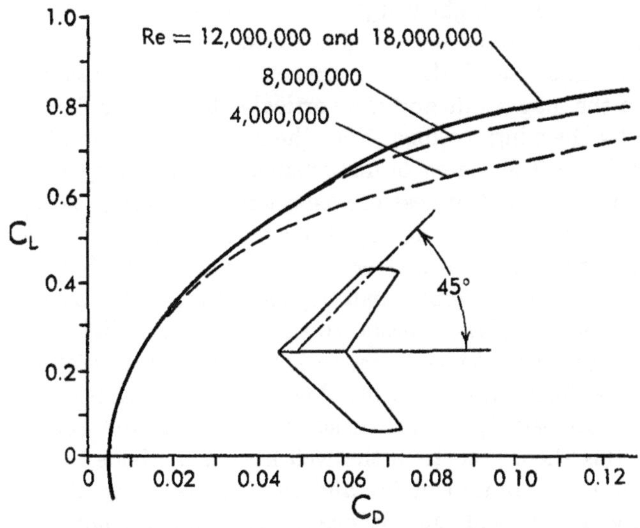

Fig. A,4l. Effect of Reynolds number on drag of swept wing, from [*41*].

up to a lift coefficient of approximately 0.3. Beyond this point the drag curve bends over and begins to follow the curve defined by

$$C_D = C_{D_{min}} + C_L \tan \alpha \tag{4-11}$$

which is the variation expected when the resultant force is at right angles to the chord plane of the model and when no leading edge suction force is developed. Evidently flow separation occurs at about $C_L = 0.3$. This value corresponds to a lift coefficient of 1.2 with respect to the normal component of the stream velocity and hence is about equal to the normal stalling lift coefficient in straight flow. The effect of Reynolds number in delaying the onset of separation is shown in Fig. A,4l.

Further evidence of a change in type of flow is given by the moment curve shown on the right of Fig. A,4k. Although the lift continues to rise after flow separation occurs, there is some doubt that the higher lift coefficients can actually be used in practice because of the large increase in drag and the undesirable shift in center of pressure.

CHAPTER 2. AERODYNAMICS OF THIN WINGS AT SUBSONIC SPEEDS

A,5. Extension of Thin Airfoil Theory to Three Dimensions.

Necessity for more exact treatment of three-dimensional flow. The transition in high speed design to wings of sweptback planform and, in certain cases to wings of low aspect ratio, makes the necessity for three-dimensional treatment of high speed wing problems obvious. However, even in the case of straight wings, the approximation of two-dimensional flow becomes inadequate when higher speeds are considered. As explained in Art. 3, an increase of Mach number in the subsonic range leads to an increase in the magnitude and the vertical extent of the flow disturbance produced by the wing sections. Near the critical Mach number the vertical dimension of the region of disturbance becomes large relative to the chord of the wing, and, unless the aspect ratio is very great, the scale of magnitude of the section flows will become appreciable in terms of the wing span. As will be shown later, in the vicinity of sonic speed the flow over a lifting surface of finite span begins to take on the aspect of a two-dimensional flow in planes *normal* to the direction of flight.

Role of small disturbance theory in practical wing calculations. At the present time, calculations of wing flows in three dimensions are carried out almost exclusively on the basis of the linearized theory. Certain results of this theory, such as the well-known methods for determining the spanwise distribution of lift and the induced drag of wings at low speeds, have gained wide acceptance in aeronautical engineering.

The small disturbance theory represents mathematically the limiting case of infinitesmal thickness and angle of attack. The question of its applicability in practice depends on the magnitudes of these quantities as determined by practical considerations. At low speeds, increases of thickness or angle of attack are not necessarily accompanied by a pressure drag so long as flow separation is avoided. Hence in the design of airplanes for low speeds the thickness ratios are determined, broadly speaking, by the necessity for avoiding flow separation. Usually it is found that the resulting shapes are too thick to allow determination of the pressure distribution in detail by the small disturbance theory, although the lift curve slope, the center of pressure, and the induced drag are ordinarily given with sufficient accuracy by the linear relations.

At high subsonic speeds and at supersonic speeds, on the other hand, there arises a pressure drag or a wave drag which increases rapidly with the thickness ratio. For efficient flight at these speeds, therefore, it appears necessary to restrict the thickness ratio, or the slopes of the surfaces, to low values. In practice this will usually mean extending the dimensions of the wing or body in the flight direction. Of course this process leads to more unfavorable ratios of exposed surface area to span, or of surface area to volume, and hence increases the magnitude of the skin friction. Rough calculations can be made, however, which show that, in order to obtain reasonably small values of the total drag, extremely slender shapes must be used.

As an illustration of the change in proportions brought about by transition to higher speeds it is instructive to consider the case of a body of revolution designed to contain a given volume with minimum total drag. As is well known, at low speeds the minimum drag with friction occurs at a fineness ratio of about 3 to 1. At supersonic speeds, however, conservative estimates show that for the total drag to be a minimum the fineness ratio must be in excess of 20 to 1. As indicated in Art. 3 and in Fig. A,3c and A,3d, similar proportions must apply to the wing sections as well. Hence it seems reasonable to suppose that the small disturbance theory will provide useful engineering approximations for such bodies and wings as are actually well adapted to flight at high speeds.

Extension of the Prandtl rule to three-dimensional flows. The linearized treatment of wing flows in three dimensions makes use of (1) the approximate boundary condition of the thin airfoil theory and (2) the Prandtl transformation, which reduces the subsonic compressible flow to an equivalent incompressible flow. According to the Prandtl transformation, a distribution of φ's satisfying the differential equation of the compressible flow (3-8) can be made to satisfy Laplace's equation by merely lengthening or stretching the distribution along x by the factor $1/\sqrt{1 - M_\infty^2}$. It was seen that in two dimensions the flow over a thin airfoil section of unit chord length could be made to correspond to the incompressible flow over a geometrically similar section with its chord length increased by $1/\sqrt{1 - M_\infty^2}$. In this relation the spacing of the φ's in the vertical direction is not changed, so that $\partial\varphi/\partial z$ and hence the slopes of the streamlines at corresponding points remain unchanged. The crowding of the φ's in the x direction, however, leads to an increase in the perturbation velocity $u = \partial\varphi/\partial x$.

These same considerations are directly applicable to the three-dimensional wing flow.[5] Although the airfoil sections of the corresponding wings remain geometrically similar, the aspect ratios of the wings differ because of the change in chord length. The compressible flow over a wing of aspect ratio $\mathcal{R}$ at the Mach number M_∞ is related to the incompressible flow over a wing of aspect ratio $\mathcal{R}\sqrt{1 - M_\infty^2}$. This relation between the wing shapes is illustrated by Fig. A,5. A study of the changes in the field of velocities and pressures around a given wing with increasing Mach number is thus reduced to the investigation of the incompressible flow around each of a series of wings of progressively reduced aspect ratio. Corresponding points in the compressible and incompressible flows are points of equal values of φ. These points occur at equal percentages of the chord lengths in the x-wise direction and at the same absolute values of y and z. The values of u, v, and w calculated for points in the field surrounding the extended wing are transferred to the corresponding points in the field

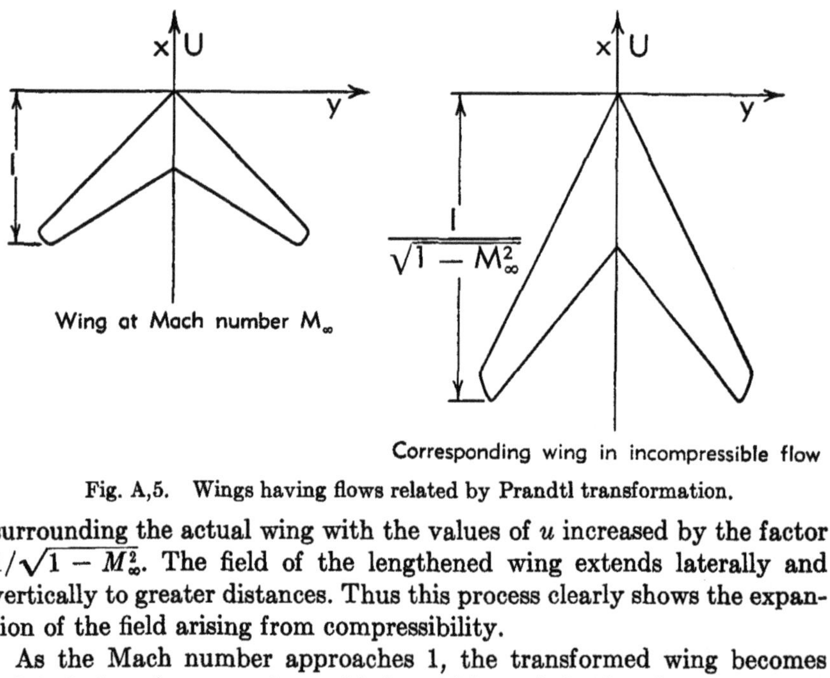

Fig. A,5. Wings having flows related by Prandtl transformation.

surrounding the actual wing with the values of u increased by the factor $1/\sqrt{1 - M_\infty^2}$. The field of the lengthened wing extends laterally and vertically to greater distances. Thus this process clearly shows the expansion of the field arising from compressibility.

As the Mach number approaches 1, the transformed wing becomes infinitely long in comparison with its width, and the flow becomes essentially two-dimensional in planes at right angles to the direction of flight. This consideration has lead to the adoption of the so-called "slender wing theory" (Art. 8) as a theory for sonic speeds.

Linearization of the boundary conditions in three-dimensional flows. As in the case of two-dimensional motion, the expression of the boundary

[5] For the treatment of more general bodies in three-dimensional flow, however, the reader should consult Göthert's paper [*42*].

condition in the form

$$\frac{dz}{dx} = \left(\frac{w}{U}\right)_{z\to 0}$$

results in a linear relation between the perturbation velocity field and the ordinates of the wing shape, so that again any wing shape may be divided into components in an arbitrary manner and its velocity flow fields may be determined as the sum of those arising from the individual components of the shape. Since problems associated with the camber and angle of attack of the mean surface present a somewhat different mathematical aspect from those associated with the thickness distribution, it is convenient to divide the wing into a thin inclined cambered or twisted "lifting surface" and a distribution of thickness which will be symmetrical above and below the plane $z = 0$.

As in the two-dimensional case, the distribution of thickness is characterized by equal and opposite values of the vertical velocity w on the upper and lower sides of the chord plane within the boundary of the wing planform, and can be represented by the action of a distribution of sources and sinks over the chord plane. Because of symmetry, the value of w is zero at all points of the plane $z = 0$ beyond the edges of the planform. The distribution of the velocities u and w extends continuously throughout the field. The thickness distribution alone gives rise to no lift and, in steady motion at subsonic speeds, to no drag other than that arising from the skin friction.

The flow associated with the lifting surface in three dimensions shows a discontinuity in v as well as in u across the chord plane. In steady motion the lifting pressure, and hence the discontinuity in u, vanishes outside the wing planform. The whole field of motion around the lifting surface can be obtained from the action of a suitable distribution of vortex filaments over the chord plane, and these filaments will extend behind the wing, forming the trailing vortex sheet. Since the trailing vortices lie parallel to the direction of flight, they represent a discontinuity in the lateral velocity v. The downwash velocity w remains continuous throughout the field.

It is important to note that the approximate boundary condition of the thin airfoil theory is limited, in three dimensions, to bodies that are thin and flat in all vertical cross sections. In such cases the flow, if continued to the interior of the body, will show a discontinuity of the velocities over an approximately plane mean surface which extends nearly to the edges of the planform. In the vicinity of such a planar distribution, the velocities and the slopes of the streamlines do not change rapidly with distance except possibly near the edges, and the condition of tangential flow can be applied without specification of the exact position of the solid boundary. Such a linearized boundary condition cannot be applied to other forms of slender bodies, however, even though the linearized flow equation may remain valid. Thus in the case of a body of revolution it is found that the streamlines of the internal flow emanate from a line of singularities along the axis of the body with an infinite radial velocity. The slopes of the streamlines diminish rapidly with the radial distance from the line, becoming equal to the slope of the surface only at the radial position of the surface. The range of physical validity of the flow equation is of course generally limited to the external flow field.

A,6. Lifting Surfaces.

General Properties. Before proceeding to a discussion of specific wing forms it is desirable to consider certain results of the theory of lifting surfaces which are of a general nature. It is found that the linearized frictionless flow theory provides at once certain theorems relating to the lift and drag of thin wings. These theorems are independent of the particular planform of the wing and in certain cases they are also independent of the state of motion, i.e. the motion may be steady or unsteady and the speed may be subsonic or supersonic.

Drag of a given distribution of lift. In general it is found that the lift of a wing is given quite accurately by considerations of frictionless flow. The drag, on the other hand, depends substantially on both the friction stresses and the distribution of normal pressures over the wing. However, for the determination of that part of the drag arising from the normal pressures, the frictionless flow theory may be used and an estimate of the total drag may be formed by adding an appropriate skin friction coefficient.

For the derivation of the theorems it is necessary to examine the process of computing the drag in the linearized theory. First consider a lifting surface having a planform such as shown in Fig. A,6a and a specified distribution of lift. The given distribution of lift will induce a certain distribution of downwash $w(x, y)$ over the plane of the wing, which results in a slope of the stream surface $dz/dx = w/U$. The form taken by the stream surface inside the boundary of the wing is identified with the form of the camber and twist of the surface required to produce the given distribution of lift. Since the drag with which we are concerned arises only from the normal pressures, it can be computed by multiplying the lift of each point of the planform by the rearward inclination of the surface at that point and integrating the resultant product over the entire wing[6] (see Fig. A,6a).

Computation of the drag of a given distribution of lift thus depends first on the computation of the downwash at each point. In a linearized

[6] If the given distribution of lift shows infinite values around the leading edge, the calculation must be performed over a stream surface slightly above the chord plane and the integration must extend a small distance beyond the wing boundary, as explained in Art. 2.

theory this may be accomplished by dividing the wing up into elementary areas and considering the field of downwash induced by each lifting element. The pattern of the downwash field surrounding an element of lift will be determined by an elementary solution of the flow equation, which may be elliptic or hyperbolic. In any event the downwash pattern will be the same for every element except for a constant factor of strength proportional to the lift of the element.

Determination of the downwash at any given point of the planform requires an integration of the effect at this point of all lifting elements of

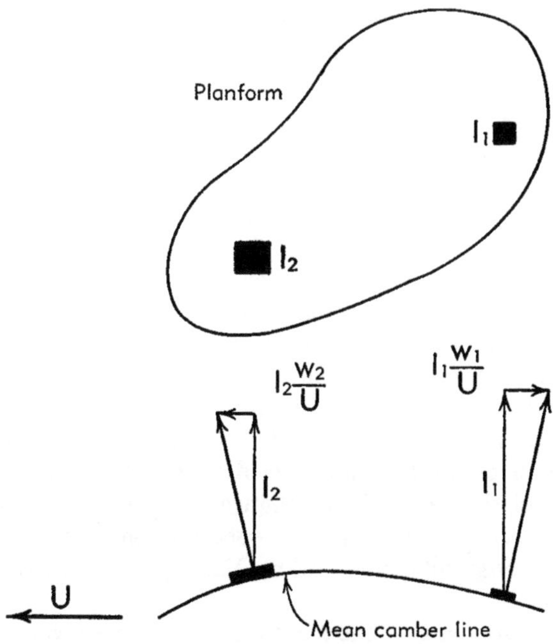

Fig. A,6a. Drag arising from inclination of lifting surface.

the wing. One integration over the surface is required to yield the drag of one element of lift, and a second surface integration will be required to yield the total drag.

EQUALITY OF DRAG IN REVERSED MOTION. A consideration of the elements of this double integral leads to the conclusion that *the drag of a given distribution of lift is unchanged by a reversal of the direction of motion.* Each element of the drag involves the product of the lift of one element l_1 and the downwash induced at its position by another element l_2. This downwash may be written

$$w_{12} = W_{12} l_2$$

where W_{12} denotes the downwash at position 1 due to a unit lift at position 2. The element of drag is then

$$l_1 W_{12} l_2$$

Now, if the direction of motion is reversed, element 2 will lie in the same relation to element 1 as was formerly occupied by l_1 relative to l_2. In other words the influence function W_{21} in reversed motion will be equal to W_{12} in forward motion. The drag of l_1 in the downwash of l_2 in forward motion is then equal to the drag of l_2 in the downwash of l_1 with the motion reversed. In symbols,

$$l_1 W_{12} l_2 = l_2 W_{21} l_1$$

For every element of drag there is thus a corresponding equal element for motion in the reverse direction. Since the correspondence includes all elements of the double integrals, the total drags are equal (see Fig. A,6b).

The invariance of drag evidently applies under very general conditions. The elements of lift need not lie in the same, nor in parallel, planes. At supersonic speeds or in unsteady motion at subsonic speeds, the zones of interaction between elements may be limited in extent. Reversal of an unsteady motion implies reversal of the velocity vector $\mathbf{q}(t)$ at each corresponding instant. For the case of steady motion at subsonic speeds the equality of forward and reverse drags may be shown by Munk's stagger theorem [*7*]. Von Kármán [*43*] and Hayes [*44*] have demonstrated the theorem for the case of steady supersonic motion.

Distribution of Lift for Minimum Drag.

Mutual drag in combined flow field. Certain necessary conditions on the distribution of lift for minimum drag can be derived through the consideration of the fields of disturbance velocities produced by a given distribution of lift in forward and reversed motion. The derivation makes use of the idea, originally advanced by Munk [*45*], of superimposing these velocity fields.

Let us first suppose that the planform of a wing is given, together with the distribution of lift. If the same distribution of lift is considered in reversed flow, it will in general be found that the downwash distribution is different from the original one, and that the wing must therefore be given a different distribution of camber and twist to maintain the same distribution of lift. In spite of this difference in shape, however, the total drag must remain unchanged by the reversal. Since the drag is obtained by integrating the product of the lift and the downwash over the planform, we have

$$D = \frac{1}{U}\iint l(x, y)w_f(x, y)dxdy = \frac{1}{U}\iint_S l(x, y)w_r(x, y)dxdy \quad (6\text{-}1)$$

where w_f and w_r are the downwash distributions in forward and reversed flow respectively.

The drag may be computed by considering the given distribution of lift to be placed in a "combined flow field," obtained by superimposing the perturbation velocities for forward and reversed motion. It is evident that for the same distribution of lift to appear in each field the horizontal perturbation velocities u must be exactly equal and opposite at each point

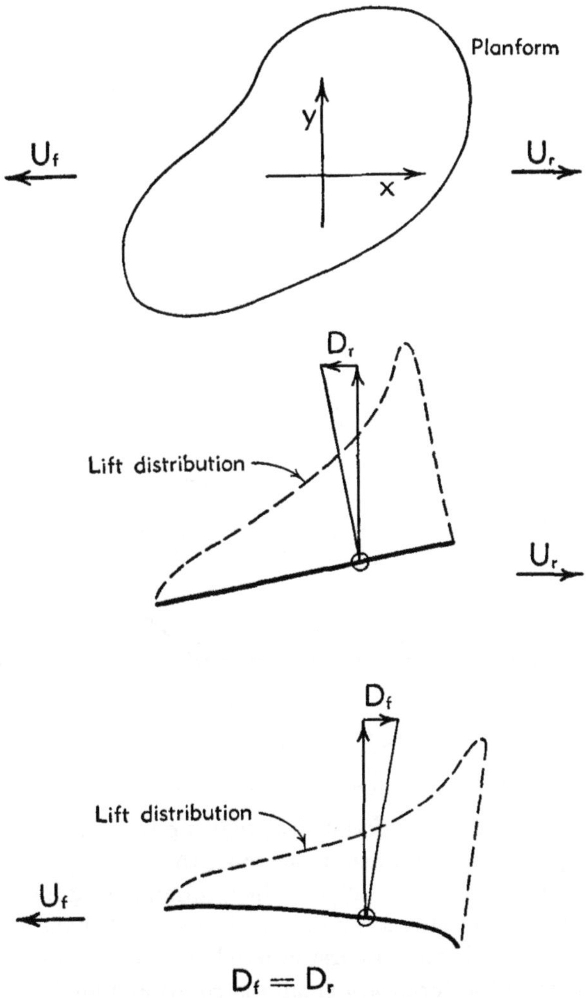

Fig. A,6b. Drag of a given distribution of lift in forward and reversed motion.

of the wing surface. Hence the discontinuity in u vanishes in the combined field. A continuous distribution of downwash remains, however. If its value is denoted by

$$\bar{w} = w_f + w_r$$

the drag will be given by

$$D = \frac{1}{2U} \iint l(x, y)\bar{w}(x, y)dxdy \tag{6-2}$$

The quantity $\overline{w}/2$ may be interpreted as the mean value of the downwash for the two directions of motion and reduces to the induced downwash of the Prandtl wing theory in the case of steady motion at subsonic speeds.

It may now be shown that *the mutual interference drags of two distributions of lift in the combined flow field are equal.* Consider two elements of lift such as l_1 and l_2 in Fig. A,6c. The drag of l_1 caused by the downwash of l_2 in the combined flow may be written

$$l_1\overline{W}_{12}l_2$$

where the symbol $\overline{W}$ denotes the combined downwash arising from a unit element of lift. Similarly, the drag of l_2 caused by the interference of l_1 is given by

$$l_2\overline{W}_{21}l_1$$

From the symmetry of the combined flow field for each element it is evident that

$$\overline{W}_{12} = \overline{W}_{21}$$

and hence the interference drags are equal. This equality of mutual drags holds for groups of elements and hence for any two distributions of lift.

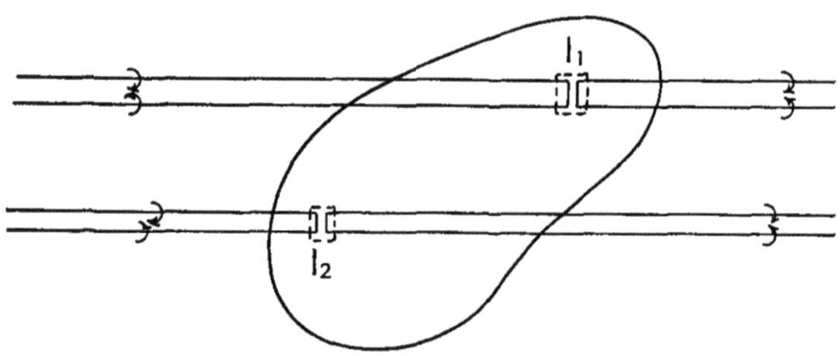

Fig. A,6c. Mutual interference of elements in combined flow field.

Distribution of lift for minimum drag with a given planform and a given total lift. Suppose the planform of the wing to be given, together with a distribution of lift designed to achieve the minimum drag for a given total lift L. If the drag is actually a minimum, then a small change in the shape of the distribution will cause no first order change in the drag. The change in shape may take the form of a small distribution added to the original lift; the condition of a fixed total lift will be met by admitting only those variations for which the total lift is zero. Such restricted variations may be divided into pairs of equal, opposite elements of lift, of which one pair may be taken as representative. Following the law of superposition, the drag added by such a pair will be composed of three parts:

1. the drag of the added elements acting alone
2. the interference drag of the added elements caused by the downwash of the original distribution

3. the interference drag of the original distribution caused by the downwash of the added elements.

The first item will be of second order in terms of the small magnitude of the added distribution. (The fact that this second order term is always positive insures that a stationary value of the drag will be a minimum and not a maximum.) According to the mutual drag theorem, the second and third items will be equal in the combined flow field. The first order variation in drag is then equal to twice the drag of the added lift in the downwash of the original distribution. For a pair of equal and opposite lifting elements the drag will be zero if the downwash is the same at each element. Furthermore, the drag added by every such pair will be zero if the downwash has the same constant value at each point of the planform.

Hence if a distribution of lift is to have the minimum drag consistent with a given total lift the "combined" downwash $\overline{w}$ *must be constant over the entire planform* (see Fig. A,6d).

Planform for minimum drag. The foregoing condition for minimum drag may be generalized to include variations in planform of the wing, provided the meaning of the term "additional lift distribution" is broadened to cover small changes in the outline shape of the planform. It will be apparent from the foregoing discussion that the drag for a specified total lift will be a minimum with respect to extensions of the wing outline in a given direction if the downwash $\overline{w}$ is not only constant over the planform but maintains this value without first order change for a small distance beyond the border of the planform. If the distribution of lift over the planform is plotted as a surface above the chord plane, there will result a three-dimensional figure whose volume is the total lift. *Then for the drag to have a stationary value relative to small changes in shape of this figure,* $\overline{w}$ *must be a constant over the planform and if* s *denotes a distance in the* x, y *plane measured from the edge of the planform* $d\overline{w}/ds$ *must equal zero.*

Such general variations in shape of the lift distribution do not ordinarily lead to the determination of a definite minimum of drag since at subsonic speeds, with the total lift fixed, the drag decreases monotonically with increases of span, while at supersonic speeds it decreases indefinitely with increases of both span and chord. The foregoing conditions are useful, however, in determining the minimum drag consistent with certain limitations on the planform, as later examples will show.

Relation to theory of induced drag. For the case of steady motion at subsonic speeds the foregoing theory agrees with the classical theory [*7,8*] of induced drag. In steady motion no wave drag arises and the whole pressure drag can be associated with the downwash produced by the trailing vortex wake. In the superposition of the forward and reversed flow fields for a fixed distribution of lift, wakes having identical distributions of vorticity will extend both ahead of and behind the wing. Since the

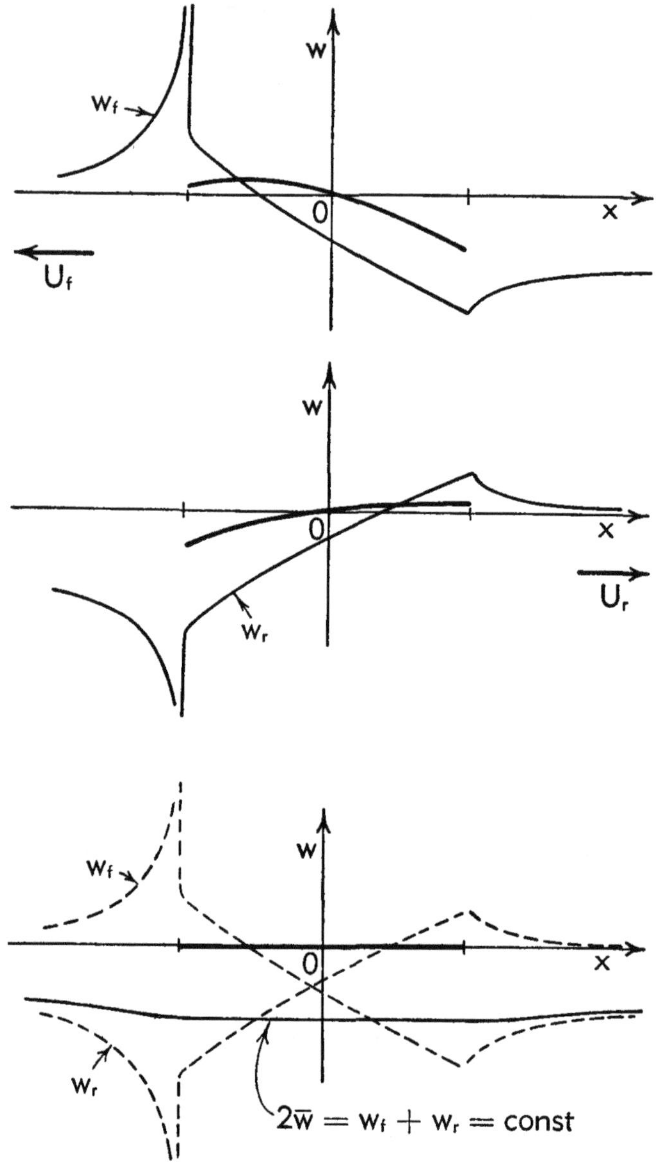

Fig. A,6d. Condition for minimum drag with a given total lift.

longitudinal component u of the perturbation velocity is cancelled at each point within the planform boundary, the vortices in the superimposed flow will continue straight across the wing in a direction parallel to the stream velocity (see Fig. A,6c). In the subsonic case the superimposed flows form a two-dimensional field of motion associated with a vortex ribbon or wake of width equal to the span and extending from $+\infty$ to

$-\infty$ in the stream direction. The distribution of vorticity in the combined wake is identical with the distribution in each of the single semi-infinite wakes and bears the same relation to the spanwise distribution of lift as given by the classical wing theory. The minimum drag for a given planform and a given total lift is achieved when the spanwise distribution of lift is elliptical—since in this case the downwash $\bar{w}$ induced by the combined wakes is constant—and is given by the well-known formula

$$D = \frac{L^2}{\pi \frac{1}{2}\rho_\infty U^2 b^2}$$

where b is the span of the lifting surface. Inasmuch as the constant value of $\bar{w}$ extends not only over the wing planform, but also over the whole infinite vortex ribbon ahead of and behind the wing, the drag remains unchanged for extensions of the planform in the chordwise direction.

According to Munk's stagger theorem, the induced drag is unaffected by the form of the distribution of lift in the chordwise direction, even though the spanwise distribution does not correspond to the minimum drag. The present considerations can be seen to yield this result by noting that in the combined flow field, which is a two-dimensional motion, $\bar{w}$ remains independent of x even though a spanwise variation is admitted. Variations which amount to a redistribution of lift along any line in the flight direction thus cannot change the drag since the downwash $\bar{w}$ has the same value at all points along such a line. (This stationary character of the drag is, of course, not that of a minimum value since even finite changes of chordwise lift distribution produce no change of drag.)

Effect of Compressibility on Drag at Subsonic Speeds. An important result of the linearized wing theory is the prediction that *the drag associated with a given distribution of lift is unchanged by changes of the Mach number in the subsonic range.* This result can be verified by applying the Prandtl transformation to the combined flow field. In this field the potential φ has the same value at all points of any line parallel to the x axis. A change in the scale of x thus produces no change in the gradient of the potential and hence does not alter the perturbation velocities. The drag which, it will be remembered, results from an integration of the product of the given lift and the downwash of the combined field at each point is thus also unchanged.

If the geometry and angle of attack of the wing are given rather than the lift distribution, then the spanwise distribution of lift and the induced drag will be affected by the Mach number, though the change may be slight. According to the Prandtl rule (Art. 5), the flow pattern at increased Mach numbers can be related geometrically to the incompressible flow around a derived wing of increased chordwise dimensions or reduced aspect ratio. The computation of induced drag at varying Mach numbers thus involves first of all the computation of the varying lift distribution.

In many cases the lift distribution is nearly elliptical and, as will be shown later, tends to become more so at higher subsonic Mach numbers so that the variation of drag with a given total lift is slight.

REVERSAL THEOREM FOR LIFT CURVE SLOPE. The combination flow, obtained by superimposing the disturbances arising in forward and reversed motion, has been used by Brown [*46*] to show that the lift curve slope of a wing in steady motion is unchanged by a reversal of the planform. Here it is sufficient to show that the lift of a flat surface at an angle of attack α is unchanged (except for sign) by a reversal of the flow direction, since, in the linear theory, the lift curve slope is unaffected by a small initial camber or thickness. The distribution of lift over the planform will, in general, be different for the two directions of motion. The flow in forward motion must satisfy the boundary condition $\overrightarrow{w} = U\alpha$ at every point of the planform. With the wing fixed and the flow reversed, the boundary condition requires that the downwash at each point be equal and opposite to that in forward flow; i.e. $\overleftarrow{w} = -\overrightarrow{w}$. Superimposing the two flows then results in a complete cancellation of downwash at each point of the planform, and the combined flow field thus satisfies the boundary condition for a flat plate at zero angle of attack.

The proof of the reversibility of the lift requires consideration of the mechanism of the drag, although the drag itself does not in general remain unchanged by the reversal of the planform. The drag in forward motion will be given by

$$\overrightarrow{D} = \overrightarrow{L}\alpha - \overrightarrow{T}$$

where the quantity $\overrightarrow{T}$ is introduced to take into account the leading edge thrust and is proportional to the strength of the singularity in the velocity around the leading edge. The drag of the airfoil can also be obtained by a different consideration, which involves a characteristic feature of the three-dimensional wing flow. In such flows the disturbance vanishes at a great distance ahead of the wing; on the downstream side, however, the disturbance remains, in the form of the wake, undiminished by distance. At supersonic speeds a downstream disturbance appears also in the Mach waves emanating from the wing. In any event the drag appears as the rate of increase of momentum involved in the continual extension of this wake at the flight velocity and can be obtained from the configuration of the flow at a great distance behind the wing. From this reasoning it will be apparent that when forward and reversed flows are superimposed, the two wakes, which now lie on opposite sides of the wing, will not interfere, and the combination flow at a great distance from the wing in either direction will be identical with either one or the other of the single flows. If the uniform stream velocity U is now added to the combined flow and the change in momentum introduced at the wing is considered, it can be seen that the drag arising from the wake extending forward is simply the negative of the drag for reversed flow, while the drag of the wing in forward flow appears in the rear wake with its normal sign. The drag in the combination flow is thus the difference of the two drags; i.e.

$$\overline{D} = \overrightarrow{D} - \overleftarrow{D}$$

Now a second independent determination of the drag can be obtained from a consideration of the inclination of the pressure forces over the surface of the wing itself. This method of calculation leads to the expressions

$$\overrightarrow{D} = \overrightarrow{L}\alpha - \overrightarrow{T}$$

$$\overleftarrow{D} = \overleftarrow{L}\alpha - \overleftarrow{T}$$

where the terms $\overrightarrow{T}$ and $\overleftarrow{T}$ are again introduced to include leading edge thrust forces. Inasmuch as the combined flow corresponds to a flat horizontal wing surface, the only drag arising in this flow must be the resultant of the suction forces around the edge of the wing or, in other words,

$$\overline{D} = \overleftarrow{T} - \overrightarrow{T}$$

provided the suction forces T are not changed in value by the superposition. The latter condition is found to be satisfied if the Kutta condition is imposed on each flow, for then the infinite edge velocities of the two flows involve completely separated regions of the planform; the trailing edge for one direction of motion becomes the leading edge for the other direction and the tangential trailing edge velocities can be added to the higher order leading edge velocities without affecting the suction force. The theorem can now be proved by writing

$$\overrightarrow{D} - \overleftarrow{D} = (\overrightarrow{L}\alpha - \overrightarrow{T}) - (\overleftarrow{L}\alpha - \overleftarrow{T}) = \overleftarrow{T} - \overrightarrow{T}$$

and hence

$$\overrightarrow{L} - \overleftarrow{L} = 0$$

The reversibility of the lift curve slope applies to steady motion at subsonic or supersonic speeds without restriction on the planform. It obviously applies also to airfoils in combination, such as wing and tail configurations.

Reciprocal Relations in Thin Airfoil Theory. In spite of the variety of form in three-dimensional wing flows, there thus exist certain remarkable and simple relations between the disturbance fields in forward and reversed motion. Consider, for example, the application of Brown's reverse flow theorem to the lift of a narrow triangular wing at supersonic speed. With the triangle moving point foremost, the flow in the vicinity of the wing is similar (Art. 8) to a solution of Laplace's equation in two dimensions, expanding smoothly in the downstream direction in proportion to the width of the triangle. With the triangle moving base foremost, the pressure distribution over the surface and the flow pattern in the vicinity of the wing have a decidedly different appearance. The flow is deflected at the leading edge by discontinuous wave fronts and the pressure distribution over the surface following is characterized by a crossed array of Mach lines along which ridges or discontinuous slopes in the pressure distribution appear. Nevertheless, according to Brown's theorem the total lift must be the same in the two cases.

A more general reverse flow theorem which includes the lift theorem as well as a number of other interesting relations is the one due to Ursell and Ward [*47*]. This theorem has recently been generalized by Heaslet and Spreiter [*48*] and by Flax [*49*], who include time-dependent as well as steady motions. In [*48*], the theorem is related to the classical reciprocal theorems in dynamics associated with the names of Maxwell, Helmholtz, and Rayleigh.

In the form given by Ursell and Ward, the theorem applies specifically to steady motion and may be stated in the following way: Consider a thin wing of planform S and let the slope of the surface, which is given arbitrarily, be α_1. These specifications together with the assumptions of the thin airfoil theory then determine a corresponding distribution of pressure p_1. Both α_1 and p_1 may vary from point to point over the surface and are not necessarily the same on the upper and lower sides of the wing. Now let the motion of the planform be reversed and consider a second independent distribution of slope α_2 together with its corresponding distribution of pressure p_2. The theorem is:

$$\int_S \alpha_1 p_2 dS = \int_S \alpha_2 p_1 dS \tag{6-3}$$

where the integration extends over both upper and lower sides of the wing. In order to make the relations connecting α_1 to p_1 and α_2 to p_2 definite, Ursell and Ward assumed that the Kutta condition was applied.

It is instructive to derive the relation (6-3) by the method previously used to demonstrate Brown's reversal theorem for the lift.[7] Supposing the range of integration to be limited strictly to the surface S, so that terms contributing to the leading edge thrust do not arise in the integration, we have for the drags in forward and in reversed motion,

$$D_1 = \int_S \alpha_1 p_1 dS + T_1 \tag{6-4}$$

$$D_2 = \int_S \alpha_2 p_2 dS + T_2 \tag{6-5}$$

[7] This general application of the method is contained in an unpublished work of W. D. Hayes.

where T_1 and T_2 denote the leading edge forces. Following Munk [45] we now superimpose the two disturbance fields and consider the combined field to be placed in a stream of velocity U. The angle of deflection in the resulting flow over the surface S will be $\alpha_1 + \alpha_2$, while the pressure will be $p_1 - p_2$. We then have for the drag in the combined disturbance field

$$D = \int_S (\alpha_1 + \alpha_2)(p_1 - p_2)dS + T_1 - T_2 \tag{6-6}$$

Eq. 6-6 is based on the assumption that the edge forces T_1 and T_2 are additive, and this will be the case if the Kutta condition is applied, since the portions of the edge on which T_1 and T_2 may arise are then completely separated in the forward and reversed flow.

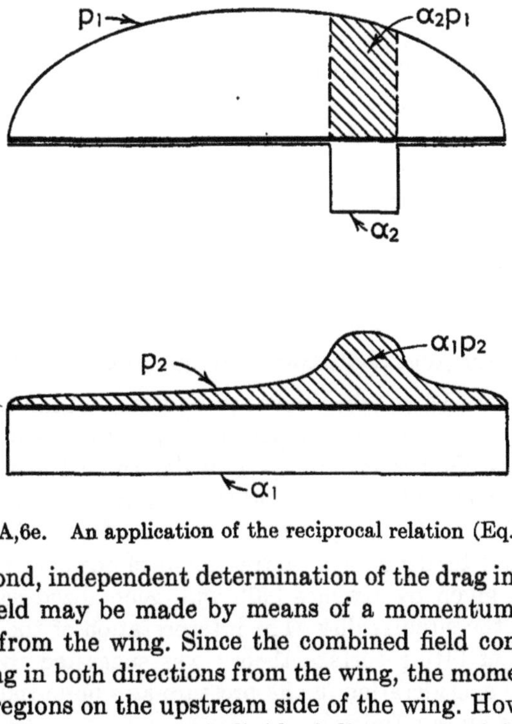

Fig. A,6e. An application of the reciprocal relation (Eq. 6-3).

Now a second, independent determination of the drag in the combined disturbance field may be made by means of a momentum survey taken at a distance from the wing. Since the combined field contains disturbances extending in both directions from the wing, the momentum integral must include regions on the upstream side of the wing. However, because of the one-sided nature of each individual disturbance field, they do not interfere, and the superposition of the two fields produces no change in either of the components at a sufficient distance from the wing. The balance of momentum between regions far upstream and far downstream in the combined field then shows the drag to be simply the difference of the drags of the individual flows, or

$$D = D_1 - D_2$$

After expressing D_1 and D_2 by means of Eq. 6-4 and 6-5 and equating the result to Eq. 6-6 we obtain

$$\int_S (\alpha_2 p_1 - \alpha_1 p_2) dS = 0$$

which is the same as Eq. 6-3.

An important result of the foregoing relation, which commits itself easily to memory, may be stated as follows (see Fig. A,6e): *The lift on a wing due to deflection of a flap is equal to the lift on the flap due to deflection of the wing with the flow reversed.* The term flap may here be taken to mean any small area of the wing surface. The statement may be verified by taking for α_1 a unit value over the whole surface, and for α_2 a value equal to unity over the region of the flap but equal to zero over the remainder of the wing.

The foregoing relation shows immediately those regions of a wing on which a flap is most effective in producing lift. The regions of greatest effectiveness are just those regions on which the lift density is greatest for the flat wing with the flow reversed. Thus in the case of a flat triangular wing, moving point foremost at an angle of attack, it is known that the lift density approaches a high peak along the leading edge. If the triangle is reversed so as to travel base foremost, the leading edge is turned into a trailing edge and it follows that the effectiveness per unit area of a narrow flap along this edge approaches a similar high peak value.

A,7. Lifting Surfaces. Solutions for Specified Planforms.

THE ELLIPTIC WING AT SUBSONIC SPEEDS. Analytic solution of Laplace's equation in three dimensions is practicable only for simple geometric forms. Of the few completely determined solutions in three dimensions the most interesting from the standpoint of wing theory are the classical solutions for the potential of ellipsoids [*1*], which include the elliptic disk as a limiting form. A treatment of the elliptic wing on this basis has been given by Krienes [*50*], who generalized earlier work by Kinner [*51*] on the circular disk. The interpretation of the classical solutions in terms of lifting surface theory was suggested by Prandtl and depends on the consideration of the pressure as a potential field. Accordingly, Prandtl defines a function

$$\Psi = -\frac{\Delta p}{\frac{1}{2}\rho_\infty U^2} = -\frac{p - p_\infty}{\frac{1}{2}\rho_\infty U^2} \tag{7-1}$$

The function Ψ is a solution of Laplace's equation, and has been termed the "acceleration potential." In steady motion the pressure disturbance Δp is related to the horizontal perturbation velocity by $\Delta p/\frac{1}{2}\rho_\infty U^2 = -2u/U$; hence in this case Ψ becomes simply $2u/U$. The corresponding vertical component w/U, whose value in the plane of the wing determines the camber and twist of the lifting surface, is found by an integration of the vertical acceleration experienced by a particle, beginning with the parallel flow at infinity ahead of the wing and ending at the point where the downwash is desired. The integral to be evaluated is

$$w = \int_{-\infty}^{x} \frac{\partial u}{\partial z}\, dx \tag{7-2}$$

Following the classical potential theory, solutions are obtained in terms of the orthogonal ellipsoidal coordinates λ, μ, and ν. The surfaces λ-const are confocal ellipsoids, while the surfaces defined by μ and ν are

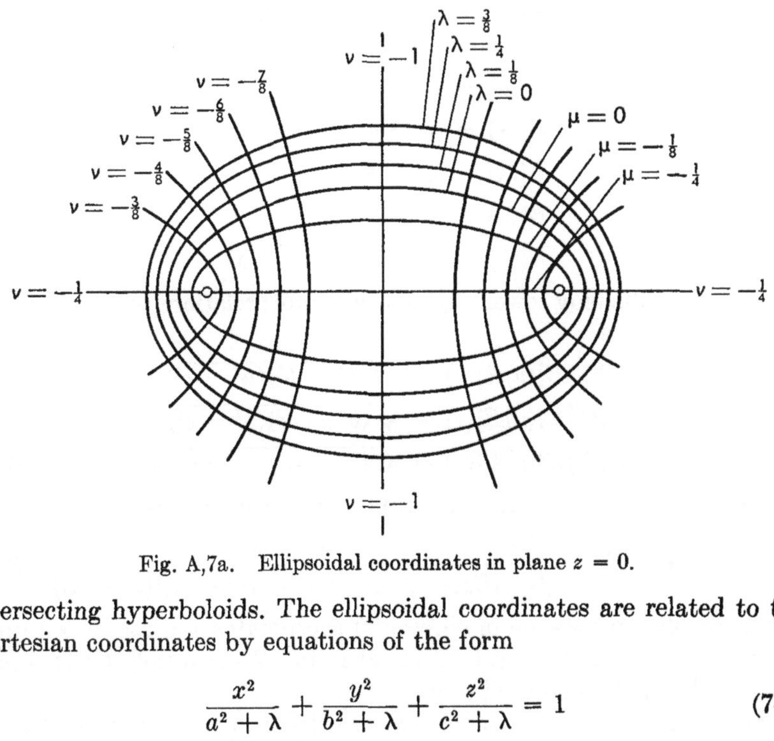

Fig. A,7a. Ellipsoidal coordinates in plane $z = 0$.

intersecting hyperboloids. The ellipsoidal coordinates are related to the Cartesian coordinates by equations of the form

$$\frac{x^2}{a^2 + \lambda} + \frac{y^2}{b^2 + \lambda} + \frac{z^2}{c^2 + \lambda} = 1 \tag{7-3}$$

where a, b, and c are the semiaxes of the base ellipsoid. Thus the base ellipsoid is given by $\lambda = 0$ and a thin ellipsoid is obtained by letting the semiaxis c approach zero. Setting $\lambda = -c^2$ gives the focal disk, which lies in the chord plane and forms the surface of discontinuity of the velocities u or w when the velocity field is continued into the interior of the base ellipsoid. The intersections with the disk of the surfaces of constant μ form a family of confocal ellipses; lines of constant ν appear as a family of hyperbolas (see Fig. A,7a).

The class of solutions considered is that expressible as the product of three functions of the three ellipsoidal coordinates. For the elliptic wing, solutions are given by products of the form

$$2\frac{u_n^m}{U} = E_n^m(\mu)E_n^m(\nu)F_n^m(\lambda) \tag{7-4}$$

where E_n^m and F_n^m are Lamé's functions of the first and second kind, respectively, of order n and class m. The use of the function of the second kind F_n^m is dictated by the requirement that the pressure vanish at infinity. The value of F is of course constant on the wing surface, and the distribution of surface pressure is determined in shape by the product $E(\mu)E(\nu)$. The degree n of the function determines the number of alternations in sign of the pressure distribution over the wing, while the class of the function determines the orientation of the isobars relative to the x, y, and z directions. Examples of the first few Lamé functions are

$$\begin{aligned} E_0 &= 1 \\ E_1^1(\lambda) &= \sqrt{a^2 + \lambda} \\ E_1^2(\lambda) &= \sqrt{b^2 + \lambda} \\ E_1^3(\lambda) &= \sqrt{c^2 + \lambda} \\ E_2^1(\lambda) &= \sqrt{(b^2 + \lambda)(c^2 + \lambda)} \end{aligned}$$

etc. Lamé products of lower degree bear simple relations to the Cartesian coordinates. Thus

$$E_1^1(\lambda)E_1^1(\mu)E_1^1(\nu) = \sqrt{(a^2 - b^2)(a^2 - c^2)}\,x \tag{7-5}$$

$$E_1^2(\lambda)E_1^2(\mu)E_1^2(\nu) = \sqrt{(b^2 - a^2)(b^2 - c^2)}\,y \tag{7-6}$$

$$E_1^3(\lambda)E_1^3(\mu)E_1^3(\nu) = \sqrt{(c^2 - a^2)(c^2 - b^2)}\,z \tag{7-7}$$

Functions of the second kind are obtained by the integration

$$F_n^m(\lambda) = E_n^m(\lambda)\int_\lambda^\infty \frac{d\lambda_1}{[E_n^m(\lambda_1)]^2\sqrt{(a^2 + \lambda_1)(b^2 + \lambda_1)(c^2 + \lambda_1)}} \tag{7-8}$$

For a complete discussion of ellipsoidal coordinates and associated functions, the reader is referred to Whittaker and Watson [*52*], Hobson [*53*], or to textbooks on potential theory.

As in the case of the complex velocity functions (Table A,2), a distinction must be made between functions for which u is discontinuous across the chord plane ($\lambda = -c^2$) and those for which w is discontinuous. Functions of the latter type yield various distributions of thickness over the base disk. Of the functions yielding distributions of lift (u discontinuous), the simplest is (Fig. A,7b)

$$2\frac{u_1^3}{U} = E_1^3(\mu)E_1^3(\nu)F_1^3(\lambda) \tag{7-9}$$

which reduces to

$$\frac{u_1^3}{U} = zf(\lambda) \tag{7-10}$$

For an ellipsoid of vanishing thickness it is found that

$$\frac{u_1^3}{U} = \frac{z_{\text{ell}}}{c} = \sqrt{1 - \frac{x^2}{a^2} - \frac{y^2}{b^2}} \tag{7-11}$$

The distribution of lifting pressure over the wing in this case can be represented by the ordinates of an ellipsoid constructed over the basic disk. By

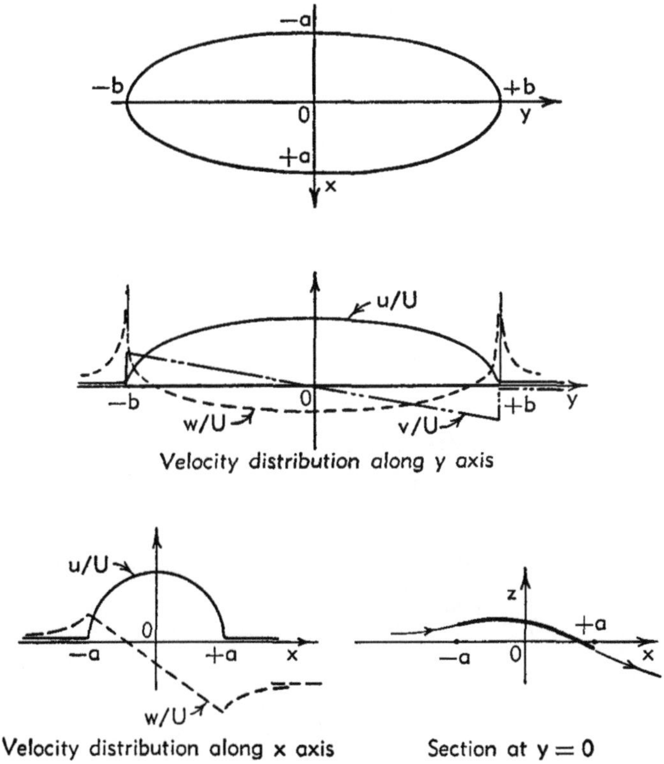

Fig. A,7b. Velocities given by first potential function for elliptic disk.

making use of the relation (7-11), the total lift of the distribution is found to be

$$L = \frac{8}{3}\pi ab \frac{1}{2}\rho_\infty U^2 \tag{7-12}$$

or, in coefficient form,

$$C_L = \frac{8}{3} \tag{7-13}$$

If the surface distributions of pressure, corresponding to the Lamé functions, are integrated in the chordwise direction to obtain the spanwise distributions of lift or circulation it is found that the latter are given by simple algebraic curves in the form of Legendre polynomials. Thus, if the total circulation around a section is expressed as $\Gamma(\eta)$, where $\eta = y/b$, it is found that

$$\Gamma_1^3(\eta) = 2\pi Ua \frac{P_0(\eta) - P_2(\eta)}{3} = \pi Ua(1 - \eta^2) \tag{7-14}$$

and the spanwise loading is parabolic.

For the determination of the downwash it is found convenient to use the general form [*52*, p. 390] of the three-dimensional potential

$$2\frac{u}{U} = \oint F[(\alpha x + \beta y + \gamma z), t]dt \tag{7-15}$$

Then the expression for w/U becomes

$$\frac{w}{U} = \oint \frac{\gamma}{\alpha} F[(\alpha x + \beta y + \gamma z), t]dt \tag{7-16}$$

For the solutions represented by Lamé functions it has been shown [*52*, pp. 567–570] that

$$2\frac{u_n^m}{U} = \frac{1}{2\pi i}\oint Q_n(\alpha x + \beta y + \gamma z)E_n^m(t)dt \tag{7-17}$$

with

$$\left.\begin{aligned} \alpha &= \frac{\sqrt{k^2 - t^2}}{ka} \\ \beta &= \frac{t}{kb} \\ \gamma &= \frac{i\sqrt{1 - t^2}}{a} \end{aligned}\right\} (\alpha^2 + \beta^2 + \gamma^2 = 0)$$

and

$$k = \sqrt{1 - \frac{a^2}{b^2}}$$

Here Q_n is the Legendre function of the second kind. The contour c extends along the real axis with loops around the points $\pm k$. With $t/k = \sin\theta$, the integration (7-17) may be interpreted geometrically as the superposition of two-dimensional fields of the form given by formula 7 in Table A,2 at varying angles θ around the z axis. The downwash function is given by (see Eq. 7-16)

$$\frac{w_n^m}{U} = \frac{1}{4\pi}\oint k\sqrt{\frac{1 - t^2}{k^2 - t^2}}\, Q_n(\alpha x + \beta y + \gamma z)E_n^m(t)dt \tag{7-18}$$

For the potential function of the first degree, evaluation of the integral yields (see [*50*])

$$\frac{w_1^3}{U} = \frac{a}{b} Q_1\left(\frac{y}{b}\right) - E(k)\frac{x}{a} \tag{7-19}$$

within the boundary of the disk. Here $E(k)$ is the complete elliptic integral of the second kind, equal to the ratio of the semiperimeter of the elliptic disk to the span $2b$. The term $E(k)(x/a)$ yields a parabolic camber shape for the sections taken parallel to the flight direction, while the remaining term shows a progressive spanwise twist of the sections along the span. As the aspect ratio $\mathcal{R} = 4b/\pi a$ approaches infinity, the spanwise variation $(a/b)Q_1(y/b)$ disappears, and the flow over the sections approaches in form the two-dimensional velocity function given by formula 5 of Table A,2.

Differentiation of Eq. 7-14 with respect to y yields the spanwise gradient of the circulation and hence the discontinuity of the lateral velocity v in the wake:

$$\frac{1}{U}\frac{d\Gamma_1^3}{dy} = 2\frac{v}{U} = -2\pi\frac{a}{b}P_1(\eta) = -2\pi\frac{a}{b}P_1\left(\frac{y}{b}\right) \tag{7-20}$$

At a great distance behind the wing, the wake forms a two-dimensional field of motion, with the axis parallel to the flight direction. Substituting v for u in formula 9 of Table A,2 shows that the velocity field of the wake is given by

$$\frac{v - iw}{U} = 2i\frac{a}{b}Q_1\left(\frac{y}{b} + i\frac{z}{b}\right) \tag{7-21}$$

The downwash along the centerline of the wing (Eq. 7-19, $x = 0$) is found to be equal to one-half the final value induced by the trailing vortices in the wake. As may be shown by considerations of symmetry, this statement must apply to every distribution of lift which is symmetrical about a line at right angles to the flight direction.

It is of interest to note that the foregoing relations for the ellipsoidal pressure distribution agree qualitatively with the results that would be obtained by application of the two-dimensional wing section theory together with the Prandtl lifting line theory. For the same camber and twist, however, the more complete calculation [*54*] shows a reduction of the lift in the ratio $1/E$.

Lift distributions of more general form may be obtained by expansion in a series of ellipsoidal harmonics, making use of the orthogonal properties of the Lamé functions. The distributions obtained in this way, however, fall to zero all around the edge of the wing, with tangential flow at both leading edge and trailing edge. In order to show the effect of changes in angle of attack of a rigid wing surface it is necessary to obtain solutions with flow around the leading edge; it is to be expected that such solutions will show infinite velocities around the edges of the focal disk.

Solutions having the desired properties may be obtained by applying various differential operators to the potential functions (7-4). The simplest of these is obtained by differentiating u_1^3/U and w_1^3/U with respect to x:

$$\mathrm{D}_x \frac{u_1^3}{U} \equiv a \frac{\partial}{\partial x} \frac{u_1^3}{U} = - \frac{x/a}{\sqrt{1 - (x^2/a^2) - (y^2/b^2)}} \tag{7-22}$$

and

$$\mathrm{D}_x \frac{w_1^3}{U} \equiv a \frac{\partial}{\partial x} \frac{w_1^3}{U} = -E(k) \tag{7-23}$$

Since the downwash is constant over the entire disk, Eq. 7-22 is a solution for a flat elliptic plate at an angle of attack. However, the flow shows infinite velocities around the trailing edge and corresponds to the noncirculatory flow depicted in part 1 of Fig. A,1a. The distribution of pressure over the sections is in fact identical with that given by formula 2 of Table A,2 for the noncirculatory two-dimensional flow over a flat plate at an angle of attack lower by the factor $1/E(k)$.

A circulatory flow over the disk is obtained by applying the operator

$$\mathrm{D}_{ab} = 1 + a \frac{\partial}{\partial x} + b \frac{\partial}{\partial y} \tag{7-24}$$

to u_1/U. Geometrically, this operation amounts to a uniform infinitesimal expansion of the scale of dimensions, followed by subtraction of the original field. Application of the operator to Eq. 7-11 and 7-19 produces

$$\mathrm{D}_{ab} \frac{u_1^3}{U} = \frac{1}{\sqrt{1 - (x^2/a^2) - (y^2/b^2)}} \tag{7-25}$$

which is seen to be analogous to formula 1 of Table A,2, and

$$\mathrm{D}_{ab} \frac{w_1^3}{U} = - \frac{a}{b} \frac{dQ_0(\eta)}{d\eta} = - \frac{a}{2} \left(\frac{1}{b - y} + \frac{1}{b + y} \right) \tag{7-26}$$

Unfortunately this latter solution cannot be combined with Eq. 7-23 in such a way as to satisfy the Kutta condition all around the trailing edge. Eq. 7-23 shows a uniform (infinite) velocity around the trailing edge at all sections, while Eq. 7-25 and 7-26 show increasing velocities and lift coefficients toward the tips. The spanwise lift distribution is rectangular (i.e. $\Gamma(\eta) \sim P_0(\eta)$) and, although each section of the wing is flat, the wing is twisted to conform to the downwash induced along the center line by a pair of finite vortices emanating from the tip (see Fig. A,7c).

The determination of the flow over the flat inclined elliptic plate in such a way as to satisfy the Kutta condition at all points of the trailing edge would require the superposition of infinitely many potential functions of the type (7-25), together with the noncirculatory flow (7-22). The total lift of the wing, however, requires only the determination of the first term of such an expansion. This calculation has been carried out by Krienes for wings of several different aspect ratios and his results are reproduced in Fig. A,7d.

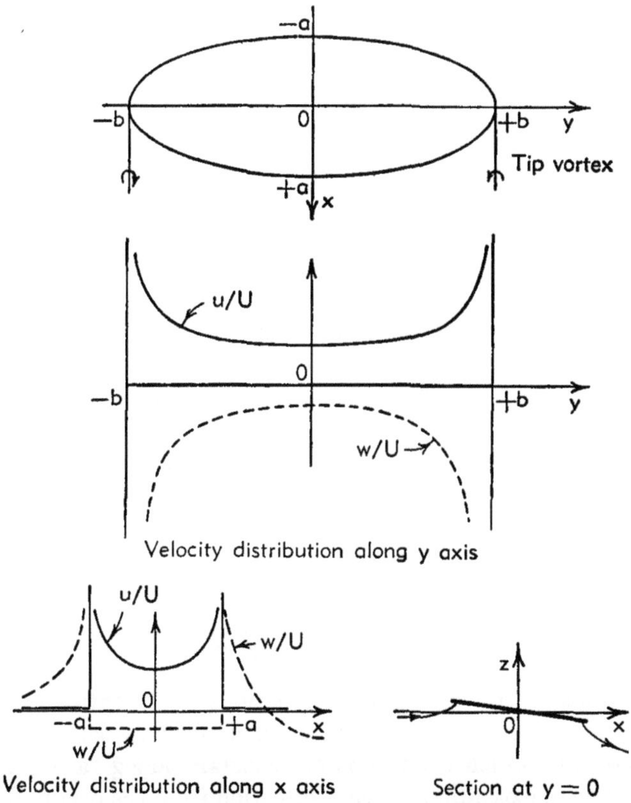

Fig. A,7c. Circulatory flow over elliptic disk.

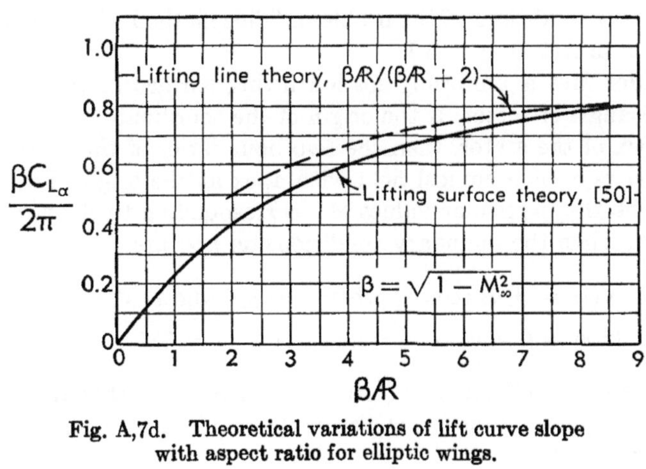

Fig. A,7d. Theoretical variations of lift curve slope with aspect ratio for elliptic wings.

Wings with Uniform Loading. Another family of completely determined solutions [55,56] gives the camber and twist of straight-sided wings with uniform loading. In principle, such solutions may be found by the straightforward integration of formulas for the downwash in terms of doublet or vortex strength to be given later in this article, but in practice the integrations are not always simple.

For straight-sided planforms it is possible to write down immediately some elementary solutions which may be combined through the use of the superposition principle of linearized theory to give the desired wing outline. The elementary solutions are found by the subsonic analogue of the conical flow theory developed by Busemann [57] in connection with supersonic wing problems. Its mathematical basis is the general solution of Laplace's equation of zero degree first given by Donkin (see [34, p. 357]), namely

$$F\left(\frac{y + iz}{x + \sqrt{x^2 + y^2 + z^2}}\right) + G\left(\frac{y - iz}{x + \sqrt{x^2 + y^2 + z^2}}\right) \tag{7-27}$$

where F and G are arbitrary analytic functions. The variable

$$\epsilon = \frac{y + iz}{x + \sqrt{x^2 + y^2 + z^2}} \tag{7-28}$$

and its conjugate are expressible as functions of the ratios y/x and z/x, so that the functions $F(\epsilon)$ and $G(\bar{\epsilon})$ may be used to describe a flow field of infinite extent in which conditions are constant along rays $y/x = \text{const}$ and $z/x = \text{const}$, extending in every direction from the origin. Such fields are known as conical flow fields, and the conical flow method consists in the superposition of a number of such fields, oriented as needed to satisfy the boundary conditions of the problem. (This method will receive further application in Art. 9 and 13.)

Consider first a uniformly loaded lifting surface in the form of an infinite triangle or sector. If the origin of the coordinate system is placed at the apex of the sector, then the two boundaries of the sector may be considered rays of a conical field with apex at the origin. Let ϵ_1 and ϵ_2 ($\epsilon_1 < \epsilon_2$) be the particular values of ϵ corresponding to the two edges of the sector. Then the boundary conditions on $u(\epsilon)$ are

$$\begin{aligned} &\text{(i)} \quad u(\epsilon) = \pm u_0 \text{ (const) when } z = \pm 0 \text{ and } \epsilon_1 < \epsilon < \epsilon_2 \\ &\text{(ii)} \quad u(\epsilon) = 0 \text{ when } \epsilon < \epsilon_1 \text{ and } \epsilon > \epsilon_2 \ (z = 0) \end{aligned} \tag{7-29}$$

By reference to Table A,2 (formula 8) it is found that the imaginary part of the Legendre function Q_0 shows the required variation, from which may be obtained

$$u = -\text{R.P.} \frac{iu_0}{\pi} \ln \frac{\epsilon - \epsilon_2}{\epsilon - \epsilon_1} \tag{7-30}$$

In order to find the corresponding vertical velocity we write the condition for irrotationality

$$\frac{\partial w}{\partial x} = \frac{\partial u}{\partial z} \tag{7-31}$$

or

$$\frac{dw}{d\epsilon} \cdot \frac{\partial \epsilon}{\partial x} = \frac{du}{d\epsilon} \cdot \frac{\partial \epsilon}{\partial z} \tag{7-32}$$

From this we find

$$dw = -\frac{i}{2}\left(\epsilon + \frac{1}{\epsilon}\right) du \tag{7-33}$$

so that

$$w = -\frac{i}{2} \int \left(\epsilon + \frac{1}{\epsilon}\right) \frac{du}{d\epsilon}\, d\epsilon \tag{7-34}$$

Applying Eq. 7-34 to the function (7-30) for the uniformly loaded sector and setting z equal to zero, we obtain the vertical component of the velocity in the plane of the wing:

$$w_{z=0} = -\frac{u_0}{2\pi}\left[\frac{\sqrt{1+m_2^2}}{m_2} \ln (M_2 - T)^2 - \frac{\sqrt{1+m_1^2}}{m_1} \ln (T - M_1)^2 \right.$$
$$\left. + \left|\frac{1}{M_2} - \frac{1}{M_1}\right| \sinh^{-1} \frac{1}{|t|}\right] + \text{const} \tag{7-35}$$

where $t = y/x$, m_1 and m_2 are the slopes of the two boundaries of the sector, and the functions

$$T(t) = \frac{t}{1 + \sqrt{1+t^2}}, \quad M_1 = T(m_1), \quad M_2 = T(m_2) \tag{7-36}$$

are the expressions to which ϵ, ϵ_1, and ϵ_2 reduce in the $z = 0$ plane. In the superposition of two or more infinite fields to produce a wing of finite chord, the constant of integration will be cancelled.

The contour of the surface, except for an arbitrary variation $z_0(y)$ of dihedral, is obtained by an integration of the slope w/U with respect to x:

$$z = -\frac{u_0 x}{2\pi U}\left\{(t - m_1)\frac{\sqrt{1+m_1^2}}{m_1^2} \ln (T - M_1)^2 \right.$$
$$+ (m_2 - t)\frac{\sqrt{1+m_2^2}}{m_2^2} \ln (M_2 - T)^2$$
$$+ \left[\left(\frac{\sqrt{1+m_2^2}}{m_2} - \frac{\sqrt{1+m_1^2}}{m_1}\right) t + \left(\frac{1}{M_2} - \frac{1}{M_1}\right)\right] \sinh^{-1}\frac{1}{|t|}$$
$$\left. + \left(\frac{1}{m_1} - \frac{1}{m_2}\right)\sqrt{1+t^2} + \text{const}\right\} + z_0(y) \tag{7-37}$$

In the simplest case, z_0 will be proportional to y and the surface will be a portion of a very shallow cone about the origin, since z/x is then a function of y/x.

The process by which wings of finite dimensions are derived is illustrated in Fig. A,7e. Starting with an infinite sector having its apex at the foremost corner of the wing planform, we superpose additional sectors of positive and negative lift in such a way as to satisfy the condition of zero lift over all areas of the plane *outside* the boundary of the planform. Since solutions satisfying the conditions (7-29) do not introduce any lift outside their own uniformly loaded sectors, the superposition of such

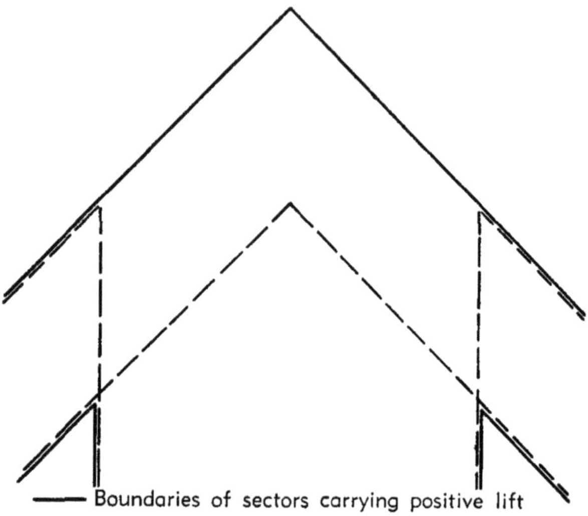

Fig. A,7e. Construction of a constant chord sweptback wing by the superposition of conical areas.

conical fields does not interfere with the condition of uniform loading on the wing.

It is found that this process requires one sector at each corner of the planform boundary. The shape of the wing $z(x, y)$ is then obtained by adding the ordinates of the stream surfaces given by Eq. 7-37, with the appropriate values of m_1 and m_2 for each superposition, and the origin displaced to the wing tips or trailing edge, as the case may be. If the stream surface associated with each sector is a conical figure, the resultant shape of the wing surface can be obtained most readily by laying out the elements (rays) of each conical surface and adding the ordinates at the points of intersection of the rays from the various fields. The dihedral may be adjusted afterwards to give a more practical shape.

The chief difficulty in the calculation of surfaces of constant loading arises in a failure of the linearized boundary condition along the root section of the triangular or sweptback wing. It will be noted on examining the form of the elementary solution for a sector of uniform loading that the downwash shows a logarithmic infinity along the x axis. The streamlines in the x, z plane therefore cross the x axis with a vertical slope. Since the integration of w to obtain the surface shape was performed, not along a streamline, but in the $z = 0$ plane, the result, along the x axis, is an apparently infinite displacement of the surface. It can be shown that for any small value of the lift coefficient the streamline passing through the

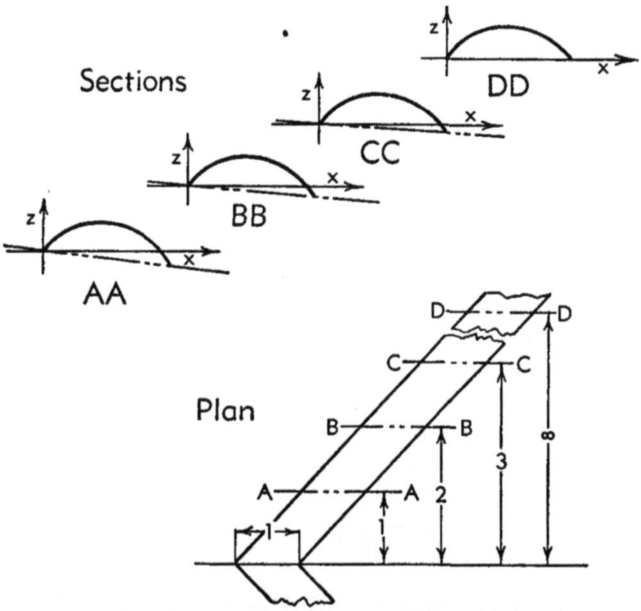

Fig. A,7f. Camber and twist required to support a uniform distribution of lift; $\Lambda = 45°$, $M_\infty = 0$.

origin is actually displaced only a small angle below the x axis. The variation in magnitude of this angle with lift coefficient is not even approximately linear, however, the initial rate of change being infinite. Moreover, the streamlines show a physically unacceptable singularity in slope (along the x axis) *above* the wing surface. While this situation invalidates a precise determination of the shape of the wing surface at the root section, it is nevertheless highly localized and may probably be disregarded if the center of the wing is to be enclosed in a fuselage. The singular behavior is removed entirely if the condition of absolute uniformity of the loading is relaxed to permit a small decrease in loading at the center section.

Lampert [*55,56*] has worked out several examples of uniformly loaded wings. From the practical standpoint the most interesting is the shape of a uniformly loaded, constant chord wing with 45° of sweep. This result is shown in Fig. A,7f.

If additional sectors were superposed at the tips, as in Fig. A,7e, to reduce the wing to finite aspect ratio, a prohibitive increase of twist near the tip and an infinite induced drag would result. In this case, the singularity in the downwash is of a higher order than the singularity following a bend in the leading edge. This effect is avoided if the tips are pointed or raked.

ARBITRARY LOADING; VORTEX SHEET THEORY. We proceed, from the foregoing quite special cases, to a consideration of the general problem, with loading and planform arbitrary. Because it is basic to most of the numerical methods of lifting surface theory and offers the advantage of being physically intuitive, we will discuss first the vortex sheet representation of the lifting surface. The relation between the lift and the vorticity is derived from the definition of the circulation as the line integral

$$\Gamma = \int_C \Delta\mathbf{q} \cdot d\mathbf{s} \tag{7-38}$$

where

$$\Delta\mathbf{q} = \mathbf{i}u + \mathbf{j}v + \mathbf{k}w$$

represents the disturbance velocity due to the airfoil in an otherwise irrotational stream. Now let the path be a small flat rectangle lying in a plane y = const and enclosing a portion of the wing surface. If the top and bottom of the rectangle are parallel to the wing (assumed in the $z = 0$ plane), Eq. 7-38 reduces to

$$\Gamma = \int_C (udx + wdz) \tag{7-39}$$

Since w is continuous across the wing (except in the neighborhood of the leading edge) the second term can be made to vanish by allowing the rectangle to become vanishingly thin. In the limit, the singularity at the leading edge can also be shown to contribute nothing to the integral. The horizontal velocity u will approach equal and oppositely-directed values at the surface of discontinuity, so that it is possible to write for the total circulation ahead of a point x, y

$$\Gamma(x, y) = 2 \int_{x_1(y)}^{x} u_{z=0}\, dx \tag{7-40}$$

where $x = x_1(y)$ is the equation of the leading edge. The function $\Gamma(x, y)$ is related to the vorticity γ by the equation $\gamma = d\Gamma/dx$, and reduces to the more familiar span loading $\Gamma(y)$ at the trailing edge.

A familiar representation of the vortex sheet is as a planar arrangement of horseshoe vortices of infinitesimal spanwise extent dy and of infinitesimal strength, which will be in the present notation $\gamma\, dx$. Then by the Biot-Savart law the vertical component of the velocity at any point

x_0, y_0 in the plane of the wing may be found as

$$w(x_0, y_0) = \frac{1}{4\pi} \iint\limits_S \gamma \left\{ \frac{x - x_0}{r^3} - \frac{\partial}{\partial y} \left[\frac{1}{y - y_0} \left(1 - \frac{x - x_0}{r} \right) \right] \right\} dxdy$$

in which $r = \sqrt{(x - x_0)^2 + (y - y_0)^2}$. The expression in square brackets is the downwash of a single trailing vortex. Taking the indicated derivative gives

$$w(x_0, y_0) = \frac{1}{4\pi} \iint\limits_S \frac{\gamma(x, y)}{(y - y_0)^2} \left(1 - \frac{x - x_0}{r} \right) dxdy \tag{7-41}$$

Eq. 7-41 cannot generally be evaluated analytically for useful distributions of γ. Numerical evaluation is complicated by the strong singularity in the integrand. An alternate formulation of the downwash suggested in [*58*] facilitates the integration to a considerable extent. In this approach, the vortices are of uniform infinitesimal strength and arranged along lines of equal Γ on the wing, a new vortex element being required for every change in circulation. In the wake there are no new sources of vorticity, so that the lines of equal Γ appear as the usual trailing vortices. Unless the surface is cambered to a vertical slope at the trailing edge the bound vortices of the wing must merge smoothly into the trailing vortices at the trailing edge of the wing in order to conform to the Kutta condition. Fig. A,7g shows an example of such a vortex representation.

By the introduction of polar coordinates r, ϑ centered in each calculation about the point x_0, y_0 at which the downwash is to be found, Eq. 7-41 is replaced by

$$w(x_0, y_0) = -\frac{1}{4\pi} \int_0^\infty \int_0^{2\pi} \frac{1}{r} \frac{\partial \Gamma}{\partial r} d\vartheta dr \tag{7-42}$$

or

$$w(x_0, y_0) = -\frac{1}{2} \int_0^\infty \frac{1}{r} \frac{d\bar{\Gamma}}{dr} dr \tag{7-43}$$

where $\bar{\Gamma}$ is the average value of Γ around the circle with center at x_0, y_0 and radius r. Eq. 7-43 is essentially the equation for the downwash at the initial point of a semi-infinite lifting line, which is readily evaluated by a numerical method presented in [*58*]. Thus the slope of the surface to support a given distribution of lift may be determined at an arbitrary number of points.

Since the determination of $\bar{\Gamma}$ is expected to be performed graphically (from a vortex pattern such as is shown in Fig. A,7g), it is not necessary to specify any analytic form for the loading. However, the accuracy will be considerably improved if the spanwise and chordwise variations of Γ can be fitted, in the neighborhood of each point x_0, y_0, with analytic expressions, to provide the initial variation of $\bar{\Gamma}$ with r.

Some Results from Lifting Surface Theory. Although the particular wings studied in [*58*] were elliptical in chord distribution, the results, interpreted in general terms, summarize most of what is presently known regarding the theoretical distribution of lift on wings of finite planform.

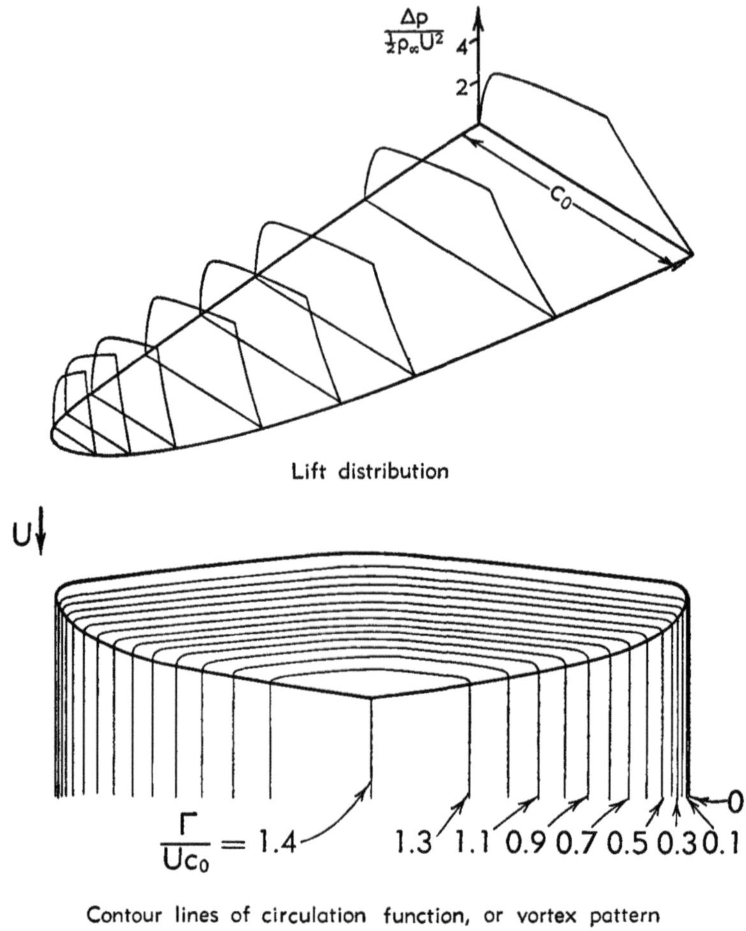

Fig. A,7g. Arbitrary distribution of lifting pressure and corresponding vortex-sheet representation.

Effects of aspect ratio. In Fig. A,7h and A,7i are shown two unswept elliptical wings, Æ = 6 and 3, and the vortex patterns corresponding to the following distribution of lift: The span loading in each case is the elliptic loading predicted for an untwisted elliptic wing by the lifting line theory, and the chordwise loading at each section is that on an infinite flat plate at an angle of attack (formula 3, Table A,2).

The exact value of the downwash along the 50 per cent chord line for this loading may be obtained without calculation from the following considerations: As was pointed out earlier, in connection with the solution given by Eq. 7-22 and 7-23, the nonlifting component of the two-dimensional solution for the flat plate at an angle of attack is also a solution for the section flow on a flat elliptic disk. The downwash in the latter

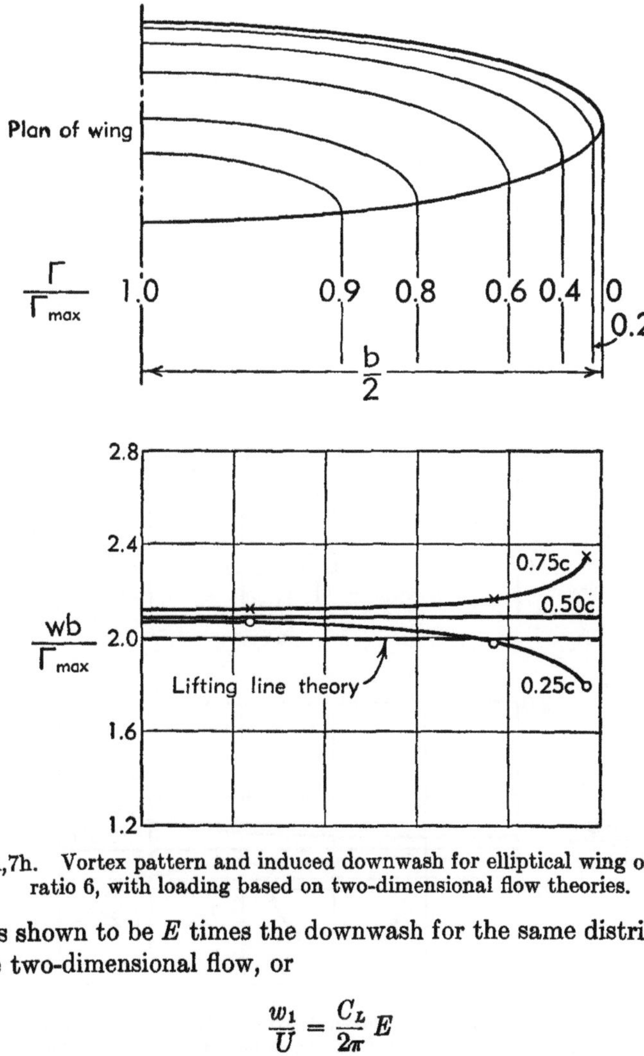

Fig. A,7h. Vortex pattern and induced downwash for elliptical wing of aspect ratio 6, with loading based on two-dimensional flow theories.

case was shown to be E times the downwash for the same distribution of u in the two-dimensional flow, or

$$\frac{w_1}{U} = \frac{C_L}{2\pi} E \tag{7-44}$$

where E is the ratio of the semiperimeter of the ellipse to its diameter.

The lifting component of the assumed two-dimensional velocity distribution corresponds to a circulatory flow symmetrical about the 50 per cent chord line. This component cannot be related to the circulatory flow over the flat elliptic disk described by Eq. 7-25 and 7-26 because the span loading in that solution is rectangular, rather than elliptical as assumed here. However, because of the fore-and-aft symmetry of the lift distribution, the downwash along the 50 per cent chord line will be exactly half the downwash induced by the wake at infinity. For the elliptic loading, this value is

$$\frac{w_2}{U} = \frac{C_L}{\pi \mathit{A\!R}} \tag{7-45}$$

as given by lifting line theory. Thus the total downwash at the 50 per

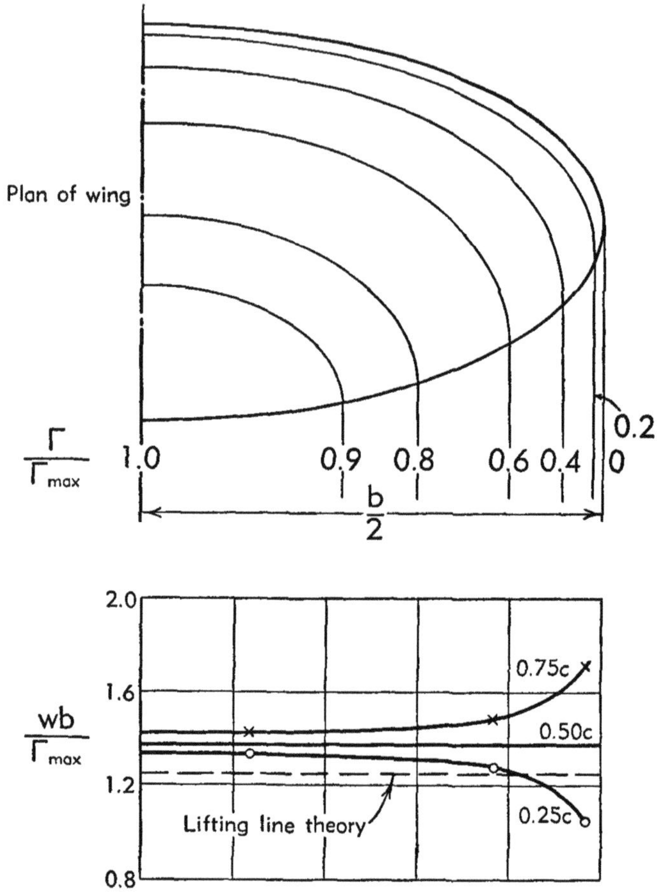

Fig. A,7i. Vortex pattern and induced downwash for elliptical wing of aspect ratio 3, with loading based on two-dimensional flow theories.

cent chord line is given by

$$\frac{w}{U} = \frac{C_L}{2\pi} \cdot \frac{E\mathit{A\!R} + 2}{\mathit{A\!R}} \tag{7-46}$$

This value and the downwash along the quarter chord and three-quarter chord lines calculated by the graphical integration are shown in the lower parts of Fig. A,7h and A,7i. The deviation of the curves from the straight lines derived by lifting line theory indicates the error in the two-dimensional theories as applied to wings of finite aspect ratio. The chief effect is seen to be an approximately symmetrical curvature of

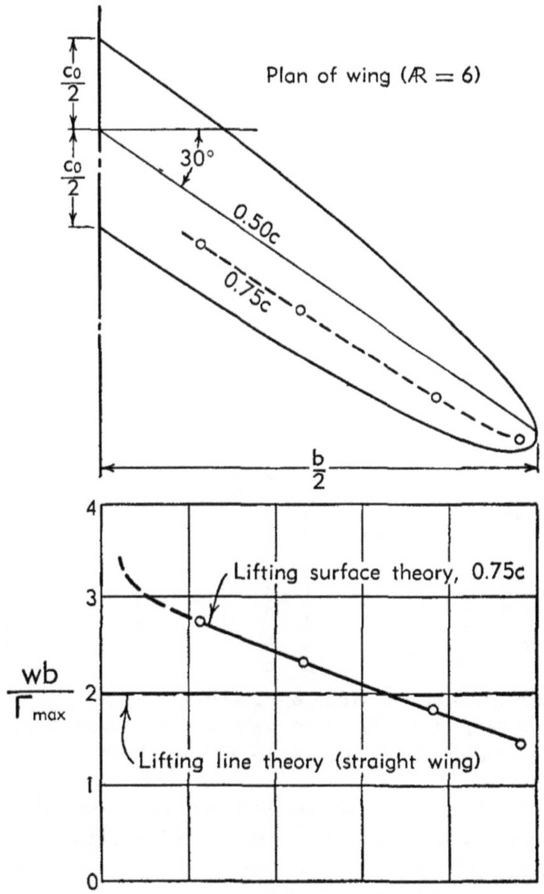

Fig. A,7j. Planform and induced downwash at three-quarter chord line of sweptback elliptical wing with loading based on two-dimensional flow theories.

the flow, in a direction requiring positive camber of the wing. The curvature is small at the center but increases sharply near the tips. The effect is of course more marked on the wing of lower aspect ratio.

Effect of sweepback. The same problem was worked for a wing with the 50 per cent chord line swept back 30°, the chord distribution remaining elliptical and the sections remaining parallel to the stream. The downwash was found only at the three-quarter chord line. The results (Fig. A,7j) indicate a linear twist for most of the wing, with more than 50 per cent reduction in angle of attack in going from a section near the root to one near the tip. Along the root section, the downwash goes to infinity, showing the assumed bending of the vortices to be physically incompatible with a continuous surface.

For the reverse problem—finding the lift distribution on a flat wing of the same planform—the vortices were rounded out at the center line to eliminate the infinite downwash of the preceding example, and the remainder of the pattern adjusted so that the downwash was approximately constant at 16 points, 8 on the quarter chord and 8 on the three-quarter chord lines.

The resulting load distribution (Fig. A,7k) shows the tendency of the load along the root section of a sweptback wing to be distributed rear-

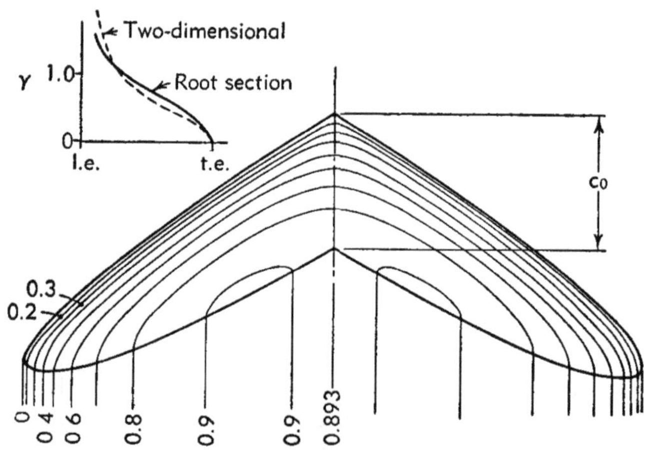

Fig. A,7k. Distribution of circulation on flat wing of aspect ratio 6, with elliptical chord distribution swept back 30° along midchord line. Contours of the circulation function Γ, from [*58*], and chordwise loading at the root section.

ward compared with that predicted by two-dimensional theory. Near the tips, the opposite effect is observed, i.e. the lift tends to concentrate near the leading edge and the section aerodynamic centers are forward of the quarter chord points. The drop in lift in the center and concentration of lift at the tips are characteristic of a sweptback wing whose sweepback is not offset by washout of the angle of attack [*59*]. These effects are more clearly seen in Fig. A,7l, where comparison of the span loading is made with that for the unswept elliptic wing, in one case at equal lifts and in the other case at equal angles of attack. In the latter case, the introduction of sweepback results in about 14 per cent loss in total lift.

Electromagnetic Analogues. Closely related to the foregoing investigation is the work of Swanson and Crandall [*60*], based on the correspondence between the velocity field of a vortex filament and the magnetic field induced by an electric current. Starting with a physical model of the vortex pattern, with the vortices replaced by wires, they find the corresponding downwash by measuring the vertical component of the magnetic field. The advantage of this method is that once the model has been constructed, the "downwash" may be measured at a very large number of points without undue expenditure of time. This advantage is partly offset by the disadvantage that numerous measurements must be made before useful results can be obtained, because of the discontinuities in field strength associated with the use of a system of discrete wires. Fig. A,7m shows a typical chordwise distribution of downwash for a model consisting of 50 wires. It should be noted also that the measurements were necessarily made at a finite height above the plane of the wing. To

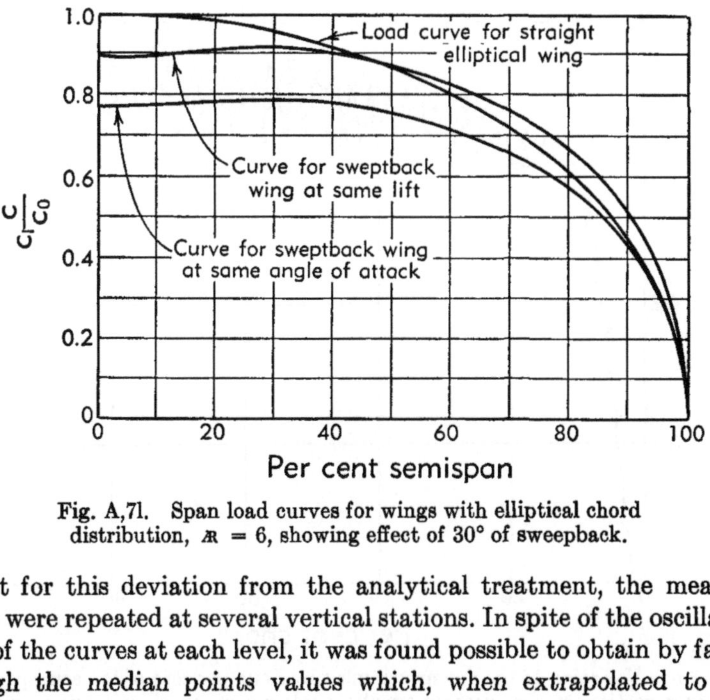

Fig. A,7l. Span load curves for wings with elliptical chord distribution, Æ = 6, showing effect of 30° of sweepback.

correct for this deviation from the analytical treatment, the measurements were repeated at several vertical stations. In spite of the oscillatory form of the curves at each level, it was found possible to obtain by fairing through the median points values which, when extrapolated to zero height, showed very smooth and consistent variations, both spanwise and chordwise.

Unpublished data obtained for the vortex configurations of Fig. A,7h and A,7i show linear chordwise variation of w, substantiating the indication of the previously obtained numerical results that the induced camber of the assumed loading is parabolic in form. Extension of the investigation to the problem of the straight elliptical wing in steady rolling motion [*61*] yielded a similar result, with the same sort of increase in camber toward the tip as with the angle-of-attack loading.

It may be appropriate to mention here a somewhat different electrical analogue—the electrolytic tank. Apparently successful tests have been made by Malavard [*62*] (see also IX,H,2), replacing the wing with a large number of small conducting plates, and the wake with strips of the same, in order to generate an electric potential to correspond to the velocity potential of the wing. The resistances in series with the plates are variable. The quantity measured is the current through the plate, which, being proportional to the gradient of the potential, corresponds to the downwash in the aerodynamic problem. This method is discussed further below.

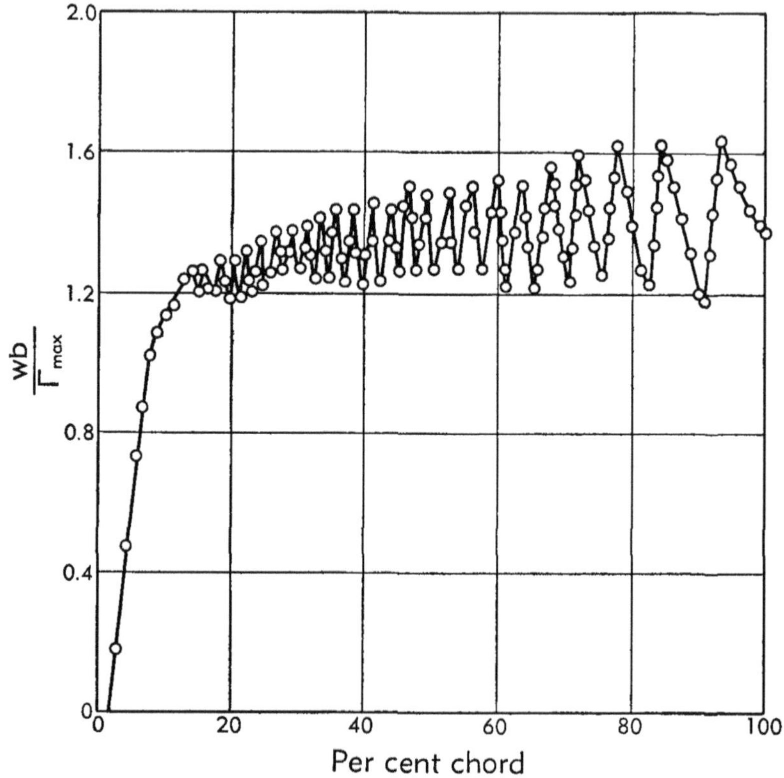

Fig. A,7m. Variation of the downwash measured one per cent of the semispan above the center section of the electromagnetic analogue of an elliptic wing, with two-dimensional flat plate loading approximated by 50 wires [*60*].

The "Direct Problem"—The Wing Completely Specified. In the foregoing, methods were presented for solving the so-called "indirect problem" of aerodynamics—that is, finding the shape of the body consistent with an assumed distribution of pressure. This type of problem is inherently easier to solve for the following reason: beyond the boundaries of the lifting surface the pressure, or u component of the perturbation velocity, must be zero. If the distribution of pressure is specified on the surface also, then the function u is described all over the plane and the related quantities φ and w can in principle be found by straightforward integration. Moreover, new solutions can be found by the superposition of simple solutions, as was done in deriving the uniformly loaded wing, since the component solutions introduce no new lift beyond their own surfaces.

However, the majority of the designer's problems are of the reverse type: he is required to predict the effect on the lift distribution of control-surface deflection, of angle of attack, of specified twist, and so on. In these problems, the boundary condition of tangential flow prescribes the variation of w on the wing, but elsewhere the condition $u = 0$ must still be satisfied. Thus we have a problem with mixed boundary conditions leading, in the case of a finite wing, to an integral equation for which no analytical method of solution has as yet been developed.

The previously mentioned "rheoelectric" method of Malavard (see IX,H,2) furnishes one method of solution through the adjustment of the resistances until the required variation of current (proportional to the downwash) is obtained. By this method span loadings and center-of-pressure locations have been found [*62*] for wings with various degrees of sweepback, including the effects of flaps and ailerons. Some work has also been done on the development of relaxation methods [*63,64*] and satisfactory results obtained.

While the detailed solutions obtainable by such methods would be of value as a basis for boundary layer studies, for flap-effect calculations, and for checking the accuracy of more approximate methods, the special equipment or lengthy calculations involved make the methods not entirely suitable for general use. Instead, the most usual procedure has been to superimpose a number of solutions to the indirect problem, in such proportions as to satisfy the conditions on the vertical component of the velocity at a finite number of points on the wing.

The methods of approximation used to reduce the computing to practical proportions generally take one of two forms: either (1) assuming the chordwise load distribution to be adequately expressed by a small number of terms of a trigonometric series, or (2) replacing the continuous vortex sheet with a finite number of discrete vortices.

Lifting surface methods. If a continuous distribution of vorticity is retained, it is generally assumed, after Birnbaum's treatment of the two-dimensional problem [*65*], to have along each chordwise section the form

$$\gamma = a_0 \cot \frac{\theta}{2} + \sum_{n=1}^{N} a_n \sin n\theta \tag{7-47}$$

$$\theta = \cos^{-1} \frac{x}{c/2} \tag{7-48}$$

where c is the chord of the section and x is the streamwise distance from the midpoint. The first term of this expression for γ is the trigonometric form of formula 3 of Table A,2 and corresponds to the two-dimensional solution for a flat plate at an angle of attack; the second term ($n=1$) is the solution for parabolic camber (formula 5). For a wing of finite span, the coefficients a_n are expressed as functions of the spanwise coordinate y. This approach was first used by Blenk [*66*] and has been the basis of most of the subsequent lifting surface work.

With the insertion of the first terms of Eq. 7-47 for γ, the chordwise integration of Eq. 7-41 may be performed analytically (in terms of elliptic integrals [*67*] or numerically). Blenk chose to expand the integrand in a series which is unfortunately not convergent at the wing tips or at the root of a sweptback wing. Very recently Multhopp has prepared some charts [*68,69*] giving numerical values for results of the first integration. With these values, the equation for w may be put into the form of the Prandtl lifting line equation and treated similarly, using Multhopp's well-established method [*70*] for that problem.

Like the form of the vorticity distribution, the control points at which the boundary condition of tangential flow is to be satisfied are also selected on the basis of two-dimensional theory. If, for example, the downwash is to be fixed at only one point of a given section, the three-quarter chord point is indicated as most representative for the determination of the section lift. The reason for this choice is the fact that, in two-dimensional flow, a flat plate and a wing with parabolic camber, if they coincide in slope at the three-quarter chord line, are found to have the same lift coefficient. Thus, whether the wing is uncambered or is cambered along a parabolic arc, the slope at the three-quarter chord will suffice to indicate the lift coefficient, in two-dimensional flow. Where chordwise discontinuities in slope are involved, as in the case of deflected flaps, the three-quarter chord line, of course, has no special significance, but some adjustment is made, usually in the form of the use of "equivalent slopes" for the sections, so that the three-quarter chord points can still be used.

If more than one control point per section is to be used, a formula due to Multhopp gives the location of the jth point,

$$\theta_j = \frac{2\pi j}{2p+1}, \qquad j = 1, 2, \ldots, p \tag{7-49}$$

where p is the total number of control points for the section.

It is generally concluded, in investigations based on the foregoing method, that one term in the series (7-47) (that is, the infinite flat plate solution) and one set of downwash points along the three-quarter chord line are sufficient for the determination of the lift and span loading of most wings. For the calculation of the aerodynamic center, at least two terms and two downwash points are required at each section.

Presumably, with more terms in the series and more control points, the surface load distribution could be worked out in greater detail. It is clear, however, that such a procedure is not really adapted to the investigation of the loading in regions of rapidly changing boundary conditions, because of the reliance on two-dimensional solutions. Moreover, it can be shown [*71*] that the form of loading assumed (Eq. 7-47 and 7-48) is fundamentally unsuitable for any wing containing discontinuities in the direction of the lines of constant per cent chord. In such cases (i.e. swept wings), the use of the per cent chord as a variable necessarily introduces inadmissible infinities in the vertical velocity, due to the assumed angularity of the vortices, with the result that the boundary conditions cannot be satisfied by any finite number of solutions.

Various methods of dealing with the failure of the two-dimensional approach at the root of the swept wing have been proposed. Multhopp [*68*] relies on the smoothing effect of the interpolation functions used in performing the spanwise integrations. Küchemann [*72*] has undertaken a more basic consideration of the problem, attempting to isolate the effect of the bend in the wing outline. His results, though interesting, do not lead to a complete lifting surface theory and appear to be of limited value because of the numerous assumptions introduced. Garner has attempted to solve the problem by a more suitable choice of variable, one which while conforming to the planform boundaries has, as its intermediate contours, lines of continuous curvature. Unfortunately, other mathematical difficulties (see the brief discussion in [*73*]) prevent a satisfactory resolution of the center-section problem.

Thus there remains as yet unsolved the question of the precise theoretical form of the lift distribution at the root of a swept wing.[8] Detailed solutions for the neighborhood of the tips, or of deflected flaps, are likewise beyond the reach of facile computing routines. Such failures of the theory need not, however, have any practical significance in determining such over-all characteristics of the wing as lift curve slope, span loading, and induced drag, since the effect of details of the loading at any one station dies away very rapidly. The chief difficulty appears when one of a small number of "control points" is located on the center section of a swept wing, since the inaccuracies introduced there then affect the entire solution.

Methods using discrete vortices. It is logical to attempt to simplify the computations by the substitution of a number of discrete vortices for the continuous distribution of circulation of the vortex sheet. The spanwise continuity of the loading is generally retained, but the lift is gathered chordwise into a finite number of lifting lines, each extending over the span of the wing. The trailing vortices leave each line in the same way as in the Prandtl lifting line model. When the wings are tapered or swept,

[8] An interesting comparison of the span loadings calculated for an infinite swept wing by various methods is given in [*74*].

the lifting lines are made to pass through points of constant per cent chord.

The formula for the downwash induced by a single lifting line with variable circulation $\Gamma(y)$, swept back through an angle Λ at the center, is

$$w(x_0, y_0) = \frac{1}{4\pi}\left[\int_{-b/2}^{b/2} \frac{\Gamma'(y)}{y - y_0}\left(1 + \frac{x_0 - |y| \tan \Lambda}{r}\right) dy \right.$$
$$\left. - (x_0 + y_0 \tan \Lambda) \int_{-b/2}^{0} \frac{\Gamma(y)}{r^3}\, dy - (x_0 - y_0 \tan \Lambda) \int_{0}^{b/2} \frac{\Gamma(y)}{r^3}\, dy \right] \quad (7\text{-}50)$$

with $r = [(x_0 - |y| \tan \Lambda)^2 + (y_0 - y)^2]^{\frac{1}{2}}$ and $y_0 \geqq 0$.

In methods employing discrete vortices, two-dimensional theory is used to determine the most representative location of the vortices as well as of the control points. If only one vortex line is used, it is placed along the center-of-pressure line in two-dimensional flow, which for a flat plate at an angle of attack is the quarter chord line.

The downwash then varies inversely as the distance behind the quarter chord line; at the position of the three-quarter chord line it just equals in magnitude the vertical component of the flow tangential to the flat plate having the same circulation. Conversely, if the condition of tangential flow is satisfied at the three-quarter chord line, the strength of the concentrated vortex will indicate the lift on the wing due to angle of attack.

The foregoing approach to the lifting surface problem—a first correction to the Prandtl lifting line theory—was initially suggested by Pistolesi [*75*]. Mutterperl [*76*] and Weissinger [*77*] developed the procedure further and also extended the formulas to cover sweptback wings. A side investigation by Weissinger indicated that the span loadings obtained might be expected to be in good agreement with the results of the vortex sheet calculations when the latter included only one term in the series (7-47) for the chordwise distribution of lift. Using Weissinger's formulas, De Young and Harper [*78*] have developed a rapid computing procedure for the span loading of wings with quarter chord lines composed of two straight-line segments and have prepared charts from which the span loading of straight-tapered wings may be obtained directly. A comparison of their results with the limited results of lifting surface methods now available[9] shows the angle-of-attack span loadings obtained by the Weissinger method to be in fact almost indistinguishable from those obtained with any of the more elaborate methods.[10]

[9] Covering 0 to 45° of sweep, aspect ratios from 3 to 8, and taper ratios from 0 to 1.

[10] An approximate integration formula for obtaining the total lift appears to be less reliable, since it does not gives results approaching the correct limit for infinite aspect ratio when applied to sweptback wings. However, this and other difficulties at high aspect ratios may be remedied by increasing the number of control points over the seven used in [*78*]. (See, for example, [*79*].)

Concentration of the lift in one vortex line makes it impossible, of course, to determine anything about the chordwise distribution of the load. For more detailed information, two or more vortex lines are required. Various arrangements of the vortices have been proposed. One, applied by Weighardt [*80*] with considerable success to rectangular wings of low aspect ratio, appears to have the advantage of simplicity of concept and application. Schlichting [*81*] and Scholz [*82*] have shown that it has also a certain physical justification in two-dimensional flow. The essence of the method is as follows:

The wing is first divided into N strips of equal depth extending from tip to tip. (In the case of a tapered wing, the strips would taper in the same ratio.) A lifting line is placed along the quarter chord line of each strip and the condition of tangential flow satisfied at M points on each three-quarter chord line. Thus, $M \cdot N$ equations are obtained for the circulation on each lifting line, which may then be expressed in the form of M terms of a Fourier series. For the physical justification of this strip method, we again turn to the two-dimensional solutions. It can be shown [*83*] that for any N, the sum of the circulations obtained for each of the vortices by satisfying the condition $\alpha = \text{const}$ at each of the three-quarter chord lines on an infinite wing is exactly equal to the circulation of the flat plate at that angle of attack, and the pitching moment is also reproduced. Moreover, the lift of a wing with parabolic camber is duplicated by the same system of vortices if the corresponding boundary conditions are inserted. (It is possible only to approximate the moment of the cambered wing, and this requires three or four strips.)

The discrete vortices are introduced only for the purpose of computing the downwash. The aerodynamic characteristics of the wing are computed from a continuous loading in the form of Eq. 7-47, related to the loading of the discrete vortices by equating section lifts, moments, etc. on the assumption of two-dimensional flow at each section. Of course, this method of obtaining correspondence between the lifting lines and the vortex sheet loses its validity near the corners of the wing, where the flow differs sharply from the two-dimensional.

This method has been applied by Scholz to rectangular wings, both flat and cambered. The results for flat wings at an angle of attack (Fig. A,7n) show a tendency of the span loadings toward the elliptical as the aspect ratio is reduced, and a tendency for the lift to be concentrated at the leading edge. However, the previously noted inaccuracy of the solution near the tips causes the results for the lowest aspect ratio to be only qualitatively correct, as may be seen by the plots (Fig. A,7n, bottom) of aerodynamic center location. The use of more vortices for low aspect ratios (here $N = 3$) would be expected to show a more marked forward movement of the aerodynamic center near the tips for the case $A\!\!\!R = 1$, in conformity with the other curves. In the exact potential-flow solution the

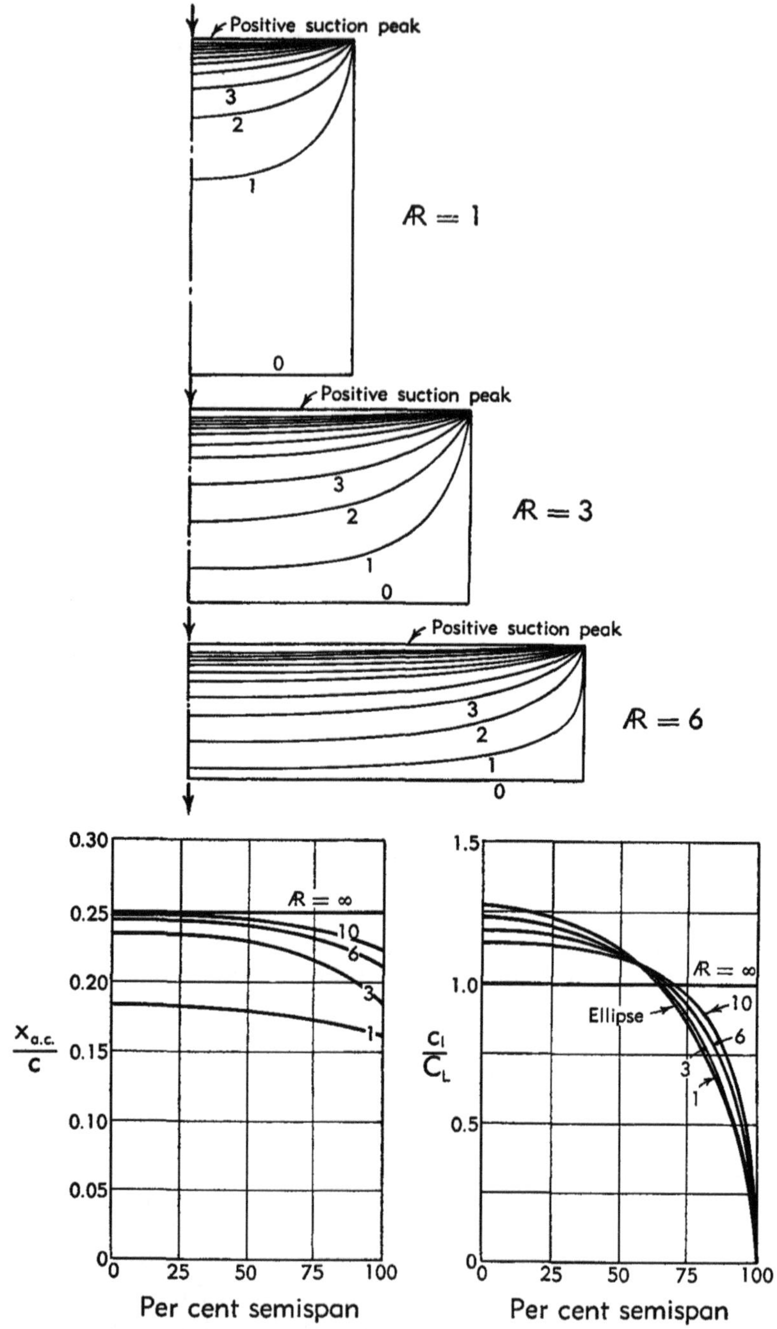

Fig. A,7n. Effect of aspect ratio on loading of flat rectangular wings, calculated by the strip method of [82]. Top, contour lines of load distribution; bottom left, aerodynamic center location aft of leading edge; bottom right, span loading.

aerodynamic center of the tip section on rectangular wings would be at the leading edge.

The contour lines of the load distribution for a rectangular wing of aspect ratio 6 with parabolic camber is shown in Fig. A,7o. It is seen that in this case, even at so large an aspect ratio, the departure from two-dimensional conditions affects the entire flow, so that the method of solution is itself invalidated.

The strip method just described has been applied to sweptback wings by Schlichting [81]. Other methods of distributing the lifting lines and the points for computing the downwash are investigated in a paper by Holme [84]. Using two lifting lines and fifteen control sections (effectively thirty control points), Holme obtains, for a wing with 40° sweepback and aspect

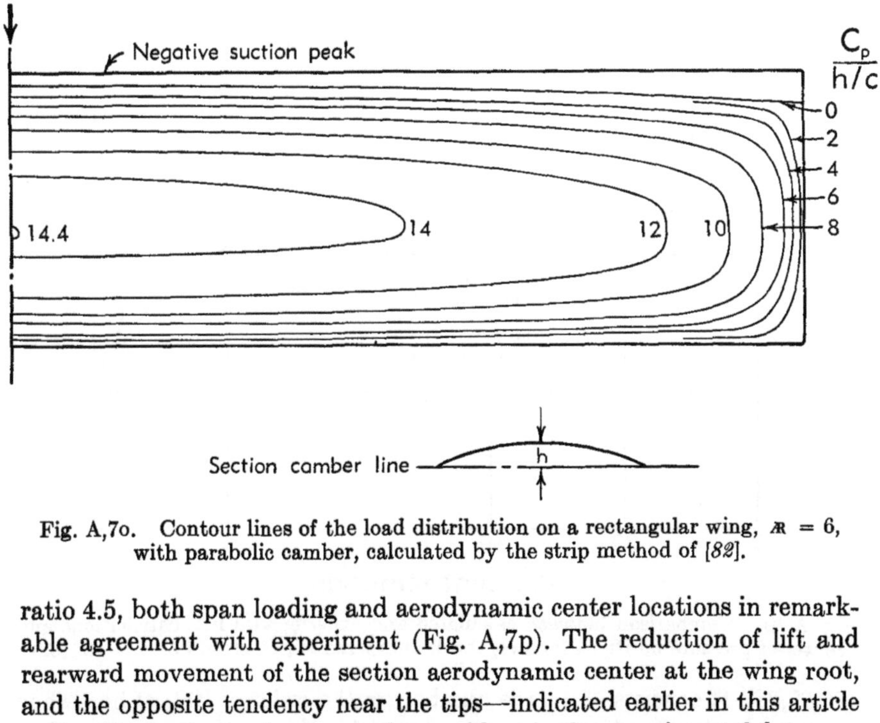

Fig. A,7o. Contour lines of the load distribution on a rectangular wing, Æ = 6, with parabolic camber, calculated by the strip method of [82].

ratio 4.5, both span loading and aerodynamic center locations in remarkable agreement with experiment (Fig. A,7p). The reduction of lift and rearward movement of the section aerodynamic center at the wing root, and the opposite tendency near the tips—indicated earlier in this article as the effects of sweepback—are here evident in the experimental data.

The foregoing procedures are essentially extensions of the lifting line theory, with the formulas modified to admit one bend in the lifting lines. It is apparent that the range of planforms to which this representation can be applied is somewhat limited, since the introduction of additional corners, or curves, would overcomplicate the formulas. Thwaites [85] has suggested the use of broken-line vortices adaptable to any planform, with the compensating simplification of a quadratic approximation to the variation of vorticity along each segment. By adjusting the constants, the discontinuities in the direction of the vortices and in the magnitude of the loading can be made to counteract each other so that no singularities occur in the downwash. This procedure would appear to avoid many of the pitfalls of other discrete-vortex methods but does not seem to have been developed to any extent.

Vortex lattice method. An arrangement designed to permit greater flexibility in the range of planforms treated has been developed by Falkner

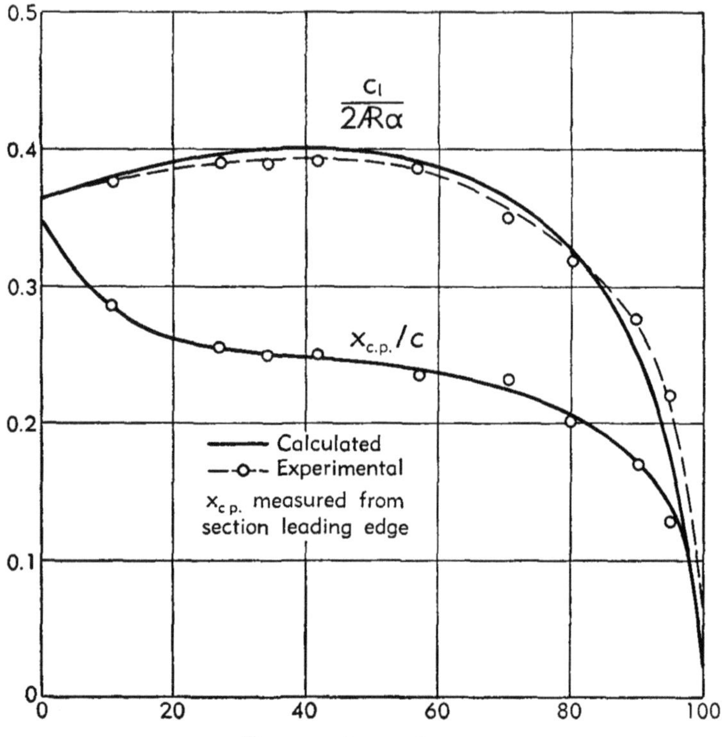

Fig. A,7p. Comparison between calculated and experimental lift distributions and center-of-pressure positions for sweptback untapered wing; $\mathcal{R}$ = 4.5, Λ = 40° [*84*].

[*86,87,88*]. It consists of a large number (of the order of 100) of horseshoe vortices, each extending over only a small fraction of the span, placed so as to form a rectangular lattice, of which Fig. A,7q shows a typical example. By this method the loading along the span as well as along the chord is made stepwise discontinuous. Use is made of extensive tables of the downwash field of a horseshoe vortex computed by the National Physics Laboratory [*89*]. The procedure is similar in other respects to the previously discussed discrete-vortex methods.

In any method employing discrete vortices, the relative location of vortices and control points is of critical importance, since the downwash tends toward infinity in the neighborhood of each vortex. When only a single vortex line is used and the control points are located on the three-quarter chord line, as in the preceding methods, their distance from the vortex is large enough so that the variation of the downwash with location is slow. When the number of vortices is large, the downwash varies from point to point in a manner similar to that shown in Fig. A,7m. It is not surprising, then, that in Falkner's method the solution appears to depend on the vortex pattern and on the control points employed.

It should be noted that Multhopp's lifting surface method, which is equally flexible as regards the planforms that can be treated, is not subject

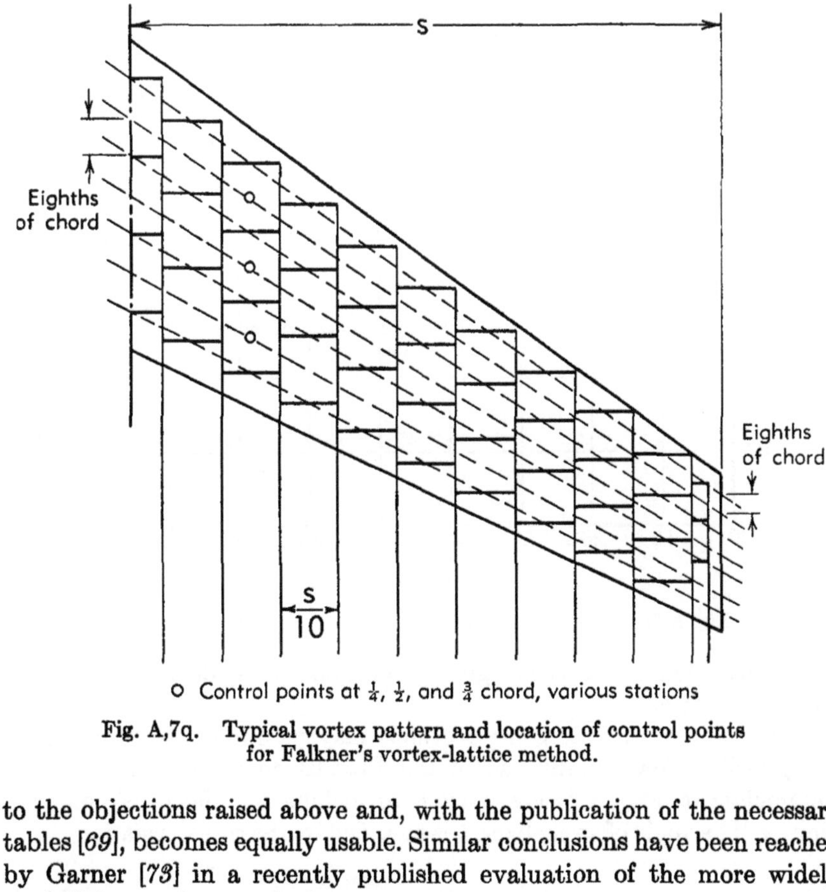

Fig. A,7q. Typical vortex pattern and location of control points for Falkner's vortex-lattice method.

to the objections raised above and, with the publication of the necessary tables [*69*], becomes equally usable. Similar conclusions have been reached by Garner [*73*] in a recently published evaluation of the more widely used lifting surface methods.

Effect of Planform on Lift Due to Angle of Attack. In Fig. A,7r are shown the effects of taper and sweep on the theoretical span loading of wings of aspect ratio 4. The curves are derived from Falkner's results and the charts of [*78*], which are consistent for the cases shown. The values of $C_{L\alpha}$ are those given by Falkner. It can be seen that at low angles of attack the effects of taper and sweepback tend to counteract each other, both with respect to the total lift and its distribution along the span. However, since the lift per unit chord length is greater near the tip of the tapered wing, the undesirable stalling characteristics of a swept-back wing are worsened by the use of taper.

In Fig. A,7s is shown the variation of lift curve slope C_{L_α} with aspect ratio according to various lifting surface theories, for untapered wings with 0, 30, and 45 degrees of sweepback. Included for comparison are the results of the Prandtl lifting line theory, which is of course applicable only to straight wings, and the low aspect ratio theory to be discussed later in

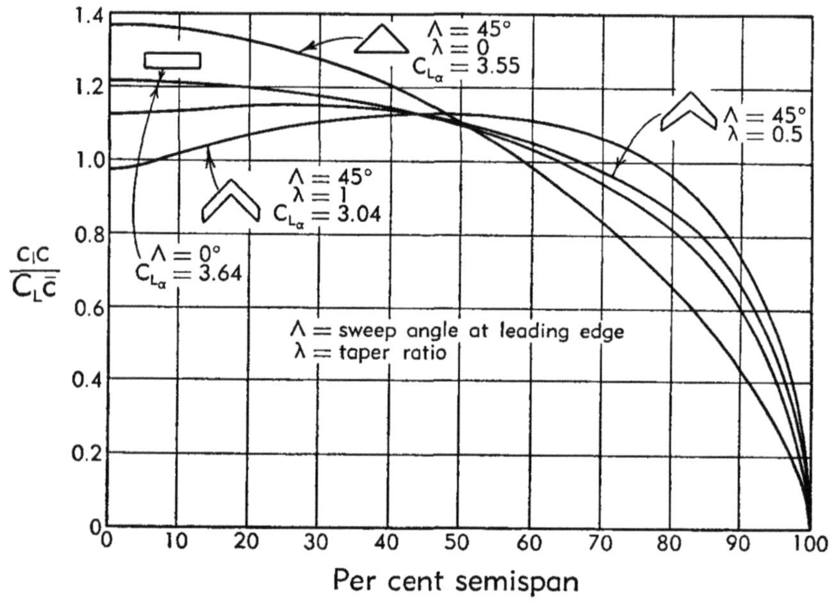

Fig. A,7r. Theoretical span loadings for four wings of Æ = 4.

this article. The latter theory gives a formula for C_{L_α} dependent only on the aspect ratio.

Since any of the foregoing methods may be applied to problems in compressible flow by the use of the Prandtl correction (Art. 5), the parameters of Fig. A,7s are written to include the effects of compressibility. The angle of sweepback Λ is modified by the transformation

$$\tan \Lambda_{comp} = \frac{\tan \Lambda}{\sqrt{1 - M_\infty^2}} \tag{7-51}$$

The lift curve slope and aspect ratio are simply multiplied by $\sqrt{1 - M_\infty^2}$.

The curve for zero sweep is drawn through values taken from [*66,81, 82,84,87*]. While the results of these various lifting surface methods are in excellent agreement as regards the simple rectangular wing, the indication with respect to the lift of sweptback wings is not quite so clear. Curves from [76] and [87] for $\Lambda_{comp} = 30°$ and 45° are in good agreement, but Schlichting's curve for the latter case is somewhat higher and Weissinger's (from [78]) considerably lower. It has already been mentioned that Weissinger's results fail to approach the correct asymptotes (Eq. 4-8) for infinite span, and are therefore not reliable for higher aspect ratios. However, the asymptotic slope at zero aspect ratio is correct. Mutterperl's formula approaches the correct values at either extreme. No further values from lifting surface theory are available for the particular case of the untapered wing. Fig. A,7s includes, however, the results given

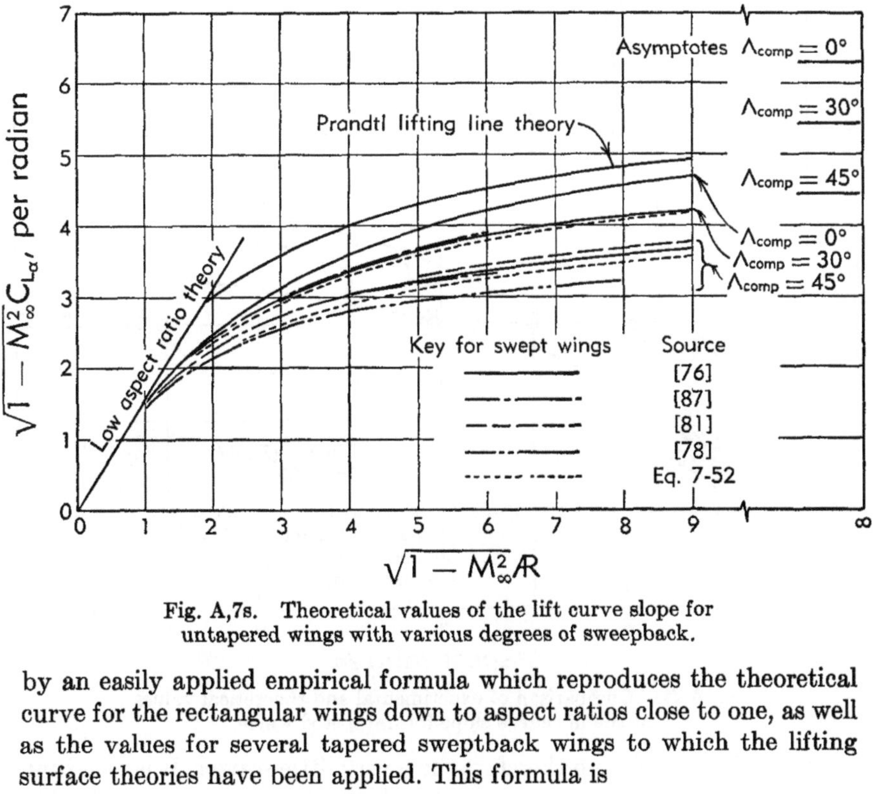

Fig. A,7s. Theoretical values of the lift curve slope for untapered wings with various degrees of sweepback.

by an easily applied empirical formula which reproduces the theoretical curve for the rectangular wings down to aspect ratios close to one, as well as the values for several tapered sweptback wings to which the lifting surface theories have been applied. This formula is

$$C_{L_\alpha} \cong \frac{2\pi Æ}{pÆ + 2} \tag{7-52}$$

where p is the ratio of the semiperimeter of the wing to its span, and results from the generalization of the first order correction to the lifting line theory derived in [54] for elliptical wings (cf. also Eq. 7-46). Applied to sweptback untapered wings, Eq. 7-52 gives values somewhat lower than the lifting surface values, but perhaps close enough for engineering requirements.

Experimental values of the lift curve slope of untapered wings, obtained from a number of sources, are plotted in Fig. A,7t and compared with the theoretical curves from [*76*], extended below Æ = 3 by the use of Falkner's results. In all the theoretical work discussed, 2π has been tacitly assumed as the value of C_{L_α} in two-dimensional flow, whereas it is well known that higher or lower values have been obtained in wind tunnel tests, largely depending on the trailing edge angle of the section. The data presented in Fig. A,7t are almost wholly for conventional 9 or 12

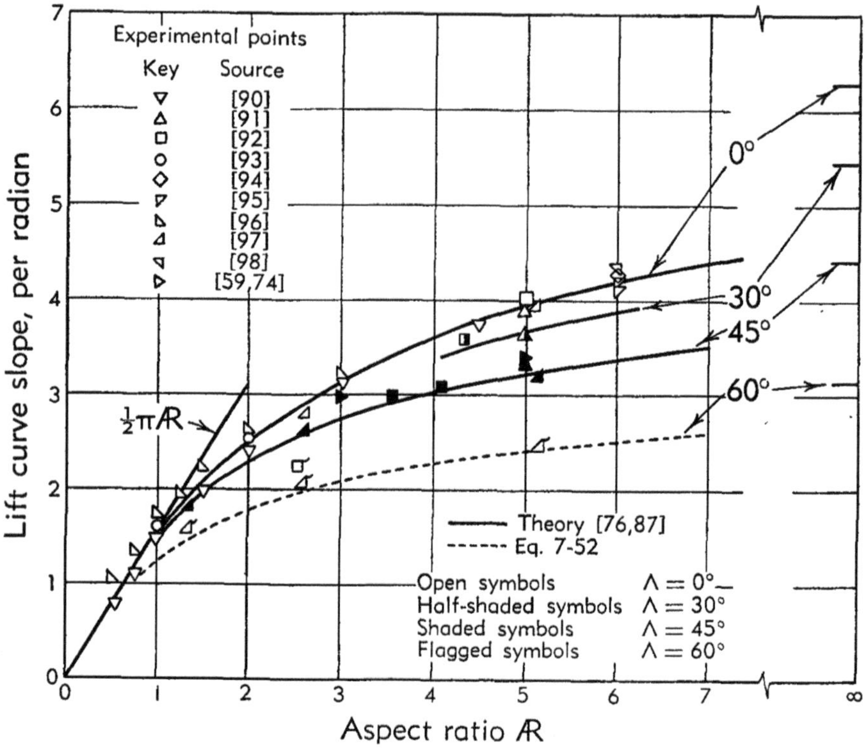

Fig. A,7t. Comparison of experimental and theoretical values of the lift curve slope for untapered wings.

per cent thick airfoils and therefore show very little scatter. It is seen that the points are in very good agreement with the theoretical curves. On this basis the approximate formula (7-52) may also be considered a practical means of obtaining values of the lift curve slope for other angles of sweepback and for tapered wings of moderate-to-high aspect ratio.

Fig. A,7u shows a number of experimental results for triangular wings as well as the theoretical curve calculated for such wings in [*78*] on the basis of Weissinger's method. This curve is seen to represent the average variation of the experimental values quite well. Two results from Falkner's work indicate a slightly higher curve. Application of Eq. 7-52 to triangular wings gives a curve which, although approaching zero along the correct tangent, appears to be perhaps ten per cent too high at higher aspect ratios. The subject of triangular wings will be treated in more detail in the following article.

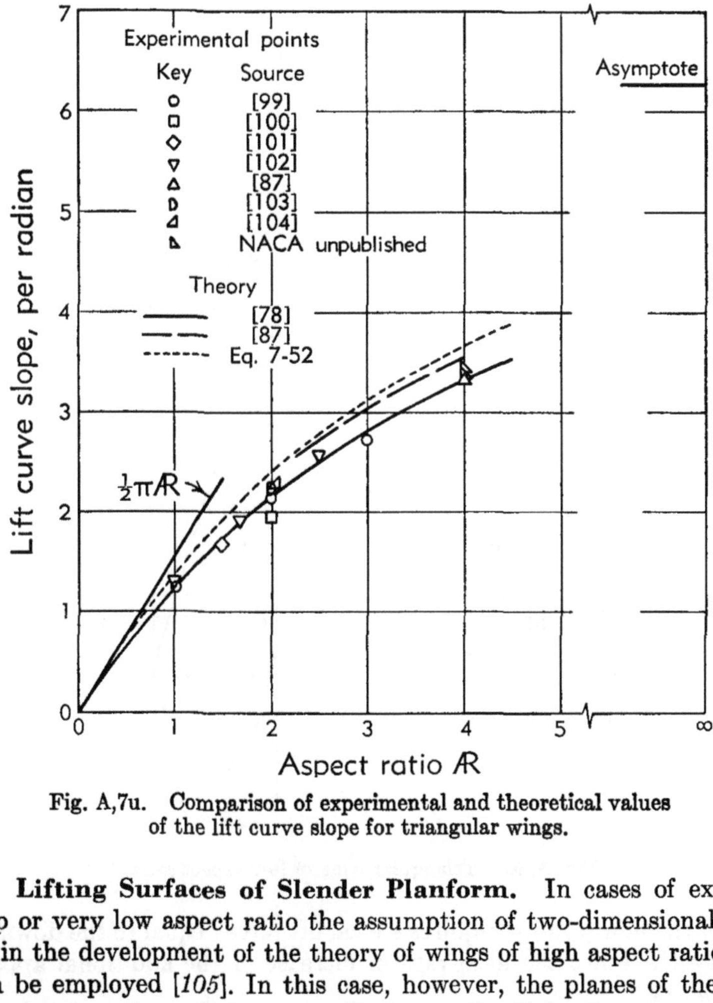

Fig. A,7u. Comparison of experimental and theoretical values of the lift curve slope for triangular wings.

A,8. Lifting Surfaces of Slender Planform. In cases of extreme sweep or very low aspect ratio the assumption of two-dimensional flow, used in the development of the theory of wings of high aspect ratio, can again be employed [*105*]. In this case, however, the planes of the two-dimensional motion will be perpendicular to the flight direction and the origin of the lift appears in a somewhat different aspect than in the case of the wing of large span and narrow chord. The formulas obtained resemble in some respects those derived by Munk [*106*] and Tsien [*107*] for the lift of an elongated body of revolution.

Theory for wings of low aspect ratio. With the idealization of two-dimensional flow, the treatment of the low aspect ratio wing becomes exceedingly simple. The case lending itself most readily to the analysis is that of the long flat triangular airfoil, traveling point-foremost at a small angle of attack. Viewed from a reference system at rest in the undisturbed fluid, the flow pattern in a plane cutting the airfoil at a distance x from the nose is the familiar two-dimensional flow caused by a flat plate having the normal velocity $U\alpha$ (see Fig. A,8a). Observed in this plane, the width of the plate and hence the scale of the flow pattern continually increase as the airfoil progresses through the plane. This increase in the

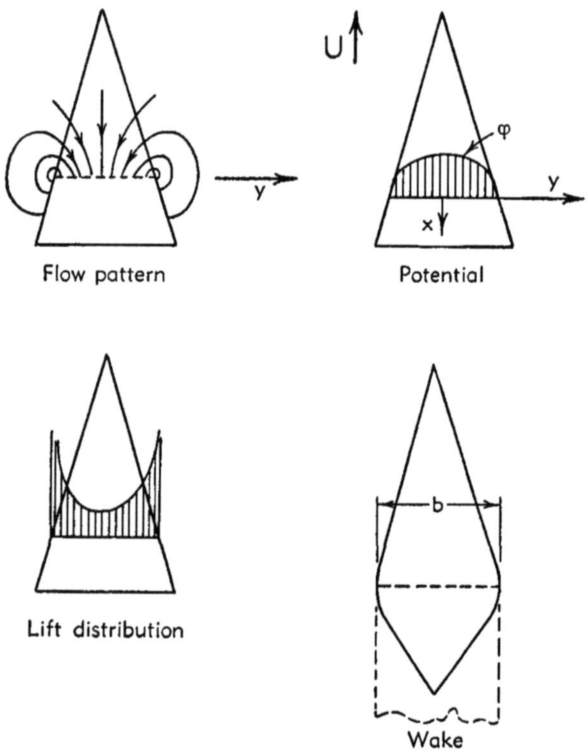

Fig. A,8a. Triangular wing of low aspect ratio.

scale of the flow pattern requires a local lift force l equal to the downward velocity $U\alpha$ times the local rate of increase of the additional apparent mass m', or

$$l = U\alpha \frac{dm'}{dt} = U^2\alpha \frac{dm'}{dx}$$

since $U = dx/dt$.

By a well-known formula from two-dimensional flow theory,

$$m' = \pi y_1^2 \rho_\infty$$

where $y_1(x)$ is the y coordinate of the leading edge and $2y_1$ is the local width of the plate. Hence

$$\frac{dm'}{dx} = 2\pi\rho_\infty y_1 \frac{dy_1}{dx}$$

and the lift l per unit length will be given by the expression

$$l = 4\pi\alpha \frac{1}{2} \rho U^2 y_1 \frac{dy_1}{dx}$$

Dividing by $\frac{1}{2}\rho_\infty U^2$ and by the span $2y_1$ gives the local lift coefficient

$$c_l = 2\pi\alpha \frac{dy_1}{dx} \tag{8-1}$$

When this flow is considered in more detail, it is found from the two-dimensional theory that the surface potential φ is distributed spanwise according to the ordinates of an ellipse, i.e.

$$\varphi = \pm U\alpha \sqrt{y_1^2 - y^2} \tag{8-2}$$

(see Fig. A,8a). An instant later, in the same plane, the ordinates are those of a slightly larger ellipse, corresponding to an increase of φ. The local pressure difference is given by the local rate of increase of φ,

$$\begin{aligned} p_{\text{lower}} - p_{\text{upper}} = \Delta p &= 2\rho_\infty \frac{\partial \varphi}{\partial t} \\ &= 2\rho_\infty U \frac{\partial \varphi}{\partial x} \\ &= 2\rho_\infty U \frac{\partial \varphi}{\partial y_1} \frac{dy_1}{dx} \end{aligned} \tag{8-3}$$

where $\partial\varphi/\partial y_1$ is a function of y. Differentiation of φ yields the equation

$$\Delta p = 2\rho_\infty U^2 \alpha \frac{y_1}{\sqrt{y_1^2 - y^2}} \frac{dy_1}{dx} \tag{8-4}$$

The pressure distribution thus shows an infinite peak along the sloping sides of the airfoil similar to the pressure peak at the leading edge of a conventional airfoil. The distribution along radial lines passing through the vertex of the triangle (lines of constant y/y_1) is uniform, however, and the center of pressure coincides with the center of area. The maintenance of lift right up to the trailing edge is an effect associated solely with the limiting case of zero width and will not exist in the case of a wing of finite aspect ratio.

Eq. 8-1 and 8-4, although derived for triangular wings, actually apply to more general planforms, since dy_1/dx is not necessarily a constant, and show that the development of lift by a long slender airfoil depends on an expansion of the sections in a downstream direction. Hence a part of the surface having parallel sides would develop no lift. Furthermore, a decreasing width would, according to Eq. 8-4, require negative lift with infinite negative pressure peaks along the edges of the narrower sections. In the actual flow, however, the edge behind the maximum cross section will lie in the viscous or turbulent wake formed over the surface ahead; for this reason it will be assumed that the infinite pressure difference indicated by Eq. 8-4 cannot be developed across these edges. It is this assumption, corresponding to the Kutta condition, that gives the plate the properties of an airfoil as distinct from another type of body such as a body of revolution.

With the aid of the Kutta condition, it may easily be shown that sections of the airfoil behind the section of greatest width develop no lift. A potential flow satisfying both the boundary condition and the Kutta condition may be obtained by the introduction of a free surface of discontinuity behind the widest section. This surface of discontinuity (Fig. A,8a) is composed of parallel vortices extending downstream from the widest section of the airfoil as prolongations of the vortices representing the discontinuity of potential over the forward part of the airfoil. Such a sheet, although possibly wider than the downstream sections of the airfoil, still satisfies their boundary condition, since the lateral arrangement of the vortices is such as to give uniform downward velocity equal to $U\alpha$ over the entire width of the sheet, including the rearward portion of the airfoil. The Kutta condition will be satisfied within the region covered by the sheet of parallel vortices regardless of the shape of the trailing edge. Since the pressure difference across the airfoil is proportional to $\partial\varphi/\partial x$, and since this gradient disappears as soon as the vortices become parallel to the stream, no lift is developed on the rearward sections.

Integration of the lifting pressures in a chordwise direction from the leading edge downstream to the widest section will give the span load distribution and the induced drag. Since, from Eq. 8-3,

$$\int \Delta p dx = 2\rho_\infty U\varphi$$

the span load distribution is

$$\frac{\partial L}{\partial y} = 2\rho_\infty U^2\alpha\sqrt{\left(\frac{b}{2}\right)^2 - y^2} \tag{8-5}$$

where b is the span of the wing. Hence $\partial L/\partial y$ is elliptical, and is furthermore independent of the planform in these simple cases.

With the elliptical span load the induced drag is equal to its minimum value

$$D_i = \frac{L^2}{\pi q b^2} \tag{8-6}$$

A second integration of $(\partial L/\partial y)dy$ across the widest section gives the total lift, which is

$$L = \frac{\pi}{4} \rho_\infty U^2 \alpha b^2 \tag{8-7}$$

The lift of the slender airfoil therefore depends only on the width and not on the area.

Dividing by $\frac{1}{2}\rho_\infty U^2 S$, we obtain the lift coefficient

$$C_L = \frac{\pi}{2} Ʀ\alpha, \qquad Ʀ = \frac{b^2}{S} \tag{8-8}$$

and, from Eq. 8-6, the induced-drag coefficient

$$C_{D_i} = \frac{C_L^2}{\pi Ʀ} = C_L \frac{\alpha}{2} \tag{8-9}$$

From Eq. 8-9 it appears that the resultant force lies halfway between the normal to the surface and the normal to the air stream.

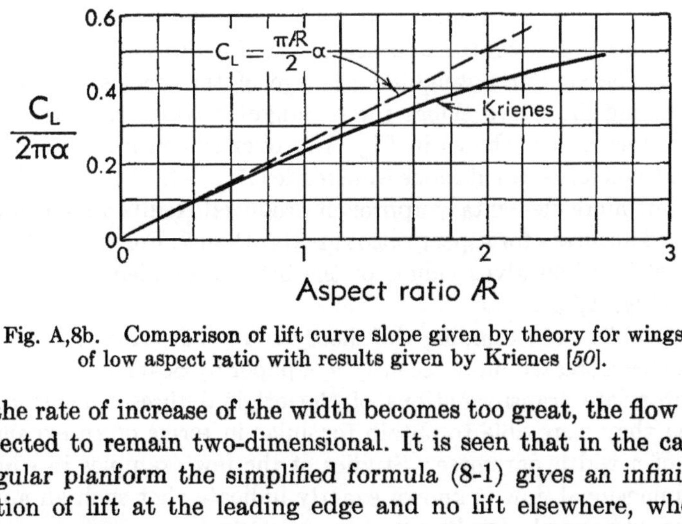

Fig. A,8b. Comparison of lift curve slope given by theory for wings of low aspect ratio with results given by Krienes [*50*].

If the rate of increase of the width becomes too great, the flow cannot be expected to remain two-dimensional. It is seen that in the case of a rectangular planform the simplified formula (8-1) gives an infinite concentration of lift at the leading edge and no lift elsewhere, whereas a more accurate theory would show some distribution of the lift rearward. However, it can be shown by examination of the three-dimensional (nonlifting) potential flow around an elliptic disk (Eq. 7-22 and 7-23), that the two-dimensional theory gives a good approximation in the case of an elliptical leading edge, which indicates that the theory is applicable over a large range of nose shapes. In Fig. A,8b is shown a comparison of the lift calculated by the present theory for elliptical wings of low aspect ratio with the results of the more accurate three-dimensional potential flow calculations of Krienes (Art.7). It is seen that the simple theory approximates the exact value very closely up to an aspect ratio of 1. Comparison with the lifting surface values of the lift curve slope for rectangular and sweptback wings (Fig. A,7s) as well as experimental data (Fig. A,7t) suggests the same range of validity for untapered wings.

Extensions of low aspect ratio theory. The essentially elliptic form of the span loading on low aspect ratio wings was surmised by Weighardt [*80*] on empirical grounds and used to reduce the lifting surface integral equation (7-41) to a one-dimensional one, which he then solved numerically for rectangular wings by means of an expansion chordwise in a Birnbaum series, satisfying the downwash condition at points along the axis of the wing.

A development which shows the transition from the theory of lifting surfaces of high aspect ratio to those of low aspect ratio has been given by Lawrence [*108*], and results in a more rigorous proof that the ellipse is the limiting form of the span loading on a flat rectangular plate of infinite depth. Applying the slender wing relations between w and φ derived for the rectangular wing (not necessarily flat) to simplify the lifting surface integral equation (7-41), Lawrence is able to write a first order integral equation in the integrated lift per unit chord, which is then solved numerically through expansion in series. In effect, the procedure involves equating a weighted average of the downwash at each spanwise section to the similarly weighted average of the local angle of attack. The resulting lift curve slopes for rectangular wings coincide with the lifting surface values shown in Fig. A,7s over the entire range of aspect ratios for which the calculations were made ($\mathcal{R} \leqq 4$), whereas Weighardt's similar but more elementary approach produced results which cannot be considered accurate for aspect ratios greater than 2. For triangular wings, Lawrence's method gives values of the lift almost identical with Weissinger's (Fig. A,7u).

The problem of extending the theory to include wings of wider planform has been treated more recently in a paper by Sears and Adams [*109*]. By appropriate transformations of the original three-dimensional flow equation they were able to obtain formulas in terms of an expansion in powers of a width parameter. Studies of the few examples in which the three-dimensional flow is known exactly indicate that such an approach should yield accurate results.

Slender sweptback wings. The consideration of wings with sweptback trailing edges introduces additional difficulties which even a return to the two-dimensional approach fails to eliminate if the trailing edge cutout extends forward of the tip region, i.e. into the lifting part of the wing (see Fig. A,8c). This problem has been studied by Heaslet and Lomax [*110*] and Robinson [*111*], and subsequently by Mangler [*112*], Legendre [*113*] and Mirels [*114*].[11]

[11] Also, in a very recent paper [*115*], Eichelbrenner has indicated how the method of Adams and Sears [*109*] can be extended to this problem.

In cross stream sections the portion of the wing between sections AA and BB in Fig. A,8c appears as two parallel flat plates, and the form of the complex velocity function can be written down from a consideration of its singularities, except that the magnitude of the circulatory term will depend on the vorticity in the wake, which in turn depends on the spanwise variation of lift at upstream sections. The expression of this dependency on upstream conditions takes the form of an integral equation. Except for a special case, which will be discussed later, this equation has been solved only numerically.

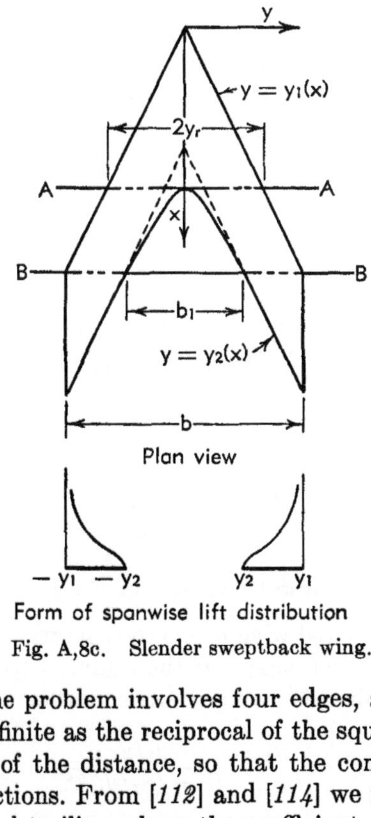

Fig. A,8c. Slender sweptback wing.

We note that the problem involves four edges, at each of which the pressure either is infinite as the reciprocal of the square root, or vanishes as the square root of the distance, so that the complete solution must involve elliptic functions. From [*112*] and [*114*] we find, for a wing with arbitrary leading and trailing edges, the coefficient of lifting pressure for the flat wing at an angle of attack α to be

$$\frac{\Delta p}{\frac{1}{2}\rho_\infty U^2} = 4\alpha S(x)\frac{dy_1}{dx}\left[Z(\psi, k) + \frac{y}{y_1}\sqrt{\frac{y^2 - y_2^2}{y_1^2 - y^2}}\right] \qquad \text{(8-10)}$$

where Z is the Jacobian zeta function of argument

$$\psi = \sin^{-1}\sqrt{\frac{y_1^2 - y^2}{y_1^2 - y_2^2}}$$

and modulus

$$k = \sqrt{1 - \left(\frac{y_2}{y_1}\right)^2}$$

and where $S(x)$ is a function to be determined from the integral equation.

Mirels has calculated values of $S(x)$ for untapered sweptback wings, and Mangler, for both tapered and untapered. Since the results of slender wing theory are of particular interest in their application to sonic speed, as explained in a later section, these results will be discussed further in connection with supersonic lift (Art. 14).

An example that lends itself to solution in closed form is provided by the condition that no vortices exist in the V-shaped region between the two wing panels. Ordinarily this condition would require a certain variation of the camber and twist of the wing surface. However, it is found

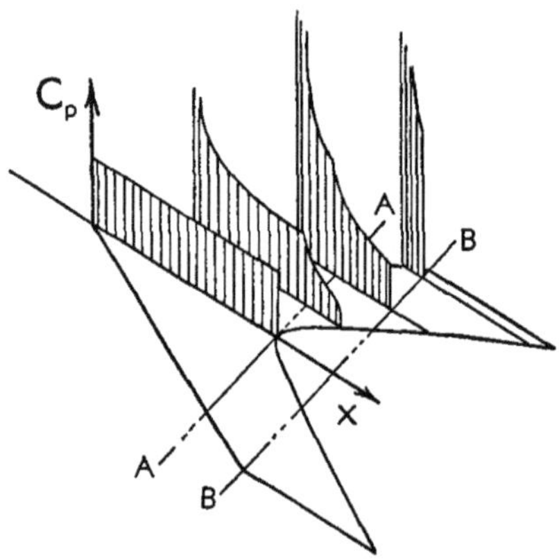

Fig. A,8d. Lift distribution on wing of Fig. A,8c (from [*110*]).

[*110*] that a flat untwisted wing meets the condition, provided the trailing edge is given a particular shape. If the leading edge is straight, the trailing edge shape required is given by the equation

$$\frac{y_2}{y_r} = \frac{y_2/y_1}{E - (y_2/y_1)^2 K} \tag{8-11}$$

where E and K are the complete elliptic integrals with the modulus $\sqrt{1 - (y_2/y_1)^2}$ and y_r is the semispan of the section AA through the trailing edge of the root chord. With the substitution of $y_1 \sim x$, Eq. 8-11 gives the equation of the trailing edge in parametric form. The shape is the one used for illustration in Fig. A,8c. At large aspect ratios, the planform approaches that of a swept wing of constant chord; there is, however, a rounding or filletting of the trailing edge at the root section.

The distribution of lift over the surface of wings having this form is illustrated in Fig. A,8d. It will be noted that a sharp drop in the lifting pressure occurs behind section AA and that no lift occurs behind the section BB at which the maximum span is reached. These discontinuities in the calculated results must of course be associated with the limiting case of zero width, and a more complete treatment, taking into account the three-dimensional nature of the flow, would show some redistribution of lift in these regions on wings of finite span.

As in the first example (Fig. A,8a), the exact shape of the tip and trailing edge behind the section of maximum span has no effect on the flow, nor on the total lift. The formulas for the lift and induced drag are

$$\frac{L}{\frac{1}{2}\rho_\infty U^2\alpha} = \frac{\pi}{2}(b^2 - b_1^2) \tag{8-12}$$

$$\frac{D_i}{\frac{1}{2}\rho_\infty U^2\alpha^2} = \frac{\pi}{4}(b^2 - b_1^2) - 2\frac{y_1}{b}[b^2E(k) - (b^2 - b_1^2)K(k)] \tag{8-13}$$

where b is the span of the wing (Fig. A,8c), b_1 is the span of the wake at section BB, and $k = b_1/b$.

It is of interest to note that Eq. 8-12 is the result that would be obtained by applying the theory of oblique cylindrical flows (Art. 4) to an equivalent swept wing having a straight trailing edge (i.e. one which lacks the fillet area indicated on Fig. A,8c), at the same time taking into account the absence of lift in the tip region indicated by the slender wing theory. The area added by the curved trailing edge is evidently just sufficient to compensate for the interference between the two sides of the wing for every ratio of b_1/b.

Increasing the aspect ratio of a wing having a large angle of sweep does not necessarily increase the slope of the lift coefficient curve. For example, if one begins with a slender triangular wing, the aspect ratio may be made as large as desired by cutting out V-shaped areas, leaving a sweptback wing of constant chord. According to the foregoing formulas the lift curve slope would at first decrease, but would eventually, as the aspect ratio approached infinity, return to the value given by the original triangle. However, the ratio of lift to drag continually increases with aspect ratio.

Effect of compressibility. Application of the Prandtl transformation to the preceding formulas shows that in the limiting condition approached by the slender wing there is no effect of compressibility on the pressures. As explained in Art. 5, the incompressible flow corresponding to the compressible flow over a wing at Mach number M_∞ is obtained by a calculation based on a fictitious wing whose lengthwise dimensions have been increased in the ratio $1/\sqrt{1 - M_\infty^2}$. The pressures at corresponding points on the actual wing are then obtained by increasing the values calculated for the fictitious wing in this same ratio. If the wing is sufficiently slender, however, the pressures will vary inversely as the length, and hence the

Prandtl transformation merely restores the pressures to their value at $M_\infty = 0$. As pointed out in [*105*], these statements apply to supersonic as well as subsonic speeds so long as the wing lies near the center of the Mach cone originating at its apex.

Application to the problem of flow in the vicinity of sonic speed. The velocity potential of the flow disturbance caused by a slender lifting surface of arbitrary form traveling at the velocity U may be represented by the function

$$\varphi = F(y + iz, x - Ut) \tag{8-14}$$

which satisfies the two-dimensional equation

$$\varphi_{yy} + \varphi_{zz} = 0 \tag{8-15}$$

in planes at right angles to the velocity of flight. In this form the pressure is given by

$$\Delta p = \rho_\infty \frac{\partial \varphi}{\partial t}$$

Eq. 8-14 represents a complete three-dimensional flow only in the sense of the approximate slender wing theory. When the velocity U is set equal to the velocity of sound a_∞, however, Eq. 8-14 becomes an exact solution of the sound wave equation in three dimensions (3-9). Thus the formulas developed for slender wings become applicable to wings of finite aspect ratio at speeds near the speed of sound [*116,117*]. This fact may be verified in another way by considering the flow to be steady, and applying the Prandtl-Glauert correction. The correction yields an equivalent wing whose aspect ratio is continually reduced as the Mach number increases toward 1.0.

At $U = a_\infty$, Eq. 8-14 represents a kind of sound wave, with lateral and vertical motions satisfying the boundary condition of the wing. The condition that the divergence of the component velocity field in the y, z planes be zero is analogous to the condition imposed by the one-dimensional flow equation (3-4), which prescribes no change in the cross-sectional area of the stream tubes at sonic velocity. Streamlines at sonic velocity are thus laterally incompressible.

An example of Eq. 8-14 is the family of solutions for wings of triangular planform, given by functions of the type

$$\frac{\varphi}{a_\infty} = (x - a_\infty t)F\left(\frac{y + iz}{x - a_\infty t}\right) \tag{8-16}$$

If the real solution is chosen and the substitution

$$\xi = \frac{y + iz}{x - a_\infty t}$$

is made, the expression

$$\frac{v - iw}{a_\infty} = F'(\xi) \tag{8-17}$$

for the lateral and vertical velocities is obtained. The function F' may be selected with the aid of examples given previously (Table A,2). The pressures are determined from the acoustic formula

$$\frac{\Delta p}{\frac{1}{2}\rho_\infty a_\infty^2} = \frac{2}{a_\infty^2}\frac{\partial \varphi}{\partial t} = 2[\xi F'(\xi) - F(\xi)] \tag{8-18}$$

The isobars are straight lines radiating from the point $x = a_\infty t$, as shown in Fig. A,8e.

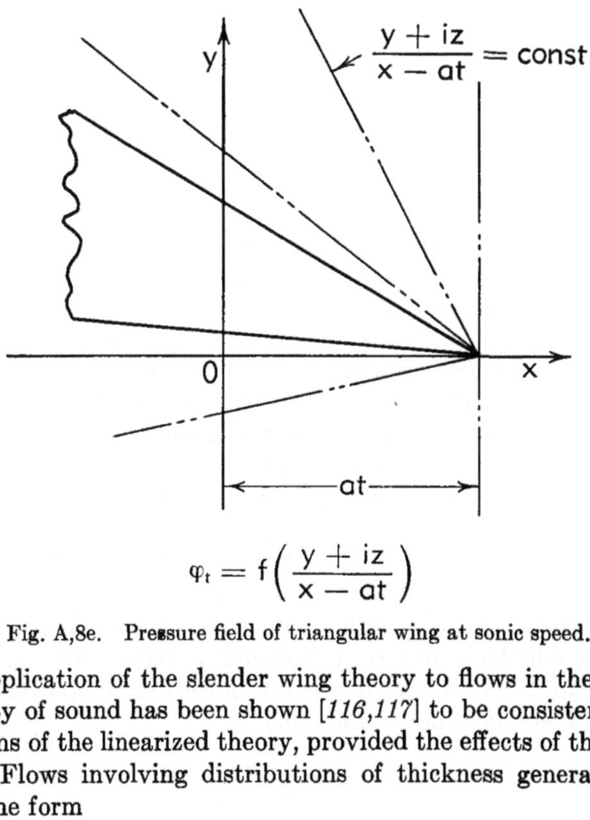

Fig. A,8e. Pressure field of triangular wing at sonic speed.

The application of the slender wing theory to flows in the vicinity of the velocity of sound has been shown [*116,117*] to be consistent with the assumptions of the linearized theory, provided the effects of thickness are excluded. Flows involving distributions of thickness generally include terms of the form

$$\varphi = \ln(y + iz)f(x - a_\infty t) \tag{8-19}$$

which lead to infinite pressure at infinity. The failure of the linearized theory in this case may be attributed to the presence of sources, which violates the condition of zero divergence. In the special case of the oblique cylindrical wing (Eq. 4-2) the total strength of the sources in each plane cross section is zero.

Nonlinear characteristics of low aspect ratio wings. Application of the linearized frictionless-flow theory to airfoils of low aspect ratio is limited more strictly to small angles of attack and small lift coefficients than is the case with airfoils of more usual proportions. As explained in Art. 4, the effect of yaw or sweep on a long narrow wing of cylindrical form is to reduce the lift coefficient for flow separation by the factor $\cos^2 \beta$. With relatively thin sections, such flow separation usually takes place at the leading edge as the result of the large pressure gradients associated with the flow around the nose of the section. With wings of low aspect ratio the flow in the vicinity of the leading edge remains locally similar to the flow around the edge of an oblique cylindrical wing, and hence flow separation can be expected at progressively smaller lift coefficients as the angle of sweep of the leading edge is increased. The linearized theory developed in the preceding sections is of course based on the assumption that the flow remains attached around the leading edge, and its validity is thus limited to a progressively smaller range of angles of attack as the leading edge is made more oblique. In the case of a sharp leading edge or a side edge parallel to the stream, separation can be expected at the smallest angles of attack.

Experimental observations of the flow after separation indicate the presence of a vortex surface which leaves the leading edge and rolls up in the form of a spiral sheet above and behind the edge. The radius of the spiral is small near the front of the wing and becomes larger at downstream sections. Partly as a result of the diffusion of vorticity, the spiral resembles a single vortex with a finite core. In this regime the lift continues to increase with angle of attack—frequently up to an angle of 45 degrees—but the variation is nonlinear and the lifting pressures are no longer divided equally between the upper and lower surfaces. Of the greatest practical consequence is the rapid increase of drag with angle of attack. After the flow becomes detached from the edge, the forward suction force no longer increases in proportion to the lift, with the result that the theoretical formulas for drag no longer apply and the resultant force on the wing falls back toward a direction at right angles to the chord plane. Prior to the occurrence of separation the drag is observed to follow roughly the theoretical minimum value

$$C_D = C_{D_0} + \frac{C_L^2}{\pi A\!\!\!R} \tag{8-20}$$

but at higher angles of attack the value

$$C_D = C_{D_0} + C_L \tan \alpha \tag{8-21}$$

is approached.

Fig. A,8f, taken from data given in [*99*], illustrates the behavior observed experimentally in the case of low aspect ratio wings having different angles of obliquity of the leading edge. In each case the aspect ratio is $\frac{4}{3}$ and the airfoil section is the NACA 0012.

It will be noted that with wings of low aspect ratio the transition from one law of drag variation to the other is more gradual than in the case of the slender sweptback wing (Fig. A,4k).

A satisfactory quantitative treatment of the forces experienced by a wing in the nonlinear range is not yet available, though the behavior illustrated in Fig. A,4k and A,8f may be used as a guide. A treatment of the limiting case of a rectangular wing of small aspect ratio was given by Bollay [*118*]. Bollay assumes that a vortex sheet leaves the side edge of the wing, forming a region of constant negative pressure above the upper

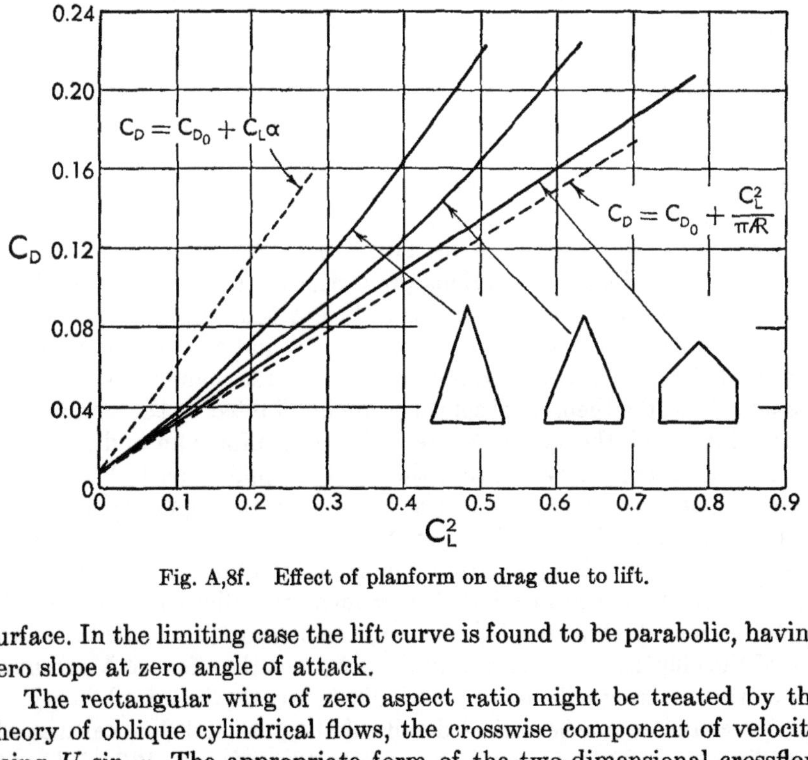

Fig. A,8f. Effect of planform on drag due to lift.

surface. In the limiting case the lift curve is found to be parabolic, having zero slope at zero angle of attack.

The rectangular wing of zero aspect ratio might be treated by the theory of oblique cylindrical flows, the crosswise component of velocity being $U \sin \alpha$. The appropriate form of the two-dimensional crossflow in this case would be of the discontinuous type, such as the Helmholtz flow, satisfying the condition of tangential velocity at the side edges. Unfortunately there exists an infinite number of solutions of this type depending on the pressure assumed in the region above the wing, and there seems to be no physical criterion for selecting a certain value of this pressure. In Helmholtz' configuration the pressure is zero, but experimentally a considerable negative pressure is observed. A result in fair agreement with experiment in the case of a sufficiently long narrow plate can be obtained by calculating the force normal to the plate from an experimentally determined drag in a perpendicular air stream, corresponding to the component $U \sin \alpha$ [*119,120*]. The crosswise drag coefficient of such a plate has approximately the value 2.0, which leads to the expression for the normal force coefficient

$$C_n = 2 \sin^2 \alpha \tag{8-22}$$

A,9. Thickness Distributions. In addition to the effects of camber and twist of the mean surface, consideration of the thickness distribution is required for the complete determination of the pressures and forces on the wing. In the linearized theory the thickness modifies the pressure distribution by the addition of equal pressures above and below the chord plane and hence does not affect the lift. Furthermore, in steady potential flow at subsonic speeds the thickness does not introduce drag. The additional flow velocities and pressure gradients arising from the thickness are decisive, however, in determining the Mach number at which the breakdown of smooth regular flow occurs and the subsequent rise in drag on transition to supersonic speeds.

AERODYNAMIC PROPERTIES OF THIN, FLAT ELLIPSOIDS. The potential flow over ellipsoids of various proportions can be completely determined by relatively simple formulas. Since a thin flat ellipsoid bears a close resemblance to a wing, a study of the flow and pressure distributions over such forms is of considerable interest in aerodynamics. With incompressible flow, the theoretical solutions are not limited by the assumptions of the thin airfoil theory and hence their study affords an understanding of certain errors involved in that theory. For application to high speeds the study of the incompressible case is made on the supposition that the aspect ratio and the velocity distribution are to be corrected by the Prandtl rule. In this application the approximations of the linearized theory must be borne in mind. Under most conditions the greatest departures from the linearized theory will occur near the rounded nose and tail of the elliptical sections and in these regions the effect of Mach number on the pressures will be less than indicated by the linearized theory.

For complete mathematical details of the potential flow calculations for ellipsoids, reference should be made to Lamb [*1*, pp. 139–156], or to [*121*, Vol. 1, pp. 293ff.] However, the results of the theory are of such simplicity that they can be completely summarized with the aid of a few elementary concepts.

As in the case of the lifting elliptic disk, the flow around nonlifting ellipsoids is studied with the aid of the ellipsoidal coordinates λ, μ, ν (Art. 7) and the associated Lamé functions. Since the vortex wake is absent, no simplification results from the use of the acceleration potential and hence the solutions are expressed in terms of the velocity potential

φ. Accordingly,

$$\varphi_n^m = E_n^m(\mu)E_n^m(\nu)F_n^m(\lambda) \tag{9-1}$$

where E_n^m and F_n^m are defined as in Art. 7.

The various solutions represented by Eq. 9-1 may be thought of as flows caused by various distributions of velocity around the surface of the base ellipsoid $\lambda = 0$. The solutions for uniform translation of the solid ellipsoid in the directions of x, y, and z are given by

$$\left.\begin{aligned} \varphi_1^1 &= A_1^1E_1^1(\mu)E_1^1(\nu)F_1^1(\lambda) = xf_1^1(\lambda) \\ \varphi_1^2 &= A_1^2E_1^2(\mu)E_1^2(\nu)F_1^2(\lambda) = yf_1^2(\lambda) \\ \varphi_1^3 &= A_1^3E_1^3(\mu)E_1^3(\nu)F_1^3(\lambda) = zf_1^3(\lambda) \end{aligned}\right\} \tag{9-2}$$

On the surface of the base ellipsoid ($\lambda = 0$), $f(\lambda) =$ const, and it is seen that in each case the potential is proportional to the coordinate of the surface in the direction of motion. Hence the equipotential lines on the surface are ellipses formed by the intersections of a series of equidistant parallel planes with the surface $\lambda = 0$. For motion in any of the three principal directions, the planes of the equipotential lines are perpendicular to the direction of motion.

The form of the surface potential (9-1) leads to a simple rule for computing the velocity distribution over the surface of any ellipsoid fixed in an incompressible stream [*9*]. For this purpose it is necessary to know the magnitude and direction of the maximum velocity. The velocity at every point on the surface can then be obtained simply by taking the projection of the maximum velocity on the plane parallel to the surface at the point. For flow parallel to a principal axis, the maximum velocity is parallel to the direction of flow at infinity and appears all around the widest section of the ellipsoid. In oblique flow the maximum velocity will be inclined to the main flow direction; it is usually more convenient to treat oblique motion by resolving the velocity at infinity into components along the principal axes of the ellipsoid and superimposing the resulting velocity distributions.

The maximum velocity can be found directly from the coefficient of x, y, or z in the expression for φ, or in other words from the constants $f_n^m(0)$ in Eq. 9-2. In a stream of velocity U along any one of the three directions, the maximum velocity on the surface of the ellipsoid will be given by

$$[1 + f_n^m(0)]U$$

Hence

$$f_n^m(0) = \left(\frac{u}{U}\right)_{\max}$$

The evaluation of the constants $f_n^m(\lambda)$ has been given in several text books [*1*; *121*, pp. 224–293] and need not be reproduced here.

Fig. A,9a, plotted from data given in [122], shows values of $(u/U)_{max}$ for ellipsoids of various proportions. It will be noted that for every elliptic cylinder the maximum additional velocity is exactly equal to the thickness-chord ratio. Thus, for a thickness-chord ratio of 10 per cent and flow parallel to the chord plane, the velocity is $1.10U$, while for flow perpendicular to the chord plane the velocity around the nose and tail of the section is $11.0U$.

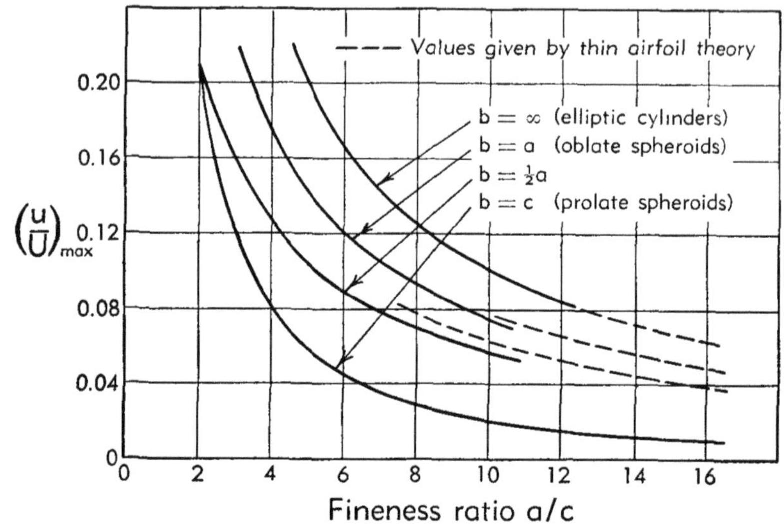

Fig. A,9a. Maximum velocity increments for ellipsoids; $\alpha = 0$, $M_\infty = 0$.

Of particular interest in the thin wing theory are the limiting values of $(u/U)_{max}$ for small thickness, i.e. $c/a \to 0$ and $c/b \to 0$. These limiting values are found to be

$$\left(\frac{u}{U}\right)_{max} = \frac{c}{a}\,\frac{ab(K - E)}{a^2 - b^2} \qquad \text{for } a > b \tag{9-3}$$

$$\left(\frac{u}{U}\right)_{max} = \frac{c}{b}\,\frac{a^2E - b^2K}{a^2 - b^2} = \frac{c}{a}\,\frac{b^2E - a^2K}{b^2 - a^2} \qquad \text{for } b > a \tag{9-4}$$

$$\left(\frac{w}{U}\right)_{max} = \frac{a}{c}\,\frac{1}{E} \tag{9-5}$$

for flow in the direction of c normal to the chord plane. Here K and E are the complete elliptic integrals with modulus $k = \sqrt{1 - (b/a)^2}$ or $k = \sqrt{1 - (a/b)^2}$. Eq. 9-5 yields an infinite velocity around the edge of a perfectly thin disk ($c = 0$), but the value differs from that for two-dimensional motion by the factor $1/E$, which is the ratio of the span $2b$ ($b > a$) to its semiperimeter.

In flow parallel to the chord plane, as indicated in Fig. A,9a, the values of the maximum velocity given by the thin airfoil approximation are exact for the case of infinite aspect ratio ($b = \infty$), but are larger than the actual surface velocities for ellipsoids of finite proportions. The thin airfoil approximation cannot of course be used for a body of revolution since the velocities along the axis of singularities are infinitely greater than the surface velocities.

Fig. A,9b, taken from data given in [*123*], shows the effect of Mach number on $(u/U)_{max}$ for ellipsoids of various aspect ratios ($\mathit{AR} = 4a/\pi b$) having thickness-chord ratios of 10 per cent. The calculations, which were based on the thin airfoil theory, show that the effect of Mach number is greatest in two-dimensional flow. Also indicated on the figure are values of the Mach number for which the linearized theory predicts the occurrence of a local supersonic region in the flow. The occurrence of such a region usually precedes the appearance of a shock, although no definite significance can be attached to such a calculation in three-dimensional flow. For the elliptic cylinder placed at an angle of yaw β, the designation of the abscissa may be replaced by $M_\infty \cos \beta$ and the quantities u_{max} and U will refer to the component velocities at right angles to the axis of the cylinder.

FLOW OVER WINGS OF CONICAL OR CYLINDRICAL FORM. The pressure field of thin nonlifting airfoils of rectilinear planform and simple thickness distribution can be obtained by direct integration of the effect of a distribution of sources over the chord plane. Calculations based on this method have been given in [*124*] and more recently in [*125,126*].

The method of [*124*] involves the use of the acceleration potential (Art. 7). In these terms a source represents a singularity in the pressure field or in what amounts to the same thing, the field of horizontal perturbation velocities $u(x, y, z)$. Thus the fundamental solution

$$\frac{u}{U} = -\frac{1}{r} = -\frac{1}{\sqrt{x^2 + y^2 + z^2}} \tag{9-6}$$

represents a point source of pressure rather than a source of volume.

To obtain the effect of a row of such sources—or a line source—along the x axis between the points a and b, it is necessary to integrate Eq. 9-6 as follows:

$$\begin{aligned} \frac{u}{U} &= -\int_a^b \frac{d\xi}{\sqrt{(x-\xi)^2 + y^2 + z^2}} \\ &= -\sinh^{-1}\frac{x-b}{\sqrt{y^2+z^2}} + \sinh^{-1}\frac{x-a}{\sqrt{y^2+z^2}} \end{aligned} \tag{9-7}$$

The pressure field of the finite line source thus consists of the sum of two conical pressure fields radiating from the ends of the line source as in Fig. A,9c. Such a conical field with rays emanating from the origin can also be

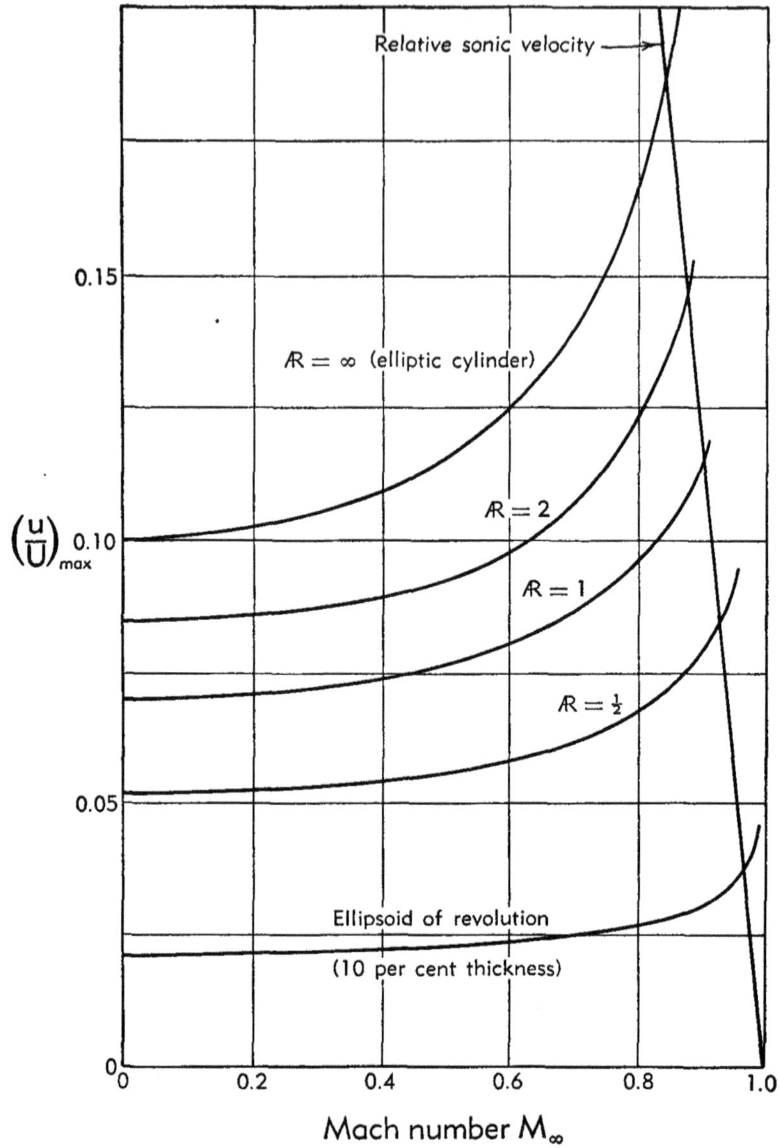

Fig. A,9b. Velocity increments vs. Mach number for thin ellipsoids; $\alpha = 0$.

represented in terms of the general solution of Laplace's equation of zero degree (Eq. 7-27); i.e.

$$\frac{u}{U} = \text{R.P.} \ln \frac{y + iz}{x + \sqrt{x^2 + y^2 + z^2}} = -\sinh^{-1} \frac{x}{\sqrt{y^2 + z^2}} \qquad (9\text{-}8)$$

If the velocity of flight is along the axis of the source, the flow will satisfy the boundary condition for a body of revolution. However, if the line source is turned out to a position oblique to the stream, the boundary shape will be distorted, and if the angle of obliquity is large enough to place the line source well outside the diameter of the original body the figure formed will be an oblique wedge. The nose angle of the wedge is formed where the streamlines of the main flow cross the line source.

The turning of the line source may be accomplished by the transformation

$$\left.\begin{aligned} x' &= x + my \\ y' &= y - mx \\ z' &= z\sqrt{1 + m^2} \end{aligned}\right\} \tag{9-9}$$

where m is the slope of the new axes relative to the old.

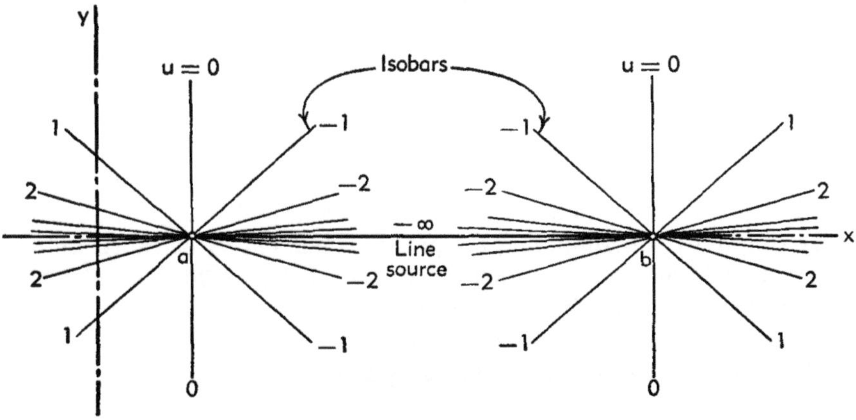

$$u = \sinh^{-1}\frac{x - a}{\sqrt{y^2 + z^2}} - \sinh^{-1}\frac{x - b}{\sqrt{y^2 + z^2}}$$

Fig. A,9c. Radial isobars from ends of line source

The vertical velocity w near $z = 0$, which determines the shape of the boundary, may be found by integrating u with respect to x and then differentiating the resulting velocity potential with respect to z:

$$w = \frac{\partial\varphi}{\partial z} = \frac{\partial}{\partial z}\int_{-\infty}^{x} u dx \tag{9-10}$$

Evaluation of this integral for the overlapping fields from the two ends of a line source gives

$$\frac{w}{U} = \pm 2\pi \frac{\sqrt{1 + m^2}}{|m|} \tag{9-11}$$

for the region of the x, y plane behind the line source. The plus and minus signs refer to the upper and lower surfaces of the chord plane, respectively. The figure formed by the streamlines crossing a line source is thus a wedge-shaped body having an oblique leading edge and extending indefinitely downstream.

By making use of the relation

$$\frac{\partial z}{\partial x} = \frac{w}{U}$$

together with Eq. 9-7 and 9-11, the pressure coefficient near the plane $z = 0$ may be expressed in terms of the slope as follows:

$$\frac{\Delta p}{\frac{1}{2}\rho_\infty U^2} = \frac{1}{\pi} \frac{|m|}{\sqrt{1 + m^2}} \frac{dz}{dx} \left(\sinh^{-1} \frac{x' - b'}{|y'|} - \sinh^{-1} \frac{x' - a'}{|y'|} \right) \tag{9-12}$$

Airfoils bounded by plane surfaces. It has been seen that the effect of a line source in the pressure field is to cause a deflection of the streamlines crossing the source in such a way as to spread the streamlines apart. If the source is followed by a sink of equal strength, an equal and opposite deflection of the streamlines will occur as they cross over the sink. The figure formed by the streamlines near the plane $z = 0$ will thus be a plate of uniform thickness with a beveled leading edge, as shown in Fig. A,9d.

The pressure distribution over such a beveled edge may be obtained very simply by superimposing the pressures marked off on radial isobars originating from the four corners of the bevel. Fig. A,9d illustrates this process for a bevel having a square planform. Only isobars from one tip are shown because of the symmetry of the figure. With the corners at ± 1 and source and sink lines parallel to the y axis, we have

$$u = -\sinh^{-1} \frac{y+1}{|x+1|} + \sinh^{-1} \frac{y-1}{|x+1|} + \sinh^{-1} \frac{y+1}{|x-1|} - \sinh^{-1} \frac{y-1}{|x-1|} \tag{9-13}$$

If the aspect ratio of the figure is increased, the isobars in intermediate regions approach parallel straight lines and the flow field thus approaches a cylindrical or two-dimensional form. Suppose that the tips are at $y = \pm\eta$, with η very much larger than 1. Then for the center of the wing, the terms $\sinh^{-1} (y \pm \eta)/(|x \pm 1|)$ in the expression for the pressure approach the limiting values

$$\sinh^{-1} \frac{y \pm \eta}{|x \pm 1|} \to \pm \ln 2 \left| \frac{y \pm \eta}{x \pm 1} \right|, \qquad y \ll \eta \tag{9-14}$$

with the sign of the logarithm determined by the sign of $y \pm \eta$, and the right-hand side of Eq. 9-13 is found to approach the Legendre function Q_0, i.e.

$$u \xrightarrow{=} -2 \ln \left| \frac{x-1}{x+1} \right| = 4Q_0(x) \tag{9-15}$$

This expression when combined with Eq. 9-11 agrees with the two-dimensional complex velocity function for the wedge

$$(u - iw) = 4Q_0(x) \mp 2\pi i P_0(x) \tag{9-16}$$

(see formula 19, Table A,2).

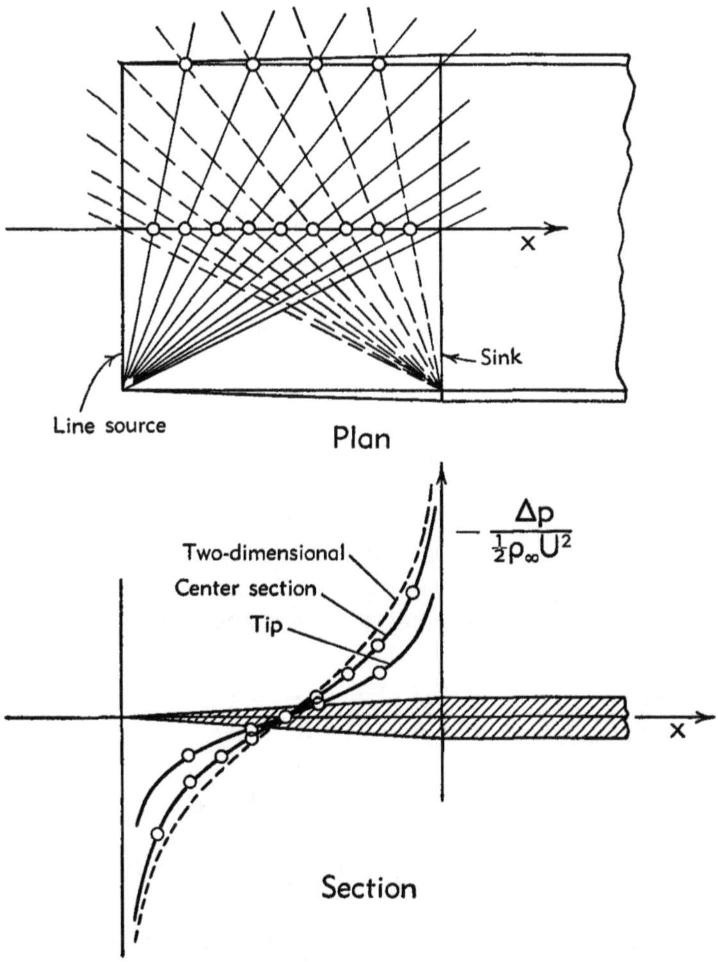

Fig. A,9d. Construction for the pressure distribution over a beveled leading edge.

Eq. 9-15 applies over the middle portion of a long narrow wedge. At the extreme tips it may be shown that the pressure falls to one half this value. Although the thin airfoil theory indicates that the thickness of the figure ends abruptly at the ends of the source distribution, a more exact consideration would be expected to show some rounding of the tip.

Airfoils of biconvex section. The use of a finite number of source-sink lines results in airfoil sections composed of straight segments. Cylindrical surfaces having continuous curvature require a continuous distribution of sources and sinks aligned with the generators of the surface. The simplest of these is the biconvex profile in which the upper and lower surfaces are parabolic arcs of constant curvature. Such a profile requires line sources of finite strength to form the desired angle of intersection of the arcs at the leading and trailing edges, together with a uniform distribution of sinks between the two sources.

The pressure field for a uniform sheet of line sources is obtained by integrating in the x direction the pressure field of a single line source. The result is

$$\int \sinh^{-1} \frac{x'}{|y'|} \, dx = \frac{\sqrt{1+m^2}}{m} y \sinh^{-1} \frac{x}{|y|} - \frac{1}{m} y' \sinh^{-1} \frac{x'}{|y'|} \tag{9-17}$$

For a bilaterally symmetrical arrangement,

$$\int \left(\sinh^{-1} \frac{x'}{|y'|} + \sinh^{-1} \frac{\bar{x}'}{|\bar{y}'|} \right) dx = \frac{1}{m} \left(\bar{y}' \sinh^{-1} \frac{\bar{x}'}{|\bar{y}'|} - y' \sinh^{-1} \frac{x'}{|y'|} \right) \tag{9-18}$$

where $\bar{x}'$ and $\bar{y}'$ have been used to denote the "conjugate" coordinates $x - my$ and $y + mx$, respectively.

To obtain a complete sweptback wing it is necessary to superimpose a number of component pressure fields, as explained in [*126*] and [*127*]. For a wing of infinite span with leading and trailing edges at $y' = +m$ and $-m$, respectively, on one side and $\bar{y}' = +m$ and $-m$ on the other side, the result is

$$\frac{\Delta p}{\frac{1}{2}\rho_\infty U^2} = \frac{2}{\pi} \left(\frac{t}{c}\right)_{\max} \frac{m}{\sqrt{1+m^2}} \left[\frac{y'}{m} \left(\sinh^{-1} \frac{x'+1}{|y'-m|} - \sinh^{-1} \frac{x'-1}{|y'+m|} \right) + \frac{\bar{y}'}{m} \left(\sinh^{-1} \frac{\bar{x}'-1}{|\bar{y}'-m|} - \sinh^{-1} \frac{\bar{x}'+1}{|\bar{y}'+m|} \right) + 2Q_1\left(\frac{y'}{m}\right) + 2Q_1\left(\frac{\bar{y}'}{m}\right) \right] \tag{9-19}$$

where $(t/c)_{\max}$ is the thickness-chord ratio of the biconvex profile. The terms in Q_1 represent the pressure distribution on the airfoil in two-dimensional flow (see formula 20, Table A,2). The appearance of these terms is the result of the assumption that the tips are removed to a great distance.

Fig. A,9e shows the pressure distributions at various stations along the span for a biconvex wing having 60° of sweepback. The curves assume the two-dimensional form at a relatively short distance ($y > c/2$) from the root section; similar behavior is to be expected near the tips. Hence the assumption of infinite aspect ratio should apply very nearly at any section situated more than one-half chord length from either root or tip.

At the root section ($y = 0$), Eq. 9-19 reduces to

$$\frac{\Delta p}{\frac{1}{2}\rho_\infty U^2} = \frac{2}{\pi}\left(\frac{t}{c}\right)_{\max} \frac{m}{\sqrt{1+m^2}}\left[4Q_1(x) - 4P_1(x)\sinh^{-1}\frac{1}{m}\right] \tag{9-20}$$

Since $dz/dx = -2(t/c)_{\max}P_1(x)$, the second term in the bracket denotes a component of the pressure which is proportional to the local slope of the

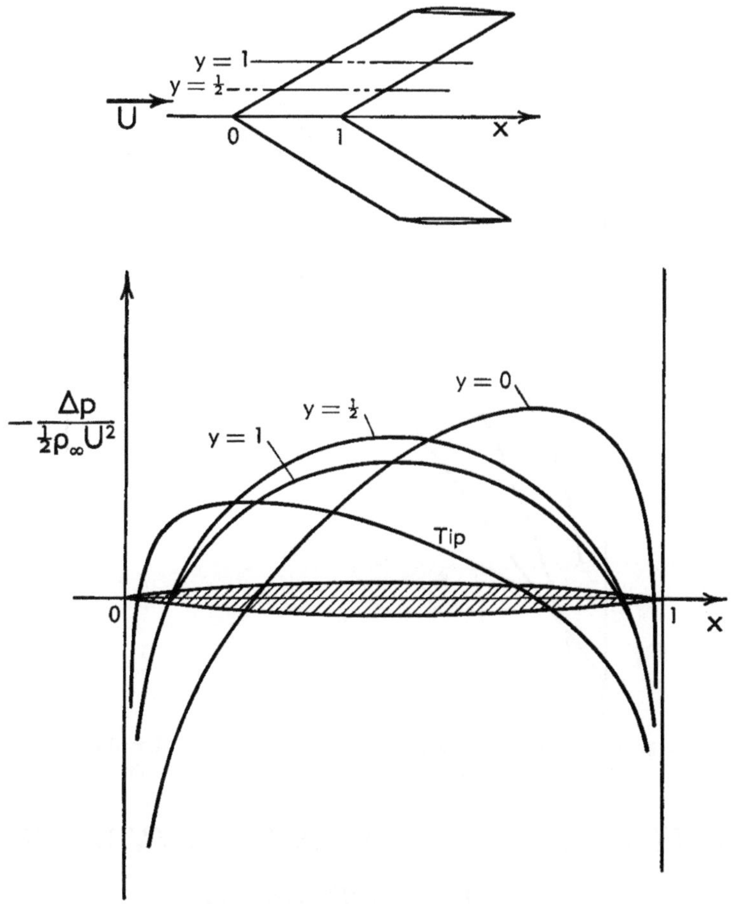

Fig. A,9e. Pressure distribution at various spanwise stations on a sweptback wing; $\Lambda = 60°$; $M_\infty = 0$.

airfoil surface. Such a component appears for a section of any shape and is a characteristic of the supersonic flow over a section. At subsonic speeds its magnitude increases progressively with the Mach number.

Fig. A,9f shows the effect of Mach number on the pressures over the root section as obtained by applying the Prandtl-Glauert transformation to the foregoing formula. It will be noted that an increase of the Mach number causes a progressive transition toward the supersonic, or Ackeret, type of pressure distribution ($M_\infty = 1.05$) to be discussed in a later section. Sections farther out along the span retain the subsonic type of pressure distribution. In the case of the straight wing, a more rapid change in the pressure distribution occurs on transition to supersonic speeds.

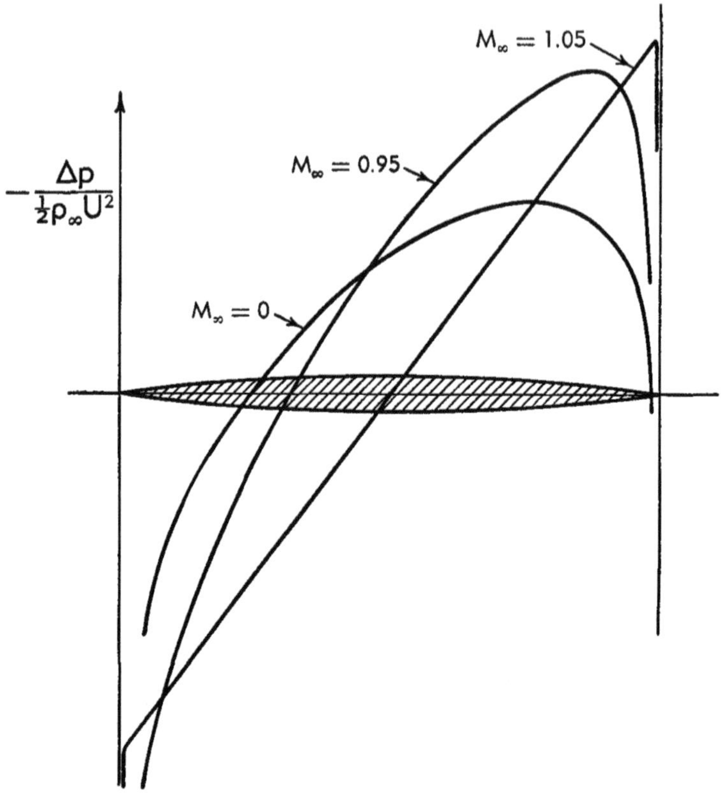

Fig. A,9f. Effect of Mach number on the pressure distribution over the root section of a sweptback wing; $\Lambda = 60°$.

CHAPTER 3. AERODYNAMICS OF THIN WINGS AT SUPERSONIC SPEEDS

A,10. Fundamental Considerations.

Geometric character of supersonic disturbances. As the speed is increased in the subsonic range the disturbance field of the moving wing becomes progressively modified, the disturbance extending to greater distances both laterally and vertically. For steady motion this distortion is shown by the Prandtl-Glauert transformation, which prescribes a change in the scale of the lateral and vertical dimensions relative to the lengthwise dimension. Except for this continuous change in proportions, however, there remains an essential similarity to incompressible flow. The features of the incompressible flow, such as the increase of speed at the sides of a body, the recovery of pressure at the rear, and the fore-and-aft balance of the pressure forces (D'Alembert's paradox) remain. Hence the intuitive concepts acquired in the study of incompressible flow suffice almost without modification for the whole range of steady subsonic potential flow.

On transition to supersonic speeds a real change in the geometric character of the disturbance field takes place. The fore-and-aft symmetry of the subsonic flow disappears and the disturbance is limited to the region behind the wing bounded by Mach waves or cones. Because of the asymmetry of the flow, all outgoing disturbances result in a drag, so that the thickness of the body or wing creates a drag in addition to the drag associated with the lift. At higher supersonic speeds the lifting efficiency of the wing begins to be diminished by the restriction on the volume of air involved in the production of the lift. As the speed is increased in the supersonic range the angle of the Mach waves becomes more acute and the volume of air coming under the influence of the wing becomes progressively smaller until, finally, it includes only that mass of air actually encountered by the frontal projection of the wing area.

The additional drag associated with the supersonic speed has been termed the "wave drag" since it appears in the energy required to extend the wave system emanating from the body. From the standpoint of an observer at rest, the streamlines of the disturbance field originate at the fronts of these waves. The streamlines always intersect the wave fronts at right angles, with the result that the motion at the wave front is parallel to the direction of propagation, in accordance with the simplified concept of a sound wave as a "longitudinal" disturbance. In the vicinity of the wave fronts, discontinuous changes in pressure and velocity may occur, but in three-dimensional supersonic flows there also appear large regions in which the motion is continuous, and the streamlines are indistinguishable from the patterns of an incompressible flow.

According to the thin airfoil theory the wave drag is zero at all subsonic speeds and rises suddenly at $M_\infty = 1.0$. As shown by Busemann [*128*] this discontinuous behavior results from the idealization of steady flow. In nonstationary or curvilinear motion, such as the motion of a propeller blade, the wave drag begins to appear at subsonic speeds.

From the mathematical standpoint, the change in flow geometry on going to supersonic speed appears in the change of the disturbance equation from the elliptic to the hyperbolic type. The steady flow equation

$$(1 - M_\infty^2)\varphi_{xx} + \varphi_{yy} + \varphi_{zz} = 0 \tag{10-1}$$

becomes on transition to supersonic speeds

$$(M_\infty^2 - 1)\varphi_{xx} - \varphi_{yy} - \varphi_{zz} = 0 \tag{10-2}$$

The disturbance fields admitted as solutions of the latter partial differential equation show a remarkable variety of pattern. In three dimensions these patterns range from the straight and discontinuous wave fronts associated with the hyperbolic equation, to the intricate curvilinear geometry associated with solutions of Laplace's equation. This variety can be foreseen by examining arrangements of the three terms of the differential equation in pairs.

First let us normalize the equation by the transformation

$$x_1 = \frac{x}{\sqrt{M_\infty^2 - 1}}; \quad y_1 = y; \quad z_1 = z \tag{10-3}$$

so that

$$\varphi_{x_1x_1} - \varphi_{y_1y_1} - \varphi_{z_1z_1} = 0 \tag{10-4}$$

Then, dropping the subscripts, we have for one pair of terms

$$\varphi_{xx} - \varphi_{zz} = 0 \tag{10-5}$$

with solutions of the form

$$\varphi = F(x \pm z) \tag{10-6}$$

corresponding to the flows considered in the Ackeret theory (see Art. 11). Such expressions can represent the wave disturbances in the vicinity of the sections of a long straight wing.

On the other hand, considering the last two terms of the equation, we have

$$\varphi_{yy} + \varphi_{zz} = 0 \tag{10-7}$$

and the solutions have the form

$$\varphi = F(y \pm iz) \tag{10-8}$$

i.e. the form of an incompressible potential flow in two dimensions. This is the form approached by the supersonic disturbance field in the vicinity of the vortex wake at a great distance behind the airfoil. In the case of a large angle of sweep, Eq. 10-7 may also apply in the vicinity of the wing itself.

Further illustration of the basic difference in geometry is shown by the fundamental solutions of the two flow equations, viz.

$$\varphi = \frac{1}{\sqrt{x^2 + (1 - M_\infty^2)(y^2 + z^2)}} \quad \text{(subsonic)} \tag{10-9}$$

and

$$\varphi = \frac{1}{\sqrt{x^2 - (M_\infty^2 - 1)(y^2 + z^2)}} \quad \text{(supersonic)} \tag{10-10}$$

These solutions, which represent the velocity potential arising from a point source, are illustrated in Fig. A,10a. In the subsonic case the equipotential surfaces are ellipsoids which can be reduced to spheres by the

Prandtl-Glauert transformation. In the supersonic case the surfaces are hyperboloids of two sheets bounded by the Mach cone, which extends both ahead of and behind the point source. Outside this cone the potential is imaginary. It is evident that the differential equation makes no distinction between forward and rearward zones of influence of the source, and the disturbance in either branch of the Mach cone complies with the differential equation. The upstream branch of the Mach cone is ordinarily to be discarded for physical reasons. Discarding the forward branch of the Mach cone, however, requires that the strength of the disturbance be doubled to provide the same flux from the source. The velocity distribution at the center of the Mach cone, along the positive branch of the x axis, is exactly twice that for a subsonic source.

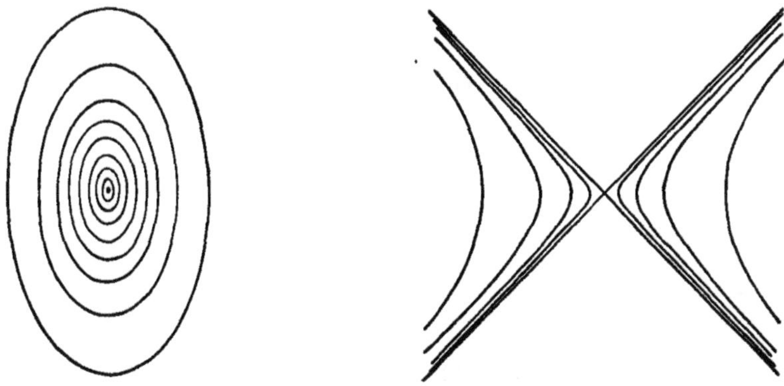

$$\varphi = \frac{1}{\sqrt{x^2 + (1 - M_\infty^2)(y^2 + z^2)}} \qquad \varphi = \frac{1}{\sqrt{x^2 - (M_\infty^2 - 1)(y^2 + z^2)}}$$

(subsonic) (supersonic)

Fig. A,10a. Equipotential surfaces for a subsonic and a supersonic source.

A physical explanation of this difference in the action of subsonic and supersonic sources may be adapted from the explanation given by Heaviside [*129*] for the case of a point charge moving at a speed less than or greater than the speed of light. A diagram similar to Heaviside's is given in Fig. A,10b. The field of the steadily moving source may be obtained by the superposition of the nonstationary fields of a series of pulses which go on and off in such a sequence as to represent the motion of the source along the line. Each pulse sends out a spherical wave which expands with the radial velocity a. In the subsonic case the expanding spheres fill the whole space without overlapping so that an observer at any point sees (or hears) the pulse traveling along in a retarded position. In the supersonic case, however, each point within the Mach cone is characterized by two intersecting spheres, one having its center along the line of motion ahead of the observer (P) and the other behind. Thus in the supersonic case the observer sees two sources whose apparent positions recede from him in opposite directions.

Wing flows related by the Prandtl-Glauert transformation. Steady flows at different Mach numbers in the supersonic range may be related geometrically by a transformation of coordinates similar to the Prandtl-Glauert transformation, but with the factor $1/\sqrt{1 - M_\infty^2}$ replaced by

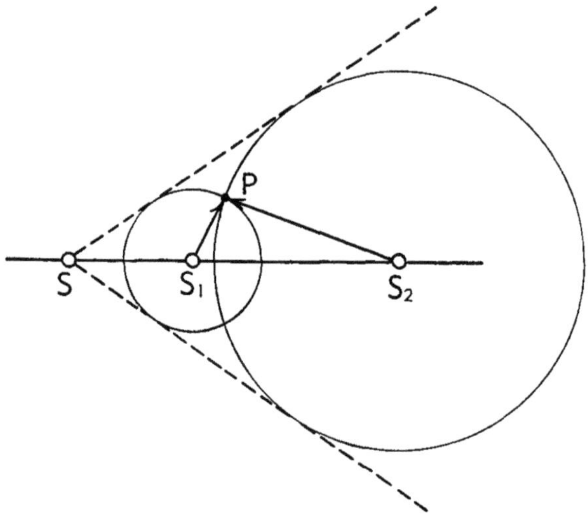

Fig. A,10b. Apparent positions S_1 and S_2 of the disturbance S traveling at supersonic speed. Observer is at P.

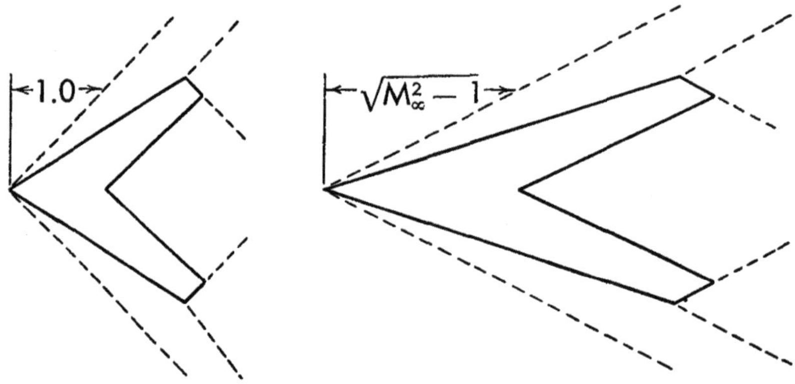

Fig. A,10c. Wings having flows related by the Prandtl-Glauert transformation.

$1/\sqrt{M_\infty^2 - 1}$. This is the transformation already introduced (Eq. 10-3). As illustrated in Fig. A,10c, it shortens or lengthens the potential field in accordance with the change in inclination of the Mach lines. In general, increases of the Mach number in the supersonic range decrease the ratio of the longitudinal to the vertical and lateral perturbation velocities by $1/\sqrt{M_\infty^2 - 1}$.

At the wave front the resultant disturbance velocity becomes normal to the wave in all cases, so that at such points

$$\frac{u}{\sqrt{v^2 + w^2}} = \frac{1}{\sqrt{M_\infty^2 - 1}} \tag{10-11}$$

At other points behind the wave front and within the zone of influence of the wing, the disturbance field presents a more-or-less complex three-dimensional pattern, tending generally toward the pattern of an incompressible flow near the center of the Mach cone as the zone of disturbance widens.

Subsonic and supersonic edges. Wherever the outline of the wing forms a continuous curve in plan view it is possible to mark off a sufficiently narrow strip around the edge such that the flow in the vicinity of this strip is locally cylindrical or two-dimensional in form. Hence the nature of the flow in the vicinity of the wing edges may be deduced from the theory of oblique cylindrical flows previously discussed (Art. 4). On this basis it is seen that an important distinction arises between cases in which the component velocity normal to the edge is subsonic and those in which this velocity is supersonic. In the former case the rules of subsonic flows will apply and hence, for brevity, it is said that the edge is subsonic. Similarly, whenever the normal component of velocity is supersonic, the edge is said to be supersonic.

Thus, in the case of a lifting wing having a rounded subsonic leading edge, there will arise a flow around the edge and a suction force will be developed, as in the Kutta-Joukowski flow (Art. 2). The effect of a symmetrical distribution of thickness, on the other hand, is to develop a positive pressure in this region, the maximum value of which will be equal to the stagnation pressure corresponding to the transverse velocity component at the edge.

If the leading edge is supersonic the flow will be locally similar to the Ackeret flow, provided the edge is sharp. In this case the flow is characterized by a curved shock wave attached to the edge. Here there will be no flow around the edge and the surface of the airfoil divides the flow into independent upper and lower regions. As in the Ackeret theory, it is no longer necessary to distinguish between pressures arising from the lift and those arising from the thickness, the pressures on each surface being determined by the shape of that surface alone. If the supersonic edge is blunt or rounded, the shock wave will be detached and the pressures cannot be calculated by the Ackeret theory. A satisfactory extension of the thin airfoil theory to deal with examples of this kind has not yet been devised.

In the case of a supersonic trailing edge, the independent pressures

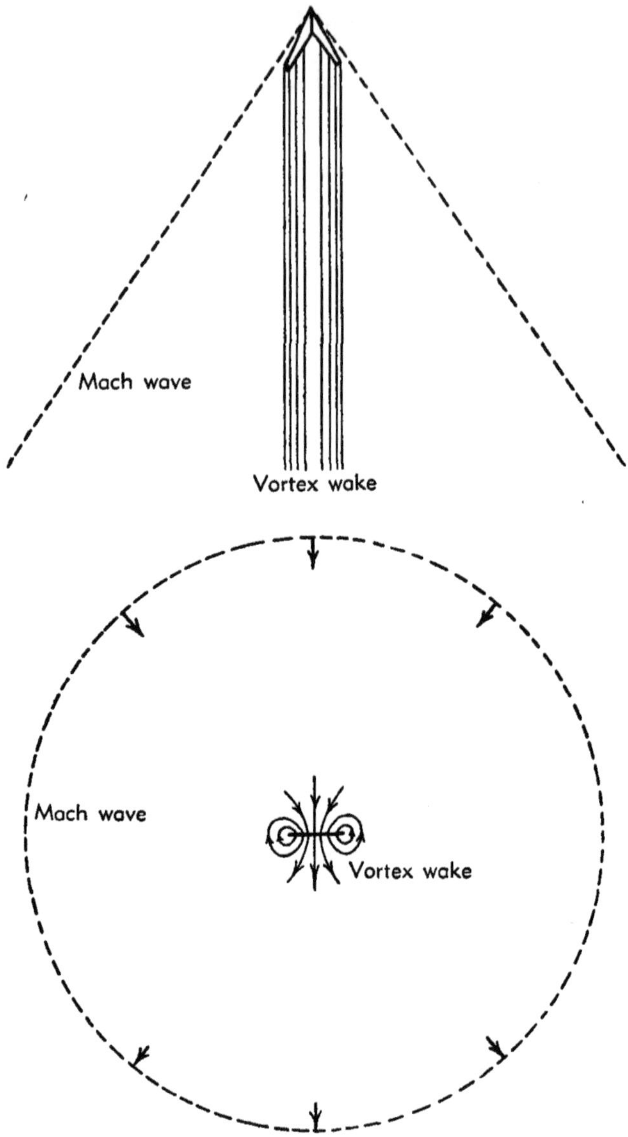

Fig. A,10d. Disturbance produced by wing at supersonic speeds.

on the upper and lower surfaces will be maintained up to a point so near the trailing edge that equalization of the pressures takes place through the boundary layer. In theory it is usually considered sufficient to satisfy the condition that the lift fall to zero behind the wing by discontinuous pressure changes across oblique shock waves attached to the edge. In practice, because of the existence of the viscous wake, the attachment of the waves to the trailing edge is never complete and the result is that the pressure changes in this region occur somewhat more gradually than the theory would indicate.

Wave drag; vortex drag. At supersonic speeds, drag arises both from the thickness of the wing and from the lift. The drag arising from the lift may be identified partly as vortex drag and partly as wave drag, though both effects appear in the rearward inclination of the lift vector. In the theory of wings at subsonic speeds it is shown that the production of lift by a wing of finite span is accompanied by a drag force that depends on the magnitude of the lift and its distribution over the span. In subsonic theory, this drag is termed the "induced drag" and may be calculated from consideration of the vorticity in the wake. It is evident that the total drag coefficient of a wing in supersonic flow cannot fall below the value of the induced drag coefficient corresponding to the same spanwise distribution of lift at subsonic speeds, since the vortex wake is similar and the motion induced by this wake at a great distance is similar to that in subsonic flow. At a great distance behind the wing, as pointed out by Hayes [*44*], the wave disturbance becomes concentrated in the vicinity of the Mach cone and does not modify the purely subsonic field of disturbance in the vicinity of the vortex wake, which occupies a relatively small region near the center of the cone (see Fig. A,10d). Hence the wave drag and the vortex drag may be separated if the calculation is made on the basis of the momentum of the flow at a great distance behind the wing.

For many purposes it is convenient to determine the drag from the pressures acting directly on the wing. In this way it is readily shown that the drag arising from the thickness is directly proportional to the square of the thickness-chord ratio, while the drag arising from the lift is proportional to the square of the lift coefficient. Changing the angle of attack or camber does not change the drag arising from the thickness, however, and hence the components of drag due to lift and thickness are additive in the approximation of the linear theory, just as in subsonic flow. These other components of the drag will be treated separately in the articles to follow.

A,11. Two-Dimensional Flow.

Ackeret theory. The first theoretical treatment of airfoils traveling at supersonic speed was given by Ackeret [*10*]. Ackeret considered the two-dimensional flow produced by a cylindrical wing traveling at right angles to its long axis. The sections were assumed to be thin, with sharp leading and trailing edges, so that the linearized equation of motion would apply. The disturbance field produced by the wing in this case turns out to be kinematically extremely simple. As illustrated in Fig. A,11a it consists merely of two plane waves of sound, traveling in the direction normal to the wave fronts and intersecting the x, y plane at the Mach angle μ

$= \sin^{-1}(1/M_\infty)$. Two waves are needed, one for the upper surface of the airfoil and one for the lower. The velocity with which the intercept travels along the x axis is greater than the actual velocity of the wave and is equal to the velocity of flight U. The equation of motion (10-6), which is the equation for plane sound waves, admits an arbitrary variation of velocity within the wave, but requires that the resultant disturbance velocity be at right angles to the wave front. Hence the horizontal and

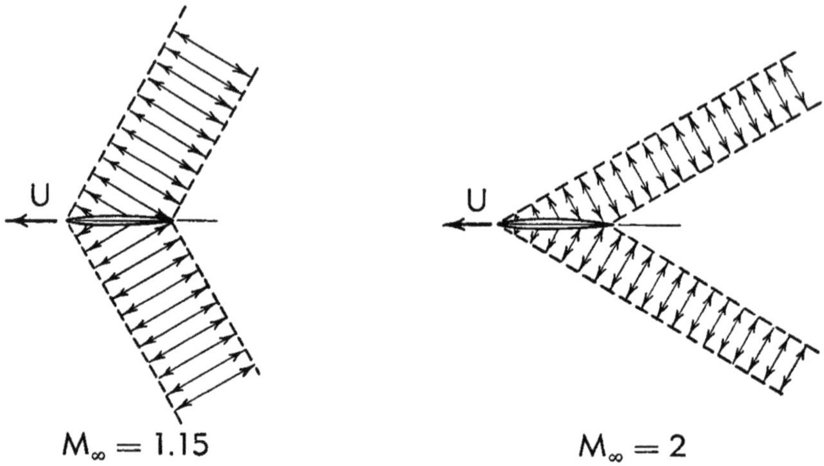

Fig. A,11a. Ackeret flow, showing disturbance velocities produced by wing.

vertical components u and w remain in a fixed ratio, which depends on the Mach angle. A simple resolution of velocities shows that

$$u = w \cot \mu = \frac{w}{\sqrt{M_\infty^2 - 1}} \tag{11-1}$$

The variation of velocity within the waves is determined by the shape of the airfoil surface through the relation

$$\frac{w}{U} = \frac{dz}{dx} \tag{11-2}$$

as in the thin airfoil theory for subsonic speeds. The pressure disturbance is again given to the first order by the relation

$$\frac{\Delta p}{\frac{1}{2}\rho_\infty U^2} = -\frac{2u}{U} \tag{11-3}$$

The latter formula agrees with the pressure relation used in acoustical theory. Combination of the three equations yields an expression for the pressure coefficient in terms of the slope of the airfoil surface:

$$\frac{\Delta p}{\frac{1}{2}\rho_\infty U^2} = \frac{2(dz/dx)}{\sqrt{M_\infty^2 - 1}} \tag{11-4}$$

The pressure is constant at all points of any plane parallel to the wave front. The pressure at any point on the airfoil surface depends solely on the slope of the surface at that point and there is no influence of one portion of the airfoil on another.

With the aid of Eq. 11-4 it is easily shown that the lift and pressure drag per unit span of an airfoil are given by

$$L = -\tfrac{1}{2}\rho_\infty U^2 \frac{2}{\sqrt{M_\infty^2 - 1}} \int \frac{dz}{dx}\, dx \tag{11-5}$$

$$D = \tfrac{1}{2}\rho_\infty U^2 \frac{2}{\sqrt{M_\infty^2 - 1}} \int \left(\frac{dz}{dx}\right)^2 dx \tag{11-6}$$

where the integrations extend over both upper and lower surfaces of the airfoil. The highest ratio of lift to drag is achieved when dz/dx is a constant, i.e. when the airfoil is a perfectly thin flat plate. In this case the formulas, in coefficient form, are

$$C_L = \frac{4\alpha}{\sqrt{M_\infty^2 - 1}} \tag{11-7}$$

$$C_D = \frac{4\alpha^2}{\sqrt{M_\infty^2 - 1}} \tag{11-8}$$

As the latter formula shows, the lift is inclined backward at an angle equal to the angle of attack, and no suction force appears at the leading edge.

Higher approximations in the calculation of two-dimensional flows. At finite values of the thickness or angle of attack, departures from the linearized theory arise. Such departures make their appearance partly as an alteration in the geometry of the disturbance pattern, and partly through the nonlinear relation between the local velocity and the pressure. The former effect may be visualized as the result of the changes in the local inclination of the Mach lines in the field of varying local velocities. The nonlinear relation between the velocity and the pressure arises in part from the variation of density. The effect of a given pressure gradient in decelerating the stream is diminished by the accompanying increase of density along the streamlines. Similarly, the acceleration, in flowing toward a region of reduced pressure, is magnified by the reduction of density.

In two-dimensional flow, calculations can be made which show the nonlinear behavior of a perfect gas with a high degree of accuracy. The method of calculation, known as the "shock-expansion method" [*130*] makes use of the relations governing finite shock and expansion waves given originally by Hugoniot [*131*] and Meyer [*132*]. For a full explanation of this method, reference should be made to Ferri's book [*133*].

For airfoils of reasonable thickness the shock-expansion method shows an essentially single-valued, though nonlinear relation between the slope of the surface and the pressure. This relation is illustrated in Fig. A,11b for two values of the stream Mach number. It is seen that positive inclinations of the surface have a larger effect than predicted by linearized theory while negative inclinations have a smaller effect. In computations of the over-all lift or drag these differences tend to compensate each other.

Fig. A,11c shows pressure distributions calculated by the shock-expansion method for biconvex circular arc profiles of varying thickness ratio at a Mach number of 2.0. The comparison with linearized theory

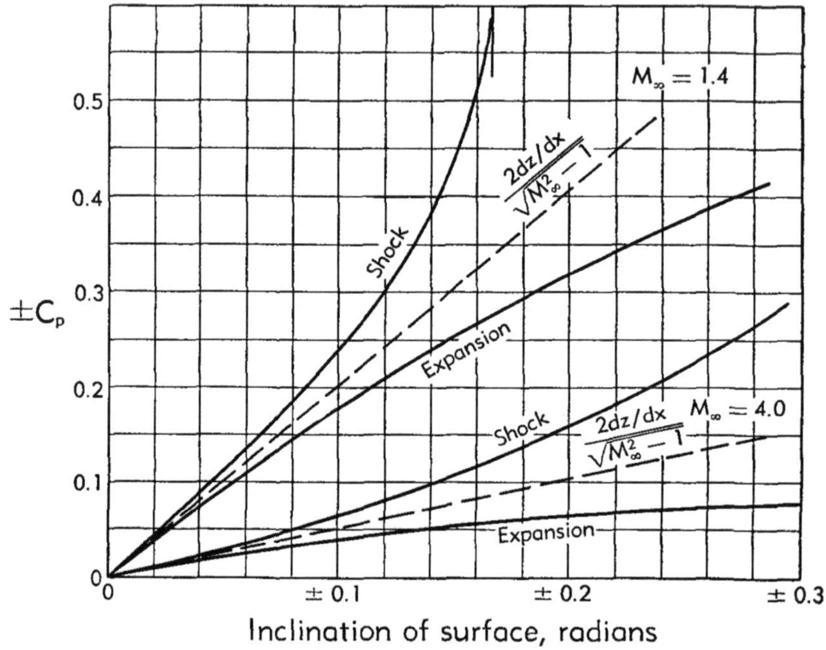

Fig. A,11b. Pressure coefficient $C_p = (p - p_\infty)/\frac{1}{2}\rho_\infty U^2$ vs. surface inclination in two-dimensional flow.

shows good agreement at $t/c = 5$ per cent, but increasing departures for 10 per cent and 15 per cent thickness.

In a series of experiments made in the supersonic wind tunnel at Guidonia, Ferri [*134*] measured the pressure distributions for a variety of airfoil sections. The observed pressure distributions were in good agreement with those calculated by the shock-expansion method over the forward parts of the sections but the pressures near the trailing edge were considerably modified by flow separation. A typical result of Ferri's experiments is shown in Fig. A,11d. Here the airfoil was of biconvex section. Evidently in the case of 10 per cent thickness the pressure rise at the trailing edge of the airfoil was sufficiently great to cause a reversal of the boundary layer flow at the trailing edge. In the experiments quoted, the Reynolds number was rather lower than full size values, and the

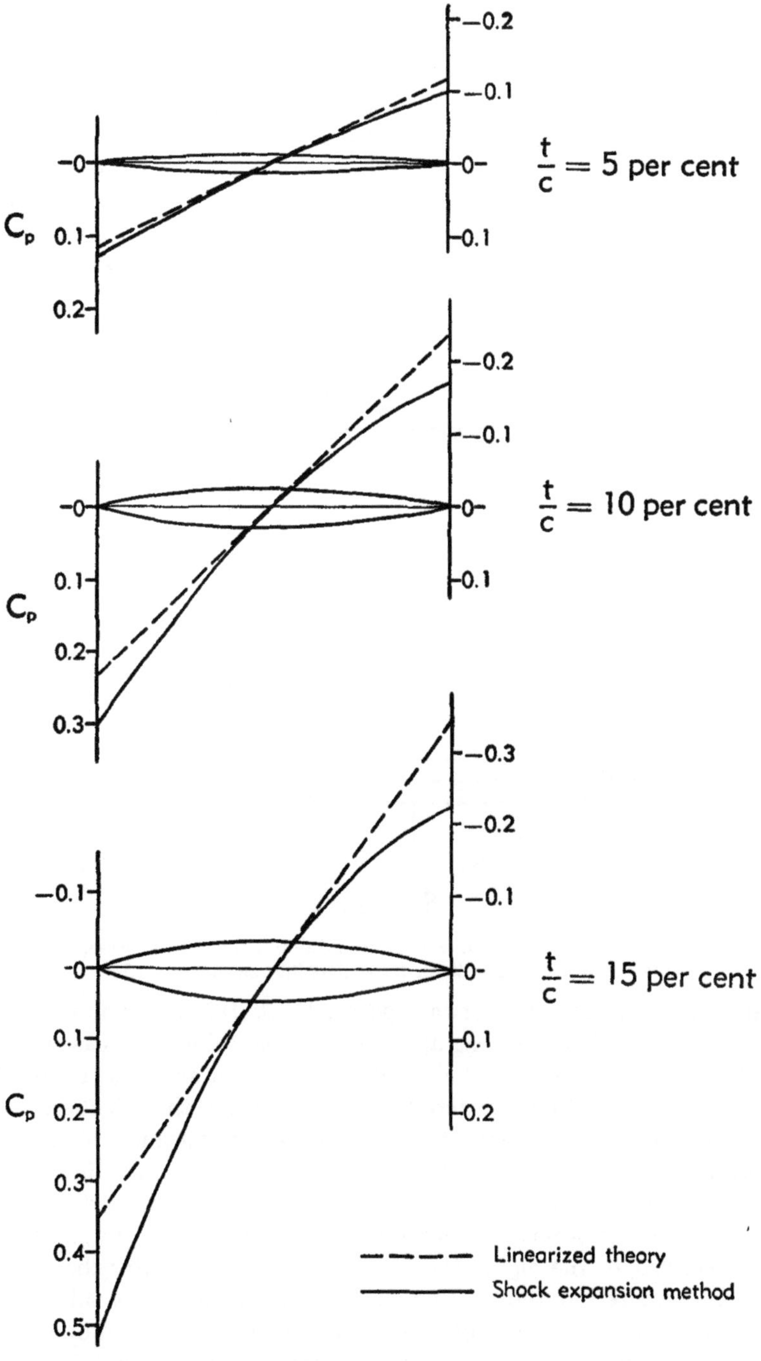

Fig. A,11c. Comparison of pressures given by linearized theory and shock-expansion method; two-dimensional flow, $M_\infty = 2.0$.

boundary layer over the airfoil was laminar. At higher Reynolds numbers the boundary layer is probably turbulent and flow separation is less likely to occur. In spite of the difference in pressure distribution, the relation between airfoil thickness and flow separation appears to be much the same at supersonic speeds as at low subsonic speeds.

Optimum profiles. Because of the magnitude of the theoretical pressure drag in two-dimensional supersonic flow, the occurrence of separation actually may have the effect of diminishing the drag in certain cases. In such cases it is desirable to thicken the airfoil at the trailing edge and to

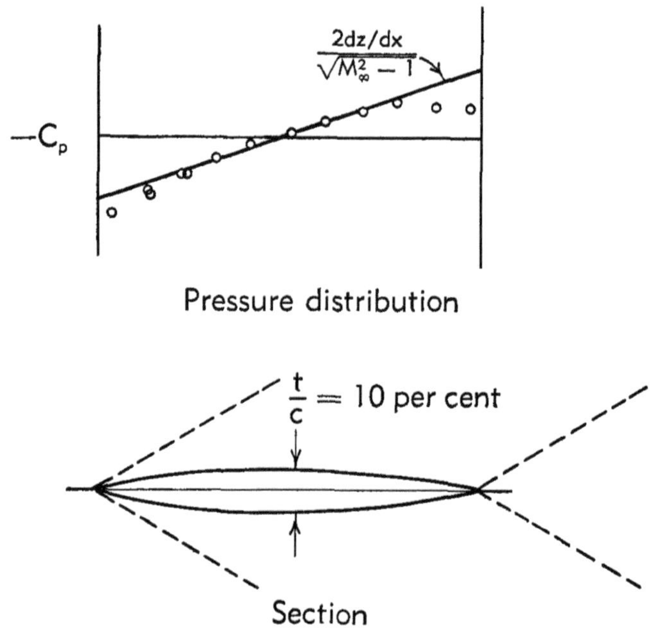

Fig. A,11d. Comparison of experimental and theoretical pressure distributions; $M_\infty = 2.13$, $Re = 640{,}000$. Data from [*134*].

diminish the slope over the forward part [*135,136,137,138*]. The wave drag is thus lessened while the drag appearing in the wake is increased.

The optimum shape of the profile, taking into account the trailing edge thickness, is not readily determined. However, if the thickness-chord ratio and one other (e.g. structural) condition are specified, a shape for minimum pressure drag can be calculated [*137,138*], provided that a value for the base pressure can be estimated. Using experimentally determined values for the base pressure, Chapman [*137,138*] found that the greater the thickness ratio, or the higher the Mach number, the greater the relative bluntness required at the trailing edge, although the exact thickness is more critical in the case of thin airfoils and moderate Mach numbers. It is interesting to note that essentially the same results were obtained both by linear and shock-expansion methods.

Oblique cylindrical flows at supersonic speed. In view of the continuous decrease of efficiency of the Ackeret type of flow with increased Mach number, Busemann [*32*] suggested that the wings be swept back so as to reduce the component Mach number of the wing sections to a lower value. Busemann showed that a sufficiently long wing, swept so that its crosswise velocity was only slightly supersonic, was aerodynamically more efficient than a straight wing.

In Busemann's analysis, the flow over the oblique wing panels was assumed to be two-dimensional or cylindrical in form, so that the solution could be obtained by superimposing an ineffective axial component of velocity on a flow of the Ackeret type. The field of disturbance in this case may be visualized as a formation of plane waves inclined both vertically and laterally in such a way that the lines of intersection with the x, y plane travel at the crosswise velocity $U \cos \Lambda$, while the point of intersection with the x axis travels at the flight velocity U. The relations governing the lift and drag remain the same as in straight flow (Eq. 11-5 and 11-6), provided the reduced values of Mach number and velocity are employed. It is important to note, however, that the pressure drag determined in this way is directed at right angles to the leading edge and hence its effectiveness in opposing motion in the direction of flight is reduced. Since the effectiveness of the lift force is not similarly reduced, it is evident that this inclination of the pressure drag is advantageous, although the advantage is partly offset by the fact that the friction drag force is still essentially governed by the resultant flight velocity U.

Later calculations [*33*] have indicated that a still greater improvement in aerodynamic efficiency may be obtained by increasing the angle of sweep, so that the wing lies inside the Mach cone and the crosswise component of velocity becomes subsonic. In the case of the infinite wing, the Ackeret type of flow is then replaced by a flow of the Kutta-Joukowski type and the wave drag disappears. The relations determining the flow and the forces on the two-dimensional airfoil with a subsonic normal velocity are given in Chap. 2. For a wing of finite span, however, the wave drag does not disappear entirely when the wing is swept behind the Mach cone. Determination of the wave drag requires consideration of the three-dimensional flow field, as described in the following articles.

A,12. The Drag of Lifting Surfaces in Three-Dimensional Flow.

Calculation from lift distribution. At supersonic speeds, the form of the vortex wake following a given distribution of lift, as well as the induced field of motion in the vicinity of the wake far behind the wing, is the same as at subsonic speeds. Hence, the vortex drag (Art. 10) of a wing traveling at supersonic speed can be identified with the induced drag which would result from a similar distribution of lift at subsonic speeds. Following Munk's "stagger theorem," this component of the drag will depend only on the spanwise distribution of lift and will be independent of the distribution of lift along the chord of the wing. However, the wave drag—i.e. the component of drag associated with the energy in the system of waves continuously being extended by the airfoil cannot be related simply to spanwise or lengthwise loading at the wing, but depends in a more complex way on the actual distribution of lift over the surface. The total drag may be found by integrating the product of the lift and the downward inclination of the streamlines over the airfoil surface; this method is usually most convenient when the geometry of the airfoil is simple (e.g. in the case of a flat wing).

In computing the total drag by integrating around the surface, care must be taken to include the alleviating effect of the suction force along any subsonic leading edges (see Art. 10). The general procedure described in Art. 2, applied to the three-dimensional problem, will reduce in the limit of zero thickness to the consideration, for each infinitesimal segment of the leading edge, of local conditions in the plane normal to the edge.

Thus, if the incremental velocity over the surface, normal to the leading edge, includes a singularity in the form $C\Delta^{-\frac{1}{2}}$, where Δ is the length of the normal to the leading edge, the leading edge suction force will be given by Eq. 2-14.

Eq. 2-14 is derived for incompressible flow. For supersonic speeds it will be necessary to take into account the effect of compressibility by means of the Prandtl-Glauert transformation, and in addition, for practical convenience, it will be desirable to rewrite the formula in terms of the usual flight-oriented coordinates. The quantity of interest is the component of the suction force in the flight direction, or the thrust. If M_∞ is the free stream Mach number and ρ_∞ the density, and m is the ratio of the local inclination of the leading edge to the tangent of the Mach angle ($\sqrt{M_\infty^2 - 1}$/tangent of the angle of sweep), the thrust per unit streamwise length of leading edge may be written

$$\frac{dT}{dx} = \pi\rho_\infty \sqrt{1 - m^2}\, C_x^2 \tag{12-1}$$

where C_x is the value at the leading edge $x = x_{\text{l.e.}}$ of $u\sqrt{x - x_{\text{l.e.}}}$, x and u being measured in the streamwise direction.

In the following subarticle an alternate method of calculating the drag arising from a given distribution of lift at supersonic speeds will be described. This method has the advantage of showing explicitly the type of lift distribution that leads to small values and, in certain cases, to the minimum value of the drag.

At subsonic speeds the drag is a minimum, for a given lift and a given span, when the induced downwash (in the special sense of the lifting line theory) has the same value at all points of the span. In Art. 6 it was shown that the minimum drag consistent with a given planform and a given total lift occurs when the superposition of the forward and reversed disturbance fields results in a constant value of the downwash at all points of the planform. In the more general theory, which includes both subsonic and supersonic speeds, the "combined downwash" thus takes the place of the induced downwash.

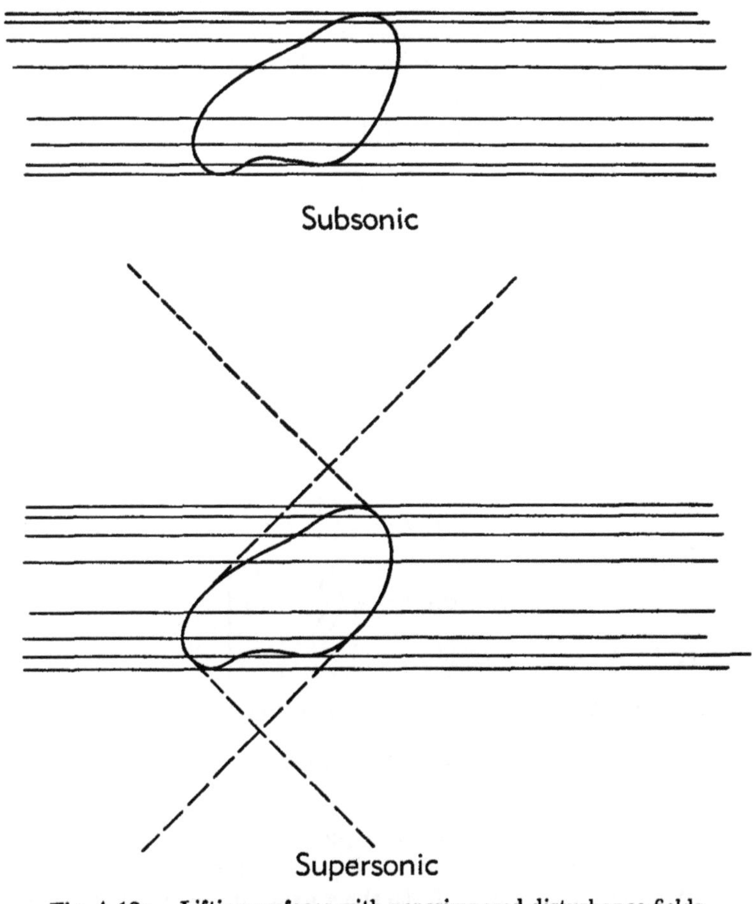

Fig. A,12a. Lifting surfaces with superimposed disturbance fields.

The superposition of the forward and reversed disturbance fields, corresponding to a given distribution of lift, results in a purely two-dimensional field of motion if the flight speed is subsonic. As illustrated in Fig. A,12a, the combined disturbance field in this case is obtained simply by extending the trailing vortex wake straight across the wing into the region ahead. Since the problem is two-dimensional, the determination of the optimum distribution of lift is relatively simple.

At supersonic speeds the combined disturbance field is again characterized by a vortex ribbon, identical in form with the trailing vortex wake, and extending across the wing both ahead of and behind the wing. Here, however, the field of motion is three-dimensional in form and is bounded by two overlapping zones of influence, as illustrated in the lower part of the figure. The determination of the ideal distributions, which yield a constant value of the downwash $\bar{w}$ at each point of the planform, thus leads to a considerably more difficult calculation if the speed is supersonic.

Plane-waves method. The method to be presented for the calculation of the drag makes use of forms similar to Whittaker's solution of Laplace's equation [*52*] and is related to a more general class of integral operators studied by Bergman [*139*]. Starting with the potential function

$$\varphi = F(X) \tag{12-2}$$

where

$$X = \alpha x - \beta y - \gamma z \tag{12-3}$$

we find that φ is a solution of the normalized potential equation (10-4) if F is a differentiable function and if α, β, and γ are parameters determined so that

$$\alpha^2 - \beta^2 - \gamma^2 = 0 \tag{12-4}$$

Setting

$$\left.\begin{aligned} \alpha &= 1 \\ \beta &= \cos\theta = \frac{1}{2}\left(\frac{1}{\lambda} + \lambda\right) \\ \gamma &= \sin\theta = \frac{i}{2}\left(\frac{1}{\lambda} - \lambda\right) \\ \lambda &= e^{i\theta} \end{aligned}\right\} \tag{12-5}$$

we obtain by superposition the general solution

$$\varphi = \int_C F(X, \lambda) d\lambda \tag{12-6}$$

where C is a suitable contour in the λ plane.

With $X = x - y \cos\theta - z \sin\theta$, Eq. 12-2 is seen to be the potential of a plane wave at an angle of 45° to the x axis ($M_\infty = \sqrt{2}$) and at a variable angle θ to the negative z axis. Eq. 12-6 then represents the three-dimensional potential as the result of the superposition of plane waves whose form and intensity are functions of the angle θ.

Real values of θ correspond to points λ on the unit circle. On the other hand, for values of λ near the origin we have

$$X \mathrel{\overset{\rightarrow}{=}} -\frac{1}{2\lambda}(y + iz) \tag{12-7}$$

and Eq. 12-2 represents a cylindrical, or two-dimensional, complex potential function which is a solution of

$$-\varphi_{yy} - \varphi_{zz} = 0$$

with $\varphi_{xx} = 0$ separately.

The form of Eq. 12-6 for the supersonic point source may be shown to be

$$\varphi_s = \frac{1}{4\pi^2} \int_C \frac{d\lambda}{i\lambda X} \tag{12-8}$$

To evaluate this integral by the method of residues it is necessary to determine the poles of the integrand. The denominator λX may be factored into

$$\lambda X = -\tfrac{1}{2}(y - iz)(\lambda - \epsilon_1)(\lambda - \epsilon_2) \tag{12-9}$$

where

$$\left.\begin{aligned} \epsilon_1 &= \frac{y + iz}{x + R} \\ \epsilon_2 &= \frac{y + iz}{x - R} \end{aligned}\right\} \tag{12-10}$$

and

$$R = \sqrt{x^2 - y^2 - z^2}$$

The integral in Eq. 12-8 may now be written

$$\varphi_s = -\frac{1}{4\pi^2} \int_C \frac{2d\lambda}{i(y - iz)(\lambda - \epsilon_1)(\lambda - \epsilon_2)} \tag{12-11}$$

Now consider the evaluation of Eq. 12-11 for a closed contour drawn just inside the unit circle on the λ plane. If the real part of R is taken as positive, then for values of x, y, z in the downstream branch of the Mach cone ($x^2 > y^2 + z^2$, $x > 0$), ϵ_1 will lie inside the unit circle while ϵ_2 lies outside (see Fig. A,12b). The value of the integral is then

$$\varphi_s = -\frac{1}{\pi(y - iz)(\epsilon_1 - \epsilon_2)} = \frac{1}{2\pi R} \tag{12-12}$$

For values of x, y, z in the upstream branch of the Mach cone, ϵ_1 and ϵ_2 are interchanged and ϵ_2 lies inside the unit circle. Hence we have

$$\varphi_s = -\frac{1}{\pi(y - iz)(\epsilon_2 - \epsilon_1)} = -\frac{1}{2\pi R} \tag{12-13}$$

Investigation of Eq. 12-10 shows that, for every point x, y, z in the "zone of silence" between the two branches of the Mach cone, both ϵ_1 and ϵ_2 lie on the unit circle and the value of the integral in Eq. 12-11 is then zero.

Exactly the same determination of φ results from a contour drawn in the opposite direction around the remaining portion of the λ surface outside the unit circle. It is found that integration and differentiation of the expression for the source are simplified if the complete contour C is speci-

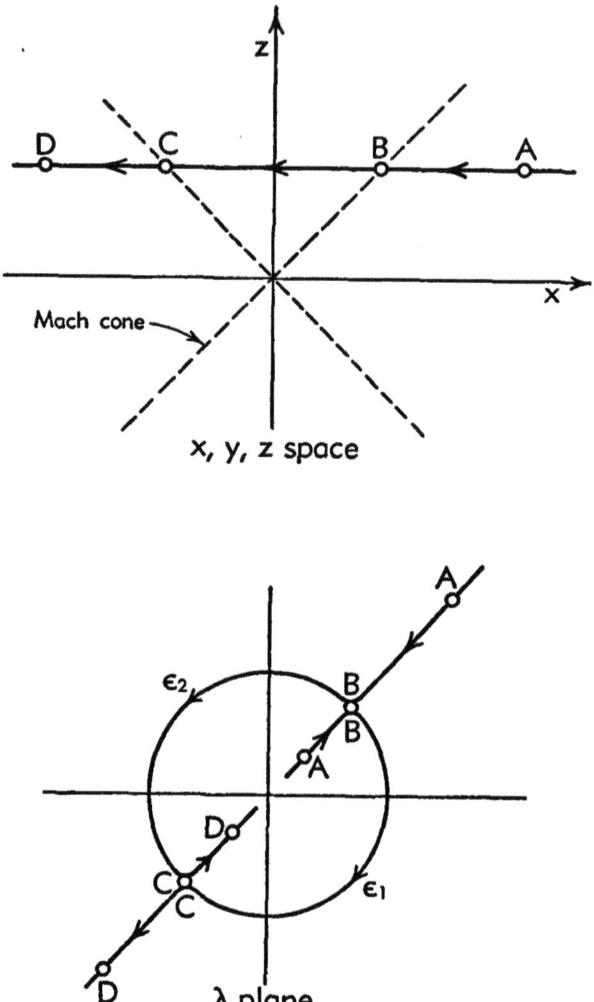

Fig. A,12b. Course of ϵ_1 and ϵ_2 during variation of point x, y, z.

fied so as to enclose the two portions of the λ surface $|\lambda| < 1$ and $|\lambda| > 1$ in opposite directions and to exclude the points $|\lambda| = 0$ and $|\lambda| = \infty$ (see Fig. A,12c).

The solution for a horseshoe vortex, which corresponds to the disturbance produced by an element of lift, may be obtained by the familiar process of integrating the expression for the source along x and then differ-

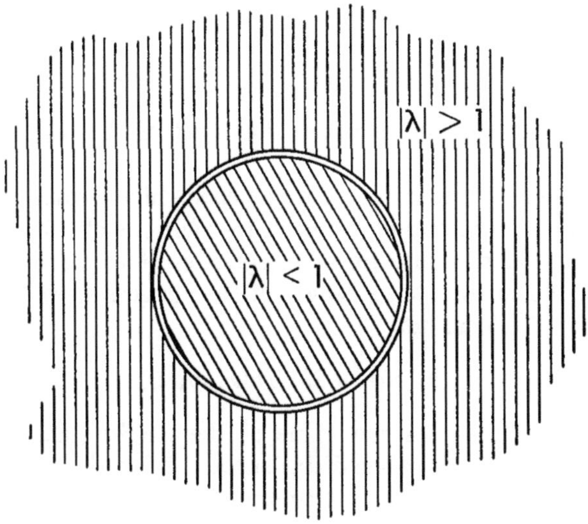

λ surface, showing cut

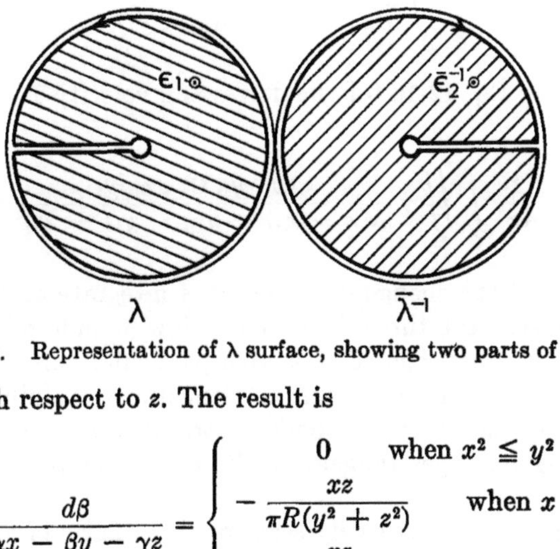

Fig. A,12c. Representation of λ surface, showing two parts of contour C.

entiating with respect to z. The result is

$$\varphi_v = \frac{1}{4\pi^2}\int_C \frac{d\beta}{\alpha x - \beta y - \gamma z} = \begin{cases} 0 & \text{when } x^2 \leq y^2 + z^2 \\ -\dfrac{xz}{\pi R(y^2+z^2)} & \text{when } x > 0 \\ +\dfrac{xz}{\pi R(y^2+z^2)} & \text{when } x < 0 \end{cases} \tag{12-14}$$

The contour C, transformed according to the relation

$$\beta = \frac{1}{2}\left(\lambda + \frac{1}{\lambda}\right)$$

encloses the poles

$$\nu_1 = \frac{xy - iRz}{y^2 + z^2} = \frac{1}{2}\left(\epsilon_1 + \frac{1}{\epsilon_1}\right)$$

and

$$\nu_2 = \frac{xy + iRz}{y^2 + z^2} = \frac{1}{2}\left(\epsilon_2 + \frac{1}{\epsilon_2}\right)$$

in opposite directions, and excludes that part of the real axis between ± 1.

The combined disturbance field for an entire lifting surface S may now be obtained by superimposing elementary solutions of the form (12-14). This superposition amounts to an integration of elementary horseshoe vortices over the surface, the strength of the vortices at each point being determined by the local lift

$$l(x_1, y_1) = 2\rho_\infty U u(x_1, y_1, 0) \tag{12-15}$$

where $u(x_1, y_1, 0)$ denotes the value of the longitudinal perturbation velocity on the upper side of the lifting surface. After the order of integration has been changed so that the contour integration is performed last, the expression for the combined potential $\bar{\varphi}$ of the lifting surface becomes

$$\bar{\varphi} = \frac{1}{4\pi^2} \int_C \iint_S \frac{u(x_1, y_1, 0)\, dx_1 dy_1}{\alpha(x - x_1) - \beta(y - y_1) - \gamma z} d\beta \tag{12-16}$$

By differentiating Eq. 12-16 for the downwash velocity $\bar{w} = \partial\bar{\varphi}/\partial z$ there is obtained

$$\bar{w} = \frac{1}{4\pi^2} \int_C \iint_S \gamma \frac{u(x_1, y_1, 0)\, dx_1 dy_1}{[\alpha(x - x_1) - \beta(y - y_1) - \gamma z]^2} d\beta \tag{12-17}$$

As the result of the change in the order of integration, Eq. 12-16 and 12-17 now represent the three-dimensional flow as infinitely many two-dimensional fields superimposed. For each particular value of λ or β, the integration over the surface S yields an elementary two-dimensional disturbance field associated with an oblique strip drawn in the plane of the wing and having its edges tangent to the outline of the planform (see Fig. A,12d).

These relations are most easily seen for those parts of the contour C near $\lambda = 0$ and $1/\lambda = 0$. Considering first the loop around $\lambda = 0$, we have

$$\beta \overset{\rightarrow}{=} \frac{1}{2\lambda}$$

$$\gamma \overset{\rightarrow}{=} \frac{i}{2\lambda}$$

and

$$\alpha(x - x_1) - \beta(y - y_1) - \gamma z \overset{\rightarrow}{=} \frac{1}{2\lambda}[y_1 - (y + iz)]$$

since $\alpha(x - x_1)$ is negligible by comparison. With these substitutions, Eq. 12-17 becomes,

$$\overline{w} = \frac{1}{4\pi^2} \int_C \iint_S \frac{u(x_1, y_1)\, dx_1 dy_1}{[y_1 - (y + iz)]^2} \frac{i}{\lambda}\, d\lambda \tag{12-18}$$

The integration along x_1 may now be performed directly, and yields a

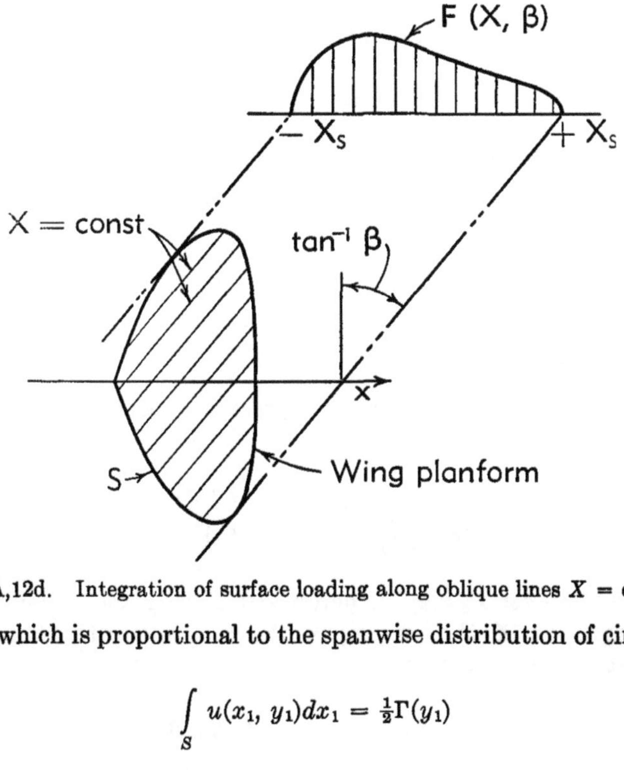

Fig. A,12d. Integration of surface loading along oblique lines X = const.

quantity which is proportional to the spanwise distribution of circulation, i.e.

$$\int_S u(x_1, y_1) dx_1 = \tfrac{1}{2}\Gamma(y_1) \tag{12-19}$$

The integral over the surface S reduces to

$$\int_S \frac{\frac{1}{2}\Gamma(y_1)}{[y_1 - (y + iz)]^2} dy_1 = \frac{1}{2} \int_S \frac{\Gamma'(y_1)}{y_1 - (y + iz)} dy_1 \tag{12-20}$$

which is the familiar formula for the complex velocity function associated with the wake of trailing vortices.

Values of λ near $1/\lambda = 0$ yield an expression similar to Eq. 12-20 except that $y - iz$ appears instead of $y + iz$. The addition of the contributions of the two parts of the contour $|\lambda| \rightrightarrows 0$ and $1/|\lambda| \rightrightarrows 0$ thus yields a real value of $\overline{w}$ which corresponds to the downwash induced by the vortex wake. In particular, it is known that this component of the downwash will be constant if the integrated spanwise loading of the wing is elliptical.

Those portions of the contour C, extending around the unit circle in the λ plane or around the unit strip $(-1 < \beta < +1)$ in the β plane, yield a second component of $\bar{w}$ which corresponds to the "wave drag" of the lifting surface. Again, the surface integral in Eq. 12-17 may be shown to reduce to the expression for a two-dimensional velocity function. The form of the velocity function in this case depends, for each particular value of β, on an integration of the surface loading of the wing in an oblique direction.

To illustrate this relation we replace $\alpha x - \beta y - \gamma z$ by the single variable

$$X = \alpha x - \beta y - \gamma z$$

and for the variable point on the wing we introduce

$$X_1 = \alpha x_1 - \beta y_1 \tag{12-21}$$

together with the orthogonal variable

$$Y_1 = \frac{\beta x_1 + \alpha y_1}{\alpha^2 + \beta^2} \tag{12-22}$$

The factor $1/(\alpha^2 + \beta^2)$ in the latter expression preserves the elementary area. Eq. 12-17 may now be written

$$\bar{w} = \frac{1}{4\pi^2} \int_C \iint_S \frac{u(X_1, Y_1)dX_1dY_1}{(X - X_1)^2} \gamma d\beta \tag{12-23}$$

The integration over Y_1 may be performed directly and amounts simply to an integration of the loading of the wing along the oblique lines $X_1 = \text{const.}$ Thus we may write

$$\int_S u(X_1, Y_1)\, dY_1 = F(X_1, \beta) \tag{12-24}$$

where $F(X_1, \beta)$ is the loading of the oblique lifting line, comparable to Γ in Eq. 12-19. The equation for $\bar{w}$ now becomes

$$\bar{w} = \frac{1}{4\pi^2} \int_C \int_{X-}^{X+} \frac{F(X_1, \beta)\, dX_1}{(X - X_1)^2} \gamma d\beta \tag{12-25}$$

The expression under the integral is the familiar formula for the downwash of a lifting line having the loading $F(X_1, \beta)$. Eq. 12-25 thus represents the downwash of the wing as the sum of values contributed by an infinite number of lifting lines.

Drag of a given distribution of lift. The final value of the wave drag will be obtained by integrating the product of the lift $l(x, y)$ and the downwash angle $\overline{w}/U$ over the surface S. In this way we obtain

$$D_w = -\rho_\infty \iint_S u\overline{w}\,dxdy \tag{12-26}$$

where u corresponds to the specified distribution of lift and $\overline{w}$ is given by Eq. 12-25. Making use of Eq. 12-21, 12-22, and 12-24, we obtain

$$D_w = -\frac{\rho_\infty}{4\pi^2}\int_C \int_{X-}^{X+}\int_{X-}^{X+} \frac{F(X,\beta)F(X_1,\beta)dXdX_1}{(X-X_1)^2}\gamma d\beta \tag{12-27}$$

Eq. 12-27 represents the wave drag of the lifting surface as the sum of the drags of the oblique lifting lines. Following a well-known procedure in subsonic wing theory, we may represent each of the oblique loadings of the wing by a Fourier series. Thus if

$$X + \text{const} = X_s \cos\delta$$

and

$$F(X,\beta) = \sum_1^\infty A_n \sin n\delta$$

the drag of each oblique lifting line will be proportional to $\sum_1^\infty nA_n^2$. Since the total lift L depends only on A_1, we obtain finally

$$D_w = \frac{1}{\pi}\int_{-1}^{+1} \frac{L^2}{\pi\frac{1}{2}\rho_\infty U^2(2X_s)^2}\left(1 + \sum_2^\infty na_n^2\right)\sqrt{1-\beta^2}\,d\beta \tag{12-28}$$

where $a_n = A_n/A_1$ and the integration around the contour C has been replaced by four times the real integral over -1 to $+1$.

In Eq. 12-28 the quantity $2X_s$, which is a function of β, represents the "oblique span" of the lifting surface (see Fig. A,12d). For the particular value $\beta = 0$, $2X_s$ is simply the over-all length of the wing. The coefficients a_n, which will also generally vary with β, represent the departures of the loading curves from the elliptic form. As the formula shows, for a wing of given dimensions, departures from elliptic loadings have the effect of increasing the wave drag.

Eq. 12-28 is entirely analogous to the well-known formula [*140*] for the induced drag or vortex drag

$$D_v = \frac{L^2}{\pi\frac{1}{2}\rho_\infty U^2 b^2}(1 + 2a_2^2 + 3a_3^2 + \cdots)$$

where the a_n's are now the coefficients occurring in the spanwise load distribution. It is evident that, in order to minimize both the wave drag and the vortex drag, the span loading and the oblique loadings for the range $-1 < \beta < 1$ should be as nearly elliptical as possible.

In the case of a long narrow wing, the preceding formulas may be simplified by the assumption that the projected dimension of the planform does not differ significantly from the over-all length c measured along the x axis. If we further assume that both the lengthwise and spanwise loadings are elliptical, so as to obtain the minimum drag, there results the formula

$$D_v + D_w = \frac{L^2}{\pi \frac{1}{2}\rho_\infty U^2 b^2} + \frac{M_\infty^2 - 1}{2} \frac{L^2}{\pi \frac{1}{2}\rho_\infty U^2 c^2} \tag{12-29}$$

As pointed out by E. W. Graham, in certain cases the optimum distribution of lift will not cover the whole area of the allotted planform.

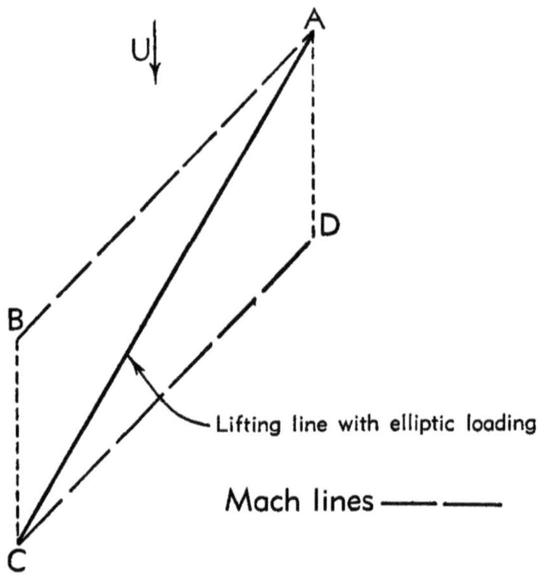

Fig. A,12e. Ideal distribution of lift for parallelogram $ABCD$.

Thus if we describe a parallelogram bounded by Mach lines and lines parallel to the direction of flight, as shown in Fig. A,12e, the ideal distribution will place all the lift along a narrow lifting line extending diagonally across the area. Within the region shown in Fig. A,12f, formed by Mach-line and streamwise tangents to an ellipse, a uniform concentration of lift over the ellipse, with zero lift elsewhere, will result in the least drag.

The minimum drag of a wing having an elliptic planform. The foregoing discussion indicates that the minimum drag of a wing of elliptic planform will occur when the lift is distributed uniformly over the surface, since the integrated loading in every oblique direction is then elliptical.

The downwash $\overline{w}$ over the elliptic wing may be calculated directly from Eq. 12-17. To evaluate the surface integral we write

$$\iint\limits_S \frac{u(x_1, y_1)dx_1dy_1}{[\alpha(x - x_1) - \beta(y - y_1) - \gamma z]^2} = \pi i F'(X, \beta) \tag{12-30}$$

After introducing $u = u_0$ and $X = \alpha x - \beta y - \gamma z$, and integrating over x_1, we obtain

$$F'(X, \beta) = \frac{1}{\pi i} \int_{-y_s}^{+y_s} \frac{1}{\alpha} \left[\frac{u_0 dy_1}{X - \alpha x_1 + \beta y_1} \right]_{-x_s}^{+x_s} \tag{12-31}$$

where x_s and y_s correspond to the edges of the planform. For the ellipse

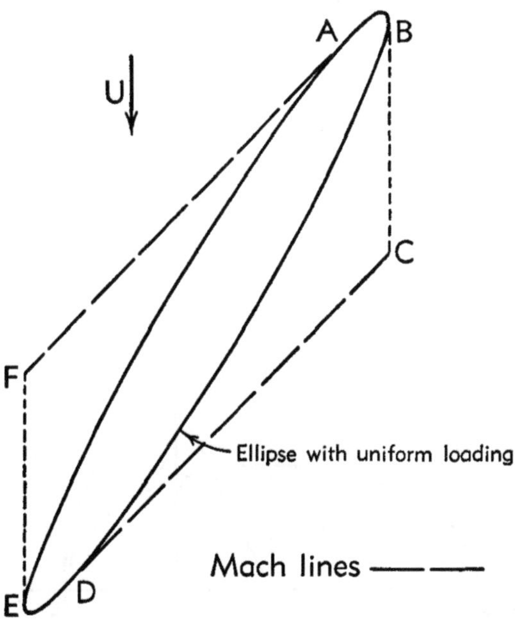

Fig. A,12f. Ideal distribution of lift for area *ABCDEF*.

with semiaxes a and b (Fig. A,12g) we have in parametric form

$$y_1 = b \cos \vartheta$$

$$x_s = -a \sin \vartheta$$

The integral in Eq. 12-31 now becomes

$$F'(X, \beta) = \frac{u_0 b}{\pi i \alpha} \int_0^{2\pi} \frac{\sin \vartheta d\vartheta}{X + b\beta \cos \vartheta + a\alpha \sin \vartheta} \tag{12-32}$$

The evaluation of integrals of this form is given in [*53*]. There are two determinations, depending on the location of the poles of the integrand

$$e_1 = \frac{-X - \sqrt{X^2 - (b\beta)^2 - (a\alpha)^2}}{b\beta - i a\alpha} \tag{12-33}$$

and

$$e_2 = \frac{-X + \sqrt{X^2 - (b\beta)^2 - (a\alpha)^2}}{b\beta - ia\alpha} \tag{12-34}$$

For points x, y inside the elliptic disk we have

$$\frac{x^2}{a^2} + \frac{y^2}{b^2} < 1$$

and in this case the general formula

$$\int_0^{2\pi} \frac{\cos n\vartheta d\vartheta}{X + b\beta \cos \vartheta + a\alpha \sin \vartheta} = \int_0^{2\pi} \frac{i \sin n\vartheta d\vartheta}{X + b\beta \cos \vartheta + a\alpha \sin \vartheta}$$

$$= \pi \frac{e_1^n - e_2^n}{\sqrt{X^2 - (b\beta)^2 - (a\alpha)^2}} \tag{12-35}$$

applies. Eq. 12-32 corresponds to $n = 1$, and the formulas give

$$F'(X, \beta) = \frac{2u_0 b}{\alpha} \frac{1}{b\beta - ia\alpha} \tag{12-36}$$

Higher values of n correspond to nonuniform loadings over the wing. It is evident that in the present case $F'(X, \beta)$ is independent of X over

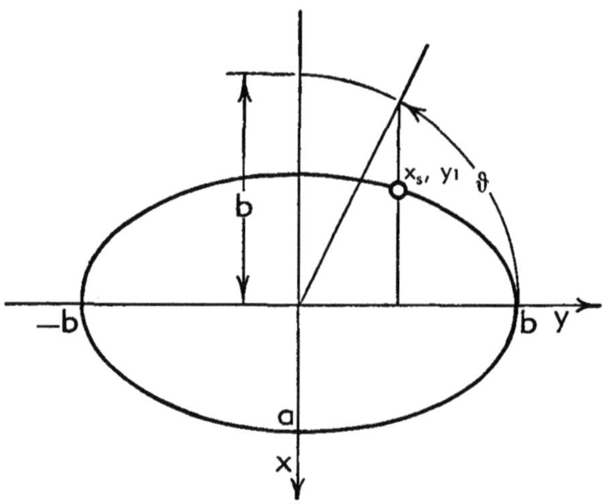

Fig. A,12g. Notation for elliptic wing.

the surface of the ellipse and hence the downwash will be constant at these points.

The value of the downwash is now obtained by introducing the value (12-36) into (12-30) and substituting in Eq. 12-17, making use of the relations

$$\alpha = 1$$
$$\gamma = \sqrt{1 - \beta^2}$$

Thus

$$\frac{\bar{w}}{U} = \frac{u_0}{2\pi i U} \int_C \frac{\sqrt{1-\beta^2}}{\beta - i(a/b)} = \frac{2u_0}{U}\sqrt{1+\frac{a^2}{b^2}} \tag{12-37}$$

Since the value $\beta = i(a/b)$ corresponds to two distinct poles, one on each portion of the λ surface (or, one pole on each sheet of the β surface), the value of the integral is twice the residue at this point.

The drag is given by

$$D_{\min} = \frac{1}{2}\frac{\bar{w}}{U}L$$

where L is the total lift, equal to

$$L = 2\rho_\infty u_0 US$$

for a wing of area S. The formula for the minimum drag of the elliptic wing then becomes

$$D_{\min} = \frac{L^2}{4(\frac{1}{2}\rho_\infty U^2)S}\sqrt{1+\frac{a^2}{b^2}} \tag{12-38}$$

or, in terms of the coefficients C_L and C_D,

$$C_{D_{\min}} = C_L^2\sqrt{\left(\frac{1}{4}\right)^2 + \left(\frac{1}{\pi Æ}\right)^2} \tag{12-39}$$

In the latter form it can be seen that the first term under the radical sign corresponds to the value given by the Ackeret theory for a wing of very high aspect ratio, while the second term corresponds to the value given by the Munk theory. The latter value is approached at small values of the aspect ratio Æ, for which the vortex drag predominates over the wave drag.

Eq. 12-37 and 12-39 apply at $M_\infty = \sqrt{2}$. The variation with Mach number may be deduced, however, by correcting the downwash $\bar{w}/U$ by the factor $\sqrt{M_\infty^2 - 1}$ and multiplying the ratio a/b by the factor $1/\sqrt{M_\infty^2 - 1}$.

For the elliptic wing in straight flow, the drag is invariably greater than that given by the Ackeret theory for a two-dimensional flat plate. It is to be expected that lower values could be obtained if a long narrow ellipse were yawed behind the Mach cone.

The treatment of the yawed ellipse follows the preceding analysis closely. Again the minimum drag occurs when the lift density is constant over the whole surface. If the equation of the ellipse is written in the form

$$x_s = m'y_1 \pm a'\sqrt{1-\frac{y_1^2}{b'^2}}$$

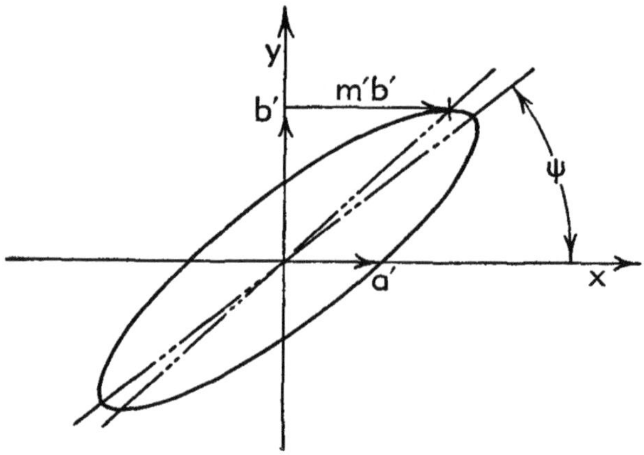

Fig. A,12h. Oblique ellipse.

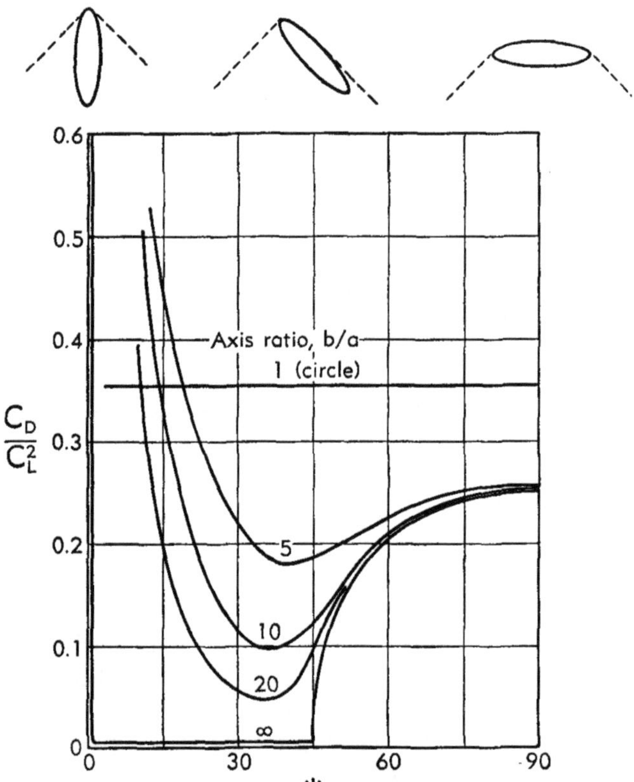

Fig. A,12i. Minimum drag of elliptic wings at various angles of yaw.

(see Fig. A,12h), the expression for the drag reduces to

$$D_{\min} = \frac{L^2}{4(\frac{1}{2}\rho_\infty U^2)S}\,\text{R.P.}\sqrt{1 - \left(m' + i\frac{a'}{b'}\right)^2} \tag{12-40}$$

The limit $a'/b' \to 0$ corresponds to infinite aspect ratio and the expression for the drag then involves

$$\text{R.P.}\ \sqrt{1 - m'^2}$$

The change of this quantity from real to imaginary, as m' passes through 1, shows the disappearance of the wave drag when the wing of infinite aspect ratio is yawed behind the Mach cone. The variation of drag coefficient with angle of yaw ψ for ellipses of finite proportions is shown in Fig. A,12i.

A,13. Determination of Lift in Three-Dimensional Flow. Methods and Formulas.

Introduction. The preceding article describes a general method for finding the drag when the distribution of lift on a surface is given. Finding the shape of the surface under the same circumstances is generally a direct mathematical procedure; this problem is adequately discussed in VI,D,11. The problem of finding the lift distribution when the wing geometry is given is more difficult. We are first of all confronted with the problem of mixed boundary conditions, the pressure being specified off the wing surface and the vertical component of velocity on it. Consequently the known relations take the form of an integral equation, solution of which is possible only under certain limited conditions. The solutions are not determinate without a knowledge of the singularities. This latter information is obtained from physical considerations and is based on the previously discussed classification of the edges (Art. 10). It is obvious that this classification is not so much a matter of the wing planform as it is of the speed range. Every sweptback wing, for example, on entering the supersonic regime, has subsonic leading edges, and in most cases, subsonic trailing edges. At a higher Mach number the same planform may have subsonic leading edges and supersonic trailing edges. Finally, if the Mach number is increased sufficiently, both leading and trailing edges will be supersonic. Since the interference effects also vary radically with the flight speed because of the dependence of the extent of the disturbance fields on the Mach angle, it is clear that no single concise formula can describe the flow over a given wing throughout the supersonic speed range.

Nevertheless, we are in a far better position with regard to the lifting surface problem at supersonic speeds than we were (Art. 7) with the wing at subsonic speeds. Just because the interference effects (in the first order theory) are confined to the interior of the Mach cones from the various points of disturbance, it is possible to perform far more complete calculations of the flow over most planforms than in the corresponding subsonic problems.[12] It is no longer necessary to resort to the device of the lifting line approximation, and for many planforms the entire lifting surface solution can be obtained with no simplifying assumptions beyond those used in linearizing the flow equation.

The methods presently used fall into two broad classifications which may be termed, from their elements, conical flow methods and source-distribution methods. The choice between them depends largely on the planform and Mach number under consideration. In many cases, it will be more effective to treat part of the problem by one method and part by the other, than to use either alone [*141*]. When the planform is a simple polygon, the major part of the flow field can usually be described immediately in terms of one or more conical flow solutions. For this reason the conical flow method is presented first.

Conical Flow Theory. As originally conceived by Busemann [*57*], a conical flow field is defined as a field in which all the velocity components, and hence the pressure and the slope of the streamlines, are constant along rays emanating from a single point. An application of conical solutions in subsonic flow has been described in Art. 7 and another indicated in Art. 9. However, the chief use of conical disturbance fields has been in connection with supersonic flow, where the Mach cone bounding the field of a point disturbance is itself one of the conical surfaces of constant pressure of the field. Busemann found that by restricting the problem to small disturbances, solutions of very general and elegant form could be obtained. We further confine ourselves here to essentially planar disturbances. For these, the work of Lagerstrom [*142*], Hayes [*143*], Germain [*144*], and others has made available an extensive and useful theory, of which we will present, with a minimum of development, only the most directly applicable results.

Solutions of conical form constitute part of a more general class of theoretical flows having considerable interest from the standpoint of practical wing shapes. These flows are described by homogeneous functions of the coordinates x, y, and z, and contain the conical fields as the special class of degree zero. Solutions of higher degree are sometimes called "quasi-conical," and the pressure, instead of being constant, varies according to some power of x, y, or z along each ray. An example of such a field has already been encountered (Art. 9) through the integration of the conical field of a simple wedge to obtain a parabolically curved surface. Another method of obtaining solutions of quasi-conical form is through application of the easily proved theorem that, if u, v, and w are the velocity components of any disturbance field satisfying the linearized flow

[12] The wing of elliptic planform must be noted as an exception to this statement.

equation, then

$$f = xu + yv + zw$$

also satisfies the equation. If u, v, and w are components of a conical velocity field—or more generally, are homogeneous functions—the solution f will obviously be of one higher degree in the coordinates.

The general subject of quasi-conical fields has been treated by Heaslet and Lomax in this series (VI,D,13), Germain [*144*], Robinson [*145*], Squire [*146*], and Multhopp [*147*]. Applications to rolling and pitching wings may be found in [*148,149,150*]. The reader will have no difficulty in generalizing much of what follows to similar problems in quasi-conical flow.

Returning to the consideration of the conical solutions, comparison of the three-dimensional Laplace equation and the linearized equation for supersonic speeds in the normalized form (10-4) suggests that solutions of the latter may be obtained by replacing y by $\pm iy$ and z by $\pm iz$ in the general solution (7-27) of the former. The principal result is the replacement of the radial distance $\sqrt{x^2 + y^2 + z^2}$ by the so-called hyperbolic distance $\sqrt{x^2 - y^2 - z^2}$. The numerator of the argument becomes $\mp(z \pm iy)$, but because y and z are interchangeable in Eq. 10-4, we may choose a form for our variable more closely resembling the subsonic one, namely

$$\epsilon = \frac{y + iz}{x + \sqrt{x^2 - y^2 - z^2}} \tag{13-1}$$

It can be shown [*142*] that Eq. 10-4, written in terms of the real and imaginary parts of ϵ, reduces identically to the two-dimensional Laplace equation. However, since the new coordinates bear no direct relation to the velocity components (such as Eq. 2-7), this fact does not offer as great an advantage as might be expected. Instead it is convenient in the supersonic case to make use of a further transformation of variables, writing

$$\tau = \frac{2\epsilon}{1 + \epsilon^2} \tag{13-2}$$

The variable τ has the property of mapping the interior of the Mach cone into the whole of the complex plane. The Mach cone itself and the region outside it are transformed by Eq. 13-2 into the positive and negative branches of the real axis outside ± 1 and the circle at infinity (see Fig. A,13a). In the neighborhood of the x, y plane, the variable τ approaches the value $(y + iz)/x$, so that the contour of a thin airfoil is mapped undistorted into the neighborhood of the real axis of the τ plane. Thus the expression of the boundary conditions in terms of τ is very simple.

The relations between the components of the velocity, obtained from

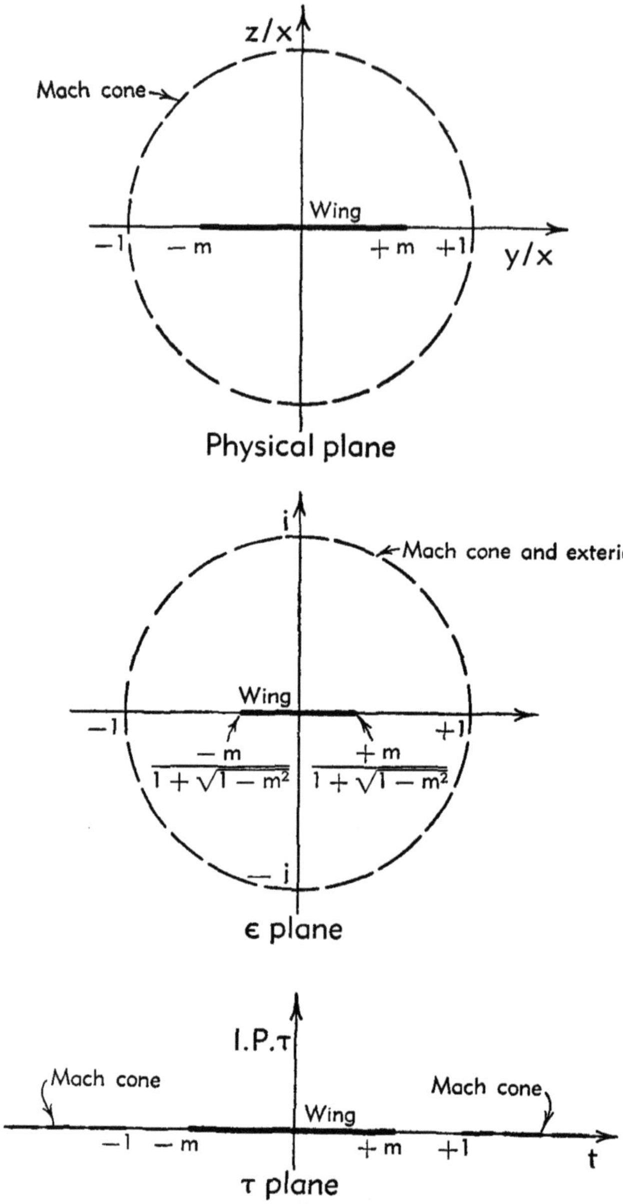

Fig. A,13a. Transformation of flow field to the plane of the conical variable τ.

the irrotationality conditions, are

$$d\hat{W} = -i\frac{\sqrt{1-\tau^2}}{\tau}\,d\hat{U} \tag{13-3}$$

and

$$d\hat{V} = -\frac{1}{\tau}\,dU \tag{13-4}$$

where $\hat{U}$, $\hat{V}$, and $\hat{W}$ are the analytic functions of τ of which u, v, and w are the real parts.

If the given boundary conditions are sufficient to determine either $\hat{U}$ or $\hat{W}$, the remaining components can be found by integrating. The conditions to be satisfied on the Mach cone determine the constant of integration, except for an imaginary part of no significance in the final application. When the boundary conditions are "mixed," that is $\hat{U}$ is conditioned in some regions and $\hat{W}$ in others, Eq. 13-3 yields some general conclusions useful in translating the conditions on $\hat{W}$ into terms of $\hat{U}$. First, it is seen that, in the plane of the wing ($z = 0$) and within the Mach cone ($|\tau| < 1$), a necessary condition for a flat airfoil surface, $w =$ const, is that the imaginary part of $\hat{U}$ be zero. However, this condition is not sufficient, for if $\hat{U}' = d\hat{U}/d\tau$ is continuous in the neighborhood of $\tau = 0$, integrating across the singularity in Eq. 13-3 will result in a discontinuity in $\hat{W}$ of the magnitude $-\pi\hat{U}'(0)$, corresponding to a step in the downwash, unless R.P. $\hat{U}'(0)$ is equal to zero.[13] Thus for a continuous surface of asymmetric planform including the $\tau = 0$ axis, two terms are generally required to express the pressure distribution, so that the lateral gradient may be zero through the axis.

We are further aided in the construction of solutions by the following considerations concerning the flow at the edges:

1. If the edge is a leading edge at an angle of attack, and the flow normal to it is subsonic, there will be a $\frac{1}{2}$-power singularity in the pressure on the wing and in the downwash just ahead of the leading edge.
2. If the normal component of the stream velocity is supersonic, both u and w will be stepwise discontinuous on crossing the edge.
3. If the edge is trailing and the normal flow is subsonic, application of the Kutta condition determines that the discontinuity in u at the plane of the wing will vanish at the edge, and that w will remain finite on crossing the edge.
4. If the edge is trailing, but the normal flow is supersonic, the Kutta condition may be replaced by a discontinuous expansion to zero pressure. (Nonlifting wings are discussed elsewhere.)

[13] If I.P. $\hat{U}'(0)$ is not equal to zero there will be a logarithmic infinity in the downwash along the ray $\tau = 0$.

The logarithmic functions have been found most useful in setting up solutions to fit simple boundary conditions. Fig. A,13b illustrates the behavior of these functions. If $\zeta(\tau)$ is any complex function of τ which on the plane of the wing takes on real values, then the real part of $i/\pi \ln \zeta(t)$ will provide a unit step wherever $\zeta(t)$ goes through zero. The function $\zeta(t)$ may be otherwise constructed to cause the disturbance field to vanish at infinity ($\tau \to \infty$) and to satisfy the edge conditions where desired.

Conversely, the inverse trigonometric functions (see Fig. A,13b) can be set up to satisfy the Kutta condition at a trailing edge and provide

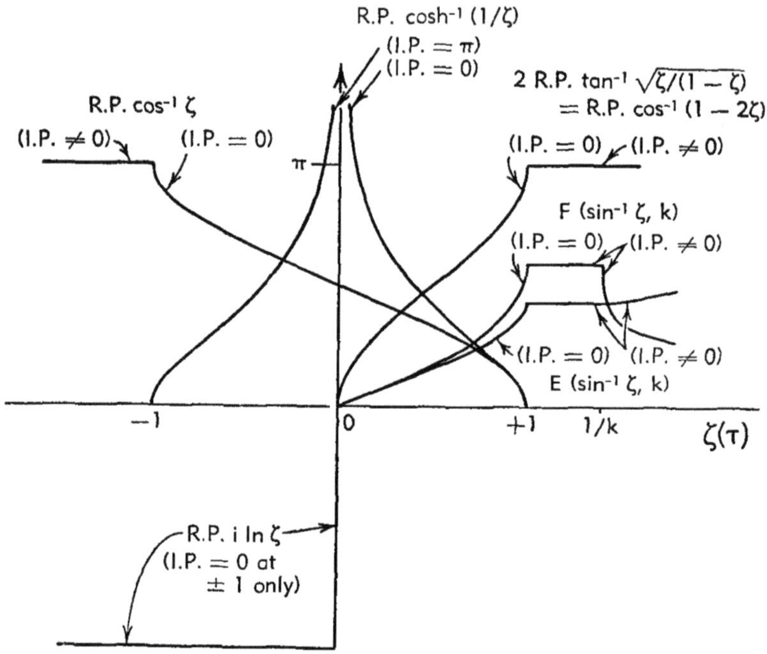

Fig. A,13b. Sketch showing real parts of functions useful as conical flow solutions.

constant pressure (or downwash) over some other region of the field; it will follow that in the region in which the real part of the function varies, the downwash (or pressure) from Eq. 13-3, will either be constant or step-shaped. The elliptic integrals $F(\vartheta, k)$ and $E(\vartheta, k)$ show essentially the same pattern, as $\sin \vartheta$ varies from zero, through one, to $1/k$ [*151*]. Square-root terms will be needed to provide the singularities at leading edges and also give constant or step-shaped downwash or pressure curves. The inverse hyperbolic functions result from assuming one velocity component in logarithmic form.

A table of conical flow solutions[14] follows with sketches of the bound-

[14] Most of these solutions are derived in [*142*]. A few solutions have been derived especially for this table. References for the others are as follows: solutions 10 and 11, [*152*]; solution 12, [*146*]; solution 18, [*148*]; solutions 13 and 14, [*153*].

ary conditions they are designed to satisfy. Only the u and w components of the flows are given, but since these are given in complex form, the sidewash v, if desired, can be obtained by the use of Eq. 13-4. The velocities are plotted against the (real) values of τ in the plane $z = 0$. The symbol m usually denotes the value of τ along the leading or trailing edge of the wing, that is, $\sqrt{M_\infty^2 - 1}$ times the inclination of the edge to the stream. However, it may merely refer to the ray separating two regions where different boundary conditions apply, and in the superposition procedures to be described later, m will often be replaced by a variable.

The double signs in the designation of u and w refer to values on the upper and lower surfaces and indicate discontinuities in their values across the plane of the wing. A discontinuity in w indicates divergence of the upper and lower surfaces of the wing, or thickness; a discontinuity in u corresponds to lift. However, when there is no possibility of interaction between the upper and lower surfaces, as in the case of a wing with supersonic edges, or of a control surface at the rear of a wing, any thickness solution may be used to obtain the lift on a thin surface of similar inclination by reversing the signs of the solution to take account of the actual slopes of the upper and lower surfaces (cf. solutions 2 and 3). The lifting solutions may be used in the same way to determine the thickness distribution corresponding to a specified pressure distribution.

Table A,13a is of course not exhaustive; other solutions are to be found in [*142,152*] and elsewhere. On the other hand, several of the solutions presented are readily derivable one from another, but are included because of their individual usefulness or to indicate methods of devising additional solutions. For example, the solution (2) for the bilaterally symmetrical nonlifting wedge is obtained from the solution (1) for the semi-infinite wedge by the superposition of the latter and its reflection in the x axis. Solutions 2 and 10 are related in that the vertical component in the former is required to satisfy the same boundary conditions as the streamwise component in the latter, so that the same function will do for both, the remaining components in each case being determined from Eq. 13-3 and 13-4. Some of the solutions are simply special cases of others, but conversely some may be regarded as generalizations of others, obtainable (e.g. 7 from 6) by an oblique transformation, discussed immediately below. It should be possible to satisfy any simple combination of conical boundary conditions with a modified form of the solutions given.

THE OBLIQUE TRANSFORMATION APPLIED TO CONICAL FLOWS. It may be helpful to discuss briefly the subject of the oblique transformation. The subsonic form of this transformation was used (except for a scale factor which has no significance in a conical field) in Art. 9 (Eq. 9-9) and corresponds to a simple rotation of the axis. Again omitting the scale

Table A,13a. Conical flow functions.

Flow field; u and w components	Values in plane of wing	Plan view and boundary conditions
1. Wedge, supersonic leading edge, streamwise side edge $u = \text{R.P.}\, \frac{-m}{\pi\sqrt{m^2-1}} \cos^{-1} \frac{1-m\tau}{m-\tau}$ $w = \text{R.P.}\, \frac{i}{\pi} \ln \frac{\tau}{\tau - m}$	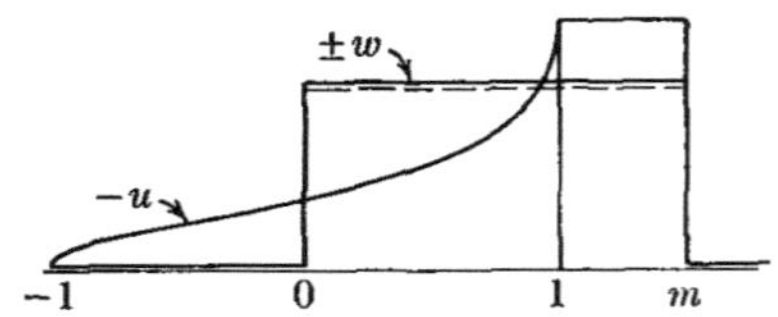	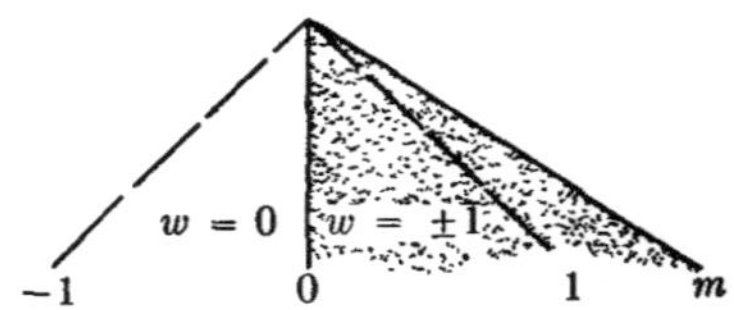
2. Unyawed wing, wedge section, $m > 1$ $u = \text{R.P.}\, \frac{-m}{\pi\sqrt{m^2-1}} \left(\cos^{-1} \frac{1-m\tau}{m-\tau} + \cos^{-1} \frac{1+m\tau}{m+\tau}\right)$ $w = \text{R.P.}\, \frac{i}{\pi} \ln \frac{\tau + m}{\tau - m}$	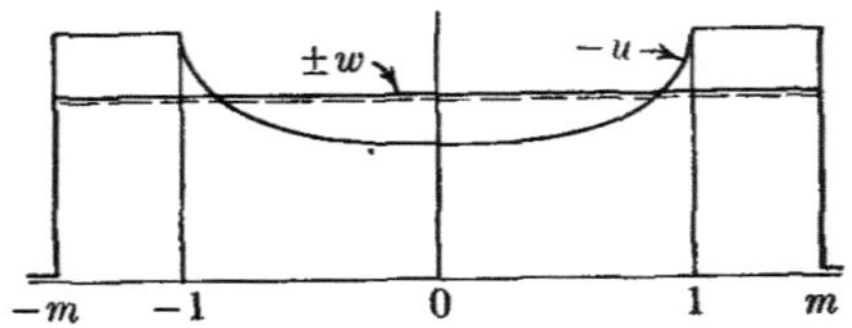	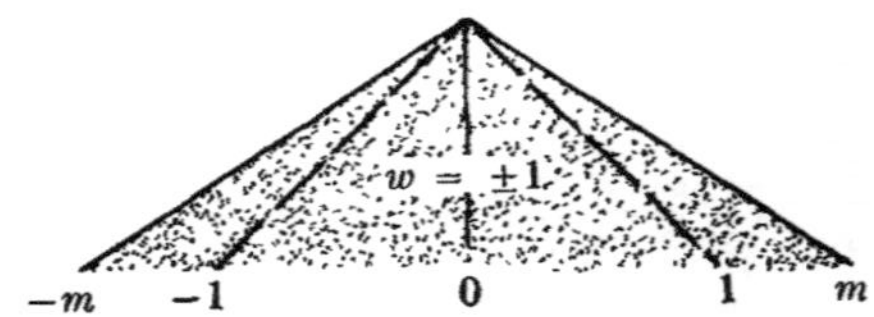
3. Flat lifting triangle, $m > 1$ $u = \pm\text{R.P.}\, \frac{\alpha U}{\beta\pi} \frac{m}{\sqrt{m^2-1}} \left(\cos^{-1} \frac{1-m\tau}{m-\tau} + \cos^{-1} \frac{1+m\tau}{m+\tau}\right)$ $w = \mp\text{R.P.}\, i \frac{\alpha U}{\beta\pi} \ln \frac{\tau + m}{\tau - m}, \qquad \beta = \sqrt{M_\infty^2 - 1}$ (Two independent flows of form 2)	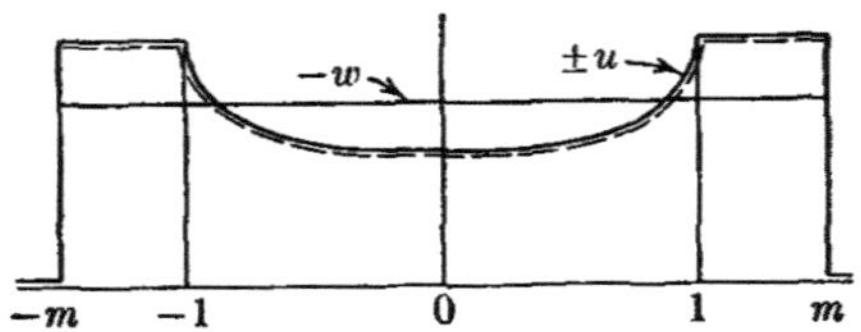	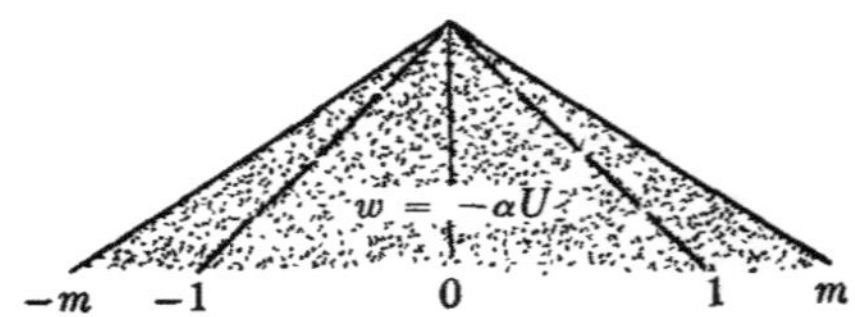

4. Yawed triangle, both leading edges supersonic

$$u = \pm \text{R.P.}\, \frac{\alpha U}{\beta\pi}\left(\frac{m_1}{\sqrt{m_1^2 - 1}}\cos^{-1}\frac{1 - m_1\tau}{m_1 - \tau} + \frac{m_2}{\sqrt{m_2^2 - 1}}\cos^{-1}\frac{1 + m_2\tau}{m_2 + \tau}\right)$$

$$w = \mp \text{R.P.}\, i\frac{\alpha U}{\beta\pi}\ln\frac{\tau + m_2}{\tau - m_1}$$

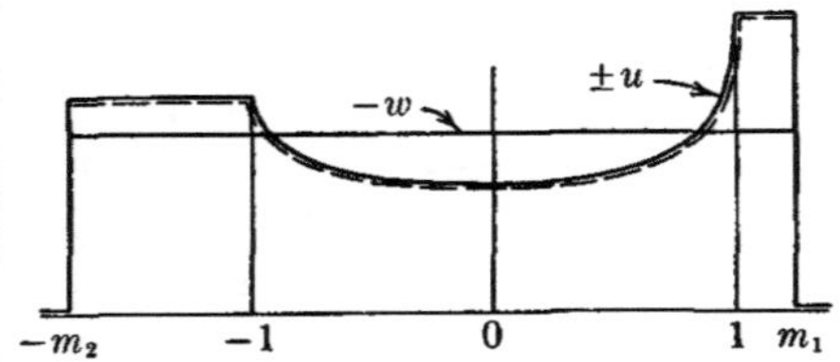

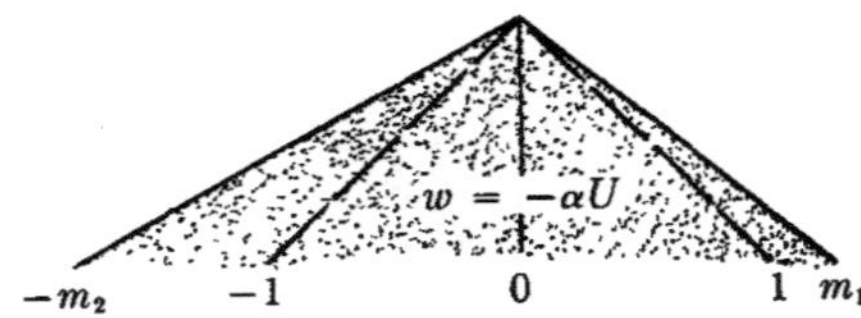

5. Wedge, subsonic leading edge, streamwise side edge

$$u = \text{R.P.}\, \frac{-m}{\pi\sqrt{1 - m^2}}\cosh^{-1}\frac{1 - m\tau}{m - \tau}$$

$$w = \text{R.P.}\, \frac{i}{\pi}\ln\frac{\tau}{\tau - m}$$

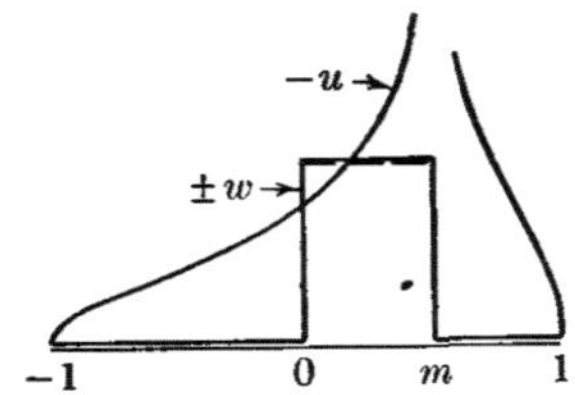

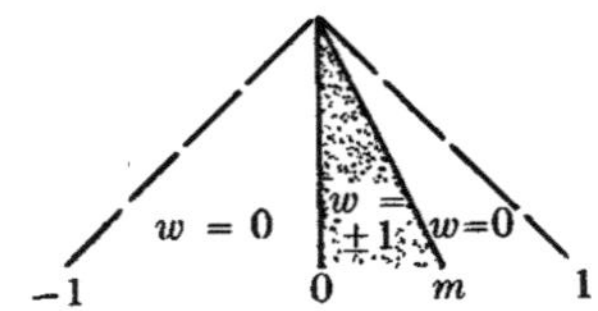

6. Flat lifting triangle, $m < 1$

$$u = \text{R.P.}\, \frac{m^2\alpha U}{\beta E(k)}\frac{1}{\sqrt{m^2 - \tau^2}}$$

$$w = \text{R.P.}\, \frac{\alpha U}{\beta E(k)}\left[\tau\sqrt{\frac{1 - \tau^2}{\tau^2 - m^2}} - E(\vartheta, k)\right]$$

$$\vartheta = \sin^{-1}\sqrt{\frac{1 - \tau^2}{1 - m^2}}, \qquad k = \sqrt{1 - m^2}$$

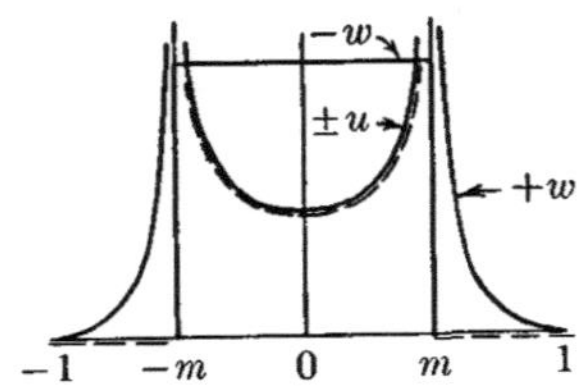

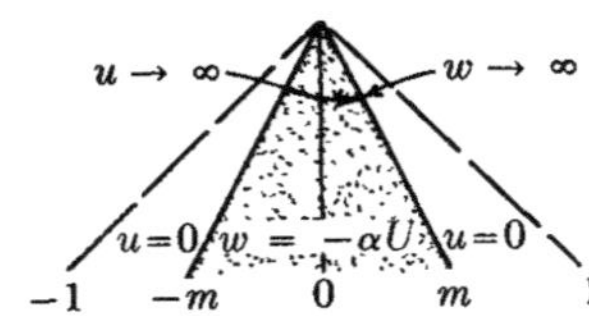

Table A,13a (continued)

Flow field; u and w components	Values in plane of wing	Plan view and boundary conditions

7. Yawed triangle, both leading edges subsonic

$$u = \text{R.P.}\ \frac{\alpha U}{\beta A E(k)}\left[m_1\sqrt{\frac{m_2+\tau}{m_1-\tau}} + m_2\sqrt{\frac{m_1-\tau}{m_2+\tau}}\right]$$

$$w = \text{R.P.}\ \frac{\alpha U}{\beta A E(k)}\left\{(1+m_1)\sqrt{\frac{(\tau+m_2)(1-\tau)}{(\tau-m_1)(1+\tau)}}\right.$$

$$+ (1-m_2)\sqrt{\frac{(\tau-m_1)(1-\tau)}{(\tau+m_2)(1+\tau)}}$$

$$\left. + A[(1-\sqrt{1-k^2\sin^2\vartheta})\cot\vartheta - E(\vartheta, k)]\right\}$$

$$A = \sqrt{(1+m_1)(1+m_2)} + \sqrt{(1-m_1)(1-m_2)}$$

$$k = \frac{2}{A}[(1-m_1^2)(1-m_2^2)]^{\frac{1}{4}},$$

$$\vartheta = \tan^{-1}\frac{A}{2}\sqrt{\frac{1-\tau^2}{(\tau-m_1)(\tau+m_2)}}$$

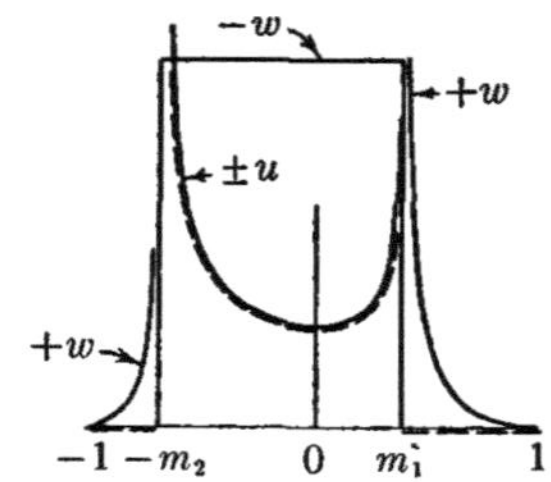

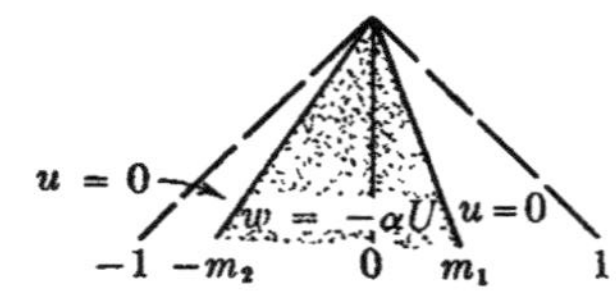

8. Yawed triangle, one edge subsonic, one supersonic

$$u = \text{R.P.}\ \frac{2U\alpha}{\pi\beta}\left[\frac{m_1}{1+m_1}\sqrt{\frac{(m_1+m_2)(1+\tau)}{(1+m_2)(m_1-\tau)}}\right.$$

$$\left. + \frac{m_2}{\sqrt{m_2^2-1}}\tan^{-1}\sqrt{\frac{(m_2-1)(m_1-\tau)}{(m_1+m_2)(1+\tau)}}\right]$$

$$w = \text{R.P.}\ \frac{2U\alpha}{\pi\beta}\left[\sqrt{\frac{(m_1+m_2)(1-\tau)}{(1+m_2)(\tau-m_1)}}\right.$$

$$\left. - \tan^{-1}\sqrt{\frac{(m_1+m_2)(1-\tau)}{(1+m_2)(\tau-m_1)}}\right]$$

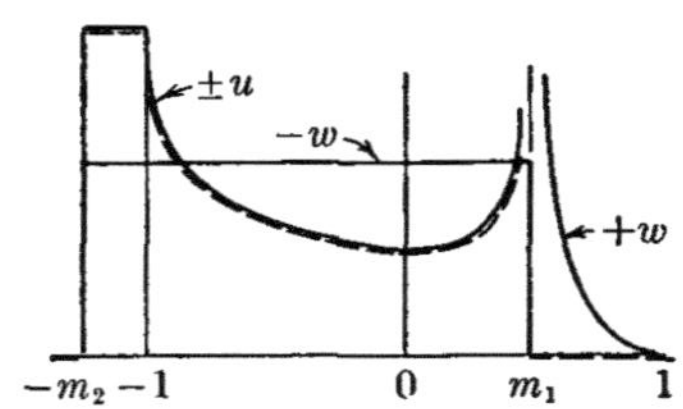

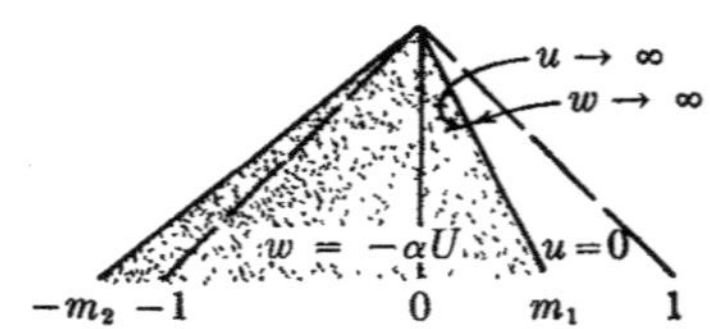

9. Yawed triangle, one edge trailing

$$u = \text{R.P.}\,\frac{2m_2\alpha U}{\pi\sqrt{m_2^2-1}}\tan^{-1}\sqrt{\frac{(m_2-1)(\tau-m_1)}{(m_2-m_1)(1-\tau)}}$$

$$w = \text{R.P.}\,\frac{2\alpha U}{\pi\sqrt{m_2+1}}\left[\sqrt{\frac{m_2-m_1}{m_1}}\times\sinh^{-1}\sqrt{\frac{m_1-\tau}{(1+m_1)\tau}} - \sqrt{m_2+1}\tan^{-1}\sqrt{\frac{(m_2-m_1)(1+\tau)}{(m_2+1)(m_1-\tau)}}\right]$$

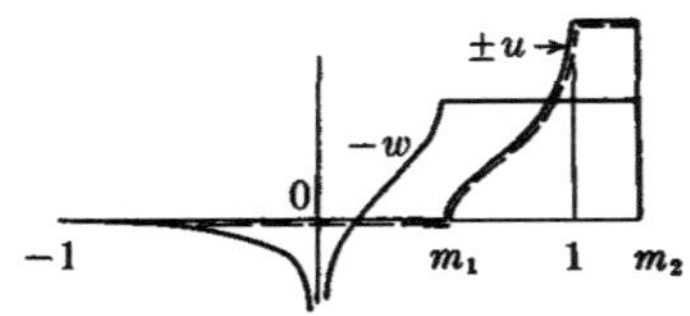

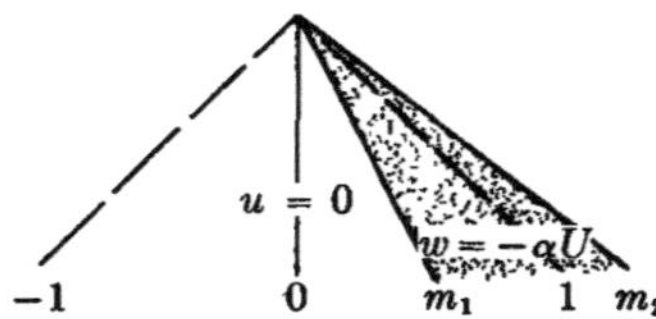

10. Uniformly loaded triangular wing

$$u = \text{R.P.}\,\frac{i}{\pi}\ln\frac{\tau+m}{\tau-m}$$

$$w = \text{R.P.}\,\frac{1}{\pi}\left[\frac{\sqrt{1-m^2}}{m}\left(\cosh^{-1}\frac{1+m\tau}{\tau+m} + \cosh^{-1}\frac{1-m\tau}{\tau-m}\right) - \frac{2}{m}\cosh^{-1}\frac{1}{\tau}\right]$$

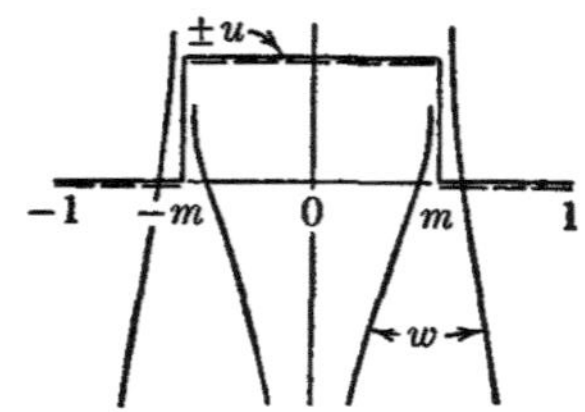

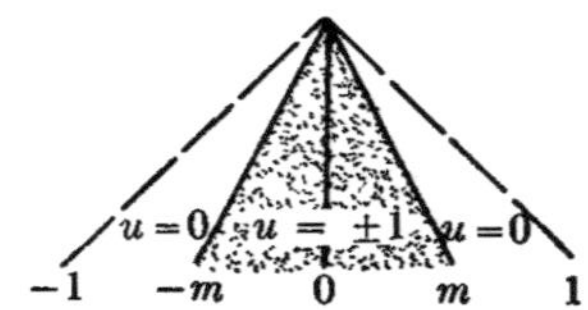

11. Elliptically loaded triangular wing

$$u = \text{R.P.}\,i\left(\tau - \sqrt{\tau^2-m^2}\right)$$

$$w = \text{R.P.}\left[\sqrt{1-\tau^2} - \cosh^{-1}\frac{1}{\tau} + F(\vartheta, k) - E(\vartheta, k)\right]$$

$$k = \sqrt{1-m^2}, \qquad \vartheta = \sin^{-1}\sqrt{\frac{1-\tau^2}{1-m^2}}$$

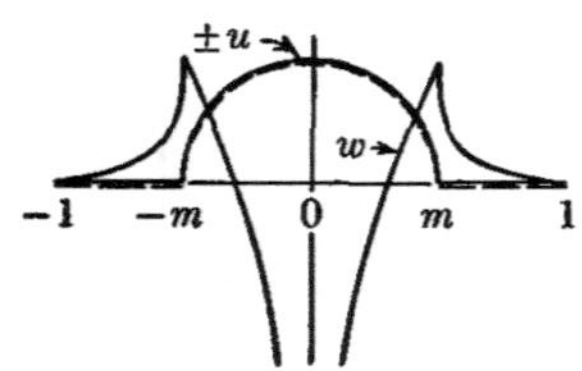

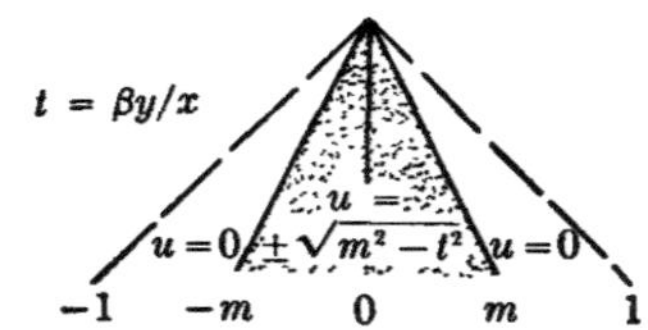

Table A,13a (continued)

Flow field; u and w components	Values in plane of wing	Plan view and boundary conditions
12. Cone with elliptic cross section $u = \text{R.P.}\,\dfrac{-m^2}{1-m^2}\,[\tau \tan\vartheta + F(\vartheta, k) - E(\vartheta, k)]$ $k = \sqrt{1-m^2}, \qquad \vartheta = \sin^{-1}\sqrt{\dfrac{1-\tau^2}{1-m^2}}$ $w = \text{R.P.}\,\dfrac{-m^2}{\sqrt{m^2-\tau^2}}$	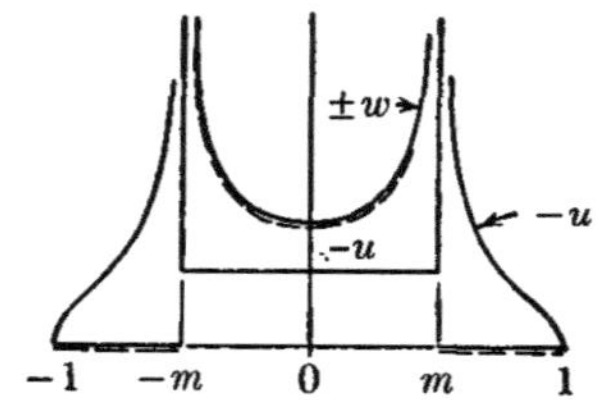	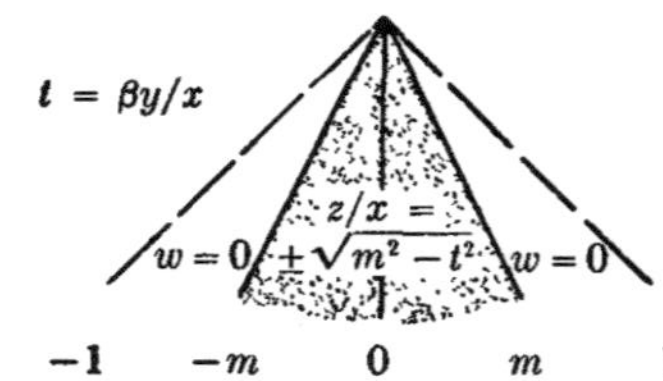
13. Symmetrical flap deflection at leading edge $u = \text{R.P.}\,\cosh^{-1}\sqrt{\dfrac{m_1^2-m_2^2}{\tau^2-m_2^2}}$ $w = \text{R.P.}\,\sqrt{m_1^2-m_2^2}\,[\Pi(\vartheta, n, k) - F(\vartheta, k)]$ $w_1 = \sqrt{m_1^2-m_2^2}\left[\Pi\left(\dfrac{\pi}{2}, n, k\right) - K(k)\right]$ $w_2 = w_1 - \dfrac{\pi}{2m_2}\sqrt{1-m_2^2}$ $\vartheta = \sin^{-1}\sqrt{\dfrac{1-\tau^2}{1-m_1^2}}, \quad n = -\dfrac{1-m_1^2}{1-m_2^2}$ $k = \sqrt{1-m_1^2},$ $\Pi = \displaystyle\int_0^{\vartheta} \frac{d\theta}{(1+n\sin^2\theta)\sqrt{1-k^2\sin^2\theta}}$	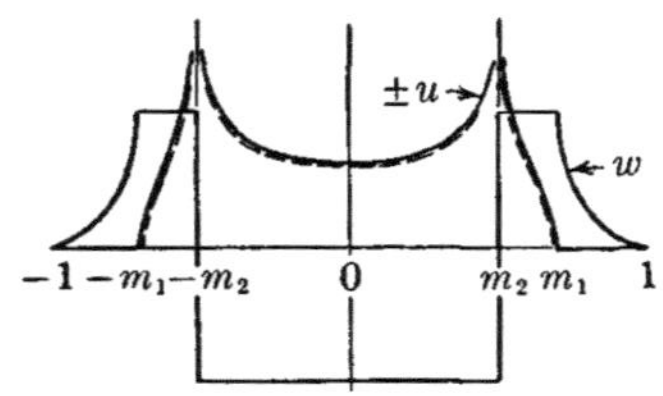	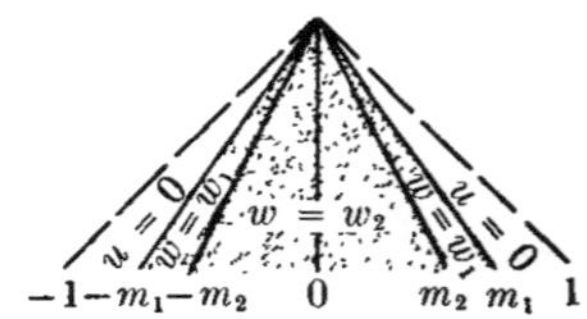

$$u = \text{R.P.}\ \frac{2w_1}{\pi\sqrt{1 - m_2^2}}\left[m_2 \cosh^{-1} \frac{m_2\sqrt{m_1^2 - \tau^2}}{m_1\sqrt{m_2^2 - \tau^2}} - \tau\sqrt{\frac{m_1^2 - m_2^2}{m_1^2 - \tau^2}}\right]$$

$$w = \text{R.P.}\ \frac{2w_1}{\pi}\left[\tan^{-1}\sqrt{\frac{(m_1^2 - m_2^2)(1 - \tau^2)}{(1 - m_2^2)(\tau^2 - m_1^2)}} - \sqrt{\frac{(m_1^2 - m_2^2)(1 - \tau^2)}{(1 - m_2^2)(\tau^2 - m_1^2)}}\right]$$

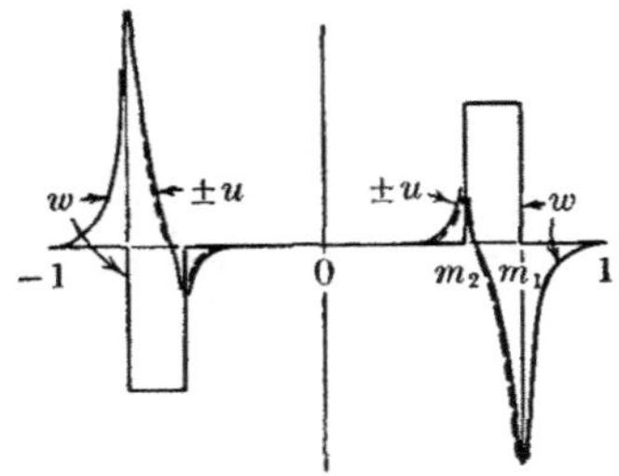

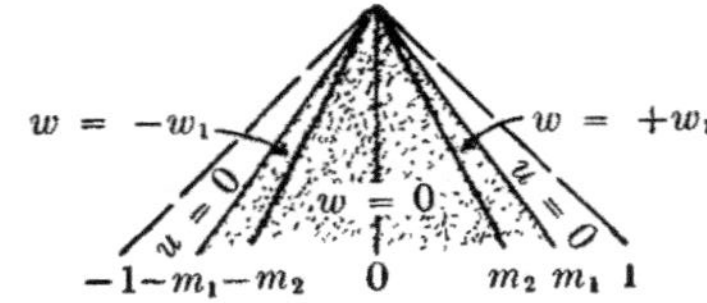

15. Symmetrical flap deflection at trailing edge

$$u = \text{R.P.}\left[m_1\sqrt{\frac{m_1^2 - m_2^2}{1 - m_1^2}}\,\Pi(\vartheta, n, k) + Z(\psi, k)F(\vartheta, k)\right]$$

$$w = \text{R.P.}\left\{\frac{1}{m_2}\left[-Z(\psi, k) + \frac{1}{m_1}\sqrt{(1 - m_1^2)(m_1^2 - m_2^2)}\right]\sinh^{-1}\sqrt{\frac{m_2^2 - \tau^2}{\tau^2}} - \frac{\sqrt{1 - m_1^2}}{m_1}\tan^{-1}\sqrt{\frac{m_1^2 - m_2^2}{m_2^2 - \tau^2}}\right\}$$

See 13 for Π

$$\vartheta = \sin^{-1}\sqrt{\frac{1 - \tau^2}{1 - m_2^2}}, \qquad n = -\frac{1 - m_2^2}{1 - m_1^2}$$

$$k = \sqrt{1 - m_2^2}, \quad \psi = \sin^{-1}\sqrt{\frac{1}{-n}}, \quad w_1 = \frac{-\pi\sqrt{1 - m_1^2}}{2m_1}$$

$$Z(\psi, k) = E(\psi, k) - \frac{E(k)\,F(\psi, k)}{K(k)}$$

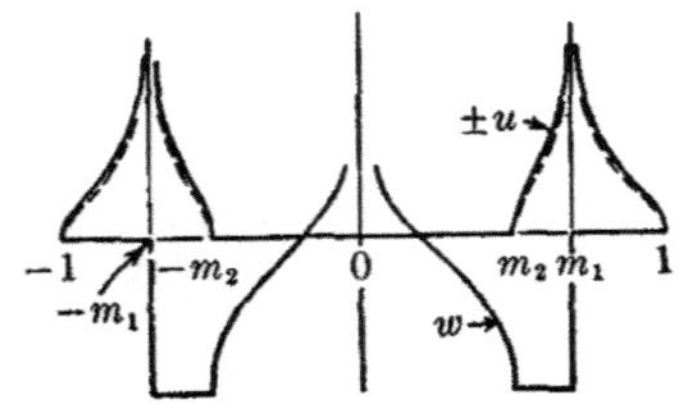

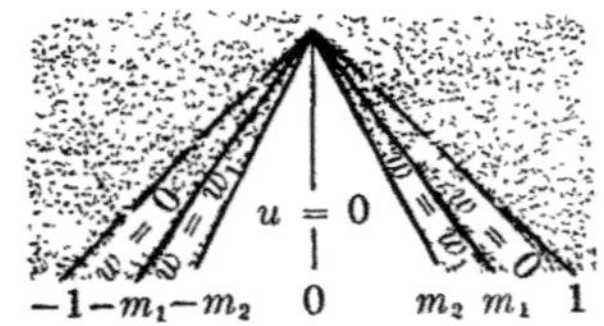

Table A,13a (continued)

Flow field; u and w components | **Values in plane of wing** | **Plan view and boundary conditions**

16. Antisymmetrical flap deflection at trailing edge

$$u = \text{R.P.} \cosh^{-1} \sqrt{\frac{(1 - m_1^2)(\tau^2 - m_2^2)}{(1 - m_2^2)(\tau^2 - m_1^2)}}$$

$$w = \text{R.P.} \frac{\sqrt{1 - m_1^2}}{m_1} \tan^{-1} \frac{\tau}{m_1} \sqrt{\frac{m_1^2 - m_2^2}{m_2^2 - \tau^2}}$$

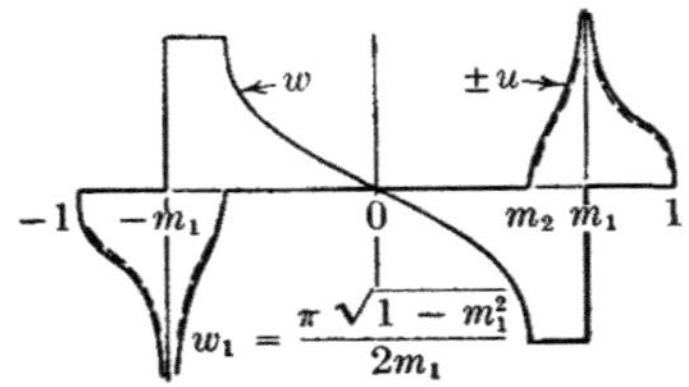

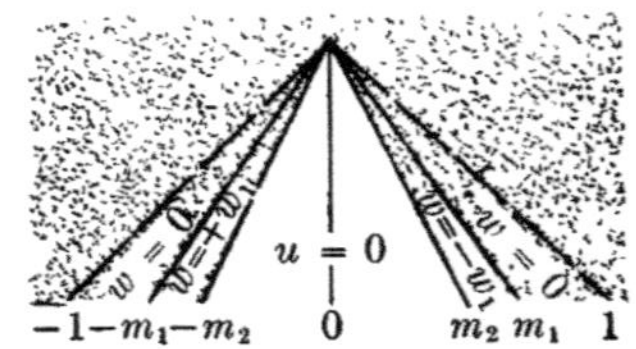

17. Side edge, wing with supersonic leading edge

$$u = \text{R.P.} \frac{2m\alpha U}{\beta\pi \sqrt{m^2 - 1}} \tan^{-1} \sqrt{\frac{-(1 + m)\tau}{m(1 + \tau)}}, \qquad m > 1$$

$$w = \text{R.P.} \frac{2\alpha}{\beta\pi} U \left[\sqrt{\frac{m(1 - \tau)}{\tau(m - 1)}} - \tan^{-1} \sqrt{\frac{m(1 - \tau)}{\tau(m - 1)}} \right]$$

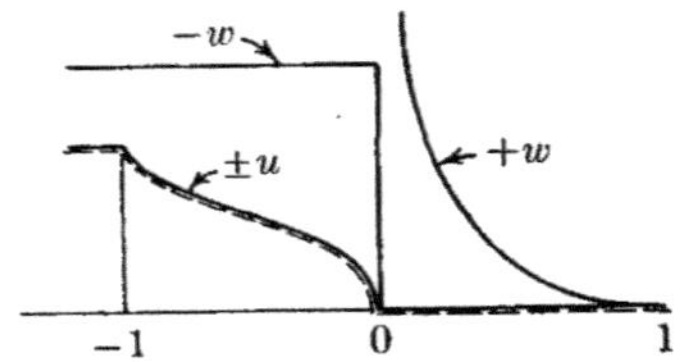

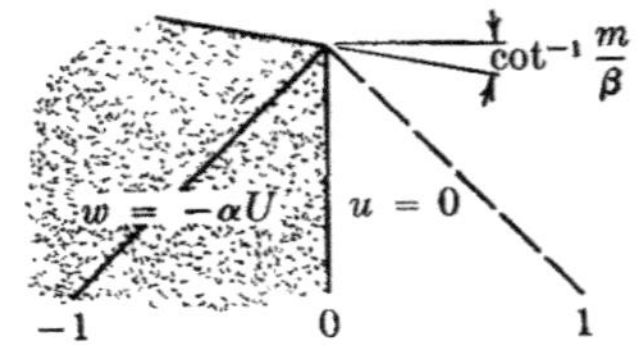

18. Symmetrical lift-cancellation field

$$u = \text{R.P.} \pm \frac{F(\vartheta, k)}{K(k)}$$

$$w = \text{R.P.} \frac{+1}{mK(k)} \cosh^{-1} \frac{m}{\tau}$$

$$\vartheta = \sin^{-1} \sqrt{\frac{1 - \tau^2}{1 - m^2}}, \qquad k = \sqrt{1 - m^2}$$

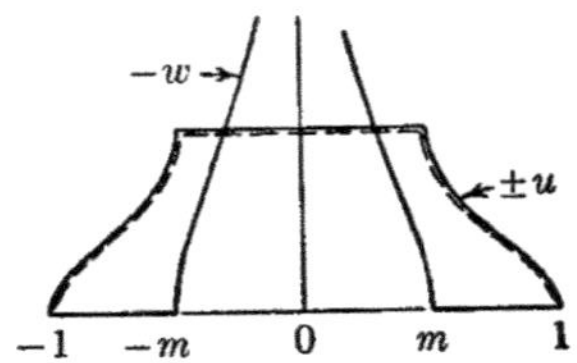

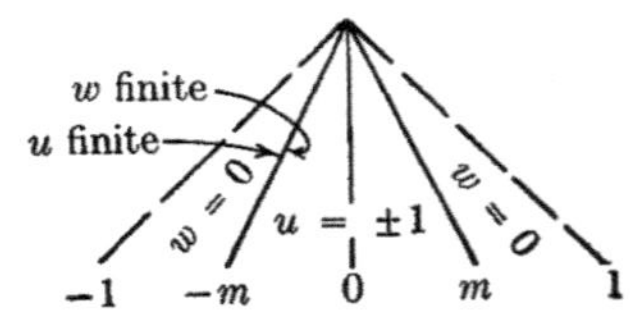

19.

$$u = \text{R.P. } \frac{1}{E(k)} [E(\vartheta, k) - \tau \tan \vartheta]$$

$$w = \text{R.P. } \frac{-1}{mE(k)} \left[\cosh^{-1} \frac{m}{\tau} - \frac{m(1 - m^2)}{\sqrt{m^2 - \tau^2}} \right]$$

$$\vartheta = \sin^{-1} \sqrt{\frac{1 - \tau^2}{1 - m^2}}, \qquad k = \sqrt{1 - m^2}$$

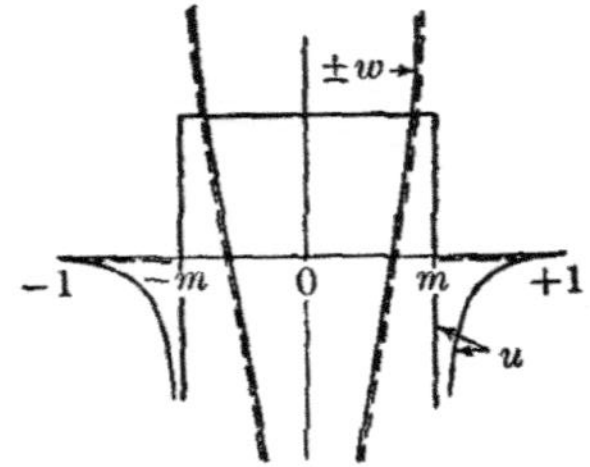

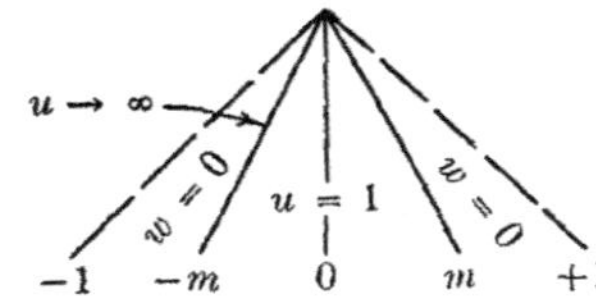

20.

$$u = \text{R.P. } \cosh^{-1} \frac{m}{\tau}$$

$$w = \text{R.P. } \left[mF(\vartheta, k) - \frac{1}{m} E(\vartheta, k) + \frac{1}{m\tau} \sqrt{(1 - \tau^2)(\tau^2 - m^2)} \right]$$

$$\vartheta = \sin^{-1} \sqrt{\frac{1 - \tau^2}{1 - m^2}}, \qquad k = \sqrt{1 - m^2}$$

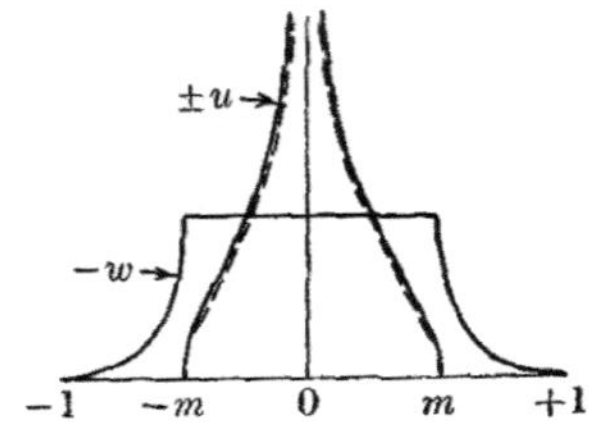

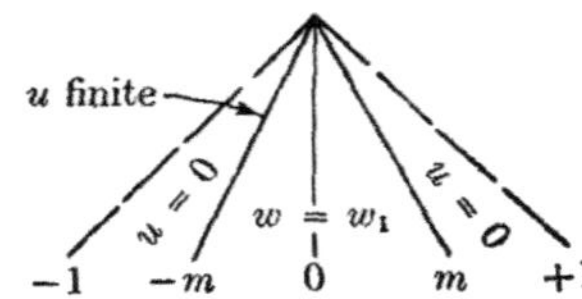

$$w_1 = mK(k) - \frac{1}{m} E(k)$$

21.

$$u = \text{R.P. } \frac{2}{\pi} \tan^{-1} \sqrt{\frac{;m(1 + \tau)}{-(1 + m)\tau}}$$

$$w = \text{R.P. } \frac{2}{\pi} \sqrt{\frac{1 + m}{m}} \left[\sqrt{\frac{1 - m}{m}} \cosh^{-1} \sqrt{\frac{(1 - m)\tau}{\tau - m}} - \sqrt{\frac{1 - \tau}{\tau}} \right]$$

$$m \leqslant 1$$

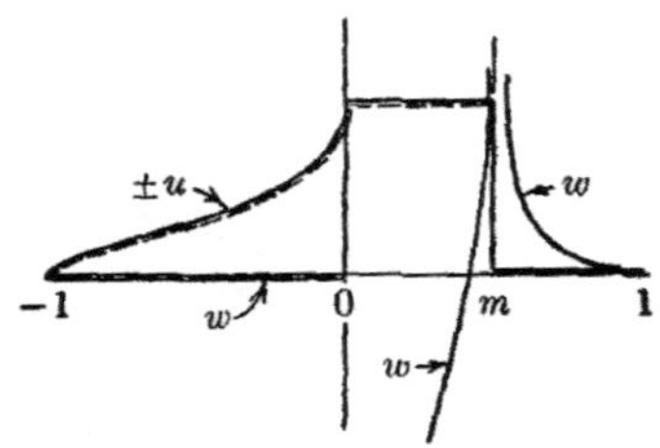

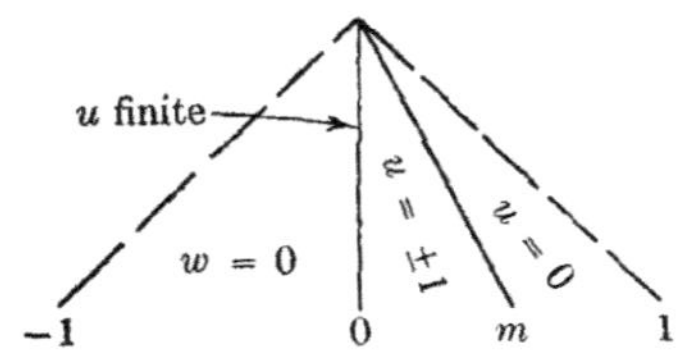

Table A,13a (continued)

Flow field; u and w components	Values in plane of wing	Plan view and boundary conditions
22. $u = \text{R.P.}\,\frac{2}{\pi}\tan^{-1}\sqrt{\frac{(m_2 - m_1)(1 - \tau)}{(1 - m_1)(\tau - m_2)}}$ $w = \text{R.P.}\,\frac{2}{\pi}\frac{\sqrt{1 - m_1}}{m_1}\left[\sqrt{\frac{m_2 - m_1}{m_2}} \times \cosh^{-1}\sqrt{\frac{m_2 - \tau}{-(1 + m_2)\tau}} - \sqrt{1 + m_1}\cosh^{-1}\sqrt{\frac{(1 + m_1)(m_2 - \tau)}{(1 + m_2)(m_1 - \tau)}}\right]$	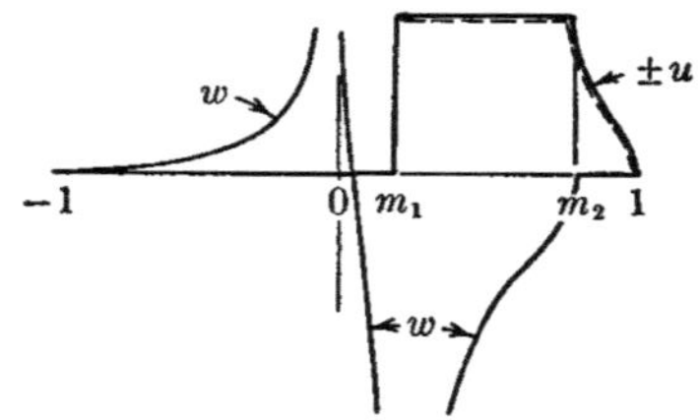	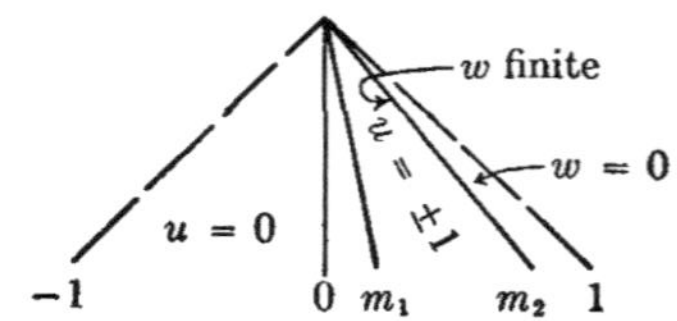
23. $u = \text{R.P.}\,\frac{i}{\pi}\ln\frac{m_1 - \tau}{m_2 - \tau}$ $w = \frac{1}{\pi}\,\text{R.P.}\left[\frac{m_2 - m_1}{m_1 m_2}\cosh^{-1}\frac{1}{\tau} - \frac{\sqrt{1 - m_1^2}}{m_1}\cosh^{-1}\frac{1 - m_1\tau}{m_1 - \tau} - \frac{2}{m_2}\sqrt{m_2^2 - 1}\tan^{-1}\sqrt{\frac{(m_2 - 1)(1 + \tau)}{(m_2 + 1)(1 - \tau)}}\right]$	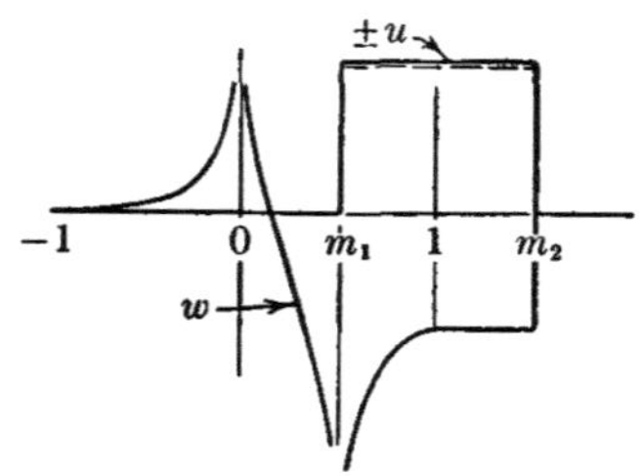	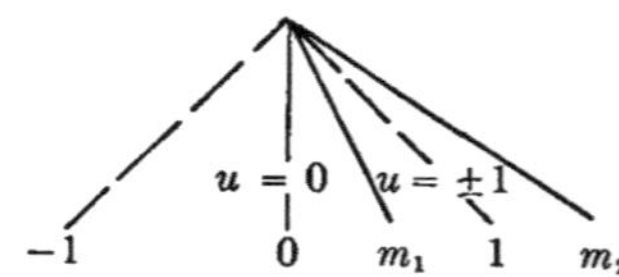
24. $u = \frac{2m^{\frac{1}{2}}}{\pi(1 + m)}\,\text{R.P.}\sqrt{\frac{1 + \tau}{m - \tau}}$ $w = \frac{-2}{\pi}\,\text{R.P.}\left[\tan^{-1}\sqrt{\frac{m(1 - \tau)}{\tau - m}} - \sqrt{\frac{m(1 - \tau)}{\tau - m}}\right]$	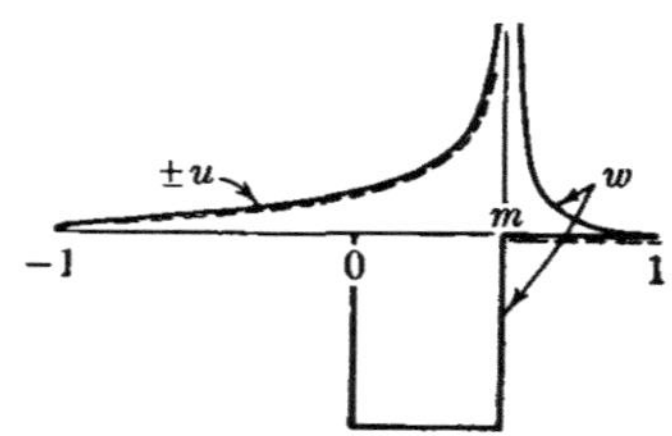	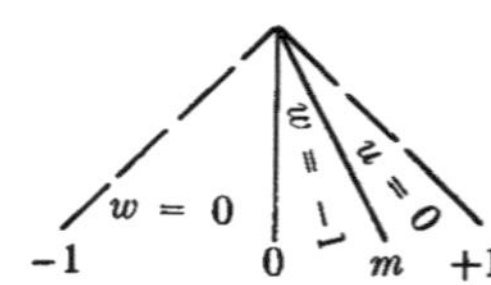

25.

$$u = \text{R.P.}\,\frac{2}{\pi}\left[\tan^{-1}\sqrt{\frac{(m_2 - m_1)(1 - \tau)}{(1 - m_1)(\tau - m_2)}} - \frac{m_2(1 - m_1)}{m_1(1 - m_2)}\sqrt{\frac{(m_2 - m_1)(1 - \tau)}{(1 - m_1)(\tau - m_2)}}\right]$$

$$w = \text{R.P.}\,\frac{2}{\pi}\,\frac{\sqrt{1 - m_1}}{m_1}\left[\sqrt{m_2 - m_1}\sqrt{\frac{1 + \tau}{m_2 - \tau}} - \sqrt{1 + m_1}\cosh^{-1}\sqrt{\frac{(1 + m_1)(m_2 - \tau)}{(1 + m_2)(m_1 - \tau)}}\right]$$

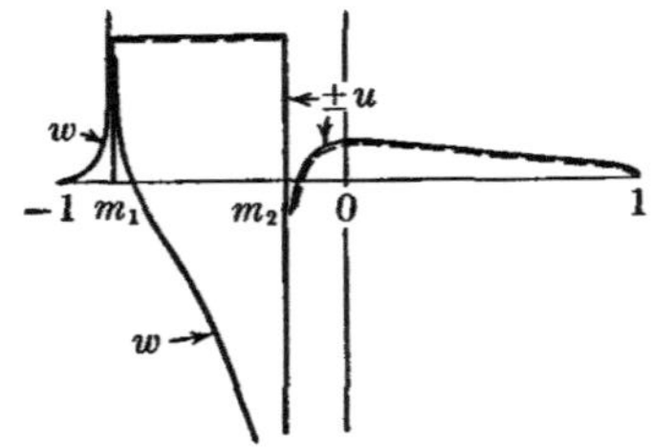

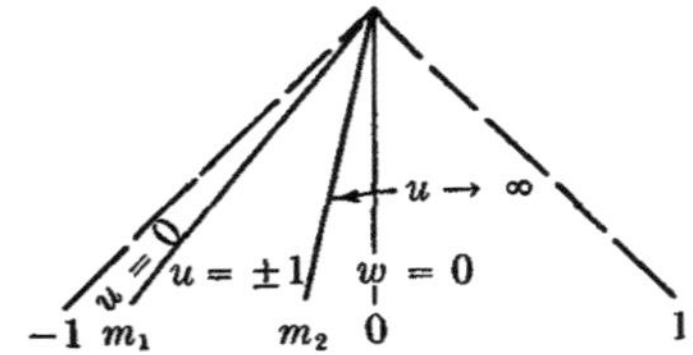

26.

$$u = \text{R.P.}\,\frac{2}{\pi}\left[\frac{m_1}{\sqrt{1 - m_1^2}}\cosh^{-1}\sqrt{\frac{(1 + m_1)(m_2 - \tau)}{(1 + m_2)(m_1 - \tau)}} - \frac{m_2}{1 + m_2}\sqrt{\frac{(m_2 - m_1)(1 + \tau)}{(1 - m_1)(m_2 - \tau)}}\right]$$

$$w = \text{R.P.}\,\frac{2}{\pi}\left[\tan^{-1}\sqrt{\frac{(m_2 - m_1)(1 - \tau)}{(1 - m_1)(\tau - m_2)}} - \sqrt{\frac{(m_2 - m_1)(1 - \tau)}{(1 - m_1)(\tau - m_2)}}\right]$$

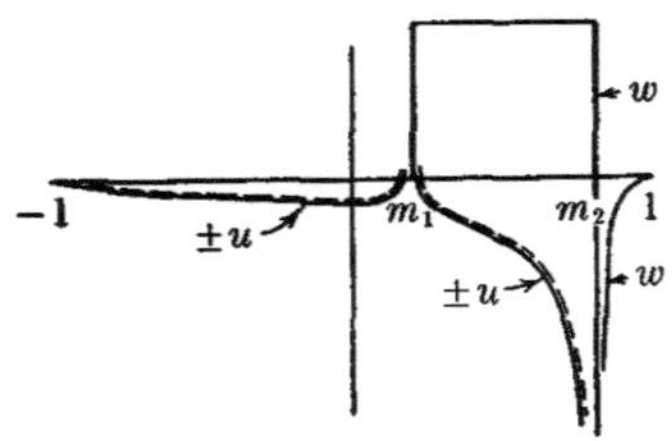

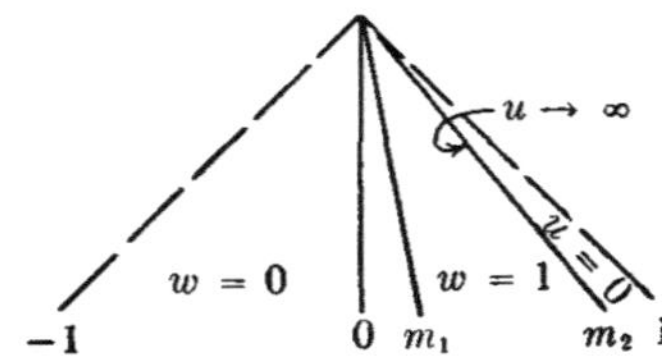

27.

$$u = \text{R.P.}\,\frac{2}{\pi}\,\frac{m_1}{\sqrt{1 - m_1^2}}\cosh^{-1}\sqrt{\frac{(1 - m_1)(\tau - m_2)}{(1 - m_2)(\tau - m_1)}}$$

$$w = \text{R.P.}\,\frac{2}{\pi}\left[\cos^{-1}\sqrt{\frac{(1 + m_1)(m_2 - \tau)}{(1 + m_2)(m_1 - \tau)}} - \sqrt{\frac{m_1 - m_2}{m_2(1 + m_1)}}\cosh^{-1}\sqrt{\frac{m_2 - \tau}{-(1 + m_2)\tau}}\right]$$

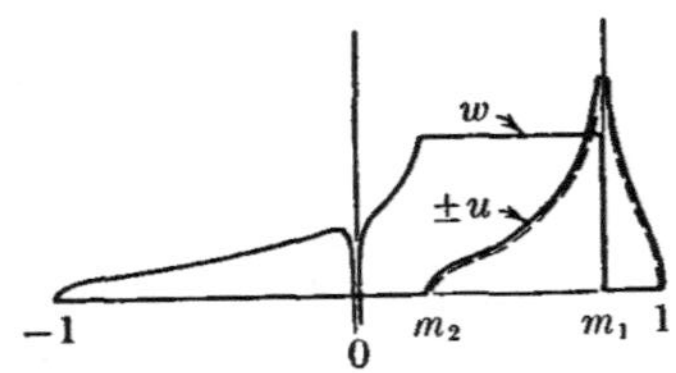

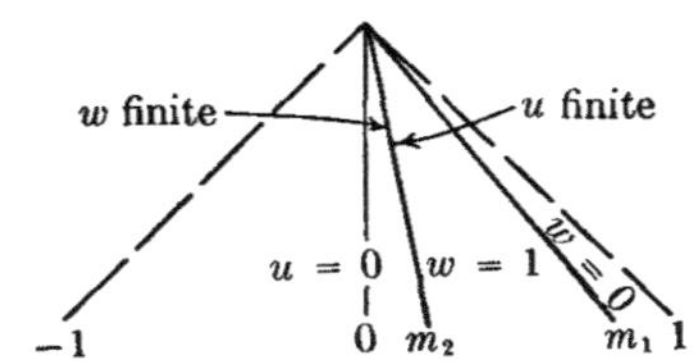

factor, the supersonic form may be written

$$\left.\begin{aligned} x' &= x - \mu y \\ y' &= y - \mu x \\ z' &= \sqrt{1 - \mu^2}\, z \end{aligned}\right\} \tag{13-5}$$

where μ is the tangent of the angle between the axes (Fig. A,13c). This transformation is a special application of the well-known Lorentz transformation, and has the following properties:

1. If $f(x, y, z)$ is a solution of the steady flow equation (Eq. 10-4 or its equivalent expression in terms of one of the velocity components), then $f(x', y', z')$ is also a solution.
2. The Mach cone $x^2 = y^2 + z^2$ remains invariant under the transformation.
3. The x' and y' axes are oblique to each other and only the Mach angle is preserved.

In the plane of the wing, the conical variable τ becomes

$$\tau(x', y', 0) = \frac{t - \mu}{1 - \mu} \tag{13-6}$$

where $t = \tau(x, y, 0)$.

If the variable τ is replaced by $\tau(x', y', z')$ in any expression for u or w in Table A,13a, it follows from (1) above that a new solution will have been generated. However, the transformation will hold only for the velocity component to which it was applied, since, from property (3), the relations (13-3 and 13-4) between the velocity components are not invariant under the transformation. Thus if the boundary conditions in u are to be satisfied in regions skewed according to Eq. 13-5, the new expression for w must be derived by the reapplication of Eq. 13-3, and vice versa if the boundary conditions are specified for w. For a more complete discussion, see [*142*].

Method of Application of Conical Flow Theory. The simplest use of the conical flow method is in finding the flow over wings of essentially conical form. Thus, following a preliminary application of the Prandtl-Glauert transformation (10-3), the pressure over a thin flat triangular wing at an angle of attack is given directly by solution 6 if the wing has subsonic leading edges, or by solution 3 if the leading edge is supersonic, when τ is replaced by y/x, its value in the $z = 0$ plane. Almost as direct is the solution for the flat rectangular wing of high aspect ratio (Fig. A,13d), in which each tip may be considered independently as long as the Mach cones from the extremities of the leading edge do not overlap.

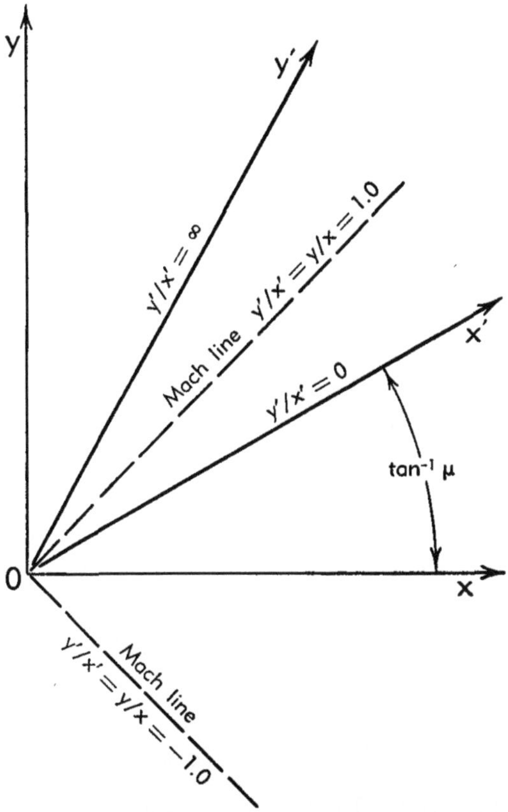

Fig. A,13c. Oblique coordinates.

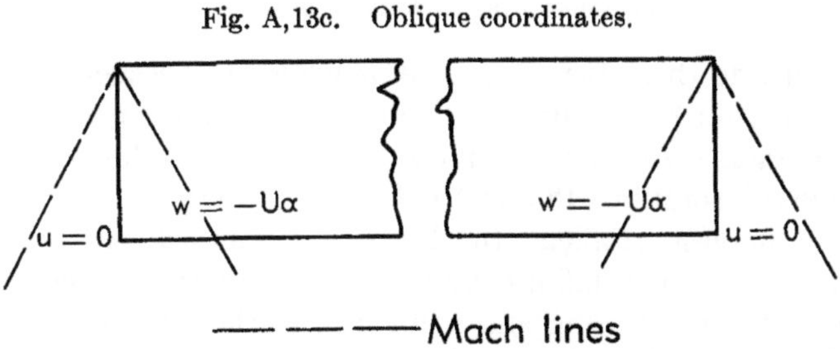

Fig. A,13d. Rectangular wing as composite of two independent conical fields (solution 17, Table A,13a, with $m \to \infty$).

Between the tip regions, of course, the pressure is given by the two-dimensional (Ackeret) theory.

In either of the cases cited, the fact that the entire wing lies ahead of the Mach lines from points on the trailing edge makes it unnecessary to consider the effect of terminating the infinite conical solution along that edge, if one is interested only in the lift distribution. If the flow behind the wing is also to be investigated, it becomes necessary to take into account the fact that the wing is finite. By the method of superposition of conical flows the entire flow field behind such finite wings of simple planform can be found [*154*]. The procedure is the same as that by which the lift on finite wings of more general form may be derived from the infinite conical or two-dimensional solutions. A description of this procedure follows.[15]

Cancellation-of-lift procedure for finite wings. The basic consideration in determining the flow around the finite wing is that (except across a shock) there can exist no discontinuities in the pressure unsupported by a surface. Then the problem becomes the cancellation of such discontinuities introduced by the infinite conical flow field beyond the planform of the finite wing. This must be done without disturbing the boundary conditions elsewhere. In the case of a flat wing of specified camber, the latter provision means that the canceling flow must (i) induce no downwash within the boundaries of the finite wing and (ii) introduce no new lifting

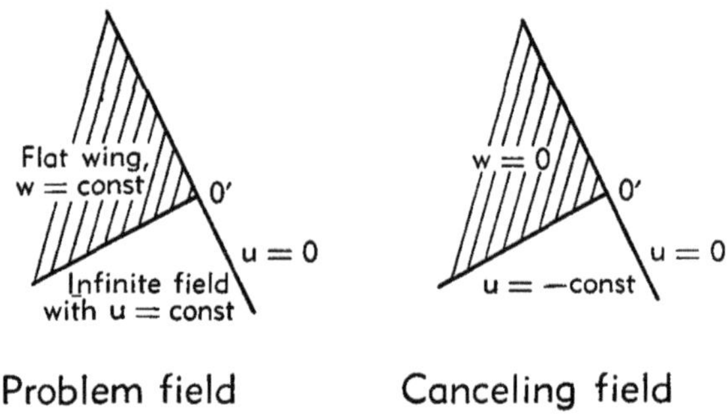

Fig. A,13e. Boundary conditions for cancellation of uniform lift to derive a finite wing.

pressure outside these boundaries. It is generally necessary to consider also the character of the edge at which cancellation occurs—whether it is leading or trailing, subsonic or supersonic—since the solution is otherwise not unique (cf. solutions 18 and 19 of Table A,13a).

In the simplest problem, the lift to be canceled will be uniformly distributed over a semi-infinite region bounded by two straight lines, one the edge of the finite wing and one a ray of the infinite field (see Fig. A,13e). With conditions (i) and (ii) listed above, the boundary conditions of the cancellation field are then conical with respect to the intersection $0'$ of the two lines. Thus the flow field due to angle of attack of the wing shown in Fig. A,13f may be derived from that of a triangular wing (after application of the Prandtl-Glauert transformation) by the superposition, with its origin at $0'$, of the field given by solution 21, and at $0''$ of the mir-

[15] The procedure presented here is superficially different from that first suggested in [*142*], and employed in [*154*] and subsequent papers using conical flow theory. However, the reader will see that either reduces to the other through an integration by parts.

ror image of such a field. These fields are characterized by a region of zero vertical velocity adjacent to a region of constant streamwise velocity and a finite value of u along the ray between. Since from solution 3 the pressure difference to be canceled is $2\rho_\infty U$ times

$$u_1 = \frac{mU\alpha}{\sqrt{m^2 - 1}} \tag{13-7}$$

where α is the angle of attack of the wing and U the free stream velocity, the decrement in lifting pressure induced on the wing by the cancellation

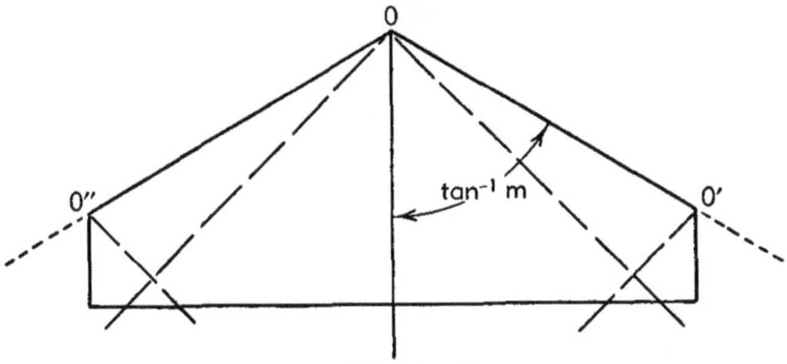

Fig. A,13f. Wing with sweptback leading edge, straight trailing edge, and streamwise tips, at a high supersonic Mach number. (Plane transformed by Prandtl-Glauert transformation to $M_\infty = \sqrt{2}$; $m \gg 1$.)

process is, within the Mach cone of the right tip, $2\rho_\infty U$ times

$$\Delta u(x, y) = -\frac{2mU\alpha}{\pi\sqrt{m^2 - 1}} \tan^{-1} \sqrt{\frac{m(1 + t')}{-(1 + m)t'}}$$

where t' is the conical coordinate of $(x, y, 0)$ referred to $0'$ as origin. With a similar expression for the correction to the velocity in the left-hand tip cone, the problem of the lift on the depicted wing is completely solved.

However, the pressure to be canceled is not generally a constant, but varies, for example, in the case of a sweptback wing, with $t = \tau(x, y, 0)$ according to the streamwise component u in solution 3 or 6, depending on whether the leading edge is supersonic or subsonic. We therefore require an infinitesimal region of uniform lift (or u) with zero lift on either side, the lifting strip being bounded by two adjacent rays of slope t and $t + dt$ of the original conical field (Fig. A,13g) and an infinitesimal segment of the tip or trailing edge of the wing. Such a field may be obtained as the differential of a conical field with the appropriate boundary conditions. For the tip problem shown in Fig. A,13g, solution 21 is again employed, with the parameter m replaced by t, and the variable τ referred to the intersection $0'$ of the ray t with the tip as origin. Differentiating the resulting expression for u with respect to t would give the lifting pressure on the wing due to an infinitesimal strip with unit lift. The total decrement in lift at any point x, y within the Mach cone from the right tip would be $2\rho_\infty U$ times

$$\Delta u(x, y) = \frac{2\rho_\infty U}{\pi} \int_{t'=-1}^{t=m} u_\Delta(t) \left[\frac{\partial}{\partial t} + \frac{\partial}{\partial t'}\frac{dt'}{dt}\right] \tan^{-1} \sqrt{\frac{t(1+t')}{-(1+t)t'}}\, dt \quad (13\text{-}8)$$

in which $t' = \tau'(x, y, 0) = (y - s)/[x - (s/t)]$ (see Fig. A,13g), u_Δ is given by solution 3 or 6, and the lower limit of integration corresponds to the value of t such that the Mach cone from 0′ passes through the point x, y.

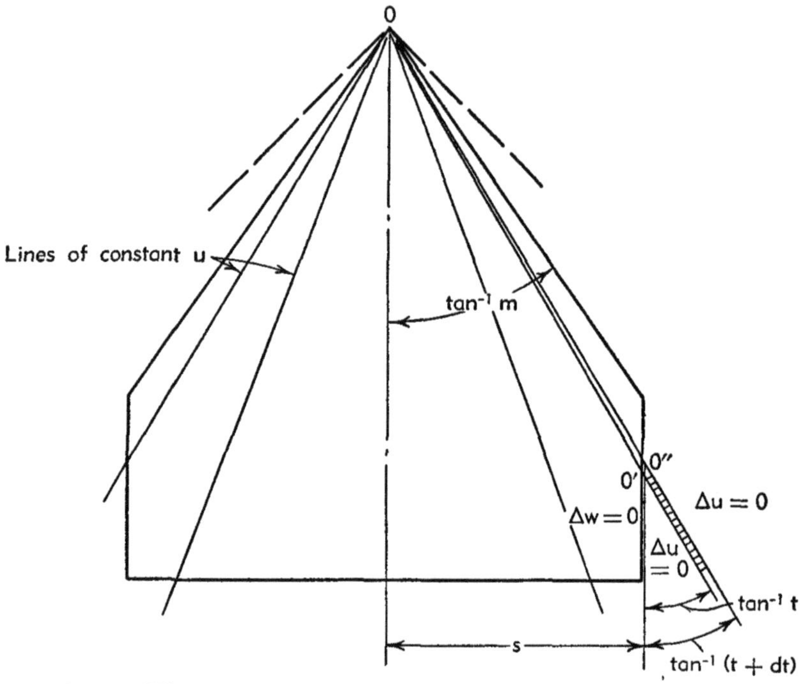

Fig. A,13g. Wing of Fig. A,13f at a lower Mach number. (Plane transformed by Prandtl-Glauert transformation to $M_\infty = \sqrt{2}$; $m < 1$.)

If the leading edge of the wing is subsonic ($m < 1$), Eq. 13-8 is readily integrated.[16] If the leading edge is supersonic, the integral is not as tractable, and the source-sink methods to be described later appear more suited to the problem. In either case, the regions influenced by the two tips may overlap. Where this occurs, the effects of the cancellation of lift are of course additive.

To find the flow behind a straight or swept trailing edge, a similar procedure would be followed, employing solution 23 (see Fig. A,13h). If the angle of sweep is greater than the sweep of the Mach lines (Fig. A,13i), a portion of the wing will lie within the Mach cones of the cancel-

[16] The result is contained in a more general formula (13-37a) to be derived later.

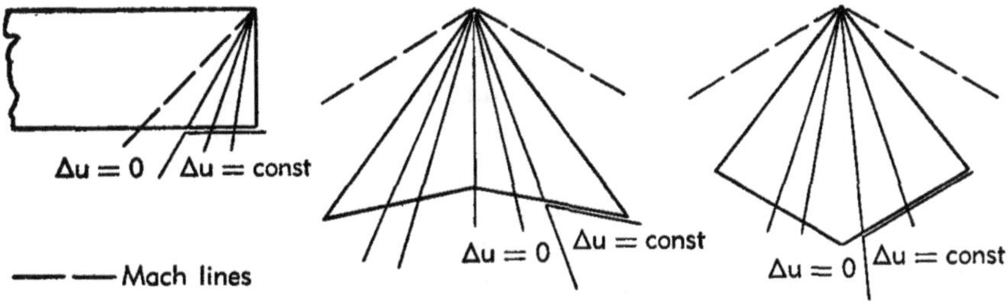

Fig. A,13h. Wings with supersonic trailing edges, and the conical flow fields required to cancel lift in the stream.

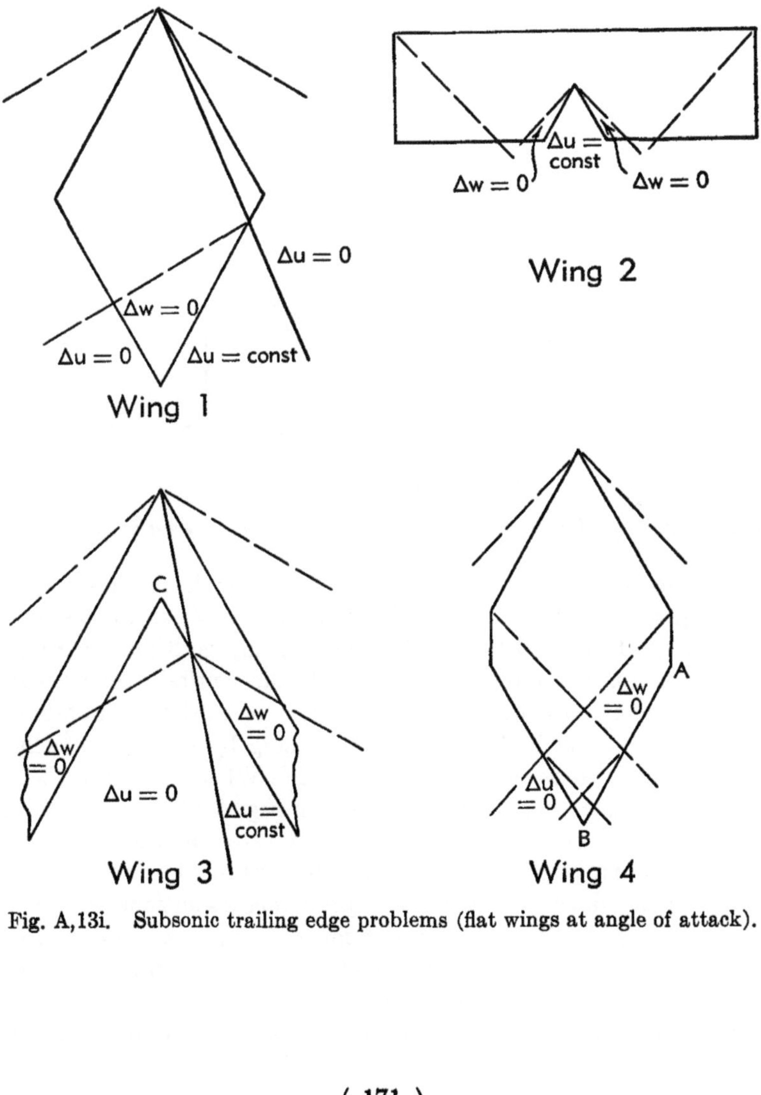

Fig. A,13i. Subsonic trailing edge problems (flat wings at angle of attack).

ing fields, complicating the boundary conditions so that solution 23 will no longer apply. A uniform distribution of lift (wing 2) could be canceled behind a subsonically sweptback trailing edge without altering the slope of the streamlines ahead, by the application of solution 18 (or an oblique transformation of the same field) with opposite signs on upper and lower surfaces. The lift distribution on the wing would then be modified in such a way as to satisfy the Kutta condition. Cancellation of a nonuniform lift, however, would generally require displacement of the origin along the trailing edge, disrupting the conical form of the boundary conditions (wing 3). If the trailing edge is swept forward (wings 1 and 4) a similar problem arises or, in general, whenever another edge of the wing cuts

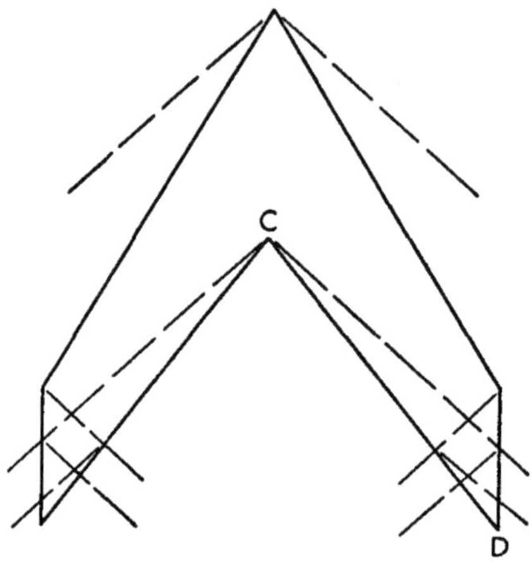

Fig. A,13j. Interference regions on a typical sweptback wing at $M_\infty = 1.5$.

across the Mach cone of a canceling field. In principle the resulting non-conical regions can be treated by the same superposition process as was the original problem, but in practice the resulting multiple integrals can seldom be evaluated analytically. If one edge lies entirely downstream of the other, as at *A* on wing 4, the lift distribution in the neighborhood of their intersection can be found in a single step by the lift cancellation method of [*141*] (discussed later in this article). When this is not the case, as at *B* on the same wing, and at *C* and *D* in Fig. A,13j, any cancellation of nonuniform lift at one edge will introduce errors at the other, so that such problems require an infinite number of steps for their complete solution by presently developed methods. Approximate solutions for the problem shown in Fig. A,13j and for the still more complex case (Fig. A,13k) in which both the tip and part of the leading edge lie within the region of influence of the trailing edge are discussed in [*155*] and later in this article.

Other applications. The use of the conical flow solutions is not, of course, restricted to the cancellation of lift. In a later article, nonlifting components will be superimposed to derive finite wings of specified thickness distribution. The lifting solutions characterized by constant w over part of the x, y plane are useful in finding the effect of twist or of control surface deflection [*156,157*]. An interesting example of their application is

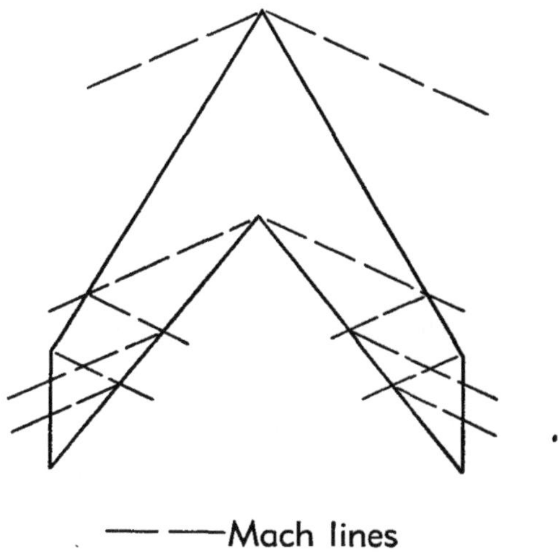

Fig. A,13k. Interference regions on the wing of Fig. A,13j at $M_\infty = 1.2$.

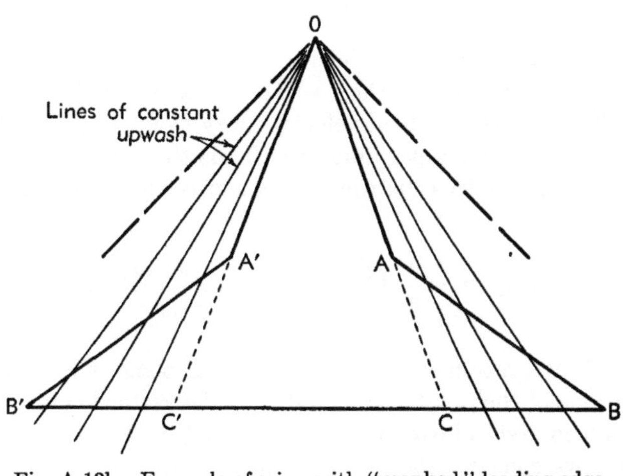

Fig. A,13l. Example of wing with "cranked" leading edge.

the wing with so-called "cranked" leading edge shown in Fig. A,13l, where the outer panels ABC and A'B'C' are operating in a conically varying upwash field. The solutions required [*158*] are of the same type as in the case of flap deflection.

Calculation of integrated forces and moments. If the planform, twist, and camber of a wing are derivable by the superposition of conical flows, the calculation of the total lift and pitching or rolling moment should also be found by consideration of the component fields. In each conical field, the pressure or u varies only with t, so that a single integration suffices to determine the lift or moment associated with that field. The resulting expressions may then be combined in exactly the same way as the velocity fields would be for the same wing. Since the result of the first integration is usually a simple algebraic expression, the conical flow method is often the most practical for the calculation of these integrated effects (see [*155*]).

Source-Distribution Methods. Basic Formulas. It is obvious that the conical flow method, though providing in certain cases very complete solutions in just one or two steps, must be restricted in its application to wings with rectilinear planforms and a minimum of interference between the edge flows. Somewhat more flexible tools are provided by the methods based on the solution (10-10) for a supersonic source, and related quantities.

For convenience of presentation, it will be assumed that the flow field has been normalized by the application of the Prandtl-Glauert transformation (10-3), so that the solution for the source takes the form

$$\varphi \sim \frac{1}{\sqrt{x^2 - y^2 - z^2}} \tag{13-9}$$

Puckett [*159*] has shown that the vertical velocity associated with this solution is zero everywhere in the $z = 0$ plane except at the location of the source itself, where the vertical velocity away from the plane is exactly π times the source strength per unit area. Thus the velocity potential φ at any point x, y, z, due to a distribution of sources in the horizontal plane, is

$$\varphi(x, y, z) = -\frac{1}{\pi} \iint \frac{w_u(\xi, \eta) d\xi d\eta}{\sqrt{(x - \xi)^2 - (y - \eta)^2 - z^2}} \tag{13-10}$$

The area of integration is the portion of the plane intercepted by the Mach forecone of the point x, y, z, and w_u is the vertical component of the velocity immediately above any source-point ξ, η, 0.

If, now, we consider the flow around a vertically symmetric wing—that is, an uncambered wing at zero angle of attack—the vertical velocity will be zero in the plane of symmetry everywhere except in the region occupied by the wing, where it will be determined by the shape of the wing surface. Thus the simple "thickness problem" is solved immediately by the integration of Eq. 13-10.

The expression for the streamwise component is obtained by the differentiation of Eq. 13-10 in the x direction and can be shown (VI,D,10) to reduce to

$$u(x, y, z) = \frac{1}{\pi} \int d\eta \, \displaystyle{\not}\!\!\int \frac{(x - \xi)w_u(\xi, \eta)}{[(x - \xi)^2 - (y - \eta)^2 - z^2]^{\frac{3}{2}}} d\xi \quad (13\text{-}11)$$

in which the bar on the inner integral sign signifies that the finite part is to be used (see VI,D,14.1). When the value of w_u outside of the wing is zero, the area of integration may be reduced to the part of the wing plan intercepted by the Mach forecone from $P(x, y, z)$ (see Fig. A,13m). Then integration of Eq. 13-11 by parts gives

$$u(x, y, z) = -\frac{1}{\pi}\left[\int_{\widehat{AB}} \frac{w_u(\xi, \eta)d\eta}{\sqrt{(x - \xi)^2 - (y - \eta)^2 - z^2}} + \iint_S \frac{\partial w_u/\partial \xi}{\sqrt{(x - \xi)^2 - (y - \eta)^2 - z^2}} d\xi d\eta \right] \quad (13\text{-}12)$$

The first term of Eq. 13-12 is a line integral along the part of the leading edge intercepted by the Mach forecone from P. When the slope of the wing

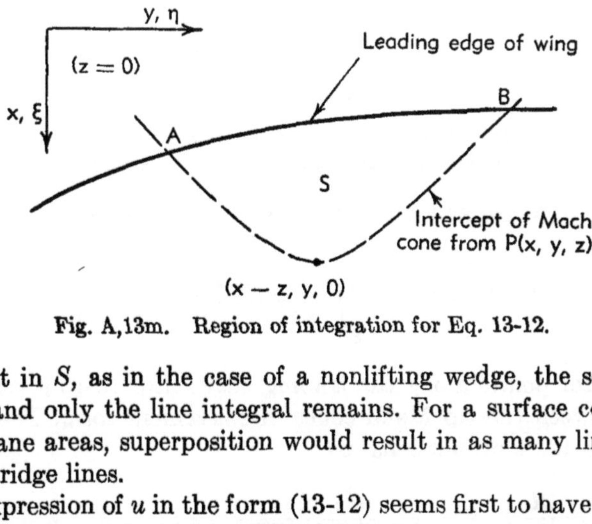

Fig. A,13m. Region of integration for Eq. 13-12.

is constant in S, as in the case of a nonlifting wedge, the second term vanishes and only the line integral remains. For a surface consisting of several plane areas, superposition would result in as many line integrals along the ridge lines.

The expression of u in the form (13-12) seems first to have been noted by Evvard [*160*], and will be seen to have great practical value.

Source Distribution Formulas Applied to Thin Lifting Wings with Supersonic Leading Edges. If the boundary of the wing is supersonic within the area of integration, as for point x, y in Fig. A,13n, Eq. 13-10 may be evaluated directly and Eq. 13-12 will apply even in the case of a nonsymmetrical airfoil, because the airfoil can cause no disturbance in the field ahead of such a boundary, and w is consequently zero there. The upper and lower surfaces may be considered separately, since the only possibility of interaction is around a subsonic edge. Thus, outside the regions influenced by subsonic edges, we may write, for points on the upper surface,

$$\varphi(x, y, +0) = \frac{U}{\pi} \iint_S \frac{\alpha_u(\xi, \eta) d\xi d\eta}{\sqrt{(x - \xi)^2 - (y - \eta)^2}} \tag{13-13}$$

and for the lower surface,

$$\varphi(x, y, -0) = -\frac{U}{\pi} \iint_S \frac{\alpha_l(\xi, \eta) d\xi d\eta}{\sqrt{(x - \xi)^2 - (y - \eta)^2}} \tag{13-14}$$

where α_u and α_l are the local slopes of the upper and lower surfaces in the stream direction, and U is the free stream velocity. For a thin cam-

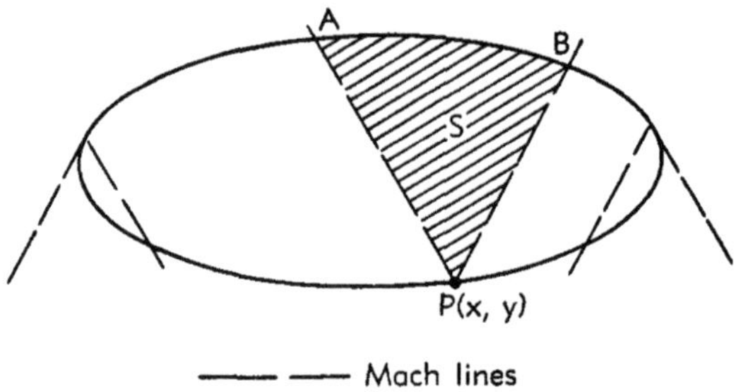

Fig. A,13n. Example of the application of Eq. 13-12 to a finite lifting wing (arc AB entirely supersonic).

bered wing, $\alpha_u = \alpha_l$, and the difference in potential across the surface is simply

$$\Delta\varphi(x, y) = \frac{2U}{\pi} \iint_S \frac{\alpha(\xi, \eta) d\xi d\eta}{\sqrt{(x - \xi)^2 - (y - \eta)^2}} \tag{13-15}$$

Evaluated for points along the trailing edges, Eq. 13-15 gives the span loading between the tip Mach cones in convenient form. The local lift, from Eq. 13-12, is proportional to

$$\Delta u(x, y) = -\frac{2U}{\pi} \left[\int_{\widehat{AB}} \frac{\alpha(\xi, \eta) d\eta}{\sqrt{(x - \xi)^2 - (y - \eta)^2}} + \iint_S \frac{d\alpha/d\xi}{\sqrt{(x - \xi)^2 - (y - \eta)^2}} d\xi d\eta \right] \tag{13-16}$$

Leading edge partially subsonic. If the boundary of the wing is subsonic anywhere within the Mach forecone of the point $x, y, \varphi(x, y)$ cannot in general be found by direct integration of Eq. 13-10. However, it may be (as in Fig. A,13l, with AB and $A'B'$ supersonic) that the upwash ahead of the wing is known, either from conical flow theory or otherwise. In such a case one may imagine the source sheet to cover the entire area of disturbance—that is, the entire region included by the Mach lines from O—and thus reduce the problem to the preceding case.

Using a variation of the above idea, Evvard [*160*] and Krasilshchikova [*161*] independently arrived at the general solution for a thin lifting wing of arbitrary planform, subject only to the condition that the foremost portion of the planform boundary be supersonic for some distance. Briefly, the development in this case is as follows:

Consider a point $P(x, y)$ near the right-hand tip of a wing of which the leading edge is largely supersonic. Let O (Fig. A,13o) be the point of the leading edge at which the leading edge becomes subsonic. Then the

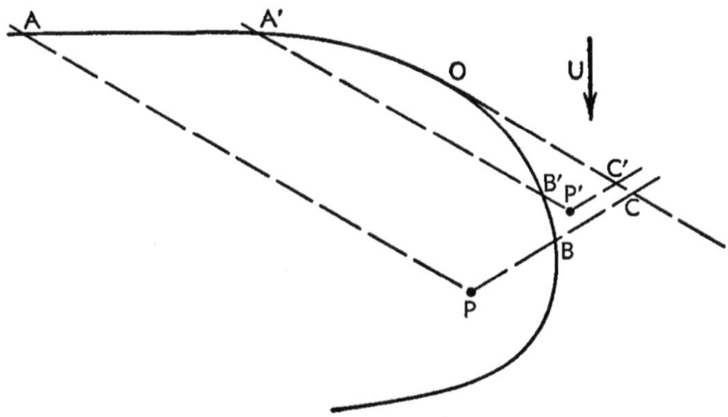

Fig. A,13o. Region of integration for point near tip of wing with supersonic leading edge.

evaluation of the velocity potential at P by Eq. 13-10 includes integration over the area OCB of a function involving a vertical velocity generally undetermined in that region.

However, we may also write the expression for φ at a point P' of region OCB:

$$\varphi(x', y') = -\frac{1}{\pi} \iint_{A'OC'P'} \frac{w(\xi, \eta)d\xi d\eta}{\sqrt{(x' - \xi)^2 - (y' - \eta)^2}} \tag{13-17}$$

and this must equal zero if the point P' is ahead of the wing, or at any rate not in its wake. (When P' is in the wake, a different development is required.) If the integral equation obtained by setting $\varphi(x', y')$ equal to zero can be solved for $w(x', y')$, this quantity will be known both off and on the wing and Eq. 13-10 can be evaluated for point P.

The solution of the integral equation for w is made feasible by a transformation to a characteristic coordinate system, i.e. one with its axes in the direction of the Mach lines of the flow. The new variables r and s

(Fig. A,13p) are related to x and y by the equations[17]

$$\begin{aligned} r &= (x - x_0) - (y - y_0) \\ s &= (x - x_0) + (y - y_0) \end{aligned} \tag{13-18a}$$

where x_0, y_0 are the coordinates of the point O. The coordinates of the source-point will again be designated by the Greek counterparts

$$\begin{aligned} \rho &= (\xi - x_0) - (\eta - y_0) \\ \sigma &= (\xi - x_0) + (\eta - y_0) \end{aligned} \tag{13-18b}$$

With this transformation, the hyperbolic distance $\sqrt{(x - \xi)^2 - (y - \eta)^2}$ becomes $\sqrt{(r - \rho)(s - \sigma)}$, and the variables are then separable. The

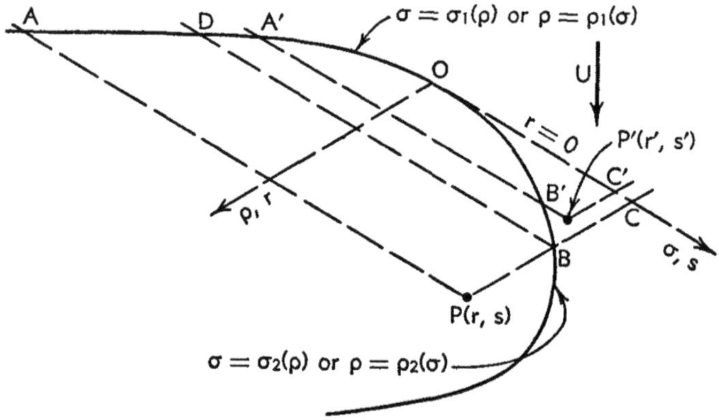

Fig. A,13p. Characteristic coordinate system and equations of boundary curves.

element of area $d\xi d\eta$ transforms into $\frac{1}{2}d\rho d\sigma$. For point P' we obtain

$$\int_0^{r'} \int_{\sigma_1(\rho)}^{s'} \frac{w(\rho, \sigma) d\sigma d\rho}{\sqrt{r' - \rho}\sqrt{s' - \sigma}} = 0 \tag{13-19}$$

where $\sigma = \sigma_1(\rho)$ is the equation of the supersonic portion of the leading edge.

Eq. 13-19 may be considered an Abel equation of the form

$$\int_0^{r'} \frac{F(\rho) d\rho}{\sqrt{r' - \rho}} = 0 \tag{13-20}$$

for which the solution is $F(\rho) = 0$ provided that $F(\rho)$ is known to have no singularities in the region of integration. In the integrand of

$$F(\rho) = \int_{\sigma_1(\rho)}^{s'} \frac{w(\rho, \sigma) d\sigma}{\sqrt{s' - \sigma}} \tag{13-21}$$

[17] It may be well to recall at this time that the x, y coordinate system already includes a Prandtl-Glauert transformation, so that the Mach lines, and the new axes are at 45° to the stream.

the only singularities are known to be of integrable order, so that $F(\rho)$ is itself nonsingular in the range $0 < \rho < r'$.

The integral equation $F(\rho) = 0$ can, in turn, be solved for w (which is *not* continuous, and is therefore not necessarily zero). The solution is of interest in itself as giving the upwash off the tip of a thin wing of arbitrary camber, but the explicit value is not needed for the calculation of the potential at P. Instead we write

$$F(\rho) = \int_{\sigma_1(\rho)}^{\sigma_2(\rho)} \frac{w(\rho, \sigma)d\sigma}{\sqrt{s' - \sigma}} + \int_{\sigma_2(\rho)}^{s'} \frac{w(\rho, \sigma)d\sigma}{\sqrt{s' - \sigma}} \tag{13-22}$$

$\sigma_2(\rho)$ being the subsonic portion of the wing boundary (to the right of O). Then

$$\int_{\sigma_2(\rho)}^{s'} \frac{w(\rho, \sigma)d\sigma}{\sqrt{s' - \sigma}} = - \int_{\sigma_1(\rho)}^{\sigma_2(\rho)} \frac{w(\rho, \sigma)d\sigma}{\sqrt{s' - \sigma}} \tag{13-23}$$

In the characteristic coordinates the velocity potential at P becomes (Fig. A,13p)

$$-\frac{1}{2\pi} \iint_{AOCP} \frac{w(\rho, \sigma)d\rho d\sigma}{\sqrt{(r - \rho)(s - \sigma)}} = -\frac{1}{2\pi} \left[\int_0^{\rho_2(s)} \frac{d\rho}{\sqrt{r - \rho}} \int_{\sigma_1(\rho)}^{\sigma_2(\rho)} \frac{w(\rho, \sigma)}{\sqrt{s - \sigma}} d\sigma \right.$$
$$+ \int_{\rho_2(s)}^{r} \frac{d\rho}{\sqrt{r - \rho}} \int_{\sigma_1(\rho)}^{s} \frac{w(\rho, \sigma)}{\sqrt{s - \sigma}} d\sigma$$
$$\left. + \int_0^{\rho_2(s)} \frac{d\rho}{\sqrt{r - \rho}} \int_{\sigma_2(\rho)}^{s} \frac{w(\rho, \sigma)}{\sqrt{s - \sigma}} d\sigma \right] \tag{13-24}$$

The first two terms in the expanded expression for $\varphi(x, y)$ are obtainable from the specification of the wing. The last term is obtainable from Eq. 13-23, with $s' = s$ (P' on BC, Fig. A,13p) and has the effect of canceling exactly the first term, so that

$$\varphi(x, y) = -\frac{1}{2\pi} \iint_{ADBP} \frac{w(\rho, \sigma)d\rho d\sigma}{\sqrt{r - \rho}\sqrt{s - \sigma}} \tag{13-25}$$

This result is identical with that obtained as Eq. 12-39 of VI,D, but is rederived here using the Evvard-Krasilshchikova approach as being more intuitive, and therefore more readily extended.

Eq. 13-25 leads to the interesting conclusion that the shape or existence of the surface ahead of the "reflected" Mach line BD has no effect on the flow at point P. Physically interpreted, it leads to the conjecture that the influence of a negative point disturbance on the under surface of a thin wing travels around the subsonic tip and somehow cancels the influence of the corresponding positive disturbance on the upper side within a region which may be thought of as resulting from the wrapping of the Mach cone from the under side around the side edge.

If the Mach forecone from P includes a second subsonic edge, as $O'A$ in Fig. A,13q, the upwash behind the Mach line from O' will cancel the influence of the region ahead of the "reflected" Mach line AE in the same way; and if the lines AE and BD intersect on the wing, the negative effects

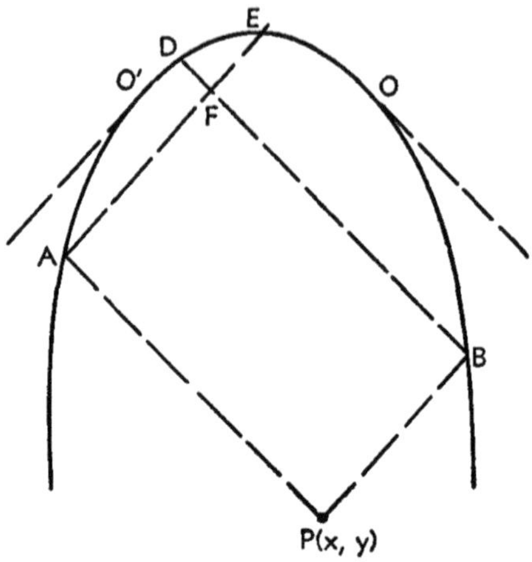

Fig. A,13q. Location of point and regions of integration for Eq. 13-26 and 13-27.

will be superimposed in the region ahead of the intersection. The net effect is to give, for a point located as in Fig. A,13q,

$$\varphi(x, y) = -\frac{1}{2\pi}\left[\iint\limits_{AFBP}\frac{w(\rho, \sigma)d\rho d\sigma}{\sqrt{r-\rho}\sqrt{s-\sigma}} - \iint\limits_{DEF}\frac{w(\rho, \sigma)d\rho d\sigma}{\sqrt{r-\rho}\sqrt{s-\sigma}}\right] \tag{13-26}$$

Eq. 13-26 may be differentiated to obtain the component of velocity proportional to the pressure. If $\sigma = \sigma_1(\rho)$ describes the portion of the planform boundary containing A, and $\rho = \rho_2(\sigma)$ the portion containing B, the expression for $\partial\varphi/\partial x$ may be written

$$u(x, y) = -\frac{1}{2\pi}\left\{\iint\limits_{AFBP}\frac{\partial w/\partial x}{\sqrt{r-\rho}\sqrt{s-\sigma}}d\rho d\sigma\right.$$

$$-\iint\limits_{DEF}\frac{\partial w/\partial x}{\sqrt{r-\rho}\sqrt{s-\sigma}}d\rho d\sigma + \int\limits_{\widehat{DE}}\frac{w(d\rho - d\sigma)}{\sqrt{r-\rho}\sqrt{s-\sigma}}$$

$$+\frac{1}{\sqrt{s-\sigma_1(r)}}\left[1-\left(\frac{d\sigma_1}{d\rho}\right)_{\rho=r}\right]\int\limits_{\overrightarrow{EA}}\frac{w}{\sqrt{r-\rho}}d\rho$$

$$\left.+\frac{1}{\sqrt{r-\rho_2(s)}}\left[1-\left(\frac{d\rho_2}{d\sigma}\right)_{\sigma=s}\right]\int\limits_{\overrightarrow{DB}}\frac{w}{\sqrt{s-\sigma}}d\sigma\right\} \tag{13-27}$$

in which $\sigma_1(r)$ is the value of σ at the left-hand intersection A of the Mach forecone from P with the planform boundary and $\rho_2(s)$ is the value of ρ at the right-hand intersection (point B). The evaluation of the double integrals is facilitated by reference to Fig. A,13r, where the limits of integration are expressed in terms of the characteristic coordinates. If the location of P relative to the wing boundary is such that the lines AE and DB do not intersect on the wing, the second area integral is of course equal to zero, and the line integral along the arc $\widehat{DE}$ changes sign.

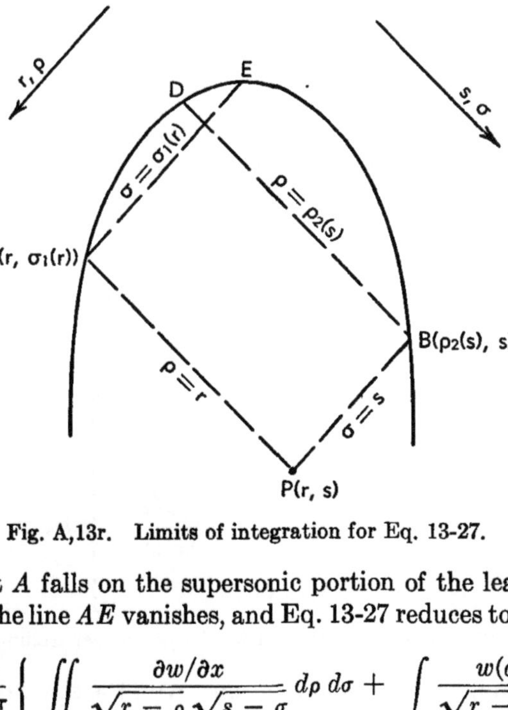

Fig. A,13r. Limits of integration for Eq. 13-27.

If the point A falls on the supersonic portion of the leading edge, as in Fig. A,13p, the line AE vanishes, and Eq. 13-27 reduces to

$$u(x, y) = -\frac{1}{2\pi}\left\{\iint_{ADBP} \frac{\partial w/\partial x}{\sqrt{r-\rho}\sqrt{s-\sigma}}\, d\rho\, d\sigma + \int_{\widehat{DA}} \frac{w(d\rho - d\sigma)}{\sqrt{r-\rho}\sqrt{s-\sigma}}\right.$$

$$\left. + \frac{1}{\sqrt{r-\rho_2(s)}}\left[1 - \left(\frac{d\rho_2}{d\sigma}\right)_{\sigma=s}\right] \int_{\overrightarrow{DB}} \frac{w d\sigma}{\sqrt{r-\rho}\sqrt{s-\sigma}}\right\} \qquad (13\text{-}28)$$

If A or B is on a streamwise edge (tip), the corresponding one of the last two terms in Eq. 13-27 again drops out, since the coefficient becomes zero. In the case of a flat plate wing, the first two terms vanish and only line integrals remain to be calculated.

Pressure at point near subsonic trailing edge. If the Mach forecone from P includes a portion of the wake, as in Fig. A,13s, the foregoing development is no longer applicable. However, it may be applied to eliminate the region $OB'C$, where $\varphi = 0$, as before. It is then found that the effect of the upwash in this region cancels the influence of any downwash in region $D'OB'$, so that Eq. 13-27 may be applied as if the wing included the region of trailing vorticity behind it. The value of w in the region GBB' is now the only unknown quantity in the expression for $u(r, s)$. We may write Eq. 13-27 for point B, the intersection of the line $\sigma = s$ with the trailing edge, in terms of the downwash in the same region, and, according to the Kutta condition, set the resulting expression equal to zero. The equation thus obtained can then be written in the form of Eq. 13-20 and, following the procedure used in deriving Eq. 13-23, solved for

$$\int_{\sigma_t(\rho)}^{s} \frac{\partial w/\partial x}{\sqrt{s-\sigma}}\, d\sigma$$

in which $\sigma_t(\rho)$ is the value of σ along the trailing edge, in terms of ρ. This solution is all that is required to complete the expression of $u(r, s)$ in

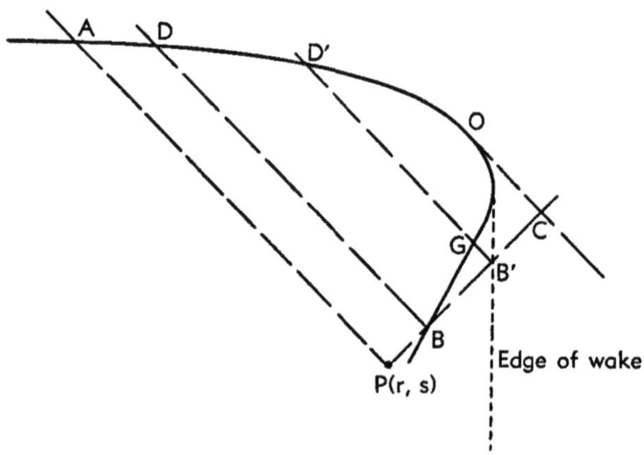

Fig. A,13s. Regions of integration for point near trailing edge.

terms of known quantities, and has, moreover, the effect of canceling the contribution of the region $BDD'G$ to that expression, so that the final formula is simply (for P of Fig. A,13s)

$$u = -\frac{1}{2\pi}\left[\iint_{ADBP} \frac{\partial w/\partial x}{\sqrt{r-\rho}\sqrt{s-\sigma}}\, d\rho d\sigma + \int_{\widehat{DA}} \frac{w(d\rho - d\sigma)}{\sqrt{r-\rho}\sqrt{s-\sigma}}\right] \tag{13-29}$$

If the leading edge at A is subsonic, or if the line $\rho = r$ from P intersects the wing tip, as in Fig. A,13r, a similar derivation will yield Eq. 13-27 with the last term omitted. If the point A, as well as B, lies on a subsonic trailing edge, both the last and the next-to-last terms in Eq. 13-27 will drop out.

Eq. 13-29 is directly derivable from Eq. 13-53 of VI,D but it is hoped that the present derivation will lend itself more readily to extension by the reader to other cases. Evvard [*160*] has carried through an even more general derivation, in that the slopes of the upper and lower surfaces of the wing are allowed to vary independently, so that any distribution of thickness, as well as camber and twist, may be taken into account. In the present treatment, however, thickness effects will be considered separately (see Art. 14), in line with usual linear theory practice.

Table A,13b summarizes the formulas derived in the foregoing section and indicates the conditions under which they apply. Also indicated is a situation (region VII) for which further analysis is required. While the Evvard-Krasilshchikova method can be extended to cover such regions, it is probable that the full range of its usefulness per se has been explored. However, several of the basic ideas have been incorporated into a highly practical cancellation method first suggested by Goodman [*162*] and developed by Mirels [*141*]. This method is particularly useful for the treatment of the part of the wing influenced by a subsonic trailing edge. Therefore we will not consider further extensions of the preceding method, but will assume that the formulas given will be supplemented by the use of the Goodman-Mirels cancellation technique, a discussion of which follows.

Cancellation of Lift by Source Distributions. The basic concept of the cancellation method has already been discussed, and in connection with the present problem of finding the lift distribution over a wing of given shape may be summarized as follows:

1. The lift distribution must be known over some wing of which the specified wing forms a part.
2. A flow must be formulated which (a) coincides in lift distribution with that calculated over the regions of the plane outside of the plan area of the specified wing, (b) introduces no vertical velocity in the area occupied by the specified wing, and (c) introduces no lift outside the boundaries of the large wing. Note that the original calculation may be made by any method—conical flows, Evvard-Krasilshchikova, or any other—so that the present method is no longer limited to wings with supersonic leading edges.

The general formula for a flow in which the downwash is specified over part of the plane and the lift over the remainder has been derived in VI,D. If the dividing line is a subsonic trailing edge of the wing of which the pressure distribution is being sought, so that the continuity of pressure across it is a boundary condition of the problem, the formula for the local lift is given explicitly. From VI,D (Eq. 12-40) we obtain, by setting w equal to zero on the specified wing,

$$\Delta u(r, s) = -\frac{\sqrt{r - \rho_2(s)}}{\pi} \int_{\rho_1(s)}^{\rho_2(s)} \frac{u(\rho, s)d\rho}{(r - \rho)\sqrt{\rho_2(s) - \rho}} \tag{13-30}$$

Table A,13b. Formulas for lifting pressure on a wing with the foremost portion of the leading edge supersonic. (Formulas derived by the Evvard-Krasilshchikova method.)

Points in region	*Mach forecone intersects*[1]	*Formula for u*
I	supersonic edges only[2]	13-16
II	1 subsonic leading edge, 1 supersonic leading edge	13-28
III	2 subsonic leading edges[3]	13-27
IV	1 subsonic trailing edge,[3] 1 supersonic leading edge	13-29
V	1 subsonic trailing edge,[3] 1 subsonic leading edge	13-27, without last term
VI	2 subsonic trailing edges[3]	13-27, without last two terms
VII	2 interacting subsonic edges	Not presented. See discussion on p. 188.

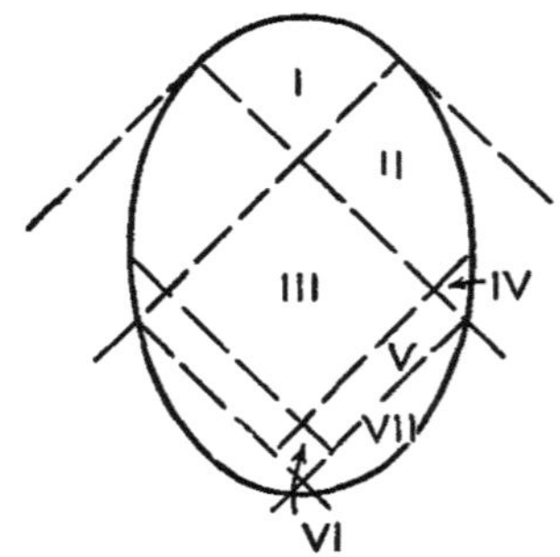

[1] Number of intersections limited to two.
[2] Leading edge assumed supersonic throughout region I.
[3] or streamwise edge.

where $\Delta u(r, s)$ is the incremental streamwise velocity, proportional to the loss in lift, induced at the point r, s on the specified wing by the cancellation flow; $\rho_1(s)$ (see Fig. A,13t) is the value of ρ at the intersection A of the line $\sigma = s$ with the leading edge of the wing for which the lift is known; and $\rho_2(s)$ is the value of ρ at the intersection B of $\sigma = s$ with the trailing edge of the specified wing. (The lift to be canceled is assumed to occur to the right of the specified wing. If not, r and s are simply interchanged in Eq. 13-30 and ρ replaced by σ.)

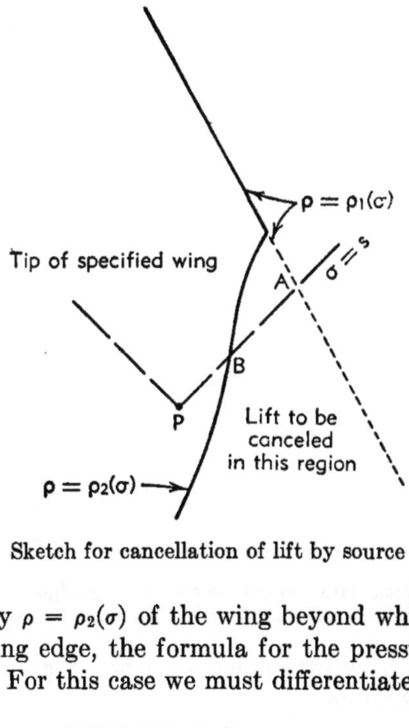

Fig. A,13t. Sketch for cancellation of lift by source distribution.

If the boundary $\rho = \rho_2(\sigma)$ of the wing beyond which the lift is to be canceled is a leading edge, the formula for the pressure is considerably more complicated. For this case we must differentiate

$$\Delta\varphi = -\frac{\sqrt{r - \rho_2(s)}}{\pi} \int_{\rho_1(s)}^{\rho_2(s)} \frac{\varphi d\rho}{(r - \rho)\sqrt{\rho_2(s) - \rho}} \tag{13-31}$$

which is Eq. 12-41 of VI,D with the condition of zero vertical velocity on the specified wing inserted. The result (see Appendix of [*141*]) may be written in the form

$$\Delta u = -\frac{1}{2\pi}\left[\sqrt{r - \rho_2(s)} \int_{\rho_1(s)}^{\rho_2(s)} \frac{2u(\rho, s)d\rho}{(r - \rho)\sqrt{\rho_2(s) - \rho}} - \frac{1 - (d\rho_2/d\sigma)_{\sigma=s}}{\sqrt{r - \rho_2(s)}} \int_{\rho_1(s)}^{\rho_2(s)} \frac{(u - v)d\rho}{\sqrt{\rho_2(s) - \rho}}\right] \tag{13-32}$$

in which v is the sidewash associated with the original wing flow, obtainable from the streamwise component of velocity u by the relation

$$v = \frac{\partial}{\partial y} \int_{x_1(y)}^{x} u(x, y)dx \tag{13-33}$$

where $x = x_1(y)$ is the leading edge of the original wing. The second term of Eq. 13-32 introduces the expected suction peak at the leading edge into the pressure distribution, and is important in computing the drag (Art. 2).

As an illustration of the use of the Goodman-Mirels method, yielding a result also of considerable interest in itself, consider the derivation of the formula for the lifting pressure at a point near the arbitrarily curved

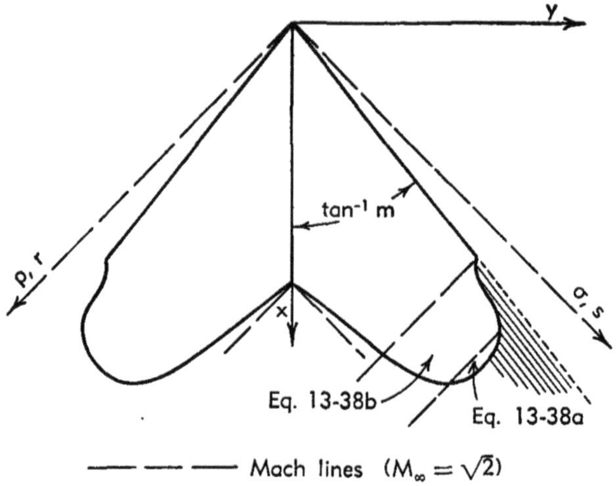

Fig. A,13u. Wing with straight swept leading edge and curved tips, for which lift distribution is given by Eq. 13-34 and 13-38.

tip of a flat wing with sweptback leading edge (Fig. A,13u), subject only to the following restrictions:

1. At the root, the leading edge of the wing consists of two straight line segments lying within the Mach cone from their juncture.
2. The remainder of the leading edge and the tips must also be subsonic at all points and lie within the infinite triangle formed by the rearward extension of the straight portions of the leading edge.
3. The Mach forecone from the point at which the pressure is to be calculated intersects only, on the left, the straight portion of the leading edge and, on the right, the nearby tip (or curved portion of the leading edge). If the point of intersection with the tip is one at which the wing is entering the stream, there is the further restriction that it is not directly downstream of some other part of the wing (as in a re-entrant side edge).

The basic solution is that for the infinite triangular wing at an angle of attack. From solution 6, Table A,13a, we obtain for the value of the u component of the velocity in the plane of the wing (after applying the Prandtl-Glauert transformation),

$$u_\Delta = \frac{mu_0 x}{\sqrt{m^2x^2 - y^2}} \tag{13-34}$$

in which m is the tangent of the semiapex angle,

$$u_0 = \frac{mU\alpha}{E'(m)} \tag{13-35}$$

is the value of u_Δ along the axis of symmetry, $E'(m)$ is the elliptic integral, U is the stream velocity, and α the angle of attack. The velocity u_Δ is to be canceled in the region shown shaded in Fig. A,13u.

The expression for u_Δ, rewritten in terms of ρ and σ, with the origin at the apex of the wing, is

$$u_\Delta = \frac{mu_0(\sigma + \rho)}{\sqrt{m^2(\sigma+\rho)^2 - (\sigma-\rho)^2}} \tag{13-36}$$

and the right-hand leading edge of the triangular wing is given by

$$\rho = \rho_1(\sigma) = \frac{1-m}{1+m}\sigma \tag{13-37}$$

Substituting these expressions into Eq. 13-30 and integrating gives [*152*]

$$\Delta u(r, s) = -u_\Delta \Lambda_0(\psi, k) + \frac{u_0}{\pi}\sqrt{\frac{m[r - \rho_2(s)]}{s}}\,K(k) \tag{13-38a}$$

$$k = \frac{1}{2}\sqrt{\frac{1-m}{ms}[(1+m)\rho_2(s) - (1-m)s]}$$

$$\psi = \tan^{-1}\sqrt{\frac{(1+m)s - (1-m)r}{(1-m)[r - \rho_2(s)]}}$$

$$\Lambda_0(\psi, k)^{18} = \frac{2}{\pi}\{K(k)E(\psi, k') - [K(k) - E(k)]F(\psi, k')\}$$

for a point of which the Mach forecone cuts a trailing edge; that is, the slope $\rho_2'(s)$ exceeds or is equal to one. If $\rho_2'(s) < 1$, Eq. 13-32 applies, and the following *additional* term results:

$$\frac{u_0}{\pi}[1 - \rho_2'(s)][2E(k) - K(k)]\sqrt{\frac{s}{m[r - \rho_2(s)]}} \tag{13-38b}$$

When the point $\rho_2(s)$, s lies within the vortex sheet originating at some portion of the wing ahead, the sidewash v in Eq. 13-32 is affected so that Eq. 13-38 no longer holds. However, the problem may still be solved with

[18] Tabulated in [*151,163*].

the aid of Eq. 13-37. A simple example of such a case is worked out in [*141*].

The Goodman-Mirels method is particularly useful in finding the pressure near the trailing edge when the variation of lift ahead is already somewhat complex. For example, let us consider region VII of the wing shown in Table A,13b. The new element in region VII is the effect of the reflection, at the trailing edge ahead of the region, of the Mach wave bounding the region of influence of the far tip. This influence is expressed by the difference between the velocity in region V and that in region IV (or equally between $u(x,y)$ in region III and $u(x,y)$ in region II), and is to be canceled in the region between the trailing edge and the extension of the Mach line separating those regions. Thus the expressions for u, ρ_1, and ρ_2 can be determined for substitution into Eq. 13-30 and the increment u found for extending the solution from region V into region VII.

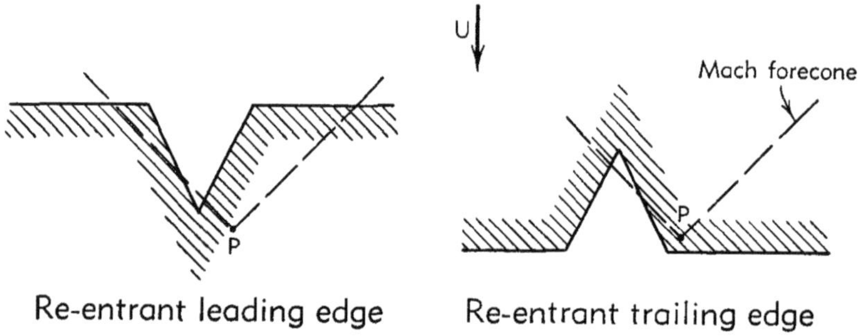

Fig. A,13v. Problems not readily solved by lift-cancellation methods.

As indicated in the discussion of the conical flow method, the foregoing procedure may be applied any number of times on alternate sides of a wing, so that in principle the pressure on any wing, of no matter how low aspect ratio, may be completely determined, providing only that the trailing edge eventually has a supersonic segment. Where the wing terminates in an angle formed by two subsonic edges (Fig. A,13i) the process has no end, but may be carried to a point where the region of undetermined pressures is arbitrarily small. A solution in closed form would require the inversion of an integral equation involving more than the two regions considered in the derivation of Eq. 13-30, and this has not as yet been attempted for the general case (however, see [*148,164*]).

If two subsonic leading or trailing edges meet in a re-entrant angle, the Mach forecones of points on the wing are again divided into a number of disconnected or irregular regions (see Fig. A,13v). However, whereas in the preceding problem the errors introduced by ignoring the more remote regions took the form of extraneous lift, which could be eliminated by successive applications of the same lift-cancellation procedure, in the problems of Fig. A,13v the errors would involve unwanted variations of downwash on the wing, so that a somewhat different procedure must be followed.

For the problem of re-entrant trailing edge, the lift cancellation procedure may be carried out in a reverse direction, so to speak, to obtain an expression for the pressure at any point in the region of the trailing edge. Again an infinite number of terms is required for a complete solution, but an arbitrary degree of accuracy may be obtained with a finite number. Thus (Fig. A,13w) the lift at a point P on one side of the wing is expressed by Eq. 13-30 in terms of the known value of u in the region to be cut out and of the as-yet-undetermined value of u at points P_1 on the other side of the cutout. The value of u at P_1 is, in turn, expressed in terms of u at

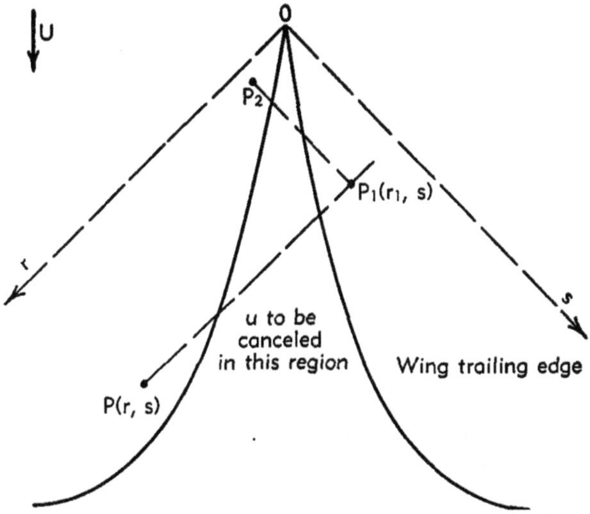

Fig. A,13w. Successive terms in pressure at point within Mach cone from apex of sweptback trailing edge.

the points P_2 on the first side of the wing, and so on. At any step in this procedure it may be assumed that the effect of the as-yet-undetermined pressures will be negligible and the series terminated. This procedure has the advantages that even in the first approximation, obtained by assuming Δu to be zero on the opposite wing panel, the Kutta condition will be satisfied at the trailing edge, and that the only violation of the boundary conditions occurs at points relatively remote from the point at which the pressure is being determined. If the trailing edge cutout occupies only a small portion of the Mach cone from its apex, a second or third step may be required for satisfactory accuracy.

It is apparent that, in the important case of a flat sweptback wing with straight leading edges, applying the foregoing procedure to Eq. 13-34 gives, even in the first approximation, incomplete elliptic integrals of the third kind, functions which have not as yet been tabulated. For engineering purposes, further simplification will generally be desirable. Actually, the lift to be canceled between the branches of the trailing edge differs very little from $2\rho_\infty U u_0$, the value along the axis of symmetry. In the following approximation, the Kutta condition is satisfied by taking into account the actual value of the lift to be canceled at the trailing edge near the point at which the pressure is to be calculated. Furthermore, the variation of lift is closely approximated along the forward Mach line ($\rho = r$, Fig. A,13x) up to a point r, s_0 at which the approximate value equals $2\rho_\infty U u_0$. The latter value is then assumed for the remainder of the

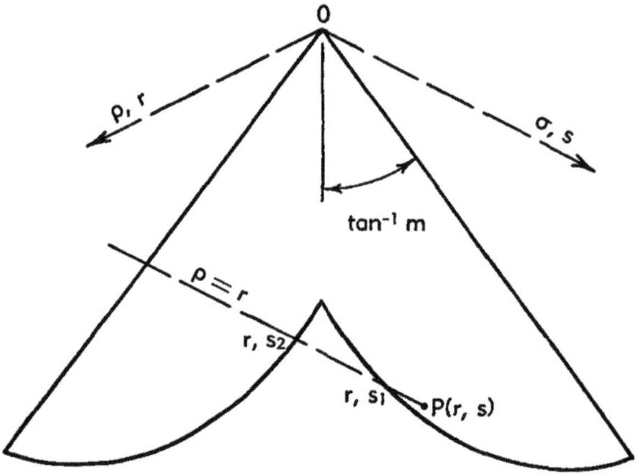

Fig. A,13x. Sketch for Eq. 13-39.

wake region. This procedure results in the following expression for the trailing edge effect on u:

$$\Delta u(r, s) = -\frac{2u_0}{\pi}\left[\tan^{-1}\sqrt{\frac{s_1 - s_2}{s - s_1}} - \tan^{-1}\sqrt{\frac{s_1 - s_0}{s - s_1}} + \sqrt{\frac{(1+m)r - (1-m)s_0}{(1+m)r - (1-m)s}}\tan^{-1}\sqrt{\frac{(1+m)r - (1-m)s}{(1+m)r - (1-m)s_0}\left(\frac{s_1 - s_0}{s - s_1}\right)}\right] \tag{13-39}$$

Eq. 13-39 is written for a point on the right-hand wing, as shown in Fig A,13x. The origin of the coordinate system is at the leading edge apex. The quantities s_1 and s_2 are the σ coordinates (see Eq. 13-18) of the near and far intersections of the line $\rho = r$ with the trailing edge of the wing, and s_0 has the value

$$s_0 = \frac{(1 - m^2)rs_1 - (r - ms_1)^2}{(1 - m)[(1 + m)s_1 - (1 - m)r]}$$

Eq. 13-39 applies to an arbitrarily curved trailing edge. If the trailing edge is composed of straight lines, the first term in the equation, which approximates the effect of canceling the uniform lift $2\rho_\infty U u_0$ throughout the wake region, may be replaced by the exact value given by solution 18 of Table A,13a. Thus, for a simple V-shaped wing with subsonic leading and trailing edges, the reduction in u due to the trailing edge is

$$\Delta u(r, s) = -u_0 \left\{ \frac{F(\phi, \sqrt{1 - m_t^2})}{K(\sqrt{1 - m_t^2})} - \frac{2}{\pi}\left[\tan^{-1}\sqrt{\frac{s_1 - s_0}{s - s_1}} \right.\right.$$

$$\left.\left. - \sqrt{\frac{(1 + m)r - (1 - m)s_0}{(1 + m)r - (1 - m)s}} \tan^{-1}\sqrt{\frac{(1 + m)r - (1 - m)s}{(1 + m)r - (1 - m)s_0}\left(\frac{s_1 - s_0}{s - s_1}\right)} \right] \right\}$$

$$\phi = \sin^{-1}\frac{1}{x - c_0}\sqrt{\frac{(r - c_0)(s - c_0)}{1 - m_t^2}} \tag{13-40}$$

where m_t is the inclination of the trailing edge to the stream and c_0 is the length of the root chord.

Eq. 13-40 gives accuracy varying from complete agreement at the trailing edge to a maximum error of 1.5 per cent when compared with the trailing edge corrections given in [*155*]. Eq. 13-39 gives results within 8 per cent of the correct values.

A,14. Specific Planforms. Lift Distribution, Lift and Drag Due to Angle of Attack. This article will present results of the application of the foregoing procedures to specific planforms. The planforms considered will be limited to those bounded by straight lines, although the trailing edges, if supersonic, may be of almost any shape. Only the effect of angle of attack (the flat plate problem) and unyawed wings will be considered. Calculated loadings will be shown in plots of $C_p = (p_{\text{lower}} - p_{\text{upper}})/\frac{1}{2}\rho_\infty U^2 = 4u/U$, and formulas for the lift, drag, and aerodynamic center location will be given. The formulas will be written in terms of the physical dimensions of the wing—root chord c_0, span b or semispan s, and aspect ratio $\mathcal{R}$—so that the Prandtl-Glauert transformation factor $1/\beta$ ($\beta = \sqrt{M_\infty^2 - 1}$) will now appear with all the streamwise quantities in the previously derived expressions. Then, since the pressure is proportional to the streamwise gradient of φ, the formulas for lift will contain an additional factor of $1/\beta$.

Rectangular wings. The solution for the high aspect ratio rectangular wing is most readily obtained by the superposition of two conical fields (solution 21, Table A,13a) on a cylindrical field (Ackeret's solution for a flat plate). The resulting form of the lift distribution is shown in spanwise sections in Fig. A,14a. Also shown are experimental values obtained [*165*] on the flat upper surface of a wing with half-diamond profile (see figure) in tests at a Mach number of 2.

If either the Mach number or aspect ratio is low enough to cause the tip Mach cones to overlap, the negative effects are additive in the region of overlapping (region III in sketch, Fig. A,14b). The tip effects add up to a constant all along the line $x = 2s$ joining the points of intersection of the Mach lines with the opposite tips, and combine to reduce the lift to zero at that streamwise station (a distance $1/\beta$ times the span downstream of the leading edge). Behind this line the lift is negative.

Reflection of the tip Mach waves at the opposite side edges has the

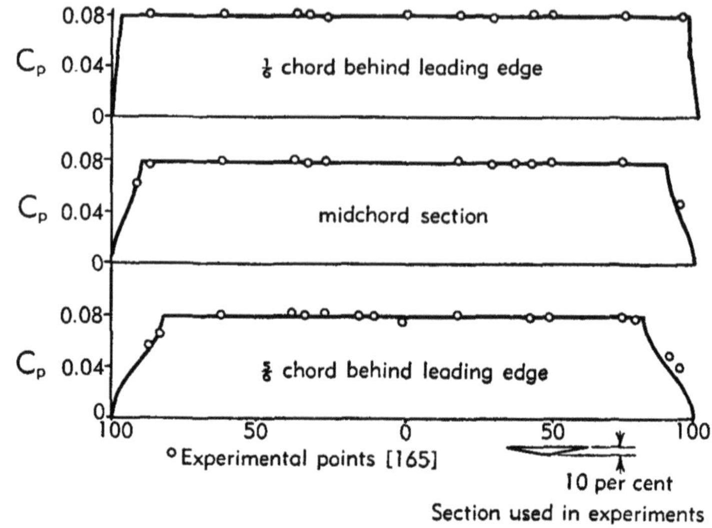

Fig. A,14a. Distribution of lift on a rectangular flat plate (Æ = 5.7) at 4° angle of attack and Mach number of 2.0.

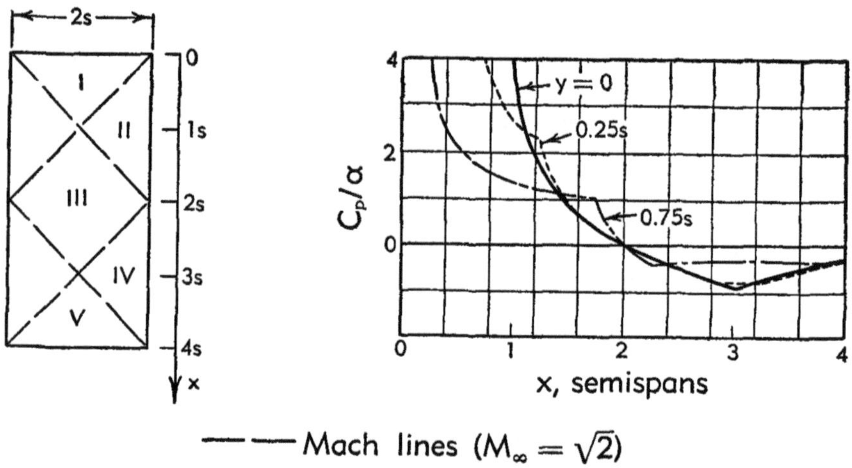

Fig. A,14b. Chordwise lift distribution at various spanwise stations of rectangular wings [*167*].

effect of restoring some of the lift. The magnitude of this effect (according to the simplified theory being applied) may be calculated about equally well by conical-flow [*168*] or source-distribution [*167*] methods. It should be noted, however, that early separation of the flow and the formation of vortices around the side edges (Art. 1) make the flow near the tips rather different from that assumed, so that the results of this simplified theory are of limited interest.

The loading on rectangular wings has been calculated as far downstream as the intersections of the reflected Mach waves with the side edges. The results are presented for three chordwise sections in Fig. A,14b. Discontinuities in slope mark the location of the waves and reflected waves from the tips. It is seen that at the farthest station downstream the lift is still negative.

The total lift on the rectangular wing with aspect ratio $A\!\!R \geqq 1/\beta$ is readily found by subtracting the integrated tip losses from the lift for the infinite wing. In coefficient form we get

$$C_{L_\alpha} = \frac{4}{\beta}\left(1 - \frac{1}{2\beta A\!\!R}\right), \qquad \text{for } \beta A\!\!R \geqq 1 \tag{14-1a}$$

For $\beta A\!\!R < 1$, application of the conical flow method gives [*168*]

$$C_{L_\alpha} = \frac{4}{\pi\beta}\left[\left(2 - \frac{1}{\beta A\!\!R}\right)\sin^{-1}\beta A\!\!R + (\beta A\!\!R - 2)\cosh^{-1}\frac{1}{\beta A\!\!R} + \left(1 + \frac{1}{\beta A\!\!R}\right)\sqrt{1 - \beta^2 A\!\!R^2}\right] \tag{14-1b}$$

$$\tfrac{1}{2} \leqq \beta A\!\!R \leqq 1$$

The variation of lift curve slope with Mach number for rectangular wings of various aspect ratios is shown in Fig. A,14c. In Fig. A,14d, the variation of aerodynamic center location, $x_{a.c.}$, with Mach number is shown, obtained from the same source [*168*]. The drag due to angle of attack, as of any flat wing with supersonic leading edge, is merely equal to the lift times the angle of attack.

Trapezoidal wings. The high aspect ratio trapezoidal wing is treated very similarly to the rectangular. On the two-dimensional field we superimpose, in the case of subsonically raked-in tips, a field obtained from solution 22, Table A,13a, by letting $m_1 \to -\infty$. If the tips are raked out, the conditions for a subsonic edge will be satisfied by solution 8; a solution for supersonic edges can be derived from solution 1 by subtracting the given field from one of the same form, with infinite m. With supersonically raked-in tips, the pressure has the two-dimensional value all over the wing.

If the side edges are subsonic, reflection of the tip Mach waves will again occur at low Mach numbers, on low aspect ratio wings. If the edges are raked in, or trailing, Eq. 13-30 provides a means of calculating the incremental lift. For raked-out edges, the solution has been obtained by a source-distribution method in the previously cited work [*167*] by Behrbohm.

Section lift distributions for this last case are reproduced in Fig.

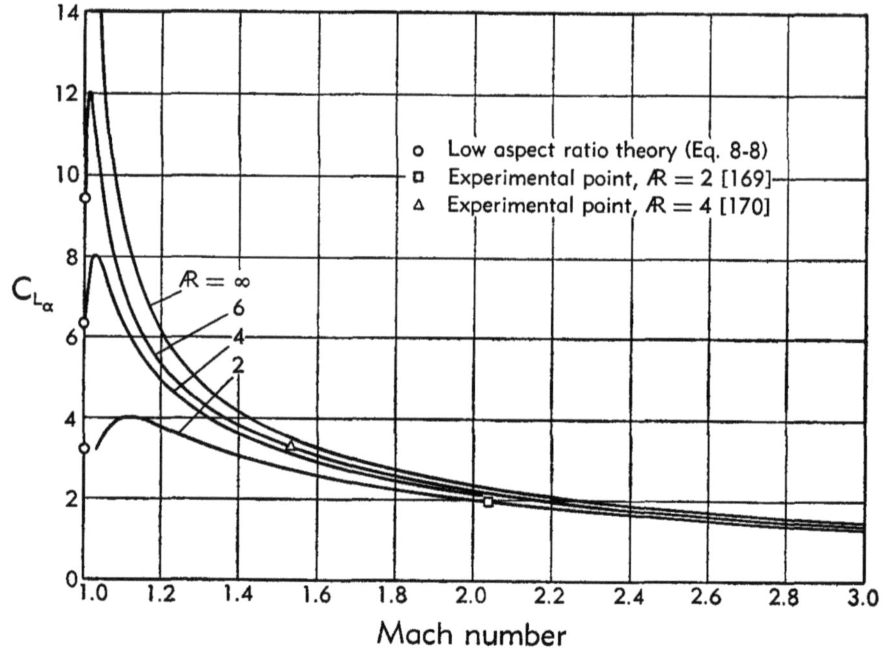

Fig. A,14c. Theoretical variation with Mach number of the lift curve slope of rectangular wings.

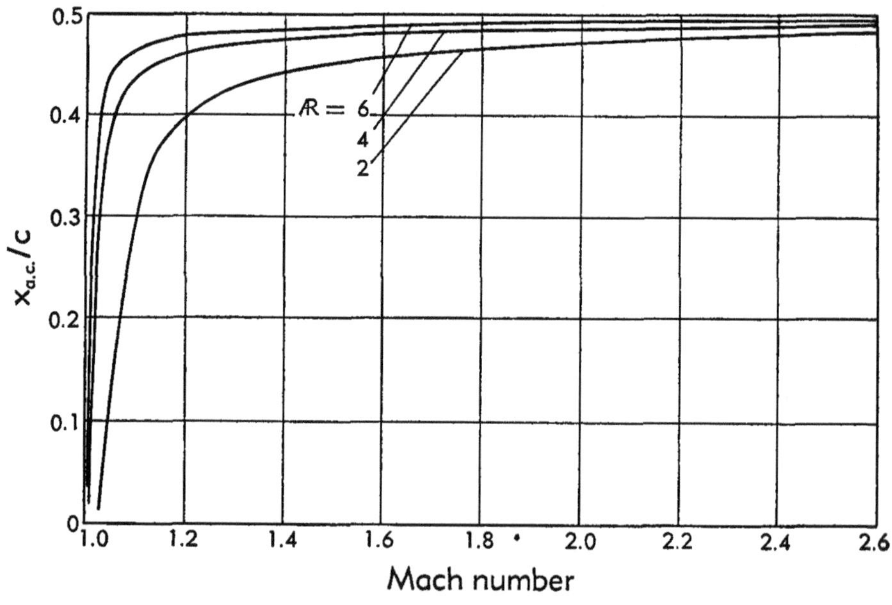

Fig. A,14d. Theoretical variation with Mach number of the aerodynamic center location of rectangular wings.

A,14e. In addition to showing the expected suction peaks at the raked-out side edges, the loading differs from that on the rectangular wing by remaining positive at all points. Discontinuities in pressure gradients again indicate the influence of the tips. If the side edges were extended beyond the tip Mach lines, the suction peaks would be replaced by plateaus in the loading, covering the regions between the tip Mach lines and the side edges.

As a result of the reversibility property of the total lift (Art. 6), only two cases need be considered in determining the lift curve slopes for

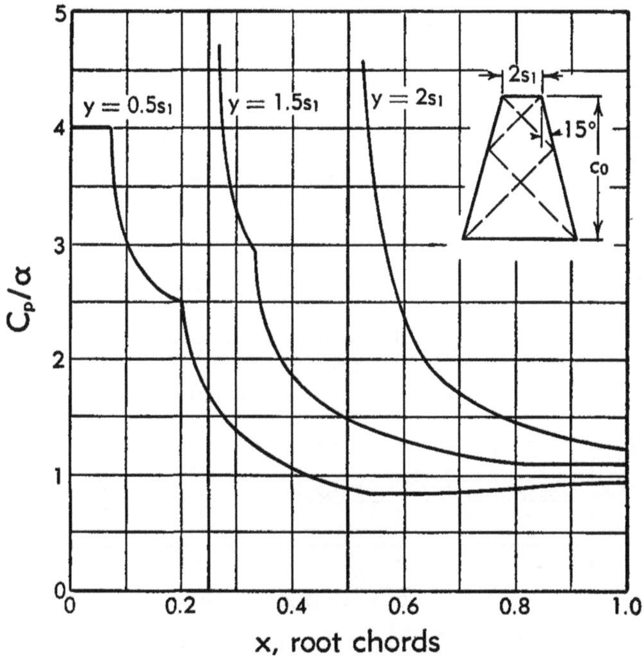

Fig. A,14e. Section lift distributions on a low aspect ratio trapezoidal wing with raked-out subsonic tips [167].

trapezoidal wings: (1) subsonically raked and (2) supersonically raked tips. In both cases, the raked-in tip presents the simpler approach. If the side edge lies ahead of the tip Mach cone, the lift is uniform over the entire wing and the lift coefficient is identical with that for infinite aspect ratio, so that

$$C_{L_\alpha} = \frac{4}{\beta}$$

If the side edge is subsonic, subtracting the tip-induced losses in lifting pressure from the two-dimensional lift gives

$$C_{L_\alpha} = \frac{4}{\beta}\left[1 - \frac{(1 - \beta \tan \delta)c_0}{2\beta\bar{b}}\right], \qquad \bar{b} \geq \frac{c_0}{\beta} \tag{14-2}$$

in which δ is the angle of inclination of the side edge to the stream and $\bar{b}$ is the width of the wing at midchord. For aspect ratios and Mach numbers such that $\beta\bar{b} < c_0$, no closed expression for C_{L_α} has been derived. Numerical values for two values of $\beta \tan \delta$ have been calculated by Behrbohm [*167*] and are shown in Fig. A,14f, together with the curve for $\beta \tan \delta = 0$, the rectangular wing.

In Fig. A,14g, aerodynamic-center locations are presented for the same families of wings, flying with the longer base trailing. Numerical values for $\beta\bar{b} < c_0$ are again taken from [*167*]. If $\beta\bar{b} \geqq c_0$, the aerodynamic center is readily located by resolving the loading into its component

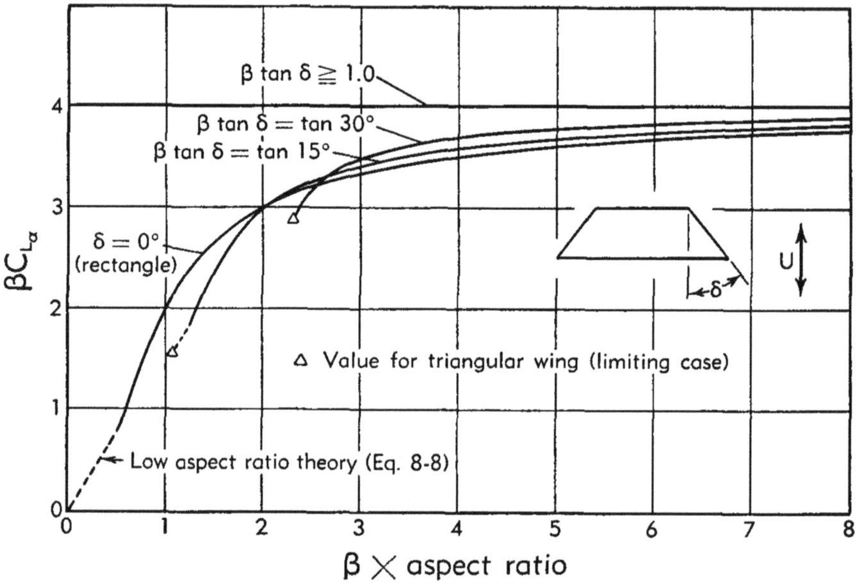

Fig. A,14f. Lift curve slope of trapezoidal wings.

parts: the uniform two-dimensional loading and the conical tip corrections, the latter having their centroids at the centroids of area of the tip Mach cones. Results for both raked-in and raked-out trapezoids in such cases are included in the figure.

The drag due to lift of the trapezoidal wing with raked-in tips is simply equal to the lift times the angle of attack, since there are no subsonic leading edges. The same is true if the tips are raked out through more than the Mach angle. However, if the side edges are raked out but lie within the Mach cones from the leading edge, the drag will be reduced by a suction force calculated by means of Eq. 12-1 as follows:

From solution 8, Table A,13a, the streamwise component of velocity u approaches infinity at the tip as

$$\frac{2U\alpha m}{\pi(1 + m)\beta}\sqrt{\frac{x + \beta y}{mx - \beta y}} \qquad (14\text{-}3)$$

where m is β times the inclination of the side edge to the stream, and the origin of the coordinate system is at the tip of the leading edge. Then

$$C_x = \frac{2U\alpha \sqrt{mx}}{\pi\beta \sqrt{1+m}} \tag{14-4}$$

$$\frac{dT}{dx} = \frac{4\rho_\infty U^2\alpha^2 mx}{\pi\beta^2} \sqrt{\frac{1-m}{1+m}} \tag{14-5}$$

and the total thrust is

$$T = \frac{4\rho_\infty U^2\alpha^2 m}{\pi\beta^2} c_0^2 \sqrt{\frac{1-m}{1+m}} \tag{14-6}$$

as long as the tip Mach cones do not intersect the opposite side edges. If this condition is violated ($c_0/\beta > \bar{b}$, the average span), the strength of

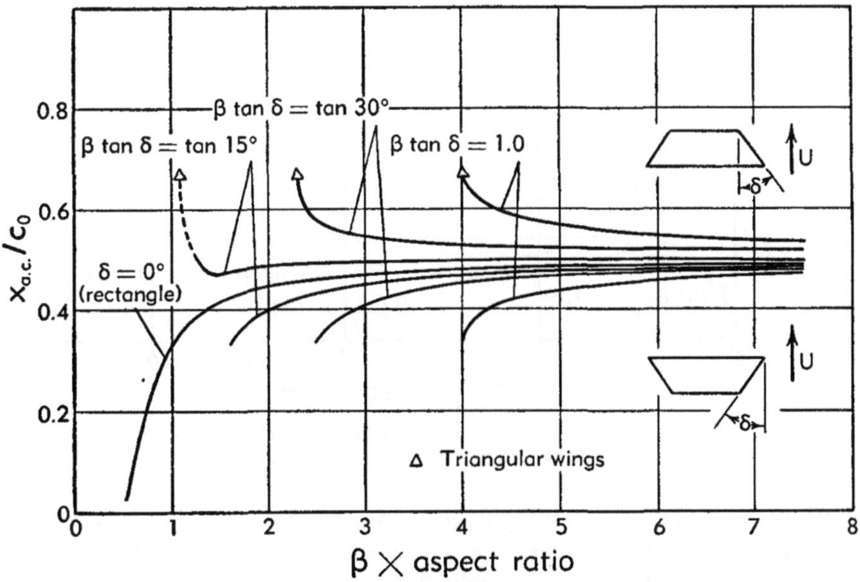

Fig. A,14g. Aerodynamic center location of trapezoidal wings.

the singularity along the edge will be modified by the interfering tip flow. The effect of the interference would not be difficult to calculate, but is not of great practical interest, being somewhat unrealistic. The variation of drag rise with aspect ratio and rake angle is shown in Fig. A,14h.

Pointed wings (triangular, diamond-shaped and swallowtail planforms). The loading on any wing with supersonic trailing edges and tips tapered to a point can be represented by a single conical expression depending only on the relative sweep of the leading edge and the Mach lines. The expression is, of course, that for the corresponding triangular wing, which is merely a special example of such wings. At low Mach numbers, such that the foremost Mach lines lie ahead of the leading edge, the loading is proportional to the incremental velocity given by Eq. 13-34 and shown as solution 6 in Table A,13a. In chordwise sections the lift distribution has the form shown in Fig. A,14i, where section lift distributions are plotted for a slender triangle. It is of interest to note that the span loading on the triangular wing is elliptic when the leading edge is subsonic.

Experiments on triangular wings at supersonic speeds have given pressure distributions in good agreement with the linear theory at very low angles of attack. At higher angles of attack, particularly when the aspect ratio is very low, the flow separates at the leading edge and the simple inviscid fluid theory no longer applies. The data shown in Fig. A,14i were obtained at $M_\infty = 1.37$ with a wing having a 40° apex angle

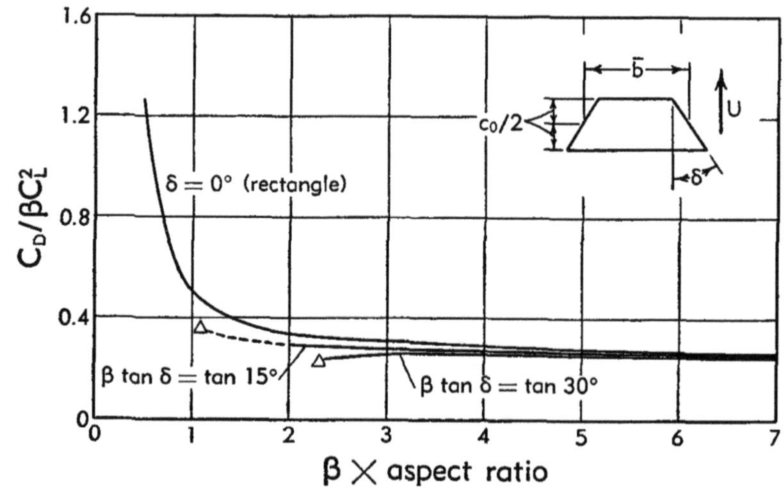

Fig. A,14h. Variation of drag rise factor with aspect ratio and rake angle for trapezoidal wings.

(70° sweepback of the leading edge) and 6 per cent thick, biconvex sections with a rounded leading edge. The effects of separation at the leading edge are evident even at a 3° angle of attack.

At higher Mach numbers, the flow normal to the leading edge becomes supersonic and the suction peaks observed in Fig. A,14i are replaced by the two-dimensional plateau in the loading (solution 3, Table A,13a). Some experimental lift distributions have been published *[171]* for triangular wings at such speeds, but the thickness of the wings (10 per cent) combined with the high Mach number, resulted in prominent second order effects.

General formulas for the total lift of pointed wings have been derived by Puckett and Stewart *[172]*. If m is β times the cotangent of the angle of sweepback of the leading edge and ξ the ratio of the cotangent of the angle of sweepback of the leading edge to that of the trailing edge, the lift curve slope is[19]

$$C_{L_\alpha} = \frac{4m}{\beta E'(m)}\left[\frac{\xi}{1+\xi} + \frac{1-\xi}{(1-\xi^2)^{\frac{1}{2}}}\cos^{-1}(-\xi)\right] \qquad (14\text{-}7)$$

when $m \leqq 1$, and

$$C_{L_\alpha} = \frac{8m}{\beta\pi(1+\xi)}\left[\frac{1}{\sqrt{m^2-\xi^2}}\cos^{-1}\left(-\frac{\xi}{m}\right) + \frac{\xi}{\sqrt{m^2-1}}\cos^{-1}\frac{1}{m}\right] \qquad (14\text{-}8)$$

when $m \geqq 1$. Diamond-shaped wings are included by permitting ξ to take on negative values.

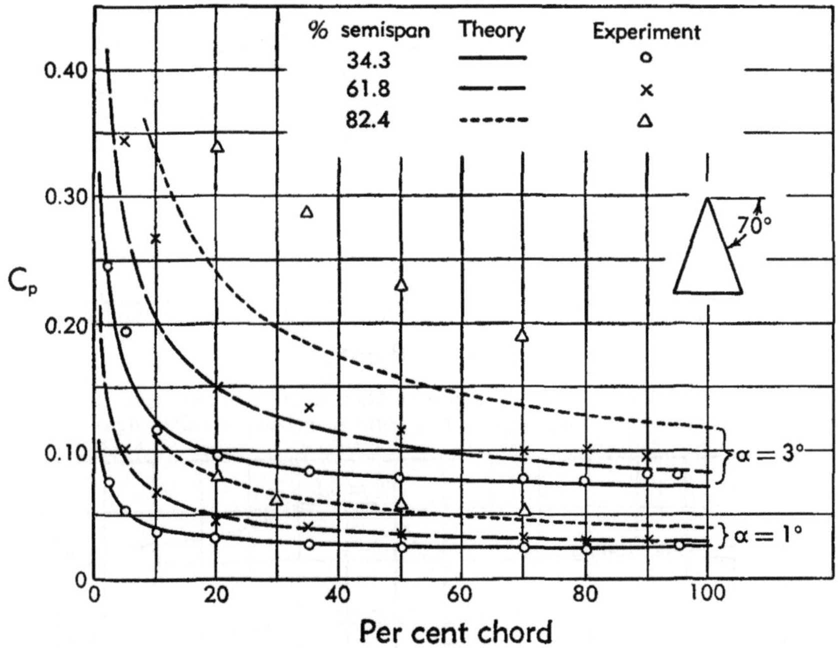

Fig. A,14i. Variation of lifting pressure along three streamwise sections of a triangular wing with 40° apex angle, at $M_\infty = 1.37$; theory and experiment [*173*].

Values of βC_{L_α} are plotted against m for various values of ξ in Fig. A,14j, and the corresponding center-of-pressure locations in Fig. A,14k. Positive values of ξ correspond to wings with cutouts at the relatively inefficient rear center of the planform, so that the lift curve slope inceases sharply with this parameter.

For the triangular wing ($\xi = 0$), the formulas for the lift curve slope reduce to

$$C_{L_\alpha} = \frac{2\pi m}{\beta E'(m)}, \qquad m \leqq 1 \qquad (14\text{-}9)$$

[19] $E'(m)$ is the complete elliptic integral of the second kind with modulus $\sqrt{1-m^2}$.

and

$$C_{L_\alpha} = \frac{4}{\beta}, \qquad m \geqq 1 \tag{14-10}$$

Figure A,14l shows this variation, together with experimental data taken from various sources. We have already seen that the local pressures are accurately predicted by the theory when the leading edge is well within

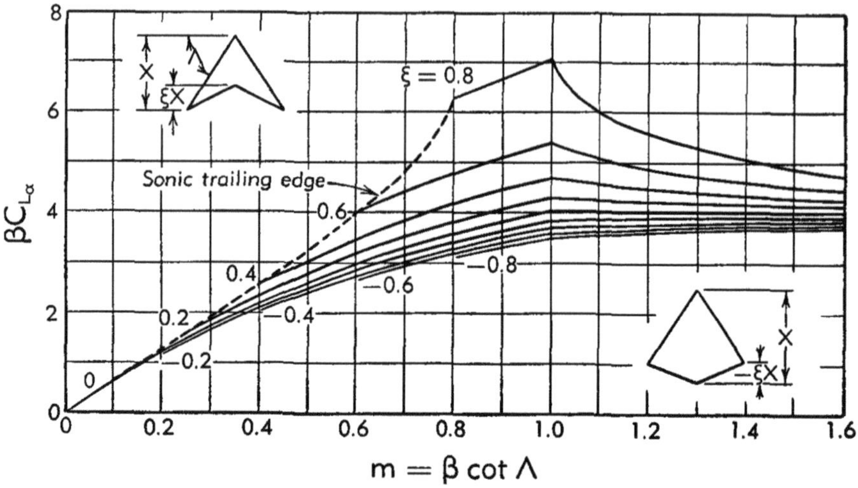

Fig. A,14j. Variation of lift curve slope with Mach number, sweep angle, and trailing edge cutout for wings tapered to a point (supersonic trailing edges).

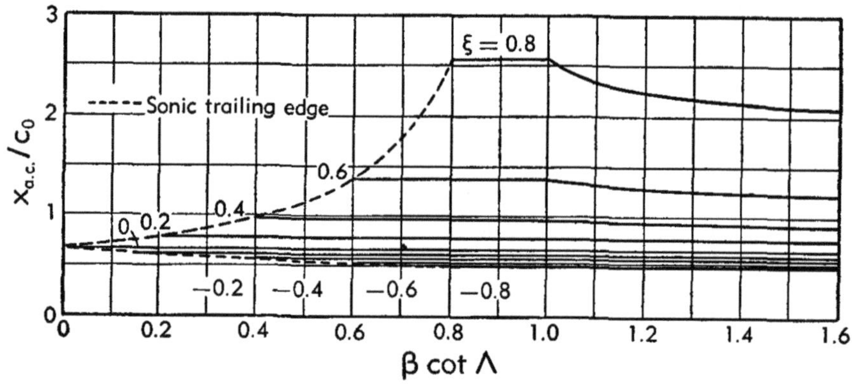

Fig. A,14k. Variation of aerodynamic center location for the wings of Fig. A,14j.

the Mach cone from its apex. Accordingly, the total lift also follows the theoretical curve closely when the Mach number and/or aspect ratio is low. As the Mach line approaches the leading edge ($m \to 1$; $\beta A\!\!R \to 4$), the experimental values appear to fall below the theoretical ones. The reason for this failure to realize the theoretical lift in this case is not understood at present.

In computing the drag due to lift, it is necessary, when m is less than 1, to take into account the suction force at the leading edge. From Eq. 13-34 we obtain as the strength of the leading edge singularity

$$C_\Delta = \lim_{x \to \frac{\beta y}{m}} u_\Delta \sqrt{x - \frac{\beta y}{m}} = u_0 \sqrt{\frac{x}{2}} \tag{14-11}$$

with

$$u_0 = \frac{mU\alpha}{\beta E'(m)} \tag{14-12}$$

Then substituting in Eq. 12-1 and integrating gives the thrust

$$T = \frac{\pi}{2} \rho_\infty U^2 \alpha^2 \frac{s^2}{[E'(m)]^2} \sqrt{1 - m^2} \tag{14-13}$$

Subtracting this quantity, which depends only on the Mach number, sweep, and spanwise extent of the leading edge, from the product of the

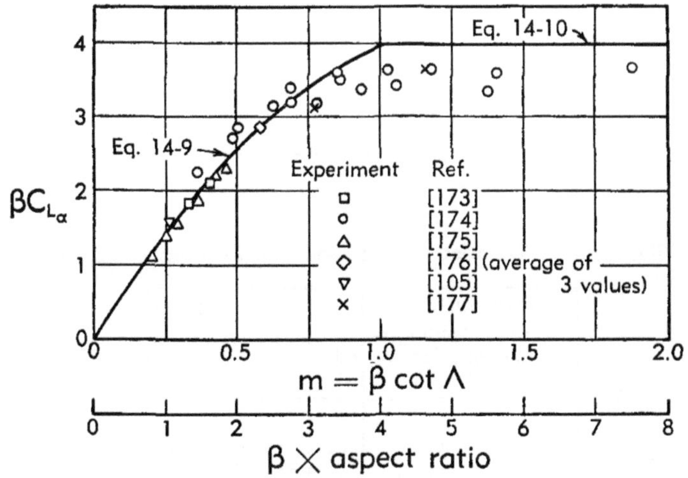

Fig. A,14l. Lift curve slope of triangular wings at supersonic speeds; theory and experiment.

lift and the angle of attack gives the drag due to angle of attack for any wing with straight subsonic leading edges, provided that there is no disturbance of the oncoming stream. The ratio C_D/C_L^2 for pointed-tip wings is plotted against m for various values of ξ in Fig. A,14m.

In Fig. A,14n, comparison is made of the experimental drag of two wings differing only in the rounding of the leading edge. The wings were of triangular planform, with the sweep and Mach number such that m had the value of 0.58. It is seen that the sharp-nosed airfoil showed a drag rise almost equal to the theoretical value for the wing in the absence of leading edge suction. A slight rounding of the nose reduced the drag to values approaching those predicted by the more complete theory. (The effect is of course not expected to be maintained beyond a small range of angle of attack.)

Wings of hexagonal planform with supersonic trailing edges. We will consider next wings having sweptback leading edges, blunt tips, and

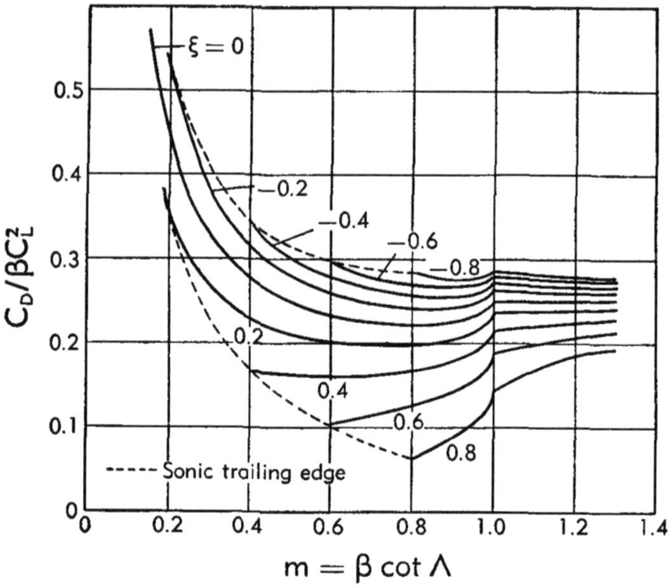

Fig. A,14m. Variation of drag rise factor for wings of Fig. A,14j.

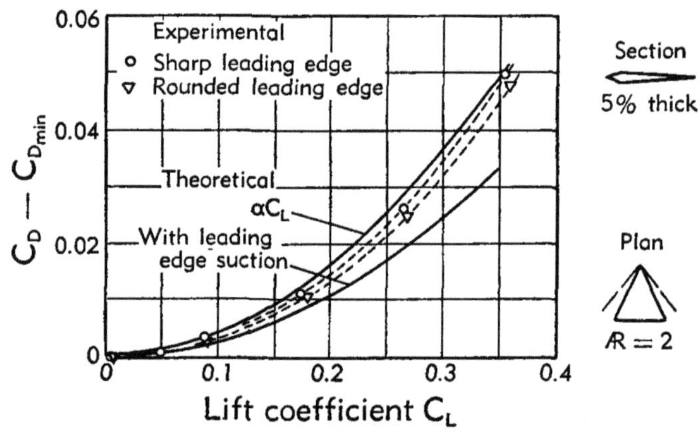

Fig. A,14n. Effect on drag rise of rounding the leading edge on a sweptback wing (Double wedge section rounded to radius of 0.25% chord.) $M_\infty = 1.53$. From [*176*].

trailing edges swept either forward or back, but not so much as to be subsonic. Except in the neighborhood of the tips, the lift distributions on such wings are the same as for the pointed wings (of the preceding discussion) having the same sweep of the leading and trailing edges. If the leading edge is supersonic, the Evvard-Krasilshchikova method gives the lift distribution within the tip Mach cones immediately (see Table A,13b). Formulas and numerical results obtained by this method for the loading and lift curve slopes of a large family of wing planforms are available in [*178*]. The pressure distributions are typified by that shown in Fig. A,14o. In the region between the leading edge and the Mach cones from the apex and tips, the loading is constant. The discontinuities in the planform outline appear in the loading as discontinuities in the pressure gradient at the Mach lines from the corners of the wing.

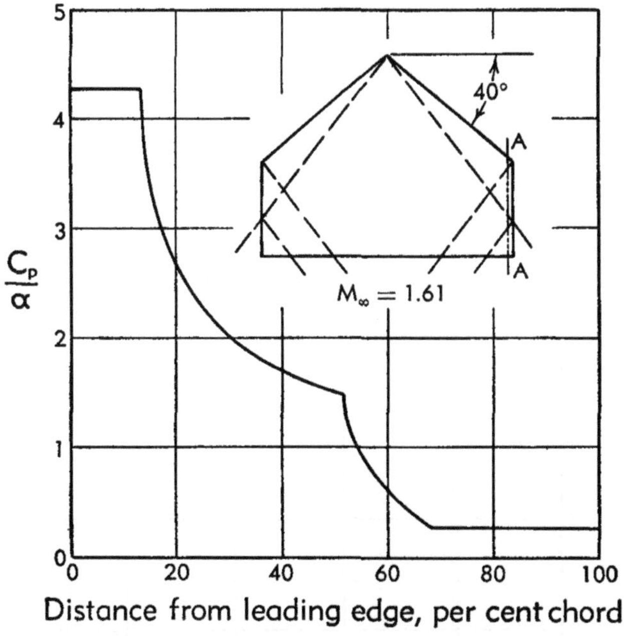

Fig. A,14o. Distribution of lifting pressure near the tip of a sweptback wing with supersonic leading and trailing edges and streamwise tips (section AA in sketch).

The lift distribution on wings with subsonic sweptback leading edges has been treated in considerable detail in connection with the lift-cancellation methods. Fig. A,14p shows the same wing as that of Fig. A,14o at a somewhat lower Mach number. The section lift distribution was calculated by means of Eq. 13-34 and 13-38a. Two basic differences are immediately noted: in addition to the appearance of the familiar leading edge suction peak, because of the now subsonic normal component of the oncoming stream, there also occurs a discontinuity in the *magnitude* of the pressure, rather than in the gradient, at the edge of the Mach cone from the tip. The latter effect is of course the consequence of the former and occurs whenever an abrupt change in conditions is introduced where the velocities are (theoretically) infinite.

The magnitude of the jump in the local lift at a point on the tip Mach line is readily calculated, being equal to

$$\Delta p = -\frac{\rho_\infty U u_0 \sqrt{b}}{\sqrt{(1+m)d}} \tag{14-14}$$

in which $2\rho_\infty u_0 U$ (Eq. 14-12) is the magnitude of the lifting pressure along the center line, m is β times the tangent of the semiapex angle, b is the

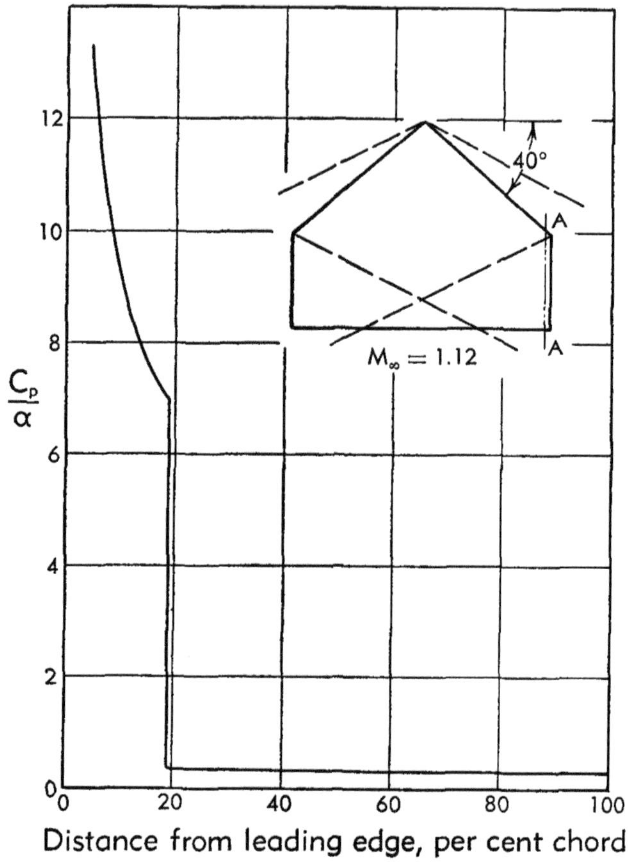

Fig. A,14p. Lift distribution at same spanwise station of the wing shown in Fig. A,14o, at a Mach number such that the leading edge is subsonic.

wing span, and d the spanwise distance of the point from the tip. The jump, which always results in the loss of more than half the local lift on crossing the tip Mach line, is nicely illustrated in a set of isobar diagrams calculated by Gilles [*179*], two of which are reproduced in Fig. A,14q. It is interesting to note that the isobars in the tip regions are, for practical purposes, straight lines. Since these lines are also almost parallel to the stream, it will usually be sufficient merely to calculate the pressure at points along the tip Mach lines with and without the tip effect, and to assume the lift to be constant at the former values rearward from these points (see also Fig. A,14p).

At the lower Mach number of the two for which the isobars are shown, the tip Mach cones overlap. Superposition of the tip losses in the region of

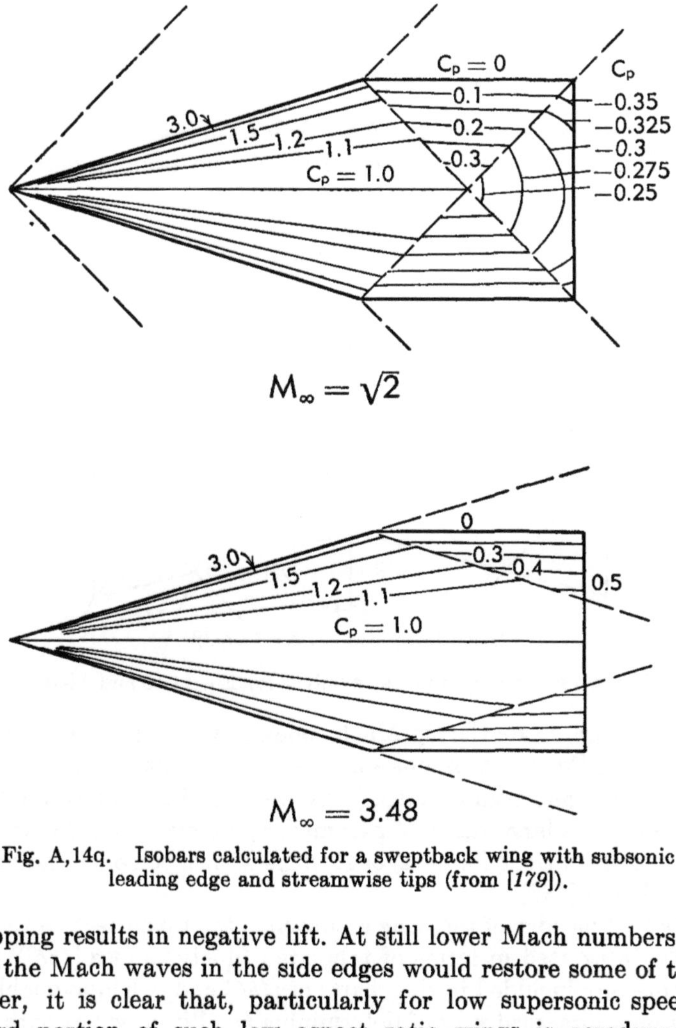

Fig. A,14q. Isobars calculated for a sweptback wing with subsonic leading edge and streamwise tips (from [*179*]).

overlapping results in negative lift. At still lower Mach numbers, reflection of the Mach waves in the side edges would restore some of this lift. However, it is clear that, particularly for low supersonic speeds, the rearward portion of such low aspect ratio wings is aerodynamically inefficient.[20]

The integrated lift of hexagonal wings with supersonic leading and trailing edges and streamwise tips is, in coefficient form,

[20] Note that, for $M_\infty = 1$, linear theory indicates *no* lift in this region.

$$C_L = \frac{2m\mathcal{R}\alpha}{\pi a_t^2\sqrt{m^2-1}}\left\{m\frac{(m_t-a_t)^2}{m_t^2-m^2}\left(\frac{m}{m_t}\cos^{-1}\frac{1}{m}-\frac{m_t}{m}\sqrt{\frac{m^2-1}{m_t^2-1}}\cos^{-1}\frac{1}{m_t}\right)\right.$$
$$+\left(\frac{m_t}{m}\right)^{\frac{3}{2}}\frac{1}{\sqrt{m_t+1}}\left[\frac{(m+a_t)^2\sqrt{m-1}}{m_t+m}\cos^{-1}\sqrt{\frac{(1+m)a_t}{m+a_t}}\right.$$
$$\left.-\frac{(m-a_t)^2\sqrt{m+1}}{m_t-m}\cos^{-1}\sqrt{\frac{m(1-a_t)}{m+a_t}}\right]$$
$$\left.+2m_t\frac{(m_t-a_t)^2}{m_t^2-m^2}\sqrt{\frac{m^2-1}{m_t^2-1}}\cos^{-1}\sqrt{\frac{m_t(1-a_t)}{m_t-a_t}}\right\} \quad (14\text{-}15)$$

The symbols used in Eq. 14-15 are defined in Fig. A,14r. The formula presented covers only those cases in which the trailing edge does not

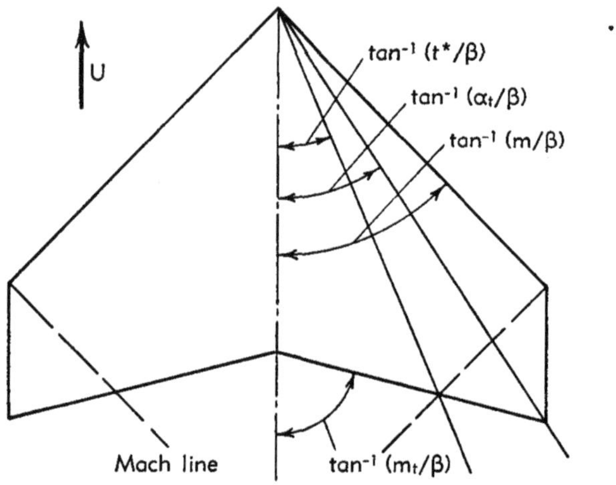

Fig. A,14r. Sketch showing symbols for Eq. 14-15 and 14-16.

change direction inside the tip Mach cones and $a_t \leqq 1$. Special forms for cases in which Eq. 14-15 becomes indeterminate (untapered wings, sonic edges, and straight trailing edges), as well as charts of computed lift curve slopes for a large range of parameters, are given in [*178*] and [*180*]. Span load distributions for hexagonal wings with supersonic leading edges are available in [*181*].

If the leading edge of the wing is subsonic, the tip correction to the lift cannot be integrated in terms of tabulated functions (see [*155*]). Calculated values are included in the charts of [*180*] and, when available, provide the most practical way of determining the lift on wings with supersonic trailing edges. The following approximate formula is based on the previously suggested assumption of uniform loading along each section downstream from the tip Mach line:

$$C_L = \frac{\alpha\mathcal{R}}{E'(m)}\left\{\frac{(1+m)^2(m_t-t^*)^2}{(m_t^2-m^2)(1+t^*)^2}\left[\frac{m_t}{\sqrt{m_t^2-m^2}}\left(\cos^{-1}\frac{m^2-m_tt^*}{m(m_t-t^*)}\right.\right.\right.$$

$$
\begin{aligned}
& \left. - \cos^{-1}\frac{m}{m_t}\right) - \frac{\sqrt{m^2 - t^{*2}}}{m_t - t^*} + \frac{m}{m_t}\Bigg] \\
& + \frac{1+m}{1-m}\left[\frac{1}{\sqrt{1-m^2}}\cos^{-1}\frac{t^* + m^2}{m(1+t^*)} - \frac{\sqrt{m^2 - t^{*2}}}{1+t^*}\right] \\
& + \frac{1+m_t}{m_t(1-m)^2}\Bigg[\frac{\sqrt{m^2 - t^{*2}}}{1+t^*}\left(2 + \frac{t^* + m^2}{1+t^*}\right) \\
& - \frac{1}{\sqrt{1-m^2}}\left(m^2 + 2\frac{t^* + m^2}{1+t^*}\right)\cos^{-1}\frac{t^* + m^2}{m(1+t^*)}\Bigg] \\
& \qquad\qquad - \left.\frac{4(1+m_t)\sqrt{2}}{3m_t\sqrt{m(1+m)}}\left(\frac{m - t^*}{1+t^*}\right)^{3/2}\right\} \qquad (14\text{-}16)
\end{aligned}
$$

The formula is derived for wings with supersonic trailing edges that do not change direction within the tip Mach cones. The symbols are defined in Fig. A,14r.

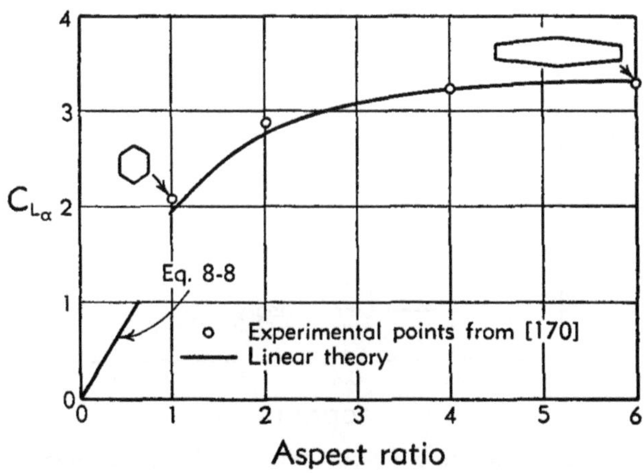

Fig. A,14s. Variation of lift curve slope with aspect ratio for tapered wings with unswept 50 per cent chord line; $\lambda = \frac{1}{3}$, $M_\infty = 1.53$.

A typical variation of lift curve slope with aspect ratio is shown in Fig. A,14s, taken from [*170*]. The curve is for wings with fore-and-aft symmetry (unswept midchord line) and a fixed taper ratio of $\frac{1}{3}$, so that there is considerable variation of the leading edge sweep. Fixing the leading edge sweep and reducing the aspect ratio by cutting off area at the tips give a slightly sharper variation of lift curve slope. When the fifty per cent chord line is unswept there is very little variation of lift curve slope with taper ratio (Fig. A,14t), but it will be seen later (Fig. A,14bb) that this is not generally the case.

Since, on the wings being considered in this section, the leading edge is always outside the zone of influence of the tips, the drag due to angle of attack is calculated exactly as for the pointed wings. If the leading edge is subsonic, the reduction in drag is given by Eq. 14-13; otherwise the drag is merely the product of the lift and the angle of attack. Span load distributions for these wings can be found in [*182*].

Sweptback wings. The preceding paragraphs dealt with hexagonal wings with supersonic trailing edges. At lower speeds the trailing edges of these wings will be subsonic and, as previously pointed out, the mathematical treatment of the problem remains incomplete. Because of the special interest attached to highly swept wings, the inquiry into their characteristics has nevertheless been pushed as far as possible. As a result, the theoretical distribution of lift can be ascertained with sufficient accuracy and completeness for any practical purpose when the planform and Mach number are such that the Mach lines from the trailing edge apex intersect the tips, but not the leading edge (Fig. A,13j). Although the

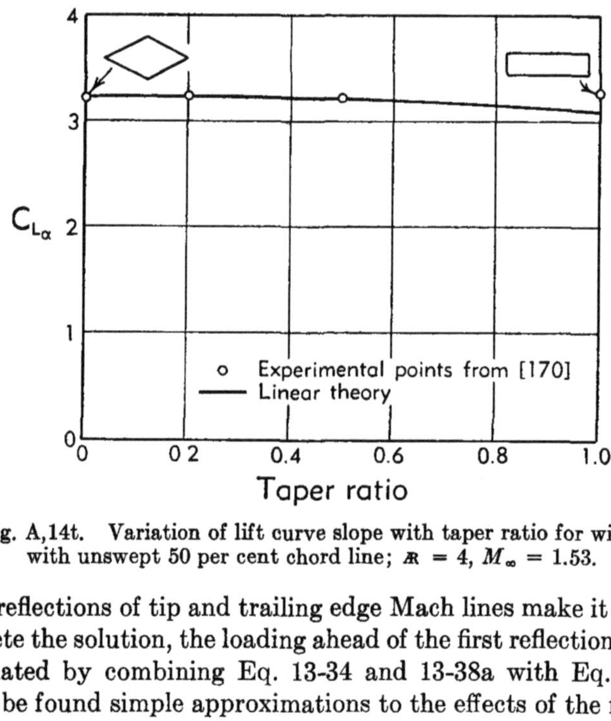

Fig. A,14t. Variation of lift curve slope with taper ratio for wings with unswept 50 per cent chord line; $\mathcal{R}$ = 4, M_∞ = 1.53.

multiple reflections of tip and trailing edge Mach lines make it impossible to complete the solution, the loading ahead of the first reflections is closely approximated by combining Eq. 13-34 and 13-38a with Eq. 13-40. In [*155*] can be found simple approximations to the effects of the first reflections of the Mach lines. Farther to the rear the exact solution would appear to be only of academic interest, in view of the limitations of the theory being employed.

A typical load distribution in the cases described is shown in Fig. A,14u. The distribution is given in more detail for the section at 75 per cent semispan (Fig. A,14v) in order to show the magnitude of the various edge effects. Component 1 corresponds to the first term in Eq. 13-40, the effect of canceling a uniform lift equal to the lift at the root section, all along the trailing edge. Component 2 is the approximate effect of

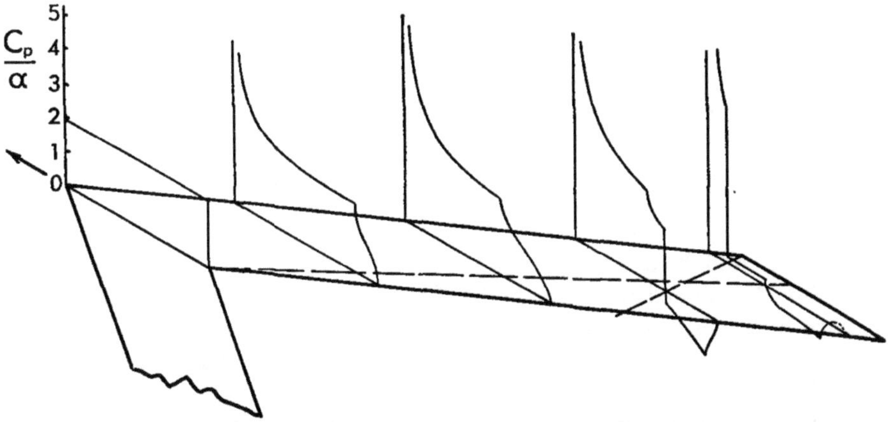

Fig. A,14u. Isometric sketch of the theoretical load distribution on an untapered, sweptback wing with subsonic leading and trailing edges. $\beta \cot \Lambda = 0.6$, $\beta Æ = 1.92$.

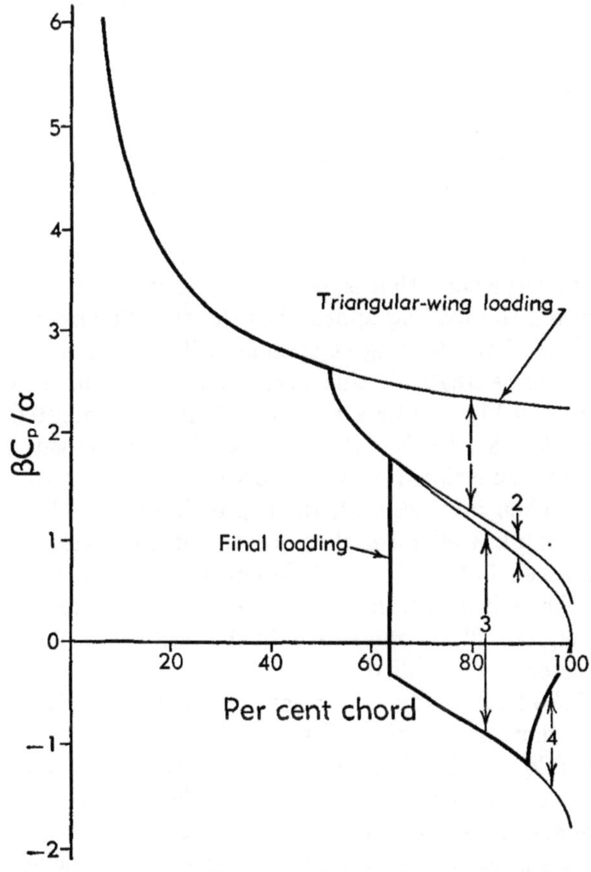

Fig. A,14v. Details of load distribution calculated by linearized theory for section at 75 per cent semispan of wing of the preceding figure.

canceling the remaining lift behind the wing. Component 3 is the primary tip effect (Eq. 13-38a), and component 4 the "reflection" of this effect in the trailing edge. It may be seen that the major portion of the trailing edge effect is given by a single conical field (solution 18 of Table A,13a). The corresponding decrement in total lift is readily found:

$$\frac{\Delta L}{\frac{1}{2}\rho_\infty U^2} = -\frac{4\alpha s^2 m}{m_t E'(m)}\left[1 - \frac{\pi/2}{K'(m_t)}\right] \tag{14-17}$$

where E' and K' are the complete elliptic integrals with complementary moduli. Combined with Eq. 14-16, Eq. 14-17 provides a fairly good estimate of the lift coefficient for a tapered sweptback wing with trailing edge Mach cone intersecting the wing tips. For more complete formulas, see [*155*].

If the trailing edge Mach lines cross the leading edge as in Fig. A,13k, one must begin to look toward the asymptotic cases. Two limiting solutions are available, one for Mach numbers approaching unity, and the other for aspect ratios approaching infinity. Calculated lift distributions show (Fig. A,14w) that, behind the reflection of the trailing edge Mach wave at the leading edge, the loading already resembles very closely that on the two-dimensional flat plate in incompressible flow. If the section is taken spanwise (the chordwise distributions shown in Fig. A,14w may then be considered to be crossplots from spanwise section curves), and a correction factor is introduced to make the two loadings agree in the strength of the singularity at the leading edge, it is found that the agreement is good everywhere; that is, on untapered wings extending infinitely far downstream, the loading approaches the two-dimensional essentially as the strength of the leading edge singularity. Even the tapered wing (Fig. A,14x) may be treated in this way, if the taper is not extreme.

Also shown in Fig. A,14w are lift distributions computed by slender wing theory (Art. 8). In the latter theory, the flow is assumed from the start to be two-dimensional in spanwise sections. Since both branches of the wing are taken into account, the expression for the lift distribution has a more complicated form than for the single plate. However, the calculations made indicate very little effect of the presence of the far wing on the shape of the lift distribution, so that the simpler expression is probably adequate. There remains only the problem of determining the correction factor to be applied.

Ahead of the trailing edge Mach cone, the strength of the leading edge singularity is that on the triangular wing having the same sweep of the leading edge (Eq. 14-11). Behind the shock wave from the trailing edge, the flow is of course modified. The effect on the leading edge singularity may be calculated for some distance downstream of the shock by formulas given in [*155*]. A good approximation to the modified coefficient is

$$C_\Delta(x) + \Delta C(x) = \dot{u}_0\left[\frac{1}{\sqrt{2}} - \frac{4mK(k)Z(\psi, k)}{\pi m_t K'(m_t)\sqrt{1+m}}\right]\sqrt{x} \tag{14-18}$$

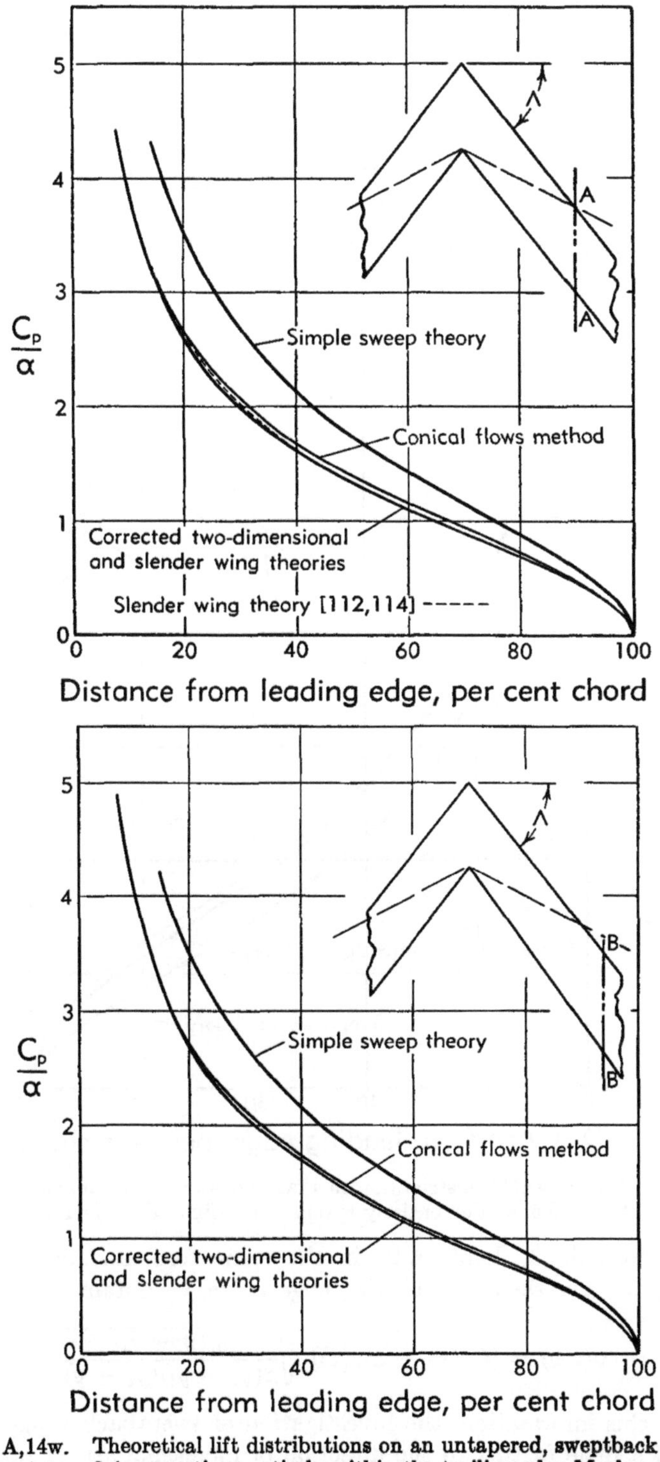

Fig. A,14w. Theoretical lift distributions on an untapered, sweptback wing, $\beta \cot \Lambda = 0.4$, on sections entirely within the trailing edge Mach cone.

in which m_t is β times the cotangent of the angle of sweep of the trailing edge, and, in the elliptic functions K and Z [151],

$$k = \sqrt{\frac{2(y_2 - m_t y_1)}{(1 - m_t)(y_1 + y_2)}}, \qquad \psi = \sin^{-1}\sqrt{\frac{y_1 + y_2}{2y_1}}$$

(see Fig. A,14y for y_1 and y_2). Expressing the two-dimensional subsonic

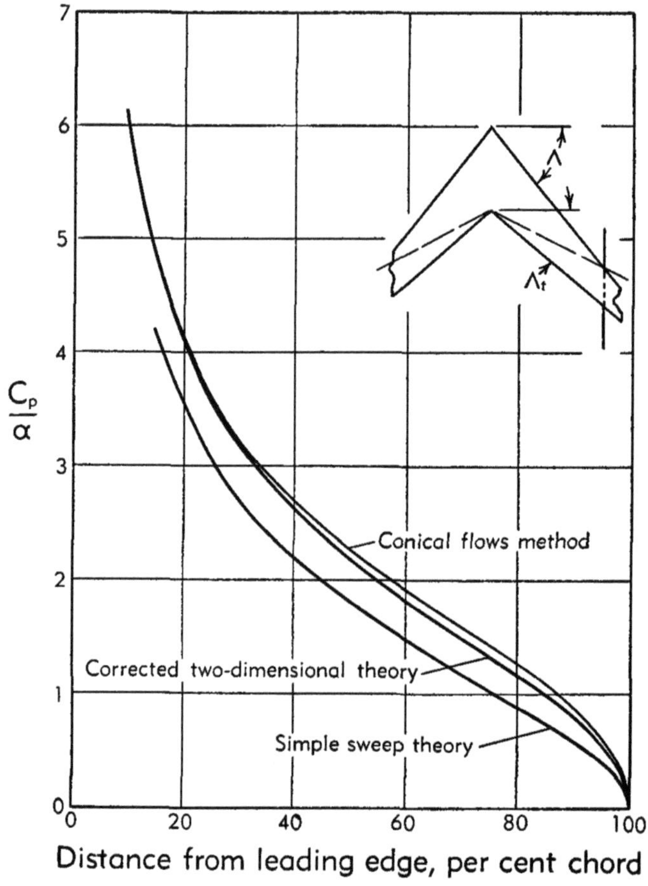

Fig. A,14x. Theoretical lift distribution on a tapered, sweptback wing; $\beta \cot \Lambda = 0.4$, $\beta \cot \Lambda_t = 0.6$; section entirely within the trailing edge Mach cone.

loading (formula 3, Table A,2) in the same form and adjusting the coefficient to give the above leading edge value, we obtain

$$u(x,y) = [C_\Delta(x) + \Delta C(x)]\sqrt{\frac{m(y - y_2)}{\beta(y_1 - y_2)(y_1 - y)}} \qquad (14\text{-}19)$$

With this information, the investigation of sweptback wings can be extended a little further in the direction of increasing aspect ratio. Approximate formulas [*155*] for the loss of lift in the tip region and for the total lift may be derived. Since the strength of the leading edge singularity is known, the leading edge thrust and hence the drag due to lift may be calculated. However, if the wing shown in Fig. A,13k were of somewhat greater aspect ratio, or if the Mach number were slightly lower so that the relative sweep of the wing were greater, the second reflection of the trailing edge Mach wave, now shown intersecting the tip, would again cross the leading edge, and further modification of the leading edge

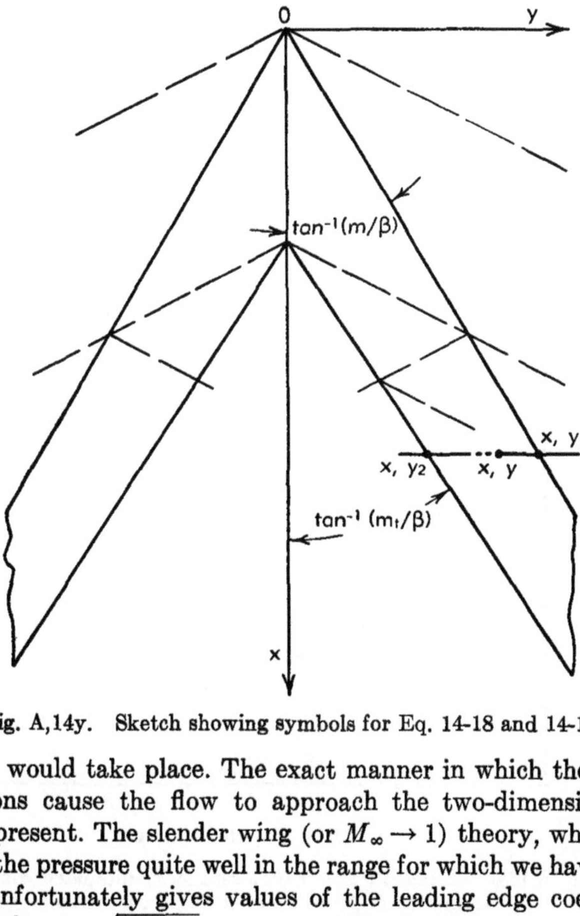

Fig. A,14y. Sketch showing symbols for Eq. 14-18 and 14-19.

singularity would take place. The exact manner in which the successive modifications cause the flow to approach the two-dimensional is not known at present. The slender wing (or $M_\infty \to 1$) theory, which appears to predict the pressure quite well in the range for which we have the exact solution, unfortunately gives values of the leading edge coefficient too low by the factor $\sqrt{1 - m^2}$ as $x \to \infty$. Thus there remains, at the high aspect ratio end of the range, a considerable gap in the theory, as shown by Fig. A,14z and A,14aa, where the lift curve slope and the drag due to lift are plotted against aspect ratio for untapered wings. In these figures, the calculations for $m = 0.2$ and 0.4, from [*155*], have been carried out as far as the current state of the theory permits. The asymptotic values for infinite aspect ratio—that is, $C_D = 0$ and, from simple sweep theory,

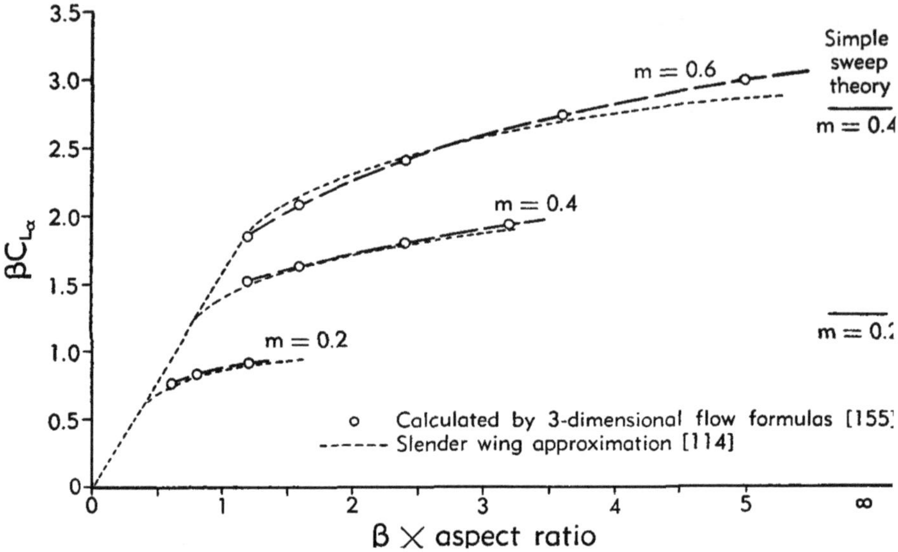

Fig. A,14z. Variation of lift curve slope of sweptback, untapered wings with aspect ratio and Mach number.

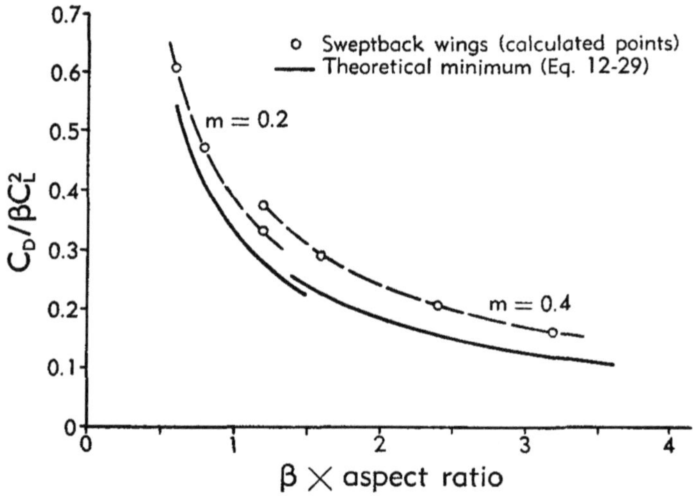

Fig. A,14aa. Drag rise factor for sweptback, untapered wings, as a function of Mach number and aspect ratio.

$\beta C_{L_\alpha} = 2\pi m/\sqrt{1 - m^2}$—are also shown, as well as the slender wing theory values for C_{L_α}. The latter appear to be in remarkable agreement with the results of the three-dimensional theory over a considerable range of aspect ratio, Mach number, and sweep angle. Results for two tapered wings, presented in [*155*], were likewise duplicated within the accuracy of the computations by slender wing theory. However, it should be recognized that these coincidences are not to be generally expected, but are the result of compensating errors, as indicated by the deviation of the curve of C_{L_α} for triangular wings (Fig. A,14l) from the straight line $C_L = \frac{1}{2}\pi A\!\!\!R$.

In the low aspect ratio range, interference between the disturbance fields of the two tips further complicates the calculations and puts a practical limit on the use of the three-dimensional theories. If the leading edge of the wing is highly swept (relative to the Mach lines), the slender wing theories of [*110–114*] will provide solutions of reasonable accuracy. The chief discrepancies will occur in the tip regions, where the slender wing theory result of zero lift is approached by the three-dimensional solution only as M_∞ approaches 1. Thus, if such regions form a large part of the wing planform, or if the trailing edge is swept forward, as in a diamond, a more accurate procedure such as that of [*115*] will be required.

The dotted curves in Fig. A,14z, being for untapered wings, were computed using Mirels' numerical results [*114*] (as were the pressure distributions of Fig. A,14x). Results for other taper ratios have been given by Mangler [*112*] and by Legendre and Eichelbrenner [*113*]. Legendre has fitted the numerical results with a simple formula which, if λ is the taper ratio and μ is the sweep ratio m/m_t, may be written

$$C_{L_\alpha} = \frac{\pi}{2} A\!\!\!R \frac{(1+\lambda)(1-\mu)}{1-\lambda\mu}\left[1 - 0.004(1+6\lambda^2)\frac{\mu-\lambda}{1-\mu}\right], \qquad \lambda < \mu \leqq 1 \tag{14-20a}$$

For untapered wings ($\lambda = \mu = 1$) the formula becomes[21]

$$C_{L_\alpha} = \frac{2\pi A\!\!\!R}{A\!\!\!R + 2\cot\Lambda}(1.028\cot\Lambda - 0.014 A\!\!\!R) \tag{14-20b}$$

$$A\!\!\!R \geqq 2\cot\Lambda$$

Comparison with the results of [*114*], shows Eq. 14-19b to hold through $A\!\!\!R = 6\cot\Lambda$, with a deviation of about 3 per cent at $A\!\!\!R = 8\cot\Lambda$. The cases excluded from Eq. 14-20a and 14-20b are those in which the trailing edge lies entirely in the nonlifting region of the wing, when the formula is simply (Art. 8)

$$C_{L_\alpha} = \frac{\pi}{2} A\!\!\!R$$

Making use of slender wing values (for $M_\infty = 1$) and conical flow formulas, Eichelbrenner [*183*] has computed charts from which the lift

[21] The slender wing formulas, which are in a sense derived for sonic speed and do not show the effect of Mach number, are generalized to Mach numbers slightly different from 1 by writing βC_{L_α} for the left-hand side and substituting for cot Λ and $A\!\!\!R$, $m = \beta\cot\Lambda$ and $\beta A\!\!\!R$, the similarity parameters suggested by the Prandtl-Glauert transformation.

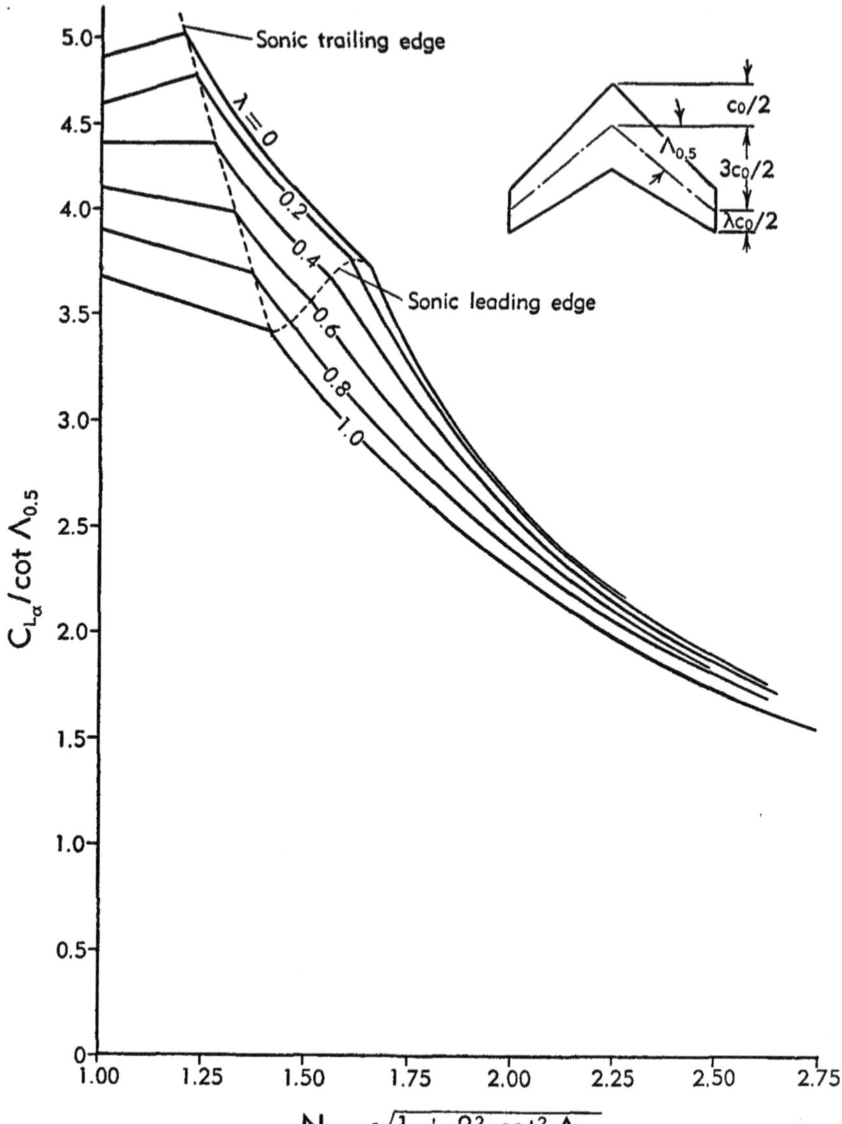

Fig. A,14bb. Chart showing effect of taper on lift curve slope of a family of swept-back wings with fixed depth-to-chord ratio as defined in sketch. One of a set of design charts given in [*183*].

curve slope for a given wing may be determined over the whole Mach number range. In the form shown in Fig. A,14bb, the charts show the effect of taper ratio and Mach number for selected values of a planform parameter involving the aspect ratio and sweep. The departure from the horizontal near $M_\infty = 1$ indicates the error in using slender wing theory for any particular case. However, the compressed scales in M_∞ should be noted.

The drag, according to slender wing theory, is computed from the span loading, plus a term of higher order in β, representing the wave drag, (see Art. 12). From [*113*], the span loading (Fig. A,14cc) is seen to have the form associated with the sweptback wing in subsonic flight, as contrasted with the elliptical span loading predicted by the same theory for wings of convex planform. The curve plotted in Fig. A,14aa as the theo-

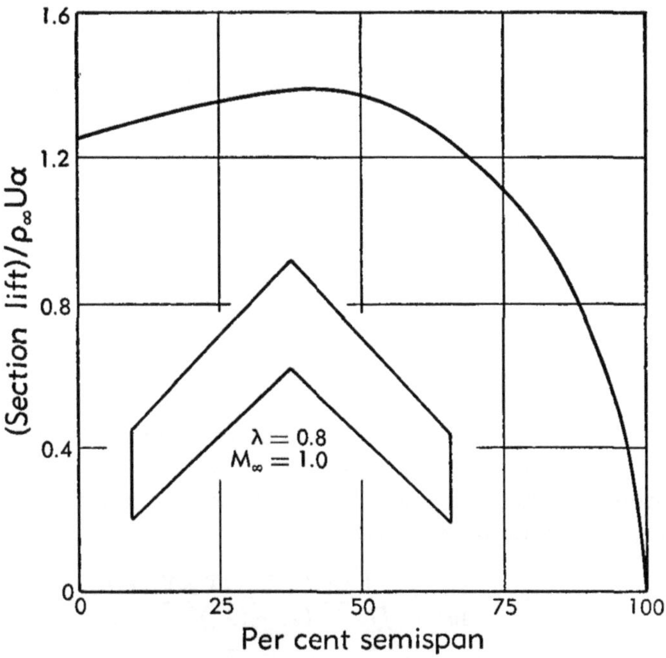

Fig. A,14cc. Theoretical span loading on tapered sweptback wing at a Mach number of 1 [*113*].

retical minimum is based on an elliptical span loading, but includes the wave drag.

A,15. Wave Drag and Pressure Distribution Due to Thickness.

General formulas for wave drag due to thickness. To obtain general expressions for the wave drag due to thickness we make use of the fact that the drag is unchanged by a reversal of the direction of motion (see VI,D,31). After superimposing the perturbation velocities in forward and reversed motion, the drag is obtained from the distribution of pressure in the combined field.

The effect of a symmetrical (nonlifting) distribution of thickness may be obtained by superimposing fundamental solutions of the form which corresponds to the combined disturbance field of a source. The local source strength is proportional to the vertical perturbation velocity w, which is related to the prescribed thickness distribution $t(x_1, y_1)$ by the equation

$$w = \frac{1}{2}\frac{\partial t}{\partial x_1} U \tag{15-1}$$

By following steps similar to those outlined in the lifting case we obtain for the longitudinal and vertical perturbation velocities in the combined flow field:

$$\bar{u} = -\frac{1}{4\pi^2}\int_C \iint_S \frac{w(x_1, y_1)dx_1dy_1}{[\alpha(x - x_1) - \beta(y - y_1) - \gamma z]^2}\frac{\alpha d\lambda}{i\lambda} \tag{15-2}$$

$$\overline{w} = \frac{1}{4\pi^2}\int_C \iint_S \frac{w(x_1, y_1)dx_1dy_1}{[\alpha(x - x_1) - \beta(y - y_1) - \gamma z]^2}\frac{\gamma d\lambda}{i\lambda} \tag{15-3}$$

The value of $\overline{w}$ should of course vanish at each point of the given planform S. The drag will be given by

$$D = \rho_\infty \iint_S \bar{u}w dx dy \tag{15-4}$$

Eq. 15-3 leads directly to the concept of equivalent linear source distributions originally introduced by Hayes [44].

Following the steps outlined in Art. 12, we introduce the variables X, X_1, Y_1, and obtain for Eq. 15-2

$$\bar{u} = -\frac{1}{4\pi^2}\int_C \iint_S \frac{w(X_1, Y_1)dX_1dY_1}{(X - X_1)^2}\frac{\alpha d\lambda}{i\lambda} \tag{15-5}$$

For $\lambda = e^{i\theta}$, the lines X = const correspond to the intercepts of a system of plane waves at 45° to the x axis ($M_\infty = \sqrt{2}$) and at an angle θ in the y, z plane. Eq. 15-5 then yields the three-dimensional disturbance field by the superposition of elementary plane waves at varying angles θ around the x axis. Inspection of Eq. 15-5 shows that at a particular value of θ the influence of an element of thickness w depends only on the relative position of its intersecting plane wave front, and is independent of its position Y_1 in the plane of the wave. An integration over Y_1 thus corresponds to the collection of sources at a single point in each plane wave, and, for the whole system, results in an equivalent linear source distribution of strength

$$\sigma'(X_1, \theta) = \frac{2}{U}\int w(X_1, Y_1)dY_1 \tag{15-6}$$

for each angle θ (see Fig. A,15a). Then the double integral over S in Eq.

15-5 may be written

$$\frac{U}{2}\int_{-X_s}^{+X_s}\frac{\sigma'(X_1,\theta)dX_1}{(X-X_1)^2} \tag{15-7}$$

which is the form of the integral for the induced downwash in the Prandtl wing theory. Following a procedure employed in that connection [*140*, pp. 135–140], we may expand the function $\sigma'(X, \theta)$ in the Fourier series

$$\sigma'(X,\theta) = \sum_1^\infty A_n \sin n\varphi \tag{15-8}$$

where $\varphi = \cos^{-1}(X/X_s)$ and the coefficients A_n are functions of θ. Then, as in lifting line theory, we may introduce Eq. 15-8 into 15-7 and obtain

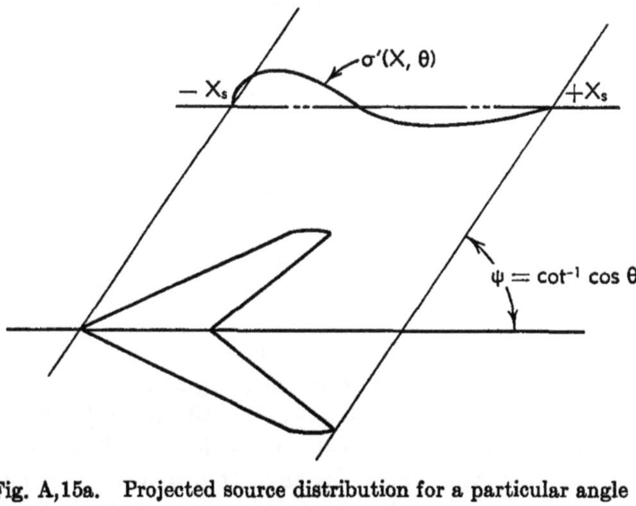

Fig. A,15a. **Projected source distribution for a particular angle θ.**

for the drag of each elementary linear source distribution

$$\frac{dD}{d\theta} = \int_{-X_s}^{X_s}\int_{-X_s}^{X_s}\frac{\sigma'(X,\theta)\sigma'(X_1,\theta)dXdX_1}{(X-X_1)^2} = \frac{\pi^2}{2}\sum_1^\infty n[A_n(\theta)]^2 \tag{15-9}$$

Finally after replacing $\alpha d\lambda/i\lambda$ by $d\theta$ in Eq. 15-5 we obtain

$$D = \frac{\rho_\infty U^2}{8}\int_0^\pi \sum_1^\infty n[A_n(\theta)]^2 d\theta \tag{15-10}$$

The coefficients A_n may be found from Eq. 15-8. Integration shows that the coefficient A_1 is given by

$$A_1 = \frac{2}{\pi}\frac{\sigma(X_s)}{X_s} \tag{15-11}$$

where X_s is of course a function of θ and the quantity

$$\sigma(X_s) = \int_{-X_s}^{X_s} \sigma' dX$$

is the "base area" of the wing. Hence A_1 will be zero if the wing is a finite closed body. Assuming $A_1 = 0$, we obtain for A_2

$$A_2 = -\frac{4}{\pi}\frac{\text{volume}}{X_s^2} \tag{15-12}$$

Higher terms of the series contribute nothing to the volume, but do contribute to the drag. We may thus obtain a lower bound for the drag with a given volume by suppressing higher terms of the series.

Relation of wave drag to area distribution. The foregoing equations represent the wave drag of a planar source distribution in terms of a series of equivalent linear source distributions. Thus, by making use of the well-known relation between source strength and cross-sectional area, the drag associated with the thickness of a wing may be obtained as the average of the values for a series of equivalent bodies of revolution.

As the Mach number approaches 1 the planes X = const approach a single set of planes perpendicular to the flight direction. As remarked by Hayes [*44*], the drag of a wing, or of a wing-body system, then approaches the drag of a single equivalent body of revolution having the same longitudinal distribution of area.

It is interesting to note that this formal result of linear theory has been verified in a striking way by recent developments in transonic theory and experiment. On the theoretical side, Oswatitsch [*184*] has shown that the nonlinear transonic potential field of a slender wing-body combination approaches, with increasing radial distance, that of an equivalent body of revolution. In the vicinity of the wing or body the potential fields differ by a two-dimensional, laterally incompressible flow potential, as in the theory of Jones [*105*] or Ward [*185*].

Experiments made by Whitcomb [*186*] at transonic speeds show that the incremental drag of a wing-body combination follows rather closely the drag of the equivalent body of revolution, and furthermore that the drag of the complete airplane can be reduced considerably by re-shaping the fuselage so that the whole system presents a smooth area distribution along the longitudinal axis.

Wing shapes of minimum drag. Examples of wing shapes for which the wave drag due to thickness is a minimum are naturally of considerable interest to the wing designer. Such examples provide an indication of efficient wing shapes, and at the same time yield approximate engineering formulas for the drag of a variety of shapes which do not differ too greatly from the optimum.

The term minimum drag as used herein refers of course to the drag arising from pressure forces, exclusive of the friction. In practice, a balance must be achieved between the pressure drag and the friction drag. For present purposes, the friction drag is taken to be proportional to the wetted area of the wing surface. With a wing of given depth or thickness, increasing the chord diminishes the slope of the surface, and hence, the pressure drag; but the increased area results in a greater friction drag. Therefore the problem of minimizing the pressure drag for a wing having the smallest possible area is of interest.

Conditions for the drag to have the minimum value under certain auxiliary assumptions may be stated in terms of the pressure distribution in the combined flow field, given by

$$\overline{\Delta p} = \rho_\infty \bar{u} U \tag{15-13}$$

The demonstration of these conditions involves a repetition of the reasoning employed in connection with the minimum drag of lifting surfaces (Art. 6 and [*187*]).

If the volume of the wing is specified, then the thickness distribution having the smallest wave drag must yield

$$\frac{\partial \overline{\Delta p}}{\partial x} \quad \text{or} \quad \frac{\partial \bar{u}}{\partial x} = \text{const} \tag{15-14}$$

over the entire planform S. Since the pressure gradient $\partial \overline{\Delta p}/\partial x$ is the drag per unit volume, minimum drag requires that the drag per unit volume be constant.

Similarly, if a line is drawn on the wing planform and if the thickness distribution along this line is required to have a specified projected or frontal area, then it is found that $\overline{\Delta p}$ must have the same value at all points of the planform ahead of this line and again a constant value (perhaps different) at all points behind this line. In this case it will be observed that the drag per unit frontal area is constant in the superimposed disturbance field.

If the auxiliary assumptions are made more restrictive, then the condition for the drag to be a minimum becomes, of necessity, less restrictive. Thus, if the volume of the wing is specified but all sections along the span are required to have the same shape, then the streamwise gradient of the pressure must be permitted to vary along the span. The condition for minimum drag is then

$$\frac{\partial}{\partial x} \int_S \bar{u} dy = \text{const} \tag{15-15}$$

Minimum wave drag for elliptic planform. If the planform of the wing is elliptical, the formulas (15-8) and (15-10) for the drag are considerably simplified. In particular, if the sections are assumed to be parabolic arcs such that

$$\frac{\partial t}{\partial x} = \text{const} \cdot x$$

over the whole planform, then the equivalent source distribution will be given by Eq. 15-18, with $n = 2$. This particular thickness distribution—which gives what might be termed an "elliptic lens"—thus yields the minimum drag consistent with a given volume. A similar parabolic biconvex section shape also yields the minimum drag in two-dimensional flow. With the elliptical planform, however, the thickness-chord ratio becomes smaller near the tips. The drag coefficient of the elliptic lens at $M_\infty = \sqrt{2}$ is given by

$$C_{D_{\min}} = \frac{t_{\max}^2}{a^2} \frac{1}{\sqrt{1 + \frac{a^2}{b^2}}} \left(1 + \frac{\frac{a^2}{b^2}}{1 + \frac{a^2}{b^2}} \right) \tag{15-16}$$

where a and b are the semiaxes of the ellipse.

Fig. A,15b shows the variation of drag coefficient with Mach number for elliptic wings of three different axis ratios. The values were obtained from Eq. 15-16 after applying the Prandtl-Glauert transformation. With the lowest aspect ratio the variation of drag with Mach number is almost completely suppressed.

It is of interest to compare the foregoing results with the results for a body of revolution designed to contain the same volume within the same specified length. The shape of the body and the minimum value of the drag coefficient in this case have been worked out by Sears [*188*] and Haack [*189*]. It may be imagined that the elliptic figure is derived by flattening the body of revolution. At first the flattening produces only a second order effect on the drag, but eventually the wave drag diminishes appreciably. If the body is flattened into a circular planform the wave drag (at $M_\infty = \sqrt{2}$) for the given volume and length is diminished to one half. Of course the wetted area and the skin friction are greatly increased by this process.

The drag for a given volume could be further reduced by increasing the value of X_s. Since X_s is one half the projected maximum dimension of the wing planform, this process would involve spreading the volume out over a larger area. However, since the projections involve only those angles lying between the forward and rearward Mach characteristic angles, it is clear that no value of $X_s(\theta)$ can be increased by additions to the planform if such additions are restricted to the interior of the characteristic envelope of the wing (see Fig. A,15c). Thus the drag of the elliptic lens, or of any system of sources for which Eq. 15-10 holds, cannot be diminished by any redistribution of volume elements within its characteristic envelope.

Calculated wave drags. It is of course possible to calculate the wave drag of a particular wing by the integration of the theoretical pressures on the surface. The determination of these pressures, in the case of a thin symmetrical airfoil at zero lift, is a problem that lends itself readily

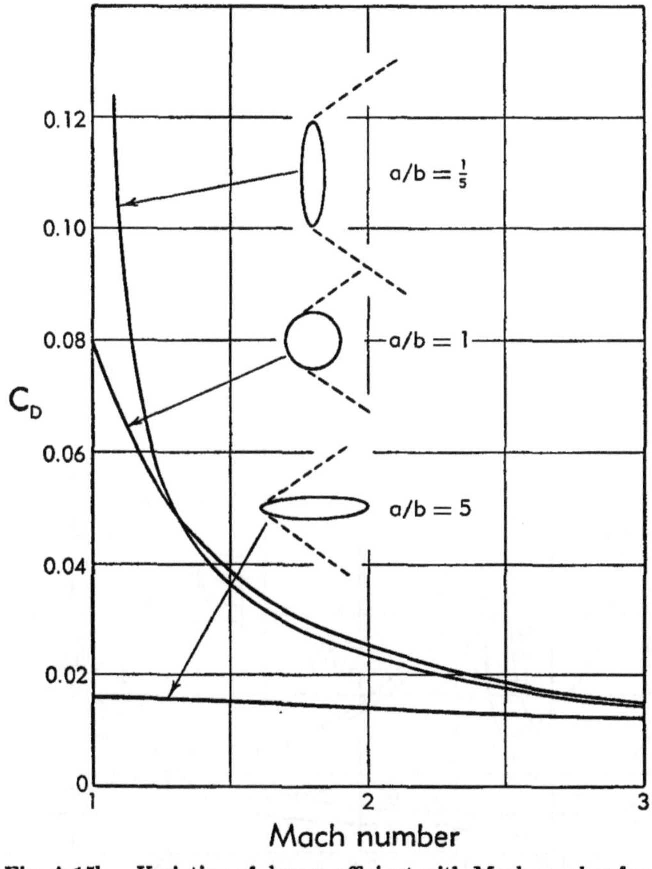

Fig. A,15b. Variation of drag coefficient with Mach number for elliptic wings with biconvex sections, $t/c = 0.1$, at zero lift.

to analysis by the linearized theory. For many simple wing shapes the entire process can be carried out analytically. For example, complete results for wings of triangular planform with double-wedge sections were early given by Puckett and Stewart [*172*]. Swanson and Harmon [*190,191*] present formulas and charts for untapered wings of biconvex section. Bismut [*192*] treats the "swallowtail" wing, including results for wings with parabolic arc section, Chang [*193*] and Nielsen [*194*] consider tapered wings with diamond profile (that is, maximum thickness at 50 per

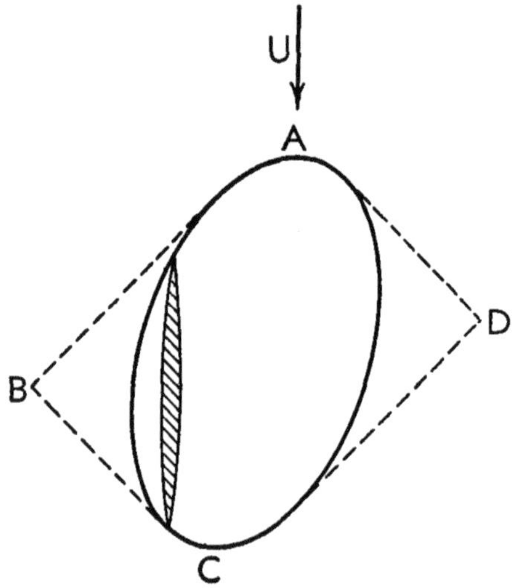

Fig. A,15c. Ideal distribution of thickness for area *ABCD*, given volume.

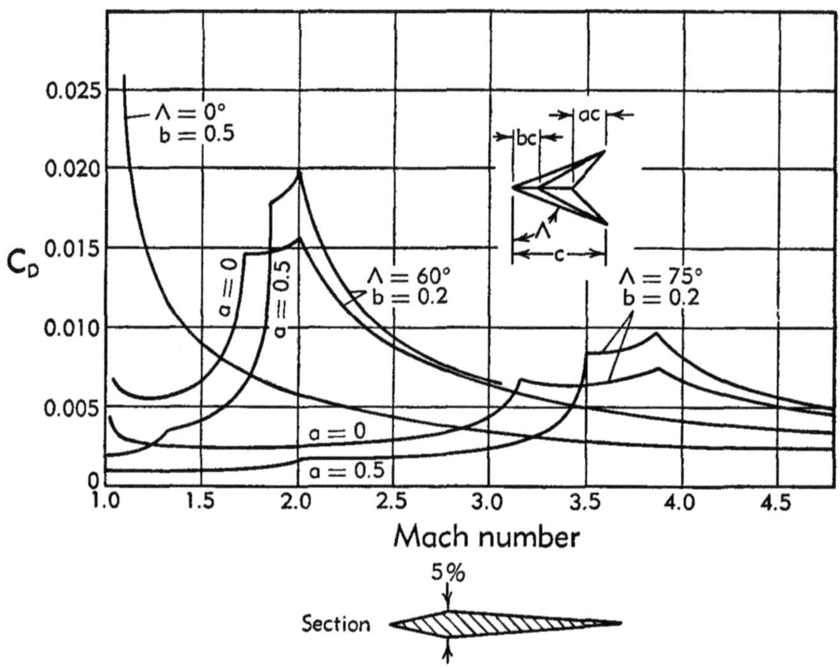

Fig. A,15d. Variation of drag coefficient with Mach number for "swallowtail" wings with double-wedge sections, at zero lift [*172*].

cent chord), and Margolis [*195,196,197*] covers the double-wedge profile with varying location of the maximum thickness line. Most of these results have been summarized by Lawrence [*198*] in the form of design charts.

Some typical results of these calculations are reproduced in Fig. A,15d, A,15e, and A,15f. A characteristic feature of all such calculations for wings of rectilinear planform is the occurrence of sharp peaks in the variation of wave drag with Mach number. Such peaks arise when the Mach lines coincide with bends in the stream surface, which ordinarily

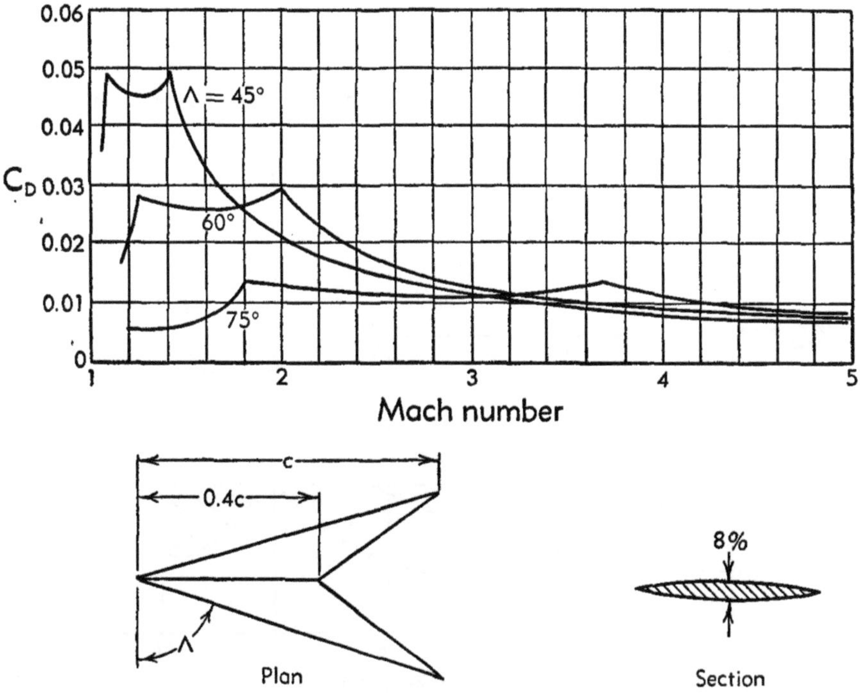

Fig. A,15e. Variation of drag coefficient with Mach number for "swallowtail" wings with biconvex sections, at zero lift [*192*].

occur at the leading and trailing edges of the wing. The linearized theory, of course, displays the Mach lines in a perfectly rectilinear pattern, undistorted by superimposed variations in the local relative velocity. In the vicinity of a leading edge or a trailing edge the disturbance velocity becomes very large—of the order of the stream velocity—so that this coincidence of the Mach lines can hardly be expected to occur in practice. Furthermore, the effect of friction will lead to a smoothing of the effective contour of the airfoil at the trailing edge, so that a sharp bend of the stream surface in this region will not occur.

A situation similar to the foregoing arises in the prediction of the wave resistance of ships [*199*]. Here, calculations based on a linearized theory, disregarding friction, show a succession of peaks in the variation of wave resistance with speed, the peaks occurring when multiples of the wave length coincide with the length of the ship. By supposing a cusp-shaped extension added to the ship, smoothing out the lines of the stern to represent the probable effect of the boundary layer, the peaks in the wave resistance can be eliminated from the calculations and the results can be brought into better agreement with experimental values [*199*].

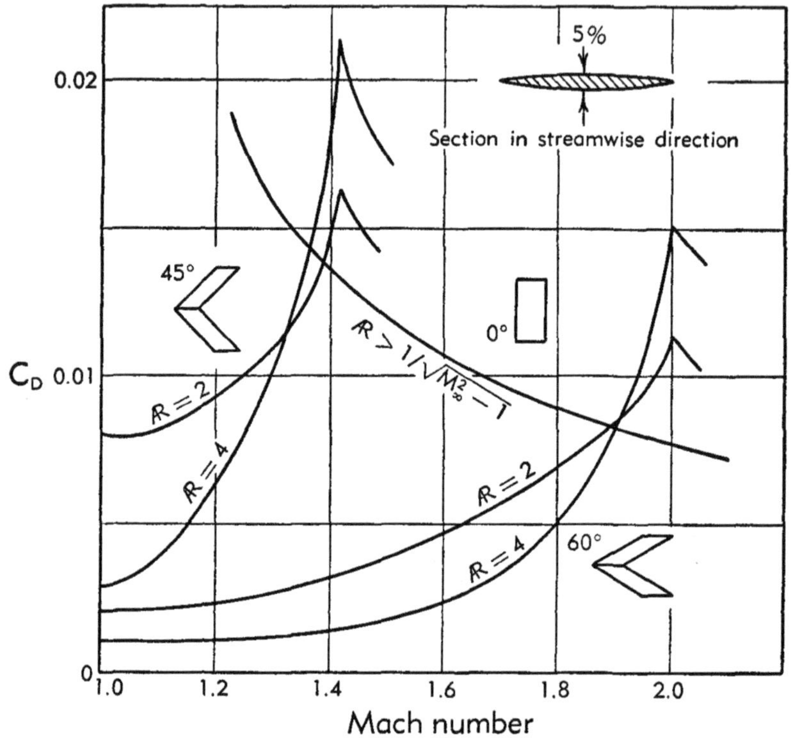

Fig. A,15f. Variation of drag coefficient with Mach number for straight and sweptback wings with biconvex sections, at zero lift [*190*].

Calculated pressure distributions for sweptback wings. Perhaps the most direct approach to the problem of determining the pressure distribution over symmetrical nonlifting airfoils is the method of source distributions (see Art. 13) suggested by von Kármán in his 1935 Volta Congress paper [*200*]. For airfoils composed of cylindrical or conical surfaces it is convenient to consider the sources to be distributed along the generating elements of the surface. If the source strength is held constant along each line, the distribution of source strength can then be represented in terms of a single variable. This simplification makes possible the treatment of fairly general section shapes.

A similar procedure was employed in Art. 9 in treating the corresponding problem in subsonic flow. The application to supersonic flow is found in [*127*]. As in Art. 9, the fundamental solution (10-10) may be used to represent the longitudinal velocity component u rather than the velocity potential φ. Then to get the effect of a line of sources of uniform density along the x axis we merely integrate the expression for the fundamental solution:

$$\frac{u_0}{U}(x, y, z) = -\int_0^{\xi} \frac{dx_1}{\sqrt{(x - x_1)^2 - y^2 - z^2}}$$

$$= -\cosh^{-1}\frac{x}{\sqrt{y^2 + z^2}} \tag{15-17}$$

letting ξ represent the location of the last source whose Mach cone includes the points x, y, z, i.e.

$$\xi = x - \sqrt{x^2 + z^2}$$

Such a distribution of sources along the x axis satisfies the boundary condition for a circular cone, and corresponds to the solution for that case given by von Kármán and Moore [*201*]. The solution for an oblique line source may be derived by integrating to obtain the effect of a row of sources along the oblique line $y = mx$. It is more convenient, however, to make use of the transformation to oblique coordinates discussed in Art. 13, that is,

$$\left.\begin{aligned} x' &= x - my \\ y' &= y - mx \\ z' &= \sqrt{1 - m^2}\, z \end{aligned}\right\} \tag{15-18}$$

The expression for the oblique line source then becomes, for $m < 1.0$,

$$\frac{u_0}{U} = -\cosh^{-1}\frac{x'}{\sqrt{y'^2 + z'^2}} \tag{15-19}$$

and for $m > 1.0$,

$$\frac{u_0}{U} = -\cos^{-1}\frac{x'}{\sqrt{y'^2 + z'^2}} \tag{15-20}$$

Since u is proportional to the pressure through the relation

$$\frac{\Delta p}{\frac{1}{2}\rho_\infty U^2} = -2\frac{u}{U}$$

these equations describe the pressure field associated with the source line. The vertical velocity w near $z = 0$ determines the shape of the boundary and may be obtained by integrating u with respect to x and then differentiating the resulting velocity potential with respect to z. Thus

$$w = \frac{\partial \varphi}{\partial z} = \frac{\partial}{\partial z} \int u dx \tag{15-21}$$

When this operation is carried out for either Eq. 15-19 or 15-20 it is found that w near $z = 0$ is zero everywhere except over the radial sector of the plane in which $y' < 0 < y$ (see Fig. A,13c). Over this area w has a constant value, given by

$$\frac{w_0}{U} = \pm \pi \frac{\sqrt{1 - m^2}}{|m|} \tag{15-22}$$

The constant value of w/U implies a constant value of $\partial z/\partial x$, which in this case takes on equal and opposite values on the upper and lower sides of the chord plane. The solutions (15-19) and (15-20) thus satisfy the boundary condition for a thin oblique wedgelike body having its leading edge along the line $y' = 0$. The relation between the thickness and the magnitude of the pressure disturbance may be obtained by introducing the relation

$$\frac{\partial z}{\partial x} = \frac{w}{U}$$

together with Eq. 15-22. The result is

$$\frac{\Delta p}{\frac{1}{2}\rho_\infty U^2} = \begin{cases} \dfrac{2}{\pi} \dfrac{\partial z}{\partial x} \dfrac{m}{\sqrt{1 - m^2}} \cosh^{-1} \dfrac{x'}{\sqrt{y'^2 + z'^2}}; & \text{for } |m| < 1 \\ \dfrac{2}{\pi} \dfrac{\partial z}{\partial x} \dfrac{m}{\sqrt{m^2 - 1}} \cos^{-1} \dfrac{x'}{\sqrt{y'^2 + z'^2}}; & \text{for } |m| > 1 \end{cases} \tag{15-23}$$

For a bilaterally symmetrical wedge, we may superimpose the foregoing solution and its mirror image in the x axis, since in each solution w is zero in the portion of the x, y plane occupied by the other. The terms for the left-hand wedge are obtained by replacing m by $-m$ in Eq. 15-18 and 15-23.

Since the flows produced by the wedge-shaped figures are conical in form, they may also be derived by the conical flow method described earlier. To satisfy the boundary condition for a bilaterally symmetrical wedge we write

$$w_0 = \text{R.P.} \frac{i}{\pi} \ln \frac{\tau + m}{\tau - m} \tag{15-24}$$

where τ is the conical variable defined by Eq. 13-1 and 13-2. The real part of this function (Fig. A,13b) shows the desired behavior, yielding the discontinuous value

$$w = \pm 1$$

between the points $\tau = -m$ and $+m$.

The corresponding values of u_0 and v_0 are found by integrating the relations (13-3) and (13-4). The integration yields, for $m < 1.0$,

$$u_0 = \text{R.P.} \frac{-m}{\pi \sqrt{1 - m^2}} \left(\cosh^{-1} \frac{1 - m\tau}{m - \tau} + \cosh^{-1} \frac{1 + m\tau}{m + \tau} \right) \quad (15\text{-}25)$$

$$v_0 = \text{R.P.} \frac{1}{\pi \sqrt{1 - m^2}} \left(\cosh^{-1} \frac{1 - m\tau}{m - \tau} - \cosh^{-1} \frac{1 + m\tau}{m + \tau} \right) \quad (15\text{-}26)$$

The expression for u_0 is seen to correspond to the superposition of two (right- and left-hand) disturbance fields of the form 5, Table A,13a; the

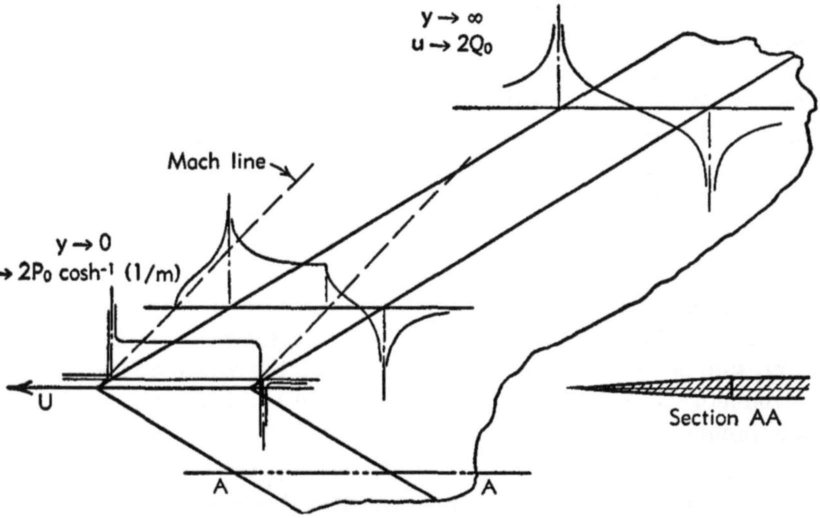

Fig. A,15g. Variation of pressure distribution along beveled edge. $M_\infty = \sqrt{2}$; $m = \tan 30°$ [*127*].

latter can, in turn, be shown to reduce to Eq. 15-19, except for the constant coefficient. The corresponding expression for $m > 1.0$ is solution 3 of Table A,13a.

Airfoils of rectilinear planform bounded by a finite number of sloping plane surfaces, can be obtained by superimposing a finite number of solutions of the form (15-19) or (15-20). This process is described in detail in [*127*] and follows essentially the procedure described in Art. 9.

Fig. A,15g illustrates the result of superimposing two solutions of the form (15-25) with the origins displaced along the x axis. The figure obtained is the beveled leading edge of a plate of uniform thickness. The sweep of the leading edge in this example is 60° while the Mach number is $\sqrt{2}$, so that the normal component of the Mach number is less than 1.0. As shown in the figure, the pressure distribution over the sections approaches the subsonic form (see formula 19 of Table A,2) with increasing distance from the root section. At the root section the pressure is constant along the chord, as given by the Ackeret theory for a straight wing with wedge sections. Inspection of Eq. 15-25 at the limit $y = 0$ shows that the effect of sweep is to reduce the pressures at the root section by the factor

$$\frac{2}{\pi}\frac{m}{\sqrt{1-m^2}}\cosh^{-1}\frac{1}{m} \tag{15-27}$$

relative to the value given by the Ackeret theory.

From the practical standpoint the solutions of most interest are those for airfoils having continuously curved surfaces. Curved surfaces require a continuous distribution of line sources or sinks parallel to the generators of the wing surface. As in the analogous development for subsonic speeds, the simplest analytic example is the untapered wing having upper and lower surfaces formed by parabolic arcs. As a starting point, a streamline surface of parabolic shape may be obtained by integrating the solution (15-25) along x. To obtain the integral of Eq. 15-25, we make use of the general relation (Euler's formula)

$$(n+1)\varphi = xu + yv + zw \tag{15-28}$$

valid for any homogeneous potential field of order n in the velocities u, v, and w. For the conical field, $n = 0$ and, since $\varphi = \int u dx$, the right-hand side of Eq. 15-28 yields the integral of the solution (15-25). After introducing the expressions (15-25, 15-26, and 15-24) for u, v, and w there is obtained

$$u_1 = \int u_0 dx = \frac{1}{\pi\sqrt{1-m^2}}\left[(y - mx)\cosh^{-1}\frac{1-m\tau}{m-\tau} - (y + mx)\cosh^{-1}\frac{1+m\tau}{m+\tau} - i\sqrt{1-m^2}\, z \ln\frac{\tau - m}{\tau + m}\right] \tag{15-29}$$

Eq. 15-29 yields the horizontal perturbation velocity (and hence the pressure) for a parabolically curved surface whose ordinates are obtained by integrating the ordinates of the wedge along the x direction.

The pressure field for the sweptback wing with biconvex sections may now be obtained by superimposing the solutions (15-25) and (15-29) in the following way: If the root chord of the wing lies between $x = -1$ and $x = +1$ and if we use $u(\pm 1)$ to denote a disturbance field with its origin displaced to the point $x = \pm 1$, then the nose angle of the wing section is obtained by a solution $u_0(-1)$, and the curvature of the surface by $u_1(-1)$. Two additional solutions are required at $x = +1$ to terminate the wing. The resulting disturbance field may be represented by the sum

$$u_0(-1) - u_1(-1) + u_1(+1) + u_0(+1) \tag{15-30}$$

This summation yields a relatively simple expression for the pressure coefficient in the plane of the wing:

$$\frac{\Delta p}{\frac{1}{2}\rho_\infty U^2} = \frac{4}{\pi}\frac{t}{c}\frac{m}{\sqrt{1-m^2}}\left[\frac{y'}{m}\left(\cosh^{-1}\frac{x'+1}{|y'-m|} - \cosh^{-1}\frac{x'-1}{|y'+m|}\right.\right.$$
$$\left.\left. + \frac{\bar{y}'}{m}\cosh^{-1}\frac{\bar{x}'-1}{y'-m} - \cosh^{-1}\frac{\bar{x}'+1}{|y'+m|}\right)\right] \quad (15\text{-}31)$$

The pressure distribution over the same wing at subsonic speeds has been given in Art. 9, Eq. 9-19.

Fig. A,15h shows the distribution of pressure as given by Eq. 15-31 at various sections along the span for the case of a wing with 60° sweep at a Mach number $M_\infty = \sqrt{2}$ ($m = 0.577$). These calculations illustrate the nature of the transition from the supersonic, or Ackeret, type of pressure distribution at the root to the subsonic form at sections farther outboard. Comparison with Fig. A,9e shows that the same trends exist in a qualitative sense for the same wing at subsonic speeds. At supersonic speeds, however, the pressure distribution at the root section is identical in form with that given by the Ackeret theory except that the pressures are reduced in the ratio (15-27). This statement holds for the root section of any sweptback wing formed by joining two cylindrical surfaces, provided the root section remains ahead of the zone of influence of the tip sections.

An interesting feature of these calculations is the characteristic sharp compression, or pressure recovery, which occurs along the wave emanating from the trailing edge of the root section. The linearized theory shows an infinite gradient in the pressure behind this wave, but of course in practice the wave will form a shock of finite strength. As in the lifting case (Art. 14), transition to the two-dimensional subsonic type of pressure distribution is essentially complete at the point where this compression wave crosses the leading edge of the wing. At sections beyond that point, the pressure distribution is approximated very closely by the velocity function 20 of Table A,2 after correction for the transverse velocity.

Spanwise distribution of wave drag of sweptback wings. It is apparent from the form of the pressure distributions that the drag of the untapered sweptback wing is concentrated near the root sections, or the foremost sections. The variation of the drag of the individual sections with distance from the root is shown in Fig. A,15i. In the case shown ($m = 0.577$) the drag of the root section is about half the value for a similar section of a straight wing. The section drag falls to zero at the point where the compression wave from the trailing edge crosses the line of maximum thickness. Sections beyond this point have essentially zero pressure drag, so that additions to the span of a wing beyond this point can be made without additional pressure drag.

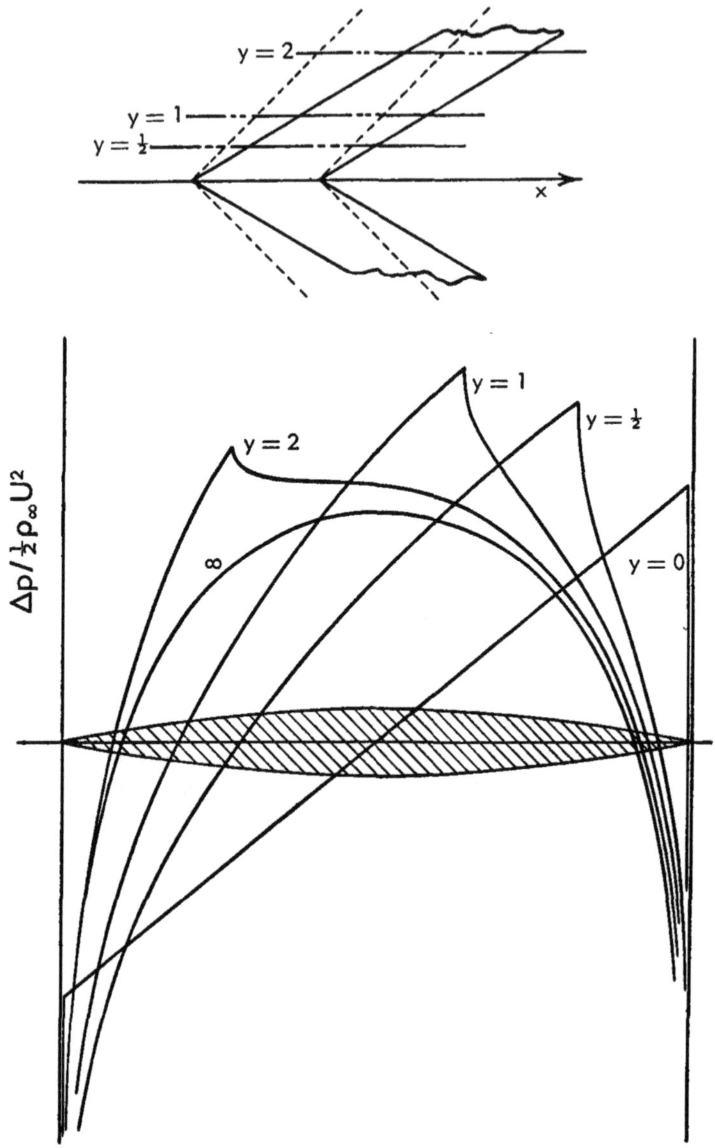

Fig. A,15h. Distribution of pressure at different sections along the span of infinite wing; $M_\infty = 1.4$; $\Lambda = 60°$; biconvex sections.

Although the discussion has been based on the results for an infinite wing, the effect of cutting the wing off at some point to form a tip can be foreseen from the forms of the elements used in the superposition. Thus it can be seen that the pressure distribution at the extreme tip will be composed of two curves (assuming that the wing is long enough to avoid multiple interaction between root and tip), one of which will be the pressure on the infinite wing and the other equal to one half the pressure of the root section but with the sign reversed. The extreme tip section thus has a negative drag, although it may be shown [*190*] that the total drag increment associated with the tip region is zero for an untapered wing.

Fig. A,15j, taken from [*190*], shows the variation of section drag coefficient along the span for three different values of the Mach number

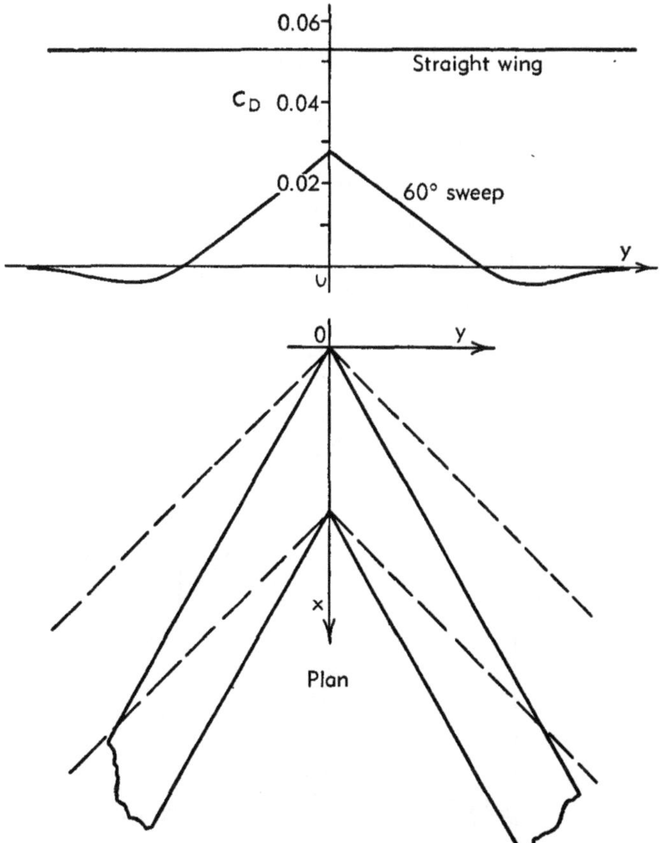

Fig. A,15i. Spanwise distribution of drag of straight and sweptback wings, 10 per cent thick biconvex sections.

in the case of a wing with biconvex sections and 45 degrees of sweep. At the lowest value of the Mach number ($M_\infty = 1.1$) the wing is well inside the Mach cone and the pressure drag falls to zero rapidly on proceeding away from the center section. At $M_\infty = 1.34$ the Mach line lies close to the leading edge and the drag is then nearly constant over the portion of the wing shown. At $M_\infty = 1.41$ the Mach lines coincide with the generators of the wing surface, corresponding to a transverse Mach number of 1. In this case the section drag coefficient approaches infinity as the flow approaches the two-dimensional form at a great distance from the root section. As illustrated by the figure, the actual rate of increase is quite slow, the drag coefficient being ultimately proportional to the square root of the distance from the apex of the wing.

Influence of section shape on the wave drag of swept wings. Since the wave drag of a long narrow wing, swept behind the Mach cone, is concentrated in the vicinity of the root section, von Kármán [*43*] has proposed that the wave drag of an infinite swept wing be used as an approximation to the drag of a long narrow wing of finite dimensions.

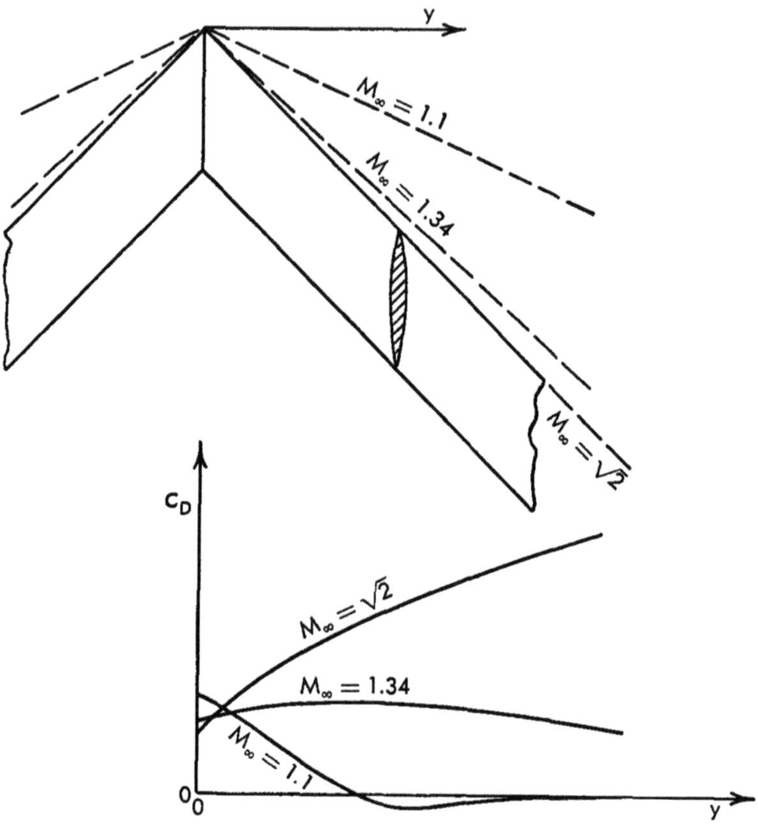

Fig. A,15j. Spanwise distribution of drag at different Mach numbers [*190*].

Assuming the wing to consist of two cylindrical pieces joined at the root section with the sweepback angle Λ, we find that the equivalent linear source distribution $\sigma'(X, \theta)$ (see Art. 15) has the same form as the distribution of thickness along any streamwise section (see Fig. A,15k). While remaining similar to the section shape, the amplitude of the equivalent source distribution varies with θ in a manner dependent on the sweep angle.

In this case the coefficients A_n in Eq. 15-8 will maintain the same ratio at all values of θ and the drag can be expressed directly in terms of the wing section shape. Supposing a streamwise section of the wing to lie between the points $-c/2$ and $+c/2$ and setting

$$\frac{x}{c/2} = \cos\phi$$

we may write

$$\frac{t(x)}{t(0)} = \sum a_n \sin n\phi \tag{15-32}$$

where $t(x)$ is twice the ordinate of the (symmetrical) wing section at x. Eq. 15-10 yields for the drag

$$D = \pi\tfrac{1}{2}\rho_\infty U^2[t(0)]^2 f(M_\infty, \Lambda) \sum_1^\infty n a_n^2 \tag{15-33}$$

where

$$f(M_\infty, \Lambda) = \frac{\cos^2\Lambda}{\sin\Lambda}\,\frac{1 + \sin^2\Lambda - M_\infty^2\cos^2\Lambda}{(1 - M_\infty^2\cos^2\Lambda)^{\frac{3}{2}}} \tag{15-34}$$

Evaluation of Eq. 15-10 for the root section of the sweptback wing alone yields half the value given above. However, as pointed out by Hayes

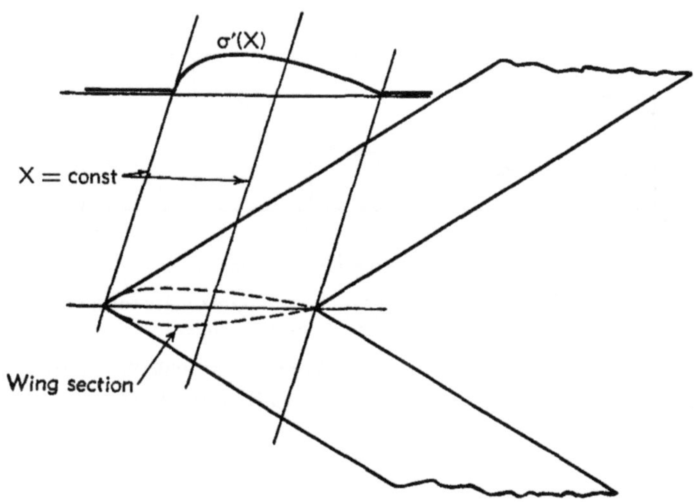

Fig. A,15k. Equivalent source distribution for sweptback wing.

[*44*] an allowance must be made for the tips of the wing in obtaining the actual pressure drag. It appears that a suitable approximation to the drag of a wing having a large but finite aspect ratio is obtained when the contribution of the tips to the drag is made equal to the contribution from the root section.

Fig. A,15l shows the drag as calculated by exact formulas from the linearized theory [*190*] compared to the values given by the approximate equation (15-33). The approximation is good for wings having a sufficiently high aspect ratio and lying well inside the Mach cone from the apex.

Eq. 15-33 is of interest in that it shows the effect of section shape on the drag of wings swept behind the Mach angle. As remarked by Germain

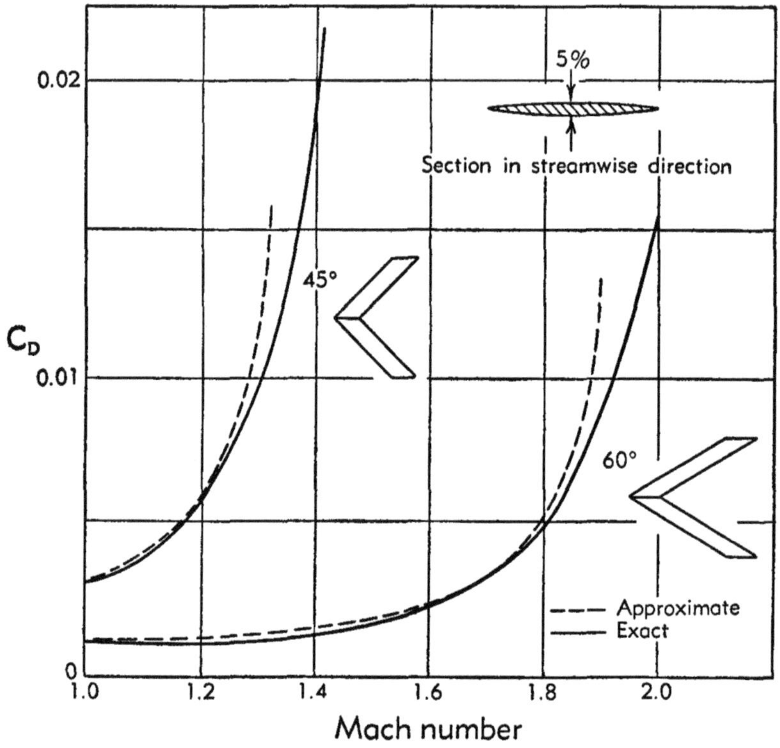

Fig. A,15l. Comparison of exact and approximate formulas for the wave drag of swept wings.

in [*144*], the minimum drag for a given area of the section occurs when the section shape is elliptical. In this case Eq. 15-32 reduces to

$$\frac{t(x)}{t(0)} = \sin \phi \tag{15-35}$$

so that the coefficient a_1 is unity and the remaining coefficients are zero.

A,16. Cited References and Bibliography.

Cited References

1. Lamb, H. *Hydrodynamics*, 6th ed. Dover, 1945.
2. Kutta, W. Auftriebskräfte in strömenden Flüssigkeiten. *Illustrierte Aeronaut. Mitt.*, July 1902.
3. Joukowski, N. De la chute dans l'aire de corps légers de forme allongée, animés d'un mouvement rotatoire. *Bull. de l'Inst. Aeronaut. Koutchino, Fasc. I*, St. Petersburg, 1906.

4. Jacobs, E. N. Preliminary report on laminar flow airfoils and new methods adopted for airfoil and boundary layer investigations. *NACA Advance Confid. Rept.*, June 1939.
5. Abbott, I. H., Von Doenhoff, A. E., and Stivers, L. S., Jr. Summary of airfoil data. *NACA Rept. 824*, 1945.
6. Lanchester, F. W. *Aerodynamics.* London, 1907.
7. Munk, M. M. The minimum induced drag of aerofoils. *NACA Rept. 121*, 1921.
8. Prandtl, L., and Betz, A. *Vier Abhandl. zur Hydrodynamik und Aerodynamik*, Göttingen, 1927. Reprinted by Edwards Brothers, 1943.
9. Munk, M. M. *Fluid Dynamics for Aircraft Designers.* Ronald Press, 1929.
10. Ackeret, J. Air forces on airfoils moving faster than sound. *NACA Tech. Mem. 317*, 1925.
11. McCullough, G. B., and Gault, D. E. Boundary-layer and stalling characteristics of the NACA 64A006 airfoil section. *NACA Tech. Note 1923*, 1949.
12. Liepmann, H. W., and Puckett, A. E. *Introduction to Aerodynamics of a Compressible Fluid.* Wiley, 1947.
13. Busemann, A. Drag near sonic velocities. *Third Intern. Congr. of Mechanics*, Stockholm, 1930.
14. Prandtl, L. General considerations on the flow of compressible fluids. *NACA Tech. Mem. 805*, 1936.
15. Glauert, H. The effect of compressibility on the lift of an airfoil. *Aeronaut. Research Council Repts. and Mem. 1135*, 1927.
16. Molenbroeck, P. *Arch. Mat. Phys. 9*, 157–195 (1890).
17. Chaplygin, S. A. Gas jets. *NACA Tech. Mem. 1063*, 1944.
18. von Kármán, Th. *J. Aeronaut. Sci. 8*, 337 (1941).
19. Kaplan, C. Effect of compressibility at high subsonic velocities on the lifting force acting on an elliptic cylinder. *NACA Rept. 834*, 1946.
20. Hantzsche, W., and Wendt, H. *Z. angew. Math. u. Mech. 22*, 72–86 (1942). Transl. as *R.T.P. Transl. 2198, British Ministry of Aircraft Production.*
21. von Kármán, Th. *J. Math. and Phys., 26(3)*, 182–190 (1947).
22. Guderley, G. On the transition from a transonic potential flow to a flow with shocks. Transl. as *U.S. Air Force Air Materiel Command Tech. Rept. F-TR-2160-ND*, 1947.
23. Oswatitsch, K., and Weighardt, K. Theoretical analysis of stationary potential flows and boundary layers at high speed. *NACA Tech. Mem. 1189*, 1948.
24. Oswatitsch, K. *Z. angew. Math. u. Mech. 30*, 17–24 (1950).
25. Stack, J., and Lindsey, W. F. Characteristics of low-aspect-ratio wings at supercritical Mach numbers. *NACA Rept. 922*, 1949.
26. Spreiter, J. R. Similarity laws for transonic flow about wings of finite span. *NACA Tech. Note 2273*, 1951.
27. Berndt, S. B. Similarity laws for transonic flow around wings of finite aspect ratio. *Roy. Inst. Technol. Stockholm KTH-Aero. TN 14*, 1950.
28. Kantrowitz, A. R. The formation and stability of normal shock waves in channel flows. *NACA Tech. Note 1225*, 1948.
29. Busemann, A. *J. Aeronaut. Sci. 16*, 337–344 (1949).
30. Kuo, Y. H. *J. Aeronaut. Sci. 18*, 1–6 (1951).
31. Feldman, F. K. Untersuchung von Symmetrischen Tragflugelprofilen bei hohen Unterschallgeschwindigkeiten in einem geschlossenen Windkanal. *Mitt. 14, Inst. für Aerodynamik.* A. G. Gebr. Leeman and Co., Zurich, 1948.
32. Busemann, A. Aerodynamischer Auftrieb bei Überschallgeschwindigkeit. *Volta Congr.*, 328–360 (1935). Also *Luftfahrtforschung 12*, 210–220 (1935).
33. Jones, R. T. Wing plan forms for high-speed flight. *NACA Tech. 863*, 1947.
34. Bateman, H. *Partial Differential Equations.* Dover, 1944.
35. Mathews, C. W., and Thompson, J. R. Comparative drag measurements at transonic speeds of rectangular and sweptback NACA 65-009 airfoils mounted on a freely falling body. *NACA Tech. Note 1969*, 1949.
36. Busemann, A. Pfeilflügel bei Hochgeschwindigkeit. *Lilienthal-Gesellschaft, Bericht 164*, 1943.

37. Struminsky, V. V. *Doklady Akad. Nauk. S.S.S.R. 54*, 765–768 (1946).
38. Jones, R. T. Effects of sweepback on boundary layer and separation. *NACA Rept. 884*, 1947.
39. Sears, W. R. *J. Aeronaut. Sci. 15*, 49 (1948).
40. Relf, E. H., and Powell, C. H. Tests on smooth and stranded wires inclined to the wind direction and a comparison of the results on stranded wires in air and water. *Aeronaut. Research Council Repts. and Mem. 307*, 1917.
41. Kolbe, C. D., and Boltz, F. W. The forces and pressure distribution at subsonic speeds on a plane wing having 45° of sweepback, an aspect ratio of 3, and a taper ratio of 0.5. *NACA Research Mem. A51G31*, 1951.
42. Göthert, B. Ebene und räumliche Strömung bei hohen Unterschallgeschwindigkeiten (Erweiterung der Prandtlschen Regel). *Lilienthal-Gesellschaft, Bericht 127*, 97–101 (1940); *NACA Tech Mem. 1105*, 1946.
43. von Kármán, Th. *J. Aeronaut. Sci. 14*, 373–409 (1947).
44. Hayes, W. D. Linearized supersonic flow. *North Amer. Aviation Co., Rept. AL-222*, Los Angeles, 1947.
45. Munk, M. M. *J. Appl. Phys. 21*, 159–161 (1950).
46. Brown, C. E. The reversibility theorem for thin airfoils in subsonic and supersonic flow. *NACA Tech. Note 1944*, 1949.
47. Ursell, F., and Ward, G. N. *Quart. J. Mech. and Appl. Math. 3*, 326–348 (1950).
48. Heaslet, M. A., and Spreiter, J. R. Reciprocity relations in aerodynamics. *NACA Rept. 1119*, 1953.
49. Flax, A. H. *J. Aeronaut. Sci. 19*, 361–374 (1952).
50. Krienes, K. The elliptic wing based on the potential theory. *NACA Tech. Mem. 971*, 1941.
51. Kinner, W. *Ing.-Arch. 8*, 47 (1937).
52. Whittaker, E. T., and Watson, G. N. *A Course of Modern Analysis*, 4th ed. Cambridge Univ. Press, 1940.
53. Hobson, E. W. *The Theory of Spherical and Ellipsoidal Harmonics.* Cambridge Univ. Press, 1931.
54. Jones, R. T. Correction of the lifting-line theory for the effect of the chord. *NACA Tech. Note 817*, 1941.
55. Lampert, S. *J. Aeronaut. Sci. 18*, 107–114 (1951).
56. Lampert, S. Some applications of conical-flow methods to subsonic lifting-surface problems. *NACA Tech. Note 2262*, 1950.
57. Busemann, A. *Jahrbuch deut. Luftfahrtforschung 7B(3)*, 105–121 (1943); *NACA Tech. Mem. 1100*, 1947.
58. Cohen, D. A method for determining the camber and twist of a surface to support a given distribution of lift, with applications to the load over a sweptback wing. *NACA Rept. 826*, 1945.
59. Falkner, V. M., and Lehrian, D. Low-speed measurements of the pressure distribution at the surface of a swept-back wing. *Aeronaut. Research Council Repts. and Mem. 2741*, 1953.
60. Swanson, R. S., and Crandall, S. M. An electromagnetic analogy method of solving lifting-surface-theory problems. *NACA Advance Restricted Rept. L5D23*, 1945. (Issued as Wartime Rept.)
61. Swanson, R. S., and Priddy, E. L. Lifting-surface-theory values of the damping in roll and of the parameters used in estimating aileron stick forces. *NACA Advance Restricted Rept. L5F23*, 1945. (Issued as Wartime Rept.)
62. Malavard, L., and Duquenne, R. *Recherche Aéronaut. 23*, 3–22 (1951).
63. Allen, D. N. deG., and Dennis, S. C. R. *Quart. J. Mech. and Appl. Math. 4*, 199–208 (1951); *6*, 81–100 (1953).
64. Tyler, C. M., Jr. *Solution of the Lifting-line and Lifting-Surface Integral Equations by the Relaxation Method.* Ph. D. Thesis, Univ. Pittsburgh, 1949.
65. Birnbaum, W. *Z. angew. Math. u. Mech. 3*, 290–297 (1923).
66. Blenk, H. *Z. angew Math. u. Mech. 5*, 36–47 (1925); *NACA Tech. Mem. 1111*, 1947.
67. Jones, W. P. Theoretical determination of the pressure distribution on a finite wing in steady motion. *Aeronaut. Research Council Repts. and Mem. 2145*, 1943.

68. Multhopp, H. Methods for calculating the lift distribution of wings (subsonic lifting surface theory). *Roy. Aircraft Establishment Rept. Aero. 2353*, 1950.
69. Curtis, A. R. Tables of Multhopp's influence functions. *Brit. Natl. Phys. Lab. Math. Div. Rept. MA/21/0505*, 1952.
70. Multhopp, H. *Luftfahrtforschung 15*, 153–169 (1938).
71. Garner, H. C. Methods of approaching an accurate three-dimensional potential solution for a wing. *Aeronaut. Research Council Rept. 12,222*, 1948.
72. Küchemann, D. *Aeronaut. Quart. 4*, 261–278 (1953).
73. Garner, H. C. Swept-wing loading. A critical comparison of four subsonic vortex sheet theories. *Aeronaut. Research Council, C.P. 102*, 1952.
74. Küchemann, D. A simple method for calculating the span loadings on thin swept wings. *Roy. Aircraft Establishment Rept. Aero. 2392*, 1950.
75. Pistolesi, E. Betrachtungen über die gegenseitige Beeinflussung von Tragflügelsystemen. *Lilienthal-Gesellschaft*, 214–219, 1937.
76. Mutterperl, W. The calculation of span load distributions on swept-back wings. *NACA Tech. Note 834*, 1941.
77. Weissinger, J. Ueber die Auftriebsverteilung von Pfeilflugeln. *FB 1553*, Berlin-Adlershof, 1942; *NACA Tech. Mem. 1120*, 1947.
78. De Young, J., and Harper, C. W. Theoretical symmetric span loading at subsonic speeds for wings having arbitrary plan form. *NACA Rept. 921*, 1948.
79. Schneider, W. C. A comparison of the spanwise loading calculated by various methods with experimental loadings obtained on a 45° sweptback wing of aspect ratio 8 at a Reynolds number of 4.0×10^6. *NACA Rept. 1208*, 1954.
80. Weighardt, K. *Z. angew. Math. u. Mech. 19*, 257 (1939); *NACA Tech. Mem. 963*, 1940.
81. Schlichting, H., and Kahlert, W. Calculation of lift distribution of swept wings. *Roy. Aircraft Establishment Rept. Aero. 2297*, 1948.
82. Scholz, N. *Ing.-Arch. 18*, 84–105 (1950).
83. Byrd, P. F. *Ing.-Arch. 19*, 321–323 (1951).
84. Holme, O. On the approximate solution of the lifting-surface problem with the aid of discrete vortices. *Roy. Inst. Technol. Stockholm, KTH-Aero TN 6*, 1949.
85. Thwaites, B. A continuous vortex line method for the calculation of lift on wings of arbitrary plan form. *Aeronaut. Research Council Rept. 12,082*, 1949; Addendum, *Aeronaut. Research Council Rept. 12,389*, 1949.
86. Falkner, V. M. The solution of lifting plane problems by vortex lattice theory. *Aeronaut. Research Council Repts. and Mem. 2591*, 1947.
87. Falkner, V. M. Calculated loadings due to incidence of a number of straight and swept-back wings. *Aeronaut. Research Council Repts. and Mem. 2596*, 1948.
88. Falkner, V. M. *Aircraft Eng. 22*, 296 (1950).
89. Staff of the Mathematics Division, N. P. L. Tables of complete downwash due to a rectangular vortex. *Aeronaut. Research Council Repts. and Mem. 2461*, 1953. Lehrian, D. E. A calculation of the complete downwash in three dimensions due to a rectangular vortex. *Aeronaut. Research Council Repts. and Mem. 2771*, 1953.
90. Holme, O. A. M. Measurements of the pressure distribution on rectangular wings of different aspect ratios. *FFA Meddelande 37*, Stockholm-Elvsunda, 1950.
91. Jacobs, W. *Ing.-Arch. 18*, 344 (1950).
92. Letko, W., and Goodman, A. Preliminary wind-tunnel investigation at low speed of stability and control characteristics of swept-back wings. *NACA Tech. Note 1046*, 1946.
93. Winter, H. Flow phenomena on plates and airfoils of short span. *NACA Tech. Mem. 798*, 1936.
94. Jacobs, E. N., and Clay, W. C. Characteristics of the NACA 23012 airfoil from tests in the full-scale and variable-density tunnels. *NACA Rept. 530*, 1935.
95. Goett, H. J., and Bullivant, W. K. Tests of NACA 0009, 0012, and 0018 airfoils in the full-scale tunnel. *NACA Rept. 647*, 1939.
96. Zimmerman, C. H. Characteristics of Clark Y airfoils of small aspect ratios. *NACA Rept. 431*, 1932.
97. Goodman, A., and Brewer, J. D. Investigation at low speeds of the effect of

aspect ratio and sweep on static and yawing stability derivatives of untapered wings. *NACA Tech. Note 1669*, 1948.
98. Jacobs, E. N., Ward, K. E., and Pinkerton, R. M. The characteristics of 78 related airfoil sections from tests in the variable-density wind tunnel. *NACA Rept. 460*, 1933.
99. Lange, and Wacke. Pruefbericht über 3- und 6- Komponenten-messungen an der Zuspitzungsreihe von Fluegeln kleiner Streckung. *UM 1023, Berlin-Adlershof*, 1943. (Partial report on triangular wings translated as *NACA Tech. Mem. 1176*, 1948.)
100. Wick, B. H. Chordwise and spanwise loadings measured at low speed on a triangular wing having an aspect ratio of two and an NACA 0012 airfoil section. *NACA Tech. Note 1650*, 1948.
101. Staff of Navy Aerodynamical Lab. Air force and moment for gliding wing. *Department of Aeronautics Rept. 677*, 1943.
102. Berndt, S. B. Three component measurement and flow investigation of plane delta wings at low speeds and zero yaw. *Roy. Inst. Technol. Stockholm KTH-Aero TN 4*, 1948.
103. Graham, D. Chordwise and spanwise loadings measured at low speeds on a large triangular wing having an aspect ratio of 2 and a thin, subsonic-type airfoil section. *NACA Research Mem. A50A04a*, 1950.
104. Anderson, A. Chordwise and spanwise loadings measured at low speed on large triangular wings. *NACA Research Mem. A9B17*, 1949.
105. Jones, R. T. Properties of low-aspect-ratio pointed wings at speeds below and above the speed of sound. *NACA Rept. 835*, 1946.
106. Munk, M. M. The aerodynamic forces on airship hulls. *NACA Rept. 184*, 1924.
107. Tsien, H. S. *J. Aeronaut. Sci. 5*, 480 (1938).
108. Lawrence, H. R. The lift distribution on low-aspect-ratio wings at subsonic speeds. *Inst. Aeronaut. Sci. Preprint 313*, 1951.
109. Adams, M. C., and Sears, W. R. *J. Aeronaut. Sci. 20*, 85–98 (1953).
110. Lomax, H., and Heaslet, M. A. Linearized lifting-surface theory for swept-back wings with slender plan forms. *NACA Tech. Note 1992*, 1949.
111. Robinson, A. *Aeronaut. Quart. 4*, 69–82 (1952).
112. Mangler, K. W. Calculation of the pressure distribution over a wing at sonic speeds. *Roy. Aircraft Establishment Rept. Aero. 2439*, 1951.
113. Legendre, R., Eichelbrenner, E. A., and von Baranoff, A. Écoulement transsonique autour d'ailes a forte flèche. *Office natl. Études et Recherches aéronaut 53*, 1952.
114. Mirels, H. Aerodynamics of slender wings and wing-body combinations having swept trailing edges. *NACA Tech. Note 3105*, 1954.
115. Eichelbrenner, E. A. Théorie des corps élancés. *Office natl. Études et Recherches aéronaut. 68*, 1954.
116. Heaslet, M. A., Lomax, H., and Spreiter, J. R. Linearized compressible-flow theory for sonic flight speeds. *NACA Rept. 956*, 1950.
117. Robinson, A., and Young, A. D. Note on the application of the linearized theory for compressible flow to transonic speeds. *College of Aeronautics, Cranfield, Rept. 2*, 1947.
118. Bollay, W. *Z. angew. Math. u. Mech. 19*, 21-35 (1939).
119. Kriesis, P. *Z. angew. Math. u. Mech. 24*, 1–5 (1944).
120. Flax, A. H., and Lawrence, H. R. The aerodynamics of low-aspect-ratio wings and wing-body combinations. *Third Anglo-American Aeronaut. Conf.*, 1951.
121. Durand, W. F., ed. *Aerodynamic Theory.* Springer, Berlin, 1934–1936.
122. Zahm, A. F. Flow and force equations for a body revolving in a fluid. *NACA Rept. 323*, 1929.
123. Hess, R. V., and Gardner, C. S. Study by the Prandtl-Glauert method of compressibility effects and critical Mach number for ellipsoids of various aspect ratio and thickness ratios. *NACA Tech. Note 1792*, 1949.
124. Jones, R. T. Subsonic flow over thin oblique airfoils at zero lift. *NACA Tech. Note 1340*, 1947. Republished as *NACA Rept. 902*, 1948.

125. Neumark, S. Velocity distribution on straight and swept-back wings of small thickness and infinite aspect ratio at zero incidence. *Roy. Aircraft Establishment Rept. Aero. 2200. ARC 10907*, 1947.
126. Küchemann, D. Wing junction, fuselage and nacelles for swept-back wings. *Roy Aircraft Establishment Rept. Aero. 2219. ARC 11035*, 1947.
127. Jones, R. T. Thin oblique airfoils at supersonic speed. *NACA Tech. Rept. 851*, 1946.
128. Busemann, A. *Schriften. deut. Akad. für Luftfahrtforschung 30*, 17–36 (1940). Transl. as *R.T.P. Trans. 1830, British Ministry of Aircraft Production.*
129. Heaviside, O. *Electromagnetic Theory*. Dover, 1950.
130. Epstein, P. S. *Proc. Natl. Acad. Sci. 17*, 532–547 (1931).
131. Hugoniot, H. *J. École Polytech. 57*, 3–97 (1887).
132. Meyer, T. The two-dimensional phenomena of motion in a gas flowing at supersonic velocity. *Dissertation*, Göttingen, 1908. (Reprinted in Carrier, G. F. *Foundations of High Speed Aerodynamics*. Dover, 1951.)
133. Ferri, A. *Elements of Aerodynamics of Supersonic Flows*. Macmillan, 1949.
134. Ferri, A. Investigations and experiments in the Guidonia supersonic wind tunnel. *NACA Tech. Mem. 901*, 1939.
135. Busemann, A., and Walchner, O. *Forsch. Gebiete Ingenieurw. 4A*, 87–92 (1933).
136. Sanger, E. *Raketen-Flugtechnik*. Oldenbourg, Munich and Berlin, 1933.
137. Chapman, D. R. Airfoil profiles for minimum pressure drag at supersonic velocities—General analysis with application to linearized supersonic flow. *NACA Tech. Rept. 1063*, 1952.
138. Chapman, D. R. Airfoil profiles for minimum pressure drag at supersonic velocities—Application of shock-expansion theory including consideration of the hypersonic range. *NACA Tech. Note 2787*, 1952.
139. Bergman, S. On solutions with algebraic character of linear partial differential equations. *Trans. Am. Math. Soc. 68*, 461–507 (1950).
140. Glauert, H. *Aerofoil and Airscrew Theory*. Macmillan, 1943.
141. Mirels, H. Lift-cancellation technique in linearized supersonic wing theory. *NACA Rept. 1004*, 1950.
142. Lagerstrom, P. A. Linearized supersonic theory of conical wings. *NACA Tech. Note 1685*, 1948.
143. Hayes, W. D., Browne, S. H., and Lew, R. J. Linearized theory of conical supersonic flow with application to triangular wings. *North Amer. Aviation Rept. NA-46-818*, 1946.
144. Germain, P. La théorie générale des mouvements coniques et ses applications à l'aérodynamique supersoniques. *Office natl. Études et Recherches aéronaut. 34*, 1949. Transl. as *NACA Tech. Mem. 1354*, 1955.
145. Robinson, A. Aerofoil theory of a flat delta wing at supersonic speeds. *Brit. Aeronaut. Research Council Repts. and Mem. 2548*, 1946.
146. Squire, H. B. An example in wing theory at supersonic speed. *Aeronaut. Research Council Repts. and Mem. 2549*, 1951.
147. Multhopp, H. A unified theory of supersonic wing flow, employing conical fields. *Roy. Aircraft Establishment Rept. Aero. 2415*, 1951.
148. Ribner, H. S. Some conical and quasi-conical flows in linearized supersonic wing theory. *NACA Tech. Note 2147*, 1950.
149. Walker, H. J., and Ballantyne, M. B. Pressure distribution and damping in steady roll at supersonic Mach numbers of flat sweptback wings with subsonic edges. *NACA Tech. Note 2047*, 1950.
150. Walker, H. J., and Ballantyne, M. B. Pressure distribution and damping in steady pitch at supersonic Mach numbers of flat sweptback wings having all edges subsonic. *NACA Tech. Note 2197*, 1950.
151. Byrd, P. F., and Friedman, M. D. *Handbook of Elliptic Integrals for Engineers and Physicists*. Grundlehren der Mathematischen Wissenschaften, Band LXVIII. Springer-Verlag, Heidelberg, 1953.
152. Baldwin, B. S., Jr. Triangular wings cambered and twisted to support specified distributions of lift at supersonic speeds. *NACA Tech. Note 1816*, 1949.

153. Bolton-Shaw, B. W. Nose controls on delta wings, at supersonic speeds. *College of Aeronautics, Cranfield, Rept. 36*, 1950.
154. Lagerstrom, P. A., and Graham, M. E. Downwash and sidewash induced by three-dimensional lifting wings in supersonic flow. *Douglas Aircraft Co. Rept. SM13007*, Apr. 1947.
155. Cohen, D. Formulas for the supersonic loading, lift, and drag of flat sweptback wings with leading edges behind the Mach lines. *NACA Rept. 1050*, 1951.
156. Lagerstrom, P. A., and Graham, M. E. Linearized theory of supersonic control surfaces. *Douglas Aircraft Co. Rept. SM13060*, 1947.
157. Frick, C. W., Jr. Application of the linearized theory of supersonic flow to the estimation of control-surface characteristics. *NACA Tech. Note 1554*, 1948.
158. Cohen, D., and Friedman, M. D. Theoretical investigation of the supersonic lift and drag of thin, sweptback wings with increased sweep near the root. *NACA Tech. Note 2959*, 1953.
159. Puckett, A. E. *J. Aeronaut. Sci. 13*, 475–484 (1946).
160. Evvard, J. C. Use of source distributions for evaluating theoretical aerodynamics of thin finite wings at supersonic speeds. *NACA Rept. 951*, 1950.
161. Krasilshchikova, E. A. *Uchenye Zapiski 154, Mekhanika 4*, 181–239 (1951). Transl. *NACA Tech. Mem. 1383*, 1956.
162. Goodman, T. R. The lift distribution on non-conical flow regions of thin finite wings in a supersonic stream. *Cornell Aeronaut. Lab. Rept.*, Oct. 1948.
The problem of edges in the doublet distribution method of obtaining supersonic lift. *Cornell Aeronaut. Lab. Rept.*, Feb. 1950.
163. Heumann, C. *J. Math. and Phys. 20*, 127–206 (1941).
164. Behrbohm, H., and Oswatitsch, K. *Ing.-Arch. 18*, 370–377 (1950).
165. Cramer, R. H. Interference between wing and body at supersonic speeds, Part V, Phase 1. *Cornell Aeronaut. Lab. Rept. CAL/CM 597*, 1950.
166. Coleman, T. F. Supersonic lift solutions obtained by extending the simple linearized conical flow theory. *North Amer. Aviation Co. Rept. CM440*, 1948.
167. Behrbohm, H. The lifting trapezoidal wing with small aspect ratio at supersonic speed. *SAAB Aircraft Co. Tech. Note 10*, Linköping, Sweden, 1952.
168. Lagerstrom, P. A., and Graham, M. E. Low aspect ratio rectangular wings in supersonic flow. *Douglas Aircraft Co. Rept. SM13110*, 1947.
169. Picard, C., and Chevalier, J. P. *Recherche aéronaut. 26*, 3–12 (1952).
170. Nielsen, J. N., Matteson, F. H., and Vincenti, W. G. Investigation of wing characteristics at a Mach number of 1.53. III: Unswept wings of differing aspect ratio and taper ratio. *NACA Research Mem. A8E06*, 1948.
171. Dye, F. E., Jr. Interference between wing and body at supersonic speeds. Part IX: Pressure distribution tests of wing-body interference models at Mach No. of 2.0. Phase V tests of Aug. 1951. *Cornell Aeronaut. Lab. Rept. CAL/CF 1684*, Dec. 1951.
172. Puckett, A. E., and Stewart, H. J. *J. Aeronaut. Sci. 14*, 567–578 (1947).
173. Drougge, G., and Larsson, P. O. *Report of the Aeronautical Research Inst. of Sweden (FFA)*. To be published.
174. Ellis, M. C., Jr., and Hasel, L. E. Preliminary tests at supersonic speeds of triangular and swept-back wings. *NACA Research Mem. 26L17*, 1947.
175. Drougge, G. Some measurements at low supersonic speeds by a method for continuous variation of the Mach number. *FFA Rept. 42*, Stockholm-Elvsunda, 1952.
176. Vincenti, W. G., Nielson, J. N., and Matteson, F. H. Investigation of wing characteristics at a Mach number of 1.53. I: Triangular wings of aspect ratio 2. *NACA Research Mem. A7I10*, 1947.
177. Chang, C. C., and Houser, J. E. Correlation of test data and comparison with theoretical results for thin wings in supersonic flow. *Glenn L. Martin Co. Eng. Rept. 2729*, 1947.
178. Harmon, S. M., and Jeffreys, I. Theoretical lift and damping in roll of thin wings with arbitrary sweep and taper at supersonic speeds. Supersonic leading and trailing edges. *NACA Tech. Note 2114*, 1950.

179. Gilles, A. *Recherche aéronaut. 19*, 3-10 (1951).
180. Lapin, E. Charts for the computation of lift and drag of finite wings at supersonic speeds. *Douglas Aircraft Co. Rept. SM-13480*, 1949.
181. Martin, J. C., and Jeffreys, I. Span load distributions resulting from angle of attack, rolling and pitching for tapered swept-back wings with streamwise tips. Supersonic leading and trailing edges. *NACA Tech. Note 2643*, 1952.
182. Hannah, M. E., and Margolis, K. Span load distributions resulting from constant angle of attack, steady rolling velocity, steady pitching velocity, and constant vertical acceleration for tapered sweptback wings with streamwise tips. Subsonic leading edges and supersonic trailing edges. *NACA Tech. Note 2831*, 1952.
183. Eichelbrenner, E. A. *Recherche aéronaut. 25*, 19–20 (1952).
184. Oswatitsch, K. Die theoretischen Arbeiten über schallnahe Strömung am Flugtechnischen Inst. der Kungl. Tekniska Högskolan, Stockholm. *Proc. Eighth Intern. Congress on Theoretical and Appl. Mech.*, 1953.
185. Ward, G. N. *Quart. J. Mech. and Appl. Math. 2*, 75–97 (1949).
186. Whitcomb, R. T. A study of the zero-lift drag-rise characteristics of wing-body combinations near the speed of sound. *NACA Research Mem. L52H08*, 1952.
187. Jones, R. T. *J. Aeronaut. Sci. 19*, 813–822 (1952).
188. Sears, W. R. *Quart. Appl. Math. 5*, 361–366 (1947).
189. Haack, W. Geschossformen kleinsten Wellenwiderstandes. Lilienthal-Gesellschaft, *Bericht 139*, 1941.
190. Harmon, S. M., and Swanson, M. D. Calculation of the supersonic wave drag of non-lifting wings with arbitrary sweepback and aspect ratio. *NACA Tech. Note 1319*, 1947.
191. Harmon, S. M. Theoretical supersonic wave drag of untapered sweptback and rectangular wings at zero lift. *NACA Tech. Note 1449*, 1947.
192. Bismut, M. *Recherche aéronaut. 21*, 9–18 (1951).
193. Chang, C. C. Applications of von Kármán's integral method in supersonic wing theory. *NACA Tech. Note 2317*, 1951.
194. Nielsen, J. N. The effect of aspect ratio and taper on the pressure drag at supersonic speeds of unswept wings at zero lift. *NACA Tech. Note 1487*, 1947.
195. Margolis, K. Supersonic wave drag of sweptback tapered wings at zero lift. *NACA Tech. Note 1448*, 1947.
196. Margolis, K. Effect of chordwise location of maximum thickness on the supersonic wave drag of sweptback wings. *NACA Tech. Note 1543*, 1948.
197. Margolis, K. Supersonic wave drag of nonlifting sweptback tapered wing with Mach lines behind the line of maximum thickness. *NACA Tech. Note 1672*, 1948.
198. Lawrence, T. Charts of the wave drag of wings at zero lift. *Roy. Aircraft Establishment Tech. Note Aero. 2139*, 1952.
199. Havelock, T. H. Calculations illustrating the effect of boundary layer on wave resistance. *Trans. Inst. Nav. Arch. 90*, 259–271 (1948).
200. von Kármán, Th. The problems of resistance in compressible fluids. *Proc. 5th Volta Congr.* 255–64 (1935).
201. von Kármán, Th., and Moore, N. B. *Trans. Am. Soc. Mech. Engrs. 54*, 303–310 (1932).

Bibliography for Table A,2

Allen, H. J. General theory of airfoil sections having arbitrary shape or pressure distribution. *NACA Rept. 833*, 1945.
Glauert, H. *The Elements of Aerofoil and Airscrew Theory.* Cambridge Univ. Press, 1943.
Munk, M. M. *Fundamentals of Fluid Dynamics for Aircraft Designers.* Ronald Press, 1929.

SECTION B

AERODYNAMICS OF BODIES AT HIGH SPEEDS

CLINTON E. BROWN

B,1. Drag of Bodies. The drag of bodies moving in a fluid is a problem which has confronted engineers and physicists for many years. Newton, d'Alembert, and other great scientists devoted considerable effort to explain the resistance of common observation. Prior to World War II the body drag problem at supersonic speeds was only of practical interest to the ballistician and the meteorologist; recently, however, as the speeds of aircraft have increased and radio-controlled projectiles have appeared, the drag problem at supersonic speeds has become one for the aeronautical engineer. The ballistician has previously been concerned with two factors contributing to the drag of projectiles, namely, the form drag or drag caused by wave formation, and the base drag which is intimately connected with the effects of viscosity and form. The drag produced by the viscous shearing in boundary layer was of secondary importance since the form and base drags were of a higher order of magnitude. This situation was the result of the high body diameter-to-length ratios required by projectiles for stability reasons, it having been found that slender bodies could not be spin-stabilized. However, with the introduction of fin-stabilized missiles and the need for low drag fuselages for supersonic and transonic aircraft, the body diameter-to-length ratios have been reduced so that the practical drag problem requires the knowledge of form drag, base drag, and viscous drag. Fortunately these three factors may be treated in normal cases so that each constitutes a separate problem.

The perfect fluid pressure drag or form drag may be defined as the drag the body would have in a fluid devoid of viscosity, with the exception that for bodies having bases the pressures acting on the base area are considered base drag and hence are not to be included in the form drag. In pure subsonic flow over closed bodies the pressure drag is then zero, whereas in transonic flow and supersonic flow it is a drag which shows up downstream in the fluid as a momentum loss caused directly by shock waves produced by the body. In transonic flow, the necessity for solving a nonlinear differential equation of mixed elliptic and hyperbolic type makes the drag calculation extremely difficult, and no generally applicable methods have as yet been found which are free of criticism of their range of convergence or of their basic assumptions. In pure supersonic flow, the calculation of the form drag has reached its highest point of development. Various methods are available of which the linearized theory affords the greatest ease and applicability but also the least accuracy, although in many cases the accuracy is sufficient for engineering purposes and to give some insight into the real flow conditions. Iteration methods have appeared which lead to increased computation time. However, the use of high speed computing machines has brought the exact computations into reach, especially for the case of bodies possessing axial symmetry.

The base drag is simply the product of base area and base pressure, since in most cases the pressures acting over the base are nearly constant due to the relatively small velocities in the dead air wake just behind the base. The base pressures are, of course, not the same as those which the fluid would produce in the absence of viscosity. In fact, the streamline picture at the base is completely altered by the viscous action so that the body streamlines leave the body at the base shoulder and enclose a dead air wake. Thus the calculation of the base pressure is not simple since it involves an unknown streamline along which there is a mixing zone of high and low velocity fluid. Therefore, the prediction of the base pressure depends on experimental observations and measurements.

The viscous drag of slender bodies may be estimated quite well, provided the position of transition from a laminar to a turbulent boundary layer is known. Unfortunately the transition point is highly dependent on Reynolds number, surface roughness, surface heat transfer, and surface pressure distribution. The transition point is also sensitive to the turbulence of the air stream and hence wind tunnel results often give unreliable predictions. The interrelation between wind tunnel turbulence and transition is an extremely difficult problem, which together with the fundamental problem of transition in free flight continues to be one of the most difficult confronting today's research in aeronautics. When the transition point is not known, the best thing to be done is to bracket the true viscous drag by computing the pure laminar and pure turbulent values. In this regard the reader is referred to the volume on viscous fluids. However, in the following pages is found a discussion of the effect of applying two-dimensional skin friction factors to three-dimensional bodies.

In recent years flight tests incorporating rocket drives have been invaluable for obtaining aerodynamic information. In many cases the measurements are made in flight under varying amounts of acceleration which may range from plus to minus 20g. The effect of the acceleration on the drag must be determined, and it is possible to show at least in the flight region away from $M_\infty = 1$ that the acceleration introduces an additional force which is negligible compared with the steady state drag [*1*].

Another important factor required for the estimation of the drag of bodies is the effect of yaw. At small angles of yaw the flow can be calculated in several ways, i.e. linearized theory, linearized characteristic theory, or full characteristic theory; in general the viscous drag is relatively constant. However, at large angles the flow is complicated by separation of the boundary layer and in this case the perfect fluid theories are inaccurate.

Pressure Drag in a Perfect Fluid. The form drag of a closed body in a completely subsonic steady flow is zero. This fact was demonstrated for incompressible flow many years ago by D'Alembert and resulted in the now famous "D'Alembert's paradox." It is easily shown, also, that in a compressible medium devoid of viscosity and containing no surfaces of discontinuous velocity there can be no drag. The proof follows from Kelvin's theorem which implies that along any stream tube the stagnation pressure in addition to the entropy will be a constant. Since all stream tubes about a body originate and end in fluid either at rest or at constant conditions, it follows that the energy of the stream tubes is unchanged when the fluid returns to the ambient pressure; since a drag on the body would require a continual transfer of energy to the fluid, there can be no drag. The restriction to cases in which there are no sheets of discontinuity implies the absence of trailing vortices, and it is well established that these vortices constitute the only mechanism in pure subsonic flow which can produce energy dissipation and hence drag.

The problem of the perfect fluid drag is therefore of primary interest in the transonic and supersonic speed ranges, and since little progress has been made on the transonic problem the remaining portions of this article are devoted to the supersonic flight condition.

Slender Bodies of General Cross Section. For the bodies of general cross section the most useful result is given by linear theory. The following development is due largely to Ward [2]. The differential equation to be solved can be written in cylindrical coordinates (see VI,C).

$$(M_\infty^2 - 1)\varphi_{xx} - \varphi_{rr} - \frac{1}{r}\varphi_r - \frac{1}{r^2}\varphi_{\theta\theta} = 0 \qquad (1\text{-}1)$$

where φ is the disturbance velocity potential and M_∞ the Mach number of the undisturbed stream. The pressure is given by the expansion in a Taylor series of Bernoulli's equation and is

$$p = p_\infty - \rho_\infty\left[\varphi_x U + \frac{1}{2}\left(\varphi_r^2 + \frac{1}{r^2}\varphi_\theta^2\right)\right] \qquad (1\text{-}2)$$

where ∞ refers to free stream conditions, and U is the free stream velocity. Generally, only the first few terms are used since the disturbance velocities are assumed to be small. Writing Eq. 1-2 in the form of a pressure coefficient we obtain

$$\frac{\Delta p}{\frac{1}{2}\rho_\infty U^2} = \frac{\Delta p}{q} = \frac{-2\varphi_x}{U} - \frac{\left(\varphi_r^2 + \frac{1}{r^2}\varphi_\theta^2\right)}{U^2} \tag{1-3}$$

In the preceding equations it is convenient for the treatment of slender bodies to measure r and x in terms of the body length; thus r and x can be assumed to be nondimensional quantities and for slender bodies r can be said to be small ($r \ll 1$) near the body surface. The general solution of Eq. 1-1 for diverging waves is as given by Lamb [*3*, p. 527], with suitable change in notation,

$$\varphi = U \sum_s (Q_s r^s \cos s\theta + P_s r^s \sin s\theta) \tag{1-4}$$

where

$$Q_s = \left(\frac{\partial}{r\partial r}\right)^s Q_0$$

$$P_s = \left(\frac{\partial}{r\partial r}\right)^s P_0$$

and

$$Q_0 = -\frac{1}{2\pi}\int_0^\infty f(x - \sqrt{M_\infty^2 - 1}\, r \cosh u) du$$

$$P_0 = -\frac{1}{2\pi}\int_0^\infty g(x - \sqrt{M_\infty^2 - 1}\, r \cosh u) du$$

The portion of Lamb's general solution which corresponds to converging waves does not apply to the external flow over slender bodies since all disturbances originate on the body and diverge into the flow field along Mach surfaces (characteristics), $(x - r\sqrt{M_\infty^2 - 1}) = \text{const.}$

For the case of rotational symmetry only the first term of Eq. 1-4 is required, thus the potential and its derivatives may be obtained as

$$\varphi = -\frac{U}{2\pi}\int_0^\infty f_0(x - \beta r \cosh u) du \tag{1-5}$$

$$\varphi_r = +\frac{\beta U}{2\pi}\int_0^\infty f_0'(x - \beta r \cosh u) \cosh u du \tag{1-6}$$

$$\varphi_x = -\frac{U}{2\pi}\int_0^\infty f_0'(x - \beta r \cosh u) du \tag{1-7}$$

where $\beta = \sqrt{M_\infty^2 - 1}$ for convenience.

Substituting $\xi = (x - \beta r \cosh u)$ in Eq. 1-5, 1-6, and 1-7, there are obtained the alternative equations

$$\varphi = -\frac{U}{2\pi}\int_0^{x-\beta r} \frac{f_0(\xi)d\xi}{\sqrt{(x-\xi)^2-\beta^2r^2}} \tag{1-8}$$

$$\varphi_r = \frac{U}{2\pi r}\int_0^{(x-\beta r)} \frac{f_0'(\xi)(x-\xi)d\xi}{\sqrt{(x-\xi)^2-\beta^2r^2}} \tag{1-9}$$

$$\varphi_x = -\frac{U}{2\pi}\int_0^{(x-\beta r)} \frac{f_0'(\xi)d\xi}{\sqrt{(x-\xi)^2-\beta^2r^2}} \tag{1-10}$$

Prandtl [4] has given a very clear derivation of Eq. 1-8 (see also Ferri [5]) in which the integral is shown to represent a distribution of point sources along the x axis beginning at $x = 0$ and in which the strength is given by the arbitrary function $f_0(\xi)$. Since $f_0(\xi)$ is arbitrary, it is evident that the boundary conditions for any slender body of revolution at zero angle of attack can be met by such a distribution of sources. The additional terms of Eq. 1-4 can be found to correspond to distributions of dipoles, $s = 1$, quadripoles, $s = 2$, etc. which do not produce fluid, but merely act to direct the flow outward along some axes and inward along others. Thus by suitable distributions of these singularities, the flow over any slender body of arbitrary cross section can be approximated.

Eq. 1-4 can be considerably simplified if we consider only the region where r is small compared with unity, i.e. in the vicinity of the body boundary. Thus, putting the body nose at $x = 0$ and integrating by parts, the first term, Eq. 1-8, can be reduced to

$$\varphi_0 = -\frac{U}{2\pi}\left[f_0(0)\cosh^{-1}\frac{x}{\beta r} + \int_0^{x-\beta r} f_0'(\xi)\cosh^{-1}\frac{x-\xi}{\beta r}\,d\xi\right] \tag{1-11}$$

for small r,

$$\cosh^{-1}\frac{(x-\xi)}{\beta r} \to \ln\frac{2(x-\xi)}{\beta r} + 0\left(\frac{\beta r}{x-\xi}\right)^2$$

hence

$$\varphi_0 \underset{r\to 0}{=} +\frac{U}{2\pi}\left[f_0(x)\ln\frac{\beta r}{2} - \int_0^x f_0'(\xi)\ln(x-\xi)d\xi\right] \tag{1-12}$$

Here $f_0(0)$ is assumed zero since the body starts at $x = 0$ and thus there are no sources ahead of $x = 0$. It will be noticed that Eq. 1-12 is correct only when $f'(\xi)$ is of the same order as $f(\xi)$ since for large $f'(\xi)$ it would be possible for neglected terms in the $f_0'(\xi)\cosh^{-1}(x-\xi)/\beta r$ term of Eq. 1-11 to become of the same order as those retained. It will be shown that for the body of revolution this requires the body contour to be smooth and free from abrupt changes in slope. In like manner the remaining terms of Eq. 1-4 may be simplified; thus, for example, the cosine term of $s = 2$.

$$\varphi_2 = -\frac{U\cos 2\theta}{2\pi r^2}\int_0^{x-\beta r} \frac{f_2'(\xi)(x-\xi) + f_2''(\xi)(x-\xi)^2}{\sqrt{(x-\xi)^2-\beta^2r^2}}\,d\xi \tag{1-13}$$

for small r

$$\varphi_2 = -\frac{U \cos 2\theta}{\pi r^2} f_2(x) \tag{1-14}$$

Again it is necessary that $f_2''(\xi)$ and $f_2'(\xi)$ be of the order of $f_2(\xi)$ so as to insure the accuracy of Eq. 1-14.

Following the same process for each term of Eq. 1-4 thus yields for small r

$$\varphi = +\frac{U}{2\pi}\left\{f_0(x) \ln \frac{\beta r}{2} - \int_0^x f_0'(\xi) \ln (x - \xi) d\xi - \sum_1^\infty (-1)^s \left[\frac{s \cos s\theta}{r^s} f_s(x) + \frac{s \sin s\theta}{r^s} g_s(x)\right]\right\} \tag{1-15}$$

The velocities are now

$$\varphi_x = +\frac{U}{2\pi}\left\{f_0'(x) \ln \frac{\beta r}{2} - f_0'(0) \ln x - \int_0^x f_0''(\xi) \ln (x - \xi) d\xi - \sum_1^\infty (-1)^s \left[\frac{s \cos s\theta}{r^s} f_s'(x) + \frac{s \sin s\theta}{r^s} g_s'(x)\right]\right\} \tag{1-16}$$

$$\varphi_r = +\frac{U}{2\pi}\left\{\frac{f_0(x)}{r} + \sum_1^\infty (-1)^s \left[\frac{s^2 \cos s\theta}{r^{s+1}} f_s(x) + \frac{s^2 \sin s\theta}{r^{s+1}} g_s(x)\right]\right\} \tag{1-17}$$

$$\frac{1}{r}\varphi_\theta = \frac{U}{2\pi}\sum_1^\infty (-1)^s \left[\frac{s^2 \sin s\theta}{r^{(s+1)}} f_s(x) - \frac{s^2 \cos s\theta}{r^{(s+1)}} g_s(x)\right] \tag{1-18}$$

Term-by-term inspection of Eq. 1-15, 1-16, and 1-17 indicates that for any term the φ_r and $(1/r)\varphi_\theta$ terms are of the same order whereas the φ_x terms are of lower order. However, it is not possible to determine the true order of the velocities without solving the given problem. Nevertheless, experience with the body of revolution problem and consideration of the physical phenomenon indicate that if the body surface is smooth and free from rapid changes in curvature the φ_x terms at any point will be of lower order of smallness than the cross velocities. Thus in the pressure equation (Eq. 1-3) the cross-velocity terms should be retained.

Inspection of Eq. 1-15 shows that at any cross plane x = const the potential is a solution of the Laplace equation in two dimensions r and θ; thus Eq. 1-15 satisfies

$$\varphi_{rr} + \frac{1}{r}\varphi_r + \frac{1}{r^2}\varphi_{\theta\theta} = 0 \tag{1-19}$$

This result will be of importance in the following development.

Boundary conditions. Consider Fig. B,1a. The two contours shown are assumed to be body cross sections in planes normal to x and a distance Δx apart. If Δn represents the normal distance between the two contours, the conditions of zero flow through the body surface then requires

$$\frac{\partial \varphi}{\partial n} = (U + \varphi_x)\frac{\Delta n}{\Delta x} \tag{1-20}$$

Since we are considering only first order terms and φ_x is of lower order than U, we obtain

$$\frac{\partial \varphi}{\partial n} = U\frac{dn}{dx} \tag{1-21}$$

Furthermore if τ is a distance measured along the contour it can be seen that the increase in cross-sectional area ΔS, which occurs as we pass from

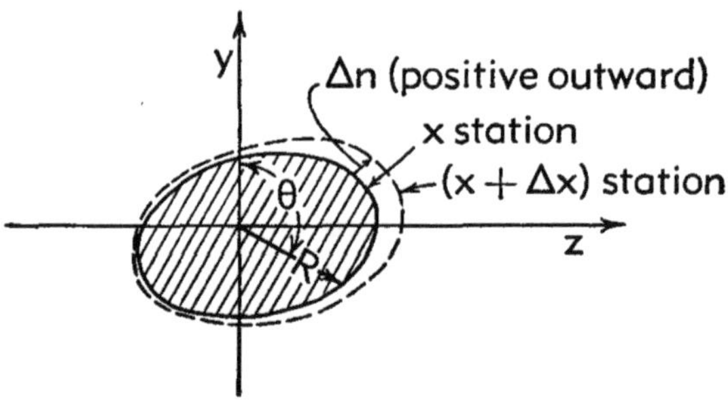

Fig. B,1a. Coordinate notation.

x to $x + \Delta x$, is now the area between the two contours and

$$\Delta S = \int_C \Delta n d\tau \tag{1-22}$$

or from Eq. 1-21 and 1-22

$$\frac{\Delta S}{\Delta x} = S'(x) = \frac{1}{U}\int_C \frac{\partial \varphi}{\partial n} d\tau \tag{1-23}$$

where the prime indicates differentiation with respect to x. Taking the solution, Eq. 1-15, and remembering that φ is a solution of the Laplace equation in r and θ in any plane x = const, the symmetrical form of Green's theorem shows that the line integral of Eq. 1-23 is independent of the contour C; hence we may choose a circle about the origin of any radius r_i. Making use of Eq. 1-15 in Eq. 1-23 for $C = r_i$ thus indicates that only the source term which contains no cosine or sine terms contributes to the integral, hence

$$S'(x) = \frac{1}{U}\int_0^{2\pi} \frac{\partial \varphi}{\partial r} r_i d\theta = f_0(x) \tag{1-24}$$

The additional terms of the series for the potential φ are needed only to account for the change in shape of the section and a translational motion of the sections with respect to the x axis.

The drag integral. The drag of the general body can be conveniently obtained by computing the momentum change of the fluid passing through a cylinder of radius r_0 drawn about the body (see Fig. B,1b). The summation of forces on the cylinder yields

$$D = \pi r_0^2 \rho_\infty U^2 - \int_{\beta r_0}^{1} dx \int_0^{2\pi} \rho\varphi_r(U + \varphi_x) r_0 d\theta - \int_{R(1,\theta)}^{r_0} rdr \int_0^{2\pi} \rho(U + \varphi_x)^2 d\theta - \int_0^{r_0} rdr \int_0^{2\pi} \Delta p(1, r, \theta) d\theta \quad (1\text{-}25)$$

Here, $R(r, \theta)$ represents the body radius at the base and Δp $(1, r, \theta)$ the difference between the local pressure and the stream pressure at the base.

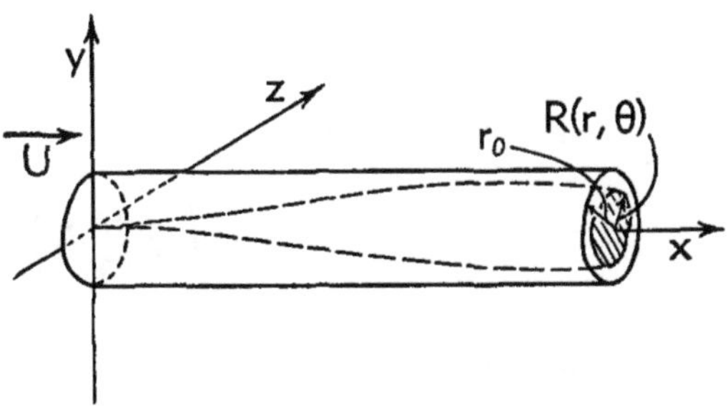

Fig. B,1b. Hypothetical cylinder enclosing body.

The lower limit, βr_0, of the first integral represents the point where the first disturbance from the nose ($x = 0$) reaches the control cylinder. Continuity of mass requires that

$$\pi r_0^2 \rho_\infty U - \int_{\beta r_0}^{1} dx \int_0^{2\pi} \rho\varphi_r r_0 d\theta - \int_{R(1,\theta)}^{r_0} rdr \int_0^{2\pi} \rho(U + \varphi_x) d\theta = 0 \quad (1\text{-}26)$$

Multiplying U into Eq. 1-26 and substituting in Eq. 1-25 gives

$$D = - \int_{\beta r_0}^{1} dx \int_0^{2\pi} \rho\varphi_r\varphi_x r_0 d\theta - \int_{R(1,\theta)}^{r_0} rdr \int_0^{2\pi} \rho(U\varphi_x + \varphi_x^2) d\theta - \int_0^{r_0} rdr \int_0^{2\pi} \Delta p(1, r, \theta) d\theta \quad (1\text{-}27)$$

When r_0 is small it is generally true under the stated restrictions that the φ_x terms are of lower order than the φ_r and $(1/r)\varphi_\theta$ terms, hence in Eq. 1-27 it is permissible to drop the φ_x^2 terms. Since the density differs from the stream value by a factor of the order of φ_x, the density in Eq. 1-27 can be replaced by the stream value. Substituting Eq. 1-3 in Eq. 1-27 and dropping terms of the order of φ_x^2 there is obtained, for r_0 approaching zero,

$$D = -\rho_\infty \int_0^1 dx \int_0^{2\pi} \varphi_x \varphi_r r_0 d\theta + \rho_\infty \int_{R(1,\theta)}^{r_0} r dr \int_0^{2\pi} \frac{1}{2}\left(\varphi_r^2 + \frac{1}{r^2}\varphi_\theta^2\right) d\theta + D_{\text{base}} \quad (1\text{-}28)$$

In Eq. 1-28 the pressure integral over the body base area is written as base drag since this contribution is to be treated separately. Thus the remaining terms represent the pressure drag of the body up to the base. The second integral of Eq. 1-28 may be seen to be similar to that representing the kinetic energy of a two-dimensional flow. Thus remembering that φ satisfies Laplace's equation in the two dimensions r and θ, application of Green's theorem for the plane converts the surface integral to two line integrals about the boundary contours. Hence Eq. 1-28 becomes

$$D = -\rho_\infty \int_0^1 dx \int_0^{2\pi} \varphi_x \varphi_r r_0 d\theta + \frac{\rho_\infty}{2}\left(\int_0^{2\pi} \varphi \frac{\partial \varphi}{\partial r} r_0 d\theta - \int_C \varphi \frac{\partial \varphi}{\partial n} d\tau\right) + D_{\text{base}} \quad (1\text{-}29)$$

where n and τ are the normal and tangent to the body cross section at the plane $x = 1$.

Substitution of Eq. 1-15 into Eq. 1-29 gives now the final relation

$$D = \frac{-\rho_\infty U^2}{4\pi} \int_0^1 \int_0^1 S''(x)S''(\xi) \ln (x - \xi) dx d\xi + \frac{\rho_\infty U^2}{4\pi} [S'(1)] \int_0^1 S''(\xi) \ln (1 - \xi) d\xi - \frac{\rho_\infty}{2} \int_C \varphi \frac{\partial \varphi}{\partial n} d\tau + D_{\text{base}} \quad (1\text{-}30)$$

In this computation the following integrals must be used:

$$\int_0^{2\pi} \cos s\theta \cos \theta d\theta = \int_0^{2\pi} \sin s\theta \sin \theta d\theta = 0 \qquad s \neq 1$$

and

$$\int_0^1 S''(x)dx \int_0^x S''(\xi) \ln (x - \xi) d\xi = \tfrac{1}{2} \int_0^1 \int_0^1 S''(x)S''(\xi) \ln (x - \xi) d\xi$$

Several important considerations[1] result from Eq. 1-30. When the body ends in a cylinder whose elements are parallel to the stream, $S'(x)$ is zero on the cylinder and since there is no lateral motion of the boundary,

[1] These results are, of course, subject to the restrictions requiring smooth and faired body shapes.

$\partial\varphi/\partial n$ is zero on the cylinder. It is clear then that regardless of the angle or shape of the body forward of the cylindrical portion, the drag is that of an equivalent body of revolution having the same distribution of cross-sectional area with x. This drag is given by the first two terms of Eq. 1-30. Further, since $S'(x)$ vanishes if the body is closed, the drag is again the same as that of the equivalent body of revolution; this result is independent of the cross section and angle of attack of the body. Lastly, it will be seen for both the above cases that the drag is independent of the Mach number. It can be seen that, for other cases in which there is some boattailing or variation in cross-section shape or an angle of attack, the drag problem consists in computing an integral involving the lengthwise cross-sectional area distribution and a term which represents the kinetic energy of a two-dimensional flow; in particular the two-dimensional flow resulting from the movement of the boundary of the base section. This flow may therefore consist of a translation of the section as a whole, or general distortion of the boundary including a growth or contraction of section area. For the general case this two-dimensional flow is not simple to calculate. However, all the well-developed techniques of two-dimensional incompressible flow may be applied.

Drag of Bodies of Revolution. For a body of revolution at zero angle of attack only the source term of Eq. 1-15 is needed. The contour integration of Eq. 1-30 is easily performed and thus the drag is given by the relation

$$D = -\frac{\rho_\infty U^2}{4\pi}\int_0^1\int_0^1 S''(x)S''(\xi)\ln(x-\xi)dxd\xi + \frac{\rho_\infty U^2}{2\pi}[S'(1)]\int_0^1 S''(\xi)\ln(1-\xi)d\xi - \frac{\rho_\infty U^2}{4\pi}[S'(1)]^2\ln\frac{\beta R(1)}{2} + D_{\text{base}} \quad (1\text{-}31)$$

When the body has shape such that there is some angle of boattailing, the drag is seen to be dependent on the Mach number, the drag for a given body decreasing slightly with an increase in Mach number.

In comparing drag predictions with experiment it is best to compare pressure distributions or to compare calculated drag with the integrated pressures acting on the bodies. In this way the unknown experimental viscous forces can be eliminated. The expression for the pressure coefficient along a body of revolution subject to the restrictions of the preceding analysis is obtained from Eq. 1-3, 1-12, and 1-24. The expression for φ_x is required, i.e.

$$\varphi_x = \frac{U}{2\pi}\left\{f_0'(x)\ln\frac{\beta R(x)}{2} - \lim_{\epsilon\to 0}\left[f_0'(x)\ln\epsilon + \int_0^{x-\epsilon}\frac{f_0'(\xi)d\xi}{x-\xi}\right]\right\} \quad (1\text{-}36)$$

Hence the pressure coefficient is

$$\frac{\Delta p}{\frac{1}{2}\rho_\infty U^2} = -\frac{2\varphi_x}{U} - \frac{\varphi_r^2}{U^2} = \frac{1}{\pi}\left\{S''(x)\ln\frac{2}{\beta R(x)} + \lim_{\epsilon\to 0}\left[S''(x)\ln\epsilon + \int_0^{x-\epsilon}\frac{S''(\xi)d\xi}{x-\xi}\right] - [R'(x)]^2\right\} \tag{1-37}$$

Eq. 1-37 has been used to determine the theoretical pressure distribution for two slender bodies for which experimental data are available. One of

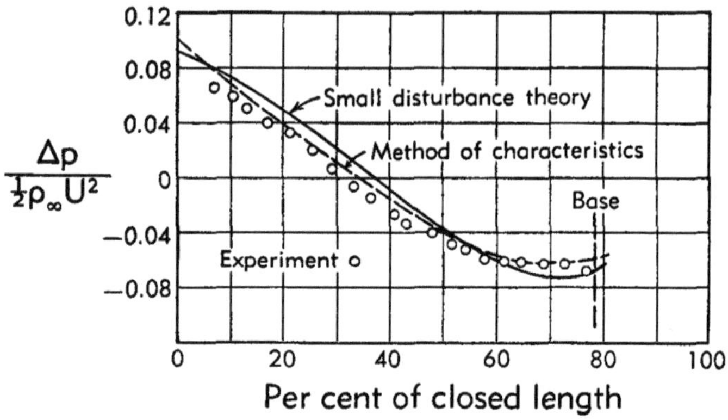

Fig. B,1c. Experimental and theoretical pressures on a parabolic body of revolution, $M_\infty = 1.93$. Parabolic profile $l/D_{max} = 11$.

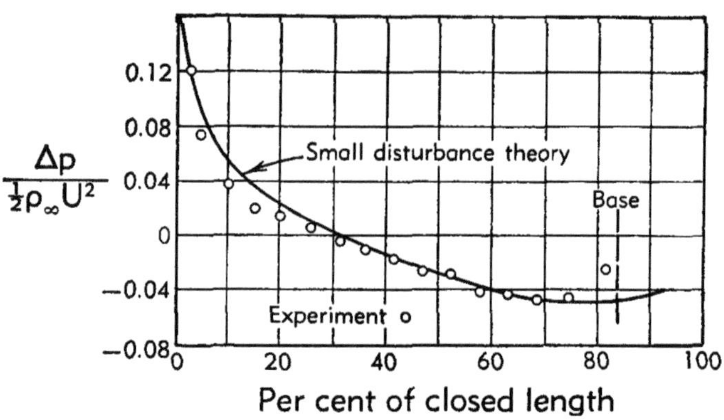

Fig. B,1d. Experimental and theoretical pressure over a slender body of revolution, $M_\infty = 1.93$. Profile $R/l = (R_{max}/l)[(x/l) - (x^2/l^2)]^{\frac{3}{4}}$; $R_{max}/l = 0.04$.

the bodies is formed by rotating a parabolic arc about the flight axis, the other is the closed body shape given by Haack [6] having a minimum drag for a given volume and length. The comparison of theory and experiment is given in Fig. B,1c and B,1d.[2] On Fig. B,1c the pressure obtained by means of the characteristics method is also presented for com-

[2] Tests conducted in NACA Langley 9-inch supersonic tunnel.

parison.[3] In the experiments the pressures near the base were influenced by the base conditions, hence the comparison of theory and experiment should not be made near the base. It is seen that the linear theory gives reasonably good estimates of the pressure distribution, but that the estimated drag will be somewhat too high. In Fig. B,1e is shown the variation of drag coefficient with Mach number for the minimum-drag body of Fig. B,1d. The experimental wave drag values for this figure were obtained by subtracting an appropriate skin friction drag to account for the laminar boundary layer over the body. The values used were those for a flat plate multiplied by the ratio of body wetted area to frontal area. It is seen that the agreement is excellent, even in the prediction of the Mach number effect. Clearly, the linear theory results will become less accurate as the bodies become blunter. However, for fine bodies such as those needed for efficient flight it would appear that the linear theory is adequate.

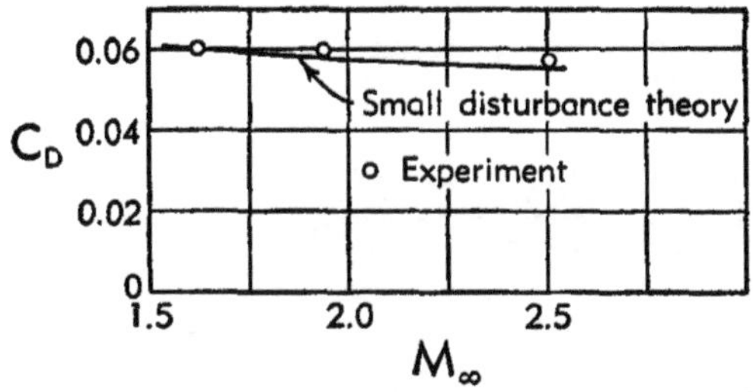

Fig. B,1e. Experimental and theoretical wave drag of body of Fig. B,1d.

The body pressure drag may be computed more accurately by the use of the method of characteristics or by iteration (VI,G and VI,E, respectively). Both of these methods require greater labor but the latter method has been systemized by Van Dyke [*7*] so that the calculation time is considerably less than that for the characteristics method.

Bodies with Shoulders or Open Noses. In the preceding analysis it was found that the flow over a body of revolution at zero angle of attack could be approximated by a distribution of sources along the axis. The source strength for slender but smoothly contoured bodies was shown to be proportional to the rate of change of cross-sectional area. When the body has corners or shoulders, the stated result does not hold and the boundary conditions must be reexamined to determine the source distribution in the vicinity of the shoulder. The following material is essentially due to Lighthill [*8*].

Let it be assumed that the body has one shoulder (extension to more than one is easily made) and that at all other points the body is smooth

[3] The author is indebted to E. S. Love of the NACA for this calculation. For details of this method see VI,G,11.

and free from rapid changes in curvature. The boundary condition can be written

$$\frac{\varphi_r}{U} = R'(x) \tag{1-38}$$

on the boundary. Hence from Eq. 1-9 and 1-38

$$\int_0^{x-\beta R} \frac{f'(\xi)(x-\xi)d\xi}{\sqrt{(x-\xi)^2 - \beta^2 R^2}} = 2\pi R R' = S'(x) \tag{1-39}$$

Now the rate of change of cross-sectional area, $S'(x)$, is a discontinuous function in our problem at say $x = a$. Hence it can be written

$$S'(x) = \int_0^x S''(y)dy + \Delta S'(a)1(x-a) \tag{1-40}$$

where $\Delta S'(a) = 2\pi R'(a)R(a)$ and is the jump in $S'(x)$ at $x = a$. The source distribution can be assumed to be the sum of two functions $F_1(x) + F_2(x)$ such that

$$\int_0^{x-\beta R} \frac{F_1'(\xi)(x-\xi)d\xi}{\sqrt{(x-\xi)^2 - \beta^2 R^2}} = \int_0^x S''(y)dy \tag{1-41}$$

and

$$\int_0^{x-\beta R} \frac{F_2'(\xi)(x-\xi)d\xi}{\sqrt{(x-\xi)^2 - \beta^2 R^2}} = \Delta S'(a)1(x-a) \tag{1-42}$$

Now for slender bodies such that $\beta R/x$ is small, Eq. 1-41 is to the first order

$$F_1'(x) = S''(x) \tag{1-43}$$

This simplification is valid so long as the function $F_1'(\xi)$ is not large, and since $F_1'(\xi)$ is near $S''(\xi)$ it appears that Eq. 1-43 is valid when the curvature of the body is not too rapid. Since the influence of the function $F_2'(\xi)$ must be zero for points ahead of $x = a$ the function itself must be zero ahead of the point $a - \beta R(a)$. Furthermore it is evident from the same simplification used to obtain Eq. 1-43, that $F_2(\xi)$ becomes constant and independent of R for small values of $\beta R/(x - a)$. Hence the dependence of the function $F_2(x)$ on R exists only for points near $x = a$ where R is approximately $R(a)$. To the first order then the R in Eq. 1-42 may be replaced by $R(a)$ with no loss of generality. Eq. 1-42 is now

$$\int_{a-\beta R(a)}^{x-\beta R(a)} \frac{F_2'(\xi)(x-\xi)d\xi}{\sqrt{(x-\xi)^2 - \beta^2 R^2(a)}} = \Delta S'(a)1(x-a) \tag{1-44}$$

This integral equation has been solved by Lighthill [8] in terms of the velocity φ_x associated with the source distribution $F_2(x)$. Hence from

Eq. 1-10 and Lighthill's solution to Eq. 1-44

$$\varphi_{2x} = -\frac{U}{2\pi}\int_{a-\beta R(a)}^{x-\beta R(a)} \frac{F_2'(\xi)d\xi}{\sqrt{(x-\xi)^2 - \beta^2 R^2(a)}}$$

$$= -\frac{\Delta S'(a)U}{2\pi\beta R(a)} U_1\left(\frac{x-a}{\beta R(a)}\right) \quad (1\text{-}45)$$

The function $U_1[(x-a)/\beta R(a)]$ is plotted in Fig. B,1f. It will be seen that φ_{2x} for $(x-a)/\beta R(a)$ approaching zero gives the two-dimensional jump at the corner which is expected from dimensional considerations and that for large $(x-a)/\beta R(a)$ the pressure becomes $\Delta S'(a)/2\pi(x-a)$, independent of R. Thus the function $U_1[(x-a)/\beta R(a)]$ asymptotes the

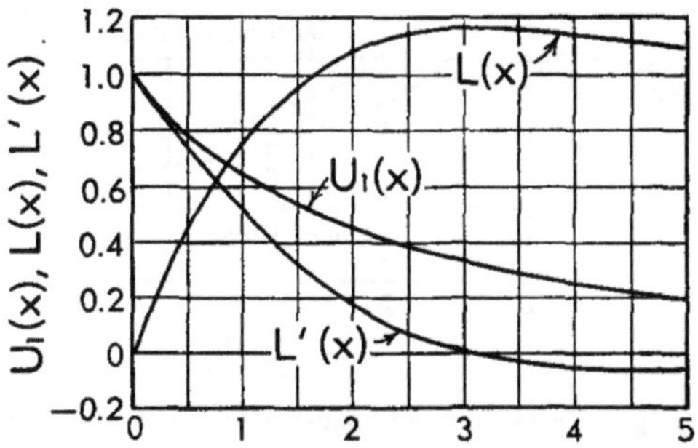

Fig. B,1f. Functions $U_1(x)$, $L(x)$, and $L'(x)$.

function $\beta R(a)/(x-a)$ far downstream of the shoulder. For the complete pressure distribution we need φ_x which is now, from Eq. 1-10,

$$\varphi_x = -\frac{U}{2\pi}\int_0^{x-\beta R} \frac{S''(\xi)d\xi}{\sqrt{(x-\xi)^2 - \beta^2 R^2}} - \frac{\Delta S'(a)U}{2\pi\beta R(a)} U_1\left(\frac{x-a}{\beta R(a)}\right) \quad (1\text{-}46)$$

This expression may be further simplified by integrating the first term by parts and expanding the result in series form, retaining only first order terms. Thus

$$\varphi_x = -\frac{U}{2\pi}\left\{-S''(0)\ln\frac{2x}{\beta R} + \Delta S''(a)\ln\frac{2(x-a)}{\beta R}1(x-a) - \int_0^x S'''(\xi)\ln\frac{2(x-\xi)}{\beta R}d\xi + \frac{\Delta S'(a)}{\beta R(a)} U_1\left(\frac{x-a}{\beta R(a)}\right)\right\} \quad (1\text{-}47)$$

The pressure distribution is now given from Eq. 1-47 and 1-3:

$$\frac{\Delta p}{\frac{1}{2}\rho_\infty U^2} = -\frac{S''(0)}{\pi}\ln\frac{2x}{\beta R} + \frac{\Delta S''(a)}{\pi}\ln\frac{2(x-a)}{\beta R}1(x-a) - \frac{1}{\pi}\int_0^x S'''(\xi)\ln\frac{2(x-\xi)}{\beta R}d\xi + \frac{\Delta S'(a)}{\pi\beta R(a)} U_1\left(\frac{x-a}{\beta R(a)}\right) - [R'(x)]^2 \quad (1\text{-}48)$$

Clearly the second term is not accurate near $x = a$ and should not be retained in this region.

Eq. 1-48 may be applied to open-nosed bodies and bodies with shoulders since an open-nosed body can be considered to be a portion of an infinitely long body starting with a cylinder. To illustrate the type of pressure distribution on such open-nosed bodies, Eq. 1-48 has been applied to a truncated cone, and the results are given in Fig. B,1g together with pressures calculated by Ferri [*9*], using the method of characteristics.

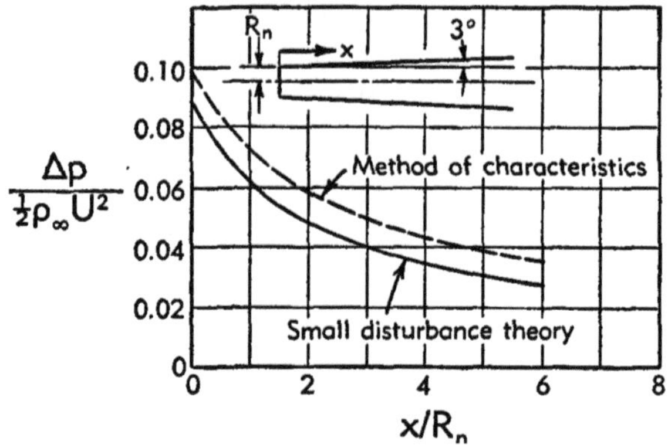

Fig. B,1g. Pressure over open-nosed body, $M_\infty = 1.525$.

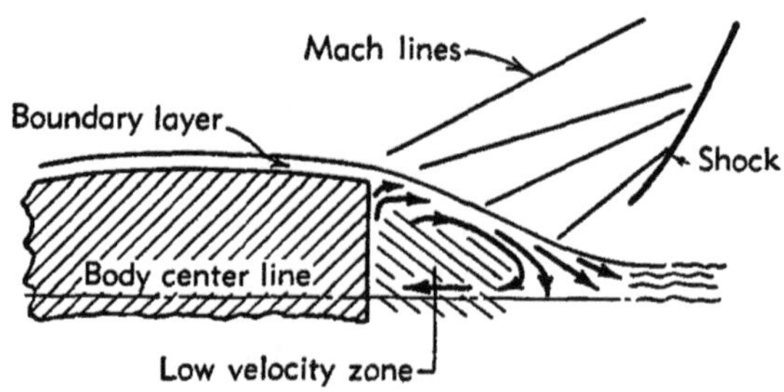

Fig. B,1h. Base flow conditions at supersonic speeds.

Base Drag. The flow conditions at a blunt base of a body of revolution are quite complex and at present no theory exists which can quantitatively predict the variations in base drag with Reynolds number, Mach number, and body shape. Reasonably good estimates can be made for certain cases by using available experimental data and correlation of the experimental results with suitable parameters such as boundary layer thickness. To understand the variations of base pressure under various flight conditions it is necessary to visualize the type of flow and flow action at the body base. A typical flow condition at a supersonic speed is shown in Fig. B,1h. Behind the base area the outer streamlines tend to converge toward the axis; the initial turning tends to reduce the pressure on the streamline and hence in the base region. The straightening out of the streamlines as the axis is approached requires a pressure rise, the pressure rise being greater the more rapid the change in curvature of the outer streamline. The ability of the boundary layer to accommodate this pressure rise is one of the main factors in influencing the base pressure. Consider the action in detail as follows: Along the boundary, fluid is removed from the base region by jet-pump action, and thus some fluid must be returned at the downstream convergence point. If the convergence is too great, the pumping length is shortened, while at the same time the pressure rise downstream is increased causing more fluid to be returned to the base region. Thus the fluid entrained tends to increase, which in turn reduces the convergence of the streamlines and the base drag. It is clear that a turbulent boundary layer will provide increased jet-pumping action and greater capacity for overcoming the final adverse pressure gradient, hence the turbulent flow should generally yield lower base pressures and higher drag than laminar flow. Another consideration would be, of course, the ratio of boundary layer thickness to base height. An increase in relative boundary thickness would be expected to reduce the base drag since in the limit of very large boundary layer thickness the base pressure should approach the stream value. From considerations similar to those just discussed, Chapman [*10*] obtained a reasonable correlation of base pressures with boundary layer thickness and type of boundary layer. It was assumed that the relative laminar boundary layer thickness at the base was proportional to the length-to-diameter ratio, L/h, and inversely proportional to the square root of the Reynolds number. For the turbulent boundary layer the relative thickness was taken proportional to L/h and inversely proportional to the fifth root of the Reynolds number. In both cases, the variation of boundary layer thickness with body shape (see the discussion of viscous drag below) was neglected, although a small correction was applied for pressure on the body just ahead of the base. The test results are shown in Fig. B,1i.

It can be seen that the data appears to correlate well with a straight line; however, the slope of the line is a function of Mach number and type of boundary layer, laminar or turbulent. The results indicate that for a turbulent boundary layer the base pressure varies very slowly with the boundary layer thickness; this effect is also verified by tests in the Naval Ordnance Laboratory 40 × 40-cm supersonic wind tunnel as reported by Kurzweg [*11*].

Some general effects of Mach number and body shape can be reasoned from the physics of the flow. The pressures acting in the base region must depend largely on the shape of the streamline just outside the boundary layer. Now in subsonic flow it is known that the pressures on the rear of the body tend to produce a thrust; hence the base pressures in subsonic flow under certain conditions may be above stream pressure. At supersonic speeds, however, the streamlines aft of the base generally form a shape along which the pressure coefficients are negative; thus it is clear that base drag at subsonic speeds should generally be less than at supersonic speeds. The pressure distribution for closed bodies of revolution is characterized by a stagnation point at the rear and a peak negative pressure at some point behind the section of maximum thickness. As the Mach number changes from subsonic to supersonic, the pressure distribution curve is sheared backward so that the minimum pressure point is farther aft; the same general trend with Mach number occurs for bodies

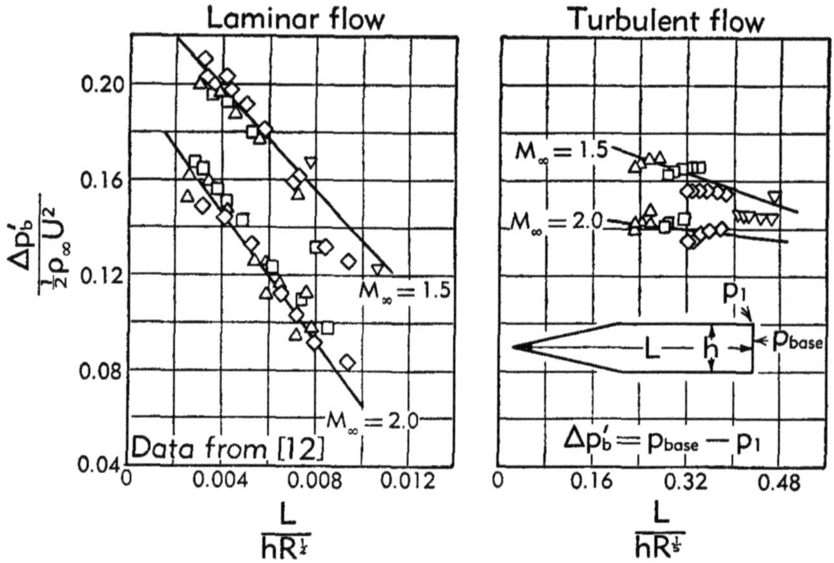

Fig. B,1i. Correlation of base pressure with relative boundary layer thickness for various bodies without boattailing.

ending as a cylinder. For small ratios of base diameter to maximum diameter, the base region may lie in a positive pressure position of the field, although at high supersonic speeds it is unlikely that the base pressures would become positive. It is clear then that the base pressures will be dependent on the type of boundary layer, the Mach number, and the body shape. In practical cases the body may also have tail fins, rotation bands, and rocket tube openings, and hence strong boundary layer disturbances may be present together with the pressure fields produced by the fins. Indeed, the size and position of the fins relative to the base should influence the base pressure. Clearly, base pressure estimation requires a knowledge of the influence of all of these variables, but at present the experimental information is quite restricted. The range of base pressures which occur for some typical bodies with and without fins can be seen in Fig. B,1j on which test data from free flight [*12*], rocket firings

[*13*], and wind tunnel tests [*10,11*] are given. The data are presented for models without boattailing on which the boundary layer was turbulent. The absolute vacuum conditions at the base represent the simplest limiting pressures and are given by

$$\frac{\Delta p_{\text{base}}}{\frac{1}{2}\rho_\infty U^2} = -\frac{\gamma M_\infty^2}{2} \tag{1-49}$$

where $\gamma = 1.4$ for air. The curve is presented for reference.

It can be seen that the fins used on the rocket models produced a considerable increase in base drag in the moderate supersonic range. The magnitude of the fin effect will be expected to vary from one fin configuration to another [*14*], and certainly extrapolation of the data for other fin sizes and shapes will not be possible until further information is available on the effects of fin sweep, thickness, location, etc.

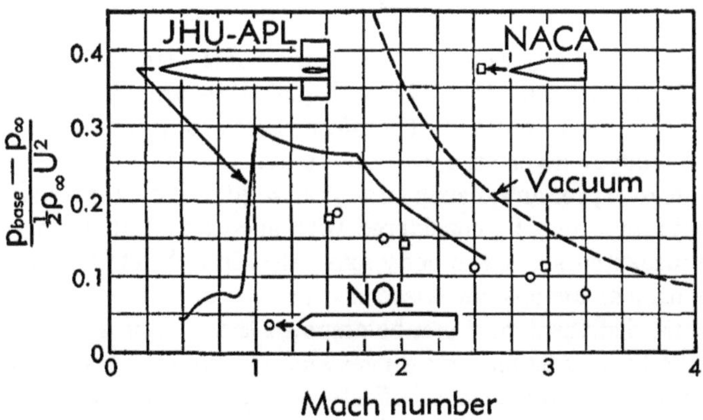

Fig. B,1j. Typical base pressures for finned and finless bodies of revolution.

Viscous Drag. It has been found for thin two-dimensional wings at zero angle of attack that the viscous drag can be estimated with reasonably good accuracy by simply using the skin friction coefficient for a flat plate of the same Reynolds number. Thus it is found that small pressure gradients do not affect the viscous drag to the first order even though the skin friction distribution may be somewhat altered. Of course, the pressure distribution can produce significant effects near the critical Reynolds numbers. However, for fully laminar or turbulent boundary layers the above statements hold approximately true. For slender bodies it is again commonly assumed that the viscous drag may be computed using the skin friction coefficient for a flat plate. In this case the justification is not physically clear since the boundary layer on the nose of a body of revolution will certainly be thinner than on a corresponding flat plate. For laminar boundary layers Mangler [*15,16*] has given an interesting method by which the boundary layer on bodies of revolution can be related to a two-dimensional boundary layer; thus the validity of the use of two-dimensional skin friction factors for bodies of revolution can be checked. Briefly, the work of Mangler shows that the boundary layer equations for a body of revolution expressed in coordinates measured along and normal to a meridian contour can be transformed by a suitable transformation of coordinates into the boundary layer equations for two-dimensional flow. Thus, if s is the length along the body meridian contour, n the normal to s, and all barred quantities denote those pertaining to two-dimensional flow, the following equations transform the boundary layer equations for rotational symmetry to those of two-dimensional flow:

$$\bar{s} = \left(\frac{c}{L}\right)^2 \int_0^s R^2(s)ds \tag{1-50}$$

$$\bar{n} = \left(\frac{c}{L}\right) R(s)n \tag{1-51}$$

where $R(s)$ is the body radius, L is a characteristic length, and c is a constant scale factor. All other physical quantities are the same at corresponding points s, n and $\bar{s}$, $\bar{n}$. Therefore there is a similarity between the laminar boundary layer on a given body and that on a flat plate subject to an analogous pressure distribution $\bar{p}(s) = p(s)$. Now for slender bodies the coordinate s may be considered with good approximation to be the distance measured along the body axis. Also, in light of the fact that for two-dimensional airfoils the viscous drag is reasonably independent of the pressure distribution, it can be assumed that the pressure is constant over the body. In this manner we get an approximate idea of the magnitude of the three-dimensional effects on the skin friction coefficients. Therefore consider a body of length L having a surface of revolution given by $r = R(s)$. The viscous drag of such a body having constant pressure distribution will be obtained directly from the boundary layer momentum thickness Θ, at the base of the body. Thus

$$D = \rho_\infty U^2 \Theta_{\text{base}} 2\pi R_{\text{base}} \tag{1-52}$$

since the boundary layer is considered to be thin compared with R. The momentum thickness is expressible by the relation:

$$\Theta = \int_0^\infty \frac{\rho u}{\rho_\infty U}\left(1 - \frac{u}{U}\right) dn \tag{1-53}$$

or from Eq. 1-51:

$$\Theta_{\text{base}} = \int_0^\infty \frac{\rho u}{\rho_\infty U}\left(1 - \frac{u}{U}\right) \frac{L}{cR_{\text{base}}}\, d\bar{n} = \bar{\Theta}\, \frac{L}{cR_{\text{base}}} \tag{1-54}$$

since the density ρ and the velocity u are unchanged at corresponding transformed points. The quantity $\bar{\Theta}$ can now be referred to a flat plate length L by adjusting the constant c. Thus, from Eq. 1-50, setting $\bar{s} = s = L$

$$\frac{L}{c} = \sqrt{\frac{\int_0^L R^2(s)ds}{L}} \tag{1-55}$$

and hence

$$\Theta(L) = \sqrt{\frac{\int_0^L R^2(s)ds}{LR^2_{\text{base}}}}\,\bar{\Theta}(L) \tag{1-56}$$

The expression (Eq. 1-56) now relates the momentum thickness of the body base $s = L$ to the momentum thickness of a boundary layer on a flat plate of length L. It will be seen that for a slender cone the result is obtained that the three-dimensional momentum thickness is smaller, by a factor $1/\sqrt{3}$, than the momentum thickness over a two-dimensional flat plate of the same length and in the same air stream. This result has also been shown by Hantzsche and Wendt [17].

For the general body the commonly used method of drag estimation would yield

$$D_e = \frac{\rho_\infty U^2 C_f S_w}{2} = \frac{\rho_\infty U^2 \bar{\Theta} S_w}{L} \tag{1-57}$$

where C_f is the two-dimensional flat plate skin friction coefficient for a plate of length L, in the same air stream as that of the body, and S_w is the body wetted area. Making use of Eq. 1-57, 1-56, and 1-52, the ratio of true drag of the body to that computed from the rule of thumb, Eq. 1-57 becomes

$$\frac{D}{D_e} = \frac{\sqrt{\int_0^L \frac{R^2}{L}\,ds}}{\int_0^L \frac{R}{L}\,ds} \tag{1-58}$$

Thus from Eq. 1-58 the drag of a cone would be about 15 per cent greater than that computed by Eq. 1-57, while for a parabolic body the drag increase would be about 9 per cent. As would be expected the ratio for an open-nosed cylinder of constant radius would be unity. It is interesting to note that the ratio of Eq. 1-58 is simply the ratio of root-mean-squared body radius to the mean body radius.

For turbulent flow the above relations do not hold. However, if the Prandtl mixing length concept of turbulence is adopted, a transformation similar to that for laminar flow can be formulated. The transformation for turbulent flow rests on less rigorous concepts than that for laminar flow, but should give an approximate indication of the magnitude of the three-dimensional effects. Mangler gives the following equation for zero pressure gradients along the body in an incompressible flow:

$$\bar{s} = \left[\int_0^s \frac{c}{L}\,(R(s))^{\frac{5}{4}}ds\right]^{\frac{4}{5}} \tag{1-59}$$

$$= \frac{c}{L}\,R(s)n \tag{1-60}$$

Thus the ratio o drag computed for a body of revolution to that computed for a plate of equal wetted area and length is given by the relation

$$\frac{D}{D_c} = \frac{\left[\int_0^L \frac{(R)^{\frac{5}{4}}}{L}\,ds\right]^{\frac{4}{5}}}{\int_0^L \frac{R}{L}\,ds} \tag{1-61}$$

It is easily checked that the simple rule of thumb for calculating the drag is very good in the case of turbulent flow, since, for example, the expression (Eq. 1-61) yields a value of about 1.04 for cones and 1.02 for parabolic bodies.

It should be realized that the values of Eq. 1-61 are so close to unity that the differences are not within the probable accuracy of the method. Thus it appears that for the turbulent flow over bodies of revolution, the drag should be given closely by multiplying the two-dimensional skin friction factor by the wetted area and dynamic pressure $\frac{1}{2}\rho_\infty U^2$.

B,2. Lift and Pitching Moment.

Linearized treatment. The lift and side forces acting on a slender body are obtained most conveniently from momentum considerations. Let the body, which is assumed to be at a small angle of attack, be enclosed by a cylindrical box having its axis parallel to the stream as shown in Fig. B,2a. The box is drawn so that the radius is that at which the Mach cone from the nose intersects a plane drawn through the base and normal to the free stream. The lift and side force are now given by the change in lateral momentum of the stream in passing from infinity upstream to the control surface. Thus the force in the y direction is

$$Y = -\int_{A_1} \rho_\infty(U + \varphi_x)\varphi_y dA_1 \tag{2-1}$$

and the force in the z direction is

$$Z = -\int_{A_1} \rho_\infty(U + \varphi_x)\varphi_z dA_1 \tag{2-2}$$

where the area A_1 is shown in Fig. B,2a and does not include the base area of the body. Clearly Eq. 2-1 and 2-2 contain terms of second and third order which cannot be retained, hence Eq. 2-1 becomes

$$Y = -\int_{A_1} \rho_\infty U \varphi_y dA_1 \tag{2-3}$$

Integrating once with respect to y, Eq. 2-3 gives the line integrals

$$Y = \rho_\infty U \int_{c_1} \varphi dz - \rho_\infty U \int_{c_2} \varphi dz$$

On the contour c_1 (the trace of the Mach cone in the base control plane) the potential is zero, hence the force in the y direction is simply

$$Y = -\rho_\infty U \int_{c_2} \varphi dz \tag{2-4}$$

where the integral is taken counterclockwise when looking upstream at the base. In the same manner the force in the z direction becomes

$$Z = \rho_\infty U \int_{c_2} \varphi dy \tag{2-5}$$

The magnitude and direction of the lateral force can be expressed in vector notation; thus if $\mathbf{j}$ and $\mathbf{k}$ are unit vectors in the y and z directions respectively, the vector lateral force is

$$\mathbf{j}Y + \mathbf{k}Z = -\rho_\infty U \int_{c_2} \varphi[\mathbf{j}dz - \mathbf{k}dy] = U \int_{c_2} \rho_\infty \varphi \bar{\mathbf{n}} d\tau \tag{2-6}$$

where $\bar{\mathbf{n}}$ is the outward directed unit vector normal to the contour c_2, and τ as before is the distance along c_2. The problem of calculating the

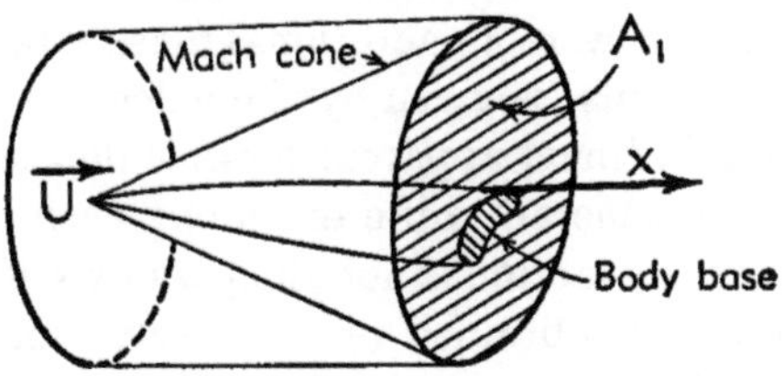

Fig. B,2a. Sketch showing control surface for calculating lateral forces.

lift of a slender body is now resolved to finding the potential function at the body base. For slender bodies having smooth contours we may again use the simplified potential given by Eq. 1-15, and since Eq. 1-15 is a solution of Laplace's equation in the variables r and θ, the calculation of lift depends on a two-dimensional flow involving the contraction, distortion, and translation of a cylindrical body of the shape of the body base. The motion of the boundary of the two-dimensional flow is given by the boundary condition (1-21)

$$\frac{\partial \varphi}{\partial n} = U \frac{dn}{dx}$$

Hence, boattailing has the effect of contracting the boundary, the angle of attack produces a translational velocity of the contour equal to $U\alpha$, and general changes of cross-sectional shape produce corresponding normal velocities of the boundary for the two-dimensional problem. This result is essentially that given by application of Munk's airship theory [*18*] to slender bodies of arbitrary shape in incompressible flow. The term $\rho_\infty \varphi$ of Eq. 2-6 is the so-called impulsive pressure of the incompressible flow, and its integral over the boundary results in the vector fluid impulse necessary to create the given fluid motion from a state of rest. In hydrodynamics textbooks, the problem is generally not complicated by the growth or contraction of the contour cross-sectional area; however, this trouble in the present problem can be circumvented by considering φ to be the superposition of two potentials φ_A and φ_B, φ_A containing the source flow but contributing zero lateral force and φ_B representing the additional potential necessary to create the lift. The potential φ_A is complete at the attitude of zero lift and the resultant normal velocity at the boundary accounts for the growth and distortion of the cross section. The lifting potential φ_B, therefore, contributes only to a lateral translation of the boundary of magnitude $U\alpha$, where α is measured as the change in body inclination to the stream from the position of zero lift. The lateral force is now related to the fluid impulse of a two-dimensional incompressible flow produced by the translation of a two-dimensional cylinder of the same cross section as the body base. When this impulse happens to be directed along the line of motion, the body is moving along one of its two perpendicular "axes of permanent translation" and for this condition the fluid acts as though it were an additional mass attached rigidly to the body. Thus the fluid impulse is given by the velocity times an additional apparent mass. The additional apparent mass is dependent on the cross-sectional shape, and is generally different along each axis of permanent translation [*18*]. When the motion is not along an axis of permanent translation the fluid impulse is obtained by the vector sum of the apparent momentum along the lines of permanent translation. Let the direction of motion coincide with the y axis (no loss of generality); then in our hypothetical two-dimensional flow the impulsive pressures $\rho_\infty \varphi_B$ acting over the contour produce a fluid impulse which is now

$$-\int_{C_2} \rho_\infty \varphi_B \bar{\mathbf{n}} d\tau = -\mathbf{j} U\alpha(m_a \cos^2 \beta + m_b \sin^2 \beta) + \mathbf{k} U\alpha(m_a \cos \beta \sin \beta - m_b \sin \beta \cos \beta) \quad (2\text{-}7)$$

where m_a and m_b are the additional apparent masses along the axes of permanent translation, a and b, which make an angle β with the Cartesian coordinates as shown in Fig. B,2b. The values of m_a and m_b have been computed for ellipses and are given as

$$m_b = \frac{\rho_\infty \pi a^2}{4} \tag{2-8}$$

$$m_a = \frac{\rho_\infty \pi b^2}{4} \tag{2-9}$$

where a and b are used here to denote the major and minor axes of the

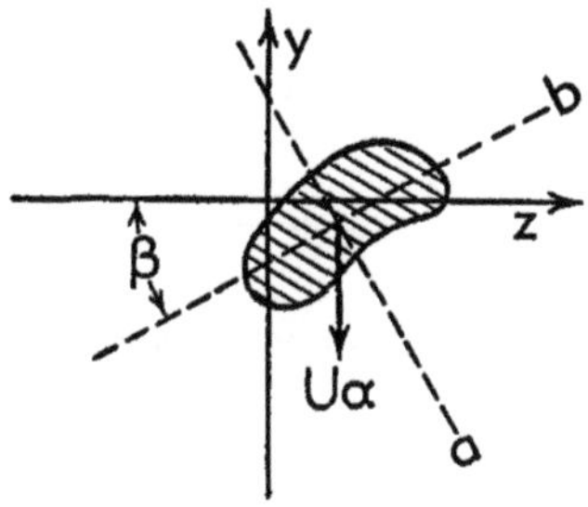

Fig. B,2b. Axes of permanent translation.

ellipse. From Eq. 2-6 and 2-7 the lateral forces for the arbitrary body become

$$Y = \alpha U^2(m_a \cos^2 \beta + m_b \sin^2 \beta) \tag{2-10}$$

$$Z = \alpha U^2(m_b - m_a) \sin \beta \cos \beta \tag{2-11}$$

and for the body having an elliptical base contour,

$$Y = \frac{\rho_\infty U^2 \alpha \pi}{4} (b^2 \cos^2 \beta + a^2 \sin^2 \beta) \tag{2-12}$$

$$Z = \frac{\rho_\infty U^2 \alpha}{4} \pi (a^2 - b^2) \sin \beta \cos \beta \tag{2-13}$$

Munk [*18*] has pointed out that for contours not greatly unlike ellipses the additional apparent masses are only slightly changed and hence the above results will yield very good estimates of the lift.

Since the above results yield the lateral forces acting ahead of any section, the local forces per unit length are given by differentiating Eq. 2-10 and 2-11 with respect to x. Multiplying by a moment arm x, and integrating, we obtain the moments about the z and y axes respectively:

$$M_z = -\alpha U^2 \int_0^1 \frac{\partial(m_a \cos^2 \beta + m_b \sin^2 \beta)}{\partial x} x dx \tag{2-14}$$

$$M_y = -\alpha U^2 \int_0^1 \frac{\partial[\sin \beta \cos \beta(m_b - m_a)]}{\partial x} x dx \tag{2-15}$$

For the body with an elliptical section a particular expression can be obtained from Eq. 2-8, 2-9, 2-14, and 2-15.

For the body of revolution there is obtained as a special case of Eq. 2-12:

$$y = \frac{\pi \rho_\infty U^2 \alpha d_{\text{base}}^2}{4}$$

where d is the diameter. The lift coefficient based on the base area is

$$C_L = \frac{y}{(\frac{1}{2}\rho_\infty U^2)(\pi d^2/4)} = 2\alpha \tag{2-16}$$

The moment about the nose is from Eq. 2-14 and 2-8,

$$M = -\frac{\rho_\infty U^2 \alpha \pi}{4} \int_0^1 \frac{dd^2}{dx} x dx = -\rho_\infty U^2 \alpha \left[S_{\text{base}} - \int_0^1 S(x) dx \right]$$

or the moment coefficient, based on base area and body length is,

$$C_M = -2\alpha \left(1 - \frac{S_{\text{av}}}{S_{\text{base}}}\right) \tag{2-17}$$

The drag produced by the lift of the slender and smooth but otherwise arbitrary body can be shown to be equal to the lift times one half the angle of attack. The proof follows: Consider again the potential φ_A and φ_B defined in the previous discussion. At zero lift the potential φ_B is zero and, as will be seen, the drag at this condition is the minimum drag. The addition of the lifting potential φ_B gives rise to three additional terms in the drag equation (1-30). These terms constitute the additional drag due to lift ΔD,

$$\Delta D = -\frac{\rho_\infty}{2} \left(\int_{C_2} \varphi_B \frac{\partial \varphi_B}{\partial n} d\tau + \int_{C_2} \varphi_B \frac{\partial \varphi_A}{\partial n} d\tau + \int_{C_2} \varphi_A \frac{\partial \varphi_B}{\partial n} d\tau \right) \tag{2-18}$$

Since in the present approximation φ_A and φ_B are both solutions of the two-dimensional Laplace equation, the latter two terms of Eq. 2-18 are equal since from the symmetrical form of Green's theorem

$$\int_{C_2} \varphi_A \frac{\partial \varphi_B}{\partial n} d\tau - \int_{C_2} \varphi_B \frac{\partial \varphi_A}{\partial n} d\tau = \iint (\nabla^2 \varphi_A + \nabla^2 \varphi_B) dA = 0$$

Again considering the angle of attack to be in the direction of the negative y axis, the quantity $(\partial \varphi_B / \partial n) d\tau$ in the last term of Eq. 2-18 can be resolved on the contour to $U\alpha dz$; thus

$$\Delta D = -\frac{\rho_\infty U \alpha}{2} \int_{C_2} \varphi_B dz - \rho_\infty U \alpha \int_{C_2} \varphi_A dz \tag{2-19}$$

but the integrals shown are equal to the lift times $\alpha/2$ since the latter term is zero by the definition of φ_A and the first integral term is the lift in the y direction times $\alpha/2$ from Eq. 2-4. We now can write

$$\Delta D = \frac{L\alpha}{2}$$

or

$$\Delta C_D = \frac{C_L\alpha}{2} \tag{2-20}$$

Bodies of revolution. In the previous article the simplified solution (1-15) was used. However, for open-nosed bodies and for bodies having slopes which are discontinuous, Eq. 1-15 is no longer valid and the form of the potential near the nose must be examined more closely. The problem becomes extremely difficult for bodies of arbitrary cross section, hence the discussion will be limited to bodies of revolution. In Art. 1, the body of revolution at zero angle of attack was replaced by a suitable distribution of sources along the flight axis (x axis). For treating the case of inclined bodies of revolution the boundary conditions make it desirable to work with coordinates fixed to the body. It is then found that the required potential can be obtained as a distribution of sources and doublets along the body axis. The velocity potential for the flow in terms of the body coordinates x_1, r_1, θ_1 can be obtained from Eq. 1-4 by simply replacing $x\ r\ \theta$ with x_1, r_1, θ_1. That this is a permissible operation is easily proved by considering the differential equation for which Eq. 1-4 is the general solution. In the proof it is most convenient to use the corresponding Cartesian sets of coordinates x, y, z, and x_1, y_1, z_1, the latter again being fixed to the body and therefore rotated about the z axis through the small angle α. The differential equation of motion is

$$\beta^2\varphi_{xx} - \varphi_{yy} - \varphi_{zz} = 0 \tag{2-21}$$

but

$$\begin{aligned} x_1 &= x\cos\alpha - y\sin\alpha \\ y_1 &= y\cos\alpha + x\sin\alpha \\ z_1 &= z \end{aligned} \tag{2-22}$$

Substituting in Eq. 2-21 and setting $\cos\alpha = 1$ and $\sin\alpha = \alpha$, there is obtained

$$\beta^2\varphi_{x_1x_1} - \varphi_{y_1y_1} - \varphi_{z_1z_1} = 0(2\alpha M_\infty^2\varphi_{x_1y_1}) = 0 \tag{2-23}$$

The term on the right is apparently of the same order of magnitude as terms already dropped in linearizing the equation of motion; hence the differential equation for tilted coordinates is still Eq. 2-21, and its general solution is Eq. 1-4 with x, r, θ, replaced by x_1, r_1, θ_1.

The equation for the pressure in terms of the new coordinates is affected by the rotation; thus, using Eq. 2-22, 1-22, and the relation $y_1 = r_1 \cos \theta_1$, the pressure coefficient becomes

$$\frac{\Delta p}{\frac{1}{2}\rho_\infty U^2} = -\left[\frac{2\varphi_{x_1}}{U} + \frac{2\alpha}{U}\left(\varphi_{r_1} \cos \theta_1 - \varphi_{\theta_1} \frac{\sin \theta_1}{r_1}\right) + \frac{\varphi_{r_1}^2}{U^2} + \frac{1}{r_1^2}\frac{\varphi_{\theta_1}^2}{U^2}\right] \quad (2\text{-}24)$$

For bodies of revolution the boundary conditions of zero flow through the surface give

$$U\frac{dR_1}{dx_1} - U\alpha \cos \theta_1 = \left(\frac{\partial \varphi}{\partial r_1}\right)_{R_1} \quad (2\text{-}25)$$

Here only quantities of highest order of magnitude are retained, and R_1 is the body radius. The necessary solution is now the first two terms of Eq. 1-4. Performing the indicated operations and making use of the relation $\xi = x_1 - \beta r_1 \cosh u$, we have

$$\varphi = -\frac{U}{2\pi}\int_{-\infty}^{x_1-\beta r_1} \frac{f_0(\xi)d\xi}{\sqrt{(x_1-\xi)^2 - \beta^2 r_1^2}} + \frac{U \cos \theta_1}{2\pi r_1}\int_{-\infty}^{x_1-\beta r_1} \frac{f_1'(\xi)(x_1-\xi)d\xi}{\sqrt{(x_1-\xi)^2 - \beta^2 r_1^2}} \quad (2\text{-}26)$$

The first term (source term) satisfies the first term of Eq. 2-25 as discussed in Art. 1; hence the second (doublet) term of Eq. 2-26 is required to satisfy the second condition of Eq. 2-25. We have then an integral equation for $f_1(\xi)$, from Eq. 2-26 and 2-25,

$$U\alpha \cos \theta_1 = \frac{U \cos \theta_1}{2\pi R_1^2}\int_{-\infty}^{x_1-\beta R_1} \frac{f_1''(\xi)(x_1-\xi)^2 d\xi}{\sqrt{(x_1-\xi)^2 - \beta^2 R_1^2}} \quad (2\text{-}27)$$

The differentiation of the second term of Eq. 2-26 is best carried out before changing the variable from u to ξ. If the expression $f''(\xi)$ is bounded and of the same order of magnitude as $f'(\xi)$, the integral may be expanded in a series involving powers of βR_1 and only terms of highest order of magnitude retained. For this case Eq. 2-27 becomes

$$2\alpha S(x_1) = \int_{-\infty}^{x_1} f_1''(\xi)(x_1-\xi)d\xi = f_1(x_1) \quad (2\text{-}28)$$

The doublet distribution is therefore proportional to the cross-sectional area and the angle of attack. It is seen that the condition that $f''(x)$ be bounded and of the order of $f'(x)$ requires the body to be smooth and without regions of rapid curvature. If φ_0 is defined as the contribution of the source distribution to the potential, the velocity components on the slender body are

$$\varphi_{x_1} = (\varphi_0)_{x_1} + 2U\alpha \cos\theta_1 \frac{dR_1}{dx_1}$$

$$\varphi_{r_1} = U\frac{dR_1}{dx_1} - U\alpha\cos\theta_1 \tag{2-29}$$

$$\frac{1}{R_1}\varphi_{\theta_1} = -U\alpha\sin\theta_1$$

The pressure distribution can now be obtained by substitution of Eq. 2-29 into Eq. 2-24. Hence the contribution to the pressure coefficient due to angle of attack may be written as

$$\frac{\Delta p}{\frac{1}{2}\rho_\infty U^2} - \left(\frac{\Delta p}{\frac{1}{2}\rho_\infty U^2}\right)_{\alpha=0} = -4\alpha\cos\theta_1\frac{dR_1}{dx_1} + \alpha^2(1 - 4\sin^2\theta_1) \tag{2-30}$$

Following the development by Lighthill [*8*], in essence, we consider a body having discontinuous slopes. It is clear that such bodies have infinite curvature and hence Eq. 2-28 is not the correct result. Let us assume an open-nosed body having one shoulder at some point $x = a$. For this case we must reexamine Eq. 2-27 to find the correct doublet distribution. Rewriting Eq. 2-27, the integral equation for $f_1(\xi)$ becomes

$$2\alpha S(x_1) = \int_{-\infty}^{x_1-\beta R_1} \frac{f''(\xi)(x_1-\xi)d\xi}{\sqrt{(x_1-\xi)^2-\beta^2R_1^2}} \tag{2-31}$$

The function $S(x)$ has a discontinuity at the nose, $x_1 = 0$, and a discontinuous derivative at $x_1 = a$; hence we may write the identity

$$S(x_1) = S(0)1(x_1) + x_1S'(0)1(x_1) + (x_1 - a)\Delta S'(a)1(x_1 - a) + \int_0^{x_1} d\lambda \int_0^{\lambda} S''(\xi)d\xi \tag{2-32}$$

where the integral term represents an integrative process neglecting the singular points $x_1 = 0$ and $x_1 = a$, and represents a function having a continuous derivative. We also assume that $S''(x)$ is of the order of the body thickness ratio squared, thus restricting the regions of high curvature to the points $x_1 = 0$ and $x_1 = a$. Proceeding as in the zero angle-of-attack case (see Art. 1) we assume the doublet distribution to be given by the sum of four functions, $G_1(\xi)$, $G_2(\xi)$, $G_3(\xi)$, and $G_4(\xi)$, each function satisfying one of the components of $S(x)$ in Eq. 2-32. Four integral equations result; the first,

$$2\alpha\int_0^{x_1} d\lambda \int_0^{\lambda} S''(\xi)d\xi = \int_{-\infty}^{x_1-\beta R_1}\frac{G_1''(\xi)(x_1-\xi)^2d\xi}{\sqrt{(x_1-\xi)^2-\beta^2R_1^2}} \tag{2-33}$$

has for its solution $G_1''(\xi) = +2\alpha S''(\xi)$ as previously discussed for the case of smooth bodies (Eq. 2-27 and 2-28). The other integral equations are

$$2\alpha S(0)1(x_1) = \int_{-\beta R_1(0)}^{x_1-\beta R_1} \frac{G_2''(\xi)(x_1-\xi)^2 d\xi}{\sqrt{(x_1-\xi)^2-\beta^2R_1^2}} \tag{2-34}$$

$$2\alpha S'(0)1(x_1) = \int_{-\beta R_1(0)}^{x_1-\beta R_1} \frac{G_3''(\xi)(x_1-\xi)^2 d\xi}{\sqrt{(x_1-\xi)^2-\beta^2R_1^2}} \tag{2-35}$$

$$2\alpha(x_1-a)\Delta S'(a)1(x_1-a) = \int_{a-\beta R(a)}^{x_1-\beta R_1} \frac{G_4''(\xi)(x_1-\xi) d\xi}{\sqrt{(x_1-\xi)^2-\beta^2R_1^2}} \tag{2-36}$$

For large values of $\beta R_1/x_1$ and $\beta R_1/(x_1 - a)$, whichever is appropriate, the dependence of the above integrals on βR_1 becomes of a low order of magnitude, and the dependence on βR_1 is only first order near $x_1 = 0$ or $x_1 = a$. Hence the values of βR_1 may be set equal to $\beta R_1(0)$ in Eq. 2-34, and 2-35 and $\beta R_1(a)$ in Eq. 2-36, with no loss in accuracy. The resulting simplified integral equations have been solved by Ward [*19*] in terms of the corresponding x_1-wise velocity which is needed to determine the pressure. Ward gives the following result for φ_{x_1} resulting from the doublet distributions $G_2(\xi)$ and $G_3(\xi)$ and $G_4(\xi)$:

$$(\varphi_{x_1})_2 = \frac{U\cos\theta_1}{2\pi R_1}\int_{-\beta R_1(0)}^{x_1-\beta R_1(0)} \frac{G_2''(\xi)(x_1-\xi)d\xi}{\sqrt{(x_1-\xi)^2-\beta^2R_1(0)^2}}$$

$$= \frac{U\alpha\cos\theta_1 S(0)}{\pi\beta R_1(0)R_1(x)} L'\left(\frac{x_1}{\beta R_1(0)}\right) \tag{2-37}$$

$$(\varphi_{x_1})_3 = \frac{U\alpha\cos\theta_1 S'(0)}{\pi R_1} L\left(\frac{x_1}{\beta R_1(0)}\right) \tag{2-38}$$

$$(\varphi_{x_1})_4 = \frac{U\alpha\cos\theta_1 \Delta S'(a)}{\pi R_1} L\left(\frac{x_1-a}{\beta R_1(a)}\right) \tag{2-39}$$

The functions L and L' are given in Fig. B,1f. Several important features of these solutions must be noted; first, the solution near the open nose (Eq. 2-37) indicates a pressure jump from stream pressure to the two-dimensional value at the 0 and 180° meridians, and therefore the pressure is of the order of the angle of attack α. In the terms arising from the slope discontinuity at $x_1 = 0$ and $x_1 = a$, however, the resulting φ_{x_1} is of the order of α times the change in body slope. Now the pressure jump at these points due to the zero-angle-of-attack flow is of the order of the slope change alone; hence it is questionable whether the results, Eq. 2-38 and 2-39, are any more accurate than the result given by Eq. 2-30. In addition, the $\beta^2\varphi_{x_1x_1}$ term of Eq. 2-23 which arises from the doublet flow is (from Fig. B,1f and Eq. 2-38 and 2-39) of the order of α on the body. Since the neglected right-hand side of Eq. 2-23 contained terms of the order of α near the shoulders, the use of Eq. 2-23 is to be questioned, and it is clear that the calculation of the potential due to lift in this case can be no more accurate than that obtained by use of Laplace's equation in the cross flow variables. It is concluded therefore that the small disturbance theory cannot give the lift of a *pointed* slender body to any greater accuracy than that calculated by use of Eq. 2-30. Integration of the pressures acting thus yields for the lift of the *pointed* body

$$C_L = 2\alpha \tag{2-40}$$

in which the lift coefficient is based on the base area.[4] For the case of the open-nosed body, the doublet-produced pressures in the region near the nose are of the same order as those of the zero angle flow, and hence the results of Eq. 2-37 are correct. For points close to the nose the only first order pressure term contributed by the lift is then

$$\left[\frac{\Delta p - (\Delta p)_{\alpha=0}}{\frac{1}{2}\rho_\infty U^2}\right] = -\frac{2\alpha \cos \theta_1 S(0)}{\pi\beta R_1(0)R_1(x_1)} L'\left(\frac{x_1}{\beta R_1(0)}\right) \tag{2-41}$$

Downstream of the nose the L' function falls rapidly and the pressure must then include the terms of the order of α^2 and $\alpha(dR/dx)$. For these downstream points the function L in Eq. 2-38 approaches 1; hence Eq. 2-38 becomes simply

$$(\varphi_{x_1})_3 = \frac{U\alpha \cos \theta_1 S'(0)}{\pi R_1}$$

The contribution of $G_1(\xi)$ to the axial velocity is

$$(\varphi_{x_1})_1 = \frac{U\alpha \cos \theta_1}{\pi R_1} \int_0^x S''(\xi)d\xi \tag{2-42}$$

where the integral ignores the singularity at $x = 0$. Combining the preceding expressions with Eq. 2-24 the pressure becomes

$$\left[\frac{\Delta p - \Delta p_{\alpha=0}}{\frac{1}{2}\rho_\infty U^2}\right] = -\frac{2\alpha \cos \theta_1 S(0)}{\pi\beta R_1(x)R_1(0)} L'\left(\frac{x_1}{\beta R_1(0)}\right) - \frac{2\alpha \cos \theta_1 S'(x_1)}{\pi R_1} + 4\alpha^2(1 - 4 \sin^2 \theta) \tag{2-43}$$

and the last two terms must be dropped close to $x = 0$. The total lift is now

$$\int_0^{2\pi} \int_0^1 \Delta p \cos \theta_1 R_1 d\theta_1 dx_1 = \rho_\infty U^2\alpha \left[S(0)L\left(\frac{x_1}{\beta R_1(0)}\right) + S(1) - S(0)\right] \tag{2-44}$$

[4] Lighthill [8] has examined the order of the neglected terms near the shoulder and concludes that Eq. 2-39 may be retained in calculating the lifting pressure, but that the total lift is still no more accurate than that given by Eq. 2-40.

For large $x_1/\beta R_1(0)$ the function L approaches 1, hence the lift is the same as for the pointed body (Eq. 2-40) and independent of Mach number. For short open-nosed bodies, however, the lift is a function of Mach number, which is obvious since in the limit as $x_1 \to 0$, the body approaches a ring airfoil which, from elementary analysis, would have a lift curve slope inversely proportional to β as does a two-dimensional wing.

Limitations of small disturbance theory. In the development of the small disturbance theory for bodies of revolution, some controversy has arisen among contributors [*8,20,21*] as to what procedure must be followed to insure the greatest accuracy. We should really like to know the point at which we must break off the calculations and say, "This is all the information the small disturbance theory is capable of giving." The problem is important because the theory will be invariably used for cases unwarranted by the assumptions of the theory, which imply vanishingly thin bodies at angles of attack approaching zero. Briefly the disagreement concerns the following: On one hand it is felt that, since the small disturbance theory neglects supposedly second order terms in the differential equation, it is inconsistent to use the exact boundary conditions and pressure relation. Rather it is felt that only terms arising which are of the first order should be retained. Of course, first order terms can be composed of squares of the velocity components if these terms are the highest order terms which appear in the particular equation. Thus it is argued that the neglected terms of the differential equation may yield terms of the second order about which nothing is known, and since these terms are unknown even as to sign, there can be no additional accuracy claimed for including second order terms arising from use of the exact pressure relation and exact boundary conditions. On the other hand, another school of thought is that the inclusion of the second order terms can do no harm and may give increased accuracy. Van Dyke [*20*] has brought forward the latter view and shown that for cones at least, improved accuracy is obtained by using the exact equations for the boundary conditions and pressure. In two-dimensional linearized flow, however, it is easily proved that the use of exact boundary conditions yields a second order term which is of opposite sign and different magnitude than that given by exact nonlinear theory; thus the only justification for using exact boundary conditions and pressure relations in the case of three-dimensional flow must come from examination of the terms dropped in the exact differential equation, or from sufficient checks of the method against exact nonlinear solutions. Clearly it is permissible to use the exact boundary conditions and pressure equation for cases where experience has shown that such a procedure allows a better approximation to the true flow; however, in other cases it is difficult to see how information obtained from the second order terms can be trusted.

Effects of viscosity. It has been found that the fluid viscosity can produce a considerable change in the lifts and moments of slender bodies. The boundary layer is found to accumulate on the lee side of the body, and as the angle of attack is increased the flow separates from the body. When this occurs to any reasonable extent it is clear that the potential flow previously discussed cannot represent the true flow picture. The fundamental reason for the early separation on slender bodies at an angle of attack has been demonstrated in a paper by Jones [*22*] (see also Sec. A) in which it is shown that for laminar flow over infinite sweptback wings the boundary layer characteristics are only dependent on the pressure distribution, and the Reynolds number based on the component of the stream velocity normal to the generators of the cylindrical surface. Hence the boundary layer behaves essentially as though the flow component parallel to the elements of the surface were zero. Applying the result to a cylindrical body at an angle of attack, it is seen that back on the body where longitudinal pressure gradients have fallen to zero the initial separation will be determined by the pressure distribution and the Reynolds number of the cross flow component of stream velocity. For the case of a circular cylinder the data obtained on two-dimensional flow over such cylinders can be used to roughly estimate the normal forces produced by the separation. Thus the normal force per unit length of an inclined cylinder of diameter d is

$$N = \frac{C'_D \pi d^2 \rho_\infty U^2 \sin^2 \alpha}{8} \tag{2-45}$$

where C'_D is the drag coefficient of a unit length of an infinite circular cylinder based on a cross-sectional area $\pi d^2/4$. In Eq. 2-45 the drag coefficient must be obtained from experiments and the correct cross flow parameters must be used. That is, the test Reynolds numbers and Mach numbers should be $\rho_\infty U \sin \alpha d/\mu_\infty$ and $M_\infty \sin \alpha$, respectively. Application of the cross flow concept to a general body of revolution is not rigorously possible; however, it is clear from Eq. 2-30 that there will be regions along the body where the pressure gradients are very similar to those on a cylinder and hence similar separation phenomena are to be expected. Eq. 2-30 indicates that near the body nose the pressure gradients will be such that very little separation will occur. In regions where the body is nearly cylindrical the pressure distribution is the same as that of the two-dimensional cylinder and hence a corresponding separation is to be expected. The net result of the separation is to increase the lift of the body by producing more negative pressures on the upper surface at points some distance downstream of the nose. Clearly, bodies of the projectile type will be less influenced by separation than closed bodies since the boundary layers will be thinner and the adverse pressure gradients producing separation will be smaller. An experimental verification of the phenomena just discussed is given in Fig. B,2c. The figure shows a comparison of experimental and theoretical lift distribution for a parabolic body of revolution together with experimental and theoretical pressure distributions at two stations along the body. In the top graph it is seen that the experimental lift is higher than that of theory over the portion of the body behind the station $x/l = 0.2$. The bottom graphs show clearly what is happening; that is, the flow over the rear portion of the body is separating and preventing the pressure recovery at $\theta = \pi$ which is predicted by the

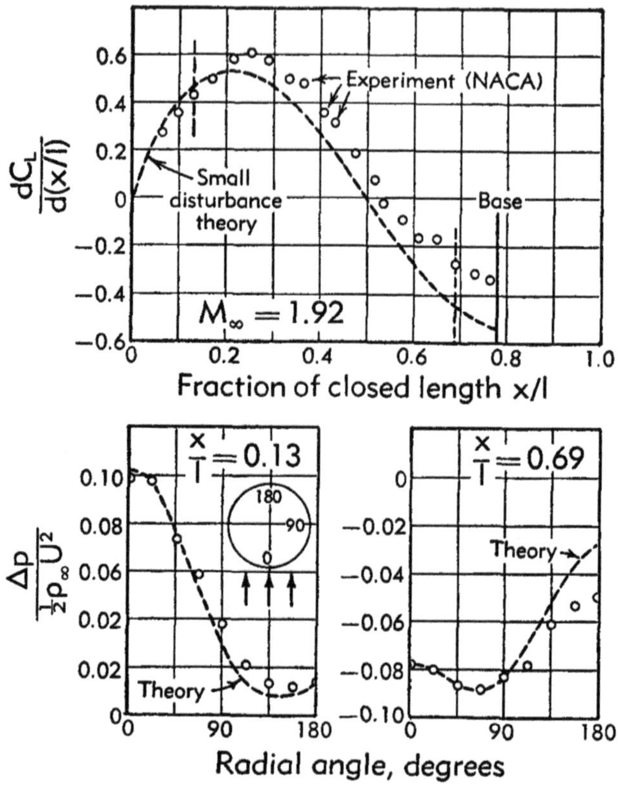

Fig. B,2c. Comparison of experimental and theoretical lift distribution over a parabolic body of revolution.

theory. On the nose sections, however, the theory indicates a favorable pressure gradient (negative along the streamlines), and hence there is no separation and the theory and experiment are in reasonably good agreement. It should be noted that the lifting area produced by the separation (Fig. B,2c, top) is quite large when compared with the theoretical lift and hence the lift at only 5 degrees angle of attack for this body is about 83 per cent higher than the theoretical value. It is interesting that the lift of the body up to the maximum diameter station would be only about 17 per cent higher than the corresponding theoretical result. Thus it is clear that for small angles of attack the agreement between theory and experiment will depend greatly on the ratio of maximum diameter to base diameter. It is readily apparent that from a practical standpoint the use of exact perfect fluid solutions would be a waste of time, since the viscous effects alter the lift of a body of revolution to a degree much greater than the differences between a first order theory and an exact perfect fluid theory.

B,3. Stability Derivatives for Bodies of Revolution. The derivatives of general interest in stability calculations are the lift curve slope C_{L_α}, pitching moment curve slope C_{m_α}, and the damping derivatives C_{m_q} and $C_{m_{\dot\alpha}}$. Since the lifts and moments acting on bodies have already been discussed, the following will deal with the latter two quantities. The derivative C_{m_q} is a measure of the pitching moment produced by rotational motion of the body in space and is defined as

$$C_{m_q} = \frac{\partial C_m}{\partial(ql/2U)} \tag{3-1}$$

where C_m is the moment coefficient, q is the angular velocity of the body axis relative to a reference fixed in space, and l is the body length. $C_{m_{\dot\alpha}}$ is a measure of the pitching moment produced by the rate of change of angle of attack α, where the angle of attack is measured as the angle between the body axis and the flight path. The definition of $C_{m_{\dot\alpha}}$ is

$$C_{m_{\dot\alpha}} = \frac{\partial C_m}{\partial(\dot\alpha l/2U)} \tag{3-2}$$

where $\dot\alpha$ is the time rate of change of the angle of attack.

For the rotary motions described we need the time dependent differential equation of motion (see VI, F):

$$\frac{1}{a_\infty}\varphi_{tt} + \frac{2U}{a_\infty}\varphi_{tx} + \beta^2\varphi_{xx} = \varphi_{yy} + \varphi_{zz} = 0 \tag{3-3}$$

Here the coordinates are moving through the air parallel to the x axis with velocity U; the speed of sound c is that in the undisturbed stream. As in Art. 2 it is convenient to work in coordinates fixed to the body; hence we assume the body coordinate system x_1, y_1, z_1 to be displaced by a small distance h along the y axis, and rotated through a small angle θ. The new relations are now, to first order in θ and h,

$$\begin{aligned} x_1 &= x - \theta y \\ y_1 &= y + \theta x - h \\ z_1 &= z \\ t_1 &= t \end{aligned} \tag{3-4}$$

For small rotation and displacement and small rates of rotation and displacement, the differential equation is unchanged since the transformation of coordinates introduces only new terms expected to be of the same order of magnitude as those previously dropped under the assumptions of small disturbances and slow motions. The pressure relation is

$$\frac{\Delta p}{\frac{1}{2}\rho_\infty U^2} = -\frac{2\varphi_x}{U} - \frac{2\varphi_t}{U^2} - \frac{\varphi_y^2 + \varphi_z^2}{U^2} \tag{3-5}$$

and with Eq. 3-4

$$\frac{\Delta p}{\frac{1}{2}\rho_\infty U^2} = -2\left(\frac{\varphi_{x_1}}{U} + \theta\frac{\varphi_{y_1}}{U}\right) - \frac{2}{U^2}\left[\varphi_{t_1} - y_1\dot{\theta}\varphi_{x_1} + (\dot{\theta}x_1 - \dot{h})\varphi_{y_1}\right] - \frac{\varphi_{y_1}^2 + \varphi_{z_1}^2}{U^2} \tag{3-6}$$

where the dot over the symbol denotes differentiations with respect to time. From the definition of θ, α, and h it can be seen that $\dot{h}/U$ equals $\theta - \alpha$, and $\dot{\theta} = q$; hence Eq. 3-6 becomes

$$\frac{\Delta p}{\frac{1}{2}\rho_\infty U^2} = -\frac{2}{U}(\varphi_{x_1} + \alpha\varphi_{y_1}) - \frac{2}{U^2}(\varphi_{t_1} - y_1 q\varphi_{x_1} + qx_1\varphi_{y_1}) - \frac{\varphi_{y_1}^2 + \varphi_{z_1}^2}{U^2} \tag{3-7}$$

The boundary condition for a slender body of revolution is most conveniently expressed in cylindrical body coordinates x_1, r_1, θ_1 as

$$(\varphi_{r_1})_{r_1=R_1} = U\frac{dR_1}{dx_1} - U\alpha\cos\theta_1 - qx_1\cos\theta_1 \tag{3-8}$$

In the above expression the first term on the right is independent of time for a fixed body, and the last two terms may be functions of time.

In the region close to the body surface the above equations can be considerably simplified, since for slender bodies performing slow motions of small amplitude it is found that the cross flow perturbations are much larger than the axial disturbances. Thus $\varphi_{t_1t_1}$, $\varphi_{x_1t_1}$, and $\varphi_{x_1x_1}$ can be dropped in the differential equation and Eq. 3-3 in body coordinates becomes

$$\varphi_{y_1y_1} + \varphi_{z_1z_1} = 0 \tag{3-9}$$

In Eq. 3-7 the terms φ_{x_1} and $(1/U)\varphi_t$ are found to be of the same order of magnitude as the square of the cross velocities φ_r and $(1/r)\varphi_\theta$; hence Eq. 3-7 becomes on transforming to cylindrical coordinates

$$\frac{\Delta p}{\frac{1}{2}\rho_\infty U^2} = -\frac{2\varphi_{x_1}}{U} - \frac{2\varphi_{t_1}}{U^2} - \frac{2(\alpha + qx_1/U)}{U}\left(\cos\theta_1\varphi_{r_1} - \frac{\sin\theta_1\varphi_{\theta_1}}{r_1}\right) - \frac{\varphi_{r_1}^2 + (1/r^2)\varphi_{\theta_1}}{U^2} \tag{3-10}$$

The problem is therefore similar to that of the steady motion except that the boundary conditions are now functions of time. The disturbance potential may now be taken in three parts: one for the thickness distribution (first term of Eq. 3-8), one for the angle of attack (second part of Eq. 3-8), and one satisfying the boundary conditions for the rotation q (third term of Eq. 3-8). That part of the potential function satisfying the thickness condition is a steady flow potential and is that given in Art. 1. The solution for the lateral motion of each body section is (as given in Art. 2) that of a doublet distribution on the body axis; hence

$$\varphi = \varphi_0 + \frac{\cos\theta_1 S(x_1)}{\pi r_1}\left[U\alpha(t_1) + q(t_1)x_1\right] \tag{3-11}$$

where φ_0 is the thickness solution, and $S(x)$ is the body cross-sectional area. The validity of the original simplifications may be checked by substitution of Eq. 3-11 into the differential equation (3-3). The pressure distribution on the body due to the α and q motion is now

$$\frac{(\Delta p - \Delta p_{\alpha=0})}{\frac{1}{2}\rho_\infty U^2} = -4\cos\theta_1 \frac{dR_1}{dx_1}\left(\alpha + \frac{qx_1}{U}\right) - \frac{2\cos\theta_1 R_1 q}{U} + \left(\alpha + \frac{qx_1}{U}\right)^2 (1 - 4\sin^2\theta) - \frac{2\cos\theta R_1}{U}\left(\dot\alpha + \frac{\dot q x_1}{U}\right) \tag{3-12}$$

Integration of the pressure given by Eq. 3-12 over the body of length l yields the following expressions for the lift and moment coefficients:

$$C_L = 2\alpha - \frac{2lq}{U}\left(\frac{x_{\text{base}}}{l}\right) + \frac{2\dot\alpha l}{U}\left(\frac{\text{volume}}{ls_{\text{base}}}\right) + \frac{2\dot q l}{U^2}\left(\frac{\text{moment of volume}}{l^2 s_{\text{base}}}\right) \tag{3-13}$$

$$C_m = \frac{\text{moment}}{\frac{1}{2}\rho_\infty U^2 l s_{\text{base}}} = -2\alpha\left(\frac{x_{\text{base}}}{l} - \frac{\text{volume}}{ls_{\text{base}}}\right) - \frac{2lq}{U}\left(\frac{x^2_{\text{base}}}{l^2} - \frac{\text{volume moment}}{l^2 s_{\text{base}}}\right) - \frac{2\dot\alpha l}{U}\left(\frac{\text{volume moment}}{l^2 s_{\text{base}}}\right) - \frac{2\dot q l^2}{U}\left(\frac{\text{volume moment of inertia}}{l^3 s_{\text{base}}}\right) \tag{3-14}$$

The stability derivatives C_{m_q} and $C_{m_{\dot\alpha}}$ are now obtained from Eq. 3-13 and 3-14:

$$C_{m_q} = -4\left[\left(\frac{x_{\text{base}}}{l}\right)^2 - \frac{\text{volume moment}}{l^2 s_{\text{base}}}\right] \tag{3-15}$$

$$C_{m_{\dot\alpha}} = -4\left(\frac{\text{volume moment}}{l^2 s_{\text{base}}}\right) \tag{3-16}$$

It should be noted that the moments given by the above equations are to be taken about the point of rotation or moment reference $x = 0$. The foregoing analysis is restricted to slender bodies performing slow motions. Mathematically less-restricted results require a study of the nonlinear equations of motion; however, such a study is probably not warranted in view of the large effects of viscosity on the forces acting over slender bodies.

B,4. Cited References.

1. Frankl, F. I. Effect of the acceleration of elongated bodies of revolution upon the resistance in a compressible flow. *NACA Tech. Note 1230*, 1949.
2. Ward, G. N. Supersonic flow past slender pointed bodies. *Quart. J. Mech. and Appl. Math. 2, Part 1*, 1949.
3. Lamb, H. *Hydrodynamics*, 6th ed. Cambridge Univ. Press, 1932.
4. Prandtl, L. Theorie des Flugzeugtragflugels im zusammendruckbaren medium. *Luftfahrtforschung 13*, 1936.
5. Ferri, A. *Elements of Aerodynamics of Supersonic Flows*. Macmillan, 1949.
6. Haack, W. Geschossformen kleinsten wellenwiderstandes. *Lilienthal-Gesellschaft, Bericht 139*, 14–28 (1941).
7. Van Dyke, M. D. Practical calculation of second order supersonic flow past nonlifting bodies of revolution. *NACA Tech. Note 2744*, 1951.
8. Lighthill, M. J. Supersonic flow past slender bodies of revolution the slope of whose meridian section is discontinuous. *Quart. J. Mech. and Appl. Math. 1, Part 1*, 1948.
9. Ferri, A. Application of the method of characteristics to supersonic rotational flow. *NACA Tech. Note 1135*, 1946.
10. Chapman, D. R. An analysis of base pressure at supersonic velocities and comparison with experiment. *NACA Tech. Note 2137*, 1950.
11. Kurzweg, H. H. New experimental investigations on base pressure in the NOL supersonic wind tunnels at Mach numbers 1.2 to 4.24. *Nav. Ord. Lab. Rept. MA0113*, 1950.
12. Charters, A. K., and Turetsky, R. A. Determination of base pressure from free-flight data. *Aberdeen Ballist. Research Lab. Rept. 653*, 1948.
13. Faro, I. D. V. Experimental determination of base pressures at supersonic velocities. *Johns Hopkins Univ. Appl. Phys. Lab. Rept. 106*, Nov. 1949.
14. Spahr, J. R., and Dickey, R. R. Effect of tail surfaces on the base drag of a body of revolution at Mach numbers of 1.5 and 2.0. *NACA Tech. Note 2360*, 1951.
15. Mangler, W. Boundary layers with symmetrical airflow about bodies of revolution. *Goodyear Aircraft Corp. Rept. R-30-18, Part 20*, 1946. (Transl.)
16. Mangler, W. Compressible boundary layers on bodies of revolution. *MAP-Völkenrode 83–47 T*, 1946.
17. Hantzsche, W., and Wendt, H. Laminar boundary layer at a circular cone with zero angle of incidence in supersonic flow. *Jahrbuch deut. Luftfahrtforschung 1*, 76, 77 (1941).
18. Munk, M. M. The aerodynamic forces on airship hulls. *NACA Rept. 184*, 1924.
19. Ward, G. N. The approximate external and internal flow past a quasi-cylindrical tube moving at supersonic speeds. *Quart. J. Mech. and Appl. Math. 1*, 225 (1948).
20. Van Dyke, M. D. First order and second order theory of supersonic flow past bodies of revolution. *Rand Corp. Rept. P-125*, 1949.
21. Laitone, E. V. The linearized subsonic and supersonic flow about inclined slender bodies of revolution. *J. Aeronaut. Sci. 14*, 631–642 (1947).
22. Jones, R. T. Effects of sweepback on boundary layer and separation. *NACA Rept. 884*, 1947.

SECTION C

INTERACTION PROBLEMS

C. FERRARI

Aerodynamics interference, in a restricted sense, is the change of the fluid dynamic properties of an object immersed in a flowing stream due to the presence of one or more other objects introduced into the same field of flow. Typical cases are: the mutual interaction between the wings of a biplane; the interaction of an infinite sequence of airfoils forming a cascade; and the effect of a solid or fluid surface bounding a wind tunnel jet upon the aerodynamic field around a wing inserted in the tunnel.

In a broader sense, which is the way the term is used in this discussion, *interference effects* are the changes in the above-mentioned properties of the flow, not only when the several objects are separate and distinct from one another but also when they are so closely bound together that they actually form one single configuration. Thus a conventional airplane, whether of either subsonic or supersonic design, may be visualized as an aggregate of the following parts: (1) a streamlined body called the fuselage, (2) a lifting surface (the wing), usually displaying a plane of symmetry, (3) a horizontal stabilizer called the horizontal tailplane, (4) a vertical stabilizer (the vertical tailplane), and finally, (5) some kind of propulsion device (either a propeller or a jet). Now, although the wings (or tailplanes) and the fuselage interpenetrate and are faired into each other, it is convenient in most cases to express the value of each individual aerodynamic characteristic pertaining to any one of the separate parts of the airplane as the sum obtained by totalling two distinct contributions: the value of the particular characteristic when that part of the airplane under consideration is operating as though isolated in free stream flow, and the increment to this quantity due to the presence of the other appendages. This increment is precisely the measure of the amount of interference present.

This allows us:

1. To derive quantitative laws, or at least some qualitative ones, to predict the aerodynamic behavior that can be expected from a configuration composed of any combination of fundamental elements whatsoever on the basis of knowledge of the aerodynamic action and properties of each of these elements (the wing, the fuselage), when considered to be operating alone and not influenced by the rest of the

ensemble of which it is a part. Of course, the isolated components can be analyzed more completely and with less effort than the composite configuration.

2. To bring out clearly what effect the juxtaposition of the component parts has upon the above-mentioned aerodynamic properties, so as to furnish insight into the ways in which such fundamental elements may be combined in order to produce desired optimum results.

If the aerodynamically interacting parts (as exemplified by a wing-fuselage configuration) are joined together tightly, the determination of the interference effects is rendered much more complicated because of the confluence of the boundary layers enveloping the respective component parts of the configuration. Discussion of the results of such boundary layer interactions is not generally dealt with in this section.

Referring to the conventional airplane mentioned above, the following problems are considered here:

1. Interference of the wing on the tailplanes.
2. Mutual interference between the wing and the fuselage.
3. Mutual interference of the means of propulsion upon the lifting surfaces (the wings or the tailplanes).

The interference of the tailplanes on the wing is not examined, because it is nil in the case of supersonic flow and negligibly small when the flow is subsonic in the conventional airplane.

Moreover, if the wing design incorporates nacelles, these appendages are to be considered as forming integral sections of the wing in the sense that, if infinitesimally thin slices of the nacelles are imagined to be marked out by a sheaf of planes running parallel to the plane of symmetry of the airplane, then such sections may be regarded as equivalent to airfoils [*1*; *2*, Vol. 4, p. 158]. Based on this concept, the determination of the lift distribution and the drag induced by such a nacelle-wing configuration becomes the same problem as that encountered when one treats the analogous case of a wing made up of a sequence of airfoil sections with changing properties as one progresses out along the span, and which has an entirely arbitrary planform shape. (See Sec. A for a discussion of this problem.)

Finally, because of the intrinsic differences between subsonic and supersonic flow, it is apposite to treat all the enumerated interference effects in distinctly separate ways, depending upon whether $M_\infty \lessgtr 1$. In both cases it is assumed that the magnitude of the perturbations produced by the several component parts of the configurations is small enough to make it possible to linearize the equation of motion governing the flow. Because of this assumption it is possible to treat each of the various interference effects as a separate problem.

CHAPTER 1. WING-TAILPLANE INTERFERENCE AT SUBSONIC SPEEDS

C,1. Statement of the Problem. It is assumed that the perturbations due to the wing and the tail are sufficiently small to permit the linearization of the equation governing the potential function of the perturbed flow. Under these conditions the actual equation defining the flow may be converted into that which applies for an incompressible fluid by means of the Prandtl-Glauert transformation (Sec. A):

$$x_1 = \frac{x}{\beta}, \qquad y_1 = y, \qquad z_1 = z \tag{1-1}$$

where

$$\beta = \sqrt{1 - M_\infty^2} \tag{1-2}$$

In this way it is permissible to solve fluid flow problems, and in particular the one of immediate interest, by working in the transformed space, obtained from the real physical space through means of Eq. 1-1, and by proceeding thereafter to determine the solution just as though the fluid were incompressible. Thus the effect of compressibility is interpreted as a distortion undergone by the physical space according to Eq. 1-1.

On the basis of the previous reasoning, the influence of the tail surface upon the wing is disregarded, and the wing is replaced by an equivalent system of bound vortices and the trailing vortices which are shed from it. Then the fluid stream impinging on the tail is equal to that resulting from the superposition of the undisturbed flow having the velocity **U** and of the incremental flow induced by the vortical system Σ_w.

It is now essential to distinguish between the two following practical setups:

1. A wing with its high-lift flap retracted, and operating in an unstalled condition. For this type of flight attitude one may assume that the wake is of zero thickness; consequently the trailing vortices form a simple vortex sheet. Within the degree of approximation assumed in this investigation it is then sufficient to consider merely the w component of the velocity increments arising from the Σ_w system; the w component is, of course, the component which is perpendicular to the free stream velocity vector **U** and which lies in a direction parallel to the plane of symmetry of the lifting system.
2. A wing with the high lift flap extended, or in a condition where the flow has separated (stalled wing). In this case there is an appreciable thickness of the wake and, within this wake, a not inconsiderable variation in the magnitude of the component of velocity lying in the

same direction as U. Therefore, whenever the tailplane penetrates into the wake, it is necessary to take account not only of the transverse component w, but also of the longitudinal component u that is induced by the vortex system Σ_w.

The small perturbations assumption is controverted, to be sure, when such conditions of flight are encountered, but, on the other hand, whenever this state of affairs arises, the corresponding Mach numbers are found to be so small that it continues to be legitimate to consider the equation of motion as remaining linear.

In any case, once the induced velocities produced by the vortex system Σ_w are known at all points of the stabilizing surface, the determination of the field of flow around the tail is accomplished by resorting to the same hypotheses and exactly equivalent procedures as those brought to bear when the solution for the flow around any isolated wing surface is carried out. In this way the question of determining the influence exerted on a tail by an interfering wing is effectively reduced to the calculation of the aforementioned induced velocities.

With these conventions in mind, examination of case 1 is undertaken first. A prime consideration is the determination of the w component induced by wings of not too small aspect ratio and having any amount of sweep whatsoever. The effect of the wake is discussed in Art. 4.

C,2. Calculation of the *w* Component of Induced Velocity in Instances Where Thickness of the Wake Is Zero.

Case of a Wing Having a Moderately Large Aspect Ratio and a Small Amount of Sweep. The Multhopp Method.

Computation of the induced velocity generated at any point of the trailing vortex sheet. Under the conditions stipulated above, or to be more definite, for wings of aspect ratio $Æ \geqq 4$ and with angle of sweep $\Lambda \leqq 30°$, the wing's system of bound vortices may be taken to be compressed into one single vortex, the axis of which is normal to the plane of symmetry and passes through the mean aerodynamic center of the lifting surface. The single vortex has a circulation Γ that is variable in strength along the span; a trailing vortex sheet Σ is shed across the whole breadth of this bound vortex core. The actual shape of the vortex sheet is rather difficult to specify inasmuch as it depends upon the particular velocities induced by the vortex system at the points occupied by the sheet Σ.

The problem is complicated even more seriously because the trailing band of vortices is unstable. As a result of this instability the band tends to roll up at the edges in such a way that two distinct vortical corelike nuclei of appreciable size are formed, which are characterized by equal but opposite amounts of circulation.

When dealing with wings of high enough aspect ratios such as those considered here, the coiling-up phenomenon may be ignored without entailing any sizeable error. Kaden has shown on the basis of his investigations [3] that the complete rolling up of the trailing vortex surface is accomplished only at a certain distance L behind the rear edge of the wing, which may be calculated from the relationship

$$\frac{2L}{b} = 0.56 \frac{Æ}{C_L} \tag{2-1}$$

where b is the wing span.

For a lift coefficient amounting to $C_L = 1.3$, and for an aspect ratio taken to be $Æ = 4$ at worst, Eq. 2-1 predicts that $2L/b = 1.72$; i.e. the distance required for total roll-up is more than double the distance aft of the wing at which the tailplane is normally placed. For this reason it

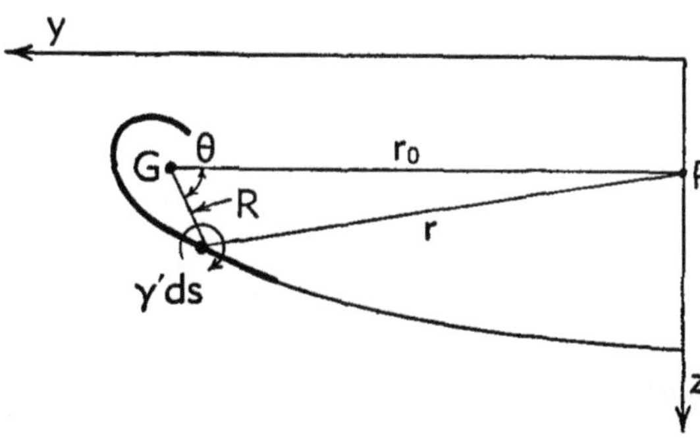

Fig. C,2a. Geometric relationships useful in computing effect of a partially rolled-up vortex sheet on the downwash behind a wing.

appears that, even when the lift coefficient is uncommonly large, the region of the vortex sheet Σ which stands nearest to the stabilizer, and which thus contributes most strongly to the induction effects exerted upon it, is not yet involved in the roll-up phenomenon. At the same time one may be certain that those regions of the coiled-up vortex sheet lying farther downstream do not appreciably affect the induced downwash velocities acting on the tail.[1]

[1] Let s denote the part of the vortex sheet which is involved in the roll-up phenomenon (Fig. C,2a), and imagine that the vortices all have straight-line cores of infinite length. For the sake of simplicity, attention is centered on what happens at a point P lying on the z axis (refer to Fig. C,2a). The component of velocity, directed parallel to the z axis that is induced at the point P by action of the vortices composing the ribbonlike band s, is given by the integral

$$w = \frac{1}{2\pi} \int_s \frac{\gamma' y ds}{r^2} = \frac{1}{2\pi} \int_s \frac{\gamma' y ds}{r_0^2 + R^2 - 2r_0 R \cos\theta} \tag{2-2}$$

where γ' represents the local strength of circulation per unit width of the vortex sheet. Furthermore let G indicate the barycenter of the distribution of vorticity, and

Since, at first, attention is focused on obtaining the induced velocities arising from the vortex sheet Σ, no provision is made in the initial calculations to account for the effect of the velocities upon the form and character of the sheet itself. From this standpoint it is logical to assume for the purposes of the initial calculation that the axes of the trailing vortices are all straight lines lying parallel to the free stream velocity vector.

In a second step, however, the results obtained from the initial, admittedly rough, calculations are used to obtain a more realistic approximation for the description of the real form of the vortex sheet, as well as to determine its location more precisely with respect to the geometry of the airplane. The necessity to improve the first crude computations is quite impelling when one seeks to determine the value of the downwash w existing at the tail because, as brought out forcefully in what follows, the values of w do not pass through a maximum at the level of the vortex sheet Σ, but rather exhibit a sharp change in slope there.

A system of coordinate axes is selected which is linked to the oncoming

---

then the distance PG may be denoted by the symbol r_0. The significance of the other symbols may be gleaned most easily by referring to Fig. C,2a. If y_0 and z_0 are now taken to represent the coordinates of the point G, and if z_P is the value of z which corresponds to the vertical height of P, then one may write

$$\cos\theta = \frac{y_0(y_0 - y) + (z_0 - z_P)(z_0 - z)}{r_0 R}$$

Thus the denominator of Eq. 2-2 may be rearranged to give:

$$r_0^2 + R^2 - 2r_0 R\cos\theta = r_0^2\left(1 + \frac{R^2}{r_0^2} - 2\frac{R}{r_0}\cos\theta\right)$$

$$= r_0^2\left[1 + \frac{R^2}{r_0^2} - \frac{2y_0(y_0 - y)}{r_0^2} - \frac{2(z_0 - z_P)(z_0 - z)}{r_0^2}\right]$$

Now if this expression is inserted into Eq. 2-2 and the reciprocal of the term enclosed in brackets is expanded by means of the binomial theorem, one finds that

$$w = \frac{1}{2\pi r_0^3}\left[r_0\int_s \gamma' y\,ds + 2\int_s \gamma'\frac{y}{r_0}[y_0(y_0 - y) + (z_0 - z_P)(z_0 - z)]ds - \int_s \gamma'\frac{y}{r_0}R^2 ds + \cdots\right]$$

A simplifying substitution is now made in the second and third of these integrals for the quantity y/r_0, because if r_0 is sufficiently large, it appears to be legitimate to make the approximation that $y/r_0 \cong r/r_0 \cong 1$. Upon using this approximation it results that the second integral is zero, because it is proportional to the moments of the elements of the vorticity distribution s taken with respect to the set of axes passing through the barycenter. At the same time the first integral remains constant during the roll-up process, while the third can also be taken to be constant to good approximation [*4*]. It follows directly from these considerations that what was asserted above concerning the effect of the roll-up phenomenon is surely within reason.

flow in such a way as to be orientated and to have its origin located in conformity with the arrangement sketched in Fig. C,2 b.

Then it results that the induced velocity w_1, created at any point of the x, y plane by action of the bound vortices obviously is parallel to

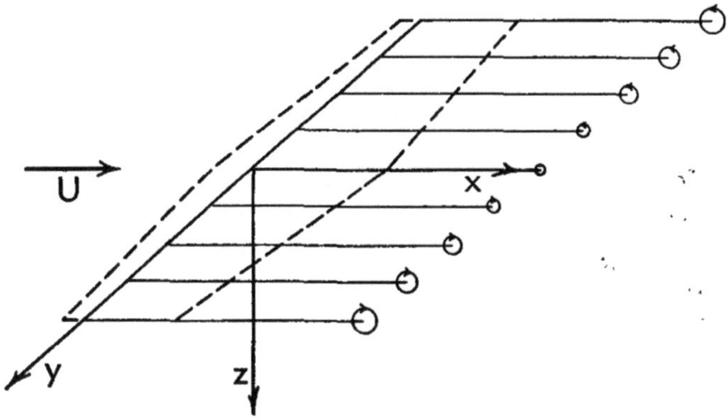

Fig. C,2b. Vortex system equivalent to the wing.

the z axis. The magnitude of the induced velocity is computed by means of the Biot-Savart formula,

$$\frac{w_1}{U} = \frac{1}{2\pi} \int_{-1}^{1} \frac{\gamma \xi_1 d\eta'}{[\xi_1^2 + (\eta - \eta')^2]^{\frac{3}{2}}} = \epsilon_1 \tag{2-3}$$

wherein

$$\xi_1 = \frac{2x_1}{b}, \qquad \eta = \frac{2y}{b}, \qquad \gamma = \frac{\Gamma}{bU} \tag{2-4}$$

and ϵ_1 is the associated contribution to the downwash angle.

The system of trailing vortices induces a velocity w_2 at the same point, which is likewise directed parallel to the z axis, and of a magnitude given by the relation

$$\frac{w_2}{U} = \frac{1}{2\pi} \int_{-1}^{1} \frac{d\gamma}{d\eta'} \left[1 + \frac{\xi_1}{\sqrt{\xi_1^2 + (\eta - \eta')^2}}\right] \frac{d\eta'}{\eta - \eta'} = \epsilon_2 \tag{2-5}$$

to produce a total contribution to the downwash of

$$\epsilon = \epsilon_1 + \epsilon_2 \tag{2-6}$$

From Eq. 2-5 and 2-3 one obtains

$$\begin{aligned} \epsilon &= \frac{1}{2\pi} \int_{-1}^{1} \left\{ \frac{d\gamma}{d\eta'} \left[1 + \frac{\xi_1}{\sqrt{\xi_1^2 + (\eta - \eta')^2}}\right] \frac{1}{\eta - \eta'} + \frac{\gamma \xi_1}{[\xi_1^2 + (\eta - \eta')^2]^{\frac{3}{2}}} \right\} d\eta' \\ &= \frac{1}{2\pi} \int_{-1}^{1} \frac{d}{d\eta'} \left\{ \gamma \left[1 + \frac{\xi_1}{\sqrt{\xi_1^2 + (\eta - \eta')^2}}\right]\right\} \frac{d\eta'}{\eta - \eta'} \end{aligned} \tag{2-7}$$

Using the shorthand notation that

$$\tilde{\gamma} = \gamma\left[1 + \frac{\xi_1}{\sqrt{\xi_1^2 + (\eta - \eta')^2}}\right] \tag{2-8}$$

the expression for the downwash angle may be written then as

$$\epsilon = \frac{1}{2\pi}\int_{-1}^{1}\frac{d\tilde{\gamma}}{d\eta'}\frac{d\eta'}{\eta - \eta'} \tag{2-9}$$

which is formally identical with the equation giving the induced downwash angle ϵ_w related to the bound vortex on the wing. According to Multhopp's method [*5*], one may now proceed by setting

$$\gamma = \frac{2}{m+1}\sum_{n=1}^{m}\gamma_n\sum_{\mu=1}^{m}\sin\mu\theta_n\sin\mu\theta \tag{2-10}$$

in which γ_n is the value of γ when $\eta = \eta_n$, the meaning of the latter being given by $\eta_n = \cos[n\pi/(m+1)]$.

When this is done (see Sec. A), it results that

$$\epsilon_w(\eta_\nu) = b_{\nu\nu}\gamma_\nu - \sum_{n=1}^{m}{}' b_{\nu n}\gamma_n \tag{2-11}$$

where the use of the prime signifies that terms with repeated subscripts are excluded from the summation; i.e. whenever $n = \nu$. In addition, one finds that

$$b_{\nu\nu} = \frac{m+1}{4\sin\theta_\nu} \tag{2-12}$$

while

$$b_{\nu n} = \frac{\sin\theta_n}{(m+1)(\cos\theta_n - \cos\theta_\nu)^2} \quad \text{for } |n-\nu| = 1, 3, 5, \cdots$$

and

$$b_{\nu n} = 0 \quad \text{for } |n-\nu| = 2, 4, 6, \cdots$$

On the basis of these preparatory developments, and with the assumption that it is also possible to express even $\tilde{\gamma}$ by means of such a trigonometric series in $\sin\theta$, just as was done for γ in Eq. 2-10, the following solution may be written as

$$\epsilon = b_{\nu\nu}\tilde{\gamma}_\nu - \sum_{n=1}^{m}{}' b_{\nu n}\tilde{\gamma}_n = 2b_{\nu\nu}\gamma_\nu - \sum_{n=1}^{m}{}' b_{\nu n}\left[1 + \frac{\xi_1}{\sqrt{\xi_1^2 + (\eta_n - \eta_\nu)^2}}\right]\gamma_n \tag{2-13}$$

or upon transferring back to the physical space coordinates (ξ, η, ζ), one obtains finally that

$$\epsilon(\xi,\eta_\nu) = 2b_{\nu\nu}\gamma_\nu - \sum_{n=1}^{m}{}' b_{\nu n}\left[1 + \frac{\xi}{\sqrt{\xi^2 + \beta^2(\eta_n - \eta_\nu)^2}}\right]\gamma_n \quad (2\text{-}13')$$

which provides the means of determining the values of ϵ at the points ξ, η_ν, provided that the pattern of variation of the circulation distributed along the wing span is known.

Determination of the shape and location of the trailing vortex sheet. The shape and location of the trailing vortex sheet can be determined more precisely now that the above results are available. This calculation is based on the assumption that the inclination at an arbitrary point P of this vortex sheet is equal in magnitude to the downwash angle ϵ, calculated by means of Eq. 2-13′. Employing a system of axes traveling with the impinging flow, one has that $dz/dt = w$, and thus $dz = \epsilon U dt = \epsilon dx$.

It follows thus that

$$\zeta_1 = \zeta_0 + \int_{\xi_0}^{\xi} \epsilon d\xi \quad (2\text{-}14)$$

where ζ_0 is the vertical coordinate, measured at its trailing edge, of the airfoil lying in an arbitrary vertical plane, $\eta = \text{const}$. Inasmuch as ϵ may be considered a function of ξ and of η through means of Eq. 2-13′, Eq. 2-14 constitutes the analytic expression for the discontinuity surface composed of the trailing vortices.

Calculation of the angle of downwash, ϵ, pertaining to points located on the tailplane. The values of ϵ obtained by evaluation of Eq. 2-13′ apply to those points which belong to the surface defined by Eq. 2-14. For points located off this surface, but still lying nearby, one may set up an expression for ϵ by carrying out a Taylor series development in terms of the variable ζ. The expression for ϵ may thus be written as

$$\epsilon(\xi, \eta, \zeta) = \epsilon(\xi, \eta, \zeta_1) + (\zeta - \zeta_1)\left(\frac{\partial \epsilon}{\partial \zeta}\right)_{\zeta=\zeta_1} + \cdots \quad (2\text{-}15)$$

Now the motion in an arbitrary $\xi = \text{const}$ plane may be considered as two-dimensional, at least for those values of ξ which are likely to enter into the calculation of the induced velocities that exist in the region of the tail. Consequently the continuity equation may be written simply as

$$\frac{\partial \epsilon}{\partial \zeta} + \frac{\partial \sigma}{\partial \eta} = 0$$

where the spanwise v component has been nondimensionalized by setting $\sigma = v/U$.

The calculation of the coefficient $(\partial\epsilon/\partial\zeta)_{\zeta=\zeta_1}$ may then be carried out quite simply, on the basis of the justifiable assumption just mentioned, since it now follows immediately that

$$\left(\frac{\partial\epsilon}{\partial\zeta}\right)_{\zeta=\zeta_1} = -\left(\frac{\partial\sigma}{\partial\eta}\right)_{\zeta=\zeta_1} \tag{2-16}$$

If one ignores the rolling-up phenomena, then the relationship between the local circulation and the spanwise components of velocity is given by

$$\sigma(\zeta_1 + 0) - \sigma(\zeta_1 - 0) = -2\frac{d\gamma}{d\eta}$$

where also

$$\sigma(\zeta_1 + 0) = -\sigma(\zeta_1 - 0) \tag{2-17}$$

so that when inserted in Eq. 2-16 these expressions give the significant result

$$\left(\frac{\partial\epsilon}{\partial\zeta}\right)_{\zeta=\zeta_1+0} = -\left(\frac{\partial\epsilon}{\partial\zeta}\right)_{\zeta=\zeta_1-0} = \frac{d^2\gamma}{d\eta^2} \tag{2-18}$$

Of course, it is also true, as a result of the relationship given in Eq. 2-10, that

$$\frac{d^2\gamma}{d\eta^2} = -\frac{2}{m+1}\cdot\frac{1}{\sin^2\theta}\left(\sum_{n=1}^{m}\gamma_n\sum_{\mu=1}^{m}\mu\sin\mu\theta_n\frac{\cos\mu\theta\cos\theta}{\sin\theta}\right.$$
$$\left.+\sum_{n=1}^{m}\gamma_n\sum_{\mu=1}^{m}\mu^2\sin\mu\theta_n\sin\mu\theta\right) \tag{2-19}$$

The expressions given in Eq. 2-18 show that the derivative $\partial\epsilon/\partial\zeta$ changes abruptly in going from one side of the vortex sheet to the other, which in other words means that ϵ does not have a maximum at the height $\zeta = \zeta_1$ but that it exhibits a sharp change in slope there.

The determination of the quantity $d^2\gamma/d\eta^2$ is almost immediate, once the values of

$$A_n = \sum_{\mu=1}^{m}\mu^2\sin\mu\theta_n\sin\mu\theta$$

and of

$$B_n = \cot\theta\sum_{\mu=1}^{m}\mu\sin\mu\theta_n\cos\mu\theta$$

have been computed. The values of A_n and B_n have been evaluated for the two selected cases where $\eta = 0$ and $\eta = 0.3827 = \cos(6\pi/16)$; the results are listed in Table C,2. These computations afford a quick evaluation of $d^2\gamma/d\eta^2$ for any given distribution of γ_n and for $m = 15$.

Table C,2

n	A_n $\eta = 0$	B_n $\eta = 0.3827$	A_n $\eta = 0.3827$
1	−3.245	−1.081	−8.061
2	7.173	2.343	19.315
3	−12.859	−4.102	−40.776
4	22.628	7.222	99.311
5	−43.102	−15.937	−411.20
6	100.94	−0.686	679.31
7	−412.31	17.325	−411.98
8	680	−8.659	100.94
9	−412.31	5.625	−43.429
10	100.94	−4.000	23.312
11	−43.102	2.937	−13.961
12	22.628	−2.150	8.802
13	−12.859	1.516	−5.572
14	7.173	−0.971	3.313
15	−3.245	0.474	−1.552

In some instances it is found that the value of $\zeta - \zeta_1$, which corresponds to the point at which it is desired to compute the value of ϵ, is not small enough to make it legitimate to halt the Taylor series development at the linear term in $\zeta - \zeta_1$. When this happens, rather than continue with the series expansion, it will be found more convenient in general to calculate the value of $\epsilon(\xi, \eta, \zeta)$ directly by use of the Biot-Savart formula:

$$\begin{aligned}\epsilon(\xi, \eta, \zeta) &= \frac{1}{2\pi} \int_{-1}^{1} \frac{\gamma(\eta')[(\zeta - \zeta_1)^2 - (\eta - \eta')^2]}{[(\zeta - \zeta_1)^2 + (\eta - \eta')^2]^2} \\ &\times \left\{1 + \frac{\xi}{\sqrt{\xi^2 + \beta^2[(\eta - \eta')^2 + (\zeta - \zeta_1)^2]}}\right\} d\eta' + \frac{\beta^2 \xi (\zeta - \zeta_1)^2}{2\pi} \\ &\times \int_{-1}^{1} \frac{\gamma(\eta') d\eta'}{[(\zeta - \zeta_1)^2 + (\eta - \eta')^2][\xi^2 + \beta^2(\eta - \eta')^2 + \beta^2(\zeta - \zeta_1)^2]^{\frac{3}{2}}}\end{aligned} \qquad (2\text{-}20)$$

Influence of the Mach number on the induced downwash angle. Fig. C,2e, C,2f, and C,2g show the values of ϵ/α at the trailing vortex sheet as they result from the application of the method developed in the preceding paragraphs for the untwisted wing represented in Fig. C,2c. The spanwise distribution of the circulation for Mach numbers $M_\infty = 0.3$, 0.6, and 0.8 is shown in Fig. C,2d for this wing. The values of $\zeta_1 - \zeta_0$ and hence of the downward displacement of the points of the vortex sheet Σ afforded by

Eq. 2-14 are given in Fig. C,2h; C,2i; and C,2j. From these figures it is apparent that for a given α, *an increase in the Mach number results in an increase of the downwash angle* ϵ *on* Σ *as well as of the displacement* ζ_1.

Owing to the fact that the lift curve slopes of the wing in Fig. C,2c for the aforementioned Mach numbers are $C_{L_\alpha} = 4.38, 4.96, 5.97$ for $M_\infty = 0.3, 0.6$, and 0.8 respectively, one may infer that, again at the level of the vortex sheet Σ, *the downwash angle* ϵ *for a given lift coefficient*

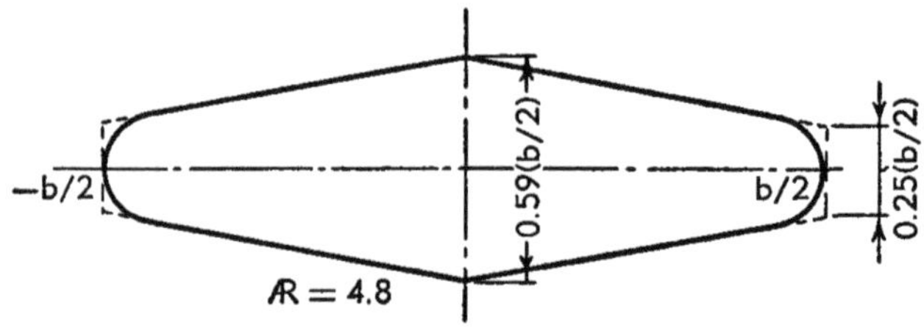

Fig. C,2c. Geometric characteristics of typical wing employed in illustrations of wing-linked interference phenomena.

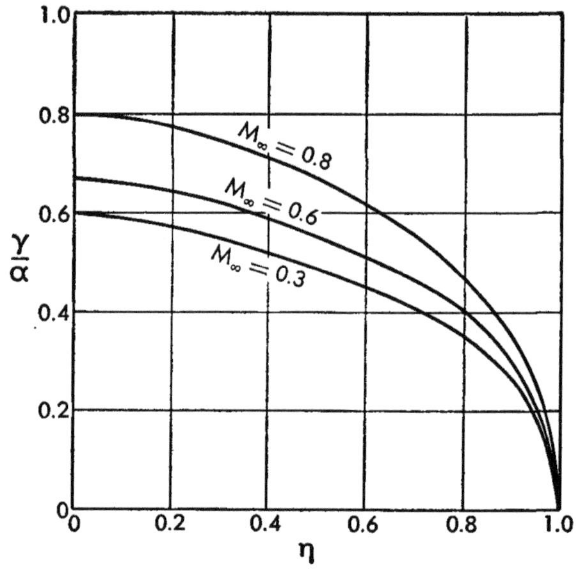

Fig. C,2d. Spanwise distribution of the circulation for the wing shown in Fig. C,2c.

C_L *decreases as* M_∞ *increases and the downward displacement* $\zeta_1 - \zeta_0$ *decreases as well* (see Fig. C,2l and C,2m). The way ϵ/α and ϵ/C_L vary as functions of the distance off of the vortex sheet is shown in Fig. C,2k and C,2n respectively, for the same three Mach numbers selected previously and for the wing shown in Fig. C,2c. The results apply solely to the specific location where $\xi = 1$.

$\eta = 0$. From these figures it may be seen that the rate of decrease in the value of ϵ/C_L as $\zeta - \zeta_1$ increases remains unchanged to good approximation, regardless of any changes taking place in M_∞. It is also true that $(\epsilon/C_L)_{\zeta-\zeta_1=0}$ decreases with increase in the Mach number, but at the

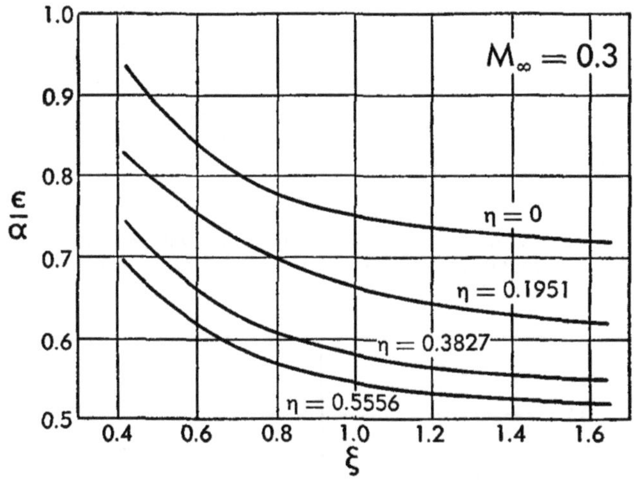

Fig. C,2e. Distribution of the downwash over the trailing vortex sheet for the case of $M_\infty = 0.3$.

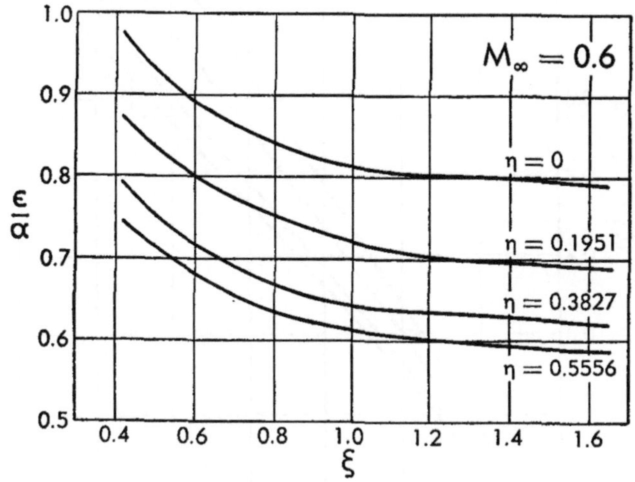

Fig. C,2f. Distribution of the downwash over the trailing vortex sheet for the case of $M_\infty = 0.6$.

same time an increase in M_∞ produces a decrease in $\zeta - \zeta_1$ so it may be deduced that, for a given point bearing a fixed geometric relationship to the airplane, the induced angle of flow inclination ϵ is hardly influenced at all by Mach number changes, provided the lift coefficient C_L remains

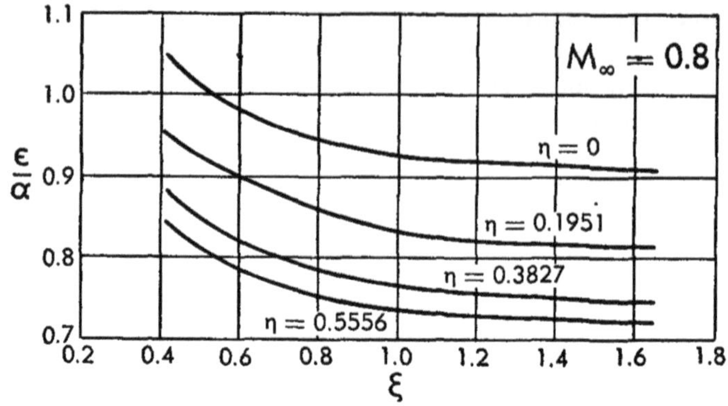

Fig. C,2g. Distribution of the downwash over the trailing vortex sheet for the case of $M_\infty = 0.8$.

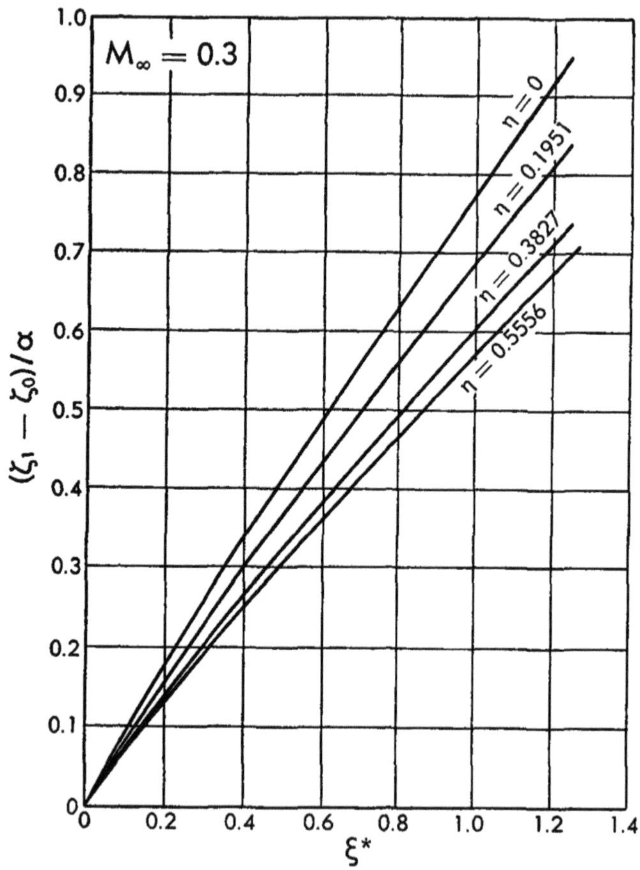

Fig. C,2h. Vertical displacement of trailing vortices at a distance ξ^* behind the trailing edge.

unaltered. Thus in the example under discussion, for the condition $C_L = 1$ and for a point which lies in the plane of the wing, the value of ϵ diminishes about 3 per cent in going from $M_\infty = 0.3$ to $M_\infty = 0.6$, while it diminishes about 6 per cent in going from $M_\infty = 0.3$ to $M_\infty = 0.8$.

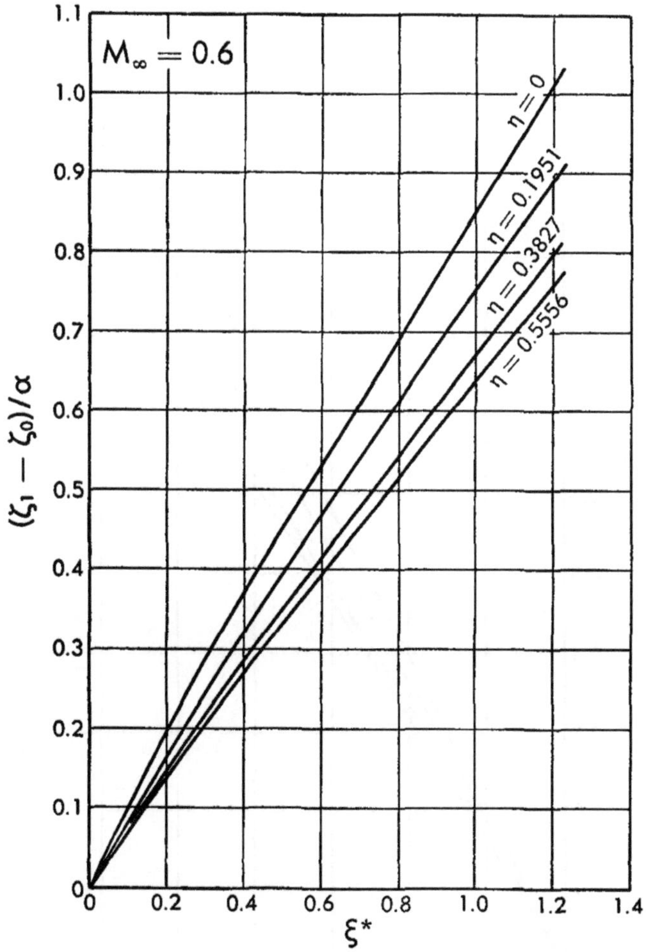

Fig. C,2i. Vertical displacement of trailing vortices at a distance ξ^* behind the trailing edge.

Case of a Wing of Small Aspect Ratio or with a Large Amount of Sweep.

Physical and mathematical foundation for evaluation of the downwash. As ordinarily found to be true for conventional wings, it is taken for granted that the line joining the aerodynamic centers of the various airfoil sections making up the wing is a straight line; the angle that this line makes with the normal to the plane of symmetry of the lifting surface is denoted by Λ (the angle of sweep).

Weissinger's theory [6] for the determination of the distribution of circulation created along the span of wings having any shape and aspect ratio whatsoever is effective here. Application of his theory, even in cases

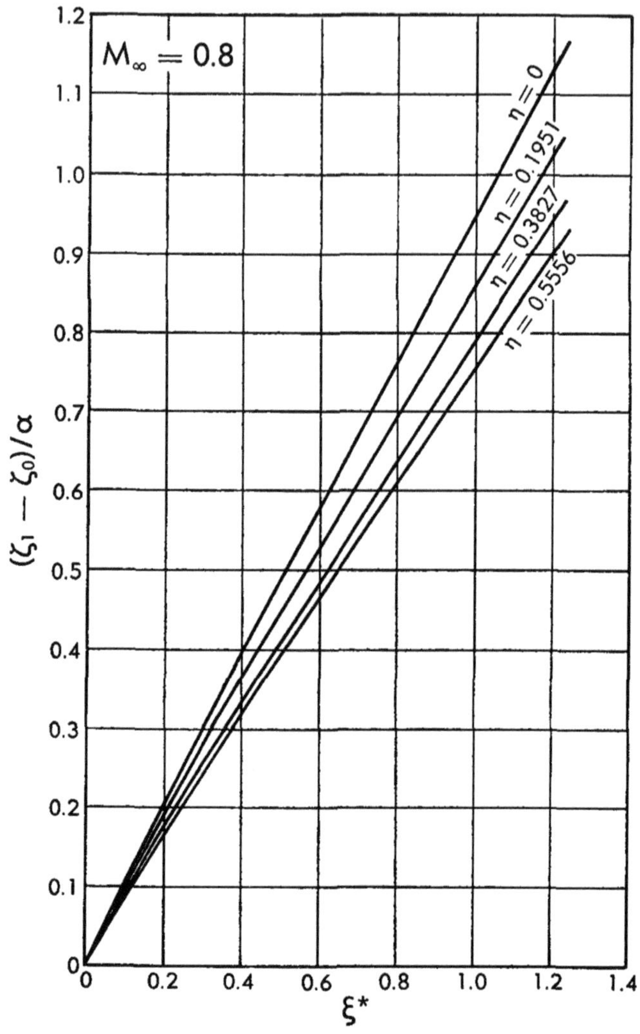

Fig. C,2j. Vertical displacement of trailing vortices at a distance ξ^* behind the trailing edge.

where the aspect ratio is small [7], has shown that a close approximation is obtained in the calculation of the induced velocities produced by the vortex system which replaces the wing, provided merely that the bound vortices on the wing are treated as though they have been compressed

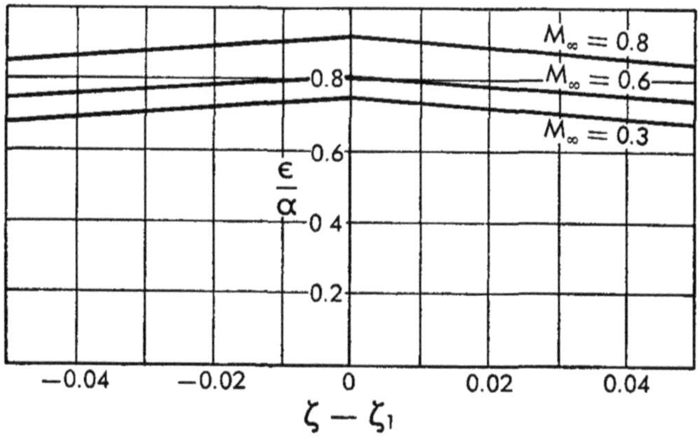

Fig. C,2k. Downwash variation normal to the trailing vortex sheet, for $\xi = 1$ and $\eta = 0$.

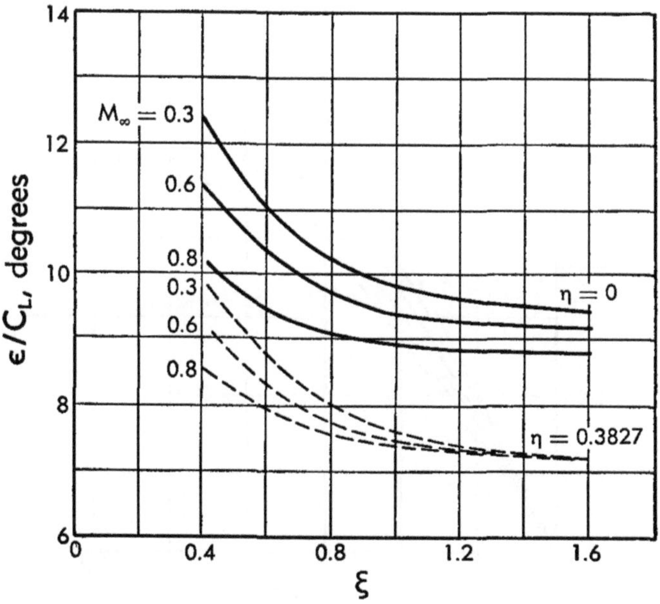

Fig. C,2l. Distribution of the downwash, in degrees, over the trailing vortex sheet for the case where the lift coefficient is maintained constant as the Mach number is varied.

into a single vortex whose axis lies along the locus of the aerodynamic centers of the airfoil sections. By hypothesis, this locus is to be taken as a straight line, and the induced velocities computed at the three-quarter chord line must satisfy the boundary conditions, i.e align themselves with the airfoil surface. On the basis of these assuring results it appears that the same method of attack ought to be suitable for calculation of w in the region of the tail.

In order to determine the downwash w at points of the trailing vortex sheet Σ, this system of vortices is again taken to be composed of a continuum of vortices having straight line axes which lie parallel to the free

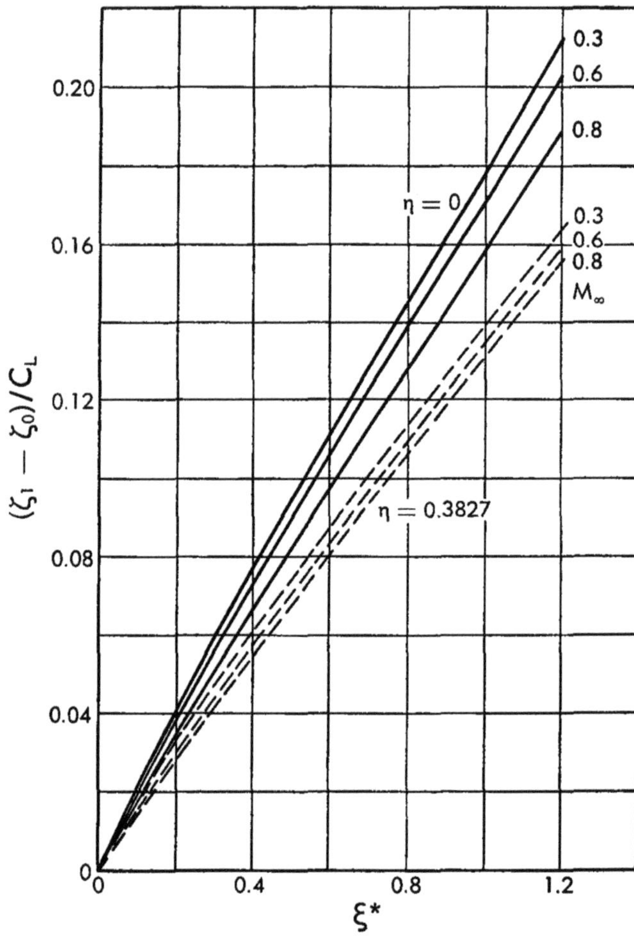

Fig. C,2m. Vertical displacement of the trailing vortices at a distance ξ^* behind the trailing edge for the case where the lift coefficient is maintained constant as the Mach number is varied.

stream velocity vector **U** (see Fig. C,2o). This assumption is still legitimate, even for wings of small aspect ratio, as long as sufficiently small values of the wing lift coefficient C_L are used (Eq. 2-1).

Under these conditions, the induced velocity generated at an arbitrary general point P lying on the vortex sheet Σ is given, according to the Biot-Savart formula, by the expression:

$$\epsilon = \frac{w}{U} = \frac{1}{2\pi}\int_{-1}^{1}\frac{1}{\eta-\eta'}\left[\frac{1+\xi_1-|\eta'|\tan\Lambda_1}{(\xi_1-|\eta'|\tan\Lambda_1)^2+(\eta-\eta')^2}\right]\frac{d\gamma}{d\eta'}\,d\eta'$$

$$+\frac{1}{2\pi}\int_{-1}^{1}\frac{\left(\xi_1-\eta\frac{\eta'}{|\eta'|}\tan\Lambda_1\right)\gamma(\eta')d\eta'}{[(\xi_1-|\eta'|\tan\Lambda_1)^2+(\eta-\eta')^2]^{\frac{3}{2}}}$$

$$=\frac{1}{2\pi}\int_{-1}^{1}\frac{1}{\eta-\eta'}\frac{\partial}{\partial\eta'}(\gamma H)d\eta' = \frac{1}{2\pi}\int_{-1}^{1}\frac{1}{\eta-\eta'}\frac{d\tilde{\gamma}}{d\eta'}\,d\eta' \qquad (2\text{-}21)$$

where, because of the transformation (Eq. 1-1), the relationship

$$\tan\Lambda_1 = \frac{1}{\beta}\tan\Lambda \qquad (2\text{-}22)$$

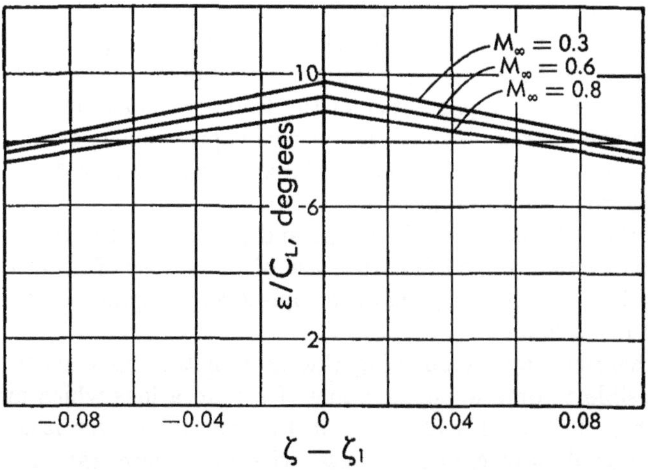

Fig. C,2n. Downwash variation, in degrees, normal to the trailing vortex sheet, for $\xi = 1$ and $\eta = 0$, for the case where the lift coefficient is maintained constant as the Mach number is varied.

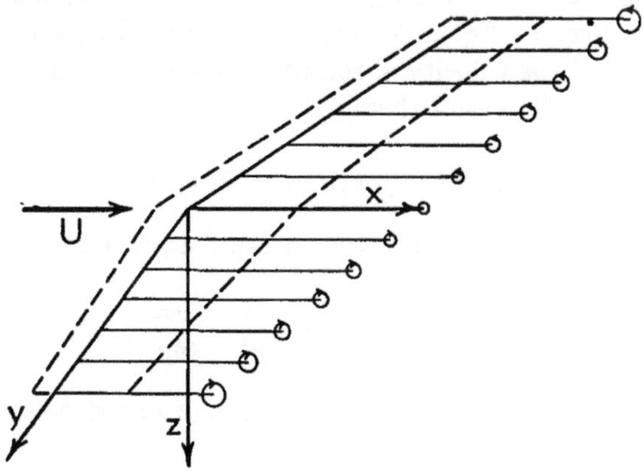

Fig. C,2o. Equivalent vortex system replacing the wing exhibiting an appreciable amount of sweep.

holds true, and

$$H = 1 + \frac{\xi_1 - |\eta'| \tan \Lambda_1}{(\xi_1 - |\eta'| \tan \Lambda_1)^2 + (\eta - \eta')^2} \tag{2-23}$$

$$\tilde{\gamma} = \gamma H \tag{2-24}$$

Consequently upon expressing γ once again by means of Eq. 2-10 it follows that

$$\epsilon(\xi, \eta_\nu) = 2b_{\nu\nu}\gamma_\nu - \sum_{n=1}^{m}{}' b_{\nu n} \left[1 + \frac{\xi - |\eta_n| \tan \Lambda}{\sqrt{(\xi - |\eta_n| \tan \Lambda)^2 + \beta^2(\eta_n - \eta_\nu)^2}}\right] \gamma_n \tag{2-25}$$

The values of $\gamma(\eta)$ which apply for various wing planforms and aspect ratios have been computed by De Young and Harper [7], and their results may be taken to be valid for any Mach number whatsoever, within the limits of applicability of the Prandtl-Glauert transformation. Although the restriction that γH must be expressible as a trigonometric series of the form of Eq. 2-10 is stipulated, this procedure does not differ otherwise from what has already been discussed.

It is then obvious that the shape and position of the trailing vortex sheet may be determined from the use of the values of ϵ, calculated by means of Eq. 2-25, in a way that is completely analogous to the procedure followed on p. 289.

Furthermore, by reemploying the same method used on p. 289 it is then possible to determine the value of ϵ at locations which are off the vortex sheet, but which still lie near it. In general, however, it is necessary to make due allowance, in those cases where the aspect ratio is small, for the rate of rolling up of that part of the vortex sheet which is in proximity to the plane of symmetry.

Because of the rate of rolling up of the vortex sheet, the second of the relationships given as Eq. 2-17 must be modified. Upon letting v_r denote the component of the velocity of roll-up which is directed along the y axis, and then by nondimensionalizing it to $\sigma_r = v_r/U$, it follows that Eq. 2-17 now becomes

$$\sigma(\zeta_1 + 0) + \sigma(\zeta_1 - 0) = 2\sigma_r \tag{2-26}$$

and, therefore, the expression given as Eq. 2-18 now takes the form

$$\left(\frac{\partial \epsilon}{\partial \zeta}\right)_{\zeta=\zeta_1} = \pm \frac{d^2\gamma}{d\eta^2} - \frac{\partial \sigma_r}{\partial \eta} \tag{2-27}$$

This result discloses that the distribution of ϵ on either side of Σ is not symmetric (see Fig. C,2p).

For want of any reliable experimental data, one may proceed to evaluate the derivative $\partial\sigma_r/\sigma\eta$ in the following manner according to Multhopp's procedure [5]:

The velocity v_r existing at any arbitrary point of the vortex sheet Σ is assumed to be the same as the induced velocity produced at the point in question by that part of the vortex sheet which has already been rolled up. This part of the sheet is imagined to be an infinite straight-line vortex with a circulation strength of Γ_r. Upon evaluation of the derivative $\partial\sigma_r/\partial\eta$ at the plane of symmetry, $\eta = 0$, a close approximation to

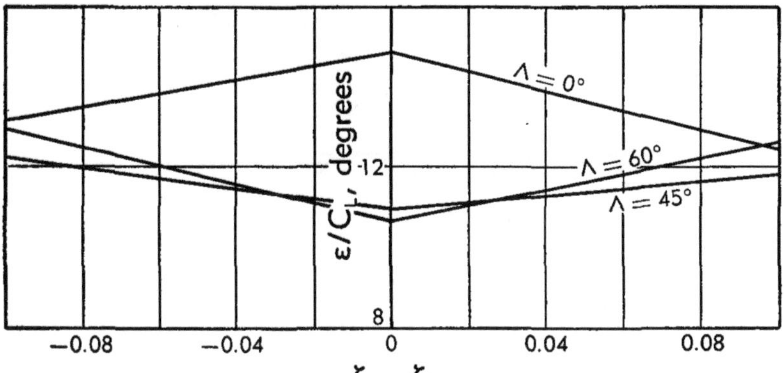

Fig. C,2p. Downwash variation, in degrees, normal to the trailing vortex sheet, for $\xi = 1$ and $\eta = 0$, for the case where the lift coefficient is maintained constant as the sweep angle is varied.

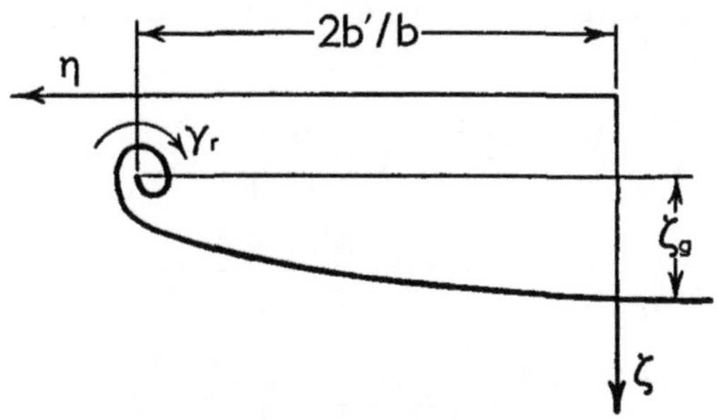

Fig. C,2q. Rolling up of the trailing vortices.

the effect of the two vortex cores, produced by the coiling-up tendency of the vortex sheet, is given by the following expression:

$$\frac{\partial\sigma_r}{\partial\eta} \cong \frac{4}{\pi} \frac{\gamma_r \zeta_g \frac{2b'}{b}}{\left(\frac{4b'^2}{b^2} + \zeta_g^2\right)^2} \tag{2-28}$$

The meanings of the symbols used in Eq. 2-28 may be understood most readily by referring to Fig. C,2q. Lacking more specific experimental data for insertion into this equation, it may be assumed that $2b'/b = 1$, that ζ_0 is equal to the value of ζ_1 given by Eq. 2-14, and that γ_r is equal to γ_{max}.

If the value of $\zeta - \zeta_1$ corresponding to the location at which it is desired to compute ϵ is too large, then it is no longer permissible to use Eq. 2-15. In this event, however, the value of ϵ may be obtained by a method analogous to that used earlier; i.e. by applying the Biot-Savart

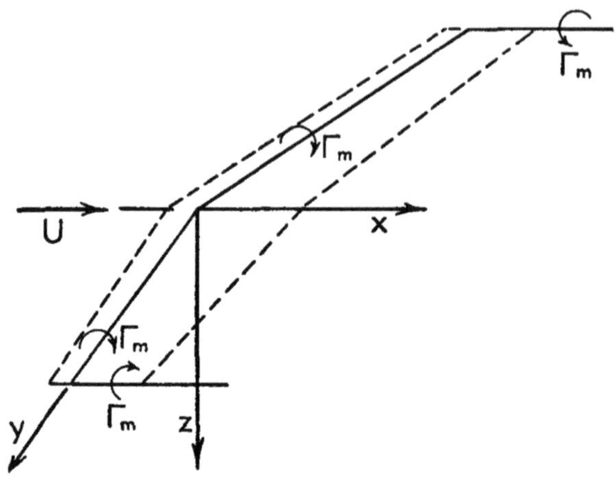

Fig. C,2r. Schematic diagram illustrating how rolled-up trailing vortices are imagined take disposed.

formula directly to the vortex system illustrated in Fig. C,2o. The pertinent formula is

$$\epsilon(\xi, \eta, \zeta) = \frac{1}{2\pi}\int_{-1}^{1} \frac{(\zeta - \zeta_1)^2 - (\eta - \eta')^2}{[(\eta - \eta')^2 + (\zeta - \zeta_1)^2]^2}$$

$$\times \left[1 + \frac{\xi - |\eta'| \tan \Lambda}{\sqrt{(\xi - |\eta'| \tan \Lambda)^2 + \beta^2(\eta - \eta')^2 + \beta^2(\zeta - \zeta_1)^2}}\right] \gamma d\eta'$$

$$+ \frac{\beta^2}{2\pi}(\zeta - \zeta_1)^2 \int_{-1}^{1} \frac{\xi - |\eta'| \tan \Lambda}{(\eta - \eta')^2 + (\zeta - \zeta_1)^2}$$

$$\times \frac{\gamma d\eta'}{[(\xi - |\eta'| \tan \Lambda)^2 + \beta^2(\eta - \eta')^2 + \beta^2(\zeta - \zeta_1)^2]^{\frac{3}{2}}} \qquad (2\text{-}29)$$

Of course, the effect of the roll-up is not accounted for in the least by this equation (2-29).

For small values of the aspect ratio Æ, and for the larger values of the wing lift coefficient C_L, one is confronted with the necessity of making better provision for the roll-up effects than is afforded by the relationships given in Eq. 2-27 and 2-28. Under the conditions just mentioned, an alternative approach to the calculation of ϵ must thus be essayed, in place of the procedure suggested previously. Lacking any systematic experi-

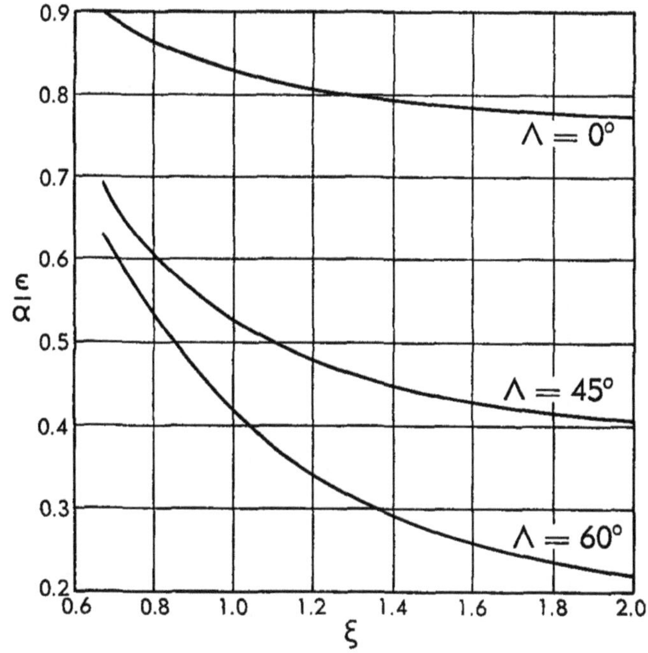

Fig. C,2s. Distribution of the downwash on the trailing vortex sheet, for $\eta = 0$.

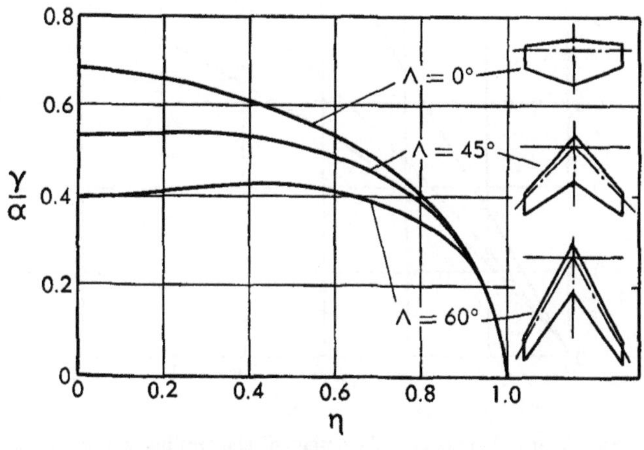

Fig. C,2t. Spanwise distribution of circulation of three swept wings having equal aspect ratios and taper ratios.

mental evidence on this question, one is constrained to rely on the following proposal.

The actual vortex system is replaced by a much simpler one made up of a single vortex according to the scheme illustrated in Fig. C,2r. According to this point of view the rolling up of the band of trailing vortices is imagined to have taken place completely by the time that they have detached themselves from the wing. This way of idealizing the aerodynamic situation corresponds to a limiting representation for the vortex system responsible for the induction effects to be studied. This idealization is exactly opposite to the one used previously, but since the real

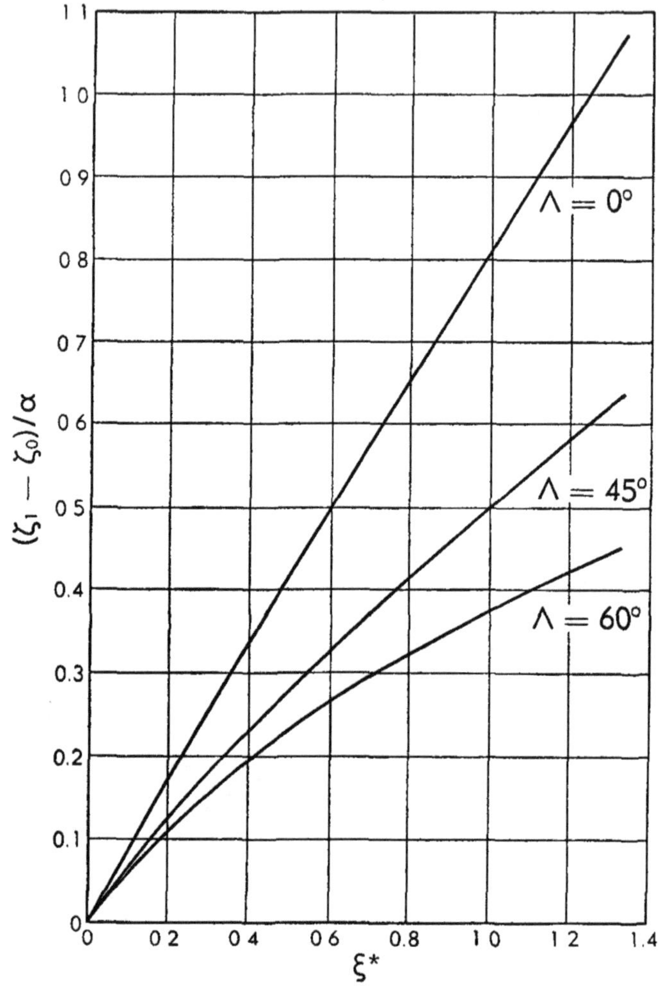

Fig. C,2u. Vertical displacement of the trailing vortex sheet at a distance ξ^* behind the trailing edge.

physical picture must actually lie somewhere in between the two extremes, it is legitimate to presume that the true values of ϵ ought to fall at some intermediate point between the results which would be calculated on the basis of the two limiting cases.

Along the single vortex Σ_w now being dealt with, the circulation is constant and has the value Γ_m which coincides with the value of the circulation found at the airfoil section located in the plane of symmetry. Now the two trailing vortices are assumed to have straight-line axes lying parallel to the free stream velocity vector **U**, and these axes constitute the side legs of the vortex system Σ_w. The distance between these sides is given by the expression

$$\eta_0 = -\frac{1}{\gamma_m}\int_0^1 \eta \frac{d\gamma}{d\eta} d\eta = \frac{1}{\gamma_m}\int_0^1 \gamma d\eta \tag{2-30}$$

which is derived from the theorems of vortex motion.

The calculation of ϵ at any point P whatsoever may now be obtained in a straightforward manner by application of the Biot-Savart formula.

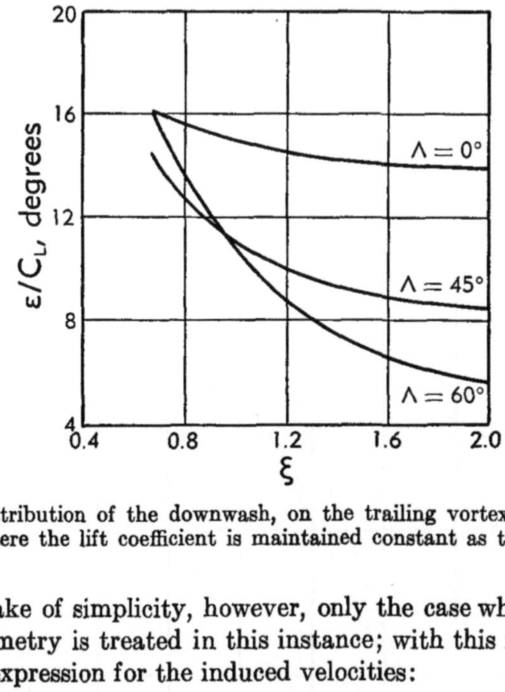

Fig. C,2v. Distribution of the downwash, on the trailing vortex sheet, for $\eta = 0$, for the case where the lift coefficient is maintained constant as the sweep angle is varied.

For the sake of simplicity, however, only the case where P lies in the plane of symmetry is treated in this instance; with this reservation one obtains this expression for the induced velocities:

$$\begin{aligned}\epsilon(\xi, 0, \zeta) = \frac{\gamma_m}{\pi}\Bigg\{&\frac{1}{\sqrt{\xi_1^2\cos^2\Lambda_1 + \zeta^2}}\Bigg[\frac{\xi_1 \sin\Lambda_1}{\sqrt{\xi_1^2 + \zeta^2}}\\ &+ \frac{1}{\cos\Lambda_1}\cdot\frac{\eta_0 - \xi_1\sin\Lambda_1\cos\Lambda_1}{\sqrt{(\xi_1 - \eta_0\tan\Lambda_1)^2 + \eta_0^2 + \zeta^2}}\Bigg]\\ &+ \frac{1}{\eta_0}\Bigg[1 + \frac{\xi_1 - \eta_0\tan\Lambda_1}{\sqrt{(\xi_1 - \eta_0\tan\Lambda_1)^2 + \eta_0^2 + \zeta^2}}\Bigg]\Bigg\}\end{aligned} \tag{2-31}$$

where again

$$\xi_1 = \frac{\xi}{\beta} \quad \text{and} \quad \tan\Lambda_1 = \frac{1}{\beta}\tan\Lambda$$

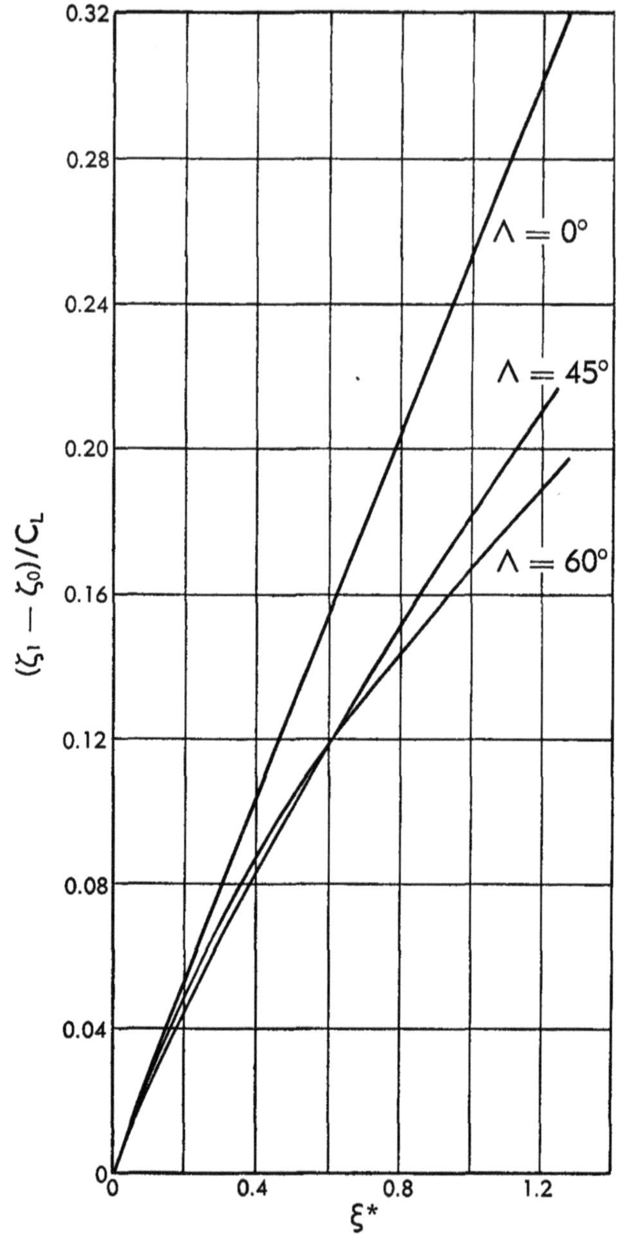

Fig. C,2w. Vertical displacement of the trailing vortex sheet, for the case where the lift coefficient is maintained constant as the sweep angle is varied at a distance ξ^* behind the trailing edge.

It is worthy of note that ϵ varies somewhat less rapidly as a function of ζ according to the relationship just derived in Eq. 2-31 than it does in response to the dictates of either Eq. 2-15 or Eq. 2-29. Consequently, in view of the approximate nature of the present analysis, one may safely ignore the downward displacement of the trailing vortices when proceeding to evaluate ϵ at any arbitrary point P.

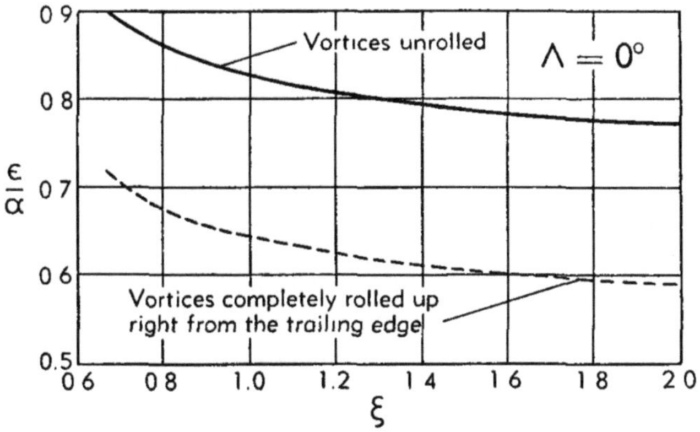

Fig. C,2x. Downwash for $\eta = 0$ with vortices unrolled and with vortices completely rolled up right from the trailing edge.

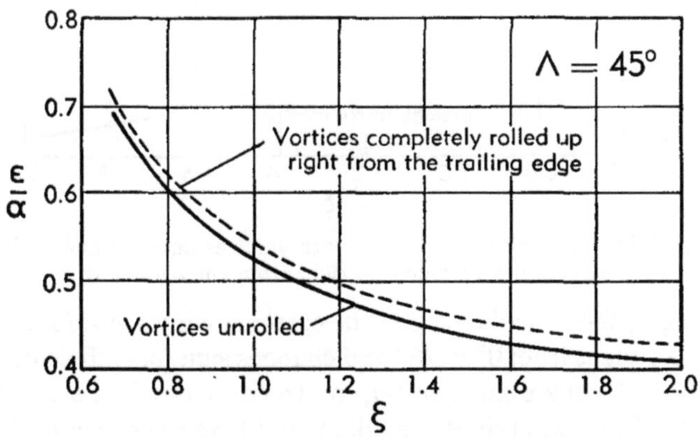

Fig. C,2y. Downwash for $\eta = 0$ with vortices unrolled and with vortices completely rolled up right from the trailing edge.

Effect of the sweep and the roll-up phenomenon on the downwash. In Fig. C,2s the results of the application of Eq. 2-25 are plotted for $\eta_v = 0$ and for three tapered wings, each having a taper ratio of 0.5, $\AR = 3$, and sweepback angles of respectively $\Lambda = 0°$, $\Lambda = 45°$, and $\Lambda = 60°$, for which pertinent distributions of the circulation γ in terms of η are given in Fig. C,2t. It appears that, for a given α, as Λ increases ϵ decreases. From Fig. C,2u, on which the corresponding values of the downward displacement $(\zeta_1 - \zeta_0)/\alpha$ are plotted, one deduces that the same property holds also for this parameter. If a comparison of the effects of sweep on the downwash is made on the basis that the lift coefficient C_L must remain fixed, rather than α, the relationship just noted still holds, but only for those locations lying aft of a certain fixed value of ξ; this state of affairs is shown in Fig. C,2v, and the corresponding data pertaining to the vertical displacement of the trailing vortex sheet, in Fig. C,2w.

Fig. C,2n shows the dependence of ϵ/C_L upon $\zeta - \zeta_1$ for the previously mentioned values of Λ and for $\xi = 1$. As $\zeta - \zeta_1$ increases, the induction decreases for $\Lambda = 0$ and increases for Λ sufficiently large. Furthermore the rate of increase is larger the greater Λ becomes; therefore, as $\zeta - \zeta_1$

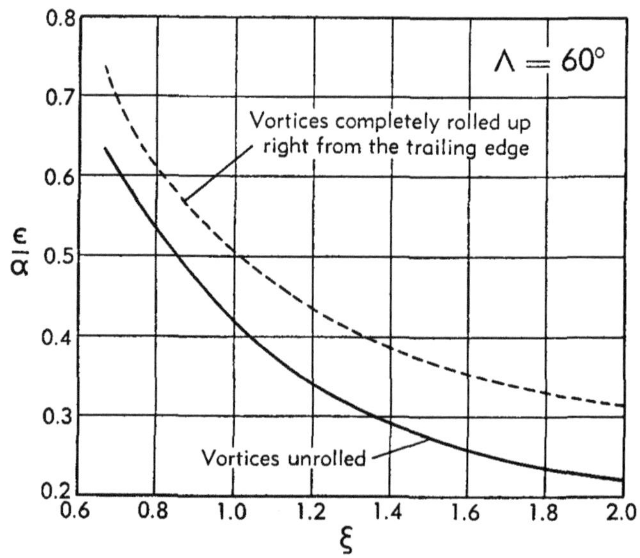

Fig. C,2z. Downwash for $\eta = 0$ with vortices unrolled and with vortices completely rolled up right from the trailing edge.

becomes larger, the spread between the extreme values of ϵ, for a given C_L, narrows as Λ ranges from 0° to 45° and changes sign for sufficiently large Λ.

The effect of vortex sheet roll-up on the downwash is evinced by the curves in Fig. C,2p, wherein it is evident that here the slopes of the graphs are different depending on whether $\zeta - \zeta_1$ is positive or negative. In addition to what was remarked above concerning the maximum possible influence of the roll-up, it may be noted from the curves plotted in Fig. C,2x, C,2y, and C,2z that (1) the influence of roll-up is so weak that if one arbitrarily selects any values whatsoever to represent the downwash ϵ, as long as they are bracketed by the extreme values predicted by either Eq. 2-25 or 2-31, the amount of error actually introduced will be inconsequential even at fairly large values of C_L, and that (2) the effect of roll-up upon the downwash changes in sign as one progress from small to large values of sweep.

C,3. Determination of the Aerodynamic Characteristics of the Tailplane. Under the hypotheses already stipulated in Art. 1, the values computed for the downwash ϵ may be directly substituted with impunity for the induced angle of flow inclination. Consequently the first step in the determination of the aerodynamic properties of the tail when it is operating in the disturbed flow emanating from the wing is to compute the downwash values ϵ, all along the span of the stabilizer at the three-quarter chord locations of the airfoil sections constituting the surface. Once this is accomplished, the rest of the computation is carried out in just the same way that the isolated wing was treated in Sec. A.

By applying Multhopp's procedure [*8*] for handling this case, the distribution of circulation due to the interference effect may be written down simply as

$$b_\nu \Delta\gamma_\nu = -\epsilon_\nu + \sum_{n=1}^{m}{}' b_{\nu n}\Delta\gamma_n \tag{3-1}$$

where

$$b_\nu = b_{\nu\nu} + \frac{b}{c_\nu (c_{l\alpha})_\nu}$$

The meaning of the symbols, though fairly obvious, is explained as follows: $b_{\nu\nu}$ and $b_{\nu n}$ are the Multhopp coefficients defined earlier, c_ν is the chord of the airfoil section of the tail at the spanwise position given by $\eta = \eta_\nu$, while $(c_{l\alpha})_\nu$ is the slope of the lift curve of a two-dimensional wing possessing the same airfoil section.

The changes in the values of the local lift coefficient and local drag coefficient are furnished by the following expressions, provided one has already obtained the results for the distribution of the incremental circulation $\Delta\gamma_n$ due to the interference:

$$(\Delta C_L)_t = \frac{\pi(\mathit{AR})_t}{m+1}\sum_{n=1}^{m}\Delta\gamma_n \sin\theta_n$$

$$(\Delta C_{D_i})_t = \frac{\pi(\mathit{AR})_t}{m+1}\left[\sum_{n=1}^{m}\Delta\gamma_n(\epsilon_n + \alpha_i + \Delta\alpha_i) + \sum_{n=1}^{m}\gamma_n(\epsilon_n + \Delta\alpha_i)\sin\theta_n\right] \tag{3-2}$$

where α_i is the self-induced angle of attack created by the circulation distribution γ_n.

The most frequently occurring type of tail surface encountered in practical applications is that for which the absolute geometric angle of attack of all the airfoil sections making up the tail surface is constant over the whole span. Consider only this kind of tail surface, therefore, and take it for granted that an average constant value of the wing-induced flow inclination angle ϵ may be used to represent adequately the actual angles ϵ, which as a matter of fact do not vary over a wide range of values as one moves from location to location on the stabilizing surface. Then it is obvious from comparison of the equations defining $\Delta\gamma_n$ and γ_n that the graph of $\Delta\gamma$ as a function of η is affine to the graph of γ. The constant of proportionality between these two affine curves is also found to be equal to $-\epsilon/\alpha$.

On this basis it follows that the changes in the tail lift coefficient and of the tail drag coefficient due to the interference effect may be expressed as

$$(\Delta C_L)_t = -\frac{\epsilon}{\alpha}(\overline{C}_L)_t$$

$$(\Delta C_{D_i})_t = (\overline{C}_L)_t \cdot \epsilon\left(1 - \frac{\epsilon}{\alpha}\right) - \frac{\epsilon}{\alpha}(\overline{C}_{D_i})_t\left(2 - \frac{\epsilon}{\alpha}\right) \tag{3-3}$$

Furthermore the resultant lift coefficient for the tail in the presence of the wing, $(C_L)_t$, and also the resultant induced drag coefficient, $(C_{D_i})_t$, are consequently given by the relations

$$(C_L)_t = (\overline{C}_L)_t\left(1 - \frac{\epsilon}{\alpha}\right)$$

$$(C_{D_i})_t = \epsilon(C_L)_t + \left(1 - \frac{\epsilon}{\alpha}\right)^2(\overline{C}_{D_i})_t \tag{3-4}$$

where $(\overline{C}_L)_t$ and $(\overline{C}_{D_i})_t$ are the coefficients of lift and of induced drag, respectively, applying to the isolated tail surface.

C,4. Effect of the Wake on the Interference. Experimental Results of Silverstein and Katzoff. The interference effect of the wing upon the tail is of a much more complex character if the wing is provided with an extended high lift flap or if the flow is stalled over even the narrowest of spanwise regions. In both of these cases the lift distribution is altered from that produced at a small angle of attack. In the event that flow separation is present, it is necessary to distinguish between the cases of tip stalling and root stalling; i.e. whether the separation first takes place at the tips or over the central portions of the wing. If one is confronted with the case of a tip stall, the spanwise distribution of circulation over the wing bears qualitative similarity to the distribution arising from a wing with a partial-span flap (see Fig. C,4a). In such an instance the induced angles of attack in the region occupied by the tail are increased over what would have been found to hold true there if no separation had occurred and if the lift coefficient were the same. On the other hand, the opposite situation exists in regard to the distribution of circulation when the separation is in the nature of a root stall; thus the induced angles of attack at the tail are reduced under this condition.

However, in addition to the change in the form of the lift distribution, regardless of whether it is due to the presence of the flap or to stalling, the separated flow conditions produce a wake whose thickness is of such magnitude that it can no longer be neglected. When the tail surface lies within the wake or in close proximity to its borders, it will be necessary to add to the field of velocities directed normal to the free stream velocity **U** (considered in the preceding articles) the field of interference velocities generated by the wake itself.

Such wake interference is characterized by a decrease of velocity within the wake and in a change in the induced angles of flow inclination

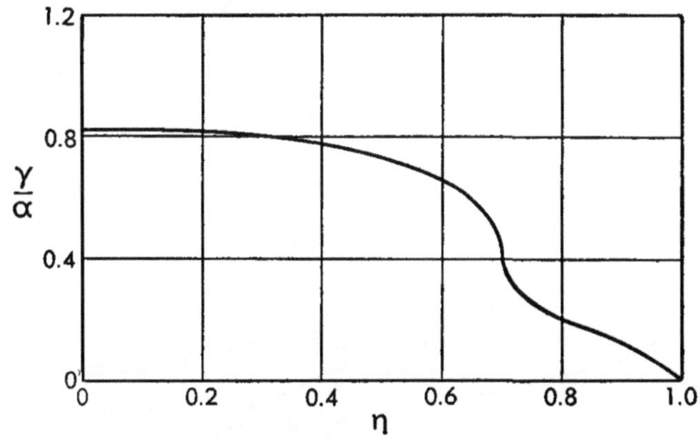

Fig. C,4a. Distribution of the circulation along the span of a tip-stalled wing.

compared to what would have been found to hold true in the absence of the wake. These changes are due to the convergence of the streamlines in the flow field occupied by the wake. The effect of such a convergence is to increase the amount of induction occurring in the upper part of the wake and to decrease the induction in the lower part. Because of the present inadequate state of the theory, the quantitative determination of such effects cannot be reckoned from purely theoretical considerations. Satisfactory results may be obtained, nevertheless, by using a procedure offered by Silverstein and Katzoff [*9*]; this method can be substantiated both on theoretical grounds and on the basis of systematic experimental research results.

It may be assumed that the surface constituting the center of the wake is coincident with the trailing vortex sheet already described above and that the location of this sheet is determined approximately by means of Eq. 2-14. However, it is no longer sufficient to rely completely upon this formula to furnish an adequate approximation to the wake position, because it is obvious that for wings with extended flaps the line along which the trailing vortex sheet or the wake may be considered to detach itself from the wing does not coincide with the trailing edge of the wing. Instead, these lines of detachment will have to lie somewhere below the trailing edge of the wing in this case (see Fig. C,4b). On the other hand, when a wing exhibits flow separation, the lines of detachment are found to lie at some distance above the trailing edge of the wing (see Fig. C,4b, middle and bottom).

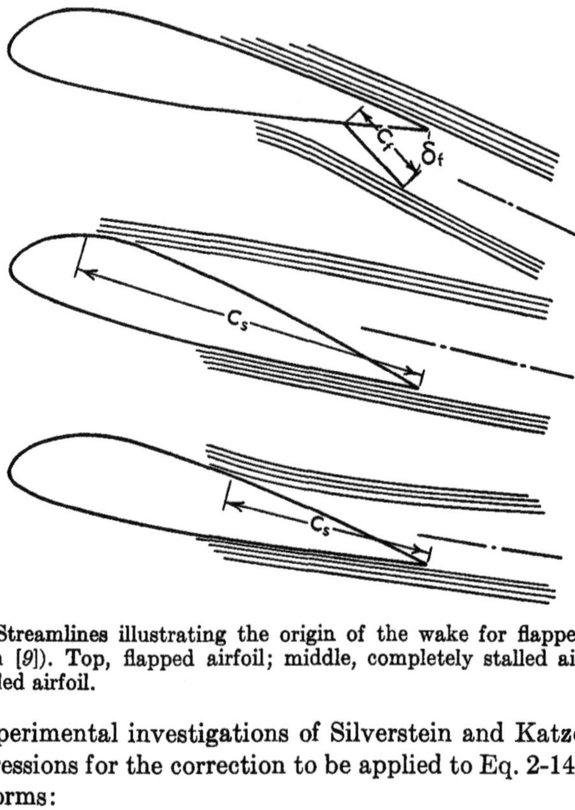

Fig. C,4b. Streamlines illustrating the origin of the wake for flapped and stalled airfoils (from [9]). Top, flapped airfoil; middle, completely stalled airfoil; bottom, partially stalled airfoil.

The experimental investigations of Silverstein and Katzoff have led to two expressions for the correction to be applied to Eq. 2-14, having the following forms:

(i) $$\Delta\zeta_1 = \tfrac{1}{2}c_f \sin \delta_f + kc \qquad \text{for a flapped wing} \qquad (4\text{-}1)$$

wherein c_f denotes the chord length of the flap, δ_f is the angle through which the flap is deflected, c represents the wing chord, and k is an empirical factor whose value is plotted in Fig. C,4c as a function of the flap deflection δ_f.

(ii) $$\Delta\zeta_1 = -\tfrac{1}{2}c_s \sin \alpha \qquad \text{for a stalled wing} \qquad (4\text{-}2)$$

wherein c_s denotes the abscissa of the point from which the detachment takes place; this distance is measured back along the wing chord, starting at the trailing edge.

The thickness of the wake and the distribution of the velocities within the wake may be determined by applying the rules derived by Prandtl [*2*, Vol. 3, pp. 162–178] and by Schlichting [*10*]. These laws hold true independent of the Mach number, at least to the same order of approximation as that implied when these laws were originally formulated for the case of $M_\infty = 0$. The constants which appear in these relationships will

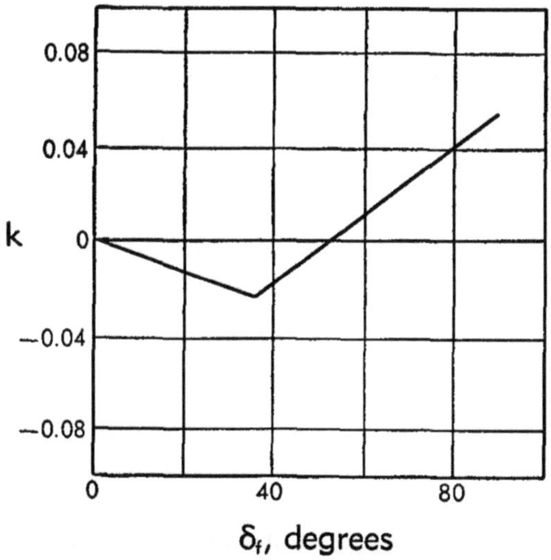

Fig. C,4c. Plot of factor k for correction to the vertical position of the wake origin at the trailing edge of a wing with a flap (from [*9*]).

be assumed to take on the values that have been found to fit the experimental results most closely. On the basis of these understandings, therefore, the following formula may be utilized for computing the wake semiwidth:

$$\zeta_w = 0.68 c_{d_0}^{\frac{1}{2}} (\xi_w + 0.15)^{\frac{1}{2}} \tag{4-3}$$

where c_{d_0} is the profile drag coefficient applying to the airfoil section located at the spanwise position corresponding to the plane $\eta = \text{const}$, in which the wake's behavior is being investigated. The definitions of the other two symbols in this formula are simply

$$\zeta_w = \frac{1}{2} \frac{\text{wake thickness}}{\text{chord length of the airfoil}}$$

and

$$\xi_w = \frac{\text{distance downstream from the trailing edge of the airfoil}}{\text{chord length of the airfoil}} \tag{4-4}$$

Likewise the velocity distribution in the wake is given by the formula

$$\frac{\mu'}{\mu} = \left[1 - \left(\frac{\zeta'_w}{\zeta_w}\right)^{1.75}\right]^2 \cong \cos^2 \frac{\pi \zeta'_w}{\zeta_w} \tag{4-5}$$

in which

$$\left.\begin{aligned} \mu &= 1 - \frac{u_0^2}{U^2} \\ &\text{and} \\ \mu' &= 1 - \frac{u^2}{U^2} \end{aligned}\right\} \tag{4-6}$$

where u_0 denotes the velocity at the center of the wake, and u stands for the value of the velocity at a point in the wake which lies at some vertical distance off the centerline; the vertical distance in question, when non-dimensionalized by reference to the chord length of the wing, is represented by the symbol ζ'_w.

Finally on the basis of the experimental studies mentioned above, one may assume that μ is given by the relation

$$\mu = \frac{2.42 c_{d_0}^{\frac{1}{2}}}{\xi_w + 0.3} \tag{4-7}$$

which only differs from the expression proposed by Prandtl [*2*, Vol. 3, pp. 162–178] in the manner in which it depends upon ξ_w, the fore and aft location in the wake.

Consequently one is now in a position to determine whether the stabilizer lies within or outside of the wake, since the wake's position in any artibrary plane η = const is given by means of Eq. 4-1 or Eq. 4-2 and 2-14 while knowledge of its thickness at different distances behind the trailing edge of the airfoil section lying in the same η = const plane is given by means of Eq. 4-3. Whenever the tailplane lies within the wake it is essential to make allowance for the fact that, when proceeding to compute the aerodynamic forces exerted upon this surface, the dynamic pressure applying to the region occupied by the tail is no longer as much as $\frac{1}{2}\rho_\infty U^2$, but is reduced approximately to

$$q = \left(\frac{1}{1 + \frac{\gamma - 1}{2} M_\infty^2 \mu'}\right)(1 - \mu') \frac{1}{2} \rho_\infty U^2 \tag{4-8}$$

where γ is the adiabatic exponent. The term in Eq. 4-8, which represents the influence of compressibility, is so small for the values of M_∞ that are of general interest that it may be dropped in most instances.

As already mentioned, however, the presence of the wake not only alters the longitudinal components of the field of flow but in addition creates further interference effects by altering the transverse components of the flow, and these alterations will have a reaction upon the tail surface even though the surface is not actually in the wake but merely lies near its boundaries.

The quantitative determination of this effect due to the transverse velocity components may be considered as follows: In the plane of symmetry ($\eta = 0$), for example, the expression for the flux density, or the amount of fluid mass transported per unit of time, across the streamtube comprised of the streamline which passes through the mid-point of the wake (ξ_w, 0) and the streamline which passes through the arbitrary point $P(\xi_w, \zeta'_w)$, may be written as:

$$\begin{aligned}\Psi &= c\int_0^{\zeta'_w} \rho u d\zeta'_w \\ &\cong c\rho_\infty U\left[\int_0^{\zeta'_w}\sqrt{1-\mu\cos^2\left(\frac{\pi}{2}\frac{\zeta'_w}{\zeta_w}\right)}\,d\zeta'_w \right.\\ &\qquad \left. - \frac{\gamma-1}{2}M_\infty^2\int_0^{\zeta'_w}\frac{u}{U}\mu' d\zeta'_w\right] \qquad (4\text{-}9)\end{aligned}$$

But since the term which represents the influence of compressibility in this expression is always very small, it is permissible to replace the ratio u/U by unity, approximately, without any serious consequences. It follows then that the explicit formula for the flux density is

$$\begin{aligned}\Psi = c\rho_\infty U\frac{2}{\pi}\zeta_w\Bigg[\mathsf{E}(\sqrt{\mu}) - E(\sqrt{\mu}, \Theta)' \\ - \frac{\gamma-1}{2}M_\infty^2\left(\frac{\Theta}{2}+\frac{1}{4}\sin 2\Theta\right)\Bigg] \qquad (4\text{-}10)\end{aligned}$$

wherein E denotes the complete elliptic integral of the second kind with modulus $\sqrt{\mu}$, while E denotes the incomplete elliptic integral of the second kind with modulus $\sqrt{\mu}$ and argument Θ'. This argument and the other parameter appearing in Eq. 8-10 are defined by the relations

$$\Theta' = \frac{\pi}{2}\left(1-\frac{\zeta'_w}{\zeta_w}\right) \quad \text{and} \quad \Theta = \frac{\pi}{2}\frac{\zeta'_w}{\zeta_w}$$

The value of the velocity w', normal to the free stream velocity vector **U**, that is created by presence of the wake may then be calculated directly from the relationship

$$\Delta\epsilon = \frac{w'}{U} = \frac{1}{\rho_\infty U}\frac{\partial\Psi}{c\,\partial\xi_w} \tag{4-11}$$

The values of $\Delta\epsilon$ have been calculated by means of Eq. 4-11 by Silverstein and Katzoff, and their results are reproduced here as Fig.C,4d, the effect of compressibility being neglected.

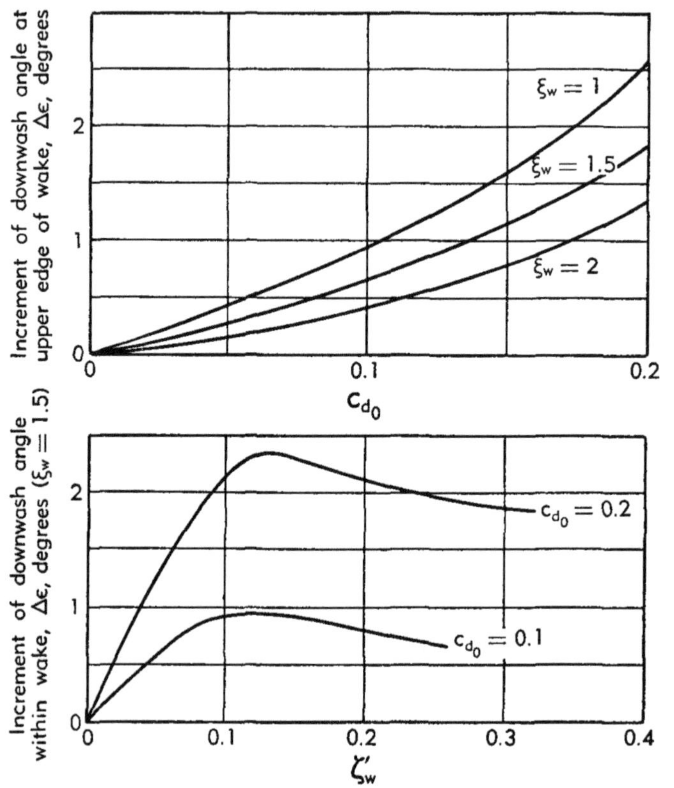

Fig. C,4d. Effect of wake on downwash. Top, at the upper edge of the wake; bottom, within the wake. (The effects are of equal magnitude, but of opposite sign, below the wake center.)

CHAPTER 2. WING-BODY INTERFERENCE AT SUBSONIC SPEEDS

C,5. General Considerations. The mutual interferences acting between a wing and an adjoining body may be considered to arise because of circumstances that are usually interpreted as follows. The velocities that are induced upon the surface of the fuselage, by the action of the vortex system which represents the wing, will have components normal to this body which are in general not zero. Consequently it is necessary to create an additional flow so that its effect upon the fuselage will just counteract the above-mentioned non-null components. On the other hand, the additional flow, together with the basic flow that is produced because of the impingement upon the body of the stream which originally was undisturbed far ahead of the body, will modify the velocity field surrounding the lifting surface, thus making it also necessary to alter the system of vortices representing the wing.

Since an exact solution of the general problem presents almost insurmountable difficulties, it is therefore prudent to narrow the discussion to certain particular cases which may be given an analytic treatment not excessively complex. The simplified cases may be handled by making suitable opportune assumptions in order to reduce the flow phenomena down to their elementary features. When this is done, the essential aspects of the interference effects are highlighted. Two such cases worthy of intensive study are (1) a wing whose aspect ratio is not very small, incorporating a small amount of sweep, and mounted on a long fuselage, and (2) a wing of small aspect ratio, with pronounced sweep, and also mounted on an elongated body.

CASE 1: USE OF MULTHOPP'S PROCEDURE

C,6. Distribution of Lift Along the Wing Span and on the Fuselage. The solution of the problem now under examination may be accomplished by using the method proposed by Multhopp [*11*]. This method is based on the following fundamental hypotheses: (1) The idealized fuselage is supposed to be an infinitely long cylinder whose diameter is equal to that of the actual fuselage at the location corresponding to the three-quarter chord position on the wing; its angle of attack with respect to the free stream velocity vector **U** is denoted by α_f. (2) The wing should have a chord of sufficiently small length to make it legitimate to replace the actual wing by a single bound vortex with its axis lying perpendicular to the plane of symmetry and passing through the aerodynamic center; the angle of attack of this wing with respect to **U** is denoted by α_w. (3) The trailing vortices should have straight-line axes lying parallel to the generatrices of the cylindrical fuselage.

The justification for making assumption 1 lies in the following fact: In the case of a fuselage having the form of a body of revolution, when considering the flow at the point of maximum width of this type of body, it will be found that its fineness ratio exerts practically no influence on the generalized upwash distribution out along the wing, as may be seen by reference to the plot of the quantity $H = u_f w_f/UW$ in Fig. C,6a. The meaning of these symbols is almost obvious; i.e. u_f denotes the axial component of the velocity generated about the fuselage by action of the longitudinal component of the undisturbed flow, $U \cos \alpha_f \cong U_f$. Likewise, w_f denotes the component of the flow about the fuselage which is normal to the horizontal axis and which is produced by the action of the transverse component, W, of the oncoming undisturbed stream. Of course it is exactly this H-ratio which has the most pronounced and direct influence upon the form of the circulation distribution that is found to exist along the span of the wing.

Assumption 2 is acceptable, provided that the chord of the wing is quite small with respect to its span and to the length of the body on which

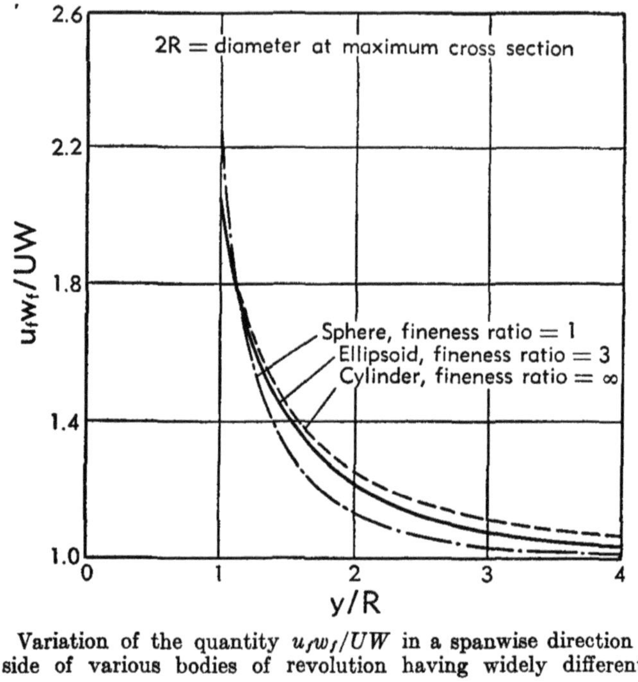

Fig. C,6a. Variation of the quantity $u_f w_f/UW$ in a spanwise direction out away from the side of various bodies of revolution having widely different fineness ratios.

it is mounted, but these are just the same stipulations which must be satisfied if one intends to make use of the procedure developed by Multhopp.

Finally the third assumption may be justified on the grounds that the trailing vortex sheet actually aligns itself with the fuselage for a considerable portion of its full length and for quite a wide range of angles of incidence. This is due to the action of the self-induced velocities in producing displacements in the flow behind the lifting surface that are of just sufficient size to maintain the kind of alignment being premised.

Under the dictates of these assumptions the vortex system Σ_w, which effectively replaces the wing, has the form and position shown in Fig.

C,6b and the system of axes used in this investigation is also shown in the same figure. It may be observed that the x axis is taken to coincide with the axis of the fuselage, while the y axis is normal to the x axis and is likewise normal to the plane of symmetry of the configuration. Now let the symbol Γ represent the strength of the circulation about the airfoil section lying in any arbitrary plane y = const. Then the potential function describing the field of flow created about the wing-body system, by

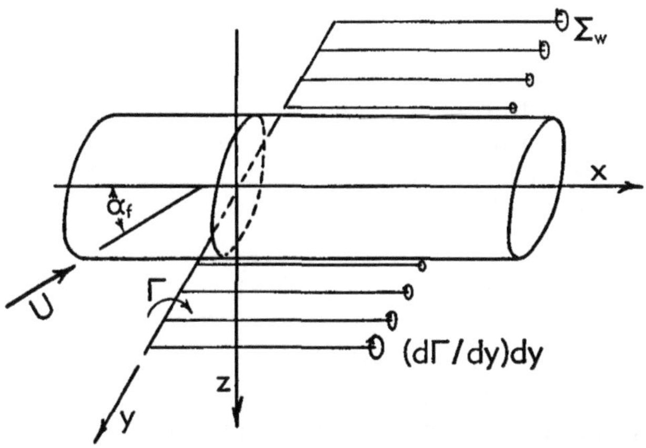

Fig. C,6b. Schematic arrangement of vortices and body geometry utilized in the study of mutual interference between a wing and a body, according to the Multhopp method.

action of the impinging flow with a uniform velocity of U far upstream, has a discontinuity at the surface of the vortex sheet Σ_w of an amount given by the relationship

$$\phi_{z=z_w^-} - \phi_{z=z_w^+} = \Gamma(y) \tag{6-1}$$

where $z = z_w$ is the equation of the horizontal plane in which Σ_w lies.

This potential for the flow about the configuration may be expressed in the more explicit form

$$\phi(x, y, z) = U[x - \alpha_f z - \alpha_f \phi_1(y, z) + \phi_2(x, y, z)] \tag{6-2}$$

where $\phi_1(y, z)$ is the potential for the additional flow which must be superimposed on the cylindrical body to counteract the cross flow component which is composed of a uniform stream in the y, z plane having a velocity of magnitude $W = 1$, while $U\phi_2$ denotes a harmonic function in the coordinates x_1, y, z which represents the potential describing the field of flow that is created by the vortex system Σ_w in the presence of the fuselage. Of course the coordinate x_1 is again related to the physical axial distance x by means of the relation $x_1 = x/\beta = x/\sqrt{1 - M_\infty^2}$.

In view of the discontinuity relationship given in Eq. 6-1, it holds true on the vortex sheet Σ_w that

$$(\phi_2)_{z=z_w^-} - (\phi_2)_{z=z_w^+} = \frac{\Gamma(y)}{U} \tag{6-3}$$

while throughout the rest of the flow field the potential ϕ_2 is continuous, together with its derivatives.

On the other hand, the expression for the circulation may be put in the form

$$\Gamma(y) = \tfrac{1}{2}c_{l_\alpha}(\alpha_w)_{\text{eff}}Uc \tag{6-4}$$

where $(\alpha_w)_{\text{eff}}$ denotes the effective absolute angle of incidence of the airfoil section measured with respect to the zero-lift line, and where the chord length is represented by c, while c_{l_α} stands for the slope of the lift curve for a wing of this profile having an infinite aspect ratio. Now the effective absolute angle of attack is given by the expression

$$(\alpha_w)_{\text{eff}} = \alpha_w + \alpha_f\left(\frac{\partial\phi_1}{\partial z}\right)_{z=z_w} - \epsilon \tag{6-5}$$

where ϵ is the induced angle of attack due to the flow generated by the trailing vortex system operating in the presence of the fuselage.

The flow approaching the fuselage in a direction perpendicular to the axis is called the cross flow and it is assumed for the present purpose that the magnitude of its uniform velocity is unity. If the cross flow is subjected to a conformal transformation in such a way as to convert the cross section of this body into a repeated segment of a line lying along the $\hat{z}$ axis (see Fig. C,6c), the determination of the derivative $(\partial\phi_1/\partial z)_{z=z_w}$ then comes about quite easily. Because the shape of the cross section is always rather close to being an ellipse or a circle, the transformation function can be readily set down; for example, one may use the Kármán-Trefftz method [12] or the procedure due to Theodorsen [13].

If this transformation function is symbolically represented in the form

$$\hat{t} = \hat{t}(t) \tag{6-6}$$

where t and the transformed point $\hat{t}$ are defined by the relationships

$$t = z + iy \qquad \text{and} \qquad \hat{t} = \hat{z} + i\hat{y} \tag{6-7}$$

then the expression for the desired derivative is given by

$$\left(\frac{\partial\phi_1}{\partial z}\right)_{z=z_w} = \left[\text{R.P.}\left(\frac{d\hat{t}}{dt}\right) - 1\right]_{t=z_w+iy} \tag{6-8}$$

where R.P. $(d\hat{t}/dt)$ is the real part of the derivative of $\hat{t}$ with respect to t.

Now the determination of the induced angle ϵ due to the influence of the bound vortex lying along the wing span may be carried out as follows.

First of all, it is noted that ϵ may be written as

$$\epsilon = \frac{1}{2} \lim_{x \to \infty} \left(\frac{\partial \phi_2}{\partial z}\right)_{z=z_w} = \frac{1}{2} \left(\frac{\partial \phi_{2\infty}}{\partial z}\right)_{z=z_w} \tag{6-9}$$

provided one takes $\phi_{2\infty}$ to be defined by the relation

$$\phi_{2\infty} = \lim_{x \to \infty} \phi_2(x, y, z) \tag{6-10}$$

but this is immediately seen to hold true if one imagines that the system of trailing vortices is prolonged through the semiplane $z = z_w$, $x < 0$.

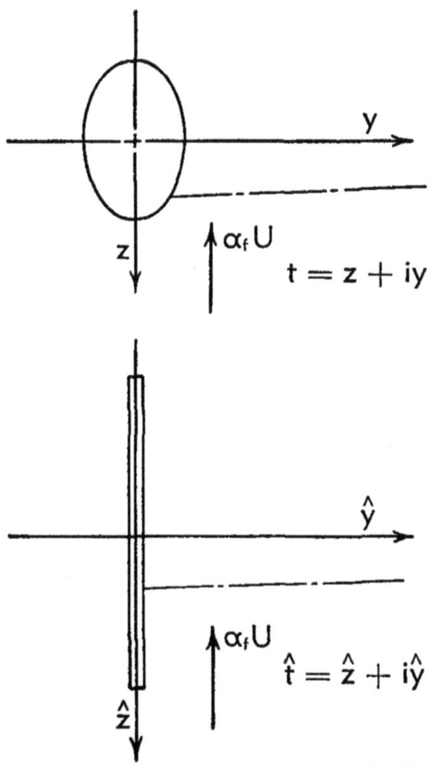

Fig. C, 6c. Fuselage cross section and the repeated segment of the $\hat{z}$ axis into which it is transformed conformally.

On the basis of the three assumptions already made, it is seen that $U\phi_{2\infty}$ is the potential which describes the two-dimensional flow that is generated around the cross section of the fuselage by action of the system of trailing vortices, which are now considered to be extended infinitely in both directions. The strength of these vortices per unit span distance is represented by the symbol $d\Gamma/dy$, and thus if one is still transforming the flow in the plane $x = \infty$ by means of the function given as Eq. 6-6, the corresponding potential in the transformed plane will appear as

$$\phi_{2\infty}(\hat{z}, \hat{y}) = \frac{1}{2\pi U} \text{R.P.} \left[\oint_s \frac{d\Gamma}{d\hat{s}} \ln (\hat{t} - \hat{t}')d\hat{s} \right] \tag{6-11}$$

where the symbol $\hat{t}'$ represents the complex coordinate, in the $\hat{t}$ plane, that gives the location of the element d_s of the line corresponding to the segments $t = z_w + iy$, along which the free vortices in the t plane are located. In fact, on account of the flow symmetry with respect to the $\hat{y} = 0$ plane, the boundary condition, relative to the repeated segment corresponding to the body cross section, is automatically satisfied.

The expression given above as Eq. 6-11 may be simplified somewhat inasmuch as one may generally replace $d\hat{s}$ by $d\hat{y}'$ within close approximation; in this manner the expression for $\phi_{2\infty}$ becomes

$$\phi_{2\infty}(\hat{z}, \hat{y}) = \frac{1}{2\pi U} \text{R.P.} \left[\int_{-\hat{b}/2}^{\hat{b}/2} \frac{d\Gamma}{d\hat{y}'} \ln (\hat{t} - \hat{t}')d\hat{y}' \right] \tag{6-12}$$

where $\hat{b}/2$ is the semispan of the wing in the $\hat{t}$ plane, which is related to the span of the physical wing by the formula (see Fig. C,6c)

$$\hat{b} = 2 \left\{ \text{I.P.} \left[\hat{t} \left(z_w + i\frac{b}{2} \right) \right] \right\} \tag{6-13}$$

Relying upon the accuracy of the above-mentioned approximation, one may write finally that

$$2\epsilon = \left(\frac{\partial \phi_{2\infty}}{\partial z} \right)_{z=z_w} = \left(\frac{\partial \phi_{2\infty}}{\partial \hat{z}} \right)_{\hat{z}=\hat{z}_w} \cdot \text{R.P.} \left(\frac{d\hat{t}}{dt} \right)_{t=z_w+iy}$$

$$= \frac{1}{2\pi U} \text{R.P.} \left(\frac{d\hat{t}}{dt} \right)_{t=z_w+iy} \left[\int_{-\hat{b}/2}^{\hat{b}/2} \frac{d\Gamma}{d\hat{y}'} \frac{d\hat{y}'}{\hat{y} - \hat{y}'} \right] \tag{6-14}$$

It follows, therefore, by inserting Eq. 6-5, 6-8 and 6-14 into Eq. 6-4 that the expression for the circulation is

$$\Gamma(\hat{y}) = \frac{1}{2} c_{l_\alpha} Uc \left\{ \alpha_w + \alpha_f \left[\text{R.P.} \left(\frac{d\hat{t}}{dt} \right) - 1 \right] - \frac{1}{4\pi U} \left[\text{R.P.} \left(\frac{d\hat{t}}{dt} \right) \right] \int_{-\hat{b}/2}^{\hat{b}/2} \frac{d\Gamma}{d\hat{y}'} \frac{d\hat{y}'}{\hat{y} - \hat{y}'} \right\} \tag{6-15}$$

provided that the derivative $d\hat{t}/dt$ is evaluated at the location where $t = z_w + iy$.

A further improvement in notation may be made by setting

$$\hat{\gamma} = \gamma \frac{b}{\hat{b}} = \frac{\Gamma}{\hat{b}U} \quad \text{and} \quad \hat{\eta} = \frac{2\hat{y}}{\hat{b}} \tag{6-16}$$

from which it follows that the converted circulation may be written as

$$\hat{\gamma} = \frac{1}{2} c_{l_\alpha} \frac{c}{\hat{b}} \left[\text{R.P.}\left(\frac{d\hat{t}}{dt}\right)\right] \left\{ \frac{\alpha_w - \alpha_f}{\left[\text{R.P.}\left(\frac{d\hat{t}}{dt}\right)\right]} + \alpha_f - \frac{1}{2\pi} \int_{-1}^{1} \frac{d\hat{\gamma}}{d\hat{\eta}'} \frac{d\hat{\eta}'}{\hat{\eta} - \hat{\eta}'} \right\} \tag{6-17}$$

It is convenient to have the portion of $\hat{\gamma}$ which is due to the incidence setting, $\alpha_w - \alpha_f$, separated from the part that is produced by the angle of attack α_f. If, therefore, the converted circulation is written as the sum of two terms

$$\hat{\gamma} = \hat{\gamma}_0 + \alpha_f \hat{\gamma}_f \tag{6-18}$$

one obtains the following expressions for the two contributions:

$$\hat{\gamma}_0 = \frac{1}{2} c_{l_\alpha} \frac{c}{\hat{b}} \left[\text{R.P.}\left(\frac{d\hat{t}}{dt}\right)\right] \left\{ \frac{\alpha_w - \alpha_f}{\text{R.P.}\left(\frac{d\hat{t}}{dt}\right)} - \frac{1}{2\pi} \int_{-1}^{1} \frac{d\hat{\gamma}_0}{d\hat{\eta}'} \frac{d\hat{\eta}'}{\hat{\eta} - \hat{\eta}'} \right\} \tag{6-19}$$

and

$$\hat{\gamma}_f = \frac{1}{2} c_{l_\alpha} \frac{c}{\hat{b}} \left[\text{R.P.}\left(\frac{d\hat{t}}{dt}\right)\right] \left(1 - \frac{1}{2\pi} \int_{-1}^{1} \frac{d\hat{\gamma}_f}{d\hat{\eta}'} \frac{d\hat{\eta}'}{\hat{\eta} - \hat{\eta}'} \right) \tag{6-20}$$

As far as the determination of the converted circulation $\hat{\gamma}$ is concerned, it may be observed by examination of the form of Eq. 6-19 and 6-20 that the wing behaves in the presence of the fuselage just as though it were an isolated wing, differing from the actual wing merely in that the chord lengths of the transformed wing are obtained by multiplying the chord lengths of the real wing by the factor R.P. $(d\hat{t}/dt)$. The incidence setting, $\alpha_w - \alpha_f$, of the transformed wing is obtained by dividing the setting for the real wing by this same quantity.[2]

[2] If the cross section of the fuselage is circular or elliptic, the transformation function $\hat{t}(t)$, as well as the other expressions which depend on knowledge of this transformation, and in which one is primarily interested when trying to compute the circulation distribution $\hat{\gamma}$, are quite easily found. To be specific, one finds for the case of a fuselage with circular cross section the following: The transformation function is

$$\hat{t} = t + (h^2b^2/t) \tag{1}$$

where hb is the radius of the circular cross section; and from this expression it follows that

$$\hat{z} = z\left(1 + \frac{h^2b^2}{z^2 + y^2}\right)$$

$$\hat{y} = y\left(1 - \frac{h^2b^2}{z^2 + y^2}\right) \tag{2}$$

and thus

$$\hat{b} = b\left[1 - \frac{h^2b^2}{z_w^2 + (b^2/4)}\right] \tag{3}$$

so that finally

$$\text{R.P.}\left(\frac{d\hat{t}}{dt}\right)_{t=z_w+iy} = 1 - \frac{z_w^2 - y^2}{(z_w^2 + y^2)^2} h^2b^2 \tag{4}$$

For the case of a fuselage with elliptic cross section the following applies: The

C,7. Lift of the Wing and the Fuselage. Once the values of γ have been obtained, the lift coefficient applying to the wing is computed from the expression

$$
\begin{aligned}
(C_L)_w &= \frac{2}{AU}\left[\int_{-b/2}^{-y_w} \Gamma dy + \int_{y_w}^{b/2} \Gamma dy\right] \\
&= \mathcal{R}\left[\int_{-1}^{-l_0} \gamma d\eta + \int_{l_0}^{1} \gamma d\eta\right]
\end{aligned}
\tag{7-1}
$$

where $l_0 = 2y_w/b$.

In order to compute the total lift of the entire configuration it is necessary to add the lift acting on the fuselage to the above-mentioned wing lift. The fuselage lift is obtained by noting first that Bernoulli's theorem gives the pressure change along a streamline as

$$\int_{-\infty}^{\infty} (p - p_\infty)dx = -\rho_\infty U^2 \phi_{2\infty} \tag{7-2}$$

provided that one makes the usual approximation of dropping the squared terms of the superimposed velocities. It is true, of course, that

$$\int_{-\infty}^{\infty} (p - p_\infty)dx = \frac{dL_f}{dy} \tag{7-3}$$

---

transformation function is

$$\hat{t} = \frac{1}{h_2 - h_1}\left[h_2 t - h_1\sqrt{t^2 - b^2(h_2^2 - h_1^2)}\right] \tag{1'}$$

where $h_1 b$ and $h_2 b$ are the minor and major semiaxes, respectively, of the elliptical cross section, which are assumed to extend along the y and z axes.

The expressions connecting the corresponding real and transformed coordinates y and $\hat{y}$ and also $\hat{z}$ and z are most easily derived by making the substitution $t = z + iy = A\cos\theta + iB\sin\theta$, where A and B are now the major and minor semiaxes, respectively, of an ellipse which has the same foci as the elliptical cross section of the fuselage and which passes through the point P, with complex coordinate t. Consequently A is the arithmetic mean of the distances from the point P to the foci, while B is given by the relation $B = \sqrt{A^2 - E^2}$ where $E = \sqrt{h_2^2 - h_1^2}\, b$. One then finds that

$$\hat{y} = \frac{y}{h_2 - h_1}\left(h_2 - h_1\frac{A}{\sqrt{A^2 - E^2}}\right), \quad \hat{z} = \frac{z}{h_2 - h_1}\left(h_2 - h_1\frac{\sqrt{A^2 - E^2}}{A}\right) \tag{2'}$$

and therefore

$$\hat{b} = \frac{b}{h_2 - h_1}\left(h_2 - h_1\frac{A_0}{\sqrt{A_0^2 - E^2}}\right) \tag{3'}$$

where A_0 signifies the value of A that represents the wing tip location.

So that finally

$$\text{R.P.}\left(\frac{d\hat{t}}{dt}\right)_{t=z_w+iy} = \frac{1}{h_2 - h_1}\left(h_2 - h_1\frac{\dfrac{A}{\sqrt{A^2 - E^2}}}{1 + \dfrac{E^2y^2}{(A^2 - E^2)^2}}\right) \tag{4'}$$

Consequently it is evident that the way in which the potential $\phi_{2\infty}$ is distributed over the region occupied by the body in the plane $x = \infty$ determines how the lift is distributed over the fuselage.

Now, still working in this plane at infinity, let the following Taylor series development represent the potential, $\phi_{2\infty}$, along the edges of the straight-line segment which corresponds to the actual fuselage cross section:

$$\phi_{2\infty}(\hat{z}) = \phi_{2\infty}(\hat{z}_w) + (\hat{z} - \hat{z}_w)\left(\frac{\partial \phi_{2\infty}}{\partial \hat{z}}\right)_{\hat{z}=\hat{z}_w} + \cdots \qquad (7\text{-}4)$$

It is usually satisfactory to terminate this development after the first two terms, while in addition the indicated coefficients may be seen to have

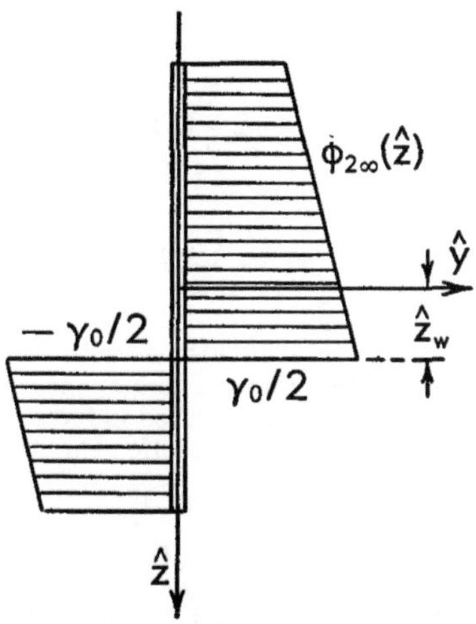

Fig. C,7. Variation of the potential $\phi_{2\infty}(\hat{z})$ along the repeated segment of vertical axis into which the fuselage cross section has been transformed.

the following explicit formulations:

$$\left.\begin{aligned} \phi_{2\infty}(\hat{z}_w) &= \frac{1}{2}\frac{\Gamma_{(y=0)}}{U} \qquad \text{wherever } \hat{z} - \hat{z}_w < 0 \\ &\text{or} \\ \phi_{2\infty}(\hat{z}_w) &= -\frac{1}{2}\frac{\Gamma_{(y=0)}}{U} \qquad \text{wherever } \hat{z} - \hat{z}_w > 0 \end{aligned}\right\} \qquad (7\text{-}5)$$

whereas in both cases the coefficient multiplying the linear term is

$$\left(\frac{\partial \phi_{2\infty}}{\partial \hat{z}}\right)_{\hat{z}=\hat{z}_w} = 2\hat{\epsilon} = \frac{2\epsilon}{\left[\text{R.P.}\left(\dfrac{d\hat{t}}{dt}\right)_{t=z_w+iy_w}\right]} \qquad (7\text{-}6)$$

The coordinates y_w, z_w appearing here represent the point at which the trailing vortex sheet becomes detached from the fuselage cross-sectional contour. One concludes from these facts that the distribution of $\phi_{2\infty}$ over the fuselage may be depicted in the $\hat{t}$ plane by means of a graph like the one drawn in Fig. C,7. Since the way y is related to $\hat{z}$ is known, one may immediately deduce from this distribution the variation of $\phi_{2\infty}$ or, that is, of dL_f/dy relative to the real body.

It follows that the lift coefficient applying to the fuselage is thus

$$(C_L)_f = 2\mathcal{R}\frac{\hat{b}}{b}\left[-\int_{\hat{\zeta}_1}^{\hat{\zeta}_2}\hat{\phi}_{2\infty}(\hat{\zeta})\cdot\frac{d\eta}{d\hat{\zeta}}d\hat{\zeta}\right] \tag{7-7}$$

where $\hat{\phi}_{2\infty} = \phi_{2\infty}/\hat{b}$, and where the symbols $\hat{\zeta}_1$ and $\hat{\zeta}_2$ represent the values of $\hat{\zeta}$ that correspond to the end points of the slot into which the fuselage is transformed in the $\hat{t}$ plane.

Upon substitution of Eq. 7-4 and 7-5 for the $\hat{\phi}_{2\infty}$ function appearing in the above integral, the fuselage lift coefficient becomes[3]

$$(C_L)_f = \mathcal{R}\left[2\gamma_0 l_0 - 2\frac{\hat{b}}{b}\hat{\epsilon}\int_{\hat{\zeta}_1}^{\hat{\zeta}_2}\frac{d\eta}{d\hat{\zeta}}(\hat{\zeta} - \hat{\zeta}_w)d\hat{\zeta}\right] \tag{7-8}$$

where

$$\gamma_0 = \frac{\Gamma_{(y=0)}}{bU} \tag{7-9}$$

[3] In the case of a fuselage with a circular or with an elliptic cross section, the determination of the lift distribution and the corresponding integrated lift coefficient may be carried out by use of the formulas derived in footnote 2. In fact one finds that for the case of a fuselage with circular cross section the lift distribution is given by

$$\frac{1}{\frac{1}{2}\rho_\infty U^2 A}\frac{dL_f}{d\eta} = \begin{cases} \mathcal{R}\left[\pm\frac{1}{2}\gamma_0 - 2\hat{\epsilon}\cdot 2h\left(\sqrt{1-\frac{\eta^2}{4h^2}} - \sqrt{1-\frac{l_0^2}{4h^2}}\right)\right] \\ \mathcal{R}\left[\pm\frac{1}{2}\gamma_0 - 2\hat{\epsilon}\cdot 2h\left(-\sqrt{1-\frac{\eta^2}{4h^2}} - \sqrt{1-\frac{l_0^2}{4h^2}}\right)\right] \end{cases}$$

in which the upper expression is to be used when $\zeta > 0$, while the lower one is valid whenever $\zeta < 0$. Consequently the fuselage lift coefficient is

$$(C_L)_f = 2\mathcal{R}(\gamma_0 l_0 - 4\hat{\epsilon}\pi h^2)$$

For the case of a fuselage with elliptic cross section the lift distribution is given by

$$\frac{1}{\frac{1}{2}\rho_\infty U^2 A}\frac{dL_f}{d\eta} = \mathcal{R}\left[\pm\frac{1}{2}\gamma_0 - \frac{\hat{\epsilon}}{h_2 - h_1}\left(h_2 - \frac{h_1^2}{h_2}\right)2h_1\left(\sqrt{1-\frac{\eta^2}{4h_2^2}} - \sqrt{1-\frac{l_0^2}{4h_2^2}}\right)\right] \quad \text{for } \zeta > 0$$

and

$$\frac{1}{\frac{1}{2}\rho_\infty U^2 A}\frac{dL_f}{d\eta} = \mathcal{R}\left[\pm\frac{1}{2}\gamma_0 - \frac{\hat{\epsilon}}{h_2 - h_1}\left(h_2 - \frac{h_1^2}{h_2}\right)2h_1\left(-\sqrt{1-\frac{\eta^2}{4h_2^2}} - \sqrt{1-\frac{l_0^2}{4h_2^2}}\right)\right] \quad \text{for } \zeta < 0$$

Consequently the fuselage lift coefficient is

$$(C_L)_f = 2\mathcal{R}[\gamma_0 l_0 - 2\hat{\epsilon}(h_2 + h_1)\pi h_1]$$

If Eq. 7-8 is now compared with Eq. 7-1 it will be seen that there is a deficiency of lift created on the fuselage in comparison with the amount which would be produced by an extension of the wing corresponding to the spanwise section, $2y_w$ in extent, actually covered by the fuselage; this lift deficiency is exactly the second term in Eq. 7-8, or

$$2\frac{\hat{b}}{b}\hat{\epsilon}\mathcal{R}\int_{\hat{\xi}_1}^{\hat{\xi}_2}\frac{d\eta}{d\hat{\xi}}(\hat{\xi}-\hat{\xi}_w)d\hat{\xi} \tag{7-10}$$

C,8. Moment Coefficient of the Fuselage in Presence of the Wing. The determination of the moment created by the resultant aerodynamic forces acting upon the fuselage in the presence of the wing (where the axis about which the moments are taken is to be the y axis) requires knowledge of the distribution of these forces along the axis of the fuselage; i.e. it is necessary to know the potential ϕ_2, not merely the Trefftz value $\phi_{2\infty} = \lim_{x\to\infty}\phi_2$. Further, even though the procedure resorted to in Art. 6 is quite acceptable for calculation of the lift, such a treatment is no longer justifiable when one wants to compute the moment, because in the latter event it is obvious that the actual length of the body is of utmost importance.

Fortunately there exists an approximate method due to Multhopp [*11*], which gives quite good agreement with experiment, despite the fact that it is founded on patently crude assumptions. The fundamental idea underlying this method is to imagine that the distribution of forces exerted along the length of the body is separated into two parts; one part is represented by a resultant force acting at the aerodynamic center of the wing's root-chord airfoil section, while the other part is visualized as giving rise to a resultant couple. The genesis of the former part of the total moment is very likely a result of the concentrative action of the adjacent wing. As a matter of fact, the system of bound vortices on the wing really ought to be thought of as extending over into the region occupied by the fuselage so as to fill out the region of the wing span normally covered by the body. In this way, the continuity of the vortex filaments would not be destroyed and therefore Helmholtz's theorem would be upheld. Of course, fundamentally the body lift is created directly by action of these vortices.

Contrariwise, it is the regions of the body lying farthest from the wing which play the most important role in the formation of the other contribution to the total moment. In order to obtain a quantitative estimate of this effect one may now proceed as follows. Because of the assumption of the high fineness ratio of the body, it may be taken for granted that the flow is two-dimensional in any plane perpendicular to the body axis; i.e. it is assumed that the pattern of flow in such a plane would be identical to the velocity field produced by a stream approaching the body in a direction perpendicular to its axis and impinging on an infinite cylinder whose directrix has a shape identical to that of the fuselage at the cross section in question. It is then possible to compute the force exerted on each cross-sectional element dx along the body by the impulse method, in the same way that Munk [14] applied it for determination of the aerodynamic behavior of dirigibles. In the present case, though, one must remember that the stream flowing normal to the body axis is formed from two separate sources which must be summed; one is the transverse component of the free stream flow at infinity, while the other is generated by the system of vortices associated with the flow about the wing.

According to the impulse concept, the slice of fuselage lying between the planes x = const and $x + dx$ = const has a velocity along the direction of the z axis, relative to the mass of air contained between the partitioning planes, of an amount given by

$$W = U(\alpha_t + \epsilon) = \delta U \tag{8-1}$$

where

$$\delta = \alpha_t + \epsilon$$

If attention is now focused upon the air mass itself, it is seen that its reaction produces an amount of momentum whose component in the z direction is given by the relation

$$dQ = \rho_\infty \delta U A^* dx$$

where the factor $\rho_\infty A^* dx$ stands for the so-called apparent mass of this slice of fuselage.

In the case of a body having an elliptic cross section with semihorizontal axis (in the y direction) equal to $h_1 \cdot b$, the apparent area is

$$A^* = \pi h_1 b^2 \tag{8-2}$$

Since this formula holds regardless of what the size of the semiaxis lying in the z direction happens to be, and since it even applies in the two limiting cases when $h_2 = h_1$ (circular body section) as well as when $h_2 = 0$ (flat plate body), one may always employ Eq. 8-2 as the definition of A^*, irrespective of the fuselage cross section. On the basis of the momentum theorem, therefore, one may write the expression for the force exerted on this thin slice of body, dx in length, in the following form

$$dF_L = \frac{d(dQ)}{dt} \cong \rho_\infty U^2 dx \frac{d(\delta A^*)}{dx}$$

where

$$\frac{d}{dt} \cong U \frac{d}{dx}$$

The summation of the total force acting on the body from this source is thus given by the integral

$$F_{\mathrm{L}} = \rho_\infty U^2 \int_0^L \frac{d}{dx}(\delta A^*)dx = 0$$

provided the origin from which distances along the x axis are measured is taken to be placed at the tip of the nose. The system of local forces dF_{L} when exerted on the individual elements dx of the fuselage reduces to a couple, and the magnitude of this couple is represented by the formula

$$M_{\mathrm{f}} = \rho_\infty U^2 \int_0^L x \frac{d(\delta A^*)}{dx} dx = -\rho_\infty U^2 \int_0^L \delta A^* dx \tag{8-3}$$

In order to improve the accuracy of the approximation (8-3) one may insert a correction factor f, so that the more universal formula becomes

$$M_{\mathrm{f}} = -\rho_\infty U^2 f \int_0^L \delta A^* dx \tag{8-4}$$

The correction factor is to be selected in such a way that the value obtained by evaluating the right-hand side of Eq. 8-4, in the case of a body having the shape of an ellipsoid with three unequal axes, is identical with the result obtained for the same type of ellipsoid when the problem is solved by means of the exact theory [*15*, p. 152] and δ is taken to be the angle of attack of the fuselage, α_{f}, in both cases. In particular, if the three semiaxes of such an ellipsoid are denoted by h_3b, h_1b, and h_2b to differentiate between the axes lying along the x, y, and z coordinate directions respectively, then the correction factor may be dissociated into the separate factors

$$f = \frac{h_2}{h_1} K \tag{8-5}$$

The values for the correction coefficient K have been computed over a range of values for the ratios h_2/h_3 and h_1/h_2 by Zahm [*16*]. The results of his computations are summarized in Fig. C,8a.

Having obtained the moment, the expressions for the moment coefficient may be written immediately as

$$(C_M)_{\mathrm{f}} = -2 \int_0^{L/c_{\mathrm{av}}} \delta \frac{A^*}{A} d\left(\frac{x}{c_{\mathrm{av}}}\right) \tag{8-6}$$

where c_{av} is the average chord of the wing ($c_{\mathrm{av}} = A/b$).

The stability factor, or the derivative of the moment coefficient with respect to the angle of attack, which accounts for the effect of the wing upon the fuselage, is a parameter of prime importance in determining the location of the aerodynamic center of the complete airplane. This quantity is obtained from the formula

$$\frac{d(C_m)_t}{d\alpha} = -2\int_0^{L/c_{av}} \frac{d\delta}{d\alpha}\frac{A^*}{A}\, d\left(\frac{x}{c_{av}}\right) \tag{8-7}$$

provided the way $d\delta/d\alpha$ varies along with x/c_{av} is known.

In order to see how $d\delta/d\alpha$ varies, one need merely observe that along that portion of the body where the wing is located the flow may be considered to follow the direction of the wing chord, and thus for practical purposes $d\delta/d\alpha = 0$. Downstream of the wing, at the rear end of the fuselage and thus in the neighborhood of the tail, the value of δ is $\alpha_t - \epsilon_t$, where ϵ_t is the angle of attack induced upon the tail itself; consequently one obtains the expression $d\delta/d\alpha = 1 - d\epsilon_t/d\alpha$.

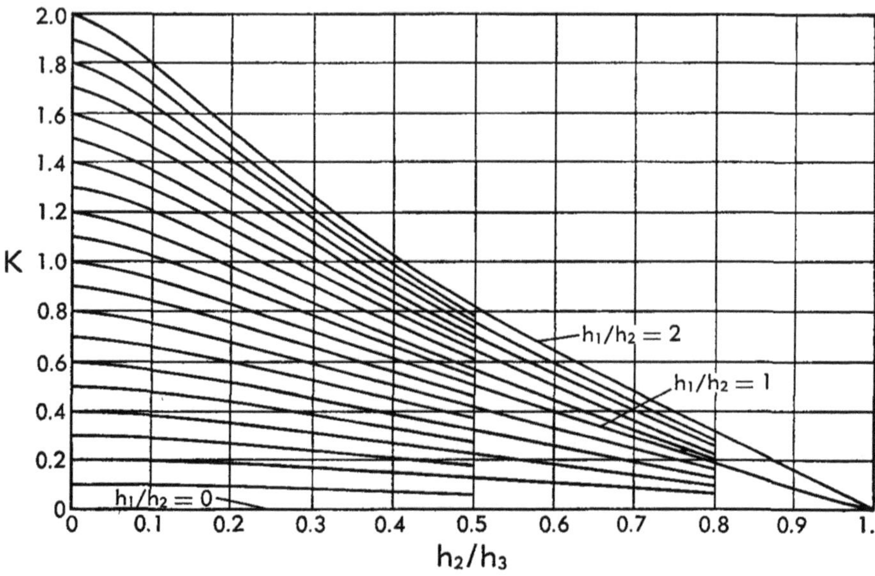

Fig. C,8a. Correction coefficients for the pitching moment due to the body.

In view of the rather small amount of variation exhibited by $d\delta/d\alpha$ throughout the above-mentioned region, Multhopp substitutes a linear relationship for the actual variation of $d\delta/d\alpha$ along the rear of the body; the constants in this linear formula are adjusted so as to make the values obtained at the beginning and end of the interval in question coincide with the actual values that this derivative has at these endpoints. Finally, in regard to the behavior of $d\delta/d\alpha$ ahead of the wing, it may be noted that its value is always greater than 1, and that it grows rapidly larger as one moves closer to the wing itself. The graph of $d\delta/d\alpha$ as a function of x/c_{av} thus has the shape of the curve shown qualitatively in Fig. C,8b.

In order to determine $d\delta/d\alpha$ quantitatively in the region ahead of the wing one may replace the vortex system associated with the wing by a simple horseshoe vortex with a span equal to $\pi/4$ times the wing span (this represents an elliptic lift distribution), having a circulation strength given by $\Gamma = (2/\pi)C_L c_{av} U$. If the x axis is still taken to be orientated as shown previously in Fig. C,6b, then the sought quantitative expression is

$$\frac{d\delta}{d\alpha} = 1 + C_{L\alpha}\left[-\frac{8}{\pi^3}\cdot\frac{1}{\mathcal{R}}\left(1 + \frac{2\xi^*}{\sqrt{4\xi^{*2} + \beta^2(\pi^2/16)\mathcal{R}^2}}\right) - \frac{\beta^2}{4\pi\xi^*}\frac{\mathcal{R}}{\sqrt{4\xi^{*2} + \beta^2(\pi^2/16)\mathcal{R}^2}}\right] \quad (8\text{-}8)$$

where the substitution $\xi^* = x/c_{av}$ has been made for sake of simplicity.

The values of the quantity enclosed within square brackets in Eq.

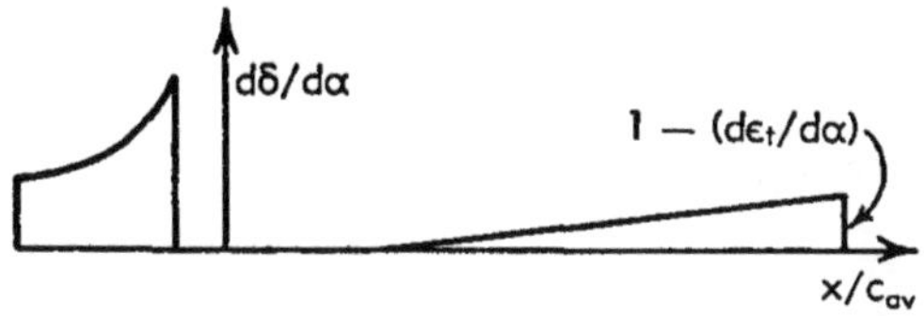

Fig. C,8b. Variation of $d\delta/d\alpha$ along the axis of the body.

8-8, i.e. $[(d\delta/d\alpha) - 1]/C_{L\alpha}$, are listed in Table C,8 as a function of the parameter $\xi = \xi^*/\beta$ for three values of the aspect ratio.

Table C,8. Incremental values to upwash produced at several locations ahead of the wing, for wings of three different aspect ratios.

ξ	Aspect ratio $\mathcal{R}$		
	4	6	8
−0.25	0.3629	0.3813	0.3909
−0.3	0.2795	0.2973	0.3069
−0.5	0.1483	0.1640	0.1732
−1.0	0.0554	0.0668	0.0738
−2.5	0.0115	0.0153	0.0197

C,9. Influences Attributable to the Variation of the Mach Number and the h_1 Parameter. The way the circulation per unit span distance, γ, is distributed along the wing span and over the fuselage is depicted in Fig. C,9a and C,9b for the two fairly standard configurations schematically represented on these same graphs. The geometrical characteristics of the configurations shown are: the wing is trapezoidal with taper ratio = 0.42 and $\mathcal{R}$ = 4.83; it is mounted in the mid-wing position on a fuselage of circular cross section at zero angle of incidence; i.e. $\alpha_w = \alpha_t = \alpha$ (the wing is identical with that shown in Fig. C,2c). The ratio that the body radius bears to the wing span is $h_1 = 0.07$ for the configuration having the kind of circulation distribution displayed in Fig. C,9a, while this parameter has been increased in value to $h_1 = 0.1$ for the curves in Fig. C,9b. In both figures the three representative values $M_\infty = 0.3$, 0.6, and 0.8 have been selected for purposes of illustrating the effect of Mach number on the size and shape of the γ/α distribution.

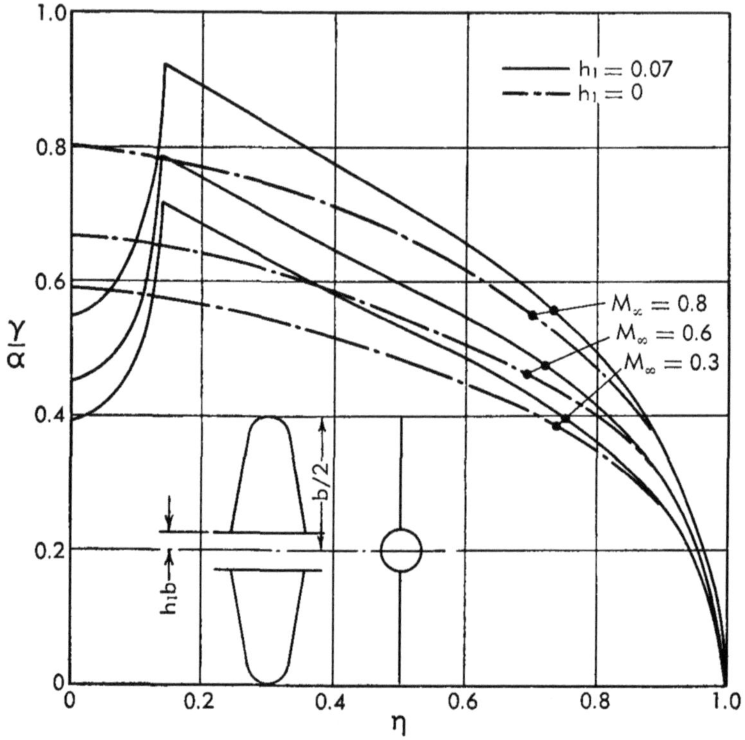

Fig. C,9a. Spanwise circulation distribution for the wing-body combination and for the isolated wing.

It is worthy of note that the curves corresponding to the different Mach numbers are roughly affine to each other, at least for those values of the spanwise location coordinate η corresponding to the portion of the total span occupied by the wing itself. The factor of proportionality relating this set of curves is then found to be

$$\frac{(\gamma)_{M_\infty}}{(\gamma)_{M_\infty=0}} = \frac{0.965}{\sqrt{1 - M_\infty^2} + \frac{1}{\pi A\!\!R}(C_{L\alpha})_{M_\infty=0}(1 - \sqrt{1 - M_\infty^2})} \tag{9-1}$$

where the denominator of the right-hand side represents the ratio of the slope of the lift curve in the incompressible case to that obtained for the

Mach number in question, M_∞; i.e. the denominator is equivalent to the ratio $(C_{L\alpha})_{M_\infty=0}/(C_{L\alpha})_{M_\infty}$, provided the assumption of an elliptic lift distribution is postulated. The coefficient 0.965 is a correction factor that most likely arises because of the fact that the circulation distribution is not elliptic in this case.

Even the slopes of the lift curve applying to the complete configuration are likewise related to each other for the various Mach numbers through the same factor of proportionality as given in Eq. 9-1. This

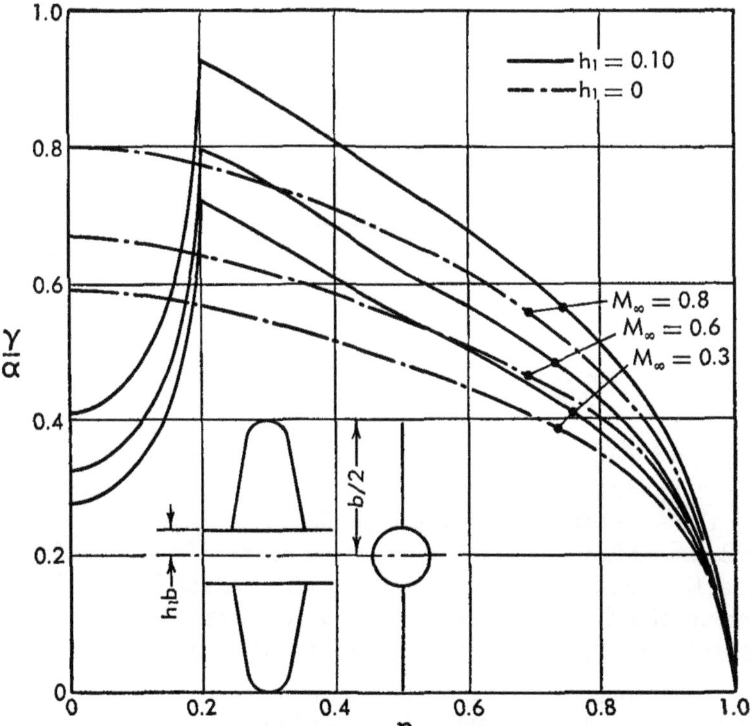

Fig. C,9b. Spanwise circulation distribution for the wing-body combination and for the isolated wing.

correspondence may be checked by comparing the values listed in Table C,9. The value of the slope of the lift curve for the two-dimensional profile ($Æ = \infty$), which was used in making the calculations, is also included in the headings of this table.

By reference to Table C,9 or to the summary curves of Fig. C,9c, it may be observed that the slope of the lift curve at first increases as h_1 increases but then after the fuselage diameter passes a certain size the $C_{L\alpha}$ values begin to grow smaller again, and, in fact, the rapidity with which $C_{L\alpha}$ diminishes at these higher values of h_1 is accentuated at the larger Mach numbers. Because of these trends it is seen that although the value of C_{L_α} is greater for $h_1 = 0.1$ than it is for $h_1 = 0$ in the case where $M_\infty = 0.3$, the reverse is true when the Mach number increases to $M_\infty = 0.8$. Consequently it may henceforth be presumed to hold true that, provided the body diameter to span ratio h_1 is large enough, then the value of the slope of the lift curve C_{L_α} for the combination of wing plus body will always be less than the figure which applies to the wing alone.

Because of the lack of any experimental curves illustrating pertinent circulation distributions, it is necessary to resort to a comparison made

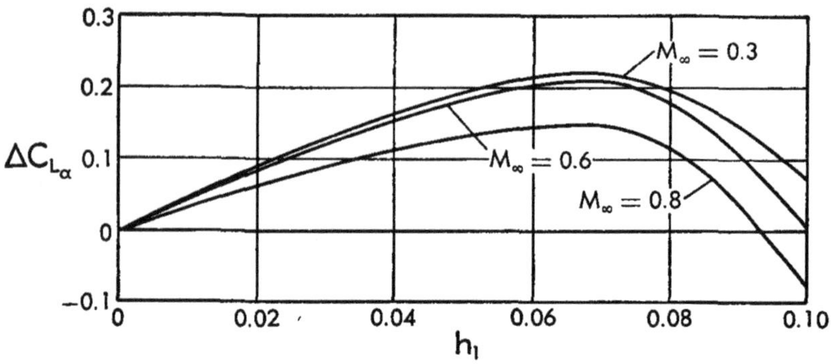

Fig. C,9c. Variation of ΔC_{L_α} vs. h_1 for the wing-body combination of Fig. C,9a and C,9b.

with derived results; i.e. an indirect comparison can still be made by determining whether the over-all values of C_{L_α}, obtained experimentally, match those resulting from the theory discussed above. Suitable experimental results have been obtained in the course of tests, made at the Aeronautical Laboratory of the Polytechnic Institute of Turin (Laboratorio di Aeronautica del Politecnico di Torino), on a wing-body combination composed of a wing with the same features as that employed in the above

Table C,9. Slope of lift curve for constituent parts and complete wing-body configuration, for three representative body radius to wing span ratios, showing the effect of Mach number.

	$M_\infty = 0.3$, $c_{l_\alpha} = 6.18$	$M_\infty = 0.6$, $c_{l_\alpha} = 7.38$	$M_\infty = 0.8$, $c_{l_\alpha} = 9.83$
$h_1 = 0$	$(C_{L_\alpha})_w = 4.39$	$(C_{L_\alpha})_w = 4.96$	$(C_{L_\alpha})_w = 5.97$
$h_1 = 0.07$	$(C_{L_\alpha})_w = 3.98$ $(C_{L_\alpha})_t = 0.626$ $C_{L_\alpha} = 4.61$	$(C_{L_\alpha})_w = 4.46$ $(C_{L_\alpha})_t = 0.707$ $C_{L_\alpha} = 5.17$	$(C_{L_\alpha})_w = 5.27$ $(C_{L_\alpha})_t = 0.85$ $C_{L_\alpha} = 6.12$
$h_1 = 0.1$	$(C_{L_\alpha})_w = 3.74$ $(C_{L_\alpha})_t = 0.716$ $C_{L_\alpha} = 4.46$	$(C_{L_\alpha})_w = 4.14$ $(C_{L_\alpha})_t = 0.819$ $C_{L_\alpha} = 4.96$	$(C_{L_\alpha})_w = 4.89$ $(C_{L_\alpha})_t = 0.999$ $C_{L_\alpha} = 5.89$

calculations, the fuselage of which has a cross section at the three-quarter chord location of the wing of such a size as to produce a value of $h_1 = 0.07$; hence this model reproduces in all respects one of the illustrative numerical examples considered above. It was determined from these tests that

$$\frac{\Delta C_{L\alpha}}{(C_{L\alpha})_w \text{ for } h_1 = 0} = 0.0475$$

while the comparable value obtained from the present theoretical computations turns out to be slightly more, 0.05 for $M_\infty = 0.3$.

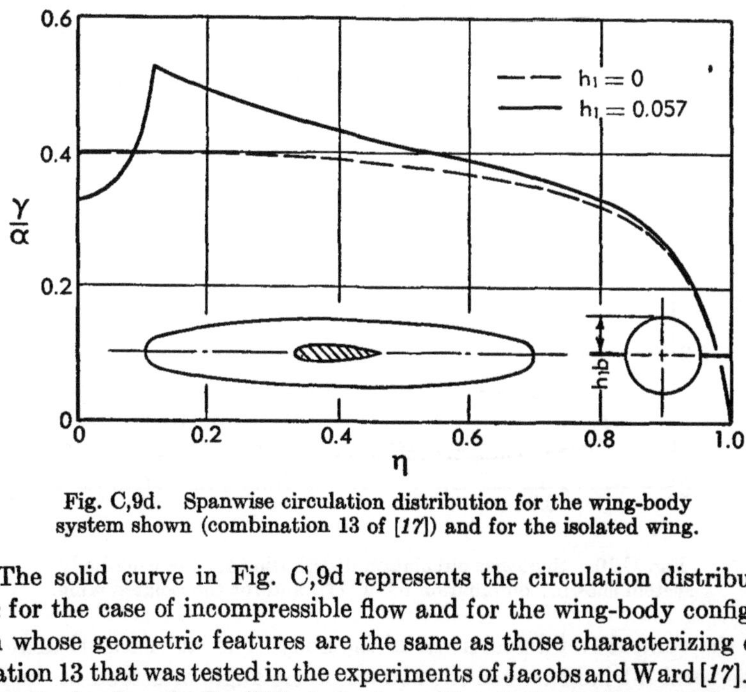

Fig. C,9d. Spanwise circulation distribution for the wing-body system shown (combination 13 of [*17*]) and for the isolated wing.

The solid curve in Fig. C,9d represents the circulation distribution γ/α for the case of incompressible flow and for the wing-body configuration whose geometric features are the same as those characterizing combination 13 that was tested in the experiments of Jacobs and Ward [*17*]. The theoretical value of $\Delta C_{L\alpha}/[(C_{L\alpha})_w \text{ for } h_1 = 0]$ applying to this configuration has been computed by Multhopp [*11*] with a result of 0.073. On the other hand the experimental value for this ratio obtained from the Jacobs and Ward investigation is only 0.065.

Although this correspondence between theory and experiment is satisfactory, it is appropriate to point out why the theory can be expected to predict a larger amount of lift than that which is actually found from experimental tests. Above all, the boundary layer must exert a pronounced effect on the actual results because in the first approximation it can be likened to a virtual increase in the body thickness parameter h_1. In addition, the vorticity connected with the trailing vortices lying close to the fuselage is exceptionally intense (see Fig. C,9a, C,9b, and C,9d), and thus a non-negligible amount of curvature is introduced into the part of the flow in which the central section of the wing operates. In order to take this curvature into account it would be necessary to compute the induced velocity arising at the three-quarter chord point, not at the quarter chord point.

On the other hand, an additional lift contribution ought to be determined and added to the theoretical amount already computed, which is due merely to the induction process from the wing. This additional lift contribution is generated by and on the surfaces of the fuselage itself, exclusive of the wing; it is ordinarily called viscous cross flow lift, inasmuch as it is due to the action of the transverse flow component of the impinging stream, which produces a dissymmetry in the boundary layer

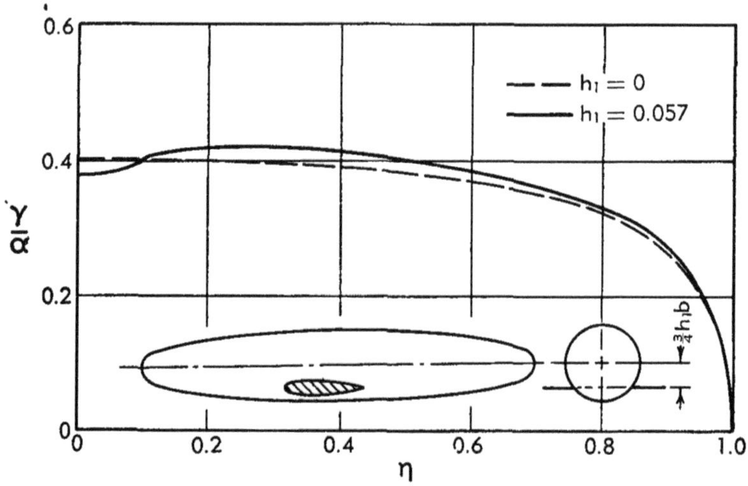

Fig. C,10. **Spanwise circulation distribution for the wing-body system shown (combination 64 of [*17*]) and for the isolated wing.**

over the ventral and dorsal areas of the body. The latter constituent part of the total lift increases as the square of the angle of attack of the fuselage α_f, although in spite of this it is usually quite small for the kinds of fuselage shape commonly met in practice.

C,10. Effect of the Wing's Vertical Position on the Fuselage. Thus far the discussion has centered on mid-wing configurations. Now, however, the results pertinent to a low-wing design, exemplified by combination 64 in the above-mentioned tests of Jacobs and Ward [*17*], are cited. The circulation distribution, γ/α, applying to this combination, is plotted in Fig. C,10, and for comparative purposes the corresponding distribution of the wing alone is also shown.

In the case of a low-wing configuration it is evident that the fuselage has a conspicuously reduced influence on the spanwise lift distribution;

thus the effect on the ratio

$$\frac{\Delta C_{L\alpha}}{(C_{L\alpha})_w \text{ for } h_1 = 0}$$

which is a measure of the integrated over-all effect, is also a good deal less in this case, amounting to merely 0.032 (calculated) or 0.039 (experimental). If one takes into account the viscous cross flow lift, not being accounted for here, it will be appreciated that, though small in absolute amount, these viscous effects are actually of the same order of magnitude in this case as the fuselage lift deficiency itself (see Eq. 7-10). Thus the fairly good correlation between the theoretical and the experimental results noted above is perhaps more remarkable than might be supposed at first glance.

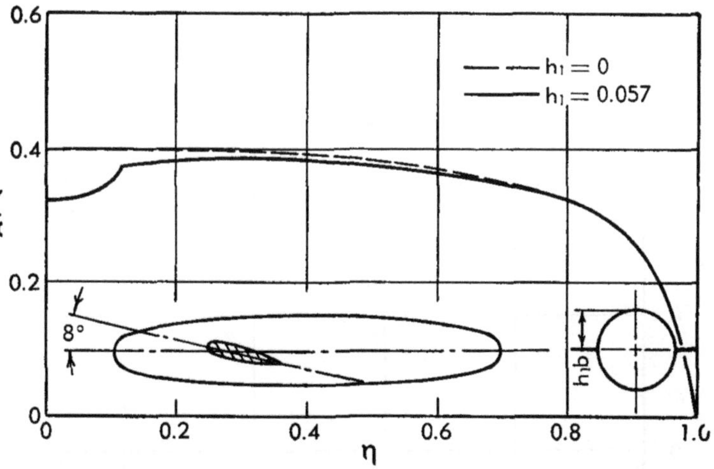

Fig. C,11. Spanwise circulation distribution for the wing-body system shown (combination 12 of [*17*]) and for the isolated wing.

C,11. Effect of the Angle of Wing Setting. The distribution of circulation γ/α, as a function of spanwise distance η, is depicted in Fig. C,11 for a wing-body configuration having characteristics like those of Combination 12 tested by Jacobs and Ward [*17*] during the course of their investigations. This model represents a mid-wing configuration in which the wing has an 8° angle of incidence setting, while $\alpha_f = 0°$. For comparative purposes the circulation distribution relative to the isolated wing is also presented.

That a decrease in lift is occasioned in this instance by the presence of the fuselage, in comparison with the amount of lift supported by the wing alone, is confirmed by the graphs of Fig. C,11; such a state of affairs is in agreement with the findings of Lennertz [*18*]. This subsidence in lift is summarized by evaluation of the performance factor

$$\frac{\Delta C_L}{(C_L)_w \text{ for } h_1 = 0} = -0.05$$

The corresponding experimentally determined value is

$$\frac{\Delta C_L}{(C_L)_w \text{ for } h_1 = 0} = -0.049$$

C,12. Effect of Interference Phenomena on the Stall Angle. Curves representing the effective angle of attack distribution occurring

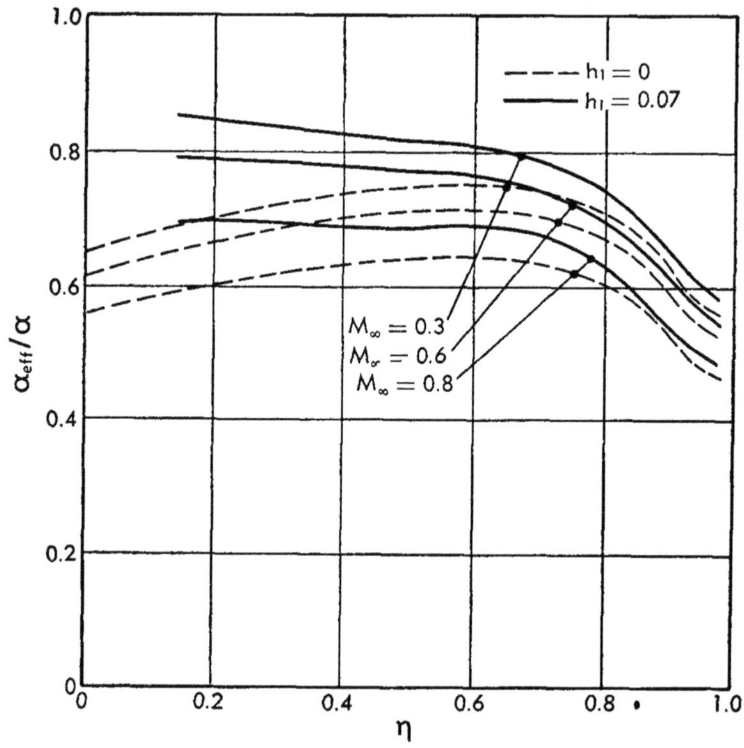

Fig. C,12. Effective angle-of-attack distribution applying to the wing-body combination shown in Fig. C,9a, and also the corresponding curves applying to the isolated wing.

along the wing span of the configuration in Fig. C,9a are given in Fig. C,12, for the usual three Mach numbers of 0.3, 0.6, and 0.8, and for a body radius to wing span ratio of 0.07. The same kinds of curves are also furnished for the case of the wing-alone configuration. It is clear from these plots that the effective angle of attack is greatest at the juncture of the wing with the fuselage and that the maximum values for the combination are larger than those which would be found to exist there if no fuselage were present. It may be concluded from this set of circumstances that the separation of the stream tubes shows up first in the juncture region, and that the flow breakdown occurs at a lesser angle of attack than would be true for the wing alone. The presence of the boundary layers tends to ameliorate this distinction between the two cases in certain respects, but it tends to aggravate it in other respects. Inasmuch as one effect of the boundary layer is to impede the attainment of as high values for the effective angle of attack as would otherwise have been induced by the fuselage upon the wing, in the neighborhood of the wing-body juncture, the boundary layer mitigates separation. On the other hand, since the boundary layer adversely affects the ability of the stream tubes to resist separation, it exerts a deleterious effect on the stall. A well-designed fillet eliminates or ameliorates the latter undesirable situation, and it thus has the beneficial effect of delaying the inception of the stall.

In the case of a low-wing (or high-wing) configuration the effective angle of attack of the section at the wing-fuselage juncture does not differ appreciably from the one occurring in the central region of an isolated wing. One of the prime causes leading to the premature stall mentioned above is thus absent in this type of design. However, the adverse effect of the boundary layer is intensified, although a suitable fairing can usually be fitted in the juncture region to eliminate the undesirable features of such designs.

Finally it may be observed that the effective angle of attack decreases as the Mach number increases, under the assumption that the geometric angle of attack remains constant. In view of the implications of this fact in regard to separation phenomena, it follows that the stall would come at a higher angle of attack at the higher Mach numbers than it would for the same configuration if it were subjected to incompressible ($M_\infty = 0$) flow.

Nevertheless the gradient of pressure existing along any chordwise section of the wing would also be steepened as the Mach number increases, and thus the fundamental determining cause of the separation would be exaggerated. Consequently it is not possible to predict which of these two contrasting trends will prevail, especially in view of the fact that the real behavior of the flow does not depend solely on the Mach number, but also on the Reynolds number.

C,13. Shift in Aerodynamic Center Location Due to Wing-Body Interference. Experimental and theoretical results are presented in Fig. C,13 to show the effect of fore and aft wing position on the longitudinal stability parameter, $dC_M/d\alpha$, which is the rate of change of the moment coefficient with angle of attack, where the moment is taken with respect to an axis through the aerodynamic center of the isolated wing in all cases. The experimental results obtained by Jacobs and Ward furnish the data for the solid curve of Fig. C,13. These results apply to Model Combinations 1, 3, 7, 11, and 13 of their tests, while the numerical computations were obtained with the aid of Eq. 8-7; notwithstanding the gross assumptions made in deriving this equation, the correspondence between theory and experiment is excellent.

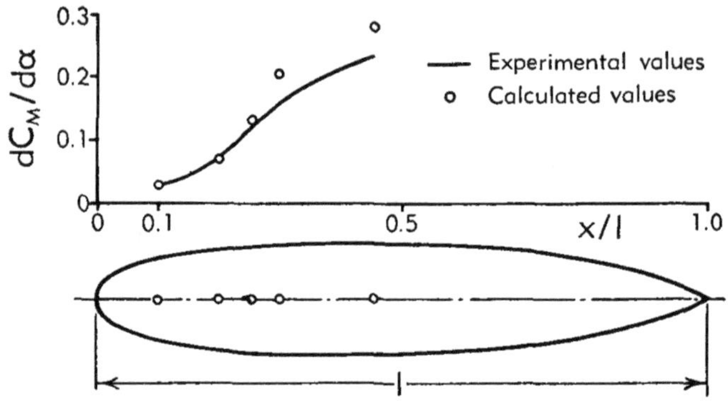

Fig. C,13. Pitching moment vs. wing position for mid-wing configurations, showing effect due to presence of fuselage.

Experimental evidence may also be adduced to confirm that the vertical positioning of the wing on the fuselage has practically no influence on the fuselage stability parameter, $dC_M/d\alpha$; this deduction is in accord with the discussion at the beginning of Art. 8.

C,14. Effect of the Presence of the Fuselage on the Rolling Moment. If the free stream uniform flow that impinges upon the wing-body combination in question has a sidewind component producing the effect of an angle of incidence to the plane of symmetry, the problem of computing the mutual wing-body interference effects is much more complex. The solution of this problem is made more difficult both because the idealizations resorted to in Art. 6 are no longer acceptable and also because it is no longer true that the induced angle of attack generated by the trailing vortices at the lifting line is one half of the angle of attack attained at infinity.

An approximate way of handling the present problem concerning the effect of the fuselage upon the rolling moment has been suggested by Multhopp [*11*]. He bases his analysis upon the assumption that the wing is elliptic, and then proceeds to find the rolling moment created by the asymmetric angle of attack distribution which is induced along the wing span due to the presence of the fuselage.

The distribution of the induced angles of attack may be ascertained by following exactly the same sort of analysis here as was used in Art. 6; in the present instance, however, the free stream flow vector **U** will be considered to impinge upon the plane of symmetry of the airplane at an angle of yaw, denoted by ψ.

Analogously to what was done previously, therefore, the function which transforms the cross section of the fuselage over into a repeated segment of line lying along the $\hat{y}$ axis (see Fig. C,14a) is denoted symbolically by the relation

$$\hat{t} = \hat{y} + i\hat{z} = \hat{t}(t) \tag{14-1}$$

where

$$t = y + iz \tag{14-2}$$

which, of course, leaves undisturbed the straight lines located at infinity in both the t and $\hat{t}$ planes. The expression for the uniform sidewind velocity in the $\hat{t}$ plane reduces to the term $U\psi\mathbf{j}$ where $\mathbf{j}$ represents the unit vector lying along either the $\hat{y}$ axis or the y axis.

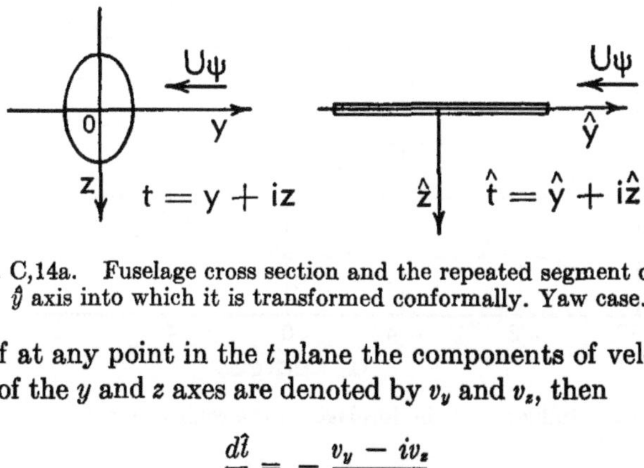

Fig. C,14a. Fuselage cross section and the repeated segment of the $\hat{y}$ axis into which it is transformed conformally. Yaw case.

Thus, if at any point in the t plane the components of velocity in the directions of the y and z axes are denoted by v_y and v_z, then

$$\frac{d\hat{t}}{dt} = -\frac{v_y - iv_z}{U\psi} \tag{14-3}$$

so that the induced angles of attack produced by the body upon the wing may be written as

$$(\alpha_i)_t = \frac{v_z}{U} = \psi\left[\text{I.P.}\left(\frac{d\hat{t}}{dt}\right)\right]_{t=y+iz_w} \tag{14-4}$$

where I.P. ( ) denotes the imaginary part of the quantity appearing in the parentheses.

Under the hypothesis that the circulation along the wing has an elliptic distribution, the expression for the rolling moment then turns out to be

$$C_l = \frac{M_l}{\frac{1}{2}\rho_\infty U^2 A(b/2)} = -C\psi \int_{-1}^{1}\left[\text{I.P.}\left(\frac{d\hat{t}}{dt}\right)\right]_{t=y+iz_w} \eta\sqrt{1-\eta^2}\,d\eta \tag{14-5}$$

wherein M_l is the rolling moment, C_l is the rolling moment coefficient, and the constant C has been substituted for the involved factor

$$C = \frac{1}{\frac{\pi\beta}{(c_{l_\alpha})_{M_\infty=0}} + \frac{2}{AR}}$$

The integrand indicated in the right-hand side of Eq. 14-5 contributes but little to the value of this integral near the ends of the interval of integration (i.e. for values of $|\eta|$ near 1) because I.P. $(d\bar{l}/dt)$ decreases very rapidly as η approaches the wing tips. No great error is committed, therefore, if the evaluation of the integral is confined merely to the interval lying between $\eta = \pm 2/\pi$ rather than ± 1, and where it is assumed that over this restricted interval it is sufficiently accurate to set $\sqrt{1 - \eta^2} \cong 1$.

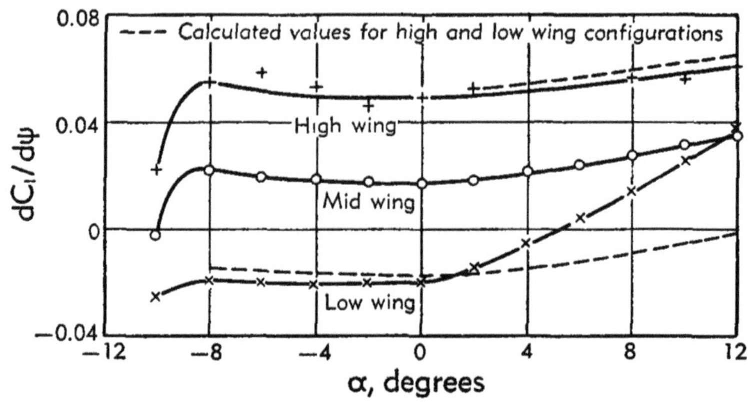

Fig. C,14b. Influence of the fuselage on the rolling moment (from [19]).

On the basis of these approximations, Eq. 14-5 reduces to the simplified expression

$$C_l = -C\psi \int_{-2/\pi}^{2/\pi} \left[\text{I.P.} \left(\frac{d\bar{l}}{dt} \right) \right]_{t=y+iz_w} \eta d\eta \qquad (14\text{-}6)$$

The easiest method of evaluating the integral on the right-hand side of this equation is to observe that

$$-\int \left[\text{I.P.} \left(\frac{d\bar{l}}{dt} \right) \right]_{t=y+iz_w} y dy = \left[\text{I.P.} \left(\int \bar{l} dt \right) \right]_{t=y+iz_w} - y[\text{I.P.}(\bar{l})]$$

In the case where the fuselage cross section is in the shape of an ellipse with semiaxes h_1b and h_2b, and thus identical with the example considered in the preceding articles, the transformation formula is given by means of Eq. 1′ appearing in footnote 2. Consequently the rolling moment for such a configuration turns out to be

$$C_l = \frac{4}{\pi} C\psi h_2(h_1 + h_2) \left[\frac{z_w}{h_2 b} \sqrt{1 - \left(\frac{z_w}{h_2 b}\right)^2} + \sin^{-1} \frac{z_w}{h_2 b} - \frac{2\pi z_w}{b} \right] \qquad (14\text{-}7)$$

for any z_w within the vertical zone $-h_2 b \leqq z_w \leqq h_2 b$.

Figure C,14b shows a series of curves [*19*] which gives the values of $dC_l/d\psi$ that were obtained experimentally on a wing-body model with a provision for placing the wing at three different vertical locations on the fuselage. In addition, comparative theoretical values obtained by use of Eq. 14-7 have been plotted on the same graph. A word of fuller explanation of how the theoretical curves are obtained follows: The final plotted values are obtained by first computing the values of the coefficient $dC_l/d\psi$ by aid of Eq. 14-7, in the two cases where the selected values of $z_w/h_2 b$ to be inserted in Eq. 14-7 correspond to the geometric features of the high-wing and low-wing configurations tested. After obtaining the wingshift factor, it is added to the experimental value of $dC_l/d\psi$, at each α value along the curve representing the test results for the wing-body combination with the wing in the midposition. The agreement between the experimental data and the computed results appears to be quite good, especially in the limited angle of attack range where $-4° \leqq \alpha \leqq 4°$, in spite of the fact that the derivations upon which the theoretical curves are based are plainly oversimplified.

CASE 2: METHOD DUE TO G. N. WARD

C,15. Fundamental Concepts and Basic Formulas. The second of the two major elementary cases into which wing-body interference problems were classified in Art. 5 will now be examined. The typical wing-body combination under consideration here is shown in Fig. C,15a, which gives the essential features of such a configuration, i.e. a long slender body supporting a very low aspect ratio wing that exhibits a quite large degree of sweep. It is imagined, furthermore, that this configuration is now to be immersed in a uniform free stream flow of velocity **U** directed along the x axis, and the ensemble is set at an angle of attack of α_w to this axis, as indicated in Fig. C,15a. Let it be agreed besides that ϕ represents the perturbation potential governing the disturbances superimposed upon the originally uniform stream, due to the presence of the wing-body combination. It can then be shown (see, for example, [*27*]) that the aerodynamic force perpendicular to the x axis that is created on the forward portion of the body, extending from the tip of the nose back to the axial station marked off by the vertical section (called the P plane) with equation x = const, is given within the degree of approximation afforded through use of the linear theory by the formula

$$F_y + iF_z = -\rho_\infty U \oint_{C_1} \phi dt \qquad (15\text{-}1)$$

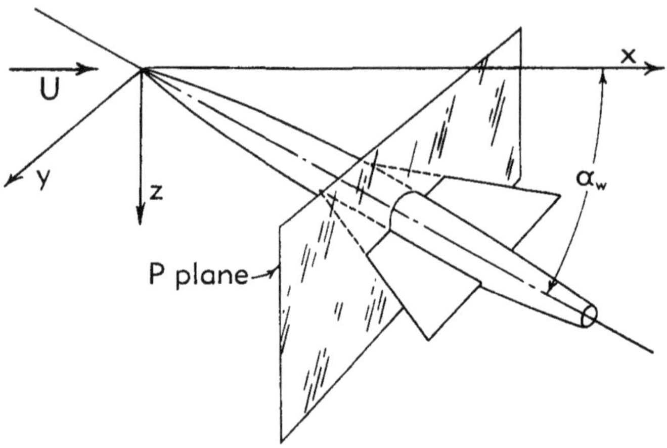

Fig. C,15a. View of the flat plate wing and slender body combination, showing orientation of the coordinate axes.

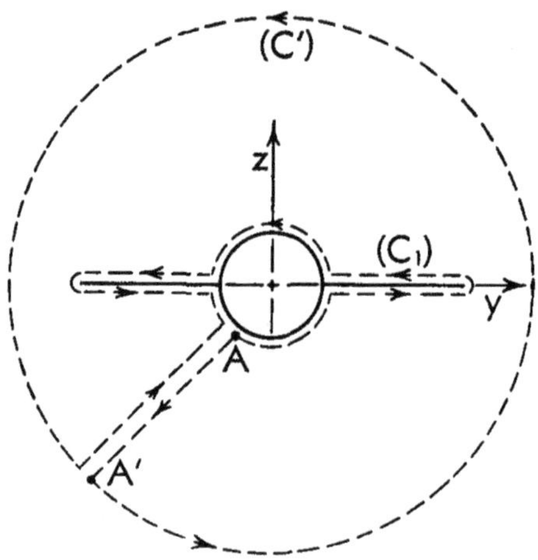

Fig. C,15b. Contour intercepted in a plane normal to the x axis by the surface of a typical slender wing-body combination and illustration of the path of integration to be followed in evaluation of the local aerodynamic force, acting at such a cross section, according to Eq. 15-8.

The indicated line integral is to be taken around the complete contour C_1, produced by intersection of the P plane with the body and wing surfaces, as indicated in Fig. C,15b. Of course, t denotes the complex variable $t = y + iz$, where the y and z axes conform to a right-handed Cartesian system, orientated with respect to the x axis and the model as shown in Fig. C,15a.

Because the form of the wing and body has been specified, above all, to possess the characteristic of slenderness, it may be observed that the field of flow generated by the cross flow component of the impinging stream will appear to be two-dimensional in any plane normal to the free stream direction, such as the P plane; i.e. it will be permissible to drop off the $1 - M_\infty^2$ term in the general equation of motion, which the potential ϕ must obey. To be more precise, let it be assumed that l^* denotes a characteristic length in the direction of the x axis (for instance, say it is the length of the body) and let b^* likewise be assigned as a characteristic length in the direction of the y axis (the semispan of the wing may be selected), so that the physical coordinates then become nondimensionalized in the form

$$x = l^*\xi, \quad y = b^*\eta, \quad \text{and} \quad z = b^*\zeta$$

and the governing equation of motion that ϕ must obey becomes

$$\beta^2 \cdot \frac{b^{*2}}{l^{*2}} \frac{\partial^2 \phi}{\partial \xi^2} + \frac{\partial^2 \phi}{\partial \eta^2} + \frac{\partial^2 \phi}{\partial \zeta^2} = 0 \tag{15-2}$$

For small values of the ratio $\beta b^*/l^*$, such as apply for the kinds of slender body configuration now under examination, the equation of motion reduces to the two-dimensional Laplace equation

$$\frac{\partial^2 \phi}{\partial \eta^2} + \frac{\partial^2 \phi}{\partial \zeta^2} = \frac{\partial^2 \phi}{\partial y^2} + \frac{\partial^2 \phi}{\partial z^2} = 0 \tag{15-3}$$

provided one abides by the restriction that the development of the dimensions of the configuration in directions normal to the x axis are to take place without abrupt variation, as one progresses in the streamwise direction at all axial locations.

The sought potential may be separated into an approximate solution and a residual term having the specific features epitomized by the relation

$$\phi(x, y, z) = \phi_0(y, z; x) + \phi_0^*(x) \tag{15-4}$$

where $\phi_0(y, z; x)$ denotes a function which is a solution of the Laplace equation

$$\nabla^2 \phi_0 = 0 \tag{15-3'}$$

as far as the y and z coordinates are concerned, and which registers merely the change in dimensions of the geometry of the system as a function of x, i.e. it accounts for the change in cross-sectional shape as one progresses from a plane x = const on, farther aft, to the next P plane. The residual term $\phi_0(x)$ has no effect on the lift, pitching moment, or rolling moment, because the related pressures are equalized on the upper and lower surfaces of the slender configuration (see VI,D,20 and 21).

The boundary conditions that this approximate potential must satisfy may be specified in the following compact manner, provided the definitions are made that $\mathbf{n}$ is the unit vector in the direction normal to the C_1

contour traced out by the wing-body combination in the P plane, while v_n represents the component of the free stream velocity **U**, taken in the direction of **n**. On the basis of this notation, then, the boundary conditions may be succinctly stated as:

On the body

$$\frac{\partial \phi_0}{\partial \eta} = -v_n \tag{15-5}$$

On the wake

$$\Delta\left(\frac{\partial \phi_0}{\partial y}\right)_0 = \left(\frac{\partial \phi_0}{\partial y}\right)_{z=0^+} - \left(\frac{\partial \phi_0}{\partial y}\right)_{z=0^-} = \frac{d\Gamma}{dy} = g(y) \tag{15-5'}$$

where Γ is the circulation pertaining to any arbitrary general one of the profiles lying out along the wing span, at successive planes y = const. This distribution of circulation is obviously not known beforehand.

In the special case where the fuselage is a body of revolution and the wing is a flat plate, the general boundary conditions become particularized to read:

On the body

$$v_n = -U\frac{dR_b}{dx} + \alpha_b U \sin\theta \tag{15-5''}$$

where α_b is the angle of attack of the body's axis with respect to the free stream flow vector **U**, and R_b is the radius of the section through the body at the axial location in question.

On the wing

$$v_n = \alpha_w U \tag{15-5'''}$$

where α_w is the angle of attack of the wing with respect to the free stream **U**.

If the fuselage is not a body of revolution, then the component of velocity v_n, which has to exist at the boundary of such a shape in order to keep the flow aligned with the surface of the body, must obey the condition that

$$v_n = -U\frac{dn}{dx} \tag{15-6}$$

where dn is an element of length on the normal through any arbitrary point of the contour C_1 of the section made by the P plane through the body; this small piece of the normal is the part intercepted on it by the C_1 contour itself and the projection upon the P plane of the contour formed when a plane parallel to the P plane is passed through the point $x + dx$ on the abscissa axis.

Irrespective of whether the relation given as Eq. 15-5'' or 15-6 is the one which is applicable in a given case, it follows that in general

$$\oint_{C_1} v_n ds = -U \oint_{C_1} \frac{dn}{dx} ds = -U \frac{dS}{dx} \tag{15-7}$$

where ds is the element of length along the C_1 contour, and S is the cross-sectional area of the fuselage intercepted by the P plane. Thus, in order to compensate for the transport of fluid taking place across the C_1 contour, corresponding to the amount of flux indicated in Eq. 15-7, it will be necessary to establish a counteracting source on the axis of the fuselage, at the x location of the P plane, with a strength of like amount, namely

$$q = U \frac{dS}{dx} \tag{15-7′}$$

It is clear from the foregoing remarks, therefore, that ϕ_0 is actually nothing more than the potential which describes the two-dimensional flow about the body in each cross-sectional P plane, in such a manner that the velocity components normal to the C_1 contour correspond to the $v_n \cdot \mathbf{n}$ distribution, provided the P plane lies upstream of the abscissa location where the maximum span falls, or, in less circumspect wording, provided it does not intersect the trailing edge. If it should happen that the P plane does intersect the trailing edge, then the potential ϕ_0 governs the two-dimensional flow corresponding to the normal velocity distribution $v_n \cdot \mathbf{n}$ taken in conjunction with the distribution of vortices lying along the segment of the wake intercepted by this P plane, where the circulation strengths of the trailing vortices composing the wake are prescribed according to the definition given as Eq. 15-5′.

By virtue of the simplifications brought about in Eq. 15-2 and thanks to the uncluttered form introduced for ϕ in Eq. 15-4, the determination of the over-all aerodynamic behavior of the sort of slender wing-body combination now under examination, as well as the delineation of the axial distribution of such forces, is reduced to a simple problem in two-dimensional flow, which may be solved by application of well-known methods which are essentially based on complex-variable techniques of the more elementary variety.

In order to make the treatment most amenable to a wide range of configurational shapes, the method employed by Ward in [*20*] may be profitably reviewed here. In order to evaluate the expression for the force acting on the ensemble, it is convenient first of all to convert Eq. 15-1 into more tractable form by introduction of the complex potential function W_0, governing the field of flow about the C_1 contour, so that $W_0 = \phi_0 + i\psi_0$ where ψ_0 is the harmonic conjugate of ϕ_0.

Now let C' denote any closed path enclosing the C_1 contour, and connect the two by means of the cut AA' passing through any point whatsoever on C_1, such as through the point A, as indicated in Fig. C,15b. Since

W_0 does not have any singularity and is single-valued throughout the region bounded by the C_1 and C' curves and along the two branches of the cut AA', it follows that the line integral of W_0 around the C_1 contour may be expressed as

$$\oint_{AA'+(C')+A'A} W_0 dt = \oint_{C_1} W_0 dt$$

On the other hand, since $\phi_0 = W_0 - i\psi_0$, it follows that Eq. 15-1 then becomes converted to

$$F_y + iF_z = -i\rho_\infty U \left[\oint_{AA'+(C')+A'A} W_0 dt - i \oint_{C_1} \psi_0 dt \right] \qquad (15\text{-}8)$$

Inasmuch as W_0 is analytic with no singular points at all along the whole path of integration $AA' + (C') + A'A$, it is permissible to express this complex potential function as a power series of the form

$$W_0 = \mathfrak{F} + \mathfrak{F}_0 \ln t + \sum_{m=1}^{\infty} \mathfrak{F}_m t^{-m} \qquad (15\text{-}9)$$

where $\mathfrak{F}$, $\mathfrak{F}_0$, and $\mathfrak{F}_m$ are constants with respect to t, although they will be, in general, functions of x. Evaluation of the first of the line integrals appearing in Eq. 15-8 gives then

$$\oint_{AA'+(C')+A'A} W_0 dt = 2\pi i(\mathfrak{F}_1 + t_0\mathfrak{F}_0)$$

Returning to consideration of the other integral in Eq. 15-8, it may be seen that an integration by parts results in the relation

$$\oint_{C_1} \psi_0 dt = (\psi_0 t)_{C_1} - \oint_{C_1} t \frac{\partial \psi_0}{\partial s} ds$$

where ds denotes the element of length along the contour C_1. By referring back to Eq. 15-5''' and 15-6 it is immediately evident that

$$(\psi_0)_{C_1} = \oint_{C_1} \frac{\partial \psi_0}{\partial s} ds = - \oint_{C_1} \frac{\partial \phi_0}{\partial n} ds = U \frac{dS}{dx}$$

while on the other hand, it follows from evaluation in Eq. 15-9 that

$$(\psi_0)_{C_1} = 2\pi\mathfrak{F}_0$$

to give finally:

$$\mathfrak{F}_0 = \frac{1}{2\pi} U \frac{dS}{dx} \qquad (15\text{-}10)$$

The remaining integral requiring evaluation gives a simple enough result, because

$$\oint_{C_1} t \frac{\partial \psi_0}{\partial s} ds = U \oint_{C_1} t \frac{dn}{dx} ds = U \frac{d}{dx} (t_0 S)$$

where $t_g = y_g + iz_g$ represents the complex coordinate of the centroid of the area S.

Consequently, the sought expression for the aerodynamic action taking place on the configuration is

$$F_y + iF_z = 2\pi\rho_\infty U\mathfrak{F}_1 + \rho_\infty U^2 \frac{d}{dx}(t_g S) \tag{15-11}$$

Likewise, the normal force per unit axial distance is analogously given as

$$\frac{dF_y}{dx} + i\frac{dF_z}{dx} = 2\pi\rho_\infty U\left[\frac{d\mathfrak{F}_1}{dx} + \frac{U}{2\pi}\frac{d^2}{dx^2}(t_g S)\right] \tag{15-11$'$}$$

By use of the above differential relation the moment with respect to the y and z axes may easily be calculated:

$$\begin{aligned} M_z + iM_y &= \int_0^x x\left(\frac{dF_y}{dx} + i\frac{dF_z}{dx}\right)dx \\ &= x(F_y + iF_z) - 2\pi\rho_\infty U\left(\int_0^x \mathfrak{F}_1 dx + \frac{U}{2\pi}t_g S\right) \end{aligned} \tag{15-12}$$

where the moment so obtained pertains to the whole forward portion of the body, extending from the tip of the nose back to the axial station marked off by the intersection with the P plane. The relations adduced here as Eq. 15-11 and 15-12 constitute the basic formulas derived by Ward in [*20*].

Implications of the "slender-body" results and their limits of applicability. It follows from examination of the formulas just derived, and from their manner of derivation under the restrictions of this approximate theory, that the compressibility of the fluid plays no part whatsoever in the genesis of the flow phenomena under consideration. Furthermore, it is worthy of note that practical experience furnishes good confirmation that this property is a fundamental feature of such flows; at least substantiation is provided for this assertion by the results obtained in the limiting case when the wing-body combination reduces to a body-alone configuration. For example, values of

$$\frac{1}{q2\alpha d_{max}}\frac{dL}{dx}$$

have been calculated with the aid of Eq. 15-11$'$ for the body shape sketched in Fig. C,15c, where d_{max} represents the maximum cross-sectional diameter, the position of which is also indicated in this diagram. These theoretically computed results are superimposed upon the graphs of the experimental results given in Fig. C,15c; the test data were obtained from experiments conducted on such a spindlelike body by Ackeret [*21*] for Mach numbers running from $M_\infty = 0$ to $M_\infty = 0.95$. Ample proof is supplied by these results to support the general conclusion cited above.

It is still a matter of paramount import to look into the question which arises at this juncture concerning just how much the wing's aspect ratio may be increased before the assumptions, upon which the treatment just described is based, cease to be valid. This question may be settled by scrutiny of the curves plotted in Fig. C,15d, applying to wing-alone configurations. The solid curves in this figure represent the results which

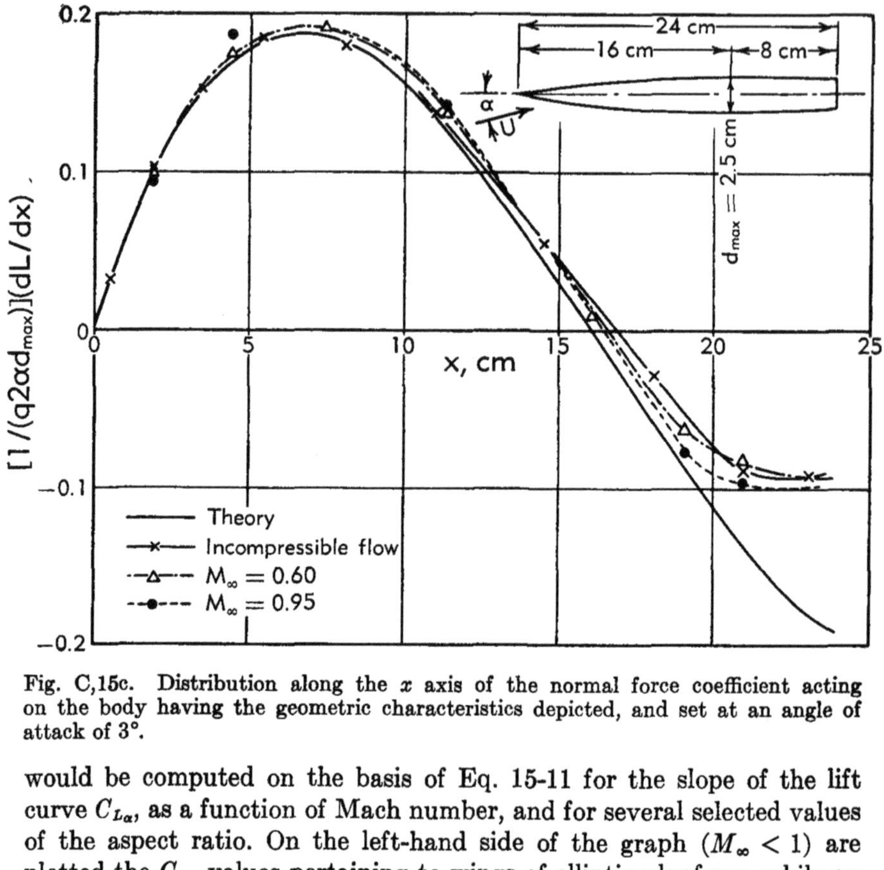

Fig. C,15c. Distribution along the x axis of the normal force coefficient acting on the body having the geometric characteristics depicted, and set at an angle of attack of 3°.

would be computed on the basis of Eq. 15-11 for the slope of the lift curve C_{L_α}, as a function of Mach number, and for several selected values of the aspect ratio. On the left-hand side of the graph ($M_\infty < 1$) are plotted the C_{L_α} values pertaining to wings of elliptic planform, while on the right-hand side ($M_\infty > 1$) appear the C_{L_α} values applying to wings of triangular planform. In addition to these horizontal lines (showing no variation with Mach number), another series of dashed-line curves are provided, for ease of comparison of these approximate findings with those obtained from more exact methods. The comparative calculations for the subsonic regime ($M_\infty < 1$) were performed according to Krienes' method [*22*], after it was modified to take into account the effect of compressibility according to the procedure recommended by Robinson and Young [*23*].

The comparative computations in the supersonic regime ($M_\infty > 1$) follow the standard methods formulated by such authors as Robinson [24], Stewart [25], etc.

It may be deduced from these plots that the degree of accuracy provided by the kind of computational approach being subscribed to in this article is certainly sufficiently good to warrant putting full confidence in this method whenever $\mathit{AR} \leqq 1.5$. It is worthwhile to bear in mind in this regard, besides, that an extension to the slender body theory has recently been presented by Adams and Sears in [26], with the avowed purpose of widening the limits of applicability and increasing the general utility of

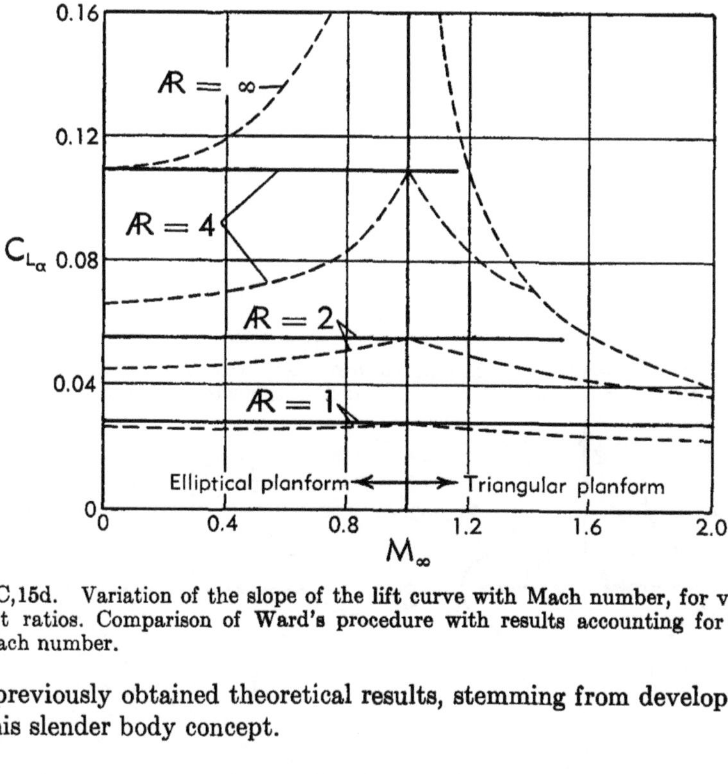

Fig. C,15d. Variation of the slope of the lift curve with Mach number, for various aspect ratios. Comparison of Ward's procedure with results accounting for effect of Mach number.

the previously obtained theoretical results, stemming from development of this slender body concept.

C,16. Applications of the Method. Now that the requisite preliminary agreements and basic relationships have been established, it is not difficult to determine the aerodynamic forces acting on winged bodies of the two different general types illustrated in Fig. C,16a and C,16b, which are characterized by the fact that in the one case (1) the configuration is a body of revolution of arbitrary axial variation in cross-sectional diameter, supporting wings having edges which are always "leading edges" upstream from the base (see Fig. C,16a), and in the other case (2) the configuration is a body of revolution with a uniform cross-sectional diameter for the afterbody, lying downstream of the wing's trailing edge (see Fig. C,16b).

Determination of the lift and moment for a finned missile. A section through the system made with a plane P having the equation x = const

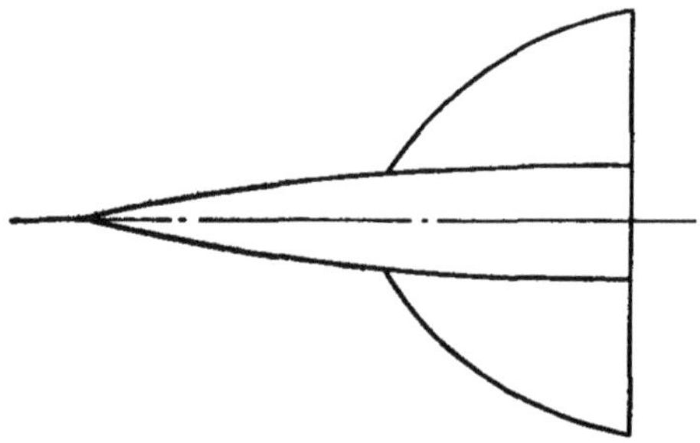

Fig. C,16a. Sketch showing general features of any wing-body configuration to be included under the classification: case 1, finned missile.

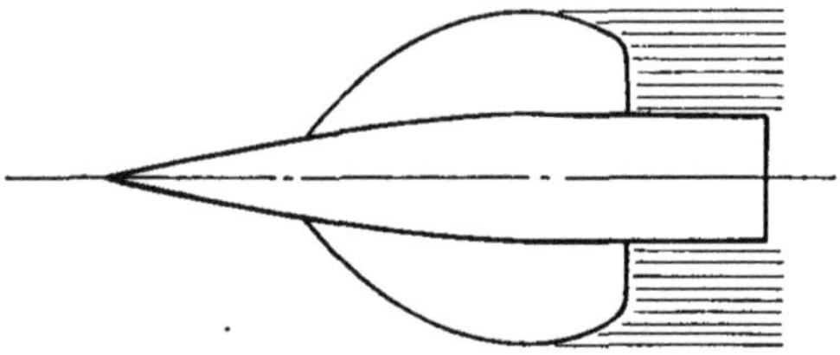

Fig. C,16b. Sketch showing general features of any wing-body configuration to be included under the classification: case 2, missile with cylindrical rump.

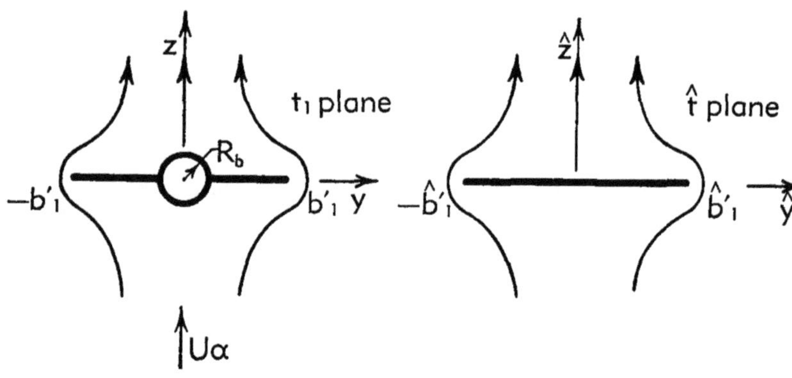

Fig. C,16c. Schematic representation of the two-dimensional flow fields generated about the actual flat plate wing plus slender body combination, and its transform in the complex $\hat{t}$ plane.

produces the outline shown in Fig. C,16c; thus the complex potential function W_0 becomes specified solely in terms of the boundary conditions given as Eq. 15-5″ and 15-5‴, so that if W_0 is sought in the form

$$W_0 = \frac{U}{2\pi}\frac{dS}{dx} \ln (t - t_g) + W_1(t) \tag{16-1}$$

where

$$W_1(t) = \phi_1 + i\psi_1$$

it follows that this $W_1(t)$ potential function pertaining to the two-dimensional motion must obey the requirements that

$$\left(\frac{\partial \phi_1}{\partial R}\right)_{R=R_b} = -\alpha_w U \sin \theta$$

$$\left(\frac{\partial \phi_1}{\partial R}\right)_{z=0} = -\alpha_w U \qquad \text{for the exposed wing surface, where } R_b \leqq |y| \leqq b_1'$$

and

$$\left(\frac{\partial \phi_1}{\partial R}\right)_{R=\infty} = 0$$

where b_1' represents the semispan of the wing intercepted by the P plane, and where the angle of incidence of the wing is zero; i.e. $\alpha_b = \alpha_w$. It is quite evident, in fact, that $W_1(t) - i\alpha_w Ut$ is the complex potential which corresponds to the two-dimensional flow created about the outline of the configuration traced out when it is intercepted by a P plane (as indicated in Fig. C,16c). The flow generated is equivalent to that existing around a rigid obstacle, having such a contour, when it is immersed in a uniform stream of velocity $\alpha_w U$, directed along and in the same sense as the positive z axis.

Having assumed that $t_1 = t - t_g$ and by applying to the t_1 plane the transformation

$$\hat{t} = \hat{y} + i\hat{z} = t_1 + \frac{R_b^2}{t_1}$$

the C_1 contour is transmuted into a flat plate $\overline{C}_1$ placed along the real axis, of length

$$2\hat{b}_1' = 2\left(b_1' + \frac{R_b^2}{b_1'}\right)$$

Inasmuch as the complex potential for the flow about the $\overline{C}_1$ contour is

$$W(\hat{t}) = -iU\alpha_w \sqrt{\hat{t}^2 - \hat{b}_1'^2}$$

it follows that the retransformed flow back in the real plane is characterized by a complex potential having the form

$$W_1(t) = -iU\alpha_w\left[\sqrt{\left(t_1 + \frac{R_b^2}{t_1}\right)^2 - \left(b_1' + \frac{R_b^2}{b_1'}\right)^2} - t_1\right] \tag{16-2}$$

Upon development of $W_1(t)$ in a Laurent series, it may readily be seen that

$$\mathfrak{F}_1 = iU\alpha_w \frac{b_1'^4 + R_b^4}{2b_1'^2} + i\alpha_w x \frac{U}{2\pi}\frac{dS}{dx} \tag{16-3}$$

provided it is recognized that

$$t_0 = -i\alpha_w x \tag{16-4}$$

The sought expression for the lift force is obtained, therefore, as

$$F_s = \pi\rho_\infty U^2\alpha_w \frac{b_1'^4 + R_b^4 - b_1'^2R_b^2}{b_1'^2} \tag{16-5}$$

through insertion of the expression just derived for $\mathfrak{F}_1$ into Eq. 15-11.

By assuming as the P plane the plane containing the maximum span $b = 2b'$ and indicating by R_0 the radius of the body section lying on this plane, F_s becomes coincident with the total lift of the system, that is

$$\frac{L_{bw}}{\frac{1}{2}\rho_\infty U^2} = 2\pi\alpha_w \frac{b'^4 + R_0^4 - b'^2R_0^2}{b'^2} \tag{16-6}$$

As a standard of comparison, it may be noted that the total lift L_w, acting on an isolated wing of the same span (and filled in across the area actually covered by the fuselage in the case of the complete configuration), is given by the formula

$$\frac{L_w}{\frac{1}{2}\rho_\infty U^2} = 2\pi\alpha_w b'^2 \tag{16-6'}$$

If an interference effectiveness factor is constructed from these separate relations by forming a ratio for the total lifts, it develops that

$$\frac{L_{bw}}{L_w} = \frac{(C_L)_{bw}}{(C_L)_w} = 1 - \frac{R_0^2}{b'^2} + \frac{R_0^4}{b'^4} \tag{16-7}$$

If the wing is mounted on a cylindrical body that is infinitely long, the lift L_1 corresponding to the fore portion of the body is given by

$$\frac{L_b}{\frac{1}{2}\rho_\infty U^2} = 2\pi\alpha_w R_0^2 \tag{16-8}$$

and, thus, if this quantity L_b is subtracted from the amount of lift created on the complete finite-length model, as determined from the use of Eq. 16-6, the residual amount of lift operating over that portion of the wing-body system confined within the fore and aft limits, marked off by intersection of the wing's leading and trailing edges with the side of the body, may be computed by evaluation in the formula

$$\frac{L_{bw} - L_b}{L_w} = \left(1 - \frac{R_0^2}{b'^2}\right)^2 \tag{16-9}$$

These results are displayed pictorially in Fig. C,16d, where the computed values of $(C_L)_{bw}/(C_L)_w$, furnished by Eq. 16-7 and 16-9, have been plotted as a function of the body diameter-to-wing span parameter, K_h, where $K_h = R_0/b'$.

In similar manner, the expression for the total moment being exerted on the wing-body combination under study may be written down immediately by reference to the general formula given as Eq. 15-12, which in the present case becomes particularized to

$$(M_y)_{bw} = l \cdot L_{bw} - \int_0^l \pi\rho_\infty U^2\alpha_w \cdot \frac{b_1'^4 + R_b^4 - b_1'^2R_b^2}{b_1'^2}\,dx \qquad (16\text{-}10)$$

where l is the length of the fuselage. If the shape of the configuration is stylized even further to become a triangular wing mounted on a conical

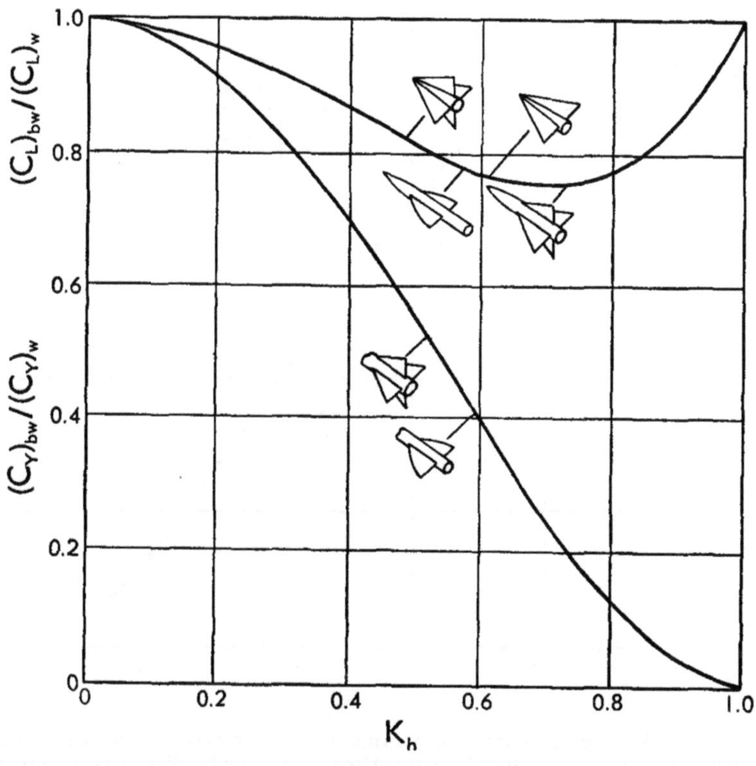

Fig. C,16d. Lift and side force ratios, for several types of wing-body combination, plotted as functions of body diameter-to-wing span parameter K_h.

body, as indicated by the small sketches set beside the corresponding curve in Fig. C,16e, then the variable ratios appearing in Eq. 16-10 become expressible as

$$\frac{R_b}{b_1'} = \frac{R_0}{b'} \quad \text{and} \quad \frac{b_1'}{b'} = \frac{x}{l}$$

so that

$$(M_y)_{bw} = \tfrac{2}{3}lL_{bw}$$

Inasmuch as the corresponding moment produced by the isolated "extended" wing is just

$$(M_y)_{\text{w}} = \tfrac{2}{3} l L_{\text{w}}$$

it follows that

$$\frac{(C_M)_{\text{bw}}}{(C_M)_{\text{w}}} = \frac{L_{\text{bw}}}{L_{\text{w}}} = 1 - \frac{R_0^2}{b'^2} + \frac{R_0^4}{b'^4} \tag{16-11}$$

If the wing is mounted on a cylindrical body that is infinite in axial extent, the moment about the y axis (which is assumed to pass through the

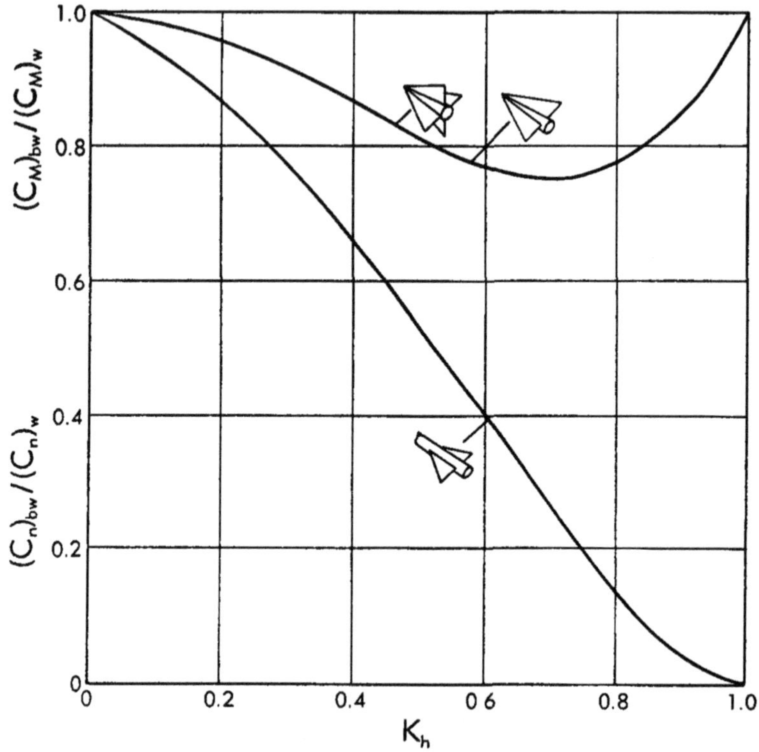

Fig. C,16e. Pitching moment and yawing moment ratios, for several types of wing-body combination, plotted as functions of the body diameter-to-wing span parameter K_h.

leading edge of the profile section forming the juncture between the wing and body), created by the aerodynamic forces being exerted on the system for the portion affected by the wing, is given by

$$(M_y)_{\text{bw}} = \pi\rho_\infty U^2 \alpha_{\text{w}} \left[l\left(b'^2 + \frac{R_0^4}{b'^2} - R_0^2\right) - \int_0^l \left(b_1'^2 + \frac{R_0^4}{b_1'^2} - R_0^2\right) dx \right]$$

$$= \pi\rho_\infty U^2 \alpha_{\text{w}} \frac{lb'^2}{3}\left(1 - \frac{R_0}{b'}\right)^2 \left(2 + 3\frac{R_0}{b'} + 3\frac{R_0^2}{b'^2}\right) \tag{16-12}$$

As a standard of comparison, it may be observed that the integrated moment $(M_y)_w$ produced by the flow about an isolated triangular "extended" wing of maximum semispan b', for an axis of rotation y, taken to be at the leading edge of the root chord, is given by the expression

$$(M_y)_w = \tfrac{2}{3}\pi\rho_\infty U^2\alpha_w l b'^2 \tag{16-13}$$

Consequently, if another interference effectiveness factor pertaining to the present situation (wherein the body is assumed to be infinitely long) is

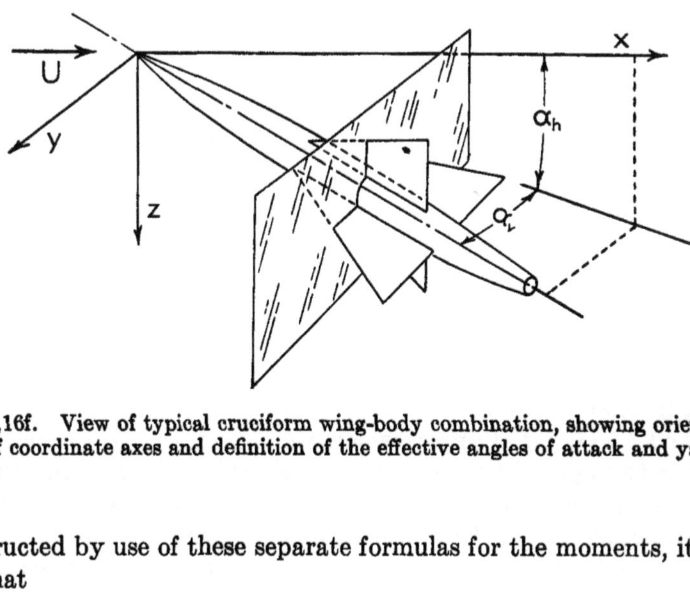

Fig. C,16f. View of typical cruciform wing-body combination, showing orientation of coordinate axes and definition of the effective angles of attack and yaw.

constructed by use of these separate formulas for the moments, it comes out that

$$\frac{(C_M)_{bw}}{(C_M)_w} = \frac{(M_y)_{bw}}{(M_y)_w} = \frac{1}{2}\left(1 - \frac{R_0}{b'}\right)^2\left(2 + 3\frac{R_0}{b'} + 3\frac{R_0^2}{b'^2}\right) \tag{16-14}$$

These results are presented graphically in Fig. C,16e, where the computed values of $(C_M)_{bw}/(C_M)_w$, furnished by Eq. 16-11 and 16-14, have been plotted as a function of the body diameter-to-wing span parameter, R_0/b', for the two different classes of nose (pointed, or infinitely-long cylindrical) treated in the above analysis.

It is also interesting to consider the case in which the body carries two wings of cruciform type and the velocity has an incidence α_h with respect to one wing plane (which is conventionally referred to as a horizontal plane) and an incidence α_v with respect to the other wing plane (Fig. C,16f). Under these conditions, the $W_1(t)$ potential is given by

$$W_1(t) = -iU\alpha_h\left[\sqrt{\left(t_1 + \frac{R_b^2}{t_1}\right)^2 - \left(b_1' + \frac{R_b^2}{b_1'}\right)^2} - t_1\right] + U\alpha_v\left[\sqrt{\left(t_1 + \frac{R_b^2}{t_1}\right)^2 + \left(b_1' - \frac{R_b^2}{b_1'}\right)^2} - t_1\right] \quad (16\text{-}15)$$

while in the present context

$$t_g = -x(\alpha_v + i\alpha_h) \quad (16\text{-}16)$$

It follows, analogously to what was found to hold for the single-wing configuration, that under the present circumstances

$$\mathfrak{F}_1 = \frac{b_1'^4 + R_b^4}{2b_1'^2}U(i\alpha_h + \alpha_v) + x\frac{U}{2\pi}\frac{dS}{dx}(i\alpha_h + \alpha_v) \quad (16\text{-}17)$$

From the structure of this result it may be seen that, when it is inserted into Eq. 15-11 and 15-12, the corresponding forces and moments (taken about the z axis) will turn out to be determined by formulas which are identical to those previously derived for the single horizontal wing (with moments taken about the y axis), with the mere substitution of α_v for α_w, and with the similar replacement of α_w by α_h in the case where the performance of the horizontal cross member of the cruciform configuration is under study. It is scarcely necessary to point out the obvious stipulation that one must use the wing and body dimensions, in the generalized force and moment formulas, as now provided by Eq. 16-7, 16-9, 16-11, and 16-14, for example, which are appropriate to the particular one of the two crossed-wing members in which interest may be centered.

Because of this similitude in the formulas for the lift and moments, acting on either one of the two perpendicular wing members of typical cruciform configurations, it follows that the previously given curves drawn in Fig. C,16d and C,16e are equally valid for the side force and yawing moment exerted on cruciform-wing models, such as sketched on these graphs. Consequently, a mere change in symbols on these curves suffices to make them apply to the coefficients

$$(C_Y)_{bw} = \frac{F_Y}{\frac{1}{2}\rho_\infty U^2 A_v}, \qquad (C_n)_{bw} = \frac{M_z}{\frac{1}{2}\rho_\infty U^2 A_v b_v'}$$

where A_v is the area of the vertical member of the crossed wings.

In the event that the two members of the cruciform synthesis are not identical in shape, a rolling moment will be impressed upon the system, and this moment may be computed by use of a general formula analogous to the one applying for the pitching moment of a single wing, so that in the present instance

$$M_x = \rho_\infty U \oint_{C_1} \phi(y\,dy + z\,dz) \quad (16\text{-}18)$$

In the special case of a configuration composed of a cylindrical body carrying triangular wings, the aforementioned rolling moment was worked out by Spreiter in [*27*], by application of a method which is quite different, however, from the one outlined above. His result turns out to be

$$C_l = \frac{M_x}{\frac{1}{2}\rho_\infty U^2 2b'_h A_h} = \frac{\alpha_h \alpha_v}{3}\left(\frac{A_v b'_v}{A_h b'_h}\nu_v - \nu_h\right) \tag{16-19}$$

where the factor ν, as a function of the body diameter-to-wing span ratio K_h, has been plotted in Fig. C,16g.

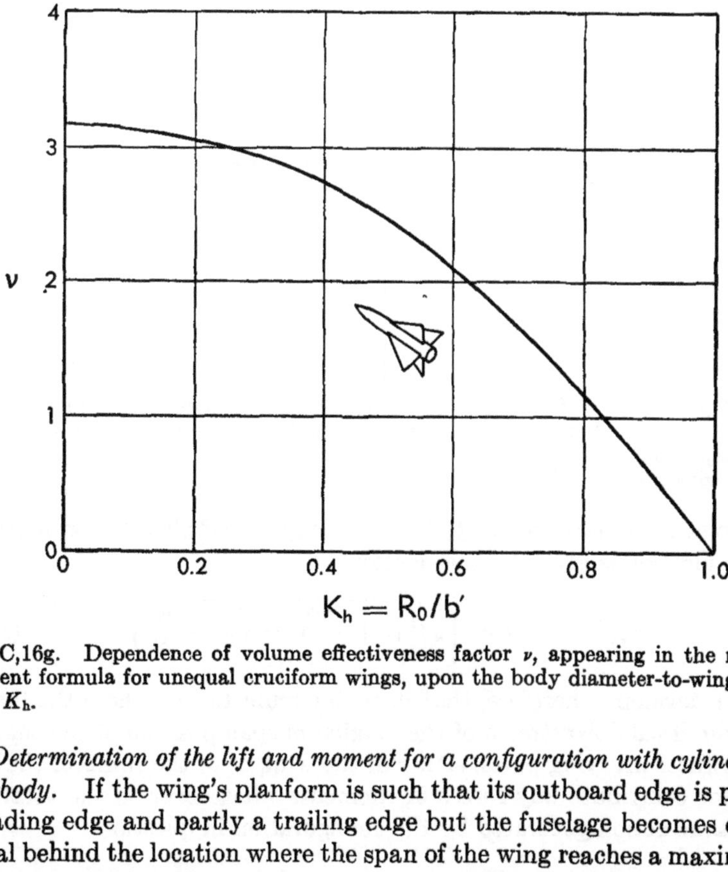

Fig. C,16g. Dependence of volume effectiveness factor ν, appearing in the rolling moment formula for unequal cruciform wings, upon the body diameter-to-wing span ratio K_h.

Determination of the lift and moment for a configuration with cylindrical afterbody. If the wing's planform is such that its outboard edge is partly a leading edge and partly a trailing edge but the fuselage becomes cylindrical behind the location where the span of the wing reaches a maximum, then the solution of the problem of determining the aerodynamic behavior of such a configuration may be seen to reduce without difficulty to essentially the same situation encountered in dealing with case 1, on the basis of the following observations. Consider a cross section made through the configuration by a plane P with equation x = const and located aft of the axial station where the wing span reaches its maximum spread; such a cross section and the indication of the shed vorticity in the wake of the wing is illustrated in Fig. C,16h. It is readily appreciated in this instance that the boundary conditions which have to be met on the surface of the configuration, as well as on the wake shed from the wing, as imposed by the stipulations written above in Eq. 15-5 and 15-5′, will be satisfied automatically provided the complex potential $W_0(t)$ is selected to be the one corresponding to the expression which applies to the two-dimensional flow in the cross-sectional P plane, located right at the particular axial station where the maximum span of the wing occurs. The particular

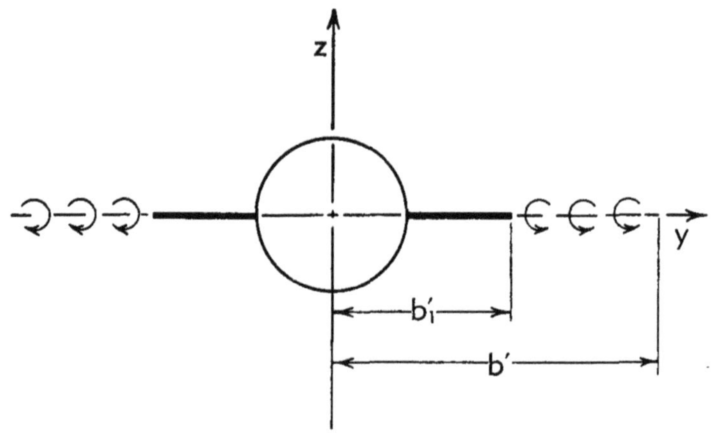

Fig. C,16h. Trace of a wing-body combination having a cylindrical rump, together with its associated sheet of trailing vortices shed from the fore part of the wing, as intercepted in a normal plane, x = const, lying downstream of the location of maximum wing span.

complex potential applying to flow in this specific P plane at the maximum span location is given by the relation

$$W_1(t) = -iU\alpha_w \left[\sqrt{\left(t_1 + \frac{R_0^2}{t_1}\right)^2 - \left(b' + \frac{R_0^2}{b'}\right)^2} - t_1\right] \quad (16\text{-}20)$$

It develops, therefore, that in such circumstances where the fuselage is cylindrical downstream of the maximum span position of the combination, these aft-lying parts, either of the wing or of the rump of the body, do not contribute any forces or moments whatsoever to the total aerodynamic action generated by the configuration in question.

C,17. Interference Effects Produced by the Body upon the Tailplane. If a body and tailplane combination may be classified within the group of lifting surface and body configurations which were examined in Art. 15 the determination of the aerodynamic characteristics of such a combination follows immediately, without further consideration, through direct application of the results already presented there. If, on the contrary, the aspect ratios pertaining to the stabilizing surfaces are too large or if their geometric features controvert the properties cited in Art. 15 as being prerequisite to the use of these methods, then the determination of the effect of the fuselage upon the tail is rendered somewhat more complicated than was true even in the case of wing-body interactions.

The principal reason for this more complex state of affairs in regard to body-tail interference is primarily because there are much more drastic effects attributable to the action of the boundary layers which develop over the long stretches of the fuselage and on the rear lifting surfaces, in contrast to what takes place farther forward near the wing. In the second place, the prospects of determining these fuselage-tail interference effects with precision are significantly worsened, because of the fact that the idealizations and concepts utilized by Multhopp appear far too remote from reality in this instance for one to expect that any satisfactory results could be elicited from the theoretical procedures founded on such inappropriate assumptions.

Notwithstanding these serious drawbacks to the use of the Multhopp procedure for analysis of the body-tail interference problem, it may be worthwhile, at least for the purpose of deducing some qualitative inferences as to what the general behavior may be, to apply the method willy-nilly to the case of the horizontal tail. If such computations are carried out, it is clear from the trends of the results obtained in Art. 9 that the slope of the lift curve for the tail is likely to be low, especially at the higher Mach numbers. In fact, this unfortunate situation is aggravated even more by the presence of the boundary layer, inasmuch as this thickened layer has the same effect, roughly speaking, as a virtual increase in the ratio of the body diameter to the span of the stabilizer.

On the other hand, however, the downwash induced by the wing in the presence of the body and that attributable to the body itself is generally less severe than that produced behind the same wing if it were operating as an isolated entity. This assertion is substantiated by the data presented in Fig. C,17a; these curves apply to the particular wing-body configuration depicted in Fig. C,9a. It is thus apparent that the two interference effects from the fuselage that make themselves felt in this instance are of opposite sign, so that the combined operation of the two contrasting mechanisms results in a residue effect that is probably small.

The state of uncertainty surrounding such computations of interference reactions is even more confounded in the case of the vertical stabilizer, if this surface is merged directly into the fuselage. From a qualitative viewpoint there can be no doubt that the fuselage constitutes a sort of reflection-plane shield to the lifting surface represented by the vertical fin, so that the fuselage tends to act as an end plate. Experimental data bearing on this question have been procured by Root [*28*], and his results are reproduced here as Fig. C,17b. This curve thus furnishes a quantitative measure of the effectiveness of the fuselage, acting

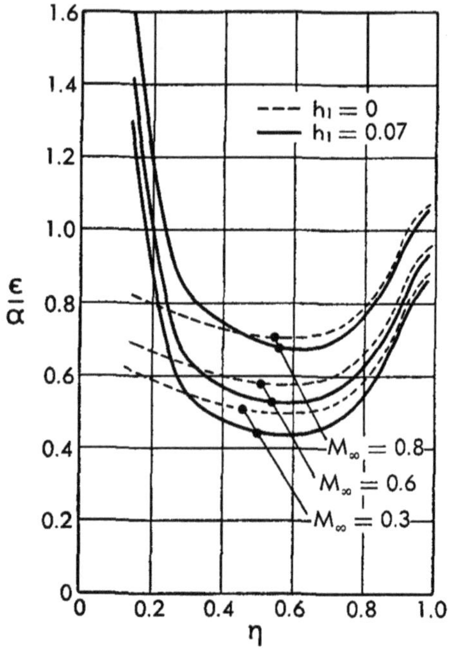

Fig. C,17a. Diagram of downwash ϵ/α in the Trefftz plane, where $x = \infty$ (wing-body combination of Fig. C,9a).

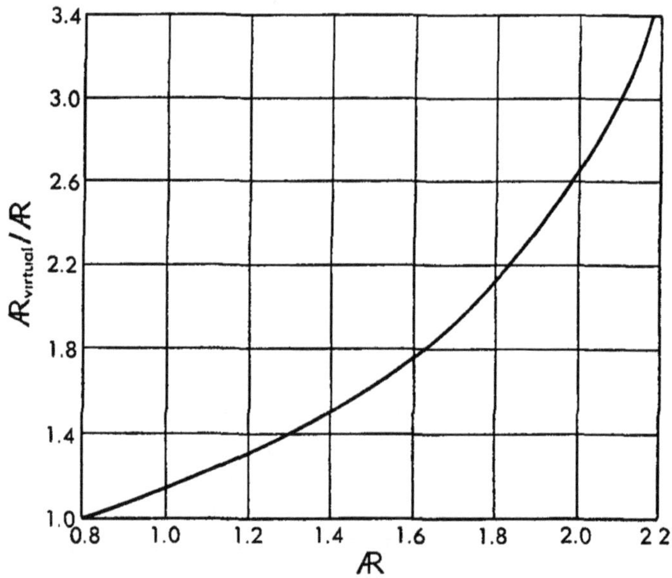

Fig. C,17b. Virtual aspect ratio $Æ_{virtual}$ of the vertical tail surface as a function of its geometric aspect ratio $Æ$ [*28*].

as an end-plate shield, in raising the geometric aspect ratio Æ to a larger virtual aspect ratio $Æ_{virtual}$.

A further stage of complexity is encountered if one attempts to account for the wing-body interference effects in determining the aerodynamic characteristics of the vertical tail. Because no significant theoretical work has been done on this subject, there is no course left but to rely on qualitative surmises.

If one considers, for instance, what happens when a high-wing airplane is yawed slightly, it is readily appreciated that the half-wing on the windward side of the fuselage experiences a higher angle of attack (because of the effect of the velocities induced by the flow about the fuselage) than the leeward half-wing is subjected to (Art. 14). It would

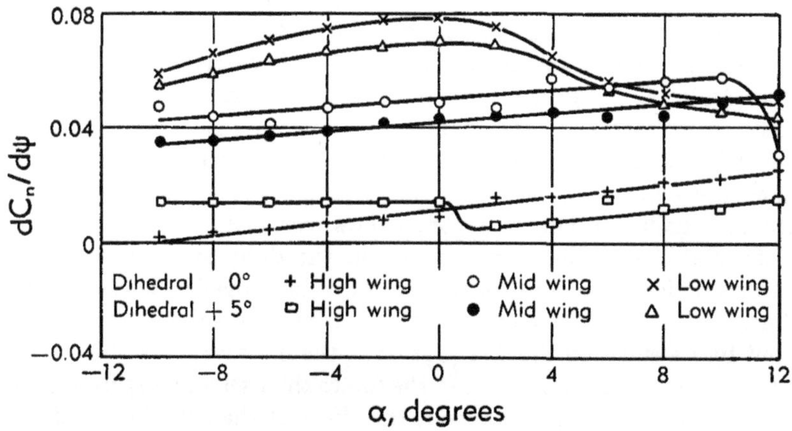

Fig. C,17c. Body's influence on the yawing moment due to yaw, produced by the vertical tailplane, operating on high-, mid-, and low-wing models (from [*19*]).

appear therefore that the system of trailing vortices from the upwind wing-half creates a stronger downwash field than would arise if the fuselage were removed; consequently, because of the dissymmetry of pressures on the wing halves, there are created, in the region above the wake, velocity components directed away from the upper half of the plane of symmetry (i.e. toward the windward side), and thus the directional stability of the airplane should decrease below that in the wing-alone case. Of course, the opposite train of events occurs when the configuration has a low wing.

These heuristic arguments are confirmed by practical experience, as exemplified by the experimental data in Fig. C,17c. The curves give the value of the yawing moment due to yaw for several angles of attack and for three different wing positions; these data were all excerpted from the experimental results described in [*19*].

CHAPTER 3. PROPELLER AND WING INTERACTIONS AT SUBSONIC SPEEDS

C,18. Introduction. It is convenient to separate the propeller-wing interactions into those effects produced separately by action of the wing upon the propeller and the effects attributable to the influence exerted by the propeller upon the lifting surface or surfaces. This dichotomous mode of attack is advocated not only because it makes the treatment simpler but also because the order of magnitude of the effects of the mutual interference is much larger in the latter case than in the former (at least for conventional placements of these two interfering parts of the airplane). Moreover, the importance of this type of interference in regard to the general behavior of the airplane is vastly different in the two cases. As far as the propeller is concerned, for instance, the mutual interference merely acts in a way that produces such a generally slight alteration of the propeller's aerodynamic characteristics that the consequences, which are evinced by the behavior of the complete configuration, are rather insignificant. On the contrary, however, it is essential in any study of the stability of the airplane and in the determination of its flying characteristics to account for the fact that the propeller exerts a profound influence upon the lift and drag of the wing; both the changes brought about by its effect on the distribution of these forces along the wing span as well as upon the magnitude of the forces themselves must be accounted for. Of even greater importance is the effect of the propeller in changing the characteristics of the tail surface, whether these changes are produced directly by the action of the propeller or indirectly by the action of the downwash from the wing, upon which interference effects from the propeller have had a chance to come into play.

It is assumed in what follows, in order to determine the character of the flow created about a wing when it is operating in the presence of a propeller, that the velocity and pressure fields produced by the propeller remain unaltered from what they would have been if the propeller were acting as an isolated entity. Likewise, in setting out to study the effect of the lifting surface upon the propeller, the flow over the wing is first considered to be that generated solely by action of the free stream (uniform at infinity) which impinges upon the wing, uninfluenced by the presence of the propeller. By proceeding in this manner it is possible, on the basis of the first approximation results thus obtained, to deduce what alterations must be applied to the first determinations in order to take into account, at least in a qualitative way, the effect of the "interference of the interference."

Consequently the present chapter is divided into three sections in which the following topics are individually considered in this order:

1. Influence of the propeller upon the wing
2. Influence of the propeller and the propeller-wing combination upon the tail surface
3. Influence of the lifting surface upon the propeller, and the effect in turn of this interference upon problems 1 and 2.

Besides the assumption just made concerning the possibility of separating out the constituent parts of the over-all interaction, some additional simplifying hypotheses must be introduced into the study of the first two problems in order to ensure that the solutions to be derived do not involve an excessive amount of labor. To be specific, it is further postulated that:

1. The increments in velocity produced by the action of the propeller, both in the direction of the free stream velocity and perpendicular thereto, are assumed to be constant over the propeller disk (and therefore at an infinite distance downstream).
2. The character of the slipstream is assumed to be uninfluenced by the presence of the wing, and likewise the trailing vortex sheet behind the wing is assumed to remain composed of straight-line vortices which lie parallel to the free stream velocity vector everywhere, and which are thus assumed to be uninfluenced by the induced velocities generated by the vortices shed from the propeller.
3. The propeller may be considered to have an infinite number of blades, and to be placed sufficiently far away from the wing so that it is possible to assume that any variations in the velocity increments taking place along the axis of the slipstream are small and thus tractable, throughout that extent of its length which is of any importance in this study.
4. The wing is supposed to have a chord length which is small enough and a sweep that is slight enough to make it legitimate to replace the actual wing by a single bound vortex, the axis of which runs perpendicular to the plane of symmetry of the airplane. This vortex may be specified in terms of the circulation created at one of its arbitrary general points. The circulation, denoted by the symbol Γ, is given by the formula

$$\Gamma = \tfrac{1}{2} c_{l_\alpha} c \alpha_{\text{eff}} U_s \tag{18-1}$$

where α_{eff} represents the effective angle of attack and where U_s is the sum of the free stream velocity, U, and the velocity increment also lying in the direction of this free stream velocity which is produced by the action of the propeller.

5. Finally, the dwindling-away of the vorticity in the slipstream, produced by the viscosity and turbulence, is not considered.

Assumption 1 appears to be quite permissible inasmuch as it simplifies the solution to the problem and does not alter the visualization of the basic flow pattern in any appreciable way.

Assumption 2 presupposes that the disturbances created by the presence of the wing are quite small, and thus it is assumed, to be more precise, that the product $c_l(c/R)$ is small at each airfoil section (where R is the radius of the propeller disk).

No serious objection can be raised in regard to the reasonableness of assumption 3, while it is merely necessary to postulate that the ratio c/R is to be small in order to make Eq. 18-1 hold true. In fact, the justification for proceeding in this way appears in the corresponding two-dimensional problem which arises when an airfoil section is placed in a jet of width $2R$, having a velocity U_s surrounded by a stream of constant velocity U, and separated from the uniform flow by vortical shear planes. In this case the airfoil section lift coefficient c_l has been shown [*29*] to be a function of the position of the airfoil section within the jet and also of the ratio, c/R, now under discussion. Assumptions 2 and 4 act in concert to establish the fact, therefore, that the investigation to be undertaken in the following paragraphs must be considered in the guise of a limiting case, which gives worthwhile results only if $c_l(c/R)$ tends toward zero. For the larger values of this parameter, however, the theoretical values obtained ought to be adjusted to take into account the results of practical experience.

At any rate, once the above-listed assumptions are premised, it is possible to divide up the field of flow into two parts S_{I} and S_{II} (the first corresponding to the flow taking place outside of the slipstream, and the second, internal to the slipstream flow), both describable by means of potentials, and separated by a vortical surface of given shape and of given strength. Consequently a different potential function is separately determined in what follows for each one of these regions.

As a result of this division of the flow, it is no longer permissible to apply the usual procedure for finding the flow about a wing, i.e. by means of superimposing an additional flow upon the uniform stream impinging upon the wing, in such a way that the lifting surface comprises a flow surface. The breakdown in the familiar process of determining the flow about the wing is due to the fact that in addition to the boundary conditions which must be satisfied at the surface of the wing, there are other conditions that the resultant flow must also obey on the vortical surfaces constituting the boundaries of the parts into which the flow field is now separated. These surfaces, dividing the regions composing the constituent parts of the whole flow, are *fluid* surfaces, and thus they must neither be surfaces of discontinuity with respect to pressure nor with respect to the velocity component which is normal to any such surface.

The effect of satisfying the requirements concerning the behavior of the flow at the surface of the slipstream makes a great deal of difference in the aerodynamic characteristics of the wing, depending on whether the wing lies behind the propeller or ahead of it. The two contrasting situations are examined separately, therefore, and the first case to be considered is that where the wing lies downstream of the propeller.

INTERFERENCE OF THE PROPELLER UPON THE WING (WING LOCATED DOWNSTREAM OF THE PROPELLER)

C,19. Flow Field Generated by an Isolated Propeller.

CASE OF A PROPELLER OPERATING AT ZERO ANGLE OF ATTACK. If, in conjunction with hypothesis 1 stated in the previous article, it is now assumed in addition that $\alpha_T = 0$, then the induced velocity field generated by the propeller is readily determined [*2*, Vol. 4, pp. 364–373; *30*, pp. 353–358] in the following ways.

Ahead of the propeller. The field in any plane through the axis is the same as that which would be produced by the action of a uniform distribution of sinks dispersed over the propeller disk, and having a strength per unit area equal to $-dE/2\pi\bar{r}d\bar{r} = -aU$ where the quantity aU signifies the increment experienced by the axial component of the velocity in going through the disk and persisting for locations lying immediately downstream, and wherein

$$a = \frac{\Delta p}{2\rho_\infty U_\infty^2} \tag{19-1}$$

Δp being the pressure variation through the propeller.

Behind the propeller. The field here is equivalent to the sum of the velocity produced by the above-mentioned uniform distribution of sinks taken together with the flow that consists of a stream having a constant velocity equal to $2a\beta^2 U$ within the slipstream and a zero velocity outside of it.

Now let v'_x and v'_r denote, respectively, the axial and radial components of velocity generated at an arbitrary general point P (situated at a distance r from the propeller axis) by the action of the uniform distribution of sinks. Then, as pointed out by Weinig [*30*, pp. 353–358], the lines

$$v'_x = \text{const} \quad \text{and} \quad \frac{1}{\beta} r v'_r = \text{const} \tag{19-2}$$

represent, respectively, the equipotential lines and the streamlines for the induced flow created by a circular vortex filament having a radius equal to R and lying on the propeller disk.

It follows therefore that v'_x and v'_r in the transformed space (x_1, y, z) are given by

$$\left.\begin{aligned} v'_x &= -aU\frac{\Omega}{2\pi} \\ \text{and}\qquad \frac{1}{\beta}rv'_r &= aU\frac{\tilde{r}_1+\tilde{r}_2}{\pi}[\mathsf{K}(k)-\mathsf{E}(k)] \end{aligned}\right\} \qquad (19\text{-}3)$$

where Ω is the solid angle subtended by the circle with radius R as seen from the point P, while K and E are the complete elliptic integrals of the first and second kinds with modulus k, where $k = (\tilde{r}_2 - \tilde{r}_1)/(\tilde{r}_1 + \tilde{r}_2)$ and

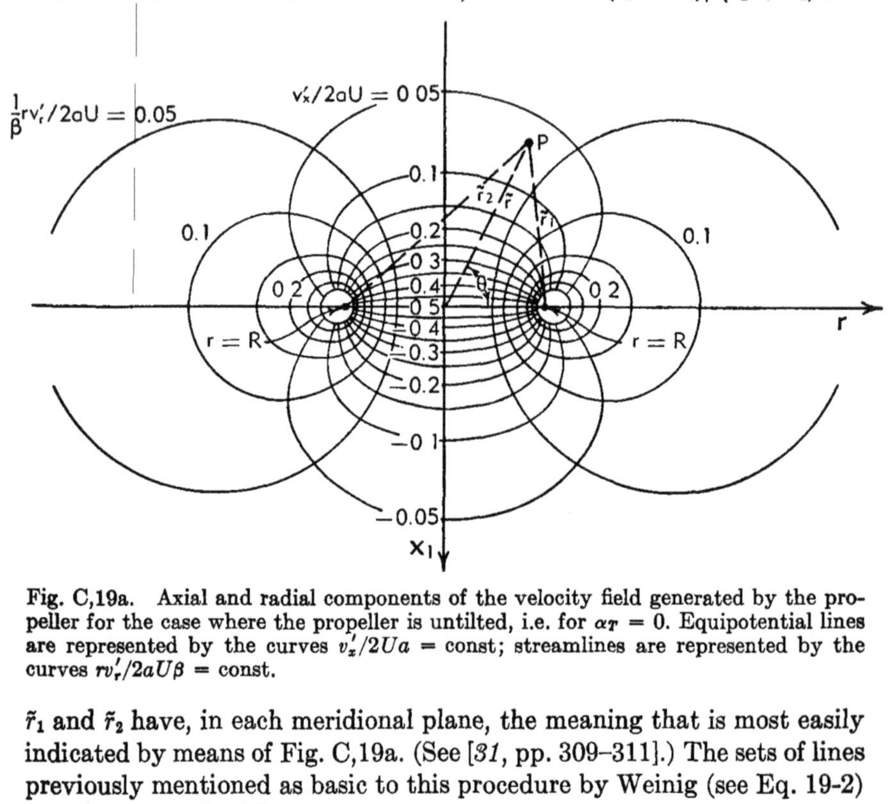

Fig. C,19a. Axial and radial components of the velocity field generated by the propeller for the case where the propeller is untilted, i.e. for $\alpha_T = 0$. Equipotential lines are represented by the curves $v'_x/2Ua$ = const; streamlines are represented by the curves $rv'_r/2aU\beta$ = const.

$\tilde{r}_1$ and $\tilde{r}_2$ have, in each meridional plane, the meaning that is most easily indicated by means of Fig. C,19a. (See [31, pp. 309–311].) The sets of lines previously mentioned as basic to this procedure by Weinig (see Eq. 19-2) are also drawn in this figure.

It now becomes necessary to add to the axial and radial components of the incremental velocity, just defined, the peripheral component of the increment in velocity that takes place within the slipstream. Under the hypothesis that the propeller is composed of N blades with a constant circulation distribution equal to Γ_p, the peripheral velocity increment is given by

$$v_\theta = \frac{N\Gamma_p}{2\pi r} = \frac{\gamma_p R}{\pi r}U$$

where the definition of γ_p is taken to be $\gamma_p = N\Gamma_p/2RU$, although if it is desired to express this nondimensionalized circulation in terms of the angular velocity of the propeller ω and the quantity aU, it takes on the representation $\gamma_p = 2a\beta^2 U\pi/\omega R$. Then the expression for the peripheral velocity increment is converted into the convenient form

$$v_\theta = 2a\beta^2 U \frac{R}{r} \tan \Phi \tag{19-4}$$

where $\tan \Phi = U/\omega R$.

For points lying aft of the propeller disk, as long as they are sufficiently far away from it, the formulas due to Koning [*2*, Vol. 4, pp. 364–373] will stand in good stead here for determination of the total increment in velocity produced by the propeller, instead of relying on formulas like Eq. 19-2. Based on the system of coordinates shown in Fig. C,19b, Koning's expressions for the velocity increments may be written as

$$\left.\begin{aligned} v_x &= aU\left(1 - 2M_\infty^2 + \frac{x_1}{\sqrt{R^2 + x_1^2}}\right) \\ \frac{1}{\beta} v_y &= +\frac{1}{2} aU \frac{R^2 y}{(R^2 + x_1^2)^{\frac{3}{2}}} - 2aU \frac{Rz}{y^2 + z^2} \tan \Phi \\ \frac{1}{\beta} v_z &= +\frac{1}{2} aU \frac{R^2 z}{(R^2 + x_1^2)^{\frac{3}{2}}} + 2aU \frac{Ry}{y^2 + z^2} \tan \Phi \end{aligned}\right\} \tag{19-5}$$

for

$$y^2 + z^2 \leqq R^2$$

and

$$\left.\begin{aligned} v_x &= -\frac{1}{2} aU \frac{R^2 x_1}{(x_1^2 + y^2 + z^2)^{\frac{3}{2}}} \\ \frac{1}{\beta} v_y &= \frac{1}{2} aU \frac{R^2 y}{(x_1^2 + y^2 + z^2)^{\frac{3}{2}}} \\ \frac{1}{\beta} v_z &= \frac{1}{2} aU \frac{R^2 z}{(x_1^2 + y^2 + z^2)^{\frac{3}{2}}} \end{aligned}\right\} \tag{19-5$'$}$$

for $y^2 + z^2 \geqq R^2$ where, as usual, $x_1 = x/\beta$.

The first of the expressions given as Eq. 19-5 is identical with the equation $v_x = v_x' + 2a\beta^2 U$, provided v_x' is taken to be given by Eq. 19-3 at all points along the x axis. Thus, within the limits of the approximation for which this relationship holds true, the boundary of the slipstream is seen to be a body of revolution whose radius R_1 at every point x is determined through the equation of continuity to be

$$\pi R_1^2 \rho_1 (U + v_x) = \pi R^2 U[1 + a(2\beta^2 - 1)]\rho_p$$

where ρ_p is the density of the fluid immediately behind the propeller.

If Eq. 19-1 is utilized for the expression of Δp, it is found that

$$\frac{\rho_1}{\rho_p} = 1 + M_\infty^2 \left[a(1 - 2M_\infty^2) - \frac{v_x}{U} \right]$$

and thus it follows that

$$\frac{R_1^2}{R^2} = \frac{1 + a - 2aM_\infty^2}{1 + M_\infty^2 \left[a(1 - 2M_\infty^2) - \frac{v_x}{U} \right]} \frac{1}{1 + \frac{v_x}{U}} \tag{19-6}$$

The formulas written above account correctly for the effect of compressibility, provided that one negelects terms of the same order of magnitude as a^2.

Case of a Propeller Operating at an Angle of Attack That Is Not Zero. If the angle of attack α_T which the axis of the propeller

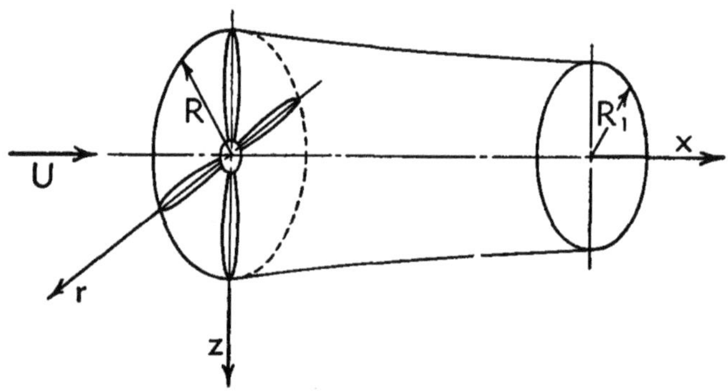

Fig. C,19b. Slipstream contraction, such as described analytically in formulas suggested by Koning [2, Vol. 4, pp. 364–373].

makes with the free stream velocity vector **U** is not zero, but yet still small enough to permit the approximations that $\cos \alpha_T \cong 1$ and $\sin \alpha_T \cong \alpha_T$, the determination of the velocity increments produced by the propeller operating alone can be obtained at a large distance from the propeller in a relatively easy manner. In fact under these stipulations the axial component of the increment in velocity is still given by the formula

$$\left.\begin{aligned} v_x &= 2a\beta^2 U && \text{for points lying within the slipstream} \\ v_x &= 0 && \text{for points lying outside the slipstream} \end{aligned}\right\} \tag{19-7}$$

Meanwhile the factor a is still related to the thrust through the following relationship

$$T = 4C_T R^2 \rho_\infty U^2 = \pi R^2 \rho_\infty U^2 (1 + a) 2a\beta^2$$

so that the value of a is given by[4]

$$a = -\frac{1}{2} + \sqrt{\frac{1}{4} + \frac{2C_T}{\pi\beta^2}} \qquad (19\text{-}7')$$

Now in addition to the increment in the axial component, there is also an increment created in the direction normal to the free stream velocity vector **U**, whose average value is denoted by $\bar{v}_{z\infty}$, where **z** is in a direction normal to **U**. This increment is also related to the thrust vector **T**, and its analytic expression is given by the following formula [*32*]

$$\pi R^2 \rho_\infty U(2 + a)\bar{v}_{z\infty} = Z + \alpha_T T = (C_z + \alpha_T C_T) \cdot 4R^2 \rho_\infty U^2$$

where Z is the force in the direction of **z**. From this it follows that

$$\frac{\bar{v}_{z\infty}}{U} = \frac{4}{(2 + a)\pi}(C_z + \alpha_T C_T) \qquad (19\text{-}8)$$

Outside of the slipstream, the field of flow produced by action of the velocity increments in the directions of the y and z axes are identical with those produced by a doublet of strength $2\pi R^2 \bar{v}_{z\infty}$ that would be located at the center of the circular cross section of the slipstream, having its axis orientated in the direction of the z axis. It follows therefore that

$$\left.\begin{aligned} v_{y\infty} &= \bar{v}_{z\infty} \frac{R^2 yz}{(y^2 + z^2)^2} \\ \text{and} \qquad v_{z\infty} &= -\bar{v}_{z\infty} \frac{R^2(y^2 - z^2)}{(y^2 + z^2)^2} \end{aligned}\right\} \qquad (19\text{-}9)$$

external to the slipstream, where $y^2 + z^2 > R^2$.

In the case where the angle of attack α_T is small, and provided that one keeps far enough away from the condition at which Φ takes on the value that renders C_T zero, then C_z is small enough in comparison with the product $\alpha_T C_T$ so that it is quite permissible to rewrite Eq. 19-8 in the form

$$\frac{\bar{v}_{z\infty}}{U} = \frac{4\alpha_T}{(2 + a)\pi} C_T = \frac{1 + a}{2 + a} 2a\alpha_T \beta^2 \qquad (19\text{-}8')$$

and, furthermore, for small values of a this can be further reduced to just

$$\frac{\bar{u}_{z\infty}}{U} = a\alpha_T \beta^2 \qquad (19\text{-}8'')$$

[4] Eq. 19-7′, 19-8, etc. are correct for $\beta = 1$; for $\beta < 1$ these formulas are correct if one neglects the terms of the same order of magnitude as a^2.

If the x coordinate of the point P, at which it is required to determine the velocity increments created by the propeller, is not sufficiently large to permit the assumption that the increments under consideration can be taken to be the same as those which occur infinitely far downstream, then the computation of these increments can be carried out by means of the analysis presented in [*32*]. Such a procedure is somewhat lengthy, however, and consequently the following method of attack is now proposed in order to deal with any point P lying in the x, y plane.

The incremental velocity components in the axial and peripheral directions are still to be obtained by the use of Eq. 19-5 and 19-4 while, in order to carry out the calculations for the v_z component, which is due to the angle of attack of the thrust axis, another useful idealization of the situation is assumed. It is premised, for instance, that the real vortex

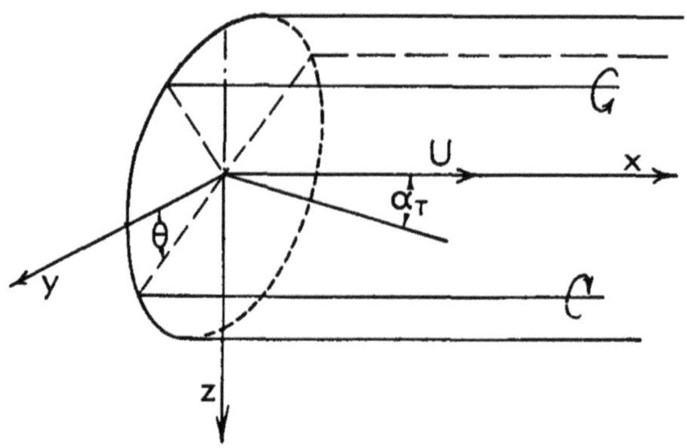

Fig. C,19c. Vortex system equivalent to the propeller, utilized in the calculation of the normal component of the increment in velocity produced when the propeller is at an angle of attack.

system associated with the propeller may be reduced simply to that of a radial system of bound vortices, lying in the plane of the propeller disk, together with a complex of shed trailing vortices with axes aligned parallel with the free stream velocity vector (see Fig. C,19c).

Now let $\gamma_p(\theta)Rd\theta$ denote the strength of the bound vortices in any sector of the circular system of vortices, where $d\theta$ represents the angular amplitude of the sector in question. Then one may decompose γ_p into two parts as follows

$$\gamma_p = \gamma_p^{(0)} + \gamma_p^{(1)} \cos \theta \tag{19-10}$$

where $\gamma_p^{(0)}$ and $\gamma_p^{(1)}$ are constants. It follows that Eq. 19-10 thus gives the pattern of variation of the strength of the trailing vortex system over the cylindrical surface that constitutes the boundary of the slipstream.

It may be deduced that

$$\gamma_p^{(1)} = 2\bar{v}_{z\infty} \tag{19-11}$$

while

$$v_z = \frac{\bar{v}_{z\infty}}{2}(1 + A) \qquad \text{for } y \leq R$$

and

$$v_z = -\frac{\bar{v}_{z\infty}}{2}\frac{R^2}{y^2}(1 - B) \qquad \text{for } y \geq R \tag{19-12}$$

where the quantities A and B may be evaluated from the following expressions:

$$\begin{aligned} A = \frac{2x_1}{\pi}\Bigg\{&\left(\frac{y^2 - 2yR - 5R^2}{4y^2m} + \frac{m'}{y}\right)\mathsf{K}(\alpha) - \frac{Rm'}{y}\mathsf{E}(\alpha) \\ &+ \frac{x_1^2 + y^2}{y^2}\frac{R}{\sqrt{x_1^2 + (y + R)^2}} \\ &\times\left[\left(\frac{R}{\sqrt{x_1^2 + y^2}} - 1\right)\frac{1}{\sqrt{x_1^2 + y^2} + y}\Pi(n, k)\right. \\ &\left.+ \left(1 + \frac{R}{\sqrt{x_1^2 + y^2}}\right)\frac{1}{\sqrt{x_1^2 + y^2} - y}\Pi(n', k)\right] \\ &- \frac{y - R}{y + R}\frac{y^2 + R^2}{2y^2R}\frac{R}{\sqrt{x_1^2 + (y + R)^2}}\Pi(n'', k)\Bigg\} \end{aligned}$$

and

$$\begin{aligned} B = \frac{2x_1}{\pi}\Bigg\{&\left(\frac{y^2 - 2yR - 5R^2}{4R^2m} + \frac{ym'}{R}\right)\mathsf{K}(\alpha) - \frac{ym'}{R}\mathsf{E}(\alpha) \\ &+ \frac{x_1^2 + y^2}{R^2}\frac{R}{\sqrt{x_1^2 + (y + R)^2}} \\ &\times\left[\left(\frac{R}{\sqrt{x_1^2 + y^2}} - 1\right)\frac{1}{\sqrt{x_1^2 + y^2} + y}\Pi(n, k)\right. \\ &\left.+ \left(1 + \frac{R}{\sqrt{x_1^2 + y^2}}\right)\frac{1}{\sqrt{x_1^2 + y^2} - y}\Pi(n', k)\right] \\ &- \frac{y - R}{y + R}\frac{y^2 + R^2}{2R^3}\frac{R}{\sqrt{x_1^2 + (y + R)^2}}\Pi(n'', k)\Bigg\} \end{aligned} \tag{19-12'}$$

where K, E, and Π denote the complete elliptic integrals of the first, second, and third kinds, respectively, while in addition the following definitions have also been inserted for the sake of simplicity:

$$m = \frac{1}{2}\sqrt{x_1^2 + (y + R)^2} \quad \text{and} \quad m' = \frac{m}{yR}$$

$$\sin^2 \alpha = k^2 = \frac{4yR}{x_1^2 + (y+R)^2}$$

$$n = -\frac{2y}{y + \sqrt{x_1^2 + y^2}}, \quad n' = \frac{2y}{\sqrt{x_1^2 + y^2} - y}, \quad \text{and} \quad n'' = \frac{-4yR}{(y+R)^2}$$

It is worthy of note that if $y = 0$, then the first of the expressions given as Eq. 19-12 reduces to the simple form

$$v_z = \frac{\bar{v}_{z\infty}}{2}\left(1 + \frac{\sqrt{R^2 + x_1^2}}{x_1}\right) \tag{19-12''}$$

C,20. Discussion of the Boundary Conditions. First of all, let the symbols $\phi_{\text{I}}U$ and $\phi_{\text{II}}U$ stand for the two potential functions that

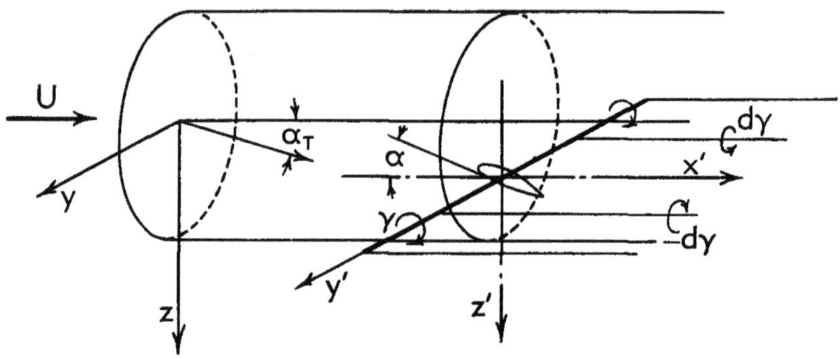

Fig. C,20a. Coordinate system and shifted location of the propeller-wing system relative to the vortex system depicted in Fig. C,19c.

describe the fields of flow, undisturbed by the presence of any wing, which exist in the region lying outside of the propeller's slipstream in the first case, and the region invested by the slipstream itself in the second case. In addition let $\varphi_{\text{I}}U$ and $\varphi_{\text{II}}U$ denote the two potential functions which represent the small velocity increments produced by the wing throughout the two aforementioned regions, respectively, into which all free space is now considered to be divided. Then, on the basis of the second hypothesis made in Art. 18, the potentials which describe the total resultant flow may be expressed as the sums of the foregoing separate constituent parts:

$$\phi_{\text{I}}^* = (\phi_{\text{I}} + \varphi_{\text{I}})U \quad \text{and} \quad \phi_{\text{II}}^* = (\phi_{\text{II}} + \varphi_{\text{II}})U \tag{20-1}$$

Furthermore let a new coordinate system be set up in which the y' axis coincides with the bound vortex (see Fig. C,20a) and which is otherwise merely shifted parallel to the system of axes depicted in Fig. C,19c. Then, relying upon the usual assumption that a lifting line may be substituted for the lifting surface, the boundary conditions which must be satisfied by the potentials φ_{I} and φ_{II}, and which thus serve to determine these potentials uniquely, are as follows:

Upon the trailing vortex sheet Σ_t. The potentials φ_{I} and φ_{II} are discontinuous at the level of the vortex sheet, and, to be precise, the mathematical expression for the jump is given by

$$\left.\begin{aligned}(\varphi_{\text{I}})_{z'=0^-} - (\varphi_{\text{I}})_{z'=0^+} &= \frac{\Gamma_{\text{I}}}{U}\\ \text{and}\qquad (\varphi_{\text{II}})_{z'=0^-} - (\varphi_{\text{II}})_{z'=0^+} &= \frac{\Gamma_{\text{II}}}{U}\end{aligned}\right\} \tag{20-2}$$

where Γ_{I} is the value of the circulation around the lifting line in any arbitrary general vertical plane $y = \text{const}$, provided that it is located beyond the region occupied by the slipstream (and thus the plane is confined within the range $R_1 < |y| < b/2$), while Γ_{II} denotes the analogous value for the circulation in the case where $|y| \leq R_1$. The circulation Γ at any point on the bound vortex may be expressed on the basis of Eq. 18-1, by the relation

$$\Gamma = \frac{1}{2} c_{l_\alpha} c U \left(1 + \frac{v_x}{U}\right) \alpha_p \tag{20-3}$$

where v_x/U (which is to be calculated at points which lie on the piece of axis constituting the lifting line) may be obtained by employing the formulas presented in Art. 19. In addition, the value of α_p is given by the relationship

$$\alpha_p = \alpha - \frac{1}{1 + v_x/U}\left[\left(\frac{\partial \phi}{\partial z'}\right)_{z'=x'=0} + \alpha_i\right] \tag{20-4}$$

where α_i is the angle of attack induced by action of the wing's trailing vortices at the location corresponding to the straight-line axis constituting the lifting line; thus its value is given by

$$\alpha_i = \frac{1}{2}\left(\frac{\partial \varphi}{\partial z'}\right)_{\substack{x'=\infty\\ z'=0}} \tag{20-4$'$}$$

Consequently Eq. 20-3 may be rewritten in the compact form

$$\Gamma = \frac{1}{2} c_{l_\alpha} c U \left[\left(1 + \frac{v_x}{U}\right)\alpha - \left(\frac{\partial \phi}{\partial z'}\right)_{z'=x'=0} - \alpha_i\right] \tag{20-3$'$}$$

wherein one must affix subscripts I or II to the symbols Γ, ϕ, and φ according to whether the part of the flow which lies outside of the propeller's slipstream, or within it, is under consideration.

Upon the vortical surface constituting the boundary of the slipstream. Because the boundary is a fluid surface, the pressure and density must be continuous in passing through it. At locations corresponding to the vortical (slipstream) surface Σ_p, therefore, it must be true that

$$p_{\mathrm{I}} = p_{\mathrm{II}} \quad \text{and} \quad \rho_{\mathrm{I}} = \rho_{\mathrm{II}} \tag{20-5}$$

where the indicated subscripts have the usual significance.

Since the boundary surface Σ_p constitutes a surface composed of a complex of individual streamtube surfaces, and since it is also a surface of revolution (Art. 19), then the component of velocity normal to the vortical

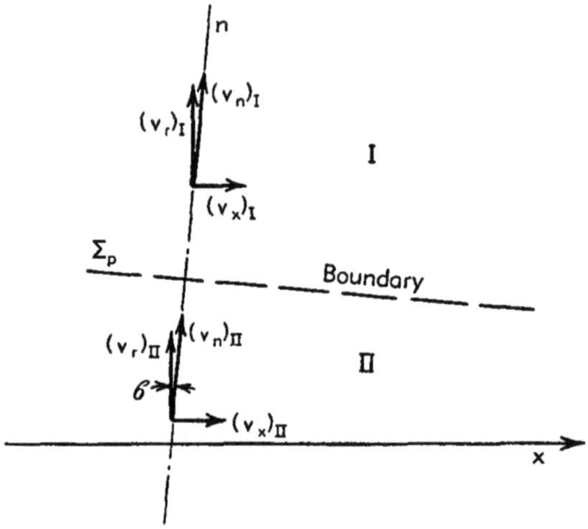

Fig. C,20b. Application of the condition of continuity at the boundary of the slipstream.

slipstream surface has to vanish. Now the normal components in the two regions on either side of this surface are given by

$$(v_n)_{\mathrm{I}} = (v_r)_{\mathrm{I}} \cos \delta + (v_x)_{\mathrm{I}} \sin \delta$$

and

$$(v_n)_{\mathrm{II}} = (v_r)_{\mathrm{II}} \cos \delta + (v_x)_{\mathrm{II}} \sin \delta$$

where δ is the angle formed between the normal to the vortical slipstream surface Σ_p and the plane passed perpendicularly to the x axis (see Fig. C,20b), and where the symbol v stands for the total value making up the complete contribution of the constituent parts of the velocity in each of the regions I or II. It follows therefore that the normality condition to be satisfied is

$$\left(\frac{v_r}{v_x}\right)_{\mathrm{I}} = \left(\frac{v_r}{v_x}\right)_{\mathrm{II}} \tag{20-6}$$

Conditions (20-5) and (20-6) may be cast into a simpler form if the wing is assumed (according to the hypothesis already explicitly stated in Art. 18) to lie at a sufficient distance from the propeller in order to ensure the validity of Eq. 19-7, 19-8, and 19-9 for predicting the values of the velocity increments created by such a propeller. In fact, under the stipulations premised here, one may write that

$$(v_x)_{\rm I} = U\left(1 + \frac{\partial \varphi_{\rm I}}{\partial x}\right) \quad \text{and} \quad (v_x)_{\rm II} = U\left(1 + 2a\beta^2 + \frac{\partial \varphi_{\rm II}}{\partial x}\right)$$

On the other hand, since Bernoulli's equation holds within each of the regions (I and II) under examination, in consequence of the fact that the flow is irrotational throughout these two regions, it may be deduced that the velocities and pressures are related in the following way, provided that the squared terms involving the perturbations produced by the wing are considered to be negligible:

$$\frac{\gamma}{\gamma - 1}\frac{p_{\rm I}}{\rho_{\rm I}} + \frac{1}{2}U^2\left(1 + 2\frac{\partial \varphi_{\rm I}}{\partial x}\right) = \frac{\gamma}{\gamma - 1}\frac{p_\infty}{\rho_\infty} + \frac{1}{2}U^2$$

and

$$\frac{\gamma}{\gamma - 1}\frac{p_{\rm II}}{\rho_{\rm II}} + \frac{1}{2}U^2(1 + 2a\beta^2)^2\left(1 + 2\frac{\partial \varphi_{\rm II}}{\partial x}\frac{1}{1 + 2a\beta^2}\right) = \frac{\gamma}{\gamma - 1}\frac{p_\infty}{\rho_\infty} + \frac{1}{2}U^2(1 + 2a\beta^2)^2$$

because at an infinite distance it is true that

$$p_{\rm I} = p_{\rm II} = p_\infty$$

and

$$\rho_{\rm I} = \rho_{\rm II} = \rho_\infty$$

Taking into account the continuity condition (20-5), therefore, one obtains that

$$\frac{\partial \varphi_{\rm I}}{\partial x} = S\frac{\partial \varphi_{\rm II}}{\partial x} \tag{20-7}$$

where S is defined by the equation

$$S = 1 + 2a\beta^2 \tag{20-7$'$}$$

It follows from the above-written connection between the two derivatives that

$$\varphi_{\rm I} = S\varphi_{\rm II} \qquad \text{on the surface of the slipstream } \Sigma_p \tag{20-8}$$

which results from integrating Eq. 20-7 with respect to x, and discarding the unessential constant of integration.

In an analogous manner it is found from the use of Eq. 20-6 that

$$S\frac{\partial \varphi_{\rm I}}{\partial r} = \frac{\partial \varphi_{\rm II}}{\partial r} \qquad \text{on the surface of the slipstream } \Sigma_p \tag{20-8$'$}$$

These conditions were first stated in this form by Koning [*2*, Vol. 4, pp. 364–373] when dealing with the same situation in the case of incompressible flow.

C,21. General Equations for Determining the Distribution of Circulation out along the Wing Span. In order to discover how the circulation is distributed along the wing span it is no longer necessary, as it was earlier for the solution of the wing-fuselage interference problem (see Chap. 2), to determine the potentials φ_{I} and φ_{II} explicitly. Instead, thanks to Eq. 20-5 and 20-3′, it suffices to compute merely

$$\varphi_\infty = \lim_{x' \to \infty} \varphi$$

in such wise as to ensure that Eq. 20-8 and 20-8′ are satisfied.

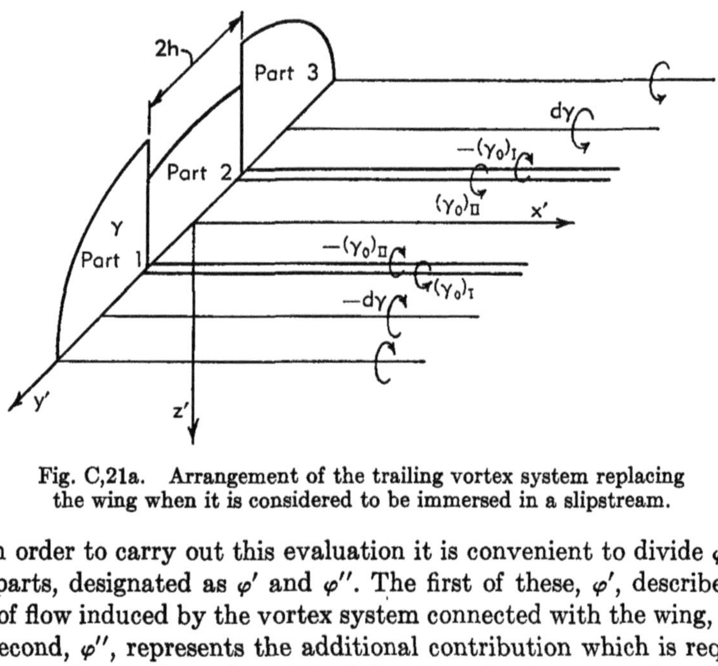

Fig. C,21a. Arrangement of the trailing vortex system replacing the wing when it is considered to be immersed in a slipstream.

In order to carry out this evaluation it is convenient to divide φ into two parts, designated as φ' and φ''. The first of these, φ', describes the field of flow induced by the vortex system connected with the wing, while the second, φ'', represents the additional contribution which is required to be superimposed on the original flow in order to ensure that the resultant flow pattern obeys the boundary conditions that must be satisfied on the slipstream surface Σ_p.

Now define a nondimensional form of the circulation as

$$\gamma(y) = \frac{\Gamma(y)}{bU}$$

and let the distribution of γ out along the wing span be divided up into the three parts indicated in Fig. C,21a. Then the potential φ' may be obtained as the sum of the three potentials which correspond to the three circulation distributions just mentioned, regardless of whether the external region, I, or the region internal to the slipstream, II, is under consideration.

As a result of proceeding in this way the potential

$$\varphi'_\infty = \lim_{x' \to \infty} \varphi'$$

is seen to be the potential representing the infinitely long straight-line vortices composing the trailing vortex sheet (see Fig. C,21a).

Furthermore let $d\Gamma$ denote the circulation of the vortex located at a point P, which may lie externally or internally to the cross section H,

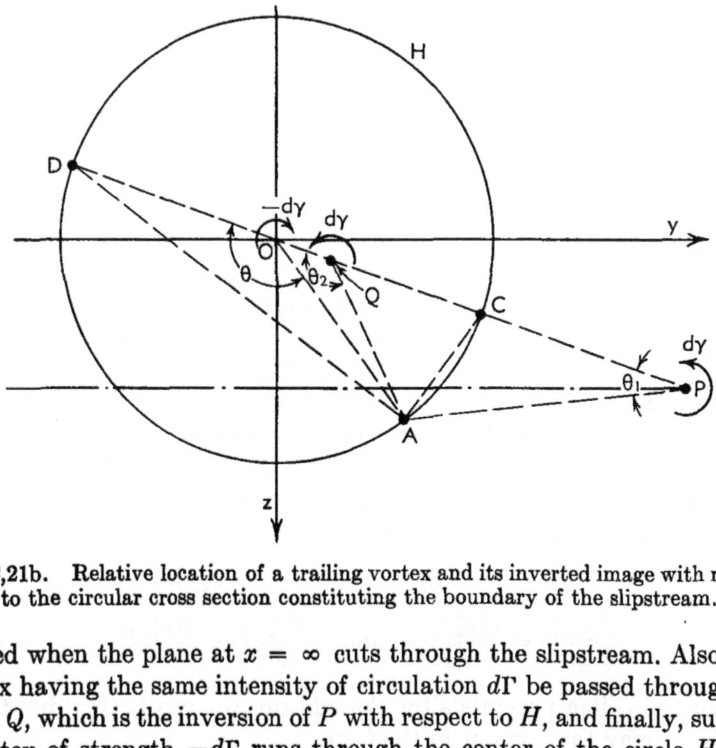

Fig. C,21b. Relative location of a trailing vortex and its inverted image with respect to the circular cross section constituting the boundary of the slipstream.

formed when the plane at $x = \infty$ cuts through the slipstream. Also let a vortex having the same intensity of circulation $d\Gamma$ be passed through the point Q, which is the inversion of P with respect to H, and finally, suppose a vortex of strength $-d\Gamma$ runs through the center of the circle H. The potential function describing the three vortices has the value

$$\frac{d\Gamma}{2\pi} (\theta_1 + \theta_2 - \theta)$$

at any arbitrary general point A. The angles θ_1, θ_2, and θ appearing in this formula are the angles depicted in Fig. C,21b.

Because the points P and Q separate the points C and D harmonically, where C and D are the points of intersection of the line PQ with the circle H, it follows that the segments AC and AD are the internal and external bisectors of $\angle PAQ$. In addition since $\theta = 2 \cdot \angle ACD$, it may be deduced that

$$\theta_1 + \theta_2 - \theta = 0$$

Consequently, for points belonging to H, it is true that

$$d\varphi'_\infty + d(\varphi'_\infty)_r = 0$$

if the notation $d(\varphi'_\infty)_r$ is used to denote the potential corresponding to the vortices running through Q and through the center O, which are related, as explained above, to the one passing through P.

It is further recognized that the vortex of strength $d\Gamma$ located at P and the one of strength $d\Gamma$ located at the inverted point, Q, generate a flow field of which H is a streamline, and thus it is true that

$$\frac{\partial d\varphi'_\infty}{\partial r} = \frac{\partial d(\varphi'_\infty)_r}{\partial r}$$

on the circumference H, since H is also a streamline corresponding to the flow generated by the vortex located at O.

Now let the potential function which corresponds to the vortices lying external to the slipstream (i.e. in region I) be denoted by $(\varphi'_\infty)_{\text{I}}$ and let the analogous potential representing the vortices lying internal to the slipstream (i.e. in region II) be denoted by $(\varphi'_\infty)_{\text{II}}$. In like manner, let the potential functions corresponding to the vortices which have been inverted with respect to H be denoted by $(\varphi'_\infty)_{\text{II}r}$ and $(\varphi'_\infty)_{\text{I}r}$, such that the subscripts II and I employed in this case also serve to indicate in which region (in or out of the slipstream) they are located.

On the basis of what has already been derived, therefore, it may be seen that the following relationships hold true on the circumference of a cross section through the slipstream:

$$(\varphi'_\infty)_{\text{I}} = -(\varphi'_\infty)_{\text{II}r}, \quad (\varphi'_\infty)_{\text{II}} = -(\varphi'_\infty)_{\text{I}r},$$

$$\frac{\partial(\varphi'_\infty)_{\text{I}}}{\partial r} = \frac{\partial(\varphi'_\infty)_{\text{II}r}}{\partial r}, \quad \text{and} \quad \frac{\partial(\varphi'_\infty)_{\text{II}}}{\partial r} = \frac{\partial(\varphi'_\infty)_{\text{I}r}}{\partial r}$$

The corresponding potentials for the additional contribution due to the slipstream may thus be put in the form

$$(\varphi''_\infty)_{\text{I}} = n_1(\varphi'_\infty)_{\text{II}} + n_2(\varphi'_\infty)_{\text{II}r}$$

and

$$(\varphi''_\infty)_{\text{II}} = n_3(\varphi'_\infty)_{\text{I}} + n_4(\varphi'_\infty)_{\text{I}r}$$

Since $(\varphi'_\infty)_{\text{II}}$ and $(\varphi'_\infty)_{\text{II}r}$ describe vortices which lie external to the slipstream, while $(\varphi'_\infty)_{\text{I}}$ and $(\varphi'_\infty)_{\text{I}r}$ correspond to vortices lying within the slipstream it appears that the boundary conditions given as Eq. 20-2, which the resultant potentials $(\varphi_\infty)_{\text{I}}$ and $(\varphi_\infty)_{\text{II}}$ must obey, are automatically satisfied by this procedure.

If the further condition is now imposed that the requirements of Eq. 20-8 and 20-8′ are also to be met, it follows that these boundary conditions can be written as

$$\left.\begin{aligned}(\varphi_\infty)_{\mathrm{I}} &= (\varphi'_\infty)_{\mathrm{I}} + (\varphi''_\infty)_{\mathrm{I}} = \varphi'_\infty - \frac{(S-1)^2}{1+S^2}(\varphi'_\infty)_{\mathrm{II}} - \frac{S^2-1}{1+S^2}(\varphi'_\infty)_{\mathrm{II}r}\\ &\text{and}\\ (\varphi_\infty)_{\mathrm{II}} &= (\varphi'_\infty)_{\mathrm{II}} + (\varphi''_\infty)_{\mathrm{II}} = \varphi'_\infty - \frac{(S-1)^2}{1+S^2}(\varphi'_\infty)_{\mathrm{I}} + \frac{S^2-1}{1+S^2}(\varphi'_\infty)_{\mathrm{I}r}\end{aligned}\right\} \tag{21-1}$$

and if it is assumed that a is small enough to make terms in a^2 negligible in their consequences, the above-stated conditions reduce to:

$$(\varphi_\infty)_{\mathrm{I}} = \varphi'_\infty - 2a\beta^2(\varphi'_\infty)_{\mathrm{II}r} \quad \text{and} \quad (\varphi_\infty)_{\mathrm{II}} = \varphi'_\infty + 2a\beta^2(\varphi'_\infty)_{\mathrm{I}r} \tag{21-1′}$$

Furthermore, from the definition given as Eq. 20-4′, it follows that

$$\left.\begin{aligned}(\alpha_i)_{\mathrm{I}} &= \frac{1}{2}\left[\frac{\partial\varphi'_\infty}{\partial z'} - \frac{(S-1)^2}{1+S^2}\frac{\partial(\varphi'_\infty)_{\mathrm{II}}}{\partial z'} - \frac{S^2-1}{1+S^2}\frac{\partial(\varphi'_\infty)_{\mathrm{II}r}}{\partial z'}\right]_{z'=0}\\ &\text{and}\\ (\alpha_i)_{\mathrm{II}} &= \frac{1}{2}\left[\frac{\partial\varphi'_\infty}{\partial z'} - \frac{(S-1)^2}{1+S^2}\frac{\partial(\varphi'_\infty)_{\mathrm{I}}}{\partial z'} + \frac{S^2-1}{1+S^2}\frac{\partial(\varphi'_\infty)_{\mathrm{I}r}}{\partial z'}\right]_{z'=0}\end{aligned}\right\} \tag{21-2}$$

and when a is assumed small enough to make the terms in a^2 negligible as before, then these expressions for the induced angles reduce to

$$\left.\begin{aligned}(\alpha_i)_{\mathrm{I}} &= \frac{1}{2}\left[\frac{\partial\varphi'_\infty}{\partial z'} - 2a\beta^2\frac{\partial(\varphi'_\infty)_{\mathrm{II}r}}{\partial z'}\right]\\ &\text{and}\\ (\alpha_i)_{\mathrm{II}} &= \frac{1}{2}\left[\frac{\partial\varphi'_\infty}{\partial z'} + 2a\beta^2\frac{\partial(\varphi'_\infty)_{\mathrm{I}r}}{\partial z'}\right]\end{aligned}\right\} \tag{21-2′}$$

C,22. Circulation Distribution for a Generic Wing Traversing the Slipstream at a Very Large Distance Aft of the Propeller. Considered now is the case in which the wing, which may be of any shape whatsoever within the limitations of the hypotheses stated in Art. 18, is assumed to traverse the slipstream at a very large distance aft of the propeller. The velocities induced by the slipstream upon the lifting line are still given by means of Eq. 19-7 and 19-9, and the radius of the circular section taken through the slipstream is obtained from Eq. 19-6.

Inasmuch as Eq. 19-7 and 19-9 satisfy the boundary conditions pertaining to the dividing wall, Σ_p, which delimits the extent of the slipstream, and since it is also likely that the factor $2a$ can take on rather sizeable values, it is fitting to undertake the solution of this problem by having recourse to the expressions for φ_∞ taken in the form of Eq. 21-1, rather than in the simplified form represented by Eq. 21-1′.

In order to reduce the complexity of the equations to be derived, however, the restriction is stipulated now that only the case of the lifting line passing diametrically through the center of the slipstream is investigated at present. The peripheral components of the velocity increments, defined according to Eq. 19-3, are not determined now. (For more information on these subjects see [33].) To avoid complicating the development to be followed immediately below, these questions are discussed in Art. 23 where the approximate solution based on Eq. 21-1′ is taken up in some detail.

By utilizing the definition (21-1), it is readily apparent that Eq. 20-3′ may be written in the form

$$\frac{2b\gamma}{c_{l_\alpha}c} = \alpha\left(1 + \frac{v_x}{U}\right) - \varepsilon_p - \alpha_i \tag{22-1}$$

wherein v_x/U assumes the values

$$\left.\begin{aligned} \frac{v_x}{U} &= 2a\beta^2 \qquad \text{for } |y| \leqq R_1 \\ \frac{v_x}{U} &= 0 \qquad \text{for } |y| \geqq R_1 \end{aligned}\right\} \tag{22-2}$$

The symbol ε_p employed here is defined for $|y| \leqq R_1$ by means of Eq. 19-8, or for $|y| \geqq R_1$ by

$$\varepsilon_p = (\varepsilon_p)_{|y| \geqq R_1} = \frac{-4}{\pi(2 + a)}\,(C_x + \alpha_T C_T)\,\frac{h^2}{\eta^2} \tag{22-3}$$

where

$$h = \frac{2R_1}{b} \quad \text{and} \quad \eta = \frac{2y}{b} \tag{22-4}$$

Of course, the value of α_i appearing in the above Eq. 22-1 is given by means of Eq. 21-2.

Because the pressure cannot suffer any discontinuity over the surface of the wing, the lift is likewise continuous, and thus γ must be discontinuous at $|\eta| = h$. Furthermore the jump in γ is determinable from Eq. 20-8, if one makes use of the relationships given as Eq. 20-2; that is, the change in γ is

$$\Delta\gamma = (\gamma_{\mathrm{I}} - \gamma_{\mathrm{II}})_{|\eta|=h} = \frac{S - 1}{S}\,(\gamma_{\mathrm{I}})_{|\eta|=h} \tag{22-5}$$

It may be seen from Eq. 20-8′ that $d\gamma/d\eta$ is also discontinuous at $|\eta| = h$, and the mathematical expression for the jump may thus be written as

$$\Delta\left(\frac{d\gamma}{d\eta}\right) = \left(\frac{d\gamma_{\mathrm{I}}}{d\eta} - \frac{d\gamma_{\mathrm{II}}}{d\eta}\right)_{|\eta|=h} = -(S - 1)\left(\frac{d\gamma_{\mathrm{I}}}{d\eta}\right)_{|\eta|=h} \tag{22-5′}$$

One may easily recognize that Eq. 22-5 and 22-5′ are the necessary and sufficient conditions which must be fulfilled in order that α_i should remain finite throughout the interval $-1 \leqq \eta \leqq 1$.

It is instructive to decompose γ into the sum of two terms, i.e.

$$\gamma = \gamma_d + \gamma_c \tag{22-6}$$

where γ_d is to incorporate the discontinuities $\Delta\gamma$ and $\Delta(d\gamma/d\eta)$, as defined by means of Eq. 22-5 and 22-5′, and where γ_c is the part of the circulation distribution which is continuous, together with its first derivative, throughout the whole interval $-1 \leqq \eta \leqq 1$.

Then one may write explicitly

$$\left.\begin{aligned} \gamma_d &= 0 && \text{for } |\eta| > h \\ \gamma_d &= -a_1 - a_2\eta^2 && \text{for } |\eta| < h \end{aligned}\right\} \tag{22-7}$$

while

$$\gamma_c = \frac{2}{m+1}\sum_{n=1}^{m}\gamma_n\sum_{\mu=1}^{m}\sin\mu\theta_n\sin\mu\theta + \alpha_k\sum_{\mu=1}^{\infty}\beta_\mu\sin\mu\theta \tag{22-7′}$$

where $\cos\theta = \eta$, $\theta_n = n\pi/(m+1)$, and a_1, a_2, α_k, and the various γ_n's are constants that are a priori unknown. The β values employed in Eq. 22-7′ are obtained by means of the relationships

$$\left.\begin{aligned} \beta_1 &= -\frac{4}{\pi}(2\theta_0 - \pi - \sin 2\theta_0) \\ \beta_\mu &= \frac{8}{\pi}\frac{1}{\mu}\left[\frac{-\sin(\mu-1)\theta_0}{\mu-1} + \frac{\sin(\mu+1)\theta_0}{\mu+1}\right] \quad \text{for } \mu = 3, 5, 7, \text{etc.} \end{aligned}\right\} \tag{22-8}$$

where it has been assumed that

$$\cos\theta_0 = h$$

The first summation on the right-hand side of Eq. 22-7′ corresponds to a continuous distribution of the induced angle of attack along the wing span, while the second of the summations corresponds to the arrangement of the induced angles of attack that exhibits zero values for $|\eta| \geqq h$ and that is equal to $2\alpha_k$ for $|\eta| \leqq h$.

It should then be noted that

$$\sum_{\mu=1}^{\infty}\beta_\mu\sin\mu\theta = \frac{4\alpha_k}{\pi}\left[(\cos\theta + \cos\theta_0)\ln\frac{\cos\dfrac{\theta-\theta_0}{2}}{\left|\cos\dfrac{\theta+\theta_0}{2}\right|} + \pi\sin\theta - (\cos\theta - \cos\theta_0)\ln\frac{\sin\dfrac{\theta+\theta_0}{2}}{\sin\left|\dfrac{\theta-\theta_0}{2}\right|}\right] \tag{22-8′}$$

It is now required that the stipulation imposed by Eq. 22-1 be satisfied at the m points having the coordinates $\eta_n = \cos \theta_n$, for n running from 1 to m. Then the following formulas may be deduced:

$$\left. \begin{aligned} &\left(A_{\nu\nu} + \frac{2b}{c_{l_\alpha} c_\nu}\right) \gamma_\nu + \frac{2b}{c_{l_\alpha} c_\nu} \alpha_k \sum_{\mu=1}^{\infty} \beta_\mu \sin \mu\theta_\nu + \sum_{n=1}^{m}{}' A_{\nu n} \gamma_n + M_\nu a_2 + M'_\nu \\ &\qquad = \alpha_\nu - (\varepsilon_p)_{\text{I}} \quad \text{for } |\eta_\nu| > h \\ &\text{and} \\ &\left(B_{\nu\nu} + \frac{2b}{c_{l_\alpha} c_\nu}\right) \gamma_\nu + \frac{2b}{c_{l_\alpha} c_\nu} \alpha_k \sum_{\mu=1}^{\infty} \beta_\mu \sin \mu\theta_\nu + \sum_{n=1}^{m}{}' B_{\nu n} \gamma_n + N_\nu a_2 + N'_\nu \\ &\qquad - \frac{2b}{c_{l_\alpha} c_\nu} (a_1 + a_2 \eta_\nu^2) + 2\alpha_k = \alpha_\nu S - (\varepsilon_p)_{\text{II}} \quad \text{for } |\eta_\nu| < h \end{aligned} \right\} \tag{22-9}$$

where the various coefficients $A_{\nu\nu}$, $A_{\nu n}$, etc., are defined below in terms of the Multhopp coefficients, $b_{\nu\nu}$ and $b_{\nu n}$, and the factors

$$C = \frac{S^2 - 1}{1 + S^2} \qquad \text{and} \qquad D = \frac{(S - 1)^2}{1 + S^2}$$

The coefficients are:

$$\left. \begin{aligned} A_{\nu\nu} &= b_{\nu\nu} + \frac{1}{\pi(m+1)} \sum_{\mu=1}^{m} \mu \sin \mu\theta_\nu [D I'_{\mu\nu} + C(J_{\mu\nu} + J''_{\mu\nu})] \\ A_{\nu n} &= -b_{\nu n} + \frac{1}{\pi(m+1)} \sum_{\mu=1}^{m} \mu \sin \mu\theta_n [D I'_{\mu\nu} + C(J_{\mu\nu} + J''_{\mu\nu})] \\ B_{\nu\nu} &= b_{\nu\nu} + \frac{1}{\pi(m+1)} \sum_{\mu=1}^{m} \mu \sin \mu\theta_\nu [D(I_{\mu\nu} + I''_{\mu\nu}) - C J'_{\mu\nu}] \\ B_{\nu n} &= -b_{\nu n} + \frac{1}{\pi(m+1)} \sum_{\mu=1}^{m} \mu \sin \mu\theta_n [D(I_{\mu\nu} + I''_{\mu\nu}) - C J'_{\mu\nu}] \\ M_\nu &= \frac{2h}{\pi} (1 - D) + \frac{\eta_\nu}{\pi} (1 - D) \ln \left| \frac{\eta_\nu - h}{\eta_\nu + h} \right| \\ M'_\nu &= \frac{\alpha_k}{2\pi} \sum_{\mu=1}^{\infty} \mu \beta_\mu [D I'_{\mu\nu} + C(J_{\mu\nu} + J''_{\mu\nu})] \\ N_\nu &= \frac{2h}{\pi} + \frac{\eta_\nu}{\pi} \ln \left| \frac{\eta_\nu - h}{\eta_\nu + h} \right| - \frac{2Ch^3}{\pi \eta_\nu^2} - \frac{Ch^4}{\pi \eta_\nu^3} \ln \left| \frac{\eta_\nu - h}{\eta_\nu + h} \right| \\ N'_\nu &= \frac{\alpha_k}{2\pi} \sum_{\mu=1}^{\infty} \mu \beta_\mu [D(I_{\mu\nu} + I''_{\mu\nu}) - C J'_{\mu\nu}] \end{aligned} \right\} \tag{22-10}$$

provided the auxiliary definitions for $I_{\mu\nu}$ and $J_{\mu\nu}$ are taken to be

$$\left.\begin{aligned} I_{\mu\nu} &= \int_0^{\theta_0} \frac{\cos \mu\theta' d\theta'}{\cos \theta_\nu - \cos \theta'} \\ \text{and} \qquad & \\ J_{\mu\nu} &= \int_0^{\theta_0} \frac{\cos \theta' \cos \mu\theta'}{\cos \theta' \cos \theta_\nu - \cos^2 \theta_0} d\theta' \end{aligned}\right\} \tag{22-11}$$

and provided that the same symbols to which primes and double primes have been added are to stand for the same integrals when taken between the limits of θ_0 and $\pi - \theta_0$ in the first instance, and between $\pi - \theta_0$ and π in the second instance.

These integrals, Eq. 22-11, may be evaluated as follows: First let the notation be simplified by setting

$$I_\mu = \int_0^{\theta_0} \frac{\cos \mu\theta' d\theta'}{\cos \theta - \cos \theta'}$$

and

$$J_\mu = \int_0^{\theta_0} \frac{\cos \theta' \cos \mu\theta'}{\cos \theta' \cos \theta - \cos^2 \theta_0} d\theta'$$

Then it is noted that the value of the first of these integrals may be obtained from the relation

$$I_\mu = F_\mu + \frac{\cos \mu\theta}{\sin \theta} \ln \frac{\cos \theta_0 - \cos \theta}{1 - \cos (\theta_0 + \theta)} \tag{22-12}$$

where the following recurrence formula may be used to determine the $F_{\mu+1}$ values:

$$F_{\mu+1} + F_{\mu-1} - 2 \cos \theta F_\mu = -\frac{2}{\mu} \sin \mu\theta_0 \tag{22-12$'$}$$

in which $F_0 = 0$ and $F_1 = -\theta_0$.

Likewise it is noted that

$$2J_\mu = \frac{2}{\cos \theta} \frac{\sin \mu\theta_0}{\mu} + \frac{2 \cos^2 \theta_0}{\cos \theta} K_\mu \tag{22-13}$$

where

$$K_\mu = H_\mu + \frac{\cos \mu\psi}{\cos \theta \sin \psi} \ln \frac{\cos \theta[1 - \cos (\theta_0 + \psi)]}{\cos \theta_0(\cos \theta - \cos \theta_0)} \tag{22-14}$$

in which the value of ψ is obtained from

$$\cos \psi = \frac{\cos^2 \theta_0}{\cos \theta} \tag{22-14$'$}$$

and the first term on the right-hand side of Eq. 22-14 may be evaluated by means of the following recurrence formula:

$$H_{\mu+1} + H_{\mu-1} - \frac{2\cos^2\theta_0}{\cos\theta} H_\mu = \frac{2}{\cos\theta}\frac{\sin\mu\theta_0}{\mu} \qquad (22\text{-}14'')$$

in which $H_0 = 0$ and $H_1 = \theta_0/\cos\theta$.

Thus in like manner it is found that

$$I'_\mu = F'_\mu - \frac{\cos\mu\theta}{\sin\theta}\ln\frac{(\cos\theta - \cos\theta_0)(\cos\theta + \cos\theta_0)}{\sin^2(\theta_0 + \theta)} \qquad (22\text{-}15)$$

where the first term on the right-hand side of Eq. 22-15 is evaluated through the use of the following recurrence formula:

$$F'_{\mu+1} + F'_{\mu-1} - 2\cos\theta F'_\mu = \begin{Bmatrix} \frac{4}{\mu}\sin\mu\theta_0 & \text{for } \mu \text{ even} \\ 0 & \text{for } \mu \text{ odd} \end{Bmatrix} \qquad (22\text{-}15')$$

in which $F'_0 = 0$ and $F'_1 = -\pi + 2\theta_0$.

Likewise, provided that it is agreed that $\cos^2\theta_0 \leqq \cos\theta \leqq 1$, so that it is possible to set $\cos\psi = \cos^2\theta_0/\cos\theta$, then the value of J'_μ is given by the relation

$$2J'_\mu = \frac{2}{\cos\theta}\frac{1}{\mu}\begin{Bmatrix} -2\sin\mu\theta_0 \\ 0 \end{Bmatrix} + \frac{2\cos^2\theta_0}{\cos\theta}K'_\mu \qquad (22\text{-}16)$$

where the expression utilizing the $-2\sin\mu\theta_0$ factor appearing between the brackets is the one which holds if μ is even, or else the term is zero if μ is odd. In addition, the K'_μ factor is defined by the expression

$$K'_\mu = H'_\mu + \frac{\cos\mu\psi}{\cos\theta\sin\psi}\ln\frac{\cos^2\theta_0(\cos\theta_0 + \cos\theta)(\cos\theta_0 - \cos\theta)}{\cos^2\theta\sin^2(\psi + \theta_0)} \qquad (22\text{-}17)$$

where the first term on the right-hand side of this expression is evaluated through means of the following recurrence formula:

$$H'_{\mu+1} + H'_{\mu-1} = \frac{2}{\cos\theta}\frac{1}{\mu}\begin{Bmatrix} -2\sin\mu\theta_0 \\ 0 \end{Bmatrix} + \frac{2\cos^2\theta_0}{\cos\theta}H'_\mu$$

in which $H'_0 = 0$ and $H'_1 = (\pi - 2\theta_0)/\cos\theta$, and in which the expression utilizing the $-2\sin\mu\theta_0$ factor appearing in brackets is the one which holds if μ is even; the term is zero if μ is odd.

If, on the other hand, it is required that $0 \leqq \cos\theta \leqq \cos^2\theta_0$, then it is useful to define a quantity ψ' by means of the relation

$$\cos\psi' = \frac{\cos\theta}{\cos^2\theta_0}$$

and the evaluation of J'_μ is obtained by having recourse to the following formula:

$$2J'_\mu = K'_{\mu+1} + K'_{\mu-1} = \frac{2}{\cos^2\theta_0 \cos\psi'}\left\{\begin{matrix}\dfrac{-2\sin\mu\theta_0}{\mu}\\ 0\end{matrix}\right\} + \frac{2}{\cos\psi'}K'_\mu \quad (22\text{-}18)$$

in which the expression utilizing the $-2\sin\mu\theta_0/\mu$ factor appearing in brackets is that which holds if μ is even; the term is zero if μ is odd. In addition, the definition of K'_μ is given, in the case of $\mu = 0$ and $\mu = 1$ by

$$\left.\begin{aligned} K'_0 &= -\frac{2}{\cos^2\theta_0\sin\psi'}\tan^{-1}\left(\frac{\sin\psi'}{\tan\theta_0}\right)\\ \text{and}\qquad & \\ K'_1 &= \frac{1}{\cos^2\theta_0\cos\psi'}(\pi - 2\theta_0) + \frac{1}{\cos\psi'}K'_0\end{aligned}\right\} \quad (22\text{-}18')$$

Finally for the case where $\pi - \theta_0 \leqq \theta' \leqq \pi$, the evaluation of the integrals given as Eq. 22-11, where the double primes now apply, is carried out by means of the expression

$$I''_\mu = F''_\mu + \frac{\cos\mu\theta}{\sin\theta}\ln\frac{\cos\theta + \cos\theta_0}{1 + \cos(\theta_0 + \theta)} \quad (22\text{-}19)$$

in which the value of $F''_{\mu+1}$ is obtained by making use of the following recurrence formula:

$$F''_{\mu+1} + F''_{\mu-1} - 2\cos\theta F''_\mu = \mp\frac{2}{\mu}\sin\mu\theta_0$$

Here, again, the negative sign is employed if μ is even, while the positive sign is employed if μ is odd; the starting values of F''_μ are

$$F''_0 = 0 \qquad \text{and} \qquad F''_1 = \pi - 2\theta_0$$

Meanwhile the evaluation of the second of the two integrals given as Eq. 22-11 becomes

$$2J''_\mu = \frac{2}{\cos\theta}\left(\pm\frac{\sin\mu\theta_0}{\mu}\right) + \frac{2\cos^2\theta_0}{\cos\theta}K''_\mu \quad (22\text{-}20)$$

where, this time, the positive sign is used in front of the term in brackets if μ is even and the negative sign if μ is odd, and where the final factor on the right-hand side is given by

$$K''_\mu = H''_\mu - \frac{\cos\mu\psi}{\cos\theta\sin\psi}\ln\frac{\cos\theta_0(\cos\theta_0 + \cos\theta)}{\cos\theta[1 + \cos(\theta_0 + \psi)]} \quad (22\text{-}20')$$

The first term on the right-hand side may be valuated by means of the following recurrence formula:

$$H''_{\mu+1} + H''_{\mu-1} = \frac{2}{\cos\theta}\left(\pm\frac{\sin\mu\theta_0}{\mu}\right) + \frac{2\cos^2\theta_0}{\cos\theta}H''_\mu \quad (22\text{-}20'')$$

where, as usual, the upper (plus) sign is used if μ is even and the lower (negative) sign is used if μ is odd, and where the starting values are

$$H_0'' = 0 \qquad \text{and} \qquad H_1'' = \frac{\theta_0}{\cos\theta_0}$$

while it is again taken for granted that $\cos\psi = \cos^2\theta_0/\cos\theta$.

All the formulas set forth above for evaluating the two integrals, I_μ and J_μ, hold for the first quadrant where $\pi/2 \leqq \theta \leqq 0$. Since the distribution of the circulation γ is symmetric with respect to $\theta = \pi/2$, however, there is no loss in generality by limiting oneself to consideration of merely those values of θ_ν which lie within the first quadrant.

The system of Eq. 22-9 constitutes a set of $(m + 1)/2$ equations in the unknowns γ_ν (of which there are $(m + 1)/2$), together with the three unknown constants a_1, a_2, and α_k. If the size of h is such that it happens to fall exactly at one of the η_ν values, the two Eq. 22-9 have to be satisfied by the same value of γ_ν existing at the point $\eta_\nu = h$, and thus it is required that

$$\frac{1}{\pi(m+1)}\sum_{n=1}^{m}\gamma_n\sum_{\mu=1}^{m}\mu\sin\mu\theta_n[D(I^*_{\mu,0} - I'^*_{\mu,0} + I''^*_{\mu,0})$$
$$- C(J^*_{\mu,0} + J'^*_{\mu,0} + J''^*_{\mu,0})] + a_2[N^*(h) - M^*(h)]$$
$$+ \frac{\alpha_k}{2\pi}\sum_{\mu=1}^{\infty}\mu\beta_\mu[D(I^*_{\mu,0} - I'^*_{\mu,0} + I''^*_{\mu,0}) - C(J^*_{\mu,0} + J'^*_{\mu,0} + J''^*_{\mu,0})]$$
$$+ 2\alpha_k - \frac{2b}{c_{l_\alpha}c_0}(a_1 + a_2h^2) = \alpha_0(S - 1) - [(\varepsilon_p)_{II} - (\varepsilon_p)_I]_0 \qquad (22\text{-}21)$$

wherein the asterisk is placed on the I, J, M, and N symbols to denote that only the *finite part* of the quantities represented by these symbols is to be evaluated for the case where $\theta = \theta_0$. Thus, for instance, in the evaluation of the quantity $I^*_{1,0}$, if one returns to Eq. 22-12 and inserts the concrete values now specified, it is seen that the finite part of the integral is just

$$I^*_{1,0} = -\theta_0 - \cot\theta_0 \ln(1 - \cos 2\theta_0)$$

In the case of the N^* and M^* symbols the following relationship holds:

$$N^*(h) - M^*(h) = \frac{2h}{\pi}(D - C) + \frac{h}{\pi}(C - A)\ln 2h$$

As a matter of fact, because of the definition of the jump (22-5′), the sum $L(\eta)$ of the coefficients of the $\ln|\eta - h|$ quantity, which appears in the series of terms in Eq. 22-21, tends to zero as a limit for $\eta = h$, when the *complete values* of I_μ, J'_μ, etc. are employed therein, and in fact it is true that

$$\lim_{\eta \to h} L(\eta) \ln |\eta - h| = 0$$

The rest of the notation is made to conform with the style of the symbolism just utilized, i.e. it is assumed that

$$\alpha_0 = \alpha(h), \qquad (\varepsilon_p)_{I,0} = (\varepsilon_p)_I(h), \qquad \text{etc.}$$

Finally, on the basis of Eq. 22-5 and 22-5′ it is deduced that

$$\left.\begin{aligned} &(a_1 + a_2h^2)\frac{S}{S-1} \\ &\qquad = \frac{2}{m+1}\sum_{n=1}^{m}\gamma_n\sum_{\mu=1}^{m}\sin\mu\theta_n\sin\mu\theta_0 + \alpha_k\sum_{\mu=1}^{\infty}\beta_\mu\sin\mu\theta_0 \\ &\text{and} \\ &\frac{2a_2h}{S-1} \\ &\qquad = \frac{2}{m+1}\sum_{n=1}^{m}\gamma_n\sum_{\mu=1}^{m}\mu\sin\mu\theta_n\frac{\cos\mu\theta_0}{\sin\theta_0} + \alpha_k\sum_{\mu=1}^{\infty}\mu\beta_\mu\frac{\cos\mu\theta_0}{\sin\theta_0} \end{aligned}\right\} \quad (22\text{-}22)$$

In this manner, it evolves that Eq. 22-9, 22-21, and 22-22 constitute a system of $[(m + 1)/2] + 3$ equations in the $[(m + 1)/2] + 3$ unknowns composed of the quantities γ_n, a_1, a_2, and α_k. Consequently this setup of the problem formally solves the question of what is the distribution of circulation out along a wing, which is operating in the presence of a propeller.

C,23. Determination of the Distribution of Circulation in the Case of a Wing Placed at Any Distance from the Propeller. An approximation is now sought for a wing traversing the slipstream at a relatively short distance aft of the propeller. The increments of velocity produced by the propeller in this case were explicitly derived in Art. 19, and it should be understood that they satisfy only the boundary conditions with respect to the slipstream surface Σ_p to the order of terms which run up no higher than those in a^2. It would serve no purpose, therefore, to try to compute the distribution of circulation out along the wing to any better approximation than this. Thus if one requires only an approximation of the magnitude represented by terms involving a^2 or less, a great simplification to the problem now under study comes about. In order to carry out the required computations, the induced angle of attack, α_i, is taken to be defined by Eq. 21-2′. In conformity with the only justifiable degree of approximation just mentioned, it is seen that the quantities $(\varphi'_\infty)_{Ir}$ and $(\varphi'_\infty)_{IIr}$ (see Eq. 21-2′) represent the potential existing far downstream where $x = \infty$ for the flow generated by the inverted vortices that are related to the trailing vortices of the wing when it is considered to be operating in an isolated condition.

Now let the distribution of circulation existing along the isolated wing be represented by

$$\gamma_0 = \frac{2}{m+1}\sum_{n=1}^{m}\gamma_n^{(0)}\sum_{\mu=1}^{m}\sin\mu\theta_n\sin\mu\theta$$

and also let the change in the circulation values owing to propeller interference be denoted by $\Delta\gamma = \Delta\gamma_d + \Delta\gamma_e$, where $\Delta\gamma_d$ is still to be expressed in accordance with the method used in Eq. 22-7.

Consequently it may be set down at once that Eq. 22-22 are now transformed into[5]

$$\left.\begin{aligned} a_1 + a_2h^2 &= (S-1)\frac{2}{m+1}\sum_{n=1}^{m}\gamma_n^{(0)}\sum_{\mu=1}^{m}\sin\mu\theta_n\sin\mu\theta_0 \\ a_2h &= (S-1)\frac{1}{m+1}\sum_{n=1}^{m}\gamma_n^{(0)}\sum_{\mu=1}^{m}\mu\sin\mu\theta_n\frac{\cos\mu\theta_0}{\sin\theta_0}\end{aligned}\right\} \quad (23\text{-}1)$$

By making use of the convenient definitions that

$$(S_x)_I = 1 + \frac{(v_x)_I}{U} \qquad \text{and} \qquad (S_x)_{II} = 1 + \frac{(v_x)_{II}}{U}$$

it may then be readily deduced that Eq. 22-21 takes on the simplified form

$$2\alpha_k = \alpha_0[(S_x)_{II} - (S_x)_I] - [(\varepsilon_p)_{II} - (\varepsilon_p)_I]_0 + \frac{2b}{c_{l_\alpha}c_0}(a_1 + a_2h^2)$$
$$+ \frac{S-1}{\pi(m+1)}\sum_{n=1}^{m}\gamma_n^{(0)}\sum_{\mu=1}^{m}\mu\sin\mu\theta_n(J'^*_{\mu,0} + J^*_{\mu,0} + J''^*_{\mu,0}) \quad (23\text{-}2)$$

Finally, by letting $\Delta\gamma_e$ be written in the form

$$\Delta\gamma_e = \frac{2}{m+1}\sum_{n=1}^{m}\Delta\gamma_n\sum_{\mu=1}^{m}\sin\mu\theta_n\sin\mu\theta + \alpha_k\sum_{\mu=1}^{\infty}\beta_\mu\sin\mu\theta \quad (23\text{-}3)$$

[5] As substitutes for Eq. 23-1 one may use the relationships which are derivable from these equations by replacing S with the quantity $(S_x)_{II}$. The symbol $(S_x)_{II}$ has the following significance: $(S_x)_{II} = 1 + (v_x)_{II}/U$, where $(v_x)_{II}$ is the increment in the axial component of the velocity, evaluated at the location of the lifting line and within the slipstream of the propeller.

it can be determined that the $\Delta\gamma_n$ appearing here is given by the following expressions:

$$\left(b_{\nu\nu} + \frac{2b}{c_{l_\alpha}c_\nu}\right)\Delta\gamma_\nu + \frac{2b}{c_{l_\alpha}c_\nu}\alpha_k \sum_{\mu=1}^{\infty} \beta_\mu \sin \mu\theta_\nu - \sum_{n=1}^{m}{}' b_{\nu n}\Delta\gamma_n$$

$$= \alpha[(S_x)_\mathrm{I} - 1] - (\varepsilon_p)_\mathrm{I} - M_\nu a_2$$

$$- \frac{S-1}{(m+1)\pi}\sum_{n=1}^{m}\gamma_n^{(0)}\sum_{\mu=1}^{m}\mu \sin \mu\theta_n \,(J_{\mu\nu} + J''_{\mu\nu}) \quad \text{for } |\eta_\nu| > h \qquad (23\text{-}4)$$

and

$$\left(b_{\nu\nu} + \frac{2b}{c_{l_\alpha}c_\nu}\right)\Delta\gamma_\nu + \frac{2b}{c_{l_\alpha}c_\nu}\alpha_k \sum_{\mu=1}^{\infty} \beta_\mu \sin \mu\theta_\nu - \sum_{n=1}^{m}{}' b_{\nu n}\Delta\gamma_n$$

$$= (\alpha_\nu - \alpha_0)(S_x)_\mathrm{II} + \alpha_0(S_x)_\mathrm{I} - \alpha_\nu - [(\varepsilon_p)_\mathrm{II} - (\varepsilon_p)_{\mathrm{II},0}]$$

$$- (\varepsilon_p)_{\mathrm{I},0} - M_\nu a_2 + \frac{2b}{c_{l_\alpha}c_\nu}a_2(\eta_\nu^2 - h^2)$$

$$- \frac{S-1}{(m+1)\pi}\sum_{n=1}^{m}\gamma_n^{(0)}\sum_{\mu=1}^{m}\mu \sin \mu\theta_n(J^*_{\mu,0} + J'^*_{\mu,0} + J''^*_{\mu,0} - J'_{\mu\nu})$$

$$\text{for } |\eta_\nu| < h \qquad (23\text{-}4')$$

If $|\eta_\nu| = h$ in Eq. 23-4 and 23-4′, the right-hand sides of the expressions remain finite, because of the relationships given above as Eq. 23-1; what is more, they are equal. Eq. 23-4 and 23-4′, therefore, are suitable for determining the $(m + 1)/2$ unknowns, $\Delta\gamma_n$, appearing in the formal expansion for $\Delta\gamma_o$ which gives rise to the continuous part of the distribution of the induced angles of attack.

As noted in the case of Eq. 23-1, equivalent alternate expressions to the ones given here as Eq. 23-2, 23-4, and 23-4′ may be produced by substituting the quantity $(S_x)_\mathrm{II}$ for the symbol S which appears in the above-cited equations.

The inclusion in these determinations of the computation of the peripheral or rotational component of the incremental velocity induced by the propeller does not present any great complication to the method already set forth. To carry out this additional phase of the investigation it is merely necessary to substitute the expression

$$\bar{v}_\theta = 2aU\beta^2 \tan \Phi$$

in place of the similar expression for this same peripheral component of the increment in the velocity that was given previously in Eq. 19-4. In this new expression the symbol $\bar{v}_\theta$ stands for the weighted average of the v_θ values, i.e.

$$\bar{v}_\theta = \frac{1}{\pi R^2} \int_0^R 2\pi r v_\theta dr$$

Under these conditions, the peripheral component of the velocity increment produces a change in the angle of attack, which is induced along the wing span by action of the propeller, in the amount of

$$\left.\begin{aligned} \Delta\varepsilon &= 2a\beta^2 \tan \Phi && \text{for } 0 \leqq \eta \leqq h \\ \Delta\varepsilon &= -2a\beta^2 \tan \Phi && \text{for } 0 \geqq \eta \geqq -h \\ \Delta\varepsilon &= 0 && \text{for } |\eta| \geqq h \end{aligned}\right\} \tag{23-5}$$

It is obvious that these changes effected by the propeller are exactly analogous to the kind of angle-of-attack distribution that is produced by a deflection of a pair of inboard ailerons. Thus, in order to account for the peripheral components, it suffices merely to add the $\Delta\gamma'$ distribution of circulation corresponding to the angle-of-attack distribution given in Eq. 23-5 to the $\Delta\gamma$ circulation distribution that has already been obtained by proceeding according to the method previously outlined. The usual Multhopp procedure, which is valid for any wing whatsoever, is employed here, of course, in order to obtain the $\Delta\gamma'$ distribution.

Once the circulation distribution $\gamma(\eta)$ has been calculated, the changes in the lift and drag coefficients due to the interference from the propeller can easily be obtained. To the degree of approximation consistent with the derivations provided in this article, the changes in the lift and drag coefficients are given by the following formulas:

$$\begin{aligned} \Delta C_L &= A\!\!R \left[\int_{-1}^{1} \Delta\gamma d\eta + \int_{-1}^{1} \frac{v_x}{U} \gamma^{(0)} d\eta \right] \\ &= A\!\!R \left\{ -2h(S-1) \left[\gamma^{(0)}(h) + \frac{h}{3} \left(\frac{d\gamma_0}{d\eta} \right)_{\eta=h} \right] \right. \\ &\qquad + \frac{\pi}{m+1} \sum_{n=1}^{m} \Delta\gamma_n \sin\theta_n - 2\alpha_k(2\theta_0 - \pi - \sin 2\theta_0) \\ &\qquad \left. + \int_{-1}^{1} \frac{v_x}{U} \gamma^{(0)} d\eta \right\} \end{aligned} \tag{23-6}$$

and

$$\Delta C_{D_i} = A\!\!R \left[\int_{-1}^{1} \Delta\alpha_i \gamma^{(0)} d\eta + \int_{-1}^{1} \alpha_i^{(0)} \Delta\gamma d\eta \right] \tag{23-6'}$$

where the induced angles of attack are related to the circulation at the m points, η_n, by means of the following expressions:

$$\Delta\alpha_i = (\varepsilon_p)_I + b_{\nu\nu}\Delta\gamma_\nu - \sum_{n=1}^{m}{}' b_{\nu n}\Delta\gamma_n + M_\nu a_2$$
$$+ \frac{S-1}{(m+1)\pi}\sum_{n=1}^{m}\gamma_n^{(0)}\sum_{\mu=1}^{m}\mu \sin \mu\theta_n(J_{\mu\nu} + J''_{\mu\nu})$$
$$- \alpha_i^{(0)}[(S_x)_I - 1] \quad \text{for } |\eta| > h \qquad (23\text{-}7)$$

and

$$\Delta\alpha_i = (\varepsilon_p)_{II} + b_{\nu\nu}\Delta\gamma_\nu - \sum_{n=1}^{m}{}' b_{\nu n}\Delta\gamma_n + M_\nu a_2 + 2\alpha_k$$
$$- \frac{S-1}{(m+1)\pi}\sum_{n=1}^{m}\gamma_n^{(0)}\sum_{\mu=1}^{m}\mu \sin \mu\theta_n J'_{\mu\nu} - \alpha_i^{(0)}[(S_x)_{II} - 1] \quad \text{for } |\eta| < h$$
$$(23\text{-}7')$$

while the evaluation of $\alpha_i^{(0)}$ evolves from the usual Multhopp formula, i.e.

$$\alpha_i^{(0)} = b_{\nu\nu}\gamma_\nu^{(0)} - \sum_{n=1}^{\infty}{}' \gamma_n^{(0)} b_{\nu n}$$

C,24. Interference When the Wing Is Elliptic. If the wing is elliptic in planform it is possible to derive formulas for the calculation of the incremental circulation distribution $\Delta\gamma(\eta)$ and the related ΔC_L and ΔC_{D_i} coefficients that are significantly more simple than those encountered so far, provided that it is again further stipulated that the case to be treated is the one where the wing is located in the middle of the propeller slipstream. In order to obtain a concise analysis under these circumstances, the hypothesis is made that h is to be so small that it is perfectly legitimate to neglect terms of the order of magnitude of h^2. In addition, it is assumed that the wing is located sufficiently far downstream from the propeller so that it is permissible to set

$$v_x = 0 \qquad \text{for } |\eta| \geqq h \qquad (24\text{-}1)$$

while v_x is taken to be given by the relation stated as the first of the set of Eq. 19-5 when $|\eta| \leqq h$.

In addition one also has that

$$\left.\begin{aligned} v_z &= \bar{v}_{z\infty}(1 + A_0) && \text{for } |\eta| \leqq h \\ \text{and} \quad v_z &= -\bar{v}_{z\infty}(1 - A_0)\frac{h^2}{\eta^2} && \text{for } |\eta| \geqq h \end{aligned}\right\} \qquad (24\text{-}2)$$

wherein the quantity A_0 is determined from the relation

$$A_0 = \frac{1}{2} B^2 \frac{R^2}{x^2}$$

The expressions written here as Eq. 24-2 are derived from those given as Eq. 19-12, when the value of A appearing therein is assumed to be given by Eq. 19-12″ and the value of B is taken as $B = -1 + A_0$.

Now let $\gamma^{(0)}$ be written in the form

$$\gamma^{(0)} = \gamma_0^{(0)} \sin \theta$$

in which the symbol $\gamma_0^{(0)}$ obviously stands for the value that $\gamma^{(0)}$ takes at the root chord in the case where the wing is operating as an isolated entity. Then the continuous part of the angle of attack distribution due to the interference, ε_e (taken to be positive if in the same direction as α), may be separated into two parts arising from the operation of signally different causes. Thus, if one writes

$$\varepsilon_e = \varepsilon_{e,1} + \varepsilon_{e,2} \tag{24-3}$$

the first term on the right-hand side, $\varepsilon_{e,1}$, may be considered to be that contribution to the total value of ε_e which is due to the vortices that are the inversions of the corresponding trailing vortices being shed from the isolated wing, and which also includes the influence of the discontinuous part of the vorticity distribution, denoted by $\Delta\gamma_d$, that is linked to the $-a_2\eta^2$ term. On the other hand, the second term on the right-hand side of Eq. 24-3 represents the contribution to ε_e that is due to the velocities which are induced by action of the propeller. It follows, therefore, that Eq. 22-9 and the associated formulas that follow them may be recast into the compressed forms

$$\left.\begin{aligned}
\varepsilon_{e,1} &= \frac{-\gamma_0^{(0)}(S_x - 1)}{2\pi \cos \theta}\left[2\theta_0 \cos \psi + \frac{\cos^2 \psi}{\sin \psi} \ln \frac{\sin (\psi + \theta_0)}{\sin (\psi - \theta_0)}\right] \\
&\qquad - Ma_2 \quad \text{for } \theta_0 \geqq \theta \geqq 0 \\
&\text{or} \\
\varepsilon_{e,1} &= \frac{-\gamma_0^{(0)}(S_x - 1)}{2\pi \cos \theta}\left[(2\theta_0 - \pi) \cos \psi + \frac{\cos^2 \psi}{\sin \psi} \ln \frac{\sin (\theta_0 + \psi)}{\sin (\theta_0 - \psi)}\right] \\
&\qquad - \frac{\gamma_0^{(0)}}{2} (S_x - 1) - Ma_2 \quad \text{for } \cos^2 \theta_0 \leqq \cos \theta \leqq \cos \theta_0 \\
&\text{or} \\
\varepsilon_{e,1} &= \frac{-\gamma_0^{(0)}(S_x - 1)}{2\pi \cos^2 \theta_0}\left[\frac{2\theta_0 - \pi}{\cos^2 \psi'} + \frac{2}{\sin \psi'} \frac{\tan^{-1} (\cot \theta_0 \sin \psi')}{\cos^2 \psi'}\right] \\
&\qquad - \frac{S_x - 1}{2} \gamma_0^{(0)} - Ma_2 \quad \text{for } 0 \leqq \cos \theta \leqq \cos^2 \theta_0
\end{aligned}\right\} \tag{24-4}$$

where the value of a_2 employed here is given by

$$a_2 = \frac{S_x - 1}{2} \gamma_0^{(0)}$$

It is recognized, of course, that $\varepsilon_{e,1}$ is an even function, i.e.

$$\varepsilon_{e,1}(\theta) = \varepsilon_{e,1}(\pi - \theta)$$

For the other constituent distribution of the angles of attack, it is true then that

$$\left.\begin{aligned} \varepsilon_{e,2} &= -(\varepsilon_p)_I = \frac{\bar{v}_{z\infty}}{U}(1 - A_0)\frac{h^2}{\eta^2} \qquad \text{for } |\cos\theta| \geqq h \\ &\text{or} \\ \varepsilon_{e,2} &= -[(\varepsilon_p)_I]_{\theta=\theta_0} = \frac{\bar{v}_{z\infty}}{U}(1 - A_0) \qquad \text{for } |\cos\theta| \leqq h \end{aligned}\right\} \tag{24-5}$$

Because ε_e is continuous and finite throughout the whole interval $0 \leqq \theta \leqq \pi$, the related incremental circulation distribution may be computed from the integral relationship

$$\Delta\gamma_0'(\theta) = \frac{4}{\mathcal{R}} \int_{\bar{\theta}=-\pi}^{\bar{\theta}=\pi} \varepsilon_e(\bar{\theta}) \sin\bar{\theta} \frac{dV_{\mathcal{R}}(\bar{\theta} - \theta)}{d\bar{\theta}} d\bar{\theta} \tag{24-6}$$

To this end, the function $\varepsilon_e(\bar{\theta})$ has been extended beyond the $0 \leqq \bar{\theta} \leqq \pi$ interval by making $\varepsilon_e(\bar{\theta})$ an even function with respect to the π location, i.e. the definition of $\varepsilon_e(\bar{\theta})$ here includes the fact that $\varepsilon_e(\bar{\theta}) = \varepsilon_e(2\pi - \bar{\theta})$. In addition, the new notation $dV_{\mathcal{R}}(\bar{\theta} - \theta)/d\bar{\theta}$ is introduced here to represent the influence function for an elliptic wing, the values of which have been listed by Rossner [*34*]. These values are reproduced here in Table C,24 for values of θ lying within the interval $0 \leqq \theta \leqq \pi$, and for various values of λ. The meaning of the latter parameter is given by

$$\lambda = \frac{\pi}{c_{l_\alpha}} \mathcal{R}$$

If θ lies outside of the interval from 0 to π, the following connections between the known values of $V_{\mathcal{R}}(\theta)$ in this interval and the sought ones may be utilized in computing $V_{\mathcal{R}}(\theta)$ everywhere; i.e. one may have recourse to the relations

$$V_{\mathcal{R}}(\theta) = -V_{\mathcal{R}}(-\theta) \qquad \text{and} \qquad V_{\mathcal{R}}(\theta + 2n\pi) = V_{\mathcal{R}}(\theta) + n$$

where n is an integer.

In order to compute the value of the incremental lift and drag coefficients due to this interference, it is convenient to first put $\varepsilon_{e,1}$ in the form

$$\varepsilon_{e,1} = \sum_{m=1}^{\infty} \frac{\alpha_m \sin m\theta}{\sin\theta} \tag{24-7}$$

Table C,24. **Influence function applying to an elliptic wing.**

θ	$\lambda = \frac{\pi}{c_{l_\alpha}} Æ$							
	0	$\frac{1}{2}$	1	$\frac{3}{2}$	2	3	4	8
0	0	0	0	0	0	0	0	0
$\pi/72$	0.00694	0.03149	0.05090	0.06737	0.08184	0.10661	0.18751	0.18905
$\pi/36$	0.01389	0.05358	0.08342	0.10772	0.12734	0.16544	0.19160	0.26372
$\pi/18$	0.02778	0.08877	0.13151	0.16446	0.19117	0.23252	0.26368	0.33835
$\pi/12$	0.04167	0.11750	0.16793	0.20523	0.23444	0.27785	0.30894	0.37806
$\pi/9$	0.05556	0.14226	0.19750	0.23699	0.26706	0.31028	0.34009	0.40269
$\pi/8$	0.08333	0.18408	0.24396	0.28451	0.31403	0.35436	0.38066	0.43166
$2\pi/9$	0.11111	0.21909	0.27980	0.31919	0.34691	0.38335	0.40617	0.44817
$\pi/3$	0.16667	0.27659	0.33333	0.36788	0.39100	0.41984	0.43695	0.46648
$4\pi/9$	0.22222	0.32377	0.37304	0.40166	0.42008	0.44246	0.45545	0.47673
$\pi/2$	0.25000	0.34483	0.38968	0.41532	0.43169	0.45117	0.46221	0.48040
$5\pi/9$	0.27778	0.36464	0.40486	0.42749	0.44180	0.45862	0.46809	0.48352
$2\pi/3$	0.33333	0.40143	0.43191	0.44867	0.45910	0.47117	0.47778	0.48864
$7\pi/9$	0.38889	0.43559	0.45602	0.46709	0.47389	0.48168	0.48598	0.49282
$8\pi/9$	0.44444	0.46817	0.47841	0.48391	0.48270	0.49110	0.49319	0.49652
$17\pi/18$	0.47222	0.48413	0.48926	0.49260	0.49367	0.49558	0.49662	0.49828
π	0.50000	0.50000	0.50000	0.50000	0.50000	0.50000	0.50000	0.50000

where

$$\alpha_m = \frac{2}{\pi} \int_0^\pi \varepsilon_{e,1} \sin\theta \sin m\theta d\theta \tag{24-7'}$$

To the degree of approximation consistent with what has previously been permitted in these derivations, under the assumption that h is small, it is true that

$$\alpha_1 = \frac{4}{\pi} \gamma_0^{(0)} h(S - 1)$$

Consequently it follows that the expression for the incremental circulation $\Delta\gamma_{e,1}$, which is related to the angle-of-attack distribution $\varepsilon_{e,1}$, is given by the formula

$$\Delta\gamma_{e,1} = -2 \sum_{m=1}^{\infty} \alpha_m \frac{\sin m\theta}{m + (\pi Æ/c_{l_\alpha})}$$

and likewise, the induced angles of attack are given by

$$\Delta\alpha_{i,1} = \sum_{m=1}^{\infty} \alpha_m \frac{m}{m + (\pi Æ/c_{l_\alpha})} \frac{\sin m\theta}{\sin\theta} \tag{24-8}$$

It may be deduced, therefore, that the change in the total lift coefficient is expressible as

$$(\Delta C_L)_1 = -(C_{L\alpha})_0 \frac{4}{\pi} \gamma_0^{(0)} h(S_x - 1)$$

$$= -(C_{L\alpha})_0 \frac{A_p}{A} \frac{2(C_L)_0}{\pi \mathcal{R}} (S_x - 1) \tag{24-9}$$

wherein $(C_{L\alpha})_0$ denotes the slope of the lift curve for the entire isolated wing, and where A_p is the portion of the total wing area which is immersed in the slipstream from the propeller.

The change in the over-all induced drag coefficient then turns out to be

$$(\Delta C_{D_i})_1 = \frac{2A_p}{A} (S_x - 1)(C_{D_i})_0 \left[1 - \frac{2}{\pi \mathcal{R}} (C_{L\alpha})_0\right] \tag{24-10}$$

If one also casts $\varepsilon_{e,2}$ into the form

$$\varepsilon_{e,2} = \sum_{m=1}^{\infty} \frac{\alpha'_m \sin m\theta}{\sin \theta} \tag{24-11}$$

it is seen that, in analogy with the expressions related to $\varepsilon_{e,1}$, the following similar formulas may be set down immediately:

$$\Delta\gamma_{e,2} = 2 \sum_{m=1}^{\infty} \alpha'_m \frac{\sin m\theta}{m + (\pi \mathcal{R}/c_{l\alpha})}$$

and

$$\Delta\alpha_{i,2} = \sum_{m=1}^{\infty} \alpha'_m \frac{m}{m + (\pi \mathcal{R}/c_{l\alpha})} \frac{\sin m\theta}{\sin \theta}$$

The angle of tilt caused by the slipstream is here given by

$$\alpha'_1 = \frac{4}{\pi} \frac{\bar{v}_{z\infty}}{U} (1 - A_0) 2h$$

and the change in the total lift coefficient corresponding to the angular change is

$$(\Delta C_L)_2 = \frac{4}{\pi} \frac{A_p}{A} (C_{L\alpha})_0 (1 - A_0)(C_z + \alpha_T C_T) \tag{24-12}$$

Thus a change in the drag coefficient is brought about in the amount of

$$(\Delta C_{D_i})_2 = -\frac{A_p}{A} \frac{2}{\pi} (1 - A_0)(C_z + \alpha_T C_T)(C_L)_0 + \frac{2}{\pi \mathcal{R}} (C_L)_0 (\Delta C_L)_2 \tag{24-12'}$$

The discontinuous part of the distribution of induced angles of attack which are due to the propeller's interference are obtainable on the basis of Eq. 23-2 in the form:

$$\left.\begin{aligned} \varepsilon_d &= 0 && \text{for } |\eta| \geqq h \\ \varepsilon_d &= (S_x - 1)\alpha - \frac{2\bar{v}_{z\infty}}{U} + \frac{S_x - 1}{2}\gamma_0^{(0)} + \frac{\pi}{2}\frac{\mathcal{R}}{c_{l_\alpha}}a_1 && \text{for } |\eta| \leqq h \end{aligned}\right\} \tag{24-13}$$

where, through reference to Eq. 22-5, it is seen that the definition of a_1 is given by

$$a_1 = (S_x - 1)\gamma_0^{(0)} \tag{24-14}$$

The corresponding distribution of circulation $\Delta\gamma_{e,3}$, related to these angles of attack, is expressible in the form

$$\Delta\gamma_{e,3} = \frac{\varepsilon_d}{\pi}\left\{\frac{8h}{1 + (\pi\mathcal{R}/c_{l_\alpha})}\sin\theta + 4\sum_{n=3}^{\infty}\frac{\sin n\theta}{n + (\pi\mathcal{R}/c_{l_\alpha})}\left[-\frac{\sin(n-1)\theta_0}{n-1} + \frac{\sin(n+1)\theta_0}{n+1}\right]\right\} \quad \text{for } n = 3, 5, \cdots \tag{24-15}$$

For any rational value of λ whatsoever, with the exception of the case when $\lambda = 1$, it is possible to sum the series appearing in Eq. 24-15, and one finds that

$$\Delta\gamma_{e,3} = \varepsilon_d F(\lambda, \theta) \tag{24-15'}$$

where

$$\begin{aligned} F = \frac{1}{\pi}\frac{1}{\lambda^2 - 1}\Bigg\{&\lambda\left(\frac{\pi - \theta_0 - \theta}{|\pi - \theta_0 - \theta|} - \frac{\theta_0 - \theta}{|\theta_0 - \theta|}\right)(\sin\theta - \sin\theta_0) \\ &- \frac{2}{\pi}\left[(\pi - 2\theta_0)\sin\theta - \cos\theta\ln\left|\frac{\sin(\theta_0 + \theta)}{\sin(\theta_0 - \theta)}\right|\right] \\ &- 2\lambda\sin\theta_0[V_{\mathcal{R}}(\theta + \theta_0 - \pi) + V_{\mathcal{R}}(\theta - \theta_0 + \pi) - V_{\mathcal{R}}(\theta - \theta_0) \\ &- V_{\mathcal{R}}(\theta + \theta_0)] + \frac{2}{\lambda}\cos\theta_0\frac{\partial}{\partial\theta}[V_{\mathcal{R}}(\theta + \theta_0 - \pi) \\ &+ V_{\mathcal{R}}(\theta - \theta_0 + \pi) + V_{\mathcal{R}}(\theta - \theta_0) + V_{\mathcal{R}}(\theta + \theta_0)]\Bigg\} \end{aligned}$$

Thus the induced angles of attack, produced by the incremental amount of circulation $\Delta\gamma_{e,3}$, are given by

$$\left.\begin{aligned}\Delta\alpha_{i,3} &= -\frac{\pi A\!\!R}{2c_{l_\alpha}}\frac{\Delta\gamma_{o,3}}{\sin\theta} && \text{for } \pi \geqq \theta \geqq \pi - \theta_0 \\ &= \varepsilon_d - \frac{\pi A\!\!R}{2c_{l_\alpha}}\frac{\Delta\gamma_{o,3}}{\sin\theta} && \text{for } \pi - \theta_0 \geqq \theta \geqq \theta_0 \\ &= \frac{-\pi A\!\!R}{2c_{l_\alpha}}\frac{\Delta\gamma_{o,3}}{\sin\theta} && \text{for } \theta_0 \geqq \theta \geqq 0\end{aligned}\right\} \tag{24-15''}$$

In consequence it may be seen that the change in the total lift coefficient occasioned by these angular changes is computable from the relation

$$(\Delta C_L)_3 = (S_x - 1)\frac{A_p}{A}(C_L)_0 - \frac{4}{\pi}\frac{A_p}{A}(C_{L\alpha})_0(C_s + \alpha_T C_T) + (C_{L\alpha})_0\frac{A_p}{A}(S_x - 1)\frac{(C_L)_0}{\pi A\!\!R} + (C_{L\alpha})_0\frac{A_p}{A}\frac{S_x - 1}{c_{l_\alpha}}(C_L)_0 \tag{24-16}$$

while the corresponding change in the over-all induced drag coefficient is obtainable from evaluation of the following formula:

$$\begin{aligned}(\Delta C_{D_i})_3 &= (S_x - 1)(C_{D_i})_0\frac{A_p}{A} - 2(C_L)_0\frac{A_p}{A}\frac{(C_{L\alpha})_0}{\pi A\!\!R}\frac{4}{\pi}(C_s + \alpha_T C_T) \\ &\quad + \frac{4}{\pi}(C_s + \alpha_T C_T)(C_L)_0\frac{A_p}{A} + 2(S_x - 1)(C_{D_i})_0\frac{A_p}{A}\frac{(C_{L\alpha})_0}{\pi A\!\!R} \\ &\quad - (S_x - 1)(C_{D_i})_0\frac{A_p}{A} + \frac{2(S_x - 1)}{c_{l_\alpha}}(C_L)_0^2\frac{A_p}{A}\frac{(C_{L\alpha})_0}{\pi A\!\!R}\end{aligned} \tag{24-16'}$$

Finally, in order to obtain the value of the total change in C_L and C_{D_i} caused by the propeller interference, the following effects must be added together:

1. The changes $(\Delta C_L)_4$ and $(\Delta C_{D_i})_4$ that are due to the discontinuous part, $\Delta\gamma_d$, of the incremental circulation $\Delta\gamma$. Now, within the limits of approximation set forth at the beginning of this article, it is possible to assume for this increment merely the following form:

$$\begin{aligned}\Delta\gamma_d &= 0 && \text{for } |\eta| \geqq h \\ \Delta\gamma_d &= -a_1 = -(S_x - 1)\gamma_0^{(0)} && \text{for } |\eta| \leqq h\end{aligned}$$

The increments in the lift and drag are then given by

$$\left.\begin{aligned}(\Delta C_L)_4 &= -(S_x - 1)(C_L)_0\frac{A_p}{A} \\ \text{and} \qquad (\Delta C_{D_i})_4 &= -\frac{1}{\pi A\!\!R}(C_L)_0^2(S_x - 1)\frac{A_p}{A}\end{aligned}\right\} \tag{24-17}$$

and

2. The changes $(\Delta C_L)_5$ and $(\Delta C_{D_i})_5$ due to the fact that a relatively higher speed flow is impinging upon that portion of the wing which lies within the slipstream. These increments are given by the expressions

$$\left.\begin{aligned}(\Delta C_L)_5 &= (S_x - 1)(C_L)_0 \frac{A_P}{A}\\(\Delta C_{D_i})_5 &= (S_x - 1)(C_{D_i})_0 \frac{A_p}{A}\end{aligned}\right\} \tag{24-18}$$

Consequently the total changes may be combined into the summary equations:

$$\Delta C_L = \sum_{n=1}^{5} (\Delta C_L)_n = 2(S_x - 1) \frac{A_P}{A} (C_L)_0 \left[1 - \frac{(C_{L\alpha})_0}{\pi Æ}\right] - \frac{4}{\pi} \frac{A_P}{A} A_0 (C_z + \alpha_T C_T)(C_{L\alpha})_0 \tag{24-19}$$

and

$$\Delta C_{D_i} = 4 \frac{A_P}{A} (S_x - 1)(C_{D_i})_0 \left[1 - \frac{(C_{L\alpha})_0}{\pi Æ}\right] + \frac{2}{\pi} (1 + A_0) \frac{A_P}{A} (C_L)_0 (C_z + \alpha_T C_T) - \frac{8}{\pi^2 Æ} (C_L)_0 \frac{A_P}{A} (C_z + \alpha_T C_T) A_0 \tag{24-19'}$$

In addition, the form drag of each airfoil section along the span of the wing changes as a result of the change in the magnitude of the velocity of the impinging flow and because of the change of direction experienced by this same flow. The latter change can also be calculated on the basis of the results just elicited, as follows:

Let Ƅ denote the downward directed angle by which the flow is deflected by the action of the propeller interference effects and also because of the influence of the induction effects created along the wing, and divide Ƅ into two parts

$$Ƅ = Ƅ' + Ƅ''$$

where Ƅ' is the contribution to Ƅ coming from the *continuous* distribution of angles of attack produced from the interference actions, and Ƅ'' the part which is left over. Then one has for the continuous part that

$$\begin{aligned}Ƅ' &= \frac{2\pi}{c_{l_\alpha}} \frac{1}{\sin\theta} \left[(S_x - 1) \frac{2}{\pi Æ} (C_L)_0 \frac{E_1}{2\pi} - \frac{\bar{v}_{z\infty}}{U} (1 - A_0) E_2\right]\\&= -\frac{\pi}{2} \frac{Æ}{c_{l_\alpha}} \frac{\Delta\gamma_0'}{\sin\theta}\end{aligned} \tag{24-20}$$

in which

$$\left.\begin{aligned} E_1 &= \int_{\bar\theta=-\pi}^{\bar\theta=\pi} \varepsilon^*_{c,1}(\bar\theta)\sin\bar\theta dV_{\mathcal{R}}(\bar\theta-\theta) \\ \text{and} \qquad E_2 &= \int_{\bar\theta=-\pi}^{\bar\theta=\pi} \varepsilon^*_{c,2}(\bar\theta)\sin\bar\theta dV_{\mathcal{R}}(\bar\theta-\theta) \end{aligned}\right\} \tag{24-21}$$

where the simplifying definitions

$$\varepsilon^*_{c,1} = -\varepsilon_{c,1}\frac{2\pi}{\gamma_0^{(0)}(S_x-1)} \quad \text{and} \quad \varepsilon^*_{c,2} = \varepsilon_{c,2}\frac{U}{\bar v_{x\infty}}\frac{1}{1-A_0}$$

have been utilized for the sake of convenience.

The part which is left over may then be written

$$\left.\begin{aligned} ь'' &= (S_x-1)\alpha - \frac{\pi\mathcal{R}}{2c_{l_\alpha}}\frac{\Delta\gamma_{c,3}+\Delta\gamma_d}{\sin\theta} - \frac{1}{\pi\mathcal{R}}(C_L)_0(S_x-1) \\ &= (S_x-1)\alpha - \frac{\pi\mathcal{R}}{2c_{l_\alpha}}\frac{\varepsilon_d F(\lambda,\theta)}{\sin\theta} + \frac{(S_x-1)(C_L)_0}{c_{l_\alpha}\sin\theta} \\ &\qquad - \frac{1}{\pi\mathcal{R}}(C_L)_0(S_x-1) \quad \text{for } |\eta| \leqq h \\ ь'' &= -\frac{\pi\mathcal{R}}{2c_{l_\alpha}}\frac{\Delta\gamma_{c,3}}{\sin\theta} = -\frac{\pi\mathcal{R}}{2c_{l_\alpha}}\frac{\varepsilon_d F(\lambda,\theta)}{\sin\theta} \quad \text{for } |\eta| \geqq h \end{aligned}\right\} \tag{24-22}$$

Therefore it is clearly evident that the increment in form drag, experienced by an arbitrary general one of the airfoil sections located out along the wing span, is just

$$\Delta c_{d_0} = (c_{d_0})_{\alpha-ь}(2S_x-1) - (c_{d_0})_\alpha \tag{24-23}$$

where $(c_{d_0})_\alpha$ represents the profile drag of the airfoil section when it is at an angle of attack denoted by α, and where $(c_{d_0})_{\alpha-ь}$ is then the form drag of the airfoil when it is at an angle of attack $\alpha - ь$.

C,25. Rolling Moment of the Elliptic Wing Produced by the Induced Rotational Components Generated by the Propeller.

The determination of the rolling moment induced by the rotation of the propeller is very easily and quickly arrived at in the case of an elliptic wing, provided that one makes use of the previously discussed expressions, Eq. 23-5, for the distribution of the increment in the peripheral component of the induced velocities. If $\Delta\gamma_r(\theta)$ is used to denote the circulation distribution existing out along the wing in this case, its explicit formulation is

$$\begin{aligned} \Delta\gamma_r(\theta) = \frac{8a\beta^2\tan\Phi}{\pi}&\left\{\sum_{n=2}^{\infty}\frac{\sin n\theta}{n+\lambda}\left[\frac{(-1)^{n/2}+\sin(n-1)\theta_0}{n-1}\right.\right. \\ &\left.\left.+\frac{(-1)^{n/2}-\sin(n+1)\theta_0}{n+1}\right] \quad \text{for } n = 2, 4, 6, \cdots\right\} \end{aligned} \tag{25-1}$$

and the corresponding rolling moment produced on the wing (for a right-handed propeller) is given by

$$C_l = \frac{M_r}{\frac{1}{2}\rho_\infty U^2 A b} = -\frac{\text{Æ}\pi}{8}\frac{4 \cdot 2a\beta^2 \cdot \tan\Phi}{\pi}\frac{1}{2+\lambda}\left(-\frac{4}{3} + \sin\theta_0 - \frac{\sin 3\theta_0}{3}\right)$$

$$\cong \frac{\text{Æ}}{2+\lambda} a\beta^2 \tan\Phi \, 2h^2 \qquad (25\text{-}2)$$

C,26. Comparison of Results Obtained from Theory with Experimental Tests. Influence of the Mach Number on Such Results. The comparisons cited below are limited to confrontation of the computed values of the over-all effects, embodied in the increments ΔC_L and ΔC_D, derivable from the formulas presented in Art. 23, with the corresponding experimentally determined increments in lift and drag. No such comparisons are afforded on the basis of the circulation distributions themselves, because there is a scarcity of experimental results showing the variation of $\Delta\gamma$ with η, and what results are available are of questionable value.

The experimentally determined values of ΔC_L and ΔC_D that were obtained by Possio [*35*], working with a wing having a trapezoidal planform and an aspect ratio of 6.7 and operating behind a propeller set at zero angle of attack, $\alpha_T = 0$, are recorded in Fig. C,26a and C,26b. The other pertinent geometric features which characterize this particular propeller-wing combination are also shown in the same figures.

Superimposed on these experimental data are the curves obtained from the appropriate calculations made on the basis of Eq. 24-19, 24-19′, and 24-23. The latter formula for the incremental form drag has been replaced by the simpler relationship expressed as

$$\Delta c_{d_0} = 2(c_{d_0})_\alpha(S_x - 1)$$

which is obtained on the assumption that it is permissible to neglect the small changes produced in the profile drag because of any local changes in the angle of attack. This simplified formula is also equivalent to a change in the form drag of the entire wing in question, of the amount

$$\Delta C_{D_0} = 2(c_{d_0})_\alpha(S_x - 1)\frac{A_p}{A}$$

From a comparison of the calculations and the experimental data it is apparent that the correspondence between them is quite satisfactory in the case of the ΔC_D values, but that the ΔC_L computed values are somewhat too high to fit the actual test results. The latter trend had already been noticed in connection with the case of wings of infinite aspect ratio which were studied by Franke and Weinig [*30*]. These investigators attributed the discrepancy to the fact that, in spite of the basic stipulation regarding the smallness of the ratio c/R that is made in establishing the theoretical background of this propeller interference problem (see the statement of hypothesis 4 and the writing of Eq. 18-1 in Art. 18), nevertheless, in reality the value of this ratio is greater than unity.

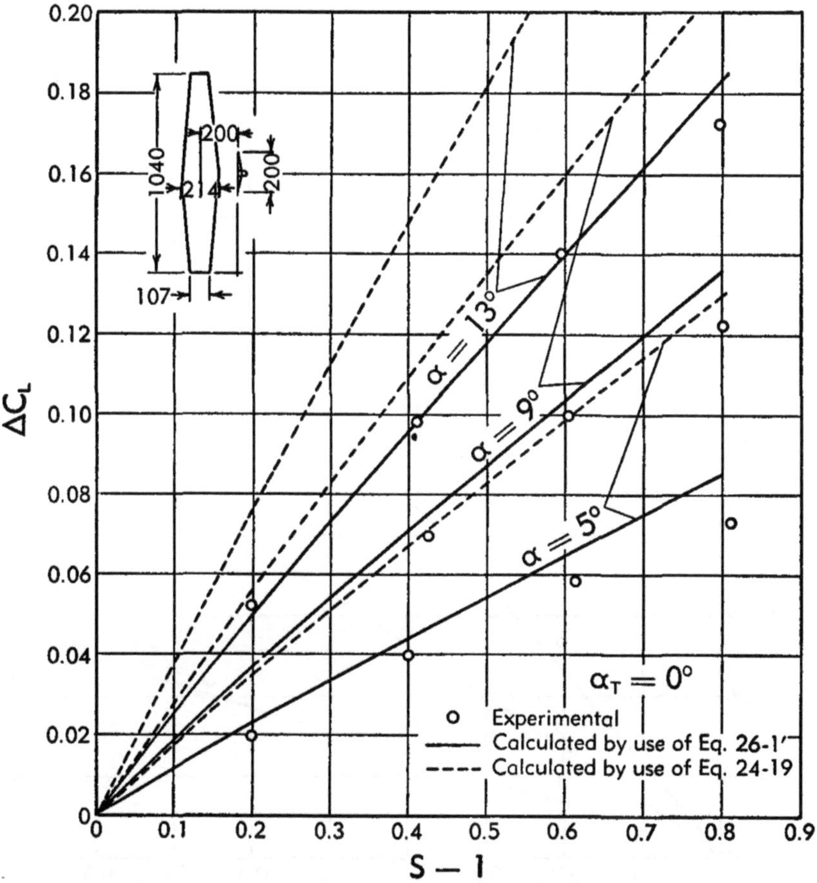

Fig. C,26a. Experimental and calculated values of ΔC_L, produced on configuration sketched, by action of an untilted propeller.

It has been proposed by Weinig that a suitable correction factor be assumed to apply to the circulation distribution $\Delta\gamma$, and consequently it would also multiply the ΔC_L values corresponding to the circulation so treated. This factor is taken to be given by the expression

$$K = \frac{1}{3} + \frac{2}{3}\frac{R}{c}\tanh\frac{c}{R} \tag{26-1}$$

When $c/R = 2$, the value of K is 0.654. The values of

$$\langle \Delta C_L \rangle = K \Delta C_L \tag{26-1'}$$

have also been plotted in the same Fig. C,26a. It may be seen from such plots that the values of $\langle \Delta C_L \rangle$ computed in this manner are in much better agreement with the experimental results than the former unadjusted ones.

Experimental results have also been obtained by Possio [35] in the case where the same wing-propeller combination as previously used is now constrained to give values of $\alpha_w = 12°$ and $\alpha_T = 10°$ for various values of

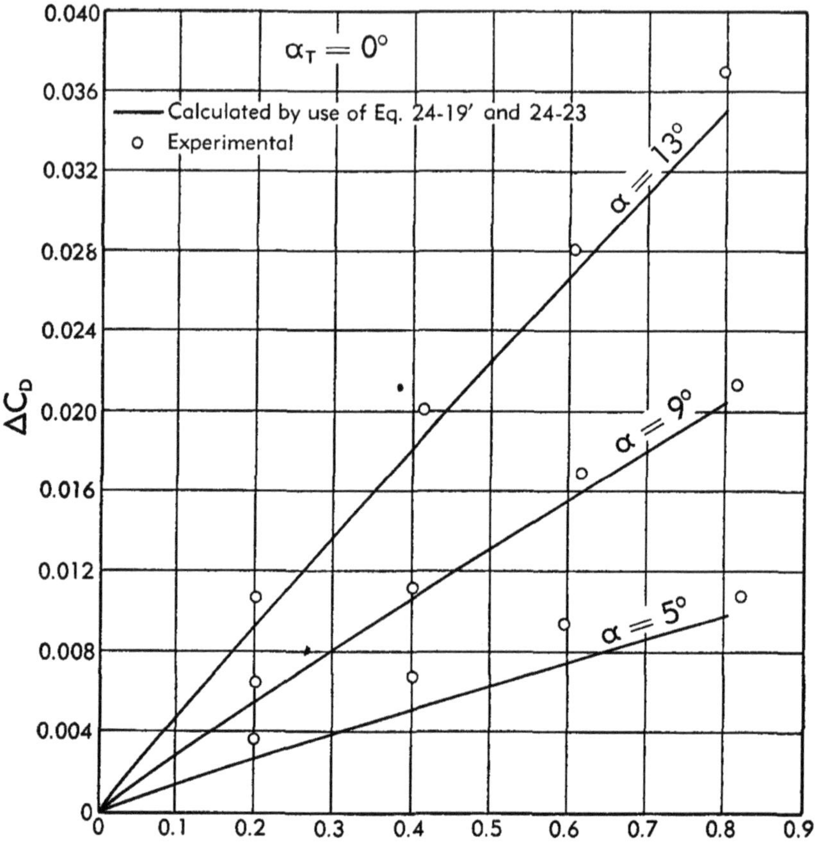

Fig. C,26b. Experimental and calculated values of ΔC_D, produced by an untilted propeller acting on the configuration sketched in Fig. C,26a.

$S - 1$. The effects on the ΔC_L and ΔC_D coefficients are recorded here in Fig. C,26c.

From a comparison of the data presented in this figure with the results plotted in the preceding figures it is evident that the angle of attack α_T of the propeller axis has a remarkably strong influence, causing a decrease in ΔC_L. In contrast, however, it may be seen from Eq. 24-19 that the value of α_T has a small effect on ΔC_L values. The changes which are brought about in ΔC_L as a result of an increase in the angle of attack of the propeller are significantly amplified over what would be obtained from calculation with Eq. 24-19, provided that one takes into account additional factors in computation of these lift increments, which have so far been neglected. For instance, one would obtain larger values if the quadratic terms in $h = 2R/b$ were included, and if the increased velocity existing in the slipstream flow were more adequately represented, which means that a quadratic term in $(S_x - 1)$ should also be inserted into the equations.

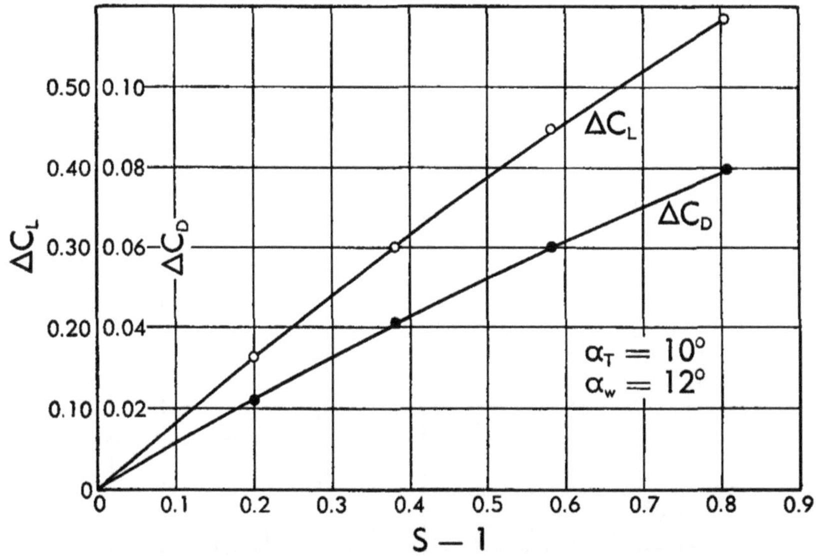

Fig. C,26c. Experimental values of ΔC_L and ΔC_D, produced on the propeller-wing system shown in Fig. C,26a, for the case where the wing is at an angle of attack of 12° while the propeller is tilted up 10°.

If the above considerations are added, the expression previously listed as Eq. 24-12 becomes modified to

$$(\Delta C_L)_2 = \frac{4}{\pi}\frac{A_p}{A}(C_{L\alpha})_0\left(1 - A_0 - \frac{\pi}{4}h\right)(C_s + \alpha_T C_T) \qquad (26\text{-}2)$$

and at the same time the second term on the right-hand side of Eq. 24-16 is now altered to

$$-\frac{4}{\pi}\frac{A_p}{A}(C_{L\alpha})_0(C_s + \alpha_T C_T)\left[1 + (S_x - 1)\frac{4h}{\pi}\right] \qquad (26\text{-}3)$$

In this way, the change in the total lift coefficient ΔC_L that is attributable to α_T, and which is now denoted by $(\Delta C_L)_T$, takes on the expanded formulation

$$(\Delta C_L)_T = -\frac{4}{\pi}\frac{A_p}{A}(C_s + \alpha_T C_T)(C_{L\alpha})_0\left[(S_x - 1)\frac{4h}{\pi} + \frac{\pi h}{4} + A_0\right] \qquad (26\text{-}4)$$

A quantitative comparison between the experimental results presented in Fig. C,26c and the theoretical results corrected in the manner just indicated cannot, unfortunately, be carried out because the additional aerodynamic characteristics of the propeller used in these experiments are not available, i.e. it is not known what C_s and C_T values apply to this particular propeller.

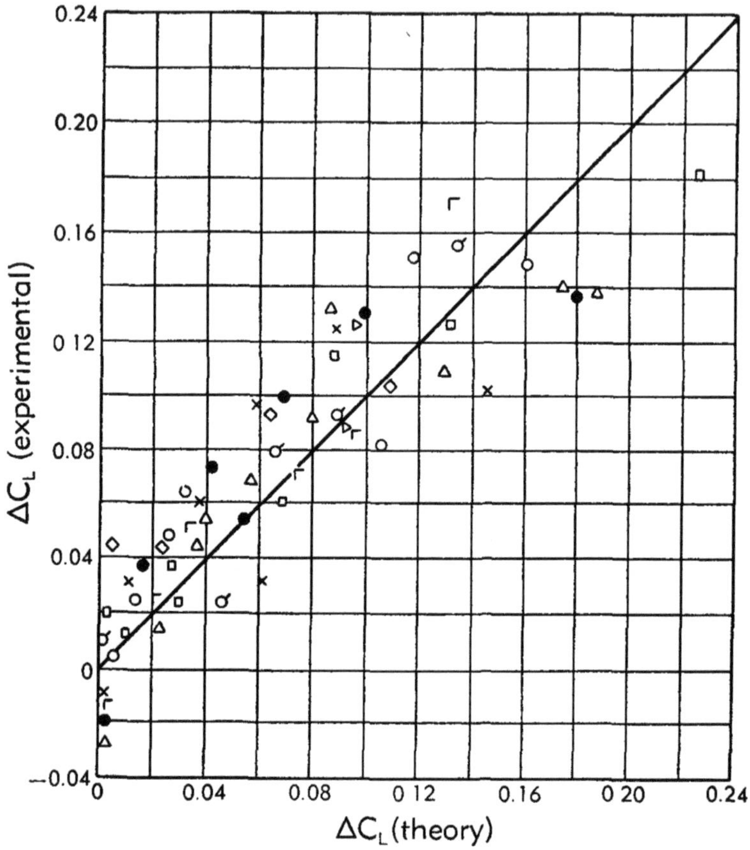

Fig. C,26d. Comparison of ΔC_L values, computed by use of Eq. 26-5, with the ΔC_L values obtained by experimental testing (from [*36*]).

It is also to be regretted that no quantitative comparison can be made between the theoretical results deducible from the above-presented formulas and the experimental results reported by Weil and Sleeman [*36*]. Their results are replotted here as Fig. C,26d. The $(\Delta C_L)_{\text{exptl}}$ increments given in this figure are listed in [*36*] as a function of the product $C_T(C_L)_0$. Of course it is possible to correlate C_T with the values of $(C_L)_0$ through knowledge of the aerodynamic characteristics of the airplane model employed in the experiments, but the listing of these characteristics is not included in the data reported.

Nevertheless one may still make a worthwhile check on these results in the following manner. The above-mentioned authors give the experimentally determined relation between ΔC_L and C_T in the form

$$\Delta C_L = 0.57 C_T (C_L)_0 \frac{c_w}{\bar{c}_w} \frac{4R^2}{A} \tag{26-5}$$

wherein c_w represents the length of the wing chord located at a distance of $0.75R$ from the root chord, and where $\bar{c}_w$ stands for the average chord length. This relationship may be converted, in the case of an elliptic wing, into the specialized equation

$$\Delta C_L = 1.45 C_T (C_L)_0 \frac{A_p}{A} \frac{R}{c_0} \tag{26-5$'$}$$

If it is further particularized to the case where $R/c_0 = \frac{1}{2}$, then one has

$$\Delta C_L = 0.725 C_T (C_L)_0 \frac{A_p}{A}$$

Now, by returning to Eq. 24-19, and taking into account the correction factor given as Eq. 26-1, and still assuming that the value of R/c_0 is one half, the corresponding theoretical treatment stemming from the kind of analysis given above leads to the analogous expression

$$\Delta C_L = 0.995 C_T (C_L)_0 \frac{A_p}{A}$$

It must be remembered, however, that the latter formula holds only for the case where $\alpha_T = 0$, while the effects of a nonzero value of α_T are included in the empirical formula first cited. It is worthy of note that the ratio between the two different ΔC_L values is in agreement with the results given in Fig. C,26a and C,26c.

Finally, in regard to what effects the Mach number exerts on these interference phenomena, it should be mentioned that these influences are exhibited in the following manner:

1. There is an increase in the ratio R_1/R, as is evident from reference to Eq. 19-6′, and thus there is an increase in the ratio A_p/A and in turn in $\Delta C_L/(C_L)_0$, as the Mach number increases.
2. There is an increase in the size of the $(C_{L\alpha})_0$ value, and this particular augmentation brings with it a decrease in the $\Delta C_L/(C_L)_0$ ratio, as the Mach number increases.

These two Mach number effects are thus counteracting and it is probably safe to assume that the magnitude of M_∞ (to at least the degree of approximation that is contingent upon acceptance of the basic hypotheses underlying the analyses followed above) has but a trivial effect upon the interference action produced on the wing by the propeller.

INTERFERENCE OF THE PROPELLER UPON THE WING (WING LOCATED AHEAD OF THE PROPELLER)

C,27. Simplifications Possible by Discarding the Slipstream Boundary Conditions. If the wing is located ahead of the pusher-propeller, provided it is not too close to it, the interference problem which arises exhibits markedly simplified features over what has been encountered up to now in the tractor-propeller case. This occurs because it is possible to drop out of the analysis any consideration of the boundary conditions on the propeller slipstream, as previously required when the wing was behind the propeller; thus the computation of the circulation distribution out along the wing span is vastly condensed. In fact, the problem is merely reduced to the determination of how the distribution of the circulation increments $\Delta\gamma$ varies along the wing as a function of η, in the idealized case where the given wing is supposed to be operating in a non-uniform, but nonetheless definitely specified flow field. The known flow field is composed of the free stream flow plus the flow created by the action of the propeller. Such a situation can be handled by the usual methods.

INTERFERENCE EFFECTS EXERTED UPON THE TAILPLANE, ARISING THROUGH ACTION OF THE PROPELLER AND THE PROPELLER-WING COMBINATION

C,28. Direct Effect of the Propeller upon the Stabilizer. The study of how the propeller *directly* influences the tail surface, that is, how the propeller affects the tail surface when the wing is not present, may be carried out in exactly the same way that the problem of propeller-wing interference was dealt with in the foregoing articles. The only real difference which arises between the situation encountered in Art. 24, and that under consideration, lies in the fact that the ratio between the span of the tail b_t, and the diameter of the propeller $2R$, now comes out to be close to unity.

With regard to configurations for which the tail plane may be considered as entirely immersed within the slipstream it should be realized that the induced angles of attack due to interference are generated not only by action of the trailing vortices streaming from the propeller, but also by induction from the inverted images of the trailing vortices shed from the tail plane. Consequently, if the circulation distribution $\gamma_0^{(0)}$ may be assumed elliptic, then the induced angle of attack at the tail location is given by

$$(\varepsilon_0)_t = -\frac{\bar{v}_{z\infty}}{U} - \frac{\gamma_0^{(0)}}{2}(S-1)\frac{1}{\cos^2\theta}\left(-1 + \frac{1}{\sin\theta}\right) \qquad (28\text{-}1)$$

wherein it is now agreed that $\cos\theta = 2y/b_t$.

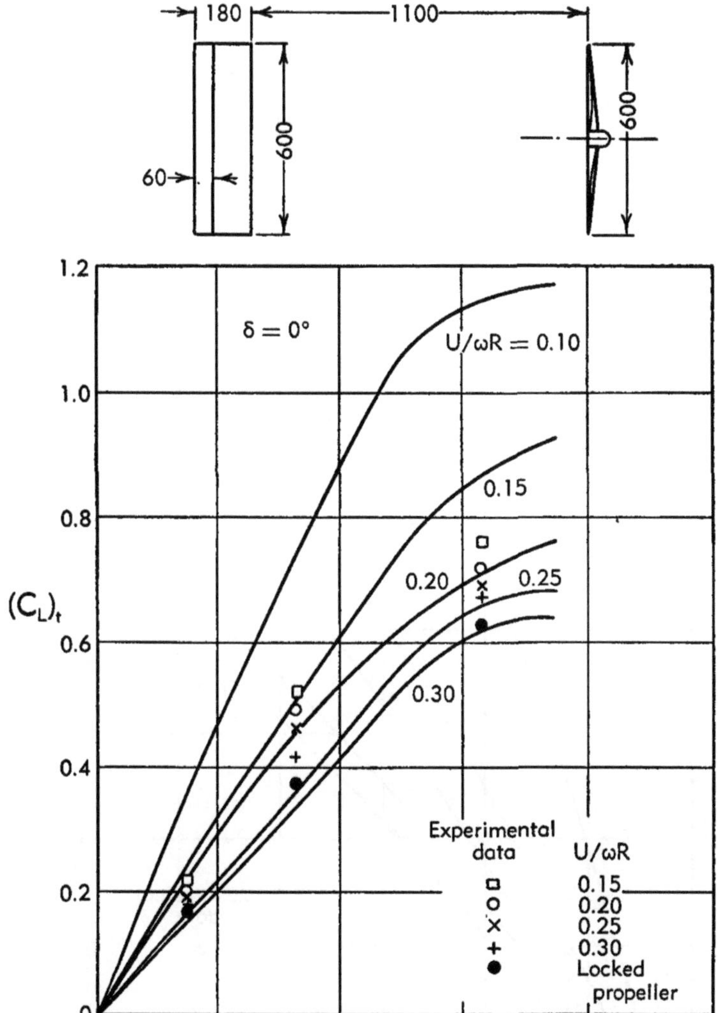

Fig. C,28a. Computed and experimental values of C_L applying to the case where the stabilizer lies entirely within the slipstream. Condition of zero-setting for the elevator angle.

The formula for the lift coefficient of the stabilizer in presence of the propeller thus comes out in the form

$$(C_L)_t = (C_{L_0})_t \left[S^2 - \frac{2}{\pi} \frac{\pi - 2}{1 + (\pi \mathit{Æ}_t / c_{l\alpha})} (S - 1) - (C_{L_\alpha})_{t,0} \frac{\bar{v}_{s\infty}}{U} S \right] \quad (28\text{-}2)$$

The first term on the right-hand side of this equation, namely $S^2(C_{L_0})_t$, represents the influence on the tail lift of the increment in the slipstream

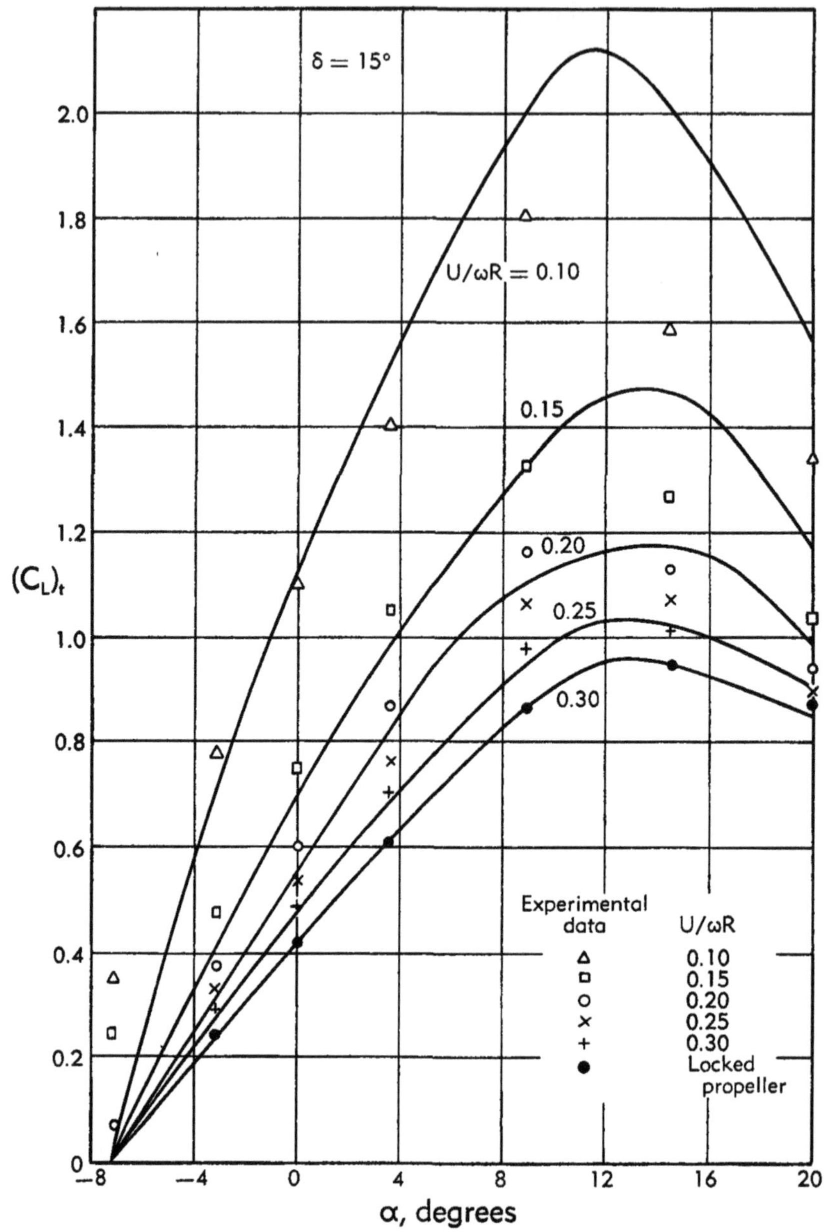

Fig. C,28b. Computed and experimental values of C_L applying to the case where the stabilizer lies entirely within the slipstream. Condition of elevator deflection of 15°.

velocity produced by the propeller. The remainder of the quantity in brackets represents the decrease in tail lift occasioned by the presence of the induced angles of attack arising from the image vortices. Finally the last term on the right corresponds to the deflection produced in the flow by action of the tilted propeller.

Theoretical and experimental results [*37*] are plotted in Fig. C,28a and C,28b to illustrate what happens in the case where a tailplane of elliptic planform is located entirely within the slipstream and is operated under the conditions of no angle of deflection on the elevator, and then with the elevator deflection set at 15°. The pertinent geometric characteristics and the position occupied by the stabilizer with respect to the propeller producing the interference phenomena are sketched in Fig. C,28a.

First of all, it should be pointed out that the correction factor K, defined in Art. 26, is practically equal to unity if the value of c/R is quite small, as it is in these examples of tail-propeller interference. It is not surprising therefore to find that the agreement between theoretical and test results is more than satisfactory at small angles of attack. The matching of the experimentally obtained results is particularly good when the theoretical value for the change in the angle of attack for zero lift, corresponding to the given angle of elevator deflection under consideration, is computed by aid of Eq. 28-2.

The discrepancy between theory and experiment grows worse, however, when the high angle of attack region is entered. However, this not uncommon situation is easily explained in the present instance, because as the angle of attack becomes larger and larger, the part of the tail which emerges from the slipstream grows more and more extensive.

C,29. Effect of the Propeller-Wing Combination upon the Tail Surface. In order to take into account the wing's presence when investigating the interference pheomena that come into play at the tail, it is necessary to add to the *direct* interference effects producing induced angles of attack at the tail (according to Eq. 28-1) those additional increments to the induced angles of attack that arise from the interference actions coming from the wing. The determination of these further contributions to the induced angles may be obtained on the basis of the results that have already been compiled in the preceding articles, if it is assumed that the stabilizer lies at a sufficient distance from the wing so that one can consider the induced angle of attack to have the same value that it has at infinity. Relying on this simplifying idealization, it is seen that the downward deflection of the flow caused by the wing may be expressed by means of Eq. 24-20 and 24-22, and thus it follows that

$$(\varepsilon_t)_w = 2(\text{Ь}' + \text{Ь}'') - \frac{\bar{v}_{z\infty}}{U}(3 + A_0) \qquad (29\text{-}1)$$

Of course the value of $(\varepsilon_t)_w$ changes from point to point along the span of the tail surface, but, if one replaces this variable value by a constant average value $(\bar{\varepsilon}_t)_w$, the expression given previously as Eq. 28-2 will be altered by the addition of this term, so that it now reads

$$(C_L)_t = (C_{L_0})_t \left[S^2 - \frac{2}{\pi} \frac{\pi - 2}{1 + (\pi \mathit{A\!R}_t / c_{l_\alpha})} (S - 1) \right] - (C_{L_\alpha})_{t,0} S \left[\frac{\bar{v}_{z\infty}}{U} + (\bar{\varepsilon}_t)_w \right] \qquad (29\text{-}2)$$

Even if the wing planform is not elliptic, it is still possible to calculate the lift coefficient of the stabilizer by means of Eq. 29-2. All that is required fundamentally is the determination of the Ƅ′ and Ƅ″ values by

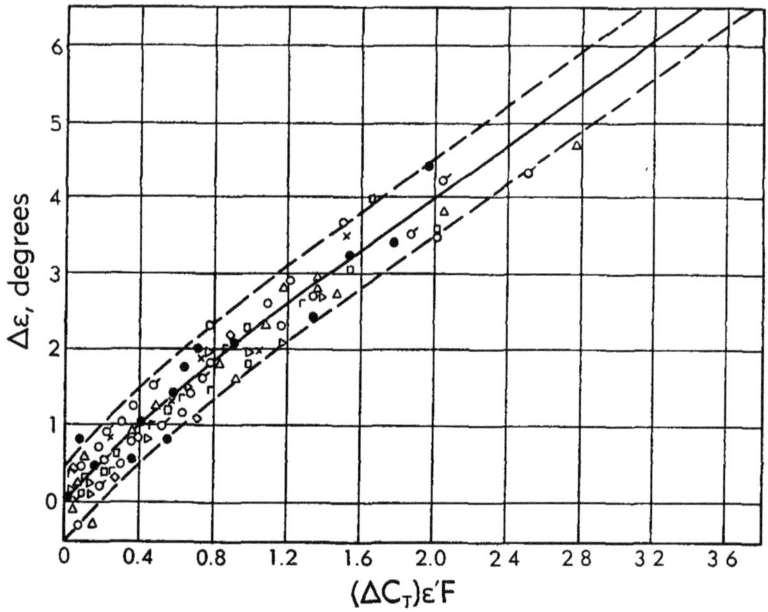

Fig. C,29a. Experimental data and suggested design curve, showing variation of the increment of downwash angle due to power, with the parameter $(\Delta C_T)\varepsilon' F$.

having recourse to the methods expounded in Art. 23 for calculating the values of $\Delta\gamma_e$ and $\Delta\gamma_d$, which are the quantities upon which the calculation of Ƅ′ and Ƅ″ is based.

The influence of the propeller upon the aerodynamic characteristics of the tail surface is very strong, and it also depends heavily upon the kind of wing planform involved. Keeping this in mind, it is instructive to see what light can be shed on these very significant phenomena by the experimental results of Weil and Sleeman [*36*]. For illustration one of their curves incorporating the results of tests on 28 model configurations investigated by these authors is reproduced here as Fig. C,29a. The ordinates in this plot are the familiar angle of deflection (downwash) increments caused by the presence of the propeller, or $\Delta\varepsilon$, but the abscissas need a further word of explanation. The three multiplicative factors making up this parameter have the following meanings: ΔC_T is the increment in the propeller thrust coefficient which occurs in going from the condition of a windmilling propeller up to a series of various power settings, ε' is the increment in the induced angle of attack produced at the tail surface by action of the windmilling propeller in comparison with the condition existing there when $(C_L)_w = 0$, and, lastly, F is an empirical factor which is related to the alterations brought about by the taper of the wing. The taper may be expressed in terms of a single "taper ratio"

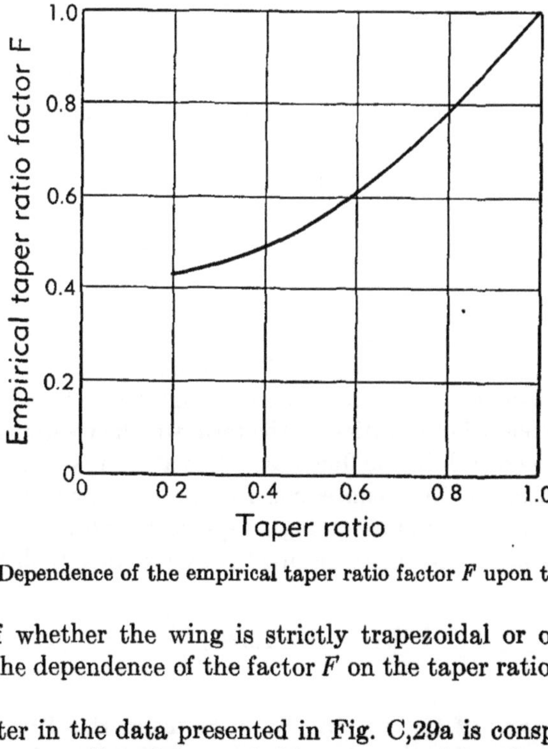

Fig. C,29b. Dependence of the empirical taper ratio factor F upon the taper ratio.

regardless of whether the wing is strictly trapezoidal or only approximately so. The dependence of the factor F on the taper ratio is plotted in Fig. C,29b.

The scatter in the data presented in Fig. C,29a is conspicuously evident. These severe disparities probably are caused by the fact that, as may be seen by reference to Eq. 29-2, $\Delta\varepsilon$ is not solely a function of the parameter $(\Delta C_T)\varepsilon' F$. For one thing, the effective downwash angle affecting the tail depends strongly on the aspect ratio in particular, but it also varies with its position relative to the wing and propeller, as well as upon the connection between the lift coefficient $(C_L)_w$ of the wing and the thrust coefficient C_T, etc. Nevertheless, the trend of variation between the downwash and the parameter $(\Delta C_T)\varepsilon' F$ is a very useful tool to the designer, because it gives him a rough statistical measure of the general value of

$\Delta\varepsilon$ which can be expected to result from a particular configuration, without necessitating detailed calculations; the side-stepping of these somewhat laborious computational chores can often be sanctioned in preliminary design.

Further interesting conclusions may also be drawn from these experimental data. For instance, if the quantity dC_M/di_T is defined as the efficiency factor of the tail, it can be related to the thrust coefficient. Of course the C_M appearing in this differential quotient is the pitching moment due to the aerodynamic forces acting on the tail surface under consideration, while the i_T represents the angle of attack for zero lift of the wing measured with respect to the propeller axis. The sign of this angle is considered to be positive when the trailing edge of the airfoil is deflected downward. Now, form the ratio of such an efficiency factor, obtained when the airplane is in condition of flight corresponding to a given power setting of the propeller, to the same factor obtained when the propeller is windmilling, and call it R_T. It is at once apparent, on the basis of these experimental data, that the ratio can be adequately expressed, on the average, by means of the relation

$$R_T = 1 + 2.1\left[\Delta C_T \frac{2R}{b_t}\left(1 - \frac{|d_T|}{2R}\right)\right] \tag{29-3}$$

where d_T is the perpendicular distance from the hinge line, about which the elevator rotates, up to the axis of the propeller.

It is worthwhile to note the following significant fact in connection with such efficiency factor ratios. If the term which accounts for the influence of the downwash in the flow upon the value of $(C_L)_t$ is dropped out of Eq. 29-2, the resulting expression for the efficiency factor ratio R_T, obtained on this basis and assuming a representative value of the aspect ratio Æ to be 3.5, turns out to be practically identical with Eq. 29-3, i.e. in this case

$$R_T = 1 + 2.22 C_T \frac{2R}{b_t}$$

C,30. Influence of the Lifting Surface upon the Propeller. The interference exerted by the wing upon the propeller is of much less importance than the effect of the propeller upon the wing. The effect of the wing upon the propeller is due to the following two paramount causes.

Change of direction of the airflow. The change produced in the direction of the airflow entering the propeller is caused by the action of the system of vortices Σ_w, which are associated with the flow over the wing. If the wing is located far enough away from the propeller so that the velocities induced by the vortex system Σ_w exhibit but a small amount of variation over the propeller disk region, it is permissible to assume a constant mean value for the induced velocity which will hold over the whole disk area, and the value that the induced velocity attains at the center of the propeller disk may be taken as an adequate representative value for the constant velocity increment. If the notation $\delta - \alpha_w$ is now used to denote the change in direction experienced by the velocity as a result of the action of the vortex system Σ_w, the actual angle of attack of the flow that is impinging upon the propeller, measured with respect to the propeller axis, may be expressed by the formula

$$(\alpha_T)_p = \alpha_T + (\delta - \alpha_w) \tag{30-1}$$

Then, if the propeller is located ahead of the wing, by following the same procedure as used in Art. 8, it is found that

$$\begin{aligned} \delta - \alpha_w = (C_L)_w \Bigg[\frac{-8}{\pi^3} \frac{1}{Æ} \left(1 + \frac{2\xi^*}{\sqrt{4\xi^* + \beta^2(\pi^2/16) Æ^2}}\right) \\ - \frac{\beta^2}{4\xi^* \pi} \frac{Æ}{\sqrt{4\xi^{*2} + \beta^2(\pi^2/16) Æ^2}} \Bigg] \\ = (C_L)_w H \end{aligned} \tag{30-2}$$

wherein $\xi^* = x/c_{av}$, while the abscissa value x is measured positively, starting from the aerodynamic center of the wing and proceeding rearward toward the tail. It follows that the new angle of attack of the propeller is related to the wing lift coefficient by the formula

$$(\alpha_T)_p = \alpha_T + (C_L)_w H \tag{30-3}$$

Because H is a positive quantity (see Art. 8), it follows that, for a positive value of lift on the wing, a positive increment in the angle of attack of the air stream impinging upon the propeller axis is created. To this increment in angle of approach there corresponds an increase in the lift force coefficient C_z, the thrust coefficient C_T, and the torque produced by the propeller.

If, on the other hand, the propeller happens to lie behind the wing, provided it is sufficiently close to it, a simpler formula for the effective angle of flow inclination that exists ahead of the propeller can be substituted for Eq. 30-3; this formula is now seen to be merely (refer to Art. 8)

$$(\alpha_T)_p = \alpha_T - \alpha_w$$

It may be deduced from this fact that there is a decrease produced in the angle of attack of the approaching flow with respect to the direction of the propeller axis.

Change of velocity. Changes are produced in the magnitude of the velocities which enter the area of the propeller disk. In addition, the velocities are now distributed over the face of the propeller disk in a nonuniform manner. For the usual configurations employed in wing-propeller combinations the approach velocity is augmented over the upper part of the disk and diminished over the lower part. It turns out therefore that a pitching moment is created which should be added to that due to the angle of attack of the propeller, $(\alpha_T)_p$.

Of the two effects due to the workings of the factors cited above, the results associated with the change of direction are the salient ones. Therefore it would appear that, at least from a qualitative point of view, it is possible to take into account the "interference of the interference" by merely assuming that $(\alpha_T)_p > \alpha_T$ or $(\alpha_T)_p < \alpha_T$ according to whether the propeller lies ahead of or behind the wing, and by subsequently adjusting the values of the coefficients C_s and C_T accordingly.

CHAPTER 4. WING-TAIL PLANE INTERFERENCE AT SUPERSONIC SPEEDS

C,31. Introduction. The determination of the interference effects produced on the horizontal tailplane by the action of the neighboring wing may be broken down into two subproblems which may be handled in sequence according to the method used when a similar problem was met in the subsonic regime. The solution of the fundamental question thus resolves itself into (1) finding the field of induced angles of attack produced by the wing, and then (2) calculating from knowledge of the induced angles of attack how the aerodynamic behavior of the stabilizer is altered from what it would otherwise be when operating as an isolated entity.

In taking up the study of the first subproblem just mentioned, several modes of attack are expounded, but they may all be classified under one of the two general headings:

1. Methods based on the idea of considering the wing to be a lifting surface.
2. Methods stemming from replacement of the actual wing by a simple lifting line idealization.

From the standpoint of the basic concepts involved, the methods that are founded on the lifting line theory are quite obviously analogous to the procedures discussed in Chap. 1, although from the analytic point of view, to be sure, this latter class of methods of approach exhibits quite noticeable differences in development according to whether the flow is subsonic or supersonic.

The principal advantage accruing from use of such lifting line treatment of the problem lies in the relatively simple form of solution to which they lead. However, they afford a sufficiently close approximation for purposes of calculating the induced angle of attack ϵ quite accurately, so long as the point in question, where the value of the downwash is desired, falls a good distance downstream of the trailing edge of the wing (the precise statement of just how far downstream one needs to go in order to be assured of a given degree of accuracy is discussed more fully later on).

On the other hand, the methods which arise from the use of lifting surface theory lead, in general, to formulas which are a good deal more complicated when it comes to actual numerical computations, although they do possess the highly desirable attribute of furnishing the means for determining ϵ at any point whatsoever, located anywhere at all without restriction with respect to the wing. Furthermore, they afford a check on the degree of approximation that one can obtain by working with the simpler lifting line procedures.

The value of the upwash ϵ, here too, depends upon just what is assumed concerning the rolling-up process which takes place in the vortex sheet formed by the continuum of trailing vortices that are shed from the rear of the wing. The effect of such a roll-up phenomenon upon the downwash is accounted for in an approximate manner in what follows by proceeding along the same lines adopted in Chap. 1 for handling the same situation.

If one wishes to have an estimate of how far downstream from the trailing edge of the wing it is possible to go, before arriving at the location where the vortex sheet is actually completely rolled up, the approximate formula given in Chap. 1 as Eq. 2-1 is still applicable in supersonic flow, while the exact formula for the distance between the legs of the single horseshoe vortex which results from the rolling-up of the trailing vortex sheet may also be extracted directly from Eq. 2-30, for application without change to the supersonic problems now arising.

Unfortunately it is no longer possible, however, to take into account what effect a deflected high lift flap or a separated flow condition may have upon the interference phenomena in supersonic flow, because the equivalent experimental data which permitted such estimates to be made in the subsonic case as yet have no suitable counterpart in the supersonic regime. Thus, lacking even a semiempirical development such as that presented in Chap. 1 for treatment of cases of flapped or stalled wings, one is constrained to consider only those more normal situations wherein the thickness of the wake may be assumed to be infinitesimally small, and where merely that component of the velocities induced by the wing lying perpendicular to the free stream velocity vector **U** need be computed.

In regard to the solution of the second subproblem mentioned in the first paragraph, it suffices at this time to note merely that the solution to the second part of the over-all question posed originally resolves itself into nothing more than the determination of the aerodynamic behavior of a lifting surface when immersed in a nonuniform flow field. Of course this problem can be solved in a standard straightforward manner, as explained in detail in VI,D. For this reason no further description of the calculational procedures is given here.

C,32. Calculation of the Induced Velocities Produced by the Wing. Lifting Surface Theory.

Classification of the Methods Used in Computing the Field of Velocities Induced by the Wing. The methods available for handling the problem of how to compute the downwash behind a wing operating in a supersonic stream may be divided into two categories: (1) that based on the conical flow theory, and (2) that using the method of singularities.

Although the second mode of approach furnishes more general techniques for arriving at the required determinations of the induced velocities, nevertheless a brief explanation of the first type of treatment is set forth here, both because the conical flow methods were the first to be

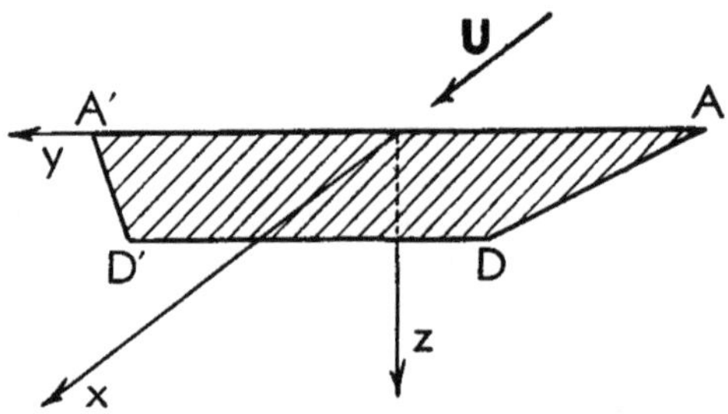

Fig. C,32a. Orientation of the coordinate axes, and sketch of the typical trapezoidal wing studied in Art. 32.

employed historically, and also because in certain special cases of appropriate wing planforms they furnish a rapid and practical means of arriving at the numerical results in most convenient form.

In both cases, the wing is assumed to have an incidence small enough with respect to **U** so that, having referred the flow to the system of axes x, y, z shown in Fig. C,32a, the projection of the wing on the x, y plane can be considered to be coincident with the wing itself.

The Method Proposed by Lagerstrom and Graham. If the outline of the wing's planform when compressed into the x, y plane is such as to be polygonal (composed of pieces of straight lines), it is found of particular advantage to make use of the method of attack proposed by Lagerstrom and Graham [*38*]. The cases considered here are those which fall within the basic and simple category of trapezoidal wings with leading and trailing edges aligned normally to the impinging free stream flow vector **U** (see Fig. C,32a). In the limiting cases the general trapezoidal wing becomes either rectangular or triangular in planform, so that this shape encompasses a great variety of actual configurations.

Case of the semi-infinite trapezoidal wing. Before going on to consideration of actual wings it is worthwhile to give consideration first to the case of a semi-infinite trapezoidal wing such as depicted in Fig. C,32b. The flow in the region I shown in this figure is examined first. Region I is that space which falls within the Mach cone emanating from point A (the tip of the leading edge) and which likewise lies external (upstream) from the Mach cone which emanates from point D (the tip of the trailing edge). The presence of the wing creates a field of flow in region I which may be obtained from the superposition of the constituent parts of the total flow resulting from (1) the undisturbed free stream flow having a velocity in the x direction of magnitude U and inclined to the wing surface at an

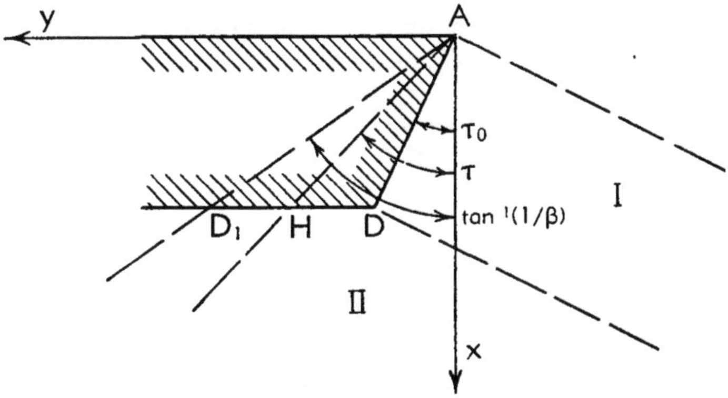

Fig. C,32b. Definition of the regions of flow about a semi-infinite trapezoidal wing wherein the expressions for the downwash take on different formulations.

angle of attack denoted by α and (2) the additional conical velocity field which emanates from point A.

The conical flow generated at A produces at any point P, lying within the above-defined region I, a velocity whose component in the z direction has, according to the dictates of the conical flow theory, a magnitude which may be expressed in the case under consideration as

$$\frac{w}{U} = \frac{2\alpha}{\pi} \int_{t_0}^{1} \frac{1}{\sqrt{1-t}} \frac{1}{\sqrt{t-t_0}} G(t, r, \theta) dt \tag{32-1}$$

wherein the following auxiliary notation has been employed:

$$\left.\begin{aligned} t &= \beta \tan \tau \\ r &= \frac{\beta \sqrt{y^2 + z^2}}{x} = \beta \tan \sigma \\ \theta &= \tan^{-1} \frac{z}{y} \\ t_0 &= \beta \tan \tau_0 \end{aligned}\right\} \tag{32-2}$$

and as usual

$$\beta = \sqrt{M_\infty^2 - 1} = \cot \text{ (Mach angle)}$$

The symbols employed in these definitions are most easily illustrated in terms of the geometry sketched in Fig. C,32b, where it may be seen that τ represents the angle that any arbitrary general ray, originating in point A and lying in the x, y plane, makes with the x axis, and where the convention is adopted that the sense of this angle is positive if it corresponds to the direction in which the positive x axis must be rotated to become aligned with the positive y axis by sweeping out only 90°. Likewise the angle denoted by τ_0 is the angle that the wing tip (line segment AD) makes with the x axis, while σ is used to denote the angle formed between the x axis and any ray whatsoever in space so long as it passes through A.

The explicit formula for G is then given by

$$G = \frac{1}{\pi}\left(\frac{1}{t} \ln |R| + \tan^{-1} \frac{2R \cos \theta}{1 - R^2} + \frac{C^2 - 1}{4C} \ln \left| \frac{R^2 + C^2 - 2CR \cos \theta}{1 + C^2R^2 - 2CR \cos \theta} \right| + \frac{\pi}{2}\right) \tag{32-3}$$

For convenience the quantities R and C have been employed herein to stand for the expressions

$$\left.\begin{aligned} R &= \frac{1 - \sqrt{1 - r^2}}{r} \\ \text{and} \qquad C &= \frac{1 - \sqrt{1 - t^2}}{t} \end{aligned}\right\} \tag{32-4}$$

For points lying in the plane of the wing where $z = 0^-$, (the x, y plane) let $\theta = 2\pi$ for $y > 0$ and $\theta = \pi$ for $y < 0$, and stipulating further that

$$t' = \frac{\beta y}{x} \quad \text{and} \quad T = \frac{1 - \sqrt{1 - t'^2}}{t'} \tag{32-5}$$

then Eq. 32-3 may be rewritten more simply as

$$G(t, t') = \frac{1}{\pi}\left(\frac{1}{t} \ln |T| + 2 \tan^{-1} T + \frac{C^2 - 1}{2C} \ln \left| \frac{T - C}{1 - CT} \right| + \frac{\pi}{2}\right) \tag{32-3'}$$

Finally the expression for the induced downwash along a ray t' reduces to

$$\left(\frac{w}{U}\right)_{z=0} = \frac{2\alpha}{\pi} \int_{t_0}^{1} \frac{1}{\sqrt{1 - t}} \frac{1}{\sqrt{t - t_0}} G(t, t')dt \tag{32-1'}$$

in this special instance.

Now consider the flow that is created in region II of Fig. C,32b. Region II is the space lying downstream behind the Mach cone emanating from point A and also downstream of the envelope of Mach cones coming from each one of the representative points H which lie along the stretch of wing trailing edge running from D inboard to D_1, where the Mach cone from the foretip A intersects the trailing edge. The field of flow created in region II by the presence of the type of wing now under consideration is made up of the sum of three contributing parts, viz. the flow listed above as the basic field produced by the free stream (1), the additional disturbance due to the conical flow from vertex A, which is field (2), and the further perturbations (3) to the flow which are due to the action of an infinite set of individual elemental conical fields emanating from the set of points located along the strip DD_1 such that each of these points, like H, is the vertex of such a conical flow.

Each one of these individual conical flows constituting the perturbations (3), such as the one generated at point H, has the following essential property: within the region of the x, y plane marked out by the piece of trailing edge running from D_1 to H and within the area bounded by the ray AH which goes through A and makes an angle τ with the x axis, the field with vertex at H must produce a velocity whose component in the x direction (on the upper surface of the wing where $z = 0^-$) is constant and equal to $-(du/dt)dt$. The u value entering this expression is the x component of the velocity produced by the field of flow (2) discussed above. In addition, the conical flows emanating from the H points must produce a zero x component of velocity throughout the part of the x, y plane that lies external to this latter-defined region.

Therefore, it follows that if the two fields of flow now defined as perturbations (2) and (3) are summed, the x component of the combined velocity field induced about the wing by these flows turns out to be zero throughout all of region II, and thus the x component of induced velocity is zero in the whole x, y plane which falls outside the region occupied by the wing. Consequently the boundary conditions defining the type of flow which is required to take place around the wing are now fulfilled at all required locations.

In addition, since it is true on the basis of trapezoidal wing theory that

$$\frac{u}{U} = \frac{2\alpha}{\pi\beta} \sin^{-1} \sqrt{\frac{t - t_0}{1 - t_0}} \tag{32-6}$$

it may be seen that the z component of the velocities induced in region II is given by

$$\frac{w}{U} = \frac{2\alpha}{\pi} \int_{t_0}^{1} \frac{1}{\sqrt{1 - t}} \frac{1}{\sqrt{t - t_0}} [G(t, r, \theta) - G(t, r^*, \theta^*)]dt \tag{32-7}$$

where the new symbols employing the asterisk superscript are defined according to the formulas

$$\left.\begin{aligned} r^* &= \frac{\sqrt{\left(\dfrac{\beta y}{c_w} + t\right)^2 + \dfrac{\beta^2 z^2}{c_w^2}}}{\dfrac{x}{c_w} - 1} \\ \text{and} \qquad \theta^* &= \tan^{-1}\left(\frac{\dfrac{\beta z}{c_w}}{\dfrac{\beta y}{c_w} + t}\right) \end{aligned}\right\} \tag{32-8}$$

For the special instance of points lying in the x, y plane, the formula for the z component of the induced velocities reduces simply to

$$\left(\frac{w}{U}\right)_{z=0} = \frac{2\alpha}{\pi}\int_{t_0}^{1} \frac{1}{\sqrt{1-t}}\frac{1}{\sqrt{t-t_0}}[G(t, t') - G(t, t'^*)]dt \tag{32-7'}$$

wherein the special definition for t'^* is introduced here, having the meaning

$$t'^* = \frac{\dfrac{\beta y}{c_w} + t}{\dfrac{x}{c_w} - 1} \tag{32-9}$$

Furthermore, if one defines

$$T^* = \frac{1 - \sqrt{1 - t'^{*2}}}{t'^*} \tag{32-10}$$

then the expression for $G(t, t'^*)$ may be obtained from the corresponding formula for $G(t, t')$ by merely substituting the quantity T^* for T in the latter formula.

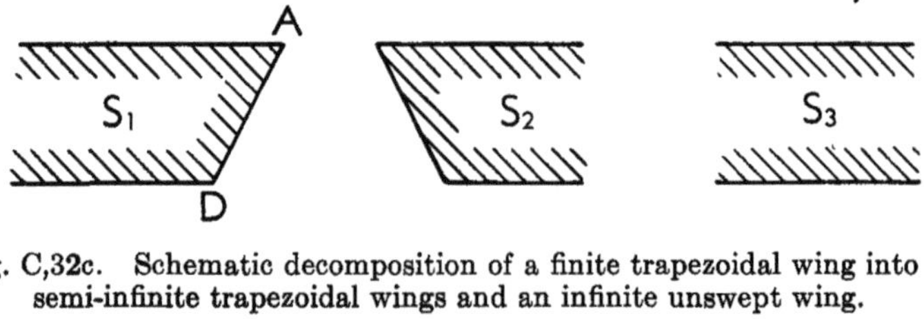

Fig. C,32c. Schematic decomposition of a finite trapezoidal wing into two semi-infinite trapezoidal wings and an infinite unswept wing.

Case of the complete trapezoidal wing. The expression giving the values of the downwash w in the case of a complete trapezoidal wing may now be readily obtained by making use of the results discussed above. All that is required is to sum up the effects produced by the semi-infinite (right and left) wing halves S_1 and S_2 and to subtract the effects due to the infinite (two-dimensional) wing S_3 having the same chord, as indicated in Fig. C,32c. The vertical velocity component w which is generated by these three wings acting in concert thus takes on values, in the various locations diagrammed in Fig. C,32d, according to the following schedule:

Region I

$$\frac{w}{U} = \left(\frac{w}{U}\right)_r$$

Region I'

$$\frac{w}{U} = \left(\frac{w}{U}\right)_l$$

Region II

$$\frac{w}{U} = \left(\frac{w}{U}\right)_r$$

Region II'

$$\frac{w}{U} = \left(\frac{w}{U}\right)_l$$

Region III

$$\frac{w}{U} = 0$$

Region IV

$$\frac{w}{U} = \left(\frac{w}{U}\right)_r + \left(\frac{w}{U}\right)_l$$

where $(w/U)_r$ stands for the downwash distribution described above in the previous subarticle for the various regions there defined, while the

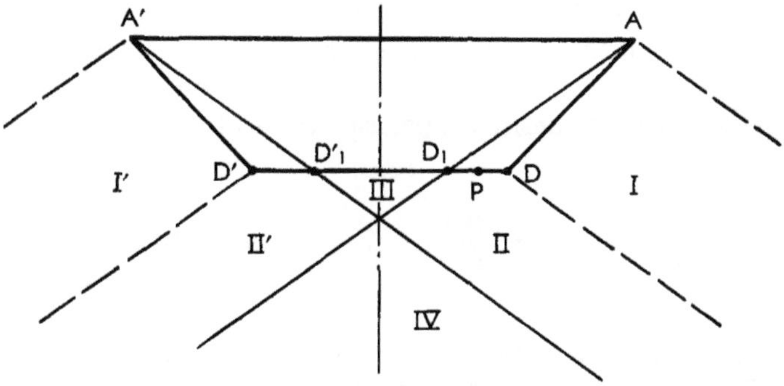

Fig. C,32d. View of the typical finite trapezoidal wing, showing the various regions in which alternate expressions for the downwash become valid.

$(w/U)_l$ notation refers to the analogous distribution that one would obtain when the system of conical coordinates t, r, θ is shifted so that the origin lies at point A' (and where the system t'^*, r^*, θ^* has its origin shifted likewise, so as to coincide with point D').

The evaluation of the integrals which appear in the above expressions for w/U is apt to be troublesome, because of difficulties involving the singular points exhibited by the integrand. There are two places, however, in which the calculations become much more simple than in the general case. These two locations are at the trailing edge of the wing, and at the Trefftz plane, i.e. for $x = \infty$.

INDUCED VELOCITIES AT THE LOCATION OF THE WING TRAILING EDGE. To examine the situation in detail, consider the arbitrary general point P which lies at the level of the zero-thickness wake streaming behind the wing, but restricted to only those locations lying immediately aft of the trailing edge. To each such point there corresponds a certain assigned value of t'; it is then noted that for locations at which $t > |t'|$ P falls entirely outside of the whole field of flow composed of the elemental conical fields which have their vertices at the points H whose coordinate is denoted by t. Consequently it follows from this consideration that in the case where $t > |t'|$ the value of $G(t, t'^*)$ at P is zero.

On the other hand, for the case where $t < |t'|$ the point P in question belongs to the region of the x, y plane where the flow due to each of the individual elemental conical fields, having its vertex in one of the H points, is two-dimensional; thus in this instance $G(t, t'^*) = 1$.

Consequently it follows that the z component of the velocity which is induced at the arbitrary point P by the action of the continuum of elemental conical flows, having their vertices at the various H locations along the trailing edge of the wing, is given by

$$-\beta \int_{t_0}^{t'} \frac{du}{dt}\, dt = -\beta u(t')$$

On the other hand the conical field emanating from the vertex A generates, at any arbitrary point P of the wing, a velocity whose component in the z direction is given simply by the term αU. Therefore, it is quite apparent that the desired expression for w at the trailing edge of the wing is

$$\left.\begin{aligned} \left(\frac{w}{U}\right)_P &= \alpha - \beta \frac{u(t')}{U} && \text{if } P \text{ lies in region II} \\ &= 0 && \text{if } P \text{ lies in region III} \end{aligned}\right\} \qquad (32\text{-}11)$$

INDUCED VELOCITIES AT THE LOCATION CORRESPONDING TO THE TREFFTZ PLANE. A general rule for the determination of the w values in the Trefftz plane is formulated, after a fuller discussion of this question, at the end of the present article. At this point it suffices to confine oneself to the straightforward determination of what the expression for w/U reduces to in the case of points lying along the line $z = 0$, $x = \infty$. This is merely a matter of substituting the limiting value of x into Eq. 32-7′ and determining the limiting value that the integrand takes on under this condition. One thus finds that

$$\left(\frac{w}{U}\right)_{\substack{z=0\\x=\infty}} = \frac{2\alpha}{\pi^2} \int_{t_0}^{1} \frac{1}{t\sqrt{1-t}} \frac{1}{\sqrt{t-t_0}} \ln \left| \frac{\beta \dfrac{y}{c_w}}{\dfrac{\beta y}{c_w} + t} \right| dt \qquad (32\text{-}12)$$

In the special case of a rectangular wing for which $t_0 = 0$, the above-written relationship simplifies to

$$\left(\frac{w}{U}\right)_{\substack{z=0\\x=\infty}} = \frac{2\alpha}{\pi}\left(1 - \sqrt{\frac{\beta\frac{y}{c_w} - 1}{\beta\frac{y}{c_w}}}\right) \quad \begin{array}{l}\text{for } \beta\frac{y}{c_w} < 0\\ \text{or for } \frac{\beta y}{c_w} > 1\end{array}$$
$$= \frac{2\alpha}{\pi} \qquad \text{for } 0 \leqq \frac{\beta y}{c_w} \leqq 1 \tag{32-13}$$

Applications in the Special Case of Downwash Produced by a Rectangular Wing. A determination of the induced angle of attack $\epsilon = w/U$ has been carried out according to the above method due to Lagerstrom [*38*] for the case of a semi-infinite rectangular wing, and the results have been plotted here in Fig. C,32e for various downstream locations x/c_w in the plane of the wing, for which $z = 0$. Likewise, similar calculations have been made to determine ϵ for the same wing but in the Trefftz plane where $x = \infty$; the results are depicted in Fig. C,32f at various horizontal levels above the plane of the wing, i.e. for various values of $\beta z/c_w$.

On the basis of these diagrams for the semi-infinite rectangular wing it is a simple matter to determine the value of ϵ which applies for the downwash behind any rectangular complete wing by proceeding in a manner identical with the process illustrated above under the heading *Case of the complete trapezoidal wing*. By application of this method, the values of ϵ/α and of ϵ/C_L have been computed for the illustrative case of a rectangular wing having an aspect ratio of 4, and the plotted data for the $z = 0$ level and for various distances behind the wing are given in Fig. C,32g and C,32h, respectively. It appears that, at equal values of α, with an increase in M_∞, ϵ decreases, whereas at equal values of C_L, with an increase of M_∞, ϵ increases or decreases depending on the values of x/c_w and y/c_w. It should be further observed that, for rectangular wings, the dependence of ϵ on y/c_w is felt somewhat more strongly at supersonic speeds than at subsonic speeds; such dependence is considerably influenced by the values of x/c_w and M_∞. The dependence of ϵ on z/c_w, for $x = \infty$, is finally shown in Fig. C,32i.

The Method of Singularities. The method of singularities has been applied to the determination of the downwash behind a lifting wing in supersonic flow in various ways. One way of proceeding is to employ pressure doublets as the singularities and to distribute them over the wing surface according to the method propounded by Martin [*39*]. Another way of handling the problem is to distribute a set of velocity doublets over the wing and back on the wake region as well, according to the plan suggested by Lomax and Heaslet in [40]. A third method, suggested by Robinson and Hunter-Tod [41], is to use vortices which are distributed over both the wing and its trailing wake region. However, these methods of applying the concept of singularities for the solution at hand are really equivalent,

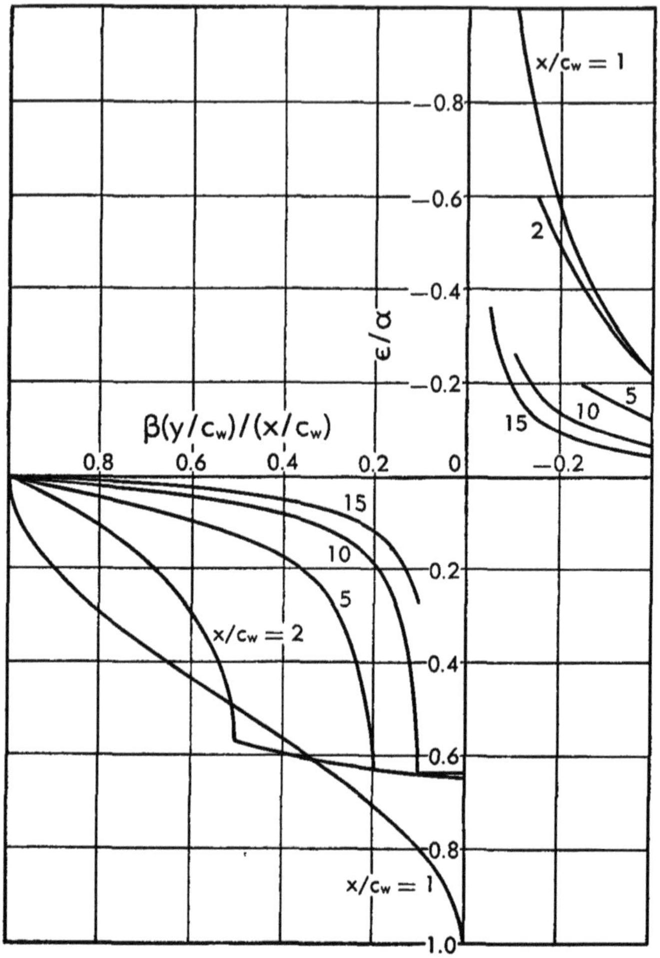

Fig. C,32e. Spanwise distribution of downwash behind a semi-infinite rectangular wing, for various downstream survey locations, x/c_w, lying in the $z = 0$ plane.

because the formulas derived on the basis of any one of the proposed procedures may be easily converted into the formulas holding for the other treatments.

As a matter of fact the interdependence of the three methods may be shown analytically by noting the following connections between the three versions mentioned.

On the basis of the method which uses pressure doublets in its analytic formulation, the potential which describes the perturbation field for the flow about the wing in question is given by

$$\phi = -\frac{z}{2\pi}\iint_{\tau} \frac{\Delta\phi_{x_1}(x - x_1)dx_1dy_1}{[(y - y_1)^2 + z^2]R(x)} \tag{32-14}$$

where

$$\Delta\phi_{x_1} = \left(\frac{\partial\phi}{\partial x}\right)_{z=0^-} - \left(\frac{\partial\phi}{\partial x}\right)_{z=0^+}$$

and where

$$R(x) = \sqrt{(x - x_1)^2 - \beta^2[(y - y_1)^2 + z^2]}$$

The subscript τ on the area integral employed here is used to denote the region marked out on the plane of the wing by the intersection of this

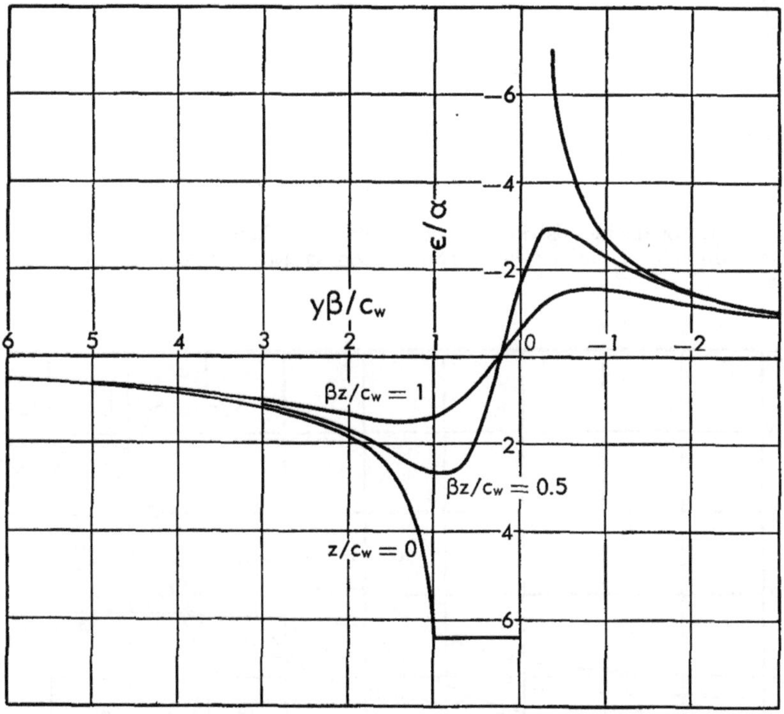

Fig. C,32f. Spanwise distribution of downwash behind a semi-infinite rectangular wing, far downstream in the Trefftz plane, at several selected horizontal levels.

plane with the forward facing Mach cone issuing from the point $P(x, y, z)$.

Now it is true, of course, that

$$-\frac{x - x_1}{[(y - y_1)^2 + z^2]R(x)} = \beta^2 \int_{x_1}^{\infty} \frac{d\xi}{[(x - \xi)^2 - \beta^2(y - y_1)^2 - \beta^2z^2]^{\frac{3}{2}}} \tag{32-15}$$

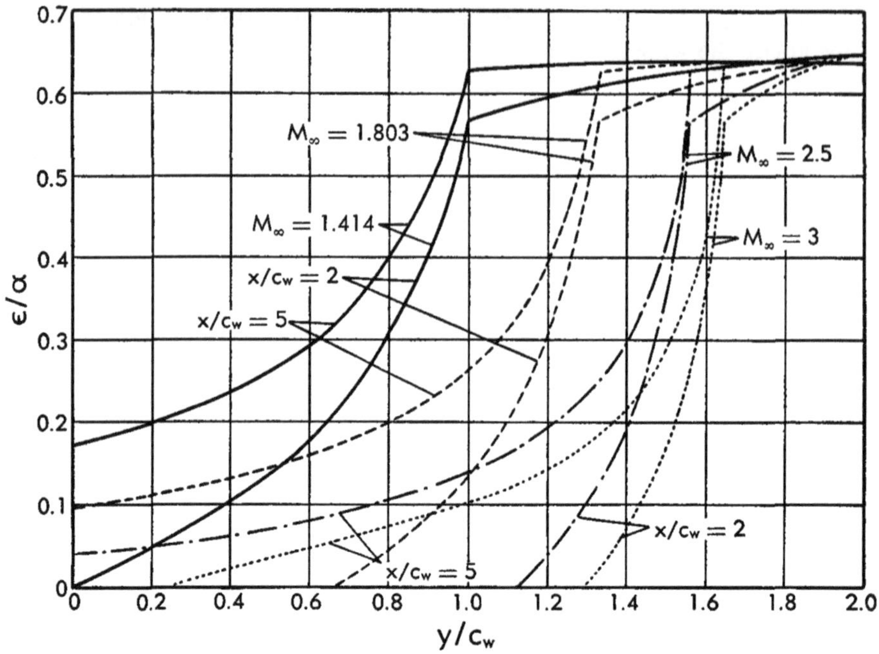

Fig. C,32g. Spanwise distribution of downwash behind a rectangular wing of aspect ratio 4.0, for various downstream survey locations, x/c_w, lying in the $z = 0$ plane.

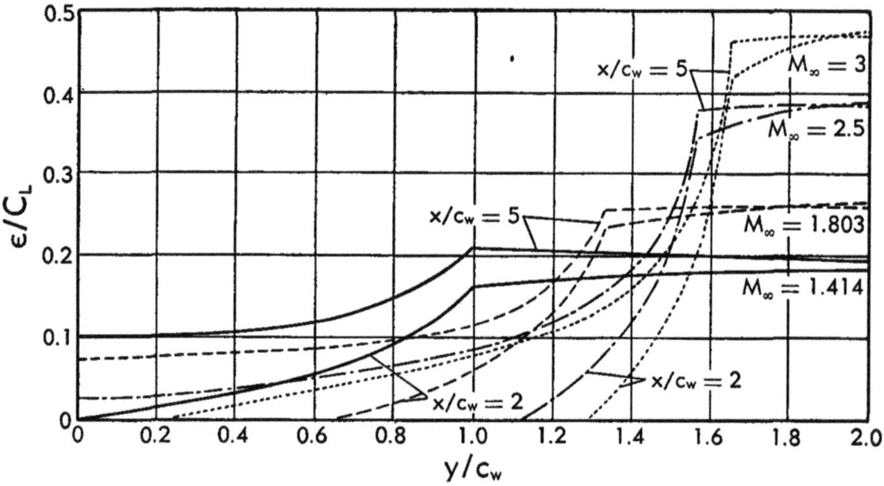

Fig. C,32h. Spanwise distribution of downwash for same situation for which plots of Fig. C,32g are valid, but comparison made in this instance on basis of a constant lift coefficient.

where the crossbar on the integral sign means that only the finite part of the indicated integral is to be counted (in the sense of Hadamard [*42*, pp. 184ff.]). Consequently, if one inserts the expression just given as Eq. 32-15 into the definition of the potential (32-14), and if the order of inte-

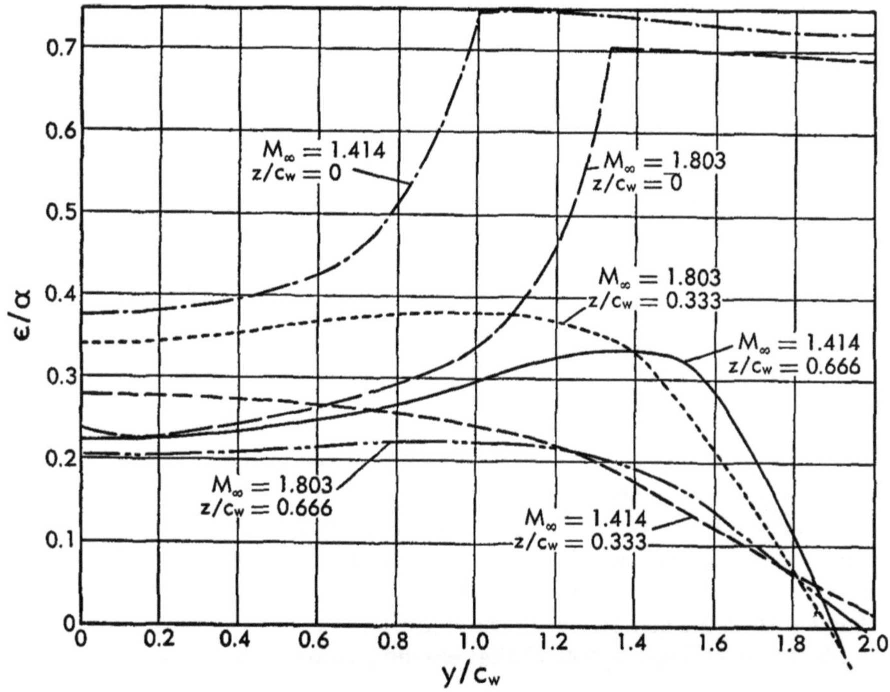

Fig. C,32i. Spanwise distribution of downwash behind a rectangular wing of aspect ratio 4.0, far downstream in the Trefftz plane, at several selected horizontal levels and for two different Mach numbers.

gration is then interchanged, it turns out that the one can perform the quadrature

$$I = \int_{-\infty}^{\xi} \Delta\phi_{x_1}(x_1\, y_1)dx_1$$

simply, to obtain $I = \Delta\phi$, and thus the expression for the desired potential is obtained in the form

$$\phi = \frac{\beta^2 z}{2\pi} \iint_{\tau'} \frac{\Delta\phi dx_1 dy_1}{[(x - x_1)^2 - \beta^2(y - y_1)^2 - \beta^2 z^2]^{\frac{3}{2}}} \tag{32-16}$$

where now the τ' affix on the double integral is used to denote that part of the region composed of the plane of the wing and including the wake surface, which is restricted to that upstream portion for which $R > 0$.

Now it may be observed that Eq. 32-16 is nothing other than the potential which describes a distribution of supersonic doublets of velocity, the strength of which is equal to the jump in the value of the potential as one passes from one side of the wing to the other (or from the bottom to the top of the wake), and which are so orientated that the doublet axes are aligned with the z direction.

On the other hand, Eq. 32-16 also represents the potential which describes a field of flow produced by a distribution of vortices disposed over the region τ' as defined above. All that is necessary to make this correspondence evident is to note that if the x and y components of the vortex distributions are denoted by Γ_x and Γ_y respectively, then these components may be linked to the potential difference $\Delta\phi$ which appears in Eq. 32-16 through means of the very same relationships which hold when $M_\infty < 1$, i.e.

$$\left.\begin{aligned} \Gamma_x &= \frac{\partial \Delta\phi}{\partial y} \quad \text{and} \quad \Gamma_y = -\frac{\partial \Delta\phi}{\partial x} \qquad \text{on the wing surface} \\ \Gamma_x &= \frac{\partial (\Delta\Gamma_{\text{t.e.}})}{\partial y} \quad \text{and} \quad \Gamma_y = 0 \qquad \text{on the wake surface} \end{aligned}\right\} \tag{32-17}$$

wherein the notation $\Delta\phi_{\text{t.e.}}$ is used to denote the jump experienced by the potential Γ at the trailing edge of the arbitrary general airfoil section lying in any one of the y = const planes.

Determination of the Expression Giving the Induced Velocities. Provided it is assumed that the pressure distribution over the wing surface is known, which means that one may take for granted that ϕ is known, then the component in the z direction of the velocities (compared with the free stream velocity) that are induced by such a wing is given by

$$\frac{w}{U} = \frac{1}{U}\frac{\partial \phi}{\partial z} \tag{32-18}$$

wherein one may employ either of the expressions presented above as Eq. 32-14 or Eq. 32-16 for the potential ϕ.

If one now proceeds to obtain w/U explicitly it may be noted first that

$$\int \frac{z(x - x_1)dx_1}{[(y - y_1)^2 + z^2]R} = \tan^{-1}\left[\frac{zR}{(x - x_1)(y - y_1)}\right]$$

Thus by integrating Eq. 32-14 by parts it follows that

$$\begin{aligned} \phi = \frac{1}{2\pi} \oint_s \Delta\phi_{x_1} \tan^{-1}\left[\frac{zR}{(x - x_1)(y - y_1)}\right] dx_1 \\ + \frac{1}{2\pi} \iint_\tau \Delta\phi_{x_1y_1} \tan^{-1}\left[\frac{zR}{(x - x_1)(y - y_1)}\right] dx_1dy_1 \end{aligned} \tag{32-19}$$

where the s affix on the line integral denotes that the path of integration is to be carried around the boundary of the region τ previously described.

Upon taking the derivative of this expression for ϕ with respect to the z coordinate, one obtains the formula for the induced downwash angles as

$$\frac{w}{U} = \frac{1}{2\pi U} \iint_{\tau} \Delta\phi_{x_1y_1} K dx_1 dy_1 + \frac{1}{2\pi U} \oint_{s} \Delta\phi_{x_1} K^* dx_1 \qquad (32\text{-}20)$$

where

$$K = \frac{(x - x_1)(y - y_1)(R^2 - \beta^2 z^2)}{[(x - x_1)^2 - \beta^2 z^2][(y - y_1)^2 + z^2]R} \qquad (32\text{-}21)$$

and likewise K^* is the same expression except that the values of x_1 and y_1 are restricted to those belonging to the boundary of the τ region.

For points which lie in the wake (for $z = 0$), Eq. 32-20 has a simpler form, i.e.

$$\left(\frac{w}{U}\right)_{z=0} = \frac{1}{2\pi U} \iint_{\tau} \frac{\Delta\phi_{x_1y_1} R}{(x - x_1)(y - y_1)} dx_1 dy_1 + \frac{1}{2\pi U} \oint_{s} \Delta\phi_{x_1} \frac{\sqrt{(x - x_1^*)^2 - \beta^2(y - y_1^*)^2}}{(x - x_1^*)(y - y_1^*)} dx_1^* \qquad (32\text{-}22)$$

where here too the points confined to the contour line s have been distinguished from the general x_1, y_1 coordinates by using the asteristic notation x_1^*, y_1^*.

If one again performs an integration by parts on the surface integral appearing on the right-hand side of Eq. 32-22 it is found that it takes on the form

$$\left(\frac{w}{U}\right)_{z=0} = \frac{1}{2\pi U} \oint_{s} (\Delta\phi_{y_1} dy_1^* + \Delta\phi_{x_1} dx_1^*) \frac{R^*}{(x - x_1^*)(y - y_1^*)} - \frac{1}{2\pi U} \iint_{\tau} \Delta\phi_{y_1} \frac{\partial}{\partial x_1} \left[\frac{R}{(x - x_1)(y - y_1)}\right] dx_1 dy_1 \qquad (32\text{-}22')$$

Evaluation of the Induced Velocities Under the Special Circumstances that the Survey Region Is Either the Trefftz Plane or the Trailing Edge of the Wing. The calculation of the induced downwash is materially simplified in those special cases where one is interested in the values of the velocities which pertain exclusively to the Trefftz plane location or to the trailing edge of the wing. The summary formulas applying to these instances are presented below.

Induced velocity obtained at the Trefftz plane. For the case when $x = \infty$, Eq. 32-14 becomes modified in form to simply

$$\phi(\infty, y, z) = \frac{-z}{2\pi} \iint_{S_W} \frac{\Delta\phi_{x_1}}{(y - y_1)^2 + z^2} dx_1 dy_1 \qquad (32\text{-}23)$$

where the previously encountered area of integration τ is now replaced by the whole region of the wing surface S_w. Since it is also true that

$$\int_{\text{l.e.}}^{\text{t.e.}} \Delta\phi_{x_1} dx_1 = \Delta\phi_{\text{t.e.}}$$

it follows that Eq. 32-23 may be reduced further to just

$$\phi(\infty, y, z) = \frac{-z}{2\pi} \int_{-b/2}^{b/2} \frac{\Delta\phi_{\text{t.e.}}}{(y - y_1)^2 + z^2} dy_1 \qquad (32\text{-}24)$$

which is completely identical to the formula giving ϕ in the Trefftz plane that holds for incompressible flows.

By use of this definition for ϕ, one finds that the corresponding formula for the induced downwash angles is

$$\left(\frac{w}{U}\right)_{x=\infty} = -\frac{1}{2\pi U} \int_{-b/2}^{b/2} \frac{\Delta\phi_{\text{t.e.}}}{(y - y_1)^2 + z^2} dy_1 + \frac{z}{2\pi U} \int_{-b/2}^{b/2} \frac{2z\Delta\phi_{\text{t.e.}}}{[(y - y_1)^2 + z^2]^2} dy_1$$

$$= \frac{1}{2\pi U} \int_{-b/2}^{b/2} \Delta\phi_{\text{t.e.}} \frac{z^2 - (y - y_1)^2}{[z^2 + (y - y_1)^2]^2} dy_1$$

This expression may be further simplified by noting that

$$\Delta\phi_{\text{t.e.}} = \Gamma$$

and that

$$\Gamma\left(\frac{b}{2}\right) = \Gamma\left(-\frac{b}{2}\right) = 0$$

so that one obtains, through an integration by parts, that the Trefftz plane downwash values may be written in the compact form

$$\left(\frac{w}{U}\right)_{x=\infty} = \frac{1}{2\pi U} \int_{-b/2}^{b/2} \frac{y - y_1}{(y - y_1)^2 + z^2} \frac{d\Gamma}{dy_1} dy_1 \qquad (32\text{-}25)$$

It appears from the foregoing formula, therefore, that there is no Mach number influence exerted upon the values of the induced velocity provided that the circulation distribution Γ out along the wing span does not become altered as the Mach number varies. Consequently, in order to determine the downwash, one may have direct recourse to the calculational methods already dealt with in Chap. 1. Lagerstrom and Graham have proposed [*43*] a method of computing these values of $(w/U)_{x=\infty}$, starting from the actual circulation distribution Γ, which exists out along the wing span, and they have given the results for triangular and trapezoidal wings when treated according to their theory.

Induced velocity obtained at the trailing edge location by use of the Lagerstrom formula [*38*]. If the trailing edge is supersonic, the change in the u component of the velocity, which takes place at any arbitrary point P lying along this trailing edge (passing from the value $u \neq 0$ on the wing to the value $u = 0$ in the wake), is produced by a Mach wave which is the envelope of the Mach cones whose vertices lie at the aforementioned points P. The plane tangent at P to the said envelope has the following equation:

$$(x - x_P) \tan \upsilon + (y - y_P) - \frac{\sqrt{\beta^2 - \cot^2 \upsilon}}{\cot \upsilon} z = 0$$

where υ is the angle that the tangent line to the trailing edge of the wing, at the point P, makes with the x axis. Of course, the above tangent plane contains the straight line l_1 which is tangent to the trailing edge of the planar wing at the point P; the equation of this line is

$$z = 0, \quad y - y_P + (x - x_P) \tan \upsilon = 0$$

The tangent plane also contains the straight line l_2, which is a generatrix of the Mach cone through the point P. In parametric form,

$$x - x_P = L \cos \mu, \quad y - y_P = -L \frac{\sin^2 \mu}{\cos \mu} \cot \upsilon,$$
$$z = L \sqrt{1 - \cot^2 \upsilon \tan^2 \mu} \sin \mu$$

where $\beta = \cot \mu$, and μ is the Mach angle.

Consequently, since the Mach wave produces merely a change in the component of velocity that is normal to this Mach wave when a streamline of the flow passes through it, the following relationships may be written down to express the fact that the components of the velocity taken in both the direction of the line l_1, and of the line l_2, upstream and downstream of the wave, are equal:

$$(U + u_1) \cos \upsilon - v_1 \sin \upsilon = U \cos \upsilon - v_2 \sin \upsilon$$

and

$$(U + u_1) \cos \mu - v_1 \frac{\sin^2 \mu}{\cos \mu} \cot \upsilon + w_1 \sqrt{1 - \cot^2 \upsilon \tan^2 \mu} \sin \mu$$
$$= U \cos \mu - v_2 \frac{\sin^2 \mu}{\cos \mu} \cot \upsilon + w_2 \sqrt{1 - \cot^2 \upsilon \tan^2 \mu} \sin \mu$$

where the subscripts 1 and 2 serve to indicate that the value of the quantity to which they are affixed is that which holds just upstream and just downstream of the Mach wave, respectively.

From the above relationships it may be seen that the following connections between the velocity components upstream and downstream of the Mach wave must hold

$$v_2 = v_1 - u_1 \cot \upsilon$$

and

$$w_2 = w_1 + u_1\sqrt{\frac{\beta^2}{\cot^2 v} - 1}\cot v$$

Now since the additional simple relationships

$$w_1 = \alpha U$$

and

$$u_1 = -\tfrac{1}{2}(\Delta\phi_x)_{\text{t.e.}}$$

are easily seen to be valid, and if one makes the further simplification in notation that

$$\cot v = m$$

then it follows directly that the sought expression for the downwash is just

$$\left(\frac{w}{U}\right)_{\text{t.e.}} = \alpha - \frac{1}{2}\frac{(\Delta\phi_x)_{\text{t.e.}}}{mU}\sqrt{m^2\beta^2 - 1} \qquad (32\text{-}26)$$

If the leading edge of the wing is subsonic, then in this instance $(\Delta\phi_x)_{\text{t.e.}} = 0$, and thus no sudden variation in w takes place at the trailing edge.

Applications of the Method of Singularities.

Case of the semi-infinite trapezoidal wing. This is the same as the problem investigated by the method of conical flows under the heading The Method Proposed by Lagerstrom and Graham. If one also avails himself here of the advantages accruing from use of the conical coordinate system for which

$$t = \frac{\beta y_1}{x_1} \quad \text{and} \quad t' = \frac{\beta y}{x}$$

and if one then expresses $\Delta\phi_{xy}$ by means of the relationship given as Eq. 32-6, and makes the further simplification that $\xi = x_1/x$, it is readily apparent that Eq. 32-22 becomes converted into the expression

$$\left(\frac{w}{U}\right)_{z=0} = \frac{2\alpha}{\pi}\int_{t_0}^{1}\frac{dt}{\sqrt{(t - t_0)(1 - t)}}\int_0^{c_w/x}\frac{\sqrt{(1 - \xi)^2 - (t' - t\xi)^2}}{(1 - \xi)(t' - t\xi)}\,d\xi \qquad (32\text{-}27)$$

If the convenient definition is made that

$$I(t, t') = \frac{1}{\pi}\int_0^{c_w/x}\frac{\sqrt{(1 - \xi)^2 - (t' - t\xi)^2}}{(1 - \xi)(t' - t\xi)}\,d\xi$$

then Eq. 32-27 becomes simply

$$\left(\frac{w}{U}\right)_{z=0} = \frac{2\alpha}{\pi^2}\int_0^1\frac{I(t, t')dt}{\sqrt{(t - t_0)(1 - t)}} \qquad (32\text{-}28)$$

Now, by using the notation previously introduced in this article it is recognized at once that

$$I(t, t') = G(t, t') - G(t, t'^*)$$

Thus it is apparent that Eq. 32-28 just adduced above is identical to the formula derived previously by the use of the conical flow method suggested by Lagerstrom, which culminated in the Eq. 32-7′ of the foregoing analysis.

Case of the triangular wing with subsonic leading edges. Under these conditions the potential describing the flow at the upper surface of the wing is given by

$$(\phi)_{z=0^-} = \frac{U\alpha}{\mathsf{E}_0}\sqrt{x_1^2 \tan^2 \delta - y_1^2} = \frac{U\alpha}{\mathsf{E}_0\beta}\sqrt{\theta_0^2 x_1^2 - \beta^2 y_1^2}$$

wherein $\theta_0 = \beta \tan \delta$, and δ is the half angle at the vertex of this triangular wing, while the symbol E_0 stands for the complete elliptic integral of the second kind with modulus $k = \sqrt{1 - \theta_0^2}$.

From this expression for the governing potential, it may be deduced that

$$\Delta\phi_y = 2(\phi_y)_{z=0^-} = \frac{-2U\alpha\beta}{\mathsf{E}_0} \frac{y_1}{\sqrt{\theta_0^2 x_1^2 - \beta^2 y_1^2}} \tag{32-29}$$

and thus by inserting this expression into Eq. 32-22′ it follows that the downwash at the level of the wing surface is given by

$$\left(\frac{w}{U}\right)_{z=0} = -\frac{\alpha\beta}{\pi\mathsf{E}_0}\left\{\oint_s \frac{y_1^* R^*}{\sqrt{\theta_0^2 x_1^{*2} - \beta^2 y_1^{*2}}} \frac{dy_1^*}{(x - x_1^*)(y - y_1^*)} - \iint_T \frac{y_1}{\sqrt{\theta_0^2 x_1^2 - \beta^2 y_1^2}} \frac{\partial}{\partial x_1}\left[\frac{R}{(x - x_1)(y - y_1)}\right] dx_1 dy_1\right\} \tag{32-30}$$

The evaluations of these integrals take on especially convenient formulations in certain regions lying behind the triangular wing (see Fig. C32,j). For instance, for points lying along the centerline of the wake region (i.e. along the line $z = 0$, $y = 0$) and confined to region I, for which the range of x is limited to the stretch for which $x - c_{w_0} \leq (b/2)\beta = c_{w_0}\theta_0$, it turns out that

$$\begin{aligned} H_1 &= \oint_s \frac{y_1^* dy_1^*}{\sqrt{\theta_0^2 x_1^{*2} - \beta^2 y_1^{*2}}} \cdot \frac{R^*}{(x - x_1^*)(y - y_1^*)} \\ &= -\int_{-\frac{x - c_{w_0}}{\beta}}^{\frac{x - c_{w_0}}{\beta}} \frac{\sqrt{(x - c_{w_0})^2 - \beta^2 y_1^{*2}}}{\sqrt{\theta_0^2 c_{w_0}^2 - \beta^2 y_1^{*2}}} \cdot \frac{dy_1^*}{x - c_{w_0}} \end{aligned} \tag{32-31}$$

If one now makes use of the auxiliary definitions that

$$t = \frac{\beta y_1^*}{x - c_{w_0}} \qquad \text{and} \qquad t' = \frac{x - c_{w_0}}{\theta_0 c_{w_0}} \tag{32-32}$$

then the expression for H_1 becomes simply

$$H_1 = -\frac{1}{\beta}\int_{-1}^{1} \frac{\sqrt{1-t^2}}{\sqrt{1-t'^2t^2}}\, t'dt = -\frac{2}{\pi}\,\frac{\mathsf{E}(t') - (1 - t'^2)\mathsf{K}(t')}{t'} \tag{32-33}$$

wherein $\mathsf{K}(t')$ and $\mathsf{E}(t')$ are the complete elliptic integrals of the first and second kind, respectively, having the common modulus t'.

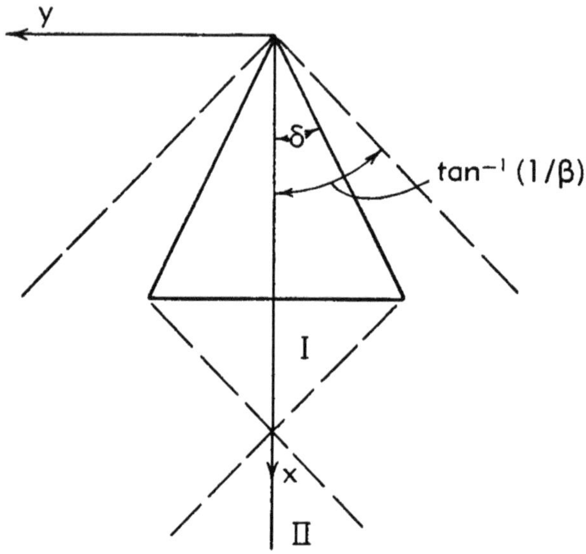

Fig. C,32j. View of a typical triangular wing with subsonic leading edges, including the definitions of wake regions I and II wherein different formulations for the downwash hold.

For those points lying farther downstream along this centerline, but in the wake region II shown in Fig. C,32j, where the range of x is restricted by a lower bound in the form $x - c_{w_0} > c_{w_0}\theta_0$, one finds that a new, but still simple formulation holds, i.e. it now turns out that

$$H_1 = -\int_{-\frac{c_{w_0}\theta_0}{\beta}}^{\frac{c_{w_0}\theta_0}{\beta}} \frac{\sqrt{(x - c_{w_0})^2 - \beta^2 y_1^{*2}}}{\sqrt{\theta_0^2 c_{w_0}^2 - \beta^2 y_1^{*2}}} \cdot \frac{dy_1^*}{x - c_{w_0}} \tag{32-31'}$$

Upon making further auxiliary definitions that in this case

$$t = \frac{\beta y_1^*}{\theta_0 c_{w_0}} \qquad \text{and} \qquad t' = \frac{\theta_0 c_{w_0}}{x - c_{w_0}} \tag{32-32'}$$

then the expression for H_1 becomes

$$H_1 = -\frac{1}{\beta}\int_{-1}^{1}\sqrt{\frac{1-t'^2t^2}{1-t^2}}\,dt = -\frac{2}{\beta}\,\mathsf{E}(t') \tag{32-33'}$$

In an analogous manner, the second integral appearing in Eq. 32-30 may be evaluated in especially convenient forms for the same stretches along the centerline as discussed above; for instance, for the x range in which $x - c_{w_0} \leqq c_{w_0}\theta_0$ holds, one finds that

$$H_2 = \iint_\tau \frac{y_1}{\sqrt{\theta_0^2x_1^2 - \beta^2y_1^2}}\frac{\partial}{\partial x_1}\left[\frac{R}{(x-x_1)(y-y_1)}\right]dx_1dy_1$$

$$= \int_0^{\frac{x}{1+\theta_0}}\frac{dx}{x-x_1}\int_{-\frac{\theta_0x_1}{\beta}}^{\frac{\theta_0x_1}{\beta}}\left\{\frac{1}{\sqrt{1-[\beta y_1/(x-x_1)]^2}} - \sqrt{1-\left(\frac{\beta y_1}{x-x_1}\right)^2}\right\}$$

$$\times\frac{dx_1dy_1}{\sqrt{\theta_0^2x_1^2-\beta^2y_1^2}} + \int_{\frac{x}{1+\theta_0}}^{c_{w_0}}\frac{dx}{x-x_1}\int_{-\frac{x-x_1}{\beta}}^{\frac{x-x_1}{\beta}}\left\{\frac{1}{\sqrt{1-[\beta y_1/(x-x_1)]^2}}\right.$$

$$\left. - \sqrt{1-\left(\frac{\beta y_1}{x-x_1}\right)^2}\right\}\frac{dx_1dy_1}{\sqrt{\theta_0^2x_1^2-\beta^2y_1^2}}$$

and if the following set of auxiliary definitions is introduced

and

$$\left.\begin{aligned} t &= \frac{\beta y_1}{x-x_1}; \quad t' = \frac{x-x_1}{\theta_0x_1} \\ \tau' &= \frac{\theta_0x_1}{x-x_1}; \quad \tau = \frac{\beta y_1}{\theta_0x_1}\end{aligned}\right\} \tag{32-34}$$

the value of H_2 comes out to be

$$H_2 = \frac{1}{\beta}\int_0^1\frac{d\tau'}{\theta_0+\tau'}\int_{-1}^{1}\left[\frac{1}{\sqrt{1-\tau^2\tau'^2}} - \sqrt{1-\tau'^2\tau^2}\right]\frac{d\tau}{\sqrt{1-\tau^2}}$$

$$+\frac{1}{\beta}\int_{\frac{x-c_{w_0}}{\theta_0c_{w_0}}}^{1}\frac{dt'}{1+\theta_0t'}\int_{-1}^{1}\left[\frac{1}{\sqrt{1-t^2}} - \sqrt{1-t^2}\right]\frac{dt}{\sqrt{1-t'^2t^2}}$$

$$= \frac{2}{\beta}\left[\int_0^1\frac{\mathsf{K}_1-\mathsf{E}_1}{\theta_0+\tau'}\,d\tau' + \int_{\frac{x-c_{w_0}}{\theta_0c_{w_0}}}^{1}\frac{\mathsf{K}_2-\mathsf{E}_2}{t'^2(1+\theta_0t')}\,dt'\right] \tag{32-35}$$

wherein the symbols K_1 and E_1 stand for the complete elliptic integrals of the first and second kind, respectively, with common modulus τ', while K_2 and E_2 are the same kinds of elliptic integrals with common modulus t'.

For points lying farther downstream on the centerline for which

$x - c_{w_0} > c_{w_0}\theta_0$ holds, the expression for H_2 takes on the alternate formulation

$$H_2 = \frac{2}{\beta}\int_0^{\frac{x - c_{w_0}}{\theta_0 c_{w_0}}} \frac{\mathsf{K}_1 - \mathsf{E}_1}{\theta_0 + \tau'}\, d\tau' \tag{32-36}$$

Consequently the expression for the induced velocity may be assuredly calculated from the following formulas, resulting from the recombination of the above H evaluations:

(i) *In the region where* $x - c_{w_0} \leqq c_{w_0}\theta_0$

$$\left(\frac{w}{U}\right)_{y=z=0} = \frac{2\alpha}{\pi \mathsf{E}_0}\left[\frac{\mathsf{E}(t') - (1 - t'^2)\mathsf{K}(t')}{t'} + \int_0^1 \frac{\mathsf{K}_1 - \mathsf{E}_1}{\theta_0 + \tau'}\, d\tau' + \int_{\frac{x - c_{w_0}}{\theta_0 c_{w_0}}}^1 \frac{\mathsf{K}_2 - \mathsf{E}_2}{t'^2(1 + \theta_0 t')}\, dt'\right] \tag{32-37}$$

while,

(ii) *In the region where* $x - c_{w_0} > c_{w_0}\theta_0$

$$\left(\frac{w}{U}\right)_{y=z=0} = \frac{2\alpha}{\pi \mathsf{E}_0}\left[\mathsf{E}(t') + \int_0^{\frac{x - c_{w_0}}{\theta_0 c_{w_0}}} \frac{\mathsf{K}_1 - \mathsf{E}_1}{\theta_0 + \tau'}\, d\tau'\right] \tag{32-37'}$$

Analogous but somewhat more complex formulas were arrived at by Lomax, Sluder, and Heaslet [*40*] for the determination of the downwash values at points lying in the plane of the wake, i.e. for $z = 0$, and also in the case of points lying in the plane of symmetry, i.e. for $y = 0$. The numerical results obtained by the aforementioned authors through application of their formulas for the case of points lying in the plane of the wake, $z = 0$, are reproduced here in the form of the curves shown in Fig. C,32k for several values of the subsonic leading edge sweep parameter θ_0 (the discrete values selected for illustration are $\theta_0 = 0.4, 0.6$, and 0.8) and for several appropriately chosen values of the spanwise location parameter, $y_0 = \beta y/c_{w_0}$, as noted on the figures themselves. Likewise, the spanwise variation of the induced downwash angles (for a given value of α), which appertains to the same configuration, is depicted graphically in Fig. C,32l for the cases where the spanwise survey locations lying downstream from the trailing edge have been selected at certain discrete values noted on the corresponding plotted data; the latter curves have been constructed solely for the case where the sweep parameter has assumed the value $\theta_0 = 0.6$.

On the basis of the information supplied by these charts it becomes quite evident that the values of ϵ/α evince but an extremely small amount of change even though the spanwise location parameter y_0 runs through

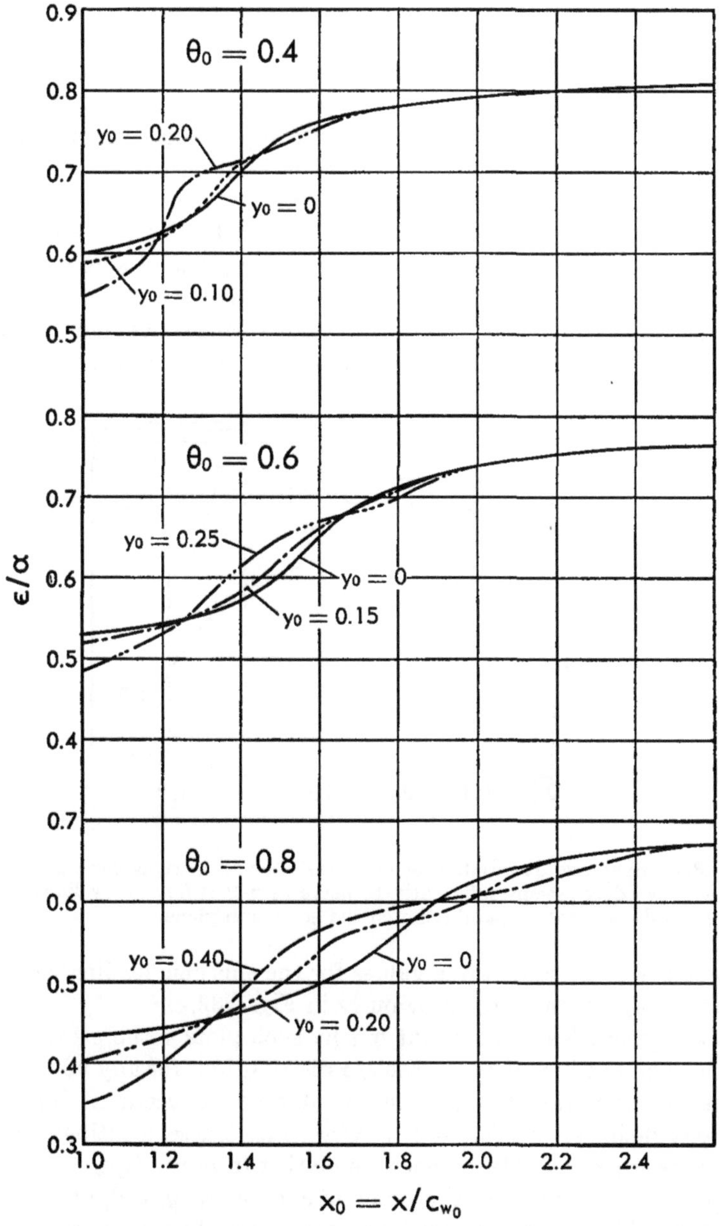

Fig. C,32k. Chordwise variation in the downwash behind a triangular wing with subsonic leading edges (three different values of sweep angle selected) for survey lines spaced at several spanwise locations, and lying on the horizontal plane.

its total possible range of values, *provided that one has moved downstream from the trailing edge far enough to assure that* $x/c_{w_0} \geqq 1 + \theta_0$. It may likewise be concluded from examination of Fig. C,32k that, provided $y_0 \leqq \theta_0/3$, then the values of ϵ/α, for any x/c_{w_0} value, i.e. for any fore-and-aft location behind the wing whatsoever, exhibit scarcely any influence at all coming from variation in the y_0 locations.

The way the downwash angle (for constant value of the angle of attack) varies as a function of distance behind the trailing edge, i.e. for increase in $x_0 = x/c_{w_0}$ for points lying in the plane of symmetry where $y = 0$, is depicted in Fig. C,32m, where ϵ/α is plotted vs. x_0 for three discrete values of height above the wake surface, denoted by $z_0/\theta_0 = \beta z/(\theta_0 c_{w_0})$. Three separate sets of curves are drawn on this graph to indicate

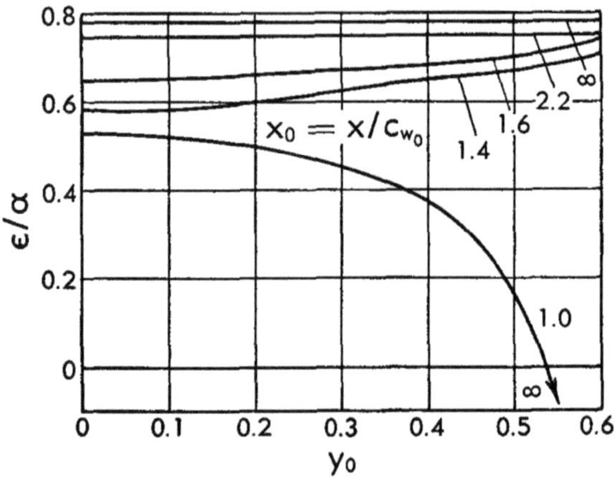

Fig. C,32l. Spanwise distribution of downwash behind a triangular wing with subsonic leading edges (swept sufficiently to make $\theta_0 = 0.6$) for survey lines spaced at several chordwise locations, and lying on the horizontal plane.

the trends for the same three values for the subsonic leading edge sweep parameter θ_0, as employed previously in Fig. C32,k.

A significant feature brought out by such plots is the presence of the singular points which occur in the ϵ/α distributions for any nonzero value of the height parameter z_0. The singular points occur at fore-and-aft locations defined by $x/c_{w_0} = 1 + \sqrt{\theta_0^2 + z_0^2}$. These locations represent the places where the Mach cones emanating from the tips of the trailing edge of the wing intersect the plane of symmetry, $y = 0$, at the various heights at which the survey levels of interest have been selected.

Such singularities are caused by the fact that the radial velocity has a logarithmic infinity on the surface of the tip Mach cones; the nature of this infinity thus explains why the singularity in question does not produce an infinite value for ϵ/α right at the plane of the wing where $z = 0$. This somewhat peculiar behavior of the singularity is due to the special circumstances existing about the wing, which is considered to be idealized to the case of a lifting surface which supports an infinite (in absolute value) pressure differential at its leading edge, according to the dictates of Eq. 32-29, and then exhibits a sudden falling off to zero at the trailing

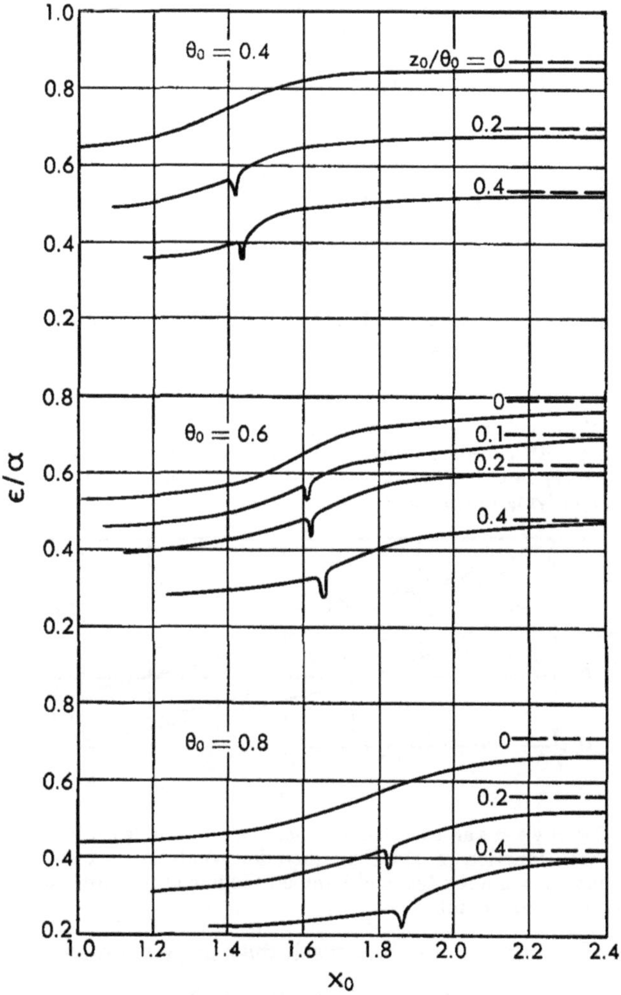

Fig. C,32m. Chordwise variations in the downwash behind a triangular wing with subsonic leading edges (three different values of sweep angle selected) for survey lines located at several vertical height spacings, and lying in the plane of symmetry.

edge. Owing to the occurrence of this singularity, the linear theory can no longer be applied. Due to the nature of the singularity, however, the region where the linear theory is not valid can be considered to be confined within a quite narrow region close to the surface of the Mach cones coming from the wing tips. Excluding this small region of nonapplicability, the rest of the downwash field may be calculated by using the same formula encountered previously in Chap. 1 provided the level below (or above) the plane of the wing at which the values of $\epsilon = w/U$ are to be determined is not too great. For the wing under study, this extension of

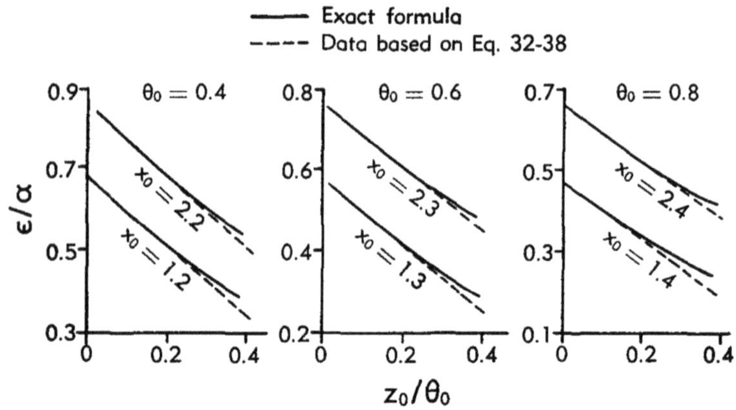

Fig. C,32n. Variation in the downwash as one makes a vertical traverse behind a triangular wing with subsonic leading edges (three different values of sweep angle selected) for survey lines located at various discrete downstream positions behind the trailing edge, and lying in the plane of symmetry.

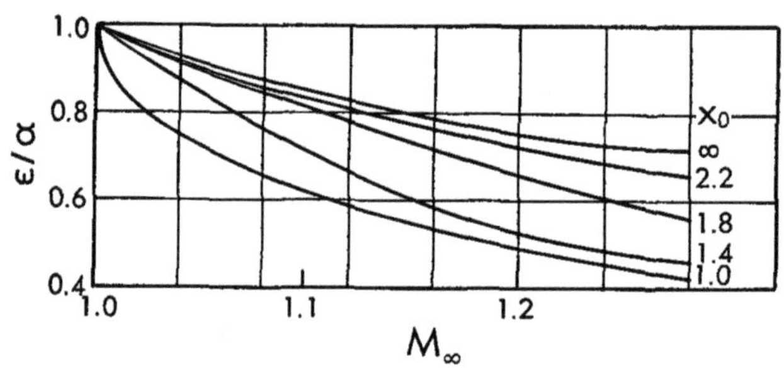

Fig. C,32o. Downwash variation behind a triangular wing with leading edge sweep angle of 45°, as a function of Mach number, for points located at certain selected distances behind the trailing edge, and lying on the line of intersection of the plane of symmetry with the horizontal.

the familiar formula for the downwash may be written as

$$\frac{\epsilon(x, 0, z)}{\alpha} = \frac{\epsilon(x, 0, 0)}{\alpha} - \frac{\beta z}{E_0 \theta_0 c_{w_0}} \frac{z}{|z|} \tag{32-38}$$

which holds approximately for small values of z/c_{w_0}.

In order to show how well Eq. 32-38 approximates the actual downwash, the results of computations utilizing Eq. 32-38 are compared in Fig. C,32n with the exact values in the cases of several discrete values of $x/c_{w_0} = x_0$ and for a sufficiently large range of vertical heights, in order to demonstrate that if $\beta z/\theta_0 c_{w_0} = z_0/\theta_0 \leqq 0.4$ then there is no practical difference between the two sets of curves, regardless of the value of x_0 under consideration. These curves have been prepared for the cases representing the same three values of the leading edge sweep parameter θ_0 as used previously; the general conclusion just drawn holds in all three cases.

Fig. C,32o shows the influence of the Mach number on the value of ϵ/α for a triangular wing with a sweep of 45°, for the different values of x/c_w.

C,32.1. Lifting Line Theory.

CONVERSION OF A LIFTING SURFACE INTO AN EQUIVALENT LIFTING LINE. The method proposed on p. 431 possesses the distinct advantage of being applicable to the determination of the downwash behind a wing of any planform whatsoever, but its actual execution is apt to lead to a great deal of labor. As pointed out in Art. 31 a very substantial simplification in the calculational complexities comes about if, for the calculation of the downwash, one replaces the lifting surface concept with that of a lifting line, just as is usually done in the subsonic case. Such a lifting line approach to the problem was formulated and applied to some practical examples by Mirels and Haefeli [*44*].

The determination of just where to place the equivalent lifting line, so as to render the error a minimum when calculating the downwash by this method is an essential point in such a process, and fairly easy to effect. It is merely necessary to observe that if one uses the symbol $\phi_s(x, y, z)$ to represent the potential for the flow when using lifting surface theory, and if the symbol ϕ_l is assumed to represent the flow for the equivalent lifting line that replaces the wing, then the difference of these two potentials is given by

$$\phi_s - \phi_l = \frac{\beta^2 z}{2\pi}\left[\iint\limits_{S} \frac{\Delta\phi(\xi, \eta)d\xi d\eta}{R^3} - \iint\limits_{S_1} \frac{\Delta\phi(\eta)d\xi d\eta}{R^3}\right] \qquad (32.1\text{-}1)$$

where the following glosses explain the notation:

1. The part of the lifting surface which is bounded by the forward-facing Mach cone having its vertex at the point $P(x, y, z)$, where one wishes to obtain the value of ϕ, is to be denoted by S.
2. The part of the same surface described above which lies within the forward-facing Mach cone but which is further restricted to just the region bounded upstream by the lifting line is to be denoted by S_1.
3. If ξ_2 is taken to stand for the abscissa value of the trailing edge location in any arbitrary one of the airfoil sections cut out by the general plane η = const, then the definition of $\Delta\phi(\eta)$ is given by

$$\Delta\phi(\eta) = \Delta\phi(\xi_2, \eta)$$

Now if ξ_1 is likewise taken to represent the abscissa value of the leading edge location for any arbitrary general one of the airfoils so cut out by the planes η = const, then for ξ locations for which it is true that $\xi \leqq \xi_1$ it is recognized that

$$(\phi)_{z=0^-} - (\phi)_{z=0^+} = 0$$

holds for any value of η whatsoever. Furthermore, provided that one is concerned only with obtaining the downwash at the points $P(x, y, z)$ which have downstream x coordinates that are so large that the forward-facing Mach cone having its vertex at P contains the whole wing, then it follows that the difference in the two potentials turns out to be expressible in the form

$$\phi_s - \phi_l = \frac{z\beta^2}{2\pi}\left[\int_{-b/2}^{b/2} d\eta \int_{\xi_1}^{\xi_2} \frac{d\xi}{R^3} \int_{\xi_1}^{\xi} m(\xi', \eta) d\xi' \right.$$

$$\left. - \int_{-b/2}^{b/2} d\eta \int_{\xi_0}^{\xi_2} \frac{d\xi}{R^3} \int_{\xi_1}^{\xi_2} m(\xi', \eta) d\xi' \right]$$

The symbol $m(\xi, \eta)$ has been substituted for the local strength of the doublet distribution, i.e. the weighting function is given by

$$m(\xi, \eta) = \frac{\partial}{\partial \xi} \Delta\phi(\xi, \eta) \qquad (32.1\text{-}2)$$

and $\xi_0(\eta)$ has been used as a lower limit on the integral sign above to signify the abscissa value of the point of intersection of any one of the arbitrary general planes y = const with the most representative lifting line being sought.

If one now inverts the indicated order of integration with respect to the variables ξ and ξ' appearing in the above expression for $\phi_s - \phi_l$, and if it is further recognized that

$$\int \frac{d\xi}{R^3} = \frac{1}{\beta^2} \frac{x - \xi}{(y - \eta)^2 + z^2} \frac{1}{R}$$

then the difference expression becomes simply

$$\phi_s - \phi_l = \frac{z\beta^2}{2\pi} \int_{-b/2}^{b/2} \frac{d\eta}{\beta^2[(y - \eta)^2 + z^2]}$$

$$\times \int_{\xi_1}^{\xi_2} \left[\frac{x - \xi'}{\sqrt{(x - \xi')^2 - \beta^2(y - \eta)^2 - \beta^2 z^2}} \right.$$

$$\left. - \frac{x - \xi_0}{\sqrt{(x - \xi_0)^2 - \beta^2(y - \eta)^2 - \beta^2 z^2}} \right] m(\xi', \eta) d\xi' \qquad (32.1\text{-}3)$$

Now consider more minutely the terms appearing under the radical signs in Eq. 32.1-3. If these quantities are developed in series, then for sufficiently large values of x it is true that

$$\int_{\xi_1}^{\xi_2} [\quad] m(\xi', \eta) d\xi' = \frac{\beta^2}{2} [(y - \eta)^2 + z^2] \left\{ \frac{2}{(x - \xi_0)^3} \int_{\xi_1}^{\xi_2} (\xi' - \xi_0) m d\xi' + \frac{3}{(x - \xi_0)^4} \int_{\xi_1}^{\xi_2} (\xi' - \xi_0)^2 m d\xi' + \cdots \right\}$$

out as far as the terms which contain $1/(x - \xi_0)$ raised to no higher than the fourth power.

Consequently, regardless of what values of y and z are involved, so long as the value of ξ is selected to satisfy the relation that

$$2(x - \xi_0) \int_{\xi_1}^{\xi_2} (\xi' - \xi_0) m d\xi' + 3 \int_{\xi_1}^{\xi_2} (\xi' - \xi_0)^2 m d\xi' = 0 \qquad (32.1\text{-}4)$$

the difference $\phi_s - \phi_l$ is zero, to the degree of accuracy of terms in $1/(x - \xi_0)$ raised to no higher than the fourth power.

If one now lets the symbol x_g stand for the location coordinate for the barycenter of the mass density that is distributed along the chord of each one of the arbitrary general airfoils under consideration, according to the dictates of the m function (which means that x_g coincides with the center of pressure of the airfoil or is the "center of gravity" of the circulation distribution over such an airfoil), and furthermore, if the symbol x_0 is taken to stand for the barycenter ("center of gravity") of the moments of the mass density distribution, where the moments are taken with respect to the y axis (the line $z = 0$, $x = 0$), then it is found from substitutions of these definitions for x_g and x_g into Eq. 32.1-4 that

$$\xi_0 \cong x_g \left(1 + \frac{3}{2} \frac{x_0 - x_g}{x}\right) \qquad (32.1\text{-}5)$$

out to terms which contain $1/x$ to no higher power than the first.

Because it is always going to be true that $x_0 > x_g$, it becomes clear at once that in order to keep the error to a minimum, in replacing the lifting vortical surface by a single lifting vortical line when one is attempting to obtain the downwash ϵ, at a goodly distance downstream from the wing, one must select for the point at which the total amount of vorticity is assumed to be concentrated on each of the arbitrary general airfoil sections over which the vorticity is spread in actuality, that location which falls somewhere in between the center of pressure and the trailing edge, according to the dictates of Eq. 32.1-5. It is seen, besides, that the point giving minimum error lies closer and closer to the center of pressure as the x location at which the evaluation of the downwash is required retreats farther and farther away from the wing.

Because the ξ_0 coordinate derived on the basis of Eq. 32.1-5 is independent of both y and z, the very same rule holds for the calculation of the induced velocities everywhere in the arbitrary general plane $x = \text{const.}$

COMPUTATION OF THE INDUCED DOWNWASH VELOCITIES. The expression for the downwash produced at a point $P(x, y, z)$ lying downstream from such a lifting line is then

$$\frac{w}{U} = \frac{\beta^2}{2\pi U} \int_{y_1}^{y_2} \frac{\Gamma[(y-\eta) - \nu(x-f)]d\eta}{\nu R^3} + \frac{1}{2\pi U} \int_{y_1}^{y_2} \frac{(x-f)(y-\eta)}{R[(y-\eta)^2 + z^2]} \frac{d\Gamma}{d\eta} d\eta \quad (32.1\text{-}6)$$

where the following conventions in notation have been employed:

1. The equation defining the location in a fore-and-aft direction of the most representative lifting line is given by

$$\xi_0 = f(\eta) \quad (32.1\text{-}7)$$

2. $\nu = d\eta/d\xi_0$.
3. The symbols y_1 and y_2 are the values of the spanwise coordinate η at which the forward-facing Mach cone emanating from the point P intersects the most advantageous lifting line (call it the line L).
4. Finally $R = \sqrt{(x-f)^2 - \beta^2(y-\eta)^2 - \beta^2 z^2}$.

If it so happens that the lifting line in question can be adequately approximated by two equal pieces of a straight-line vortical axis which are inclined to the x axis by the same angle of sweep δ^*, then one may represent the slope of these two segments with respect to the x axis by the symbols ν_0^+ and ν_0^- where $\nu_0^- = -\nu_0^+$ and $\nu_0^+ = \tan \delta^*$. Therefore, upon making use of this notation in Eq. 32.1-6, and upon carrying out an integration by parts in order to evaluate the first of the integrals appearing on the right-hand side of this equation, it reduces, in this special case of straight-line swept vortical axes, to

$$\frac{w}{U} = \frac{\Gamma(0)}{2\pi} [G(\nu_0^-) - G(\nu_0^+)]_{\eta=0} - \frac{1}{2\pi} \int_{y_1}^{y_2} [G(\nu) - G(0)] \frac{d\Gamma}{d\eta} d\eta \quad (32.1\text{-}8)$$

where

$$G(\nu) = -\frac{[y - \eta - \nu(x-f)][\beta^2\nu(y-\eta) - (x-f)]}{R\{[(y-\eta) - \nu(x-f)]^2 + (1-\beta^2\nu^2)z^2\}} \quad (32.1\text{-}9)$$

in which the f now stands for the ratio y/ν, and the constants $\Gamma(0)$ and $G(0)$ represent the values that Γ and G take on when $\eta = 0$ and $\nu = 0$, respectively.

Note that one has to insert the value ν_0^+ for ν in this expression for $G(\nu)$ if η lies within the region running from 0 to $b/2$, while if the interval of integration encompasses the region running from $-b/2$ to 0 then the ν_0^- value of ν is to be used. It should be further pointed out that Eq.

32.1-8 for the downwash holds good only when the forward-facing Mach cone originating at the point P cuts across both sides of the lifting vortical axis pair now under scrutiny. If it turns out that only one of the straight-line vortical axes lies inside the forward-facing Mach cone emanating from P, one must make use of an alternate expression for w/U obtained by dropping out of Eq. 32.1-8 the term involving $\Gamma(0)$.

If the ultimate in simple wings is to be handled by this method, where the two swept lifting straight-line axes are now assumed to coalesce into one straight line which is parallel to the y axis, then Eq. 32.1-8 becomes even more elementary in form, reducing to

$$\frac{w}{U} = \frac{1}{2\pi U}\int_{y_1}^{y_2} \frac{x(y-\eta)(R^2-\beta^2 z^2)}{R(x^2-\beta^2 z^2)[(y-\eta)^2+z^2]}\frac{d\Gamma}{d\eta}\,d\eta \tag{32.1-10}$$

For the calculation of the values of w/U which hold for points lying on the plane of the undisplaced wake behind the wing, it is readily apparent that Eq. 32.1-10 simplifies still further to

$$\left(\frac{w}{U}\right)_{z=0} = \frac{1}{2\pi U}\int_{y_1}^{y_2} \frac{R}{x(y-\eta)}\frac{d\Gamma}{d\eta}\,d\eta \tag{32.1-10'}$$

which is, in fact, merely an extension to the supersonic regime of the formula which was deduced in Chap. 1.

Now if the point $P(x, y, z)$ at which the downwash is to be calculated lies in region II of the plane $z = 0$, which is defined as the region comprising all points which lie inside of both Mach cones emanating from the extremities of the lifting straight-line axis under consideration, then evaluation of Eq. 32.1-10′ may be made in the form of

$$\frac{1}{2\pi U}\int_{-b/2}^{b/2} \frac{d\Gamma}{d\eta}\frac{R}{x(y-\eta)}\,d\eta = \frac{1}{2\pi U}\int_{-b/2}^{b/2}\frac{d\Gamma^*}{d\eta}\frac{d\eta}{y-\eta} \tag{32.1-11}$$

where the circulation symbol carrying the asterisk superscript is defined as

$$\Gamma^* = \Gamma(\eta)\frac{x}{\sqrt{x^2-\beta^2(y-\eta)^2}} \tag{32.1-12}$$

Thus, by inserting this relationship into Eq. 32.1-10′ it follows that the formula for the downwash in this special case becomes identical with that which was found to hold in the subsonic regime. In fact, in order to obtain the identical formula derived in Chap. 1 for the subsonic case it is merely necessary to represent the circulation distribution in the usual manner, employing the m discrete values of Γ_n as follows:

$$\Gamma = \frac{2}{m+1}\sum_{n=1}^{m}\Gamma_n\sum_{\mu=1}^{m}\sin(\mu\theta_n)\sin(\mu\theta)$$

where Γ_n is the value of Γ at the location θ_n, and where the spanwise locations have been converted into angular measure according to the scheme

$$\theta = \cos^{-1}\frac{2y}{b}, \quad \text{so that} \quad \theta_n = \cos^{-1}\frac{n\pi}{m+1}$$

Once this convention is set up, then the downwash is given in the already familiar form of Eq. 2-13′

$$(\epsilon)_{z=0} = 2b_{\nu\nu}\frac{\Gamma(y_\nu)}{bU} - 2\sum_{n=1}^{m}{}' b_{n\nu}\frac{\Gamma(y_n)}{bU}\frac{x}{\sqrt{x^2 - \beta^2(y_\nu - y_n)^2}} \tag{32.1-13}$$

where the symbols $b_{\nu\nu}$ and $b_{\nu n}$ are the well-known Multhopp coefficients and the prime superscript on the summation symbol denotes that the term with repeated subscripts, i.e. when $n = \nu$, must be omitted.

If, on the other hand, the point P at which the downwash is to be calculated does not lie in region II, but is situated instead in region I, which embraces all of the area which is external to the region covered by both of the Mach cones emanating from the extremities of the lifting line L, then new variables should be introduced obeying the requirements that

$$\eta_1 = y - \frac{x}{\beta} \quad \text{and} \quad \eta_2 = y + \frac{x}{\beta}$$

so that the expression for the downwash at such a point P may be written as

$$(\epsilon)_{z=0} = \frac{1}{2\pi U}\int_{\eta_1}^{\eta_2}\frac{R}{x(y-\eta)}\frac{d\Gamma}{d\eta}\,d\eta \tag{32.1-14}$$

Now let the further notation be introduced that

$$\beta\frac{y-\eta}{x} = t = \cos\theta$$

and let Γ_1 and Γ_2 denote the values that Γ takes on at the locations where $\eta = \eta_1$ and $\eta = \eta_2$, respectively. It is then possible to rearrange the circulation distribution into the form

$$\Gamma = \frac{\Gamma_1 + \Gamma_2}{2} + \frac{\Gamma_1 - \Gamma_2}{2}\cos\theta + \Gamma^*(\eta)$$

provided the end conditions are imposed that

$$\Gamma^*(\eta_1) = \Gamma^*(\eta_2) = 0 \tag{32.1-15}$$

Now it is also evident that

$$\Gamma^*(\eta) = \Gamma^*(y - \beta x\cos\theta) = \Gamma^*(\theta)$$

and thus it follows immediately that the expression for the downwash given as Eq. 32.1-14 reduces to

$$(\epsilon)_{z=0} = \frac{1}{2\pi U}\frac{\beta}{x}\int_0^\pi \frac{d\Gamma^*}{d\theta}\tan\theta d\theta \qquad (32.1\text{-}16)$$

Continuing with this manipulation, let the new converted circulation distribution $\Gamma^*(\theta)$ be represented once again by means of a trigonometric series (employing only terms in $\sin\theta$) and having the explicit form

$$\Gamma^*(\theta) = \frac{2}{m+1}\sum_{n=1}^{m}\Gamma_n^*\sum_{\mu=1}^{m}\sin(\mu\theta_n)\sin(\mu\theta)$$

where the angular notation employed here is based on the following equivalences:

$$\eta_n = y - \beta x\cos\frac{n\pi}{m+1}$$

and

$$\cos\theta_n = \cos\frac{n\pi}{m+1} = t_n = \beta\frac{y-\eta_n}{x}$$

and where the value of Γ^* at the location for which $t = t_n$ or, that is, for $\eta = \eta_n$ is denoted by Γ_n^*. Of course, by use of such a sine series arrangement for Γ^*, the conditions stated as Eq. 32.1-15 are automatically satisfied.

On the basis of such a series representation for the converted circulation, the integral appearing in Eq. 32.1-16 may be simplified to give the downwash angle in the plane of the wake, as

$$(\epsilon)_{z=0} = \frac{\beta}{2\pi U}\frac{2}{m+1}\sum_{n=1}^{m}\Gamma_n^*\sum_{\mu=1}^{m}\mu\sin(\mu\theta_n)\int_0^\pi\tan\theta\cos(\mu\theta)d\theta$$

The evaluation of the remaining integral in this expression for the induced vertical velocities (as compared with the free stream velocity) comes about in a straightforward manner; thus the following is found:

(i) *For odd values of* μ

$$A_\mu = \int_0^\pi \tan\theta\cos(\mu\theta)d\theta = -\frac{2}{\mu(\mu-2)} - 2\binom{\mu-1}{1}\frac{2}{(\mu-2;2;2)}$$
$$+2\binom{\mu-1}{3}\frac{2^2\cdot 2!}{(\mu-4;2;3)} - 2\binom{\mu-1}{5}\frac{2^3\cdot 3!}{(\mu-6;2;4)} + \cdots$$

where the quantities represented by the parentheses appearing in the numerators of the terms are the binomial coefficients

$$\binom{g}{h} = \frac{g!}{h!(g-h)!}$$

and where the quantities appearing in the denominators of these terms represent the series defined in general terms according to the scheme

$$(c; d; f) = c(c + d)(c + 2d) \cdots [c + (f - 1)d]$$

(ii) *For even values of* μ

$$\int_0^\pi \tan\theta \cos(\mu\theta)d\theta = 0$$

With the aid of such series computations, therefore, it results that the expression for the downwash reduces in this case to just

$$(\epsilon)_{z=0} = \frac{\beta}{2\pi Ux}\left(C_{\nu\nu}\Gamma_\nu^* + \sum_{n=1}^{m}{}' C_{\nu n}\Gamma_n^*\right) \tag{32.1-17}$$

where the new symbols are defined as

$$C_{\nu\nu} = \frac{2}{m+1}\sum_{\mu=1}^{m} \mu A_\mu \sin\left(\mu\frac{\pi}{2}\right)$$

$$C_{\nu n} = \frac{2}{m+1}\sum_{\mu=1}^{m} \mu A_\mu \sin(\mu\theta_n)$$

$$\Gamma_\nu^* = \Gamma^*\left(\frac{\pi}{2}\right) = (\Gamma^*)_{\eta=y}$$

The use of the prime on the summation symbol has its usual significance.

The coefficients $C_{\nu n}$ and $C_{\nu\nu}$ have values which are independent of the x and y coordinates of the point $P(x, y, z)$ at which the downwash angle ϵ is to be computed. Consequently it is possible to work out a table of values for these coefficients once and for all. Such evaluations have been carried out and the results are listed here in Table C,32.1 in the case where

Table C,32.1. List of $C_{\nu n}$ coefficients, including values of μA_μ

μ	μA_μ	n	$C_{\nu n}$	n	$C_{\nu n}$
1	2.0000	1	− 0.27918	9	−10.22665
3	− 9.9999	2	− 0.12257	10	− 0.09376
5	15.3333	3	− 0.45546	11	− 1.17431
7	−22.2667	4	− 0.10418	12	− 0.10418
9	28.0572	5	− 1.17431	13	− 0.45546
11	−34.7364	6	− 0.09376	14	− 0.12257
13	40.6886	7	−10.22665	15	− 0.27918
15	−47.2557				
		$C_{\nu\nu} = C_{88} = 25.04222$			

15 discrete points have been assumed sufficient for describing the circulation distribution, that is, m is selected equal to 15, and thus $\nu = 8$.

The realization of how great a simplification is brought about in these computations for the induced downwash angle ϵ prompts one to attempt to replace the actual, most advantageous lifting line L, as defined and determined on p. 445, by a representative straight-line piece of vortical axis. By proceeding in a manner which is entirely analogous to that just mentioned, such a "most representative" straight-line vortical axis may be derived from the "most representative" lifting line L by setting up the criterion that the lifting straight-line axis must give the smallest deviation from the result obtained by use of the lifting line L under the condition that one wishes to calculate the downwash at a goodly distance downstream of such lines. Upon carrying out such a determination, it is found, as previously, that the error committed in computing the downwash by use of a lifting straight-line axis is a minimum, for sufficiently large x values, if the x coordinate for such a straight-line vortical axis, normal to the x axis, is chosen to be

$$\xi_0^* \cong X_g\left(1 + \frac{3}{2}\frac{X_0 - X_g}{x}\right) \tag{32.1-18}$$

where, in analogy to the definitions made previously, X_g stands for the abscissa location of the barycenter of the masses that are assumed to be distributed out along the most representative lifting line L according to the rule that the local mass density is proportional to $\Gamma(d\eta/ds)$ (where ds is the infinitesimal linear element of length along the line L). In addition, X_0 is the barycenter of the moments of the mass distribution where the moments are taken about the y axis.

It may be observed by inspection from Eq. 32.1-18 that the "best" bound vortex axis, sought as a valid replacement for the wing, lies somewhere in between the center of pressure of the wing and its trailing edge. It is further evident that it lies closer and closer to the center of pressure as the point at which one wishes to compute the downwash retreats farther and farther downstream from the wing [*45*].

Applications Wherein the Lifting Line Theory Is Utilized. The results computed for the values of the downwash angle ϵ along the line of intersection of the plane of symmetry with the plane of the wake, obtained by applying the method of the lifting line, reported by Mirels and Haefeli [*44*], are reproduced here in graphic form for the case of two triangular (subsonic leading edge) wings having aspect ratios of 1.6 and 3.2, and for the normalizing Mach number = $\sqrt{2}$. The same kind of calculations have been made for the case of two rectangular wings with aspect ratio of 2.0 and 4.0, again for a Mach number = $\sqrt{2}$. In addition to the values of ϵ/α presented in the curves in Fig. C,32.1a, C,32.1b, and

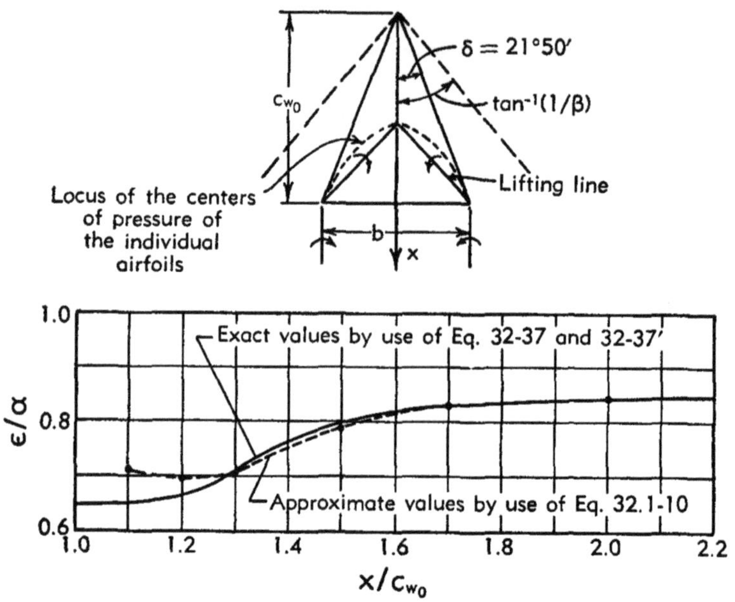

Fig. C,32.1a. Chordwise variation in the downwash behind the trailing edge of a triangular wing of aspect ratio 1.6 in supersonic flow at the normalizing Mach number of $\sqrt{2}$, for a survey line along the juncture of the plane of symmetry and the horizontal.

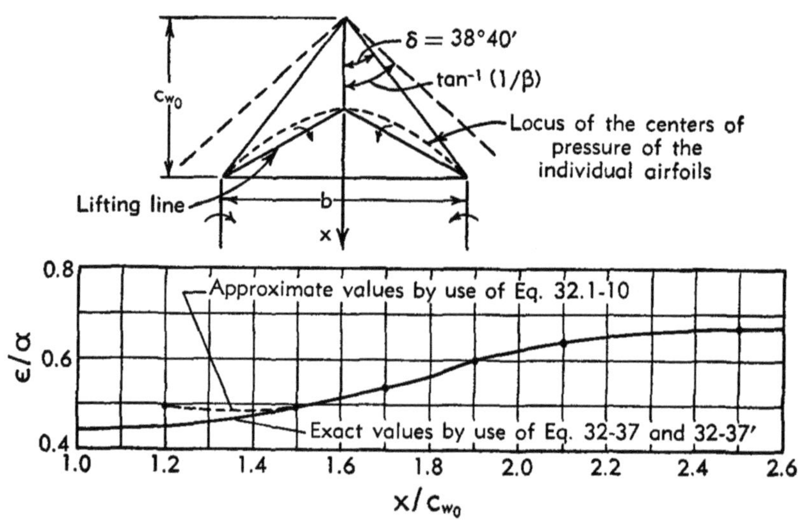

Fig. C,32.1b. Chordwise variation in the downwash behind the trailing edge of a triangular wing of aspect ratio 3.2 in supersonic flow at the normalizing Mach number of $\sqrt{2}$, for a survey line along the juncture of the plane of symmetry and the horizontal.

C,32.1c, a sketch is provided to show how the lifting line is situated with respect to the geometry of the wing planform. A comparison is also included on the plots of the results of employing the more exact formulas given on p. 434.

It appears from a comparison of the results obtained by the method of singularities and those obtained by use of the lifting line concept that the agreement between the exact and approximate values is quite satisfactory, provided that one moves downstream behind the trailing edge of the wing

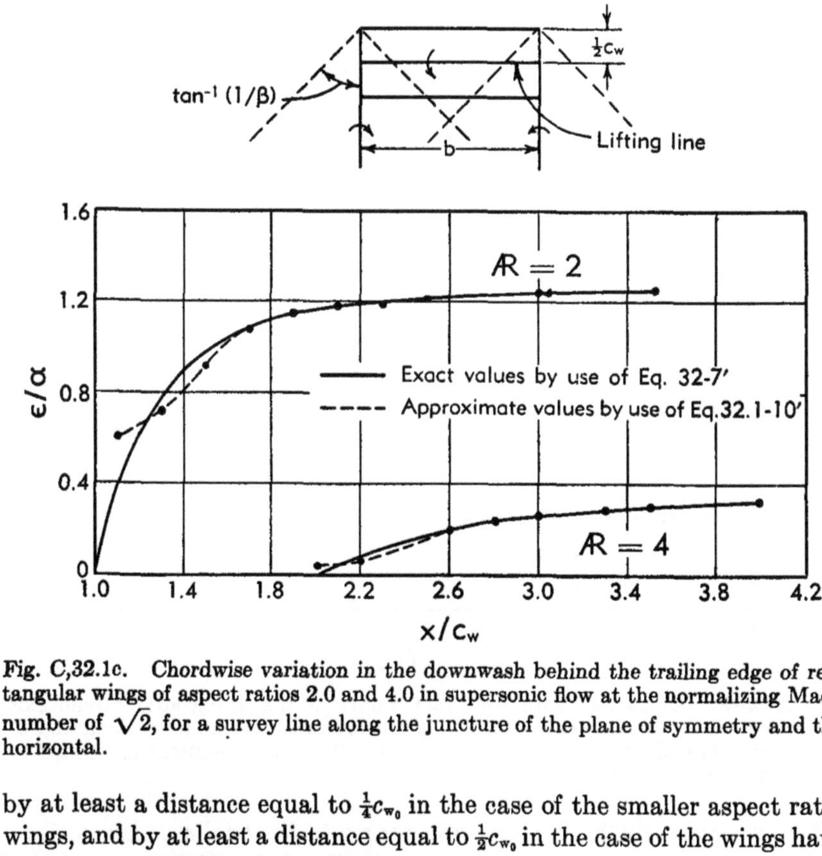

Fig. C,32.1c. Chordwise variation in the downwash behind the trailing edge of rectangular wings of aspect ratios 2.0 and 4.0 in supersonic flow at the normalizing Mach number of $\sqrt{2}$, for a survey line along the juncture of the plane of symmetry and the horizontal.

by at least a distance equal to $\frac{1}{4}c_{w_0}$ in the case of the smaller aspect ratio wings, and by at least a distance equal to $\frac{1}{2}c_{w_0}$ in the case of the wings having an aspect ratio twice as large.

For the calculation of the approximate value curves drawn in on Fig. C,32.1d, the triangular wings (of Æ = 1.6 and 3.2 again) were assumed to be replaced by the best lifting straight-line vortical axis (represented schematically for either wing in the same figure). The location selected for the placement of this bound vortex is at a distance of $\frac{3}{4}c_{w_0}$ from the apex of the triangle, while of course it is true that the center of pressure for such a wing lies farther forward, at only $\frac{2}{3}c_{w_0}$ from the vertex. It appears that once again the agreement between the exact and the approximate results is highly satisfactory, provided that the comparison is restricted to those locations which lie far enough downstream by the same amounts noted as applying to the wings illustrated in Fig. C,32.1a and C,32.1b (i.e. at least farther downstream than $\frac{1}{4}c_{w_0}$ behind the trailing edge for Æ = 1.6, and at least $\frac{1}{2}c_{w_0}$ behind the trailing edge for Æ = 3.2). In the present instance, however, the value of ϵ falls off quite precipitously

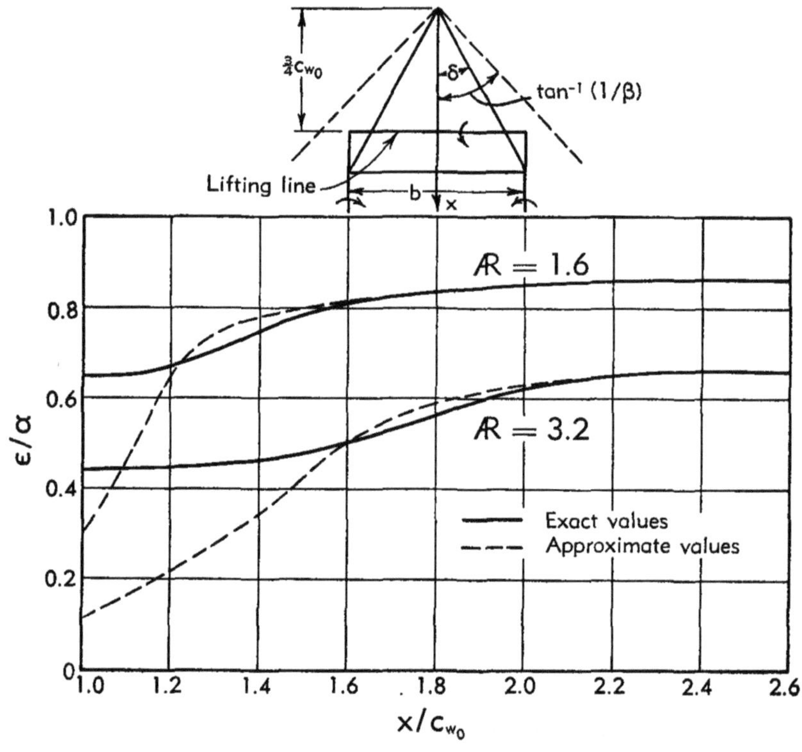

Fig. C,32.1d. Comparison with the exact values of the lifting straight-line axis approximation for the chordwise variation in the downwash behind the trailing edge of triangular wings of aspect ratio 1.6 and 3.2, for the same conditions illustrated in Fig. C,32.1a and C,32.1b.

at the locations which lie nearer and nearer to the trailing edge, in contradistinction to the gentle rise which took place formerly for points lying close behind the trailing edge, as evinced by the approximate value curves shown in Fig. C,32.1a and C,32.1b.

The corresponding results, which apply for rectangular wings (of Æ = 2 and 4) when treated according to the method of the lifting surface theory as opposed to the approximate lifting line theory (see Fig. C,32.1c), also supply confirming evidence that the approximations involved do not lead to serious error in the downwash values provided that one selects the location of the lifting straight-line axis to be at the $\frac{1}{2}c_{w_0}$ position be-

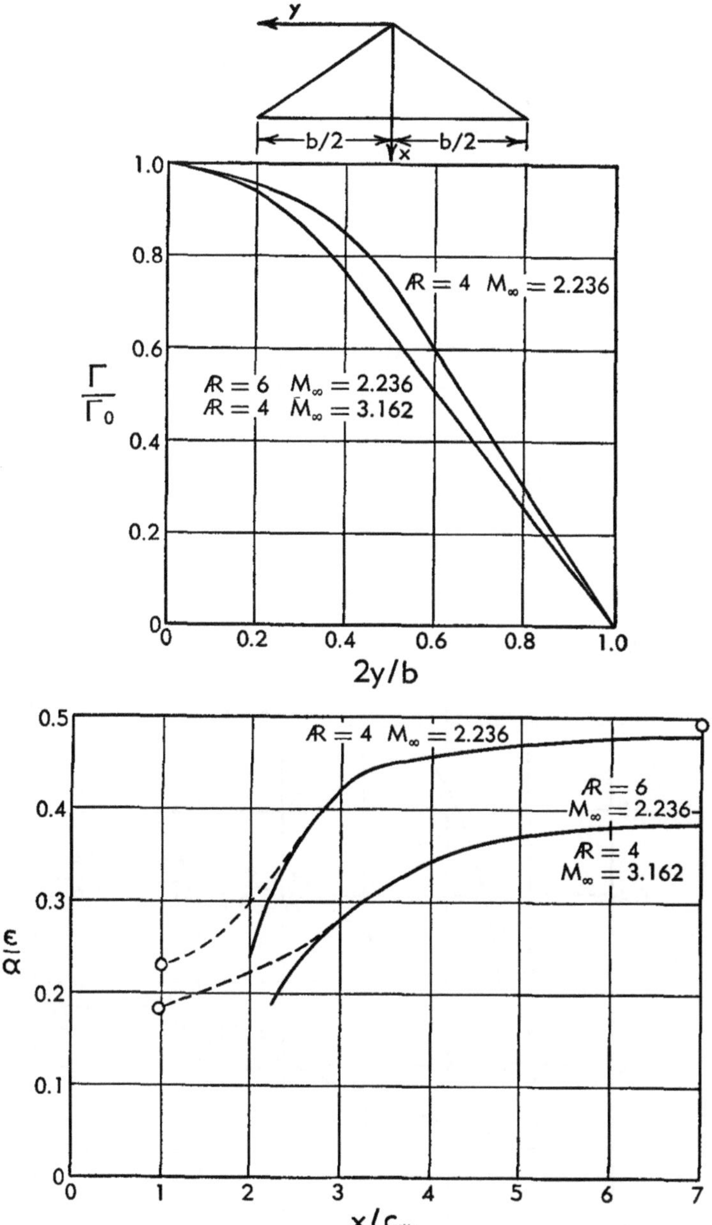

Fig. C,32.1e. Circulation distribution along the span and the chordwise variation in the downwash behind the trailing edge of two large aspect ratio (AR = 4.0 and 6.0) triangular wings having supersonic leading edges (at selected values of the Mach number) for a survey line along the juncture of the plane of symmetry and the horizontal, by use of the lifting straight-line axis approximation.

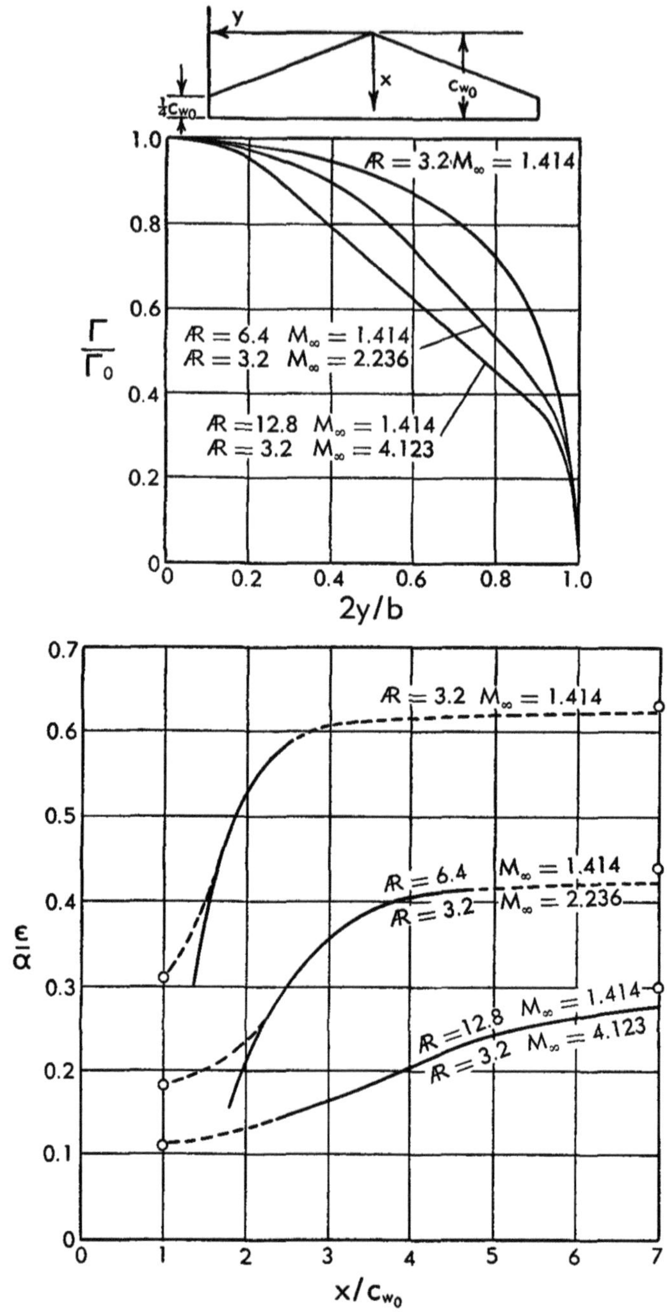

Fig. C,32.1f. Circulation distribution along the span and the chordwise variation in the downwash behind the trailing edge of several representative trapezoidal wings having supersonic leading edges (at selected values of the Mach number) for a survey line along the juncture of the plane of symmetry and the horizontal, by use of the lifting straight-line axis approximation.

hind the leading edge. Of course the center of pressure locations for such wings is at $\frac{4}{9}c_{w_0}$ downstream from the leading edge for the Æ = 2 wing, while it is somewhat closer to the wing axis for the Æ = 4 wing, the precise value being 10/21 chord lengths downstream from the leading edge.

The induced downwash has also been computed by Haefeli, Mirels, and Cummings [*46*], using the method of the lifting line, for various wings of rectangular, triangular, and trapezoidal planform, which possess solely supersonic leading and trailing edges in all cases. Their results, applying to triangular and trapezoidal wings, are reproduced here as Fig. C,32.1e and C,32.1f. It is evident from these plots how changes in the aspect ratio and in the Mach number of the flow about the triangular and trapezoidal wings affect the distribution in downwash produced by such surfaces at points lying behind them on the x axis.

It is worthwhile to point out in connection with these latter curves that, as far as the influence of the Mach number is concerned, its effect is produced on the downwash angle ϵ in a way opposite to that in the subsonic case. It may be remembered that in the subsonic regime the values of ϵ decreased as one progressed farther downstream, i.e. as x/c_{w_0} was allowed to increase but, in the present instance, where $M_\infty > 1$, the values of ϵ increase as one reaches larger values of x/c_{w_0} (regardless of whether the comparison is made on the basis of equivalent C_L values or with a fixed angle of attack α).

C,33. Computation of the Downwash Angles Induced in the Region Occupied by the Horizontal Tail. The procedure expounded in Chap. 1 may be utilized here to determine the downwash induced in the region occupied by the horizontal tailplane, under the condition that the rolling-up of the trailing vortex sheet Σ may be neglected at this stage of the computations. In following out the steps required in this procedure, it is first assumed that the values of $(\epsilon)_{z=0}$ have been obtained by the use of the various methods available for such determinations; one then proceeds to find the actual displaced position of the trailing vortex sheet Σ. Consequently if the notation is employed that $z_{\text{t.e.}}$ and $x_{\text{t.e.}}$ are the vertical and horizontal coordinates of the trailing edge of the wing in question, and if $(\epsilon)_{z=0}$ is the induced angle of downwash at the level of the wake, as computed previously, then the actual displaced position of points lying on the trailing vortex sheet may be obtained from the relation

$$z_1 = z_{\text{t.e.}} + \int_{x_{\text{t.e.}}}^{x} (\epsilon)_{z=0}\, dx \tag{33-1}$$

where the system of axes being defined here is now linked to the streamlines of the flow.

With knowledge of the location of the vortex sheet (Eq. 33-1), one may determine immediately how far away from the sheet it is necessary to go before encountering the actual location of the stabilizer. Having measured the distance away from the Σ sheet up to where the tail plane is located, one may go on to compute the value of the downwash angle ϵ at levels off of the actual vortex sheet itself, by having recourse to the same Taylor expansion method (Chap. 1) which is valid for values of $(z - z_1)$

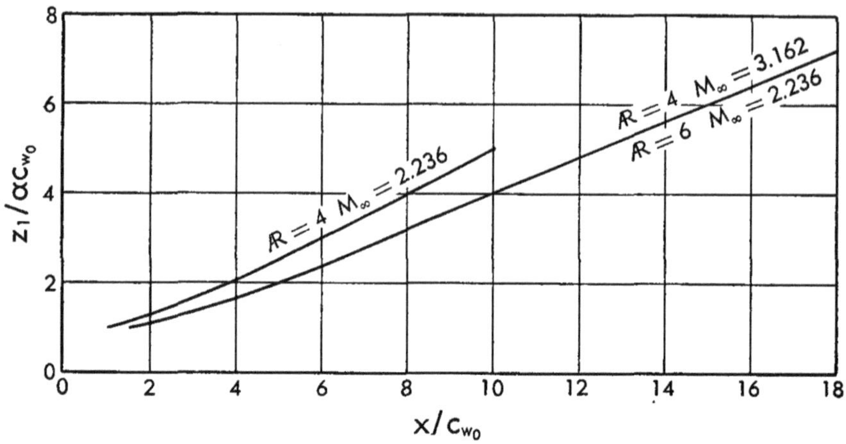

Fig. C,33a. Vertical displacement of the vortex sheet behind the trailing edge (for points confined to the x axis) in the case of the triangular wings illustrated in Fig. C,32.1e.

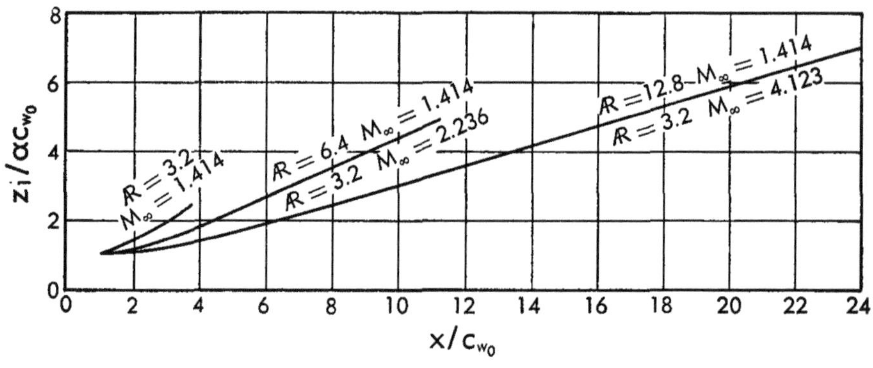

Fig. C,33b. Vertical displacement of the vortex sheet behind the trailing edge (for points confined to the x axis) in the case of the trapezoidal wings illustrated in Fig. C,32.1f.

which are not immoderately large. In fact, as pointed out earlier for small distances away from the vortex sheet, the use of the approximate method formulated in Chap. 1 produces results which are in very close agreement with the exact values (see Fig. C,32n). If the stabilizer happens to be located at a goodly distance above (or below) the vortex sheet, then the values of $z - z_1$ are too large to use the approximate method with any assurance of valid results. If this difficulty is encountered in practice, then one is forced to revert to use of the original formulas developed for prediction of the w/U values, where one now merely inserts into the appropriate expressions the values of $z - z_1$ in question, as substitutes for the previously employed z values.

The calculated values for some typical displacements of the trailing vortex sheet are shown in Fig. C,33a and C,33b for the triangular and trapezoidal wings mentioned in Art. 32 and sketched in Fig. C,32.1e and C,32.1f [*46*]. These curves of $z_1/\alpha c_{w_0}$ vs. x/c_{w_0} show the progressive effect of the aspect ratio and the Mach number upon the displacement of the vortex sheet as one proceeds downstream from the trailing edge.

C,34. Calculation of the Downwash in the Case Where the Trailing Vortex Sheet Is Considered Completely Rolled Up. As pointed

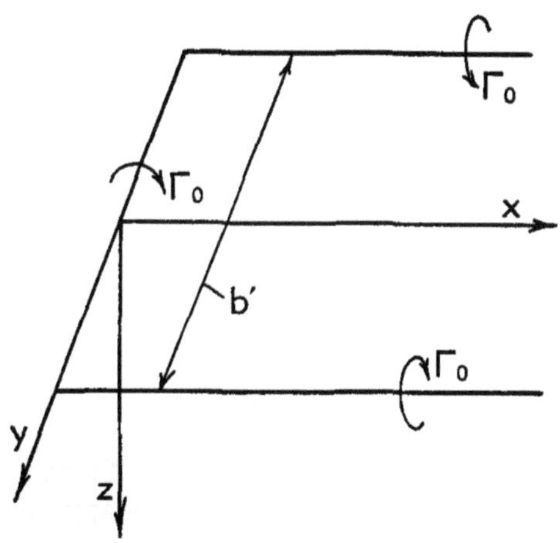

Fig. C,34. Schematic diagram of the horseshoe vortex configuration which is assumed to represent the idealization of a completely rolled-up trailing vortex sheet.

out in Art. 31, an alternative idealization for which the computation of the downwash angles can be carried out, without excessive complexity, arises under the condition where one stipulates that the trailing vortices are to be assumed completely rolled up, starting at the position where they are shed from the lifting line, in complete contrast to what has been premised in all the calculations discussed so far in this chapter.

Since the circumstances now being postulated are not truly representative of what actually happens unless the x coordinate is quite large for any point $P(x, y, z)$ at which it is desired to determine the downwash, under the limiting assumption of a fully rolled-up vortex sheet, it is quite permissible to take it for granted that the lifting line may be legitimately replaced by a simple straight-line vortical axis L^*, normal to the x axis. Consequently a vortex system which applies adequately in the present situation for explanation of the induced velocities may be taken as the austere idealization of a single horseshoe vortex as shown in Fig. C,34.

The distance between the side legs of the horseshoe vortex may be determined by the use of the same formula as in Chap. 1 (Eq. 2-30) and likewise the circulation which characterizes such trailing vortices is given by the same expression as employed in the subsonic regime. Let the spanwise location coordinate of the trailing vortices which represent the completely rolled-up vortex sheet be denoted by $y = \pm b'/2$, and assume that the lifting straight-line vortical axis L^* is located in the plane $z = 0$. The expression for the downwash produced at an arbitrary general point $P(x, y, z)$ as a result of induction from the horseshoe vortex system so orientated is then

$$\frac{w}{U} = \frac{x\Gamma_0}{2\pi U(x^2 - \beta^2 z^2)} \left(\frac{\left\{x^2 - \beta^2\left[\left(y - \frac{b'}{2}\right)^2 + 2z^2\right]\right\}\left(y - \frac{b'}{2}\right)}{\left[\left(y - \frac{b'}{2}\right)^2 + z^2\right]\sqrt{x^2 - \beta^2\left[\left(y - \frac{b'}{2}\right)^2 + z^2\right]}} - \frac{\left\{x^2 - \beta^2\left[\left(y + \frac{b'}{2}\right)^2 + 2z^2\right]\right\}\left(y + \frac{b'}{2}\right)}{\left[\left(y + \frac{b'}{2}\right)^2 + z^2\right]\sqrt{x^2 - \beta^2\left[\left(y + \frac{b'}{2}\right)^2 + z^2\right]}} \right) \tag{34-1}$$

wherein the circulation at the plain of symmetry is denoted by Γ_0, i.e.

$$\Gamma_0 = (\Gamma)_{y=0}$$

If one is interested in what happens at the Trefftz plane for such a rolled-up vortex sheet configuration, it is readily apparent that if $x \to \infty$ in Eq. 34-1 it reduces to

$$\left(\frac{w}{U}\right)_{x=\infty} = \frac{\Gamma_0}{2\pi U}\left[\frac{y - \frac{b'}{2}}{\left(y - \frac{b'}{2}\right)^2 + z^2} + \frac{y + \frac{b'}{2}}{\left(y + \frac{b'}{2}\right)^2 + z^2}\right] \tag{34-2}$$

The vertical displacement experienced by the two actual trailing vortices of such a horseshoe vortex system may be determined approximately by the use of Eq. 34-2 under the assumption that an induced velocity of constant amount is produced along the axis of each of the legs of the horseshoe vortex as a result of the downwash created by action of an equivalent strength vortex at the axis of the other leg. The vertical height of the cores of the trailing vortices may thus be referred to a coordinate system linked to the streamlines of the flow according to the relation

$$z_1 = z_{\text{t.e.}} + \frac{\Gamma_0}{2\pi U}\frac{x - x_{\text{t.e.}}}{b'} \tag{34-3}$$

Table C,34. Ratio of distance between legs of horseshoe vortex representing completely rolled-up vortex sheet and the actual span distance, for several wings previously illustrated in Fig. C,32.1e and C,32.1f.

	For triangular wings	
Mach no. = 2.24	Aspect ratio = 4	$b'/b = 0.64$
Mach no. = 2.24	Aspect ratio = 6	$b'/b = 0.60$
Mach no. = 3.16	Aspect ratio = 4	$b'/b = 0.60$
	For trapezoidal wings	
Mach no. = $\sqrt{2}$	Aspect ratio = 3.2	$b'/b = 0.84$
Mach no. = $\sqrt{2}$	Aspect ratio = 6.4	$b'/b = 0.75$
Mach no. = $\sqrt{2}$	Aspect ratio = 12.8	$b'/b = 0.68$
Mach no. = 2.24	Aspect ratio = 3.2	$b'/b = 0.75$
Mach no. = 4.12	Aspect ratio = 3.2	$b'/b = 0.68$

In Table C,34 are listed the values of the distance between the legs of the horseshoe vortex used as an idealization for the completely rolled-up vortex sheet, as compared to the actual wing span, for the wing planforms of which the downwash has already been computed by Haefeli, Mirels, and Cummings, and reported here in Fig. C,32.1e and C,32.1f. The b' values have been determined to agree with the actual circulation distribution which holds along the span of these specific wings for the various aspect ratios and Mach numbers as selected for the cases shown in the previous plots.

CHAPTER 5. WING-BODY INTERFERENCE AT SUPERSONIC SPEEDS

C,35. Introduction. The fundamental characteristic which distinguishes treatments of the wing-body interference problem at supersonic speeds from the analogous problem at subsonic speeds is rooted in the fact that it is no longer legitimate in the high speed range to replace the wing conceptually by a simple lifting line, even for purposes of finding the circulation (and thus the local lift) supported by an airfoil at an arbitrary location along the wing span.

Only if the wing is of quite short span with a leading edge which is slightly inclined to the free stream vector **U** and the body upon which it is mounted is spindlelike will the mode of attack on the problem of determining their mutual interference remain unaltered regardless of whether $M_\infty > 1$ or $M_\infty < 1$, provided that the leading edge of the wing is notably subsonic.

Therefore, two cases are considered herein, namely (1) the leading and trailing edges of the wing are supersonic, or (2) the leading edge is subsonic (to be generalized ultimately to include the possibility of a subsonic trailing edge), with the implicit understanding that the size and planform of the wing and the shape of the body are such as to preclude the application of slender body theory if accurate results are to be attained. The problem of determining the lift carried by a wing-body configuration is the only one considered in this chapter; it is sufficient for this purpose to assume that the wing is of zero thickness. The question of determining the drag, as well as the proper selection and placement of the components, in order to reduce the interference drag between the wing and the body, is not treated in this section.

In final analysis, the solution of the problem now set, whether of type 1 or of type 2, merely resolves itself into a detailed working-out of

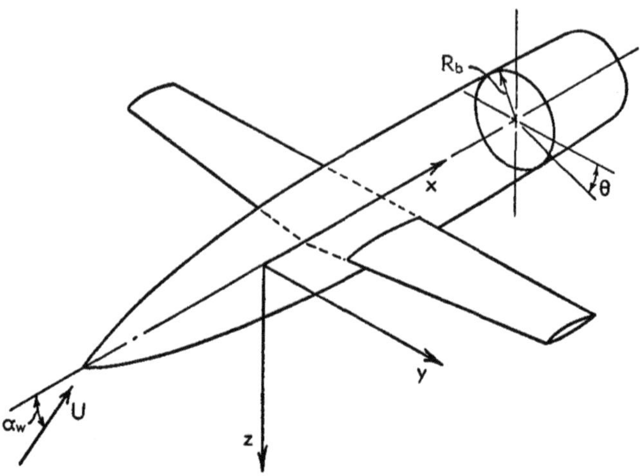

Fig. C,36. Orientation of the coordinate axes and a typical wing body combination.

the familiar method of singularities, although the exact form that this procedure will assume can be somewhat different from the conventional one.

Besides the exact methods (in the limits of approximation afforded by the linear theory), a brief account is given of simpler procedures which, under certain conditions and with a degree of approximation sufficient in most cases for technical purposes, can be applied to the calculation of the over-all aerodynamic actions.

C,36. Configurations Having Wings with Supersonic Leading and Trailing Edges. Statement of the Problem. The case is first examined for which the wing of the interacting ensemble has a supersonic leading edge of any arbitrary shape, and a trailing edge perpendicular to the plane of symmetry (see Fig. C,36). It is to be taken for granted, besides, that the wing is infinitesimally thin, that the fuselage is a body of revolution, and that the wing is centered on the body in the midwing position without incidence, so that the plane of the wing contains the axis of the body.

For simplicity of notation it will be convenient to assume that $\beta = \sqrt{M_\infty^2 - 1} = 1$. In order to obtain the solution which applies for a Mach number $M_\infty \neq \sqrt{2}$ it is merely necessary to convert the x, y, z coordinates employed herein into the physical coordinates x_1, y_1, z_1 by use of the Prandtl-Glauert transformation:

$$x = x_1/\beta, \quad y = y_1, \quad \text{and} \quad z = z_1$$

Now let U represent the magnitude of the free stream velocity, and let α_w denote the angle of attack between **U** and the x axis, in the reference system indicated in Fig. C,36.

The flow about such a wing-body configuration may be described by means of a velocity potential that takes the following specific form:

$$\phi = U\frac{b}{2}\{\xi + (\varphi_b)_1 + \alpha_w[(\varphi_b)_2 - \zeta + \varphi_w + \varphi^{(w)} + \varphi^{(b)}]\} \qquad (36\text{-}1)$$

wherein

$$\xi = 2x/b \quad \text{and} \quad \zeta = 2z/b \qquad (36\text{-}2)$$

provided the tip-to-tip wing span is taken to be of length b.

The other quantities appearing in Eq. 36-1 have the following meanings:

1. $U\frac{b}{2}(\xi - \alpha_w\zeta)$ stands for the potential which governs the free stream flow.
2. $U\frac{b}{2}[(\varphi_b)_1 + \alpha_w(\varphi_b)_2]$ stands for the incremental potential that corresponds to the flow perturbations produced by the body when it alone is inserted into the free stream.
3. $U\frac{b}{2}\alpha_w\varphi_w$ stands for the similar incremental potential which corresponds to the flow perturbations produced by the presence of the isolated wing in the otherwise undisturbed stream; the wing meant in this case is that resulting from an extension of the real exposed planform across the area normally covered by the fuselage, to result in an "extended" wing of the sort indicated in Fig. C,36.
4. $U\frac{b}{2}\alpha_w\varphi^{(w)}$ stands for the incremental potential representing the interference effect of the body upon the wing.
5. $U\frac{b}{2}\alpha_w\varphi^{(b)}$ stands for the analogous incremental potential representing the interference effect of the wing upon the body.

The potentials referring to the isolated components are assumed to be readily available by use of standard treatments, and it is obvious that the boundary condition at the surface of the wing alone is thus satisfied by the $\alpha_w\varphi_w$ contribution to the flow field, while the boundary condition at the surface of the body alone is satisfied by the $(\varphi_b)_1 + \alpha_w(\varphi_b)_2$ contribution. Consequently, upon invoking the physical fact that the velocity vectors representing the composite field of flow about the wing-body combination must be tangent to the surfaces of the configuration, it follows that the interference potentials must satisfy the following boundary conditions:

$$\left.\begin{aligned} &\left[\frac{\partial\varphi^{(w)}}{\partial\zeta}\right]_{\zeta=0} = -\left[\frac{\partial(\varphi_b)_2}{\partial\zeta}\right]_{\zeta=0} - \left[\frac{\partial\varphi^{(b)}}{\partial\zeta}\right]_{\zeta=0} \quad \text{over that part of the } \zeta = 0 \text{ plane which lies within the outline of the ``extended'' wing} \\ &\left[\frac{\partial\varphi^{(b)}}{\partial r}\right]_{r=r_b} = -\left[\frac{\partial\varphi_w}{\partial r} + \frac{\partial\varphi^{(w)}}{\partial r}\right]_{r=r_b} \quad \text{on the body} \end{aligned}\right\} \tag{36-3}$$

where the nondimensionalized radial coordinate r is defined by

$$r = \frac{2R}{b} \quad \text{and} \quad r_b = \frac{2R_b}{b} \tag{36-4}$$

when a cylindrical system of coordinates x, R, θ is employed in studying the flow about the body of revolution. The orientation of this coordinate system with respect to the body is illustrated in Fig. C,36 and, of course, it is plain that R_b denotes the radius of a circular cross section produced by a plane passed normal to the axis of the body at the generic axial location x = const.

C,37. Demonstration that Interference Potentials May Be Determined Independently of Each Other. Because the edges of the wing planform are supersonic, it is clear that the flow in the zone Σ, which is that region of space marked out by the envelopes of the Mach cones which pass through the leading and trailing edges of the wing (considered to be extended across the fuselage), may be analyzed independently in the half space $z > 0$ because there is no mechanism available for propagation, above the $z = 0$ plane, of disturbances arising in the half space $z < 0$ and vice versa. Since it is merely necessary to study the flow in one of these infinite half spaces, the following investigation concentrates on searching out the interference effects occurring solely in the region $z > 0$, at least for that part of total space encompassed in the Σ region. It is obvious that the potential $\varphi^{(w)}$ must be an odd function of z under these conditions, so that the determination of $\varphi^{(w)}$ for $z < 0$ may be obtained from the corresponding potential for $z > 0$ through the relation

$$\varphi^{(w)}(\xi, \eta, \zeta) = -\varphi^{(w)}(\xi, \eta, -\zeta) \tag{37-1}$$

where $\eta = 2y/b$.

Now in this infinite half space, $\zeta > 0$, the value of $\varphi^{(b)}$ takes on certain unique and definite values as θ varies over the range $0 \leqq \theta \leqq \pi$, but in consequence of what has just been noted about the independence of the flows in the upper and lower half spaces, it is evident that this potential may be extended in any manner whatsoever for values of θ in the interval from π to 2π. Thus, if this extension is made assuming that $\varphi^{(b)}$ must be an *even* function of θ, the $\varphi^{(b)}$ can be expressed by a Fourier series composed solely of cosine terms, namely

$$\varphi^{(b)} = \sum_n \varphi_n^{(b)}(\xi, r)r^n \cos n\theta \tag{37-2}$$

Here it is necessary to include only the even values of n, inasmuch as the field of interference velocities about the body, produced by the presence of the wing, will certainly exhibit mirrorlike symmetry with respect to the vertical plane $\theta = \pi/2$.

It follows[6] that

$$\left[\frac{\partial \varphi^{(b)}}{\partial \zeta}\right]_{\zeta=0} = \frac{1}{r}\left[\frac{\partial \varphi^{(b)}}{\partial \theta}\right]_{\theta=0 \text{ or } \pi} = 0 \tag{37-3}$$

and consequently the first of the boundary conditions given as Eq. 36-3 reduces to just

$$\left[\frac{\partial \varphi^{(w)}}{\partial \zeta}\right]_{\zeta=0} = -\left[\frac{\partial (\varphi_b)_2}{\partial \zeta}\right]_{\zeta=0} \quad \text{on the wing surface} \tag{37-3'}$$

so that the determination of $\varphi^{(w)}$ may be carried out without first having to discover the $\varphi^{(b)}$ potential. In this manner it has been demonstrated that one may proceed to find the unknown potential functions separately, because $(\varphi_b)_2$ can be considered as known.

For the determination of $\varphi^{(w)}$, it is found most effective to consider the wing as extending across the region actually covered by the body, and since $[\partial(\varphi_b)_2/\partial\zeta]_{\zeta=0}$ is not defined inside the body, it is necessary to

[6] It was first observed by Lagerstrom and Van Dyke [*47*], in a discussion of an investigation of wing-body interference made earlier by Ferrari, that it would be of advantage to suppress any sin $m\theta$ terms in the Fourier representation of the body-interference potential $\varphi^{(b)}$ so that only terms in cos $m\theta$ would appear, and then these even functions would possess the convenient property of assuring that $\partial\varphi^{(b)}/\partial\zeta = 0$ on the wing's surface. The subsequent treatment given in Art. 38–49 is actually based on work that has been expounded at length by Ferrari in [*48,49,50,51*] and by Nielsen and Pitts [*52*] but the suggestion made by Lagerstrom and Van Dyke has now been incorporated into these studies to produce a more attractive and compact development of the fundamental procedures.

assign the value of the derivative in this portion of the wing. The arbitrariness thus resulting is soon eliminated, because in the next step when $\varphi^{(b)}$ is determined it is required that the velocity field generated by this potential must negate *all* the velocities which are normal to the surface of the body regardless of whether they arise from the φ_w contribution to the total field or whether they come from the $\varphi^{(w)}$ interference potential just concocted. The ϕ thus obtained fulfills all the boundary conditions, and, on the other hand, the uniqueness theorem ensures that such ϕ, if it exists, is really the sought solution.

The salient result of the foregoing considerations is the reduction of the problem of determining $\varphi^{(w)}$ to the simple task of finding the governing potential corresponding to the perturbation field of velocities surrounding a given wing, for which the upwash velocities normal to its surface have been assigned. This elementary problem may be solved, of course, by employing any one of the pertinent methods which have been discussed in Vol. VI; for example, the standard method of singularities would be quite suitable. For present purposes, however, in order to give a more well-rounded treatment of the whole question and to gain certain advantages in ease of derivation and coherence between the various steps of the development, the formulas sought are obtained here by a procedure which differs formally from the more standard mode of attack, but which reduces to it without any real difficulty.

C,38. Determination of the Potential Corresponding to the Interference Effect of the Body upon the Wing in the Restricted Portion of Space Σ Not Influenced by the Wing Tips. The calculation of $\varphi^{(w)}$ will now be made in region Σ, as defined in Art. 37, and more precisely in that part of Σ which lies outside of the Mach cones emanating from the tips of the leading edge of the wing; let Σ' denote this inboard portion of the region Σ. The determination of $\varphi^{(w)}$ in the region of Σ which is inside of these cones is an independent problem which can be solved by applying the methods given in Vol. VI.

Define $(w_b^*)_2/(\alpha_w U)$ as a function which coincides with $[\partial(\varphi_b)_2/\partial\zeta]_{\zeta=0}$ on the wing, whereas upstream of the wing it is null. This distribution of relative angles of attack may be represented by the series

$$\frac{(w_b^*)_2}{\alpha_w U} = \sum_m G_m(\xi)\cos\left(\frac{\pi m\eta}{2}\right) \qquad \text{for } m = 1, 3, 5, \cdots \tag{38-1}$$

The possibility of expressing the $(w_b^*)_2/\alpha_w U$ by a Fourier series instead of by a Fourier integral derives from the stipulated supersonic nature of the entire wing now being considered. Because of this property it is easy to understand that the flow field about such a wing shape, at least within the region Σ', does not undergo any change if, instead of considering the wing-body system as an isolated entity, one imagines the fundamental system to be joined together with the infinite number of other systems which are mirror images of the first one with respect to the planes $y = \pm mb/2$.

This series consists only of cosine terms, owing to the symmetry of the field with respect to the plane $y = 0$; on the other hand, it is perfectly legitimate to stipulate that only the odd terms are to be included in this representation because this means merely that the $(w_b^*)_2/(\alpha_w U)$ values are being extended antisymmetrically into the $1 \leqq \eta \leqq 2$ region, which adjoins the basic interval $0 \leqq \eta \leqq 1$.

According to Eq. 38-1, it is natural to look for $\varphi^{(w)}$ in the form

$$\varphi^{(w)} = \sum_m \varphi_m^{(w)}(\xi, \zeta) \cos\left(\frac{\pi m \eta}{2}\right) \tag{38-2}$$

Because this sought potential must satisfy the linearized equation of motion governing the flow it will be necessary for each of the general harmonics making up Eq. 38-2 to satisfy the relation

$$\frac{\partial^2 \varphi_m^{(w)}}{\partial \zeta^2} - \frac{\partial^2 \varphi_m^{(w)}}{\partial \xi^2} = k^2 \varphi_m^{(w)} \tag{38-3}$$

where k is a shorthand notation for $\pi m/2$.

The imposition of the boundary conditions to be satisfied requires obviously that

$$\left[\frac{\partial \varphi_m^{(w)}}{\partial \zeta}\right]_{\zeta=0} = -G_m(\xi) \tag{38-4}$$

The solution of this differential equation, with attendant boundary conditions, is given in the infinite half space $\zeta > 0$ by the expression

$$\varphi_m^{(w)} = -\pi h_m(\xi^* - \zeta) + \pi k \zeta \int_0^{\xi^*-\zeta} h_m(\xi^{*\prime}) J_1[k\sqrt{(\xi^* - \xi^{*\prime})^2 - \zeta^2}] \frac{d\xi^{*\prime}}{\sqrt{(\xi^* - \xi^{*\prime})^2 - \zeta^2}} \tag{38-5}$$

where a shift in the origin of coordinates has been introduced by letting $\xi^* = \xi - r_b$. The other symbols not yet explained are J_n, which is the cylindrical Bessel function of nth order, and $h_m(\xi^*)$, which is an arbitrary function that is determinable immediately, however, through invoking the boundary condition (Eq. 38-4). It must be true according to this condition that

$$\pi \dot{h}_m(\xi^*) + \frac{\pi k^2}{2} \int_0^{\xi^*} h_m(\xi^{*\prime})(J_0 + J_2) d\xi^{*\prime} = -G_m(\xi^*) \tag{38-6}$$

where $\dot{h}_m(\xi^*) = dh_m/d\xi^*$.

It may then be shown that an explicit relationship for h_m may be obtained in terms of the G_m by means of the formula

$$h_m = -\frac{1}{\pi}\int_0^{\xi^*} G_m(\xi^{*\prime})J_0[k(\xi^* - \xi^{*\prime})]d\xi^{*\prime} \qquad (38\text{-}7)$$

Consequently, when this result is inserted into Eq. 38-5 and the evaluated expression is employed in turn in Eq. 38-2, the problem of determining $\varphi^{(w)}$ has been formally solved.

C,39. Determination of the Influence Exerted by the Wing Interference Potential upon the Body. By substituting Eq. 38-7 into Eq. 38-5, according to a formula deduced by Goodman [*53*] for $G_m(\xi^*)$ constant, the expression for $\varphi_m^{(w)}$ reduces to

$$\varphi_m^{(w)} = \frac{2}{k}\int_0^{\xi^*-\zeta} L_m(\xi^*, \zeta; \xi^{*\prime})dG_m \qquad (39\text{-}1)$$

where

$$L_m(\xi^*, \zeta; \xi^{*\prime}) = \sum_{j=1,3,5,\cdots} \left(\frac{\xi^* - \xi^{*\prime} - \zeta}{\xi^* - \xi^{*\prime} + \zeta}\right)^{j/2} J_j[k\sqrt{(\xi^* - \xi^{*\prime})^2 - \zeta^2}] \qquad (39\text{-}2)$$

and where the integral indicated in Eq. 39-1 is intended to be of the Stieltjes variety.

It has likewise been shown [*53*] that

$$\frac{\partial\varphi_m^{(w)}}{\partial\zeta} = -\int_0^{\xi^*-\zeta} L_m'(\xi^*, \zeta; \xi^{*\prime})dG_m \qquad (39\text{-}3)$$

where

$$L_m'(\xi^*, \zeta; \xi^{*\prime}) = \sum_{n=0,2,4,\cdots} \epsilon_n \left(\frac{\xi^* - \xi^{*\prime} - \zeta}{\xi^* - \xi^{*\prime} + \zeta}\right)^{n/2} J_n[k\sqrt{(\xi^* - \xi^{*\prime})^2 - \zeta^2}] \qquad (39\text{-}4)$$

in which ϵ_n stands for the Neumann factor, defined according to the scheme that

$$\epsilon_n = 1 \text{ for } n = 0$$

while

$$\epsilon_n = 2 \text{ for } n \neq 0$$

The component of the induced velocities, arising from each of the separate harmonics, and directed radially outward from the body axis in the system of cylindrical coordinates ξ, r, θ, may be written down in general terms as

$$\frac{\partial}{\partial r}[\varphi_m^{(w)} \cos k\eta] = \frac{\partial}{\partial \eta}[\varphi_m^{(w)} \cos k\eta] \cos\theta + \frac{\partial}{\partial \zeta}[\varphi_m^{(w)} \cos k\eta] \sin\theta$$

$$= -2 \cos\theta \sin k\eta \int_0^{\xi^*} L_m dG_m$$

$$- \sin\theta \cos k\eta \int_0^{\xi^*} L'_m dG_m \tag{39-5}$$

Consequently the component of velocity normal to the surface of the body of revolution in question, which is induced by action of the wing interference potential $\varphi^{(w)}$, is just

$$\left[\frac{v_{rb}^{(w)}}{\alpha_w U}\right]_{r=r_b} = \left[\frac{\partial \varphi^{(w)}}{\partial r}\right]_{r=r_b}$$

$$= -2\cos\theta \left(\sum_m \sin k\eta \int_0^{\xi^*} L_m dG_m\right)_{\substack{\eta=r_b\cos\theta\\ \zeta=r_b\sin\theta}}$$

$$- \sin\theta \left(\sum_m \cos k\eta \int_0^{\xi^*} L'_m dG_m\right)_{\substack{\eta=r_b\cos\theta\\ \zeta=r_b\sin\theta}} \tag{39-6}$$

In the event that one can be content with the calculation of the induced radial velocity components not actually at the surface of the body at the points P, but at the projections P', of such actual points of interest into the $\zeta = 0$ plane, the above-given formulas simplify remarkably. The projection process should be carried out by following along the inclined system of parallel lines, in the plane of symmetry $\eta = 0$, which constitute the intersections there of Mach planes normal to the plane of symmetry and passing through the body points in question, so that each equivalent point P' would be shifted upstream an amount proportional to the distance at which the body surface point stands up off of the $\zeta = 0$ plane.

If this very utilitarian hypothesis is accepted, then the expression for L_m becomes greatly simplified to read

$$L_m = \tfrac{1}{2} \int_0^{k(\xi^* - \xi^{*\prime})} J_0(x) dx$$

which is a well-known tabulated function; while, on the other hand, the value of L'_m reduces still further to $L'_m = 1$. In this especially abbreviated case, therefore, the description of the induced normal velocities is given by

$$\left[\frac{v_{rb}^{(w)}}{\alpha_w U}\right]_{r=r_b} = -2\cos\theta \left(\sum_m \sin k\eta \int_0^{\xi^*} L_m dG_m\right)_{\substack{\eta=r_b\cos\theta\\ \zeta=0}} - \sin\theta \frac{(w_b^*)_2}{U} \tag{39-6'}$$

At any rate, regardless of whether Eq. 39-6 or Eq. 39-6′ is considered to be most appropriate for a given situation, the values of $v_{rb}^{(w)}/\alpha_w U$ may now be assumed to be determined at all points on the body. Once having obtained the evaluation of these normal velocities at the various locations of interest on the body, it is then assumed that, at each one of the generic parallels (a cross section taken through the body at right angles to its axis) selected at an arbitrarily chosen series of axial positions along the length of the body, one may represent the distribution of normal velocities by means of a trigonometric series of the form

$$\left(\frac{v_{rb}^{(w)}}{\alpha_w U}\right)_{r=r_b} = \sum_n F_n(\xi) \cos n\theta \qquad \text{for } n = 0, 2, 4, \cdots \tag{39-7}$$

in the interval $0 \leqq \theta \leqq \pi$.

It is especially convenient to proceed to find the harmonic coefficients $F_n(\xi)$ of this typical description of the distribution of the normal velocities around a generic parallel of the body by appeal to the compact formula due to de la Vallée-Poussin. Thus, if the interval over which the function is to be represented, running from 0 to π in this instance, is subdivided into N equal parts, then the sought harmonic coefficients are given by the formulas

$$\left.\begin{aligned} F_s &= \frac{2}{N} \sum_{r=1}^{N-1} \frac{v_{rb}^{(w)}\left(\dfrac{r\pi}{N}\right)}{\alpha_w U} \cos\left(\frac{sr\pi}{N}\right) \\ &\qquad + \frac{\left[\dfrac{v_{rb}^{(w)}}{\alpha_w U}\right]_0 + \left[\dfrac{v_{rb}^{(w)}}{\alpha_w U}\right]_\pi}{N} \qquad \text{for } s = 0, 2, 4, \cdots N-2 \\ \text{while} \\ F_N &= \frac{1}{2}\left\{\frac{2}{N} \sum_{r=1}^{N-1} (-1)^r \frac{v_{rb}^{(w)}\left(\dfrac{r\pi}{N}\right)}{\alpha_w U} + \frac{\left[\dfrac{v_{rb}^{(w)}}{\alpha_w U}\right]_0 + \left[\dfrac{v_{rb}^{(w)}}{\alpha_w U}\right]_\pi}{N}\right\} \end{aligned}\right\} \tag{39-8}$$

where N is even in this particular case because of the symmetry of the flow with respect to the plane $\theta = \pi/2$. The notation $v_{rb}^{(w)}(r\pi/N)/\alpha_w U$ used here is practically self-explanatory, inasmuch as this expression represents the value of $v_{rb}^{(w)}/\alpha_w U$ which exists at the angular location given by $\theta = r\pi/N$, while $[v_{rb}^{(w)}/\alpha_w U]_0$ stands for the value of this normal velocity at the right side of the body (looking upstream and head down) where $\theta = 0$, and similarly the π subscript refers to the left-side location where $\theta = \pi$.

C,40. Determination of the Potential Corresponding to the Interference Effect of the Wing upon the Body in the Restricted Portion of Space Σ Not Influenced by the Wing Tips.

The procedure of Ferrari and Cramer. Following the same procedure as above when the induced interference velocities from the wing, $\partial\varphi^{(w)}/\partial r$, were being dealt with, one may now describe the induced velocities arising from the basic wing flow, $\partial\varphi_w/\partial r$, when evaluated at the surface of the body, by a trigonometric representation of the form

$$\left(\frac{\partial\varphi_w}{\partial r}\right)_{r=r_b} = \left(\frac{v_{rb,w}}{\alpha_w U}\right)_{r=r_b} = \sum_n F'_n(\xi)\cos n\theta \quad \text{for } n = 0, 2, 4, \cdots \tag{40-1}$$

likewise in the interval $0 \leqq \theta \leqq \pi$.

Consequently the boundary condition which was stated as the second of the Eq. 36-3 may be translated over into the requirement that

$$\left[\frac{\partial\varphi^{(b)}}{\partial r}\right]_{r=r_b} = -\sum_n (F_n + F'_n)\cos n\theta = \sum_n F^*_n(\xi)\cos n\theta = -\frac{v_{rb}}{\alpha_w U} \tag{40-2}$$

If the sought total potential for this body flow is then taken to be representable in the form introduced in Eq. 37-2, it is necessary that the following differential equation be satisfied by each of the individual potentials $\varphi_n^{(b)}$:

$$\frac{\partial^2\varphi_n^{(b)}}{\partial r^2} + \frac{1+2n}{r}\frac{\partial\varphi_n^{(b)}}{\partial r} = \frac{\partial^2\varphi_n^{(b)}}{\partial\xi^2} \tag{40-3}$$

It has been shown by Cramer [*54*] that the solution of Eq. 40-3 is

$$\varphi_n^{(b)} = \frac{(-1)^n}{r^n}\int_{\cosh^{-1}\xi/r}^{0} \mathfrak{F}_n(\xi - r\cosh u)\cosh nu\, du \tag{40-4}$$

and upon imposition of the boundary condition described in Eq. 40-2 it follows that each and every one of the arbitrary functions $\mathfrak{F}_n$ may be determined from the relation

$$\begin{aligned}\left[\frac{\partial(r^n\varphi_n^{(b)})}{\partial r}\right]_{r=r_b} &= -\int_{\cosh^{-1}\xi/r_b}^{0} \dot{\mathfrak{F}}_n(\xi - r_b\cosh u)\cosh nu\cosh u\, du \\ &= -F^*_n(\xi)\end{aligned} \tag{40-5}$$

where

$$\dot{\mathfrak{F}}_n(\xi) = \frac{d\mathfrak{F}_n}{d\xi}$$

For the general case where the body radius r_b varies in some arbitrarily specified manner with respect to position along the ξ axis, one may proceed to derive the solutions for the various arbitrary functions appearing in Eq. 40-5 by having recourse to the familiar step-by-step procedure employed by von Kármán in solving the similar case of flow about a body alone. In applying this method it is necessary to divide up the interval of integration into a large number of small subintervals within each of which the functions $\dot{\mathfrak{F}}_n(\xi)$ are assumed to be essentially constant, so that they may be taken out in front of the integral sign. Then, if the constant value that each $\dot{\mathfrak{F}}_n(\xi)$ takes on in the ith interval is denoted by the symbol $\mathcal{K}_n^{(i)}$, and if the symbols ξ'_{i-1} and ξ'_i are used to represent the end points of the left- and right-hand sides of this ith small interval, it follows that at every point $P^{(m)}$ on the body contour with coordinates ξ_m, r_{b_m} the boundary conditions to be satisfied by the successive $\mathcal{K}_n^{(i)}$ values turn out to be

$$\frac{1}{n^2-1}\sum_{i=1}^{m}\mathcal{K}_n^{(i)}\left[\sqrt{\left(\frac{\xi_m-\xi'_{i-1}}{r_{b_m}}\right)^2-1}\cosh\left(n\cosh^{-1}\frac{\xi_m-\xi'_{i-1}}{r_{b_m}}\right)\right.$$
$$-\sqrt{\left(\frac{\xi_m-\xi'_i}{r_{b_m}}\right)^2-1}\cosh\left(n\cosh^{-1}\frac{\xi_m-\xi'_i}{r_{b_m}}\right)$$
$$-n\frac{\xi_m-\xi'_{i-1}}{r_{b_m}}\sinh\left(n\cosh^{-1}\frac{\xi_m-\xi'_{i-1}}{r_{b_m}}\right)$$
$$\left.+n\frac{\xi_m-\xi'_i}{r_{b_m}}\sinh\left(n\cosh^{-1}\frac{\xi_m-\xi'_i}{r_{b_m}}\right)\right]=F_n^*(\xi_m) \qquad (40\text{-}6)$$

Thus the evaluation of the constants $\mathcal{K}_n^{(i)}$ may be carried out in each successive small interval without difficulty, but at the expense of some lengthy computational effort.

C,41. Determination of the Potential $\varphi^{(b)}$ in the Case of a Cylindrical Body. Relation of Boundary Condition to a Generalized Abel Equation. If, as often the case in practice, the body is a right circular cylinder (r_b = const) in the region where the wing abuts up against it, then Eq. 40-5 can be reduced to a generalized Abel integral equation, and the solution of such an equation may be brought about by applying the usual methods. To cast Eq. 40-5 into more standard form, let the following changes in notation be introduced:

$$\cosh u=\frac{\xi-\xi'}{r_b},\quad t=\frac{\xi}{r_b}-1,\quad t'=\frac{\xi'}{r_b},\quad \dot{\mathfrak{F}}_n(\xi')=U_n(t'),$$
$$\mathcal{G}(t-t')=\frac{1+t-t'}{\sqrt{2+t-t'}}\cosh\left[n\cosh^{-1}(1+t-t')\right] \qquad (41\text{-}1)$$

so that when these quantities are inserted into Eq. 40-5 it is converted into the generalized Abel integral equation

$$\int_0^t U_n(t')\frac{\mathcal{G}(t-t')}{\sqrt{t-t'}}\,dt'=-F_n^*(t) \qquad (41\text{-}2)$$

This integral equation may be transformed into a Volterra equation of the second kind. By letting

$$Q_n(t - t') = Q_n(z) = \int_0^z \frac{\mathcal{G}(s)ds}{\sqrt{s(z-s)}} \tag{41-3}$$

one obtains

$$\frac{\pi}{\sqrt{2}} U_n(t) + \int_0^t \dot{Q}_n(t-t') U_n(t')dt' = -\frac{d}{dt}\int_0^t F_n^*(t') \frac{dt'}{\sqrt{t-t'}}$$
$$= -\int_0^t \frac{dF_n^*}{\sqrt{t-t'}} \tag{41-4}$$

where, as usual,

$$\dot{Q}_n(z) = dQ_n/dz \tag{41-3'}$$

The indicated quadrature on the right-hand side of Eq. 41-3 may be carried out to give

$$Q_n(z) = \frac{\pi}{2\sqrt{2}}\left\{\left(\frac{2+z}{2}\right)^{-\frac{1}{2}+(n-1)} F\left[\frac{1}{2} - (n-1), \frac{1}{2} - (n-1); 1; \frac{z}{2+z}\right]\right.$$
$$\left. + \left(\frac{2+z}{2}\right)^{-\frac{1}{2}+(n+1)} F\left[\frac{1}{2} - (n+1), \frac{1}{2} - (n+1); 1; \frac{z}{2+z}\right]\right\} \tag{41-5}$$

where F represents the Gaussian hypergeometric function.

By making use of the well-known equivalence that $F(\frac{1}{2}, \frac{1}{2}; 1; x) = 2\mathsf{K}/\pi$, where K is the complete elliptic integral of the first kind with modulus $\sqrt{x}$, and upon noting that

$$\frac{\Gamma(c-a+n)}{\Gamma(c-a)} x^{c-a-1}(1-x)^{a+b-c-n}F(a-n, b; c; x)$$
$$= \frac{d^n}{dx^n}[x^{c-a+n-1}(1-x)^{a+b-c}F(a, b; c; x)]$$

it will be found that Eq. 41-5 may be written in the alternate form

$$Q_n(z) = \frac{\pi}{\sqrt{2}} x^{\frac{1}{2}}\{[\Gamma(n - \tfrac{1}{2})]^{-2}(1-x)^{n-\frac{1}{2}} B_{n-1}$$
$$+ [\Gamma(n + \tfrac{3}{2})]^{-2}(1-x)^{n+\frac{3}{2}}B_{n+1}\} \text{ for } n > 0$$

provided it is agreed that the newly introduced notation shall mean

$$B_m = \frac{d^m}{dx^m}\left[x^m \frac{d^m}{dx^m}(x^{m-\frac{1}{2}}\mathsf{K})\right]$$

while $x = z/(2+z)$.

In the case for which $n = 0$, the requisite formula becomes particularized to

$$Q_0(z) = 2\sqrt{2+z}\,\mathsf{E} - \frac{2}{\sqrt{2+z}}\mathsf{K}$$

where E represents the complete elliptic integral of the second kind with modulus $\sqrt{x}$.

Consequently, the sought derivative of Q_n is given by

$$\frac{dQ_n}{dz} = \frac{\pi}{2\sqrt{2}}(1-x)^2 \left\{ \left[\Gamma\left(n-\frac{1}{2}\right)\right]^{-2} \left[x^{-\frac{1}{2}}(1-x)^{n-\frac{3}{2}}\left(\frac{1}{2}-nx\right)B_{n-1} + x^{\frac{1}{2}}(1-x)^{n-\frac{1}{2}}\frac{dB_{n-1}}{dx}\right] + \left[\Gamma\left(n+\frac{3}{2}\right)\right]^{-2} \times \left[x^{-\frac{1}{2}}(1-x)^{n+\frac{1}{2}}\left(\frac{1}{2}-nx-2x\right)B_{n+1} + x^{\frac{1}{2}}(1-x)^{n+\frac{3}{2}}\frac{dB_{n+1}}{dx}\right]\right\} \tag{41-6}$$

By assuming that

$$\frac{\sqrt{2}}{\pi}\dot{Q}_n(z) = Q_n^*(z), \qquad H_n = -\frac{\sqrt{2}}{\pi}\int_0^t \frac{dF_n^*}{\sqrt{t-t'}} \tag{41-7}$$

it follows that the resolvent kernel of this Volterra integral equation may be written as

$$S_n(z) = Q_n^*(z) - \int_0^z Q_n^*(z-y)Q_n^*(y)dy + \int_0^z Q_n^*(z-y_1)dy_1 \int_0^{y_1} Q_n^*(y_1-y_2)Q_n^*(y_2)dy_2 + \cdots \tag{41-8}$$

so that the solution of Eq. 41-4 is just

$$U_n(t) = H_n(t) - \int_0^t S_n(t-t')H_n(t')dt' \tag{41-9}$$

If the expression for $H_n(t)$ given in Eq. 41-7 is inserted into this solution, it may be written explicitly as

$$U_n(t) = -\frac{\sqrt{2}}{\pi}\left[\int_0^t \frac{dF_n^*}{\sqrt{t-t'}} - \int_0^t \int_{t''}^t \frac{S_n(t-t')}{\sqrt{t-t'}}dt'dF_n^*\right]$$

or in more compact style as

$$U_n(t) = \frac{-\sqrt{2}}{\pi}\int_0^t \left[\frac{1}{\sqrt{t-t''}} - \mathcal{S}_n(t-t'')\right]dF_n^* = \int_0^t \Phi_n(t-t'')dF_n^* \tag{41-9'}$$

provided it is agreed to introduce the following new definitions:

$$\left.\begin{aligned} \mathcal{S}_n(t-t'') &= \int_{t''}^t \frac{S_n(t-t')}{\sqrt{t-t'}}dt' = \int_0^{t-t''} \frac{S_n(t-t''-t^*)}{\sqrt{t^*}}dt^* \\ &\text{and} \\ \Phi_n(z) &= -\frac{\sqrt{2}}{\pi}\left[\frac{1}{\sqrt{z}} - \mathcal{S}_n(z)\right] \end{aligned}\right\} \tag{41-10}$$

The method of Nielsen and Pitts. Returning anew to the solution for the body-flow interference potential as given in Eq. 40-4, it is readily seen that the indicated integral appearing therein may be integrated by parts to give

$$\varphi_n^{(b)} = \frac{(-1)^{n+1}}{r^n n} \int_0^{\xi - r} \dot{\mathfrak{F}}_n(\xi') \sinh\left[n \cosh^{-1}\left(\frac{\xi - \xi'}{r}\right)\right] d\xi' \qquad \text{for } n \neq 0 \quad (41\text{-}11)$$

provided the condition that $\mathfrak{F}_n(0) = 0$ is agreed to hold true for the sort of supersonic flow field now being investigated.

On the other hand, when $n = 0$, the individual zeroeth harmonic may be expressed as

$$\varphi_0^{(b)} = -\int_0^{\xi - r} \dot{\mathfrak{F}}_0(\xi') \cosh^{-1}\frac{\xi - \xi'}{r}\, d\xi' \qquad (41\text{-}11')$$

Now, however, it may be observed that the operational form of the expression $\sinh\,[n \cosh^{-1}(\xi - \xi')/r]$ appearing in the integrand of Eq. 41-11 is

$$\sinh\left(n \cosh^{-1}\frac{\xi - \xi'}{r}\right) = \frac{n}{(2n-1)(2n-3)\cdots 1}\,\frac{1}{r^n}\,\frac{\partial^{n-1}}{\partial \xi^{n-1}}\,[(\xi - \xi')^2 - r^2]^{n-\frac{1}{2}}$$

provided $n \neq 0$, and meanwhile it is also true that

$$\cosh^{-1}\left(\frac{\xi - \xi'}{r}\right) = \int \frac{d\xi'}{\sqrt{(\xi - \xi')^2 - r^2}}$$

Consequently, if one makes the simplification in notation that

$$\mathfrak{F}_n^* = \frac{\dot{\mathfrak{F}}_n}{(2n-1)(2n-3)\cdots 1}$$

then Eq. 41–11, for the body interference potential, reduces to

$$\varphi_n^{(b)} = \frac{(-1)^{n+1}}{r^n} \int_0^{\xi - r} \frac{\mathfrak{F}_n^*(\xi')}{r^n}\,\frac{\partial^{n-1}}{\partial \xi^{n-1}}\,[(\xi - \xi')^2 - r^2]^{n-\frac{1}{2}} d\xi' \qquad \text{for } n \neq 0 \quad (41\text{-}12)$$

But even if $n = 0$, this formula may be considered to hold, provided it is understood that one should replace the derivative operation indicated in the integrand by a single integral to stand for the $\cosh^{-1}\,[(\xi - \xi')/r]$ term.

By comparison of this result with what is given by Nielsen and Pitts in [*52*] it is immediately apparent that the formula of Eq. 41-12 is identical with the form assumed for the body interference potential by these authors, so that there has now been established in this simple manner a direct correlation between what Nielsen and Pitts have done and what was developed earlier by Ferrari and subsequently improved in form by Cramer.

Continuing to investigate this mode of attack, following Nielsen and Pitts, the imposition of the boundary condition to be satisfied by the normal components of the velocities impinging on the fuselage surface, for the case of r_b = const, results in the relation

$$\left[\frac{\partial(r^n\varphi_n^{(b)})}{\partial r}\right]_{r=r_b} = (-1)^n r_b^{n-2} \int_0^{\xi-r} \dot{\mathfrak{F}}_n^* \left\{ n \frac{\partial^{n-1}}{\partial \xi^{n-1}} \left[\left(\frac{\xi-\xi'}{r_b}\right)^2 - 1\right]^{n-\frac{1}{2}} + (2n-1)\frac{\partial^{n-1}}{\partial \xi^{n-1}}\left[\left(\frac{\xi-\xi'}{r_b}\right)^2 - 1\right]^{n-\frac{3}{2}}\right\} d\xi' = -F_n^*(\xi) \quad (41\text{-}13)$$

which is comparable to Eq. 40-5.

Upon introduction here of the same notation given in Eq. 41-1 it follows that the boundary condition in question may now be written as

$$\int_{0]}^{t} U_n^*(t') \left\{ n \frac{\partial^{n-1}}{\partial t^{n-1}} [(1+t-t')^2 - 1]^{n-\frac{1}{2}} + (2n-1)\frac{\partial^{n-1}}{\partial t^{n-1}}[(1+t-t')^2-1]^{n-\frac{3}{2}}\right\} dt' = -F_n^*(t) \quad (41\text{-}13')$$

where $U_n^*(t') = \dot{\mathfrak{F}}_n(\xi')/[(2n-1)(2n-3)\cdots 1]$.

This integral equation may be solved, following the treatment given by Nielsen and Pitts, through the use of the Laplace transform technique. Let

$$\mathfrak{L}[F(t)] = \overline{F}(p)$$

Upon transforming both sides of Eq. 41-13′, it turns out that

$$\overline{U}_n^*(p)[np^{n-1}\overline{\psi}(p) + (2n-1)p^{n-1}\overline{\psi}'(p)] = -\overline{F}_n^*(p)$$

where $\psi(t) = [(1+t)^2 - 1]^{n-\frac{1}{2}}$ and $\psi'(t) = [(1+t)^2-1]^{n-\frac{3}{2}}$.

It is seen that

$$\overline{\psi}(p) = e^p \frac{\Gamma(n+\frac{1}{2})}{\sqrt{\pi}}\left(\frac{2}{p}\right)^n K_n(p)$$

and

$$\overline{\psi}'(p) = e^p \frac{\Gamma(n-\frac{1}{2})}{\sqrt{\pi}}\left(\frac{2}{p}\right)^{n-1} K_{n-1}(p)$$

where $K_m(p)$ is the modified cylindrical Bessel function of the second kind, with imaginary argument, and of order m.

Table C,41. Axial strength functions (after Nielsen and Pitts).

t	Φ_0^{**}	Φ_2^{**}	Φ_4^{**}	Φ_6^{**}
0				
0.05			−0.531	−0.716
0.1	−1.321	−0.787	0.405	1.334
0.15				1.155
0.2	−0.868	−0.166	0.902	0.735
0.25				0.279
0.3		0.131	0.831	−0.119
0.35				−0.417
0.40	−0.538	0.304	0.560	−0.595
0.45				−0.675
0.50			0.245	−0.655
0.6	−0.378	0.450	−0.040	−0.417
0.7			−2.53	−0.070
0.75				0.095
0.8	−0.292	0.457	−0.386	0.235
0.85				0.340
0.9			−0.434	0.406
0.95				0.433
1.0	−0.224	0.392	−0.412	0.420
1.1			−0.337	0.306
1.2	−0.180	0.294	−0.231	0.123
1.3			−0.115	−0.062
1.4	−0.147	0.188	0.000	−0.197
1.5			0.095	−0.256
1.6	−0.122	0.088	0.163	−0.239
1.7			0.204	−0.162
1.8	−0.102	0.005	0 211	−0.062
1.9			0.197	0.036
2.0	−0.086	−0.058	0.164	0.105
2.1				0.134
2.2	−0.073	−0.100	0.069	0.121
2.3			0.020	0.085
2.4	−0.062	−0.123	−0.022	0.036
2.5			−0.053	−0.013
2.6	−0.053	−0.126	−0.073	−0.053
2.7				−0.063
2.8	−0.045	−0.119	−0.080	−0 059
2.9			−0.071	−0.042
3.0	−0.038	−0.103	−0.060	−0.019
3.1			−0.038	0.005
3.2	−0.031	−0.083	−0.019	0.022
3.3			−0.002	0.030
3.4	−0.026	−0.061	0.011	0.028
3.5			0.021	0.019
3.6	−0.022	−0.040	0.027	0.008
3.7			0.027	−0.002
3.8	−0.018	−0.022	0.025	−0.011
3.9			0.022	−0.014
4.0	−0.016	−0.007	0.015	−0.103

By algebraic rearrangement of the term multiplying $\overline{U}_n^*(p)$ in this Laplace transform of Eq. 41-13′ it is seen that

$$np^{n-1}e^p \frac{\Gamma(n+\frac{1}{2})}{\sqrt{\pi}}\left(\frac{2}{p}\right)^n K_n(p) + (2n-1)p^{n-1}e^p \frac{\Gamma(n-\frac{1}{2})}{\sqrt{\pi}}\left(\frac{2}{p}\right)^{n-1} K_{n-1}(p)$$
$$= \frac{2^n}{\sqrt{\pi}} e^p \left(\frac{n}{p} K_n + K_{n-1}\right) \Gamma(n+\tfrac{1}{2})$$
$$= \frac{2^n}{\sqrt{\pi}} e^p \dot{K}_n(p) \Gamma(n+\tfrac{1}{2})$$

where

$$\dot{K}_n(p) = \frac{dK_n}{dp}$$

and thus the relationship giving $\overline{U}_n^*(p)$ turns out to be just

$$\overline{U}_n^*(p) = -\frac{\sqrt{\pi}}{2^n\Gamma(n+\frac{1}{2})} \frac{p\overline{F}_n^*(p)}{pe^p\dot{K}_n(p)}$$

If the inverse Laplace transform is now taken of both sides of this equation, and if the auxiliary definition is made that

$$\mathfrak{L}^{-1}\left[\frac{\sqrt{\pi}}{2^n\Gamma(n+\frac{1}{2})} \frac{1}{pe^p\dot{K}(p)}\right] = \Phi_n^*(t)$$

where $\mathfrak{L}^{-1}$ indicates the operation of applying an inverse Laplace transform to the quantity within brackets, then it follows directly that the same formula for $U_n(t)$ as given in Eq. 41-9′ is now reobtained in this alternate way.

Inasmuch as the Laplace transform treatment and the integral equation analysis have now both led to the same result, it is expedient to base subsequent practical computations on the tabulated values, already prepared by Nielsen and Pitts, of the quantity

$$\Phi_n^{**}(t) = \frac{2n!}{n!}\frac{1}{2^n}\Phi_n^*(t)$$

which they worked out for the first four terms of the approximating series, i.e. for $n = 0, 2, 4$, and 6. These evaluations are reproduced here as Table C,41.

C,42. Calculation of Pressures Produced by Interference Effect on the Wing. Now that the incremental interference potentials $\varphi^{(w)}$ and $\varphi^{(b)}$ have been determined throughout the Σ' region, it is a routine matter to employ this knowledge in the calculation of the pressures produced by the interference effect upon the wing and upon that part of the body which is confined within the limits of Σ'. In fact, according to the precepts of linear theory, the pressure coefficients on the wing and on the body are respectively

$$\left.\begin{aligned}(\Delta C_p)_w &= -2\left[\frac{\partial\varphi^{(w)}}{\partial\xi} + \frac{\partial\varphi^{(b)}}{\partial\xi}\right]_{\zeta=0^+} \\ \text{and} \qquad & \\ (\Delta C_p)_b &= -2\left[\frac{\partial\varphi^{(b)}}{\partial\xi} + \frac{\partial\varphi^{(w)}}{\partial\xi}\right]_{r=r_b} \quad \text{in the region } \zeta > 0\end{aligned}\right\} \tag{42-1}$$

By reference to Eq. 38-2 and 38-5, the first part of the contribution to the incremental wing pressures due to interference may be obtained explicitly as

$$\begin{aligned}\left[\frac{\partial\varphi^{(w)}}{\partial\xi}\right]_{\zeta=0^+} &= \sum_m \left[\frac{\partial\varphi_m^{(w)}}{\partial\xi}\right]_{\zeta=0^+} \cos\frac{(\pi m\eta)}{2} \\ &= \sum_k \cos k\eta \int_0^{\xi^*} J_0[k(\xi^* - \xi^{*\prime})]dG_m \\ &= \frac{(w_b^*)_2}{\alpha_w U} - \sum_m \cos k\eta \int_0^{\xi^*} \{1 - J_0[k(\xi^* - \xi^{*\prime})]\}dG_m\end{aligned} \tag{42-2}$$

in which the last term on the right may be interpreted as the fall-off in the pressure distribution $(\Delta C_p)_w$, which is due to the three-dimensional nature of the flow, inasmuch as simple two-dimensional strip theory would give the first term on the right-hand side of Eq. 42-2. It should be remembered that in the present instance this formula has been derived for the normalizing Mach number, making $\beta = 1$, and thus reducing the Ackeret relation between vertical and horizontal incremental velocities in two-dimensional flow from $v/\sqrt{M_\infty^2 - 1} = u$ to $v = u$, where v is the velocity component normal to the free stream direction and u is the incremental component parallel to this direction.

The second part of the contribution to the incremental wing pressures, arising from the body flow potential, is obtained by first evaluating the axial derivative of each of the individual harmonics composing this total potential, so that one obtains from the formula for an arbitrary general $\varphi_n^{(b)}$ term that

$$\left[\frac{\partial(r^n\varphi_n^{(b)})}{\partial\xi}\right]_{\zeta=0^+} = (-1)^n \int_{\cosh^{-1}(\xi/\eta)}^{0} \dot{\mathfrak{F}}_n(\xi - \eta\cosh u)\cosh nu\, du$$

which may be converted into more tractable form by making the transpositions that $\tau = r_b/\eta$, $\tau_1 = (\eta/r_b) - 1$, and $\tau_2 = (\eta/r_b) + 1$ so that, when the expression for $U_n(t)$ given as Eq. 41-9′ is made use of, one finds for the case wherein the body diameter remains constant in the region of the wing that

$$\left[\frac{\partial(r^n\varphi_n^{(b)})}{\partial\xi}\right]_{\zeta=0^+}$$

$$= (-1)^{n+1}\int_0^{t-\tau_1}\left[\frac{\cosh\,[n\cosh^{-1}\tau(1+t-t')]}{\sqrt{(t-t'-\tau_1)(t-t'+\tau_2)}}\int_0^{t'}\Phi_n(t'-t'')dF_n^*\right]dt'$$

$$= (-1)^{n+1}\int_0^{t-\tau_1}\Omega_n(t-t'')dF_n^*(t'') \tag{42-3}$$

where

$$\Omega_n(z) = \int_0^{z-\tau_1}\frac{\Phi_n(t^*)\cosh\,[n\cosh^{-1}\tau(z+1-t^*)]}{(z-\tau_1-t^*)(z+\tau_2-t^*)}dt^*$$

$$= -\frac{\sqrt{2}}{\pi}\int_0^{z-\tau_1}\frac{\cosh\,[n\cosh^{-1}\tau(z+1-t^*)]}{\sqrt{t^*}\sqrt{(z-\tau_1-t^*)(z+\tau_2-t^*)}}dt^*$$

$$+\frac{\sqrt{2}}{\pi}\int_0^{z-\tau_1}S_n(t'')dt''\int_{t''}^{z-\tau_1}\frac{\cosh\,[n\cosh^{-1}\tau(z+1-t^*)]}{\sqrt{t^*-t''}\sqrt{(z-\tau_1-t^*)(z+\tau_2-t^*)}}dt^* \tag{42-4}$$

It may be observed, however, that

$$-\frac{\sqrt{2}}{\pi}\int_0^{z-\tau_1}\frac{\cosh\,[n\cosh^{-1}\tau(z+1-t^*)]}{\sqrt{t^*}\sqrt{(z-\tau_1-t^*)(z+\tau_2-t^*)}}dt^*$$

$$= -\frac{2}{[\Gamma(\frac{1}{2}+n)]^2}\frac{1}{\tau^{*\frac{1}{2}}}x^{*\frac{1}{2}}(1-x^*)^{n+\frac{1}{2}}B_n^*$$

$$= M_n(z,\tau) \tag{42-5}$$

where $$\tau^* = \eta/r_b,\quad x^* = \frac{1+z-\tau^*}{1+z+\tau^*}$$

and $$B_n^* = \frac{d^n}{dx^{*n}}\left[x^{*n}\frac{d^n}{dx^{*n}}(x^{*n-\frac{1}{2}}\mathsf{K})\right] \tag{42-6}$$

wherein the complete elliptic integral of the first kind, K, has the modulus $\sqrt{x^*}$.

It is important to note likewise that

$$\frac{\sqrt{2}}{\pi}\int_{t''}^{z-\tau_1}\frac{\cosh\,[n\cosh^{-1}\tau(z+1-t^*)]}{\sqrt{t^*-t''}\sqrt{(z-\tau_1-t^*)(z+\tau_2-t^*)}}dt^*$$

$$= -M_n(z-t'',\tau) \tag{42-7}$$

so that one may now write more compactly that

$$\Omega_n(z) = M_n(z,\tau) - \int_0^{z-\tau_1}S_n(s)M_n(z-s,\tau)ds \tag{42-8}$$

and consequently, upon substitution of this expression into Eq. 42-3, the formula for the sought pressures becomes

$$\left[\frac{\partial(r^n\varphi_n^{(b)})}{\partial\xi}\right]_{\zeta=0^+} = (-1)^{n+2}\sqrt{\tau}\,F_n^*(t-\tau_1) + (-1)^{n+2}\int_0^{t-\tau_1} F_n^*(t'')\dot{\Omega}_n(t-t'')dt'' \quad (42\text{-}9)$$

provided it is understood that

$$\dot{\Omega}_n(z) = \frac{d\Omega_n}{dz} = \frac{\partial M_n}{\partial z} - \sqrt{\tau}S_n(z-\tau_1) - \int_0^{z-\tau_1} S_n(s)\left[\frac{\partial}{\partial z}M_n(z-s,\tau)\right]ds$$

Finally, the desired evaluation for the body interference contribution to the wing pressures is obtained as

$$\left[\frac{\partial\varphi^{(b)}}{\partial\xi}\right]_{\zeta=0^+} = \left(\frac{\bar{v}_{rb}}{\alpha_w U}\right)_{\theta=0}\sqrt{\tau} + \sum_n \int_0^{t-\tau_1} F_n^*(t'')\dot{\Omega}_n(t-t'')dt'' \quad (42\text{-}10)$$

where $(\bar{v}_{rb}/\alpha_w U)_{\theta=0}$ denotes the radial component of the induced velocity produced on the body by interference action from the wing, in that point on the wing-body juncture line in the plane $\theta = 0$ where a downstream-facing Mach line would have to originate in order to pass through the particular point $P(\xi, \eta)$ on the wing at which it is required to find the pressure.

The complete expression for the wing incremental pressures due to the total of all interference is thus

$$(\Delta C_p)_w = -2\frac{(w_b^*)_2}{\alpha_w U} - 2\sqrt{\tau}\left(\frac{\bar{v}_{rb}}{\alpha_w U}\right)_{\theta=0} + 2\sum_m \cos k\eta \int_0^{\xi^*}\{1 - J_0[k(\xi^* - \xi^{*\prime})]\}dG_m - 2\sum_n \int_0^{t-\tau_1} F_n^*(t'')\dot{\Omega}_n(t-t'')dt'' \quad (42\text{-}11)$$

In the more general case of a noncylindrical body for which r_b cannot be taken as a constant, this expression for $(\Delta C_p)_w$ should be modified by replacing the part derived from Eq. 42-9 by an equivalent expression which employs the succession of constants $\mathcal{K}_n^{(i)}$ that result from working out the stepwise procedure of evaluation embodied in Eq. 40-6. This more elementary way of solving the problem leads to the substitute relation

$$\left[\frac{\partial(r^n\varphi_n^{(b)})}{\partial\xi}\right]_{\zeta=0^+} = \frac{1}{n}\sum_{i=1}^{m}\mathcal{K}_n^{(i)}\left\{\sinh\left[n\cosh^{-1}\left(\frac{\xi_m - \xi_i'}{\eta_m}\right)\right] - \sinh\left[n\cosh^{-1}\left(\frac{\xi_m - \xi_{i-1}'}{\eta_m}\right)\right]\right\} \quad (42\text{-}12)$$

where ξ_m, η_m denotes the coordinates of any arbitrary point on the wing surface at which the pressures are to be computed, although ξ_i' denotes the successive discrete "reduced" positions along the body axis, upstream of the ξ_m location, of the sequence of points at which the boundary condition was invoked in carrying out the evaluations of the $\mathcal{K}_n^{(i)}$ as required by Eq. 40-6.

C,43. Calculation of Pressures Produced by Interference Effect on the Body. By proceeding now in a manner which is fully analogous

Table C,43. Generalized Lighthill weighting function for body slopes.

t	$-\dot{\Omega}_0$	$-\dot{\Omega}_2$	$-\dot{\Omega}_4$	$-\dot{\Omega}_6$	$-\dot{\Omega}_8$	$-\dot{\Omega}_{10}$
0	0.500	0.500	0.500	0.500	0.500	0.500
0.05						2.787
0.10				2.022	3.136	4.459
0.15					3.884	5.264
0.20	0.4319	0.7541	1.644	2.888	4.176	5.164
0.25				3.037	4.021	4.213
0.35						1.209
0.40	0.3755	0.878	1.929	2.462	1.702	−0.345
0.50				1.545	−0.215	−2.296
0.60	0.3284	0.892	1.428	0.508	−1.458	−1.986
0.70					−1.697	−0.461
0.80	0.2887	0.826	0.706	−0.940	−1.116	0.980
0.85						1.317
9.90					−0.163	1.345
0.95						1.107
1.0	0.2550	0.706	−0.024	−0.974	0.634	0.676
1.1					0.930	−0.279
1.2	0.2262	0.557	−0.464	−0.150	0.741	−0.803
1.3						0.638
1.4	0.2014	0.402	−0.558	0.457	−0.213	−0.045
1.5					−0.479	0.399
1.6	0.1801	0.255	−0.418	0.455	−0.514	0.475
1.7						0.251
1.8	0.1616	0.128	−0.142	0.073	0.022	−0.129
1.9						−0.297
2.0	0.1454	0.026	0.081	−0.222	0.276	−0.214
2.1						−0.005
2.2	0.1313	−0.048	0.195	−0.217	0.043	0.154
2.3						0.167
2.4	0.1189	−0.097	0.196	−0.035	−0.173	0.061
2.5						−0.058
2.6	0.1080	−0.122	0.120	0.106	−0.053	−0.108
2.8	0.0983	−0.128	0.025	0.098	0.075	0.007
3.0	0.0898	−0.122	−0.049	−0.001		0.058
3.2	0.0822	−0.104	−0.078	−0.043	−0.049	−0.066
3.4	0.0754	−0.082	−0.068	−0.050	−0.030	−0.017
3.6	0.0693	−0.058	−0.035	−0.008	0.011	0.022
3.8	0.0638	−0.035	−0.003	0.046		
4.0	0.0590	−0.017		0.024		

to what was just carried out in Art. 42, the calculation of the pressures produced by the interference effect on the body may now also be readily obtained. To evaluate the indicated derivatives in Eq. 42-1 which pertain to the pressures on the body, it is necessary, first of all, to make use of the formula given by Goodman [53] to obtain the relation for the axial derivative of the wing interference potential which is just as valid at points up off the surface of the wing as in the plane $\zeta = 0^+$. This formula is

$$\left[\frac{\partial\varphi^{(w)}}{\partial\xi}\right]_{r=r_b} = \sum_m \cos(kr_b\cos\theta)\left[\frac{\partial\varphi_m^{(w)}}{\partial\xi}\right]_{\zeta=r_b\sin\theta}$$

$$= \sum_m \cos(kr_b\cos\theta)\int_0^{\xi^*-r_b\sin\theta} J_0[k\sqrt{(\xi^*-\xi^{*\prime})^2-r_b^2\sin^2\theta}]dG_m \tag{43-1}$$

On the other hand, at least for the case where the body is cylindrical in the region of the wing so that r_b may be assumed constant there, the axial derivative of the body interference potential is obtained directly from Eq. 42-9 by particularizing it to apply to points on the surface of the body, through setting $\tau = 1$ and $\tau_1 = 0$. Consequently this derivative becomes

$$\left[\frac{\partial\varphi^{(b)}}{\partial\xi}\right]_{\substack{r=r_b\\ \zeta>0}} = \frac{v_{rb}}{\alpha_w U} + \sum_n \int_0^t F_n^*(t'')\dot{\Omega}(t-t'')dt''\cos n\theta \tag{43-2}$$

where, in this instance, the value of Ω which is meant is the specific result applying when the values $\tau = 1$ and $\tau_1 = 0$ are inserted into Eq. 42-8.

The values of $\dot{\Omega}_n(z)$, for this specialized case where $\tau = 1$ and $\tau_1 = 0$, have been computed by Nielsen and Pitts [52] for the first six values of n (values of significance in the case of symmetric flow patterns, i.e. for $n = 0, 2, 4, 6, 8$, and 10). Their values are found in Table C,43.

The complete expression for the body incremental pressures due to the total of all interference is thus

$$(\Delta C_p)_b = \frac{-2v_{rb}}{\alpha_w U} - 2\sum_n \cos n\theta\int_0^t F^*(t'')\dot{\Omega}_n(t-t'')dt''$$

$$+ 2\sum_k \cos(kr_b\cos\theta)\int_0^{\xi^*-r_b\sin\theta} J_0[k\sqrt{(\xi^*-\xi^{*\prime})^2-r_b^2\sin^2\theta}]dG_m \tag{43-3}$$

C,44. Determination of the Field of Flow in the Region Lying Downstream of the Trailing Edge and Outside of Region Σ. If attention is now directed to the region of space lying downstream of the wing's trailing edge and falling outside of the region Σ in which the flow is determined by following through the procedures just explained, it is clear that the potential that describes the flow in this region, call it the T region, will have to satisfy the boundary conditions:

$$\left.\begin{array}{l}\text{1. } \Delta\phi = (\phi)_{\zeta=0^+} - (\phi)_{\zeta=0^-} = \Delta\phi(\xi_w, \eta) \text{ on the plane } \zeta = 0 \text{ wherein } \xi_w \text{ is used to denote the axial location of any one of the points lying on the trailing edge of the wing.}\\ \text{2. The total induced normal velocity on the body itself is zero.}\\ \text{3. } \phi \text{ must be continuous on the surface of the Mach plane } \xi - |\zeta| = \xi_w \text{ that issues from the trailing edge of the wing, while the discontinuity of grad } \phi \text{ is perpendicular to this Mach plane.}\end{array}\right\} \quad (44\text{-}1)$$

The first and third of these conditions may be satisfied easily in T by taking the sought potential to be of the form

$$\varphi'^{(w)} = \sum_m [\varphi_m^{(w)} + \varphi_m^{*(w)}] \cos\left(\frac{\pi m \eta}{2}\right) \quad (44\text{-}2)$$

where

$$\varphi_m^{(w)} = \mp\pi h_m(\xi_w^*) + \pi k\zeta \int_0^{\xi_w^*} h_m(\xi^{*\prime}) J_1[k\sqrt{(\xi^* - \xi^{*\prime})^2 - \zeta^2}] \frac{d\xi^{*\prime}}{\sqrt{(\xi - \xi^{*\prime})^2 - \zeta^2}} \quad (44\text{-}3)$$

in which $\xi_w^* = \xi_w - r_b$, while $h_m(\xi^*)$ represents the same source strength that was obtained previously by means of Eq. 38-7. On the other hand, the $\varphi_m^{*(w)}$ must be determined in such a manner so as to cancel the discontinuity in the component $u^{(b)}/(\alpha_w U)$ of the velocity along the x axis, produced in T by the potential $\varphi^{(b)}$ obtained as set forth above. In addition the $\varphi_m^{*(w)}$ potential must become null on the Mach plane $\xi - |\zeta| = \xi_w$.

Let us define a pressure jump function as

$$D(\xi, \eta) = \frac{\Delta u^{(b)}}{\alpha_w U} = 2\left[\frac{\partial\varphi^{(b)}}{\partial\xi}\right]_{\zeta=0^+} \quad \text{for } \eta \geqq r_b$$

while $D(\xi, \eta) = D(\xi, r_b)$ for all $\eta < r_b$ at any selected axial station ξ.

If the point P at which $\Delta u^{(b)}/\alpha_w U$ is to be calculated falls within the region T', as indicated in Fig. C,44, $\partial\varphi^{(b)}/\partial\xi$ may be determined there by direct appeal to the result given earlier as Eq. 42-10. On the other hand, if the point P in question lies within the region designated as T'' in Fig. C,44, then $\varphi_n^{(b)}$ is given in this instance by means of the expression

$$\varphi_n^{(b)} = \frac{(-1)^n}{r^n} \int_{\cosh^{-1}(\xi/r)}^{\cosh^{-1} r[1 + (\xi - \xi_w)/r_b]} \mathfrak{F}_n(\xi - r\cosh u) \cosh nu \, du \quad (44\text{-}4)$$

instead of by means of Eq. 40-4. It follows, consequently, that the axial components of the induced flow at the T'' locations in question will be obtained by means of the new formulation

$$\left[\frac{\partial(r^n\varphi_n^{(b)})}{\partial\xi}\right]_{\zeta=0^+}$$

$$= \mathfrak{F}_n(\xi_w - r_b)\cosh\left[n\cosh^{-1}\left(\frac{\xi - \xi_w + r_b}{\eta}\right)\right]\frac{1}{\sqrt{(\xi - \xi_w + r_b)^2 - \eta^2}}$$

$$+ \int_{\cosh^{-1}(\xi/\eta)}^{\cosh^{-1}\tau(1+[\xi-\xi_w]/r_b)} \dot{\mathfrak{F}}_n(\xi - \eta\cosh u)\cosh nu\, du \quad (44\text{-}5)$$

The pressure jump function may now be separated into harmonic components, as previously done with the potential $\varphi^{(b)}$, so that the in-

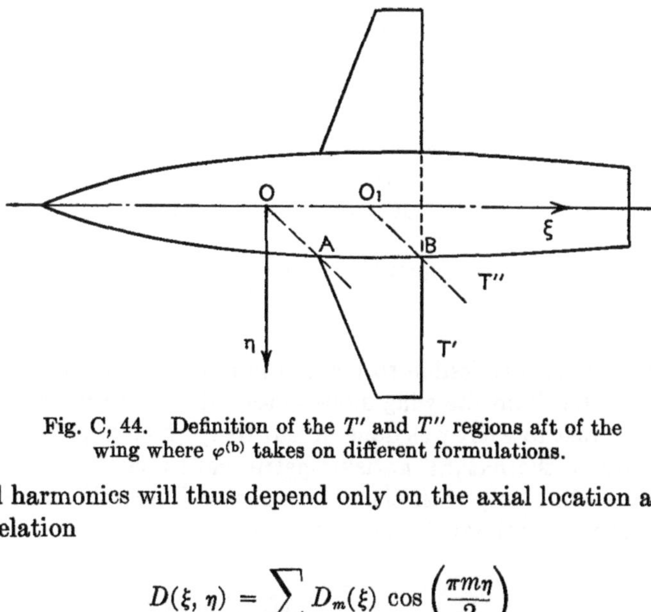

Fig. C, 44. Definition of the T' and T'' regions aft of the wing where $\varphi^{(b)}$ takes on different formulations.

dividual harmonics will thus depend only on the axial location according to the relation

$$D(\xi, \eta) = \sum_m D_m(\xi)\cos\left(\frac{\pi m\eta}{2}\right) \quad (44\text{-}6)$$

and therefore, if the corresponding source strength is taken as

$$h_m^*(\xi) = \frac{1}{2\pi}\int_{\xi_w}^{\xi} D_m(\xi')d\xi' \quad (44\text{-}7)$$

the expression sought for $\varphi_m^{*(w)}$, which fits the stipulations being imposed, has the form

$$\varphi_m^{*(w)} = \mp h_m^*(\xi - |\zeta|)$$

$$+ \pi k\zeta \int_{\xi_w}^{\xi-|\zeta|} h_m^*(\xi')J_1[k\sqrt{(\xi - \xi')^2 - \zeta^2}]\frac{d\xi'}{\sqrt{(\xi - \xi')^2 - \zeta^2}} \quad (44\text{-}8)$$

Returning to consideration of the second boundary condition (item 2 of Eq. 44-1), it should be noticed first of all that the augmented wing interference potential $\varphi'^{(w)}$, as represented by Eq. 44-2, has a radially directed derivative which will *not* vanish at the surface of the body, in general, and it is readily seen that this radial derivative will be an odd function of θ, inasmuch as $\partial\varphi'^{(w)}/\partial\zeta$ is even and $\partial\varphi'^{(w)}/\partial\eta$ is odd with respect to θ (see Eq. 39-5).

Furthermore, in this same aft-lying T region the body interference potential induces radial components of velocity that also will exhibit an odd variation with respect to the angular coordinate, since in the T region

$$\left.\begin{aligned} \varphi^{(b)} &= \sum_n \varphi_n^{(b)} \cos n\theta \qquad \text{for } 0 \leqq \theta \leqq \pi \\ &= -\sum_n \varphi_n^{(b)} \cos n\theta \quad \text{for } 0 \geqq \theta \geqq -\pi \end{aligned}\right\} \tag{44-9}$$

where the $\varphi_n^{(b)}$ are defined according to Eq. 44-4.

The total radial derivative (evaluated at the surface of the body) arising from both of these interference contributions may now be represented by an antisymmetric development of the type

$$\left[\frac{\partial(\varphi'^{(w)} + \varphi^{(b)})}{\partial r}\right]_{r=r_b} = \sum_n T_n(\xi, r_b) \sin n\theta \tag{44-10}$$

Likewise, it is permissible to recast the induced velocities normal to the body coming from the wing-alone potential φ_w in terms of a series of sines, in contrast to what was done in Eq. 40-1, which applies only in the region Σ, where the natural antisymmetric nature of the flow was suppressed by dealing only with one of the semi-infinite half planes. Consequently it is now legitimate in the T region to let

$$\left(\frac{\partial\varphi_w}{\partial r}\right)_{r=r_b} = \sum_n T_n'(\xi, r_b) \sin n\theta$$

so that the total of the induced radial velocities normal to the fuselage, aft of the wing, which are propagated from the flow about the wing alone, and from the interference interactions besides, are given in the T region by the simple formulation

$$\frac{v_r}{\alpha_w U} = \sum_n (T_n + T_n') \sin n\theta = \sum_n T_n^*(\xi, r_b) \sin n\theta$$

Thus it is clear that a counteracting potential $\varphi^{*(b)}$ should be introduced which will negate, at the surface of the afterbody, these induced radial components of the flow. This new potential must satisfy the condition in region T that

$$\left[\frac{\partial \varphi^{*(b)}}{\partial r}\right]_{r=r_b} = -\sum_n T_n^*(\xi, r_b) \sin n\theta \qquad (44\text{-}11)$$

A potential which will be satisfactory for the purposes under consideration may be taken to have the form

$$\varphi^{*(b)} = \sum_n \varphi_n^{*(b)} \sin n\theta \quad \text{for } n = 1, 3, 5, \cdots$$

where

$$\varphi_n^{*(b)} = \frac{(-1)^n}{r^n} \int_{\cosh^{-1}\{[\xi - \xi_w + (r_b)_w]/r\}}^{0} \mathfrak{F}_n^*(\xi - r\cosh u) \cosh nu\, du \qquad (44\text{-}12)$$

in which the value of the body radius at the wing's trailing edge, located at $\xi = \xi_w$ is denoted by $(r_b)_w$, and where the definition of the individual potentials has been kept quite general by allowing r_b to be variable.

Upon invoking the nullification condition given as Eq. 44-11, it follows that the $\mathfrak{F}_n^*$ may be determined from the following relation, which is entirely analogous to what was encountered in Eq. 40-5:

$$\begin{aligned} -T_n^*(\xi, r_b) &\\ = (-1)^{n+1} &\int_{\cosh^{-1}\{[\xi - \xi_w + (r_b)_w]/r_b\}}^{0} \mathfrak{F}_n^{*\prime}(\xi - r_b \cosh u) \cdot \cosh nu \cdot \cosh u \cdot du \\ &+ (-1)^{n+2}\mathfrak{F}_n^*[\xi_w - (r_b)_w] \cosh\left[n \cosh^{-1}\left(\frac{\xi - \xi_w + (r_b)_w}{r_b}\right)\right] \\ &\times \frac{\xi - \xi_w + (r_b)_w}{r_b\sqrt{[\xi - \xi_w + (r_b)_w]^2 - r_b^2}} \end{aligned} \qquad (44\text{-}13)$$

and the method of solution is identical to that indicated in Art. 41.

C,45. Determination of Harmonic Components. The solution of the problem formulated in Art. 44 requires the determination of the harmonic components T_n^* which describe the pattern of variation of the induced radial velocities normal to the body arising from the three superimposed flows governed by the potentials $\varphi^{(b)}$, $\varphi'^{(w)}$, and φ_w. The wing-alone potential φ_w contributes to the totality of induced normal components of velocity acting on the afterbody through agency of the harmonic components T_n'. These quantities may be conveniently evaluated by use of the formula of de la Vallée-Poussin or by recourse to the standard formulas for determining Fourier coefficients, particularly if the wing-alone flow field is a very simple one which would lead to easily integrable expressions for the T_n'' coefficients.

The T_n coefficients composing the rest of the contribution to the T_n^* harmonics can be further broken down into two parts, one of which corresponds to the influence exerted on the afterbody by action of the forebody interference flow, governed by the potential $\varphi^{(b)}$, and which may be designated $T_n^{*\prime}$; while the other part arises from the augmented wing interference potential $\varphi'^{(w)}$ and may be designated the $T_n^{*\prime\prime}$ components. It is a simple matter to determine such $T_n^{*\prime}$ coefficients by use of the formula of de la Vallée-Poussin, just as was done in Eq. 39-8. In order to obtain the sought $T_n^{*\prime}$ coefficients by this scheme, the radial components of velocity arising from the $\varphi_n^{(b)}$ potential need to be calculated in the T region by means of the relation

$$\left[\frac{\partial(r^n\varphi_n^{(b)})}{\partial r}\right]_{r=r_b}$$
$$= (-1)^n \int_{\cosh^{-1}(\xi/r_b)}^{\cosh^{-1}\{[\xi-\xi_w+(r_b)_w]/r_b\}} \dot{\mathfrak{F}}_n(\xi - r_b \cosh u) \cosh nu \cosh u \, du$$
$$+ (-1)^{n+1}\mathfrak{F}_n[\xi_w - (r_b)_w] \cosh\left\{n \cosh^{-1}\left[\frac{\xi - \xi_w + (r_b)_w}{r_b}\right]\right\}$$
$$\times \frac{\xi - \xi_w + (r_b)_w}{r_b\sqrt{[\xi - \xi_w + (r_b)_w]^2 - r_b^2}} \quad (45\text{-}1)$$

and then the application of the de la Vallée-Poussin formula to such a distribution of radial velocities serves to determine the $T_n^{*\prime}$ coefficients.

In regard to the augmented wing interference potential, the following plan for obtaining the induced radial velocity components at the body's surface caused by this flow may be resorted to: First, note that it is legitimate in general to make the assumption that the vertical and horizontal derivatives of $\varphi'^{(w)}$ may be represented by a Taylor series development as given below, at any arbitrary general point $P(\xi_P, \eta_P, \zeta_P)$ lying on the body, provided the value of r_b lies within reasonable bounds such as may be met in practical cases:

$$\left.\begin{aligned}\frac{\partial\varphi'^{(w)}}{\partial\eta} &= \left[\frac{\partial\varphi'^{(w)}}{\partial\eta}\right]_{\substack{\xi=\xi_P\\ \eta=\zeta=0}} + \left[\frac{\partial^2\varphi'^{(w)}}{\partial\eta^2}\right]_{\substack{\xi=\xi_P\\ \eta=\zeta=0}} \eta_P + \left[\frac{\partial^2\varphi'^{(w)}}{\partial\eta\partial\zeta}\right]_{\substack{\xi=\xi_P\\ \eta=\zeta=0}} \zeta_P\\ \frac{\partial\varphi'^{(w)}}{\partial\zeta} &= \left[\frac{\partial\varphi'^{(w)}}{\partial\zeta}\right]_{\substack{\xi=\xi_P\\ \eta=\zeta=0}} + \left[\frac{\partial^2\varphi'^{(w)}}{\partial\eta\partial\zeta}\right]_{\substack{\xi=\xi_P\\ \eta=\zeta=0}} \eta_P + \left[\frac{\partial^2\varphi'^{(w)}}{\partial\zeta^2}\right]_{\substack{\xi=\xi_P\\ \eta=\zeta=0}} \zeta_P\end{aligned}\right\} \quad (45\text{-}2)$$

Now, remembering that $\varphi^{(w)} = \sum_m \varphi_m^{(w)} \cos k\eta$, according to Eq. 38-2, it follows that

$$\left.\begin{aligned}\left[\frac{\partial\varphi^{(w)}}{\partial\eta}\right]_{\substack{\xi=\xi_P\\ \eta=\zeta=0}} &= \left[\frac{\partial^2\varphi^{(w)}}{\partial\eta\partial\zeta}\right]_{\substack{\xi=\xi_P\\ \eta=\zeta=0}} = 0\\ \left[\frac{\partial^2\varphi^{(w)}}{\partial\eta^2}\right]_{\substack{\xi=\xi_P\\ \eta=\zeta=0}} &= -\frac{1}{2}\left[\frac{d^2\Gamma^{(w)}}{d\eta^2}\right]_{\eta=0}\end{aligned}\right\} \quad (45\text{-}3)$$

where $\Gamma^{(w)}$ indicates the circulation at the root chord of the wing at the location $\eta = 0$ (normally covered by the body) that corresponds to the interference flow about the wing, characterized by the perturbation potential $\varphi^{(w)}$. This circulation is to be taken as positive if it has a direction which is opposite in sense from a rotation carrying the positive x axis into the positive z axis around the y axis.

In view of the way $\varphi^{(w)}$ has been defined (Art. 38), it is readily apparent that each of the separate individual potentials composing $\varphi^{(w)}$ will have a vertical derivative in this aft T region of the form

$$\left[\frac{\partial \varphi_m^{(w)}}{\partial \zeta}\right]_{\zeta=0} = \frac{\pi k^2}{2} \int_0^{\xi_w^*} h_m(\xi^{*\prime})\{J_0[k(\xi^* - \xi^{*\prime})] + J_2[k(\xi^* - \xi^{*\prime})]\} d\xi^{*\prime}$$

so that

$$\left[\frac{\partial \varphi^{(w)}}{\partial \zeta}\right]_{\zeta=\eta=0} = \frac{\pi}{2} \sum_m k^2 \int_0^{\xi_w^*} h_m(\xi^{*\prime})\{J_0[k(\xi^* - \xi^{*\prime})] + J_2[k(\xi^* - \xi^{*\prime})]\} d\xi^{*\prime} \tag{45-4}$$

while likewise

$$\left[\frac{\partial^2 \varphi^{(w)}}{\partial \zeta^2}\right]_{\zeta=\eta=0} = -\left[\frac{\partial^2 \varphi^{(w)}}{\partial \eta^2}\right]_{\zeta=\eta=0} = +\frac{1}{2}\left[\frac{d^2 \Gamma^{(w)}}{d\eta^2}\right]_{\eta=0} \tag{45-5}$$

inasmuch as the equation of continuity must hold in this region as elsewhere while in addition $[\partial^2 \varphi^{(w)}/\partial \xi^2]_{\eta=\zeta=0} = 0$ in the T region, because no pressure differences can be supported in the plane $\zeta = 0$ behind the wing.

In an entirely analogous manner, it becomes immediately evident that the $\varphi^{*(w)}$ potential also has derivatives in the T region of the form

$$\left[\frac{\partial \varphi^{*(w)}}{\partial \eta}\right]_{\zeta=\eta=0} = \left[\frac{\partial^2 \varphi^{*(w)}}{\partial \eta \partial \zeta}\right]_{\zeta=\eta=0} = \left[\frac{\partial^2 \varphi^{*(w)}}{\partial \eta^2}\right]_{\zeta=\eta=0} = 0 \tag{45-6}$$

where the vanishing of the second derivative with respect to η comes about in this instance because of the constancy of $D(\xi, \eta)$ in the region actually covered by the fuselage.

Likewise it is obvious that

$$\left[\frac{\partial \varphi_m^{*(w)}}{\partial \zeta}\right]_{\zeta=\eta=0} = \pi \dot{h}_m^*(\xi) + \frac{\pi k^2}{2} \int_{\xi_w}^{\xi} h_m^*(\xi')\{J_0[k(\xi - \xi')] + J_2[k(\xi - \xi')]\} d\xi'$$

$$= \pi \dot{h}_m^*(\xi) + \pi k \int_{\xi_w}^{\xi} \dot{h}_m^*(\xi') \sum_{\nu=0}^{\infty} \epsilon_\nu J_{1+2\nu}[k(\xi - \xi')] d\xi' \tag{45-7}$$

and consequently

$$\left[\frac{\partial \varphi^{*(w)}}{\partial \zeta}\right]_{\zeta=\eta=0} = \frac{1}{2}\frac{\Delta u^{(b)}}{\alpha_w U} + \frac{1}{2}\sum_m k \int_{\xi_w}^{\xi} D_m(\xi') \sum_{\nu=0}^{\infty} \epsilon_\nu J_{1+2\nu}[k(\xi - \xi')] d\xi' \tag{45-8}$$

On the other hand, by having recourse to the equation of continuity once again, it is seen that in this instance

$$\left[\frac{\partial^2\varphi^{*(w)}}{\partial\zeta^2}\right]_{\eta=\zeta=0} = \left[\frac{\partial^2\varphi^{*(w)}}{\partial\xi^2}\right]_{\eta=\zeta=0} = -\frac{1}{2}\left[\frac{d}{d\xi}\frac{\Delta u^{(b)}}{\alpha_w U}\right]_{\eta=0} \tag{45-9}$$

All the essential evaluations have thus been completed for finding $[\partial\varphi'^{(w)}/\partial r]_{r=r_b}$ because, when Eq. 45-3, 45-4, and 45-5, relating to the potential $\varphi^{(w)}$, and Eq. 45-6, 45-8, and 45-9, relating to the potential $\varphi^{*(w)}$, are inserted into Eq. 45-2, the derivatives $\partial\varphi'^{(w)}/\partial\eta$ and $\partial\varphi'^{(w)}/\partial\zeta$ are obtained, and together these provide the means for calculating the radial derivative of $\varphi'^{(w)}$. As soon as the normal velocities are computed, the standard procedure for obtaining the corresponding harmonic components $T_n^{*\prime\prime}$ may be carried out through use of the formula of de la Vallée-Poussin.

C,46. Simplified Method Applicable to Configurations Having Wings of High Aspect Ratio. If the aspect ratio of the configuration under consideration is very large, the value of ξ_w^* is quite small, and it is found under these circumstances that the foregoing procedures become much less complicated. In fact it is readily apparent from Eq. 39-6′ that the induced velocities normal to the body which are caused by the wing interference flow are so small when $\xi_w^* \ll 1$ that it is legitimate to neglect the contribution to the total flow impinging on the body arising from these $v_{rb}^{(w)}/\alpha_w U$ terms. The elimination of the $v_{rb}^{(w)}$ terms is permitted only if the values of $(w_b^*)_2/\alpha_w U$ are to be selected as being zero everywhere in the region of the wing actually covered by the fuselage (with the possible exception of locations near the wing-body juncture). Thus the relative upwash distribution to be annulled will be as shown in Fig. C,50d. If these agreements are held to, it follows that the only normal components of velocity that need to be counteracted at the surface of the body are those arising from the wing-alone potential φ_w, which is known to begin with.

It is of further advantage to represent the normal velocity components $(\partial\varphi_w/\partial r)_{r=r_b}$ by means of a series of sines, instead of by a series of cosines, as in Eq. 40-1. Thus, if the former notions about considering only the flow in the upper semi-infinite half plane are rescinded, the influence of the wing-alone potential on the fuselage may now be represented in the form

$$\left(\frac{\partial\varphi_w}{\partial r}\right)_{r=r_b} = \sum_n F_n^{\prime *}(\xi)\sin n\theta \quad \text{for } n = 1, 3, 5, \cdots \tag{46-1}$$

throughout the interval $0 \leq \theta \leq 2\pi$.

The corresponding body interference potential may then be expressed as

$$\varphi^{(b)} = \sum_n \varphi_n^{(b)} \sin n\theta \tag{46-1'}$$

where the individual harmonic components $\varphi_n^{(b)}$ are to have the same formulation given them in Eq. 40-4. The process of determining their specific values may therefore be carried out in the same way as in Art. 40, the only difference being that here the newly determined values of $F_n'^*$ are to be used.

It is necessary to recognize that under the present conditions the body interference flow produces a velocity component normal to the wing surface which, in general, is not zero. However, since it is true that these induced velocities normal to the wing are given by

$$\begin{aligned}\left[\frac{1}{r}\frac{\partial\varphi^{(b)}}{\partial\theta}\right]_{\theta=0} &= \sum_n n\frac{(-1)^n}{r}\int_{\cosh^{-1}(\xi/r)}^{0} \mathfrak{F}_n(\xi - r\cosh u)\cosh nu\,du \\ &= \left[\frac{w^{(b)}}{\alpha_w U}\right]_w\end{aligned} \tag{46-2}$$

it is evident that even these values of $[w^{(b)}/(\alpha_w U)]_w$ are small for the condition where $\xi - r_b = \xi^* \ll 1$, which is now assumed to hold.

Furthermore, these quantities exert their influence only over a rather narrowly restricted region of the wing (within the area marked out by the wing-body juncture line and the Mach line emanating from the point of intersection of the wing leading edge with the side of the body), so that they may be neglected with more and more assurance as to the closeness of the resulting approximation the smaller the given value of ξ^* happens to be.

In view of the above considerations and hypotheses it may be recognized that under the present special circumstances the boundary conditions originally stated as Eq. 36-3 now reduce to

$$\left.\begin{aligned}\left[\frac{\partial\varphi^{(w)}}{\partial\zeta}\right]_{\zeta=0} &= -\left[\frac{\partial(\varphi_b)_2}{\partial\zeta}\right]_{\zeta=0} \\ \left[\frac{\partial\varphi^{(b)}}{\partial r}\right]_{r=r_b} &= -\left[\frac{\partial\varphi_w}{\partial r}\right]_{r=r_b}\end{aligned}\right\} \tag{36-3''}$$

Thus it becomes quite obvious that the wing interference and body interference potentials, $\varphi^{(w)}$ and $\varphi^{(b)}$, may be worked out completely independently of each other in this case.

In fact, it may be observed that $\varphi^{(w)}$ is now expressed in exactly the same way that was employed in Art. 38 (see Eq. 38-2 and 38-5), while the various $\varphi_n^{(b)}$ individual potentials, making up the body interference potential, may be determined by use of the formula previously given in Art. 40 and 41, with the sole alteration, as pointed out above, that the harmonic components $F_n'^*(\xi)$ replace the $F_n^*(\xi)$ coefficients. The computation of the new $F_n'^*$ values should not cause any difficulty, however, because the calculation is just a repetition of a process now to be considered routine.

It is worth noting that if one wishes to obtain the difference in pressure acting on the top and bottom surface of the wing at any given point, it is merely necessary to substitute the pertinent coordinates into the formula for $[\partial\varphi^{(w)}/\partial\xi]$, given as Eq. 42-2, while if the evaluation of the lift acting on a given thin axial strip of the fuselage (of length $d\xi$) is required, it is sufficient merely to determine the single first harmonic of the body interference potential $\varphi_1^{(b)}$. Because of the extreme simplicity of the formula for the integrated induced lift on the body, which involves only this single first harmonic, it will consequently be of most use to give explicitly merely the corresponding formula for $U_1(t)$. The general expression for $U_1(t)$, derived as Eq. 41-9′, is still applicable, and the resolvent kernel of the Volterra integral equation S_1, which is involved here through Eq. 41-10, is determined in this instance by evaluation of the following particularized formula for Q_1 (see Eq. 41-5):

$$Q_1(z) = \frac{4}{3}(1+z)\sqrt{2+z}\,\mathsf{E}\left(\sqrt{\frac{z}{2+z}}\right) - \frac{2}{3}\frac{1+2z}{\sqrt{2+z}}\,\mathsf{K}\left(\sqrt{\frac{z}{2+z}}\right) \tag{41-5′}$$

Once the governing potentials have been determined, it is then a simple matter to compute the corresponding pressures and the integrated lift forces. In regard to the action of the interference flow on the wing, it may be seen at once that the difference in pressure supported by the wing at a generic location is computable from the relation

$$\begin{aligned}[(\Delta C_p)_w]_{\zeta=0^+} &- [(\Delta C_p)_w]_{\zeta=0^-} \\ &= -4\frac{(w_b^*)_2}{\alpha_w U} + 4\sum_m \cos k\eta \int_0^{\xi^*} \{1 - J_0[k(\xi^* - \xi^{*\prime})]\}\, dG_m \end{aligned} \tag{46-3}$$

while the corresponding integrated lift over a chordwise strip of the wing is given by

$$(\Delta c_l)_w = (\overline{\Delta c_l})_w + \frac{4}{\lambda_w}\sum_m \cos k\eta \int_0^{\xi_w^*} \{1 - J_0[k(\xi^* - \xi^{*\prime})]\}\, G_m(\xi^{*\prime})\, d\xi^{*\prime} \tag{46-4}$$

where $(\overline{\Delta c_l})_w$ is the two-dimensional lift coefficient of the strip of wing under consideration, when the lift is assumed to be distributed over this particular profile according to the Ackeret formula (relating lift on a planar element to a local angle of attack in two-dimensional flow). The nondimensional local chord length λ_w has been introduced to preserve the symmetry in the notation by making $\lambda_w = 2c_w/b$, where c_w is the actual local length of the profile at the spanwise location in question.

In regard to the interference action on that part of the body influenced by the wing, it is appropriate to list the various contributions to the total induced lift ΔL_b produced there by the wing. These separate parts of the total lift arise from:

1. The axial components of velocity, stemming from the wing-alone potential φ_w, and giving rise to a $\Delta_1 L_b$ contribution to the total.
2. The axial components of velocity, stemming from the wing interference potential $\varphi^{(w)}$, and giving rise to a $\Delta_2 L_b$ contribution.
3. The radial components of velocity, stemming from the wing-alone potential φ_w, and giving rise to a $\Delta_3 L_b$ contribution.

Symbolically,

$$\Delta L_b = \Delta_1 L_b + \Delta_2 L_b + \Delta_3 L_b$$

The second of these quantities may be computed by use of the relation

$$(\Delta_2 C_l)_b = \frac{8}{b^2}\frac{1}{\rho_\infty U^2}\frac{1}{\lambda_{w_0}}\left(\frac{d}{d\eta}\Delta_2 L_b\right) \tag{46-5}$$

wherein $\lambda_{w_0} = 2c_{w_0}/b$ and where c_{w_0} denotes the length of the wing chord at the wing-body juncture. The value of the incremental induced lift coefficient appearing in Eq. 46-5 is given by

$$(\Delta_2 C_l)_b = -\frac{8}{\lambda_{w_0}}\sum_m \frac{\cos(kr_b\cos\theta)}{k}\int_{r_b}^{\xi_w} L_m(\xi_w + r_b\sin\theta, r_b\sin\theta; \xi')dG_m \tag{46-6}$$

where

$$L_m(\xi_w + r_b\sin\theta, r_b\sin\theta; \xi')$$
$$= \frac{k}{2}\int_{\xi'}^{\xi_w} J_0[k\sqrt{(\xi_w + r_b\sin\theta - \xi'')^2 - r_b^2\sin^2\theta}]d\xi''$$

Consequently an integration by parts of Eq. 46-6 produces the summary formula

$$(\Delta_2 C_l)_b = -\frac{4}{\lambda_{w_0}}\sum_m \cos(kr_b\cos\theta)$$
$$\times\int_{r_b}^{\xi_w} G_m(\xi')J_0[k\sqrt{(\xi_w + r_b\sin\theta - \xi')^2 - r_b^2\sin^2\theta}]d\xi' \tag{46-6'}$$

The incremental induced lift coefficient $(\Delta_1 C_l)_b$ corresponding to the first of the contributions mentioned above may be defined as a function of $\Delta_1 L_b$ in a manner entirely analogous to that given for $(\Delta_2 C_l)_b$ (see Eq. 46-5). Such evaluation is almost immediate since knowledge of the potential φ_w gives the desired axial velocity components with ease.

Finally, the third contribution to the incremental lift listed above, $\Delta_3 L_b$, is given by utilizing a relationship which is almost the same as that encountered in Art. 43 (Eq. 43-2), inasmuch as $\varphi^{(b)}$ is now given by Eq. 46-1′. In fact it is evident that the only change required in order to obtain the sought expression for $[\partial\varphi^{(b)}/\partial\xi]_{r=r_b}$ is to substitute sines for cosines in the cited formula.

As mentioned above, it is only necessary to deal with the first of the harmonic components $\varphi_1^{(b)}$ when one is satisfied to determine merely the total integrated lift acting all around the periphery of a narrow strip of the circular body. Consequently, in carrying out the calculations for determining the value of $d(\Delta_3 L_b)/d\xi$ it is sufficient in this case to have at hand the explicit expression for $[\partial\varphi_1^{(b)}/\partial\xi]_{r=r_b}$. In complete analogy with what was done in Art. 42, therefore, one may proceed to write down the specific formula pertaining to the body interference flow when $n = 1$. Thus

$$\left[\frac{\partial(r_b\varphi_1^{(b)})}{\partial\xi}\right]_{r=r_b} = \int_0^t \Omega_1(t - t'')dF_1'^*(t'') \tag{46-7}$$

where

$$\Omega_1(z) = M_1'(z) - \int_0^z S_1(s)M_1'(z - s)ds$$

in which now

$$M_1'(z) = -\frac{2\sqrt{2}}{\pi}\left[\sqrt{2+z}\,\mathsf{E}\left(\sqrt{\frac{z}{2+z}}\right) - \frac{1}{\sqrt{2+z}}\,\mathsf{K}\left(\sqrt{\frac{z}{2+z}}\right)\right] \tag{46-8}$$

It follows therefore that

$$\left[\frac{\partial(r_b\varphi_1^{(b)})}{\partial\xi}\right]_{r=r_b} = -F_1'^*(t) - \int_0^t F_1'^*(t'')\dot{\Omega}_1(t - t'')dt'' \tag{46-7'}$$

and thus the incremental body lift acting at a generic parallel along the body axis is given by

$$\begin{aligned}\frac{1}{\frac{1}{2}\rho_\infty U^2}\frac{d}{d\xi}\Delta_3 L_b &= -\frac{\pi b^2}{2}\left[\frac{\partial(r_b\varphi_1^{(b)})}{\partial\xi}\right]_{r=r_b} \\ &= \frac{\pi b^2}{2}\left[F_1'^*(t) + \int_0^t F_1'^*(t'')\dot{\Omega}_1(t - t'')dt''\right]\end{aligned} \tag{46-9}$$

In addition, for this special case in which the aspect ratio of the wing is taken to be very large, it turns out that the determination of the flow in the afterbody region T, lying downstream of the wing's trailing edge, is also vastly simplified. This is found to be so because, as a result of using a sine series representation in Eq. 46-1′, no pressure-jump components, D_m, need to be annulled, and consequently the $\varphi_m^{*(w)} = 0$. Thus the sought potential describing the flow in the T region is made up of the three contributions:

1. $\sum_m \varphi_m^{(w)} \cos k\eta$, where the $\varphi_m^{(w)}$ are still defined according to Eq. 44-3,
2. φ_w,
3. $\varphi^{(b)}$, where this body interference potential continues to be described in this aft-lying T region by Eq. 46-1′, which retains its validity here.

For the sake of convenience, however, it is advantageous to continue the convention of considering the body interference potential in the Σ region to be distinct from that applying in the T region, so that the $\varphi^{(b)}$ potential is restricted to refer solely to the distribution of singularities that are confined to the segment running from 0 to $\xi_w - r_b$ on the ξ axis, while the afterbody interference potential $\varphi^{*(b)}$ is restricted to refer to the distribution of singularities which is arrayed along the ξ axis downstream of the $\xi_w - r_b$ location.

Further simplification in computation is brought about if only the axial distribution of lift acting on the body is to be found, and not the pressures. In this event, it is not necessary to determine any of the higher harmonics, but instead the desired integrated lift forces may be obtained merely through evaluation of the single first terms, $\varphi_1^{(b)}$ and $\varphi_1^{*(b)}$.

C,47. Extension of Analysis to Apply to Configurations Having Supersonic Trailing Edges that May Be Swept at Any Angle. The foregoing study of wing-body interference effects may be extended with no difficulty to cover the possibility that the trailing edge of the wing might be swept at any arbitrary angle Λ, rather than be perpendicular to the plane of symmetry as assumed so far. It is still to be taken for granted that the trailing edge remains supersonic. The kind of configuration which meets this extended classification is shown schematically in Fig. C,47 (top). The flow around such a system may be visualized as resulting from the superposition of two flow fields emanating from configurations having properties that conform with the procedures now educed. In fact, the configuration of C,47 (top) produces the same kind of disturbances in the air stream as would result from the addition of the flows created by the proper joint action of the configurations shown in the middle and bottom portions of Fig. C,47. The behavior of the middle configuration may be obtained in its entirety through application of the processes already elicited, because its trailing edge is normal to the free stream direction. Likewise, the bottom configuration also has a normal

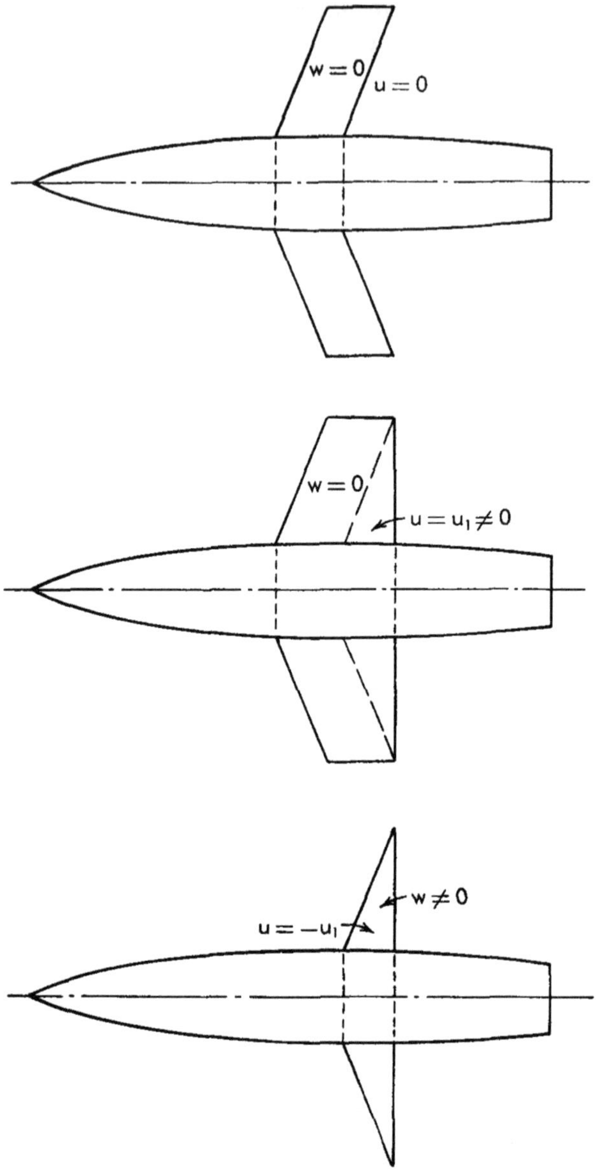

Fig. C,47. Scheme for decomposing configuration with swept supersonic trailing edge into equivalent system consisting of two wings with normal trailing edges. Top, original configuration for which flow field is sought. Middle, configuration with filled-in wing. Bottom, equalizer wing.

trailing edge and may thus be treated by means of techniques now available, provided the following special treatment is used for the determination of the cancellation flow provided by this small equalizer wing.

There will be u components of velocity induced over the "exposed" region of the wing shown in Fig. C,47 (bottom), i.e. in the region lying between its leading and trailing edges and outboard of the wing-fuselage juncture. These pressure discontinuities arise from the presence of the filled-in configuration shown in the middle sketch, and they will have to be canceled out in this subsequent step in the superposition process. It is a simple matter to bring about the elimination of the pressure jumps, however, inasmuch as the pattern of their distribution over the "exposed" region of the equalizer wing is precisely attained as a result of applying the foregoing methods to solution of the flow over the filled-in configuration. The distribution of the velocities $\Delta u/\alpha_w U$ may be continued into that region of the equalizer wing actually covered by the fuselage by making the same sort of assumption as to their constancy there as suggested in Art. 44. The rest of the steps required to determine the flow over the small equalizer wing shown in Fig. C,47 (bottom) are entirely analogous to those in Art. 44, although it should be remarked that the radial velocity components normal to the fuselage were made to vanish when the characteristics of the filled-in wing were obtained in the first step of the superposition treatment.

C,48. Iteration Method and Its Application in Extending Simplified Analysis for Cases in Which Reduced Chord Length Is Not Small. The simplified analysis given in Art. 46 may be looked upon as supplying the first step in a longer iteration process which is capable of providing a solution to the wing-body interference problem when the wing aspect ratio is no longer assumed to be very large, i.e. ξ_w^* need not be restricted to small values, as demanded in Art. 46.

It may be assumed in the present case where ξ_w^* has any value whatsoever that the first part of the procedure recommended in Art. 46 has been carried out and that the determination of the flow about the wing-body combination under examination has furnished the pertinent information about the velocity field C_I, now to be considered as a first approximation to the characteristics of the wing in the presence of the body.

A second approximation may then be obtained in the following manner. The chord AC of the profile constituting the wing-body juncture is first to be divided into small equal-sized intervals of length $\Delta\xi_{w_0}^*$; call them subdivisions AA_1, A_1A_2, A_2A_3, etc., where the magnitude of $\Delta\xi_{w_0}^*$ is very much less than unity (see Fig. C,48). Because the influence of the velocity field C_I is now propagated onto the body, the velocity components normal to the body in the narrow strip between A and A_1 must be counteracted by the effect of a set, Σ_1, of multipole supersonic sources.

These must be distributed along the piece of the ξ axis lying between O and O_1 in sufficient strength to produce the required cancellation in the AA_1 region, when the perturbations generated on the axis are propagated out along the "characteristics" to exert their combined influence ultimately over the region of the body in question. Now, the multipoles Σ_1 produce an upwash on the wing in the region ABC, which is not accounted for in the velocity field representing the first step of this procedure, denoted by C_I. Let the additional values of the relative upwash produced in the region ABC (Fig. C,48) be represented by the symbol $w_\mathrm{I}^*/\alpha_w U$. Furthermore, let the distribution of the $w_\mathrm{I}^*/\alpha_w U$ values be continued over into the region of the wing which is actually covered by the fuselage by taking them to be zero almost everywhere there, as in

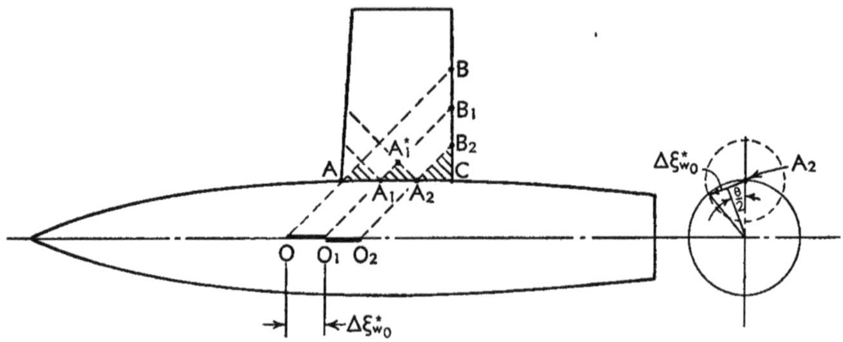

Fig. C,48. Delineation of wing and body areas affected at successive steps in the iteration procedure employed to extend the simplified analysis. The radial velocities induced upon the fuselage are not completely compensated over that arc (of the cross section through the point A_2) marked out on this parallel by the intersection with the dotted circle, of radius $\Delta\xi_{w0}^*$.

The cross-hatched areas indicate those regions in which are located the singularities, corresponding to the wing interference potential, that produce flow on the body which is not compensated completely.

Art. 46. It may be noted besides that the $w_\mathrm{I}^*/\alpha_w U$ values are zero over all of that region of the wing which lies outboard of the Mach line AB. Once these conditions are accepted, the additional relative upwash values so defined are counteracted by supplying a new wing interference potential of the proper sort to accomplish this objective, according to the procedures already explained in Art. 38.

After this operation has been carried out, an additional flow field from the wing is created in the region downstream of the point A_1; let this new field be designated the S_1 contribution. Then the S_1 field will induce further normal velocity components back upon the body in the narrow strip A_1A_2. The additional induced radial components of velocity, together with those which have also been generated there by action of the C_I field, must now be counteracted by setting up a suitable distribution of multipole sources, Σ_2, in the strip of axis lying between O_1 and O_2. The next step required is to compute the upwash components acting on the wing in the region bounded upstream by the Mach line A_1B_1, which are produced by action of the new Σ_2 distribution of multipoles. These additional relative upwash velocities on the wing may then be annulled by supplying another wing interference potential, as in the previous step.

By proceeding step by step in this way the flow over the wing-body combination is built up to give a mathematical solution which, on the wing, rigorously adheres to the physical boundary condition stipulating that the resultant velocity vector must be aligned everywhere with the surface of the wing. On the other hand, this procedure leads to a flow field which satisfies the boundary condition pertaining to the body to no worse a degree of approximation than that tolerated in the corresponding treatment of Art. 46.

As a matter of fact, however, if one considers where the induced velocities originate which reach the surface of the body within a selected one of the sequence of narrow annular rings marked out there by successive parallels, it will be recognized that it is only the distribution of singularities on the adjacent small triangle (e.g. $A_1A_2A_1^*$ for the section A_1A_2) that produces radial velocities which are not exactly compensated, and the maximum angular extent up the side of the body for which the compensation is not complete is given by $\theta = 2\sin^{-1}\Delta\xi^*_{w_0}/2r_b$ (see Fig. C,48).

By a judicious selection of the subdivision length $\Delta\xi^*_{w_0}$, therefore, it is often found to be thoroughly practical to stop such an iteration process just as soon as the second approximation is completed. (More information on this point may be found in Art. 50.)

C,49. Morikawa's Procedure. The question of wing-body interference has likewise been considered by Morikawa [*55, 56*]. The fundamental difference between his procedure and the method discussed in the preceding articles lies in the fact that Morikawa bases his mode of attack on the variant concepts that (1) the determination of the interference potentials $\varphi^{(w)}$ and $\varphi^{(b)}$ need not be carried out separately, attempting instead to discover at one time the entire field of flow represented by the sum of $\varphi_w + \varphi^{(w)} + \varphi^{(b)}$ and (2) the Laplace transform technique is used in place of procedures relying almost exclusively on Fourier transforms or the method of singularities.

In the formalization of the exact mathematical problem now to be solved, the body is considered to be idealized into a right circular cylinder of infinite length. Although the method is theoretically capable of handling the case of a swept leading edge, the application actually carried out is further restricted to determination of the behavior of a configuration for which the wing sweep angle Λ is zero, because the analytical complexities involved even in treating this case are still found to be rather great in themselves. In addition, no information is provided by this method of attack concerning the flow generated downstream of the wing's trailing edge, which is also assumed to be perpendicular to the plane of symmetry.

Let the potential describing the entire flow about the ensemble be denoted by φ so that

$$\varphi = \varphi_w + \varphi^{(w)} + \varphi^{(b)} \tag{49-1}$$

Now this potential will have to satisfy the linearized equation of motion governing any such (normalized) flow, which in cylindrical coordinates is

$$\frac{\partial^2 \varphi}{\partial \xi^2} = \frac{\partial^2 \varphi}{\partial r^2} + \frac{1}{r}\frac{\partial \varphi}{\partial r} + \frac{1}{r^2}\frac{\partial^2 \varphi}{\partial \theta^2} \tag{49-2}$$

Whereas the boundary conditions are

$$\left.\begin{aligned} &\frac{1}{r}\left(\frac{\partial \varphi}{\partial \theta}\right)_{\theta=0} = 1 + \frac{r_b^2}{r^2} \qquad \text{for } x > 0, \quad \text{and} \quad r \geqq r_b \\ &\left(\frac{\partial \varphi}{\partial r}\right)_{r=r_b} = 0 \\ &\varphi(r, \theta, 0) = \left(\frac{\partial \varphi}{\partial x}\right)_{x \leqq 0} = 0 \end{aligned}\right\} \tag{49-3}$$

The simple expression for the first of these boundary conditions (Eq. 49-3) is a consequence of the assumption made concerning the shape of the body. Then, upon application of the Laplace transform to the differential equation governing the flow (49-2), it becomes converted to

$$\nabla^2 \bar{\varphi} - s^2 \bar{\varphi} = 0 \tag{49-2'}$$

while the boundary conditions become recast as

$$\left.\begin{aligned} &\frac{1}{r}\left(\frac{\partial \bar{\varphi}}{\partial \theta}\right)_{\theta=0} = \frac{1}{s}\left(1 + \frac{r_b^2}{r^2}\right) \qquad \text{for} \quad r \geqq r_b \\ &\left(\frac{\partial \bar{\varphi}}{\partial r}\right)_{r=r_b} = 0 \end{aligned}\right\} \tag{49-3'}$$

Furthermore let the Green function related to the governing differential equation as reformulated in Eq. 49-2′ be represented by $\mathcal{G}(r, \theta; r', \theta')$, and let the boundary of the domain in which $\bar{\varphi}$ is defined be denoted by C. The actual shape of the contour in the y, z plane which delimits the region in question is shown in Fig. C,49, under the supposition that the analysis is only to apply to the infinite semiplane where $z > 0$. The mathematical description of the pertinent properties of such a Green function is as follows:

1. It must be true for any two distinct points $P \neq Q$, that $\mathcal{G}(P; Q) \equiv \mathcal{G}(r, \theta; r', \theta')$ will satisfy the differential equation (49-2′).

2. The function $\mathcal{G}(r, \theta; r', \theta')$ will have a logarithmic singularity for the case where $P = Q$.
3. $\partial\mathcal{G}/\partial n = 0$ on the contour C, where $\partial\mathcal{G}/\partial n$ denotes the directional derivative taken normal to the contour C.

If one knows the Green function, then the solution of Eq. 49-2′ which satisfies the stipulated boundary condition (Eq. 49-3′) may be written down at once as

$$\bar{\varphi}(r, \theta; s) = \frac{1}{2\pi} \oint_C \mathcal{G} \frac{\partial\bar{\varphi}}{\partial n} \, dc \tag{49-4}$$

where the notation $\partial\bar{\varphi}/\partial n$ stands for the directional derivative of $\bar{\varphi}$ taken normal to the contour C.

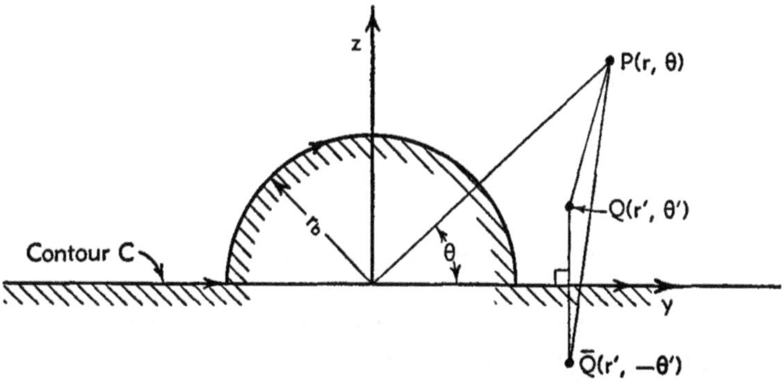

Fig. C,49. Path of integration in semiplane $z > 0$, utilized in evaluating the transformed potential appearing in Morikawa's method.

Morikawa was successful in determining the appropriate Green function, and the result he found to be applicable in the present circumstances is given by the expression

$$\mathcal{G}(P; Q) = K_0(s\bar{\rho}) + K_0(s\bar{\rho}_1) - 2\frac{\dot{I}_0(s)}{\dot{K}_0(s)} K_0(sr)K_0(sr') + 4\sum_{n=1}^{\infty} \frac{\dot{I}_n(s)}{\dot{K}_n(s)} K_n(sr)K_n(sr') \cos n\theta \cos n\theta' \tag{49-5}$$

where r, θ and r', θ' are the polar coordinates of the points P and Q lying in the y, z plane, as indicated in Fig. C,49. The other distance-measuring parameters in these functions are defined as

$$\bar{\rho} = \sqrt{r^2 + r'^2 - 2rr' \cos(\theta - \theta')} \quad \text{and} \quad \bar{\rho}_1 = \sqrt{r^2 + r'^2 - 2rr' \cos(\theta + \theta')}$$

while the usual conventional symbols have been used to designate the functions themselves, i.e. I_n and K_n are the modified cylindrical Bessel functions of the first and second kinds, respectively, with imaginary argument, and of order n. In addition, the customary convention has also been adopted that $\dot{I}_n(s) = dI_n/ds$, etc.

The sought transformed potential $\bar{\varphi}$ thus may be written out explicitly as

$$\begin{aligned}\bar{\varphi}(r, \theta; s) &= \frac{1}{2\pi s}\int_1^\infty \left(1 + \frac{r_b^2}{r'^2}\right)[\mathcal{G}(r, \theta; r', \pi) + \mathcal{G}(r, \theta; r', \theta)]dr' \\ &= \frac{1}{\pi s}\int_1^\infty \left(1 + \frac{r_b^2}{r'^2}\right)\{K_0[s\sqrt{r^2 + r'^2 - 2rr'\cos\theta}] \\ &\quad + K_0[s\sqrt{r^2 + r'^2 + 2rr'\cos\theta}]\}dr' \\ &\quad - \frac{1}{\pi s}\int_1^\infty \left(1 + \frac{r_b^2}{r'^2}\right)\left[\frac{\dot{I}_0(s)}{\dot{K}(s)}K_0(sr)K_0(sr') \right. \\ &\quad \left. + 2\sum_{n=1}^{\infty}\frac{\dot{I}_{2n}(s)}{\dot{K}_{2n}(s)}K_{2n}(sr)K(sr')\cos 2n\theta\right]dr' \\ &= \bar{\varphi}_1 + \bar{\varphi}_2 \end{aligned} \qquad (49\text{-}4')$$

where advantage has been taken of the fact that $\partial\bar{\varphi}/\partial n$ vanishes on the circular part of the contour C.

The potential φ may therefore be attained upon reversing the transformation process at this stage, so that finally

$$\varphi = \mathcal{L}^{-1}[\bar{\varphi}(r, \theta; s)] = \mathcal{L}^{-1}(\bar{\varphi}_1 + \bar{\varphi}_2)$$

Unfortunately the inversion of the indicated Laplace transforms presents a practically insurmountable difficulty, as far as the $\bar{\varphi}_2$ term is concerned. Faced with this obstacle it is expedient to resort to an approximation, following Morikawa, which simplifies the Green function in question to read

$$\begin{aligned}\mathcal{G}(r, \theta; r, \theta') &= K_0[s\sqrt{r^2 + r'^2 - 2rr'\cos(\theta - \theta')}] \\ &\quad + K_0[s\sqrt{r^2 + r'^2 + 2rr'\cos(\theta + \theta')}] \\ &\quad - \frac{K_0(sr')}{I_0(s/r')}\cdot\frac{\dot{I}_0(s)}{\dot{K}(s)}\left\{K_0\left[s\sqrt{r^2 + \frac{1}{r'^2} - 2\frac{r}{r'}\cos(\theta - \theta')}\right]\right. \\ &\quad \left. + K_0\left[s\sqrt{r^2 + \frac{1}{r'^2} - 2\frac{r}{r'}\cos(\theta + \theta')}\right]\right\} \end{aligned} \qquad (49\text{-}5')$$

which grows to be a good enough representation of the true $\mathcal{G}$ at large values of s.

As soon as this simpler expression (Eq. 49-5′) for the Green function is admitted to replace the intractable one encountered in Eq. 49-5 the inversion of the Laplace transform becomes feasible, but even so the resulting formula for the potential φ will constitute an acceptable solution only if the values of $\xi - r_b$ at which the computations are to be made are kept down to rather small size. Thus the limits of validity of this procedure scarcely differ from the severe bounds which restrict the range of applicability of the approximate method described in Art. 46, the impaired effectiveness of which was improved in Art. 48.

In addition to the restrictions on its accuracy, except for small values of $\xi - r_b$, the Morikawa procedure is not particularly suited to rapid calculation of the pressure coefficients applying to the wing and body. In fact, the task of carrying out such a computation is just as laborious as if the job of finding these pressures were accomplished by the approximate method of Art. 46. For these reasons, it is understandable that it would not serve any useful purpose to give a more extensive résumé of the Morikawa procedure here, although further reference will be made to this method when the results of numerical applications for various conventional configurations are reviewed in Art. 50.

C,50. Numerical Applications in the Case Where the Wing Is Rectangular. Comparison with Test Results. When the wing of the configuration is of rectangular planform, the numerical applications become particularly simple to carry out according to the theories developed in the preceding articles. It is worthwhile to focus attention on such special cases that are amenable to easy solution, in order to obtain easily as much useful information as possible.

Under these special circumstances, the wing-alone flow, governed by the potential φ_w, produces a component of induced velocity perpendicular to the surface of the circular body which is given by

$$\frac{v_{rb,w}}{\alpha_w U} = \left(\frac{\partial \varphi_w}{\partial r}\right)_{r=r_b} = \sin\theta \quad \text{for that part of the body intercepted within the } \Sigma \text{ region}$$

$$= 0 \quad \text{in the region of the body lying outside the } \Sigma \text{ region} \tag{50-1}$$

On the other hand, the normal components of velocity induced in the region of the $\zeta = 0$ plane which is marked out by the outline of the wing planform can be taken to be adequately described with very good approximation by

$$\left[\frac{\partial(\varphi_b)_2}{\partial \zeta}\right]_{\zeta=0} = -a\frac{r_b^2}{r^2} = \frac{(w_b^*)_2}{\alpha_w U} \tag{50-2}$$

where a never deviates very much from unity regardless of the shape of the body.

Numerical results, sanctioning the above generalization, are presented in Fig. C,50a. The curves show the spanwise distributions of relative upwash values, $[\partial(\varphi_b)_2/\partial\zeta]_{\zeta=0}$, which exist in the region of the wing in the $\zeta = 0$ plane for two bodies of quite different shape (Fig. C,50b and C,50c). The simple relationship stated as Eq. 50-2 has been plotted on

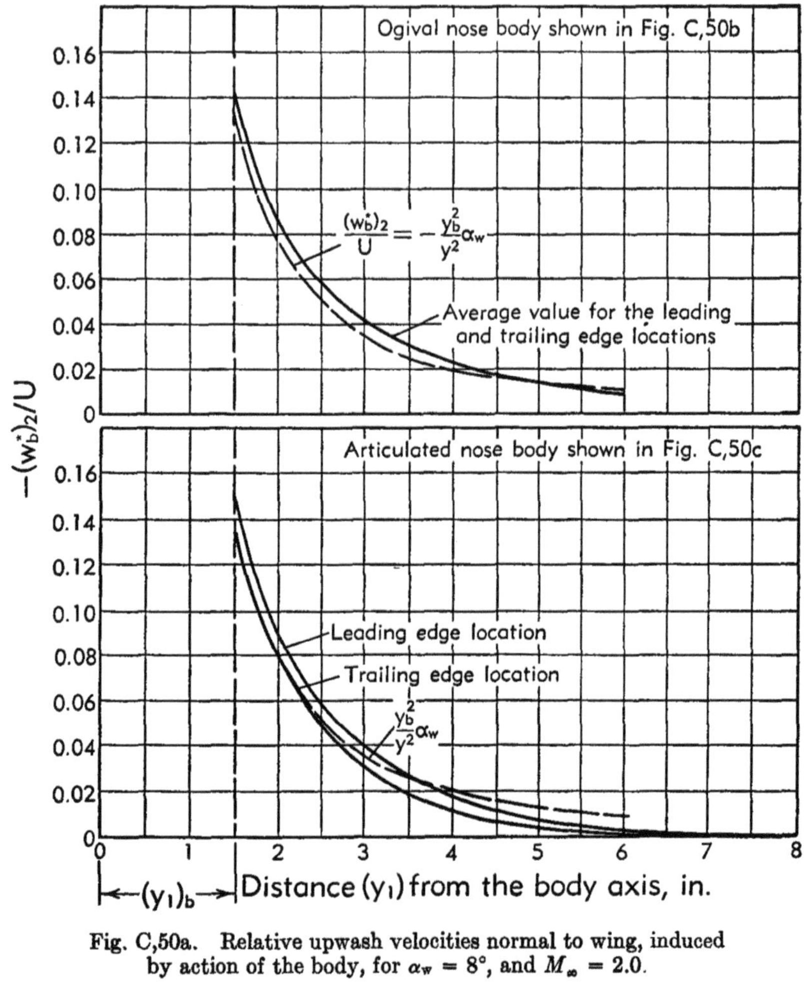

Fig. C,50a. Relative upwash velocities normal to wing, induced by action of the body, for $\alpha_w = 8°$, and $M_\infty = 2.0$.

each of the graphs of Fig. C,50a, and the accord between this idealized pattern for the upwash and the more accurate results obtained by use of a laborious step-by-step computation is so close that there can be little doubt of the complete justification for the present assumption.

The next step in the computations is to represent the upwash distribution described by Eq. 50-2, in which a is assigned the value unity, by means of a trigonometric series of cosines as required by Eq. 38-1. This

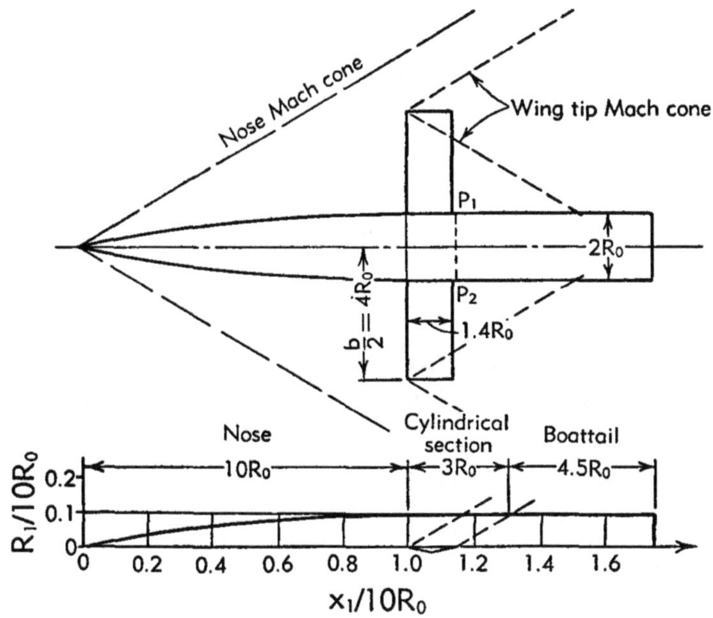

Fig. C,50b. Plan and elevation of the ogival nose body tested in Cornell Aeronautical Laboratory experiments.

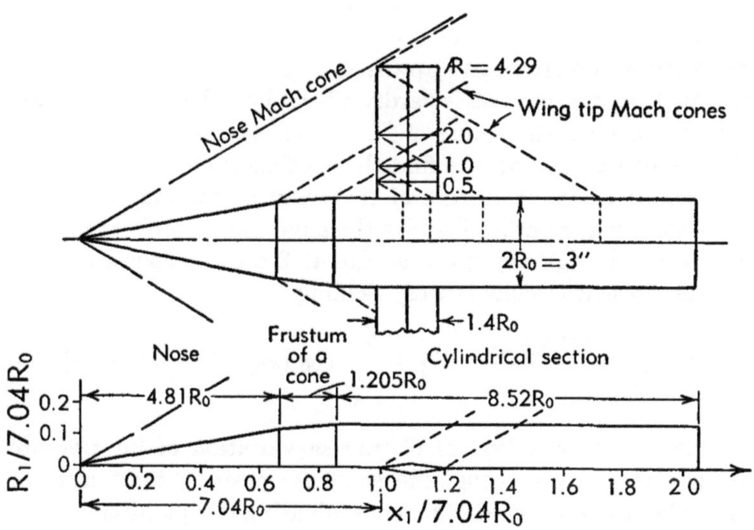

Fig. C,50c. Plan and elevation of the articulated nose body tested in Cornell Aeronautical Laboratory experiments.

process may be executed by demanding that the representing series, of the sort defined by Eq. 38-1, should pass exactly through the values of $(w_b^*)_2/\alpha_w U$ obtained from Eq. 50-2 when the following sequence of equally-spaced spanwise location coordinates is inserted into the above relation:

$$\frac{r_b}{r} = 1, \frac{3}{4}, \frac{3}{5}, \frac{3}{6}, \frac{3}{7}, \frac{3}{8}, \frac{3}{9}, \frac{3}{10}, \text{ and } \frac{3}{11}$$

while $(w_b^*)_2/\alpha_w U$ is to be assigned the value zero for

$$\frac{r}{r_b} = \frac{2}{3}, \frac{1}{3}, \text{ and } 0, \text{ where } r_b = \frac{1}{4}$$

When the numerical work has been done, it is found that the harmonic coefficients of the approximating trigonometric representation have the values:

$$G_1 = -0.35518, \quad G_3 = +0.07282, \quad G_5 = +0.18705, \quad G_7 = +0.20288$$

$$G_9 = +0.08926, \quad G_{11} = -0.01284, \quad G_{13} = -0.10629, \quad G_{15} = -0.12253$$

$$G_{17} = -0.09389, \quad G_{19} = -0.01816, \quad G_{21} = +0.05419, \quad G_{23} = +0.10267$$

and the representing curve is as shown in Fig. C,50d. It may be observed that the $(w_b^*)_2/\alpha_w U$ values are zero almost everywhere in the region of the wing covered by the body (satisfying the conditions stipulated for valid application of the approximate method given in Art. 46), while in the exposed region of the wing the approximating curve also reproduces the original distribution of relative upwash values with adequate precision.

Inasmuch as the upwash is hardly affected by changes in fore-and-aft location on the wing, the values of G_m are perforce constants with respect to ξ. Consequently, in order to find the coefficient of pressure (difference between top and bottom surface of the wing) pertaining to the induced flow over the wing, as prescribed by the governing interference potential $\varphi^{(w)}$, the formula originally given as Eq. 46-3 may be specialized to conform to the present circumstances, yielding

$$\frac{[(\Delta C_p)_w]_{\zeta=0^+} - [(\Delta C_p)_w]_{\zeta=0^-}}{4\alpha_w} = \frac{r_b^2}{r^2} + \sum_m \cos(k\eta) G_m[1 - J_0(k\xi^*)] \quad (46\text{-}3')$$

In order to show a typical chordwise variation of interference lift acting on a rectangular wing, the spanwise position right next to the fuselage (where $\eta = r_b$) is selected. On evaluating the quantity

$$\frac{[(\Delta C_p)_w]_{\zeta=0^+} - [(\Delta C_p)_w]_{\zeta=0^-}}{4\alpha_w}$$

as a function of the axial location ξ^*/r_b, the curve indicated by the dash-dot line in Fig. C,50e may be constructed.

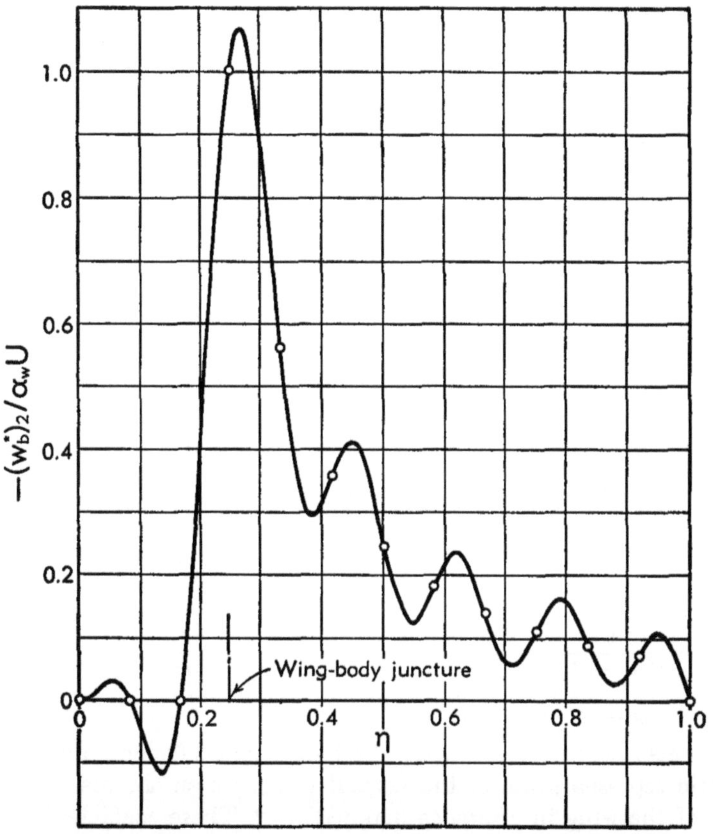

Fig. C,50d. Trigonometric representation of the upwash distribution out along the wing span, induced by an idealized fuselage.

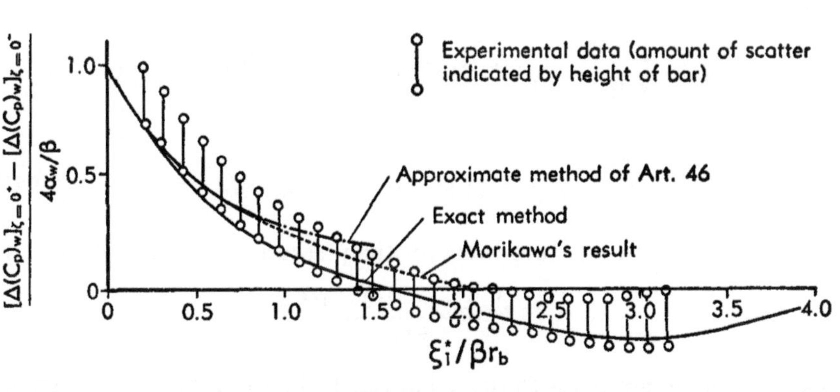

Fig. C,50e. Chordwise pressure distribution on the wing, induced by the body. Profile selected for illustrating pattern of pressures is located right at juncture between the components.

To obtain the exact result according to the detailed methods of Art. 39, 40, and 42, it is necessary to determine the contribution to the wing pressures provided by the axial derivative of the body interference potential $\partial\varphi^{(b)}/\partial\xi$. In order to take account of this source of additional pressure induced upon the wing, it is required that the harmonic coefficients F'_n, defined by Eq. 40-1, be evaluated by utilizing the simple expression (Eq. 50-1) for the normal components of the induced velocity arising from the type of wing-alone flow existing under the present circumstances. The harmonic coefficients are thus given by

$$\left.\begin{aligned} F'_0 &= \frac{2}{\pi}\left[1-\sqrt{1-\left(\frac{\xi}{r_b}-1\right)^2}\right] && \text{for } \frac{\xi}{r_b}-1 \leqq 1 \\ &= \frac{2}{\pi} && \text{for } \frac{\xi}{r_b}-1 \geqq 1 \\ F'_n &= \frac{1}{\pi}\left[\frac{2\cos(n-1)\delta}{n-1} - \frac{2\cos(n+1)\delta}{n+1} - \frac{4}{n^2-1}\right] && \text{for } \frac{\xi}{r_b}-1 \leqq 1 \\ &= -\frac{4}{\pi(n^2-1)} && \text{for } \frac{\xi}{r_b}-1 \geqq 1 \end{aligned}\right\} \quad (40\text{-}1')$$

where $\delta = \sin^{-1}[(\xi/r_b) - 1]$.

It is also necessary to obtain the harmonic coefficients for the trigonometric representation of the normal velocities on the fuselage due to action of the wing interference potential $\varphi^{(w)}$. These coefficients F_n may be determined by evaluation of Eq. 39-6 and 39-8. In the present instance, however, it is observed that a certain amount of simplification ensues because of the constancy of the G_m values. In fact, it may now be stated that expressions for $\varphi_m^{(w)}$ and $\partial\varphi_m^{(w)}/\partial\zeta$ reduce here to

$$\varphi_m^{(w)} = \frac{2}{k}\int_0^{\xi^*-\zeta} L_m(\xi^*, \zeta; \xi^{*\prime})dG_m = \frac{2G_m(0)}{k} L_m(\xi^*, \zeta; 0) \quad (39\text{-}1')$$

and

$$\frac{\partial\varphi_m^{(w)}}{\partial\zeta} = -\int_0^{\xi^*-\zeta} L'_m(\xi^*, \zeta; \xi^{*\prime})dG_m = -G_m(0)L'_m(\xi^*, \zeta; 0) \quad (39\text{-}3')$$

The corresponding exact value for the pressure differential, defined as

$$\frac{[(\Delta C_p)_w]_{\zeta=0^+} - [(\Delta C_p)_w]_{\zeta=0^-}}{4\alpha_w}$$

may then be computed with the aid of Eq. 42-11, in which τ is set equal to unity, because it is desired to obtain the chordwise pressure distribution applying specifically to the profile at the wing-body juncture where

$r = r_b$. The graph of the exact results,[7] produced by so evaluating Eq. 42-11, is also shown in Fig. C,50e, together with the experimental test data reported by Pitts, Nielsen, and Gionfriddo [*57*]. For further comparison, Morikawa's calculations are also included in this diagram.

In view of the relationships shown in Fig. C,50e, the following conclusions may be drawn:

1. The predictions of the theoretical procedures are amply confirmed by the experimental test points.
2. The approximate procedure advocated in Art. 46 yields quite good results provided the chordwise location of interest lies well forward on the wing, so that $\xi_w^*/r_b \leqq 1$. For the after portions of the wing, however, this method alone is inadequate, but still of advantage since one can use the iteration method based on this approximate method, as discussed in Art. 48. The simplest way to arrive at good results in the region of the chord running from $\xi_w^*/r_b = 1.0$ to 2.0 is to rely on the iteration process which requires that the total chord length c_{w_0} be divided into two equal segments. Not only is this choice advisable in order to cut down the work required to obtain a distribution of pressures of high accuracy in the wing-body juncture region, but it is especially advantageous to operate in this manner if, subsequently, information is required about the flow field that is generated downstream of the wing's trailing edge.
3. The Morikawa procedure gives better than rough answers, but the range over which it is really effective appears curtailed, for it can hardly be expected to be useful beyond $\xi_w^*/r_b > 1.5$.

The Cornell Aeronautical Laboratory has carried out a systematic series of comparisons of theory and experiment in order to have available an organized body of information on the subject of wing-body interference at supersonic speeds [*58,59,60,61,62,63*]. The compilations of data in question may be used to illustrate several fundamental points.

First consider the interference effects produced at a Mach number of 2.0 by a configuration having the geometric characteristics shown in Fig. C,50f. The graphs of the pressure coefficients $(C_p)_w$, which pertain to locations on the upper flat surface of such a wing at three arbitrarily selected chordwise locations for the spanwise survey lines are plotted in Fig. C,50f, C,50g, and C,50h.[8] The experimental data are from the

[7] The curve representing the exact results was also obtained by Pitts, Nielsen, and Gionfriddo [*57*], but their procedure for obtaining it formally differs entirely from the one presented in Art. 38, 39, 40, and 42 and recommended for the present application in the paragraphs immediately above; nevertheless, there is fundamental consistency between the two approaches.

[8] For reasons of clarity, the actual physical dimensions have been retained in presenting these data, rather than the normalized coordinates denoted by x, y, z. This practice has been followed not only in Fig. C,50f, C,50g, and C,50h, but also in the subsequent graphs of this article.

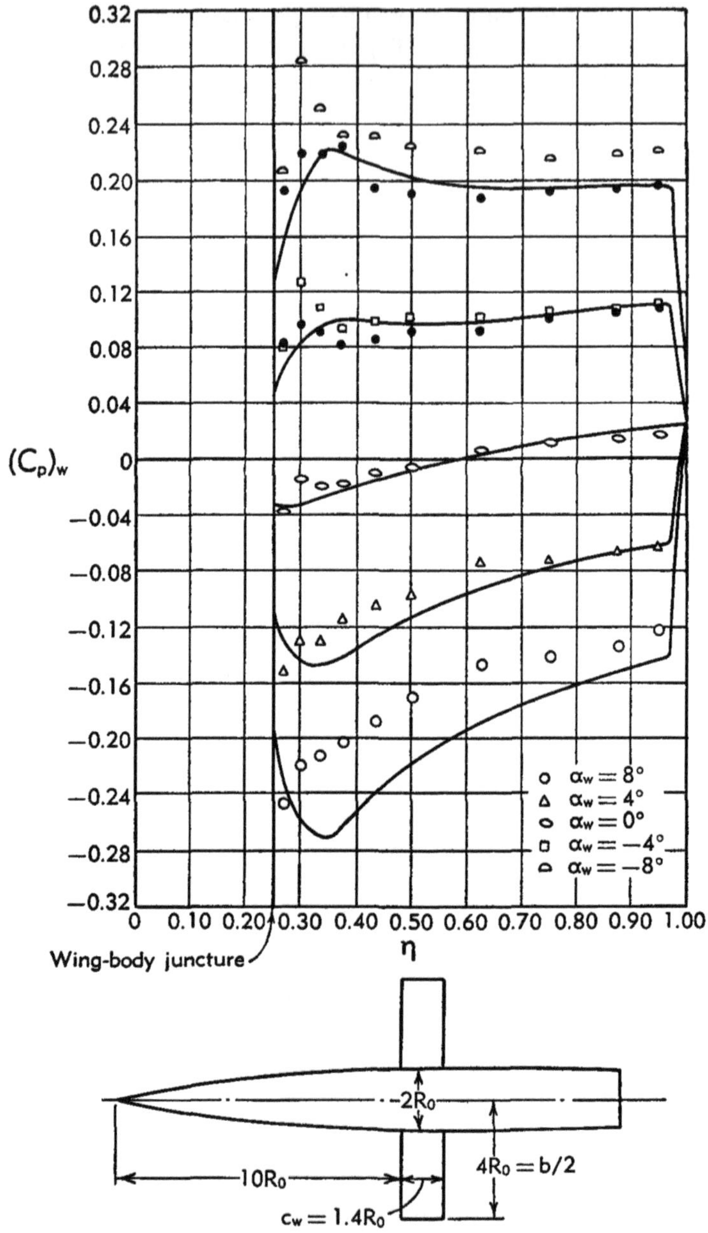

Fig. C,50f. Pressure coefficient along spanwise section at ¼ wing chord aft of wing leading edge. Comparison of experimental and theoretical results for configuration shown.

Cornell Aeronautical Laboratory and the theoretical curves were constructed by adding together the three contributions to the total pressures due to: (1) the incremental pressures, denoted by $(\Delta C_p)_w$, that arise exclusively from computation of the interference effect caused by upwash

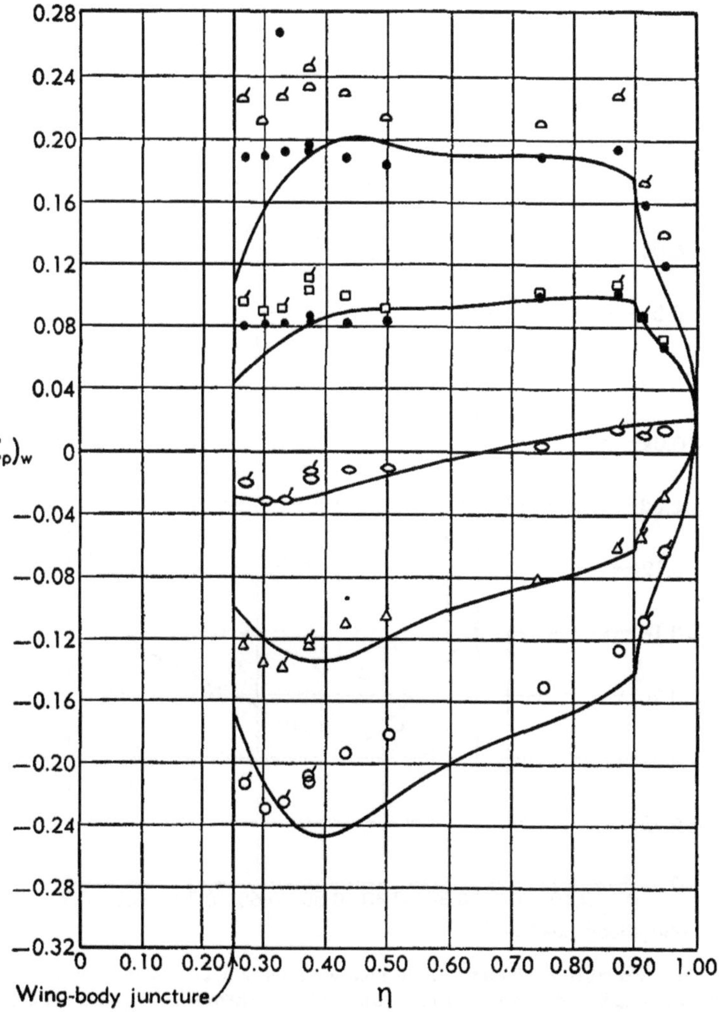

Fig. C,50g. Pressure coefficient along spanwise section at ½ wing chord aft of wing leading edge. Comparison of experimental and theoretical results for configuration shown in Fig. C,50f.

from the body impinging on the wing, according to the simplified theory discussed in Art. 46, (2) the pressures existing on the isolated wing, before interference effects are introduced, and (3) the axial components of velocity, acting equally on both sides of the wing, which are induced in this region because of the presence of the body, inasmuch as the airstream

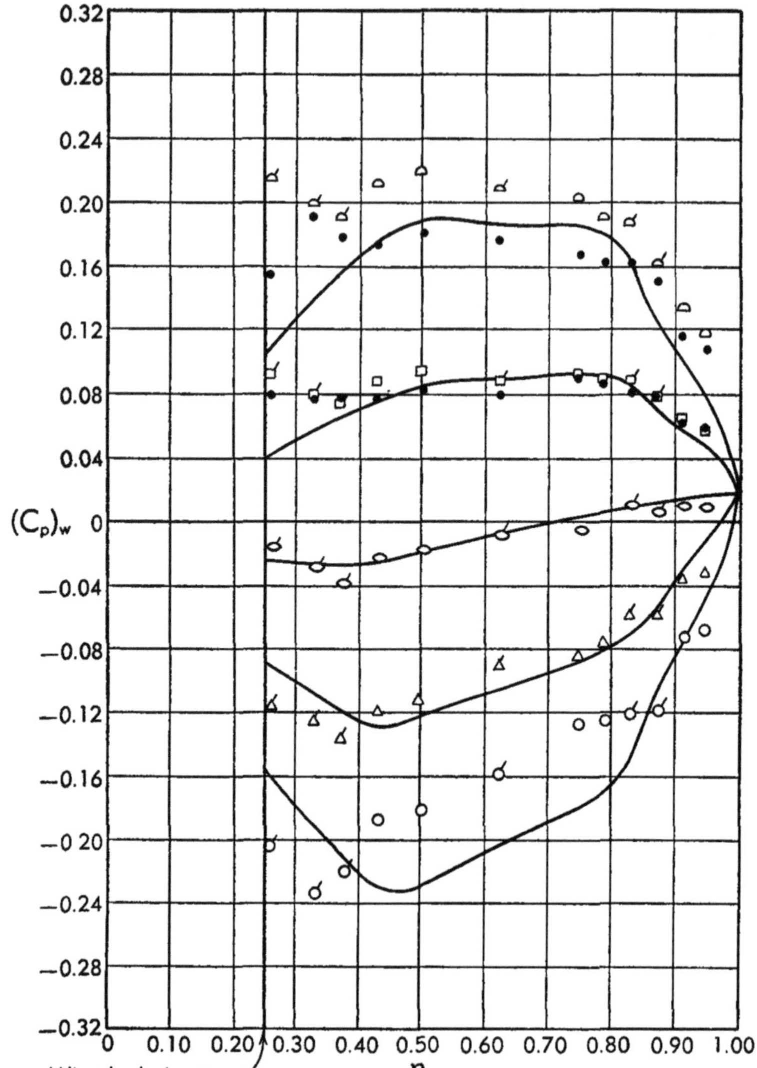

Fig. C,50h. Pressure coefficient along spanwise section at ⅟ wing chord forward of wing trailing edge. Comparison of experimental and theoretical results for configuration shown in Fig. C,50f.

must accommodate the bulk of the body even when it is at zero-degrees angle of attack.

Although the correspondence between theory and experiment for the spanwise plots may be considered generally good at all angles of attack from a qualitative point of view, it is important to note that the quantitative agreement shows up much better at the lower angles ($\alpha_w = \pm 4°$)

than at the higher ones (where $a_w = \pm 8°$). The apparent decrease in agreement between the theoretical and experimental results at the higher angles of attack may be ascribed to the fact that, in the theoretical treatment used, nonlinear terms in the expression for the pressure coefficients, which would enter into the more exact formulas required for accurate results at high angles of attack, have been ignored. In fact, it is worthy of note that an asymmetric behavior of the curves is exhibited by the experiments when passing from the region of negative angles of attack to the positive range. It is contrary to the fundamental tenets of the linearized theory for such a dissymmetry to arise. The nature of the discrepancy is exactly analogous to the nonsymmetrical behavior of the pressures on an airfoil operating in two-dimensional flow, according to the Busemann theory, which affirms that the pressures should follow the nonlinear law given by

$$\frac{p - p_\infty}{\frac{1}{2}\rho_\infty U^2} = \mp \frac{2}{\sqrt{M_\infty^2 - 1}}\alpha_w + \frac{2}{\sqrt{M_\infty^2 - 1}}\left[\frac{(\gamma + 1)M_\infty^4}{4(M_\infty^2 - 1)} - 1\right]\alpha_w^2 \quad (50\text{-}3)$$

where the negative sign applies for the pressures on the top surface of the planar element (at an angle of attack α_w to the free stream), and where the positive sign applies for the pressures on the lower surface.

Since it seems possible to surmise that much of the observed discrepancy at the higher angles of attack can be traced back to the influence of second order terms having the same general structure as the Busemann formula, it appears logical to average the experimental results obtained at positive and negative angles of attack at the same orifice, in order to arrive at a more equitable standard of comparison for judging the reliability of the results derived by use of the linear theory. The adjusted experimental values were obtained by averaging (ignoring the sign) the results for $+\alpha_w$ and $-\alpha_w$, and including the effect on the local pressures of the contribution arising from the distortion to the flow field about the body alone, due to the fact that the air must accommodate for the bulk of the body. These values have been spotted-in on the graphs of Fig. C,50f, C,50g, and C,50h as solid circles. In nearly all instances the adjusted points are in good agreement with the theoretically derived curves.

It should be recognized in addition that the prediction of *lift* according to the linear theory will also be much closer to reality than was true of the unadjusted *pressure* distribution data, because it is only the sum (of the absolute values) of the pressures from the top and bottom surfaces which contributes to the lift force. In fact, good correspondence is found to exist, even at a relatively high angle of attack of 8°, between the theory, when based on the simplified analysis which led to Eq. 46-4, and the experimental lift distribution generated by the configuration shown in Fig. C,50b. This comparison of lifts is presented in Fig. C,50i. The total integrated lift, acting on the wing at an angle of attack of 8°, thus turns out to be

$$[(C_L)_w]_{\text{theory}} = 0.3734 \quad \text{while} \quad [(C_L)_w]_{\text{experiment}} = 0.3625$$

where the reference area for the coefficients has been selected to be the exposed area of both panels.

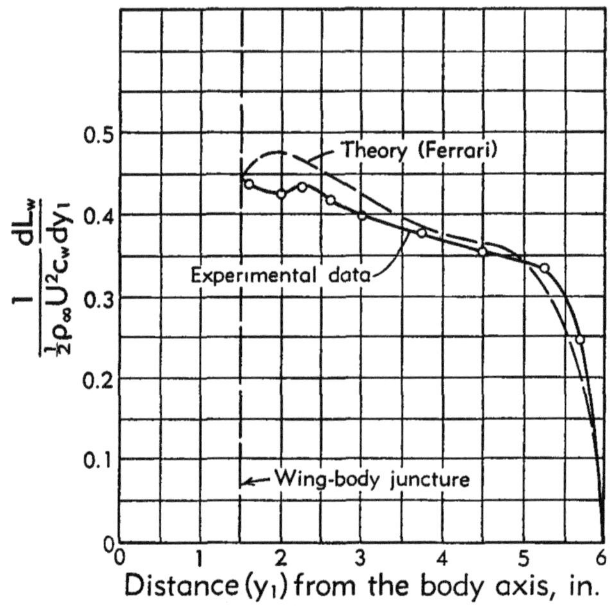

Fig. C,50i. Spanwise distribution of lift for configuration shown in Fig. C,50b. Comparison of experimental results with theory using simplified approach. Configurational angle of attack $\alpha_w = 8°$, $M_\infty = 2.0$.

Returning now to examination of the flow about the fuselage, the pertinent pressure coefficients at the surface of the body may be determined as follows, on the basis of the simplified method explained in Art. 46:

It may be observed first of all that

$$\left(\frac{\partial \varphi_w}{\partial \xi}\right)_{\zeta=0^+} = -1 \qquad \text{in the } \Sigma \text{ region, defined in Art. 37}$$

$$= 0 \qquad \text{for that part of the } \zeta = 0 \text{ plane lying outside of the } \Sigma \text{ region}$$

and consequently the incremental lift $\Delta_1 L_b$ due to interference action from the wing alone amounts to just exactly the same amount of lift as would be carried on that part of the wing, actually covered by the fuselage, when operating as a contiguous piece of the isolated total wing.

Secondly, in view of the simple way in which the normal components of the induced flow emanating from the wing may here be described, in conformity with Eq. 50-1, it follows that the corresponding harmonic components $F_n'^*$ of the representing series required according to Eq. 46-1 can be determined by evaluation of the following simple expression. (It is not necessary to employ the formula of de la Vallée-Poussin in this instance.)

$$\left.\begin{aligned} F_1'^* &= \frac{1}{\pi}(2\delta - \sin 2\delta) \\ F_n'^* &= \frac{2}{\pi}\left[\frac{\sin (n-1)\delta}{n-1} - \frac{\sin (n+1)\delta}{n+1}\right] && \text{for } n \text{ odd} \\ &= 0 && \text{for } n \text{ even} \end{aligned}\right\} \tag{50-4}$$

where $r_b \leqq \xi \leqq r_b + c_w$ and where, as before,

$$\delta = \sin^{-1}\left(\frac{\xi}{r_b} - 1\right)$$

If it so happens that $r_b + c_w \leqq \xi \leqq 2r_b$, then the required expressions are

$$\left.\begin{aligned} F_1'^* &= \frac{1}{\pi}[2(\delta - \delta') - \sin 2\delta + \sin 2\delta'] \\ F_n'^* &= \frac{2}{\pi}\left[\frac{\sin (n-1)\delta - \sin (n-1)\delta'}{n-1}\right. \\ &\qquad \left. - \frac{\sin (n+1)\delta - \sin (n+1)\delta'}{n+1}\right] && \text{for } n \text{ odd} \\ &= 0 && \text{for } n \text{ even} \end{aligned}\right\} \tag{50-4'}$$

and likewise they become altered to read

$$\left.\begin{aligned} F_1'^* &= \frac{1}{\pi}(\pi - 2\delta' + \sin 2\delta') \\ F_n'^* &= \frac{2}{\pi}\left[\frac{\sin (n-1)\delta'}{n-1} - \frac{\sin (n+1)\delta'}{n+1}\right] && \text{for } n \text{ odd} \\ &= 0 && \text{for } n \text{ even} \end{aligned}\right\} \tag{50-4''}$$

when $2r_b \leqq \xi \leqq 2r_b + c_w$, where the additional angular location coordinate

$$\delta' = \sin^{-1}\left(\frac{\xi}{r_b} - 1 - \frac{c_w}{r_b}\right)$$

is introduced to fix the lower limit to the arc on any body-parallel over which the induction action of the wing is felt directly.

The peripheral distribution of pressures acting at an axial station falling slightly behind the trailing edge of the wing (at the position P_1P_2) for the configuration shown in Fig. C,50b has been determined by use of the simple formulas above and the results have been plotted in Fig. C,50j, together with the corresponding pressures determined experimentally. The pressure coefficient represented by these curves is the total value which results from adding together the interference from the wing and the basic pressure variation existing about the isolated body when it is

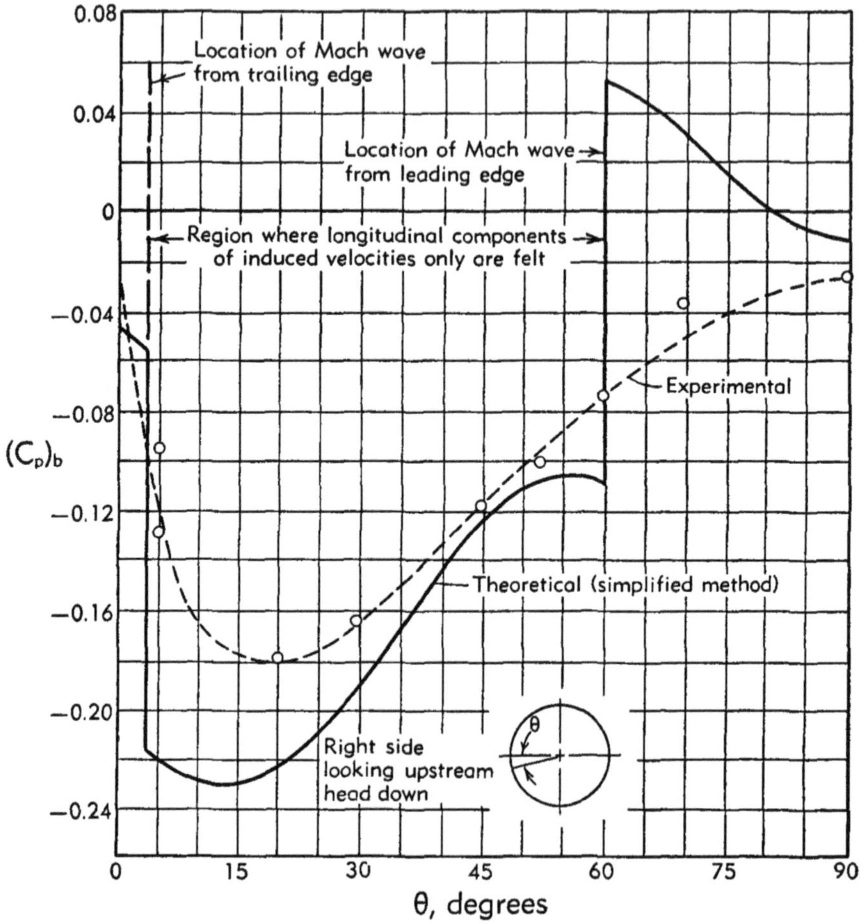

Fig. C,50j. Peripheral distribution of pressures around the body for configuration shown in Fig. C,50b at an axial location lying slightly downstream of the trailing edge of the wing, at a distance of $0.1R_b$ from it (at parallel P_1P_2, as marked on Fig. C,50b).

at an angle of attack (of 8° in this case); however, these body-alone pressures are of inconsequential size in this example. The correspondence between the theoretical and experimental results is quite satisfactory, especially over that portion of the periphery which experiences the direct effect of the longitudinal components of induced velocity from the wing. A better fit is obtained, however, when the pressures are integrated around the fuselage at each axial station and an axial distribution plot is constructed for both the theoretical and experimental data.

In order to obtain the theoretical curve for the axial distribution of the total interference lift acting along the body, it is necessary to combine the contributions to this total effect which come from the induced longitudinal components directly and from the additional field of flow which is generated because of the necessity to cancel the induced normal components of velocity at the surface of the body. Consequently, the interference lift $(\Delta C_L)_b$ will, in this simplified case, be produced by summing the effect coming solely from the isolated wing potential φ_w (for the longitudinal components) and the effect due solely to the body interference potential $\varphi^{(b)}$ (for the normal components). These two separate sources contributing to the total interference are individually plotted in Fig. C,50k, together with the distribution representing the total effect exerted by the wing upon the body.

In Fig. C,50l the theoretical curves of the axial distribution of interference lift are presented for the cases in which the angle of attack is 4° and 8° respectively; the corresponding experimentally determined distributions of interference lift are also shown. The agreement between the experimental and theoretical results is entirely acceptable, particularly because the small discrepancy between the curves is of the right sign to be readily accounted for by the terms that were not considered in this simplified theoretical treatment. It should be remembered that the approximate theory upon which these curves are based has neglected to provide for any contribution from the $\Delta_2 L_b$ term discussed in Art. 46, and, moreover, the normal velocities induced on that part of the fuselage lying downstream of the Σ region, which are produced by action of the wing interference potential $\varphi^{(w)}$, have also been dropped.

An outstanding feature of the axial plots discussed above is the persistence of the carry-over of lift upon the body for large distances downstream of the Σ region and well into the T region. At least this is true of the $d(\Delta C_L)_b/d(\xi/r_b)$ distribution referring to configurations having high aspect ratio wings, of the kind shown in Fig. C,50b and C,50c. This dominance of the $\varphi^{(b)}$ potential over the $\varphi^{(w)}$ potential in the afterbody region makes it imperative to take the $\varphi^{(b)}$ potential into account in any theory which hopes to predict body interference lift successfully.

As the aspect ratio of the wing decreases, however, the relative importance of the $\varphi^{(w)}$ potential becomes greater and greater while, in addition, the portion of the fuselage which is influenced by the Mach cones emanating from the wing tips at the leading edge also becomes larger. Because of these trends, a likely explanation is presented for the mechanisms at work in producing the progressive patterns of axial lift distribution that are evinced in the systematic collection of data given in Fig. C,50m, pertaining to a series of wing-body combinations having rectangular wings of diverse aspect ratio, as shown in the sketch accompanying these same graphs.

It is worth noting that the validity of the simplified method of Art. 46 does not depend upon the magnitude of the aspect ratio but rather

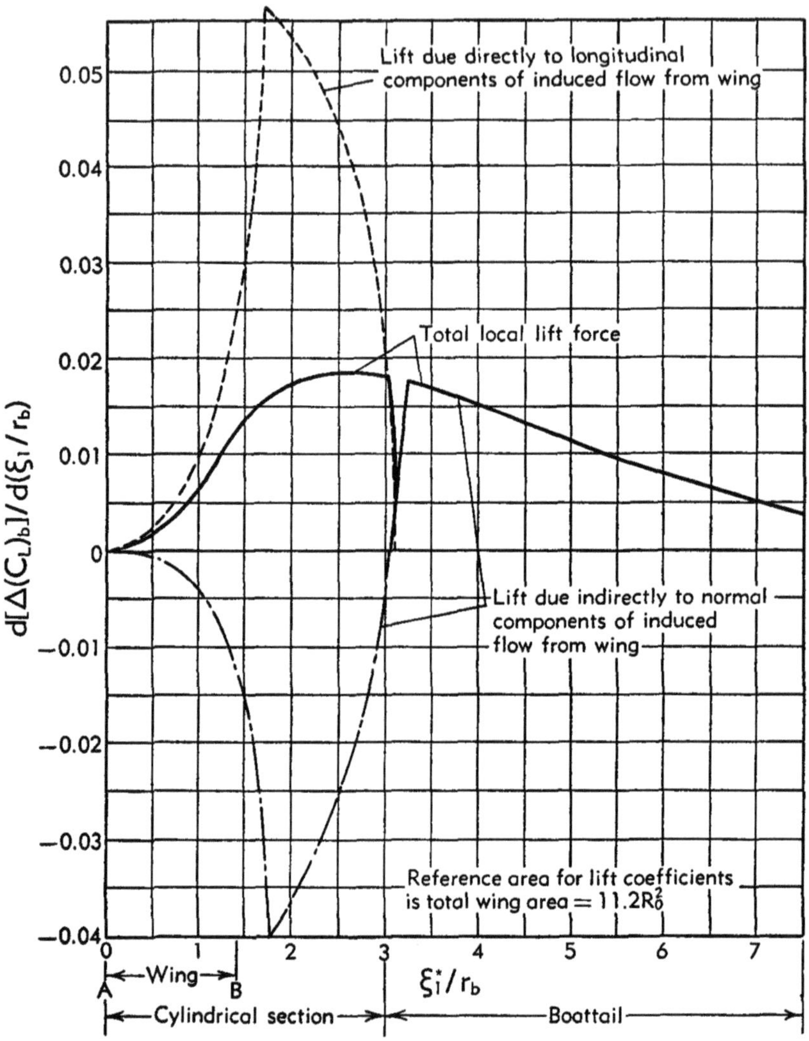

Fig. C,50k. Contributions making up total local lift force induced along the body by interference action of the wing, for configuration shown in Fig. C,50b. Configurational angle of attack $\alpha_w = 8°$, $M_\infty = 2.0$.

upon the chord-length parameter ξ_w^*/r_b. Consequently this fundamental method, though approximate, is capable of giving results of unimpaired accuracy for the aerodynamic actions taking place on the fuselage even though the wing planform is reduced to stubby proportions.

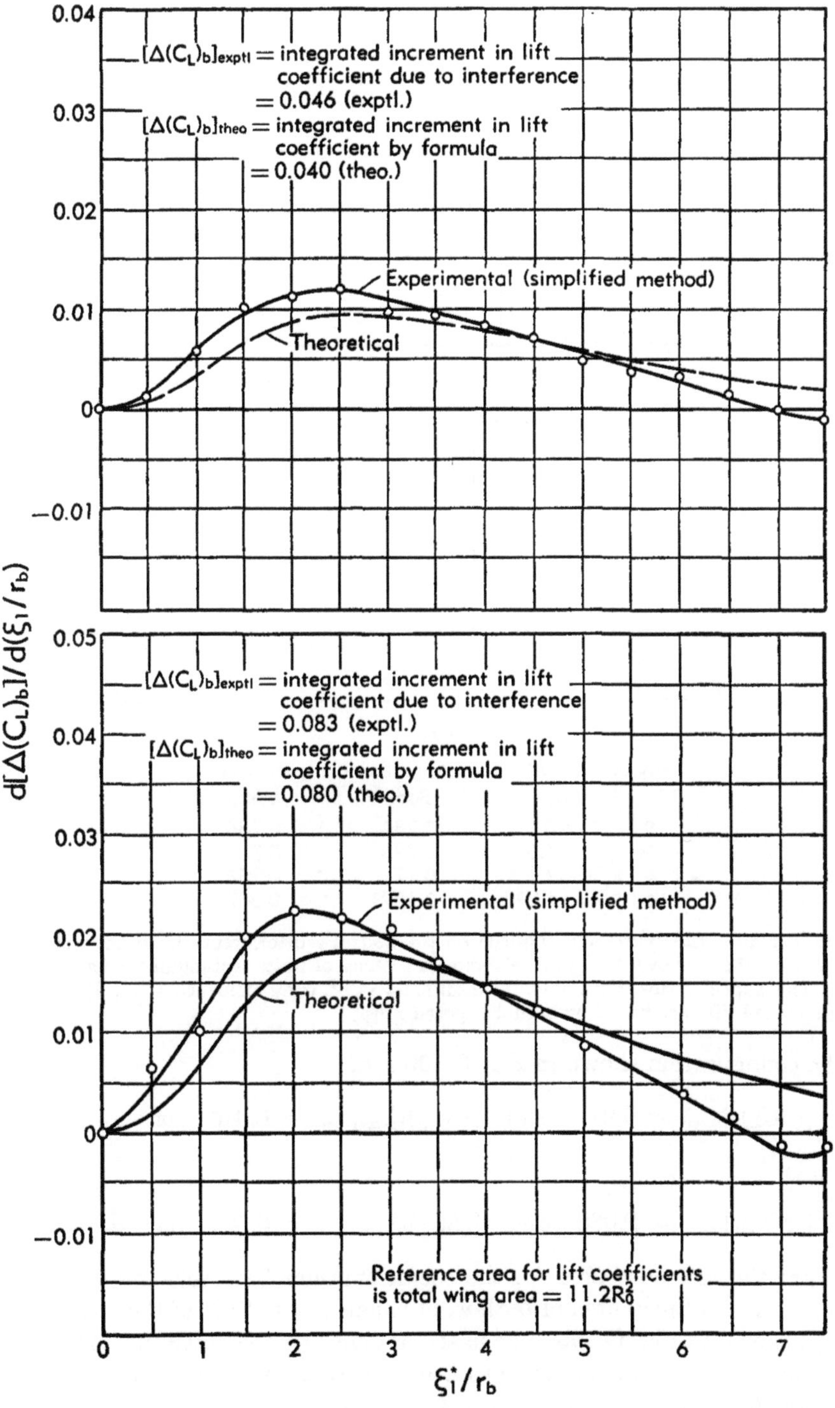

Fig. C,50l. Lift increment on body due to interference from wing, for configuration shown in Fig. C,50b. Top, $\alpha_w = 4°$; bottom, $\alpha_w = 8°$, $M_\infty = 2.0$.

In regard to the total integrated incremental lift acting over the entire length of the body, it may be observed that the approximate method provides very good estimates inasmuch as the actual values applying to

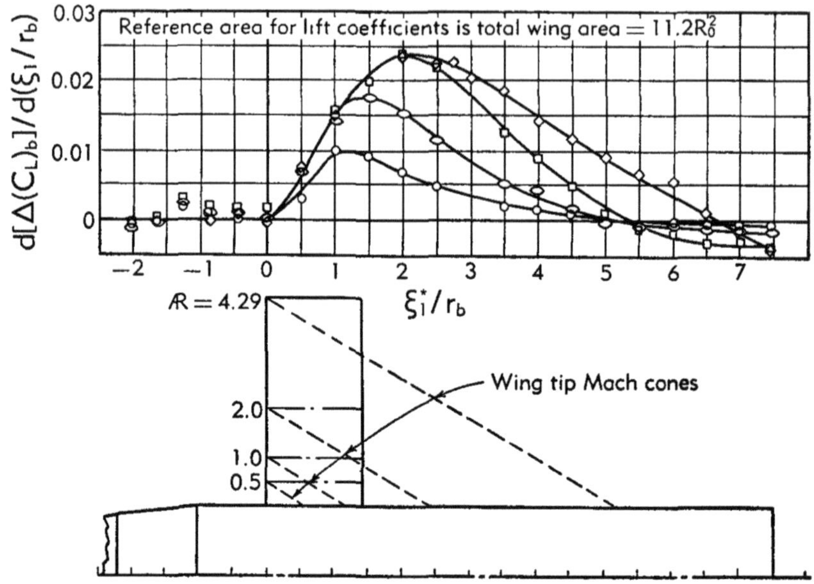

Legend

Symbol	Æ	Integrated lift coefficients: Body	Wing
○	0.5	0.234	0.286
⬭	1.0	0.226	0.363
□	2.0	0.174	0.368
◇	4.29	0 109	0.357

Fig. C,50m. Lift increment induced on the body by interference from the wing for the body shape shown in Fig. C,50c and for a series of finite rectangular wings having various aspect ratios. Experimental results. $\alpha_w = 8°$, wing incidence = 0°, $M_\infty = 2.0$. Integrated lifts are based on actual exposed area.

the configuration shown in Fig. C,50b are:

$$[(\Delta C_L)_b]_{\text{theory}} = 0.040 \quad \text{and} \quad [(\Delta C_L)_b]_{\text{experiment}} = 0.046 \quad \text{for} \quad \alpha_w = 4°$$

while

$$[(\Delta C_L)_b]_{\text{theory}} = 0.080 \quad \text{and} \quad [(\Delta C_L)_b]_{\text{experiment}} = 0.083 \quad \text{for} \quad \alpha_w = 8°$$

where the reference area for the coefficients has been selected this time to be the tip-to-tip area of both wing panels (= $11.2R_0^2$) of the configuration shown in Fig. C,50b. The usage of denoting the maximum diameter of the physical body as $2R_0$ has been reverted to in these figures, as in Chap. 2.

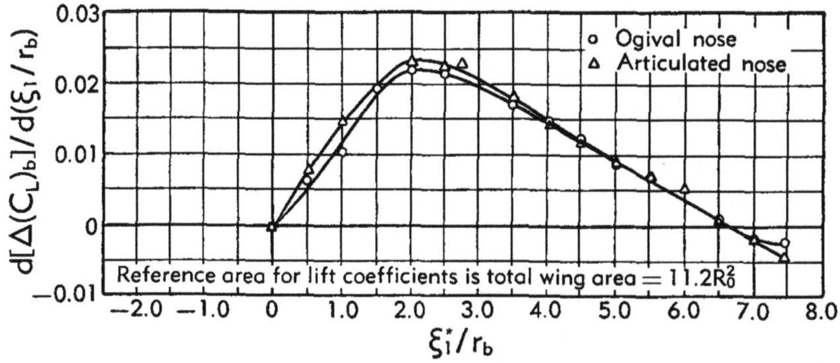

Fig. C,50n. Lift increment on body due to interference from wing. Comparison of the experimental results pertaining to the ogival nose body (Fig. C,50b) with those of the articulated nose body (Fig. C,50c). $\alpha_w = 8°$, $M_\infty = 2.0$.

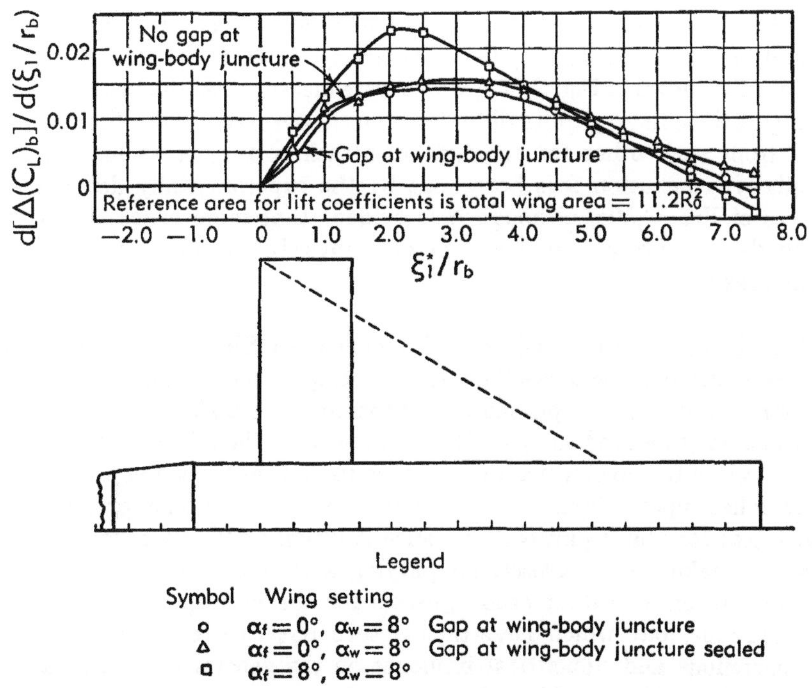

Fig. C,50o. Comparison of carry-over of lift onto the body, for configuration shown in Fig. C,50c in three different wing conditions: (1) wing at incidence, so that $\alpha_f = 0$, $\alpha_w = 8°$ with natural gap, (2) same condition but with gap sealed, and (3) configuration at angle of attack, so that $\alpha_f = \alpha_w = 8°$. $M_\infty = 2.0$.

Other general conclusions may be drawn from the theoretical results:

1. In view of the fact that the upwash distribution used in this linear theory has been generalized, so that $[\partial(\varphi_b)_2/\partial\zeta]_{\zeta=0}$ is given in simple form by Eq. 50-2, regardless of the shape of the nose of the body lying ahead of the wing, it follows that the lift induced by the wing upon the body will be independent of the missile nose shape.
2. On the basis of the linear theory given here the lift induced by the wing upon the body, according to the simplified method, will remain unaltered regardless of whether both the wing and body are operating at the same angle of attack α_w with aspect to the free stream flow or whether the wing is set on the fuselage at an angle of incidence of α_w and the fuselage is then held aligned with the free stream flow at a zero angle of attack, or, in other words, $(\Delta C_L)_b$ remains the same for a given wing condition irrespective of how its angle of incidence is varied.

The above conclusions are supported by experimental evidence presented in Fig. C,50n and C,50o, respectively. In fact, the axial distributions of lift shown in Fig. C,50n, which were measured on the bodies of two configurations having quite different nose shapes, are hardly distinguishable from each other for an angle of attack setting of as high as 8°. Similarly, conclusion 2 is borne out by the experimentally determined curves of Fig. C,50o, although it is evident that more exact numerical methods should be able to detect slight changes between the two different situations tested.

C,51. Determination of Over-All Forces by Short-Cut Methods. The estimation of the over-all lift forces acting on the wing and body may in many cases be accomplished by approximate methods.[9] Such short-cut methods may be used successfully to circumvent the labor of computing the details of the flow, which would be required if the foregoing procedures were relied upon. Such telescoping of the computational processes is legitimate for many purposes, because it usually does no harm in such cases to make certain grossly simplifying assumptions which would, to be sure, result in rather crude guesses for the corresponding pressure distributions, but which nevertheless turn out to exert a remarkably inconspicuous and noncritical influence on the general integrated aerodynamic interactions taking place.

One of the earliest examples of such an effective over-all approach was given by Kirkby and Robinson [*64*]. They found the lift induced on the wing by action of the fuselage through the simple expedient of assum-

[9] In this article the practice of using the actual dimensions has been reverted to, because the formulas for the stretched system resulting from use of the normalized coordinates are not as readily amenable to direct physical interpretation, free from distracting glosses.

ing that the force exerted on each strip of the wing would be approximately equal to the two-dimensional value of the lift which would be created on it, by virtue of the fact that the angle of attack of each such strip is effectively the variable angle of upwash created at the successive spanwise locations according to the idealized law given as Eq. 50-2. The analytical result of this elementary approach for obtaining the lift on the wing is thus expressible as

$$\Delta L_w = \tfrac{1}{2}\rho_\infty U^2 b \frac{4\alpha_w}{\beta} \int_{r_b}^{1} c_w \frac{r_b^2}{\eta^2}\, d\eta \tag{51-1}$$

which, for a rectangular wing, reduces to

$$\Delta L_w = \tfrac{1}{2}\rho_\infty U^2 b \frac{4\alpha_w}{\beta} c_w r_b (1 - r_b) \tag{51-1$'$}$$

Corresponding to this amount of lift, an appropriate total lift coefficient may be devised by basing the reference area in question on the sum of the exposed wing panel areas, and, after adding in the amount of lift on the wing alone (exposed portion only, without accounting for tip effects), the following pertinent expression is obtained

$$(C_L)_{w.\,total} = \frac{4\alpha_w}{\beta}(1 + r_b) \tag{51-2}$$

On the other hand, a rapid estimate of the lift induced on the body may be obtained by following the argument proposed by Flax [*65*]. This analysis, based on reverse-flow concepts, indicates that when the body may be considered to be infinitely long, the wing will induce upon it an amount of lift given by the relation

$$\Delta L_b = \rho_\infty U b \int_{r_b}^{1} \Gamma(\eta) \frac{r_b^2}{\eta^2}\, d\eta \tag{51-3}$$

where $\Gamma(\eta)$ is the circulation pertaining to the wing section at a generic spanwise location denoted by η.

Under the same conditions that are assumed to hold in arriving at the wing lift estimate provided by the Kirkby and Robinson formula (Eq. 51-1), the pertinent circulation distribution which applies in the present instance (for the total lift on the wing) will therefore be expressed as

$$\Gamma(\eta) = \frac{1}{2}\frac{4\alpha_w}{\beta} c_w U \left(1 + \frac{r_b^2}{\eta^2}\right)$$

and consequently the expression for the corresponding body lift is given by

$$\Delta L_b = \tfrac{1}{2}\rho_\infty U^2 \frac{4\alpha_w b}{\beta} \int_{r_b}^{1} c_w \left(1 + \frac{r_b^2}{\eta^2}\right) \frac{r_b^2}{\eta^2}\, d\eta \tag{51-3$'$}$$

If the wing is rectangular, the above formula simplifies to just

$$\Delta L_{\mathrm{b}} = \tfrac{1}{2}\rho_\infty U^2 c_{\mathrm{w}} b \frac{4\alpha_{\mathrm{w}}}{\beta} r_{\mathrm{b}}[1 - r_{\mathrm{b}} + \tfrac{1}{3}(1 - r_{\mathrm{b}}^3)] \tag{51-3''}$$

while if a corresponding lift coefficient is devised, as before, by referring this amount of lift to a reference area which is again taken to be the exposed portion of both wing panels, the pertinent formula becomes

$$(\Delta C_L)_{\mathrm{b}} = \frac{4\alpha_{\mathrm{w}}}{\beta} \frac{r_{\mathrm{b}}}{1 - r_{\mathrm{b}}} [1 - r_{\mathrm{b}} + \tfrac{1}{3}(1 - r_{\mathrm{b}}^3)] \tag{51-4}$$

Following Lawrence and Flax [66] a useful relation for the fractional part of the lift carried on the wing and body may then be concocted immediately from the foregoing, so that

$$\frac{(\Delta C_L)_{\mathrm{b}}}{(C_L)_{\mathrm{w,total}}} = r_{\mathrm{b}} \frac{1 - r_{\mathrm{b}} + \tfrac{1}{3}(1 - r_{\mathrm{b}}^3)}{1 - r_{\mathrm{b}}^2} \tag{51-5}$$

which is valid for a rectangular wing, while in the less specialized case where the wing has any planform whatsoever the general formula is

$$\frac{(\Delta C_L)_{\mathrm{b}}}{(C_L)_{\mathrm{w,total}}} = \frac{\displaystyle\int_{r_{\mathrm{b}}}^{1} c_{\mathrm{w}} \left(1 + \frac{r_{\mathrm{b}}^2}{\eta^2}\right) \frac{r_{\mathrm{b}}^2}{\eta^2}\, d\eta}{\displaystyle\int_{r_{\mathrm{b}}}^{1} c_{\mathrm{w}} \left(1 + \frac{r_{\mathrm{b}}^2}{\eta^2}\right) d\eta} \tag{51-5'}$$

Contrary to the assumption of strip theory there exist on actual wings regions of fall-off in lift at both the outboard real wing tip and in the inboard wing-body juncture region, which are ignored in arriving at Eq. 51-2. Although the validity of Eq. 51-2 is thus strictly applicable only to configurations with wings of large aspect ratio, so that the aforementioned fall-off-of-lift areas would constitute but a small fraction of the total wing surface, it has been pointed out by Lawrence and Flax that the proportionate amounts of lift supported by each component of the configuration, as specified in Eq. 51-5, has a much wider range of applicability, as attested to by the experimental evidence.

Confirming information bearing on this point is reproduced here in Fig. C,51a. In this figure, values of $\Delta L_{\mathrm{b}}/L_{\mathrm{w,total}}$, corresponding to Eq. 51-5, have been plotted as a function of the ratio of body radius to wing semispan, and corresponding experimentally determined points have been superimposed on the theoretical curve. The experimental data [67] support unequivocally the contention that Eq. 51-5 has an extensive range of usefulness, inasmuch as the aspect ratios of the wings used in the experiments include Æ = 4.29, 2.0, 1.0, and 0.5; the geometric details of the models in question are depicted in Fig. C,50m.

It is very convenient for purposes of quick estimation to have established in this manner the general validity of Eq. 51-5, and probably equal confidence can be placed in Eq. 51-5′, likewise, regardless of the size of the wing aspect ratio. Because the ratio of the lift supported by the wing and by the fuselage can be judged so closely by aid of these formulas, an effortless means is provided for obtaining either of the quantities L_b or L_w, together with their sum, just as soon as the other is known. Some other generalizations that are useful design tools may be extracted from further manipulation of the formulas now deduced. It will be illuminating, for instance, to add together the contributions to

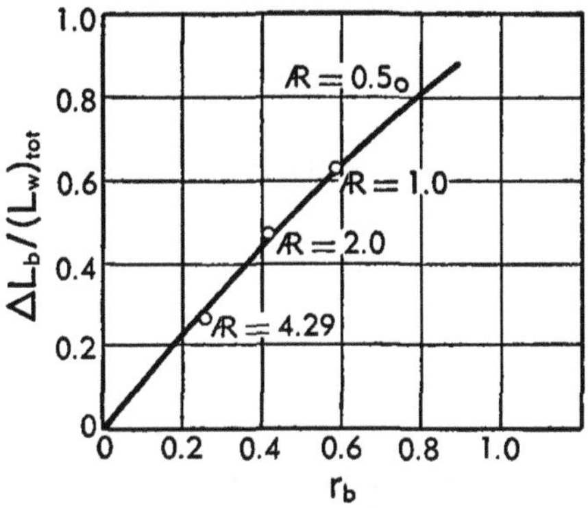

Fig. C,51a. Ratio of incremental lift induced on the body to the total lift on the wing. Comparison of experimental results with formula due to Lawrence and Flax [66]. $\alpha_w = 8°$, $M_\infty = 2.0$.

the total lift on the wing and body, as given by Eq. 51-1′ and 51-3″, and to subtract the lift on the isolated body alone, leaving a quantity to be designated as $L_{bw} - L_b$. Thus

$$L_{bw} - L_b = \Delta L_b + \Delta L_w + L_w^* = \Delta L_b + (L_w)_{\text{total}} \tag{51-6}$$
$$= \tfrac{1}{2}\rho_\infty U^2 b c_w \frac{4\alpha_w}{\beta}(1 - r_b)[1 + 2r_b + \tfrac{1}{3}r_b(1 + r_b + r_b^2)]$$

where L_w^* is the lift on the exposed portion of the wing, computed by means of strip theory, so that

$$L_w^* = \tfrac{1}{2}\rho_\infty U^2 b c_w \frac{4\alpha_w}{\beta}(1 - r_b) \tag{51-7}$$

and consequently

$$L_{bw} - L_b = L_w^*[1 + 2r_b + \tfrac{1}{3}r_b(1 + r_b + r_b^2)] \tag{51-6′}$$

Now it is noteworthy that this equation not only gives the correct result when the body shrinks to nothing, so that $r_b \to 0$, but it also gives the known exact result at the other extreme range of the r_b values, i.e. for cases where $r_b = 1$. It has been found previously, by use of slender body theory, that if the combination is of the angle-of-attack variety (for which the wing is fixed solidly at an undeflected angle of zero incidence to the body axis), and provided it is of small enough aspect ratio and swept sufficiently (with trailing edge perpendicular to the body axis) so that the stipulations of the slender body theory are met, then the corresponding expression for the lift is given as (see Chap. 2, Eq. 16-9):

$$L_{bw} - L_b = L_w\left(1 - \frac{R_0^2}{b'^2}\right)^2 = L_w(1 - r_b^2)^2$$

but since the lift on an isolated slender wing made from the exposed panels, L_w^*, is connected with the lift on the extended-area wing when operating alone, L_w, through means of the ratio

$$\frac{L_w^*}{L_w} = (1 - r_b)^2$$

it follows that

$$L_{bw} - L_b = L_w^*(1 + r_b)^2 \tag{51-6''}$$

By comparison of Eq. 51-6′ and 51-6″ it becomes quite clear that there is actually very little difference between them, over the whole range of body radius-to-wing span ratios running from 0 to 1. This similarity in the two formulas seems to intimate that Eq. 51-6″ can be used in place of Eq. 51-6′ in order to provide rapid and nearly exact answers for the lift acting on a wide range of wing-body configurations, not just for those which fall within the regulations specified in slender body theory. In making use of Eq. 51-6″ for this purpose of estimating the lift on any wing-body combination whatsoever, it is still necessary to use Eq. 51-5′ to determine in what ratio the total lift is divided between the wing and body, while the quantity L_w^* may be obtained approximately from use of Eq. 51-7 or, better yet, by referring to the more exact formulas given in Vol. VI, or through experiment.

A better estimate of the total lift experienced by the wing in presence of the fuselage than is provided by Eq. 51-2 may be derived on the basis of the following considerations. The loss of lift in the wing-body juncture region may be determined with fairly good accuracy for rectangular wings of sufficiently large aspect ratio by making use of the formula for the lift distribution which holds strictly just for the case where the wing chord c_w tends toward zero. Thus, in way of approximation, Eq. 38-7 may now be replaced by the specialized formula, applying to an infinitesimally narrow chord wing, namely:

$$h_m(\xi_1^*) = -\frac{\sqrt{2}}{\pi k} G_m(0) \sin \frac{k\xi_1^*}{\sqrt{2}\,\beta} \tag{51-8}$$

On the basis of this assumption the expression for the incremental lift induced on the wing now becomes

$$\Delta L_w = 2\rho_\infty U^2 b^2 \pi \sum_m h_m \left(\frac{2c_w}{b}\right) \int_{r_b}^{1} \cos k\eta d\eta$$

$$= -2\rho_\infty U^2 \frac{b^2}{\beta} \sum_m \int_{r_b}^{1} d\eta \int_0^{2c_w/b} G_m(0) \cos\left(\frac{k\xi_1^*}{\sqrt{2}\,\beta}\right) \cos k\eta d\xi_1^*$$

$$= -\rho_\infty U^2 \frac{b^2}{\beta} \int_0^{2c_w/b} d\xi_1^* \int_{r_b}^{1} \sum_m G_m(0) \cos k\left(\eta - \frac{\xi_1^*}{\sqrt{2}\,\beta}\right) d\eta$$

$$- \rho_\infty U^2 \frac{b^2}{\beta} \int_0^{2c_w/b} d\xi_1^* \int_{r_b}^{1} \sum_m G_m(0) \cos k\left(\eta + \frac{\xi_1^*}{\sqrt{2}\,\beta}\right) d\eta$$

while the harmonic coefficients $G_m(0)$ are to be constructed so as to represent the infinite-body upwash, in this slightly idealized situation, so that it is permissible to say for present general purposes that

$$\sum_m G_m(0) \cos k\eta = -\frac{r_b^2}{\eta^2}\alpha_w \qquad \text{for } |\eta| \geqq r_b$$

$$= 0 \qquad \text{for } |\eta| < r_b$$

and consequently

$$\int_{r_b}^{1} \sum_m G_m(0) \cos k\left(\eta - \frac{\xi_1^*}{\sqrt{2}\,\beta}\right) d\eta = \int_{r_b-(\xi_1^*/\sqrt{2}\,\beta)}^{1-(\xi_1^*/\sqrt{2}\,\beta)} \sum_m G_m(0) \cos(k\eta_1) d\eta_1$$

$$\cong -\alpha_w \int_{r_b}^{1} \frac{r_b^2}{\eta_1^2} d\eta_1$$

Likewise

$$\int_{r_b+(\xi^*/\sqrt{2}\,\beta)}^{1+(\xi_1^*/\sqrt{2}\,\beta)} \sum_m G_m(0) \cos(k\eta_2) d\eta_2 \cong -\alpha_w \int_{r_b}^{1} \frac{r_b^2}{\eta_1^2} d\eta_1 + \alpha_w \int_{r_b}^{r_b+(\xi_1^*/\sqrt{2}\,\beta)} \frac{r_b^2}{\eta_1^2} d\eta_1$$

Finally, therefore, the approximate expression for the incremental lift in this case becomes

$$\Delta L_w = 2\rho_\infty U^2 \frac{b}{\beta} c_w \alpha_w r_b (1 - r_b)$$

$$- \rho_\infty U^2 \frac{b^2}{\beta} \alpha_w \int_0^{2c_w/b} d\xi_1^* \int_{r_b}^{r_b+(\xi_1^*/\sqrt{2}\,\beta)} \frac{r_b^2}{\eta_1^2} d\eta_1 \quad (51\text{-}9)$$

The last term on the right-hand side of this expression may thus be interpreted as representing the loss of lift in the wing-body juncture region in comparison with the undiminished amount that would be predicted on the basis of strip theory according to Eq. 51-1′. More precisely, it may be stated that the part of the wing surface between the wing-body juncture and the ray emanating from the point of intersection of the wing's leading edge and the fuselage, having the equation $\eta = r_b + \xi_1^*/(\sqrt{2}\,\beta)$ constitutes a fictitious wing tip in regard to the interference flow. In this region only half as much interference lift is produced as would be created if the flow were treated according to the precepts of strip theory, for which the Kirkby and Robinson formula of Eq. 51-1′ applies. This triangular area which is half efficient in producing interference lift is shown by the shaded area of Fig. C,51b.

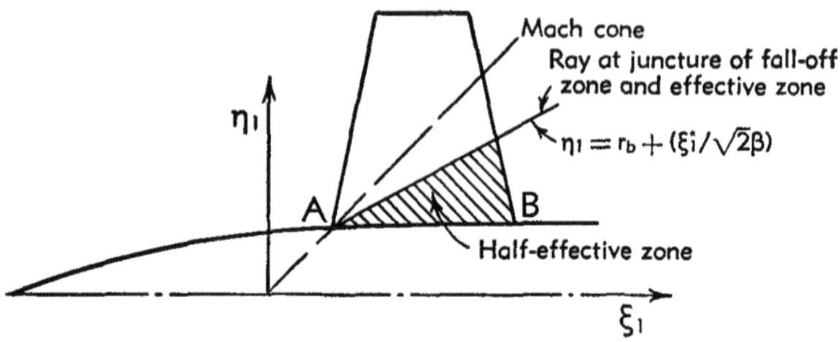

Fig. C,51b. Portion of wing area which is half effective in supporting interference lift induced on it by aerodynamic action of the fuselage (rigorously true only for very small chord wing).

In order to obtain a more explicit summary expression for the incremental lift induced on the wing, the indicated quadratures in Eq. 51-9 may be carried through, to give

$$\begin{aligned}\Delta L_w &= 2\rho_\infty U^2 \frac{b}{\beta} c_w \alpha_w r_b (1 - r_b) \\ &\quad - \rho_\infty U^2 \frac{b^2}{\beta} \alpha_w \left[r_b \frac{c_w}{b} - \sqrt{2}\,\beta r_b^2 \ln\left(1 + \frac{c_w}{\sqrt{2}\,b\beta r_b}\right)\right] \\ &= 2\rho_\infty U^2 \frac{b}{\beta} c_w \alpha_w r_b (1 - r_b) \left[1 - \frac{1 - \sqrt{2}\,\beta b \frac{r_b}{c_w} \ln\left(1 + \frac{c_w}{\sqrt{2}\,\beta b r_b}\right)}{2(1 - r_b)} \right]\end{aligned} \tag{51-9′}$$

The total lift coefficient for the entire force acting on the exposed portion of the wing, including the fall-off of lift at the real outboard wing tip besides, is thus determinable from the relation

$$(C_L)_w = \frac{4\alpha_w}{\beta}\left(1 + f r_b - \frac{1}{2 Æ \beta}\right) \tag{51-10}$$

where

$$f = 1 - \frac{1 - \sqrt{2}\,\beta b \frac{r_b}{c_w} \ln\left(1 + \frac{c_w}{\sqrt{2}\,\beta b r_b}\right)}{2(1 - r_b)} \tag{51-10′}$$

Although the concept of a half-effective inboard fictitious wing tip, as embodied in Eq. 51-9, is derived above for the specific case of a rectangular wing, it can be extended immediately to encompass the possibility that the planform may actually be of any shape whatsoever.

Coming now to the stabilizing effects produced by the distribution of lift forces acting over the body, one may intuitively appreciate that it would require extensive calculation to determine the exact moment produced by the interference forces acting on a particular configuration, but if the configuration is idealized sufficiently it is an easy matter to estimate these summary results. For purposes of arriving at a general and useful impression of what the moments will be for configurations of uncomplicated geometry it will be most profitable to consider the ultimate in model simplicity for which the body radius is allowed to approach zero and the shape of the wing is rectangular and of large aspect ratio. The moments under consideration are those which are computed by assuming the axis of rotation to be the wing axis, i.e the line joining the 50 per cent chord points on the wing. Under these circumstances, it has been shown [*48,66*] that the aerodynamic induction process will produce an alternating distribution of forces on the body which create a couple, of magnitude given by

$$\Delta_1 M_b = -\tfrac{1}{4}\pi\rho_\infty U^2 b^2 r_b^2 c_w \alpha_w \tag{51-11}$$

This formula accounts for all those contributions which arise from the potential $\varphi^{(b)}$ in the region Σ and from $\varphi^{(b)} + \varphi^{*(b)}$ in the T region, which are in each case perturbation potentials that must be introduced in order to counteract the normal components of velocity induced upon the surface of the body.

On the other hand, the part of the interference flow which is attributable to the direct action of the potential φ_w, in propagating longitudinal components of velocity over the surface of the body, will create a moment, taken about the wing axis location again, of an amount equal to exactly half that due to the couple, i.e.

$$\Delta_2 M_b = -\frac{\pi}{8}\rho_\infty U^2 b^2 r_b^2 c_w \alpha_w = \frac{\Delta_1 M_b}{2} \tag{51-11$'$}$$

Consequently the total moment acting on the *body*, other than the moment produced by the flow over the nose, is given by the expression

$$\Delta M = -\tfrac{3}{8}\pi\rho_\infty U^2 b^2 r_b^2 c_w \alpha_w \tag{51-12}$$

and it is noteworthy that this moment is stabilizing (a pitching moment tending to depress the nose).

CONFIGURATION HAVING A WING WITH SUBSONIC LEADING EDGE

C,52. Classification and Critique of Methods of Attack. The solution of the wing-body interference problem is rendered somewhat more difficult when the leading edge of the configuration is subsonic, because it is no longer possible to work out the analysis, in general, by relying on the separation of the wing interference and body interference potentials, as in Art. 37 when computing $\varphi^{(w)}$ and $\varphi^{(b)}$. It is possible to solve the problem in certain specific cases, however, and these examples may be used as guides to the general features likely to be exhibited by wing-body combinations of this sort. The combinations of particular interest are (1) a triangular wing supported on a conical body in such a manner as to place the apexes in coincidence and (2) a wing of any shape and sweep mounted on a cylindrical body.

Admittedly, case (1) constitutes a somewhat impractical configuration, but there is an advantage to be gained in studying this uninviting grouping because it is possible to obtain an exact answer for it within the limitations of linear theory without a great deal of labor. Once this

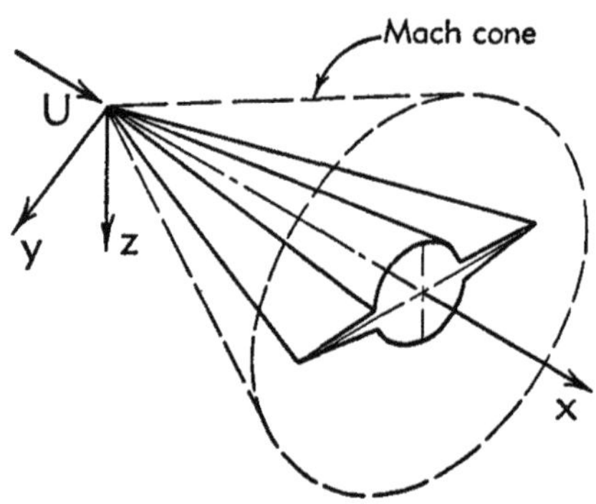

Fig. C,53a. Orientation of the coordinate axes and sketch of typical configuration.

straightforward solution has been derived, it may be used as a check or reference point against which to compare the results which would be obtained for less stylized wing-body combinations, the characteristics of which generally require treatment by introduction of certain simplifying assumptions. The procedure for case 2 is somewhat more complex, but it is valid for wings of arbitrary shape.

C,53. The Browne, Friedman, and Hodes Method for a Triangular Wing Mounted on a Cone, with Apexes Coincident. If the configuration in question is composed of a cone and a triangular wing joined with their apexes superimposed, as shown in Fig. C,53a, the flow field generated is conical and the exact method of solution proposed by Browne, Friedman, and Hodes [*68*] may be utilized. The theory of conical flows has been well advanced, so that the following properties of such fields may be applied to the solution without need of extensive verification of all details.

It is known from the theory of conical flows that when the polar angular coordinates[10] θ and ω (Fig. C,53b) are defined by

$$\theta = \tan^{-1}\frac{z}{y} \qquad \text{and} \qquad \omega = \tan^{-1}\left(\frac{\sqrt{\eta^2+\zeta^2}}{\xi}\right) \quad \text{for } 0 \leqq \omega \leqq \frac{\pi}{2}$$

then the Chaplygin transformation

$$\frac{2s}{1+s^2} = \tan\omega = t \qquad \text{for} \quad 0 \leqq s \leqq 1 \tag{53-1}$$

carries rays in the conical flow field into points in the complex $\hat{\epsilon}$ plane, where

$$\hat{\epsilon} = se^{i\theta} \qquad \text{and} \qquad \hat{\epsilon} = \hat{\epsilon}_x + i\hat{\epsilon}_y$$

Each one of the three functions u, v, and w, corresponding to the velocity components along the three Cartesian axes x, y, z is harmonic in

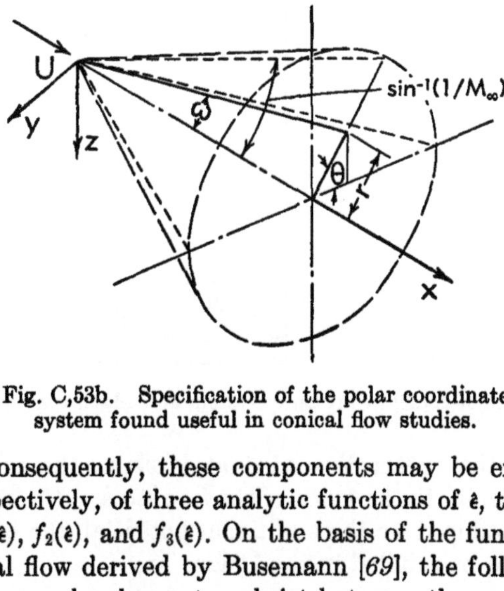

Fig. C,53b. Specification of the polar coordinate system found useful in conical flow studies.

the $\hat{\epsilon}$ plane. Consequently, these components may be expressed as the real parts, respectively, of three analytic functions of $\hat{\epsilon}$, to be designated formally as $f_1(\hat{\epsilon})$, $f_2(\hat{\epsilon})$, and $f_3(\hat{\epsilon})$. On the basis of the fundamental properties of conical flow derived by Busemann [69], the following compatibility relations may be shown to subsist between the analytic functions (see [68,70]):

$$\begin{aligned} \frac{df_2}{d\hat{\epsilon}} &= -\frac{1+\hat{\epsilon}^2}{2\hat{\epsilon}}\frac{df_1}{d\hat{\epsilon}} \\ \frac{df_3}{d\hat{\epsilon}} &= -i\frac{1-\hat{\epsilon}^2}{2\hat{\epsilon}}\frac{df_1}{d\hat{\epsilon}} \end{aligned} \tag{53-2}$$

If ω_0 is then used to denote the semivertex angle of the conical body and if the corresponding t and s values are then specialized to $t_0 = \tan\omega_0$

[10] The normalized coordinate system x, y, z is again used here for reasons of notational simplicity: conversion back to physical dimensions may be effected by means of the transformation in Art. 36.

and $s_0 = t_0/(1 + \sqrt{1 - t_0})$, it may be readily verified that the surface of the configuration under examination will be transformed into the $\hat{\epsilon}$ plane, to be represented by the circle with radius s_0 and the two segments of the real axis lying external to this circle but in juxtaposition to it, and extending radially over the distance running from $|\hat{\epsilon}_x| = s_0$ to $|\hat{\epsilon}_x| = s_1 = t_1/(1 + \sqrt{1 - t_1})$, where t_1 is the semispan of the wing in the original flow field, measured at the axial station where $\xi = 1$. The contour under consideration is depicted in Fig. C,53c. Finally, it may be observed that in this complex plane the circle with radius $s = 1$ represents the Mach cone emanating from the tip of the configuration, within the confines of which any pressure perturbations that are generated by the conical wing-body combination under study must be concentrated.

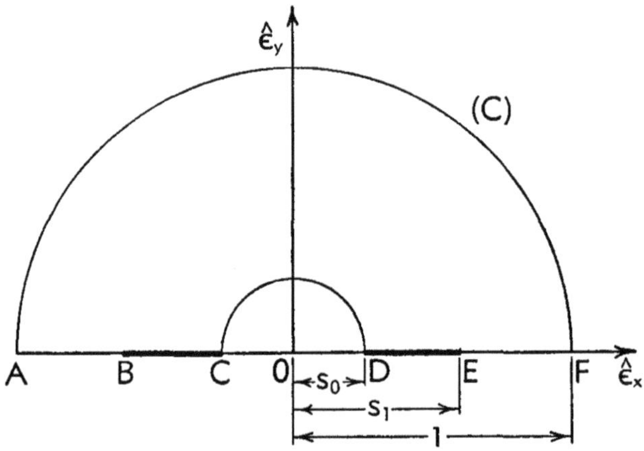

Fig. C,53c. Images of conical wing-body system and Mach cone in complex $\hat{\epsilon}$ plane.

Considering the vector $\mathbf{V} = u\mathbf{i} + v\mathbf{j} + w\mathbf{k}$ as the sum of the component perpendicular to the x axis of the free stream, $(-\alpha_w U/U)\mathbf{k}$, and of the velocities produced by the additional potential, the boundary conditions which must be fulfilled in the $\hat{\epsilon}$ plane are

(i) On the circle of unit radius, $u = v = 0$, and $w = -\alpha_w$

(ii) On the real axis within the region where

$$s_1 \leqq |\hat{\epsilon}_x| \leqq 1,\ u = 0$$

(iii) On the real axis within the region where

$$s_0 \leqq |\hat{\epsilon}_x| \leqq s_1,\ w = 0 \qquad (53\text{-}3)$$

(iv) On the image of the body, i.e. for $s = s_0$, it must be true that $v \cos\theta + w \sin\theta = 0$

Referring back to the second of the relations given as Eq. 53-2, it is evident that part (iii) of the boundary conditions just stated in Eq. 53-3 may be recast into equivalent form as

$$\frac{\partial f_1}{\partial \hat{\epsilon}_y} = \frac{\partial u}{\partial \hat{\epsilon}_y} = 0 \text{ on the real axis within the stretch limited by}$$

$$s_0 \leqq |\hat{\epsilon}_x| \leqq s_1 \quad (53\text{-}3')$$

Following the supposition made by Browne, Friedman, and Hodes, the $f_1(\hat{\epsilon})$ function is sought in the form:

$$f_1(\hat{\epsilon}) = f^{(w)}(\hat{\epsilon}) + f^*(\hat{\epsilon}) \quad (53\text{-}4)$$

where $f^{(w)}(\hat{\epsilon})$ stands for the equivalent $f_1(\hat{\epsilon})$ function describing the flow about a similar isolated extended wing (filled in right across the region otherwise occupied by the body) which is operating in the flow at an angle of attack of γ radians. The value of this angle of attack is not known

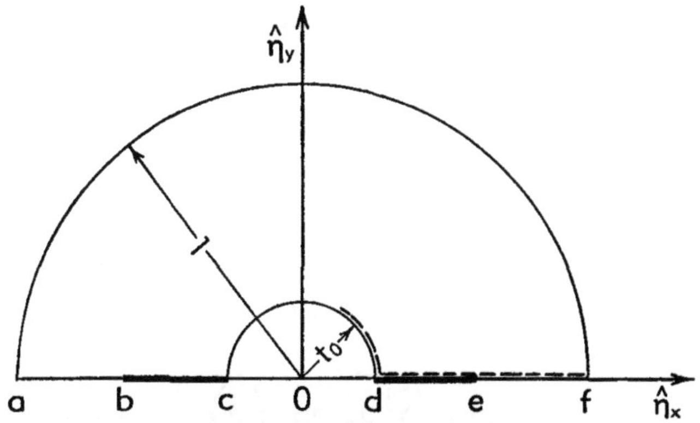

Fig. C,53d. Images of conical wing-body system and Mach cone in complex $\hat{\eta}$ plane, with indication of the path followed in evaluating integral expressions for v^* and w^*.

beforehand, but instead its magnitude must be determined in the subsequent analysis, as well as the $f^*(\hat{\epsilon})$ part of the unknown function, so that the complete set of boundary conditions (Eq. 53-3) may be satisfied exactly.

The desired solution is obtained in most convenient fashion by subjecting the $\hat{\epsilon}$ plane to the homomorphic transformation

$$\frac{1}{\hat{\eta}} = \frac{1}{2}\left(\hat{\epsilon} + \frac{1}{\hat{\epsilon}}\right) \quad (53\text{-}5)$$

where

$$\hat{\eta} = \tau e^{i\sigma} = \hat{\eta}_x + i\hat{\eta}_y$$

This transformation carries the real axis in the $\hat{\epsilon}$ plane back into the real axis in the $\hat{\eta}$ plane and on this axis, according to Eq. 53-1, $\hat{\eta}_x = t$. Corresponding to circle C, there is a circle in the $\hat{\eta}$ plane, having its radius equal to 1, whereas the image of the body can still be considered circular with a radius equal to t_0 if $t_0 \ll 1$ with $\sigma = \theta$ (see Fig. C,53d).

The expression for $f^{(w)}$ in terms of the complex variable $\hat{\eta}$ is known, having been given by Hayes [71]. In fact, the pertinent formula for this wing-alone flow is

$$f^{(w)} = \frac{\gamma t_1}{\mathsf{E}(\sqrt{1 - t_1})} \frac{1}{\sqrt{1 - (\hat{\eta}/t_1)^2}} \tag{53-6}$$

where E denotes, as usual, the complete elliptic integral of the second kind. This flow creates a w component of relative velocity at the boundary of the Mach cone which is equal to γ, while on the surface of the wing $w^{(w)} = 0$.

Under the hypothesis that the u component is to be sought in the form suggested in Eq. 53-4, the boundary conditions stated as Eq. 53-3 reduce to the following analogous ones, which apply for the function f^* in the $\hat{\eta}$ plane:

(i′) On the circle of unit radius, $u^* = v^* = 0$, and $w^* = -\alpha_w - \gamma$
(ii′) On the real axis within the region where $t_1 \leqq |\hat{\eta}_x| \leqq 1$, $u^* = 0$
(iii′) On the real axis within the region where

$$t_0 \leqq |\hat{\eta}_x| \leqq 1, \ \partial u^*/\partial \hat{\eta}_y = 0 \tag{53-7}$$

(iv′) On the image of the body, i.e for $\tau = t_0$, it must be true that

$$v^* \cos\theta + w^* \sin\theta = -v^{(w)} \cos\theta - w^{(w)} \sin\theta$$

The sought function f^* is then assumed to be represented by a series of the form

$$f^* = \sum_m A_{0,m} T_{0,m}\left(\frac{\hat{\eta}}{t_1}\right) \tag{53-8}$$

in which $A_{0,m}$ are real constants which are not assigned a priori, while the functions $T_{0,m}$ are defined according to the scheme

$$T_{0,m}(\mathfrak{k}^*) = \frac{1}{2m}(\mathfrak{k}^{*-m} - \mathfrak{k}^{*m}) \tag{53-9}$$

where

$$\frac{t_1}{\hat{\eta}} = \frac{1}{2}\left(\mathfrak{k}^* + \frac{1}{\mathfrak{k}^*}\right)$$

Expressing $\mathfrak{k}^*$ in polar form as $\mathfrak{k}^* = s^* e^{i\theta^*}$, it follows that the $T_{0,m}$ functions may be given alternatively as

$$T_{0,m} = R_{0,m}\left(\frac{t}{t_1}\right) \cos m\theta^* - \frac{i}{m} S_{0,m}\left(\frac{t}{t_1}\right) \sin m\theta^* \tag{53-9′}$$

wherein the coefficients $R_{0,m}$ and $S_{0,m}$ are defined according to Eq. 53-10.

It may be observed that the first two of the conditions (i′), and the conditions (ii′) and (iii′) of the set listed as Eq. 53-7 have been automati-

$$
\left.\begin{aligned}
R_{0,0} &= -\tanh^{-1}\sqrt{1-\frac{t^2}{t_1^2}}; & S_{0,0} &= 1\\
R_{0,1} &= \left(\frac{t}{t_1}\right)^{-1}\sqrt{1-\frac{t^2}{t_1^2}}; & S_{0,1} &= \frac{t_1}{t}\\
R_{0,2} &= \left(\frac{t}{t_1}\right)^{-2}\sqrt{1-\frac{t^2}{t_1^2}}; & S_{0,2} &= \frac{1-(t^2/t_1^2)}{t^2/t_1^2}\\
R_{0,3} &= \left(\frac{t}{t_1}\right)^{-3}\sqrt{1-\frac{t^2}{t_1^2}}\left[1+\frac{1}{3}\left(1-\frac{t^2}{t_1^2}\right)\right]; & S_{0,3} &= \frac{1+3[1-(t^2/t_1^2)]}{t^3/t_1^3}\\
R_{0,4} &= \left(\frac{t}{t_1}\right)^{-4}\sqrt{1-\frac{t^2}{t_1^2}}\left[1+\left(1-\frac{t^2}{t_1^2}\right)\right]; & S_{0,4} &= \frac{1+6[1-(t^2/t_1^2)]+[1-(t^2/t_1^2)]^2}{t^4/t_1^4}\\
R_{0,5} &= \left(\frac{t}{t_1}\right)^{-5}\sqrt{1-\frac{t^2}{t_1^2}}\left[1+2\left(1-\frac{t^2}{t_1^2}\right)+\frac{1}{5}\left(1-\frac{t^2}{t_1^2}\right)^2\right]; & S_{0,5} &= \frac{1+10[1-(t^2/t_1^2)]+5[1-(t^2/t_1^2)]^2}{t^5/t_1^5}\\
R_{0,6} &= \left(\frac{t}{t_1}\right)^{-6}\sqrt{1-\frac{t^2}{t_1^2}}\left[1+\frac{10}{3}\left(1-\frac{t^2}{t_1^2}\right)+\left(1-\frac{t^2}{t_1^2}\right)^2\right]; & S_{0,6} &= \frac{\begin{array}{r}1+15[1-(t^2/t_1^2)]+15[1-(t^2/t_1^2)]^2\\ +[1-(t^2/t_1^2)]^3\end{array}}{t^6/t_1^6}
\end{aligned}\right\}\quad(53\text{-}10)
$$

cally satisfied by taking the $T_{0,m}$ functions in the form given as Eq. 53-9′. By referring back to the compatibility conditions (Eq. 53-2), the following expressions for the u^* and w^* components of velocity may be constructed

$$\begin{aligned} v^* &= \text{R.P. } f_2(\hat{\epsilon}) = -\text{R.P.} \int_1^{\hat{\eta}} \frac{1}{\hat{\eta}} \frac{df_1}{d\hat{\eta}} \, d\hat{\eta} \\ w^* &= \text{R.P. } f_3(\hat{\epsilon}) = -\text{R.P.} \int_1^{\hat{\eta}} \frac{\sqrt{1-\hat{\eta}^2}}{\hat{\eta}} \frac{df_1}{d\hat{\eta}} \, d\hat{\eta} - \alpha_w - \gamma \end{aligned} \tag{53-2′}$$

where the constants of integration have been chosen so as to satisfy the third of the relations of part (i′), constituting the boundary conditions stated in Eq. 53-7, and where the symbol R.P. denotes that the real part of the so-designated function is meant.

In order to perform the calculations necessary to arrive at the values of v^* and w^* at any arbitrary general point lying on the circle of radius t_0 in the $\hat{\eta}$ plane (corresponding to a generatrix of the conical body in question) it is necessary to carry the integration along the path indicated by the dashed line in Fig. C,53d. For the points lying on the circle which is part of this integration path, the approximation may be resorted to that

$$T_{0,m}\left(\frac{\hat{\eta}}{t_1}\right) = t_1^m T_{0,m}(\hat{\eta}) \tag{53-11}$$

where

$$T_{0,m}(\hat{\eta}) = \frac{1}{2m} (\hat{\epsilon}^{-m} - \hat{\epsilon}^m)$$

After a good deal of manipulation, and by using the expressions for $v^{(w)}$ and $w^{(w)}$ provided by Eq. 53-6 as well as the integrated compatibility conditions represented by Eq. 53-2′, it is found, if $t_0 \ll 1$, that

$$A_{0,2k} = \left\{\frac{(2k+1)(2k-2)!}{2^{4k-3}[(k-1)!]^2}\right\} \left(\frac{t_0}{t_1}\right)^{4k} \frac{\alpha_w t_1}{\Lambda \mathsf{E}} \tag{53-12}$$

where

$$\gamma = -\frac{\alpha_w}{\Lambda}$$

provided

$$\Lambda = 1 - \frac{3}{2}\left(\frac{t_0}{t_1}\right)^4 \frac{t_1}{\mathsf{E}} I_2 - \frac{5}{16}\left(\frac{t_0}{t_1}\right)^8 \frac{t_1}{\mathsf{E}} I_4 - \frac{21}{256}\left(\frac{t_0}{t_1}\right)^{12} \frac{t_6}{\mathsf{E}} I_6 + \cdots$$

and

$$\begin{aligned} I_2 &= -\frac{t_1}{3}\mathsf{K} - \frac{1-2t_1^2}{3t_1}\mathsf{E} \\ I_4 &= \frac{16t_1^3 - 16t_1^2 + 1}{15t_1}\mathsf{E} + \frac{7-8t_1^2}{15}\mathsf{K} \\ I_6 &= \frac{-256t_1^6 + 384t_1^4 - 134t_1^2 + 3}{105t_1}\mathsf{E} + \frac{128t_1^4 - 176t_1^2 + 51}{105} t_1\mathsf{K} \end{aligned}$$

where the complete elliptic integrals of the first and second kind have been denoted, as customary, by K and E, respectively, and where the modulus for each is $\sqrt{1 - t_1^2}$.

In the half plane $\zeta > 0$, the u component of velocity may thus be determined from the formulas:

On the wing

$$u = -\frac{\alpha_w t_1}{\Lambda \mathsf{E}} \frac{1}{\sqrt{1 - (t^2/t_1^2)}} \left[1 - \frac{3}{2}\left(\frac{t_0}{t_1}\right)^4 \left(\frac{t_1^2}{t^2} - 1\right) - \frac{5}{16}\left(\frac{t_0}{t_1}\right)^8 \left(2\frac{t_1^4}{t^4} - 3\frac{t_1^2}{t^2} + 1\right) - \frac{21}{768}\left(\frac{t_0}{t_1}\right)^{12} \left(16\frac{t_1^6}{t^6} - 32\frac{t_1^4}{t^4} + 19\frac{t_1^2}{t^2} - 3\right)\right] \quad (53\text{-}13)$$

On the body

$$u = -\frac{\alpha_w t_1}{\Lambda \mathsf{E}} \left[1 - \left(\frac{t_0}{t_1}\right)^2 \cos 2\theta + \frac{1}{4}\left(\frac{t_0}{t_1}\right)^4 (3 - \cos 4\theta) + \left(\frac{t_0}{t_1}\right)^6 \left(\frac{13}{16}\cos 2\theta - \frac{1}{8}\cos 6\theta\right)\right] \quad (53\text{-}13')$$

Finally, the slope of the lift curve for the complete wing body configuration is given by

$$\frac{d(C_L)_{wb}}{d\alpha_w} = \left[\frac{2\pi t_1}{\beta}\frac{1}{\mathsf{E}}\right] \frac{1 - \frac{8}{3\pi}\left(\frac{t_0}{t_1}\right)^3 + \frac{3}{2}\left(\frac{t_0}{t_1}\right)^4 - \frac{8}{15\pi}\left(\frac{t_0}{t_1}\right)^5 + \cdots}{\Lambda} \quad (53\text{-}14)$$

where it may be recognized that the term within the square brackets is just the slope of the lift curve for the isolated wing when "extended" to fill in the central portion actually covered by the fuselage.

It must be cautioned that this result for the lift curve slope of the configuration, as derived in Eq. 53-14, has been based on the assumption that the pressure coefficients are given by the simple formula

$$C_p = \frac{p - p_\infty}{\frac{1}{2}\rho_\infty U^2} = -2u \quad (53\text{-}15)$$

without taking cognizance of the possible cross terms and squared terms which might arise if a more precise order analysis were to be carried out. This result is also restricted in applicability to those cases for which the parameter $t_0/t_1 \leqq 0.75$, which means that the spread of the conical nose must not be too large a proportion of the width of the wing planform. This hindrance to full universal use of the formulas, as now derived, arises because of the simplifications introduced into the calculation of the v and w components.

The lift curve slope $d(C_L)_{wb}/d\alpha_w$ obtained by evaluation in Eq. 53-14 is shown as a function of t_0/t_1 in Fig. C,53e. Likewise, the corresponding approximate result derived by use of slender body theory is also given in this figure. In both cases the lift was computed by relying on the formula for the pressure coefficient which contains only the single linear term $-2u$. A particularly good fit between the two curves is evident in the vicinity of $t_0/t_1 = 0.75$. Thus one may consider this value to be a convenient dividing line for deciding whether to use slender body theory (at the large values of t_0/t_1) or to use the more precise conical flow theory just presented in order to arrive at the most valid description of the velocity field generated by such a pointed configuration.

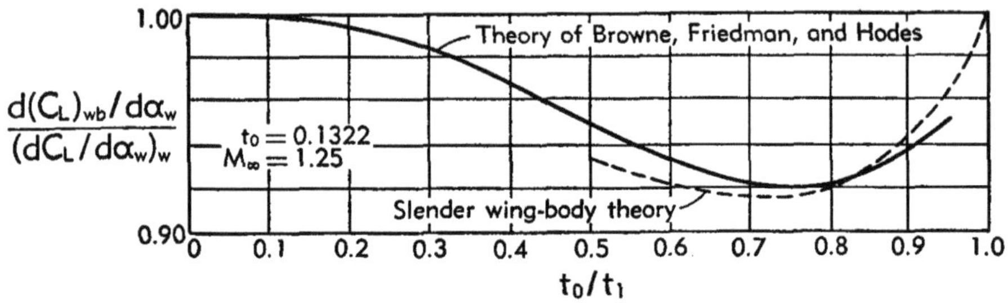

Fig. C,53e. Ratio between the slope of the lift curve for the conical configuration shown in Fig. C,53a and the slope of the lift curve pertaining to an isolated wing of the same shape (extended across the region actually occupied by the fuselage).

C,54. Configuration Composed of a Cylindrical Body Carrying a Wing Having Any Shape for the Leading Edge. Statement of the Problem. In order to attack the problem of calculating the behavior of a configuration made up of a wing of arbitrary planform mounted on a cylindrical body, it is found most apposite to break down the treatment into three subproblems, to be solved in successive stages. In the first of these steps the field of flow about the configuration is determined by applying the same general method that is described in detail in Art. 36 et seq., so that, in consequence, one arrives at expressions for the interference potentials $\varphi_{\mathrm{I}}^{(w)}$ and $\varphi_{\mathrm{I}}^{(b)}$ governing the perturbation flows about the wing and about the body, respectively, just as previously done. These potentials produce the requisite flows which are such as to satisfy the boundary conditions properly everywhere, except in region I, over which there arise some uncompensated (at this stage) potential jumps that may be denoted by

$$\Delta\varphi_{\mathrm{I}}^{(b)} = [\varphi_{\mathrm{I}}^{(b)}]_{\zeta=0^+} - [\varphi_{\mathrm{I}}^{(b)}]_{\zeta=0^-} = 2[\varphi_{\mathrm{I}}^{(b)}]_{\zeta=0^+}$$

where the I region of concern is the portion of the $\zeta = 0$ plane lying between the leading edge of the wing and the Mach lines (generatrices of the corresponding Mach cones) which are tangent to the leading edge, as shown in Fig. C,54.

In the second step, the potential jump $\Delta\varphi_{I}^{(b)}$ is canceled in region I by an additional potential $\varphi_{II}^{(w)}$, corresponding to a suitable distribution of supersonic sources in the same region I. In this manner the boundary conditions corresponding to region I and the wing surface are fulfilled, but the potential $\varphi_{II}^{(w)}$ in turn induces upon the body a field of normal velocities represented by the radial derivative $[\partial\varphi_{II}^{(w)}/\partial r]_{r=r_b}$.

In the third step of the calculation, one proceeds to negate the normal velocities, $[\partial\varphi_{II}^{(w)}/\partial r]_{r=r_b}$, by adding a potential $\varphi_{II}^{(b)}$. If this latter potential is expressed in an analogous fashion to the form used for $\varphi^{(b)}$ in Eq. 37-2, then no further normal velocities are induced back on the wing. In addition, it will be shown shortly how one may then set up a further compensating wing-interference potential, to be designated by $\varphi_{III}^{(w)}$, which

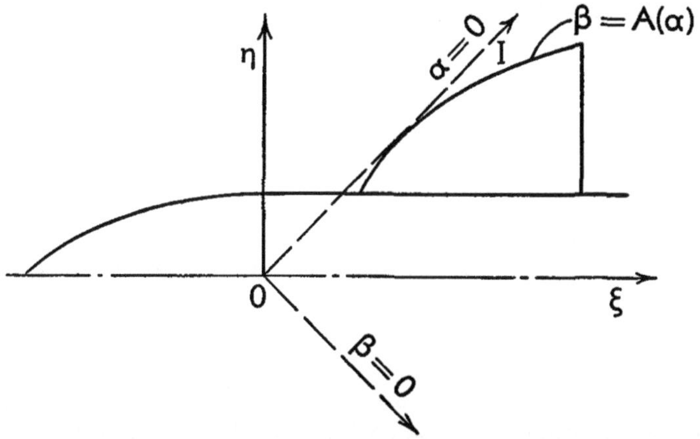

Fig. C,54. Typical configuration possessing a wing with a subsonic leading edge. (Although the wing planform may have any shape whatsoever, the shoulder adjacent to the body is taken to be supersonic.)

has the faculty of annulling the $\Delta\varphi_{II}^{(b)}$ jumps that again crop up in region I. Furthermore, the $\varphi_{III}^{(w)}$ potential also has the property of not altering the normal components of the velocity field surrounding the body, on the basis of which the body interference potential $\varphi_{II}^{(b)}$ was devised in the first place. It is thus established in the present case that the complete interference potential ought to be taken in the explicit form of

$$\varphi_i = \varphi_{I}^{(w)} + \varphi_{II}^{(w)} + \varphi_{III}^{(w)} + \varphi_{I}^{(b)} + \varphi_{II}^{(b)}$$

It should be emphasized at the outset that the determination of the $\varphi_{I}^{(w)}$ potential cannot be carried out by following the procedure used in Art. 38, because the leading edge is subsonic in the present instance. The desired solution to this first step in the 3-stage process under analysis may be obtained most readily by lumping together the $\varphi_{I}^{(w)}$ interference potential and the φ_w wing-alone potential, and by seeking their solution simultaneously by utilizing one of the two well-known methods developed in [72,73] for treating such wings. Inasmuch as the methods of Evvard and Ward are explained in sufficient detail in Vol. VI, no further exposition is felt to be necessary here, but it is worth calling attention to the fact that, once the potential $\varphi_w + \varphi_I^{(w)}$ has been found, it is most expedient to compute the values of

$$\left[\frac{\partial(\varphi_w + \varphi_I^{(w)})}{\partial r}\right]_{r=r_b}$$

by means of Taylor series expansions, analogous to those employed in Art. 45. The preliminary job of determining the first stage body interference potential $\varphi_I^{(b)}$ may also be worked out in a straightforward manner by referring back directly to the procedure recommended in Art. 40, provided one merely keeps in mind the comments to follow in Art. 57.

C,55. Determination of the Second Stage Wing Interference Potential. In carrying out the determination of the wing interference potential required in the second stage of the method now under examination, set up a system of oblique coordinates based on the characteristic lines in the $\zeta = 0$ plane, so that the rectangular system ξ, η becomes linked to the "characteristic" system α, β through the relations (see Fig. C,54)

$$\alpha = \xi - \eta \qquad \text{and} \qquad \beta = \xi + \eta \tag{55-1}$$

Then let $\Delta\varphi_I^{(b)}$ denote the discontinuities existing in the first stage body interference potential $\varphi_I^{(b)}$ in region I, and let them be described in terms of a function of the α, β location parameters, as

$$\Delta\varphi_I^{(b)} = -2Q(\alpha, \beta) \tag{55-2}$$

where, of course, $Q(\alpha, \beta)$ is a known function, in consequence of the fact that step 1 of this procedure has been assumed completed.

Furthermore, let $q(\alpha, \beta)$ denote the sought strength of supersonic sources which are to be distributed over region I, in the plane $\zeta = 0^+$ in such a way as to create the desired sort of flow (to be specified below) that is to be governed by a corresponding $\varphi_{II}^{(w)}$ potential. Then the condition that $\Delta(\varphi_I^{(b)} + \varphi_{II}^{(w)}) = 0$ becomes translated into the integral relation

$$\int_0^\alpha \frac{d\alpha'}{\sqrt{\alpha - \alpha'}} \int_{A(\alpha')}^\beta \frac{q(\alpha', \beta')}{\sqrt{\beta - \beta'}}\, d\beta' = Q(\alpha, \beta) \tag{55-3}$$

where the equation of the leading edge of the wing in question has been denoted by

$$\beta = A(\alpha) \tag{55-4}$$

If the leading edge of the wing happens to be a straight line, then in this special case

$$\beta = \beta_0 + \frac{1+m}{1-m}\alpha$$

where $\beta_0 = 2\xi_0$, and where ξ_0 is the abscissa location of the point marking the intersection of the leading edge with the side of the fuselage, while $m = \tan\delta$ is the slope of the leading edge.

Let

$$\int_{A(\alpha')}^{\beta} \frac{q(\alpha', \beta')}{\sqrt{\beta - \beta'}}\, d\beta' = \psi(\alpha', \beta) \tag{55-5}$$

so that Eq. 55-3 becomes

$$\int_0^{\alpha} \frac{\psi(\alpha', \beta)d\alpha'}{\sqrt{\alpha - \alpha'}} = Q(\alpha, \beta) \tag{55-3'}$$

The solution of this Abel integral equation is

$$\psi(\alpha, \beta) = \frac{1}{\pi}\int_0^{\alpha} \frac{\dot{Q}_{\alpha'}(\alpha', \beta)}{\sqrt{\alpha - \alpha'}}\, d\alpha' \tag{55-6}$$

where $\dot{Q}_\alpha(\alpha, \beta) = \partial Q/\partial\alpha$ and provided that $Q(0, \beta) = 0$.

Consequently Eq. 55-5 gives

$$\int_{A(\alpha)}^{\beta} \frac{q(\alpha, \beta')}{\sqrt{\beta - \beta'}}\, d\beta' = \psi(\alpha, \beta) \tag{55-5'}$$

from which it follows that

$$\begin{aligned} q(\alpha, \beta) &= \frac{1}{\pi}\frac{\partial}{\partial\beta}\int_{A(\alpha)}^{\beta} \frac{\psi(\alpha, \beta')}{\sqrt{\beta - \beta'}}\, d\beta' \\ &= \frac{1}{\pi^2}\frac{1}{\sqrt{\beta - A(\alpha)}}\int_0^{\alpha} \frac{\dot{Q}_{\alpha'}(\alpha', \beta)}{\sqrt{\alpha - \alpha'}}\, d\alpha' \\ &\quad + \frac{1}{\pi^2}\int_{A(\alpha)}^{\beta} \frac{d\beta'}{\sqrt{\beta - \beta'}}\int_0^{\alpha} \frac{\ddot{Q}_{\alpha',\beta'}(\alpha', \beta')}{\sqrt{\alpha - \alpha'}}\, d\alpha' \end{aligned} \tag{55-7}$$

where $\ddot{Q}_{\alpha',\beta'} = \partial^2 Q/\partial\alpha'\partial\beta'$.

This solution (Eq. 55-7) constitutes the formal answer to the problem originally posed of how to determine the second stage wing interference potential $\varphi_{\text{II}}^{(w)}$. It is most fitting for practical reasons, however, to further manipulate this stylized version of the solution in order to render it more amenable to routine computation.

C,56. Conversion of the Formal Solution for $\varphi_{\text{II}}^{(w)}$ into More Tractable Form. As a first step in the reduction process for rearranging Eq. 55-7 into a form which is more suitable to the needs of practical numerical analyses, the expression for the supersonic source strengths may be separated into two contributions having a different analytic character in the region I as follows:

$$q(\alpha, \beta) = \frac{1}{\pi^2} \frac{1}{\sqrt{\beta - A(\alpha)}} \int_0^\alpha \frac{\dot{Q}_{\alpha'}(\alpha', \beta) - \dot{Q}_{\alpha'}(\alpha', A(\alpha))}{\sqrt{\alpha - \alpha'}} \, d\alpha'$$

$$+ \frac{1}{\pi^2} \int_{A(\alpha)}^\beta \frac{d\beta'}{\sqrt{\beta - \beta'}} \int_0^\alpha \frac{\ddot{Q}_{\alpha',\beta'}(\alpha', \beta')}{\sqrt{\alpha - \alpha'}} \, d\alpha'$$

$$- \frac{1}{\pi^2} \frac{1}{\sqrt{\beta - A(\alpha)}} \int_0^\alpha \frac{\dot{Q}_{\alpha'}(\alpha', A(\alpha))}{\sqrt{\alpha - \alpha'}} \, d\alpha'$$

$$= q_{\mathrm{I}}(\alpha, \beta) - \frac{q_{\mathrm{II}}(\alpha)}{\sqrt{\beta - A(\alpha)}} \tag{55-7'}$$

In order to bring about the solution of the first of these contributions, $q_{\mathrm{I}}(\alpha, \beta)$, to the supersonic source strength function q, it should be recognized right away that it is permissible to assume the known distribution of potential jumps in region I in the form

$$\dot{Q}_{\alpha'}(\alpha', \beta) = \sum_n a_n(\beta) \alpha'^n \tag{56-1}$$

because $\dot{Q}_{\alpha'}$ is regular over the interval of integration involved, and because, in addition, $\dot{Q}_{\alpha'}(0, \beta) = 0$ for $\alpha' = 0$. If the truth of this observation is conceded then it follows that

$$\dot{Q}_{\alpha'}[\alpha', A(\alpha)] = \sum_n a_n[A(\alpha)] \alpha'^n$$

and likewise

$$\ddot{Q}_{\alpha',\beta'} = \sum_n \dot{a}_n(\beta') \alpha'^n \tag{56-1'}$$

where $\dot{a}_n(\beta) = da_n/d\beta$.

The integral expression in the definition of $q_{\mathrm{I}}(\alpha, \beta)$ may thus be evaluated as

$$\int_0^\alpha \frac{\dot{Q}_{\alpha'}(\alpha', \beta) - \dot{Q}_{\alpha'}(\alpha', A(\alpha'))}{\sqrt{\alpha - \alpha'}} \, d\alpha' = 2 \sum_n c_n \{a_n(\beta) - a_n[A(\alpha)]\} \alpha^{n+\frac{1}{2}}$$

where

$$c_n = 1 - \binom{n}{1}\frac{1}{3} + \binom{n}{2}\frac{1}{5} - \cdots (-1)^n \frac{1}{2n+1}$$

provided $\binom{g}{h}$ is taken to represent the binomial coefficients

$$\frac{g!}{h!(g-h)!}$$

It is immediately apparent upon substitution of these power series into Eq. 55-7′ that

$$\left.\begin{aligned} q_{\mathrm{I}}(\alpha, \beta) &= \frac{2}{\pi^2} \sum_n c_n \frac{a_n(\beta) - a_n[A(\alpha)]}{\sqrt{\beta - A(\alpha)}} \alpha^{n+\frac{1}{2}} \\ &\qquad + \frac{2}{\pi^2} \sum_n c_n \alpha^{n+\frac{1}{2}} \int_{A(\alpha)}^{\beta} \frac{\dot{a}_n(\beta') d\beta'}{\sqrt{\beta - \beta'}} \\ q_{\mathrm{II}}(\alpha) &= \frac{2}{\pi^2} \sum_n c_n a_n[A(\alpha)] \alpha^{n+\frac{1}{2}} \end{aligned}\right\} \qquad (56\text{-}2)$$

The function $q_{\mathrm{I}}(\alpha, \beta)$ turns out to be regular throughout region I, therefore, and it vanishes both on the Mach line $\alpha = 0$, which forms the upstream boundary of region I, as well as on the leading edge of the wing. Now, since

$$q(\alpha, \beta) = -\frac{1}{2\pi} \left[\frac{\partial \varphi_{\mathrm{II}}^{(w)}}{\partial \zeta} \right]_{\zeta=0} \qquad (56\text{-}3)$$

it is clear that the function $q_{\mathrm{I}}(\alpha, \beta)$ defines that part of $[\partial \varphi_{\mathrm{II}}^{(w)}/\partial \zeta]_{\zeta=0}$ which is regular in region I and consequently it may be represented by means of a trigonometric series of the sort employed in Eq. 38-1. The potential $\varphi_{\mathrm{II}}^{(w)}$ itself, as well as the corresponding radial velocity components induced at the surface of the body, may thus be obtained in a manner which is entirely analogous to the procedure illustrated in full detail in Art. 38.

Contrariwise, the second contribution to the supersonic source strength function, $q_{\mathrm{II}}(\alpha)/\sqrt{\beta - A(\alpha)}$, exhibits a row of singularities along the leading edge of the wing, although it is still true that this function also vanishes on the Mach line emanating from the point of intersection of the leading edge with the side of the fuselage. Because of the nonanalytic nature of this function on the leading edge, it is not possible to represent this second portion of the supersonic source strength by means of a trigonometric series, as for the regular portion.

It is possible in this case, however, to set down at once the required expression for the potential which describes the flow generated by the set of supersonic sources of strength $q_{\mathrm{II}}(\alpha)/\sqrt{\beta - A(\alpha)}$. One has, in fact, under these conditions, that

$$\varphi_{\mathrm{II}}^{\prime(w)} = \int_0^{\alpha^*(\alpha,\beta,\zeta)} q_{\mathrm{II}}(\alpha') d\alpha' \int_{A(\alpha')}^{\beta - [\zeta^2/(\alpha - \alpha')]} \frac{d\beta'}{\sqrt{\beta' - A(\alpha')}\sqrt{(\alpha - \alpha')(\beta - \beta') - \zeta^2}}$$

where $\alpha^*(\alpha, \beta, \zeta)$ defines the α' coordinate of the point of intersection of the leading edge of the wing with the forward-facing Mach cone emanating from the particular point at which $\varphi_{\rm II}'^{(w)}$ is to be calculated. The coordinate in question is thus obtained by evaluation in the formula

$$\beta - \frac{\zeta^2}{\alpha - \alpha'} = A(\alpha')$$

or, more simply, from

$$\beta\alpha - \zeta^2 = \beta\alpha' + (\alpha - \alpha')A(\alpha')$$

Since the latter of the two indicated quadratures in the above expression for $\varphi_{\rm II}'^{(w)}$ may be evaluated at once to give

$$\int_{A(\alpha')}^{\beta - [\zeta^2/(\alpha - \alpha')]} \frac{d\beta'}{\sqrt{\beta' - A(\alpha')}\,\sqrt{(\alpha - \alpha')(\beta - \beta') - \zeta^2}} = \frac{-\pi}{\sqrt{\alpha - \alpha'}}$$

it follows that

$$\varphi_{\rm II}'^{(w)} = -\pi \int_0^{\alpha^*(\alpha,\beta,\zeta)} \frac{q_{\rm II}(\alpha')}{\sqrt{\alpha - \alpha'}}\, d\alpha' \tag{56-4}$$

and consequently

$$\left.\begin{aligned}
\frac{\partial \varphi_{\rm II}'^{(w)}}{\partial \alpha} &= \frac{\pi q_{\rm II}(\alpha^*)}{\sqrt{\alpha - \alpha'}}\left(1 - \frac{d\alpha^*}{d\alpha}\right) - \pi \int_0^{\alpha^*} \frac{\dot q_{\rm II}(\alpha')d\alpha'}{\sqrt{\alpha - \alpha'}} \\
\frac{\partial \varphi_{\rm II}'^{(w)}}{\partial \beta} &= \frac{-\pi q_{\rm II}(\alpha^*)}{\sqrt{\alpha - \alpha^*}}\frac{d\alpha^*}{d\beta} \\
\frac{\partial \varphi_{\rm II}'^{(w)}}{\partial \zeta} &= \frac{-\pi q_{\rm II}(\alpha^*)}{\sqrt{\alpha - \alpha^*}}\frac{\partial\alpha^*}{\partial\zeta}
\end{aligned}\right\} \tag{56-5}$$

Because of the geometric relationships connecting the rectangular and characteristic coordinate systems it is evidently true that

$$\begin{aligned}
\frac{\partial \varphi_{\rm II}'^{(w)}}{\partial \xi} &= \frac{\partial \varphi_{\rm II}'^{(w)}}{\partial \alpha} + \frac{\partial \varphi_{\rm II}'^{(w)}}{\partial \beta} \\
\frac{\partial \varphi_{\rm II}'^{(w)}}{\partial \eta} &= \frac{\partial \varphi_{\rm II}'^{(w)}}{\partial \beta} - \frac{\partial \varphi_{\rm II}'^{(w)}}{\partial \alpha}
\end{aligned}$$

so that the set of derivatives given as Eq. 56-5 may be inserted without further ado into these latter formulas, in order to arrive at the expression $[\partial\varphi_{\rm II}'^{(w)}/\partial r]_{r=r_b}$ which describes the pattern of distribution of the radial components of induced velocity at the surface of the body, produced by the second part of the contribution to the total supersonic source strengths $q(\alpha, \beta)$.

C,57. Determination of Second Stage Body Interference Potential and Third Stage Wing Interference Potential. The total set of radial components of velocity induced on the body is composed not only of the contributions to the induced normal velocity components which are generated by the system of supersonic sources distributed over region I, but also of those generated by the sources lying in the symmetrically located I′ position on the other side of the body. The complete assemblage of such normal components should then be represented by a trigonometric series of the kind employed in Eq. 40-2, and the corresponding second stage of the body interference potential $\varphi_{II}^{(b)}$ may then be determined by reapplication of the method presented in Art. 40. In order to assure that the step-by-step process ends at this stage, however, it is necessary to proceed according to the following plan.

To make the analysis less cumbersome in the early phases of the discussion, it is assumed that the leading edge of the wing starts off at the side of the fuselage with a tangent that is only gently swept, so that this edge of the wing may be considered supersonic, at least over a small region adjacent to the body.

Furthermore, let H_0 denote the point of tangency where the leading edge, as now described, meets the zeroeth characteristic line, $\alpha = 0$, and let K_0 denote the point of intersection with the side of the body (the generatrix of which, lying in the plane $\zeta = 0$, is $r = r_b$) of a line through H_0 which is parallel with the $\beta = 0$ Mach line. In addition, let L_0 denote the point situated on the axis of the body at which it is met by a line passing through K_0 and lying parallel to the other characteristic axis, $\alpha = 0$. In similar fashion, let H_1 denote the outboard point of intersection of the Mach line through L_0 and K_0 with the leading edge of the wing, while K_1, L_1, K_2, L_2 are taken to represent the analogous points that are obtained in proceeding with the fabrication of a network of intersecting Mach lines in the identical way that the intermeshing lattice was begun when the similar initial locations H_0, K_0, etc. were defined. The geometric interdependencies between these points are illustrated in Fig. C,57.

By reference to this figure, it may be readily appreciated that that portion of the cylindrical fuselage which is contained between the two Mach cones emanating from the points L_0 and L_1 can only be reacted upon by the normal velocity components of the flow generated by means of the system of supersonic sources that is arrayed over the I_a portion of region I. On the other hand it is equally clear that the sequence of supersonic sources to be located along the segment L_0L_1 of the ξ axis, to produce the potential $\varphi_{II}^{(b)}$, can only exert its influence back upon that part of the region I which lies downstream of the line L_0H_1.

Similarly, the distribution of supersonic sources which is to be arrayed along the L_1L_2 segment of the ξ axis is linked directly to the strength and arrangement of the set of supersonic sources spread out over the I_b and I_c portions of the region I, but these sources located on the L_1L_2 segment cannot in turn react in any manner back upon regions I_b and I_c.

Because of the way the $\varphi_{II}^{(w)}$ potential was constructed it is evident that the discontinuities in the $\varphi_{II}^{(b)}$ potential have been properly eliminated in the regions I_a and I_b, so that there is no question of how to specify the $\varphi_{III}^{(w)}$ potential in this region, for it may now be taken for granted that $\varphi_{III}^{(w)} = 0$ there. In region I_c, however, the potential $\varphi_{III}^{(w)}$ must be assigned in such a way as to successfully counteract the jumps in the $\varphi_{II}^{(b)}$ potential resulting from the interference reactions generated by the corresponding sources distributed along the strip of axis lying between L_0 and L_1. By proceeding in this manner, therefore, it is evident that the boundary conditions are completely and exactly satisfied over that region of the

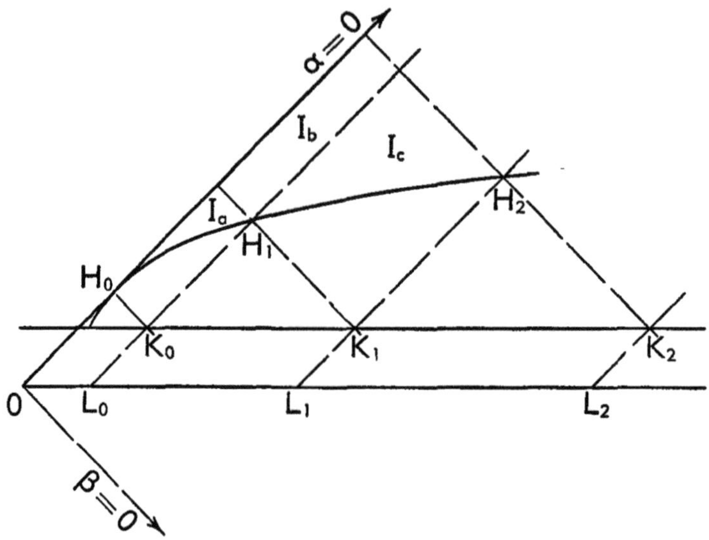

Fig. C,57. Network of intersecting Mach lines dividing the region of flow into various tessellations, which are treated differently in successive steps of the terminating iteration process employed in computing interference effects.

wing which lies upstream of the Mach lines H_2K_2 and K_2L_2. The process may be continued in similar fashion to include those areas lying farther downstream that may be of interest in any given case, by mere repetition of the same steps already used in arriving at the φ_i potential which applies in this forward area.

In actual practice it is found that the second and third steps in this method of calculating φ_i may be coalesced into just one, inasmuch as it is most convenient to compute $\varphi_{II}^{(w)}$ and $\varphi_{III}^{(w)}$ simultaneously when satisfying the boundary conditions in region I_c, and, likewise, in the other successive contiguous areas lying farther downstream, i.e. the working calculational procedure might just as well be set up to eliminate the jumps in the $\varphi_I^{(b)}$ and $\varphi_{II}^{(b)}$ potentials at one and the same time.

C,58. Modification of the Method to Remove the Restriction that the Shoulder of the Wing Must Have a Supersonic Leading Edge.

The condition imposed on the wing leading edge should not be considered too restrictive since, in any case, a very slight modification of the edge is sufficient to fulfill the above condition. Thus, if the wing has the

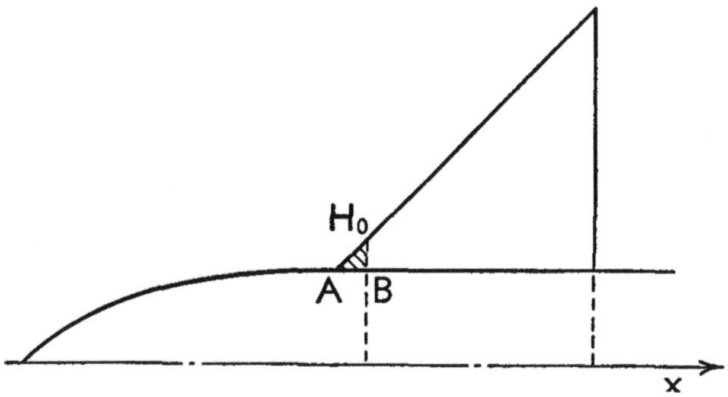

Fig. C,58a. Region to be cut from a wing having its entire leading edge swept subsonically, for application of the method given in Art. 57.

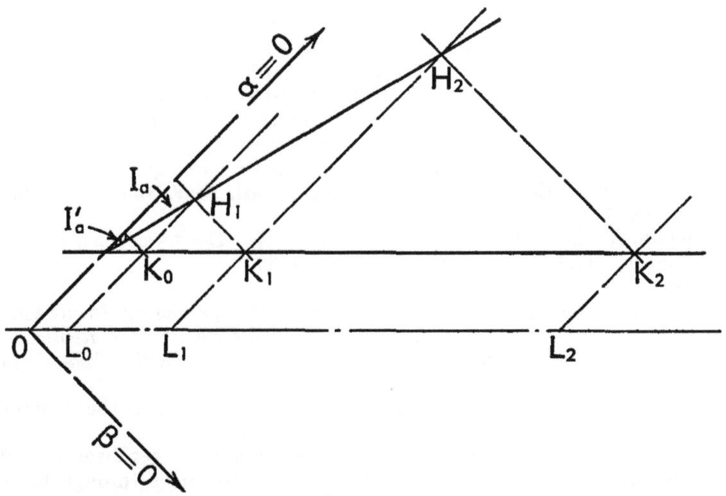

Fig. C,58b. Minute residual area in plane of wing where flow determined by application of the method developed in Art. 57 (without modification of planform in shoulder region) does not agree precisely with physical boundary conditions, for a configuration having a swept leading edge which is subsonic throughout.

shape illustrated in the plan view of Fig. C,58a, it will be sufficient to remove the portion shown in this figure to make the method given above directly applicable, and it does not appear probable that this modified wing-body system exhibits aerodynamic characteristics appreciably different from those of the original system. On the other hand, if the same procedure is applied without making any modification in the wing, the boundary conditions remain fulfilled except in a region, shown in Fig. C,58b, which may be made as small as one pleases.

C,59. Cited References.

1. Wieselsberger, C. Der Einfluss von eingebauten Motorgondeln auf die Luftkräfte eines Tragflügels (The influence of integral engine nacelles on the lift of a wing). *Intern. Eng. Congr. Paper 203*, Tokyo, 1929.
2. Durand, W. F. *Aerodynamic Theory*. Springer, Berlin, 1934–36.
3. Kaden, H. Aufwicklung einer unstabilen Unstetigkeitsfläche (The rolling-up of an unstable vortex sheet). *Ing. Arch. 2*, 140–168 (1931). Available in English transl. as *Roy. Aircraft Establishment Library Transl. 403*, Farnborough, England.
   See also:
   Ferrari, C. Determinazione dei nuclei vorticosi risultanti dall'arrotolamento della superficie di discontinuità dietro a un'ala, e calcolo della forza viva da essi indotta (Determination of the trailing vortex sheet behind a wing, and calculation of the amount of kinetic energy induced by these vortices). *Mem. reale accad. Italia*, 249–267 (1934).
4. Betz, A. Verhalten von Wirbelsystemen (Behavior of vortex systems). *Z. angew. Math. u. Mech. 12*, 164–174 (1932).
5. Multhopp, H. Die Berechnung des Abwindes hinter Tragflügeln (The calculation of downwash behind lifting surfaces). *Luftfahrtforschung 15*, 463–467 (1938).
6. Weissinger, S. Über die Auftriebsverteilung von Pfeilflügeln (The lift distribution of swept-back wings). *Zentrale für wissenschaftliches Berichtswesen der Luftfahrtforschung des Generalluftzeumeisters* (*ZWB*), *Bericht 1553*, Feb. 1942. Available in English transl. as *NACA Tech. Mem. 1120*, Mar. 1947.
7. De Young, J., and Harper, C. W. Theoretical symmetric span loading at subsonic speeds for wings having arbitrary planform. *NACA Rept. 921*, 1948.
8. Multhopp, H. Die Berechnung der Auftriebsverteilung von Tragflügeln (Calculation of the lift distribution on wings). *Luftfahrtforschung 15*, 153–169 (1938). Also available in English as *Brit. R.T.P. Transl. 2392*, 1938 or *Consolidated-Vultee Rept. 1277*, 1938.
9. Silverstein, A., Katzoff, S., and Bullivant, W. K. Downwash and wake behind plain and flapped airfoils. *NACA Rept. 651*, 1939.
   See also:
   Silverstein, A., and Katzoff, S. Design charts for predicting downwash angles and wake characteristics behind plain and flapped wings. *NACA Rept. 648*, 1939.
10. Schlichting, H. *Über das ebene Windschattenproblem* (*On the Two-Dimensional Wake Problem*). *Dissertation*, Göttingen, 1930.
11. Multhopp, H. Zur Aerodynamik des Flugzeugrumpfes (On the aerodynamics of the fuselage). *Luftfahrtforschung 18*, 52–66 (1941).
12. von Kármán, T., and Trefftz, E. Potentialströmung um gegebene Tragflächenquerschnitte (Potential flow about arbitrary cross-sections through lifting surfaces). *Z. Flugtech. u. Motorluftschiffahrt 9*, 111 (1918).
13. Theodorsen, T. Theory of wing sections of arbitrary shape. *NACA Rept. 411*, 229 (1932).
14. Munk, M. The aerodynamic forces on airship hulls. *NACA Rept. 184*, 1923.
15. Lamb, H. *Hydrodynamics*, 6th. ed. Cambridge Univ. Press, 1932.
16. Zahm, F. Flow and force equations for a body revolving in a fluid. *NACA Rept. 323*, 1928.
    See also:
    Vandrey, F. Abschätzung des Rumpfeinflusses auf das Längsmoment einer Flugzeuges (Evaluation of the body's contribution to the pitching moment of an airplane). *Jahrbuch deut. Luftfahrtforschung*, 367–370 (1940).
17. Jacobs, E. N., and Ward, K. E. Interference of wing and fuselage from tests of 209 combinations in the NACA variable density tunnel. *NACA Rept. 540*, 1935.
18. Lennertz, J. Beitrag zur theoretischen Behandlung des gegenzeitigen Einflusses von Tragfläche und Rumpf (Contribution to the theoretical treatment of

the mutual interference of wings and fuselage). *Abhandl. Aerody. Inst. Aachen 8,* 3–30 (1928). Also *Z. Flugtech. u. Motorluftschiffahrt 8,* 11–13 (1927).
19. Bamber, M. J., and House, R. O. Wind-tunnel investigation of effect of yaw on lateral-stability characteristics. II: Rectangular NACA 23012 wing with a circular fuselage and a fin. *NACA Tech. Note 730,* 1939.
20. Ward, G. N. Supersonic flow past slender pointed bodies. *Quart. J. Mech. and Appl. Math. 2, Part 1,* Mar. 1949.
See also:
Stocker, P. M. Supersonic flow past bodies of revolution with thin wings of small aspect ratio. *Aeronaut. Quart. 3, Part I,* May 1951.
21. Ackeret, J. Über die Druckverteilung an schräg angestromten Rotationskörpern bei Unterschallgeschwindigkeit (Concerning the pressure distribution on bodies of revolution placed obliquely to the impinging subsonic flow). Anniversary volume in honor of Prof. M. Panetti, published jointly by *Assoc. Ital. di Aerotecnica, Assoc. Termotecnica Ital., and Assoc. Tecnica Automobile,* 1950. Also *Aerotecnica (Special Number) 31,* 11–19 (1951).
22. Krienes, K. Die elliptische Tragfläche auf potential theoretischer Grundlage (The elliptic wing based on the potential theory). *Z. angew. Math. u. Mech. 20,* 65–88 (1940). Transl. as *NACA Tech. Mem. 971,* Mar. 1941.
23. Robinson, A., and Young, A. D. Note on the application of the linearised theory for compressible flow to transonic speeds. *College of Aeronautics, Cranfield, Rept. 2,* 1947.
24. Robinson, A. Aerofoil theory of a flat plate delta wing at supersonic speeds. *Roy. Aircraft Establishment Rept. Aero. 2151,* Farnborough, England, 1946.
25. Stewart, H. J. The lift of a delta wing at supersonic speeds. *Quart. Appl. Math. 4,* 246–254 (1946).
26. Adams, M. C., and Sears, W. R. Slender-body theory—review and extension. *J. Aeronaut. Sci. 20,* 85–98 (1953).
27. Spreiter, J. R. The aerodynamic forces on slender plane—and cruciform—wing and body combinations. *NACA Rept. 962, 1950.*
28. Root, L. E. Empennage design with single and multiple vertical surfaces. *J. Aeronaut. Sci. 6,* 353 (1939).
29. Ruden, P. Das NACA-Profil in der Nachbarschaft rechteckiger Strahl—und Windschatten-Profile (The behavior of an NACA airfoil section operating in the vicinity of a two-dimensional jet or wake having a variable velocity profile). *Jahrbuch deut. Luftfahrtforschung,* 114–138 (1939).
30. Weinig, F. *Aerodynamik der Luftschraube (Aerodynamics of the Propeller).* Springer, Berlin, 1940. Available in English transl. from *Air Documents Division, Intelligence Department, U.S. Air Materiel Command, Dayton,* 366–369 (1947).
31. Maxwell, J. C. *A Treatise on Electricity and Magnetism,* Vol. 2, Art. 704–705. Oxford Univ. Press, 1873.
32. Ferrari, C. Sul problema dell'elica con vento laterale (Note on the problem posed by a propeller operating in a side-wind). *Atti accad. sci. Torino.* Alternate title: Sul problema dell'elica in deriva. *Atti accad. sci. Torino 68,* 60–65 (1928). Also *Aerotecnica 8,* 171–176 (1928).
See also:
Ferrari, C. Il problema dell'elica con vento laterale (Study of the question concerning a propeller operating in yawed attitude). *Atti accad. sci. Torino 75,* 2–31 (1940).
33. Muggia, A. Sul calcolo dell'interferenza elica-ala (On the calculation of propeller-wing interference). *Atti. accad. nazl. Lincei, Rome, 11,* 53–57 (1951).
34. Rossner, R. Die Ermittlung der Auftriebsverteilung am elliptischen Tragflügel bei beliebig gegebener Verwindung unter Zugrundelegung des Princips der Einflussfunktion (On investigation of the lift distribution supported by an elliptic wing possessing arbitrary twist from the standpoint of the principle of the influence function). *Jahrbuch deut. Luftfahrtforschung,* 1940.
35. Possio, C. Ricerche sperimentali sull'interferenza elica-ala (Experimental investigation of propeller-wing interference). *Aerotecnica 26,* 73–77 (1946).
36. Weil, J., and Sleeman, W., Jr. Prediction of the effects of propeller operation on the static longitudinal stability of single-engine tractor monoplanes with flaps retracted. *NACA Rept. 941,* 1949.

37. Ferrari, C. Some experiments on the slipstream effect. *Rendiconti Sperimentali del Laboratorio di Aeronautica del Reale Istituto Superiore di Ingegneria di Torino, Seconda Serie, Tests on Models "Torino 30, 31, 32, and E1,"* 38–55 (1936). Also *Aerotecnica 15*, 979–1036 (1935). English transl. in *J. Roy. Aeronaut. Soc. 40*, 388–404 (1936), under title: *Test Reports of the Aeronautical Laboratory of the R. Istituto Superiore di Ingegneria of Turin.* Also *NACA Tech. Mem. 820*, 1937.
38. Lagerstrom, P., and Graham, M. Downwash and sidewash induced by three-dimensional lifting wings in supersonic flow. *Douglas Aircraft Co. Rept. SM-13007*, Apr. 1947.
39. Martin, J. The calculation of downwash behind wings of arbitrary plan form at supersonic speeds. *NACA Tech. Note 2135*, 1950.
40. Lomax, H., Sluder, L., and Heaslet, M. A. The calculation of downwash behind supersonic wings with an application to triangular plan forms. *NACA Rept. 957*, 1950.
41. Robinson, A., and Hunter-Tod, J. Bound and trailing vortices in the linearized theory of supersonic flow, and the downwash in the wake of a delta wing. *College of Aeronautics, Cranfield, Rept. 10*, 1947.
42. Hadamard, J. *Le Problème de Cauchy et les Equations aux Dérivées Partielles Linéaires Hyperboliques.* Hermann, Paris, 1932. Yale Univ. Press transl., *Lectures on Cauchy's Problem in Linear Partial Differential Equations*, 1923.
43. Lagerstrom, P., and Graham, M. Methods for calculating the flow in the Trefftz-plane behind supersonic wings. *J. Aeronaut. Sci. 18*, 179–191 (1951).
44. Mirels, H., and Haefeli, R. Line-vortex theory for calculation of supersonic downwash. *NACA Rept. 983*, 1950.
45. Ferrari, C. Calcolo della velocità indotta da un'ala in corrente supersonica linearizzata (Induced velocities generated by a wing operating in supersonic flow, on the basis of linear theory). Encomium volume of *Aerotecnica*, in honor of Prof. G. A. Crocco, *33*, 65–69 (1953). English transl. *Cornell Aeronaut. Lab. Rept. CAL-59*, July 1953.
46. Haefeli, R., Mirels, H., and Cummings, J. Charts for estimating downwash behind rectangular, trapezoidal, and triangular wings at supersonic speeds. *NACA Tech. Note 2141*, 1950.
47. Lagerstrom, P. A., and Van Dyke, M. D. General considerations about planar and non-planar lifting systems. *Douglas Aircraft Co. Rept. SM-13432*, Santa Monica, June 1949.
48. Ferrari, C. Interference between wing and body at supersonic speeds. Theory and numerical application. *J. Aeronaut. Sci. 15*, 317–336 (1948).
49. Ferrari, C. Interference between wing and body at supersonic speeds. Note on wind-tunnel results and addendum to calculations. *J. Aeronaut. Sci. 16*, 542–546 (1949).
50. Ferrari, C. Interference between wing and body at supersonic speeds. Part III: Numerical application of the linear theory. *Cornell Aeronaut. Lab. Bumblebee Rept. CAL/CM-463*, Apr. 1951.
51. Ferrari, C. Sulla determinazione di alcuni tipi di campi di corrente ipersonora. (On the determination of certain basic types of supersonic flow fields.) *Rend. reale accad. nazl. Lincei, Rome, VII (6), Ser. VIII*, 1949. Also available as *NACA Tech. Mem. 1381*, Nov. 1954.
52. Nielsen, J. N., and Pitts, W. C. Wing-body interference at supersonic speeds with an application to combinations with rectangular wings. *NACA Tech. Note 2677*, 1952.
53. Goodman, T. R. Some simplifications of Ferrari's theory of wing-body interference at supersonic speeds. *Cornell Aeronaut. Lab. Bumblebee Rept. CAL/CM-687*, Dec. 1954.
54. Cramer, R. H. Interference between wing and body at supersonic speeds. Theoretical and experimental determination of pressures on the body. *J. Aeronaut. Sci. 18*, 629–632 (1951).
55. Morikawa, G. K. The wing-body problem for linearized supersonic flow. *Calif. Inst. Technol. Jet. Propul. Lab. Progress Rept. 4-116, ORDCIT Project*, Dec. 1949.
56. Morikawa, G. K. A non-planar boundary problem for the wave equation. *Quart. Appl. Math. 10 (2)*, 129–140 (1952).

57. Pitts, W. C., Nielsen, J. N., and Gionfriddo, M. P. Comparison between theory and experiment for interference pressure field between wing and body at supersonic speeds. *NACA Tech. Note 3128*, 1954.
58. Cramer, R. H. Interference between wing and body at supersonic speeds. Part V: Phase one wind tunnel tests correlated with the linear theory. *Cornell Aeronaut. Lab. Bumblebee Rept. CAL/CM-597*, Dec. 1950.
59. Dye, F. E., Jr. A comparison of pressure predicted by exact and approximate theories with some experimental results on an ogival-nosed body at a Mach number of 2.0. *Cornell Aeronaut. Lab. Bumblebee Rept. CAL/CF-1723*, Dec. 1951.
60. Berler, I., and Nichols, S. Interference between wing and body at supersonic speeds. Part VI: Data report on pressure distribution tests of wing-body interference models at Mach number 2.0, phase two tests of June 1949. *Cornell Aeronaut. Lab. Bumblebee Rept. CAL/CF*-1569, May 1951.
61. Berler, I., and Nichols, S. Interference between wing and body at supersonic speeds. Part VII: Data report on pressure distribution tests of wing-body interference models at Mach number 2.0, phase three tests of August, 1949. *Cornell Aeronaut. Lab. Bumblebee Rept. CAL/CF*-1570, June 1951.
62. Nichols, S. Interference between wing and body at supersonic speeds. Part VIII: Data report on pressure distribution tests of wing-body interference models at Mach number 2.0, phase four tests of March 1950. *Cornell Aeronaut. Lab. Bumblebee Rept. CAL/CF-1571*, July 1951.
63. Dye, F. E., Jr. Interference between wing and body at supersonic speeds. Part IX: Data report on pressure distribution tests of wing-body interference models at Mach number 2.0, phase five tests of August 1951. *Cornell Aeronaut. Lab. Bumblebee Rept. CAL/CF-1684*, Dec. 1951.
64. Kirkby, S., and Robinson, A. Wing-body interference at supersonic speeds. *College of Aeronautics, Cranfield, Rept. 7*, Apr. 1947.
65. Flax, A. H. Integral relations in the linearized theory of wing-body interference. *J. Aeronaut. Sci. 20*, 483–490 (1953).
66. Lawrence, H. R., and Flax, A. H. Wing-body interference at subsonic and supersonic speeds. Survey and new developments. *J. Aeronaut. Sci. 21*, 289–324, 328 (1954).
67. Cramer, R. H. Wing-body interference at supersonic speeds. Recent experimental results and their bearing on advances in theory. *Proc. U.S. Navy Symposium on Aeroballistics, NAVORD Rept. 3146*, May 1952. Also available as Cornell Aeronaut. Lab. special publication.
68. Browne, S. H., Friedman, L., and Hodes, I. A wing-body problem in a supersonic conical flow. *J. Aeronaut. Sci. 15*, 443–452 (1948).
69. Busemann, A. Infinitesimale kegelige Überschallströmung. (Infinitesimal conical supersonic flow.) *Jahrbuch deut. Luftfahrtforschung*, 455–470 (1942) and *D. A. L. Jahrbuch 7B*, 105–122 (1943). Reprinted in: Carrier, G. F. (ed.) *Foundations of High-Speed Aerodynamics*. Dover, 1951. (Also available as *NACA Tech. Mem. 1100*, 1947.)
70. Stewart, H. J. The lift of a delta wing at supersonic speeds. *Quart. Appl. Math. 4 (3)*, 246–254 (1946).
71. Hayes, W. D. *Linearized Supersonic Flow. Thesis*, Calif. Inst. Technol., 1947. Also available as *North Amer. Aerophys. Lab. Rept. 222*, June 1947.
72. Evvard, J. C. Use of source distribution for evaluating theoretical aerodynamics of thin finite wings at supersonic speeds. *NACA Rept. 951*, 1950.
73. Ward, G. N. Supersonic flow past thin wings. I: General theory. *Quart. J. Mech. and Appl. Math. 2, Part 2*, 136–152 (1949).

SECTION D

PROPELLERS FOR HIGH SPEED FLIGHT

C. B. SMITH

D,1. Introduction. The high speed propeller as now envisioned is a simple airfoil type propeller having thin blade sections which operate at supersonic velocities. For reasons which are herein explained, such propellers minimize the loss in efficiency normally associated with flight Mach numbers above 0.8, and are in general applicable to high speed, long range, and general purpose aircraft in which a high propulsive efficiency is of relative importance.

The aerodynamic design and analysis of such propellers involves one of the simplest applications of the theory of supersonic flow. In principle, the procedure follows the conventional propeller strip-type analysis except for the use of supersonic section characteristics, the absence of inflow at supersonic flight speeds, and the discard of the concept of a limiting tip speed in favor of optimizing the pitch angle.

In the following sections the basic elements in the aerodynamic design of such propellers are outlined and a few of the special solutions presented.

D,2. Simple Blade Element Theory. Consider the forces acting on a single blade element such as AA in Fig. D,2a. Irrespective of the velocity at which this section is operating, the resultant force dF acting on the element is

$$dF = \frac{dL}{\cos \epsilon} \tag{2-1}$$

where

$$\epsilon = \cot^{-1} \frac{L}{D} \tag{2-2}$$

The thrustwise component of this resultant is

$$dT = dF \cos (\phi + \epsilon) = \frac{dL}{\cos \epsilon} \cos (\phi + \epsilon) \tag{2-3}$$

while the torquewise component is

$$dQ = dF \sin (\phi + \epsilon) = \frac{dL}{\cos \epsilon} \sin (\phi + \epsilon) \tag{2-4}$$

The propulsive efficiency of the two-dimensional blade element is then

$$\eta_s = \frac{U}{\omega_r}\frac{dT}{dQ} = \tan\phi\,\frac{dT}{dQ} \tag{2-5}$$

Inflow, actually present at all subsonic flight velocities, is here neglected.

Substituting Eq. 2-3 and 2-4 into Eq. 2-5 one obtains

$$\eta_s = \tan\phi\,\frac{\cos(\phi+\epsilon)}{\sin(\phi+\epsilon)} = \tan\phi\cot(\phi+\epsilon) \tag{2-6}$$

Expanding the cotangent term and substituting Eq. 2-2 for ϵ gives as a final form

$$\eta_s = \frac{L/D - \tan\phi}{L/D + \cot\phi} \tag{2-7}$$

Eq. 2-7 is entirely general, applying to blade elements operating at both subsonic and supersonic velocities. It is dictated by geometry and limited only by the assumption of zero inflow.

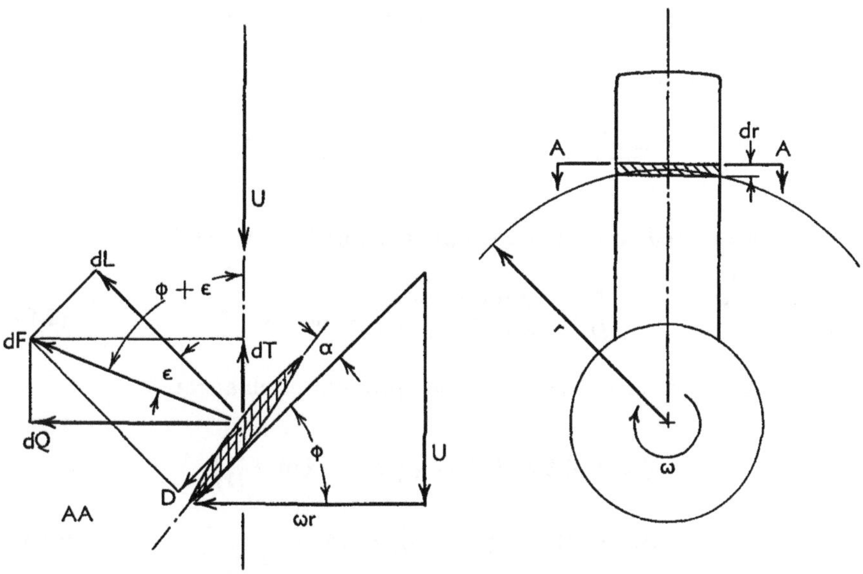

Fig. D,2a. Resolution of aerodynamic forces on two-dimensional blade element.

Eq. 2-7 indicates that blade element efficiency is fully determined by two simple quantities: (1) ϕ, the effective pitch angle at which the section is operating, and (2) L/D, the lift-drag ratio of the airfoil section. The propulsive efficiency of the two-dimensional blade element is defined by Eq. 2-7 and is plotted in Fig. D,2b for representative values of ϕ and L/D. In Fig. D,3a, values of L/D for typical supersonic sections are shown.

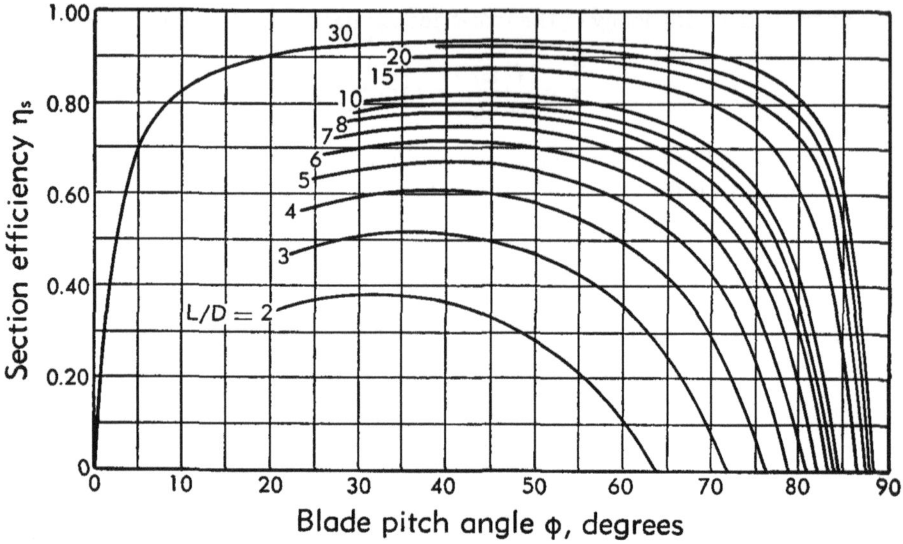

Fig. D,2b. Variation of blade element efficiency with pitch angle for various sections $(L/D)_s$.

Differentiating Eq. 2-7 with respect to ϕ, one obtains

$$\frac{\partial \eta_s}{\partial \phi} = \frac{-(L/D + \tan \phi) \sec^2 \phi + (L/D - \tan \phi) \csc^2 \phi}{(L/D + \cot \phi)^2} \tag{2-8}$$

When equated to zero and simplified, Eq. 2-8 reduces to

$$\frac{L}{D} = \frac{2 \sin \phi \cos \phi}{\cos^2 \phi - \sin^2 \phi} = \tan 2\phi \tag{2-9}$$

Solving for ϕ, one obtains for the optimum pitch angle

$$\phi_{\text{opt}} = \tfrac{1}{2} \tan^{-1} \frac{L}{D} = \tfrac{1}{2}\left(90° - \cot^{-1} \frac{L}{D}\right)$$

$$\phi_{\text{opt}} = 45° - \tfrac{1}{2} \cot^{-1} \frac{L}{D} = 45° - \frac{\epsilon}{2} \tag{2-10}$$

Substituting Eq. 2-10 into Eq. 2-7 yields the blade element efficiency at the optimum pitch angle

$$(\eta_s)_{\max} = \frac{\dfrac{1}{\epsilon} - \tan\left(45° - \dfrac{\epsilon}{2}\right)}{\dfrac{1}{\epsilon} + \cot\left(45° - \dfrac{\epsilon}{2}\right)} = \frac{\dfrac{1}{\epsilon} - \dfrac{1 - (\epsilon/2)}{1 + (\epsilon/2)}}{\dfrac{1}{\epsilon} + \dfrac{1 + (\epsilon/2)}{1 + (\epsilon/2)}}$$

Simplifying and neglecting terms of the order of ϵ^2

$$(\eta_s)_{max} = \left[\frac{1 - (\epsilon/2)}{1 + (\epsilon/2)}\right]^2 \tag{2-11}$$

which may also be written

$$(\eta_s)_{max} = \left[\frac{2(L/D) - 1}{2(L/D) + 1}\right]^2 \tag{2-12}$$

Eq. 2-7, 2-10, and 2-12 define the important features of blade element efficiency. The general approach to the design of high speed propellers can be illustrated in terms of these simple relations. The reasoning proceeds as follows.

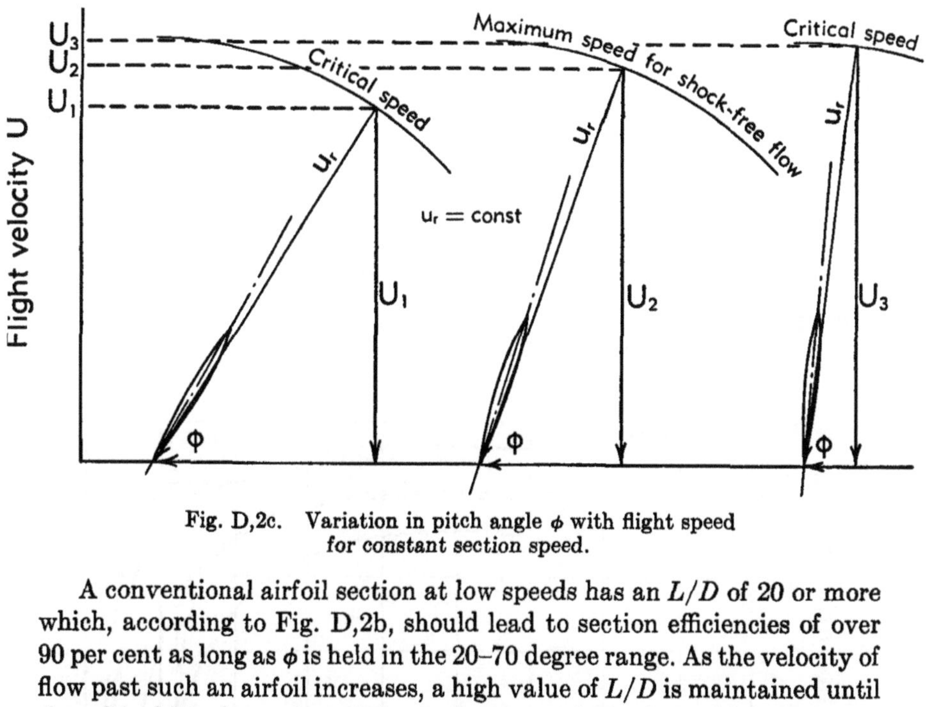

Fig. D,2c. Variation in pitch angle ϕ with flight speed for constant section speed.

A conventional airfoil section at low speeds has an L/D of 20 or more which, according to Fig. D,2b, should lead to section efficiencies of over 90 per cent as long as ϕ is held in the 20–70 degree range. As the velocity of flow past such an airfoil increases, a high value of L/D is maintained until the critical Mach number of the section is reached, after which the maximum L/D drops rapidly and low efficiencies result. The use of sweepback slightly delays but does not avoid this limit. Thus far the use of significant amounts of sweep has appeared structurally unfeasable.

It is general practice in the design of low speed propellers to limit the section speeds to values below their critical value. The fall-off in L/D is thus avoided and high efficiencies obtained; but, if the resultant section speed is kept less than its critical value, then, as flight speed increases, the pitch angle ϕ necessarily becomes larger and larger (see Fig. D,2c).

At flight Mach numbers up to about 0.8 the loss in efficiency due to high pitch angles ϕ is small and, in any event, less than that resulting

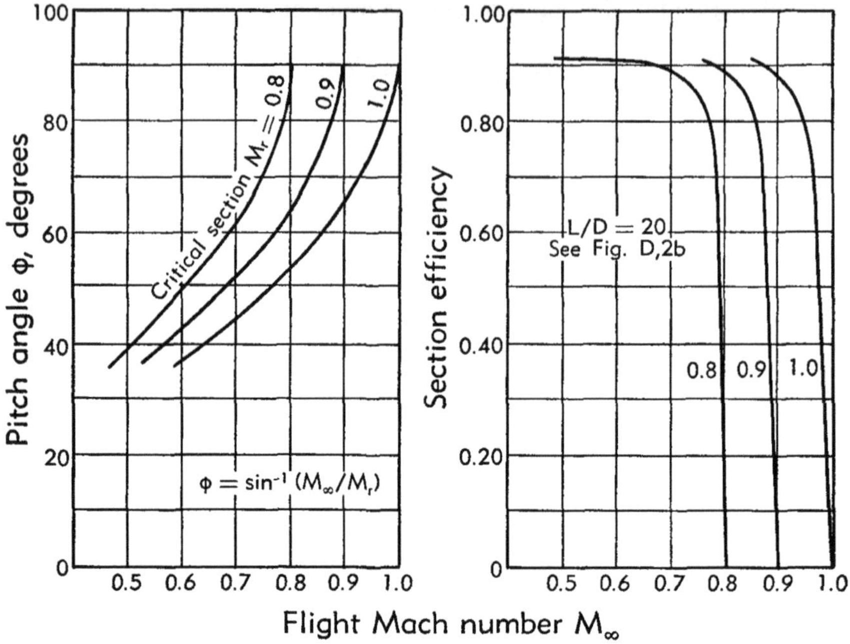

Fig. D,2d. Pitch angle vs. M_∞ for various limiting section speeds (left). Blade element efficiency vs. M_∞ for various limiting section speeds (right).

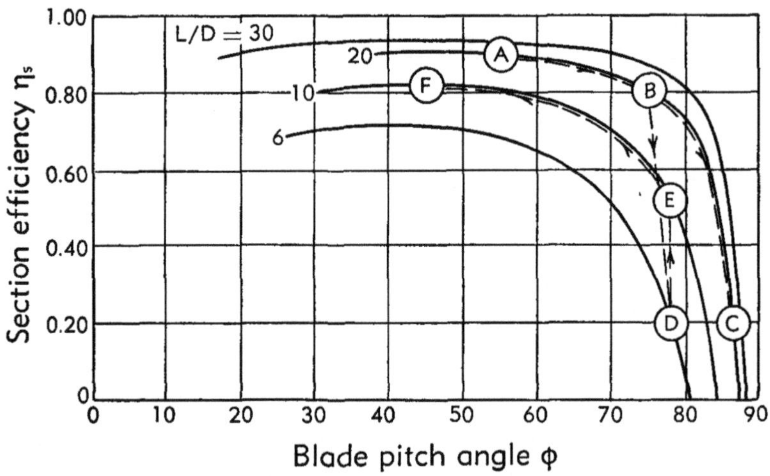

Fig. D,2e. Transition from subsonic to supersonic propeller. A representative section of usual low speed propeller operates in region *A*. Such a section may reach its critical speed at about *B*. Further increases in flight speeds bring a conventional propeller to *C* if section speeds are kept below the critical by reducing rpm (or to *D* if the critical speed is exceeded). Thinning the blade brings efficiencies up to region *E*. Further increases in rpm, such as to optimize pitch angle, bring section efficiencies into the range of region *F*. The supersonic propeller operates in this region.

when the critical section speed is exceeded. Thus in designing propellers for relatively low flight speeds, section speeds should be limited if maximum efficiency is desired.

At flight Mach numbers above 0.8, however, the loss from the excessive pitch angles resulting from attempts to limit the section speeds becomes large. In Fig. D,2d, for example, ϕ vs. flight Mach number is plotted for several critical section Mach numbers; and in the accompanying figure the corresponding blade element efficiencies are plotted (L/D assumed equal to 20). These results illustrate that, at high flight speeds, good section efficiencies cannot be obtained if section speeds are limited to the critical speeds of the usual propeller sections. It is thus clear that at flight Mach numbers above about 0.8 the concept of a limiting tip speed must be discarded and the pitch angle reduced. This is the essential change in philosophy involved in the design of high speed propellers.

Once it becomes more favorable to optimize ϕ than to limit section speed, a radical change takes place in the configuration of the propeller. The rpm increases by a factor of up to 2, resultant blade element speeds go supersonic along the blade, and the diameters required to absorb a given power are radically reduced. This transition from the essentially subsonic to the essentially supersonic propeller is traced out on the blade element efficiency diagram in Fig. D,2e.

D,3. Aerodynamic Forces on the Blade Element. In order to calculate the thrust and torque loads on the blade elements from Eq. 2-3 and 2-4 and the resulting section efficiencies from Eq. 2-7, the aerodynamic forces on the blade elements must be known. On an airfoil type propeller intended for use at flight Mach numbers of 0.9 or above, the blade elements will run at supersonic velocities, as shown in the previous article. Therefore the two-dimensional results of the small perturbation theory (Sec. A and VI,C) may in most cases be applied.

The lift on a two-dimensional section as given by the linear small perturbation theory is

$$C_L \cong C_N = \frac{4\alpha}{\beta} \tag{3-1}$$

where

α = angle of attack in radians
$\beta = \sqrt{M_r^2 - 1}$
M_r = resultant section Mach number

On the basis of the assumptions of the simple theory, the lift as given by Eq. 3-1 is independent of airfoil section.

The drag as given by the small perturbation theory is

$$C_D = C_L\alpha + \frac{K(t/c)^2}{\beta} + \frac{K(\bar{a}/c)^2}{\beta} + C_{D_f} \tag{3-2}$$

where

K = airfoil section shape constant = 4 for symmetrical double-wedge section and $\frac{16}{3}$ for biconvex section

t/c = section thickness ratio

c = airfoil chord

$\bar{a}$ = maximum displacement of mean camber line

C_{D_f} = skin friction drag coefficient

Since camber offers no advantage, the third term in Eq. 3-2 will normally be zero.

The lift-drag ratio required for Eq. 2-3, 2-4, and 2-6 is therefore given by

$$\frac{L}{D} = \frac{4\alpha/\beta}{(4\alpha^2/\beta) + [K(t/c)^2/\beta] + C_{D_f}}$$

$$= \frac{4\alpha}{4\alpha^2 + K(t/c)^2 + \beta C_{D_f}} \qquad (3\text{-}3)$$

Differentiating Eq. 3-3 with respect to α and equating to zero, one obtains for the angle of attack for $(L/D)_{max}$

$$\alpha_{(L/D)_{max}} = \tfrac{1}{2}\sqrt{K(t/c)^2 + \beta C_{D_f}} \qquad (3\text{-}4)$$

Substituting Eq. 3-4 into 3-3 and simplifying

$$\left(\frac{L}{D}\right)_{max} = \frac{1}{\sqrt{K(t/c)^2 + \beta C_{D_f}}} \qquad (3\text{-}5)$$

which is the maximum lift-drag ratio in terms of the section thickness ratio and skin friction coefficient. This expression may also be written

$$(L/D)_{max} = \frac{1}{2\alpha_{(L/D)_{max}}} \qquad (3\text{-}6)$$

or

$$\alpha_{(L/D)_{max}} = \frac{1}{2}\frac{1}{(L/D)_{max}} = \frac{\epsilon_{min}}{2} \qquad (3\text{-}7)$$

The optimum airfoil section for use near a blade tip is that having minimum drag for a given torsional stiffness, and, for the inboard section, that having minimum drag for a given bending stiffness. The biconvex section very nearly satisfies both of these conditions—for both solid and shell type construction. For this section, $K = \frac{16}{3}$.

The results as given by Eq. 3-3 to 3-7 are plotted in Fig. D,3a for a family of biconvex airfoils of different thickness ratio. The noted effects of altitude and section chord are related to the skin friction drag (laminar) and are of secondary importance.

The propulsive efficiency of any given blade element may be determined by calculating its L/D using Eq. 3-3 or Fig. D,3a if applicable, and

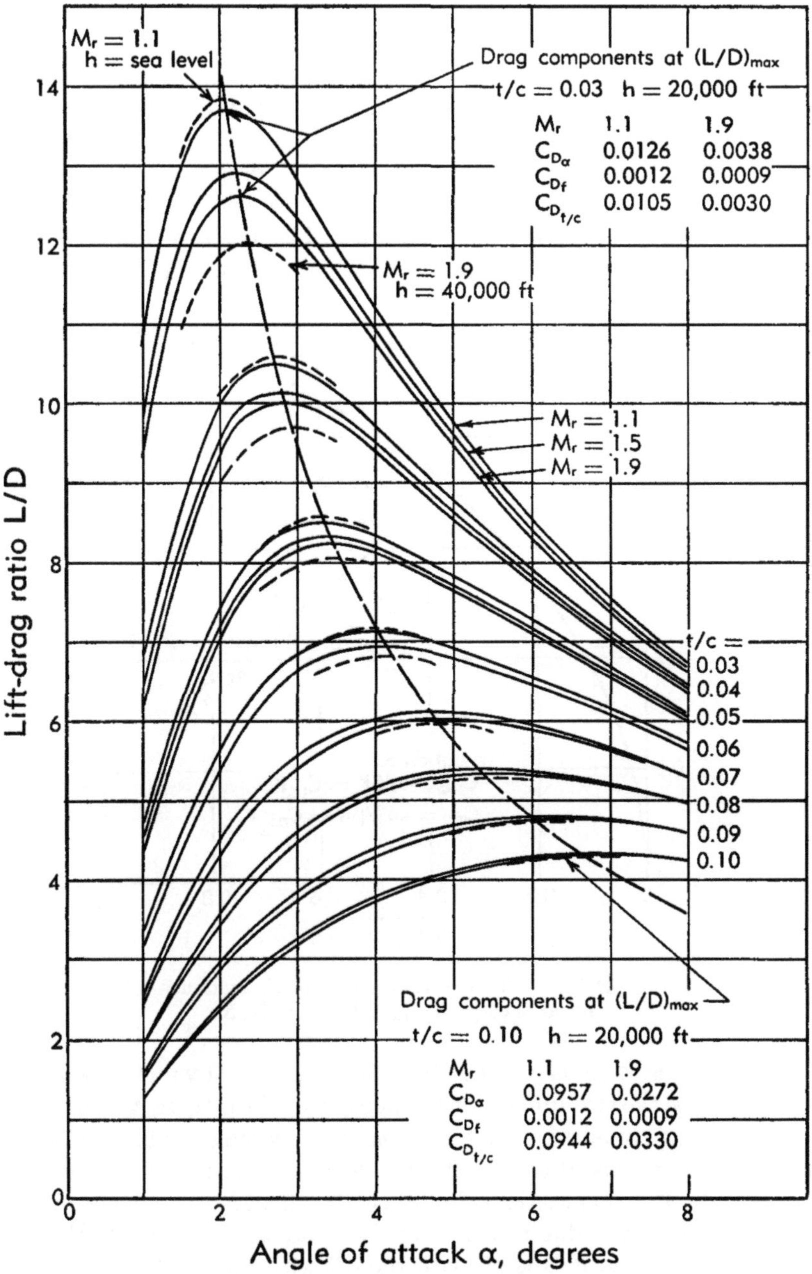

Fig. D,3a. $(L/D)_s$ for two-dimensional biconvex airfoil section (18-in. chord)—linear theory.

substituting into Eq. 2-6. The necessity for the use of thin sections is apparent.

Blade elements cut by the Mach cone from the tip of the blade do not display two-dimensional characteristics. The average lift, and drag due to lift, of such sections as given by the linear theory is $\frac{3}{4}$ the two-dimensional lift (see Sec. A); and the average wave drag, i.e. drag due to thickness, is equal to the two-dimensional wave drag. The tip-correction factor to the lift of the individual tip section is shown in Fig. D,3b. A similar section-by-section correction factor has not been worked out for the wave drag

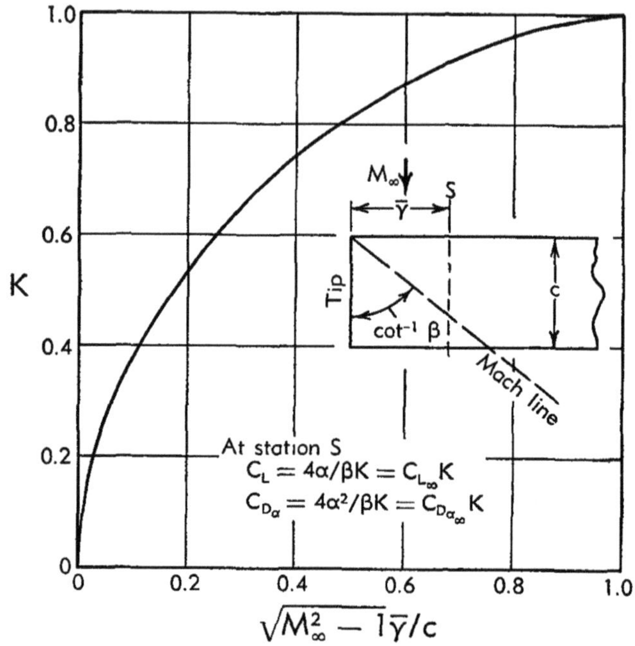

Fig. D,3b. Lift correction factor for tip sections—linear theory. $M_\infty \geqq 1.2$; $C_L = C_{L\infty}K$.

but, since the average is equal to the two-dimensional value, it is general practice to use the two-dimensional value for all of the tip sections.

The average lift-drag ratio for the tip of a rectangular blade is shown in Fig. D,3c. In general the tip effect reduces the section efficiency of the tip elements. For thin sections, however, the effect is small, since most of the resultant drag is due to the dragwise component of the lift $C_L\alpha$ (see Eq. 3-12), which itself falls off with the lift with the result that the lift-drag ratio (Eq. 3-13) is not greatly affected. The tip loss could theoretically be eliminated by raking the tip to a point.

Certain limitations must be recognized in using the steady state, small perturbation theory to predict propeller section characteristics. In visual-

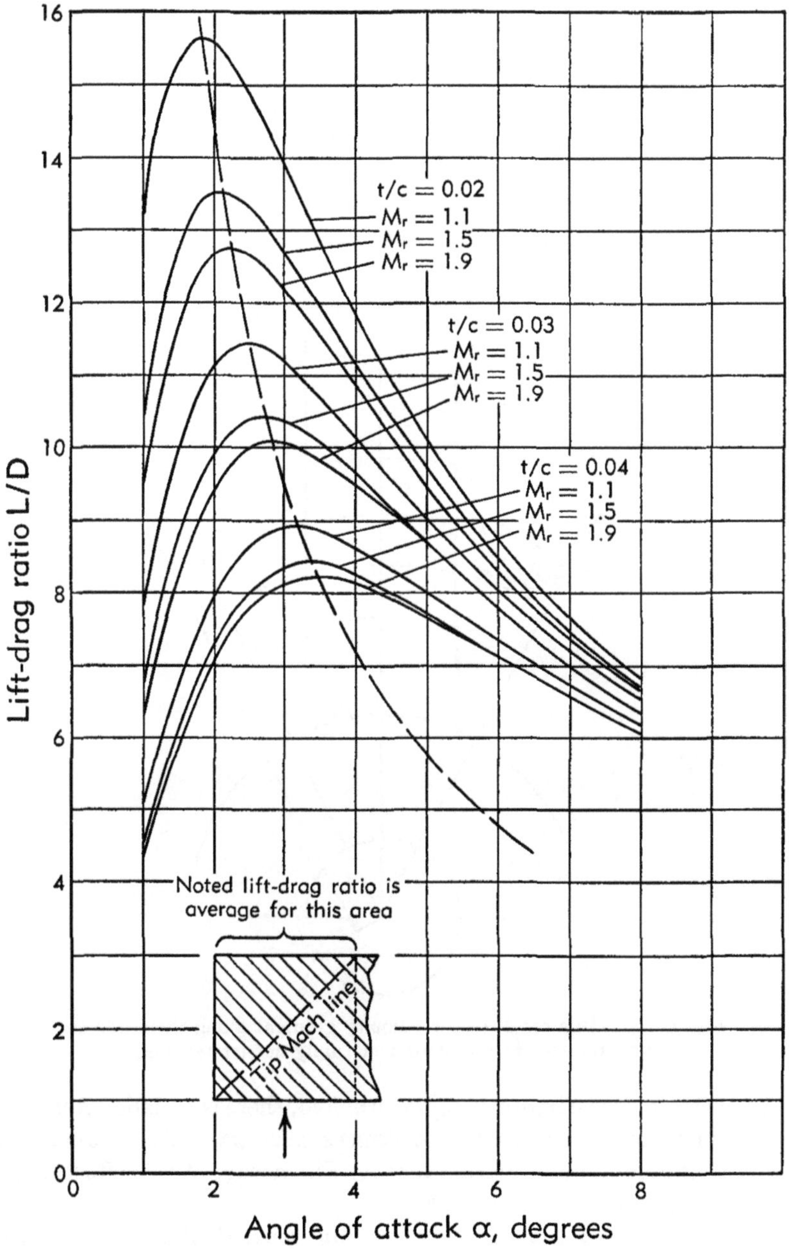

Fig. D,3c. Lift-drag ratio for tip of rectangular blade. 6-in. chord; $h = 20{,}000$ ft.

izing the flow over the blade elements as steady, one is assuming his reference axes to be fixed on the blade. But, as indicated in Fig. D,3d, a velocity gradient exists along the blade. As a result the flow relative to the blade axes is not irrotational and, strictly speaking, a velocity potential does not exist. In applying the steady state, small perturbation theory to this case, one in effect adds to the usual small perturbation assumptions the assumption that the free stream velocity increments above and below u_r at A and D are negligible relative to u_r.

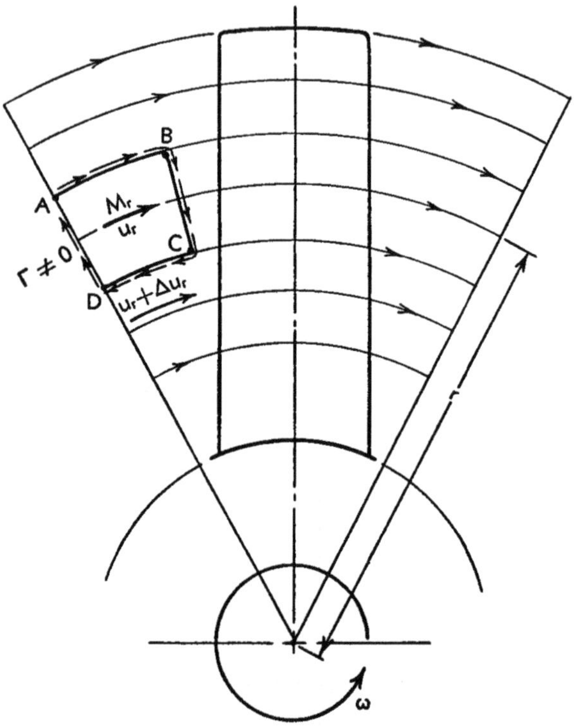

Fig. D,3d. As a result of the velocity gradient along the blade the flow relative to the blade is actually not irrotational.

In addition to the irrotationality problem, there is a blade interference effect which under certain circumstances may partially invalidate the results of the small perturbation theory. The nature of this effect may be qualitatively described as follows.

The disturbance caused by a two-dimensional blade element as it moves through the air at supersonic velocity is confined to two slotlike regions above and below the blade and lying approximately along Mach lines in the resultant flow. As long as successive blades of a propeller lie outside of the disturbance regions of the other blades in the cascade, there will be no direct interference and, as shown in Fig. D,3e, each blade will effectively operate under essentially free stream conditions. When, however, a blade lies within the disturbance region of another blade, a direct interference exists and the forces on the blade elements are not as given by the theory for the isolated airfoil. Such a condition is shown in Fig. D,3f. Here the portion of each blade lying in the disturbance region of the preceding blade is at zero angle of attack and the lift on the airfoil is

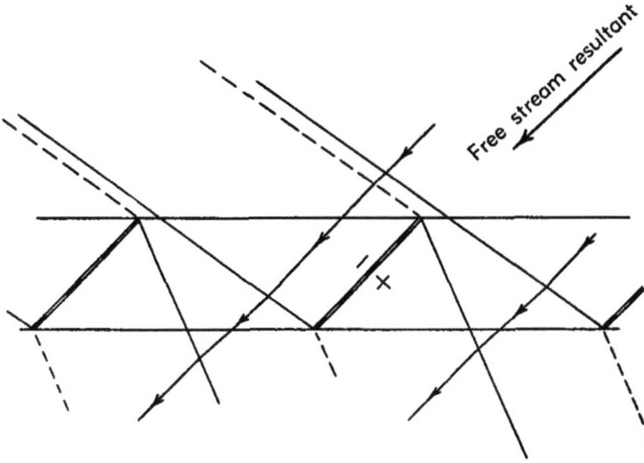

Fig. D,3e. Propeller cascade operating under interference-free condition.

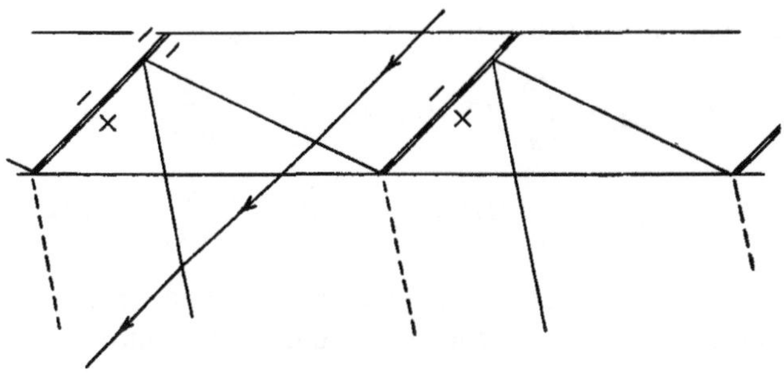

Fig. D,3f. Propeller cascade operating with direct shock interference between blades.

reduced proportionately. It is apparent that this form of interference approaches a maximum at a flight Mach number of 1.0 where the lift on the blade element would appear to be zero.

For a blade of finite length, the slotlike disturbance regions here pictured are modified and gradually dissipated within the expanding Mach cone from the tip. Usually the affected blade will lie within the tip Mach cone of the preceding blade. This fact, together with the importance of the Mach number gradient in this case, makes the actual prediction of the flow disturbance and characteristics of the affected sections a difficult and as yet unsolved problem. This situation does not arise at appreciably supersonic flight speeds, and at flight Mach numbers less than 1 it can be avoided by limiting the solidity of the propeller so as to avoid the interference condition.

In general it is apparent that the justified use of the small perturbation theory is at present restricted to the case of high aspect ratio blades operating without shock interference.

D,4. Thrust, Torque, and Efficiency of Complete Propeller. In an airfoil-type propeller the net thrust, torque, and efficiency are obtained by integrating the section forces as defined by Eq. 2-3 and 2-4 or

$$\text{Thrust} = T = B \int_H^R dT = B \int_H^R \frac{L}{\cos \epsilon} \cos (\phi + \epsilon) dr \tag{4-1}$$

$$\text{Torque} = Q = B \int_H^R r dQ = B \int_H^R \frac{rL}{\cos \epsilon} \sin (\phi + \epsilon) dr \tag{4-2}$$

$$\text{Efficiency} = \eta = \frac{\text{Thrust} \times V}{\text{Torque} \times \omega} \tag{4-3}$$

where B = number of blades. From Eq. 3-1,

$$L = qC_L c = \tfrac{1}{2}\rho_\infty (U^2 + \omega^2 r^2) \frac{4\alpha}{\beta} c \tag{4-4}$$

which when substituted into Eq. 4-1 and 4-2 gives

$$T = \tfrac{1}{2}\rho_\infty B \int_H^R (U^2 + \omega^2 r^2) \frac{4\alpha}{\beta(r)} \frac{\cos (\phi + \epsilon)}{\cos \epsilon} c dr \tag{4-5}$$

$$Q = \tfrac{1}{2}\rho_\infty B \int_H^R (U^2 + \omega^2 r^2) \frac{4\alpha}{\beta(r)} \frac{\sin (\phi + \epsilon)}{\cos \epsilon} c r dr \tag{4-6}$$

where $\beta(r)$ is β as defined by Eq. 3-1 at radius r and where c is the corresponding airfoil chord. From Fig. D,2a, neglecting inflow,

$$\phi = \tan^{-1} \frac{U}{\omega r} \tag{4-7}$$

and from Eq. 2-2 and 3-3

$$\epsilon = \cot^{-1} \frac{L}{D} = \cot^{-1} \left[\frac{4\alpha}{4\alpha^2 + K(t/c)^2 + \beta C_{D_f}} \right] \tag{4-8}$$

Thus finally,

$$\text{Thrust} \cong \tfrac{1}{2}\rho_\infty B \int_H^R (U^2 + \omega^2 r^2) \frac{4\alpha}{\beta} \cos \left\{ \tan^{-1} \frac{U}{\omega r} + \cot^{-1} \left[\frac{4\alpha}{4\alpha^2 + K(t/c)^2 + \beta C_{D_f}} \right] \right\} c dr \tag{4-9}$$

$$\text{Torque} \cong \tfrac{1}{2}\rho_\infty B \int_H^R (U^2 + \omega^2 r^2) \frac{4\alpha}{\beta} \sin\left\{\tan^{-1}\frac{U}{\omega r} + \cot^{-1}\left[\frac{4\alpha}{4\alpha^2 + K(t/c)^2 + \beta C_{D_f}}\right]\right\} crdr \quad (4\text{-}10)$$

Eq. 4-9 and 4-10 neglect inflow and assume the applicability of the two-dimensional results of the small perturbation theory. At supersonic flight speeds no inflow exists and the resultant section Mach numbers are well into the region of applicability of the small perturbation theory. For such cases Eq. 4-9 and 4-10 are clearly applicable. At subsonic flight speeds, however, an inflow exists and at least some of the sections are operating at transonic speeds. For such cases Eq. 4-1 and 4-2 are written

$$\text{Thrust} = \tfrac{1}{2}\rho_\infty B \int_H^R (U^2 + \omega^2 r^2) \left(\frac{dC_L}{d\alpha}\right) \alpha \cos\left(\phi + \cot^{-1}\frac{L}{D}\right) cdr \quad (4\text{-}11)$$

$$\text{Torque} = \tfrac{1}{2}\rho_\infty B \int_H^R (U^2 + \omega^2 r^2) \left(\frac{dC_L}{d\alpha}\right) \alpha \sin\left(\phi + \cot^{-1}\frac{L}{D}\right) crdr \quad (4\text{-}12)$$

and integrated graphically, using experimental section data to obtain $dC_L/d\alpha$ and L/D.

It is clear that at subsonic flight speeds ϕ should include an allowance for inflow, but it is not as yet clear what the magnitude of this inflow is. In a conventional low speed propeller, the bound vortices along the blades and the trailing vortices shed from the tip and shank sections combine to produce rotation and an axial slipstream velocity behind the propeller; ahead of the propeller they combine to produce an axial inflow but, since their rotary components cancel, no rotation. The actual magnitude of these induced velocities has been computed by Goldstein [*1*] and determined by Theodorsen[*2*], using an electrical analogue for a large number of practical cases. In the case of a high speed propeller, with most or all of its sections operating at supersonic section speeds, the influence of the tip vortex is confined to the tip Mach cone lying along the helix, and that of the two-dimensional blade element to the slotlike regions shown in Fig. D,3e. At supersonic flight speeds neither of these regions extends ahead of the propeller and no mechanism for the induction of an inflow exists. At subsonic flight speeds, however, both the tip Mach cone and the two-dimensional disturbance regions extend ahead of the plane of the propeller. Far ahead, these regions completely blend and, by analogy with the downwash field behind a supersonic wing, one would expect conditions to approach those of the subsonic propeller having a similar vorticity distribution. Near the face of the propeller, however, the situation is not analogous. In these regions the inflow may be largely determined by the shock interference conditions discussed in Art. 3. In general, as flight

Mach numbers approach 1.0 the shock interference effects can be expected to play a larger and larger part in determining inflow.

In the absence of direct shock interference, as described above, the inflow at the power levels associated with near-peak efficiency is expected to be small and can probably be safely neglected in making preliminary estimates of propeller performance.

In the material which follows, attention will be confined to supersonic flight speeds where the inflow problem does not exist, although it is believed that the results obtained are applicable for practical purposes at high subsonic speeds.

D,5. Nondimensional Form of Thrust and Torque Integrals. For the purpose of the solutions which follow, it is desirable to nondimensionalize the integrals in Eq. 4-5 and 4-6 and to express them in a more compact form. To accomplish this, the substitutions

$$\mathfrak{r} = \frac{r}{R} = 2\frac{r}{D} \tag{5-1}$$

and

$$J = \frac{U}{nD} = \pi\frac{U}{\omega R} = \pi \tan \phi_{\text{tip}} \tag{5-2}$$

are introduced. The major terms then reduce as follows:

$$U^2 + \omega^2 r^2 = U^2 + \pi^2 n^2 \mathfrak{r}^2 D^2 = U^2 + \pi^2 \mathfrak{r}^2 \frac{U^2}{J^2} = \frac{\pi^2 U^2}{J^2}\left[\frac{J^2}{\pi^2} + \mathfrak{r}^2\right] \tag{5-3}$$

$$\beta(r) = \sqrt{M_{\mathfrak{r}}^2 - 1} = \sqrt{\frac{U^2 + \omega^2 r^2}{a_\infty^2} - 1} = \sqrt{M_\infty^2 - 1 + \frac{\omega^2 r^2}{a_\infty^2}}$$

$$\beta(r) = \sqrt{M_\infty^2 - 1 + \frac{\pi^2 n^2 \mathfrak{r}^2 D^2}{U^2} M_\infty^2} = \sqrt{M_\infty^2 - 1 + \frac{\pi^2 M_\infty^2}{J^2}\mathfrak{r}^2}$$

$$= \frac{\pi M_\infty}{J}\sqrt{\frac{M_\infty^2 - 1}{M_\infty^2}\frac{J^2}{\pi^2} + \mathfrak{r}^2} \tag{5-4}$$

$$\cos(\phi + \epsilon)dr = (\cos\phi\cos\epsilon - \sin\phi\sin\epsilon)dr$$

$$= \left(\frac{\omega r}{\sqrt{U^2 + \omega^2 r^2}}\cos\epsilon - \frac{U}{\sqrt{U^2 + \omega^2 r^2}}\sin\epsilon\right)dr$$

$$= \frac{J}{\pi U}\frac{1}{\sqrt{(J^2/\pi^2) + \mathfrak{r}^2}}(\pi n \mathfrak{r} D\cos\epsilon - J\pi D\sin\epsilon)\frac{D}{2}d\mathfrak{r}$$

$$= \frac{\cos\epsilon}{\sqrt{(J^2/\pi^2) + \mathfrak{r}^2}}\left(\mathfrak{r} - \frac{J}{\pi}\tan\epsilon\right)\frac{D}{2}d\mathfrak{r} \tag{5-5}$$

$$\begin{aligned}
\sin(\phi + \epsilon) r\,dr &= (\sin\phi\cos\epsilon - \cos\phi\sin\epsilon) r\,dr \\
&= \left(\frac{Ur}{\sqrt{U^2 + \omega^2 r^2}}\cos\epsilon - \frac{r^2}{\sqrt{U^2 + \omega^2 r^2}}\sin\epsilon\right) dr \\
&= \frac{J}{\pi U}\frac{1}{\sqrt{(J^2/\pi^2) + \mathfrak{r}^2}}\left(U\mathfrak{r}\frac{D}{2}\cos\epsilon - \pi n \mathfrak{r}^2 \frac{D^2}{2}\sin\epsilon\right)\frac{D}{2}\,d\mathfrak{r} \\
&= \frac{J}{\pi}\frac{\cos\epsilon}{\sqrt{(J^2/\pi^2) + \mathfrak{r}^2}}\frac{D}{2}\left(\mathfrak{r} - \frac{\pi}{J}\mathfrak{r}^2\tan\epsilon\right)\frac{D}{2}\,d\mathfrak{r} \qquad (5\text{-}6)
\end{aligned}$$

Substituting Eq. 5-3, 5-4, and 5-5 into Eq. 4-5, and setting $\alpha = \sin\alpha$ one obtains

$$T = \pi\rho_\infty a_\infty n B D^3 \int_h^1 \frac{\sqrt{\frac{J^2}{\pi^2} + \mathfrak{r}^2}}{\sqrt{\frac{M_\infty^2 - 1}{M_\infty^2}\frac{J^2}{\pi^2} + \mathfrak{r}^2}} \sin\alpha\left(\mathfrak{r} - \frac{J}{\pi}\tan\epsilon\right)\frac{c}{D}\,d\mathfrak{r} \qquad (5\text{-}7)$$

where a_∞ = undisturbed speed of sound and n = revolutions per second. Similarly, substituting Eq. 5-3, 5-4, and 5-6 into Eq. 4-6, one obtains

$$Q = \frac{\pi}{2}\rho_\infty a_\infty n B D^4 \int_n^1 \frac{\sqrt{\frac{J^2}{\pi^2} + \mathfrak{r}^2}}{\sqrt{\frac{M_\infty^2 - 1}{M_\infty^2}\frac{J^2}{\pi^2} + \mathfrak{r}}} \sin\alpha\left(\frac{J}{\pi}\mathfrak{r} + \mathfrak{r}^2\tan\epsilon\right)\frac{c}{D}\,d\mathfrak{r} \qquad (5\text{-}8)$$

From Eq. 4-8

$$\begin{aligned}
\tan\epsilon &= \frac{4\sin^2\alpha + K(t/c)^2 + \beta C_l}{4\sin\alpha} \\
&= \sin\alpha + \frac{1}{\sin\alpha}\left[\frac{K(t/c)^2 + \beta C_l}{4}\right] \\
&= \sin\alpha + \frac{g(\mathfrak{r})}{\sin\alpha} \qquad (5\text{-}9)
\end{aligned}$$

where

$$g(\mathfrak{r}) = \tfrac{1}{4}[K(t/c)^2 + \beta C_l] \qquad (5\text{-}10)$$

Introducing the notation

$$I(\mathfrak{r}) = \frac{\sqrt{\frac{J^2}{\pi^2} + \mathfrak{r}^2}}{\sqrt{\frac{M_\infty^2 - 1}{M_\infty^2}\frac{J^2}{\pi^2} + \mathfrak{r}^2}} \qquad (5\text{-}11)$$

and substituting Eq. 5-9, Eq. 5-7 and 5-8 become

$$\begin{aligned}
T &= \pi\rho_\infty a_\infty n B D^3 \int_h^1 I(\mathfrak{r})\sin\alpha\left\{\mathfrak{r} - \frac{J}{\pi}\left[\sin\alpha + \frac{g(\mathfrak{r})}{\sin\alpha}\right]\right\}\frac{c}{D}\,d\mathfrak{r} \\
&= \pi\rho_\infty a_\infty n B D^3 \int_h^1 I(\mathfrak{r})\left[\mathfrak{r}\sin\alpha - \frac{J}{\pi}\sin^2\alpha - \frac{J}{\pi}g(\mathfrak{r})\right]\frac{c}{D}\,d\mathfrak{r} \qquad (5\text{-}12)
\end{aligned}$$

and

$$Q = \frac{\pi}{2} \rho_\infty a_\infty n B D^4 \int_h^1 I(\mathfrak{r}) \left[\frac{J}{\pi} \mathfrak{r} \sin \alpha + \mathfrak{r}^2 \sin^2 \alpha + \mathfrak{r}^2 g(\mathfrak{r})\right] \frac{c}{D} d\mathfrak{r} \quad (5\text{-}13)$$

The integrals in Eq. 5-12 and 5-13 are nondimensional, i.e. independent of the size or diameter of the blade.

It may be noted at this point that, as in the thin airfoil theory, the effect of thickness as introduced by the $g(\mathfrak{r})$ term appears as a simple additive term in the expressions for thrust and torque. In other words, the thrust of a given blade is equal to that of a flat plate blade *minus* a fixed increment due to thickness and skin friction; and the torque is equal to that of said flat plate blade *plus* a fixed increment due to thickness and skin friction.

D,6. Pitch Distribution for Maximum Efficiency. The radial pitch or angle-of-attack distribution

$$\alpha = f(\mathfrak{r}) \quad (6\text{-}1)$$

for maximum efficiency at a given ω and U is now determined.

The corresponding solution for subsonic propellers was obtained by Goldstein [*1*] and extended by Theodorsen [*2*]. These are potential solutions which neglect the form drag of the section and yield an optimum distribution with respect to the induced losses only. The shape and thickness of the airfoil sections and the distribution thereof do not enter the problem at all. The sections are considered to be ideal vorticity generators displaying infinite values for L/D. This is a reasonable approximation for subsonic propeller sections since the form drags are low.

At supersonic flight speeds, however, the profile drags of the blade elements are comparatively large and must be considered in determining the optimum pitch distributions. The solution [*3*] proceeds as follows.

The net thrust and torque are given by Eq. 5-12 and 5-13. The net propeller efficiency is given by

$$\eta = \frac{TU}{Q\omega} = \frac{TJD}{2\pi Q} \quad (6\text{-}2)$$

The problem is to determine the angle-of-attack distribution

$$\alpha \cong \sin \alpha = f(\mathfrak{r}) \quad (6\text{-}3)$$

which when substituted into Eq. 5-12 and 5-13 will yield the thrust and torque which maximizes the efficiency as defined by Eq. 6-2.

Assume that $\sin \alpha = f_0(\mathfrak{r})$ is the optimum distribution sought. Then replace $\sin \alpha$ in Eq. 5-12 and 5-13 by

$$\sin \alpha = f(\mathfrak{r}) = f_0(\mathfrak{r}) + \mu\varphi(\mathfrak{r}) \quad (6\text{-}4)$$

where $\varphi(\mathrm{r})$ is an arbitrary function and μ a new variable, variations of which produce a "family" of twist distributions $f(\mathrm{r})$. This is the usual substitution made in the derivation of Euler's equation in the calculus of variations.

If, as stated, $f_0(\mathrm{r})$ is the optimum distribution, then

$$\frac{d\eta}{d\mu} = 0 \qquad \text{when } \mu = 0 \tag{6-5}$$

Now

$$\left.\frac{d\eta}{d\mu}\right|_{\mu=0} = \frac{d}{d\mu}\left(\frac{TU}{Q\omega}\right)_{\mu=0} = \frac{U}{\omega}\left[\frac{Q(dT/d\mu) - T(dQ/d\mu)}{Q^2}\right]_{\mu=0}$$

$$= \frac{JD}{\pi}\left[\frac{Q(dT/d\mu) - T(dQ/d\mu)}{Q^2}\right]_{\mu=0} \tag{6-6}$$

Substituting Eq. 6-4 into Eq. 5-12 and 5-13

$$T = \pi\rho_\infty a_\infty nBD^3 \int_h^1 I(\mathrm{r}) \left\{\mathrm{r}f(\mathrm{r}) - \frac{J}{\pi}[f(\mathrm{r})]^2 - \frac{J}{\pi}g(\mathrm{r})\right\} \frac{c}{D}\, d\mathrm{r} \tag{6-7}$$

$$Q = \frac{\pi}{2}\rho_\infty a_\infty nBD^4 \int_h^1 I(\mathrm{r}) \left\{\frac{J}{\pi}\mathrm{r}f(\mathrm{r}) + \mathrm{r}^2[f(\mathrm{r})]^2 + \mathrm{r}^2 g(\mathrm{r})\right\} \frac{c}{D}\, d\mathrm{r} \tag{6-8}$$

Differentiating and evaluating at $\mu = 0$

$$\left.\frac{dT}{d\mu}\right|_{\mu=0} = \pi\rho_\infty a_\infty nBD^3 \int_h^1 I(\mathrm{r}) \left[\mathrm{r}\varphi(\mathrm{r}) - 2\frac{J}{\pi}f_0(\mathrm{r})\varphi(\mathrm{r})\right] \frac{c}{D}\, d\mathrm{r} \tag{6-9}$$

$$\left.\frac{dQ}{d\mu}\right|_{\mu=0} = \frac{\pi}{2}\rho_\infty a_\infty nBD^4 \int_h^1 I(\mathrm{r}) \left[\frac{J}{\pi}\mathrm{r}\varphi(\mathrm{r}) + 2\mathrm{r}^2 f_0(\mathrm{r})\varphi(\mathrm{r})\right] \frac{c}{D}\, d\mathrm{r} \tag{6-10}$$

Substituting Eq. 6-9 and 6-10 into Eq. 6-6, and equating to zero as indicated by Eq. 6-5,

$$\left.\frac{d\eta}{d\mu}\right|_{\mu=0} = \frac{JD}{\pi} \times \left\{\frac{\begin{array}{l} Q\pi\rho_\infty a_\infty nBD^3 \int_h^1 I(\mathrm{r})[\mathrm{r}\varphi(\mathrm{r}) - 2(J/\pi)f_0(\mathrm{r})\varphi(\mathrm{r})]\frac{c}{D}\, d\mathrm{r} \\ \quad - T(\pi/2)\rho_\infty a_\infty nBD^4 \int_h^1 I(\mathrm{r})[(J/\pi)\mathrm{r}\varphi(\mathrm{r}) + 2\mathrm{r}^2 f_0(\mathrm{r})\varphi(\mathrm{r})]\frac{c}{D}\, d\mathrm{r} \end{array}}{Q^2}\right\} = 0$$

$$\frac{J}{\pi}\left\{Q\int_h^1 I(\mathrm{r})\left[\mathrm{r}\varphi(\mathrm{r}) - 2\frac{J}{\pi}f_0(\mathrm{r})\varphi(\mathrm{r})\right]\frac{c}{D}\, d\mathrm{r} \right.$$
$$\left. - T\frac{D}{2}\int_h^1 I(\mathrm{r})\left[\frac{J}{\pi}\mathrm{r}\varphi(\mathrm{r}) + 2\mathrm{r}^2 f_0(\mathrm{r})\varphi(\mathrm{r})\right]\frac{c}{D}\, d\mathrm{r}\right\} = 0$$

$$\frac{J}{\pi}\left\{\int_h^1 I(\mathfrak{r})\left[\mathfrak{r}\varphi(\mathfrak{r}) - 2\frac{J}{\pi}f_0(\mathfrak{r})\varphi(\mathfrak{r})\right]\frac{c}{D}\,d\mathfrak{r} - \frac{T}{Q}\frac{D}{2}\int_h^1 I(\mathfrak{r})\left[\frac{J}{\pi}\mathfrak{r}\varphi(\mathfrak{r}) + 2\mathfrak{r}^2 f_0(\mathfrak{r})\varphi(\mathfrak{r})\right]\frac{c}{D}\,d\mathfrak{r}\right\} = 0$$

$$\frac{J}{\pi}\int_h^1 I(\mathfrak{r})\left[\mathfrak{r}\varphi(\mathfrak{r}) - 2\frac{J}{\pi}f_0(\mathfrak{r})\varphi(\mathfrak{r})\right]\frac{c}{D}\,d\mathfrak{r} - \eta\int_h^1 I(\mathfrak{r})\left[\frac{J}{\pi}\mathfrak{r}\varphi(\mathfrak{r}) + 2\mathfrak{r}^2 f_0(\mathfrak{r})\varphi(\mathfrak{r})\right]\frac{c}{D}\,dr = 0$$

or finally,

$$\int_h^1 I(\mathfrak{r})\left\{\frac{J}{\pi}\left[\mathfrak{r} - 2\frac{J}{\pi}f_0(\mathfrak{r})\right] - \eta\left[\frac{J}{\pi}\mathfrak{r} + 2\mathfrak{r}^2 f_0(\mathfrak{r})\right]\right\}\varphi(\mathfrak{r})\frac{c}{D}\,d\mathfrak{r} = 0 \tag{6-11}$$

Since $\varphi(\mathfrak{r})$ is arbitrary in character, the quantity in braces must itself be zero for all values of $\mathfrak{r}$. If it were not, some form of $\varphi(\mathfrak{r})$ would yield a finite value for the integral. Thus

$$\frac{J}{\pi}\left[\mathfrak{r} - 2\frac{J}{\pi}f_0(\mathfrak{r})\right] - \eta\left[\frac{J}{\pi}\mathfrak{r} + 2\mathfrak{r}^2 f_0(\mathfrak{r})\right] = 0 \tag{6-12}$$

or

$$\frac{J}{\pi}(\mathfrak{r}) - 2\frac{J^2}{\pi^2}f_0(\mathfrak{r}) - \eta\frac{J}{\pi}\mathfrak{r} - \eta 2\mathfrak{r}^2 f_0(\mathfrak{r}) = 0$$

$$f_0(\mathfrak{r}) = \sin\alpha = \frac{J}{2\pi}\frac{\mathfrak{r}(1-\eta)}{(J^2/\pi^2) + \mathfrak{r}^2\eta} \tag{6-13}$$

Eq. 6-13 indicates that the optimum angle of attack goes to zero on the axis of rotation and reaches a maximum at

$$\mathfrak{r} = \frac{J}{\eta}\sqrt{\frac{1}{\eta}} \tag{6-14}$$

For any particular case, the optimum distribution is uniquely determined by the operating J and the net propeller efficiency. Representative results are shown in Fig. D,6.

Dividing Eq. 6-13 through by $\mathfrak{r}^2$ and substituting

$$\frac{J}{\pi\mathfrak{r}} = \frac{U}{\pi n 2R\mathfrak{r}} = \frac{U}{\omega r} = \tan\phi_{\mathfrak{r}} \tag{6-15}$$

Eq. 6-13 becomes

$$\sin\alpha = \frac{1}{2}\frac{(1-\eta)\tan\phi_{\mathfrak{r}}}{\eta + \tan^2\phi_{\mathfrak{r}}} \tag{6-16}$$

At some station along the blade, the section efficiency will equal the net propeller efficiency; at this point, from Eq. 2-6,

$$\eta = \eta_s = \tan\phi_{\mathfrak{r}}\cot(\phi + \epsilon)$$

Substituting this value of η into Eq. 6-16, one obtains

$$\sin \alpha = \frac{1}{2} \frac{[1 - \tan \phi_r \cot (\phi_r + \epsilon)]\tan \phi_r}{\tan^2 \phi_r + \tan \phi \cot (\phi_r + \epsilon)}$$

which simplifies to

$$\sin \alpha = \frac{\epsilon}{2} \tag{6-17}$$

But, from Eq. 3-7, Eq. 6-17 is the angle for $(L/D)_{max}$. Thus that station along the blade, whose maximum section efficiency equals the net propeller

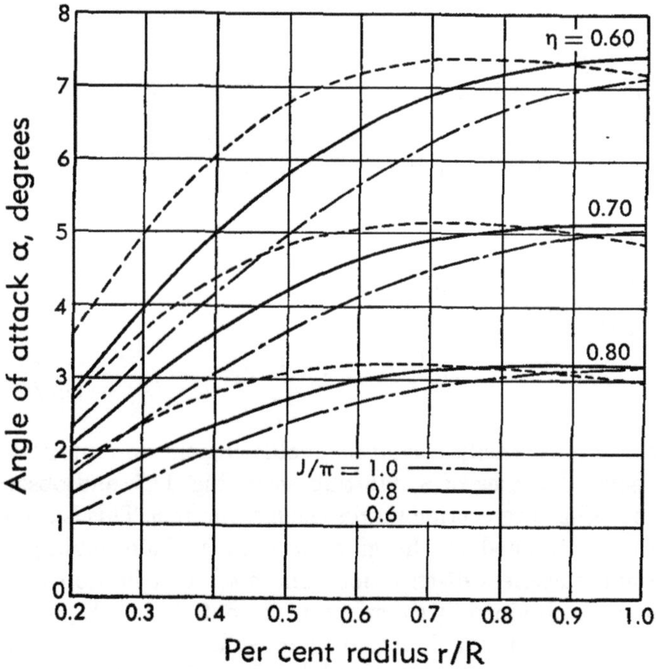

Fig. D,6. Optimum angle-of-attack distributions for various net propeller efficiencies.

efficiency, operates at $(L/D)_{max}$. Stations whose maximum section efficiency is greater than the net propeller efficiency operate at angles greater than those for $(L/D)_{max}$, and stations whose maximum section efficiency is less than the net propeller efficiency operate at angles less than that of $(L/D)_{max}$.

The above result applies to sections of arbitrary but given planform. The optimum is one having zero chord everywhere except at the station where the section efficiency as given by Eq. 2-12 is a maximum. This of course is a trivial result, the planform normally tending to favor regions of high section efficiency but being limited by structural and other practical considerations.

In general, a blade having the optimum twist distribution will be from 3 to 6 per cent more efficient than the same blade twisted so that each blade element is operating at its best L/D.

D,7. Maximum Efficiency with Optimum Pitch Distribution. The net propeller efficiency is given by Eq. 6-2 wherein T and Q are given by Eq. 6-7 and 6-8. For $f(\mathrm{r})$ in the latter, the solution (6-13) is substituted. The efficiency with the optimum pitch distribution is then

$$\eta = \frac{J}{\pi} \times$$

$$\frac{\int_h^1 I(\mathrm{r}) \left\{ \mathrm{r} \left[\frac{J}{2\pi} \frac{\mathrm{r}(1-\eta)}{(J^2/\pi^2) + \mathrm{r}^2\eta} \right] - \frac{J}{\pi} \left[\frac{\mathrm{r}}{2\pi} \frac{(1-\eta)}{(J^2/\pi^2) + \mathrm{r}^2\eta} \right]^2 - \frac{J}{\pi} g(\mathrm{r}) \right\} \frac{c}{D}\, d\mathrm{r}}{\int_h^1 I(\mathrm{r}) \left\{ \frac{J}{\pi} \mathrm{r} \left[\frac{J}{2\pi} \frac{\mathrm{r}(1-\eta)}{(J^2/\pi^2) + \mathrm{r}^2\eta} \right] + \mathrm{r}^2 \left[\frac{J}{2\pi} \frac{\mathrm{r}(1-\eta)}{(J^2/\pi^2) + \mathrm{r}^2\eta} \right]^2 + \mathrm{r}^2 g(\mathrm{r}) \right\} \frac{c}{D}\, d\mathrm{r}}$$

which simplifies to

$$4 \int_h^1 I(\mathrm{r}) \left[\frac{J^2}{\pi^2} + \mathrm{r}^2\eta \right] g(\mathrm{r}) \frac{c}{D}\, d\mathrm{r} = \frac{J^2}{\pi^2} (1-\eta)^2 \int_h^1 I(\mathrm{r}) \frac{\mathrm{r}^2}{(J^2/\pi^2) + \mathrm{r}^2\eta} \frac{c}{D}\, d\mathrm{r} \tag{7-1}$$

Eq. 7-1 is most easily solved by evaluating the two integrals graphically for several values of η, plotting as in Fig. D,7, and observing the intersection. This intersection marks the maximum η. This is the highest η which can be obtained at the given J with a blade having the given planform and thickness distribution. The pitch distribution required for this efficiency is obtained by substituting the noted η into Eq. 6-13.

To find the optimum J, Eq. 6-14 must be solved for a number of values of J and the maximum noted (see Fig. D,7). It follows from the discussion of the blade element theory contained in Art. 2 (Eq. 2-10) that the optimum J will be in the region of the J producing a pitch angle ϕ, of 45° at about the station of 85 per cent radius, i.e. $J/\pi = 0.8$.

D,8. Optimum Pitch Distribution for a Given Power Absorption. In many practical problems, it is necessary to absorb, in a given diameter, a power greater than that at peak efficiency. The optimum pitch distribution for this case is slightly different from that for peak efficiency operation.

The net thrust and torque of a propeller are given by Eq. 5-12 and 5-13. It is required to determine the angle-of-attack distribution

$$\alpha \cong \sin \alpha = f(\mathrm{r}) \tag{8-1}$$

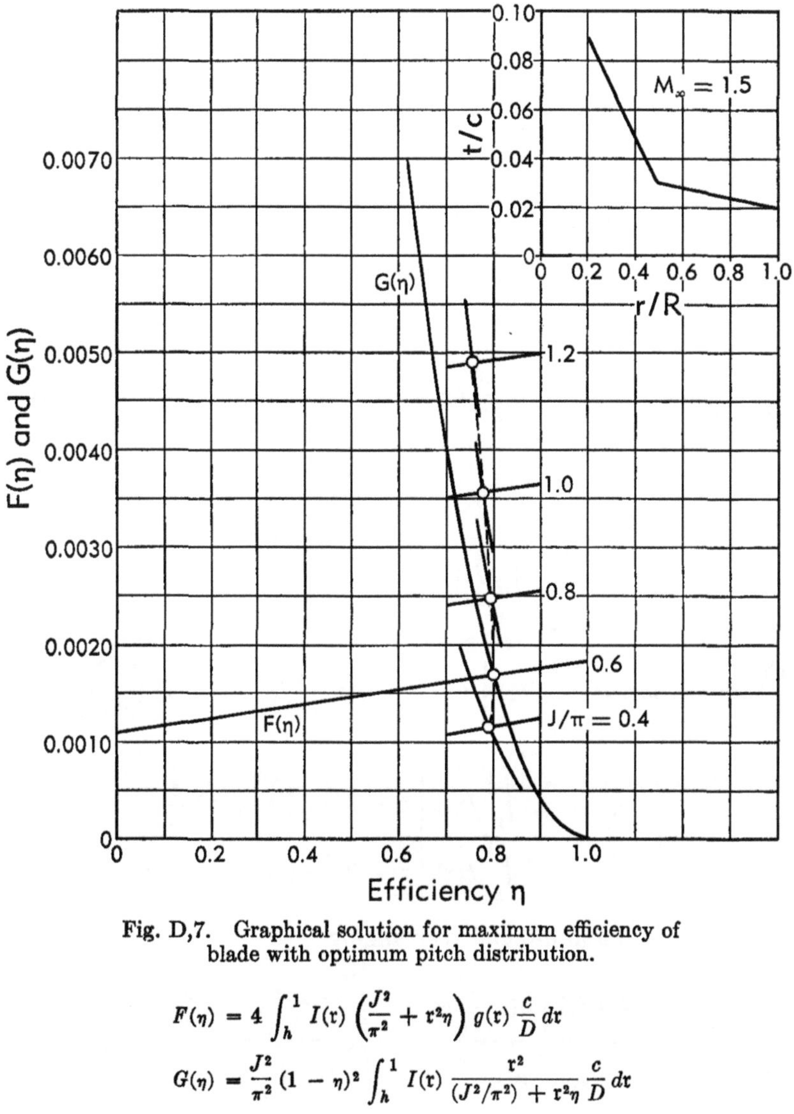

Fig. D,7. Graphical solution for maximum efficiency of blade with optimum pitch distribution.

$$F(\eta) = 4 \int_h^1 I(r) \left(\frac{J^2}{\pi^2} + r^2\eta\right) g(r) \frac{c}{D}\, dr$$

$$G(\eta) = \frac{J^2}{\pi^2} (1 - \eta)^2 \int_h^1 I(r) \frac{r^2}{(J^2/\pi^2) + r^2\eta} \frac{c}{D}\, dr$$

which maximizes the thrust for a given value of the torque. In quest of this particular distribution, the trial function

$$\sin \alpha = f(r) = \lambda f_0(r) + \mu\varphi(r) \tag{8-2}$$

is defined wherein $f_0(r)$ is the distribution sought, $\varphi(r)$ an arbitrary continuous perturbation function, and λ and μ are variables the necessary relation between which is restricted by the requirements of the problem. In particular, since $f_0(r)$ in Eq. 8-2 is the distribution sought, when $\lambda = 1$,

$\mu = 0$. The problem further requires that the torque be fixed. Thus, for all admissable values of λ, $dQ/d\lambda = 0$ and in particular $(dQ/d\lambda)_{\lambda=1} = (dQ/d\lambda)_1 = 0$. Finally, it is required that the thrust be a maximum or when $\lambda = 1$, $(dT/d\lambda)_1 = 0$.

From Eq. 8-2

$$\left[\frac{d}{d\lambda}(\sin\alpha)\right]_{\lambda=1} = f_0(\mathfrak{r}) + \varphi(\mathfrak{r})\left(\frac{d\mu}{d\lambda}\right)_1 \tag{8-3}$$

and

$$(\sin\alpha)_{\lambda=1} = f_0(\mathfrak{r}) \tag{8-4}$$

Thus with reference to Eq. 5-13

$$\left(\frac{dQ}{d\lambda}\right)_1 = \frac{\pi}{2}\rho_\infty BanD^4 \int_h^1 I(\mathfrak{r})\left[\frac{J}{\pi}\mathfrak{r} + 2\mathfrak{r}^2 f_0(\mathfrak{r})\right]\left[f_0(\mathfrak{r}) + \varphi(\mathfrak{r})\left(\frac{d\mu}{d\lambda}\right)_1\right]\frac{c}{D}\,d\mathfrak{r} = 0$$

or

$$\left(\frac{d\mu}{d\lambda}\right)_1 = -\frac{\int_h^1 I(\mathfrak{r})\left[\frac{J}{\pi}\mathfrak{r} + 2\mathfrak{r}^2 f_0(\mathfrak{r})\right] f_0(\mathfrak{r})\frac{c}{D}\,d\mathfrak{r}}{\int_h^1 I(\mathfrak{r})\left[\frac{J}{\pi}\mathfrak{r} + 2\mathfrak{r}^2 f_0(\mathfrak{r})\right]\varphi(\mathfrak{r})\frac{c}{D}\,d\mathfrak{r}} \tag{8-5}$$

Similarly from Eq. 5-12

$$\left(\frac{dT}{d\lambda}\right)_1 = \pi\rho_\infty BanD^3 \int_h^1 I(\mathfrak{r})\left[\mathfrak{r} - 2\frac{J}{\pi}f_0(\mathfrak{r})\right]\left[f_0(\mathfrak{r}) + \varphi(\mathfrak{r})\left(\frac{d\mu}{d\lambda}\right)_1\right]d\mathfrak{r} = 0$$

or

$$\left(\frac{d\mu}{d\lambda}\right)_1 = -\frac{\int_h^1 I(\mathfrak{r})\left[\mathfrak{r} - 2\frac{J}{\pi}f_0(\mathfrak{r})\right] f_0(\mathfrak{r})\frac{c}{D}\,d\mathfrak{r}}{\int_h^1 I(\mathfrak{r})\left[\mathfrak{r} - 2\frac{J}{\pi}f_0(\mathfrak{r})\right]\varphi(\mathfrak{r})\frac{c}{D}\,d\mathfrak{r}} \tag{8-6}$$

Equating Eq. 8-5 and 8-6 and rearranging

$$\frac{\int_h^1 I(\mathfrak{r})\left[\mathfrak{r} - 2\frac{J}{\pi}f_0(\mathfrak{r})\right]\varphi(\mathfrak{r})\frac{c}{D}\,d\mathfrak{r}}{\int_h^1 I(\mathfrak{r})\left[\frac{J}{\pi}\mathfrak{r} + 2\mathfrak{r}^2 f_0(\mathfrak{r})\right]\varphi(\mathfrak{r})\frac{c}{D}\,d\mathfrak{r}} = \frac{\int_h^1 I(\mathfrak{r})\left[\mathfrak{r} - 2\frac{J}{\pi}f_0(\mathfrak{r})\right] f_0(\mathfrak{r})\frac{c}{D}\,d\mathfrak{r}}{\int_h^1 I(\mathfrak{r})\left[\frac{J}{\pi}\mathfrak{r} + 2\mathfrak{r}^2 f_0(\mathfrak{r})\right] f_0(\mathfrak{r})\frac{c}{D}\,d\mathfrak{r}} \tag{8-7}$$

Since $\varphi(\mathfrak{r})$ is an arbitrary continuous function, the ratio of the left-hand integrands must, for all values of $\mathfrak{r}$ between h and 1, have the same ratio as that of the complete integrals. Thus

$$\frac{\mathfrak{r} - 2\frac{J}{\pi}f_0(\mathfrak{r})}{\frac{J}{\pi}\mathfrak{r} + 2\mathfrak{r}^2 f_0(\mathfrak{r})} = \frac{\int_h^1 I(\mathfrak{r})\left[\mathfrak{r} - 2\frac{J}{\pi}f_0(\mathfrak{r})\right] f_0(\mathfrak{r})\frac{c}{D}\,d\mathfrak{r}}{\int_h^1 I(\mathfrak{r})\left[\frac{J}{\pi}\mathfrak{r} + 2\mathfrak{r}^2 f_0(\mathfrak{r})\right] f_0(\mathfrak{r})\frac{c}{D}\,d\mathfrak{r}} \tag{8-8}$$

For any given propeller, the integrals on the right of Eq. 8-8 reduce to a single numerical value which is a function of J/π only. Thus

$$\frac{\mathrm{r} - 2(J/\pi)f_0(\mathrm{r})}{(J/\pi)\mathrm{r} + 2\mathrm{r}^2 f_0(\mathrm{r})} = f\left(\frac{J}{\pi}\right) = k\frac{\pi}{J} \tag{8-9}$$

where k is as yet an unknown function of J/π. The π/J is taken out to make the function, which follows, analogous to Eq. 6-13. Solving Eq. 8-9 for $f_0(\mathrm{r})$

$$f_0(\mathrm{r}) = \sin\alpha = \frac{J}{2\pi}\frac{\mathrm{r}(1-k)}{\mathrm{r}^2 k + (J^2/\pi^2)} \tag{8-10}$$

It will be noted that Eq. 8-10 is the same as Eq. 6-13 except that k is substituted for η. It will now be shown that, when the fixed power of the present problem approaches the power at peak efficiency, $k \to \eta$, and that in general

$$k = \frac{JD}{2\pi}\frac{dT}{dQ} = J\frac{dC_T}{dC_P} \cong \eta \tag{8-11}$$

at the power in question. C_T and C_P as here used have the usual definitions, namely

$$\begin{aligned} C_T &= \frac{T}{\rho_\infty n^2 D^4} \\ C_P &= \frac{P}{\rho_\infty n^3 D^5} \end{aligned} \tag{8-12}$$

Substituting Eq. 8-10 into Eq. 5-12 and 5-13, one obtains

$$\begin{aligned} T &= \pi\rho_\infty a_\infty nBD^3 \int_h^1 I(\mathrm{r}) \left\{ \mathrm{r}\frac{J}{2\pi}\frac{\mathrm{r}(1-k)}{\mathrm{r}^2k + (J^2/\pi^2)} - \frac{J}{\pi}\frac{J^2}{4\pi^2}\frac{\mathrm{r}^2(1-k)^2}{[\mathrm{r}^2k + (J^2/\pi^2)]^2} \right. \\ &\qquad \left. - \frac{J}{\pi}g(\mathrm{r}) \right\} \frac{c}{D}\,d\mathrm{r} \\ &= \pi\rho_\infty a_\infty nBD^3 \frac{J}{4\pi}(1-k) \int_h^1 I(\mathrm{r}) \left\{ \frac{2\mathrm{r}^2}{\mathrm{r}^2k + (J^2/\pi^2)} \right. \\ &\qquad \left. - \frac{J^2}{\pi^2}\frac{\mathrm{r}^2(1-k)}{[\mathrm{r}^2k + (J^2/\pi^2)]^2} \right\} \frac{c}{D}\,d\mathrm{r} - T_g \end{aligned} \tag{8-13}$$

where

$$T_g = \rho_\infty a_\infty nBD^3 J \int_h^1 I(\mathrm{r})g(\mathrm{r})\frac{c}{D}\,d\mathrm{r} \tag{8-14}$$

and

$$\begin{aligned} Q &= \frac{\pi}{2}\rho_\infty a_\infty nBD^4 \int_h^1 I(\mathrm{r}) \left\{ \frac{J}{\pi}\mathrm{r}\frac{J}{2\pi}\frac{\mathrm{r}(1-k)}{\mathrm{r}^2k + (J^2/\pi^2)} \right. \\ &\qquad \left. + \mathrm{r}^2\frac{J^2}{4\pi}\frac{\mathrm{r}^2(1-k)^2}{[\mathrm{r}^2k + (J^2/\pi^2)]^2} \right\} \frac{c}{D}\,d\mathrm{r} + Q_g \end{aligned} \tag{8-15}$$

where

$$Q_g = \frac{\pi}{2}\rho_\infty a_\infty nBD^4 \int_h^1 I(\mathfrak{r})\mathfrak{r}^2 g(\mathfrak{r})\frac{c}{D}\,d\mathfrak{r} \tag{8-16}$$

The second term under the integral in Eq. 8-15 may be rewritten as follows:

$$\begin{aligned}
\mathfrak{r}^2\frac{J^2}{4\pi^2}\frac{\mathfrak{r}^2(1-k)^2}{[\mathfrak{r}^2k+(J^2/\pi^2)]^2} &= \mathfrak{r}^2\frac{J^2}{4\pi^2}\frac{1}{k}\left[\mathfrak{r}^2k+\frac{J^2}{\pi^2}-\frac{J^2}{\pi^2}\right]\frac{(1-k)^2}{[\mathfrak{r}^2k+(J^2/\pi^2)]^2}\\
&= \mathfrak{r}^2\frac{J^2}{4\pi^2}\frac{1}{k}\left(\mathfrak{r}^2k+\frac{J^2}{\pi^2}\right)\frac{(1-k)^2}{[\mathfrak{r}^2k+(J^2/\pi^2)]^2}\\
&\quad - \mathfrak{r}^2\frac{J^2}{4\pi^2}\frac{1}{k}\frac{J^2}{\pi^2}\frac{(1-k)^2}{[\mathfrak{r}^2k+(J^2/\pi^2)]^2}\\
&= \mathfrak{r}^2\frac{J^2}{4\pi^2}\frac{1-k}{k}\frac{(1-k)}{[\mathfrak{r}^2k+(J^2/\pi^2)]}\\
&\quad - \frac{J^2}{\pi^2}\mathfrak{r}^2\frac{J^2}{4\pi^2}\frac{(1-k)^2}{k}\frac{1}{[\mathfrak{r}^2k+(J^2/\pi^2)]^2}
\end{aligned} \tag{8-17}$$

Substituting Eq. 8-17 into Eq. 8-15, one obtains

$$\begin{aligned}
Q &= \frac{\pi}{2}\rho_\infty a_\infty nBD^4\int_h^1 I(\mathfrak{r})\left[\left(2+\frac{1-k}{k}\right)\frac{J^2}{4\pi^2}\frac{\mathfrak{r}^2(1-k)}{\mathfrak{r}^2k+(J^2/\pi^2)}\right.\\
&\quad \left.-\frac{J^2}{\pi^2}\mathfrak{r}^2\frac{J^2}{4\pi^2}\frac{(1-k)^2}{k}\frac{1}{[\mathfrak{r}^2k+(J^2/\pi^2)]^2}\right]\frac{c}{D}\,d\mathfrak{r}+Q_g\\
&= \frac{\pi}{2}\rho_\infty a_\infty nBD^4\frac{J^2}{4\pi^2}(1-k)\int_h^1 I(\mathfrak{r})\left[\frac{1+k}{k}\frac{\mathfrak{r}^2}{\mathfrak{r}^2k+(J^2/\pi^2)}\right.\\
&\quad \left.-\frac{J^2}{\pi^2}\frac{(1-k)}{k}\frac{\mathfrak{r}^2}{[\mathfrak{r}^2k+(J^2/\pi^2)]^2}\right]\frac{c}{D}\,d\mathfrak{r}+Q_g
\end{aligned} \tag{8-18}$$

Now

$$\begin{aligned}
\frac{JD}{2\pi}T-kQ &= \frac{1}{2}\rho_\infty a_\infty nBD^4\frac{J^2}{4\pi^2}(1-k)^2\int_h^1 I(\mathfrak{r})\frac{\mathfrak{r}^2}{\mathfrak{r}^2k+(J^2/\pi^2)}\frac{c}{D}\,d\mathfrak{r}\\
&\quad -\frac{JD}{2\pi}T_g-kQ_g
\end{aligned} \tag{8-19}$$

and

$$\begin{aligned}
\frac{d}{dk}\left(\frac{JD}{2\pi}T-kQ\right) &= \frac{1}{2}\rho_\infty a_\infty nBD^4\frac{J^2}{4\pi^2}\left\{-2(1-k)\int_h^1\frac{I(\mathfrak{r})\mathfrak{r}^2}{\mathfrak{r}^2k+(J^2/\pi^2)}\right.\\
&\quad \left.-(1-k)^2\int_h^1\frac{I(\mathfrak{r})\mathfrak{r}^4d\mathfrak{r}}{[\mathfrak{r}^2k+(J^2/\pi^2)]^2}\right\}-Q_g
\end{aligned} \tag{8-20}$$

Making the same substitution in the second integral of Eq. 8-20 as made in arriving at Eq. 8-17, Eq. 8-20 reduces to

$$\frac{d}{dk}\left(\frac{JD}{2\pi}T - kQ\right)$$

$$= -\frac{1}{2}\rho_\infty a_\infty nBD^4\frac{J^2}{4\pi^2}(1-k)\left\{\frac{1+k}{k}\int_h^1 \frac{I(\mathfrak{r})\mathfrak{r}^2}{\mathfrak{r}^2k + (J^2/\pi^2)}\frac{c}{D}\,d\mathfrak{r}\right.$$

$$\left. - \frac{1-k}{k}\frac{J^2}{\pi^2}\int_h^1 \frac{I(\mathfrak{r})\mathfrak{r}^2dx}{[\mathfrak{r}^2k + (J^2/\pi^2)]^2}\right\} - Q_0 \quad (8\text{-}21)$$

Comparing Eq. 8-21 with 8-18, it is seen that

$$\frac{d}{dk}\left(\frac{JD}{2\pi}T - kQ\right) = -Q \quad (8\text{-}22)$$

But

$$\frac{d}{dk}\left(\frac{JD}{2\pi}T - kQ\right) = \frac{JD}{2\pi}\frac{dT}{dk} - Q - k\frac{dQ}{dk} \quad (8\text{-}23)$$

Equating Eq. 8-22 and 8-23 one obtains

$$\frac{JD}{2\pi}\frac{dT}{dk} = k\frac{dQ}{dk}$$

or

$$k = \frac{JD}{2\pi}\frac{dT}{dQ} = J\frac{dC_T}{dC_P} \quad (8\text{-}24)$$

Now in general,

$$\eta = \frac{V}{\omega}\frac{T}{Q} = \frac{JD}{2\pi}\frac{T}{Q}$$

or

$$T = \frac{2\pi}{JD}\eta Q$$

and

$$\frac{dT}{dQ} = \frac{2\pi}{JD}\left(\eta + Q\frac{d\eta}{dQ}\right) \quad (8\text{-}25)$$

Substituting Eq. 8-25 into 8-24 one obtains

$$k = J\frac{dC_T}{dC_P} = \eta + Q\frac{d\eta}{dQ} \quad (8\text{-}26)$$

which is the proof of Eq. 8-11.

As Q approaches the power of peak efficiency

$$\left.\begin{aligned}\frac{d\eta}{dQ} &\to 0\\ k &\to \eta\end{aligned}\right\} \quad (8\text{-}27)$$

Thus Eq. 8-10 reduces to Eq. 6-13 as it should for this special case.

For powers near that of peak efficiency, $Q(d\eta/dQ)$ is small relative to η and for practical purposes the approximation $k = \eta$ will be found acceptable.

Thus, as our final expression for the pitch distribution for optimum efficiency at a given torque, we have

$$\alpha \cong \sin\alpha = \frac{J}{2\pi}\frac{\mathfrak{r}(1-k)}{\mathfrak{r}^2 k + (J^2/\pi^2)} \cong \frac{J}{2\pi}\frac{\mathfrak{r}(1-\eta)}{\mathfrak{r}^2\eta + (J^2/\pi^2)} \tag{8-28}$$

As before, this expression is independent of planform and thickness distribution.

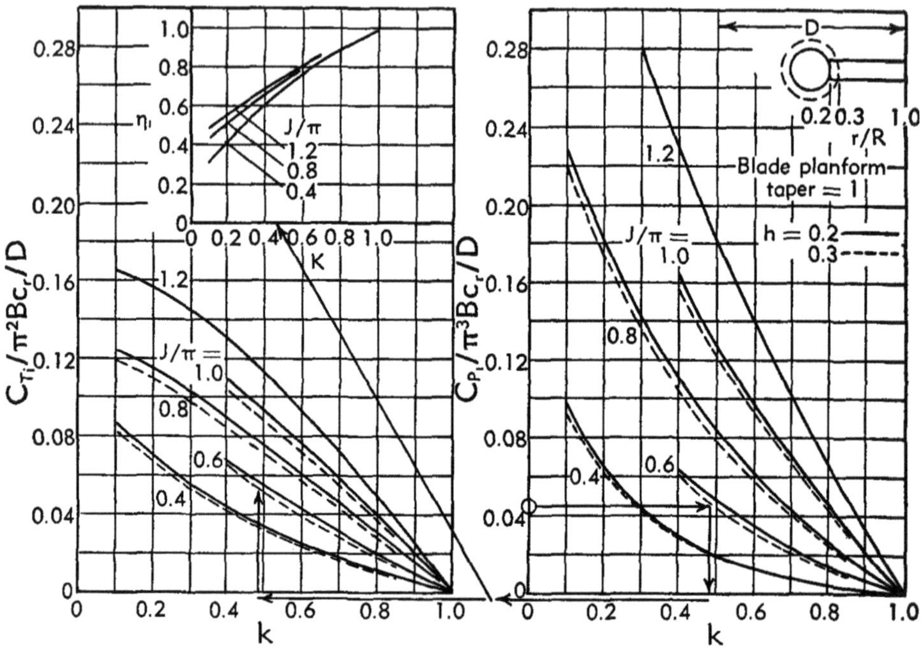

Fig. D,9a. Thrust and power coefficient per blade vs. k for ideal propeller at $M_\infty = 1.0$. Blade planform taper ratio = 1.

In Fig. D,9a, b, c, d, and e

$$\frac{C_{T_i}}{\pi^2 B(c_r/D)} = \frac{1}{M_\infty}\frac{J}{\pi}\int_h^1 I(\mathfrak{r})\left\{\mathfrak{r}\left[\frac{J}{2\pi}\frac{\mathfrak{r}(1-k)}{(J^2/\pi^2)+\mathfrak{r}^2k}\right] - \frac{J}{\pi}\left[\frac{J}{2\pi}\frac{\mathfrak{r}(1-k)}{(J^2/\pi^2)+\mathfrak{r}^2k}\right]^2\right\}d\mathfrak{r}$$

$$\frac{C_{p_i}}{\pi^3 B(c_r/D)} = \frac{1}{M_\infty}\frac{J}{\pi}\int_h^1 I(\mathfrak{r})\left\{\frac{J}{\pi}\mathfrak{r}\left[\frac{J}{2\pi}\frac{\mathfrak{r}(1-k)}{(J^2/\pi^2)+\mathfrak{r}^2k}\right] + \mathfrak{r}^2\left[\frac{J}{2\pi}\frac{\mathfrak{r}(1-k)}{(J^2/\pi^2)+\mathfrak{r}^2k}\right]^2\right\}d\mathfrak{r}$$

$$\eta_i = \frac{J}{\pi}\frac{C_{T_i}/(\pi^2Bc_r/D)}{C_{P_i}/(\pi^3Bc_r/D)}$$

D,9. Ideal Propeller. The ideal propeller is defined as that having frictionless blades of zero thickness. Such a propeller displays the maximum efficiency which can be obtained from an airfoil type propeller.

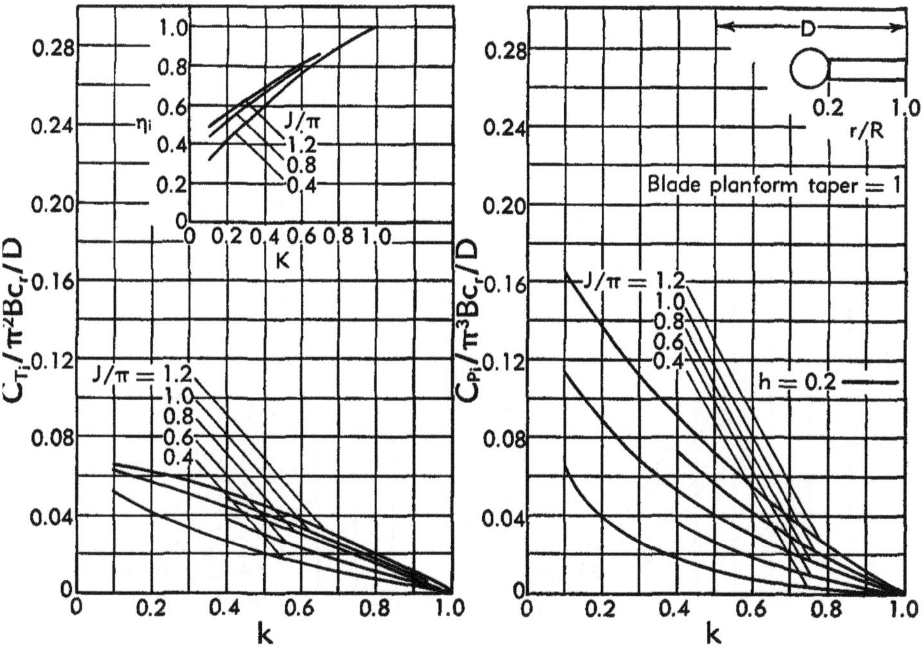

Fig. D,9b. Thrust and power coefficient per blade vs. k for ideal propeller at $M_\infty = 1.5$. Blade planform taper ratio = 1.

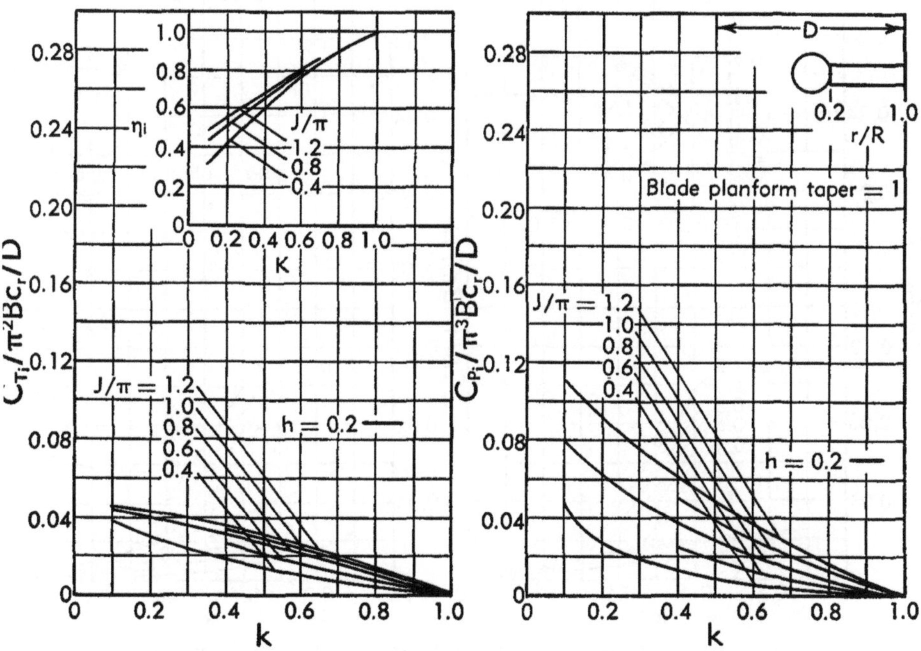

Fig. D,9c. Thrust and power coefficient per blade vs. k for ideal propeller at $M_\infty = 2.0$. Blade planform taper ratio = 1.

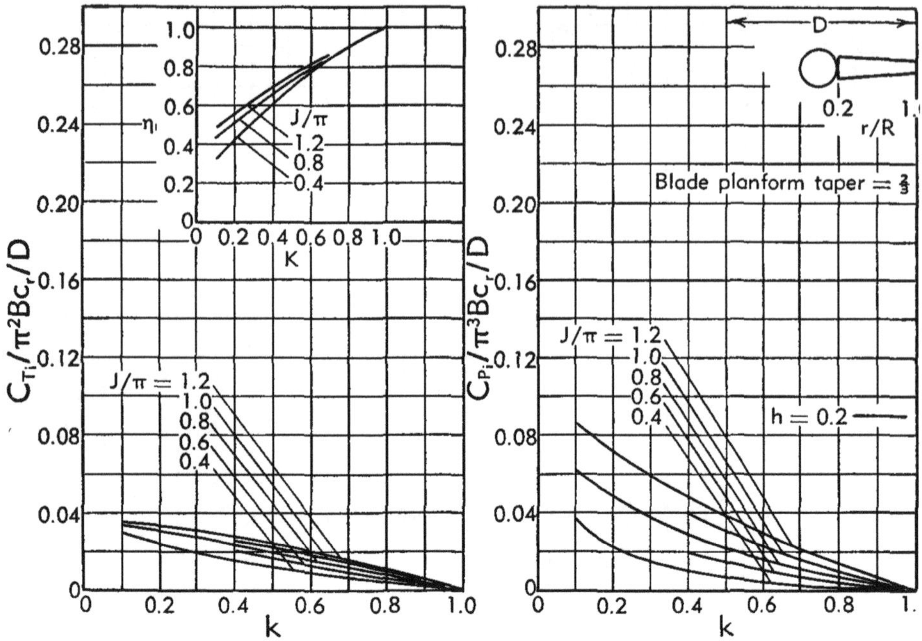

Fig. D,9d. Thrust and power coefficient per blade vs. k for ideal propeller at $M_\infty = 2.0$. Blade planform taper ratio = ⅔.

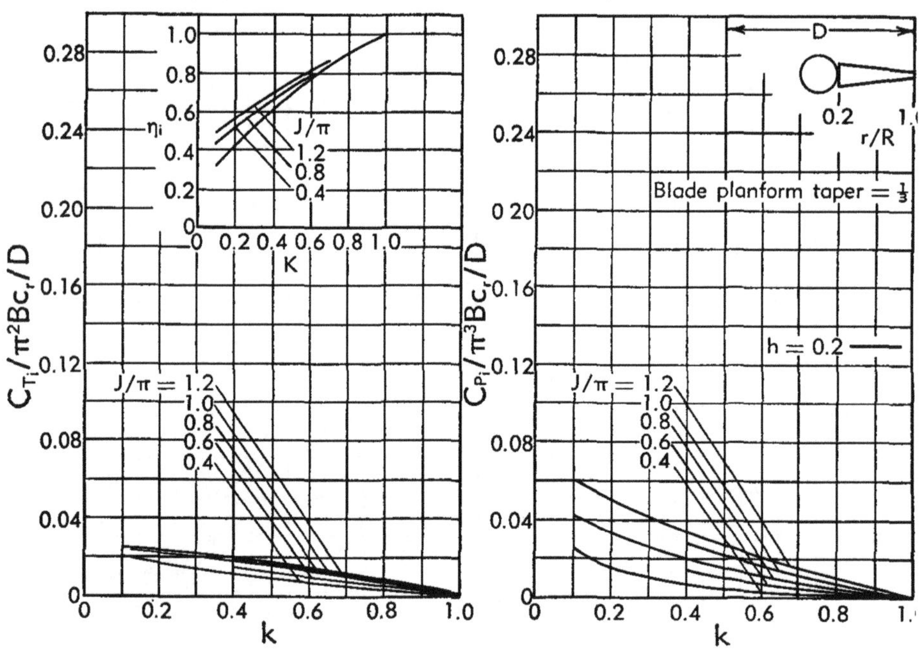

Fig. D,9e. Thrust and power coefficient per blade vs. k for ideal propeller at $M_\infty = 2.0$. Blade planform taper ratio = ⅓.

The results thus constitute an upper limit on the performance which can be expected of a propeller, the sections of which display a $C_{L}\alpha$ drag component.

The thrust and torque of an airfoil type propeller, having the optimum twist as determined in the previous articles, are given by Eq. 8-13 and 8-18. The corresponding expressions for the unit thrust and power coefficients are

$$\frac{C_T}{\pi^2 B(c_r/D)} = \frac{T}{\rho_\infty n^2 D^4} \times \frac{1}{\pi^2 B(c_r/D)} = \frac{1}{(c_r/D)} \times \frac{1}{M_\infty} \times \frac{J^2}{4\pi^2}$$
$$\times (1 - k) \int_h^1 I(\mathfrak{r}) \left\{ \frac{2\mathfrak{r}^2}{\mathfrak{r}^2 k + (J^2/\pi^2)} - \frac{J^2}{\pi^2} \frac{\mathfrak{r}^2(1 - k)}{[\mathfrak{r}^2 k + (J^2/\pi^2)]^2} \right\} \frac{c}{D}\, d\mathfrak{r}$$
$$- \frac{T_g}{\rho_\infty n^2 D^3 \pi^2 B c_r} \qquad (9\text{-}1)$$

where c_r is the root chord.

$$\frac{C_P}{\pi^3 B(c_r/D)} = \frac{Q}{\rho_\infty n^3 D^5} \times \frac{1}{\pi^3 B(c_r/D)} = \frac{1}{(c_r/D)} \times \frac{1}{M_\infty} \times \frac{J^3}{4\pi^3}(1 - k) \int_h^1 I(\mathfrak{r})$$
$$\left\{ \frac{1 + k}{k} \frac{\mathfrak{r}^2}{\mathfrak{r}^2 k + (J^2/\pi^2)} - \frac{J^2}{\pi^2} \frac{(1 - k)}{k} \frac{\mathfrak{r}^2}{[\mathfrak{r}^2 k + (J^2/\pi^2)]^2} \right\} \frac{c}{D}\, d\mathfrak{r}$$
$$+ \frac{Q_g 2\pi n}{\rho_\infty n^3 D^5 \pi^3 B(c_r/D)} \qquad (9\text{-}2)$$

For the ideal propeller, T_g and Q_g as they appear in these expressions are zero. Representative results are shown in Fig. D,9a through D,9e.

D,10. Performance of a Propeller of Given Planform and Thickness Distribution. The performance of any airfoil type propeller having a given twist distribution can be computed by evaluating Eq. 4-11 and 4-12. The performance of said propeller with the optimum pitch distribution may be determined at supersonic flight speeds by evaluating Eq. 9-1 and 9-2 for a range of values of k and plotting as in Fig. D,9a. These results automatically incorporate the optimum pitch distribution.

The performance of any propeller having one of the planforms covered in Fig. D,9a to D,9e may be rapidly determined by use of these curves. To do so, T_g and Q_g as defined by Eq. 8-14 and 8-16 are computed for the blade in question. The quantity

$$\frac{Q_g 2\pi n}{\rho_\infty n^3 D^5 \pi^3 B(c_r/D)}$$

is then subtracted from the

$$\frac{C_P}{\pi^3 B(c_r/D)}$$

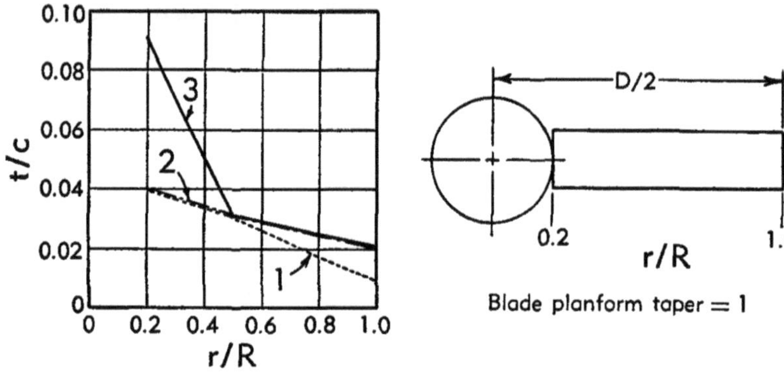

Fig. D,11a. Effect of thickness distribution on efficiency.

Distribution number	Peak efficiency	Efficiency at $\frac{C_p}{\pi^1 B(c_r/D)} = 0.44, \frac{J}{\pi} = 0.8$
1	0.83	0.805
2	0.83	0.80
3	0.775	0.755

at which the performance is sought. The result is the ideal value of

$$\frac{C_{P_i}}{\pi^3 B(c_r/D)}$$

(see Eq. 9-2). The appropriate figure is entered at this value and the corresponding k and thence

$$\frac{C_{T_i}}{\pi^2 B(c_r/D)}$$

read. For the blade in question,

$$\frac{T_g}{\rho_\infty n^2 D^4 \pi B(c_r/D)}$$

is then subtracted from

$$\frac{C_{T_i}}{\pi^2 B(c_r/D)}$$

to obtain the

$$\frac{C_T}{\pi^2 B(c_r/D)}$$

of the actual propeller corresponding to the

$$\frac{C_P}{\pi^3 B(c_r/D)}$$

in question. The efficiency is J/π times the ratio of these values.

D,11. The Effect of Planform, Thickness Distribution, and Flight Mach Number on Performance. Using the methods described in the previous articles, the performance of a series of propellers differing in thickness distribution and planform has been computed. These results are presented only to demonstrate the effect of the noted variables and are not necessarily structurally sound blades.

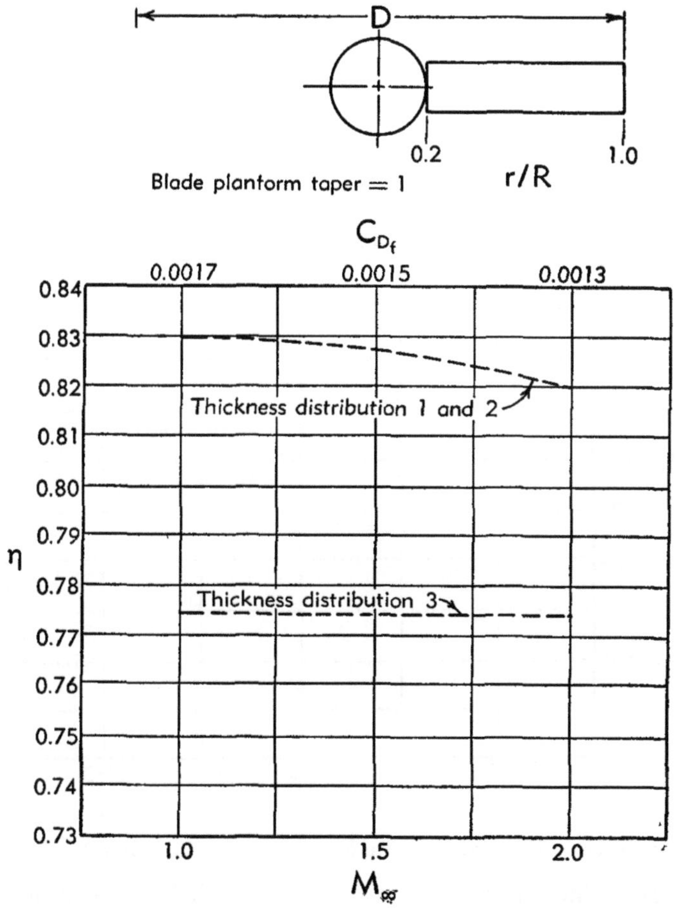

Fig. D,11b. Effect of flight Mach number on peak efficiency.

Fig. D,11a shows the effect of thickness distribution on peak efficiency and efficiency at a representative power condition. It is apparent that reduction in thickness below 2 per cent in the tip regions of the blade has little effect on performance but that appreciable thickening inboard has a definitely detrimental effect.

Fig.D,11b shows the effect of flight Mach number on peak efficiency. The gradual fall-off in efficiency is due to the reductions in section L/D

with increasing Mach number (see Fig. D,3a). Because the lift is reduced more rapidly than skin friction, L/D falls off with Mach number. This effect is less noticeable in the case of distribution 3 because, for the thicker section, skin friction is a much smaller proportion of the total drag.

Fig. D,11c shows the effect of planform taper on the efficiency of a propeller absorbing a given power in a given diameter. From the point of view of performance, tapering an overloaded blade of given thickness is

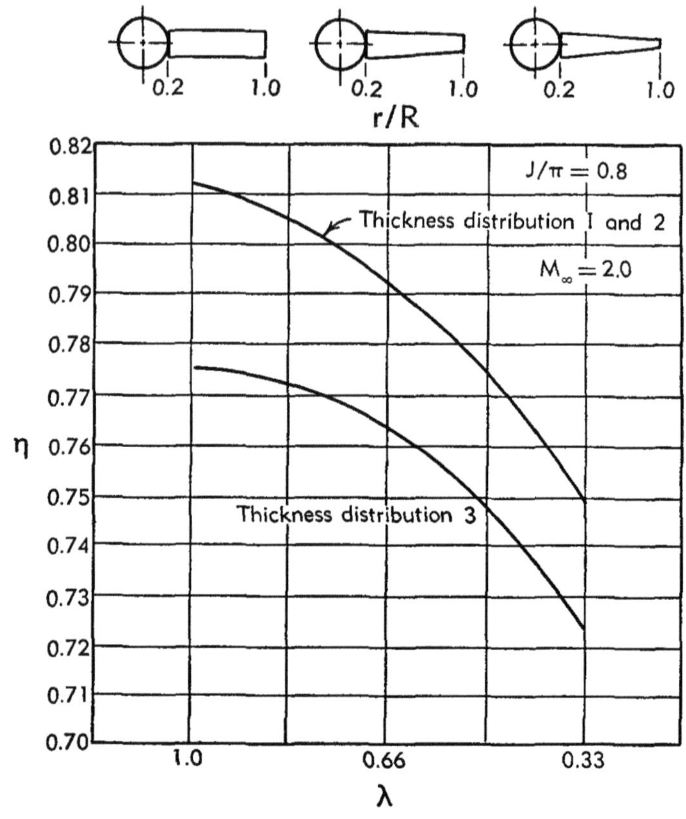

Fig. D, 11c. Effect of planform on efficiency. $C_p/\pi^3\ Bc_r/D = 0.011$.

undesirable. When structural effects are included, some form of taper may be desirable.

D,12. Practical Considerations. It is apparent from the preceding discussion that the attainment of good efficiencies depends on the use of very thin blades. The optimum thickness distribution, from an aerodynamic point of view, is zero from root to tip. Actually the thickness distribution is determined by the allowable steady and vibratory stress levels. In general the inboard thickness ratios are determined by the rigidity required to keep the natural vibratory frequencies of the blade out of the operating range, while the outboard thickness ratios are determined by flutter and durability considerations.

Planform is likewise largely determined by structural factors. At the high rpm associated with the attainment of low J at high flight Mach numbers, taper is required to limit the steady centrifugal stresses. Vibratory frequencies can also be effectively controlled by tapering the blade.

In general, aerodynamic considerations determine the over-all requirements, i.e. the necessity for thin blades and desirable rpm range; structural considerations in turn determine the required thickness ratio and planform; and aerodynamic considerations, the optimum twist distribution.

The usefulness of a propeller, no matter what its efficiency, is limited by its weight. In competition with other propulsion systems, the weight of the propeller and associated parts must be less than the weight savings resulting from the reduction in fuel load made possible by the increase in propulsive efficiency. This reduction is determined by the magnitude of the efficiency advantage, figured over all operating conditions, and the duration of flight under each of these conditions. Unless the propeller weight falls within the weight of this fuel reduction, its use cannot normally be justified. In general, from one third to one half of the propeller weight is in the blades and the remainder in the retention and pitch change mechanism, the latter being largely determined by the former. The key to a light propeller is therefore a light blade. Blade weight is determined by the type of construction employed and the ingenuity shown in meeting the requirements mentioned in the first paragraph of this article.

Finally it should be noted that no matter how efficient and how light a propeller is it must be reliable and serviceable. Reliability and serviceability are the direct result of careful design and thorough development.

Meeting all of the aforesaid requirements usually results in the most practical propeller being somewhat less efficient than that which could be built. However, by employing the aerodynamic principles outlined in the foregoing article, a practical propeller which is competitive for many high speed installations can be designed.

D,13. Cited References.

1. Goldstein, S. On the vortex theory of screw propellers. *Proc. Roy. Soc. London A123*, 440–465 (1929).
2. Theodorsen, T. *Theory of Propellers*. McGraw-Hill, 1948.
3. Nilson, E., and Smith, C. B. Optimum pitch distribution for supersonic type propellers. *Hamilton Standard Rept. HS-894*, 1951.

SECTION E

DIFFUSERS AND NOZZLES

J. C. EVVARD

CHAPTER 1. DIFFUSERS AND AIR INTAKES

E,1. One-Dimensional Equations. The process of diffusion may be defined as the conversion of velocity head or kinetic energy into pressure head or potential energy. Any geometric arrangement that accomplishes this conversion is a diffuser. The one-dimensional equations may be utilized to describe the general characteristics of a diffusion process so long as viscous effects leading to laminar or turbulent flow separation do not predominate.

In differential form and for the case without heat transfer the basic conservation equations may be summarized:

$$\text{Mass flow rate}\ \frac{d\rho}{\rho} + \frac{du}{u} + \frac{dA}{A} = 0 \tag{1-1}$$

$$\text{Momentum}\qquad dp = -\rho u du - \frac{2\rho f u^2}{D}\,dx \tag{1-2}$$

$$\text{Energy}\qquad d\left(\frac{\gamma}{\gamma - 1}\frac{p}{\rho} + \frac{1}{2}u^2\right) = 0 \tag{1-3}$$

where ρ, u, A, and p are the density, velocity, duct cross-sectional area, and static pressure, respectively, and x is the distance measured along the flow direction. The quantity $(2\rho f u^2/D)dx$ represents the skin friction axial force per unit cross-sectional area. If pipe flow exists, f represents the friction factor, and D is four times the cross-sectional area divided by the wetted perimeter, often called the hydraulic diameter. If the boundary layer is thin, f is equal to the local boundary layer skin friction coefficient. In any particularly designed flow passage, D would presumably be a known or chosen function of x.

The velocity may conveniently be nondimensionalized by means of the speed of sound a to give the Mach number

$$M = \frac{u}{a} = \frac{u}{\sqrt{\gamma p/\rho}} = \frac{u}{\sqrt{\gamma \mathcal{R} T}} \tag{1-4}$$

where a, γ, $\mathcal{R}$, and T are respectively the speed of sound, the ratio of specific heats, the specific gas constant, and the static temperature. This definition and the gas law

$$\frac{p}{\rho T} = \mathcal{R} \tag{1-5}$$

allow the basic characteristics of a one-dimensional diffusion process to be derived from Eq. 1-1, 1-2, and 1-3.

Elimination of $d\rho$ and du between Eq. 1-1 to 1-3 yields

$$\frac{dp}{p} = -\frac{\gamma \dfrac{dA}{A}}{1 - \dfrac{\gamma p}{\rho u^2}} + \frac{2(\gamma - 1)fu^2\rho\left(\dfrac{\gamma}{\gamma - 1}\dfrac{p}{\rho u^2} + 1\right)}{Dp\left(1 - \dfrac{\gamma p}{\rho u^2}\right)}dx \tag{1-6}$$

Utilizing Eq. 1-4, Eq. 1-6 becomes

$$\frac{dp}{p} = -\frac{\gamma M^2}{M^2 - 1}\frac{dA}{A} + \frac{2\gamma f M^2[1 + (\gamma - 1)M^2]}{D(M^2 - 1)}dx \tag{1-7}$$

The sign of the coefficient $-\gamma M^2/(M^2 - 1)$ is positive or negative according to whether the flow Mach number is less than or greater than unity. Thus at subsonic speeds and in the absence of friction, an increasing pressure (dp positive) must correspond to an increasing stream tube cross-sectional area, whereas at supersonic speeds a decreasing area is required to increase the pressure. The magnitude of $-\gamma M^2/(M^2 - 1)$ is very small near Mach number zero, large near unity, and asymptotically decreases to γ for values of M much larger than unity. To effect the same percentage change in pressure, a much larger relative area change is required near $M = 0$ than for any other speed. The pressure is of course most sensitive to area ratio changes near Mach number unity.

For constant area channels, friction causes a pressure drop at subsonic speeds and a pressure rise at supersonic speeds. Allowing both area changes and friction to occur simultaneously without regard to efficiency, friction promotes diffusion at supersonic speeds and demotes diffusion at subsonic speeds. If the predicted one-dimensional pressure is to remain constant (but not the velocity) the second member of Eq. 1-7 must vanish. Therefore diffusion in the sense of an increasing pressure ceases when

$$\frac{dA}{A} = \frac{2f}{D}[1 + (\gamma - 1)M^2dx] \tag{1-7'}$$

Similar results are easily derived for the differential variations of velocity, Mach number, temperature, etc., with area. Eq. 1-2 is expressed in terms of M as

$$\frac{dp}{p} = -\gamma M^2\frac{du}{u} - \frac{2\gamma M^2 f}{D}dx \tag{1-2'}$$

or, combining Eq. 1-2′ and 1-7 yields

$$\frac{du}{u} = \frac{1}{(M^2 - 1)} \frac{dA}{A} - \frac{2\gamma f M^2}{D(M^2 - 1)} dx \tag{1-8}$$

Here the trend in one-dimensional velocity with area and frictional changes is generally in the opposite sense as that for pressure. The condition for constant velocity (or Mach number from Eq. 1-11′ below) is

$$\frac{dA}{A} = \frac{2\gamma M^2 f}{D} dx \tag{1-9}$$

Integration of Eq. 1-3, utilizing the gas law and the definition of Mach number, leads directly to the relation between Mach number and static temperature:

$$T\left(1 + \frac{\gamma - 1}{2} M^2\right) = T^0 \tag{1-10}$$

where T^0, a constant, is the stagnation temperature. The temperature T from Eq. 1-10 may likewise be inserted directly into the definition of Mach number to obtain

$$\frac{M}{u\sqrt{1 + \frac{\gamma - 1}{2} M^2}} = \frac{1}{\sqrt{\gamma \mathcal{R} T^0}} = \text{const} \tag{1-11}$$

or

$$\frac{du}{u} = \frac{dM}{M\left(1 + \frac{\gamma - 1}{2} M^2\right)} \tag{1-11′}$$

If u is eliminated between Eq. 1-8 and 1-11 there results

$$\frac{dM}{M\left(1 + \frac{\gamma - 1}{2} M^2\right)} = \frac{1}{(M^2 - 1)} \frac{dA}{A} - \frac{2\gamma f M^2}{D(M^2 - 1)} dx \tag{1-12}$$

Similarly,

$$\frac{dT}{T} = -(\gamma - 1)M^2 \left[\frac{1}{(M^2 - 1)} \frac{dA}{A} - \frac{2\gamma f M^2}{D(M^2 - 1)} dx\right] \tag{1-13}$$

and

$$\frac{d\rho}{\rho} = -\frac{M^2}{(M^2 - 1)} \frac{dA}{A} + \frac{2\gamma f}{D} \frac{M^2}{(M^2 - 1)} dx \tag{1-14}$$

Another useful relation, the differential equation for the change in stagnation pressure, is obtained by combination of Eq. 1-7, 1-11, and 1-14 as

$$\frac{dp^0}{p^0} = -\frac{2\gamma f M^2}{D} dx \tag{1-15}$$

Inasmuch as all the functions in the flow process are implicitly assumed to be differentiable, which eliminates shock losses, and since losses due to flow separation are excluded by the one-dimensional assumption, the only source of stagnation pressure loss as demonstrated by Eq. 1-15 is wall friction.

The estimation of the performance of a given diffuser by one-dimensional analysis will, in general, require numerical integration. The steps in the process might be as follows:

1. Determine the relations between the cross-sectional area A, the hydraulic diameter D, and the axial distance x, and establish an independent variable, say x.
2. Estimate reasonable values of f for the particular diffuser application. These values might be taken either from turbulent pipe flow charts or from data for thin boundary layer development, depending upon the inlet flow conditions.
3. Establish the initial flow values and an increment in the independent variable.
4. The increments in p (Eq. 1-7), u (Eq. 1-8), M (Eq. 1-12), T (Eq. 1-10 or 1-13), ρ (Eq. 1-14), and p^0 (Eq. 1-15) can then be evaluated. With the exception of M, only increments of desired quantities need be evaluated.
5. Repeat in successive steps, using the new value of Mach number each time until the flow has been described.

If no losses occur in the process of diffusion, the variations of pressure, temperature, density, velocity, and area ratio with Mach number may be obtained either by integration of the appropriate equations or by writing the conservation laws in integrated form initially. The following results are obtained:

$$p^0 = p\left(1 + \frac{\gamma - 1}{2} M^2\right)^{\gamma/(\gamma-1)} \tag{1-16}$$

$$T^0 = T\left(1 + \frac{\gamma - 1}{2} M^2\right) \tag{1-10}$$

$$\rho^0 = \rho\left(1 + \frac{\gamma - 1}{2} M^2\right)^{1/(\gamma-1)} \tag{1-17}$$

$$\frac{u}{a^0} = \frac{M}{\sqrt{1 + \frac{\gamma - 1}{2} M^2}} \tag{1-18}$$

$$\frac{A}{A^*} = \frac{1}{M}\left(\frac{1 + \frac{\gamma - 1}{2} M^2}{\frac{\gamma + 1}{2}}\right)^{(\gamma+1)/2(\gamma-1)} \tag{1-19}$$

where A^* represents the duct area for Mach number unity. These functions are plotted against Mach number in Fig. E,1a. A tabulation of these and other useful relations is included in [1]. Even though the above equations were derived assuming no losses, they also represent the local isentropic conversions to stagnation conditions in a process with losses.

Another useful relation is the one-dimensional duct mass flow:

$$m = \rho u A = \frac{\gamma p}{a^0} MA \sqrt{1 + \frac{\gamma - 1}{2} M^2}$$

$$= \frac{\gamma p^0}{\sqrt{\gamma \mathcal{R} T^0}} \frac{MA}{\left(1 + \frac{\gamma - 1}{2} M^2\right)^{(\gamma+1)/2(\gamma-1)}} \tag{1-20}$$

This relation holds in a one-dimensional sense, irrespective of losses.

Losses in a diffusion process may arise from friction, from shocks, or from various other causes. These losses may be expressed in terms of the

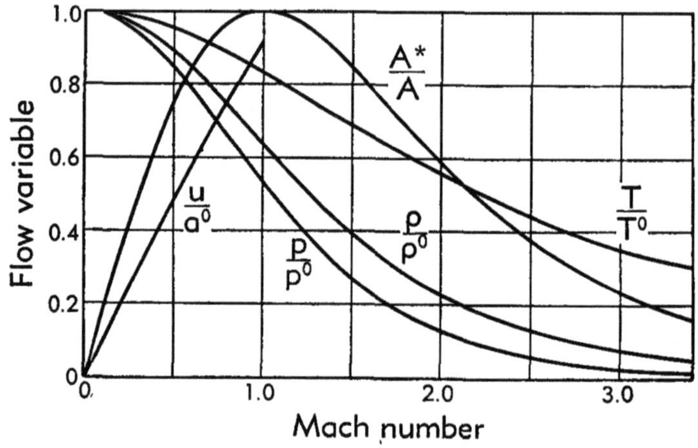

Fig. E,1a. One-dimensional flow parameters.

entrance and exit mean stagnation pressures. The general effect of decreases in stagnation pressure and any changes in stagnation temperature during diffusion may be inferred by equating the mass flow rate from Eq. 1-20 at the inlet and exit stations.

$$\frac{p_1^0}{p_2^0} \frac{A_1}{A_2} \sqrt{\frac{T_2^0}{T_1^0}} = \frac{M_2}{M_1} \left(\frac{1 + \frac{\gamma - 1}{2} M_1^2}{1 + \frac{\gamma - 1}{2} M_2^2} \right)^{(\gamma+1)/2(\gamma-1)} \tag{1-20'}$$

Thus to diffuse the flow from Mach number M_1 to M_2, the area ratio required varies inversely as the stagnation pressure ratio across the diffuser. If the local losses are due solely to friction, Eq. 1-15 may be utilized;

$$\frac{dp^0}{p^0} = -\frac{2\gamma f M^2}{D}\,dx \qquad (1\text{-}15)$$

Thus the relative skin friction pressure losses vary as the square of the local Mach number. Utilizing Eq. 1-20′ to evaluate the additional incremental area required because of total pressure losses for a given small change in Mach number there results

$$\frac{dp^0}{p^0} = -\frac{dA}{A}$$

When this expression is combined with Eq. 1-15, Eq. 1-9 again results, but this time with a broader significance;

$$\frac{dA}{A} = \frac{2\gamma f M^2}{D}\,dx \qquad (1\text{-}9)$$

Here the incremental area dA is that which is required with friction to maintain the one-dimensional potential flow velocity variation without friction. Because the equations assumed complete mixing at every station, neither Eq. 1-9 nor 1-7′ are very accurate for wall correction.

Boundary layer theory gives a somewhat different picture. For short diffusers with thin boundary layers, the geometric wall coordinates must be increased by the displacement thickness of the wall boundary layer to maintain a prescribed pressure or velocity distribution in the free stream. Because of the retardation in the velocity near the wall, the mean or one-dimensional velocity is lower than the potential flow value.

Diffuser performance parameters. Many parameters have been devised to describe the experimental performance of diffusers. Most of these parameters are based upon one-dimensional or mean flows at the inlet and exit stations of the diffuser. They must of necessity, therefore, only be approximations. For if the actual flow is converted to an equivalent one-dimensional flow by any of the usual weighting techniques, the resulting mean values of total pressure, velocity, Mach number, etc., will not simultaneously give the exact momentum, energy, mass flow, static pressure, etc., of the actual flow when these quantities are substituted into one-dimensional relations [*2,3*]. Nevertheless, the simplifications resulting from mean flow concepts greatly accelerate interpretations of experimental data, and the inaccuracies so introduced are frequently overlooked. A few of the diffuser performance parameters [*4,5*] may be listed.

1. The loss coefficient $\Delta p^0/q$. This quantity gives the loss in total pressure relative to the inlet q. As may be surmised from Eq. 1-15 this quantity is relatively independent of Mach number and density for low speed flows.

2. The diffusion factor $\Delta q/(\Delta q)_{\text{ideal}}$. This quantity measures the effectiveness of cross-sectional area changes in converting impact pressure to static pressure. Any process such as thickening of the boundary layer or flow separation from the wall reduces the diffusion factor.
3. The pressure efficiency $|\Delta p/\Delta q|$. This quantity measures how much of the impact pressure change through the diffuser is actually converted to static pressure rise.
4. The diffuser effectiveness $\Delta p/(\Delta p)_{\text{ideal}}$. This quantity describes the relative ability of a diffuser to actually obtain static pressure rise. The diffuser effectiveness is of course the product of the diffusion factor and the pressure efficiency for an adiabatic incompressible fluid. Any losses or flow separations which lead to changes in the effective area ratios as compared to the actual area ratios will decrease the diffuser effectiveness.

Most of the above parameters arise directly from the experimental measurements and therein lies their usefulness. Other quantities such as actual velocity ratio, actual-to-ideal velocity difference ratio, actual area ratio, actual-to-ideal area ratio for the same amount of diffusion, actual Mach number ratio, final-to-initial stagnation pressure ratio, etc. could be employed. The latter quantity is particularly useful to describe the internal performance of supersonic inlets. This ratio is directly related to the required area ratio for a given inlet and exit Mach number by Eq. 1-20′. When the diffuser exit Mach number is constant, as it nearly is on a turbojet engine diffuser at a given speed and altitude, the mass flow through the engine is directly proportional to the total pressure recovery ratio.

Also useful for defining diffuser performance are the various efficiencies. These are generally defined by the relation

$$\eta = 1 - \frac{\text{losses}}{\text{input}} \tag{1-21}$$

While the input and the losses may feasibly be calculated using any quantity which has dimensions of energy, only two (the over-all kinetic energy efficiencies) are discussed here.

An adiabatic diffusion process may be imagined between stations 0 and 1 (see Fig. E,1b) which entails a loss in total pressure and a change in static pressure and velocity. This flow may then be accelerated isentropically to station 2 which is at the original static pressure. The initial kinetic energy per unit mass flow rate is $u_0^2/2$. Since the static pressure is unaltered for the flow at station 2, the loss in kinetic energy, that is the decrease in the available kinetic energy per unit mass flow rate is

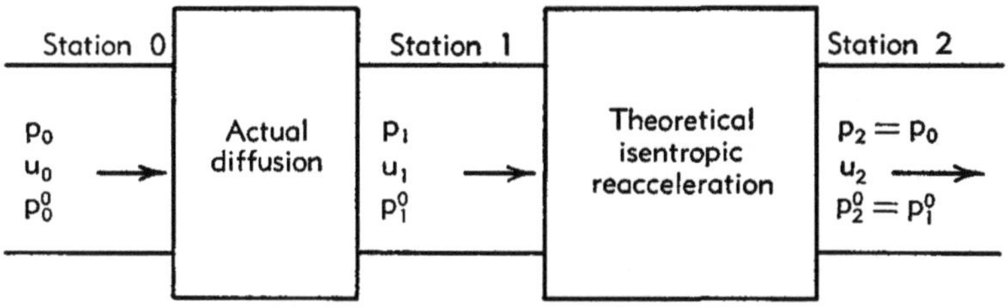

Fig. E,1b.

$(u_0^2 - u_2^2)/2$ = loss. This loss occurred in the diffuser section between stations 0 and 1. Then the kinetic energy efficiency of the diffuser [*6*] is

$$\eta_k = 1 - \frac{u_0^2 - u_2^2}{u_0^2} = \frac{u_2^2}{u_0^2} = \frac{\text{output kinetic energy available for diffusion}}{\text{input kinetic energy available for diffusion}} \tag{1-22}$$

A related efficiency has been defined by Bailey [*7*] in terms of the change of kinetic energy between stations 0 and 2 rather than of the initial value. This efficiency is

$$\eta_B = 1 - \frac{u_0^2 - u_2^2}{u_0^2 - u_1^2} \tag{1-23}$$

Both of these efficiencies may be simply related to the total pressure ratio by means of Bernoulli's equation for adiabatic flow

$$\frac{\gamma}{\gamma - 1}\frac{p_0}{\rho_0} + \frac{1}{2}u_0^2 = \frac{\gamma}{\gamma - 1}\frac{p_0^0}{\rho_0^0} = \frac{\gamma}{\gamma - 1}\frac{p_0}{\rho_2} + \frac{1}{2}u_2^2 = \frac{\gamma}{\gamma - 1}\frac{p_1^0}{\rho_2^0} \tag{1-24}$$

Utilizing the isentropic relations

$$\frac{p_0^0}{(\rho_0^0)^\gamma} = \frac{p_0}{\rho_0^\gamma}$$

and

$$\frac{p_1^0}{(\rho_2^0)^\gamma} = \frac{p_0}{\rho_2^\gamma}$$

the velocity squares become

$$\begin{aligned} u_0^2 &= \frac{2\gamma}{\gamma - 1}\frac{p_0^0}{\rho_0^0}\left[1 - \frac{1}{(p_0^0/p_0)^{(\gamma-1)/\gamma}}\right] \\ u_2^2 &= \frac{2\gamma}{\gamma - 1}\frac{p_0^0}{\rho_0^0}\left[1 - \frac{1}{(p_1^0/p_0)^{(\gamma-1)/\gamma}}\right] \end{aligned} \tag{1-25}$$

Noting that $(p_0^0/p_0)^{(\gamma-1)/\gamma} = 1 + [(\gamma - 1)/2]M_0^2$ while substituting Eq. 1-24 into Eq. 1-22, η_k becomes

$$\eta_k = 1 + \frac{2}{(\gamma - 1)M_0^2}\left[1 - \left(\frac{p_0^0}{p_1^0}\right)^{(\gamma-1)/\gamma}\right] \tag{1-22'}$$

This is the required relation between η_k, M, and p_1^0/p_0^0. Since the efficiency can never exceed unity, the stagnation pressure ratio p_1^0/p_0^0 must clearly approach unity as the initial Mach number goes to zero. This insensitivity is the reason for the general absence of stagnation pressure ratio as a measure of subsonic diffuser performance, where a poor diffuser may have stagnation pressure ratios of nearly unity. This disadvantage disappears at the higher Mach numbers.

The velocity squares of Eq. 1-24 for station 0 and a corresponding value for station 1 may be written in terms of the Mach numbers at those positions by Eq. 1-18

$$u_0^2 = (a^0)^2 \frac{M_0^2}{1 + \frac{\gamma - 1}{2} M_0^2} \qquad u_1^2 = (a^0)^2 \frac{M_1^2}{1 + \frac{\gamma - 1}{2} M_1^2}$$

so that

$$\frac{u_0^2 - u_1^2}{u_0^2} = \frac{M_0^2 - M_1^2}{M_0^2\left(1 + \frac{\gamma - 1}{2} M_1^2\right)} \tag{1-26}$$

The efficiency η_B of Eq. 1-23 becomes

$$\eta_B = 1 + \frac{2}{(\gamma - 1)M_0^2}\left[1 - \left(\frac{p_0^0}{p_1^0}\right)^{(\gamma-1)/\gamma}\right] \frac{M_0^2\left(1 + \frac{\gamma - 1}{2} M_1^2\right)}{M_0^2 - M_1^2} \tag{1-23'}$$

If M_1 is zero this reduces to η_k as it should. The relation between the two kinetic energy efficiencies is

$$\eta_B = 1 + (1 - \eta_k) \frac{M_0^2\left(1 + \frac{\gamma - 1}{2} M_1^2\right)}{M_0^2 - M_1^2} \tag{1-27}$$

Still another efficiency has been defined in differential form by Crocco [8]. His concept is that the friction term of Eq. 1-2 absorbs some of the incremental change in kinetic energy which would otherwise be converted to pressure. The quantity dp/ρ is therefore less than the incremental energy change, $-d(u^2/2)$ by a factor η, the Crocco efficiency. The correspondingly modified momentum equation then becomes

$$\frac{dp}{\rho} = -\eta u du \tag{1-28}$$

When the velocity is eliminated by means of the energy equation (1-3), there results

$$\left(\frac{\eta\gamma - \gamma + 1}{\eta\gamma}\right)\frac{dp}{p} = \frac{d\rho}{\rho} \tag{1-29}$$

If the equation is simplified by holding η constant,

$$\left(\frac{p_1}{p_0}\right)^{(\eta\gamma-\gamma+1)/\eta\gamma} = \frac{\rho_1}{\rho_0} \tag{1-30}$$

Utilizing the gas law to eliminate the density ratio gives

$$\left(\frac{p_1}{p_0}\right)^{(\gamma-1)/\eta\gamma} = \frac{T_1}{T_0} = \frac{1 + \frac{\gamma-1}{2} M_0^2}{1 + \frac{\gamma-1}{2} M_1^2}$$

by Eq. 1-10. Elimination of the static pressure by relation (1-16) leads to the equation relating stagnation pressure ratio and Crocco efficiency

$$\frac{p_1^0}{p_0^0} = \left(\frac{1 + \frac{\gamma-1}{2} M_0^2}{1 + \frac{\gamma-1}{2} M_1^2}\right)^{(\eta-1)\gamma/(\gamma-1)} \tag{1-31}$$

Liepmann and Puckett [*9*], state that a satisfactory value of η for the subsonic diffusers of supersonic wind tunnels is 0.75 (M_0 would then be the subsonic Mach number immediately downstream of the normal shock). This value is in agreement with the summary curves of Crocco [*8*].

E,2. Effects of Boundary Layer Separation. According to the one-dimensional equations summarized in Fig. E,1a, diffusion may be accomplished slowly or rapidly merely by choosing the required area variation of the walls. The inclusion of a friction factor (Eq. 1-9 and 1-15) shows that a slight increase in the potential wall boundaries is required to overcome viscous effects. No difficulty is yet apparent from the one-dimensional equations to prevent diffusion in almost arbitrary duct shapes. Yet extreme difficulties occur in practice when diffusion is too rapid or if turns and bends are required in the flow. The flow no longer follows the wall contour but chooses its own path within the duct, usually on a stream boundary of nearly constant pressure. This phenomenon is commonly referred to as separation because the main stream no longer follows the wall boundary.

Flow separation is of course related to the viscous drag which reduces the local flow velocity along the wall surface. The low energy boundary layer air adjacent to the wall may have neither sufficient inertia nor mixing to overcome the adverse pressure gradient which always accompanies a diffusion process. The fluid near the wall then becomes stationary or actually flows in a direction opposite to that of the main stream. Such reversed flow will most assuredly force the free stream away from the wall in the region of separation, and the geometric boundary then in part loses its physical significance for one-dimensional flow.

The real problem of subsonic diffuser design is to accomplish diffusion without separation. The solution to the problem depends not only on the specific design criteria but also on the uniformity and nature of the initial or entering flow. Turns, bends, and area expansions should be gradual and free of discontinuities which invite separation.

Laminar separation. Some insight into the separation problem may be gleaned from the techniques of laminar boundary layer theory. Pohlhausen [*10*] has investigated the development of boundary layers under an adverse pressure gradient corresponding to two-dimensional incompressible radial flows whose free stream velocity U varies with radial position x as shown in Fig. E,2a. In other words, $U = c/x$ where c is a constant. His technique is to assume that the boundary layer profile can be represented by a power series to fourth order terms. The coefficients of the series expansion are evaluated by wall and stream boundary conditions and the Kármán momentum equation. He obtained the noteworthy

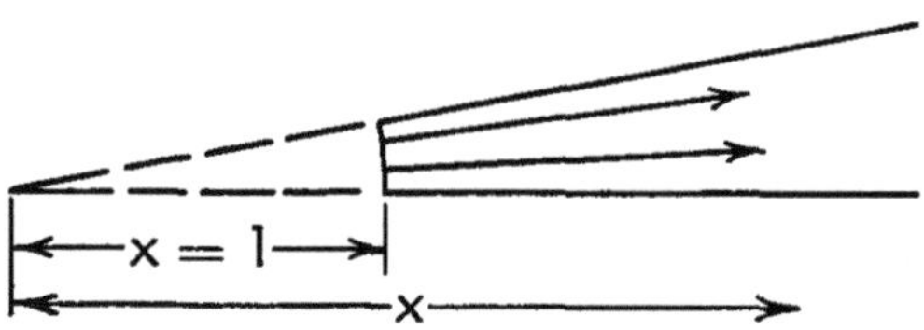

Fig. E,2a

result that even with arbitrarily small divergence angles, a separation point may always be specified. If the dimensions are normalized so that the diffuser entrance corresponds to $x = 1$, the separation point occurs at $x = 1.214$. In other words, only an area expansion of 1.214 to 1 would be permitted without separation, irrespective of the diffusion angle. This value is not physically binding because the pressure distribution deviates from the assumed value near the separation point. Even though these simplified calculations of Pohlhausen may be inaccurate, they emphasize the sensitivity of a laminar boundary layer to the adverse pressure gradients which occur during diffusion. This sensitivity is also apparent from the exact solutions of Hamel [*11*], in which reversed flow regions occur except for very small Reynolds numbers and angles of wall divergence.

Also of interest are the calculations of Stewartson [*12*] for compressible laminar boundary layers. In a decreasing velocity field,

$$U = U_1(1 - x)$$

for which the coordinate scale has been normalized to unity, the separation points are located at distances x_s and Mach number M_s in accordance with Table E,2.

Table E,2. Laminar separation points.

M_∞	0	1	2	3	4	6	8	10
x_s	0.120	0.110	0.096	0.077	0.062	0.044	0.032	0.024
M_s	0	0.87	1.69	2.48	3.18	4.51	5.76	6.96

To obtain these values Stewartson applied a method of Howarth [*13*]. The Kármán-Pohlhausen method [*10*] gives for comparison at $M_\infty = 10$, $x_s = 0.052$ and $M_s = 5.5$, which occur later in the diffusion process than the values of the table. According to Stewartson, Howarth's method should overestimate the point of separation and is therefore a closer approximation.

Turbulent separation. The calculations for laminar separation give a far more severe limitation on the ability of a pipe or channel to diffuse than is observed in practice. If the flow is initially laminar, transition or the processes associated with laminar separation soon convert the flow to turbulent. The mixing processes associated with the vortices of turbulent flow usually add enough momentum transfer to the low energy air adjacent to the wall to prevent separation, or perhaps to quickly reattach a small region of laminar separation. The practical limitation on the ability of a channel to diffuse subsonically is then established by criteria for turbulent separation.

Unfortunately, turbulent separation phenomena are so poorly understood that practically all methods of prediction depend upon simplified analyses employing the Kármán integral equation for momentum and empirical relations obtained from experimental observations. Among the common references are the work of Buri [*14*], Gruschwitz [*15*], Von Doenhoff and Tetervin [*16*], Kalikhman [*17*], Garner [*18*], and Coleman [*19*]. Each of these references proposes a different form factor as a single parameter that will describe the shape of the velocity distribution curve and the separation criteria from local conditions in the mean flow. Likewise the method of Maskell [*20*] should be included because of its simplicity. Dryden [*21*] presents an excellent summary of the earlier methods and suggests on the basis of National Bureau of Standards data that the characteristics of a turbulent layer cannot depend solely on the local mean velocity and its derivatives but must also depend on conditions upstream which influence the local fluctuating velocity components.

Although the purpose of this article is not to describe in detail turbulent separation processes, a brief summary of [*16*] is considered advisable. In that reference, criteria are presented for the prediction and avoidance of turbulent separation on wings or in channels for which thin boundary layer theory might apply.

The Kármán momentum equation for an incompressible fluid may be written

$$\frac{d\theta}{dx} + (H + 2)\frac{\theta}{U}\frac{dU}{dx} = \frac{\tau_w}{\rho U^2} \tag{2-1}$$

where

θ momentum thickness $= \int_0^\infty \frac{u}{U}\left(1 - \frac{u}{U}\right) dy$

H form factor δ^*/θ

δ^* displacement thickness $= \int_0^\infty \left(1 - \frac{u}{U}\right) dy$

U stream velocity outside the boundary layer

u local velocity at distance y from wall

τ_w wall shearing stress

ρ density of stream outside the boundary layer

The shape parameter H is determined by the velocity distribution in the boundary layer and has a value 1.286 for a $\frac{1}{7}$ power law profile. Experimentally the value of H rises as separation is approached, and at separation in an incompressible fluid has a value of about 2.6.

Von Doenhoff and Tetervin proposed the parameter $2\rho UU'\theta/\tau_w$ (almost that of Buri) to describe the effect of external forces acting on the boundary layer. The factor $\rho UU'$ is the pressure gradient so that the parameter represents the product of the pressure gradient, the momentum thickness, and the reciprocal of the wall shearing stress. From experimental data the quantity $\theta(dH/dx)$ was determined as a function of H and the pressure gradient parameter $2\rho UU'\theta/\tau_w$. The empirical equation relating these quantities is

$$\theta\frac{dH}{dx} = e^{4.680(H-2.975)}\left[-\frac{2\rho UU'\theta}{\tau_w} - 2.035(H - 1.286)\right] \tag{2-2}$$

Thus two simultaneous first order differential equations (2-1 and 2-2) are obtained for H and θ.

For purposes of designing two-dimensional diffusers, the value of H may be chosen as a constant somewhat less than the critical value, 2.6, for separation. Eq. 2-1 and 2-2 then simplify to the form

$$\frac{2\rho UU'\theta}{\tau_w} = -2.035(H - 1.280) = C_1 \tag{2-3}$$

$$\frac{d\theta}{dx} = \frac{\tau_w}{\rho U^2}\left[1 - \frac{C_1(H + 2)}{2}\right] = \frac{C_2\tau_w}{\rho U^2} \tag{2-4}$$

Elimination of τ_w and integration yield

$$\left(\frac{\theta}{\theta_1}\right)^{C_1/C_2} = \frac{U^2}{U_1^2} \tag{2-5}$$

where θ_1 and U_1 are initial values. A plot of C_1/C_2 and C_1 as functions of H is presented in Fig. E,2b.

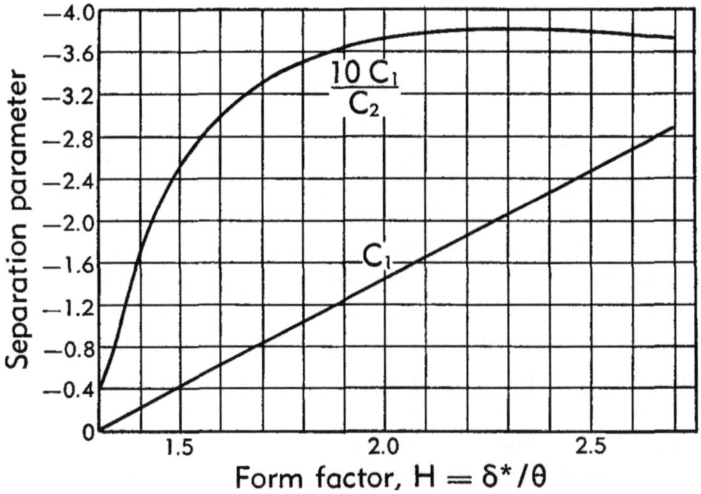

Fig. E,2b. Turbulent separation parameters.

The design of two-dimensional diffusers may be estimated from Eq. 2-3, 2-4, and 2-5. For this purpose, the simpler Falkner relation [*22*],

$$\frac{\tau_w}{\rho U^2} = 0.006534\left(\frac{\theta U}{\nu}\right)^{-\frac{1}{6}} \tag{2-6}$$

is employed herein for the shearing stress rather than the more complicated Squire-Young formula [*23*] utilized by Von Doenhoff and Tetervin. Eq. 2-5 and 2-6 then combine to give

$$\frac{\tau_w}{\rho U^2} = 0.006534\left(\frac{\theta_1 U_1}{\nu}\right)^{-\frac{1}{6}}\left(\frac{\theta}{\theta_1}\right)^{-\frac{1}{6}[1+(C_1/2C_2)]} \tag{2-6$'$}$$

Eq. 2-4 integrates to

$$\frac{\theta}{\theta_1} = \left[1 + \frac{0.006534}{\theta_1}\left(\frac{14C_2 + C_1}{12}\right)\left(\frac{\theta_1 U_1}{\nu}\right)^{-\frac{1}{6}}(x - x_1)\right]^{12C_2/(14C_2+C_1)} \tag{2-7}$$

Elimination of θ/θ_1 from Eq. 2-5 and the use of the incompressible continuity equation yields

$$\frac{A_1}{A} = \frac{U}{U_1} = \left[1 + 0.006534\left(\frac{14C_2 + C_1}{12}\right)\left(\frac{\theta_1 U_1}{\nu}\right)^{-\frac{1}{6}}\frac{(x - x_1)}{\theta_1}\right]^{6C_1/(14C_2+C_1)} \tag{2-8}$$

where A represents the cross-sectional area per unit channel width.

Eq. 2-8 clearly demonstrates that the ability of a channel to diffuse through a specified velocity ratio depends on the quality and velocity of the initial flow. The length required for diffusion $(x - x_1)$ varies directly with $\theta^{5/4}$ and $U_1^{1/4}$. Also of interest are the initial total divergence angles of diffusion permitted. These are

$$\tan \delta = \left(\frac{dA}{dx}\right)_1 = -0.003267C_1 \left(\frac{\theta_1 U_1}{\nu}\right)^{-1/4} \frac{A_1}{\theta_1} \tag{2-9}$$

Thus because $-C_1$ increases with H, the danger of turbulent separation increases with the diffusion angle δ. Likewise the smaller the initial boundary layer thickness relative to the channel width and the smaller the turbulent Reynolds number, the larger the initial divergence angle can be without separation. While Eq. 2-8 and 2-9 have presented criteria for avoiding separation, the effects of wall curvature have been neglected except as a means for creating adverse pressure gradients. If the diffuser has no initial boundary layer, the Reynolds number and initial momentum thickness of Eq. 2-9 would be the value at the transition point.

Though the applicability of the assumptions may be questioned even for an incompressible fluid, the final results nevertheless emphasize the need for evaluating the inlet flow on any scientifically performed experiments on diffusers. Examples of such evaluations are included in [*4,5,24*]. It may be added that Nitzberg and Crandall [*25*] have studied turbulent separation at transonic speeds. They employed a modified Gruschwitz-type analysis rather than extend the work of Von Doenhoff and Tetervin to compressible fluids. Also, Rubert and Persh [*26*] have employed turbulent boundary layer theory to aid in the understanding of diffusion processes.

Unfortunately, turbulent separation criteria, even if exact, would not furnish all the numbers for designing optimum diffusers. In an actual diffuser, the losses caused by small regions of separation on a short diffuser must be balanced against the increased skin friction losses on a long or gradual diffuser which has no separation. This balance in turn requires information about the point of separation, the extent of the separation region prior to flow reattachment, and the losses incurred in a separation and reattachment process. The experimental approach to this problem still predominates.

Optimum diffusion angles. Most of the connected-pipe subsonic diffuser tests reported in the literature do not include an evaluation of the inlet boundary layer profiles. To a first approximation such evaluations may not be necessary if fully developed turbulent pipe flow enters the diffuser. An excellent summary of the literature on such diffusers is given by Patterson [*27*]. He plots the pressure efficiency

$$\frac{\Delta p}{\Delta q} = \frac{p_2 - p_1}{\frac{1}{2}\rho U_1^2[1 - (A_1/A_2)^2]}$$

from the literature for a number of diffuser types against total divergence angle and concludes that the efficiencies are high only at small divergence angles. The efficiency drops rapidly with increasing divergence angles beyond the optimum, until at sudden expansion it is nearly the theoretical value given by Borda,

$$\frac{\Delta p}{\Delta q} = \frac{2}{1 + (A_2/A_1)}$$

The divergence angle for maximum efficiency depends on the shape of the channel. For conical diffusers, optimum total divergences lie in the range from 5° to 8° of expansion; square ducts would presumably have about the same adverse pressure gradients and the same range of optimum divergence angles. Because of corner effects, however, the square channel should be less efficient [*27*, Fig. 5]. The optimum angle for two-dimensional diverging channels (two walls parallel) is of course greater, and for a square entrance section is about 11°.

As demonstrated by the theory of Von Doenhoff and Tetervin (Eq. 2-8 and 2-9), initial turbulence reduces the ability of a channel to diffuse. Peters [*28*] found that a change of efficiency of approximately 7 per cent occurred between tests with initially uniform flow and initially fully developed turbulent flow. The experiments were conducted on conical diffusers with total divergence angles ranging from 5° to 30°.

Also of interest is the continued rise of pressure in the cylindrical duct following the conical section. The pressure reaches a maximum at from two to six duct diameters downstream of the conical outlet, but the position depends upon the angle of divergence and on the character of the initial inlet flow. As the divergence angle increases, or as the initial inlet boundary layer thickens, the distance downstream from the conical section outlet to the position of maximum pressure increases. (For further details see [*27*].)

Diffusers for aircraft. The installation of diffusers on aircraft is subject to compromises in design due to structural limitations, space limitations, engine location, etc. Unfortunately, each configuration has its own individual set of compromises and hence generalizations are not practical in this article. Suffice it to say that good duct designs attempt to minimize the use of high local curvature, high local pressure gradients, bends, turns, and transition sections, etc. Area changes should be gradual for sections operating near Mach number of unity, preferably with both first and second derivatives of the cross-sectional area (and also shape) with axial distance zero for duct Mach numbers of unity. The transonic portion of the duct, however, should not be so long as to incur excessive friction losses.

For subsonic flow, the performance of the diffuser is remarkably sensitive to details near the inlet station. At take-off, the inlet velocity may be many times the airplane or free stream velocity. The limiting streamline which just enters the inlet must attach itself to the cowl at a stagnation point with all the bounded flow entering the inlet. This attachment occurs with lower losses if the inlet lip (of a nose or scoop inlet) is rounded. Likewise, to minimize the external drag associated with the inlet, the external contour of the lip should also fair smoothly to the rest of the aircraft structure, the resulting inlet contour in some respects resembling the nose portion of an airfoil section. This rounding is important to capitalize on inlet "leading edge suction" and hence to reduce the airplane drag.

Proper sizing of the inlet to the engine is also of importance; if the inlet is too small, a high inlet-to-free-stream velocity ratio will be required to give the engine compressor the necessary entrance Mach number. This high velocity ratio may lead to entrance or duct losses associated with turbulent separation. Choking may also occur in the diffuser duct, and the resulting supersonic expansion and shock structure downstream of the sonic throat can precipitate poor diffuser recovery. Thus a well-designed inlet may perform poorly if it is not properly sized to the engine for optimum performance.

Since the size of an optimum inlet for a given engine varies with both flight Mach number and ambient temperature, variable geometry configurations may be required. For turbojet engines, the largest inlet size is required for the lower flight speeds. Hence the use of blow-in doors or other auxiliary air passages may be necessary for take-off. Both the matching problem and the performance penalties due to the use of sharp-lipped inlets are discussed in greater detail in Art. 3.

A number of inlet types have been utilized on the various aircraft configurations, among them the nose inlet, the scoop inlet, the wing inlet, the wing root inlet, the *NACA* submerged inlet, and others. For details in design and performance of the various types, refer to [*29,30,31,32,33,34*]. For instability problems associated with twin duct diffusers discharging into a common plenum, see [*35*].

E,3. Supersonic Diffusion Processes. There are essential differences between the perfect fluid diffusion processes of subsonic and supersonic flow. As was evident from Eq. 1-7, subsonic diffusion is accomplished by a divergence of the stream tube cross-sectional area whereas supersonic diffusion requires a convergence. The combination of a supersonic and subsonic diffuser would thus require a convergent-divergent stream passage as schematically shown in Fig. E,3a.

The amount of convergence necessary to achieve isentropic diffusion follows the curve of Fig. E,1a. Clearly, the higher the Mach number for a given inlet size, the smaller must be the diffuser throat or minimum cross section.

The diffuser may be likened to a reversed De Laval nozzle. In the attempt to establish the supersonic flow in this nozzle, lower than design Mach numbers must of course occur. At these lower Mach numbers the internal contraction (Fig. E,1a) is too great to pass the mass flow and hence spillage around the inlet lip is necessary. Associated with the spillage, a strong shock or bow wave configuration forms ahead of the inlet. Such a bow wave approximates a normal shock over a portion of the entering flow.

The normal shock discontinuously converts the supersonic free stream to subsonic, and this subsonic flow accelerates to sonic velocity at the throat. The stream tube contraction required to achieve sonic velocity, however, is less than that of the reversed De Laval nozzle in proportion to the loss in stagnation pressure across the shock; hence such a diffuser would never establish the supersonic flow at the design Mach number.

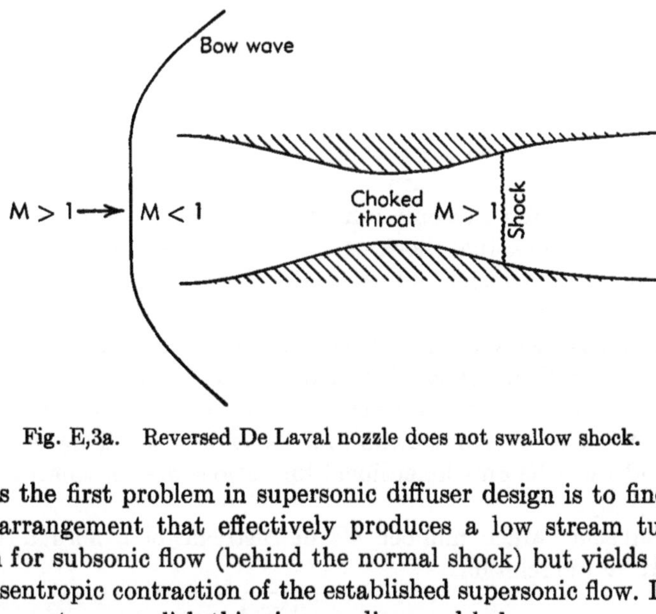

Fig. E,3a. Reversed De Laval nozzle does not swallow shock.

Thus the first problem in supersonic diffuser design is to find a geometric arrangement that effectively produces a low stream tube contraction for subsonic flow (behind the normal shock) but yields a higher nearly-isentropic contraction of the established supersonic flow. Diffusers which in part accomplish this aim are discussed below.

Normal Shock, Divergent Diffuser. This diffuser gives no internal or external contraction to the supersonic stream. The supersonic diffusion therefore occurs abruptly through a normal shock, which is followed by a divergent subsonic diffuser passage. The best pressure recovery is just that of a normal shock at the free stream Mach number, i.e.

$$\frac{p_1^0}{p_\infty^0} = \left(\frac{\frac{\gamma+1}{2}}{\gamma M_\infty^2 - \frac{\gamma-1}{2}}\right)^{1/(\gamma-1)} \left(\frac{\frac{\gamma+1}{2} M_\infty^2}{1 + \frac{\gamma-1}{2} M_\infty}\right)^{\gamma/(\gamma-1)} \tag{3-1}$$

The position of the shock ahead of the inlet is controlled by the mass flow through the diffuser until the normal shock just contacts the sharp inlet lip. In this regime the pressure recovery ratio is constant except for subsonic diffuser losses and hence the mass flow is varied, say, by adjusting some choking area downstream of the inlet. After the shock enters the diffuser, the mass flow is constant and the Mach number ahead of the shock increases above the stream value as the shock progresses down the divergent passage. The program of pressure recovery ratio p_2^0/p_∞^0 and outlet area A_2 follows the constant mass flow hyperbolic schedule,

$$\frac{p_2^0}{p_\infty^0}\frac{A_2}{A_\infty} = \sqrt{\frac{T_2^0}{T_\infty^0}}\, M_\infty \left(\frac{\dfrac{\gamma+1}{2}}{1 + \dfrac{\gamma-1}{2} M_\infty^2} \right)^{(\gamma+1)/2(\gamma-1)} \qquad (1\text{-}20'')$$

This equation applies equally well for all supersonic diffuser types in the supercritical or constant mass flow regime irrespective of the sources of pressure loss.

If the inlet lips are blunt, bow waves will remain detached from them, curling back into the inlet toward the central portion of the flow. Whether the inlet lips are blunt or sharp, the idealized normal shock, pictured as progressing down the subsonic diffuser as the outlet area is increased, does not occur. The actual shock pattern contains a complicated array of lambda shocks, expansion waves, and separated wall boundary layer regions. This complicated pattern is generated by nonuniformities in the flow field ahead of the shock and by effects of the wall boundary layer.

Convergent-Divergent Diffusers. The limitations on the contraction ratio of convergent-divergent diffusers was first discussed in the United States by Kantrowitz and Donaldson [*6*]. They and others have found that the one-dimensional limitations are reasonably verified experimentally.

The subsonic Mach number M_1 downstream of a normal shock is computed by the relation

$$M_1 = \sqrt{\frac{1 + \dfrac{\gamma-1}{2} M_\infty^2}{\gamma M_\infty^2 - \dfrac{\gamma-1}{2}}} \qquad (3\text{-}2)$$

If the maximum allowable contraction ratio is employed, all of the flow downstream of the shock at the lip enters the inlet and is accelerated to sonic velocity at the throat. From Eq. 1-19, this limiting area ratio is

$$\left(\frac{A}{A^*}\right)_{\max} = \frac{\left(\dfrac{\gamma+1}{2} M_\infty^2\right)^{(\gamma+1)/2(\gamma-1)}}{\left(1 + \dfrac{\gamma-1}{2} M_\infty^2\right)^{\frac{1}{2}} \left(\gamma M_\infty^2 - \dfrac{\gamma-1}{2}\right)^{1/(\gamma-1)}} \qquad (3\text{-}3)$$

This ratio allows all the mass flow from a free stream tube the size of the inlet to enter the diffuser. Hence the normal shock may be swallowed to establish the supersonic flow.

The shock cannot be stably located in the convergent section of the inlet. As the free stream shock enters the diffuser and moves toward the throat, the supersonic flow is decelerating, and the pressure recovery is correspondingly increasing. Thus the mass flow rate through the second throat or outlet area is increasing whereas the mass flow rate into the inlet is constant. Clearly the shock will continue to move until it has left the convergent section.

When the idealized shock enters the inlet, it immediately jumps through the throat and divergent portion of the diffuser to a station equal in cross section to the inlet area. The pressure recovery is still normal shock value until the shock is moved either upstream or downstream by controlling the back pressure, which in turn may be regulated by adjustments in the diffuser outlet area.

When the shock is swallowed, the schedule of pressure recovery and choked outlet area is given by Eq. 1-20″. If the outlet area is decreased below the size required to establish the supersonic inlet flow, the shock moves upstream in the divergent or stable portion of the diffuser to yield increased pressure recoveries. The maximum pressure is achieved when the normal shock occurs at the throat supersonic Mach number M_1, which may be computed from Eq. 3-3 and 1-20′ to give

$$\frac{\left(\frac{\gamma+1}{2}M_\infty^2\right)^{(\gamma+1)/2(\gamma-1)}}{\left(1+\frac{\gamma-1}{2}M_\infty^2\right)^{\frac{1}{2}}\left(\gamma M_\infty^2-\frac{\gamma-1}{2}\right)^{1/(\gamma-1)}} = \frac{M_1}{M_\infty}\left(\frac{1+\frac{\gamma-1}{2}M_1^2}{1+\frac{\gamma-1}{2}M_\infty^2}\right)^{(\gamma+1)/2(\gamma-1)} \tag{3-4}$$

The maximum total pressure recovery is then given by the normal shock value (Eq. 3-1), which occurs at the throat Mach number. Plots of the one-dimensional total pressure recovery and mass flow for a convergent-divergent diffuser at Mach number 2 is presented in Fig. E,3b as a function of the outlet-to-inlet area ratio.

The hysteresis in the curve and the discontinuity in the mass flow rate between swallowing and regurgitation of the shock should be noted. This discontinuity in mass flow rate and pressure recovery casts doubt on the advisability of using the convergent-divergent diffuser or any internally contracting supersonic passages on flying aircraft. The discontinuity and the nature of the starting process also imply that special

Table E,3a. Laminar separation limit pressure recovery.

Computation	Mach number			
	2	3	4	6
Normal shock recovery	0.7209	0.3283	0.1388	0.0296
Convergent-divergent recovery	0.835	0.442	0.215	0.045
Variable contraction diffuser (laminar separation limit)	0.8599	0.507	0.2811	0.091

controls would be required to achieve performance greater than normal shock recovery.

The convergent-divergent diffuser likewise suffers serious penalties in pressure recovery at the higher Mach numbers, which is shown by the values in Table E,3a. The theoretical gain in pressure recovery over the

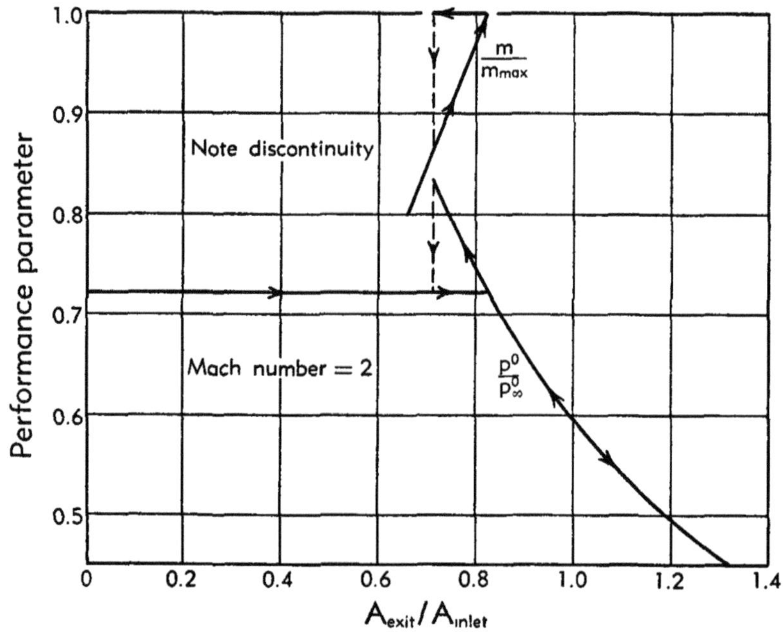

Fig. E,3b. Convergent-divergent diffuser performance.

normal shock diffuser is seldom achieved in practice. Any disturbance that oscillates the shock and hence prevents shock location close to the throat decreases the performance. Rough burning in a ramjet combustion chamber, for example, could produce such oscillations. Kantrowitz [*36*] suggested that long throat sections and smooth subsonic diffuser entries might improve the stabilization of the shock near the throat, and hence the peak pressure recovery ratio. The experiments of Wyatt and Hunczak [*37*] in general confirmed his suggestion.

The convergent-divergent diffuser is of most interest as a second throat for supersonic wind tunnels and for cascade arrangements where establishment of the supersonic flow by spillage over the inlet lip may be prevented by the adjacent diffuser sections or a boundary wall. Nevertheless the concepts must be continually kept in mind on other diffuser arrangements, because the convergent-divergent diffuser represents the maximum internal contraction that can be permitted and still establish supersonic inlet flow. Rounded lips or boundary layer growth may effectively produce this limiting or greater internal contraction ratio. Plots of the limiting contraction are included in [*1*,*6*].

Variable Contraction Convergent-Divergent Diffusers. An interesting thought suggests that the performance of the convergent-divergent diffuser could be improved if some means could be found to establish the supersonic flow. Such a means might be provided by operating the diffuser above the design speed or by designing the diffuser with variable geometry features. In a variable geometry diffuser, the throat area would first be enlarged to allow entrance and passage of the normal shock, then contracted toward the reversed De Laval nozzle value to give near-isentropic diffusion.

Experimental attempts to build such variable-geometry high performance diffusers have been relatively unsuccessful [*38*]. The usual disadvantages of the convergent-divergent diffuser persist, such as the discontinuities of mass flow rate upon regurgitation of the shock, with corresponding hysteresis in the pressure recovery curve. More important perhaps is the extreme sensitivity of the diffuser to slight disturbances which result in shock regurgitation whenever the internal contraction ratio appreciably exceeds the Kantrowitz value.

This extreme sensitivity is associated with the action of the wall boundary layer. If the diffuser is free of initial boundary layer as it is on a nose inlet, then the boundary layer growth will be laminar and hence sensitive to the diffuser adverse pressure gradient. If laminar separation occurs, the streamlines are displaced toward an increased contraction ratio. Likewise, shocks must occur which decrease the total pressure and hence increase the required throat area to pass the required mass flow rate. Under such stimulation the diffuser will probably regurgitate the normal shock unless the diffusion rate is very gradual.

Some indication of the tendency for the laminar boundary layer to separate is given by Stewartson [*12*], and the separation Mach number is summarized in Table E,2. If a variable diffuser were designed for Stewartson's assumed adverse pressure gradient, then the separation Mach number would represent the minimum throat Mach number without shock regurgitation. The corresponding total pressure recovery ratios are compared with convergent-divergent and normal shock diffuser theoretical values in Table E,3a.

Thus to achieve great advantages in performance of a variable geometry convergent-divergent diffuser over the fixed geometry type, the necessity for boundary layer suction or turbulence inducers is clearly indicated. Some success has been obtained with variable geometry diffusers, particularly at high Mach numbers, in reducing the required operating pressure ratios of hypersonic wind tunnels [*39,40*].

PERFORATED DIFFUSER. There is still another successful scheme [*41*] to avoid the Kantrowitz-Donaldson limitation on the contraction ratio of fixed geometry convergent-divergent diffusers. If the limiting contraction

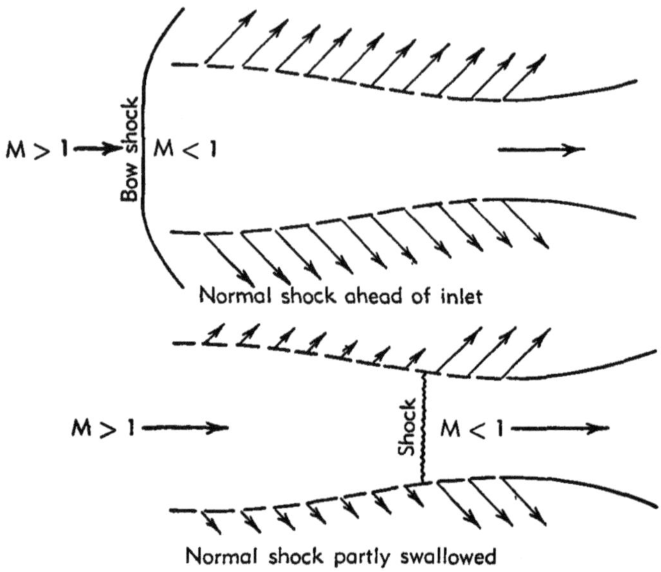

Fig. E,3c. Perforated diffuser swallows shock.

ratio is exceeded, the supersonic flow can yet be established by perforating the convergent wall of the inlet. These perforations increase the choking flow area downstream of the shock so that supersonic inlet flow may be established.

The perforations serve as automatic valves, passing high mass flow rates prior to establishing supersonic inlet flow, and low mass flow rates after the shock has been swallowed. A schematic diagram indicating this automatic valve action by the length of the flow vectors is shown in Fig. E,3c.

With the shock ahead of the inlet as in Fig. E,3c, the static pressure and density in the convergent passage will be high, and correspondingly there is a large pressure differential across each perforation. Likewise the flow is subsonic so that the effective area of the choked orifices is a reasonably large fraction of the actual area. This fraction varies with the tangential flow Mach number, but has a mean value of about 0.5. Many careful experiments have been conducted by McLafferty of the United Aircraft Corporation to evaluate more precisely these tangential flow coefficients.

As the shock passes a perforation, the local flow becomes supersonic and there is a corresponding decrease in flow density, static pressure, and pressure differential across the perforation. Likewise, the effective flow area must continually decrease with increasing tangential Mach number ahead of the holes.

This latter effect may be seen from the sketch of Fig. E,3d which shows the limiting streamline that just passes through a perforation. The initial Mach line is determined by the local supersonic flow, while the final Mach line is set by the static pressure just outside the perforation. If the

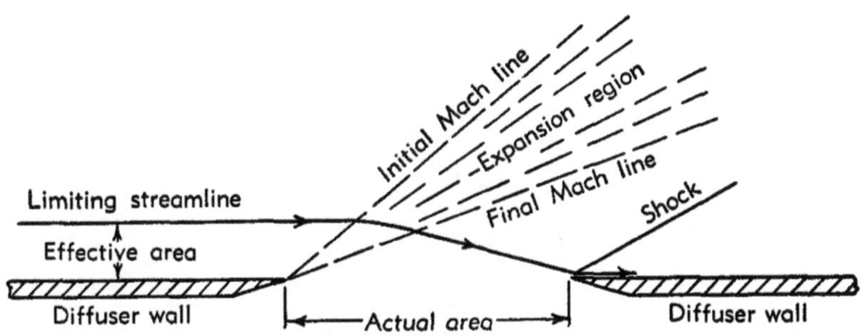

Fig. E,3d. Perforation serves as automatic valve.

final Mach line falls tangent to or below the diffuser wall, the flow coefficient of the hole is independent of the external static pressure. The angle through which the flow is turned may be determined by the Prandtl-Meyer equations for supersonic flow around a corner or from tabulated values such as are included in [*1*].

Thus, because of the automatic valve action, the subsonic mass flow entering a choked orifice may be considerably larger than the lost mass flow after the shock is swallowed. This means that the effective contraction ratio as seen by the subsonic flow downstream of the starting normal shock is much less than the contraction ratio to which the established supersonic flow is exposed. The perforated diffuser will then yield potentially high pressure recoveries without starting difficulties at the design Mach number. Perforated diffusers have been designed to give pressure recoveries as high or higher than other type diffusers for which the data are summarized in Fig. E,3e. The drag also may be higher due to the momentum loss of the air spilled through the perforations.

The shape of the pressure recovery-mass flow curve for a perforated diffuser depends somewhat on the hole distribution in the convergent portion. If the minimum hole area and proper distribution are employed, the normal shock is neutrally stable (but the system may not be) in the convergent portion and can place itself promiscuously at any position. Under such conditions the throat Mach number remains constant as the shock moves from point to point inside the diffuser. Exclusive of friction, the theoretical pressure recovery is constant with mass flow changes until the shock enters the inlet and increases proportionately to the engine mass flow as the shock moves toward the throat; then the mass flow remains constant while the pressure recovery varies as the shock moves downstream of the throat.

If the diffuser does not buzz, the pressure recovery mass flow curve may be reproducible for increasing and decreasing mass flow rates. In

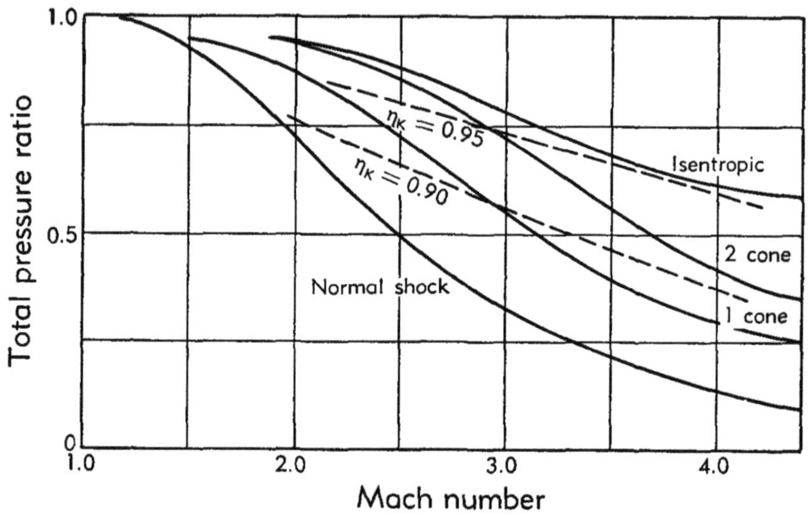

Fig. E,3e. Estimated pressure recoveries of several spike diffuser types.

other words the hysteresis of the usual convergent-divergent diffuser is eliminated. If the area of the perforations is uniformly or locally increased beyond the minimum value required for shock entrance, the shock may even be firmly located in the convergent portion. Likewise, the mass flow rate through the diffuser may be continuous and in fact, if a transient pressure disturbance were to force the shock ahead of the inlet, the shock would again be swallowed by the diffuser as soon as the disturbance ceased.

Perforated diffusers could, of course, be built with unperforated internal contraction sections which would result in the usual discontinuities in pressure recovery and mass flow rates. Likewise, buzz (pulsing flow) has been experienced (but is not necessarily the rule) on perforated diffusers such that two branches similar to a hysteresis of the pressure recovery mass flow curve may occur, one that is buzz-free, the other with

Plate E,3. Schlieren flow visualization at Mach number 2. Left, subcritical flow; right, supercritical flow.

flow oscillations. For further details concerning the design and performance of perforated diffusers see [41,42].

External Compression Diffusers. To establish supersonic flow in a diffuser, the contraction of the streamlines downstream of the entrance normal shock cannot exceed the value specified by Eq. 3-3 for the convergent-divergent diffuser. On the other hand attainment of high pressure recoveries requires that the streamlines in the established supersonic flow are submitted to a greater contraction. The use of perforations is one way to avoid this limitation. The independence of the upstream supersonic flow on the downstream geometry suggests another. If the supersonic compression could be achieved by means of a cone or spike ahead of the cowl lip, the starting difficulties associated with high supersonic contraction ratios might be overcome.

Oswatitsch [43,44] first demonstrated the effectiveness of external compression diffusers. Plate E,3 shows schlieren flow visualizations of a typical single conical shock diffuser operating at zero and maximum mass flow rates. The supersonic flow can easily be established because there is no external boundary to prevent spillage over the cowl of the subsonic air downstream of the bow wave. As the diffuser mass flow is increased, the mass flow spilled over the cowl decreases, and the shock moves downstream toward the cowl lip. In the limit or supercritical flow condition, the diffuser is passing maximum mass flow, and the supersonic flow is contracted in the ratio of the projected cowl to the annular area.

The schedule of pressure recovery with outlet area is again hyperbolic according to Eq. 1-20″ (p. 604) in the supercritical or constant mass flow regime. The variation of pressure recovery with mass flow in the subcritical or variable mass flow region depends largely on diffuser design and whether or not the flow is steady.

The theoretical total pressure recovery ratio of a spike diffuser is given as the product of the stagnation pressure ratio across each shock. As the number of shocks is increased, the average strength decreases, and also the diffuser total pressure recovery ratio improves. In the limit of a properly designed continuously curved surface, nearly isentropic flow deceleration may in principle be achieved.

These so-called isentropic spike diffusers have adverse pressure gradients on the surface so that they are sensitive to boundary layer effects. Other difficulties including sensitivity to angle of attack and high pressure drags inherent in the high cowl lip angles required for alignment with the flow, particularly at high Mach numbers, have become apparent. Although further investigation may resolve these difficulties, the simpler diffusers have continued to receive considerably more attention.

The conical spike diffuser on the other hand has no adverse pressure gradient on its surface and can be built to be rather insensitive to angle-of-attack performance. While the pressure recovery could possibly be improved by including some internal contraction, such inclusion is undesirable for flying aircraft configurations. The single shock conical diffuser without internal contraction is often called the Ferri diffuser following his important pioneering research [45].

A rough summary of the maximum pressure recovery ratios obtained on spike-type diffusers is presented as a function of Mach number in Fig. E,3e. The values correspond to optimum configurations at each design Mach number. It is interesting to note that the general trend of the curves is somewhat similar to lines of constant kinetic energy efficiency η_k from Eq. 1-22′. This similarity is apparently fortuitous.

Inasmuch as the thrust coefficient based on the free stream tube area of an idealized ramjet is $C_F = 2[\sqrt{\tau\eta_k} - 1]$, where τ is the stagnation temperature ratio across the combustion chamber, a nearly constant efficiency corresponds to a nearly constant thrust coefficient. Because η_k is relatively insensitive to pressure recovery at the higher Mach numbers, so will the thrust coefficient be insensitive. The thrust for a constant τ follows the same trend, provided that decreases in pressure recovery can be compensated by changes in outlet area so as to maintain the same mass flow through the engine. If on the other hand the exit area is fixed, a 1 per cent decrease in pressure recovery will lead to a 1 per cent decrease in mass flow through the exit. Hence the thrust will decrease by more than this proportion.

A turbojet engine operates at nearly constant Mach number at the face of the compressor for a given altitude, engine speed, and flight speed. The mass flow through the engine is then directly proportional to the total pressure recovery. For each per cent loss in pressure recovery, the thrust must decrease by more than 1 per cent. The conclusion that considerable effort may be profitably spent to improve supersonic diffuser pressure recovery is thus warranted.

An interesting question arises concerning the pressure recovery ratios that are likely to be achieved. The single shock conical diffuser gives values at Mach number 4 within 5 counts of the theoretical performance exclusive of friction. The two shock cones and the isentropic spikes on the other hand give recoveries 20 and 30 counts respectively below the usual theoretical predictions at this Mach number. Since an understanding of the nature of the losses in the latter case would aid in correcting its deficiencies and in providing concepts for better design, a tentative list of sources for pressure recovery losses follows.

Loss due to friction. According to Eq. 1-15 the percentage loss in total pressure increases approximately as the square of the surface Mach number. Such a variation could feasibly explain the downward trend in pressure recoveries at the higher Mach numbers. Numerical calculations, however, indicate that the frictional loss of total pressure amounts to only a few per cent at a Mach number of 4.

Pressure loss due to boundary layer displacement on the spike surface. Clearly, if the boundary layer corrections were improperly made, the flow would no longer be in accordance with isentropic requirements. Certainly, if flow separation occurred on the curved surfaces, the isentropic spike would effectively act as a constant pressure or conical spike with correspondingly large losses in pressure. Preliminary experiments utilizing boundary layer suction through porous spikes have been relatively unsuccessful in achieving pressure recoveries higher than those without suction. These experiments were inconclusive, however, because there was no certainty that either the amount of suction or the effective spike surface with suction was optimum.

Losses due to poor subsonic entrance flow. The isentropic spike diffusers generally have higher rates of curvature near the throat than other diffusers, and hence greater difficulty would be experienced in achieving uniform flow fields ahead of the subsonic diffuser and in placing the "normal shock" (which is actually a complicated branch shock pattern) at its optimum position. The isentropic spikes for high Mach number

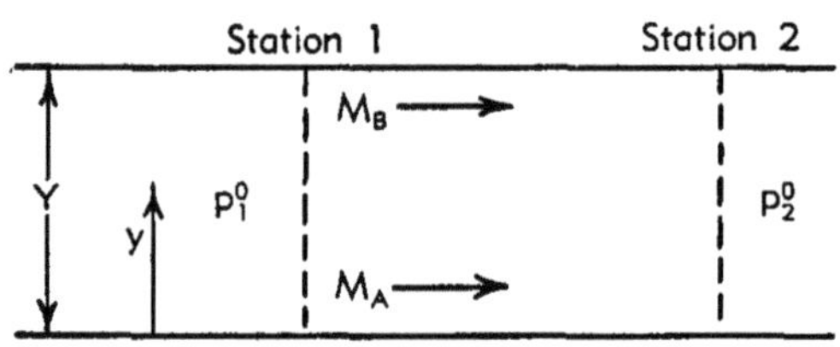

Fig. E,3f.

are usually designed to decelerate the flow only to a Mach number of about 1.5 or higher at the throat. The corresponding pressure rise across a normal shock is generally large enough at these values to separate a turbulent boundary layer [*46*]. The resulting disturbances to the subsonic entry flow, when superimposed upon the high curvatures near the throat, could easily lead to poor pressure recoveries. Pressure ratio decreases on the order of 20 per cent have been observed on side inlets at Mach number 2, due to effects from the initial boundary layer ahead of the inlet. The disturbances so created are believed to be no greater than the throat disturbances on current isentropic inlets designed for high Mach number.

Mixing losses due to nonuniform flow at the normal shock location. Because this potential loss is seldom discussed in the design of supersonic inlets, it is here considered in greater than usual detail.

If the Mach number and static pressure vary across the throat of a diffuser, it is difficult to imagine that a normal shock could span the region without mixing losses downstream and a complicated shock pattern actually occurring. To estimate the magnitude of the losses, consider that a one-dimensional frictionless duct carries a flow of uniform total pressure but varying static pressure and Mach number at station 1 (see Fig. E,3f). This flow arrives at station 2 with uniform Mach number as well as total and static pressures. If $\tilde{y} = y/Y$, the conservation equations for mass flow and momentum may be written

$$\int_0^1 \frac{M_1 d\tilde{y}}{\left(1 + \frac{\gamma - 1}{2} M_1^2\right)^{(\gamma+1)/2(\gamma-1)}} = \left(\frac{p_2^0}{p_1^0}\right) \frac{M_2}{\left(1 + \frac{\gamma - 1}{2} M_2^2\right)^{(\gamma+1)/2(\gamma-1)}} \tag{3-5}$$

$$\int_0^1 \frac{(1 + \gamma M_1^2) d\tilde{y}}{\left(1 + \frac{\gamma - 1}{2} M_1^2\right)^{\gamma/(\gamma-1)}} = \left(\frac{p_2^0}{p_1^0}\right) \frac{(1 + \gamma M_2^2)}{\left(1 + \frac{\gamma - 1}{2} M_2^2\right)^{\gamma/(\gamma-1)}} \tag{3-6}$$

When the Mach number distribution at station 1 is inserted into Eq. 3-5 and 3-6, two equations result for the unknown p_2^0/p_1^0 and M_2. In general, two solutions exist. If M_2 is supersonic, p_2^0/p_1^0 represents the mixing loss in the supersonic stream prior to placing a normal shock at the downstream Mach number M_2. The total loss will then correspond to the product of p_2^0/p_1^0 and the stagnation pressure ratio across the normal shock. The second solution gives this final pressure ratio directly, along with the Mach number downstream of the shock. Thus the combination including shock and mixing losses is independent of order in occurrence.

In the special case of $\gamma = 1.4$ and linear variation of Mach number from M_A to M_B, Eq. 3-5 and 3-6 become

$$\frac{(M_A + M_B)[1 + 0.1(M_B^2 + M_A^2)]}{2(1 + 0.2M_A^2)^2(1 + 0.2M_B^2)^2} = \left(\frac{p_2^0}{p_1^0}\right) \frac{M_2}{(1 + 0.2M_2^2)^3} \tag{3-7}$$

$$\begin{aligned} \frac{1}{(M_B - M_A)} \Bigg\{ & \left[\frac{M_B}{(1 + 0.2M_B^2)^{\frac{1}{2}}} - \frac{M_A}{(1 + 0.2M_A^2)^{\frac{1}{2}}}\right] \\ & + \frac{1}{3}\left[\frac{M_B^3}{(1 + 0.2M_B^2)^{\frac{3}{2}}} - \frac{M_A^3}{(1 + 0.2M_A^2)^{\frac{3}{2}}}\right] \\ & - \frac{6}{125}\left[\frac{M_B^5}{(1 + 0.2M_B^2)^{\frac{5}{2}}} - \frac{M_A^5}{(1 + 0.2M_A^2)^{\frac{5}{2}}}\right]\Bigg\} = \frac{p_2^0}{p_1^0} \frac{(1 + 1.4M_2^2)}{(1 + 0.2M_2^2)^{3.5}} \end{aligned} \tag{3-8}$$

Special calculations utilizing Eq. 3-7 and 3-8 yield the results in Table E,3b.

The mixing loss plus the normal shock loss thus very nearly gives the loss for a shock at the mean Mach number. That loss can still be large, indicating the desirability of reducing the average Mach number and the Mach number gradient to as low a value as possible prior to passage of the flow through the shock.

Loss due to matching the outer and inner flow fields. Connors has pointed out a fundamental limitation on the pressure recovery of current

Table E,3b. Duct mixing losses.

Initial duct Mach number		Total pressure ratio			
M_A	M_B	Mixing loss ratio	Shock loss ratio	Total	Shock at mean Mach number
1.5	2.0	0.988	0.842	0.832	0.835
1.5	3.0	0.892	0.691	0.617	0.6055
3.0	3.5	*	*	0.268	0.2645

* The supersonic Mach number solution has been excluded for this case because it leads to mixing loss ratios greater than unity in violation of the second law of thermodynamics.

isentropic diffusers. The isentropic spike is usually designed so that the Mach waves from the curved spike coalesce at a focal point, generating a strong shock in the outer flow field. The slip line trailing downstream from this focal point joins the outer and inner flow fields along a line of equal static pressure and stream direction.

The largest possible value of the static pressure in the outer flow field would be generated by a normal shock. Hence the maximum static pressure rise through the Mach wave isentropic compressions of the inner flow field is also this same value. A reflected expansion wave from the focal point cannot appreciably relieve this limitation because such an expansion increases the stream angle of the slip line and promotes detachment or forward motion of the strong shock.

Table E,3c. Maximum performance of isentropic diffusers.

Stream Mach number	2.15	2.50	3.00	3.50	4.00
Minimum isentropic diffusion Mach number	1.00	1.19	1.48	1.76	2.03
Total pressure recovery ratio	1.00	0.99	0.94	0 83	0.71

A tabulation of the minimum Mach number following isentropic diffusion and the corresponding maximum total pressure recovery is included in Table E,3c. These recoveries conform closely to the trend of maximum values observed experimentally, suggesting that the design philosophy of current isentropic diffusers must be altered before major improvements can be expected.

Losses due to unsteadiness in the flow. Maximum or peak pressure recovery usually depends on a precarious positioning of the normal shock and precise control of the mass flow rate for peak recovery. If transient low amplitude flow oscillations occur, the time-averaged pressure recovery ratio must be correspondingly decreased. Connors [47] has shown that the time-averaged maximum pressure recovery of an oscillatory flow is decreased from the steady value by approximately one half the swing of the unsteady pressure oscillation.

Diffuser Buzz. The pressure recoveries of supersonic inlets are generally steady in the supercritical or constant mass flow performance range. As the mass flow is decreased in the subcritical range, however, a point will usually be reached at which the flow abruptly starts pulsing. The resulting unsteadiness of mass flow and pressure recovery is described as diffuser buzz.

The occurrence of buzz was first noted in the historic experiments of Oswatitsch [43,44] but no explanation as to its cause was presented. An early attempt to explain buzz was given by Pearce [48]. He noted that the flow through a choked nozzle is unstable to a small disturbance if the stagnation pressure increases with mass flow; a decreasing schedule of stagnation pressure, on the other hand, is stable. From this simple rule applied at any choking station, diffuser pulsations or buzz or at least transient pressure variations would be expected if the instantaneous pressure recovery-mass flow curve for the diffuser at its choking station has a positive slope. The experimental observations on perforated and spike diffusers usually conform to this criterion but the cause of the positive slope requires further insight.

On the perforated diffuser, the positive slope of the pressure recovery-mass flow curve results from the increased supersonic contraction that is inherent as the shock moves toward the throat. Experimentally the diffuser may be buzz free with increasing mass flow until the normal shock just enters the inlet. After entrance, buzz or at least transient pressure oscillations would be expected, with amplitude that in part depends upon the perforation distribution. Though some perforated diffusers appear to be reasonably steady in this mass flow range, others begin a periodic buzzing cycle as the normal shock enters the inlet. For the latter cases, the time-averaged pressure recovery jumps discontinuously to a higher value as the diffuser buzzes. In some instances the diffuser continues to buzz even if the mass flow is reduced to zero. Thus a hysteresis [42] occurs for part of the flow range giving two values of pressure recovery, one with and one without buzz. This observation demonstrates that a nonbuzzing diffuser may feasibly buzz if the initial disturbance amplitude is large enough.

The detailed nature of a feasible buzzing cycle for a cone-type inlet is shown in Fig. E,3g. Prior to buzz, as in Fig. E,3g(1), there is a velocity discontinuity across the slipline downstream of the oblique and normal shock intersection which trails past the inlet outside the cowl. Ferri and

Nucci have shown that, as this slipline crosses the inlet lip, violent diffuser pulsations may result.

Along the slipline (assuming no mixing) the static pressure is equal for two flows of different stagnation pressure ratios. If the two flows are

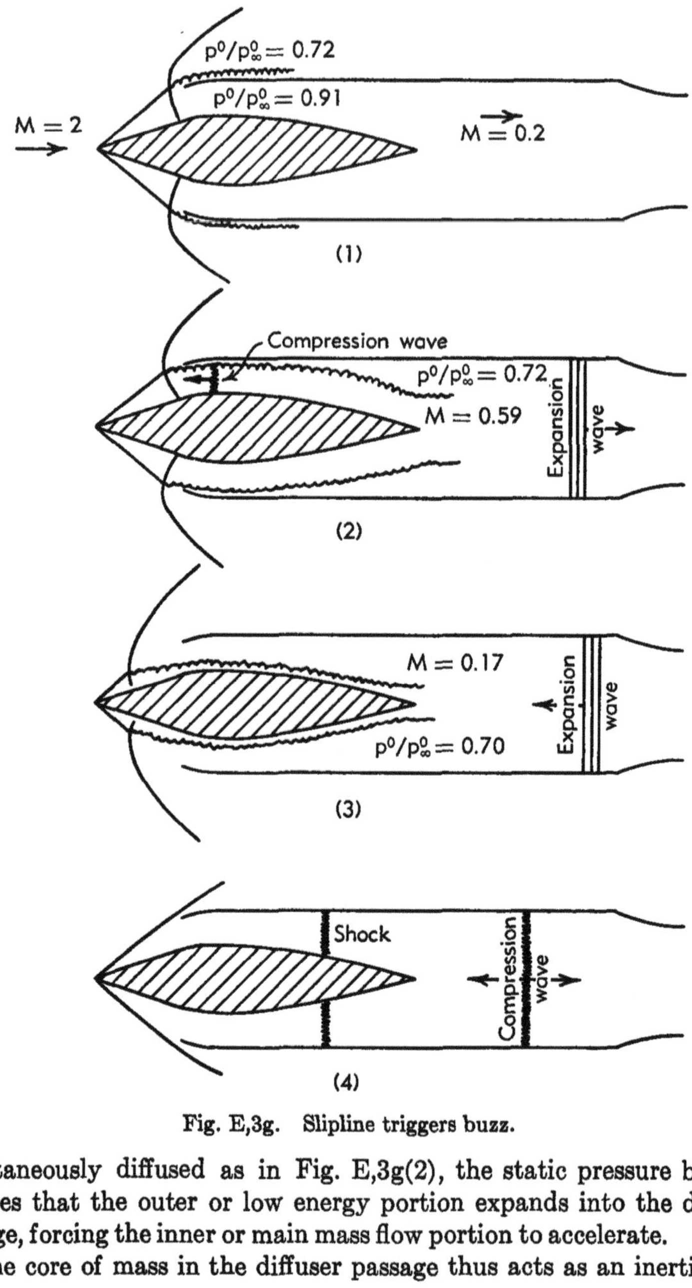

Fig. E,3g. Slipline triggers buzz.

simultaneously diffused as in Fig. E,3g(2), the static pressure balance requires that the outer or low energy portion expands into the diffuser passage, forcing the inner or main mass flow portion to accelerate.

The core of mass in the diffuser passage thus acts as an inertia plug resisting the acceleration. As a result a compression wave travels forward driving the normal shock ahead of the inlet, and an expansion wave travels rearward toward the exit. The static pressure in the combustion chamber drops with the increase in Mach number of the inner or main mass flow portion of the diffuser.

After the forward-traveling compression wave forces the shock ahead of the inlet as shown in Fig. E,3g(3), the slipline moves toward the center body and the inner portion of the flow is replaced by the outer portion as it sweeps downstream. The pressure in the combustion chamber remains at a low value because of the reduced total pressure behind the normal shock.

The rearward-moving expansion wave reflects from the exit as an expansion wave and travels forward until it meets and sucks the bow shock rapidly into the inlet. With the shock inside, the shock structure ahead of, and the mass flow into, the inlet correspond to supercritical flow conditions. The correspondingly higher mass flow rate and the potentially higher pressure recovery generate compression waves that reverberate back and forth between the exit and the normal shock, building up the pressure and forcing the shock toward the inlet to repeat the cycle.

This physical picture of longitudinally traveling waves was carefully studied by Trimpi. He has shown detailed agreement between his calculated and measured pressure variations during a buzz cycle provoked by the entrance of the slipline.

An alternative cause of buzz on spike diffusers is traceable to separation or boundary layer effects. Donaldson and Lange [*46*] have suggested that the maximum pressure rise across a shock which a boundary layer can tolerate without separation is proportional to the friction shearing stress just ahead of the shock. This concept leads to the approximate relations

$$\left.\begin{aligned}\left(\frac{\Delta p}{q}\right)_{\max} &\sim Re_x^{\frac{1}{2}} \quad \text{laminar}\\ \left(\frac{\Delta p}{q}\right)_{\max} &\sim Re_x^{-\frac{1}{5}} \quad \text{turbulent}\end{aligned}\right\} \tag{3-9}$$

Stewartson [*49*] in a detailed analysis infers that $(\Delta p/q)_{\max}$ for a laminar boundary layer is proportional to $Re_x^{-\frac{2}{5}}$ which does not differ greatly for most purposes from Eq. 3-9. Lange [*50*] supersedes his and Donaldson's earlier work with somewhat different conclusions than those of [*46*]. He shows that the surface pressure rise for turbulent separation is not necessarily the value across the shock wave because of intensive circulation in the separated region. He finds that $\Delta p/q$ evaluated from the pressure rise along the surface to the first peak is independent of Re_x and decreases with increasing Mach number.

Even though the relations for turbulent separation are only crudely defined, the resulting shock pattern may be described. The position of the bow shock ahead of the inlet cowl is largely determined by the amount of spilled mass flow, i.e. the degree to which the inlet is operating subcritically. This strong bow shock may separate the boundary layer on the compression surface generating a lambda shock configuration. The upstream leg of the lambda shock adjusts itself until the minimum required value of $\Delta p/q$ satisfies shock-induced separation criteria such as Eq. 3-9.

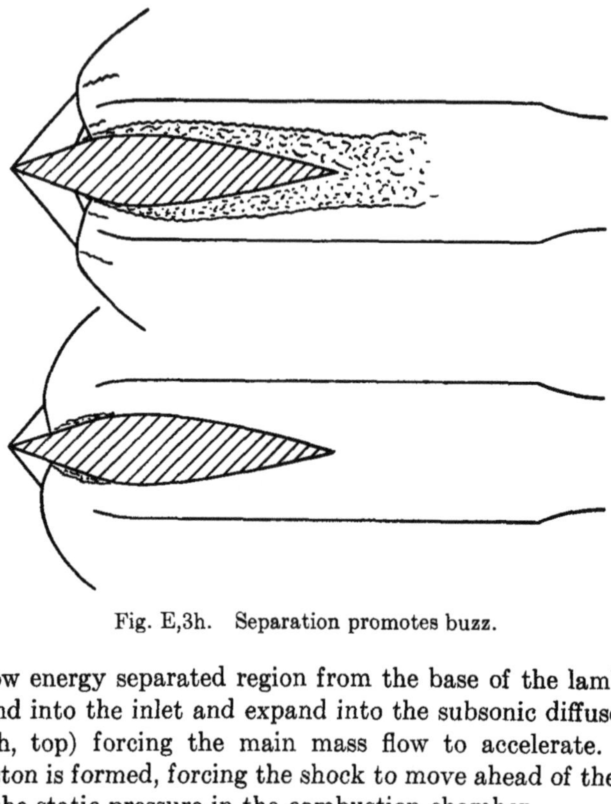

Fig. E,3h. Separation promotes buzz.

The low energy separated region from the base of the lambda shock may extend into the inlet and expand into the subsonic diffuser passage (Fig. E,3h, top) forcing the main mass flow to accelerate. Again an inertia piston is formed, forcing the shock to move ahead of the inlet and reducing the static pressure in the combustion chamber.

As the shock moves forward the flow separation region may reattach itself ahead of the throat (Fig. E,3h, bottom), or the separation may in some cases disappear entirely as is suggested by Eq. 3-9. The returning expansion wave from the exit and/or the elimination of the pressure loss due to separation moves the shock back into the inlet to repeat the process. Buzz generated by boundary layer effects may be either periodic or random, depending on the flow conditions. Buzz may also be triggered by the slipline generated at the intersection of the subcritical bow wave with the oblique shock from the separated flow region.

The discussion in connection with either Fig. E,3g or E,3h shows that the mass flow in the diffuser is alternately accelerated and decelerated during a buzzing cycle. Portions of the diffuser mass could then serve as an inertial plug resonating against the combustion chamber volume in a manner crudely analogous to a Helmholtz resonator. Utilizing this concept, Sterbentz and Evvard calculate trends of frequency with geometry changes that approximately conform to experimental results. No single theory yet proposed, however, whether it utilizes traveling longitudinal waves or Helmholtz analogy, can explain all the experimental observations for widely varying ramjet geometries.

The question next arises as to methods for eliminating buzz on particular diffuser configurations. The most important method proposed so far from the standpoint of the designer is to eliminate the possibility of the slipline crossing the cowl lip. One way to accomplish this aim is to extend the spike farther forward. To avoid expansions of the flow ahead of the cowl, and to maintain constant annular throat area and projected inlet area then requires that the cone angle be decreased. The resulting configuration is known as the low mass flow diffuser, so called because the inlet cannot capture a full stream tube on account of the supersonic spillage behind the conical shock. If the projected inlet area is allowed to increase, low mass flow diffusers may also be designed without decreasing the cone angle.

The low mass flow diffuser, while accomplishing the aim of preventing intersection of the slipline with the cowl, has compensating disadvantages. These are low mass flow rates and high drag; unfortunately the units will also buzz.

The latter effect results from boundary layer separation on the cone surface. Because of the lower cone angle, the surface Mach number outside the boundary layer is higher, and hence the normal shock is stronger. The strength of shock that is permissible may be estimated from Eq. 3-9 if the cone surface skin friction coefficients are known. In a study by Nussdorfer, criterions were presented from which he concluded that turbulent separation on cones was likely if the surface Mach number was greater than 1.33, slightly higher than the lower Mach number limit for the formation of three-shock configurations. The tendency toward separation can be alleviated by increasing the cone angle, but this again either approaches a high mass flow diffuser or increases the projected cowl area. An alternative to increasing the cone angle is to incorporate boundary layer control on the spike and other critical surfaces. The use of boundary layer control on the cone surface has been reasonably effective in extending the buzz-free subcritical range of low mass flow diffusers.

A reinspection of Fig. E,3g suggests that the difficulty on high mass flow diffusers associated with the entrance of the slipline might result from the simultaneous subsonic diffusion of two dissimilar streams. If this diffusion could be delayed by the inclusion of a long throat section so that the turbulent mixing processes could blend the two streams prior to diffusion, the acceleration forced on the inner or main mass flow portion would be eliminated. Hence the tendency toward buzz might be alleviated.

Preliminary experiments have shown that the use of long throat sections can very effectively increase the buzz-free subcritical range on high mass flow conical diffusers. The reason for the improvement, however, is not so clear. Mixing of the two streams prior to diffusion does not eliminate the fact that, as may be shown by one-dimensional analysis, entrance of the slipline corresponds to a shift of the instantaneous pressure recovery mass flow curve to a positive slope. Advocates of the positive slope criterion might say that the long throat section adds frictional damping to

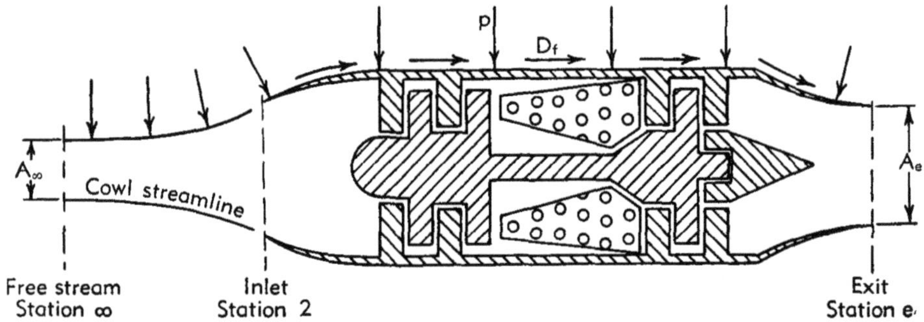

Fig. E,3i. Thrust and drag forces.

the system which eliminates the pulsations. Because friction losses generally increase with mass flow rates, such losses shift the slope toward the negative or stable values at the exit nozzle. The latter viewpoint has utility in that, if it is correct, the required frictional damping for stabilizing the inlet could be calculated. Such a calculation may not be simple, however, for the long throat section accomplishes other effects such as smoothing the entering flow to the subsonic diffuser as well.

Another way to eliminate the diffuser buzz is to incorporate in the design a control system that will prevent subcritical operation of the diffuser.

Drag. The propelling force that drives a supersonic missile or aircraft structure is simply the thrust minus the drag of the power plant. The pressure recovery has been shown to be important to achieve the highest engine mass flow rate and hence engine thrust. Minimization of the drag is equally important for obtaining an effective power plant.

The propulsive force or thrust minus drag of the power plant is simply the sum of the internal and external forces. The internal forces (Fig. E,3i) are equal to the difference in momentum of the internal flow between the exit and the inlet stations. Correspondingly, the external forces are equal to the difference in momentum of the external flow between the exit and the inlet stations. If some other station is chosen in place of the inlet, say, in the free stream ahead of the inlet, the calculated propulsive force will remain unaltered. Both internal and external forces, however, will be altered by an amount equal to the axial component of the pressure integral along the cowl stream tube ahead of the inlet.

The momentum of the internal flow at the inlet station is seldom known, but the momentum of the free stream may be clearly defined; therefore the engine thrust is conventionally defined from the free stream station.

To be consistent, the drag should also be defined from the free stream station. Drag, however, is conventionally defined as the sum of the external forces on the body. The consistent drag will then be the sum of the conventional drag and the dragwise component of the pressure integral along the cowl streamline ahead of the inlet. This latter term was defined by Ferri [*45*] as the additive drag.

In Fig. E,3i the frictional forces have been shown tangential to the body while the pressure forces are normal. Let D_f represent the axial component of the friction force on the body exterior. The thrust minus drag of the power plant is

$$(F - D) = \underbrace{[(p_e A_e + m_e u_e) - (p_\infty A_\infty + m_\infty u_\infty)]}_{\text{internal forces}} - \underbrace{\left[\int_{\text{free stream}}^{\text{inlet}} p ds + \int_{\text{inlet}}^{\text{exit}} p ds + D_f\right]}_{\text{external forces}} \tag{3-10}$$

where s represents the projected area on a plane normal to the axis with positive sign if the surface normal has a forward component. Rather than use absolute pressures, the pressure forces are usually referenced to free stream static. This transformation may be accomplished by adding the identity

$$\int_{\text{free stream}}^{\text{exit}} p_\infty ds - p_\infty(A_e - A_\infty) = 0$$

to the second member of Eq. 3-10 giving

$$(F - D) = \underbrace{(p_e - p_\infty)A_e + m_e u_e - m_\infty u_\infty}_{\text{conventional thrust}} - \underbrace{\int_{\text{free stream}}^{\text{inlet}} (p - p_\infty) ds}_{\text{additive drag}} - \underbrace{\Big[\underbrace{\int_{\text{inlet}}^{\text{exit}} (p - p_\infty) ds}_{\text{pressure}} + \underbrace{D_f}_{\text{friction}}\Big]}_{\text{conventional drag}} \tag{3-11}$$

Thus when the conventional thrust definition is employed, the consistent drag components are pressure, friction, and additive drags. The additive drag is sometimes called a subtractive thrust. It is usually not considered at subsonic speeds because of a canceling leading edge suction force on the cowl lip ([*51*] or Eq. 3-21). It should be noted that the definitions of thrust and drag can get very complicated for an aft installation for which part of the internal flow bathes the body proper [*52*].

The drag of a supersonic power plant will be strongly influenced by the amount of mass flow passing through the engine. If the engine is swallowing a full stream tube of air, there will be no deflection of the streamlines ahead of the inlet and hence no additive drag. Pressure drags can then be estimated by conventional means (cf. [*53,54,55*]). If, on the other hand, the inlet is operating subcritically, that is, with a bow wave ahead of the cowl, there will most definitely be an additive drag corresponding to the spillage. The pressure and friction drags are also altered by the mixed subsonic and supersonic flow ahead of the cowl.

Representative curves showing the experimental drag variations with mass flow ratio at Mach number 2 are presented in Fig. E,3j. The pressure drag decreases, becoming negative at the lower mass flow ratios. This results from the increased leading edge suction force around the inlet cowl associated with the subsonic flow field immediately downstream of the bow wave. The friction drag remains almost constant as would be expected because most of the body remains bathed in supersonic flow. The additive drag increases so drastically that the total drag leaps to intolerable values with decreasing mass flow ratios.

At critical mass flow, the drag on this particular model represents about four per cent of a typical afterburning turbojet engine thrust. At 40 per cent mass flow spillage, the total drag represents about 19 per cent of the critical engine thrust. These figures clearly illustrate that sub-*critical operation with air spillage through the mechanism of a bow wave* is to be avoided if possible.

Such spillage could be more economically accomplished by translating the spike forward from the cowl. Supersonic spillage by means of a conical shock then occurs with a much lower additive and total drag penalty.

An estimate of the additive drag for a normal shock inlet, schematically illustrated in Fig. E,3k, is outlined below utilizing a method due to Sibulkin [*56*] which gives approximate results for other type inlets as well.

From the engine mass flow, the corresponding free stream tube area may be calculated by Eq. 1-20. From free stream conditions, the total pressure p_1^0 and Mach number M_1 just downstream of the normal shock may be evaluated. Utilizing isentropic flow relations between the shock and the inlet and employing the inlet area allows the calculation of the inlet Mach number.

$$\frac{M_2}{\left(1+\frac{\gamma-1}{2}M_2^2\right)^{(\gamma+1)/2(\gamma-1)}} = \frac{A_\infty}{A_2}\frac{M_1}{\left(1+\frac{\gamma-1}{2}M_1^2\right)^{(\gamma+1)/2(\gamma-1)}} \quad (1\text{-}20''')$$

The inlet static pressure is given by

$$p_2 = \frac{p_1^0}{\left(1 + \frac{\gamma - 1}{2} M_2^2\right)^{\gamma/(\gamma-1)}} \tag{1-16'}$$

By definition the stream thrust is the momentum flow rate except that the pressure forces are referenced to free stream static pressure. The

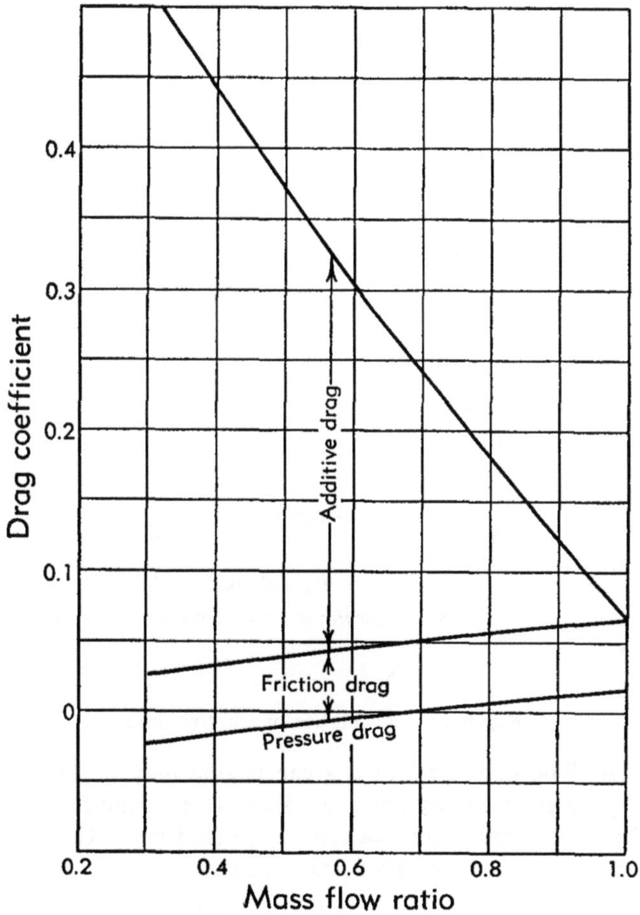

Fig. E,3j. Drag components for spike inlet.

additive drag of Eq. 3-11 is the difference in stream thrust between the inlet and the free stream, i.e.

$$D_{\text{add}} = (p_2 - p_\infty)A_2 + \gamma p_2 A_2 M_2^2 - \gamma p_\infty A_\infty M_\infty^2 \tag{3-12}$$

The unknowns in Eq. 3-12, p_2 and M_2, are evaluated by Eq. 1-16′ and 1-20‴. The results of the calculation are in fair agreement with the trends of Fig. E,3j (for refinements see [*56*]).

A knowledge of the inlet drag allows an estimate of the maximum performance point for an inlet-power plant combination. This point is generally taken to coincide with the maximum value of the quantity (engine thrust) − (airplane drag). That is when

$$\frac{d(F - D)}{d(p_3^0/p_\infty^0)} = 0 \tag{3-13}$$

The thrust on a turbojet engine depends primarily on the pressure recovery of the inlet at constant altitude, flight speed, and engine speed, while the drag increment above the minimum value varies with inlet mass flow

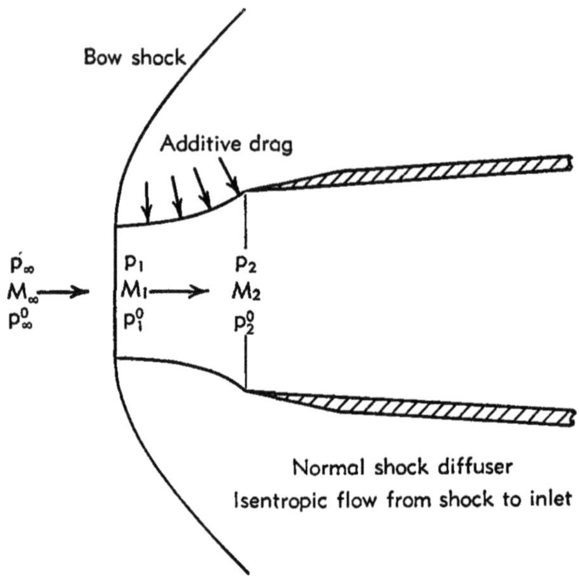

Fig. E,3k. Estimation of additive drag.

and inlet size. The inlet size for a given engine may be expressed as a function of pressure recovery and mass flow. The optimum thrust minus drag for the inlet engine combination is obtained when the slope of the pressure recovery-mass flow curve has the value

$$\frac{d(p_3^0/p_\infty^0)}{d(m/m_{cr})} = \frac{\left[\dfrac{\partial D}{\partial(m/m_{cr})}\right]}{\left[\dfrac{dF}{d(p_3^0/p_\infty^0)} - \dfrac{\partial D}{\partial(p_3^0/p_\infty^0)}\right]} \tag{3-14}$$

Two cases will be considered in order to illustrate the implications of Eq. 3-14. In the first case assume that a fixed geometry inlet of area A is equipped with a bypass system ahead of the engine so that the inlet air flow may be varied over a limited range. The engine mass flow rate m_3 is proportional to the pressure recovery; the additive drag and the bypass system drags are assumed to be proportional to the spilled mass flow rates for the respective systems. The various relations are summarized below:

$$m_3 = k_1\left(\frac{p_3^0}{p_\infty^0}\right), \qquad m_{cr} = k_2 A, \qquad D_{cr} = k_3 A$$

$$D_{add} = Ak_4(1 - m/m_{cr}), \qquad D_{bypass} = k_5\left[\left(\frac{m}{m_{cr}}\right) - \left(\frac{m_3}{m_{cr}}\right)\right]$$

D_{cr} = inlet critical drag, D_p = airplane drag excluding inlet. The quantities m and m_{cr} are the inlet and the inlet critical mass flows respectively and the k's are considered as constants. The total drag is

$$D = D_p + D_{cr} + D_{add} + D_{bypass}$$

Eq. 3-14 gives the optimum point on the inlet pressure recovery-mass flow curve as

$$\frac{d(p_3^0/p_\infty^0)}{d(m/m_{cr})} = \frac{k_5 - Ak_4}{\dfrac{dF}{d(p_3^0/p_\infty^0)} + \dfrac{k_1k_5}{k_2A}}$$

By proper design the bypass drag may be small so that k_5 is negligible. The coefficient $-Ak_4$ is $\partial D_{add}/\partial(m/m_{cr})$. Hence

$$\frac{d(p_3^0/p_\infty^0)}{d(m/m_{cr})} \cong \frac{\dfrac{1}{F_{cr}}\dfrac{\partial D_{add}}{\partial(m/m_{cr})}}{\dfrac{1}{F_{cr}}\dfrac{dF}{d(p_3^0/p_\infty^0)}}$$

The denominator of this relation has a value of about 1.2 while the numerator for the data of Fig. E,3j is about -0.4. For this case at Mach 2, the optimum performance point on the diffuser is independent of the airplane drag and occurs at that point where the nondimensional pressure recovery versus mass flow curve for the inlet has a slope of about $-\frac{1}{3}$. This point usually occurs at slightly below the critical mass flow rate.

For the second example an inlet of given design is scaled up or down so as to give optimum performance with a prescribed engine. The inlet critical mass flow and the inlet drag components will then vary with inlet size, which may be replaced in the equations by the expression

$$A = \frac{k_1(p_3^0/p_\infty^0)}{k_2(m_3/m_{cr})}$$

There is no bypass system so that the engine and inlet mass flow are identical. The optimum performance point for the inlet occurs when

$$\frac{d(p_3^0/p_\infty^0)}{d(m/m_{cr})} = -\frac{\dfrac{D_{cr}}{(m/m_{cr})} + \dfrac{D_{add}}{(m/m_{cr})[1 - (m/m_{cr})]}}{\dfrac{dF}{d(p_3^0/p_\infty^0)} - \dfrac{D_{cr}}{(p_3^0/p_\infty^0)} - \dfrac{D_{add}}{(p_3^0/p_\infty^0)}}$$

Both the numerator and the denominator of this expression are positive, showing again that the optimum diffuser operating point occurs in the negative slope portion of the subcritical region.

The discussion of supersonic inlet drags is hardly complete without a comparison of the performance of conical, two-shock, multiple-shock, and isentropic-shock configurations. Unfortunately, classification difficulties prevent such a comparison on a drag basis. Suffice it to say that because of the high cowl lip angles on the more efficient types, their drag is also higher. Whether the isentropic spike with its much higher pressure recovery is superior to the single-shock cone or not depends on the Mach number, details in the design, and the use to which the diffuser is to be applied.

Aft Inlets. The many factors essential to the design of a successful airplane may require that the inlet diffuser be mounted roughly tangent to some point on the fuselage. Such inlets have been called "side inlets," "bottom inlets," etc. depending on their location with respect to the vertical.

There are a number of reasons why aft inlets might perform differently than nose inlets. For one, most nose inlets can be designed with axial symmetry; aft inlets, on the other hand, must have modified geometries to conform to local fuselage contours. Correspondingly altered surfaces to achieve the external supersonic compression are to be expected. The more common aft inlet configurations shown in Fig. E,31 include half spikes, two-dimensional ramps, wedges normal to the fuselage surface, and continuously curved ramps, particularly on scoop inlets. Experimental research generally indicates that nearly the same peak pressure recovery may be achieved from a large variety of external compression surfaces, so that a designer has a choice of inlet types depending upon location on the fuselage and other determining factors.

A second complication arises because the aft inlet must achieve performance in the disturbed flow field of the body. First of all, the loca Mach number at the inlet station may be different from free stream. If the local Mach number is less than free stream, then improved inlet performance would be expected. The Mach number distribution, however, might be nonuniform, and therefore the inlet design should recognize such nonuniformity to yield best performance.

The most important nonuniformity in the flow field that the inlet must digest arises from the body boundary layer. The entrance of boundary layer air into a diffuser passage can cause losses far greater than the stagnation pressure deficit in the boundary layer itself. For, under the adverse pressure gradients of diffusion, this low momentum air is susceptible to flow separations and all the associated difficulties. If performance comparable to a nose inlet is to be achieved some form of boundary layer control is imperative.

The results of an experimental study by Goelzer and Cortright utilizing half of a conical spike inlet are shown in Fig. E,3m. Here the maximum

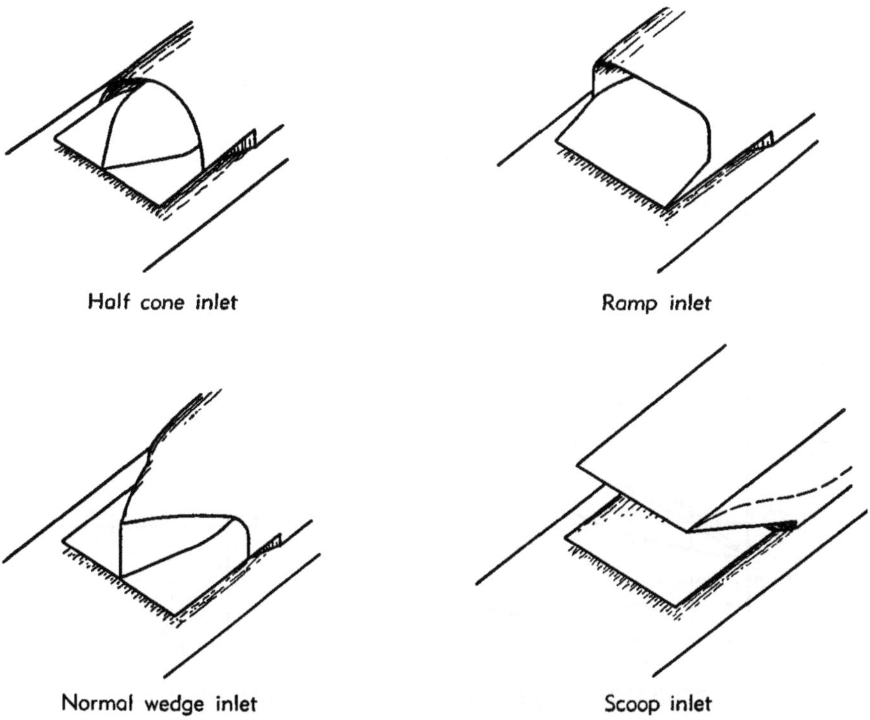

Fig. E,31. Aft inlet configurations.

total pressure recovery ratio is plotted for each value of the boundary layer scoop height with the percentage of possible mass flow through the boundary layer scoop as a parameter. With no boundary layer control ($h/\delta = 0$) the pressure recovery ratio is only 0.715, which is considerably less than normal shock recovery (0.776) for this Mach number. The available total pressure recovery ratio ahead of the inlet is, on the other hand, of the order of 0.95. Preliminary surveys of the flow field have shown that this large loss represented by the difference between 0.95 and 0.715 occurs primarily in the transonic and subsonic portions of the diffuser. The loss is triggered by the entrance of the initial or body boundary layer.

When most of the boundary layer is removed, pressure recoveries close to those observed on nose inlets can be obtained. Inasmuch as the increase in pressure recovery of about 27.5 per cent would correspond to a thrust increase for a turbojet installation of about 33 per cent, considerable penalties in drag for the boundary layer removal system could be tolerated for such a large thrust gain. Though the drag penalty varies with each configuration, a well-designed boundary layer removal system will usually cost only a few per cent in drag up to Mach 2, yielding a large net thrust gain. Indications are that the drag penalty associated with boundary layer removal increases with Mach number. In any case, a

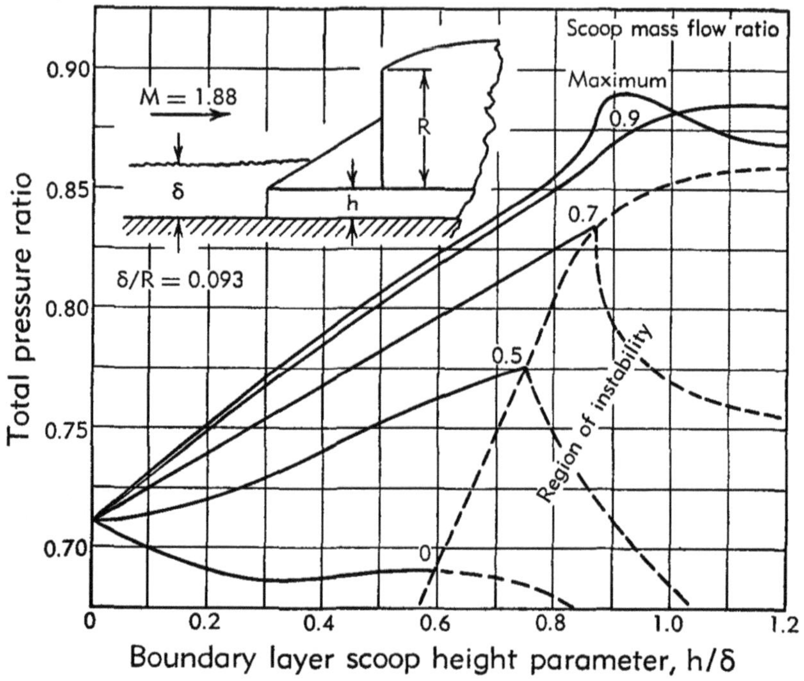

Fig. E,3m. Initial boundary layer decreases inlet performance.

designer should think twice before considering aft inlets without boundary layer control. A theoretical estimate of the maximum pressure recovery possible in the ducts of a boundary layer removal system is included in [57].

Another form of nonuniformity of the flow field ahead of an inlet arises from aircraft maneuvering. When a body of revolution is placed at angle of attack, favorable or unfavorable factors may affect inlet performance depending on location. The potential flow field on the bottom of the body decreases its surface Mach number as the angle of attack is increased. The body itself also shields an inlet so placed from angle-of-attack effects. Hence inlet mass flow and pressure recovery should improve with angle of attack for an inlet placed on the bottom of a fuselage. The local flow angle with respect to the body axis is double the angle of attack on the two sides and hence an inlet so placed should have losses equivalent to the corresponding nose inlet at twice the body angle of attack. On the top of the body there is again shielding of inlet flow angularity by the body, but the inlet Mach number is raised and therefore performance losses will occur.

When the complications due to the boundary layer are considered, the bottom inlet appears even more favorable. The secondary flows in the boundary layer due to the pressure gradients around the body tend to thin the boundary layer on the bottom and thicken it on the top of the body. Due to a tendency toward "separation of the cross flow" [*58*], low energy lobes tend to form in the top quadrants of the body. Inlets should not be located in these regions. The bottom inlet, on the other hand, requires less boundary layer control because the boundary layer is naturally thinner there. The most widely stated objection to the bottom inlet is its tendency to pick up debris from runways on take-off.

We may now return to a brief discussion of the inlets shown on Fig. E,31. The ramp inlet could be used effectively either on the side or on the bottom of the body. Because wedges are relatively insensitive to yaw, this configuration would be expected to be superior as a side inlet at angle of attack to at least some of the other models shown. Its main disadvantage is a slightly reduced performance associated with supersonic spillage around the cowl, but this difficulty could be eliminated by including side fairings. Such an inlet would make a superior bottom inlet, but the side fairings and the corner effects on the ramp cause difficulties due to cross flow at angle of attack when mounted on the side of the fuselage. The half-spike inlet experiences difficulties due to boundary layer accretion in the rectangular corners where the center body and the cowl meet the fuselage. Such difficulties do not always occur (see for example Fig. E,3m). The prime virtue of the normal wedge inlet is that it does not deflect the flow outward from the fuselage when ultimately it must be turned inward to enter the engine. Also, much of the compression surface is outside the boundary layer from the body. Hence this inlet is superior to the other configurations when no boundary layer removal is incorporated. Its potential difficulties result from the inherent instability of multiple duct passages and from possible difficulties at angle of attack if mounted on the side of the fuselage. Under such conditions only one half of the inlet is suggested, with the normal wedge compression surface on the top side. The scoop inlet (Fig. E,31) has inherent advantages of compressing on a surface free of initial boundary layer and of turning the air naturally in the direction it must ultimately take. Its big disadvantage lies in the starting difficulties associated with the effective internal contraction due to the side plates, and also due to the tendency for the boundary layer to separate off the curved throat section where the scoop compression waves coalesce. Nevertheless, promising results have been achieved in isolated experiments.

INLET-ENGINE MATCHING. From a drag study in the subcritical performance range at Mach 2, a decrease of 1 per cent in the mass flow through the inlet corresponds to a drag increase of the order of 0.4 per cent of an afterburning turbojet engine thrust. In the supercritical range, where pressure losses occur across the swallowed normal shock, each per cent loss in pressure recovery will give at least 1 per cent loss in maximum available engine thrust. It thus appears that the penalties are great if the inlet is improperly sized or mismatched to the engine mass flow

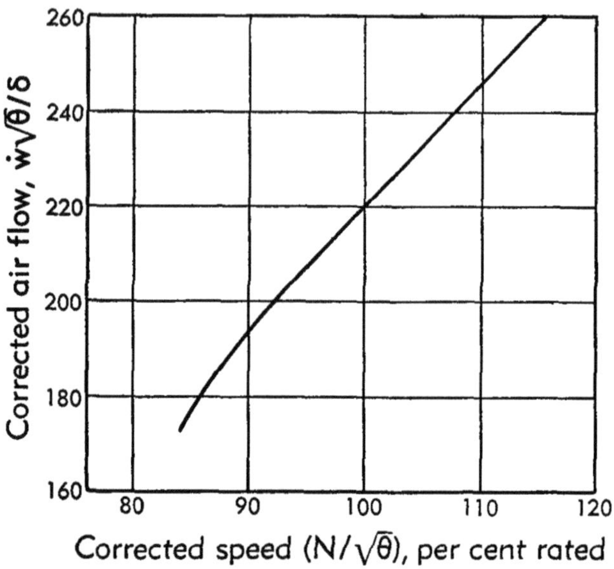

Fig. E,3n. Generalized turbojet air flow characteristics.

characteristic. A diffuser is properly matched to its engine if the inlet operates at the optimum performance mass flow rate over its range of altitudes and flight speeds. To achieve this optimization is the matching problem.

To a first order of approximation, the air flow characteristics of current turbojet engines can be generalized by plotting the reduced engine air flow $\dot{w}\sqrt{\theta}/\delta$ as a function of reduced engine speed $N/\sqrt{\theta}$, where $\dot{w}$ is the engine air flow rate, N the engine speed, and θ and δ are total temperature and pressure values at the compressor inlet respectively—nondimensionalized by dividing by sea level ambient values. Such a plot is shown in Fig. E,3n.

The reduced air flow is essentially a constant times a function of the Mach number at the engine compressor inlet. It will depend upon altitude and flight speed, for at constant engine rpm the reduced engine speed varies as $1/\sqrt{\theta}$. The value of θ in turn depends on ambient temperature, hence altitude and flight speed:

$$\theta = \frac{T^0}{519} = \frac{T^0}{T_\infty} \cdot \frac{T_\infty}{519} = \frac{T_\infty}{519}\left(1 + \frac{\gamma - 1}{2} M_\infty^2\right) \tag{3-15}$$

Thus curves of reduced air flow may be plotted as a function of flight Mach number and altitude for a given engine speed. Such curves are shown in Fig. E,3o.

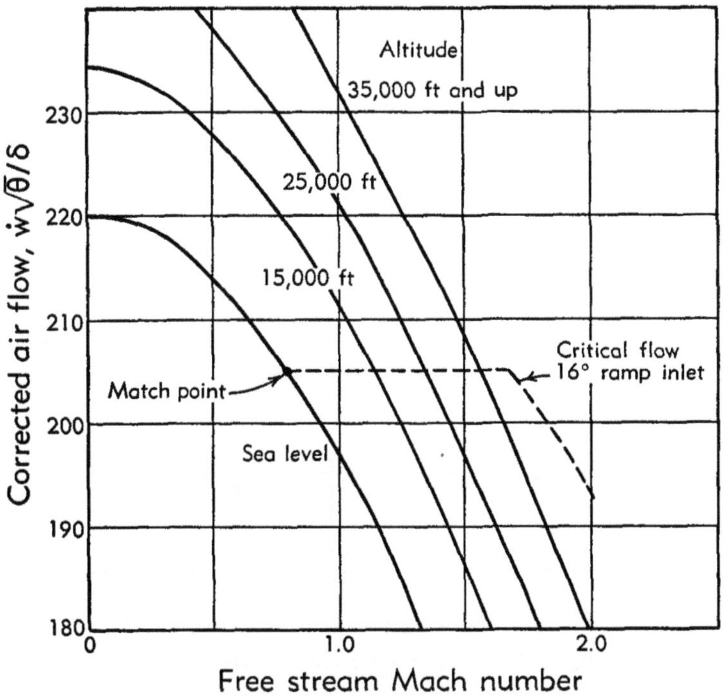

Fig. E,3o. Engine-inlet matching.

The critical values (assumed as optimum) of reduced diffuser mass flow may also be plotted as a function of flight Mach number. For a given fixed geometry diffuser, the pressure recovery curve for each flight Mach number is a function of the diffuser exit (engine inlet) Mach number, or the related quantity $\dot{w}\sqrt{\theta}/\delta A_i$ where A_i is the projected inlet area. The quantity $\dot{w}\sqrt{\theta}/\delta A_i$ for critical flow may then be determined for each flight Mach number and hence is primarily a function of flight Mach number. A representative curve for a 16° ramp inlet designed for free stream Mach number 2 is shown in Fig. E,3p. The discontinuity in slope results from shock detachment from the wedge at Mach numbers less than 1.67.

Inlet matching for any altitude and flight speed may now be accomplished simply by choosing the inlet size so that $\dot{w}\sqrt{\theta}/\delta$ for the engine and the inlet coincide. For example, matching at sea level for Mach 0.8 requires an inlet area A_i of 205/36.5 = 5.62 ft². Multiplication of the ordinate (Fig. E,3p) by this number gives the inlet corrected air flow. This curve is superimposed on Fig. E,3o. This matching method is due to Wyatt, who includes many more details in [*59*].

The altitudes and speeds for which the inlet is properly matched to the engine may now be seen from an inspection of the curve intersections. At 35,000-foot altitude and higher, the engine and inlet combination are mismatched at all Mach numbers except 1.56. For lower speeds at this

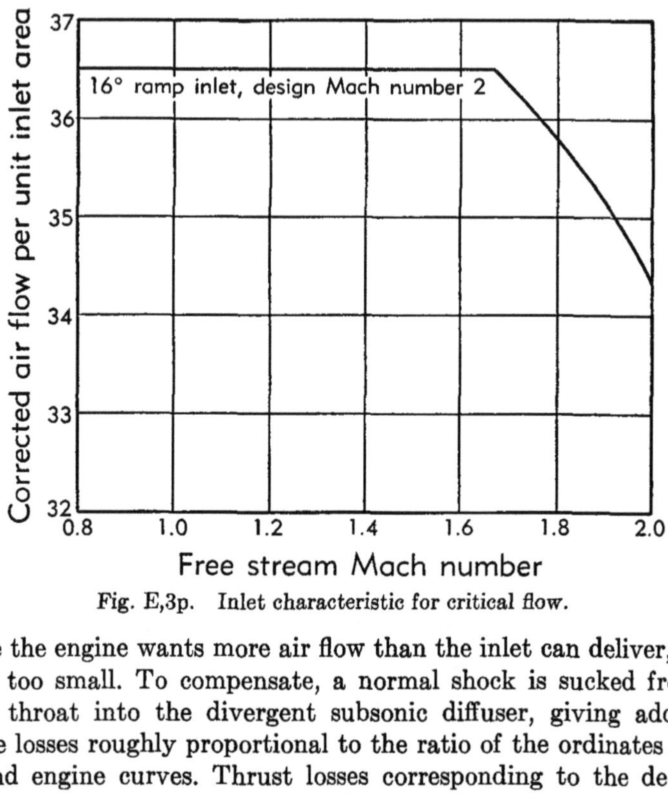

Fig. E,3p. Inlet characteristic for critical flow.

altitude the engine wants more air flow than the inlet can deliver, so the inlet is too small. To compensate, a normal shock is sucked from the choked throat into the divergent subsonic diffuser, giving additional pressure losses roughly proportional to the ratio of the ordinates for the inlet and engine curves. Thrust losses corresponding to the decreased pressure recovery would then occur. Likewise for Mach numbers higher than 1.56 the inlet is too large, so that mass flow must spill around the inlet, probably by means of a bow wave. Performance losses again occur, this time due to the additive drag increase associated with the spillage. Is it not clear from this example that inlets with variable geometry are essential for versatile performance?

A number of variable geometry diffusers have been proposed to accomplish approximate matching. First of all, blow-in doors or auxiliary inlets are required for take-off if the thrust penalty of mismatching is unacceptable. Then other means must be utilized to accomplish matching in the supersonic range.

Current suggestions for achieving high speed matching include translation of spikes or cowls on conical diffusers, translation or changing the wedge angle on ramp type inlets, and the use of a bypass system. The latter scheme is attractive because of its simplicity on some installations. A fixed geometry inlet is sized for, say Mach 0.8 at 35,000 feet; the inlet would thus be too large for Mach number 2. Rather than suffer the additive drag penalty associated with bow wave spillage, the air is taken aboard, then ejected in the axial direction through a suitable bypass nozzle ahead of the engine inlet. Experimentally, relatively small penalties compared to the additive drag result from the use of the bypass system if it is properly designed.

Supersonic Inlets at Subsonic Speeds. Subsonic inlet types generally have rounded lips on the cowl to provide smooth entry flow and to capitalize on leading edge suction. Such rounded lips on supersonic inlets generate detached bow waves and high drag; therefore the cowls on supersonic inlets usually have sharp leading edges. The question arises concerning what performance penalties will be suffered by these sharp- or thin-lipped inlets at subsonic speeds.

An estimate of the performance losses at subsonic speeds due to sharp lips may be made by means of one-dimensional equations. The limiting streamlines that just enter the inlet are shown for two cases in Fig. E,3q. Case (a) shows the inlet operating at an inlet to free stream velocity ratio less than unity whereas case (b) applies if the velocity ratio is greater than unity. The internal and external flows of both examples may be considered isentropic if the inlet lips are well rounded so that singularities in the flow do not occur.

The additive drag for case (b) may be estimated by taking the difference in the stream thrust at station 2 and in the external free stream. For ease in the calculation, station 2 is taken far enough downstream so that the flow is uniform. It will differ infinitesimally in pressure and Mach number from the free stream because of the area change in arriving at station 2. A limiting process for which A_0, the area of the external free stream, approaches infinity will be utilized. Let

$$p_2 = p_\infty + \delta p_\infty, \qquad M_2 = M_\infty + \delta M_\infty, \qquad A_2 = A_\infty + \delta A_\infty \tag{3-16}$$

where δp_∞ and δM_∞ are infinitesimal but δA_∞ is finite. From Eq. 1-16 for no loss in total pressure,

$$\frac{\delta p_\infty}{p_\infty} = -\frac{\gamma M_\infty}{1 + \frac{\gamma - 1}{2} M_\infty^2}\, \delta M_\infty \tag{3-17}$$

Likewise from Eq. 1-20 for constant stagnation temperature,

$$\delta M_\infty = \frac{M_\infty\left(1 + \frac{\gamma - 1}{2} M_\infty^2\right)}{(M_\infty^2 - 1)} \frac{\delta A_\infty}{A_\infty} \tag{3-18}$$

Combination to eliminate δM_∞ gives

$$\delta p_\infty = -\frac{\gamma p_\infty M_\infty^2}{(M_\infty^2 - 1)} \frac{\delta A_\infty}{A_\infty} \tag{3-19}$$

The additive drag is the difference in stream thrusts between stations 2 and 0.

$$D_{\text{add}} = (p_2 - p_\infty)A_2 + p_2 A_2 \gamma M_2^2 - p_\infty A_\infty \gamma M_\infty^2 \tag{3-20}$$

If Eq. 3-16, 3-18, and 3-19 are substituted into Eq. 3-20 and A_∞ is allowed to approach infinity, there results $D_{\text{add}} = 0$. The same proof also applies to case (a) except that here the additive drag plus the leading edge suction force referenced to free stream static pressure must equal the difference in external stream thrust which is again zero. Summarizing,

$$\left.\begin{array}{ll}\text{Case (a) additive drag plus leading edge force} & = 0\\ \text{Case (b) additive drag} & = 0\end{array}\right\} \tag{3-21}$$

Alternative proofs of these results are included in [*52,60,61*].

The losses may now be estimated. For velocity ratios less than unity (Fig. E,3q), there will be no large loss in pressure recovery of the flow entering the inlet. Isentropic relations may be utilized to estimate the inlet flow parameters from the mass flow and the free stream conditions. If the inlet is well rounded, the additive drag for the external flow is canceled by the leading edge suction force (plus any other nonfrictional force on the body to the cylindrical section), as long as isentropic flow equations can be utilized. If the inlet lip is sharp, thin, and for example, cylindrical, there can be no cancellation of the additive drag. The calculation breaks down in this case because the external flow can no longer be considered isentropic, but the additive drag will not be changed as it can be calculated from the internal flow. The internal flow may still be considered isentropic so that the losses due to sharp lips at velocity ratios less than unity consist of the additive drag term (Eq. 3-12).

For velocity ratios greater than unity the additive drag is zero. If the inlet lips are well rounded, the stream thrust at station 1 is then equal to the internal flow free stream thrust plus the leading edge suction force referenced to free stream static pressure. If the inlets are sharp, thin, and cylindrical, there can be no leading edge suction force and the inlet stream thrust is then equal to the free stream thrust. Corresponding losses in total pressure and maximum mass flow rates then occur because of the flow singularity at the sharp leading edge.

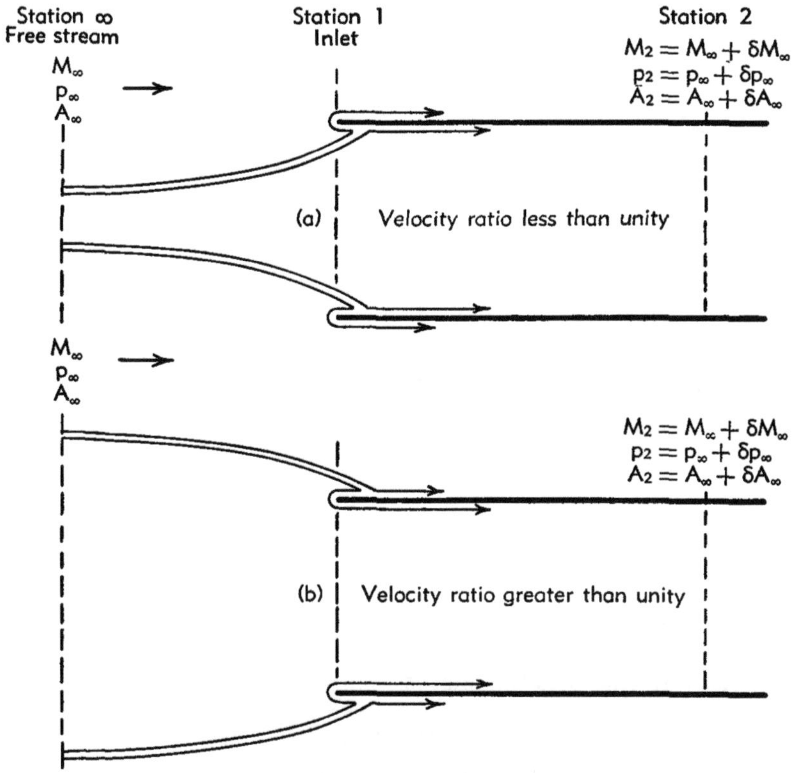

Fig. E,3q. Inlet streamlines at subsonic speeds.

The schedule of mass flow and pressure recovery for a sharp-lipped inlet with velocity ratios greater than unity may then be written from one-dimensional equations. Setting the inlet stream thrust equal to the free stream value gives

$$p_\infty(A_1 + \gamma M_\infty^2 A_\infty) = p_1(1 + \gamma M_1^2)A_1 \tag{3-22}$$

By equating the mass flow from the free stream to the inlet

$$\frac{A_1}{A_\infty} = \frac{p_\infty}{p_1} \frac{M_\infty \sqrt{1 + \frac{\gamma - 1}{2} M_\infty^2}}{M_1 \sqrt{1 + \frac{\gamma - 1}{2} M_1^2}} \tag{3-23}$$

Eliminating A_1/A_∞ between Eq. 3-22 and 3-23 gives

$$\frac{p_\infty}{p_1} = 1 + \gamma M_1^2 - \frac{\gamma M_\infty M_1 \sqrt{1 + \frac{\gamma - 1}{2} M_1^2}}{\sqrt{1 + \frac{\gamma - 1}{2} M_\infty^2}} \tag{3-24}$$

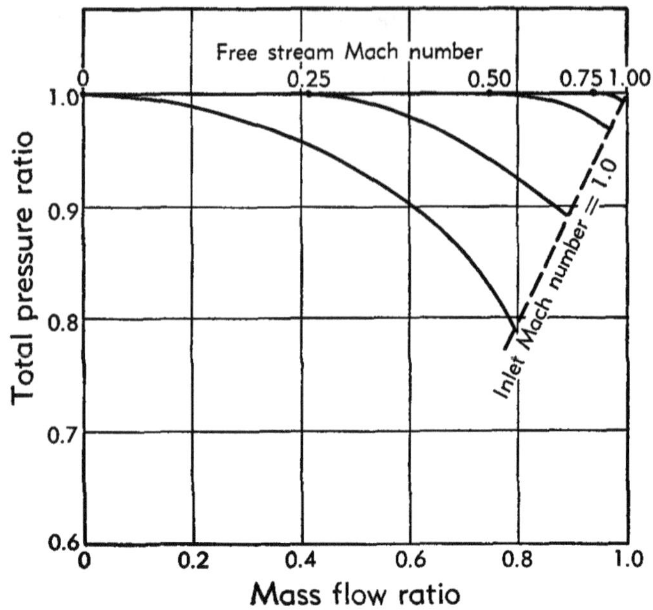

Fig. E,3r. Subsonic performance of sharp-lipped inlets.

The mass flow rate into the inlet is

$$m = \frac{\gamma p_1 A_1 M_1 \sqrt{1 + \frac{\gamma - 1}{2} M_1^2}}{\sqrt{\gamma \mathcal{R} T^0}}$$

while the maximum rate if the flow were isentropic is

$$m_{\max} = \frac{\gamma p_\infty^0 A_1}{\sqrt{\gamma \mathcal{R} T^0}\left(\frac{\gamma + 1}{2}\right)^{(\gamma+1)/2(\gamma-1)}} = \frac{\gamma p_\infty \left(1 + \frac{\gamma - 1}{2} M_\infty^2\right)^{\gamma/(\gamma-1)} A_1}{\sqrt{\gamma \mathcal{R} T^0}\left(\frac{\gamma + 1}{2}\right)^{(\gamma+1)/2(\gamma-1)}}$$

Hence

$$\frac{m}{m_{\max}} = \left(\frac{\gamma + 1}{2}\right)^{(\gamma+1)/2(\gamma-1)} \frac{M_1 \sqrt{1 + \frac{\gamma - 1}{2} M_1^2}}{\left(1 + \frac{\gamma - 1}{2} M_\infty^2\right)^{\gamma/(\gamma-1)}} \frac{p_1}{p_\infty} \tag{3-25}$$

Similarly from Eq. 1-16

$$\frac{p_1^0}{p_\infty^0} = \left(\frac{1 + \frac{\gamma - 1}{2} M_1^2}{1 + \frac{\gamma - 1}{2} M_\infty^2}\right)^{\gamma/(\gamma-1)} \frac{p_1}{p_\infty} \tag{3-26}$$

Elimination of the static pressure ratio in Eq. 3-25 and 3-26 by means of Eq. 3-24 yields two equations for the mass flow rate and the total pressure recovery ratio as a function of the inlet and free stream Mach numbers. From these parametric equations, the total pressure recovery ratio may be plotted as a function of the mass flow ratio. Such a plot is shown in Fig. E,3r. It may be noted that the equations apply only if $M_1 \geqq M_\infty$; for lower inlet Mach numbers, p_1^0/p_∞^0 remains at the limiting value of unity. Subsonic diffuser losses have not been included.

At take-off ($M_\infty = 0$) there may be a loss in maximum mass flow rate of as much as 21 per cent. If the inlet chokes, the pressure recovery loss can be even higher due to the internal "normal shock." The relief from such difficulty would come from increasing the inlet size or using auxiliary air entries.

CHAPTER 2. EXIT NOZZLES AND JETS

E,4. Thrust Estimations of Exit Nozzles by One-Dimensional Equations. A study of Eq. 3-11 shows that the engine thrust on a ramjet or turbojet engine is the difference in stream thrust between the exit and a free stream station taken far upstream of the inlet. A corresponding jet thrust may be defined as the stream thrust at the exit station,

$$F_j = m_e u_e + (p_e - p_\infty)A_e \tag{4-1}$$

The jet thrust is of course the total thrust on a rocket engine at any flight speed and, on a turbojet, at zero flight speed. It depends for its value on nozzle geometry, on stagnation-to-static pressure ratio at the exit, and on the free stream static pressure.

Sonic velocity at the exit divides the performance range into two flow regimes. If the exit Mach number is less than unity, the engine mass flow and velocity readjust until the exit static pressure equals the local external pressure, which for simplicity is assumed equal to the ambient pressure p_∞. The jet thrust is then

$$\begin{aligned} F_j &= m_e u_e = \gamma p_e M_e^2 A_e \\ &= \frac{2\gamma p_e A_e}{\gamma - 1}\left[\left(\frac{p^0}{p_e}\right)^{(\gamma-1)/\gamma} - 1\right] \end{aligned} \tag{4-2}$$

Thus for a given exit area and pressure, the subsonic jet thrust varies as the square of its Mach number which in turn varies monotonically with the jet stagnation-to-static pressure ratio.

If, on the other hand, the exit Mach number is equal to or greater than unity, the exit static pressure may be considerably different from the local static pressure again assumed to be p_∞. In this case, Eq. 4-1 can be written

$$\frac{F_j}{p_\infty A^*} = (1 + \gamma M_e^2)\left(\frac{p_e}{p^0}\right)\left(\frac{p^0}{p_\infty}\right)\left(\frac{A_e}{A^*}\right) - \frac{A_e}{A^*}$$

or

$$\frac{F_j}{p_\infty A^*} = \frac{(1 + \gamma M_e^2)}{\left(1 + \frac{\gamma - 1}{2} M_e^2\right)^{\gamma/(\gamma-1)}} \left(\frac{A_e}{A^*}\right)\left(\frac{p^0}{p_\infty}\right) - \frac{A_e}{A^*} \tag{4-3}$$

where A^* is the area of the choked nozzle throat. The exit Mach number for a choked ideal nozzle is related to the exit-to-throat area ratio A_e/A^* by Eq. 1-19. For a given exit-to-throat area ratio, the exit Mach number is constant, so that the jet thrust parameter $F_j/p_\infty A^*$ increases linearly with the total-to-free stream static pressure ratio p^0/p_∞.

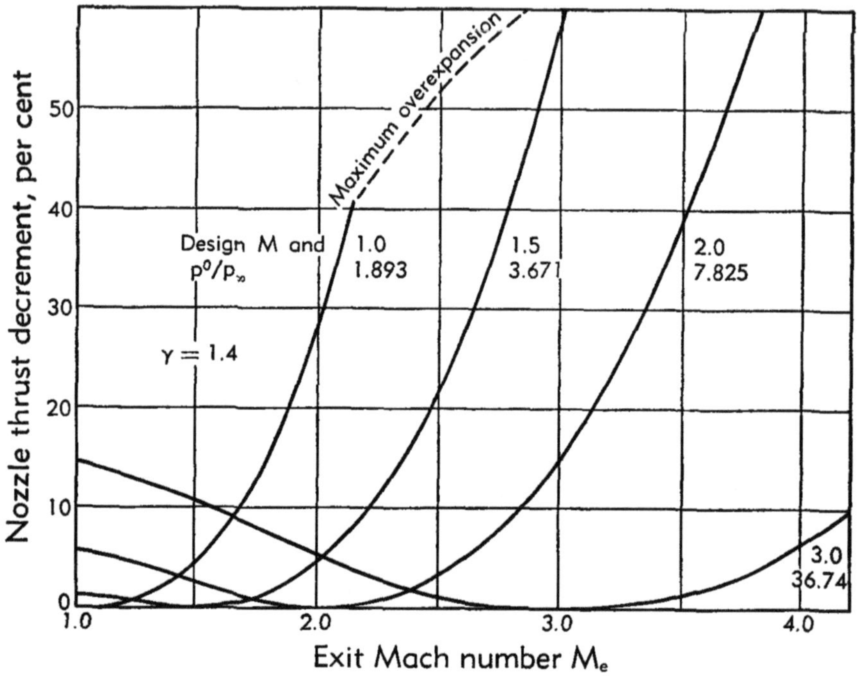

Fig. E,4a. Supersonic nozzle off-design thrust loss.

A fully expanded nozzle is one for which the exit pressure equals the free stream static pressure. This nozzle gives maximum thrust for a given pressure ratio. The jet thrust parameter for this case is

$$\frac{F_j}{p_\infty A^*} = \gamma M_e^2 \left(\frac{A_e}{A^*}\right) \tag{4-4}$$

If the nozzle is either overexpanded ($p_e < p_\infty$), or underexpanded ($p_e > p_\infty$), thrust losses occur. This thrust loss is shown for several values of pressure ratio in Fig. E,4a.

If the local flow field into which the nozzle dumps is at a static pressure other than free stream value, the optimum thrust minus drag of the aircraft configuration is obtained when the nozzle expands to this local static pressure, rather than to the free stream static pressure. In other words, the nozzle should be designed to exhaust at the static pressure of the flow field surrounding the exit.

The limit of overexpansion occurs when a normal shock at the exit raises the static pressure to free stream value. In practice, turbulent boundary layer separation usually occurs before this limit is reached. Hence the thrust losses on overexpanded nozzles, particularly for those with high divergence angles, are less than the losses shown in Fig. E,4a. In principle, the exit pressure at which separation first occurs may be calculated from shock wave boundary layer relations such as Eq. 3-9.

In rocket work "the specific impulse" is commonly used, a term denoting the jet thrust divided by the mass flow rate. For a fully expanded nozzle, the specific impulse is proportional to the exit velocity

$$I_{sp} = \frac{F_j}{mg} = \frac{u_e}{g} = \frac{\sqrt{\gamma \mathcal{R} T^0}\, M_e}{g\left(1 + \frac{\gamma - 1}{2} M_e^2\right)^{\frac{1}{2}}}$$

$$= \frac{\sqrt{\gamma \mathcal{R} T^0}\sqrt{\frac{2}{\gamma - 1}}\left[\left(\frac{p^0}{p_\infty}\right)^{(\gamma-1)/\gamma} - 1\right]^{\frac{1}{2}}}{g\left(\frac{p^0}{p_\infty}\right)^{\gamma/2(\gamma-1)}} \tag{4-5}$$

Inasmuch as all of the curves of Fig. E,4a are for constant mass flow, they also give the decrease in specific impulse due to underexpansion or overexpansion. Curves useful for rocket work or, for that matter, hot jets in general would utilize values of γ other than 1.4 as used for Fig. E,4a.

An estimate of the loss in jet thrust due to friction in the supersonic portion of the nozzle may be obtained from Eq. 1-15 and 1-20. If the nozzle is conical and of half angle α,

$$\frac{dp^0}{p^0} = \frac{-M(1 - M^2)dM}{\left(M^2 - \frac{2 \tan \alpha}{\gamma f}\right)\left(1 + \frac{\gamma - 1}{2} M^2\right)} \tag{4-6}$$

In general, $2 \tan \alpha/\gamma f$ will be much larger than M^2 and so Eq. 4-6 may be approximated

$$\frac{dp^0}{p^0} = \frac{+\gamma f}{2 \tan \alpha} \frac{M(1 - M^2)dM}{\left(1 + \frac{\gamma - 1}{2} M^2\right)} \tag{4-7}$$

or

$$\left(\frac{p^0}{p_1^0}\right)^{-2 \tan \alpha/\gamma f} = \left(\frac{1 + \frac{\gamma - 1}{2} M_e^2}{\frac{\gamma + 1}{2}}\right)^{-(\gamma+1)/(\gamma-1)^2} e^{(M_e^2-1)/(\gamma-1)} \tag{4-7'}$$

Curves of p^0/p_1^0 and M_e are plotted for several values of the friction parameter, $2 \tan \alpha/\gamma f$, as a function of the ideal exit Mach number in Fig. E,4b. The value of $2 \tan \alpha/\gamma f$ of 25 which gives a total pressure loss of 32 per cent at a Mach number of 3.25 is a rather extreme case. It corresponds approximately to a conical expansion half angle of 2°. In spite of the large loss in pressure, the jet thrust for complete expansion is only decreased about 3 per cent from ideal value. This example illustrates the insensitivity of jet thrust to losses in total pressure recovery at the higher exit Mach numbers as long as the mass flow rate is not correspondingly decreased. It should be recognized, however, that the area ratios for complete expansion may be altered considerably due to friction.

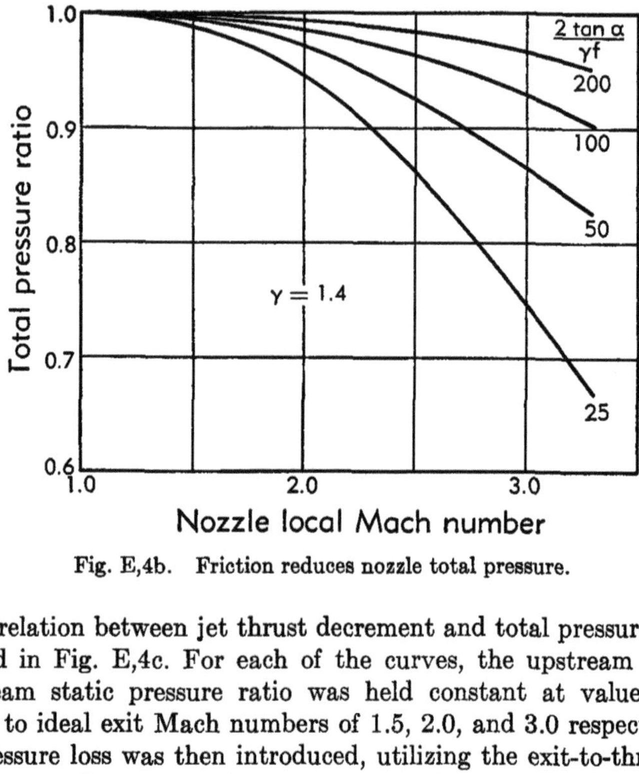

Fig. E,4b. Friction reduces nozzle total pressure.

The relation between jet thrust decrement and total pressure ratio is indicated in Fig. E,4c. For each of the curves, the upstream total-to-free stream static pressure ratio was held constant at values corresponding to ideal exit Mach numbers of 1.5, 2.0, and 3.0 respectively. A total pressure loss was then introduced, utilizing the exit-to-throat area ratio corresponding to complete expansion of the resulting flow. Had this adjustment not been made, additional losses corresponding to incomplete or overexpansion (Fig. E,4a) would have been entailed. The curves illustrate again the insensitivity of jet thrust to total pressure ratio for the higher exit Mach numbers when the area ratio is chosen so that the nozzle is fully expanded.

Another loss in jet thrust occurs because the momentum of the exit flow is not directed axially. For radial source flow used in the example, correction for stream divergence requires multiplication of the one-dimensional jet thrust by the factor $\sin^2 \alpha / 4 \sin^2 (\alpha/2)$.

In exit jets designed for rockets, ramjets, and afterburners, the choice of the proper expansion ratio is complicated by the possibility that combustion may still persist in the divergent portion of the nozzle. Heat addition in the supersonic nozzle flow causes a reduction in the exit Mach number, a decrease in total pressure, an increase in exit static pressure, and a decrease in the effective area expansion ratio of the nozzle. A nozzle of proper expansion ratio without heat addition will then be improperly expanded if heat addition occurs.

The results of heat addition in a supersonic nozzle depend in part on the rate at which the heat is added. To obtain estimates of the effects of

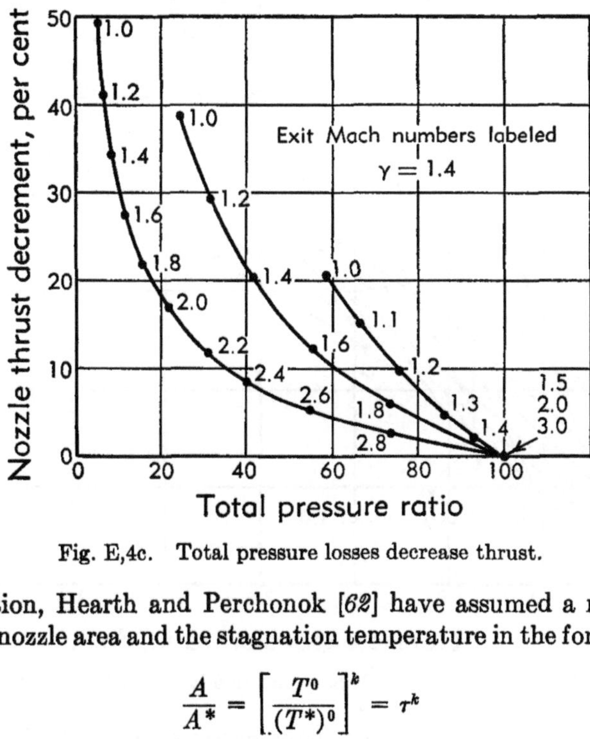

Fig. E,4c. Total pressure losses decrease thrust.

heat addition, Hearth and Perchonok [*62*] have assumed a relation between the nozzle area and the stagnation temperature in the form

$$\frac{A}{A^*} = \left[\frac{T^0}{(T^*)^0}\right]^k = \tau^k \tag{4-8}$$

Fig. E,4d taken from their results shows the variation of jet thrust with heat addition. The $\tau = 1.00$ curve represents isentropic expansion to ambient pressure, and the gain in jet thrust is due to the use of a convergent-divergent nozzle over that of a convergent-sonic exit.

The utility of Fig. E,4d may be illustrated with examples. Point *A* denotes an isentropic supersonic nozzle operating at a pressure ratio of 7.0. The corresponding jet thrust is approximately 6 per cent larger than for a converging nozzle at the same pressure ratio. If heat addition of temperature ratio 1.2 occurs in the diverging portion, the thrust increases 3.5 per cent to point *B* and the nozzle becomes underexpanded. Further thrust increases can be achieved by expanding the nozzle to point *C* lying on the pressure ratio 7.0 line.

Heat addition in the nozzle may be intentionally used to prevent overexpansion. If a fixed geometry nozzle operating at a pressure ratio of 10 (point D) has its pressure ratio reduced to 7, overexpansion may be

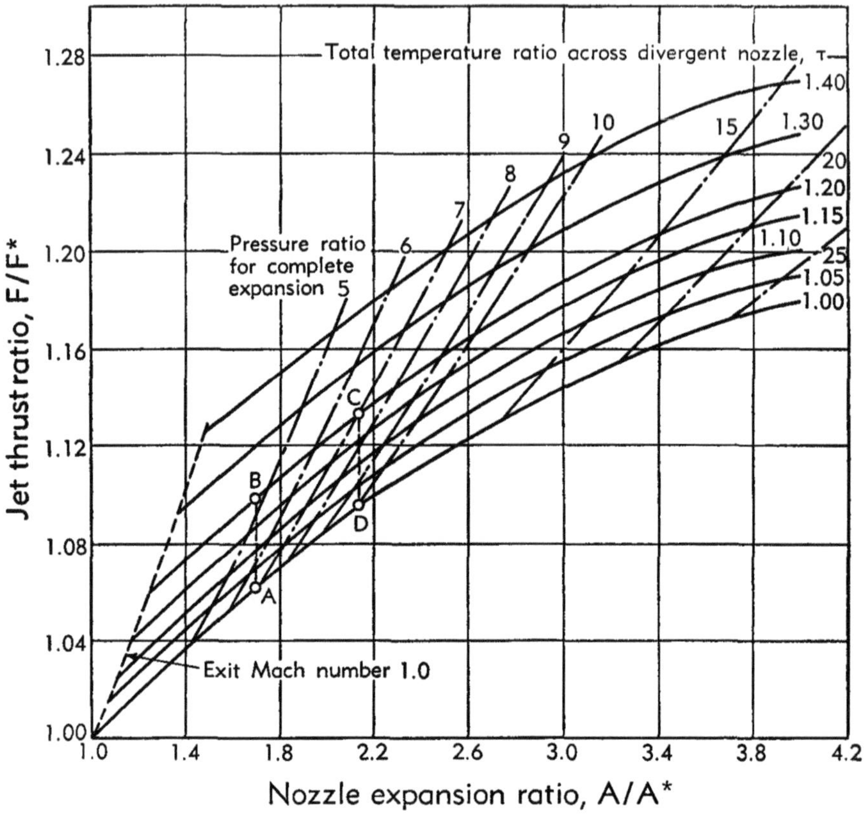

Fig. E,4d. Supersonic heat addition increases jet thrust.

prevented by heat addition to a temperature ratio of 1.2. The Mach 1.0 line is the limit to which this process may be employed. Along this line the geometric convergent-divergent nozzle has been reduced in effect to a convergent configuration by heat addition.

If combustion is delayed so that it occurs in the divergent portion of the nozzle rather than in the engine, a loss in jet thrust will occur and a corresponding decrease in the effective nozzle coefficient will then be recorded. The ordinate in Fig. E,4e [*62*] represents the ratio of the two jet thrusts, one obtained when an incremental amount of heat addition occurs in the divergent portion of the nozzle, the other when this heat addition is completed ahead of the nozzle throat. In each case the nozzle is fully expanded to free stream static pressure.

The loss in jet thrust due to delayed heat addition in the divergent portion of the nozzle depends primarily on the incremental total temperature ratio and is relatively insensitive to nozzle pressure ratio. Fig. E,4e demonstrates that the loss can be considerable. This thrust loss may be erroneously assigned as exit nozzle inefficiency yielding misleading nozzle coefficients.

Jet nozzles may be regarded as major engine components on rocket, ramjet, and turbojet engines. The nozzle throat for each of these engines

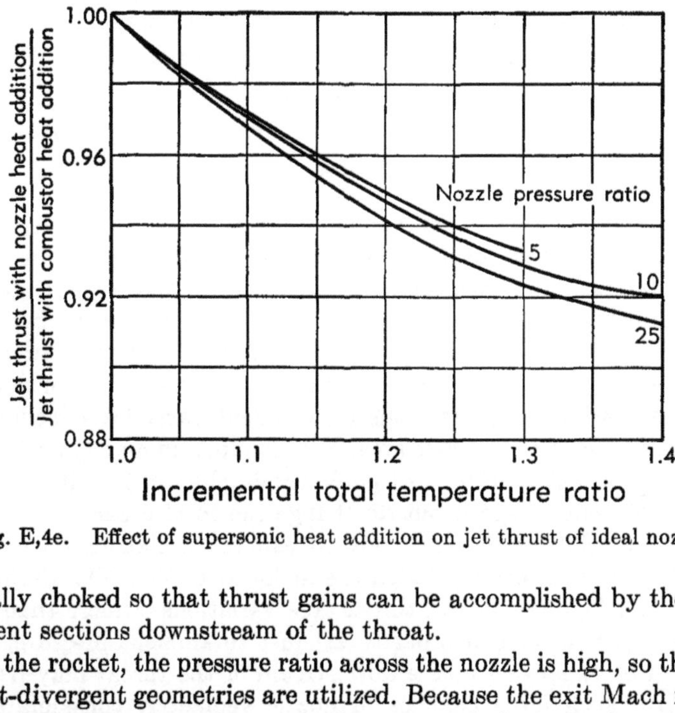

Fig. E,4e. Effect of supersonic heat addition on jet thrust of ideal nozzle.

is usually choked so that thrust gains can be accomplished by the use of divergent sections downstream of the throat.

On the rocket, the pressure ratio across the nozzle is high, so that convergent-divergent geometries are utilized. Because the exit Mach number is high, the thrust loss due to moderate nozzle imperfections is slight (Fig. E,4c). Hence the use of the method of characteristics or other means to obtain perfect shock-free flow contours is seldom warranted.

For the ramjet engine, the nozzle throat area is chosen in conjunction with fuel flow and flight plan to maintain near optimum diffuser performance. A fixed geometry exit is satisfactory if the flight Mach number and the required engine thrust coefficients are reasonably constant. As the ramjet engine is developed to a more versatile power plant, however, the need for variable exhaust throat area and supersonic expansion will increase. Variable throat area alone would substantially improve the engine efficiency and the range of thrust coefficients available at any flight speed. The gains would be at the expense of increased complications, including the variable exhaust nozzle and a control system to maintain the proper throat area for each engine fuel flow to give optimum diffuser performance.

The earlier turbojet engines incorporated fixed convergent nozzles, so that the only engine control was the fuel flow rate. The engine speed thus varied with the fuel flow. Later engines usually include variable exhaust nozzles as well as variable fuel flow rate. In particular, the variable exhaust nozzle area is essential on turbojet engines incorporating afterburning.

The exhaust throat must be of sufficient size to pass the afterburner mass flow. This quantity is in turn the product of the engine air flow and the factor $[1 + (w_f/w_a)]$ where w_f/w_a is the fuel-air ratio. Hence the outlet throat area, assumed to be choked, is

$$A_e = \left(\frac{\sqrt{T_0}}{p_0}\right)\left(\frac{\dot{w}\sqrt{\theta}}{\delta}\right)\frac{p_i^0}{p_e^0}\sqrt{\frac{T_e^0}{T_i^0}}\left(1 + \frac{w_f}{w_a}\right)\sqrt{\frac{\mathcal{R}_e}{\gamma_e}}\left(\frac{\gamma_e + 1}{2}\right)^{(\gamma_e+1)/2(\gamma_e-1)} \tag{4-9}$$

Subscripts e and i refer to exit and compressor inlet stations respectively; 0 signifies standard sea level conditions. The outlet area must thus increase with engine air flow, with temperature-rise ratio across the engine, and with the reciprocal total pressure ratio across the engine. It may at first seem strange that Eq. 4-9 continues to hold since only the outlet area is varied. The quantities in the second member must be considered as dependent variables, whereas the independent variables are engine and afterburner fuel flow and exit area. For example, in an engine for which the turbine is choked in the exit guide vanes, changes in exit area do not influence the engine speed but do change the total pressure ratio across the engine by shifting the shock downstream of the choking station. More complete details of engine design are included in Vol. XII. Suffice it to say that, for high speed jets, throat area variations of more than two to one are required—and in order to exhaust to ambient pressure, an adjustable area expansion nozzle downstream of the throat may be necessary. With such large changes in nozzle geometry, considerable care must be exercised in the design of external aerodynamic fairings to prevent large boattail or base drags when the nozzle is operating at other than its maximum size.

To obtain the throat area variations, clam shell or eyelid type nozzles, iris or finger nozzles, and translating central plugs have been employed. Independent variations of both throat area and divergent expansion ratio, however, require more ingenious designs.

The clam shell nozzle offers little promise as yet for convergent-divergent supersonic types. Double iris nozzles are being designed with independent variations of throat area and expansion ratio which may prove to be satisfactory. Internal plug nozzles can be constructed that give a scheduled variation of throat area and expansion ratio but these nozzles will seldom match all flight conditions. The plug is also hard to cool when submitted to the hot gases of an afterburner.

An aerodynamic principle may serve the designer well in his quest for variable convergent-divergent nozzle flows. Just as the use of external compression eliminates the internal contraction difficulties in a supersonic diffuser, so does external expansion eliminate the overexpansion difficulties in a supersonic nozzle. Overexpansion is traceable to jet attachment to the nozzle walls in a manner resembling a Coanda effect. If sufficient portions of the wall are eliminated, the overexpansion also disappears; yet the redesigned nozzle serves effectively as a convergent-divergent geometry to generate a supersonic jet.

For example, an external plug nozzle may be designed to resemble a spike diffuser operating in reversed flow, so that a uniform supersonic jet is obtained downstream of the plug at the design pressure ratio. Below the design pressure ratio the flow roughly follows the plug contour, generating a free stream jet boundary that is nearly at ambient pressure. Overexpansion and the associated losses in thrust are thus prevented.

Also of considerable interest is the ejector arrangement. Wilsted, Kochendorfer, and others have shown that the secondary air flow of an ejector cushions the primary stream against sudden expansion, and the unit may operate effectively as a convergent-divergent nozzle. For this reason, and because of the necessity for a pump to provide cooling air flow to afterburners at take-off condition, the ejector will find increasing application as a nozzle for high speed jet aircraft.

E,5. The Aircraft Ejector. In principle the theory of the ejector is more complicated than the turbulent mixing processes between two streams. The usual approach toward estimation of ejector performance includes the one-dimensional equations of continuity, momentum, and energy, utilizing various assumptions concerning the degree to which the primary and secondary flows are independent or mixed. Complete mixing cannot be assumed for the relatively short ejectors employed on jet-powered aircraft.

The aircraft ejector operates with high pressure ratios across the primary nozzle with choked flow in the primary and perhaps in the secondary stream. Under such conditions the ejector performance map is independent of the ambient pressure.

The criterion for flow similarity at a constant weight flow ratio is that the stagnation pressures of the primary and secondary flows are proportional. Thus at a given ejector weight flow ratio, plots of secondary against primary pressure ratio are straight-line functions through the origin. The ejector performance map in the choked regime (cold flow) can thus be reduced to a single curve of stagnation pressure ratio against corrected weight flow ratio. The shape of the curve depends on the geometry of the nozzle and the ejector shroud.

Although it is somewhat afield from the theme of this article to discuss ejector performance, Kochendorfer's theory for estimating ejector performance is included in its simplest form as a guide to physical intuition. Consider the cylindrical ejector (Fig. E,5) for which the primary and secondary fluids have the same thermodynamic properties but may be at different temperatures and pressures. The primary stream is choked at station 1 and the secondary stream accelerates to sonic speeds at station 2. As a first approximation, the primary flow is assumed to be isentropic

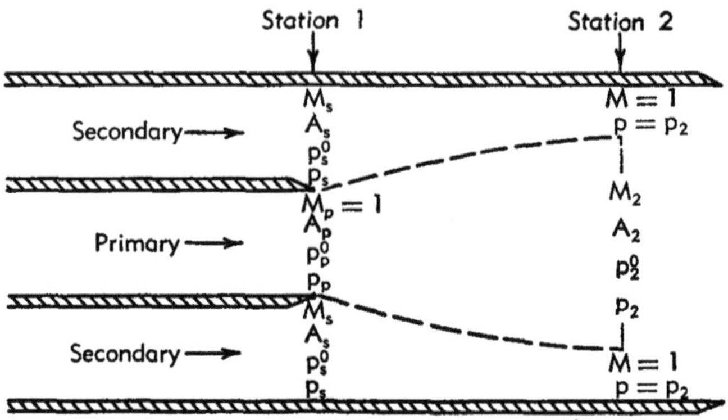

Fig. E,5. Short cylindrical high pressure ejector.

and both streams have the same static pressure at station 2. If each flow is assumed adiabatic, the one-dimensional continuity equations for mass flow and momentum then yield

$$\dot{w}_s \sqrt{T_s^0} = \sqrt{\frac{\gamma}{\mathcal{R}}} \frac{p_s^0 A_s M_s}{\left(1 + \frac{\gamma - 1}{2} M_s^2\right)^{(\gamma+1)/2(\gamma-1)}}$$

$$= \sqrt{\frac{\gamma}{\mathcal{R}}} p_s A_s M_s \sqrt{1 + \frac{\gamma - 1}{2} M_s^2} \quad (5\text{-}1)$$

$$\dot{w}_s \sqrt{T_s^0} = \sqrt{\frac{\gamma}{\mathcal{R}}} p_2(A_p + A_s - A_2) \sqrt{\frac{\gamma + 1}{2}} \quad (5\text{-}2)$$

$$\dot{w}_p \sqrt{T_p^0} = \sqrt{\frac{\gamma}{\mathcal{R}}} \frac{p_p^0 A_p}{\left(\frac{\gamma + 1}{2}\right)^{(\gamma+1)/2(\gamma-1)}} \quad (5\text{-}3)$$

$$\dot{w}_p \sqrt{T_p^0} = \sqrt{\frac{\gamma}{\mathcal{R}}} \frac{p_p^0 A_2 M_2}{\left(1 + \frac{\gamma - 1}{2} M_2^2\right)^{(\gamma+1)/2(\gamma-1)}} \quad (5\text{-}4)$$

$$p_pA_p(1 + \gamma_p) + p_sA_s(1 + \gamma M_s^2) = p_2A_2(1 + \gamma M_2^2) + p_2(A_p + A_s - A_2)(1 + \gamma) \quad (5\text{-}5)$$

From Eq. 5-1 and 5-3

$$\frac{\dot{w}_s\sqrt{T_s^0}}{\dot{w}_p\sqrt{T_p^0}} = \left(\frac{A_s}{A_p}\right)\left(\frac{p_s^0}{p_p^0}\right)\frac{M_s\left(\frac{\gamma+1}{2}\right)^{(\gamma+1)/2(\gamma-1)}}{\left(1+\frac{\gamma-1}{2}M_s^2\right)^{(\gamma+1)/2(\gamma-1)}} \quad (5\text{-}6)$$

If Eq. 5-5 is rewritten in terms of total pressures and the area ratio A_2/A_p is eliminated by means of Eq. 5-3 and 5-4 there results

$$\frac{(1+\gamma)}{\left(\frac{1+\gamma}{2}\right)^{\gamma/(\gamma-1)}} + \left(\frac{p_s^0}{p_p^0}\right)\left(\frac{A_s}{A_p}\right)\frac{(1+\gamma M_s^2)}{\left(1+\frac{\gamma-1}{2}M_s^2\right)^{\gamma/(\gamma-1)}}$$
$$= \frac{\gamma(M_2^2-1)}{\left(\frac{\gamma+1}{2}\right)^{(\gamma+1)/2(\gamma-1)}M_2\sqrt{1+\frac{\gamma-1}{2}M_2^2}} + \frac{\left(1+\frac{A_s}{A_p}\right)(1+\gamma)}{\left(1+\frac{\gamma-1}{2}M_2^2\right)^{\gamma/(\gamma-1)}} \quad (5\text{-}7)$$

Also from Eq. 5-1 and 5-2

$$\frac{p_s^0}{p_p^0}\frac{A_s}{A_p}\frac{M_s}{\left(1+\frac{\gamma-1}{2}M_s^2\right)^{(\gamma+1)/2(\gamma-1)}}$$
$$= \left[1+\frac{A_s}{A_p} - \frac{\left(1+\frac{\gamma-1}{2}M_2^2\right)^{(\gamma+1)/2(\gamma-1)}}{M_2\left(\frac{\gamma+1}{2}\right)^{(\gamma+1)/2(\gamma-1)}}\right]\frac{\sqrt{\frac{\gamma+1}{2}}}{\left(1+\frac{\gamma-1}{2}M_2^2\right)^{\gamma/(\gamma-1)}} \quad (5\text{-}8)$$

Thus for each ejector area ratio and total pressure ratio, M_s and M_2 can be obtained from Eq. 5-7 and 5-8. With M_s known, the ejector weight flow ratio $\dot{w}_s\sqrt{T_s^0}/\dot{w}_p\sqrt{T_p^0}$ may then be obtained as a function of the total pressure ratio. Of course, the weight flow ratio is a monotonically increasing function of stagnation pressure ratio and is approximately a linear function. The equations do not apply for zero weight flow ratios, however, because in this case the secondary stream does not sustain a one-dimensional static pressure difference between stations 1 and 2.

The novelty of the Kochendorfer theory lies in the realization that the secondary stream accelerates to sonic velocity between stations 1 and 2. Unlike a free jet, the static pressure varies along the boundary between the primary and secondary streams, and the primary stream responds as if it had passed through a convergent-divergent nozzle. It is this similarity that causes the thrust characteristics of an ejector configuration to resemble those of convergent-divergent nozzles.

Many refinements to this simple theory are evident. Differences in fluid properties between the primary and secondary streams as well as the interchange of mass and heat could be taken into account, but such refinements are at the expense of considerable complication and will not be discussed further.

E,6. Spreading Characteristics of Subsonic Incompressible Jets. The spreading characteristics of jets have intrigued many research investigators, and an excellent bibliography of important contributions is included in [*63*]. In general, jet spreading is relatively independent of laminar mixing processes, as laminar flow between the boundaries of two streams moving at different speeds is unstable except at extremely low

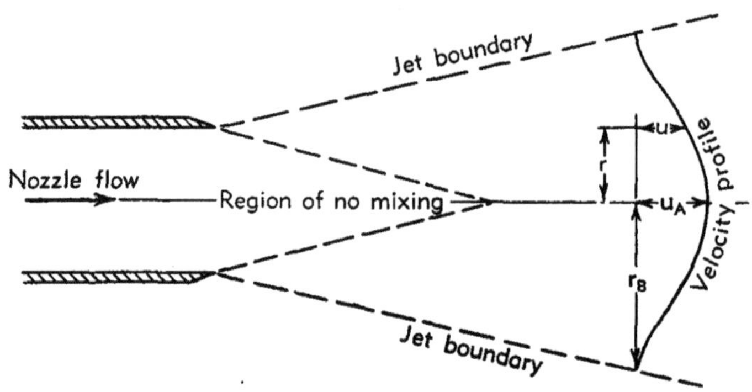

Fig. E,6. Jet spreading features.

Reynolds numbers [*64*]. Turbulent mixing processes then predominate, and as a result, in spite of many excellent contributions, much work needs to be completed before the spreading characteristics of jets are fully understood. The hot jet involves complications of heat transfer as well as the usual mass and momentum transfer due to jet entrainment of the surrounding fluid. Experimentally, both heat [*65*] and material [*63*] diffuse more rapidly than momentum in the spreading jet.

For subsonic jets exhausting into still air, the static pressure across the exit plane is nearly constant and equal to the local ambient pressure. The jet flow is then largely influenced downstream of the nozzle by the turbulent mixing processes starting at the jet rim. The features of the jet (Fig. E,6) include a conical core region of no mixing and a roughly conical jet spreading boundary, both intersecting the rim of the primary nozzle. The half angle of the no-mixing region is approximately 15°, corresponding to the usual contamination angle of turbulence spreading.

Profile similarity is clearly not possible near the nozzle exit. Somewhere downstream of the region of no mixing, similarity is approached, and the velocity profiles, for example, may be represented by

$$\frac{u}{u_A} = g\left(\frac{r}{r_B}\right) \tag{6-1}$$

where u_A is the jet speed on the axis which varies with x, and r_B is the radius of the jet boundary. The function g generally resembles an error curve although cosine or other resembling functions have been utilized in the literature.

The momentum of the jet must be conserved at all axial stations. Because the pressure is constant and the flow is assumed incompressible, conservation of momentum implies that

$$\int_0^{r_B} u^2 r dr = C_1 \text{ (a constant)} \tag{6-2}$$

Hence with Eq. 6-1, Eq. 6-2 yields

$$u_A^2 r_B^2 = \frac{C_1}{\int_0^1 \left(\frac{r}{r_B}\right) g^2 d\left(\frac{r}{r_B}\right)} = C \tag{6-3}$$

Thus the axial velocity decreases inversely with increases in the radius of the jet boundary. The quantity r_B is shown proportional to x in Fig. E,6. This proportionality has been derived from mixing length theory by Prandtl [*66*, pp. 162–178; *67*]. A summary of the various similarity arguments and their implications on free turbulent boundaries is included in [*68*].

If there is a free stream velocity, then the jet spreading characteristics are somewhat altered. Certainly if the jet and free stream velocities are equal, then no turbulent mixing would be expected. The mixing process is related to the turbulent shear which in turn depends on the velocity difference between the jet and the free stream. It seems plausible and the data seem to confirm that similarity can still be assumed for jets spreading in a free stream if the relative velocity of the two streams is used,

$$\frac{u - U}{u_A - U} = g\left(\frac{r}{r_B}\right) \tag{6-4}$$

where U is the free stream velocity. Again assuming constant static pressure throughout the jet, conservation of momentum requires that

$$\int_0^{r_B} u(U - u) r dr = \text{const} \tag{6-5}$$

At large axial distances $u \cong U$ so that Eq. 6-5 may be approximated,

$$U \int_0^{r_B} (U - u) r dr = \text{const}$$

or

$$U(U - u_A)r_B^2 \cong \text{const} \tag{6-6}$$

Since U is independent of x,

$$(U - u_A)r_B^2 \cong C \tag{6-6'}$$

Thus $(u_A - U)$ varies at large distances from the jet inversely as the jet radius squared. Jet spreading in a free stream flow field has been discussed in much greater detail in [*63,69*].

It may be remarked on physical grounds that the jet spreading angle decreases as the free stream velocity increases, reaching the laminar material diffusion angle when the jet and free stream velocities are equal.

E,7. Spreading Characteristics of Supersonic Jets. There are marked differences between the spreading characteristics of subsonic and supersonic jets in the vicinity of the jet exit. For the subsonic jet, changes in ambient pressure can cause readjustments of the internal flow so that the static pressure in the exist plane very nearly equals the ambient value. For the supersonic jet no such readjustment of the internal flow takes place, so that large changes of static pressure are permitted from the core to the jet boundary.

An example of the characteristic pattern for an over-pressure sonic jet is shown in Fig. E,7a. The jet first shows the effects of ambient pressure at the nozzle periphery, and locally expands by the Prandtl-Meyer process at supersonic speeds to ambient pressure along the jet boundary. The so-generated characteristics (see VI,G or [*70*, pp. 377–394]) crisscross through the jet, overexpanding the flow until they again reach the opposite jet boundary. Here the expansion waves must reflect as compression waves, for otherwise the jet boundary would not remain at ambient pressure. The expanding jet is thus straightened and converged, ultimately yielding the bulbous jet boundary pattern so characteristic of over-pressure sonic or supersonic nozzles.

A supersonic free stream velocity may have a pronounced effect on the jet spreading characteristics. The jet deflects the external flow, generating a shock wave at the exit and increasing the local static pressure. A supersonic free stream thus decreases the amount of jet overexpansion and correspondingly the amount of jet spreading.

The static pressure along the streamline dividing the jet and the free stream is no longer ambient as in Fig. E,7a. The pressures resulting from free stream deflections must, however, match those along the jet boundary; thus the flow field may be determined.

An example of such a two-dimensional flow field is given in Fig. E,7b. Inasmuch as the ambient pressure is the same for this and for Fig. E,7a, a comparison of the two shows the theoretical influence of an external

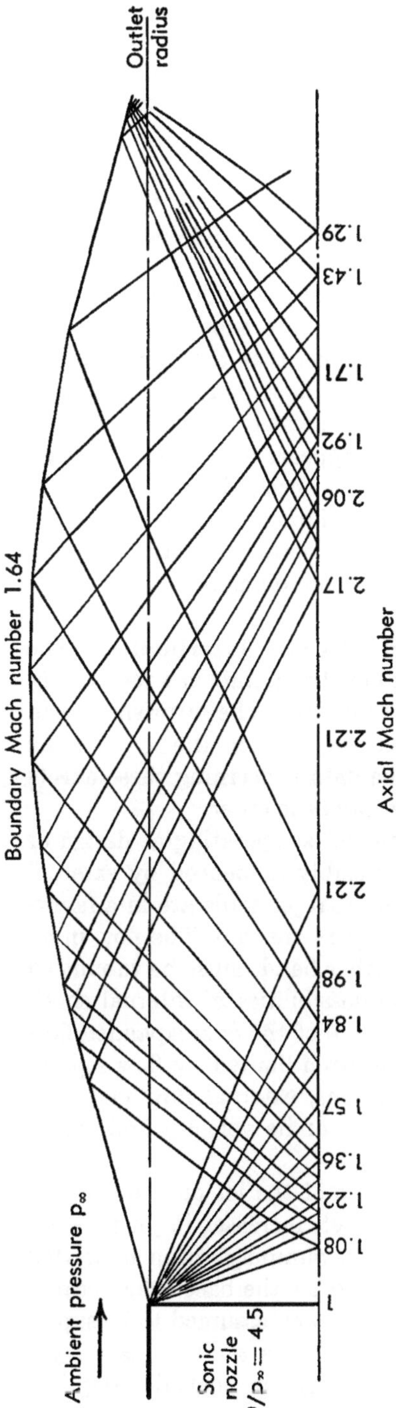

Fig. E,7a. Two-dimensional jet expansion pattern in still air.

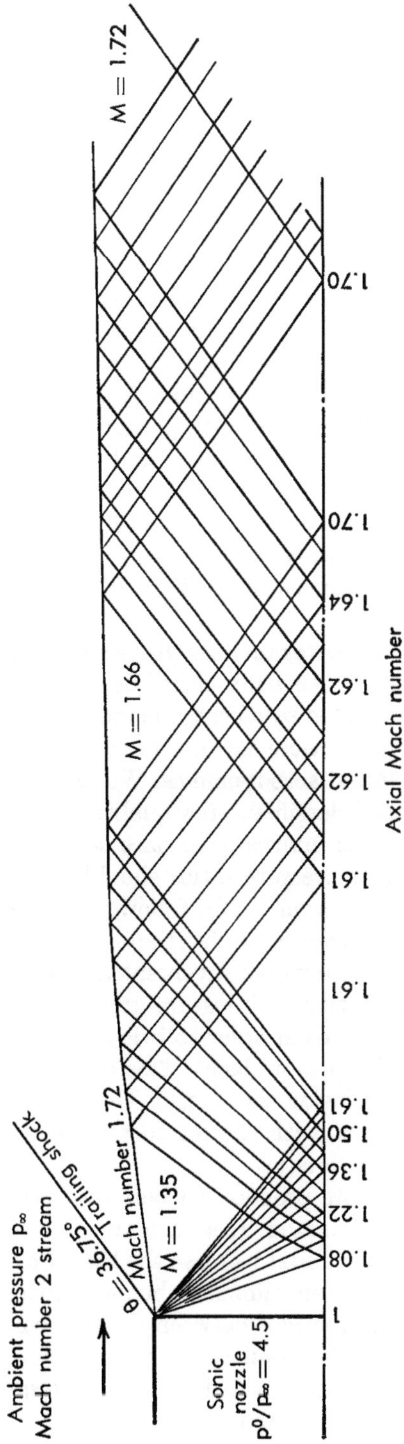

Fig. E,7b. Two-dimensional jet expansion in a Mach number 2 external stream

Mach 2 flow field on the two-dimensional potential flow jet spreading characteristics. Experimental studies of two-dimensional jets spreading into supersonic streams are described in [*71*]. The turbulent mixing processes are also discussed and formulas for predicting jet spreading are included.

The method of characteristics is effective for predicting supersonic jet spreading characteristics out to about the first maximum jet width on the bulbous flow field. Downstream therefrom, turbulent mixing processes take over, and the jet when injected into still air continues to expand linearly with distance downstream at about the subsonic jet spreading angle. Especially at the higher over-pressures, the supersonic jet at a given axial distance will hence be somewhat larger than the corresponding subsonic jet. It should be noted, perhaps that on some rocket nozzles, a supersonic core may persist downstream of the exit by as many as 10 or 12 "Mach diamonds," and the jet diameter appears to regularly increase and decrease with distance downstream of the nozzle. Even here, a subsonic jet boundary may be measured which is larger than the visible hot core, and which increases its diameter approximately linearly with distance downstream of the first maximum hot core diameter.

E,8. Interactions Between Jets and Aircraft Afterbodies. The exit jet is usually placed near the back end of the aircraft in a region contaminated by a heavy boundary layer. It would indeed be surprising, then, if there were no interferences or interactions between the jet and the external aerodynamics. While much of the data concerning these effects is still classified, a few qualitative remarks may be in order.

The internal flow on a sonic or supersonic jet operating at design or higher pressure ratios should not be appreciably influenced by external aerodynamics. This follows because the internal jet walls are in a favorable pressure gradient and the boundary layers are thin. The flow in an under-pressured divergent nozzle, on the other hand, must be submitted to an adverse pressure gradient with the possibilities of internal shock structures in order to bring the static pressure of the overexpanded flow back up to ambient. This overexpansion is often limited by flow separations inside the nozzle. Under such conditions, external flow may well interact with the jet to change its direction or its velocity and thrust characteristics.

More pronounced perhaps is the influence of the jet on the external aerodynamics. If a base area surrounds the jet, increases in jet pressure ratio will first increase the base drag due to the pumping action associated with jet entrainment, but will ultimately decrease the base drag because of interactions between the jet and the free stream, assumed to be supersonic. The feedback of base pressure through the boattail boundary layer would be expected to give corresponding changes in boattail drag, but less pronounced.

The effect of the jet on the external flow is schematically shown in Fig. E,8. Jet expansions downstream of the exit deflect the supersonic external flow, thus generating a trailing shock wave. The resulting higher pressures downstream of the trailing shock wave may feed back through the body boundary layer and reduce the boattail drag. Experimental data presented in [*72*] showed that the jet of the A4 rocket increased the body drag at subsonic speeds and decreased it at supersonic speeds.

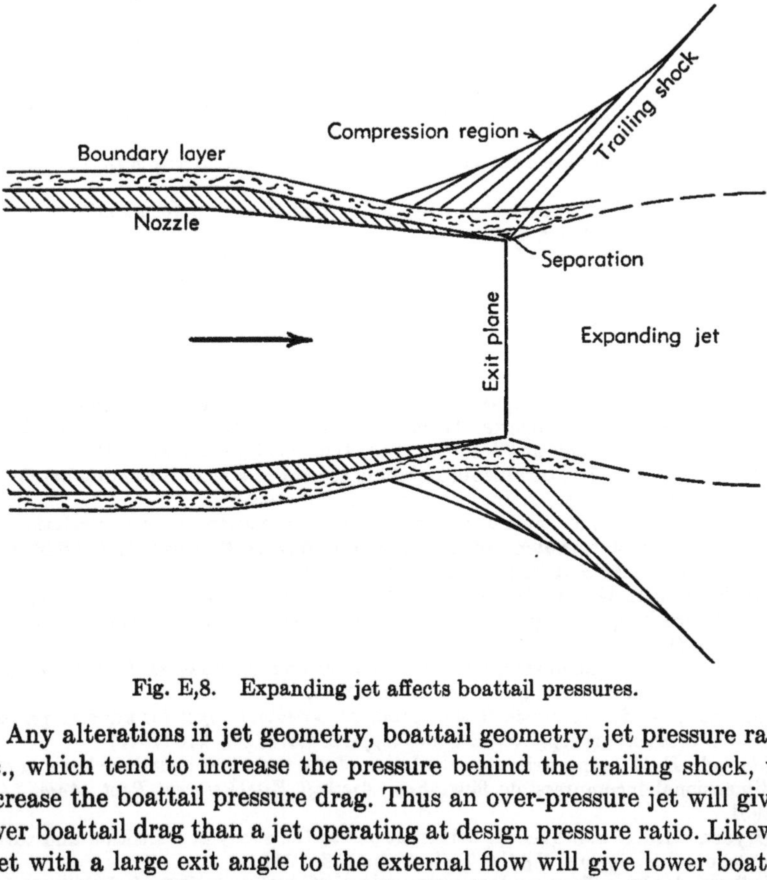

Fig. E,8. Expanding jet affects boattail pressures.

Any alterations in jet geometry, boattail geometry, jet pressure ratio, etc., which tend to increase the pressure behind the trailing shock, will decrease the boattail pressure drag. Thus an over-pressure jet will give a lower boattail drag than a jet operating at design pressure ratio. Likewise a jet with a large exit angle to the external flow will give lower boattail drags than one with a small exit angle. The lowest boattail drag does not necessarily, however, represent the optimum configuration. The optimum configuration requires simultaneous considerations of the jet thrust, the base drag, and the boattail geometry as well as the body boundary layer ahead of the base. Such considerations, however, are beyond the scope of this section.

E,9. Cited References.

1. Staff, Ames 1- by 3-Foot Supersonic Wind-Tunnel Section. Notes and tables for use in the analysis of supersonic flow. *NACA Tech. Note 1428*, 1947.

2. Wyatt, D. D. Analysis of errors introduced by several methods of weighting nonuniform duct flows. *NACA Tech. Note 3400*, 1955.
3. McLafferty, G. M. A generalized approach to the definition of average flow quantities in nonuniform streams. *United Aircraft Corp. Research Dept. Rept. R-13534-9*, July 1954.
4. Persh, J. The effect of the inlet Mach numbers and inlet boundary layer thickness on the performance of a 23° conical-diffuser-tail-pipe combination. *NACA Research Mem. L9K10*, Mar. 1950.
5. Copp, M. R., and Klevatt, P. L. Investigation of high subsonic performance characteristics of a 12°, 21-inch conical diffuser, including the effects of change in inlet-boundary layer thickness. *NACA Research Mem. L9H10*, Mar. 1950.
6. Kantrowitz, A., and Donaldson, C. duP. Preliminary investigation of supersonic diffusers. *NACA Wartime Rept. L713*, 1945. (Supersedes *NACA Advanced Confid. Rept. L5D20*).
7. Bailey, N. P. The thermodynamics of air at high velocites. *J. Aeronaut. Sci. 11*, 227–238 (1944).
8. Crocco, L. Gallerie aerodynamiche per alte velocita. *Aerotecnica 15*, 237–275, 735–778 (1935).
9. Liepmann, H. W., and Puckett, A. E. *Introduction to Aerodynamics of a Compressible Fluid.* Wiley, 1947.
10. Pohlhausen, K. Zur näherungsweisen Integration der Differentialgleichung der laminaren Grenzschicht. *Z. angew. Math. u. Mech. 1*, 252–268 (1921).
11. Hamel, G. Spiralförmige Bewegungen zäher Flüssigkeiten. *Jahresber. d. deutschen Math. Ver. 25*, 34–60 (1917). (Transl. as *NACA Tech. Mem. 1342*, 1953.)
12. Stewartson, K. *Proc. Roy. Soc. London A200*, 84–100 (1949).
13. Howarth, L. *Proc. Roy. Soc. London A194*, 16 (1948).
14. Buri, A. Eine Berechungsgrundlage für die Turbulente Grenzschicht bei Beschleunigter und Verzögerter Grundstromung. *Thesis*, E. T. H. Frey, Zurich, 1931. Transl. available as *Thesis 652*, Fed. Tech. College, Zurich, 1931.
15. Gruschwitz, E. Die turbulente Reinbungschicht in ebener Strömung bei Druckabfall und Druckansteig. *Ing.-Arch. 2*, 321–346 (1931).
16. Von Doenhoff, A. E., and Tetervin, N. Determination of general relations for the behavior of turbulent boundary layers. *NACA Wartime Rept. L382*, 1943. (Supersedes *NACA Advanced Confid. Rept. 3G13*).
17. Kalikhman, L. E. A new method for calculating the turbulent boundary layer and determining the separation point. *Compt. rend. Acad. Sci. U.R.S.S. 38*, 180–185 (1943).
18. Garner, H. C. The development of turbulent boundary layers. *British Aeronaut. Research Council Repts. and Mem. 2133 (7814)*, 1944.
19. Coleman, W. S. Analysis of the turbulent boundary layer for adverse pressure gradients involving separation. *Quart. Appl. Math. 5*, 182–216 (1947).
20. Maskell, E. C. Approximate calculation of the turbulent boundary layer in two-dimensional incompressible flow. *Roy. Aircraft Establishment Rept. Aero. 2443*, 1951.
21. Dryden, H. L. Some recent contributions to the study of transition and turbulent boundary layers. *NACA Tech. Note 1168*, 1947.
22. Falkner, V. M. The resistance of a smooth flat plate with turbulent boundary layer. *Aircraft Eng. 15*, 65–69 (1943).
23. Squire, H. B., and Young, A. D. The calculation of the profile drag of airfoils. *British Aeronaut. Research Council Repts. and Mem. 1838*, 1937.
24. Little, B. H., Jr., and Wilbur, S. High subsonic performance characteristics and boundary layer investigations of a 12°, 10-inch inlet-diameter conical diffuser. *NACA Research Mem. L50C02a*, May 1950.
25. Nitzberg, G. E., and Crandall, S. Some fundamental similarities between boundary layer flow at transonic and low speeds. *NACA Tech. Note 1623*, 1948.
26. Rubert, K. F., and Persh, J. A procedure for calculating the development of turbulent boundary layers under the influence of adverse pressure gradients. *NACA Tech. Note 2478*, 1951.
27. Patterson, G. N. Modern diffuser design (the efficient transformation of kinetic energy to pressure). *Aircraft Eng. 10*, 267–273 (1938).

28. Peters, H. Energiumsetzung im Querschnittserweiterunguen bei verschiedenen Zulaufbedingungen. *Ing.-Arch. 2*, 92–107 (1931).
29. Rogallo, F. M. Internal-flow systems for aircraft. *NACA. Rept. 713*, 1941.
30. Baals, D. D., Smith, N. F., and Wright, J. B. The development and application of high-critical-speed nose inlets. *NACA. Rept. 920*, 1948.
31. Kuchemann, D., and Weber, J. Das Einlaufproblem bei Triebwerksverkleidungen. *Schriften Deut. Akad. Luftfahrtforschung*, 1943. Transl. as *Gt. Brit. Ministry of Aircraft Production Volkenrode (VG 154-T) Repts. and Transls. 813*, 1947.
32. Roberts, H. E., and Langtry, B. D. The influence of design parameters on the performance of subsonic air inlets. *J. Aeronaut. Sci. 17*, 429–435, 456 (1950).
33. Dannenberg, R. E. A design of leading-edge inlets for unswept wings. *NACA Research Mem. A9K02b*, June 1950.
34. Frank, J. L. Pressure-distribution and ram-recovery characteristics of NACA submerged inlets at high subsonic speeds. *NACA Research Mem. A50E02*, July 1950.
35. Martin, N. J., and Holzhauser, C. A. Analysis of factors influencing the stability characteristics of symmetrical twin intake air-induction systems. *NACA Tech. Note 2049*, 1950.
36. Kantrowitz, A. The formation and stability of normal shock waves in channel flows. *NACA Tech. Note 1225*, 1947.
37. Wyatt, D. D., and Hunczak, H. R. An investigation of convergent-divergent diffusers at Mach number 1.85. *NACA Research Mem. E50K07*, Feb. 1951.
38. Luidens, R. W., and Hunczak, H. Preliminary investigation of cone-type diffusers designed for minimum spillage at inlet. *NACA Research Mem. E7K19*, May 1948.
39. Diggins, J. L., and Lange, A. H. A systematic study of a variable area diffuser for supersonic wind tunnels. *Nav. Ord. Lab. Rept. 2421*, 1952.
40. Bertram, M. H. Investigation of the pressure-ratio requirements of the Langley 11-inch hypersonic tunnel with a variable-geometry diffuser. *NACA Research Mem. L50I13*, Oct. 1950.
41. Evvard, J. C., and Blakey, J. W. The use of perforated inlets for efficient supersonic diffusion. *NACA Research Mem. E51B10*, Apr. 1951.
42. Hunczak, H. R., and Kremzier, E. J. Characteristics of perforated diffusers at free stream Mach no. 1.90. *NACA Research Mem. E50B02*, May 1950.
43. Oswatitsch, K. Der Druckrückgewinn bei Geschossen mit Rüchstossantrieb bei hohen Überschallgeschwindigkeiten (Der Wirkungsgrad Von Strossdiffusoren). *Forschungen Und Entwicklungen Des Heereswaffenamtes, Bericht. NR 1005*, 1944. (Transl. *NACA Tech. Mem. 1140*, 1947.)
44. Oswatitsch, K., and Bohm, H. Luftkräfte und Strömungsvorgänge bei angetriebenen Geschossen. *Forschungen Und Entwicklungen Des Heereswaffenamtes, Bericht. NR 1010, NR 1010/2, Kaiser Wilhelm Inst. Strömungsforschung, Göttingen, AUC*, Oct. 1944. *Douglas Aircraft Co. Transl.*, Feb. 1946.
45. Ferri, A., and Nucci, L. N. Preliminary investigation of a new type of supersonic inlet. *NACA Tech. Note 2286*, 1951.
46. Donaldson, C. duP., and Lange, R. H. Study of the pressure rise across shock waves required to separate laminar and turbulent boundary layers. *NACA Tech. Note 2770*, 1952.
47. Connors, J. F. Effect of ram-jet pressure pulsations on supersonic-diffuser performance. *NACA Research Mem. E50H22*, Nov. 1950.
48. Pearce, R. B. Causes and control of power plant surge. *Aviation Week 52*, 21–25 (1950).
49. Stewartson, K. On the interaction between shock waves and boundary layers. *Proc. Cambridge Phil. Soc. 47, Part 3*, 545–553 (1951).
50. Lange, R. H. Present status of information relative to the prediction of shock-induced boundary layer separation. *NACA Tech. Note 3065*, 1954.
51. Wyatt, D. D. Aerodynamic forces associated with inlets of turbojet installations. *Aeronaut. Eng. Rev. 10*, 20–23 (1951).
52. Klein, H. The calculation of the scoop drag for a general configuration in a supersonic stream. *Douglas Aircraft Co. Rept. SM 13744*, Apr. 1950.

53. Brown, C., and Parker, H. M. A method for the calculation of external lift, moment, and pressure drag of slender open-nose bodies of revolution at supersonic speeds. *NACA Wartime Rept. L720*, 1946. (Supersedes *NACA Advanced Confid. Rept. L5L29*).
54. Moore, F. K. Linearized supersonic axially symmetric flow about open-nosed bodies obtained by use of stream function. *NACA Tech. Note 2116*, 1950.
55. Ferri, A. Application of the method of characteristics to supersonic rotational flow. *NACA Tech. Note 1135*, 1946.
56. Sibulkin, M. Theoretical and experimental investigation of additive drag. *NACA Research Mem. E51B13*, May 1951.
57. McLafferty, G. Theoretical pressure recovery through a normal shock in a duct with initial boundary layer. *Aeronaut. Sci. 20*, 169–174 (1953).
58. Allen, H. J. Estimation of the forces and moments acting on inclined bodies of revolution of high fineness ratio. *NACA Research Mem. A9I26*, Nov. 1949.
59. Wyatt, D. D. An analysis of turbojet engine-inlet matching. *NACA Tech. Note 3012*, 1953.
60. Graham, E. W. Notes on the drag of scoops and blunt bodies. *Douglas Aircraft Co. Rept. SM* 13747, Apr. 1950.
61. Fradenburgh, E., and Wyatt, D. D. Theoretical performance characteristics of sharp-lip inlets at subsonic speeds. *NACA Tech. Note 3004*, 1953.
62. Hearth, D. P., and Perchonok, E. Analysis of heat addition in a convergent-divergent nozzle. *NACA Tech. Note 2938*, 1953.
63. Forstall, W., Jr., and Shapiro, A. H. Momentum and mass transfer in coaxial gas jets. *Mass. Inst. Technol. Guided Missiles Program Meteor Rept. 39*, July 1949.
64. Lessen, M. On stability of free laminar boundary layer between parallel streams. *NACA. Rept. 979*, 1950. (Also *NACA Tech. Note 1929*, 1949.)
65. Corrsin, S. Investigation of flow in an axially symmetrical heated jet of air. *NACA Wartime Rept. W94*, 1943. (Supersedes *NACA Advanced Confid. Rept. 3L23*).
66. Durand, W. F., ed. *Aerodynamic Theory*, Vol. III. Springer, Berlin, 1934–1936.
67. Schlichting, H. Turbulent flows. Lecture series "Boundary Layer Theory," Part II. *NACA Tech. Mem. 1218*, 1949.
68. Squire, H. B. Reconsideration of the theory of free turbulence. *Roy. Aircraft Establishment Rept. Aero. 2023*, Mar. 1945. Abstract in *Phil. Mag. 39*, 1-20 (1948).
69. Squire, H. B., and Trouncer, J. Round jets in a general stream. *British Aeronaut. Research Council Tech. Rept., Repts. and Mem. 1974*, 1944. Also *Roy. Aircraft Establishment Rept. Aero. 1904*, 1944.
70. Courant, R., and Friedrichs, K. O. *Supersonic Flow and Shock Waves. Interscience*, 1948.
71. Wilder, J. G., Jr., and Hindersinn, K. Spreading of supersonic jets in supersonic streams. *Aeronaut. Eng. Rev. 12*, 54–68 (1953).
72. Erdmann, S. Widerstandsbeiwerte für das A4VIP mit Berüchsichtigung des Strahl-und Reibungseinflusses für Unter und Überschallgeschwindigkeiten-Untersuchung der Strahlexpansion. *Peenemunde Heeresversuchsstelle Aerodynam. Inst. Archiv 66/105*, Mar. 1943. (Transl. *ATI 1899.*)

SECTION F

NONSTEADY WING CHARACTERISTICS

I. E. GARRICK

F,1. Introduction. The addition of the dimension "time" to steady aerodynamics has far-reaching effects, both practical and theoretical. There is the practical necessity for coping with many important problems involving nonsteady phenomena such as flutter, buffeting, transient flows, gusts, dynamic response in flight, maneuvers, and stability. Apart from the many applications, theoretical nonsteady aerodynamics embraces and sheds light on the realm of steady aerodynamics and introduces interesting new methods.

In the subject of nonsteady characteristics of wings, theory plays a most significant role. Some enduring parts of the edifice of aerodynamics exist in this field. Experimental work has been limited mainly to attempts at specific confirmations of theory rather than to the production of engineering data. Direct experimental measurements of the many necessary aerodynamic coefficients in the oscillating case over the required speed ranges is a most difficult, almost unending, task. Indeed the bulk of experimental evidence is of an indirect nature, as in experimental flutter research. Moreover, a great deal of the experimental effort is concerned with the development of techniques and of improved instrumentation. For these reasons, except for Art. 15, this section is devoted largely to the theoretical developments.

As in steady flow, nonsteady flow regimes may be of good or poor aerodynamic efficiency. Flows of great efficiency as strived for in good aerodynamic design may be treated by potential flow methods, whereas such flow phenomena as, for example, high angle of attack or stall flutter involve complex considerations of flow separation. There is also a merging of various flow types, such as may occur throughout the angle of attack range; and there are very significant and often profound flow changes through the subsonic, transonic, and supersonic speed ranges. Complex boundary layer phenomena are possible in transient flows, of which little is known. There may be a washing away and narrowing of the boundary layer; ineffective nonsteady motion may exist in an already separated dead flow region; buffeting flows, instabilities, and turbulence can develop. Of course, it is not possible to treat these many and complex phases of the subject. Indeed the material presented herein is limited mainly to considerations of potential flow.

As there exists a very extensive and scattered literature partly indicated in Art. 17, on nonsteady aerodynamics even for potential flows, a limited selection of the material to be included has had to be made. An attempt has been made, however, to give a unified and coherent account of well-founded phases of the subject, which it is hoped will be useful to readers interested in research and in applications.

F,2. Differential Equations for Potential Flow. In order to give a unified treatment of the theoretical methods for potential flow regardless of the speed range, it seems appropriate to briefly develop and classify the governing differential equations. (See also VI,C and VI,D.) Potential flow methods, it will be observed, are useful even for flows involving vorticity shed in specified regions of the wake. Physical ideas will be introduced along with the mathematical developments.

Velocity potential. The condition for irrotational flow is

$$\operatorname{curl} \mathbf{q} = 0$$

and this relation implies that a scalar velocity potential ϕ exists such that the velocity vector is

$$\mathbf{q} = \operatorname{grad} \phi$$

The general equation of continuity

$$\frac{\partial \rho}{\partial t} + \operatorname{div} \rho \mathbf{q} = 0$$

may be written as

$$\left(\frac{\partial}{\partial t} + \mathbf{q} \cdot \operatorname{grad}\right) \rho + \rho \operatorname{div} \operatorname{grad} \phi = 0$$

or as

$$\frac{1}{\rho} \frac{D\rho}{Dt} + \nabla^2 \phi = 0 \tag{2-1}$$

where differentiation following the particle is designated by $D/Dt = \partial/\partial t + (\mathbf{q} \cdot \nabla)$.

From Euler's equations, or from Bernoulli's general relation

$$\frac{\partial \phi}{\partial t} + \frac{q^2}{2} + \int \frac{dp}{\rho} = 0 \tag{2-2}$$

where a function, which may vary with time but which is constant in space, has been included in ϕ and where it has been assumed that p is a function of ρ only (barotropic fluid).

Introduce the acoustic relation

$$a^2 = \frac{dp}{d\rho} \tag{2-3}$$

where a is the local variable speed of sound. Then it follows from Eq. 2-2 that

$$\text{grad}\left(\frac{\partial\phi}{\partial t} + \frac{q^2}{2}\right) = -\frac{1}{\rho}\,\text{grad}\,p = -\frac{a^2}{\rho}\,\text{grad}\,\rho$$

and

$$\frac{\partial}{\partial t}\left(\frac{\partial\phi}{\partial t} + \frac{q^2}{2}\right) = -\frac{1}{\rho}\frac{\partial p}{\partial t} = -\frac{a^2}{\rho}\frac{\partial\rho}{\partial t}$$

These relations serve to eliminate ρ from Eq. 2-1 which may now be put in the form

$$\frac{1}{a^2}\left(\frac{\partial^2\phi}{\partial t^2} + \frac{\partial q^2}{\partial t} + \mathbf{q}\cdot\text{grad}\,\frac{q^2}{2}\right) = \nabla^2\phi \tag{2-4a}$$

or in the form as first given in [*1*]

$$\frac{1}{a^2}\left(\frac{\partial}{\partial t} + \mathbf{q}_c\cdot\nabla\right)^2\phi = \nabla^2\phi \tag{2-4b}$$

where the subscript c indicates that the designated variable is to be treated as constant with respect to the operations ∇ and $\partial/\partial t$.[1] With the corresponding interpretation of D_c/Dt as $\partial/\partial t + (\mathbf{q}_c\cdot\nabla)$ the equation takes on the mnemonic and classic form

$$\frac{1}{a^2}\frac{D_c^2\phi}{Dt^2} = \nabla^2\phi \tag{2-4c}$$

The resulting equation is the governing differential equation for general nonstationary, nonviscous, potential flow and is not limited to small disturbances. It has been made to exhibit the "equation of wave motion" form, showing that the disturbance represented by the velocity potential partakes of the local fluid velocity and is propagated as a wave which spreads at a rate equal to the local speed of sound.

Its invariant form permits easy conversion to any useful coordinate system whether fixed or moving. In a space-fixed Cartesian coordinate system (with $\mathbf{q} = (u, v, w)$) it becomes the formidable equation

$$\left(1-\frac{u^2}{a^2}\right)\frac{\partial^2\phi}{\partial x^2} + \left(1-\frac{v^2}{a^2}\right)\frac{\partial^2\phi}{\partial y^2} + \left(1-\frac{w^2}{a^2}\right)\frac{\partial^2\phi}{\partial z^2} - 2\frac{uv}{a^2}\frac{\partial^2\phi}{\partial x\partial y}$$
$$- 2\frac{vw}{a^2}\frac{\partial^2\phi}{\partial y\partial z} - 2\frac{wu}{a^2}\frac{\partial^2\phi}{\partial z\partial x} - 2\frac{u}{a^2}\frac{\partial^2\phi}{\partial x\partial t} - 2\frac{v}{a^2}\frac{\partial^2\phi}{\partial y\partial t} - 2\frac{w}{a^2}\frac{\partial^2\phi}{\partial z\partial t}$$
$$- \frac{1}{a^2}\frac{\partial^2\phi}{\partial t^2} = 0 \tag{2-4d}$$

[1] **Thus, using descriptive terms, the "convective" property of the material derivative is retained without its "diffusive" property.**

and this includes, for example, the classical nonsteady equation for aerial waves of finite amplitude in one dimension, and the steady flow nonlinear equations which have been the subject of various iteration and successive approximation methods of solution. A discussion and classification of this equation with regard to dimension-order and similarity conditions has been given by Lin, Reissner, and Tsien [*2*].[2]

Note that if the velocity of propagation of disturbances is assumed infinite ($a = \infty$), Eq. 2-4 reduce to Laplace's equation for the incompressible fluid, $\nabla^2\phi = 0$. It may be recalled that even this deceptively simple-looking linear equation leads to recondite matters, both of a physical and mathematical nature, for it embraces the whole of two- and three-dimensional incompressible potential flow, stationary or nonstationary, for small or large disturbances.

Perturbation velocity potential. The degree of success of small disturbance methods in aerodynamics, in that useful results should follow from such radically simplifying assumptions, is a constant source of query and gratification. A great deal of aerodynamic doctrine including nonsteady aerodynamics has been developed with such methods. There are two aspects, linearization of the differential equations and the further linearization of the boundary conditions of the problem. A distinction thus may exist in the use of small perturbations of an equilibrium flow condition for the condition of small or large disturbances in the equilibrium flow. For the condition of large disturbances which may mean inclusion of thickness effects, shock flow patterns, near-sonic flow disturbances, and stagnation points, there are methods which retain higher order terms in both the differential equation and in the boundary condition, though such methods are laden with difficulties and have not been pursued very far in nonsteady problems. The immediate developments deal with the complete combined linearization of the problem.

[2] The work of [*2*] has been extended by Miles [*3*] and Mollo-Christensen [*4*]. These references show that sufficient for linearization of the differential equations for the perturbation potential (see the following subarticle) is any one or combination of the factors:

$$|M_\infty^2 - 1| \gg O(\delta^{\frac{2}{3}})$$
$$k \gg O(\delta^{\frac{2}{3}})$$
$$\frac{1}{\mathit{R}} \gg O(\delta^{\frac{2}{3}})$$

where δ, $M_\infty\delta$, $k\delta$, and $kM_\infty\delta$ must be sufficiently small.

The perturbation equation is nonlinear only when the following conditions are jointly satisfied:

$$|M_\infty^2 - 1| = O(\delta^{\frac{2}{3}})$$
$$k = O(\delta^{\frac{2}{3}})$$
$$\frac{1}{\mathit{R}} = O(\delta^{\frac{1}{3}})$$

where M_∞ is the stream Mach number, k is a reduced frequency parameter, R is the aspect ratio, δ is a thickness, angle of attack, or camber parameter serving as a measure of the flow disturbance, and O means order of.

Linearization of Eq. 2-4 on the basis of small disturbances in a uniform stream of a compressible fluid of velocity U consists in regarding all velocity disturbances small in comparison with U, a, and $U - a$, and pressure differences and density changes small in comparison with main stream pressure and density. If the velocity potential corresponding to the main stream U is considered extracted and $dp/d\rho$ expanded around its value for the undisturbed stream, the procedure of retaining first order terms leads to the following linear equation for the *disturbance* velocity potential φ

$$\frac{1}{a_\infty^2}\left(\frac{\partial}{\partial t} + \mathbf{U}\cdot\nabla\right)^2 \varphi = \nabla^2\varphi \tag{2-5a}$$

where a_∞ is now the *constant* value of the sound speed in the uniform stream and grad φ now represents the disturbance velocity vector $\mathbf{q} - \mathbf{U}$.

The disturbance potential can be considered conveniently to correspond to that of a wing immersed in and creating small disturbances to a uniform stream flowing with velocity U in the x direction, or to correspond to the flow created by a wing moving uniformly along with the coordinate system in the negative x direction in a fluid otherwise at rest. Then Eq. 2-5a becomes in rectangular coordinates

$$\left(1 - \frac{U^2}{a_\infty^2}\right)\frac{\partial^2\varphi}{\partial x^2} + \frac{\partial^2\varphi}{\partial y^2} + \frac{\partial^2\varphi}{\partial z^2} - 2\frac{U}{a_\infty^2}\frac{\partial^2\varphi}{\partial x\partial t} - \frac{1}{a_\infty^2}\frac{\partial^2\varphi}{\partial t^2} = 0 \tag{2-5b}$$

In this form the equation has provided the theoretical basis for much of the existing background for nonsteady aerodynamics of wings both for subsonic and supersonic speeds. (The differential equation for the disturbance velocity potential for the main stream at sonic speed takes on a nonlinear form [*1*,*2*]. It is given by Eq. 2-5b with $U = a_\infty$ and with the term

$$\frac{\gamma + 1}{a_\infty}\frac{\partial\varphi}{\partial x}\frac{\partial^2\varphi}{\partial x^2}$$

added to the left-hand side of the equation, a term which it can be observed also appears in the case of steady flow; γ is the adiabatic index in the pressure-density relation $p/p_0 = (\rho/\rho_0)^\gamma$.)

For $U = 0$, Eq. 2-5a and 2-5b reduce to the classical wave equation for the propagation of sound. Moreover, in general, Eq. 2-5b can also be transformed to the wave equation by a change of coordinates converting back to a space-fixed system, involving the Galilean transformation

$$x = \xi + Ut; \quad y = \eta; \quad z = \zeta; \quad t = t \tag{2-6}$$

Then in the space-fixed coordinates, the disturbance potential created by the wing, moving in the medium otherwise at rest, satisfies the wave equation

$$\Box^2\varphi = \frac{\partial^2\varphi}{\partial\xi^2} + \frac{\partial^2\varphi}{\partial\eta^2} + \frac{\partial^2\varphi}{\partial\zeta^2} - \frac{1}{a_\infty^2}\frac{\partial^2\varphi}{\partial t^2} = 0 \tag{2-7}$$

Thus the linearized aerodynamics can be considered a branch of acoustics, and the methods available for study of wave propagation are of direct interest. For harmonic time oscillations of the form $\varphi = \varphi^* e^{i\omega t}$ the equation can be converted to the simpler harmonic wave equation form

$$\frac{\partial^2 \varphi^*}{\partial \xi^2} + \frac{\partial^2 \varphi^*}{\partial \eta^2} + \frac{\partial^2 \varphi^*}{\partial \zeta^2} + \kappa^2 \varphi^* = 0 \tag{2-8}$$

where $\kappa = \omega / a_\infty$. Fourier series, Fourier integral, or Laplace transformation methods applied to solutions of Eq. 2-8 can restore generality of solutions in time.

Acceleration potential and pressure relations. Following Prandtl's introduction of acceleration potential methods Possio, Küssner, and many others made applications to problems in nonsteady flow. A few basic formulas and relations are given here and used later. The chief usefulness of the methods of the acceleration potential has been to furnish alternative procedures to those of velocity potential methods in small disturbance compressible flow problems. The boundary value problem can be formulated in different ways, generally as integral equations with different kernels. It has sometimes been stated that the acceleration potential methods make obsolete the use of other (vortex) methods. The author of this section cannot adhere to this point of view. The methods are rather to be considered supplementary; historically it happens that one may be temporarily preferred or further advanced for certain problems.

From Euler's equation

$$\frac{D\mathbf{q}}{Dt} = -\operatorname{grad} \int \frac{dp}{\rho}$$

it follows that the acceleration vector $D\mathbf{q}/Dt$ may be expressed as the gradient of a scalar, the acceleration potential

$$\Psi = -\int \frac{dp}{\rho}$$

If $\mathbf{q} = \operatorname{grad} \phi$ there exists the following relation[3] between Ψ and the velocity potential ϕ

$$\frac{D_c \phi}{Dt} = \Psi = -\int \frac{dp}{\rho} \tag{2-9}$$

[3] The index c is to remind us that in deriving Eq. 2-9

$$\frac{D}{Dt} \operatorname{grad} \neq \operatorname{grad} \frac{D}{Dt}$$

except when the local velocity $\mathbf{q}$ in the operation

$$\frac{D}{Dt} = \frac{\partial}{\partial t} + \mathbf{q} \cdot \nabla$$

is considered constant with respect to the gradient operation. See also footnote 1.

The form of direct interest herein is, however, the one valid for *small* disturbances. There results from Eq. 2-9 using ψ instead of Ψ

$$\begin{aligned} \psi &= -\frac{p}{\rho_\infty} \\ &= \frac{\partial \varphi}{\partial t} + U\frac{\partial \varphi}{\partial x} \end{aligned} \tag{2-10}$$

where $p = p(x, y, z, t)$ denotes the disturbance pressure field (local pressure minus main stream pressure) and where ρ_∞ is the constant density in the undisturbed field. Because ψ is here directly proportional to the disturbance pressure, it is sometimes called the pressure potential; in the following ψ will be conveniently replaced by $-p/\rho_\infty$. The pressure, and the acceleration potential, are continuous everywhere except at the location of the lifting surface (and for supersonic flows except possibly at the surface of Mach cones).

The relation between the disturbance potential φ and the pressure p can be inverted to yield the relation

$$\varphi(x, y, z, t) = -\frac{1}{\rho_\infty U}\int_{-\infty}^{x} p\left(x', y, z, t - \frac{x - x'}{U}\right) dx' \tag{2-11}$$

where far ahead of the disturbance at $x = -\infty$ the potential has been assumed to be 0. It follows with the aid of Eq. 2-10 that the acceleration potential ψ or the disturbance pressure p is governed by the *same* differential equations as φ, namely Eq. 2-5 and 2-7.

The form assumed by Eq. 2-10 and 2-11 for harmonic time oscillations is of special interest. Writing

$$\varphi(x, y, z, t) = \varphi^*(x, y, z)e^{i\omega t}$$

and

$$p(x, y, z, t) = p^*(x, y, z)e^{i\omega t}$$

there is obtained

$$-\frac{p^*(x, y, z)}{\rho_\infty} = i\omega\varphi^* + U\frac{\partial \varphi^*}{\partial x} \tag{2-12}$$

$$\varphi^*(x, y, z) = -\frac{1}{\rho_\infty U}e^{-i\omega(x/U)}\int_{-\infty}^{x} p^*(x', y, z)e^{i\omega(x'/U)}dx' \tag{2-13}$$

An important consideration in acceleration potential methods, as will be observed in Art. 3, is the fact that explicit handling of the wake is not needed. In addition, for certain types of problems such as finding wing deformations producing a given load distribution, the formulation is more natural with the aid of the acceleration potential.

F,3. Boundary Conditions.

GENERAL STATEMENTS. The boundary conditions to be associated with the governing differential equations consist in the mathematical formulation of several physical or phenomenological statements. It is convenient to group here the statements chiefly intended for application of Eq. 2-5, and to elaborate specifically where later needed.

1. *Surface boundary condition.* The wing surface is impenetrable to the medium, that is the total flow is tangential to the surface.

2. *Edge conditions.* Enough viscosity remains in the "nonviscous" fluid to decisively determine the flow pattern near sharp edges. Of fundamental importance in subsonic aerodynamics is the role of the wing trailing edge in the genesis of flow pattern for the lifting surface by alteration of the primary noncirculatory pattern. The hypothesis of a condition of smooth flow at the trailing edge (Kutta condition) has been variously stated, for example, as a condition on the rear stagnation points, or on the finiteness of the velocity; however, it is generally sufficient to ascertain that the surface pressure difference remains continuous near, and vanishes at, the edge. A similar condition should be satisfied at trailing tips. Conditions to be explicitly satisfied at the leading edge have not in general been fully explored, in fact for strictly sharp edges in a compressible fluid there are unsolved difficulties. It suffices for the small disturbance applications to rounded edges in subsonic flow to require by analogy with incompressible fluid results that the total integrated forces be finite, and that infinities in the pressure distribution be of proper order.

3. *Wake conditions.* The free vorticity shed from the trailing edge in the wake is such that its circulation together with the bound circulation vanishes according to the Helmholz-Kelvin theorem. It is assumed in all small disturbance work that the shed wake remains where it is formed, floats without mutual influence along streamlines, and is considered to be a continuous sheet of discontinuity generally coplanar with the wing projection in the direction of flight. Edge effects and rolling-up of the sheet and general movements at infinity are ignored.

4. *Conditions at infinity.* The usual requirements are conditions of quiescence or of uniform flow at infinity; in addition, there are radiation conditions given explicitly by Sommerfeld [*5*, p. 25] which require waves to propagate away from sources of disturbance and to behave properly at infinity.

5. *Other conditions.* At supersonic speeds, account must be taken of appropriate zones of influence and of action; the flow pattern may be pieced together as in the method of characteristics of several regions with various edge conditions and conditions of continuity at wave fronts or Mach surfaces. There may be a reversal in the role of the leading and trailing edges as compared with subsonic conditions; the trailing edge and

the wake need play no determining role as in the subsonic case and, in fact, a compression-expansion wave mechanism beginning at the leading edge is involved in the generation of lift. In the influence of finite thickness, there are conditions necessary for an attached shock at the leading edge that require a sufficiently sharp leading edge. Otherwise, detached shocks and mixed supersonic-subsonic types of flow are involved.

The fact that some or many of the boundary conditions are implicitly or automatically satisfied in particular formulations of a problem sometimes conceals their significance. That explicit statements on wake conditions are not needed has been cited, for example, as a feature of the acceleration potential methods. In other applications some of the boundary conditions are satisfied in the elementary solutions used in the synthesis of the complete solution.

FORMULATION OF MAIN BOUNDARY CONDITIONS. If $F(x, y, z, t)$ is the equation of the boundary of the moving surface, condition 1 can be expressed [*6*, p. 7]

$$\frac{DF}{Dt} = 0 \tag{3-1}$$

To obtain the linearized form generally made use of and based on small thickness and small camber, the surface F may be utilized in the form

$$F = z - Z_{u,l}(x, y, t) = 0 \tag{3-2}$$

where the wing is assumed to lie nearly in the x, y plane, creating small disturbances to the main stream U, and where its upper and lower surfaces are distinguished respectively by u and l subscripts. Then there results, for example, for the normal fluid velocity at the upper surface:

$$w(x, y, Z_u, t) = \frac{\partial Z_u}{\partial t} + U\frac{\partial Z_u}{\partial x} \tag{3-3}$$

and by a Taylor expansion and neglect of higher order terms the left-hand side can be replaced by $w(x, y, 0, t)$. Let the mean camber surface be designated by

$$z = Z(x, y, t)$$

where $2Z = Z_u + Z_l$. Introduce the wing thickness $2Z_t = Z_u - Z_l$. Then there results

$$\begin{aligned} w(x, y, \pm 0, t) &= \frac{\partial Z}{\partial t} + U\frac{\partial Z}{\partial x} \pm \left(\frac{\partial Z_t}{\partial t} + U\frac{\partial Z_t}{\partial x}\right) \\ &= w_c \pm w_t \end{aligned} \tag{3-4}$$

where $\pm$ refers to upper and lower surfaces respectively, and where in effect the problems of camber and thickness have been separated. Assuming that the wing thickness does not change, or stated otherwise, considering that the problems of the pulsating wing and of steady flow associated with thickness have been separated out, it finally results that the profile in nonsteady flow is represented by its mean camber surface, and boundary condition 1 is

$$w(x, y, 0, t) = \frac{\partial Z}{\partial t} + U\frac{\partial Z}{\partial x} \tag{3-5}$$

where the unneeded subscript c has been dropped. It is particularly noted that the boundary condition is stated at the x, y plane, the plane of projection of the wing mean camber surface in the direction of the air stream. The two terms on the right represent the contributions to the normal velocity w (reckoned positive upwards), induced by the proper motion of the mean camber surface and by its shape. Moreover, the value of w is unchanged for the top or the bottom sides of the mean surface, and therefore w is a symmetrical or even function of z; its value at $z = 0$ is given by

$$w = \left(\frac{\partial \varphi}{\partial z}\right)_{z=0} \tag{3-6}$$

The potential φ is therefore an antisymmetric or odd function of z, and is plus on the upper surface or in the upper half plane and is minus on the lower surface or in the lower half plane. It follows also from the linearized form of the pressure (Eq. 2-10) that the disturbance pressure is also an odd function of z and is therefore zero on the x, y plane, except at the lifting surface where the *pressure difference* $P = p_u - p_l$ between top and bottom surfaces, positive downward (i.e. lift is negative) is given by

$$\begin{aligned} P(x, y, t) &= p(x, y, +0, t) - p(x, y, -0, t) \\ &= 2p(x, y, 0, t)| \\ &= -2\rho_\infty\left(\frac{\partial \varphi}{\partial t} + U\frac{\partial \varphi}{\partial x}\right) \end{aligned} \tag{3-7}$$

The mathematical problem then consists in a determination of φ, or the velocity and pressure fields, in the upper half space $z \geqq 0$ from a knowledge of the differential equations, the conditions at the boundary surface $z = 0$, and conditions at infinity. Thus on the plane $z = 0$, ahead of the location of the wing, the perturbation potential φ is zero ($\partial\varphi/\partial t$ and $\partial\varphi/\partial x$ also are zero); at the wing location, Eq. 3-5 and 3-6 hold; and aft of the wing location, P given by Eq. 3-7 is zero. The trailing edge condition or other edge conditions must be regarded as part of the boundary conditions. The further formulation of the problem is achieved in several ways, most commonly as an integral equation (obtained with the aid of the relations expressed in Eq. 2-11 and 3-6). The integral equation may relate the known normal velocity distribution w (sometimes called downwash) over the wing surface, a known kernel function or influence function K and the unknown surface distribution of potential or pressure (or "loading") to be determined. The kernel function represents a particular exact or approximately calculable normal velocity distribution associated with a unit potential difference or loading over the surface. It may assume a variety of forms depending on mathematical operations performed, and edge conditions included. Some of these forms are indicated in Art. 4. The case of a known or given loading instead of a known downwash leads to a related inverse integral equation for the wing motion or deformation.

It may be remarked that the problem of the pulsating wing or of the effects of thickness can be similarly formulated with the potential φ considered, however, as a symmetric or even function of z. Also, the small disturbances may for some problems be considered relative to a nonuniform main flow.

The reader may also note from Eq. 3-5 that $w = 0$ when the mean camber surface is of the form $z = Z(x - Ut, y, 0)$. This means that deflection waves traveling downstream with the stream velocity U do not contribute to the normal velocity and hence are aerodynamically undamped to the order of the linear approximation.

F,4. Elementary Solutions and Useful Relations. In the synthesis of solutions of boundary-value problems for linear partial differential equations, certain elementary solutions which may serve as kernel functions are important. In particular, for Laplace's equation and the wave equation, solutions appearing in relation to Green's theorem and Kirchhoff's principle are of special significance. These solutions may be recognized as primary extensions of the simple and familiar concepts of sources and doublets in hydrodynamics; and according as the velocity or acceleration potential is involved may be considered, respectively, as flow singularities or as pressure singularities. A listing and discussion of some of these solutions, together with various formulas used later, will be expedient, and moreover will serve to assist the reader in identifying the physical meaning and motivation of equations and transformations which abound in the literature of the subject.

Green's Theorem. In the theory of the potential for boundary value problems Green's theorem, a form of the divergence theorem, occurs prominently [7,8]. For any two functions φ and ψ which with their first derivatives are properly finite, single-valued, and continuous, the theorem states that

$$\int_V (\psi\nabla^2\varphi - \varphi\nabla^2\psi)dV = \int_S \left(\psi \frac{\partial\varphi}{\partial n} - \varphi \frac{\partial\psi}{\partial n}\right) dS \tag{4-1}$$

The result expresses a reduction of a volume integral into a surface integral where $\partial/\partial n$ denotes differentiation along the outward normal to the surface S. A classical application of it to the case where φ is a regular solution of Laplace's equation in the space region V, and ψ is the specially chosen function $1/r$, gives

$$\varphi_P = \frac{1}{4\pi} \int_S \left[\frac{\partial \varphi}{\partial n} \frac{1}{r} - \varphi \frac{\partial}{\partial n} \left(\frac{1}{r} \right) \right] dS \tag{4-2}$$

where φ_P the potential at a field point P is expressed in terms of the boundary surface values of φ and $\partial\varphi/\partial n$. It is noticed that solutions are here built up by integration of surface distributions of the elementary unit source solution

$$\varphi_1 = -\frac{1}{4\pi r} \tag{4-3a}$$

and of the doublet or dipole solution with axis in the direction n

$$\varphi_2 = -\frac{1}{4\pi} \frac{\partial}{\partial n} \left(\frac{1}{r} \right) \tag{4-3b}$$

where r is the geometric distance $\sqrt{(x - \xi)^2 + (y - \eta)^2 + (z - \zeta)^2}$ from the field point (x, y, z) to the source of disturbance at ξ, η, ζ. The strength of the source is determined from the total efflux of fluid of unit density through a small sphere about the source and hence the normalization factor $1/4\pi$. Similarly the moment of the doublet has been normalized to unity. In Eq. 4-2 this normalization means [*5*, p. 43] that there is a total discontinuity in the normal derivative across the surface S of magnitude $\partial\varphi/\partial n$, associated with the source distribution and a total discontinuity in the potential of magnitude φ associated with the doublets. (See also formulation for thin wings given in VI,D,3.)

Fixed Acoustic Sources and Doublets. For the wave equation, Kirchhoff's formula, the analytical statement of Huygens' principle for sound waves, appears as the basic relation corresponding to Eq. 4-2, [*5,8*]

$$\varphi_P = \frac{1}{4\pi} \int_S \left\{ \frac{1}{r} \left(\frac{\partial \varphi}{\partial n} \right)_{t-(r/a_\infty)} - \frac{\partial}{\partial n} \frac{\varphi[t - (r/a_\infty)]}{r} \right\} dS \tag{4-4}$$

or also

$$\varphi_P = \frac{1}{4\pi} \int_S G\left(t - \frac{r}{a_\infty} \right) dS$$

where

$$G(t) = \frac{1}{r} \frac{\partial \varphi}{\partial n} - \varphi \frac{\partial}{\partial n} \left(\frac{1}{r} \right) + \frac{1}{a_\infty r} \frac{\partial \varphi}{\partial t} \frac{\partial r}{\partial n}$$

The argument $t - (r/a_\infty)$ corresponds to the time of action of the disturbances whose influences reach the field point at time t, hence the designation of the potential as "retarded."

The fundamental solutions appearing in Eq. 4-4 that correspond to the source and doublet are

$$\varphi_1 = -\frac{1}{4\pi r} f\left(t - \frac{r}{a_\infty}\right) \tag{4-5a}$$

and

$$\varphi_2 = -\frac{1}{4\pi}\frac{\partial}{\partial n}\left[\frac{1}{r} f\left(t - \frac{r}{a_\infty}\right)\right] \tag{4-5b}$$

The source solution represents spherical waves issuing from a disturbance in a medium at rest. Both the strength of the source and the moment of the doublet are arbitrary and given by the function $f(t)$, which may also be considered in general to be a function of the position variables and written as $f(\xi, \eta, \zeta, t)$. In surface distributions as in Eq. 4-4, the interpretation of $f(t)$ as the total discontinuity across the surface in $\partial\varphi/\partial n$ or φ is the same as that mentioned for Eq. 4-2.

For the harmonic wave equation (2-8), Kirchhoff's formula appears as the equation of Helmholz, and the source and doublet relations (Eq. 4-5a and 4-5b) become, with the harmonic time factor $e^{i\omega t}$ suppressed, as often done (and indicated by the asterisk),

$$\varphi_1^* = -\frac{1}{4\pi r} e^{-i\kappa r} \tag{4-6a}$$

$$\varphi_2^* = -\frac{1}{4\pi}\frac{\partial}{\partial n}\left(\frac{1}{r} e^{-i\kappa r}\right) \tag{4-6b}$$

where $\kappa = \omega/a_\infty$.

In two dimensions the familiar solutions corresponding to Eq. 4-3a and 4-3b are

$$\varphi_1 = \frac{1}{2\pi}\ln r \tag{4-7a}$$

$$\varphi_2 = \frac{1}{2\pi}\frac{\partial}{\partial n}\ln r = \frac{1}{2\pi r}\frac{\partial r}{\partial n} \tag{4-7b}$$

where

$$r = \sqrt{(x-\xi)^2 + (y-\eta)^2}$$

For cylindrical wave disturbances corresponding to the plane case of Eq. 4-5a, the results are less simple because of the "infinite tail" aftereffects at any field point associated with the infinite side-extent of the plane source. There appear several forms [*6*, p. 298]:

$$\begin{aligned}\varphi_1 &= -\frac{1}{2\pi}\int_0^\infty f\left(t - \frac{r}{a_\infty}\cosh u\right) du \\ &= -\frac{1}{2\pi}\int_{-\infty}^{t - r/a_\infty} \frac{a_\infty f(\theta)d\theta}{\sqrt{a_\infty^2(t-\theta)^2 - r^2}}\end{aligned} \tag{4-8}$$

A simpler form can be obtained, however, as a source-pulse or temporary source solution by considering the source to act only during a brief interval at time $\theta = \tau$. If instead of $f(\theta)$ there is introduced $f(\tau)\delta(\theta - \tau)$, where $\delta(\theta)$ is a singular impulse function or Dirac δ function defined as zero everywhere, except at $\theta = 0$ where it is characterized by having unit area, one obtains as the source-pulse solution

$$\varphi_0 = -\frac{1}{2\pi}\frac{a_\infty f(\tau)}{\sqrt{a_\infty^2(t-\tau)^2 - r^2}} \tag{4-9}$$

The cylindrical harmonic solution corresponding to Eq. 4-6a is

$$\varphi_1^* = -\frac{1}{2\pi}\frac{\pi}{2i}H_0^{(2)}(\kappa r) = \frac{i}{4}H_0^{(2)}(\kappa r) \tag{4-10}$$

where use has been made of the following integral relations:

$$\int_0^\infty e^{-i\kappa r\cosh u}\,du = \int_1^\infty e^{-i\kappa r u}\frac{du}{\sqrt{u^2-1}} = \frac{\pi}{2i}H_0^{(2)}(\kappa r) \tag{4-11}$$

and where the Hankel function is defined as a combination of Bessel functions of the first and second (Neumann) kind,

$$H_0^{(2)}(k) = J_0(k) - iY_0(k) \tag{4-12}$$

It has the following expansions [9, pp. 73, 194]

$$H_0^{(2)}(k) = \frac{2}{i\pi}\ln k + \cdots \quad \text{for } k \to 0$$

$$= \sqrt{\frac{2}{\pi k}}\,e^{-i\left(k-\frac{\pi}{4}\right)} + \cdots \quad \text{for } k \to \infty$$

Let us also define here

$$H_1^{(2)}(k) = J_1(k) - iY_1(k) \tag{4-13}$$

and note that

$$H_1^{(2)}(k) = -\frac{d}{dk}H_0^{(2)}(k)$$

The doublet solution corresponding to Eq. 4-6b is given by

$$\varphi_2^* = \frac{i}{4}\frac{\partial}{\partial n}H_0^{(2)}(kr) = \frac{ik}{4}H_1^{(2)}(kr)\frac{\partial r}{\partial n} \tag{4-14}$$

The Helmholz form of Kirchhoff's theorem which arises in acoustics, and which is of significance for theoretical considerations of the plane harmonic case, makes use of these solutions. (See e.g. [5, p. 49], where, however, Hankel functions of the first rather than the second kind arise since the harmonic form $e^{-i\omega t}$ rather than $e^{i\omega t}$ was chosen.)

Uniformly Moving Sources and Doublets. Consider the partial differential equation (2-5b) and the corresponding fundamental solutions which may be termed moving sources and doublets. It is necessary now to distinguish between subsonic, sonic, and supersonic speeds, $U < a_\infty$, $U = a_\infty$, and $U > a_\infty$.

The results to follow can be obtained in several ways, one of which is from the classical considerations of the acoustic Doppler principle. Another more formal procedure first used by Küssner [*10*] makes use of the invariance properties of the wave equation, thus utilizing for Eq. 2-5b combined Galilean and Lorentz transformations (see also [*11*]). A less recondite procedure involves the use of product-type solutions and separation of variables, but perhaps the most direct procedure physically is that used by Prandtl in 1936 [*12*], for the case of moving constant source distributions. The method consists of the scheduling of the source strength in Eq. 4-5a in time and position to represent in effect the arbitrary source moving along a chosen path. The following derivation along these lines indicated in [*13*] for the arbitrary source in uniform rectilinear subsonic motion is believed of sufficient significance and interest to merit inclusion herein.

Let the arbitrary function $f(t)$ in Eq. 4-5a be defined as an impulse occurring at time $t = \tau$, namely as

$$f(\tau)\delta(t - \tau)$$

where, as has been defined, $\delta(\tau) = 0$ for $\tau \neq 0$, and at $\tau = 0$ it is characterized as having unit area with respect to τ. Its useful property of "sifting or selecting" the value of a function is exhibited by the following relation [*14*]:

$$\int_{-\infty}^{\infty} f(\tau - \sigma)\delta(\tau)d\tau = f(\sigma)$$

Let a succession of such impulses act, one following the other in a path, points along which are given in space-fixed coordinates by

$$\xi = \xi(\tau), \quad \eta = \eta(\tau), \quad \zeta = \zeta(\tau)$$

The effect at time t of all such impulses which act before the time t is then given as

$$\varphi = -\frac{1}{4\pi}\int_{-\infty}^{t - r/a_\infty} \frac{f(\tau)}{r}\,\delta\left(t - \tau - \frac{r}{a_\infty}\right)d\tau \tag{4-15}$$

where, as has been defined, there is a nonzero contribution to the integral only for values of τ defined by the characteristic relation

$$t - \tau = \frac{r}{a_\infty} = \frac{1}{a_\infty}\sqrt{[x - \xi(\tau)]^2 + [y - \eta(\tau)]^2 + [z - \zeta(\tau)]^2} \tag{4-16}$$

a relation which expresses the distance between the source point and the field point in terms of the time of travel of the outgoing waves.

The integral in Eq. 4-15 corresponds to a summation of temporary fixed sources. To represent the case of uniform motion with velocity U in the negative x direction, let the sources be located on the ξ axis and flow consecutively one after the other at the positions

$$\xi = -U\tau, \quad \eta = 0, \quad \zeta = 0$$

so that for $\tau = -\infty$ the source is located at $\xi = \infty$, and for $\tau = 0$ the source is located at the origin. It follows that the distance between source point and field point is

$$r = \sqrt{(x + U\tau)^2 + y^2 + z^2} \tag{4-17}$$

Replace in Eq. 4-15 the variable τ by θ where

$$t - \tau - \frac{r}{a_\infty} = -\theta \tag{4-18}$$

By the "sifting" property of the δ function the integral will have a value only for $\theta = 0$ given by

$$\varphi = -\frac{1}{4\pi}\left[\frac{f(\tau)}{r}\frac{d\tau}{d\theta}\right]_{\theta=0} \tag{4-19}$$

Eq. 4-17 and 4-18 give a quadratic equation for τ, which on choosing the solution that leads to $\tau < t$ and putting $\theta = 0$, gives

$$\tau = \frac{1}{\beta^2}\left(t + \frac{Ux}{a_\infty^2} - \frac{R}{a_\infty}\right) \tag{4-20}$$

where

$$R = \sqrt{(x + Ut)^2 + \beta^2(y^2 + z^2)}$$
$$\beta^2 = 1 - M_\infty^2$$
$$M_\infty = \frac{U}{a_\infty}$$

Eq. 4-17 and 4-18 also give

$$\frac{1}{r}\frac{d\tau}{d\theta} = \frac{a_\infty}{a_\infty r + U(x + U\tau)}$$

where on replacing r by its characteristic value $a_\infty(t - \tau)$ and replacing τ by its value given in Eq. 4-20 there is obtained for $\theta = 0$

$$\frac{1}{r}\frac{d\tau}{d\theta} = \frac{1}{R}$$

Hence Eq. 4-19 becomes

$$\varphi = -\frac{1}{4\pi}\frac{1}{R}f\left[\left(t + \frac{Ux}{a_\infty^2} - \frac{R}{a_\infty}\right)\frac{1}{\beta^2}\right] \tag{4-21}$$

This result is referred to space-fixed coordinates; it is finally desired to express this result in terms of a field point x_0, y_0, z_0 of a coordinate system moving uniformly along with the source located at ξ_0, η_0, ζ_0, so that we put

$$x + Ut = x_0 - \xi_0; \quad y = y_0 - \eta_0; \quad z = z_0 - \zeta_0$$

After performing this substitution the zero subscripts may be dropped,

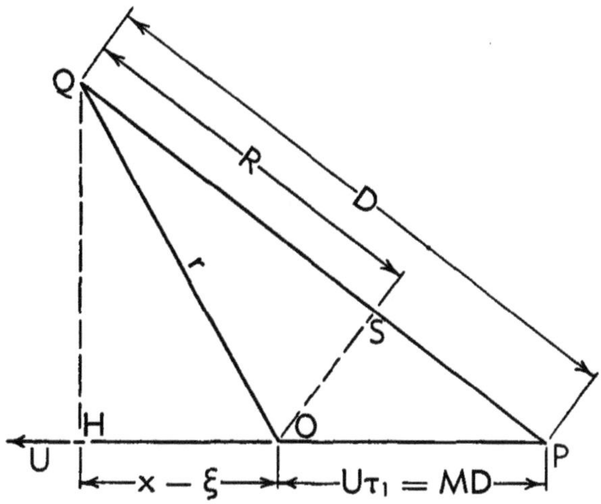

Fig. F,4a. Subsonic source O, field point Q, amplitude and phase radii R and D.

giving as the end result corresponding to Eq. 4-5a for the source and the coordinate system in uniform rectilinear motion the following relation:

Subsonic point source

$$\varphi_1 = -\frac{1}{4\pi R} f\left(t - \frac{D}{a_\infty}\right) = -\frac{1}{4\pi R} f(t - \tau_1) \tag{4-22}$$

where

$$R = \sqrt{(x - \xi)^2 + \beta^2[(y - \eta)^2 + (z - \zeta)^2]}$$

$$\tau_1 = \frac{D}{a_\infty} = \frac{-M_\infty(x - \xi) + R}{a_\infty \beta^2}$$

$$\beta^2 = 1 - M_\infty^2$$

The quantities R and D may be called "amplitude" radius and "phase" radius, respectively. Comparison of Eq. 4-22 and Eq. 4-5a leads to the mnemonic statement: To obtain the moving source solution, replace the ordinary radius r appearing in Eq. 4-5a by R in the amplitude and by D in the phase.

Of interest is a geometric interpretation of this result given with the aid of Fig. F,4a. Consider at time t a field point Q and a source point at O moving uniformly with velocity U in the negative x direction. The distance $OQ = r$, the ordinary radius. The spherical wave through Q at time t originated at the time interval τ_1 earlier, that is, at time $t - \tau_1$

when the source was at P. Thus P is determined geometrically from the fact that $\overline{OP} = M_\infty(\overline{PQ}) = U\tau_1$ and hence $\overline{PQ} = a_\infty\tau = D$, the phase radius; algebraically stated this corresponds to solving for time in the relation

$$(x - \xi - U\tau)^2 + (y - \eta)^2 + (z - \zeta)^2 = a_\infty^2\tau^2 \tag{4-23}$$

For $M_\infty < 1$ one root is positive, and a single point P physically exists. (For $M_\infty > 1$, discussed later, there are two such points.) To obtain a geometric interpretation of the amplitude radius R, drop perpendiculars $\overline{QH}$ and $\overline{OS}$. From triangles POS and PHQ, $\overline{OS}$ is determined as $M_\infty(\overline{HQ})$. Then from triangle OQS it follows that $\overline{QS} = R$.

The moving doublet corresponding to Eq. 4-5b is again given by differentiation in the direction of the axis of the doublet.

The harmonically oscillating moving source with the time factor suppressed and with R and τ_1 as in Eq. 4-22, is

$$\begin{aligned} \varphi_1^* &= -\frac{1}{4\pi R} e^{-i\omega\tau_1} \\ &= -\frac{1}{4\pi R} e^{-i\mu(x-\xi)-i\kappa R} \end{aligned} \tag{4-24}$$

where

$$\kappa = \frac{\omega}{a_\infty\beta^2}$$

$$\mu = \frac{\omega M_\infty}{a_\infty\beta^2}$$

Subsonic line source. Corresponding to the solution for cylindrical waves given in Eq. 4-8, there appears for the moving source in two dimensions

$$\varphi_1 = -\frac{1}{2\pi\beta}\int_0^\infty f\left[t + \frac{M_\infty(x-\xi)}{a_\infty\beta^2} - \frac{R\cosh U}{a_\infty\beta^2}\right] du \tag{4-25}$$

where

$$R = \sqrt{(x-\xi)^2 + \beta^2(y-\eta)^2}$$
$$\beta = |1 - M_\infty^2|^{\frac{1}{2}}$$

To the simpler result for the source-pulse (Eq. 4-9) there corresponds

$$\varphi_0 = -\frac{1}{2\pi}\frac{a_\infty f(T)}{\sqrt{a_\infty^2(t-T)^2 - [x - \xi - U(t-T)]^2 - (y-\eta)^2}} \tag{4-26}$$

For the harmonically oscillating moving source corresponding to Eq. 4-10 with the time factor $e^{i\omega\tau}$ omitted, there appears

$$\varphi_1^* = -\frac{e^{i\mu(x-\xi)}}{2\pi\beta}\int_0^\infty e^{-i\kappa R\cosh u} du$$

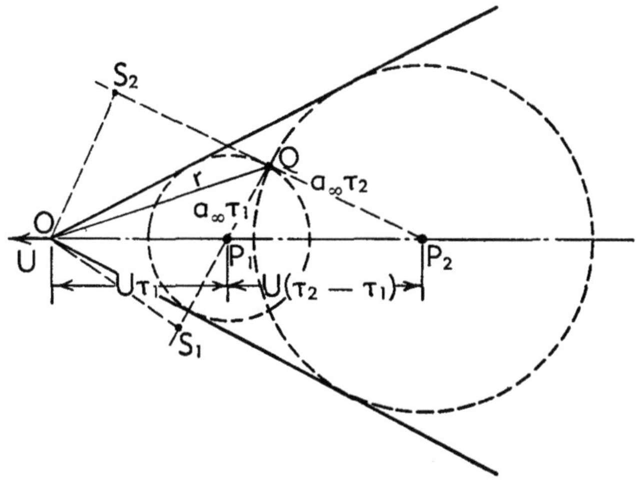

Fig. F,4b. Supersonic source O, field point Q, time intervals τ_1 and τ_2.

where

$$\kappa = \frac{\omega}{a_\infty \beta^2}$$

$$\mu = \frac{\omega M_\infty}{a_\infty \beta^2}$$

or, on integration (see Eq. 4-11),

$$\varphi_1^* = \frac{i}{4\beta} e^{-i\mu(x-\xi)} H_0^{(2)}(\kappa \sqrt{(x-\xi)^2 + \beta^2(y-\eta)^2}) \tag{4-27}$$

This result considered as an acceleration potential is made use of in Art. 9.

Supersonic point source. The solution in space corresponding to Eq. 4-5a for $M_\infty > 1$ is

$$\varphi_1 = -\frac{1}{4\pi R}[f(t-\tau_1) + f(t-\tau_2)] \tag{4-28}$$

where

$$R = \sqrt{(x-\xi)^2 - (M_\infty^2 - 1)[(y-\eta)^2 + (z-\zeta)^2]}$$

$$\tau_1 = \frac{D}{a_\infty} = \frac{1}{a_\infty}\left[\frac{M_\infty(x-\xi) - R}{M_\infty^2 - 1}\right]$$

$$\tau_2 = \frac{D'}{a_\infty} = \frac{1}{a_\infty}\left[\frac{M_\infty(x-\xi) + R}{M_\infty^2 - 1}\right]$$

A geometric interpretation of this result may be given with the aid of Fig. F,4b. Consider a source at O moving uniformly with velocity U in the negative x direction. The equation $R = 0$ now defines, besides the origin, the "Mach" cone of physical interest opening aft of the moving source. The semivertex angle or Mach angle is $\sin^{-1} 1/M_\infty$. Outside of the conical region, R is not real and the flow is undisturbed. A field point Q is chosen within the Mach cone distance r from O. The point Q at time t is influenced by two waves originating at time intervals τ_2 and τ_1 earlier, values of which are given by the roots of the quadratic equation (4-23). Two points P_1 and P_2 are thus defined for which $\overline{OP_1} = M_\infty(\overline{P_1Q})$ and $\overline{OP_2} = M_\infty(\overline{P_2Q})$, and hence $\overline{P_1Q}$ is the phase radius $a_\infty\tau_1 = D$ and $\overline{P_2Q}$ the phase radius $a_\infty\tau_2 = D'$. The amplitude radius is defined by a construction similar to that used in the subsonic case. Perpendiculars dropped from O onto the lines $\overline{QP_1}$ or $\overline{P_2Q}$ will define points S_1 and S_2, and R is then given directly by $\overline{QS_1}$ or $\overline{QS_2}$.

Another physical interpretation of Eq. 4-28 is obtained by considering a single wave pulse at different times [*11*, p. 4]. A uniformly moving field point at successive times $t + \tau_1$ and $t + \tau_2$ experiences, respectively, the effect of penetration into and emergence out of the spherical wave front of the pulse created at the origin at time t. At penetration of the wave front for a positive source there is a compression and immediate subsequent equal expansion, and at emergence an expansion and subsequent compression.

For harmonic oscillations (with the time factor omitted) the source moving at supersonic speed has the potential

$$\begin{aligned} \varphi_1^* &= -\frac{1}{4\pi R}\left(e^{-i\omega\tau_1} + e^{-i\omega\tau_2}\right) \\ &= -\frac{1}{2\pi R}\, e^{-i\mu(x-\xi)} \cos \kappa R \end{aligned} \tag{4-29}$$

where

$$\kappa = \frac{\omega}{a_\infty}\frac{1}{M_\infty^2 - 1}$$

$$\mu = \frac{\omega}{a_\infty}\frac{M_\infty}{M_\infty^2 - 1}$$

and R is defined as in Eq. 4-28.

Supersonic line source. A solution corresponding to Eq. 4-8 and 4-25 can be developed by integration of Eq. 4-28 with respect to z from $-\infty$ to ∞, or actually only over the Mach cone of influence. It is convenient to make use of the substitution $\beta z = R \sin\theta$ where $\beta = |M_\infty^2 - 1|^{\frac{1}{2}}$ and $R = \sqrt{(x-\xi)^2 - \beta^2(y-\eta)^2}$. The two integral terms that result can be combined to yield

$$\varphi_1 = -\frac{1}{2\pi\beta}\int_0^\pi f\left[t - \frac{M_\infty(x-\xi)}{a_\infty\beta^2} + \frac{R\cos\theta}{a_\infty\beta^2}\right] d\theta \tag{4-30}$$

The source-pulse solution corresponding to Eq. 4-9 is like Eq. 4-26, and with $t - T = \tau$ may be written as

$$\varphi_0 = -\frac{1}{2\pi\beta}\frac{f(t-\tau)}{\sqrt{(\tau - \tau_1)(\tau_2 - \tau)}} \tag{4-31}$$

where τ_1 and τ_2 are

$$\tau_{1,2} = \frac{M_\infty(x - \xi)}{a_\infty\beta^2} \mp \frac{R}{a_\infty\beta^2}$$

and where

$$R = \sqrt{(x - \xi)^2 - \beta^2(y - \eta)^2}$$
$$\beta = |M_\infty^2 - 1|^{\frac{1}{2}}$$

For the harmonically oscillating source (with time factor omitted) by integration of Eq. 4-31

$$\varphi_1^* = -\frac{1}{2\pi\beta}\int_{\tau_1}^{\tau_2}\frac{e^{-i\omega\tau}}{\sqrt{(\tau - \tau_1)(\tau_2 - \tau)}}\,d\tau$$

or, from Eq. 4-30,

$$\varphi_1^* = -\frac{e^{-i\mu(x-\xi)}}{2\pi\beta}\int_0^\pi e^{i\kappa R\cos\theta}d\theta$$

where

$$\kappa = \frac{\omega}{a_\infty\beta^2}$$

$$\mu = \frac{\omega M_\infty}{a_\infty\beta^2}$$

Either integral yields

$$\varphi_1^* = -\frac{1}{2\beta}e^{-i\mu(x-\xi)}J_0(\kappa\sqrt{(x - \xi)^2 - \beta^2(y - \eta)^2}) \tag{4-32}$$

a result that follows also directly by integration of Eq. 4-29, and which may be compared with the corresponding subsonic case result given in Eq. 4-27. It is made use of in Art. 10.

Sonic point source. It is of special interest also to develop results for the source moving at sonic speed. As has been indicated, difficulties arise in the use of small disturbance methods but strictly these mathematical difficulties appear, as is again noted in Art. 11, in the case of steady plane flow. Of course, one is here concerned with the problem of small oscillations superposed on an already strongly disturbed field, since, physically, near sonic speeds the flow picture is clouded by shock phenomena. Nevertheless, the availability of means of analysis of nonstationary flow at sonic speed by linearized methods should not be discounted, for this furnishes a continuous bridge in analytical results from subsonic through transonic to supersonic speeds. Moreover, the applications to physical situations sufficiently removed from the steady flow case, as for high frequency oscillations, would appear to have some degree of validity.

For the source in space moving uniformly in the negative x direction there follows from either limiting processes, or directly from Eq. 4-22 and 4-23,

$$\varphi_1 = -\frac{1}{4\pi(x-\xi)} f(t-\tau_1) \tag{4-33}$$

where

$$\tau_1 = \frac{1}{2a_\infty}\left[x-\xi+\frac{(y-\eta)^2+(z-\zeta)^2}{x-\xi}\right] = \frac{D}{a_\infty}$$

For the harmonically oscillating source the result is

$$\varphi_1^* = -\frac{1}{4\pi(x-\xi)} e^{-i\omega\tau_1} \tag{4-34}$$

Fig. F,4c illustrates the source point O and the field point Q at time t, the

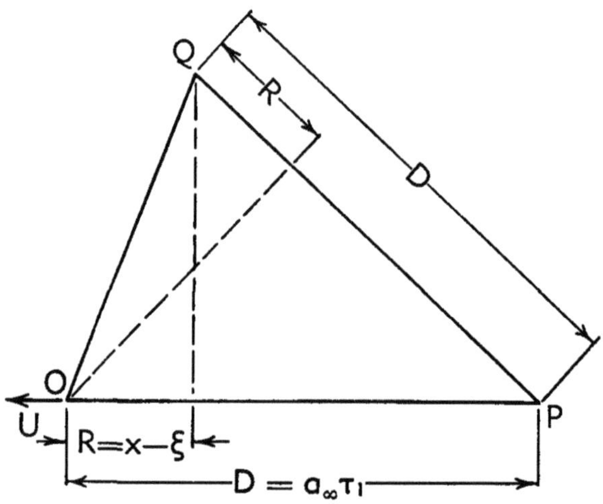

Fig. F,4c. Source O moving at sonic speed.

position P of the source at time $t - \tau_1$, and the phase and amplitude radii D and R.

Sonic line source. There results by integration of Eq. 4-33 with respect to z,

$$\varphi_1 = -\frac{1}{4\pi(x-\xi)} \int_{-\infty}^{\infty} f\left[t - \frac{(x-\xi)^2+(y-\eta)^2}{2a_\infty(x-\xi)} - \frac{z^2}{2a_\infty(x-\xi)}\right] dz \tag{4-35}$$

For the harmonically oscillating source

$$\varphi_1^* = -\frac{e^{-i\omega\tau_0}}{2\pi(x-\xi)} \int_0^\infty e^{-\lambda^2 z^2} dz \tag{4-36}$$

where

$$\tau_0 = \frac{1}{2a_\infty}\left[x - \xi + \frac{(y-\eta)^2}{x-\xi}\right]$$

$$\lambda^2 = \frac{i\omega}{2a_\infty(x-\xi)}$$

The value of the integral is $\sqrt{\pi}/2\lambda$ and hence

$$\varphi_1^* = \frac{1}{B\sqrt{x-\xi}}\, e^{-i\omega\tau_0} \tag{4-37}$$

where

$$B = -2\sqrt{2\pi}\,\sqrt{i}\,\sqrt{\frac{\omega}{a_\infty}}$$

and where $\sqrt{i} = e^{i(\pi/4)}$. This result is made use of in Art. 11.

The results for sources and doublets are to be used in subsequent developments with proper surface or line distributions to yield relevant nonsteady characteristics of wings. The treatment for the incompressible fluid case in the next article, however, proceeds along different lines.

F,5. Airfoil in Two-Dimensional Nonstationary Flow of Incompressible Fluid. It is intended to present a brief but reasonably complete development of the results for oscillating wing sections in plane incompressible flow, results which still form the basis of much current work in flutter. Contributions to the problem have been made in many places and in many different ways, starting with the different yet equivalent procedures of Birnbaum [*15*] and Wagner [*16*]. The vortex distribution method of Birnbaum was followed up particularly by Cicala [*17*] and Küssner [*18*], the conformal mapping method of Wagner, by Glauert [*19*], Duncan and Collar [*20*], Burgers [*21*], Theodorsen [*22*], and von Kármán and Sears [*23*]. In the development to be given, which is influenced by Theodorsen's work, the physical and mathematical ideas proceed in a parallel reinforcing manner. It calls for the determination of the noncirculatory flow pattern associated with the primary flow not involving the wake and the subsequent determination of the circulatory flow associated with the wake. The corresponding velocity potentials will be denoted, respectively, by φ_1 and φ_2. Supplementary remarks on Birnbaum's method and the other methods are given at the end of Art. 5.

Noncirculatory potential. The airfoil of chord $c = 2b$ is progressing uniformly with velocity U and creating small disturbances. With half chord b as reference length, let the airfoil extend from the leading edge $x = -1$ to the trailing edge $x = 1$, variable x being nondimensional, and consider an element Δx_1 of the boundary located at $x = x_1$ (Fig. F,5a). Let the normal velocity of this element be $w(x_1, t)$, given by means of

Eq. 3-5. The primary flow pattern arising from the disturbance created by this element is the so-called noncirculatory one illustrated in Fig. F,5a. Its velocity potential can be obtained as in [*22*] from the potential of a source and sink symmetrically placed on the circumference of a circle, the flow pattern of which is then mapped conformally into the flow pattern about the line. The mapping preserves the potential at corresponding

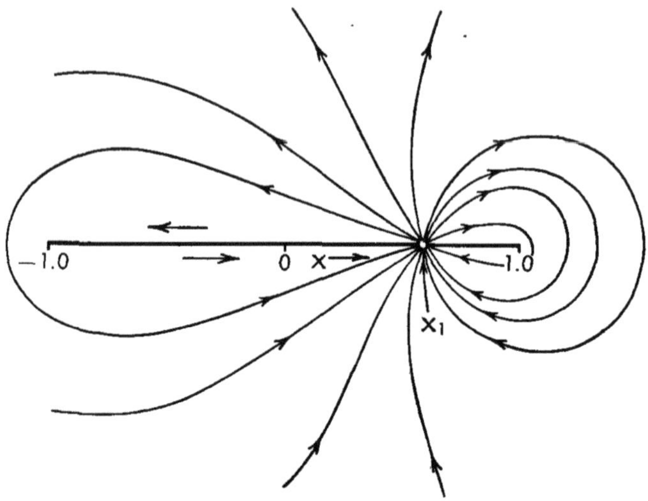

Fig. F,5a. Noncirculatory flow pattern about airfoil for normal velocity disturbance at $x = x_1$.

points and the basic result giving the (upper) surface velocity potential at any point x due to the normal velocity w at x_1 is

$$\Delta\varphi_1 = \frac{1}{2\pi} w(x_1, t)b\Delta x_1 L(x, x_1) \tag{5-1}$$

where

$$L(x, x_1) = \ln \frac{(x - x_1)^2 + (\sqrt{1 - x^2} - \sqrt{1 - x_1^2})^2}{(x - x_1)^2 + (\sqrt{1 - x^2} + \sqrt{1 - x_1^2})^2}$$

$$= 2 \ln \left| \frac{1 - xx_1 - \sqrt{1 - x^2}\sqrt{1 - x_1^2}}{x - x_1} \right|$$

Another way of arriving at Eq. 5-1 would make use of the doublet solution Eq. 4-7b and a boundary-value (Green's function or integral equation) formulation.[4] The function L is often expressed in terms of the angle parameter (between 0 and π) on a circle by means of the substitutions

$$x = -\cos\theta, \qquad x_1 = -\cos\eta \tag{5-2}$$

[4] The reader may find the classical separation-of-variable approach for the boundary-value problem given in Art. 9 of interest in this connection.

in which case the log term appears in the form

$$L(x, x_1) = 2\mathcal{L}(\theta, \eta)$$

where[5]

$$\mathcal{L}(\theta, \eta) = \frac{1}{2} \ln \frac{1 - \cos(\theta - \eta)}{1 - \cos(\theta + \eta)}$$

$$= \ln \frac{\sin \frac{|\theta - \eta|}{2}}{\sin \frac{\theta + \eta}{2}} \tag{5-3}$$

The distribution of surface pressure difference associated with the noncirculatory potential $\Delta\varphi_1$ is given by means of Eq. 3-7 as

$$\Delta P_1 = -2\rho_\infty \left(\frac{U}{b} \frac{\partial}{\partial x} + \frac{\partial}{\partial t} \right) \Delta\varphi_1$$

$$= -\frac{2\rho_\infty}{\pi} \left(\frac{U}{b} \frac{1}{x - x_1} \frac{\sqrt{1 - x_1^2}}{\sqrt{1 - x^2}} \right) bw\Delta x_1 - \frac{\rho_\infty}{\pi} L(x, x_1) b \frac{\partial w}{\partial t} \Delta x_1 \tag{5-4}$$

Circulatory potential. The resulting flow pattern gives rise, as may be noted, to an infinite velocity and pressure difference at $x = x_1$ and at the edges. The singularity at $x = x_1$ disappears in the applications by making use of continuous distributions w and Cauchy's principal part in the integrations. Because of the edge conditions, as discussed, a circulatory flow pattern is now to be superposed, of a magnitude sufficient to cancel the infinity at the trailing edge at each instant. As a consequence, a surface of discontinuity of velocity is generated at the trailing edge and free vorticity left behind in the wake. Consider an element of this surface

[5] It is of interest to indicate the general field form of the element of noncirculatory potential for which Eq. 5-1 is the boundary surface form. This will incidentally give additional insight into the character of the solution. It is convenient to introduce elliptic coordinates ξ, θ for the Cartesian field points x, y

$$x = -\cosh \xi \cos \theta, \qquad y = \sinh \xi \sin \theta$$

where these relations generalize Eq. 5-2. The boundary surface point x_1 will be denoted by $\xi = 0$, $\theta = \eta$. The general solution corresponding to Eq. 5-1 is

$$\Delta\varphi_1 = \frac{w}{2\pi} b\Delta x_1 L(x, y, x_1)$$

where, generalizing Eq. 5-3,

$$L(x, y, x_1) = 2\mathcal{L}(\xi, \theta, \eta) = \ln \frac{\cosh \xi - \cos(\theta - \eta)}{\cosh \xi - \cos(\theta + \eta)}$$

The derivative of L with respect to y can be shown to have the characteristic property of the Dirac δ function for $y = 0$ and thus assures that integration of $\Delta\varphi_1$ with respect to x_1 for all boundary surface elements will yield the required potential satisfying the given normal velocity distribution w.

at $x = x_0 > 1$, of a magnitude $E(x_0, t)$ per unit length, that is $\Delta\Gamma = Eb\Delta x_0$. The surface velocity potential of the flow pattern induced by this element at the airfoil can also be found by conformal mapping methods[6] as in [22], or otherwise, and is

$$\Delta\varphi_2' = \frac{Eb\Delta x_0}{2\pi} \tan^{-1} \frac{\sqrt{1 - x^2}\sqrt{x_0^2 - 1}}{1 - xx_0} \tag{5-5}$$

If the wake extends from $x_0 = 1$ to $x_0 = s$ (that is, the airfoil motion started at time $t = (s - 1)b/U$ earlier), the velocity potential of all elements is

$$\Delta\varphi_2 = \int_1^s \frac{\Delta\varphi_2'}{\Delta x_0} dx_0 \tag{5-6}$$

Trailing edge condition. The trailing edge condition for smooth flow (for the single element at x_1) is given by the finiteness of the expression

$$\frac{\partial}{\partial x} (\Delta\varphi_1 + \Delta\varphi_2)_{x=1}$$

Using Eq. 5-1 and 5-6, this condition becomes,

$$\frac{1}{\pi} w(x_1, t)\Delta x_1 \sqrt{\frac{1 + x_1}{1 - x_1}} = -\frac{1}{2\pi} \int_1^s E(x_0, t) \sqrt{\frac{x_0 + 1}{x_0 - 1}} dx_0 \tag{5-7}$$

and this can be regarded as an integral equation for determining the wake vorticity in terms of the normal velocity w for any element Δx_1. It is actually a form of Wagner's integral equation [16] for each element and its general solution can be made to depend on that for either a unit-step time function or a sinusoidal time function for the time variation of the function w. (As will be observed, an explicit solution of the integral equation is actually not required.)

One may note, in passing, that the property of the wake vorticity, generated at the trailing edge $x_0 = 1$, of floating with the fluid elements leads to relations of the form

$$E(x_0, t) = E(x_0 - U\tau, t - \tau) = E\left[1, t - \frac{(x_0 - 1)b}{U}\right] \tag{5-8}$$

Thus free vorticity released at the trailing edge is to be found at any other wake position x_0 at an interval of time $(x_0 - 1)b/U$ later, and E is a func-

[6] Both Eq. 5-5 and 5-1 may be shown to stem from the real and imaginary parts of the complex potential

$$\ln (z - z_1) - \ln \left(z - \frac{b^2}{z_1}\right)$$

where $z = be^{\xi + i\theta}$, $z_1 = be^{\xi_1 + i\eta}$ and where ξ, θ and ξ_1, η correspond to the coordinates referred to in the preceding footnote. The wake coordinate corresponds to $\cosh \xi_1$.

tion only of the combination $t - (x_0b/U)$ or of the distance $s - x_0$ measured back from the start of the wake.

Pressure due to circulatory potential. Before specializing the choice of the function w it is desirable to express generally in an integral form the surface pressure distribution at x as obtained from the circulatory potential. Making use of Eq. 5-4, 5-5, and 5-6, and of the fact that the time variation of the potential due to the wake arises only by virtue of the motion of the wake relative to the airfoil, i.e. because $U dt = b dx_0$ or $\partial/\partial t = (U/b)(\partial/\partial x_0)$, there results for the pressure difference associated with the circulatory potential

$$\begin{aligned} \Delta P_2 &= -2\rho_\infty \frac{U}{b}\left(\frac{\partial}{\partial x} + \frac{\partial}{\partial x_0}\right)\Delta\varphi_2 \\ &= -\frac{2\rho_\infty U}{\sqrt{1-x^2}}\int_1^s \frac{E}{2\pi}\frac{x_0 + x}{\sqrt{x_0^2 - 1}}\,dx_0 \end{aligned} \tag{5-9}$$

or making use of Eq. 5-7 and with the left-hand side of Eq. 5-7 denoted by $\Delta Q(x_1, t)$,

$$\Delta P_2 = \frac{2\rho_\infty U}{\sqrt{1-x^2}}\,\frac{\displaystyle\int_1^s E\,\frac{x_0 + x}{\sqrt{x_0^2 - 1}}\,dx_0}{\displaystyle\int_1^s E\sqrt{\frac{x_0+1}{x_0-1}}\,dx_0}\,\Delta Q(x_1, t) \tag{5-10}$$

where

$$\Delta Q(x_1, t) = \frac{1}{\pi}\,w(x_1, t)\Delta x_1\sqrt{\frac{1+x_1}{1-x_1}}$$

Harmonic motion and reduced frequency. Consider now the basis of sinusoidal motion which has been proceeding for a long time so that the wake can be regarded as extending to infinity. Let the normal velocity be written as

$$w(x_1, t) = w^*(x_1)e^{i\omega t} \tag{5-11}$$

and let the wake vorticity be written as

$$\begin{aligned} E(x_0, t) &= E_0 e^{i\omega\{t - [(x_0 - 1)b/U]\}} \\ &= E_0 e^{i[\omega t - k(x_0 - 1)]} \end{aligned} \tag{5-12}$$

where w^* and E_0 are complex amplitudes, thus also containing the phase, and where

$$k = \frac{\omega b}{U} = \frac{\omega c}{2U} \tag{5-13}$$

The reduced frequency k which appears herein combines the frequency with a reference length and a reference velocity and is an important non-dimensional parameter for characterizing unsteady flows. Here it represents the number of waves in a wake distance of 2π half chords; $2\pi/k$ is thus the wavelength in half chords. The value $k = 0$ corresponds to steady flow, values from about 0 to 0.05 occur in dynamic stability, while in flutter k may range from small values to greater than 1.0. The British literature employs the chord instead of the half chord as reference length, and the symbol λ for $2k$; the German literature often uses $-i\omega$ or $\bar{\omega}$ for k, l for b, ν for ω.

Eq. 5-10, expressing the pressure difference at x due to the wake, may be rewritten with s to approach ∞, as

$$\Delta P_2 = \frac{2\rho_\infty U}{\sqrt{1-x^2}} \frac{\displaystyle\int_1^s \frac{x_0 + x}{\sqrt{x_0^2 - 1}} e^{-ikx_0} dx_0}{\displaystyle\int_1^s \frac{x_0 + 1}{\sqrt{x_0^2 - 1}} e^{-ikx_0} dx_0} \Delta Q \tag{5-14}$$

where

$$\Delta Q = \frac{1}{\pi} w(x_1, t) \Delta x_1 \sqrt{\frac{1 + x_1}{1 - x_1}}$$

Evaluation of integral. A few considerations are involved in the explicit evaluation of the integrals for $s \to \infty$ since, if s is put equal to ∞ directly, the integrations are improper, both integrands being oscillatory at infinity. To avoid the difficulty consider, with s finite, the following expression arising in Eq. 5-14

$$\frac{\displaystyle\int_1^s \frac{x_0}{\sqrt{x_0^2 - 1}} e^{-ikx_0} dx_0}{\displaystyle\int_1^s \frac{x_0 + 1}{\sqrt{x_0^2 - 1}} e^{-ikx_0} dx_0} = 1 - \frac{\displaystyle\int_1^s \frac{e^{-ikx_0}}{\sqrt{x_0^2 - 1}} dx_0}{\displaystyle\int_1^s \frac{e^{-ikx_0}}{\sqrt{x_0^2 - 1}} dx_0 + \int_1^s \frac{x_0 e^{-ikx_0}}{\sqrt{x_0^2 - 1}} dx_0} \tag{5-15}$$

The only integral on the right side which gives difficulty when $s = \infty$ is

$$I(k) = \int_1^s \frac{x_0 e^{-ikx_0}}{\sqrt{x_0^2 - 1}} dx_0$$

Two ways to consider this integral for $s \to \infty$ follow. In one, k is considered to be complex and to give rise to a damping term in the exponential term, which is then allowed to vanish, so that from Eq. 4-11 and 4-13 there is obtained

$$\begin{aligned} I(k) &= \frac{d}{dk} \int_1^\infty \frac{ie^{-ikx_0}}{\sqrt{x_0^2 - 1}} dx_0 \\ &= \frac{d}{dk} \frac{\pi}{2} H_0^{(2)}(k) \\ &= -\frac{\pi}{2} H_1^{(2)}(k) \end{aligned}$$

This procedure involves analytic continuation or permanence of form; it is a purely mathematical device since k complex does not necessarily have the meaning physically of damped motion, and the physically proper interpretation of damped motion must follow from the superposition principle rather than from analytic extension. The improper integral can also be regarded physically as presenting a type of difficulty often appearing in electric circuit analysis in attempts to obtain steady direct current response from transient current results. From this point of view it suggests itself to add and then subtract a quantity to the integral which denotes the steady case results and which corresponds to the entire wake circulation being located *at infinity*. That is to say, if the instantaneous normal velocity at any instant is assumed to be steady the integral must be made to yield the steady case result. This type of procedure also removes the difficulty in the integration, leading to the same end results,

Table F,5

$C(k) = F(k) + iG(k)$

k	F	$-G$	k	F	$-G$	k	F	$-G$
0.000	1.00000	0.00000	0.48	0.60259	0.15347	0.98	0.54062	0.10171
0.002	0.99671	0.01258	0.50	0.59793	0.15071	1.00	0.53943	0.10027
0.02	0.96373	0.07521	0.52	0.59359	0.14800	1.1	0.53421	0.09361
0.04	0.92670	0.11600	0.54	0.58953	0.14535	1.2	0.52996	0.08771
0.06	0.89204	0.14259	0.56	0.58572	0.14277	1.3	0.52644	0.08247
0.08	0.86043	0.16040	0.58	0.58215	0.14024	1.4	0.52349	0.07777
0.10	0.83192	0.17230	0.60	0.57880	0.13779	1.5	0.52101	0.07356
0.12	0.80633	0.18007	0.62	0.57565	0.13539	1.6	0.51890	0.06977
0.14	0.78337	0.18489	0.64	0.57269	0.13305	1.7	0.51709	0.06632
0.16	0.76277	0.18757	0.66	0.56989	0.13078	1.8	0.51552	0.06318
0.18	0.74426	0.18867	0.68	0.56725	0.12857	1.9	0.51415	0.06032
0.20	0.72758	0.18862	0.70	0.56476	0.12642	2.0	0.51295	0.05769
0.22	0.71252	0.18772	0.72	0.56240	0.12433	2.5	0.50874	0.04730
0.24	0.69889	0.18619	0.74	0.56017	0.12229	3.0	0.50628	0.04000
0.26	0.68651	0.18420	0.76	0.55806	0.12031	3.5	0.50472	0.03462
0.28	0.67525	0.18188	0.78	0.55605	0.11838	4.0	0.50367	0.03050
0.30	0.66497	0.17932	0.80	0.55415	0.11660	4.5	0.50294	0.02724
0.32	0.65557	0.17659	0.82	0.55234	0.11468	5.0	0.50240	0.02460
0.34	0.64695	0.17376	0.84	0.55062	0.11290	10	0.50062	0.01245
0.36	0.63902	0.17086	0.86	0.54898	0.11117	20	0.50016	0.00625
0.38	0.63172	0.16792	0.88	0.54741	0.10949	30	0.50007	0.00417
0.40	0.62498	0.16498	0.90	0.54593	0.10785	40	0.50004	0.00312
0.42	0.61874	0.16206	0.92	0.54451	0.10626	50	0.50003	0.00250
0.44	0.61296	0.15915	0.94	0.54315	0.10470	100	0.50001	0.00125
0.46	0.60759	0.15629	0.96	0.54186	0.10319	∞	0.50000	0.00000

for $k \to 0$, and γ (Euler's const) = 0.57722,

$$C(k) \cong \left(1 - \frac{\pi k}{2}\right) + ik\left(\ln\frac{k}{2} + \gamma\right)$$

for $k \to \infty$

$$C(k) \cong \frac{1}{2}\left[\left(1 + \frac{1}{8k^2}\right) - \frac{i}{4k}\right]$$

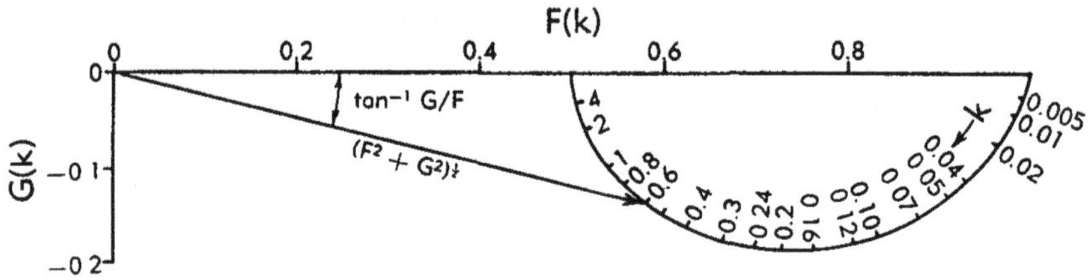

Fig. F,5b. Representation of $C(k) = F(k) + iG(k)$.

and in effect is adopted by von Kármán and Sears in [23], where the quantity added and then subtracted is denoted as "quasi-steady."

The interpretation of the expression in Eq. 5-15 for $s \to \infty$ is then

$$C(k) = \frac{\displaystyle\int_1^\infty \frac{x_0}{\sqrt{x_0^2 - 1}} e^{-ikx_0} dx_0}{\displaystyle\int_1^\infty \frac{x_0 + 1}{\sqrt{x_0^2 - 1}} e^{-ikx_0} dx_0} = 1 - \frac{H_0^{(2)}(k)}{H_0^{(2)}(k) - iH_1^{(2)}(k)} = \frac{H_1^{(2)}(k)}{H_1^{(2)}(k) + iH_0^{(2)}(k)} \tag{5-16}$$

Eq. 5-14, giving the surface pressure distribution at x due to the circulatory wake potential arising from the normal velocity of the airfoil element at x_1, may now be written as

$$\Delta P_2 = \frac{2\rho_\infty U}{\sqrt{1 - x^2}} \{C(k) + x[1 - C(k)]\} \Delta Q = \frac{2\rho_\infty U}{\pi} \sqrt{\frac{1 - x}{1 + x}} \sqrt{\frac{1 + x_1}{1 - x_1}} \left[C(k) - 1 + \frac{1}{1 - x} \right] w \Delta x_1 \tag{5-17}$$

where $C(k)$ is given in Eq. 5-16

The function $C(k)$, introduced by Theodorsen [22] appears herein as basic in the theory of the harmonically oscillating airfoil. It has been written for use in applications in the separated form

$$C(k) = F(k) + iG(k) \tag{5-18}$$

and is shown in a vector diagram in Fig. F,5b. Table F,5 gives a listing of numerical values. The significance of $C(k)$ for nonharmonic oscillations or for transient flows generally is based on the superposition principle and is introduced in Art. 6.

Total pressure difference. The sum ΔP of ΔP_1 and ΔP_2 by Eq. 5-4 and 5-17 divided by the element Δx_1 leads to the result for the rate of loading at x contributed per unit airfoil element at x_1, as

$$\frac{dP}{dx_1} = -\frac{\rho_\infty}{\pi} L(x, x_1) b \frac{\partial w}{\partial t} + \frac{2\rho_\infty U}{\pi} \sqrt{\frac{1-x}{1+x}} \sqrt{\frac{1+x_1}{1-x_1}} \left(\frac{1}{x_1 - x} + C(k) - 1 \right) w \quad (5\text{-}19)$$

The terms on the right may be classified into three important categories. The first term (see Eq. 5-1)

$$-\frac{\rho_\infty}{\pi} L(x, x_1) b \frac{\partial w}{\partial t}$$

is an apparent-mass effect which vanishes in the case of steady motion. It has been expressed with the factor $\partial w/\partial t$ instead of the harmonic factor $i\omega e^{i\omega t}$ since it holds in general. The second term

$$\frac{2\rho_\infty U}{\pi} \sqrt{\frac{1-x}{1+x}} \sqrt{\frac{1+x_1}{1-x_1}} \frac{1}{x_1 - x} w$$

may be recognized as one which appears in steady aerodynamics and can lead to all the classical results of thin airfoil theory for the steady flow case. It represents the effect that would result at each instant if the instantaneous normal velocity distribution were frozen in time. That is, it is the "quasi-steady" effect which considers the entire wake to be at infinity. The third term

$$\frac{2\rho_\infty}{\pi} \sqrt{\frac{1-x}{1+x}} \sqrt{\frac{1+x_1}{1-x_1}} [C(k) - 1] w$$

represents the effect of the harmonic wake distribution. In the steady flow case $k = 0$, $C(k) = 1$, and this term vanishes; the term with the factor $C(k)$ is discussed further in Art. 6.

For the harmonically oscillating airfoil, there follows then by proper integration of the basic relation for the density of pressure difference over the chord, Eq. 5-19, with respect to x_1 (Cauchy principal values to apply for the integrals containing the singularity at $x = x_1$) that the total pressure difference or loading at x is given by

$$P(x) = \int_{-1}^{1} \frac{dP}{dx_1} dx_1 \quad (5\text{-}20)$$

By further integration of this result with respect to x (see Art. 7), complete or partial air forces and moments as required can be obtained.

The result represented by Eq. 5-19 and 5-20 has been obtained herein by the direct means of obtaining and properly combining the elementary influence functions (Green's functions) for the noncirculatory and circulatory potentials. Eq. 5-20 was also explicitly arrived at by Küssner and Schwarz [*24*], by Schwarz [*25*], and by Söhngen [*26*]. It is to be noted that the separation into the three categories, apparent-mass, quasi-steady and wake effects, is made use of from the beginning in the procedure of von Kármán and Sears [*23*]. The different methods of derivation are worthy of the interested reader's study and comparison; brief remarks on the main concepts involved follow Eq. 5-24.

A listing here of two additional and equivalent forms in which the result appears in the literature will be useful. The first of these, due to Cicala, utilizes the variables given in Eq. 5-2, and the expression for the pressure difference $P(x, t)$ is

$$\Pi(\theta, t) = \rho_\infty U^2 e^{i\omega t}\left(2a_0 \cot\frac{\theta}{2} + 4\sum_1^\infty a_n \frac{\sin n\theta}{n}\right) \tag{5-21}$$

where

$$a_0 = C(k)(A_0 - A_1) + A_1$$

$$a_n = \frac{ik}{2} A_{n-1} - nA_n - \frac{ik}{2} A_{n+1} \qquad (n \geqq 1)$$

and where the coefficients A_n are to be regarded as known from the expansion of the normal velocity in the form

$$w = Ue^{i\omega t}\left(A_0 + 2\sum_1^\infty A_n \cos n\eta\right) \tag{5-22}$$

The form of the function representing the pressure distribution in Eq. 5-21 is a particularly appropriate one both for steady and nonsteady flows and is also useful in the subsequent developments for the compressible fluid. It may be observed that although all coefficients a_n contribute to the pressure distribution, the total forces and moments integrated over the chord depend only on the first three coefficients a_0, a_1, and a_2. Another fact that can be made use of in some developments is the observation that, for a rigid wing section undergoing translation and rotation, only the coefficients A_0 and A_1 are not zero in Eq. 5-22. It should be remarked that the German literature utilizes a function $T(k)$ instead of $C(k)$ related to it by the equation,

$$\begin{aligned} T(k) &= 2C(k) - 1 \\ &= \frac{H_0^{(2)}(k) + iH_1^{(2)}(k)}{-H_0^{(2)}(k) + iH_1^{(2)}(k)} \end{aligned} \tag{5-23}$$

Another expression for the pressure difference is

$$\Pi(\theta, t) = \frac{2}{\pi}\rho_\infty U \int_0^\pi w(-\cos\eta, t)F(\theta, \eta, k)d\eta \tag{5-24}$$

where using the notation of Eq. 5-3

$$F(\theta, \eta, k) = -ik\mathfrak{L}(\theta, \eta) \sin \eta + \frac{\sin \theta}{\cos \theta - \cos \eta}$$
$$+ \cot \frac{\theta}{2} [1 - \cos \eta)C(k) + \cos \eta]$$

Supplementary discussion. To assist the reader in identifying aspects of existing work, a few key formulas made use of in the work of Birnbaum, Küssner, Cicala, Schwarz, and others will be presented. The particular formulas exhibited have been used, not only for the two-dimensional case, but have also served as a basis of treatment in the three-dimensional unsteady-flow lifting surface theory (in particular by W. P. Jones; see Art. 13). The method makes use of various quantities measured across the lifting surface and its wake termed "bound vorticity," "free vorticity," and "total vorticity," quantities of physical significance and directly related to the pressure and to the velocity potential by the formulas indicated in the following paragraph.

Let the square brackets [ ] be used to denote differences in the enclosed quantities measured across the upper and lower sides of the lifting surface (surface as defined in Eq. 3-2). Then by Eq. 3-7 the surface pressure difference is given as

$$\rho_\infty \frac{D[\varphi]}{Dt} = \rho_\infty \left[\frac{\partial \varphi}{\partial t}\right] + \rho_\infty U \left[\frac{\partial \varphi}{\partial x}\right]$$
$$= -[p] = -P \tag{5-25}$$

Let the lift distribution be written in the classical form

$$\rho_\infty U \gamma = -P \tag{5-26}$$

where $\gamma = \gamma(x, t)$ is the so-called "bound vorticity" distribution and which is different from zero only across the lifting surface. For nonsteady flow, γ should not be confused with an element of the bound circulation; a more proper term for γ is "reduced" pressure difference or loading. From Eq. 5-26, γ may be expressed as

$$\gamma = \left[\frac{\partial \varphi}{\partial t}\right] + \frac{1}{U}\left[\frac{\partial \varphi}{\partial x}\right] \tag{5-27}$$

Introduce the useful quantity defined by the relation

$$\gamma_{tot} = \gamma + \epsilon = \left[\frac{\partial \varphi}{\partial x}\right] \tag{5-28}$$

This quantity is termed the "total vorticity"; it gives the tangential streamwise velocity discontinuity over the lifting surface or its wake. In the wake, since γ is zero there, ϵ alone gives this discontinuity and hence it is termed the "free vorticity." From Eq. 5-27 and 5-28, ϵ is also directly given as proportional to the time rate of change of the difference in potential (or circulation)

$$\epsilon(x, t) = -\frac{1}{U}\left[\frac{\partial\varphi}{\partial t}\right] \tag{5-29}$$

Differentiation of this relation with respect to x and use of Eq. 5-28 yields the result that

$$\frac{\partial}{\partial t}(\gamma_{\text{tot}}) + U\frac{\partial\epsilon}{\partial x} = 0 \tag{5-30}$$

or that

$$\frac{\partial\gamma}{\partial t} + \frac{\partial\epsilon}{\partial t} + U\frac{\partial\epsilon}{\partial x} = 0 \tag{5-31}$$

This relation can be identified as expressing both the property of persistence of vorticity and the floating property of the free vortices. It may serve, by inverting it, for example, as was done in arriving at Eq. 2-11 from Eq. 2-10 to express ϵ in terms of $\partial\gamma/\partial t$. The free vorticity ϵ may thus be eliminated from the integral equation which gives γ in terms of the known normal induced velocity w. A form in which this integral equation appears is

$$\begin{aligned} w(x, t) &= \frac{1}{2\pi}\int_{-\infty}^{\infty}\frac{\gamma + \epsilon}{x - \xi}\,d\xi \\ &= \frac{1}{2\pi}\int_{-1}^{1}\frac{\gamma(\xi, t)}{x - \xi}\,d\xi + \frac{1}{2\pi}\int_{-1}^{\infty}\frac{\epsilon(\eta, t)}{x - \eta}\,d\eta \end{aligned} \tag{5-32}$$

The solution for γ is subject to the important subsidiary condition that $\gamma = 0$ or that $\gamma_{\text{tot}} = \epsilon$ at the trailing edge.

Birnbaum [*15*] used series expansions for γ in the harmonic case, utilizing terms in the frequency ω like $\omega^n(\ln\omega)^m$, but these series were rather poorly convergent. Küssner [*18*] and Cicala [*17*] improved this procedure by utilizing the expansion for w shown in Eq. 5-22, which leads to the recurrence type solution (5-21). Later, Küssner and Schwarz [*24*] used analytic continuation to convert the recurrence solution to the form shown in Eq. 5-24. Schwarz [*25*] also obtained this result through a consistent use of the total vorticity γ_{tot} and by use of a classical inversion due to M. Munk of the integral equation which appears in the case of steady flow. Söhngen [*26*] made use of transient flow results to arrive at oscillatory flow results. The topic of transient flows and the use of superposition methods is taken up briefly in the next article.

F,6. Transient Flows and Reciprocal Relations. Before presenting integrated results for air forces and moments obtained from the pressure relations, it is appropriate to indicate some generalizations of considerable theoretical and practical interest for transient, pulsating, or accelerated flows and for effects of penetration of gustlike velocity changes in the medium.

It is to be noticed that in results for the incompressible fluid, the noncirculatory terms (arising from $\Delta\varphi_1$) act *instantaneously*. There is therefore no need to limit these results to the purely sinusoidal distribution (Eq. 5-11) of normal velocity. However, the circulatory potential is in general a function of the wake, that is, of the past history of the motion. Thus, if other than the sinusoidal case is considered, other functions arise to replace the function $C(k)$. These changes affect *only* the function $C(k)$, the other terms in the circulatory potential (sometimes referred to as quasi-steady) can also be considered to act instantaneously. (It is at this place that Küssner [*27*] prefers to introduce the function $T(k)$ to be associated with the transient terms; $T(k)$ approaches zero for $k \to \infty$, but its use actually brings in no essential differences.)

It is proposed to indicate the changes in the form of the general pressure relation for the general time function $w(x_1, t)$ or preferably $w(x_1, s)$. As already remarked following Eq. 5-7, the unit-step time function is of significance here. In fact, just as there exist in linear electric circuit theory relations between the response to steady state alternating currents and the response to indicial transient currents, analogous relations exist in the flow problem here [*28*].

Consider first the sinusoidal variation and let the total value of dP/dx_1 in Eq. 5-19 be separated into the sum of an instantaneous part $(P')_I$ and a part associated with the wake $(P')_C$, involving $C(k)$, where $(P')_C$ is given by

$$(P')_C = nC(k)w^*(x_1)e^{i\omega t} \tag{6-1}$$

and where

$$n = n(x, x_1) = \frac{2\rho_\infty U}{\pi}\sqrt{\frac{1-x}{1+x}}\sqrt{\frac{1+x_1}{1-x_1}}$$

Since $(P')_I$ is obviously the part remaining in Eq. 5-19 when $(P')_C$ is extracted, its expression need not be rewritten here. The expression for $(P')_C$ represents the contribution, to the pressure difference associated with the noninstantaneous wake terms, of the harmonically oscillating normal velocity w at x_1. The function $n(x, x_1)$ is significant and shows at once that the center of pressure acts at the quarter chord, $x = -\frac{1}{2}$. Moreover, the normal velocity distribution is weighted by the function $n(x, x_1)$ whose centroid with respect to x_1 is at the three-quarter chord position, $x_1 = \frac{1}{2}$.

Wagner's function $k_1(s)$ *and its relation to* $C(k)$. Consider now the unit step variation

$$w(x_1, t) = w_0(x_1)H(t) \tag{6-2}$$

where

$$H(t) = 0 \quad \text{for } t < 0$$
$$H(t) = 1 \quad \text{for } t > 0$$

From the principle of superposition it results that the pressure response for this unit-step normal velocity distribution can be obtained from the sinusoidal case as follows

$$(P')_H = nk_1(s)w_0(x_1)H(s) \tag{6-3}$$

where $H(s)$ replaces the exponential function $e^{i\omega t} = e^{iks}$, and where the important function $k_1(s)$, referred to as Wagner's function, appears as the indicial response function replacing $C(k)$. The relation between $k_1(s)$ and $C(k)$ has been given by Küssner [*18*] as an inverse Laplace transform, and by Garrick [*28*] in the form of a Laplace transformation in the parameter ik or as a Fourier transform in k as

$$\frac{C(k) - 1}{ik} = \int_0^\infty [k_1(s) - 1]e^{-iks}ds \tag{6-4}$$

The function $k_1(s)$ is expressed in terms of $C(k)$ by the classical inversion of the integral. Writing $C(k) = F(k) + iG(k)$ the inversion leads to the relations given in [*28*] for $s > 0$

$$\left.\begin{aligned} k_1(s) &= 1 + \frac{2}{\pi}\int_0^\infty \frac{F(k) - 1}{k} \sin ks\, dk \\ &= 1 + \frac{2}{\pi}\int_0^\infty \frac{G(k)}{k} \cos ks\, dk \end{aligned}\right\} \tag{6-5}$$

In obtaining these results, use has been made of the fact that $k_1(s) = 0$ for $s < 0$ and (as follows ab initio from the alternative convention for harmonic motion employing $e^{-i\omega t}$ instead of $e^{i\omega t}$) that $C(-k) = F(k) - iG(k)$. From the point of view of harmonic analysis and synthesis, $C(k)$ plays the dominant role; from the point of view of the superposition principle and transient flow, $k_1(s)$ plays the dominant role.

Consider briefly one of the main cases treated by Wagner, the case of $w_0(x_1) = w_0 = \text{const}$, so that the entire airfoil suddenly acquires a uniform normal velocity. Then the pressure difference is

$$P = P_H + P_I \tag{6-6}$$

where

$$\begin{aligned} P_H &= \int_{-1}^{1} (P')_H dx_1 \\ &= \pi\rho_\infty U w_0 \sqrt{\frac{1 + x}{1 - x}}\, k_1(s) \\ P_I &= \int_{-1}^{1} (P')_I dx_1 \\ &= \pi\rho_\infty b w_0 \sqrt{1 - x^2}\, \delta(s) \end{aligned}$$

where the apparent-mass term is associated with the infinite acceleration $w_0\delta(s)$ acting only for $s = 0$, and where the delta function $\delta(s)$ represents $\partial H(s)/\partial s$. The total lift per unit span may be expressed as

$$L = b\int_{-1}^{1} P dx$$

$$= 2\pi\rho_\infty bUw_0\left[k_1(s) + \frac{b}{2}\,\delta(s)\right] \tag{6-7}$$

In Art. 12 the effect of Mach number on this result is indicated (Fig. F,12). The difference $1 - k_1(s)$ is sometimes called the lift deficiency function. Table F,6a gives the function $k_1(s)$ where s, it is recalled, is measured in half chords of travel. The final lift builds up rather quickly, reaching, for example, approximately 67 per cent of its final value in 1 chord length of travel and 93 per cent in 10 chord lengths of travel. Interesting properties of $k_1(s)$ and approximations useful in applications exist; see, for example, [*24,25,26,27*]. A detailed discussion of the integral equation leading to Wagner's function is given in particular by Söhngen [*26*]. It should be mentioned that Küssner writes $T(k)$ instead of $C(k)$, and $U_1(s)$ instead of $k_1(s)$, where $U_1(s) = 2k_1(s) - 1$.

Table F,6a

s	$k_1(s)$	s	$k_1(s)$
0.0	0.5000	8.5	0.8564
0.5	0.5557	9.0	0.8631
1.0	0.6006	9.5	0.8693
1.5	0.6378	10	0.8750
2.0	0.6693	11	0.8853
2.5	0.6962	12	0.8941
3.0	0.7195	15	0.9148
3.5	0.7399	20	0.9366
4.0	0.7579	25	0.9501
4.5	0.7739	30	0.9591
5.0	0.7882	40	0.9702
5.5	0.8010	50	0.9767
6.0	0.8125	100	0.9891
6.5	0.8230	500	0.9980
7.0	0.8325	1000	0.9990
7.5	0.8412	∞	1.0000
8.0	0.8491		

Superposition principle. The application of the principle of superposition for the general transient distribution of normal velocity leads to the contribution, to the pressure term dP/dx_1, of the noninstantaneous "wake" terms. One form (Duhamel's integral form) for this result is

$$(P')_w = n(x, x_1)\left[w(x_1, 0)k_1(s) + \int_0^s k_1(s - \sigma)\frac{\partial}{\partial\sigma}w(x_1, \sigma)d\sigma\right] \tag{6-8}$$

The transform relation given in Eq. 6-4 follows itself directly from the use of the superposition principle for the appropriate exponential form of w [28,27]. The result (Eq. 6-8), together with the instantaneous terms $(P')_i$ already referred to as the part remaining in Eq. 5-19 when the term containing the factor $C(k)$ is omitted, is the complete result for the density of pressure difference or loading dP/dx_1 for the arbitrary normal velocity function $w(x_1, s)$.

Gust function $k_2(s)$ and its relation to traveling wave result. Küssner has found it convenient for certain purposes, however, to separate the normal velocity function into two parts, one w_1 associated with the motion of the airfoil, the other w_2 associated with penetration of the airfoil into a nonuniform air stream or gust. (It is necessary to consider small disturbances and to neglect the vorticity associated with the nonuniformity of the air stream.) In developing the analysis for the complete pressure distribution, there appears another significant pair of functions now to be derived. One function $k_2(s)$ corresponds to an indicial response to a unit *traveling* wave $H[(U/b)t - x_1]$ or $H(s - x_1)$, that is to the penetration effect into a sharp normal gust; and the other function corresponds to the sinusoidal velocity wave $w = e^{i(\omega t - kx_1)}$ traveling down from the leading edge toward the trailing edge. Let us obtain the lift distribution corresponding to the harmonic traveling wave and at the same time illustrate the use of Eq. 5-22 and 5-21. Put $x_1 = -\cos\eta$ in the normal velocity

$$w = w_0 e^{i(\omega t - kx_1)}$$

so that

$$w = w_0 e^{i\omega t} e^{ik\cos\eta}$$

The Fourier expansion

$$e^{ik\cos\eta} = B_0 + 2\sum_1^\infty B_n \cos n\eta$$

yields the coefficients

$$B_n = i^n J_n(k)$$

Then from Eq. 5-22

$$A_n = \frac{w_0}{U} i^n J_n(k)$$

The lift distribution is given by Eq. 5-21 where, making use of a recurrence relation for Bessel functions, it turns out rather surprisingly that all the a's except a_0 are zero and that

$$a_0 = \{C(k)[J_0(k) - iJ_1(k)] + iJ_1(k)\}\frac{w_0}{U}$$

The total lift per unit span length is given in general by

$$L = 2\pi\rho b U^2 e^{i\omega t}(a_0 + a_1) \tag{6-9}$$

and in particular by

$$L = 2\pi\rho b U w_0 e^{i\omega t}\{C(k)[J_0(k) - iJ_1(k)] + J_1(k)\} \tag{6-10}$$

The moment about the quarter-chord per unit span is given in general by

$$M = \pi\rho b^2 U^2 e^{i\omega t}\left(a_1 - \frac{a_2}{2}\right) \tag{6-11}$$

and in the present example by

$$M = 0 \tag{6-12}$$

an important observation that holds in general for penetration into an arbitrary gust [*23,27*].

Consider now the lift after penetration into the normal sharp gust $w = w_0 H(s - x_1)$. Then

$$L = 2\pi\rho b U w_0 k_2(s) \tag{6-13}$$

The function $k_2(s)$ was originally introduced by Küssner [*18*] in 1936; however, because of a sign error corresponding to a reversal in the direction of the traveling wave the function was in error, a fact pointed out and corrected by von Kármán and Sears [*23*]. Their procedure for deriving $k_2(s)$ corresponded to treatment of a broken-line airfoil with a moving hinge point.

The pair of functions

$$k_2(s) \quad \text{and} \quad C(k)[J_0(k) - iJ_1(k)] + iJ_1(k)$$

are related by means of a Laplace transformation just as are $k_1(s)$ and $C(k)$ [*27,28*]. Moreover, the "gust" function $k_2(s)$ can be expressed in terms of $k_1(s)$ by a relation of the form

$$k_2(s) = k_3(s) + k_4(s) \tag{6-14}$$

where

$$k_3(s) = \frac{1}{\pi}\int_{-1}^{1} k_1(s - \sigma)\sqrt{\frac{1+\sigma}{1-\sigma}}\,d\sigma$$

and

$$\left.\begin{aligned} k_4(s) &= \frac{1}{\pi}\sqrt{1 - s^2} \qquad |s| < 1 \\ &= 0 \qquad\qquad\qquad s > 1 \end{aligned}\right\}$$

and where it is recalled the $k_1(s) = 0$ for $s < 0$. This result is given in [*26*]; $k_3(s)$ corresponds to the wake (including the quasi-steady) effect, $k_4(s)$ to the apparent mass effect. Table F,6b presents some numerical results for $k_2(s)$, where s is measured in half chords and corresponds to the distance of penetration of the midchord into the gust, so that $s = -1$ corresponds to penetration by the leading edge.

Küssner's separated form. Küssner's result for the case of general transient flow may be put down for the sake of reference. Let the normal velocity distribution be

$$w(-\cos\eta, s) = w_1 + w_2$$

Table F,6b

s	$k_2(s)$	s	$k_2(s)$
−1.0	0.0000	7.0	0.8225
−0.8	0.1980	8.0	0.8408
−0.6	0.2757	9.0	0.8561
−0.4	0.3324	10	0.8691
−0.2	0.3782	11	0.8801
0.0	0.4167	12	0.8897
0.2	0.4500	15	0.9117
0.4	0.4794	20	0.9349
0.6	0.5057	25	0.9490
0.8	0.5293	30	0.9583
1.0	0.5508	40	0.9698
2.0	0.6351	50	0.9765
3.0	0.6945	100	0.9890
4.0	0.7388	500	0.9979
5.0	0.7731	1000	0.9990
6.0	0.8004	∞	1.0000

where w_1 is due to motion of the airfoil and w_2 is due to the penetration into the disturbed atmosphere or medium. The pressure difference expressed in notation similar to that used in Eq. 5-3 and 5-24 is given by

$$\Pi(\theta, s) = \frac{2\rho_\infty U}{\pi} \int_0^\pi [F_1(\theta, \eta, s) + F_2(\theta, \eta, s) + F_3(\theta, \eta, s)]d\eta \qquad (6\text{-}15)$$

where

$$F_1 = \left[\cos\eta \cot\frac{\theta}{2} + \frac{\sin\theta}{\cos\theta - \cos\eta} + \frac{1}{2}\sin\eta \ln\frac{1 - \cos(\theta - \eta)}{1 - \cos(\theta + \eta)}\frac{\partial}{\partial s}\right] w_1$$

$$F_2 = (1 - \cos\eta)\cot\frac{\theta}{2}\int_0^s k_1(s - \sigma)\frac{\partial w_1}{\partial\sigma}\,d\sigma$$

$$F_3 = (1 - \cos\eta)\cot\frac{\theta}{2}\int_0^{s+1} k_2(s - \sigma)\frac{\partial w_2}{\partial\sigma}\,d\sigma$$

The functions $k_1(s)$ and $k_2(s)$ correspond to those defined herein earlier. (Küssner employs different functions, $U_1(s)$ and $U_2(s)$.) F_1 corresponds to the apparent mass effect and F_2 to the wake including the quasi-steady effect for the part w_1; F_3 corresponds to the entire effect of penetration into the arbitrary gust w_2.

Additional material on transient flows is given in Art. 12.

F,7. Forces and Moments for Simple Representative Airfoil Motions. For the linearized methods, integrated results of the pressure distribution are of more direct practical interest and of greater physical validity than local results. In practical application it has been found to be very useful, and almost indispensable, to have available integrations of the local pressures to yield resultant air forces and moments for a number of simple idealized representations of the airfoil in various types of motion, sometimes referred to as degrees of freedom. With the aid of properly selected degrees of freedom it is possible by superposition to represent fairly complicated motions or combined movements of the airfoil as a linear combination of the simpler results. With a view toward applications in stability and in flutter and vibration, where it becomes necessary to represent rigid body motions such as translation and pitch, various structural modes of deformation such as bending and torsion, and control surface movements, the six types of elementary representations given in the next paragraph have been standardized. It will be noticed that the standardizations have been based on applications involving rigid body behavior of each spanwise segment or strip, a sort of aerodynamic beam representation. For platelike deformations of the wing, or for configurations in which vibration modes higher than the lowest

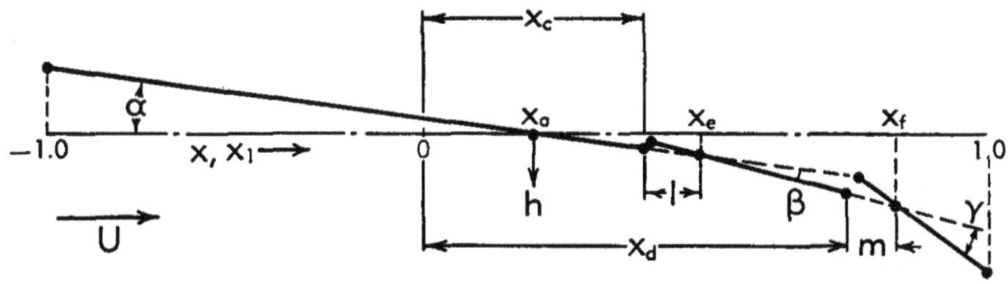

Fig. F,7. Parameters of airfoil-aileron-tab combination.

few modes, or in which chord or camber deformations play a larger role, it may become desirable to add other degrees of freedom, but these are not easily standardized. Moreover, as will be mentioned, the theoretical representations of the wing shape and motion especially for control surfaces can be chosen on a semiempirical basis with some account taken of both aerodynamic and structural experimental results.

Wing-aileron-tab system. In order to represent the harmonic oscillations of a wing-aileron-tab system including aerodynamic balance, Küssner and Schwarz [*24*] and Theodorsen and Garrick [*29*] make use of a system indicated in Fig. F,7. The figure represents a wing section with two hinges, an aileron (or elevator, or rudder) hinge at $x = x_e$ and a tab hinge at $x = x_f$. The leading edge of the wing section is again at $x = -1$ and the trailing edge at $x = 1$; the half chord $b = c/2$ is again to be used as the reference length. The leading edge of the aileron is at $x = x_c$ and the distance from the hinge to the aileron leading edge $x_e - x_c$ is denoted by l. The leading edge of the tab is at $x = x_d$ and the distance from the tab hinge to the tab leading edge $x_f - x_d$ is denoted by m. The wing is undergoing the following harmonic oscillatory motions with small amplitudes: a displacement h (velocity $\dot{h}$) in a vertical direction downward; a turning about the axis $x = x_a$, the instantaneous angle of attack being α; a rotation of the aileron about $x = x_e$, the angle of the aileron being β measured with respect to the wing (i.e. to α); and a rotation of the tab about $x = x_f$, the angle of the tab being γ measured with respect to the aileron.

The types of normal velocity distributions involved in this representation are as follows: for h and α, uniform and proportional to x over the entire chord, namely $w(x) = -[\dot{h} + U\alpha + (x - x_a)\dot{\alpha}]$; for β (and similarly for γ), zero for $x < x_c$, uniform and proportional to ξ over the aileron chord, namely $w(x) = -[U\beta + (x - x_c)\dot{\beta}]$. There is an additional term associated with the leading edge of the aileron and with the distance l, most conveniently treated as the step-shape limit $\Delta x_c \to 0$ of a distribution: $w = 0$ for $x < x_c$ and for $x > x_c + \Delta x_c$; $w = U\beta l/\Delta x_c$ for $x_c < x < x_c + \Delta x_c$ (see [24] and [29]). The velocity potential can be considered built up as a sum of terms, one associated with each term in w. Thus

$$\varphi = \sum \varphi_n \tag{7-1}$$

where subscript $n = \dot{h}, \alpha, \dot{\alpha}, \beta, \dot{\beta}, \gamma$, and $\dot{\gamma}$. In the relations for the pressure difference there also arise terms in $\ddot{h}$, $\ddot{\alpha}$, $\ddot{\beta}$, and $\ddot{\gamma}$.

Forces and moments. The total force positive downward, the total moment about $x = x_a$, and the aileron and tab hinge moments (positive moments tend to raise the leading edge and correspond to the clockwise direction in Fig. F,7), all taken per unit span length, are given by the expressions

$$\left.\begin{aligned} F &= b \int_{-1}^{1} P dx \\ M_\alpha &= b^2 \int_{-1}^{1} P(x - x_a) dx \\ M_\beta &= b^2 \int_{x_e}^{1} P(x - x_e) dx \\ M_\gamma &= b^2 \int_{x_d}^{1} P(x - x_f) dx \end{aligned}\right\} \tag{7-2}$$

where P is the pressure difference.

To indicate the pattern of the results, expressions will be given for the h and α degrees of freedom. For the fuller results it is necessary to refer to the literature [22,24,29,30] (see also, van de Vooren and Hofsommer, in Bibliography, Art. 17). The expressions which are valid for two-dimensional incompressible flow and which make use of Eq. 5-19, 5-20, and 7-2 are

$$\left.\begin{aligned} F &= -\rho_\infty b^2(\pi\ddot{h} + U\pi\dot{\alpha} - \pi b x_a \ddot{\alpha}) - 2\pi\rho_\infty UbC(k)Q \\ M_\alpha &= -\rho_\infty b^2[-\pi b x_a \ddot{h} + \pi(\tfrac{1}{2} - x_a)Ub\dot{\alpha} + \pi b^2(\tfrac{1}{8} + x_a^2)\ddot{\alpha}] \\ &\quad + 2\pi\rho_\infty Ub^2(\tfrac{1}{2} + x_a)C(k)Q \end{aligned}\right\} \tag{7-3}$$

where

$$Q = U\alpha + \dot{h} + b(\tfrac{1}{2} - x_a)\dot{\alpha}$$

One may recognize, for example, in these expressions the apparent-mass term per unit length associated with vertical acceleration of a plate $\pi\rho b^2$, the apparent moment of inertia term $\rho_\infty \pi b^4(\frac{1}{8} + x_a^2)$, the damping terms associated with the angular velocity $\dot{\alpha}$, the velocity term Q associated with $C(k)$ which here corresponds to the resultant velocity at the three-quarter chord position ($x_a = \frac{1}{2}$), and one may note the fact that the part of the moment involving $C(k)$ vanishes at $x_a = -\frac{1}{2}$, the quarter-chord position.

It is convenient to introduce section aerodynamic coefficients for lift and moment defined by

$$\left.\begin{aligned} -F &= \tfrac{1}{2}\rho_\infty U^2 c c_l \\ M_\alpha &= \tfrac{1}{2}\rho_\infty U^2 c^2 c_m \end{aligned}\right\} \qquad (7\text{-}4)$$

and to group the terms associated with vertical translation h and with rotation α in the form

$$\left.\begin{aligned} c_l &= (c_{l,h})\frac{h}{b} + (c_{l,\alpha})\alpha \\ c_m &= (c_{m,h})\frac{h}{b} + (c_{m,\alpha})\alpha \end{aligned}\right\} \qquad (7\text{-}4a)$$

For numerical purposes certain related coefficients have been introduced and used

$$\left.\begin{aligned} A_{11} &= \frac{1}{\pi}c_{l,h} \\ A_{12} &= \frac{1}{\pi}c_{l,\alpha} \\ A_{21} &= -\frac{2}{\pi}c_{m,h} \\ A_{22} &= -\frac{2}{\pi}c_{m,\alpha} \end{aligned}\right\} \qquad (7\text{-}5)$$

These coefficients are complex and may be separated into real and imaginary parts as follows:

$$\left.\begin{aligned} A_{11} &= R_{11} + iI_{11} = \frac{1}{\pi}[(c_{l,h})_r + i(c_{l,h})_i] \\ A_{12} &= R_{12} + iI_{12} = \frac{1}{\pi}[(c_{l,\alpha})_r + i(c_{l,\alpha})_i] \\ A_{21} &= R_{21} + iI_{21} = -\frac{2}{\pi}[(c_{m,h})_r + i(c_{m,h})_i] \\ A_{22} &= R_{22} + iI_{22} = -\frac{2}{\pi}[(c_{m,\alpha})_r + i(c_{m,\alpha})_i] \end{aligned}\right\} \qquad (7\text{-}6)$$

For the purpose of systematic applications the aerodynamic coefficients (the A's) can be conveniently represented in the form of a matrix, the rank of which is equal to the number of degrees of freedom considered. Each term A_{ij} can be separated as mentioned into inertia, damping, and restoring elements, or as here indicated, the aerodynamic inertia and restoring terms can be grouped into the real parts R and the damping terms into the imaginary parts I. Each complex coefficient (for example A_{22} in the expression $A_{22}\alpha$) thus represents terms in phase and 90° out of phase with the respective degree of freedom, and hence can be considered as a vector drawn from an origin rotating with frequency ω in a fixed relation to the vector representing the degree of freedom. Of course, the different degrees of freedom need not be in phase with each other.

Various notations. Some numerical results. The various coefficients are often referred to as aerodynamic derivatives, especially in the British literature. For convenience of reference to the general literature comparison of various notations is given in Table F,7a. It is especially to be noted that the moment is taken variously about spanwise axes through the quarter chord, midchord, or leading edge positions so that a conversion formula of the type $M_a = M_0 + nbF$ exists. Values for the lift and (midchord) moment coefficients for two-dimensional incompressible flow for a representative list of reduced frequencies k are given in Table F,7b. Similar tables for Mach numbers of 0.7, 1.0, 1.25, and 2.0 are given in Art. 9, 10, and 11.

With the inclusion of aileron- and tab-control surface degrees of freedom, numerous parameters occur in the aerodynamic results, functions of x_c, x_d, and k. A complete set of expressions for these is given in [*24*] and in [*29*]. (For example, [*24*] employs $-\cos\varphi$ for x_c and $-\cos\psi$ for x_d, and includes many useful numerical tables. One correction to [*24*] is noted here: the numbers listed in the table for φ_{18} should be changed in sign.)

In the theoretical treatment of the control surface with aerodynamic balance two cases have been considered, the open gap and the closed gap. The open gap has been treated in [*24*] with plausible yet somewhat arbitrary arguments in which certain terms are regarded as ineffective in altering the flow pattern. Thus the contribution of the step shape representing the foremost part of the control surface (denoted by φ_H in [*29*]) is simply dropped or not considered in the velocity potential. The closed gap idealization by means of a vertical step with its sharp edge in the air stream leads to a term having a logarithmic infinity as discussed in the references, but this term can be circumvented by not proceeding to this limit (90°), or by introducing empirical factors.

Remarks on simplified representations. The whole problem of what simplified substitutions to employ for the velocity distribution normal to the wing surface in order to properly represent at the same time the actual physical behavior of the airfoil or its control surface is a most

Table F,7a. Comparison of notation.

References	Present section	American		
		[29]	[31]	[66]
Chord	$2b$ or c	$2b$	$2b$	$2b$
Reduced frequency	k	k	k	k
Lift	$-F$	$-P$	$-L$	$-P$
Pitching moment	M_α	M_α	M	M_α
Reference axis	x_a	$x_a = a$	$x_a = -\frac{1}{2}$	$x_0 = \frac{1 + x_a}{2}$
Translation at reference axis	h	h	$[h]_{c/4}$	h
Torsional motion	α	α	$[\alpha]_{c/4}$	α
Force coefficients or derivatives	R_{11}	k^2R_{ch}	$Re\{-k^2L_h\}$	$\frac{4}{\pi}k^2L_1$
	I_{11}	k^2I_{ch}	$Im\{-k^2L_h\}$	$\frac{4}{\pi}k^2L_2$
	R_{12}	$k^2R_{c\alpha}$	$Re\{-k^2L_\alpha\}$	$\frac{4}{\pi}k^2L_3$
	I_{12}	$k^2I_{c\alpha}$	$Im\{-k^2L_\alpha\}$	$\frac{4}{\pi}k^2L_4$
Moment coefficients or derivatives	R_{21}	k^2R_{ah}	$Re\{-k^2M_h\}$	$\frac{4}{\pi}k^2M_1$
	I_{21}	k^2I_{ah}	$Im\{-k^2M_h\}$	$\frac{4}{\pi}k^2M_2$
	R_{22}	$k^2R_{a\alpha}$	$Re\{-k^2M_\alpha\}$	$\frac{4}{\pi}k^2M_3$
	I_{22}	$k^2I_{a\alpha}$	$Im\{-k^2M_\alpha\}$	$\frac{4}{\pi}k^2M_4$

British			German		Dutch
[30]	[32]	[33]	[24]	[40]	[61,62]
c	c	c	$2l$	l	$2l$
$\frac{\omega}{2}$	$\frac{\lambda}{2}$	$\frac{\omega}{2}$	$-i\omega$	ω_r	ω
$-Z$	$-Z$	$-Z$	K	$-\Delta P$	$-K$
M	M	Θ	$-M_0$	ΔM	M
$x_a = 0$	$x_a = 0$	$x_a = -1$	$x_a = -\frac{1}{2}$	$x_a = -\frac{1}{2}$	$x_a = 0$
z	z	z	$Ale^{i\nu t}$	δ	$Ale^{i\nu t}$
θ	θ	θ	$Be^{i\nu t}$	β	$Be^{i\nu t}$
Z_1	Z_1	$\frac{1}{\pi}(l_z - \omega^2 l_{\ddot{z}})$	$Re\{k_a\}$	$-(\omega_r^2 - k'_{SS})$	$-k'_a$
Z_2	Z_2	$\frac{\omega}{\pi} l_{\dot{z}}$	$Im\{k_a\}$	k''_S	$-k''_a$
$2Z_3$	$2Z_3$	$\frac{2}{\pi}(l_\theta - \omega^2 l_{\ddot{\theta}})$	$Re\{k_b\}$	$-\left(\frac{\omega_r^2}{2} - k'_{SD}\right)$	$-k'_b$
$2Z_4$	$2Z_4$	$\frac{2\omega}{\pi} l_{\dot{\theta}}$	$Im\{k_b\}$	k''_{SD}	$-k''_b$
$2M_1$	$2M_1$	$\frac{2}{\pi}(m_z - \omega^2 m_{\ddot{z}})$	$Re\{m_a\}$	$-\left(\frac{\omega_r^2}{2} - k'_{DS}\right)$	$-m'_a$
$2M_2$	$2M_2$	$\frac{2\omega}{\pi} m_{\dot{z}}$	$Im\{m_a\}$	k''_{DS}	$-m''_a$
$4M_3$	$4M_3$	$\frac{4}{\pi}(m_\theta - \omega^2 m_{\ddot{\theta}})$	$Re\{m_b\}$	$-\left(\frac{3\omega_r^2}{8} - k'_{DD}\right)$	$-m'_b$
$4M_4$	$4M_4$	$\frac{4\omega}{\pi} m_{\dot{\theta}}$	$Im\{m_b\}$	k''_{DD}	$-m''_b$

Table F,7b. Refer to Eq. 7-4, 7-5, and 7-6.

Mach number $M_\infty = 0$								
k	R_{11}	I_{11}	R_{12}	I_{12}	R_{21}	I_{21}	R_{22}	I_{22}
0	0	0	2.000	0	0	0	−1.0	0
0.05	0.0106	0.0909	1.825	−0.1658	−0.0065	−0.0454	−0.9126	0.1329
0.10	0.0245	0.1664	1.681	−0.1614	−0.0172	−0.0832	−0.8418	0.1807
0.15	0.0334	0.2318	1.574	−0.1070	−0.0280	−0.1159	−0.7896	0.2035
0.20	0.0354	0.2910	1.493	−0.03173	−0.0377	−0.1455	−0.7514	0.2159
0.30	0.0176	0.3990	1.384	+0.1409	−0.0538	−0.1995	−0.7031	0.2296
0.40	−0.0280	0.5000	1.316	0.3200	−0.0660	−0.2500	−0.6780	0.2400
0.50	−0.0993	0.5979	1.271	0.4976	−0.0753	−0.2990	−0.6669	0.2512
0.60	−0.1947	0.6946	1.240	0.6717	−0.0827	−0.3473	−0.6651	0.2641
1.00	−0.7994	1.0789	1.179	1.3389	−0.1003	−0.5394	−0.7146	0.3306

significant one. There are many reasons for supposing that the theoretical results, especially for control surfaces, may need severe modifications. This may be partly due to physical reasons as, for example, the large influence of the boundary layer on hinge moments, and partly to the simplified mathematical representation of the effective skeleton or shape. Schwarz [*34*] has proposed that, in addition to rectilinear segments, parabolic arc segments can be employed to represent the effective camber lines. Also, it has been proposed in many places that empirical or semi-empirical factors be introduced into the theoretical framework to modify various coefficients. For example, based on limited experimental information, the inertia, damping, and elastic aerodynamic terms would in certain cases each be multiplied by different factors less than unity. In other cases correction factors obtained from static measurements have been proposed for effects of thickness or for hinge moments. Such devices, though useful, are not very satisfactory in general because of the quite different behavior of the boundary layer and the flow pattern for low and high oscillation frequencies. Another important field in which the dynamics of the boundary layer and of flow separation is important, and which requires development of a healthy experimental background, is the effect of varying mean angle of attack both for unstalled and stalled conditions. Voigt (see Bibliography, Art. 17) has checked some experimental aileron flutter values by employing a reduced effective length of the aileron chord in the theoretical calculations. Indeed, in the applications of unsteady aerodynamics it often appears desirable in the absence of better procedures to maintain an attitude of mathematical experimentation to cover the likely trends of the physical situation (see also Art. 15).

F,8. Propulsion and Drag. For the sake of completeness and because of its intrinsic interest, some results on the horizontal forces are briefly given, considered of course within the framework of the theory and not including viscosity. Two approaches exist: (1) integration of local pressures to yield the drag or propulsion forces, and (2) consideration of the balance of energy in which the work required to maintain the oscillations appears in the work of propulsion or drag and as kinetic energy in the wake. Pertinent references are Birnbaum [*15*], Burgers [*21*], Garrick [*35*], and Küssner [*18,27*].

Integration of the pressure difference given in Eq. 5-20 over the surface yields for the force (per unit span length) in the flight direction

$$D_0 = b \int_{-1}^{1} P \frac{dy}{dx}\, dx \tag{8-1}$$

where dy/dx is the local slope of the representative wing surface, designated as $\partial Z/\partial x$ in Eq. 3-5. This result is incomplete, however, for as is well known for the steady case, an additional "suction" force needs to be added, representing the contribution of the sharp leading edge considered as a limit of a rounded edge. This contribution follows from momentum considerations or also readily from Blasius' theorem, and is expressed in terms of the coefficient of the factor $\sqrt{(1-x)/(1+x)}$ in the expression for P. Thus, if this expression in P is denoted as $2\rho_\infty U^2 a_0 \sqrt{(1-x)/(1+x)}$, the suction force per unit span length is

$$S = 2\pi\rho_\infty b a_0^2 \tag{8-2}$$

where the coefficient a_0 is also that given in Eq. 5-21 but, in view of the square, a_0 should here be interpreted as a real rather than a complex quantity to prevent ambiguity. The complete expression for the drag $D_0 - S$ is then

$$D = b \int_{-1}^{1} P \frac{dy}{dx}\, dx - 2\pi\rho_\infty b a_0^2 \tag{8-3}$$

For the case of a flat plate at constant small angle of attack α, suddenly started from rest to velocity U, a result is, for example, that the lift is given in terms of Wagner's function as (with $U\alpha = w_0$; see Eq. 6-7 for $s > 0$):

$$F = 2\pi\rho_\infty U w_0 b k_1(s) \tag{8-4}$$

and the drag as

$$D = 2\pi\rho_\infty w_0^2 b[k_1(s) - k_1^2(s)] \tag{8-5}$$

As additional examples, some results given in [35] for the harmonically oscillating flat plate in flapping and in pitching may be cited. For flapping motion given by $h = h_0 e^{i\omega t}$, the result for the horizontal force averaged over an oscillation cycle, is

$$\overline{D} = -\pi\rho_\infty b\omega^2 h_0^2[F^2(k) + G^2(k)] \tag{8-6}$$

where $F^2 + G^2$ represents the value of $|C(k)|^2$ and the result signifies propulsion not drag. Note that ωh_0 is the maximum vertical velocity. For

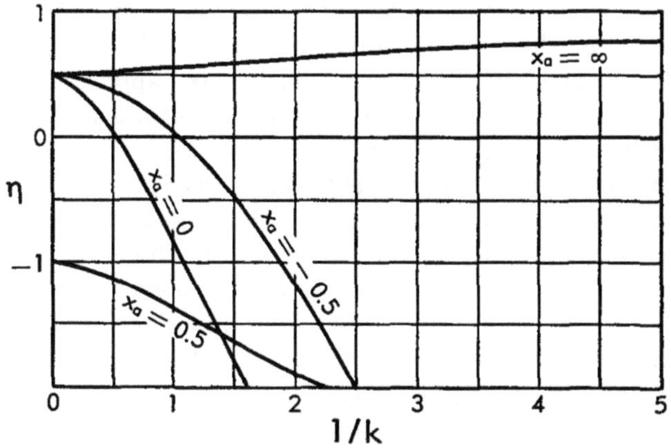

Fig. F,8. Propulsive efficiency of airfoil oscillating about axis $x = x_a$.

the case of pitching motion about a spanwise axis $x = x_a$, given by $\alpha = \alpha_0 e^{i\omega t}$, there is obtained for the average horizontal force

$$\overline{D} = -\pi\rho_\infty b^3\omega^2\alpha_0^2 \left\{ [F^2(k) + G^2(k)]\left[\frac{1}{k^2} + \left(\frac{1}{2} - x_a\right)^2\right] + \left[\frac{1}{2} - F(k)\right]\left(\frac{1}{2} - x_a\right) - \frac{F(k)}{k^2} - \left(\frac{1}{2} + x_a\right)\frac{G(k)}{k}\right\} \tag{8-7}$$

The results have fuller significance if accompanied by expressions for the work required to maintain the oscillations against the acting aerodynamic forces and moments. Corresponding respectively to Eq. 8-6 and 8-7, the following expressions for the average work per unit time are obtained:

$$\left.\begin{aligned} \overline{W} &= \pi\rho_\infty b^2\omega^3\frac{1}{k}F(k)h_0^2 = \pi\rho b^2\omega^2 UF(k)h_0^2 \\ \text{and} \\ \overline{W} &= \pi\rho_\infty b^3\omega^2 U\alpha_0^2\left\{\frac{1}{2}\left(\frac{1}{2} - x_a\right) - \left(x_a + \frac{1}{2}\right)\left[F(k)\left(\frac{1}{2} - x_a\right) + \frac{G(k)}{k}\right]\right\} \end{aligned}\right\} \tag{8-8}$$

The ratio of the average work of propulsion to the average work required yields an efficiency factor, for example, for flapping,

$$\eta_h = -\frac{\overline{D}U}{\overline{W}} = \frac{F^2(k) + G^2(k)}{F(k)}$$

Shown in Fig. F,8 are some curves of the efficiency factor η for several locations of the axis of pitching; η_h corresponds to η for $x_a = \infty$. For the ornithopter in pure flapping the theoretical efficiencies range from 50 per cent for rapid, to 100 per cent for slow, flapping. As another illustration with reference to the historic problem of bird flight, the pigeon yields values of k about 0.07 which corresponds in pure flapping to an efficiency of about 90 per cent. A large bibliography on natural flight and related aeronautics has been prepared by Fitzpatrick [*36*].

An observation of significance is that for some positions of the axis x_a and for some values of k (small), the work W required to maintain the pitching motion may become zero or negative. This corresponds to loss of damping and to the possibility of extracting energy from the air stream in the single degree of freedom. The work function for more than one degree of freedom also yields information on conditions for possible aerodynamic instability (see e.g. [*37*,*38*]).

F,9. Subsonic Oscillating Flow in Two Dimensions. The original integral equation method of Possio [*39*] for nonsteady plane subsonic flow forms the basis of numerous studies and is developed in the first part of this article. Extension of the method to finite oscillating wings is considered in Art. 13. Another treatment of subsonic oscillating flow, based on classical methods of separation of variables, is discussed in the latter part of this article.

Acceleration potential method. The differential equation satisfied by the acceleration potential or by the pressure p is given in Eq. 2-5b, where the z variable is to be omitted and y takes the role of z in the boundary conditions. The normal velocity may be written by means of Eq. 3-5, 3-6, and 2-11, as

$$w(x, 0, t) = -\frac{1}{\rho_\infty U}\int_{-\infty} \frac{\partial}{\partial y}\, p\left(x', y, t - \frac{x - x'}{U}\right) dx'$$

or, from Eq. 2-13 for harmonic oscillations $w = w^*(x)e^{i\omega t}$, as

$$w^*(x) = -\frac{1}{\rho_\infty U}\, e^{-i\omega x/U}\int_{-\infty}^{x} \frac{\partial}{\partial y}\, p^*(x', y)e^{i\omega x'/U}dx' \tag{9-1}$$

where y is to be made to approach zero in the integrals.

Satisfying the conditions of a given distribution of w requires a proper distribution over the chord of fundamental solutions involving pressure jump across the wing surface. This leads to an integral equation involving the known function w^*, an unknown distribution function, and a basic kernel function. The form of the basic kernel function, to which attention is first drawn, is determined directly from Eq. 9-1 and the fundamental doublet solution, which is obtained from the oscillating moving source solution given in Eq. 4-27 by differentiation, in this application, with respect to the position variable η. Thus, consider the doublet solution

$$\frac{-p^*(x', y)}{\rho_\infty} = -\frac{i}{4\beta}\frac{\partial}{\partial y} e^{-i\mu(x'-\xi)} H_0^{(2)}(\kappa R')e^{i\omega t} \tag{9-2}$$

where

$$\kappa = \frac{\omega M_\infty}{U\beta^2}$$

$$\mu = \frac{\omega}{U}\frac{M_\infty^2}{\beta^2}$$

$$R' = \sqrt{(x'-\xi)^2 + \beta^2(y-\eta)^2}$$

$$\beta = |1 - M_\infty^2|^{\frac{1}{2}}$$

and where use has been made of the relation $\partial/\partial\eta = -\partial/\partial y$. The doublet expression (9-2) has been normalized (see Art. 4) to represent a total jump in the pressure potential across the surface (at the point ξ, η) which is harmonic and of magnitude unity; that is, the total pressure difference by Eq. 3-7 is $P = -\rho_\infty e^{i\omega t}$.

Substitution of Eq. 9-2 into Eq. 9-1 gives a particular distribution of normal velocity w, associated with the normalized doublet and which will be used to define the basic kernel function. Thus the kernel function has the property of giving the normal velocity at any field point induced by unit harmonic loading at the loading point. The substitution yields

$$K_1(x, \xi; M_\infty, \omega) = \lim_{y\to 0} \frac{-ie^{-i(x/U)-i\mu\xi}}{4\beta U} \int_{-\infty}^{x} e^{i\omega x'/U\beta^2} \frac{\partial^2}{\partial y^2} H_0^{(2)}(\kappa R')dx' \tag{9-3}$$

The result (9-3) is a formal one in that it is in a conveniently simple form but one that is divergent for the limit $y \to 0$. Its actual interpretation, however, is readily achieved by replacing the operation $\partial^2/\partial y^2$ by its equivalent operation (obtained from the appropriate form of the wave equation satisfied by $H_0^{(2)}(\mu R)$) namely

$$\frac{\partial^2}{\partial y^2} = -\left[(1 - M_\infty^2)\frac{\partial^2}{\partial x^2} + \left(\frac{\omega}{U}\right)^2 \frac{M_\infty^2}{1 - M_\infty^2}\right]$$

Integration by parts twice, permitting η and y to approach 0, and making use of a known integral [9]

$$\int_{-\infty}^{0} e^{i\lambda} H_0^{(2)}(M_\infty|\lambda|)d\lambda = \frac{2}{\pi\beta}\ln\frac{1+\beta}{M_\infty}$$

leads to the following result for K_1:

$$K_1 = \frac{\omega}{U^2} K\left(\frac{\omega}{U}(x - \xi), M_\infty\right) \tag{9-4}$$

where K is given by

$$K\left(\frac{\omega}{U}x, M_\infty\right) = -\frac{1}{4\beta} e^{i\mu x}\left\{H_0^{(2)}(\kappa|x|) - iM_\infty \frac{x}{|x|} H_1^{(2)}(\kappa|x|) - i\beta^2 e^{-i\omega x/U\beta^2}\left(\frac{2}{\pi\beta}\ln\frac{1+\beta}{M_\infty} + \int_0^{\frac{\omega x}{U\beta^2}} e^{i\lambda} H_0^{(2)}(M_\infty|\lambda|)d\lambda\right)\right\}$$

and where each quantity enclosed by absolute value marks has the meaning of a radius. The result (9-4) represents the induced normal velocity at position x, 0 due to unit loading at ξ, 0 and depends on the difference $x - \xi$.

Integral equation of Possio's method. Introduce again the line airfoil represented by its distribution of normal velocity extending from the leading edge $x = -1$ to the trailing edge $x = 1$, the actual chord being $2b$. Hence the half chord b is the unit of length, and x and ξ in the following are nondimensional. Let the normal velocity distribution again be harmonic of the form $w^*(x)e^{i\omega t}$. Then the integral equation relating the known function $w^*(x)$, the normalized influence function K_1, and the unknown loading function $l(\xi)$ (where $l(\xi)$ represents the local intensity of the surface jump in ψ or $-P/\rho_\infty$) is

$$w^*(x) = b\int_{-1}^{1} K_1 l(\xi)d\xi$$

Expressed in another form convenient for application, the integral equation is

$$w^*(x) = k\int_{-1}^{1} \gamma^*(\xi)K(s, M_\infty)d\xi \tag{9-5}$$

where the intensity of loading, to be determined, is now given by $\gamma^*(\xi)$, defined as in Eq. 5-26 by means of the relation

$$P = -\rho_\infty U\gamma^*(\xi)e^{i\omega t} \tag{9-6}$$

and where the reduced frequency k is as before,

$$k = \frac{\omega b}{U}$$

and

$$s = k(x - \xi)$$

The kernel function $K(s, M_\infty)$ is as given in Eq. 9-4, but because x and ξ are nondimensional κ and μ are now to be interpreted as

$$\kappa = \frac{\omega b}{U}\frac{M_\infty}{\beta^2} = \frac{kM_\infty}{\beta^2}$$

$$\mu = \frac{\omega b}{U}\frac{M_\infty^2}{\beta^2} = \frac{kM_\infty^2}{\beta^2}$$

To satisfy the trailing edge condition it is necessary that $\gamma^*(1) = 0$. A convenient type distribution satisfying this requirement is the form given for the incompressible case in Eq. 5-21:

$$\gamma^*(x) = U\left[2a_0 \cot\frac{\theta}{2} + 4\sum_{1}^{\infty} a_n \frac{\sin n\theta}{n}\right] \tag{9-7}$$

where $x = -\cos\theta$ and where the a's are complex coefficients to be determined.

Before proceeding with various methods of determining the coefficients it is of interest and useful to present the kernel function $K(s, 0)$ for the incompressible flow case. It follows directly from the parallel use of the doublet potential for $M_\infty = 0$, Eq. 4-7b or, as a limiting result from Eq. 9-4, that

$$K(s, 0) = \frac{1}{2\pi s} - \frac{ie^{-is}}{2\pi}\left[\mathrm{Ci}\,(s) + i\left(\mathrm{Si}\,(s) + \frac{\pi}{2}\right)\right] \tag{9-8}$$

where

$$s = \frac{\omega b}{U}(x - \xi) = k(x - \xi)$$

and where, (see e.g. Jahnke and Emde [*222*]):

$$\mathrm{Ci}\,(s) = -\int_{-s}^{\infty} \frac{\cos u}{u}\,du = \ln e^{\gamma} s - \int_0^s \left(\frac{1 - \cos u}{u}\right) du$$

$$\mathrm{Si}\,(s) = \int_0^s \frac{\sin u}{u}\,du$$

Moreover, the limit of $kK(s, M_\infty)$ as $k \to 0$ yields as the kernel for steady flow

$$kK \to \frac{\sqrt{1 - M_\infty^2}}{2\pi(x - \xi)} \tag{9-8a}$$

Method of collocation. Whereas for the incompressible fluid case there exists the explicit solution for the coefficients a_n, discussed in Art. 5 and given in Eq. 5-21 and 5-22, for the compressible fluid case approximate numerical methods have had to be devised. These methods can now be partly circumvented by the use of tables of Mathieu functions (see end of Art. 9); nevertheless, they are of current interest since they are needed in certain extensions to finite oscillating wings. A direct, though numerically laborious, method of solving for the coefficients a_n in Eq. 9-7 is the method of collocation used originally by Possio and given in an improved form in some British papers, notably by R. A. Frazer. It has been applied mainly to the wing without aileron, undergoing vertical translation (h) and pitching motion (α). As in Art. 7, let

$$w(x, t) = e^{i\omega t}w^*(x) = -[\dot{h} + U\alpha + b(x - x_a)\dot{\alpha}] \tag{9-9}$$

in the form given in Eq. 5-22

$$w^* = U\left[A_0 + 2\sum_1^\infty A_n \cos n\theta\right]$$

Comparison with Eq. 9-9 with x replaced by $-\cos\theta$ shows at once that all coefficients A_n for $n \geqq 2$ are zero for the particular motion assumed, a fact that simplifies some of the numerical work.

The collocation method consists in setting up and solving a system of n linear equations for the first n complex coefficients $a_0 \cdots a_{n-1}$ so that the Possio integral equation (9-5) is effectively satisfied at n points of the chord $\theta_1 \cdots \theta_n$. The coefficients of the a's in this system of equations are given by definite integrals, involving the product of the known kernel K and one of the chosen aerodynamic mode shapes appearing in the relation (9-7). These integrals are obtained by numerical integrations and are functions of frequency, Mach number, and of the distance between downwash and loading points.

As mentioned in Art. 5, the total force and moment can be obtained from the first three coefficients a_0, a_1, and a_2 in Eq. 9-7. Moreover, the procedure given is one which is exact for the distribution (9-9) in the incompressible nonstationary case for boundary conditions satisfied at chosen points n equal to only 3. However, in the compressible case convenient means have not existed until recently for determining the numerical accuracy other than by repetition of the procedure with more points.

Kernel function. All numerical methods for treating Eq. 9-5 are based on some tabulations of the kernel function $K(s, M_\infty)$. Numerical values have been given in [*40,41,42*]. In the applications and tabulation it is convenient to separate singular parts of K analytically and to tabulate the remainder. In this way the sensitive parts of the numerical integrations may be exactly performed. Thus

$$K(s, M_\infty) = \frac{F(M_\infty)}{s} + iG(M_\infty) \ln s + K_r(s, M_\infty) \tag{9-10}$$

where

$$F(M_\infty) = \frac{\sqrt{1 - M_\infty^2}}{2\pi}$$

$$G(M_\infty) = \frac{1}{2\pi\sqrt{1 - M_\infty^2}}$$

and where the remainder $K_r(s, M_\infty)$ can be developed in a series in s

beginning with the constant

$$K_r(0, M_\infty) = -\frac{1}{4\sqrt{1 - M_\infty^2}} + i\left[-\frac{M_\infty^2}{2\pi\sqrt{1 - M_\infty^2}} + \frac{1}{2\pi\sqrt{1 - M_\infty^2}} \ln \frac{e^\gamma M_\infty}{2(1 - M_\infty^2)} + \frac{1}{2\pi} \ln \frac{1 + \sqrt{1 - M_\infty^2}}{M_\infty}\right]$$

where $\gamma = 0.5772 \cdots$ is Euler's constant.

Dietze's method. An iterative procedure for Eq. 9-5 has been developed by Dietze [*40*] which has some advantages over the methods of collocation. In particular, applications to include effects of aileron motion have been performed by this method. It makes essential use of the fact that the incompressible case can be treated exactly. For a given normal velocity function $w(x, t)$ the value of γ^* (or P) for the incompressible flow may be represented by subscript zero as γ_0^* (or P_0). As mentioned, this result can be obtained exactly with the aid of expansion of the pressure distribution as in Eq. 5-21 and of the normal velocity distribution as in Eq. 5-22. The solution thus obtained corresponds to solution of Eq. 9-5 with the kernel function $K(s, M_\infty)$ replaced by $K(s, 0)$. Then if the required distribution γ^* is written as

$$\gamma^* = \gamma_0^* + \Delta\gamma_0^*$$

Eq. 9-5 becomes

$$w^*(x) = k \int_{-1}^{1} [\gamma_0^*(\xi) + \Delta\gamma_0^*(\xi)]K(s, M_\infty)d\xi$$

By the definition of γ_0^*

$$w^*(x) = k \int_{-1}^{1} \gamma_0^*(\xi)K(s, 0)d\xi$$

so that it follows that

$$k \int_{-1}^{1} \Delta\gamma_0^*(\xi)K(s, M_\infty)d\xi = k \int \gamma_0^*(\xi)[K(s, 0) - K(s, M_\infty)]d\xi = w_1^*(x) \tag{9-11}$$

where $w_1^*(\xi)$ is now a *known* function being defined by γ_0^* and by the difference in the kernel functions which can be tabulated. But the additional distribution function $\Delta\gamma_0^*$ is now defined by the relation (9-11) of exactly the same form as the original equation (9-5). Hence the process can be repeated with $\Delta\gamma_0^* = \gamma_1^* + \Delta\gamma_1^*$ where γ_1^* is obtained from the solution of the incompressible case relations with normal velocity given by $w_1^*(\xi)$. The iterative process thus defined leads to the result:

$$\gamma^* = \gamma_0^* + \gamma_1^* + \gamma_2^* \cdots + \gamma_n^* + \Delta\gamma_n^* \tag{9-12}$$

where the remainder term satisfies an equation of the original type but the other terms can be obtained as exactly as desired from the simpler equations utilizing $K(s, 0)$. Theoretical justification of the convergence of the process has not been given; judged purely by numerical results, it appears to converge but the number of iterations necessary may become large and consequently difficult for values of M_∞ near 1. In the steady case the convergence of γ^* to its value $\gamma_0^*/\sqrt{1 - M_\infty^2}$ is reached in the manner of the sequence

$$\gamma_0^*(1 + \mu + \mu^2 + \cdots)$$

where

$$\mu = 1 - \sqrt{1 - M_\infty^2}$$

The iterative process can be readily systematized and reduced to routine operations. Dietze [*40*] has presented numerical results for force and moment coefficients for a wing undergoing vertical translation and pitching and has included aileron coefficients for one value (0.15) of the aileron-chord ratio. Additional results for $M_\infty = 0.7$ for aileron chord ratios 0.24, 0.33, and 0.42 are presented in [*42*].

Other work. Brief mention may be made of a few other numerical procedures for Possio's integral equation. Schade and Eichler [*43,44*] formulate the problem as a system of n simultaneous equations, and approximate the nonsingular part of the kernel function by an expansion in Legendre polynomials. In a method avoiding iteration, Fettis [*45*] has approximated the kernel function in an appropriate form to yield, in a manner similar to the incompressible flow case, exact solutions of the resulting integral equation. Numerical values obtained by Fettis include control surface derivatives and are given in [*46*].

Classical separation of variables approach for the boundary-value problem. Although the methods described thus far in this article have yielded useful numerical results for subsonic oscillating flows, the approximate or iterative character of the procedures has been a challenge for a more satisfying mathematical development. It is therefore appropriate to consider another approach, a more direct attack on the boundary-value problem which, although classical, has come later. It involves the use of infinite series of product solutions in the separated variables appropriate to the differential equation and the boundary lifting surface. In the present case of the line airfoil the appropriate variables are elliptic coordinates and the appropriate functions are Mathieu functions. The method has been considered and treated by various authors, e.g. Reissner and Sherman [*47*], Timman [*48*], Haskind [*49*], Reissner [*50*], Billington [*51*], and Timman and Van de Vooren [*52*]. Numerical results have stemmed mainly from [*52*].

A recent treatment by Küssner [*53*] is of interest in that his result takes the appearance of an immediate generalization of an explicit result available for incompressible flow. Moreover, Küssner is able to express his result with the use of a single appropriate Green function and its derivatives. The key to his method, as well as to a subsequent generalization [54], seems to be the recognition of the Dirac δ function character of some of these derivative functions on the airfoil boundary and in its wake, thus enabling the direct satisfying of the boundary conditions. Küssner indicates that he has brought the diverse treatments of Timman and Van de Vooren and of Reissner into agreement with his own.

A few main considerations of the method and of the results obtained will be indicated. The differential equation (2-5b), considered expressed in coordinates x', y', z', t', is first converted as follows for harmonic oscillations to the form of Eq. 2-8 by making use of the substitutions

$$x' = bx, \quad y' = \frac{by}{\beta}, \quad z' = \frac{bz}{\beta} \tag{9-13}$$

and dropping the z' variable. Introduce the parameters

$$k = \frac{\omega b}{U}, \quad \kappa = \frac{kM_\infty}{\beta^2}, \quad \mu = \frac{kM_\infty^2}{\beta^2}, \quad \nu = \frac{k}{\beta^2}, \quad \beta^2 = 1 - M_\infty^2 \tag{9-13a}$$

Express the velocity potential and the perturbation pressure, respectively, in the forms

$$\varphi(x', y', t') = Ue^{i(\omega t + \mu x)}\bar{\varphi}(x, y) \tag{9-14}$$

$$p(x', y', t') = \rho_\infty U^2 e^{i(\omega t + \mu x)}\bar{p}(x, y) \tag{9-15}$$

Then there is obtained

$$\left(\frac{\partial^2}{\partial x^2} + \frac{\partial^2}{\partial y^2} + \kappa^2\right)\bar{\varphi} = 0 \tag{9-16}$$

$$\left(\frac{\partial^2}{\partial x^2} + \frac{\partial^2}{\partial y^2} + \kappa^2\right)\bar{p} = 0 \tag{9-17}$$

The relation between p and φ (Eq. 2-10 or 2-12) leads also to the relation

$$\bar{p} = -\left(i\nu + \frac{\partial}{\partial x}\right)\bar{\varphi} \tag{9-18}$$

The boundary conditions for φ may be expressed for three portions of the x axis as follows

(i) for $x < -1$, $\bar{\varphi} = 0$

(ii) for $-1 \leqq x \leqq 1$

$$\frac{\partial\varphi}{\partial y} = \bar{w} = \frac{e^{-i\mu x}}{\beta}\left(ikY^* + \frac{\partial Y^*}{\partial x}\right) \tag{9-19}$$

(iii) for $x > 1$, $\bar{p} = 0$, or

$$i\nu\bar{\varphi} + \frac{\partial\bar{\varphi}}{\partial x} = 0 \tag{9-20}$$

The quantity $\bar{w}$ in (ii) is obtained by writing $Z = Y^* e^{i\omega t}$ in the boundary condition for w (Eq. 3-5). The boundary relations are to be supplemented by the condition at $x = 1$ that $\partial\bar{\varphi}/\partial x$ is finite and that, as discussed in Art. 3, there is proper behavior at $x = -1$ and at $x = \infty$.

It is convenient to separate $\bar{\varphi}$ into two parts

$$\bar{\varphi} = \bar{\varphi}_1 + \bar{\varphi}_2$$

where $\bar{\varphi}_1$ corresponds to the noncirculatory potential, stemming from the primary flow not involving the wake, and $\bar{\varphi}_2$ corresponds to the circulatory potential associated with the wake. For $y = 0$ and for $-1 \leqq x \leqq 1$, $\bar{\varphi}_1$ alone satisfies condition (ii) so that $\partial\bar{\varphi}_2/\partial y = 0$, whereas for $y = 0$ and $x > 1$, $\bar{\varphi}_2$ alone satisfies condition (iii), so that $\bar{\varphi}_1 = 0$. At $x = 1$, $\partial(\bar{\varphi}_1 + \bar{\varphi}_2)/\partial x$ is finite.

It is natural to introduce elliptic coordinates in Eq. 9-16 by means of the relations

$$\begin{aligned} x &= -\cosh\xi\cos\theta \\ y &= \sinh\xi\sin\theta \end{aligned} \tag{9-21}$$

wherein the basic line airfoil is defined by $\xi = 0$, a degenerate member of the ellipses $\xi = \text{const}$, and, as before, $\theta = 0$ and $\theta = \pi$ correspond, respectively, to the leading and trailing edges. The governing differential equation becomes

$$\frac{\partial^2\bar{\varphi}}{\partial\xi^2} + \frac{\partial^2\bar{\varphi}}{\partial\theta^2} + \kappa^2(\cosh^2\xi - \cos^2\theta)\bar{\varphi} = 0 \tag{9-22}$$

where it should be noted that the incompressible flow case $M_\infty = 0$ corresponds to $\kappa = 0$.

If series of product solutions are utilized of the form

$$\bar{\varphi} = \sum M(\xi)N(\theta)$$

$M(\xi)$ and $N(\theta)$ separately satisfy Mathieu-type differential equations and are conveniently defined by Mathieu functions. (Reference may be indicated here to McLachlan [*55*] for definitions and notations and to Bureau of Standard tables [*56*] for numerical values.)

The Mathieu functions $ce_m(\theta)$ and $se_m(\theta)$ can be made to correspond for $\kappa = 0$ to $\cos m\theta$ and $\sin m\theta$, respectively. Similarly, other modified forms can be made to correspond to hyperbolic, exponential, or Bessel or Hankel functions. Moreover, orthogonality properties exist which enable the determination of appropriate constants in the representation of given functions.

Thus the flow potential corresponding to Eq. 5-2, representing the noncirculatory potential for the element of the line airfoil, is given by

$$\Delta\bar{\varphi}_1 = \frac{1}{2\pi}\bar{w}(\eta)G(\kappa, \xi, \eta, \theta)\sin\eta\Delta\eta \tag{9-23}$$

or the whole potential is

$$\bar{\varphi}_1 = \frac{1}{2\pi}\int_0^\pi \bar{w}(\eta)G(\kappa, \xi, \eta, \theta) \sin \eta d\eta \tag{9-24}$$

where

$$G(\kappa, \xi, \eta, \theta) = \sum_1^\infty \frac{Ne_n^{(2)}(\xi)}{Ne_n^{(2)\prime}(0)} se_n(\eta)se_n(\theta) \tag{9-25}$$

and where $Ne_n^{(2)}(\xi)$ is a modified Mathieu function [55] and the prime denotes derivative with respect to ξ. The observation that the required boundary conditions for $\bar{\varphi}_1$ are properly satisfied is facilitated by noting the Dirac δ function character of $dG/d\xi$ at $\xi = 0$ (namely, $G_\xi = 0$ except at $\eta = 0$ where it is infinite but integrable).

For $\kappa = 0$ from Eq. 9-25 we have the expression

$$G(0, \xi, \eta, \theta) = -\sum_1^\infty \frac{1}{n} e^{-n\xi} \sin n\eta \sin n\theta \tag{9-26}$$

which may be recognized as the expansion of the expression

$$G(0, \xi, \eta, \theta) = \frac{1}{4} \ln \frac{\cosh \xi - \cos (\eta - \theta)}{\cosh \xi - \cos (\eta + \theta)} \tag{9-27}$$

At the boundary surface $\xi = 0$

$$G(0, 0, \eta, \theta) = \frac{1}{4} \ln \frac{1 - \cos (\eta - \theta)}{1 + \cos (\eta + \theta)} \tag{9-28}$$

which is the same as $\frac{1}{2}\mathcal{L}(\theta, \eta)$ occurring in Eq. 5-3. Thus we recover Eq. 5-2 and (as indicated in footnote 3) have in the preceding equations the extensions from the boundary values at $\xi = 0$ to the flow field ξ, θ results which can be useful for studies of dynamic stability or of oscillatory induced downwash.

The pressure field associated with the noncirculatory potential $\bar{\varphi}_1$ is

$$\bar{p}_1(\xi, \theta) = -\frac{2}{\pi}\int_0^\pi \bar{w}(0, \eta) \sin \eta[i\nu G(\kappa, \xi, \eta, \theta) + UG_\theta(\kappa, \xi, \eta, \theta)]d\eta \tag{9-29}$$

which for $\xi = 0$ corresponds through Eq. 9-15 to one half the pressure difference (Eq. 3-7) at the boundary.

The general result for the pressure associated with the total potential $\bar{\varphi}_1 + \bar{\varphi}_2$ has been given by Küssner in the form

$$\bar{p}(\xi, \theta) = -\frac{2}{\pi}\int_0^\pi \bar{w}(0, \eta) \sin \eta d\eta \left\{\left(i\nu - \frac{1}{\sin \eta}\frac{\partial}{\partial \eta}\right) G(\kappa, \xi, \eta, \theta) \right.$$
$$\left. + G_\eta(\kappa, \xi, 0, \theta)[G_\theta(\kappa, 0, \eta, \pi)T(\kappa, i\nu) - G_\theta(\kappa, 0, \eta, 0)]\right\} \tag{9-30}$$

where

$$T(\kappa, i\nu) = \frac{\int_{-\infty+i\pi/2}^{\infty-i\pi/2} e^{-i\nu\cosh\xi} G_{\eta\theta}(\kappa, \xi, 0, \pi)d\xi}{\int_{-\infty+i\pi/2}^{\infty-i\pi/2} e^{-i\nu\cosh\xi} G_{\eta\theta}(\kappa, \xi, 0, 0)d\xi} \tag{9-31}$$

The interpretation of $T(\kappa, i\nu)$ through complex contour integration (e.g. [*9*, pp. 170–180]) or its evaluation by infinite series is not gone into here. It may suffice to note that for $\kappa = 0$ the expression (9-30) can be reduced term by term to that given in Eq. 5-24, wherein $T(\kappa, i\nu)$ reduces to the quantity in Eq. 5-23.

Some expansions in the frequency parameter. Explicit presentation of some of the results is conveniently shown in the form of expansions in the frequency. Logarithmic terms, it will be noted, as well as powers of the frequency occur. These terms have figured prominently in various investigations aimed at disclosing differences in dynamic stability based on the usual quasi-steady air forces as opposed to nonsteady air forces. (For example, Goland [*57*] has discussed the incompressible flow case and Miles [*58,59*] the subsonic flow case. Runyan [*60*] has indicated effects involving single-degree-of-freedom flutter.) Such investigations disclose that the reduction in damping due to unsteady flow conditions becomes more marked with increasing Mach number and in particular near $M_\infty = 1$. Most of these comparisons, however, are incomplete, in that finite aspect ratio effects which are important especially for the low frequency range are usually neglected. The following expansions for the two-dimensional force and midchord moment coefficients of a translating and pitching wing are taken from [*61*].

$$A_{11} = -\beta\left(\nu^2\left\{2\gamma + 1 + 2\ln\frac{\nu}{2} - M_\infty^2[2 + F(M_\infty)]\right\} - (2\nu - \pi\nu^2)i\right) \tag{9-32}$$

$$A_{21} = \beta\left(\nu^2\left\{\gamma + \ln\frac{\nu}{2} - \frac{1}{2}M_\infty^2[2 + F(M_\infty)]\right\} - \frac{1}{2}(2\nu - \pi\nu^2)i\right) \tag{9-33}$$

$$\begin{aligned} A_{12} = -\frac{1}{\beta}\Bigg[&-2 + \pi\nu + \nu^2\Bigg\{-\frac{\pi^2}{2} - \frac{5}{4}M_\infty^2 + 2M_\infty^4 + \frac{1}{2}M_\infty^2\ln\frac{M_\infty}{2} \\ &+ (1 - 2M_\infty^2)\left(\gamma + \ln\frac{\nu}{2}\right) - \left(M_\infty^2 - \frac{3}{2}M_\infty^4\right)F(M_\infty) \\ &+ 2\left[\gamma + \ln\frac{\nu}{2} - \frac{1}{2}M_\infty^2 F(M_\infty)\right]^2\Bigg\} \\ &- \Bigg(\nu\left\{2 + 2\gamma + 2\ln\frac{\nu}{2} - M_\infty^2[3 + F(M_\infty)]\right\} \\ &- \pi\nu^2\left\{\frac{1}{2} + 2\nu + 2\ln\frac{\nu}{2} - M_\infty^2[1 + F(M_\infty)]\right\}\Bigg)i\Bigg] \end{aligned} \tag{9-34}$$

$$A_{22} = -\frac{1}{\beta}\left[1 - \frac{1}{2}\pi\nu + \nu^2\left\{\frac{\pi^2}{4} + \frac{1}{8} + \frac{9}{16}M_\infty^2 - \frac{5}{4}M_\infty^4\right.\right.$$

$$-\left(\frac{1}{2} - \frac{7}{4}M_\infty^2\right)\left(\gamma + \ln\frac{\nu}{2}\right) + \left(\frac{1}{2}M_\infty^2 - M_\infty^4\right)F(M_\infty)$$

$$\left.-\left[\gamma + \ln\frac{\nu}{2} + \frac{1}{2}M_\infty^2F(M_\infty)\right]^2\right\} + \left(\nu\left\{\gamma + \ln\frac{\nu}{2} - \frac{1}{2}M_\infty^2[2 + F(M_\infty)]\right\}\right.$$

$$\left.\left.- \pi\nu^2\left\{\frac{1}{4} + \gamma + \ln\frac{\nu}{2} - \frac{1}{2}M_\infty^2\left[\frac{7}{4} + F(M_\infty)\right]\right\}\right)i\right] \quad (9\text{-}35)$$

where

$$\nu = \frac{k}{\beta^2}, \quad \beta^2 = 1 - M_\infty^2, \quad \gamma = 0.57721 \cdots \text{ (Euler's constant)}$$

and where

$$F(M_\infty) = -\frac{2}{M_\infty}\ln\frac{M_\infty}{2} + \frac{2\beta}{M_\infty^2}\ln\frac{1+\beta}{M_\infty}$$

Table F,9

Mach number $M_\infty = 0.7$								
k	R_{11}	I_{11}	R_{12}	I_{12}	R_{21}	I_{21}	R_{22}	I_{22}
0	0	0	2.801	0	0	0	−1.400	0
0.02	0.0071	0.0518	2.590	−0.3295	−0.0041	−0.0258	−1.292	0.2046
0.04	0.0201	0.0952	2.392	−0.4535	−0.0120	−0.0473	−1.188	0.3038
0.06	0.0340	0.1324	2.227	−0.4985	−0.0211	−0.0653	−1.100	0.3611
0.08	0.0472	0.1651	2.093	−0.5076	−0.0307	−0.0809	−1.027	0.3992
0.10	0.0589	0.1941	1.984	−0.4639	−0.0400	−0.0945	−0.9668	0.4249
0.20	0.0945	0.3186	1.679	−0.3251	−0.0845	−0.1478	−0.7849	0.4971
0.30	0.0991	0.4332	1.573	−0.1409	−0.1282	−0.1897	−0.6973	0.5616
0.40	0.0827	0.5523	1.557	+0.0178	−0.1759	−0.2242	−0.6418	0.6407
0.50	0.0538	0.6820	1.594	0.1452	−0.2296	−0.2513	−0.5957	0.7345
0.60	0.0187	0.8229	1.661	0.2359	−0.2899	−0.2679	−0.5398	0.8385
0.70	0.0154	0.9751	1.749	0.2958	−0.3549	−0.2696	−0.4706	0.9419

(The negatives of the quantities A_{11}, A_{21}, A_{12}, and A_{22} are denoted respectively by k_a, m_a, k_b, and m_b in [*61*]; see Eq. 7-4 and notation Table F,7a.) The error in these expressions is given in [*61*] to be of the order of $\nu^3 \ln \nu$ and neglected terms are estimated to affect only the third decimal digit, provided ν is smaller than 0.05.

Miles [*59*] has also indicated various frequency expansions based on expansion of the kernel function.

Some tabulated results. In Table F,9 a tabulation for $M_\infty = 0.7$, is given corresponding to Table F,7b, valid for $M_\infty = 0$. This table was obtained by the method of Dietze [*40*] and has been checked against the tables of the National Aeronautical Research Institute of Holland [*62*], which were obtained by the use of Mathieu functions. The latter tables are among the best available, both for the wing alone and for the wing-aileron configuration in subsonic plane flow. The tables of Fettis [*46*] and those of Turner and Rabinowitz [*42*] are also useful for applications to oscillating control surfaces.

F,10. Nonsteady Flow at Supersonic Speed in Two Dimensions. The first theoretical method for linearized nonsteady flow at supersonic speeds was given briefly and tersely by Possio [*63*] and was based on the source-pulse solution (Eq. 4-31). A later treatment making use of operational or transform methods and contour integrations was given by von Borbely [*64*]. Temple and Jahn [*65*] have developed the results with the aid of Riemann's method of characteristics (in essence an application of Green's theorem). They demonstrate the uniqueness of the results and show its equivalence with [*63*] and [*64*]. A full development along Possio's lines with numerical tables for applications has been given by Garrick and Rubinow [*66*]. Some additional references are listed in the bibliography. The simple and general developments to follow make use of some results of Art. 4.

In the linearized treatment of the plane airfoil for supersonic speeds, the disturbance field representing compression or expansion of the flow created at the airfoil boundary can again be represented in general by the action of dipoles, but sources alone may also suffice. When there is independence of action and of influence of the upper and lower boundary surfaces, sources alone may be employed to represent the flow and these are simpler than dipoles. Each source affects a cylindrical angular Mach field, the wedgelike region obtained by integration of Mach cones and which opens downstream aft of the source. Each field point being influenced is affected by what takes place in the Mach angular field opening forward of the field point in the upstream direction.

Velocity Potential. It is proposed to express the total velocity potential at a field position x, y arising from a distribution of disturbances representing the airfoil located at ξ, η where $\eta = 0$. The leading edge is, for convenience, designated by $\xi = 0$. Then the total effect at x, y is the sum of effects of disturbances originating between $\xi = 0$ and the intersection $\xi = \xi_1$ of the Mach line through x, y with the ξ axis where

$$\xi_1 = x - y\sqrt{M_\infty^2 - 1} \tag{10-1}$$

since any disturbance aft of ξ_1 cannot influence x, y. Let the normal velocity distribution be given by means of Eq. 3-5 as

$$w(x, t) = \frac{\partial Z}{\partial t} + U\frac{\partial Z}{\partial x}$$

Making use of the normalized source solutions given in Art. 4, and the fact that the source distribution giving the required normal velocity at the upper surface corresponds to a total jump across the surface of $2w$, there is obtained for the velocity potential at x, y from Eq. 4-30,

$$\varphi(x, y, t) = -\frac{1}{\pi\beta}\int_0^{\xi_1}\int_0^{\pi} w\left(\xi, t - \frac{\tau_1 + \tau_2}{2} + \frac{\tau_1 - \tau_2}{2}\cos\theta\right) d\theta d\xi \tag{10-2}$$

or the representation from Eq. 4-31,

$$\varphi(x, y, t) = -\frac{a_\infty}{\pi\beta}\int_0^{\xi_1}\int_{\tau_1}^{\tau_2} \frac{w(\xi, t - \tau)}{\sqrt{a_\infty^2\tau^2 - (x - \xi - Ut)^2 - y^2}} d\tau d\xi \tag{10-3}$$

where τ_1 and τ_2 are the roots of the expression in the denominator.

The surface potential for $y = 0$ may be expressed as

$$\varphi(x, 0, t) = -\frac{1}{\pi\beta}\int_0^{x}\int_0^{\pi} w\left(\xi, t - \frac{x - \xi}{U}\frac{M_\infty^2}{\beta^2} - \frac{x - \xi}{U}\frac{M_\infty}{\beta^2}\cos\theta\right) d\theta d\xi \tag{10-4}$$

or alternatively as,

$$\varphi(x, 0, t) = -\frac{1}{\pi\beta}\int_0^{x}\int_{\tau_1'}^{\tau_2'} \frac{w(\xi, t, \tau)}{\sqrt{(\tau - \tau_1')(\tau_2' - \tau)}} d\tau d\xi \tag{10-5}$$

where

$$\tau_1' = (\tau_1)_{y=0} = \frac{x - \xi}{a_\infty}\frac{1}{M_\infty + 1}$$

$$\tau_2' = (\tau_2)_{y=0} = \frac{x - \xi}{a_\infty}\frac{1}{M_\infty - 1}$$

It is noted that the general results are valid for arbitrary time variations and that the pressure field or the surface pressure difference can be obtained directly from φ as by Eq. 3-7. The developments for the sinusoidal and the unit-step time variations are of particular interest.

APPLICATION TO THE OSCILLATING AIRFOIL. Let the normal velocity distribution be of the form

$$w(x, t) = w^*(x)e^{i\omega t} \tag{10-6}$$

Then the surface potential can be written from Eq. 10-4 or 10-5, or directly from the moving source solution Eq. 4-32, as

$$\varphi(x, t) = -\frac{1}{\beta}\int_0^x w(\xi, t)I(x - \xi)d\xi \tag{10-7}$$

where the kernel (or Green's) function is

$$I(x - \xi) = e^{-i\mu(x-\xi)}J_0[\kappa(x - \xi)] \tag{10-8}$$

and where

$$\kappa = \frac{\omega}{U}\frac{M_\infty}{\beta^2}$$

$$\mu = \frac{\omega}{U}\frac{M_\infty^2}{\beta^2}$$

A form in which the corresponding surface pressure difference may be expressed by means of Eq. 3-7 is

$$P(x, t) = \frac{2\rho_\infty U}{\beta}\left[w(0, t)I(x) + \int_0^x I(x - \xi)\left(\frac{1}{U}\frac{\partial}{\partial t} + \frac{\partial}{\partial \xi}\right)w(\xi, t)d\xi\right] \tag{10-9}$$

Consider now, for example, the motion of an airfoil involving displacement $h = h_0e^{i\omega t}$ in a vertical direction and turning about an axis at $x = x_0$ with instantaneous angle $\alpha = \alpha_0e^{i\omega t}$. Then introducing in the subsequent analysis nondimensional quantities x and ξ, obtained by dividing the old variables by the chord $c = 2b$, there is obtained for the normal velocity

$$w(x, t) = -[\dot{h} + U\alpha + c(x - x_0)\dot{\alpha}] \tag{10-10}$$

Let the velocity potential be separated into

$$\varphi = \varphi_{\dot{h}} + \varphi_\alpha + \varphi_{\dot{\alpha}} \tag{10-11}$$

where each term is associated with its respective term in Eq. 10-10. Then, with $x - \xi = u$,

$$\left.\begin{aligned} \varphi_{\dot{h}} &= \dot{h}\frac{c}{\beta}\int_0^x I(u)du \\ \varphi_\alpha &= U\alpha\frac{c}{\beta}\int_0^x I(u)du \\ \varphi_{\dot{\alpha}} &= \dot{\alpha}\frac{c^2}{\beta}\int_0^x (x - x_0 - u)I(u)du \end{aligned}\right\} \tag{10-12}$$

where

$$I(u) = e^{-i\hat{\omega}u} J_0\left(\frac{\hat{\omega}u}{M_\infty}\right)$$

and where

$$\hat{\omega} = \frac{c\omega}{U}\frac{M_\infty^2}{\beta^2} = \frac{2kM_\infty^2}{M_\infty^2 - 1}$$

$$k = \frac{\omega c}{2U}$$

(It may be observed that in contrast to the usage in the subsonic case where the airfoil extends from $x = -1$ to 1, in this case it has been chosen to extend from $x = 0$ to 1.)

The surface pressure difference is given by means of Eq. 3-7 as

$$P = -2\rho_\infty\left(\frac{\partial\varphi}{\partial t} + \frac{U}{c}\frac{\partial\varphi}{\partial x}\right)$$

The total force positive downward is given by

$$F = c\int_0^1 P dx \tag{10-13}$$

and the total moment (positive, leading edge up) about any axis at $x = x$ is

$$M_\infty = c^2 \int_0^1 (x - x_0) P dx \tag{10-14}$$

With the use of Eq. 10-12 there is obtained

$$\left.\begin{aligned} F &= -\frac{2\rho_\infty c}{\beta}\Big[U(U\alpha + \dot{h} - cx_0\dot{\alpha})r_1 \\ &\qquad + c(2U\dot{\alpha} + \ddot{h} - cx_0\ddot{\alpha})r_2 + c^2\ddot{\alpha}\frac{r_3}{2}\Big] \\ M_\alpha &= -\frac{2\rho_\infty c^2}{\beta}\Big[U(U\alpha + \dot{h} - cx_0\dot{\alpha})q_1 \\ &\qquad + c(2U\dot{\alpha} + \ddot{h} - cx_0\ddot{\alpha})\frac{q_2}{2} + c^2\ddot{\alpha}\frac{q_3}{6}\Big] - cx_0F \end{aligned}\right\} \tag{10-15}$$

The coefficients of α, $\dot{\alpha}$, and $\ddot{\alpha}$ represent respectively the aerodynamic effects of the angle-of-attack position, the angular velocity about $x = x_0$, and of the angular acceleration. The coefficients of $\dot{h}$ and $\ddot{h}$ represent the corresponding effects of the vertical velocity and acceleration. These coefficients are complex quantities each of whose real and imaginary parts defines a rotating vector which represents the harmonically oscillating force or moment in magnitude and phase associated with the particular α or h term considered. The six quantities r_1, r_2, r_3, q_1, q_2, and q_3 are functions of the stream Mach number and the frequency parameter k or $\hat{\omega}$. They are defined by certain integrals containing the kernel function $I(u)$.

All are reducible (by reduction formulas, see [*64,66*]) to forms containing the standard Bessel functions J_0 and J_1 and one additional integral function

$$f_0(M_\infty, \bar{\omega}) = \frac{1}{\bar{\omega}} \int_0^{\bar{\omega}} e^{-iu} J_0\left(\frac{u}{M_\infty}\right) du \tag{10-16}$$

This function was studied and tabulated by Schwarz [*67*]; tables are also given in [*66*].

The following list provides formulas for accomplishing the aforementioned reduction of the six functions, and hence a table of f_0 in conjunction with tables of the standard Bessel functions may be used to supply all desired numerical data for the coefficients:

$$\begin{aligned} r_1 &= f_0 \\ r_2 &= f_0 - f_1 \\ r_3 &= f_0 - 2f_1 + f_2 \\ q_1 &= f_1 \\ q_2 &= f_0 - f_2 \\ q_3 &= 2f_0 - 3f_1 + f_3 \end{aligned}$$

where

$$f_1(M_\infty, \bar{\omega}) = \frac{1}{2k}\left(-if_0 + iJ_0 e^{-i\bar{\omega}} - \frac{1}{M_\infty} J_1 e^{-i\bar{\omega}}\right)$$

$$f_2(M_\infty, \bar{\omega}) = \frac{1}{2k}\left[f_0\left(\frac{1}{\bar{\omega}} - \frac{3}{2k}\right) + J_0 e^{-i\bar{\omega}}\left(\frac{3}{2k} - \frac{1}{\bar{\omega}} + i\right) + \frac{1}{M_\infty} e^{-i\bar{\omega}} J_1\left(-1 + \frac{3i}{2k}\right)\right]$$

$$f_3(M_\infty, \bar{\omega}) = \frac{1}{2k}\left[if_0\left(\frac{15}{2k} - \frac{9}{\bar{\omega}}\right) + J_0 e^{-i\bar{\omega}}\left(\frac{5}{2k} - \frac{2}{\bar{\omega}} + i + \frac{9i}{2k\bar{\omega}} - \frac{15i}{4k^2}\right) + \frac{1}{M_\infty} e^{-i\bar{\omega}} J_1\left(-1 - \frac{4}{2k\bar{\omega}} + \frac{15}{4k^2} + \frac{5i}{2k}\right)\right]$$

and where the argument of J_0 and J_1 throughout is $\bar{\omega}/M_\infty$. It may be remarked that higher modes of normal velocity than assumed in Eq. 10-10 will introduce additional functions f_λ involving $\lambda > 3$. A table of such functions by Huckel for values of λ up to 11 is listed in the bibliography.

Some Tabulated Results. The results for F and M_α can be conveniently grouped and presented in a form as listed in Eq. 7-4. Such results with numerical tables for various frequencies and for Mach numbers $\frac{10}{9}$, $\frac{10}{8}$, $\frac{10}{7}$, $\frac{10}{6}$, $\frac{10}{5}$, $\frac{10}{4}$, $\frac{10}{3}$ and $\frac{10}{2}$ are given in [*64*]. Tables F,10a and F,10b give brief listings for $M_\infty = \frac{10}{8}$ and $M_\infty = 2.0$, corresponding to Table F,7b for $M_\infty = 0$.

Table F,10a

Mach number $M_\infty = \frac{10}{8}$								
k	R_{11}	I_{11}	R_{12}	I_{12}	R_{21}	I_{21}	R_{22}	I_{22}
0	0	0	1.698	0	0	0	0	0
0.0504	0.0076	0.0845	1.674	−0.1493	0.0025	−0.0005	−0.0105	−0.0206
0.1008	0.0289	0.1629	1.607	−0.2822	0.0091	−0.0041	−0.0398	−0.0321
0.1476	0.0577	0.2258	1.512	−0.3788	0.0171	−0.0121	−0.0791	−0.0281
0.1980	0.0934	0.2800	1.390	−0.4450	0.0248	−0.0266	−0.1265	−0.0038
0.2988	0.1559	0.3475	1.138	−0.4516	0.0249	−0.0702	−0.2016	+0.1047
0.3960	0.1779	0.3819	0.9694	−0.3364	−0.0048	−0.1127	−0.2073	0.2508
0.5040	0.1525	0.4292	0.9150	−0.1703	−0.0634	−0.1322	−0.1286	0.3888
0.5940	0.1131	0.4988	0.9500	−0.0730	−0.1103	−0.1174	−0.0305	0.4443
0.9000	0.0821	0.8183	1.031	−0.0520	−0.1214	−0.0393	+0.0884	0.4077

Table F,10b

Mach number $M_\infty = 2.0$								
k	R_{11}	I_{11}	R_{12}	I_{12}	R_{21}	I_{21}	R_{22}	I_{22}
0	0	0	0.7351	0	0	0	0	0
0.0525	0.0007	0.0385	0.7340	−0.0128	0.0002	−0.00002	−0.0004	0.0086
0.0975	0.0023	0.0714	0.7313	−0.0235	0.0008	−0.00015	−0.0015	0.0162
0.15	0.0054	0.1092	0.7261	−0.0352	0.0017	−0.00054	−0.0035	0.0253
0.2025	0.0095	0.1462	0.7190	−0.0458	0.0030	−0.00130	−0.0063	0.0351
0.30	0.0197	0.2122	0.7016	−0.0615	0.0059	−0.00405	−0.0127	0.0553
0.4125	0.0334	0.2829	0.6768	−0.0713	0.0090	−0.00975	−0.0210	0.0827
0.4875	0.0426	0.3264	0.6592	−0.0723	0.0102	−0.01506	−0.0260	0.1036
0.6000	0.0543	0.3864	0.6342	−0.0657	0.0097	−0.02486	−0.0311	0.1384
1.0125	0.0469	0.5813	0.6012	+0.0108	−0.0271	−0.05716	−0.0053	0.2748

The functions arising in the aileron case can be reduced in a similar manner. Formulas and numerical tables are given by Huckel and Durling in [*68*].

It is considered useful to list expressions of the coefficients in Eq. 10-15, valid for small values of the frequency parameter obtained from the use of a power series expansion of $I(u)$ in the integrals:

$$r_1 = 1 - i\frac{\hat{\omega}}{2}$$

$$r_2 = \frac{1}{2} - i\frac{\hat{\omega}}{6}$$

$$r_3 = \frac{1}{3} - i\frac{\hat{\omega}}{12}$$

$$q_1 = \frac{1}{2} - i\frac{\hat{\omega}}{3}$$

$$q_2 = \frac{2}{3} - i\frac{\hat{\omega}}{4}$$

$$q_3 = \frac{3}{4} - i\frac{\hat{\omega}}{5}$$

SPECIAL CASE OF HIGH MACH NUMBERS. Of interest also is a limiting result for high Mach numbers which yields a particularly simple expression. For as $M_\infty \to \infty$, it is found from Eq. 10-4, by differentiation of the definite integrals involved, that the perturbation pressure at the surface is given by

$$p = \frac{\rho_\infty U}{\beta} w(x, t) \tag{10-17}$$

This relation expresses the local pressure directly in terms of the local normal velocity function. Hence for large Mach numbers the air forces and moments approach the quasi-steady values determined by the instantaneous value of w.[7]

It has been pointed out by Lighthill [*69*], who made use of the so-called piston analogy corresponding to a one-dimensional unsteady flow, that the implications of the quasi-steady theory can, in effect, be carried beyond the linear stage by taking over higher order terms for the pressure from gas dynamics. Moreover, the theory may be applied separately to the separate surfaces of the airfoil and thus include effects of thickness as well as camber. A few relations of this theory and an illustrative application given by Lighthill follow.

[7] The result (Eq. 10-17) in essence was assumed as a suitable approximation for all $M_\infty > 1$ in R and M 2139 by A. R. Collar (1944).

Pressure and loading relations. The adiabatic pressure-density relation and the acoustic relation $p = \rho a w$ yield to the cubic approximation

$$\frac{p_l}{p_\infty} = 1 + \gamma \frac{w}{a_\infty} + \frac{\gamma(\gamma+1)}{4}\left(\frac{w}{a_\infty}\right)^2 + \frac{\gamma(\gamma+1)}{12}\left(\frac{w}{a_\infty}\right)^3 \tag{10-18}$$

where γ is the adiabatic index, $a_\infty^2 = \gamma p_\infty/\rho_\infty$ and the local excess pressure p corresponds to $(p_l - p_\infty)/p_\infty$. Eq. 10-17 is hence in accord with the linear term in Eq. 10-18 when β can be approximated by M.

Let the normal velocity be separated into effects of camber and of thickness as in Art. 3:

$$w = w_c(x, t) \pm w_t(x)$$

Using Eq. 10-18, the pressure difference or loading may be obtained from

$$\begin{aligned}\frac{P}{\gamma p_\infty} &= \frac{p_u - p_l}{\gamma p_\infty} \\ &= -\left[2 + (\gamma+1)\frac{w_t}{a_\infty} + \frac{\gamma+1}{2}\left(\frac{w_t}{a_\infty}\right)^2\right]\frac{w_c}{a_\infty} - \frac{1}{6}(\gamma+1)\left(\frac{w_c}{a_\infty}\right)^3\end{aligned} \tag{10-19}$$

The main effect of the nonlinearity is to change the coefficient of w_c/a_∞ in the first term by a factor depending on w_t/a_∞.

Application to biconvex airfoil. This main effect may be exhibited for the pitching moment of a biconvex (parabolic) airfoil. Let the airfoil be defined by

$$Z_t = 2\tau c x(1 - x) \tag{10-20}$$

where τ is the thickness-chord ratio ($M_\infty \tau \ll 1$) and let the axis of pitching be at $x = x_0$. The normal velocity due to thickness and to camber may be written as

$$\begin{aligned} w_t &= 2U\tau(1 - 2x) \\ w_c &= -[U\alpha + c\dot{\alpha}(x - x_0)] \end{aligned}$$

Then the midchord moment coefficient is given by

$$\begin{aligned}\frac{M_a}{\frac{1}{2}\rho_\infty U^2 c^2} &= c_m \\ &= [(c_{m,\alpha})_r + i(c_{m,\alpha})_i]\alpha\end{aligned} \tag{10-21}$$

where the stiffness component coefficient is

$$(c_{m,\alpha})_r = \frac{2}{M_\infty}\left\{1 - \frac{1}{3}(\gamma+1)M_\infty\tau + \frac{1}{3}(\gamma+1)M_\infty^2\tau^2 - x_0\left[2 + \frac{2}{3}(\gamma+1)M_\infty^2\tau^2\right]\right\}$$

and the damping component coefficient is

$$(c_{m,\alpha})_i = \frac{2}{M_\infty}\left\{\frac{2}{3} - \frac{1}{3}(\gamma+1)M_\infty\tau + \frac{4}{15}(\gamma+1)M_\infty^2\tau^2 - 2x_0\left[1 - \frac{1}{3}(\gamma+1)M_\infty\tau + \frac{1}{3}(\gamma+1)M_\infty^2\tau^2\right] + x_0^2\left[2 + \frac{2}{3}(\gamma+1)M_\infty^2\tau^2\right]\right\}\frac{c\omega}{U}$$

These results are used by Lighthill to discuss some center-of-pressure changes with thickness and to check some limiting values of Van Dyke [*70*] on the inclusion of higher order effects of thickness for oscillating airfoils in supersonic flow.

F,11. Nonsteady Flow at Sonic Speed in Two Dimensions. Formidable unsolved flow problems already exist for steady plane motion at sonic speeds. Questions arise embracing the origin of shocks, nonlinear flow behavior, instabilities in potential flow, and susceptibility of the flow to interaction with viscous effects. Despite these difficulties of the steady problem which have led, for example, to the search and development of transonic similarity rules, the nonsteady flow problem can be made amenable to linear analytic treatment. (See for example [*2,3,4,71*].) One recognizes that such treatment definitely becomes invalid for sufficiently low frequencies, but in such calculations there may be useful and practical frequency ranges, the validity and utility of which must ultimately be defined by experience. As mentioned in Art. 4, such treatment contains the analytical means for bridging calculations for high subsonic and low supersonic speeds. The use of such calculations in strip analysis is, of course, subject to correction for finite wing span and wing thickness, particularly for the transonic speed range.

The results to follow are developed rather similarly to the treatment for plane supersonic flow and in fact can be regarded as a mathematical limit $M_\infty \to 1$ from the supersonic side. Attention to the oscillatory sonic solution has been drawn by Rott in a brief note [*72*].

The analysis makes use of results for a plane source moving at sonic speed $U = a_\infty$ given in Eq. 4-35. The potential at x, y, corresponding to that given in Eq. 10-2 and 10-3, is

$$\varphi(x, y, t) = -\frac{1}{\pi}\int_0^x\int_0^\infty \frac{1}{x-\xi}\, w\left[\xi, 0, t - \frac{(x-\xi)^2 + y^2 + z^2}{2a_\infty(x-\xi)}\right] dz d\xi \quad (11\text{-}1)$$

where $w(\xi, 0, t)$, as before, represents the normal velocity distribution at $\eta = 0$. The surface potential is obtained by putting $y = 0$ in Eq. 11-1 and for the harmonically oscillating case, with $w(\xi, t)$ defined as in Eq. 10-6, is

$$\varphi(x, t) = -\int_0^x w(\xi, t)G(x - \xi)d\xi \tag{11-2}$$

where

$$G(x - \xi) = \frac{A}{\sqrt{\frac{\omega}{2U}(x - \xi)}} e^{-\frac{i\omega(x-\xi)}{2U}}$$

and where

$$A = \frac{1}{2\sqrt{\pi i}} = \frac{1 - i}{2\sqrt{2\pi}}$$

The result (Eq. 11-2) also follows directly from the use of Eq. 4-37.

Corresponding to the application made in Eq. 10-10, 10-11, and 10-12 there is obtained (with nondimensional x)

$$\left.\begin{aligned} \varphi_{\dot{h}} &= c\dot{h}\int_0^x G(u)du \\ \varphi_{\alpha} &= cU\alpha\int_0^x G(u)du \\ \varphi_{\dot{\alpha}} &= c^2\dot{\alpha}\int_0^x (x - x_0 - u)G(u)du \end{aligned}\right\} \tag{11-3}$$

where

$$G(u) = \frac{A}{\sqrt{ku}} e^{-iku}$$

and where

$$k = \frac{b\omega}{U} = \frac{\omega c}{2U}$$

Corresponding to Eq. 10-15 there is obtained for the total force and moment

$$\left.\begin{aligned} F &= -2\rho_\infty c\left[U(U\alpha + \dot{h} - cx_0\dot{\alpha})s_1 + c(2U\dot{\alpha} + \ddot{h} - cx_0\ddot{\alpha})s_2 + c^2\ddot{\alpha}\frac{s_3}{2}\right] \\ M_\alpha &= -2\rho_\infty c^2\left[U(U\alpha + \dot{h} - cx_0\dot{\alpha})t_1 + c(2U\dot{\alpha} + \ddot{h} - cx_0\ddot{\alpha})\frac{t_2}{2} + c^2\ddot{\alpha}\frac{t_3}{6}\right] - cx_0F \end{aligned}\right\} \tag{11-4}$$

The six quantities s_1, s_2, s_3, t_1, t_2, and t_3 are functions of the frequency parameter k and are defined by integrals containing the kernel $G(u)$. They are determined by the following set of formulas, analogous to the set in Art. 10,

$$s_1 = g_0$$
$$s_2 = g_0 - g_1$$
$$s_3 = g_0 - 2g_1 + g_2$$
$$t_1 = g_1$$
$$t_2 = g_0 - g_2$$
$$t_3 = 2g_0 - 3g_1 + g_3$$
$$g_\lambda = \int_0^1 G(u)u^\lambda du \quad (\lambda \mp 0, 1, 2, 3)$$
$$g_1 = \frac{Aie^{-ik}}{k^{\frac{3}{2}}} - \frac{i}{2k} g_0$$
$$g_2 = \frac{Aie^{-ik}}{k^{\frac{3}{2}}}\left(1 - \frac{3i}{2k}\right) - \frac{3}{4k^2} g_0$$
$$g_3 = \frac{Aie^{-ik}}{k^{\frac{3}{2}}}\left(1 - \frac{5i}{2k} - \frac{15}{4k^2}\right) + \frac{15i}{8k^3} g_0$$

Thus all quantities are given in terms of g_0, an integral related to Fresnel's integral and which can be conveniently evaluated by expansion in power series:

$$g_0 = A\left(\int_0^1 \frac{\cos ku}{\sqrt{ku}}\, du - i \int_0^1 \frac{\sin ku}{\sqrt{ku}}\, du\right)$$
$$= \frac{2A}{\sqrt{k}}\left[\sum_{n=0}^{\infty} (-1)^n \frac{k^{2n}}{(2n)!(4n+1)} - i \sum_{n=0}^{\infty} (-1)^n \frac{k^{2n+1}}{(2n+1)!(4n+3)}\right]$$

The total force and moment (Eq. 11-4) can be put in the coefficient form indicated in Eq. 7-4, 7-5, and 7-6:

$$c_l = -\frac{F}{\rho_\infty b U^2}$$
$$= \pi\left[(R_{11} + iI_{11})\frac{h}{b} + (R_{12} + iI_{12})\alpha\right] \tag{11-5}$$
$$c_m = \frac{M_\alpha}{2\rho_\infty U^2 b^2}$$
$$= -\frac{\pi}{2}\left[(R_{21} + iI_{21})\frac{h}{b} + (R_{22} + iI_{22})\alpha\right] \tag{11-6}$$

For small values of the frequency parameter k, the coefficients in Eq. 11-5 and 11-6 appear as

$$R_{11} = \frac{4k}{\sqrt{2\pi k}}(1 - k)$$

$$I_{11} = \frac{4k}{\sqrt{2\pi k}}(1 + k)$$

$$R_{12} = \frac{4}{\sqrt{2\pi k}}\left[1 + \frac{k}{3}(7 - 6x_0)\right]$$

$$I_{12} = \frac{4}{\sqrt{2\pi k}}\left[-1 + \frac{k}{3}(7 - 6x_0)\right]$$

$$R_{21} = \frac{4k}{\sqrt{2\pi k}}\left[\frac{2}{3}(1 - 3x_0) - \frac{2k}{5}(3 - 5x_0)\right]$$

$$I_{21} = \frac{4k}{\pi\sqrt{2\pi k}}\left[\frac{2}{3}(1 - 3x_0) + \frac{2k}{5}(3 - 5x_0)\right]$$

$$R_{22} = \frac{4}{\pi\sqrt{2\pi k}}\left[\frac{2}{3}(1 - 3x_0) + 2k\left(\frac{7}{5} - 3x_0 + 2x_0^2\right)\right]$$

$$I_{22} = \frac{4}{\pi\sqrt{2\pi k}}\left[-\frac{2}{3}(1 - 3x_0) + 2k\left(\frac{7}{5} - 3x_0 + 2x_0^2\right)\right]$$

Strictly speaking, these results are physically ruled out for very small k as will be noted because of their proximity to the Prandtl-Glauert type of infinity for steady flow. Note that, in the moment expressions, the factor $(1 - 3x_0)$ appears in the term which is dominant for small values of k, indicating that the center of pressure is then located one third of the chord from the leading edge, an observation made by Rott [*72*].

Some tabulated results. Another analysis for the case $M_\infty = 1$ employing Laplace transform methods has been presented along with useful numerical tables by Nelson and Berman [*73*]. A brief table corresponding to the tables given in Art. 7, 9, and 10 is given for $M_\infty = 1$ in Table F,11.

Two figures taken from [*13*] are of interest in affording an over-all view of some characteristics of the two-dimensional subsonic, sonic, and supersonic solutions. The figures indicate the oscillatory moment of an airfoil of infinite span pitching about its leading edge. The magnitude of the moment about the leading edge is indicated in Fig. F,11a as a function of Mach number and reduced frequency. Contour lines representing the phase of the moment relative to the displacement are shown in Fig. F,11b. It is of interest to observe that in this leading edge pitching case the aerodynamic damping moment can change sign, and the zero-phase contour line marks the border of the negative-phase region of M_∞ and k wherein loss of damping and instability are indicated. This loss of damping in pitch occurring as it does at low values of the reduced frequency is of special concern in dynamic stability and has been the subject of a number of studies on the effects of planform, thickness, and flow regime.

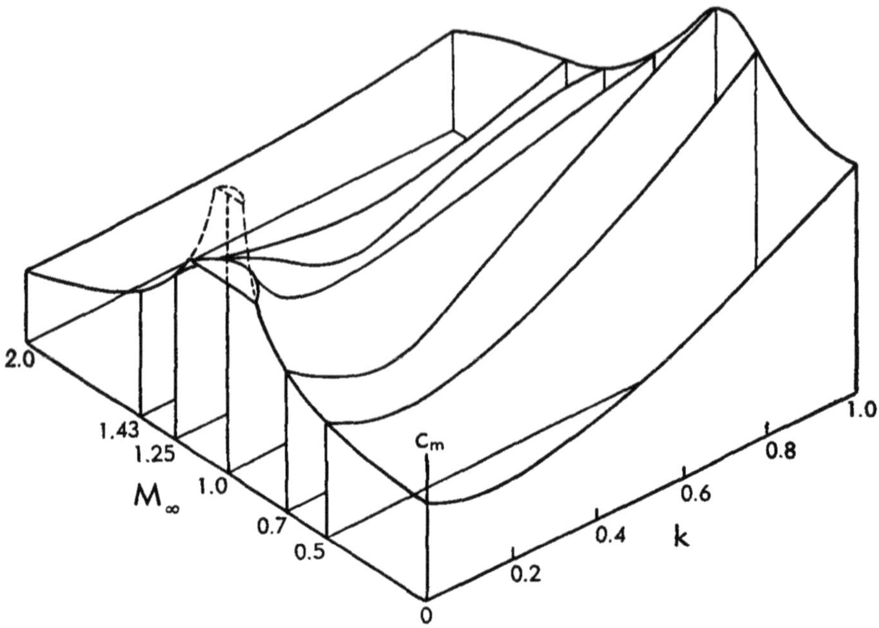

Fig. F,11a. Pitching moment of wing section oscillating about leading edge.

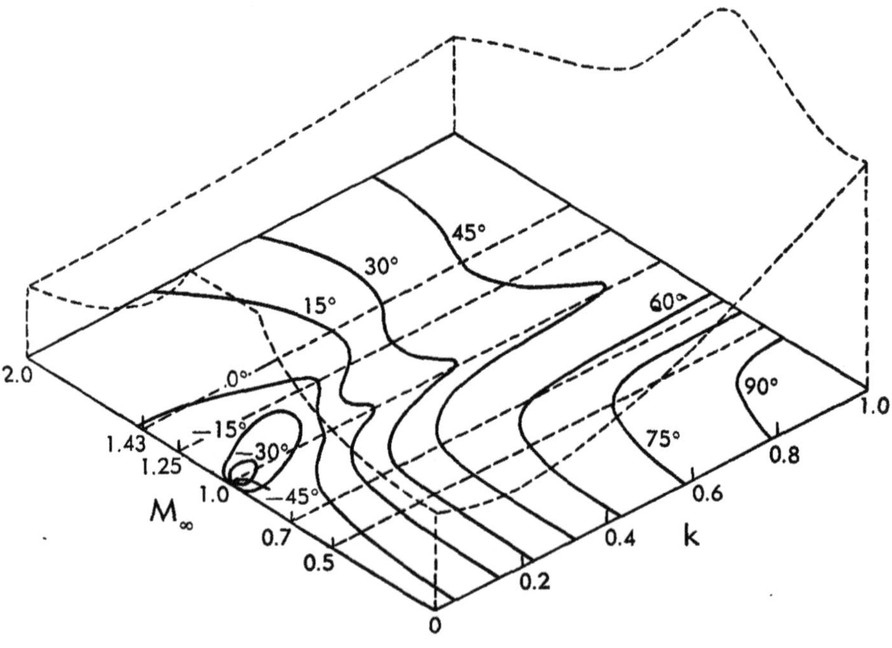

Fig. F,11b. Phase angles of moment relative to angle of pitching.

Table F,11

Mach number $M_\infty = 1.0$								
k	R_{11}	I_{11}	R_{12}	I_{12}	R_{21}	I_{21}	R_{22}	I_{22}
0	0	0			0	0		
0.05	0.1079	0.1193	2.425	−2.122	−0.0390	−0.0367	−0.6668	0.8485
0.10	0.1448	0.1769	1.826	−1.398	−0.0566	−0.0502	−0.4081	0.6650
0.16	0.1716	0.2365	1.552	−1.011	−0.0738	−0.0609	−0.2633	0.5880
0.20	0.1833	0.2740	1.454	−0.8505	−0.0841	−0.0660	−0.2004	0.5633
0.28	0.1972	0.3473	1.344	−0.6312	−0.1030	−0.0732	−0.1111	0.5397
0.40	0.2019	0.4576	1.273	−0.4268	−0.1288	−0.0781	−0.0220	0.5323
0.60	0.1835	0.6500	1.250	−0.2302	−0.1669	−0.0755	+0.0732	0.5462
0.80	0.1458	0.8576	1.272	−0.1195	−0.1998	−0.0629	0.1350	0.5714
1.00	0.0982	1.081	1.308	−0.0558	−0.2274	−0.0424	0.1771	0.6008

F,12. Transient Air Forces for Plane Flow at Supersonic, Sonic, and Subsonic Speeds. It is of interest to briefly pursue further the considerations of transient flows, extending the general developments which were given in Art. 6 for the incompressible fluid.

It may be recalled that the results for the potential given for the supersonic flow in Eq. 10-4 and for sonic flow in Eq. 11-1 are quite general as concerns the normal velocity distribution $w(x, t)$ and do not depend on harmonic motion. Hence it is appropriate to examine for these cases the special results for a choice of two basic distributions,

(i)
$$w(x, t) = w_0 H(t) \tag{12-1}$$

and

(ii)
$$w(x, t) = w_0 H\left(t - \frac{x}{U}\right) \tag{12-2}$$

where, as in Art. 6, $H(t)$ is the unit step function and (i) corresponds to the sudden acquisition of uniform downwash w_0 over the wing while (ii) corresponds to the traveling-wave type of acquisition due to penetration of a sharp-edged uniform gust.

$M_\infty > 1$. Sudden Acquisition of Uniform Downwash. **Substitution of Eq. 12-1 into Eq. 10-4 yields the result**

$$\varphi(x, t) = -\frac{w_0}{\pi\beta}\left\{x \cos^{-1}\left(\frac{M_\infty x - \beta^2 a_\infty t}{x}\right) + \beta\left[\pi - \cos^{-1}\left(\frac{x - M_\infty a_\infty t}{a_\infty t}\right)\right]\right\} \tag{12-3}$$

where $\beta = |M_\infty^2 - 1|^{\frac{1}{2}}$.

The pressure difference is given by

$$P(x, t) = \frac{2\rho_\infty U w_0}{\pi\beta}\left[\cos^{-1}\left(\frac{M_\infty x - \beta^2 a_\infty t}{x}\right) + \frac{\beta}{M_\infty}\left(\pi - \cos^{-1}\frac{x - M_\infty a_\infty t}{a_\infty t}\right)\right] \tag{12-4}$$

where the inverse trigonometric terms must be interpreted to be 0 when the arguments are $\geqq 1$ and π when the arguments are $\leqq -1$.

The results can be interpreted in terms of three distinct time phases of physical significance.

Phase 1

$$0 < t < \frac{x}{U + a_\infty}$$

This time phase corresponds to the interval from time zero to that required for the field point at x to first feel the influence of the oncoming disturbance originating at the leading edge. The pressure difference is

$$P_1 = \frac{2\rho_\infty U w_0}{M_\infty}$$

Phase 2

$$\frac{x}{U + a_\infty} < t < \frac{x}{U - a_\infty}$$

The quantity $x/(U - a_\infty)$ represents the time required for the point x to overtake the outgoing disturbance originating at the leading edge. For phase 2 the pressure P_2 is given directly by Eq. 12-4.

Phase 3

$$t > \frac{x}{U - a_\infty}$$

The pressure difference corresponds to the steady (Ackeret) result,

$$P_3 = \frac{2\rho_\infty U w_0}{\beta}$$

PENETRATION INTO A SHARP UNIFORM GUST. The use of Eq. 12-2 leads similarly to the following result for the pressure difference:

$$P = \frac{2\rho_\infty U w_0}{\pi\beta} \cos^{-1}\left(\frac{M_\infty x - \beta^2 a_\infty t}{x}\right) \tag{12-5}$$

where expressed in the same three time phases as before

Phase 1

$$P_1 = 0$$

Phase 2

P_2 given directly by Eq. 12-5

Phase 3

$$P_3 = \frac{2\rho_\infty U w_0}{\beta}$$

The foregoing results are developed by Watkins in [*74*] by the convenient intermediate use of Laplace transforms, to evaluate the integrations. Equivalent results by other methods of interest have also been given by Biot [*75*], by Heaslet and Lomax [*76*], and by Miles [*77*]. Particular mention may be made of the alternative formulation and method of solution [*75,76*] based on space-fixed coordinates and on the use of the wave equation (2-7); in this formulation the position of a given point of the airfoil boundary is associated with variable time and hence the boundary conditions are stated for time-dependent positions leading to the use of a so-called acoustic planform (as in VI,D,25). This procedure is especially useful for accelerated or nonuniform motion.

$M_\infty = 1$. For the sonic speed, only time phases 1 and 2 are involved, given by

Phase 1

$$0 < t < \frac{x}{2a_\infty}$$

Phase 2

$$\frac{x}{2a_\infty} < t < \infty$$

The results for the two conditions (i) and (ii) can be found directly from Eq. 11-1 or from taking the limit $M_\infty \to 1$ of the preceding expressions. There is obtained for (i)

$$P_1 = 2\rho_\infty U w_0$$

$$P_2 = \frac{2\rho_\infty U w_0}{\pi}\left(\sqrt{\frac{2a_\infty t}{x} - 1} + \pi - \cos^{-1}\frac{x - a_\infty t}{a_\infty t}\right)$$

and for (ii)

$$P_1 = 0$$

$$P_2 = \frac{2\rho_\infty U w_0}{\pi}\sqrt{\frac{2a_\infty t}{x} - 1}$$

$M_\infty < 1$. For the case of subsonic speeds more difficult procedures are required to evaluate the transient effects. Results corresponding to (i) and (ii) can be obtained numerically by application of the Laplace transform (reciprocal) relations already mentioned in Art. 6. Thus the numerical results arising from Art. 9 can be employed and the indicial function determined. Results of doing this for various Mach numbers have been given by Mazelsky and Drischler [*78,79*]. Since no separation of the apparent-mass term and terms due to the wake was made in this case, the transient effects would include both, and it is physically obvious that the instantaneous impulsive terms of the incompressible fluid case would correspond to action spread out over time. Indeed, the very concept of apparent mass becomes fuzzy as both inertia and damping effects are intertwined. The separation and interpretation of such terms for the subsonic plane flow case may be made with the use of the noncirculatory potentials indicated in Art. 9.

The initial value at $t = 0$ of the pressure or force turns out to be given for *all* speed ranges by a simple expression corresponding to the results shown for P_1 in the preceding cases. A direct physical interpretation may be given based on the momentum or, alternatively, on the kinetic energy of the acoustic wave system created by the starting motion. Thus, when an airfoil element of projected area ΔA suddenly pushes the adjacent air downward at a speed w_0, a rarefaction wave is formed on the top surface and a compression wave on the bottom surface. In the infinitesimal time Δt, the pressure wave travels a distance $a_\infty \Delta t$ and the mass of fluid brought to velocity w_0 is $2\rho_\infty \Delta A (a_\infty \Delta t)$. Hence, equating the impulse $F\Delta t$ to the gain in momentum, yields for the starting force:

$$F = 2\rho_\infty a_\infty w_0 \Delta A \tag{12-6}$$

Hence, the starting pressure difference is in general

$$P = -2\rho_\infty a_\infty w_0 \tag{12-7}$$

For the plane airfoil of chord $c = 2b$ acquiring vertical velocity $w = U\Delta\alpha$, corresponding to a change in angle of attack $\Delta\alpha$, the starting lift per unit span is

$$F = 4b\rho_\infty a_\infty U\Delta\alpha \tag{12-8}$$

and is thus proportional to the product of the chord, the speed of sound, and the forward velocity. This result appears to have been given originally by Possio [*80*] and, more simply, by Heaslet and Lomax [*76*]. The derivation suggests the immediate generalization to three-dimensional flow and indicates that the starting lift for a constant value of w_0 or α acts at the centroid of area.

Analytic results for initial portions of the indicial functions are given in [*71,81*] where the procedures are indicated to be based on work of

Evvard [82]. In Fig. F,12 which corresponds to a figure given in [71], there is contained a graphical summary of some of the transient results discussed. The possible role of indicial responses in the evaluation of dynamic stability effects, for example, damping in pitch under subsonic and supersonic flow conditions, has been discussed by Tobak [83] for a variety of planforms both for wing alone and for wing and tail. Other applications are indicated in (VI,D,26).

Accelerated Flows. Effects of accelerated motion in a compressible fluid are of special interest for transonic speeds. For the problem for the plane case utilizing linearized methods a few contributions have been

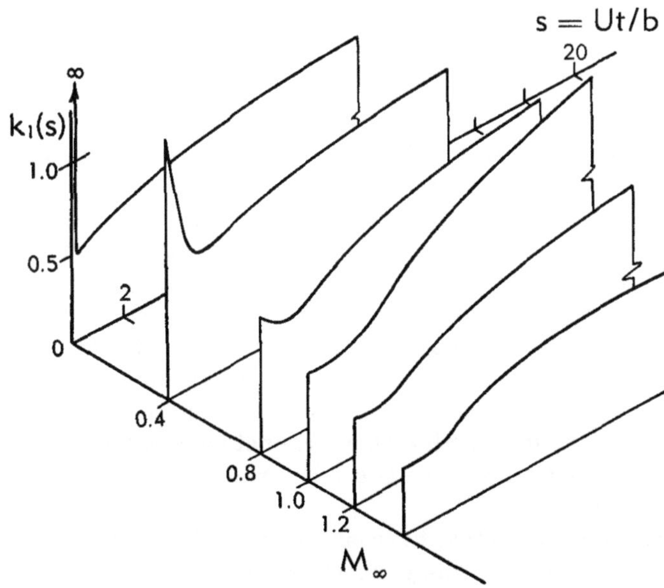

Fig. F,12. Indicial lift as functions of Mach number and distance traversed s measured in half chords.

made by Biot [84] and by Gardner and Ludloff [85]. To synthesize the solutions it is convenient to make use of the source-pulse type solutions as given in Art. 4 for problems stated either for fixed or moving coordinate systems. Solutions to the wave equation (2-7), for a space-fixed coordinate system, can be formed by making use of the pulse solution (4-9) and of boundary conditions at now time-dependent locations. It may be remarked that in general there are differences in aerodynamic characteristics for deceleration and for acceleration. For $M_\infty > 1$, the Mach wave front takes on for decelerated motion a convex character, while for accelerated motion it appears concave and cusped.

Analytical developments, as made in [84] or in [85], lead to a convenient acceleration parameter β_0, defined as twice the product of the acceleration n and the chord divided by the square of the speed of sound, that is,

$$\beta_0 = -\frac{2nc}{a_\infty^2}$$

($\beta_0 = 1$ corresponds to a deceleration from sound speed to rest in a time to travel of one chord.) The influence of deceleration for speeds not near $M_\infty = 1$ is hardly appreciable, but near $M_\infty = 1$ the pressure, lift, or drag depend strongly on the parameter β_0 (as $(\beta_0)^{-\frac{1}{3}}$). For $\beta_0 = 0$ and for $M_\infty = 1$, corresponding to the steady plane motion at sonic speed, infinite and useless results are obtained, however, as noted in [*85*], for "small" values of β_0, for example $\beta_0 = 0.02$; the linearized theory yields rational results even at sonic speeds. The general conclusions on the acceleration parameter are in accord with those on the effects of the frequency parameter as given in Art. 11.

F,13. Three-Dimensional Nonsteady Incompressible and Subsonic Potential Flow About Wings. Although a substantial development discussed in many papers exists, the problem of three-dimensional flow about wings must be considered to be in a state of continuing development. The nonsteady flow problem for wings like the steady problem, though much more so, is faced with the need for simplification of the results for ease of applicability in practice, and with the need for more detailed evaluation of the nature of the approximations and assumptions that are made, that is, for more exact treatment even on the basis of linearization. Simplified methods alone cannot be regarded as satisfactory unless compared against theoretical or experimental developments of known accuracy.

Among the methods that have been employed to treat deforming and oscillating wings are strip analysis methods, wherein the two-dimensional analyses as discussed in the preceding articles are employed (for appropriate cross sections of the wing) with or without correction factors, for example, slope of the lift curve correction factors. Other methods requiring progressively greater numerical effort are referred to as lifting line, multiple lifting line, or lifting surface methods. These methods may employ vortex distributions corresponding to discontinuities in the velocity potential or its derivatives, and hence involve both the wing and its wake explicitly, or may employ pressure or loading procedures which do not explicitly involve the wake. Some of the methods have an historical interest or an interim interest only, while others are currently in the process of development and application. In the present article, it is feasible to sketch only a few main considerations of the existing work and current developments.

Serving as a guide for some of the present methods and, in fact, a theoretical requirement of exact linearized methods are various limiting results: the exact solutions for the nonsteady two-dimensional flows (infinite aspect ratio cases), the zero aspect ratio case to be discussed later in this article, and various classical steady flow results, as for example the single line vortex results of Prandtl for wings of moderate but not small aspect ratio, or various modern developments in multiple-line methods or in wing surface theory for steady flows.

Cicala's Method. The earliest work is that of Cicala [*86,87*] for incompressible flow in which he makes use of vortex-filament methods similar to Birnbaum's for plane flow, and considers sheets of bound vorticity over the wing and free vorticity in the wake. The floating wake-vorticity with axes normal to the stream, as in the plane flow case, may be termed "shed," that, with axes in the direction of the main stream as in the steady three-dimensional flow especially near wing tips, may be

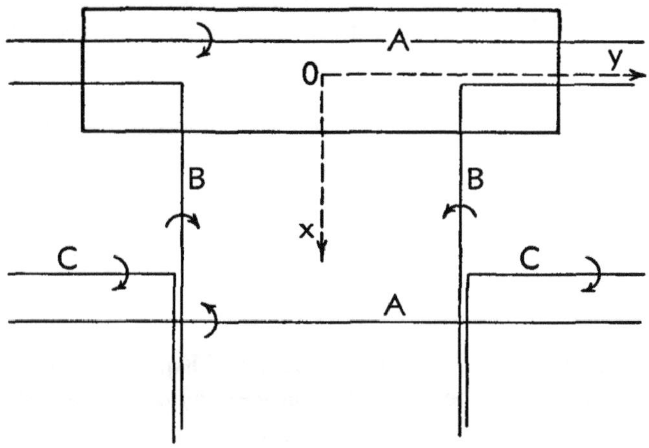

Fig. F,13a. Vortex elements in Cicala's method.

termed "trailing." Cicala introduces as a basic element the "right-angle or L" vortex line and builds up the vortex system by superposition of the three elements to be separately considered, shown in Fig. F,13a, which represent:

A. A two-dimensional system of bound and shed vortices
B. A system of trailing vortices
C. A system of angular vortices distributed in the wake

The effect of system A is treated on the basis of some results which the reader may recognize as contained in Eq. 5-21 and 5-22. Thus, Cicala writes for the normal velocity as corresponding to the representative terms in Eq. 5-22 without the time factor $e^{i\omega t}$,

$$\left.\begin{aligned} W_0 &= A \\ W_1 &= U(\tfrac{1}{2} + \cos\theta) \\ W_n &= U\cos n\theta \qquad (n \geqq 1) \end{aligned}\right\} \tag{13-1}$$

It follows from Eq. 5-21 that the associated terms of the "bound" vorticity γ^*, defined by the pressure loading $\Pi = -\rho U \gamma^* e^{i\omega t}$, are respectively

$$\left.\begin{aligned} \gamma_0^* &= 2A\left[C(k)\cot\frac{\theta}{2} + i\omega\sin\theta\right] \\ \gamma_1^* &= U\left[-\cot\frac{\theta}{2} + (2 - ik)\sin\theta - \frac{ik}{2}\sin 2\theta\right] \\ \gamma_n^* &= U\left[\frac{ik}{n-1}\sin(n-1)\theta + 2\sin n\theta \right. \\ &\qquad \left. - \frac{ik}{n+1}\sin(n+1)\theta\right] \end{aligned}\right\} \qquad (13\text{-}2)$$

Cicala assumes that only γ_0^* is affected by the finiteness of the wing, i.e. by the aspect ratio, and thus considers A as a function to be determined of the span variable y. The total section circulation, defined by the total surface difference of velocity potential over the chord, or by the equal and opposite quantity in the wake, is actually determined from γ_0^* only, the terms γ_n^* $(n \geqq 1)$ contributing nothing. Its value may be found from the following relations:

$$\Gamma = \int_{-1}^{1} \frac{\Delta\Gamma}{\Delta x_1} dx_1$$

where

$$\Delta\Gamma = \int_1^\infty bE dx_0$$

and E is as given in Eq. 5-12. By making use of Eq. 5-7, (and with care as mentioned in Art. 5 in the integration as $s \to \infty$) $\Delta\Gamma$ can be expressed as

$$\Delta\Gamma = \frac{2\pi\Delta Q}{\lambda(k)} \qquad (13\text{-}3)$$

where ΔQ is indicated in Eq. 5-14 and

$$\lambda(k) = \frac{\pi k}{2}[H_0^{(2)}(k) - iH_1^{(2)}(k)]e^{ik}$$

By making use of the substitution $x_1 = -\cos\eta$ and the trigonometric form for w (Eq. 5-22), the total circulation is

$$\Gamma = \frac{2\pi b U(A_0 - A_1)}{\lambda(k)}$$

where $U(A_0 - A_1)$ corresponds to Cicala's A in Eq. 13-1. Finally, with $c = 2b$ denoting the local chord, and with the reduced frequency k considered a function of span variable y, the total section circulation is given as

$$\Gamma(y) = \pi c \frac{A(y)}{\lambda(k)} \qquad (13\text{-}3a)$$

It may be remarked that although the use of total circulation is suggested by analogy with the steady problem, its real role in the unsteady problem is not clear. The assumptions reduce the theory in effect to a line vortex theory for oscillating flow, not a surface theory. System A is thus considered to contribute an induced velocity at a span location y given by Eq. 13-3a as

$$A(y) = \frac{\lambda(k)\Gamma(y)}{\pi c} \tag{13-4}$$

The system of trailing vortices (B) induces a velocity similar in form to that for the steady flow problem and given by

$$-\int_{-s}^{s} \frac{d\Gamma}{d\eta} \frac{d\eta}{y - \eta} \tag{13-5}$$

where s is the semispan.

It is remarked that the effect of concentrating the total circulation at the line vortex involves the greatly simplifying assumption that the wake starts at this line instead of the trailing edge. The effect of system C is found by first considering a single unit angular vortex (with vertex at ξ, η) giving an induced velocity at y of value

$$\frac{1}{4\pi}\left(\frac{1}{|y - \eta|} + \frac{1}{\xi} - \sqrt{\frac{1}{(y - \eta)^2} + \frac{1}{\xi^2}}\right) \tag{13-6}$$

To account for the whole infinite wake generated by the forward motion and of harmonic strength, Cicala introduces the function

$$F\left(\frac{\omega}{U}, |y - \eta|\right) = \int_0^\infty \left(\frac{1}{|y - \eta|} + \frac{1}{\xi} - \sqrt{\frac{1}{(y - \eta)^2} + \frac{1}{\xi^2}}\right) e^{-i\frac{\omega\xi}{U}} d\xi \tag{13-7}$$

and gives the contribution of system C as

$$\frac{i\omega}{4\pi U} \int_{-s}^{s} F \frac{d\Gamma}{d\eta} d\eta \tag{13-8}$$

Then the basic integral equation of Cicala's method is given by

$$w_0(y) = \frac{\lambda\Gamma(y)}{\pi c(y)} - \frac{1}{4\pi} \int_{-s}^{s} \left(\frac{1}{y - \eta} - \frac{i\omega F}{U}\right) \frac{d\Gamma}{d\eta} d\eta \tag{13-9}$$

where the contributions of the three systems (A, B, and C) may be recognized as the three terms on the right. In this equation $w_0(y)$, obtained from the first term in an expansion like Eq. 5-22, is to be found from the actual behavior of the wing section through y, i.e. from the normal velocity for the particular mode of oscillation under consideration. Then solution of Eq. 13-9, for example, by collocation methods will determine the spanwise distribution of Γ, and with Γ or $A(y)$ determined, the pressure distribution is expressed by the relations in Eq. 13-2 similarly to Eq. 5-21. It is observed that for $k = 0$ ($\lambda = 1$, $\omega F = 0$) Eq. 13-9 reduces to the classical induced-velocity equation of Prandtl.

Various numerical applications of Cicala's method, or extensions of it, for rectangular and elliptic wings are in the literature, e.g. [*87,88,89*]. The Cicala F function has proved useful in several investigations; values of it have been tabulated in [*87*], and more fully in [*90*]. It can be expressed for calculation purposes by the substitutions

$$X = \frac{\omega x}{U}; \qquad Y = \frac{\omega(y - \eta)}{U}$$

as

$$F(Y) = \int_0^\infty \left(\frac{1}{X} + \frac{1}{Y} - \sqrt{\frac{1}{X^2} + \frac{1}{Y^2}}\right) e^{-iX} dX \tag{13-10}$$

Küssner's Method. Küssner [*10*] utilizes the acceleration or pressure potential (Eq. 2-10) and considers the problem to be that of determining an unknown load distribution $\gamma^*(\xi, \eta)$ of harmonically oscillating pressure dipoles over the wing surface S,

$$\psi = -\frac{p}{\rho_\infty} = \frac{Ue^{i\omega t}}{4\pi} \iint_S \gamma^*(\xi, \eta) \frac{z}{r^3}\, d\xi d\eta$$

On the basis of the development in Art. 4 for the unit dipole (Eq. 4-3b) it is recognized that the surface pressure difference at x, y is, as in Eq. 9-6,

$$P = -\rho_\infty U \gamma^*(x, y) e^{i\omega t}$$

The problem of determining γ^* is set up as an integral equation by means of Eq. 3-5, 3-6, and 2-11 in terms of the normal velocity distribution w, similarly to the obtaining of Eq. 9-1, and yields

$$w^*(x, y) = \frac{1}{4\pi} \iint_S \gamma^*(\xi, \eta) \frac{\partial^2}{\partial z^2} (I)_{z=0} d\xi d\eta \tag{13-11}$$

where

$$I = \int_{-\infty}^{x} e^{i\frac{\omega(x'-x)}{U}} [(x' - \xi)^2 + (y - \eta)^2 + z^2]^{-\frac{1}{2}} dx'$$

Substitution of a new variable u' for x', defined by priming u and x in the relation

$$u = -\frac{(x - \xi)}{\sqrt{(y - \eta)^2 + z^2}}$$

yields, according to [10],

$$w^*(x, y) = \frac{1}{4\pi} \iint\limits_S \gamma^*(\xi, \eta) e^{-i\frac{\omega}{U}(x-\xi)} d\xi d\eta \left[\frac{\partial^2}{\partial z^2} \int_u^\infty e^{-i\frac{\omega}{U} u\sqrt{(y-\eta)^2+z^2}} \frac{du}{\sqrt{u^2+1}} \right]_{z=0}$$

To reduce the integrations, a significant approximation $u = 0$ is made by Küssner in the lower limit. This approximation requires that $|x - \xi| \ll |y - \eta|$, which becomes increasingly valid with increase in aspect ratio. However, since u also contains z, all the implications of this assumption and of the substitution are not clear. The equation is now written as

$$w^*(x, y) = e^{-i\omega\frac{x}{U}} \bar{w}(y) \tag{13-12}$$

where, with s as semispan,

$$\bar{w}(y) = \frac{1}{4\pi} \int_{-s}^{s} K(\eta) d\eta \left[\frac{\partial^2}{\partial z^2} \int_0^\infty e^{-i\frac{\omega}{U} u\sqrt{(y-\eta)^2+z^2}} \frac{du}{\sqrt{u^2+1}} \right]_{z=0}$$

and where, with b as semichord,

$$K(\eta) = \int_{-b}^{b} \gamma^*(\xi, \eta) e^{i\frac{\omega}{U}\xi} d\xi$$

It may be noted that the approximations involved have produced in Eq. 13-12 a very special normal velocity distribution with respect to x, one associated with harmonic traveling waves. Although this traveling-wave distribution contains the variable x, it does not appear to be more general in scope than the distribution assumed by Cicala. The value of γ^* for the assumed distribution can be found under the second main assumption, that each section of the wing behaves as part of a uniform infinite wing, hence γ^* can be considered as known for each section from any of the general pressure results (Eq. 5-20, 5-21, or 5-24), except for the factor $\bar{w}(y)$ which is to be determined. Then K is determined from Eq. 5-22 to be

$$K(\eta) = \pi c \bar{w}(\eta) f(k)$$

where $c = 2b$ is the chord, and

$$f(k) = [J_0(k) - iJ_1(k)]\{[J_0(k) - iJ_1(k)]C(k) + iJ_1(k)\}$$

Then the integral equation of Küssner is given by

$$\bar{w}(y) = \frac{K(y)}{\pi c(y) f(k)} + \frac{1}{4\pi} \int_{-s}^{s} \frac{S[(\omega/U)|y - \eta|]}{y - \eta} \frac{dK(\eta)}{d\eta} d\eta \tag{13-13}$$

where the kernel function is

$$S(\sigma) = -\sigma \int_{\sigma}^{\infty} \frac{d\tau}{\tau} \frac{d}{d\tau} \int_{0}^{\infty} \frac{e^{-i\tau u}}{\sqrt{u^2+1}} du$$

$$= \frac{i\pi}{2} \sigma \int_{0}^{\infty} G_1(u) \frac{du}{u} \tag{13-14}$$

and where $G_1(u)$ is a combination of a Struve function Z_1 and a Neumann function N_1 of the first order,

$$G_1(u) = Z_1(iu) - N_1(iu) - \frac{2}{\pi}$$

It is noted that like Eq. 13-9, Eq. 13-13 is of a form corresponding to vortex filament theory, and it also reduces for $\omega \rightarrow 0$ to the Prandtl relation for steady flow. Moreover, closer comparison of the kernel in Cicala's equation with that here reveals that these functions are in fact identical, namely

$$1 - \frac{i\omega}{U}(y-\eta)F\left(\frac{\omega}{U}|y-\eta|\right) = S\left(\frac{\omega}{U}|y-\eta|\right) \tag{13-15}$$

so that aside from interpretation ($e^{ik}\Gamma = K$) the integral equations differ formally only in one term.

As developed, Eq. 13-13 does not contain the limiting two-dimensional case except indirectly. It is necessary in strip analysis applications with general downwash distributions to interpret or to weight the special downwash distribution (Eq. 13-12) in such a manner as to make it fit the results of the plane case. Applications with the use of Multhopp's methods for span loading have been made in [*91*].

Reissner's Method. The work of Reissner [*92,93,94*] for oscillating airfoils of finite span in incompressible flow makes use of velocity potential methods, and presents a rigorous formulation of the mathematical boundary-value problem as it relates to the determination of the velocity potential in half space from the mixed boundary conditions at the dividing plane. It thus presents a framework for velocity potential methods in which approximations may be introduced.

The simplifications utilized by Reissner, however, also involve a basically two-dimensional type of analysis in strips taken streamwise, but in this method the limiting two-dimensional case is contained directly and occurs as a natural limit.

A main result of Reissner is an expression for the pressure difference [*93*, p. 32]; however, the first integral expression therein contains a sign error, herein corrected:

$$\frac{\bar{P}}{\rho_\infty U} = \frac{2}{\pi}\int_{-1}^{1}\left[\sqrt{\frac{1-x}{1+x}}\sqrt{\frac{1+\xi}{1-\xi}}\frac{1}{\xi - x} - \frac{ik}{2}L(x, \xi)\right]\bar{w}d\xi$$
$$+ \left(\frac{2}{\pi}\sqrt{\frac{1-x}{1+x}}\int_{-1}^{1}\sqrt{\frac{1+\xi}{1-\xi}}\bar{w}d\xi\right)[-1 + C(k) + \sigma] \quad (13\text{-}16)$$

where quantities having the time factor $e^{i\omega t}$ omitted have bars. The result differs from the plane flow case given in Art. 5 *solely* in a modification of the function $C(k)$ by the additive correction σ where

$$\sigma(y, k) = \left[\frac{\bar{\Omega}}{\bar{\Omega}^{(2)}} - 1\right]\left[C(k) + \frac{iJ_1(k)}{J_0(k) - iJ_1(k)}\right] \quad (13\text{-}17)$$

This function depends on the ratio of the three-dimensional "circulation" function $\bar{\Omega}$ to that for two-dimensional flow $\bar{\Omega}^{(2)}$. It is determined for a given oscillating wing mode of vibration, by means of an integral equation of the same type as Eq. 13-9 and 13-13, namely

$$\bar{\Omega}(y) + \mu(k)\frac{b}{b_0}\int_{-s}^{s}\frac{d\bar{\Omega}}{d\eta}K(k_0|y - \eta|)d\eta = \bar{\Omega}^{(2)}(y) \quad (13\text{-}18)$$

where y and η are here nondimensional span variables referred to the semichord at midspan b_0. The function $\bar{\Omega}^{(2)}$ is defined in terms of the normal velocity distribution $\bar{w}$ as

$$\bar{\Omega}^{(2)} = 4\frac{b}{b_0}e^{ik_m}\frac{\int_{-1}^{1}\sqrt{\frac{1+\xi}{1-\xi}}\bar{w}d\xi}{\pi ik[H_1^{(2)}(k) + iH_0^{(2)}(k)]}$$

where k_m is a constant which is zero for symmetrical planforms. The function $\mu(k)$ is defined by

$$\mu(k) = \frac{J_0(k) - iJ_1(k)}{\pi(k)[H_0^{(2)}(k) - iH_1^{(2)}(k)]}$$

and finally the kernel function K is defined by a relation which can be shown to make it equivalent and equal to the kernels in Eq. 13-15. The similarity of form of Eq. 13-18 and Eq. 13-9 and 13-13 is noteworthy, and suggests that the three methods have similar limitations in that they are effectively single-variable or single-line methods. There is need, in particular, for improvement of these methods in the approximations in the neighborhood of tips.

Applications of Reissner's method for rectangular and elliptic wings of aspect ratios 3 and 6 are given by Reissner and Stevens [*95*]. These applications show that the finite span effects are appreciable, that they depend on the mode shape of the vibrating wing, and that they become less marked for higher values of the reduced frequency. Other work [*96*] based on this method introduces fewer simplifying assumptions in order to treat better the effect of the wing tips, but numerical results are not given.

Remarks on Methods of R. T. Jones, Sears, and Biot and Boehnlein. As noted for the two-dimensional flow in Art. 6 and 7 there are two fundamental approaches: the harmonic based on sinusoidal motion, and the indicial, based on the unit step motion. R. T. Jones [*97,98*] examined in 1939 the problem of unsteady lift for a wing from the indicial approach and by superposition obtained results for the oscillating wing. This work deals essentially with the case of lift forces due to circulation for vertical translation of a rigid wing, and the numerical work is carried out for an elliptic wing of aspect ratios 3, 6, and ∞.

Of special interest is Jones' calculation of the starting lift for a sudden acquisition of downwash. This is done on the basis that the flow produced at the start corresponds, because of the nature of the Kutta condition, to that produced by a wing growing wider in chord at the rate U. The starting lift is calculated as a reaction to uniform vertical motion w of a wing with increasing apparent mass m'

$$L = w \frac{dm'}{dt}$$

This is calculated for an elliptic wing to be $\pi\rho Uw/2E$ where E is the ratio of the semiperimeter of the ellipse to the span, and $E \cong 1$ for high aspect ratio.

Jones then studies, by means of vortex loops and superposition, the growth of downwash, of circulation, and of lift as a function of distance traversed. The asymptotic values for large distances correspond to the steady case and approach values of Prandtl's theory. Jones utilizes, for the quantities of interest, expressions of the form

$$a + be^{-\alpha s} + ce^{-\beta s}$$

where s, the distance traversed, is the only variable for a given wing. Finite expressions of this form are especially useful in applications employing the operational or Laplace transform methods. It remains, however, rather difficult to assess the order of accuracy of the approximations except by numerical comparisons with other methods.

Sears [*99*] considered in 1938, as a possible simplification for actual wings, infinite wings with spanwise periodic downwash. This interesting type of problem was also later considered by Küssner [*10*] (see also Reissner [*100*]); however, its relation to the finite span wing problem is indirect, and it has not led to useful results for this problem.

Biot and Boehnlein [*101*] treated, by approximate vortex methods, the lift and moment on a midspan section of an elliptic wing which is translating and pitching. Their development, which cannot be described here, depends on various simplifications based on physical considerations. Their results are presented both in tabular and graphical form and in [*102*] are applied to flutter analysis. The accuracy can here, too, be judged only by numerical comparisons.

METHOD OF W. P. JONES. W. P. Jones has developed two procedures for wings of finite span in nonsteady incompressible flow. One [*89*] is essentially an extension of Cicala's work to a surface-lattice approach in which the distribution of bound vortices is expanded in a series of products of simple functions of the spanwise coordinate and functions of the chordwise coordinate taken from the two-dimensional theory with undetermined coefficients (for example, Eq. 5-21, with the a's developed as appropriate functions of the span). The discontinuity in the velocity potential over the wing is represented by a doublet distribution, and the boundary conditions yield a set of linear equations for the undetermined coefficients. In a subsequent report Jones [*103*] indicates the use of Falkner's approximate steady flow method [*104*] for systematic numerical calculations of some of the distributions required. The other procedure [*105*] adopts the inverse method of calculating the indicial functions for transient flow conditions from the results of the harmonically oscillating case. Applications to oscillating wings of various aspect ratios entailing considerable numerical effort are to be found in several British Reports and Memoranda (see [*33*]). However, the degree of compromise necessary between the labor required and the accuracy desired or obtainable is yet undecided.

A FEW COMPARISONS. Some numerical and graphical comparisons of the work of Küssner, Cicala, Possio, and R. T. Jones have been given by Katterbach [*106*] for wings of aspect ratios of 6 and 3. Qualitatively, except for Possio's results, there is fair agreement. Wasserman [*107*] has modified Cicala's circulation function by effectively changing the first term on the right-hand side of Eq. 13-9; and he has made numerical comparisons with Biot, Küssner, and Reissner, the latter two yielding identical values for the cases treated. Numerical comparisons of the methods of Biot and Reissner are given in [*95*] and [*108*], and including experimental comparison in [*109*], although the experimental scatter precluded making a choice between the theories.

The need for further resolution of the problem of the oscillating wing of small or moderate aspect ratio better to account for edge effects, and to yield the required distribution of loading for vibration modes of a given frequency is apparent. Additional procedures, some including consideration of the variation of the Mach number in subsonic flow, are given subsequently in this article. Recent surveys of the problem have been presented by Eckhaus [*110*] and by Garrick and Watkins [*111*].

WING OF CIRCULAR PLANFORM. The classical method of separation of variables in appropriate coordinates may, as mentioned in Art. 9, be considered for finite wings. A general discussion has been given by Küssner [*54*]; however, the actual carrying out of the procedures is limited theoretically by the finite choice of available orthogonal coordinate systems (see, e.g., Miles [*112*]) and also practically by the lack of mathematical details and numerical tables of the wave functions needed.

The circular planform is the only finite planform for which exact linearized procedures have been attempted. The circular wing has been treated by acceleration potential methods by Schade and Krienes, following Kinner's treatment of the steady case, and results are presented in [*113*]. This work has been limited to oscillating normal deformations of the planform of the second degree in x and y, hence to modes of oscillation including rigid body roll, pitch, and vertical translation; linear antisymmetric torsion; parabolic spanwise bending; and parabolic chordwise bending. The problem has also been considered on the basis of the velocity potential by Kochin in two papers [*114*] purporting to be an exact linearized treatment of the problem. However, inverse methods of calculation are employed in which the shape is determined for a given oscillating load.

Two approximate results given by Kochin for small frequencies may be of interest, though the numerical values seem to require checking. The first is the expression of the lift force and the moment for pitching oscillations of the form $\alpha \cos \omega t$ about the circle diameter $2a$ normal to the stream U

$$L = \rho_\infty U^2 a^2 \alpha \left(2.813 \cos \omega t - 1.766 \frac{\omega}{U} a \sin \omega t\right)$$

$$M = \rho_\infty U^2 a^2 \alpha \left(1.473 \cos \omega t + 0.867 \frac{\omega}{U} a \sin \omega t\right)$$

The second is the lift force for vertical flapping oscillations of the form $h \cos \omega t$

$$L = \rho_\infty U^2 a^2 \left(2.813 \frac{\omega}{U} h \sin \omega t + 0.301 \frac{\omega^2}{U^2} ah \cos \omega t\right)$$

An application of the theory of Schade and Krienes to a flutter calculation of an approximately circular horizontal tail surface has been indicated by Dörr [*115*]. It is of interest to mention that lifting line theory (stretched to apply to aspect ratio of unity) yields a steady flow, lift curve slope of 2.44 whereas the circular planform theory yields 1.80. Additional details including some experimental lift and moment comparisons (and results of preliminary flutter tests) may be found in a paper by Voss and Hassig [*116*].

Analysis of the general elliptic wing remains as a feasible and highly desirable objective of theoretical inquiry. It would furnish, for example, several limiting cases that could serve for comparison and evaluation of other methods.

UNSTEADY FLOW FOR WINGS OF VERY SMALL ASPECT RATIO.[8] Analytic and experimental information on nonsteady flow past wings of very small aspect ratio is now slowly being accumulated. The original concepts which have proved useful for study of water forces on planing surfaces and for forces on airships guided the developments of R. T. Jones [*118*] for steady flow about wings of small aspect ratio. These concepts rest on the assumption that the disturbance flow pattern can be considered two-dimensional for strips of the planform transverse to the main stream, and account for the effect of the change in the flow pattern on the forces. Jones considers the lift for a flat pointed triangular airfoil, at a constant angle of attack α, as resulting from the (small) rate of change in the scale of the flow pattern

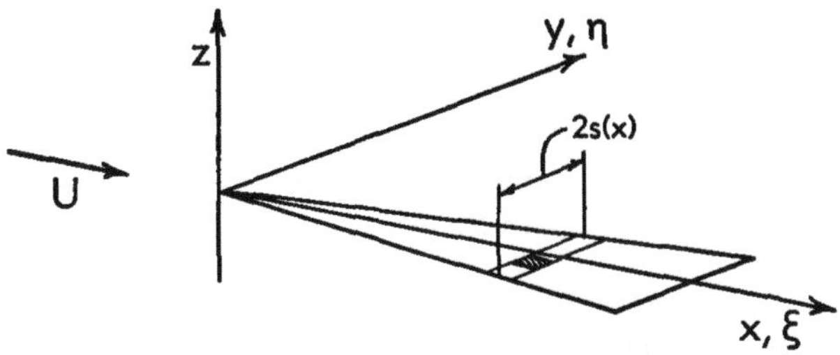

Fig. F,13b. Wing of very small aspect ratio.

transverse to the main stream U in the x direction. With $m = \pi\rho_\infty s^2 dx$ representing the apparent mass of a strip of width dx and of span $2s$, the lift per unit length streamwise is given as

$$l = U\alpha \frac{dm}{dt}$$

$$= 2\pi\alpha\rho_\infty U^2 s \frac{ds}{dx} \tag{13-19}$$

The distribution of pressure corresponding to this lift is given as

$$P = 2\alpha\rho_\infty U^2 \frac{s}{\sqrt{s^2 - y^2}} \frac{ds}{dx} \tag{13-20}$$

To generalize these results to unsteady flow, consider the flow pattern in greater detail. The velocity potential of the transverse flow pattern satisfies, in incompressible flow, Laplace's equation in two dimensions, and in compressible flow, the two-dimensional wave equation of acoustics. Consider first the incompressible flow case and the sketch in Fig. F,13b.

[8] The results given here were obtained jointly by Reissner and Garrick by different procedures during June 1948 at the Langley Aeronautical Laboratory. The exposition is that given incidentally and without references in the appendix of [*117*].

The surface velocity potential at a point y on a transverse strip of span $2s(x)$, for a disturbance w existing at the element of the strip $\Delta\eta$, is given by a function obtained from Eq. 5-1 (with interchange of the nondimensional chord variables and the dimensional span variables) as

$$\Delta\varphi = \frac{1}{2\pi} w(x, \eta, t)L(y, \eta, s)\Delta\eta \tag{13-21}$$

where

$$L(y, \eta, s) = \ln \frac{(y - \eta)^2 + (\sqrt{s^2 - y^2} - \sqrt{s^2 - \eta^2})^2}{(y - \eta)^2 + (\sqrt{s^2 - y^2} + \sqrt{s^2 - \eta^2})^2}$$

or, with $y = -s \cos \theta$, $\eta = -s \cos \psi$,

$$L = -2 \ln \frac{\sin \frac{\theta + \psi}{2}}{\sin \frac{|\theta - \psi|}{2}}$$

The rate of contribution to the pressure difference at y, of the element at η, is given by Eq. 3-7 and 13-21 as

$$\frac{dP}{d\eta} = -\frac{\rho}{\pi}\left(\frac{\partial}{\partial t} + U\frac{\partial}{\partial x}\right)(wL) \tag{13-22}$$

The total pressure difference at a point y of the strip of span $2s$ is given by the integration

$$P(x, y, t) = \int_{-s}^{s} \frac{dP}{d\eta}\, d\eta \tag{13-23}$$

Application may be made, for example, to obtain explicit results for vertical deflections of the form $h(x, t)$ for which

$$w = \frac{\partial h}{\partial t} + U\frac{\partial h}{\partial x}$$

For a nontelescoping wing (s not a function of t),

$$\frac{\partial L}{\partial t} = 0, \quad \frac{\partial L}{\partial x} = \frac{\partial L}{\partial s}\frac{ds}{dx}$$

and moreover,

$$\int_{-s}^{s} L d\eta = -2\pi\sqrt{s^2 - y^2}$$

Then

$$P = 2\rho_\infty\left(\ddot{h} + 2U\frac{\partial \dot{h}}{\partial x} + U^2\frac{\partial^2 h}{\partial x^2}\right)\sqrt{s^2 - y^2} + 2\rho U\left(\dot{h} + U\frac{\partial h}{\partial x}\right)\frac{s}{\sqrt{s^2 - y^2}}\frac{ds}{dx} \tag{13-24}$$

This result contains acceleration $\ddot{h}$, velocity $\dot{h}$, bending slope $\partial h/\partial x$, curvature $\partial^2 h/\partial x^2$, and time rate of change of slope $\partial \dot{h}/\partial x$. For constant angle of attack $(\dot{h}/U) = \alpha$, the result is that given by Jones (Eq. 13-20).

The lift per unit length streamwise is

$$l = \int_{-s}^{s} P dy$$

$$= \pi \rho_\infty s^2 \left(\ddot{h} + 2U \frac{\partial \dot{h}}{\partial x} + U^2 \frac{\partial^2 h}{\partial x^2} \right) + 2\pi \rho U s \frac{ds}{dx} \left(\dot{h} + U \frac{\partial h}{\partial x} \right) \quad (13\text{-}25)$$

Another form of interest is obtained by expressing the velocity distribution as a Fourier series

$$w = A_0 + 2 \sum_1^\infty A_n \cos n\psi \quad (13\text{-}26)$$

where the A's are functions of x and t, and ψ is a function of x through s. The lift l can then be written as

$$l = \pi \rho_\infty s^2 \left(\frac{\partial B_0}{\partial t} + U \frac{\partial B_0}{\partial x} + 2U \frac{1}{s} \frac{\partial s}{\partial x} B_0 \right) \quad (13\text{-}27)$$

and similarly the rolling moment per unit length streamwise, as

$$m = \int_{-s}^{s} P y dy$$

$$= \frac{\pi}{4} \rho_\infty s^3 \left(\frac{\partial B_1}{\partial t} + U \frac{\partial B_1}{\partial x} + 3U \frac{1}{s} \frac{\partial s}{\partial x} B_1 \right) \quad (13\text{-}28)$$

where $B_0 = A_0 - A_2$ and $B_1 = A_1 - A_3$.

The foregoing results, it is to be noted, arise from the change in the flow pattern of noncirculatory terms only, at a sufficient distance from the influence of the foremost part, and do not depend on the wake. It is thus assumed that the influence of the wake is secondary. Moreover, in the steady problem Jones assumes that the aft parts of the planform, for which the changes in width ds/dx are negative, are ineffective in producing lift. The results given here for the unsteady problem are best considered as oversimplifications requiring clarification by further work; in particular, the role of the wake and the planform edges, possible separation, and comparison with other more exact methods might be considered.

The small aspect ratio concepts have also been used and discussed by Miles [*119*], including the application of the concepts to the nonsteady flow about slender bodies. Additional independent studies have been made by Voss [*116,120*] and by Ashley [*121*].

Extension of Very Low Aspect Ratio Case to a Compressible Flow. Merbt and Landahl [*122,123*] have given the extension of the incompressible case to include Mach numbers for both subsonic and supersonic flows. The procedure is similar to that given in the preceding topic except that the two-dimensional form of Eq. 2-8 is involved in place of Laplace's equation and an equation of the form (9-24) is involved in place of Eq. 13-21, and hence the Mathieu function expansion in Eq. 9-25 occurs. The Mach number enters solely in combination with a reduced frequency based on the local span length. For sufficiently low frequencies the results reduce to the incompressible flow case regardless of the Mach number. An independent study of this topic by Mazelsky[9] makes use of the tables of [*56*], and gives detailed numerical and graphical results for the oscillating lift and moment for translation, pitching, and rolling, and also makes application to several planforms of low aspect ratio. A further study by Landahl [*124*] examines the influence of possible higher order effects on the air forces on oscillating wings of low aspect ratio at transonic speeds.

Considerations for Swept Wings. Problems of sweep encompass unswept wings of high aspect ratio on one extreme and low aspect ratio wings such as pointed triangular wings on the other, and include variably swept and tapered planforms. Progress on the wide scope of problems thus enveloped awaits the progress on the component problems, in particular on those associated with edge effects of the planform and with the induced effects of the wake. The kernel function method given later in this article offers a promising avenue of approach.

The finite swept wing of moderate to high length-to-chord ratio can be treated on the basis of a type of strip analysis, in which each strip of the wing is considered to act as though it were part of a uniform infinite yawed wing.[10] Before discussing the infinite yawed wing, however, the aerodynamic role of the vibration or deflection modes for swept wings merits attention, since this role is more explicit than for the unswept wing.

Thus consider a strip of the wing, oriented as in Fig. F,13c, undergoing vertical deflection h and turning θ with respect to an axis making an angle $90° + \Lambda$ to the forward motion. The surface normal velocity distribution is, as given in [*125*],

$$
\begin{aligned}
w &= \left(\frac{\partial}{\partial t} + U\frac{\partial}{\partial x}\right)(h + x'\theta) \\
&= \left(\frac{\partial}{\partial t} + U\cos\Lambda\frac{\partial}{\partial x'} + U\sin\Lambda\frac{\partial}{\partial y'}\right)(h + x'\theta) \\
&= h + x'\theta + U\theta\cos\Lambda + U(\sigma + x'\tau)\sin\Lambda
\end{aligned}
\tag{13-29}
$$

[9] Presented at the 1954 Annual Meeting of the Institute of the Aeronautical Sciences; *J. Aeronaut. Sci. 23*, 639–652 (1956).

[10] The flow model is in accord with that used also for the very small aspect ratio wings, the difference being in the retention of only the noncirculatory flow effects in the low aspect ratio case, whereas both noncirculatory and circulatory flow effects are retained in the swept wing case.

where the new elements that appear because of the angle of sweep Λ are $\sigma = \partial h/\partial y'$, the local bending slope of the axis of rotation (analogous to local dihedral), and $\tau = \partial\theta/\partial y'$, the local change in twist of the axis of rotation.

As in [125], this strip taken from a moderately long wing is considered to be part of a uniform infinite wing yawed at the angle Λ. The flow that establishes itself retains a vectorial character which depends on the orientation of the strip. The vectorial character of the familiar $\rho U\Gamma$ law may serve to clarify this statement. Thus the lift force per unit length of a yawed line vortex is $\mathbf{L} = \rho\mathbf{U} \times \mathbf{\Gamma}$, of magnitude $\rho|U|\,|\Gamma| \cos \Lambda$. Hence

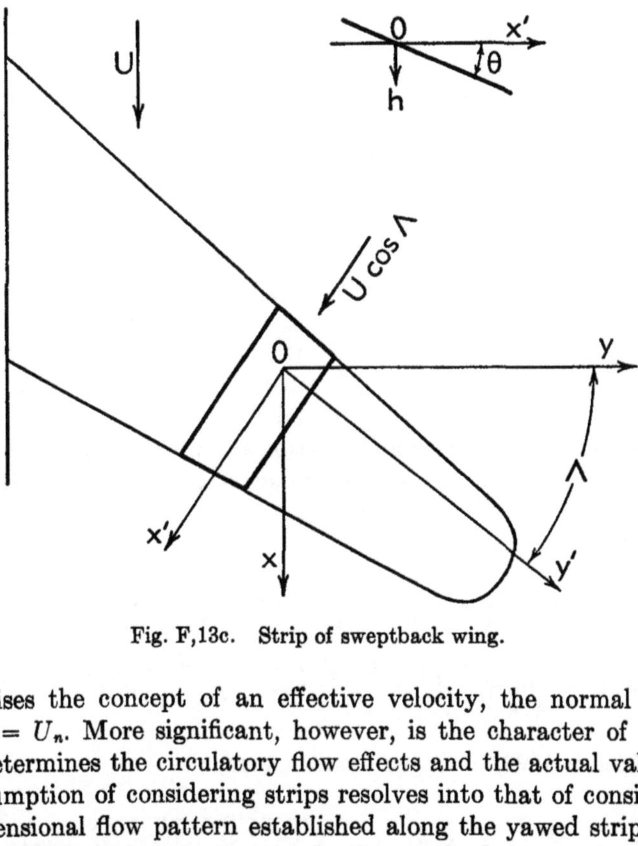

Fig. F,13c. Strip of sweptback wing.

there arises the concept of an effective velocity, the normal velocity $U \cos \Lambda = U_n$. More significant, however, is the character of the flow which determines the circulatory flow effects and the actual value of Γ. The assumption of considering strips resolves into that of considering a two-dimensional flow pattern established along the yawed strips, taken normal to the line of yaw or sweep. In the steady flow case, for angles of sweep not too near 90°, there is a rule for such flow patterns given by Busemann, which is also applicable to the oscillating flow case, as indicated by Garrick and Rubinow [11] and by Miles [126]. The rule implies that the pressures or loading associated with the uniform yawed infinite wing moving forward with velocity U in the x direction, for a given surface normal velocity distribution w, are exactly those of the same wing with the same surface normal velocity distribution, moving forward in x' direction with velocity $U \cos \Lambda$.

Development of results for the strip method are given in [*125*] for the coefficients associated with an incompressible fluid as in Art. 7. (See also [*127*].) For the subsonic case, $(U_n/a_\infty) < 1$, one may employ the coefficients of Art. 9, and for the supersonic case, those of Art. 10. In this latter case, as discussed in Art. 14, there exist explicit results which go beyond strip theory.

Turner [*128*] has developed the finite aspect-ratio, oscillating, swept-back wing along the methods of Reissner, and his results reduce for zero sweep to those given earlier in this article. It is significant that Turner introduces a kernel function, analogous to that for the right angle vortex line, as in Eq. 13-15, but which corresponds in the swept wing case to a skew angular $[(\pi/2) \pm \Lambda]$ vortex line. The method remains to be numerically and analytically evaluated. A discussion of it and the introduction of another simplified model of effects of sweep on a finite wing is contained in [*109*]. Reference to work of Van de Vooren and Eckhaus on a strip theory for oscillating swept wings in incompressible flow is made in [*110*].

Method of Lawrence and Gerber. This method [*129*] introduces a new way of reducing the surface integral equation for oscillating finite wings in incompressible flow to a single-variable integral equation, one in terms of the chordwise variable x rather than the span variable y. The method is an approximate one, limited to planforms having straight trailing edges normal to the stream, and should apply well for rigid body types of modes of translation and rotation and probably also for the lower order elastic modes. It is based on the general surface integral equation as given by Reissner in the two forms

$$w^*(x, y) = \lim_{z\to 0} \frac{1}{2\pi} \iint \frac{u^*(\xi, \eta) z d\xi d\eta}{[(x-\xi)^2 + (y-\eta)^2 + z^2]^{\frac{3}{2}}}$$

$$w^*(x, y) = \lim_{z\to 0} \frac{-1}{2\pi} \iint \frac{u^*(\xi, \eta)(x-\xi) + v^*(\xi, \eta)(y-\eta)}{[(x-\xi)^2 + (y-\eta)^2 + z^2]^{\frac{3}{2}}} d\xi d\eta$$

and modified in [*129*] to the equivalent form

$$w^*(x, y) = \frac{1}{2\pi} \frac{\partial}{\partial y} \iint \frac{u^*(\xi, \eta)}{y-\eta} \left(1 + \frac{\sqrt{(x-\xi)^2 + (y-\eta)^2}}{x-\xi}\right) d\xi d\eta \qquad (13\text{-}30)$$

where u^* is the perturbation velocity in the x direction and the integrations are to be taken over the planform and its extension in the wake.

By introducing a spanwise weighting function for symmetrical modes and various approximations, Eq. 13-30 is expressed as a single-variable integral equation

$$f(x) = \frac{1}{2}g(x) + \frac{1}{4}\int_{-b}^{b} \frac{dg}{dx}\left[1 + \frac{\sqrt{(x-\xi)^2 + \sigma^2(x)}}{x-\xi}\right] d\xi$$

$$- \frac{i\omega}{U} g(b) \int_{b}^{\infty} e^{-i\frac{\omega}{U}(x-b)} \left[1 + \frac{\sqrt{(x-\xi)^2 + s^2}}{x-\xi}\right] d\xi \qquad (13\text{-}31)$$

where b is the local semichord, $\sigma(x)$ the local semispan, s the semispan at the trailing edge, and where the known weighted downwash is

$$f(x) = \int_{-\sigma(x)}^{\sigma(x)} w(x, y) \sqrt{\sigma^2(x) - y^2}\, dy$$

and the unknown loading function is

$$g(x) = \int_{-\sigma(x)}^{\sigma(x)} \varphi^*(x, y) dy$$

The lift per unit of chord $l(x)$ is given by

$$l(x) = 2\rho U \left(\frac{\partial g}{\partial x} + \frac{i\omega}{U} g(x)\right)$$

A similar integral equation is obtained for antisymmetric modes by introducing a different weighting function, an antisymmetric one. It is stated in [*129*] that various limit results for wings of very small and very large aspect ratios are satisfactorily included in the method. Various aerodynamic coefficients are given for rectangular and triangular planforms of moderately low aspect ratio. Comparison with Reissner's method for rectangular wings of aspect ratio 3 shows differences of the order of 15 per cent.

Another study of aspect ratio effects, in which comparison of flutter calculations made with several methods is given, is that of Goland, Luke, and Dengler [*130*]. These authors also consider the use of multiple-line methods. Reissner [*131*], and also Voss and Hassig [*116*], have modified the single-variable integral equation approach to develop simultaneous integral equations which involve spanwise distribution of both the lift and the moment, and hence should lead to improved numerical results.

REMARKS ON SUBSONIC COMPRESSIBLE NONSTEADY FLOW. The chief method to be discussed for oscillating finite wings in a subsonic flow is that which generalizes the method of Possio given in Art. 9, and which involves the integral equation and its appropriate kernel function. Before doing this, brief mention may be made of other work touching on compressibility effects, the first of which is the subsonic compressibility correction for unsteady flow which may be regarded as an extension of the Göthert steady flow rule (VI,C,5), itself an extended Prandtl-Glauert rule. Miles [*132*] has made use of a series of transformations (as indicated in Art. 9, Eq. 9-13, but including also the z variable), which on neglecting terms of order of $(kM_\infty/\beta^2)^2$ reduces the governing equation for the velocity potential or for the pressure to Laplace's equation. A sequence of several operations may then be expressed [*132,116*], which relates the pressure distribution on the oscillating finite wing in subsonic flow to that of a wing of another aspect ratio oscillating at another frequency but in incompressible flow.

Starts on methods of including compressibility effects in the integral-equation approach for oscillating wings have been made in two papers by Reissner [*133,134*]. In another paper [*50*], he has also indicated the possibility of obtaining finite span corrections to the two-dimensional boundary-value approach mentioned in Art. 9, utilizing Mathieu functions. Mention should also be made again of Küssner's paper [*54*].

The Kernel Function for Oscillating Finite Wings in a Subsonic Flow. The kernel function, as introduced in Art. 9, has the property of giving the normal velocity at any field point induced by a unit harmonic loading at an isolated loading point. Consider the pressure doublet solution in space obtained from Eq. 4-24, and which is analogous to the two-dimensional form given in Eq. 9-2,

$$\frac{-p^*(x, y, z)}{\rho_\infty} = \frac{\partial}{\partial z}\frac{1}{4\pi R}e^{i\mu(x-\xi)-i\kappa R} \tag{13-32}$$

where κ, μ, and R are given in Eq. 4-24.

The induced velocity at any surface point x, y due to unit loading at ξ, η is found from the use of Eq. 2-13 to be, in analogy to Eq. 9-1 and 9-3,

$$K(x_0, y_0; M_\infty, \omega) = \lim_{z\to 0}\frac{\partial^2}{\partial z^2}e^{-i\frac{\omega x_0}{U}}\int_{-\infty}^{x_0}\frac{e^{i\mu(\lambda - M_\infty\sqrt{\lambda^2+\beta^2 y_0^2+\beta^2 z^2})}}{\sqrt{\lambda^2+\beta^2 y_0^2+\beta^2 z^2}}d\lambda \tag{13-33}$$

where $\mu = \omega U/\beta^2$, $\beta = \sqrt{1 - M_\infty^2}$, $x_0 = x - \xi$, and $y = y_0 - \eta$. A reduction of this improper integral to an explicit form has been made by Watkins, Runyan, and Woolston [*135*] and leads to a rather involved expression exhibited in the following equation:

$$\begin{aligned} K(x_0, y_0; M_\infty, \omega) &= \frac{k^2}{l^2}e^{-ikx_0}\Bigg\{-\frac{1}{k|y_0|}K_1(k|y_0|) - \frac{\pi i}{2k|y_0|}[I_1(k|y_0|) - L_1(k|y_0|)] \\ &+ \frac{iM_\infty k|y_0| + \beta}{M_\infty\beta(ky_0)^2}e^{-\frac{iM_\infty k|y_0|}{\beta}} - \int_0^{M_\infty/\beta}\sqrt{1+\tau^2}\,e^{-iky_0\tau}d\tau \\ &- \frac{M_\infty kx_0 + \sqrt{(kx_0)^2+\beta^2(ky_0)^2}}{M_\infty(ky_0)^2\sqrt{(kx_0)^2+\beta^2(ky_0)^2}}e^{\frac{i}{\beta^2}[kx_0 - M_\infty\sqrt{(kx_0)^2+\beta^2(ky_0)^2}]} \\ &+ \frac{i}{M_\infty(ky_0)^2}\int_0^{kx_0}e^{\frac{i}{\beta^2}[\lambda - M_\infty\sqrt{\lambda^2+\beta^2(ky_0)^2}]}d\lambda\Bigg\} \end{aligned} \tag{13-34}$$

where K_1 and I_1 denote modified Bessel functions, L_1 denotes the modified

Struve function, and $k = \omega l/U$, where l is an arbitrary reference length often conveniently chosen as a reference half chord. The kernel function (Eq. 13-34) has been separated for convenience, as was its two-dimensional equivalent in Eq. 9-10, into two parts. One of these contains all the singularities which in the applications can be treated analytically. The other part contains no singularities and can be numerically tabulated. An expansion of the kernel function in powers of the frequency, of possible utility in various problems of dynamic stability and low frequency flutter, is also given in [*135*].

It is shown in [*135*] that all special cases of the kernel function for incompressible flow, steady flow, and plane flow may be obtained from the general expression. Moreover, a special form for the sonic flow case suitable for use in calculations for $M_\infty = 1$ is also expressed. Another derivation of the kernel function by means of Laplace transform methods confirming [*135*] and leading to an expansion in powers of the frequency has been given by Lance [*136*]. (See also [*164*].)

Application of Integral Equation Containing the Kernel Function.[11] Consider now a procedure analogous to that used for the two-dimensional wing in Art. 9 for treating the integral equation for a finite wing of any planform. The integral equation which relates the known normal velocity distribution on the planform w^*, an unknown lift distribution or loading function L, and the kernel function K, may be expressed as

$$w^*(x, y) = \frac{1}{4\pi} \iint\limits_S L(\xi, \eta) K(x - \xi, y - \eta) d\xi d\eta \tag{13-35}$$

where the integration is to be taken over the planform surface. Let the loading L be comprised of a number of modes of pressure distribution as employed in Eq. 9-7, but involving besides the chordwise pressure terms, spanwise modifications of these terms

$$\begin{aligned} L(\xi, \eta) = \rho U \sqrt{s^2 - \eta^2} \Big[&\cot \frac{\theta}{2} (a_{00} + a_{01}\eta + a_{02}\eta^2 + \cdots) \\ &+ \sin \theta (a_{10} + a_{11}\eta + a_{12}\eta^2 + \cdots) \\ &+ \sin 2\theta (a_{20} + a_{21}\eta + a_{22}\eta^2 + \cdots) + \cdots \Big] \end{aligned} \tag{13-36}$$

where s is the semispan, $\cos \theta = (\xi_0 - \xi)/b$, and where b is the local semichord and ξ_0 is the abscissa of the local midchord (see Fig. F,13d). The first term in Eq. 13-36, for example, corresponds to the familiar $\sqrt{1 + x}/\sqrt{1 - x}$ type of loading chordwise and to the elliptic type of loading $\sqrt{s^2 - y^2}$ spanwise.

[11] The exposition of this topic is similar to that given in [*111*] and is based on work of Runyan and Woolston.

The problem is to determine the coefficients a_{ij} which appear in the expression for the loading corresponding to a specified normal velocity function $w^*(x, y)$. As in the method of collocation described for the two-dimensional case in Art. 9, the integral equation expressing w^* can be reduced to a sum of terms involving the coefficients a_{ij}; the definite integrals which occur as factors of a_{ij} involve the kernel function K and each of the assumed modes of pressure distribution, and can be evaluated numerically as accurately as one wishes. Specifications of an appropriate number of control or pivotal points x, y will lead to a desired system of simultaneous equations for determining the coefficients a_{ij}.

The procedure is thus strictly analogous to the two-dimensional one, and it also bears a resemblance to Falkner's vortex lattice method of treating finite wings in steady incompressible flow, and to the related work of W. P. Jones based on the velocity potential referred to earlier in this article. As experience develops with the use of the method, it may be possible to make better choices of aerodynamic mode shapes for the planforms or speed ranges under consideration.

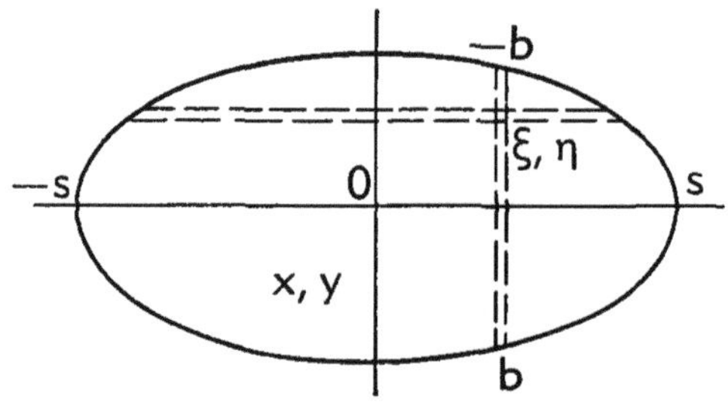

Fig. F,13d. Lifting-surface method using the kernel function.

The lifting surface methods for nonsteady flow involve great numerical labor, and for their efficient development require modern high speed computation. Systematic tabulation of the kernel function is also required. Some results of two applications obtained in preliminary numerical studies made by Runyan and Woolston are given next.

Lift and moment for a pitching wing ($M_\infty = 0$). Fig. F,13e shows the lift and moment of a rectangular wing of aspect ratio 2 oscillating in pitch about the midchord axis. The results are based on calculations involving 18 control points. The magnitude of the total lift and moment coefficients for $M_\infty = 0$ and the phase angles by which lift and moment lead the wing position are shown as a function of reduced frequency. Comparison is also made with results of Lawrence and Gerber's method and, as indicated, good agreement is found.

Variation of lift and moment with Mach number ($k = 0.22$). Some results covering a range of Mach numbers for this same case are given in Fig. F,13f for a selected value of the reduced frequency $k = 0.22$. Results

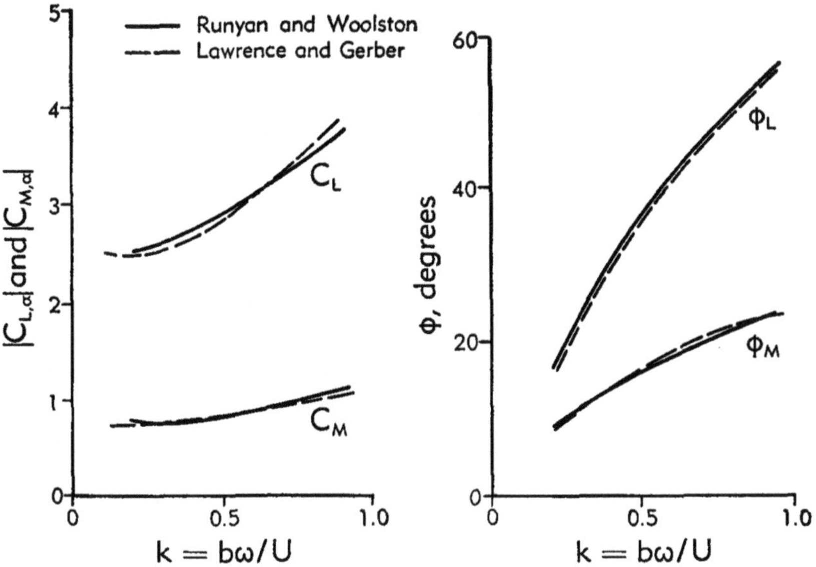

Fig. F,13e. Lift and moment for pitching rectangular wing of aspect ratio 2 for $M_\infty = 0$.

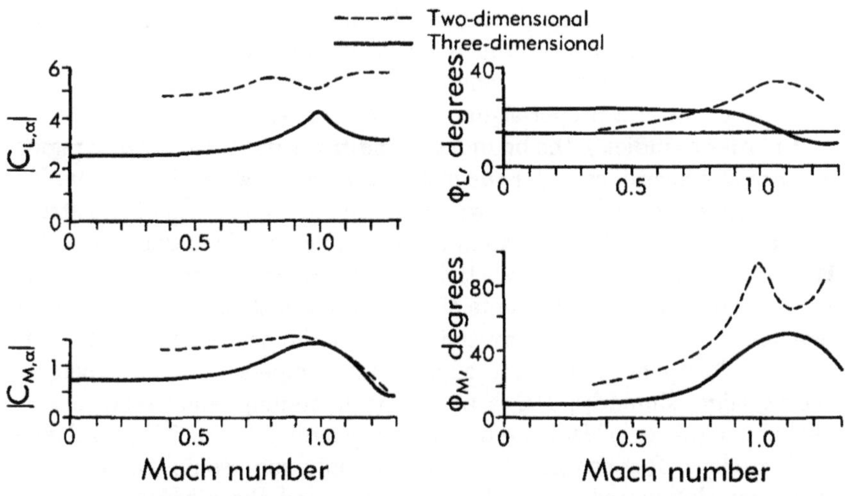

Fig. F,13f. Variation with Mach number of lift and moment for pitching rectangular wing of aspect ratio 2. (See *NACA Tech. Note 3694* for modified figure).

shown for values of M_∞ up to and including $M_\infty = 1$ were obtained by the method indicated, while some results shown for $M_\infty > 1$ were obtained from the theory of rectangular wings in supersonic flow (see Art. 14). For convenience of comparison, the results for a pitching wing based on the two-dimensional case are given in Fig. F,13f.

It is of importance to note that the numerical procedures outlined are not restricted to the rigid modes of an oscillating wing but apply equally to the elastic modes of a deformable wing. The aerodynamic forces and moments frequently appear in the applications as *generalized* forces as arise naturally in Lagranges energy methods. Some alleviation of effort numerically for the elastic wing may possibly be achieved by making use of reverse flow relations as indicated in Art. 16.

F,14. Three-Dimensional Nonsteady Supersonic Flow past Wings. The problems of finite wings in nonsteady supersonic potential flow involve a great variety of planforms and a multiplicity of boundary conditions for different ranges of supersonic speed. Considerable activity has taken place in this field[12] as portions of many of the problems are analytically tractable. However, many of these analyses are so specialized as to have only a mathematical or theoretical interest. A few of the methods being evolved show promise of development into unified procedures of general utility for arbitrary elastic wing modes. It is the purpose of this article to characterize some of the problems and methods in order to facilitate reference to, and evaluation of, current work.

The problems of supersonic flow, as observed earlier, can often be treated and synthesized in piecewise fashion. Thus the effect at a field point is found by considering what influences it in the upstream-facing Mach cone, the fore cone "zone of influence" with vertex at the point; moreover, the influence of the point of disturbance reaches only field points within the aft cone "zone of action" with vertex at the point of disturbance, after a proper signal time interval elapses.

In order to indicate the boundary conditions and the piecewise regions that are of concern, consider the oval planform shown in Fig. F,14a. The oblique characteristic lines drawn at the Mach angle to the main stream form a natural coordinate system often more convenient than any other. The tangent characteristic lines that circumscribe the planform determine several regions and arc segments of the boundary that are of significance: segment AB, characterized by the fact that the component of the main stream velocity normal to the edge remains supersonic, is a supersonic leading edge; similarly, DE is a supersonic trailing edge; segments BC and AF are subsonic leading edges; and segments CD and FE are subsonic trailing edges. Regions like S_1, involving only supersonic leading edges, have been designated as purely supersonic, and the planform, or parts of it, involving such regions alone, as simple. Regions like S_2 are influenced by upwash areas ahead of or to the side of the wing in the neighborhood of subsonic leading edges. Regions like S_3 are influenced by subsonic trailing edges and the adjacent wake, as well as possibly by upwash regions ahead of subsonic leading edges.

Purely supersonic case of simple planforms. The upper and under surfaces of purely supersonic regions such as S_1 can be treated as being

[12] An expository chapter by Temple [*137*], dealing with some analytic problems of unsteady flow, especially for supersonic speeds, has been published recently.

independent of each other, as in the two-dimensional flow treated in Art. 10, and source actions alone, based on Eq. 4-28, suffice to build up the velocity potential. A general analytical treatment for such regions as S_1 has been given by Garrick and Rubinow [*11*]. For example, the triangular wing with vertex foremost and with edges ahead of the Mach lines from the vertex (the wide delta wing) can be treated solely by such methods. The velocity potential for the boundary surface characterized as S_1 can

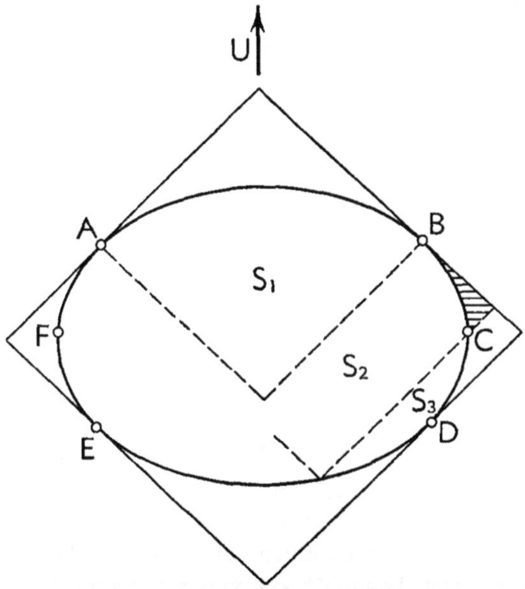

Fig. F,14a. Various boundary regions for oval planform in supersonic flow.

be expressed as a definite integration of sources (see Eq. 4-28) taken over the surface, of the form

$$\varphi(x, y, t) = \frac{-1}{2\pi\beta} \iint \frac{w(\xi, \eta, t - \tau_1) + w(\xi, \eta, t - \tau_2)}{\sqrt{(x - \xi)^2 - \beta^2(y - \eta)^2}} d\xi d\eta \qquad (14\text{-}1)$$

where

$$\tau_1 = [M_\infty(x - \xi) - R] \frac{1}{a_\infty \beta^2}$$

$$\tau_2 = [M_\infty(x - \xi) + R] \frac{1}{a_\infty \beta^2}$$

and

$$R = \sqrt{(x - \xi)^2 - \beta^2(y - \eta)^2}$$

The normal velocity w in the basic equation (14-1) is arbitrary. For harmonic oscillations the equation employs the form indicated in Eq. 4-29, and leads to the result

$$\varphi^*(x, y) = -\frac{1}{2\pi\beta} \iint w^*(\xi, \eta) e^{-i\mu(x-\xi)} \cos \kappa R d\xi d\eta \qquad (14\text{-}2)$$

where $\kappa = \omega/a_\infty\beta^2$ and $\mu = \omega M_\infty/a_\infty\beta^2$. The associated pressure relations follow directly from the use of Eq. 3-7 and these relations, therefore, constitute for this case an explicit solution of the general integral equation relating downwash and loading.

The treatment of field points in such regions of the boundary as S_2, affected by subsonic leading edges, requires consideration of the areas of upwash off the wing (shown shaded). The flow problem for such regions has been termed mixed supersonic, not to be confused with the terminology used to describe mixed supersonic-subsonic regions near shocks. No pressure difference can exist across the plane $z = 0$ off the wing. Moreover, the behavior of the pressure at the edge must be postulated. This can be chosen on the basis of analogy with subsonic flow in such a way that the pressure becomes infinite in the neighborhood of these edges as $\delta^{-\frac{1}{2}}$ where δ is a small distance normal to the edge. Any higher order infinity leads to physically unrealizable forces; a lower order will not lead to suction forces at the leading edge. In incompressible and subsonic flows about wings with rounded leading edges, a large part of this suction force is considered to exist and hence is taken into account even in thin airfoil theory. However, for actual thin airfoils and for supersonic flow, not all and, in some cases, perhaps none, of the suction force may be realized.

A relation usually in the form of an integral equation can be established between the unknown distribution of upwash off the wing and the known distribution of vertical velocity over the wing. In the steady flow case, Evvard has developed a convenient method to determine the effects of such upwash regions. The method involves the source solution to which Eq. 14-1 or 14-2 reduce for steady flow, and leads to the solution of Abel's integral equation and to a concept of "equivalent areas." A unified account of it is given in [*82*]; it applies most easily when the edges can be treated separately, that is, when the interaction of upwash regions is not involved. Evvard's method, modified to make use of Eq. 14-1 and 14-2 or their expansions, has also been applied to some unsteady flow problems, but remains essentially valid only for low frequency oscillations. Chang [*138*] has made such application for a finite swept wing with supersonic leading and trailing edges. Extensions of Evvard's concepts to apply to higher frequencies are no longer simple [*139*]. Another formulation of the problem as an integral equation appears in a Russian paper by Krasilshchikova [*140*].

In several papers dealing with nonsteady flow problems at supersonic speeds, Miles has introduced a number of interesting concepts. In [*141*], for example, after formulating the governing differential equation and the boundary conditions, Miles gives a sequence of transformations analogous to those occurring in the subsonic case (Eq. 9-13 ff.) with $\beta = |M_\infty^2 - 1|^{\frac{1}{2}}$, so that the differential equation for the reduced velocity potential for harmonic oscillations is of the form

$$\varphi^*_{yy} + \varphi^*_{zz} - \varphi^*_{xx} = \kappa^2\varphi^* \tag{14-3}$$

On neglecting terms of order κ^2, the problem can then be formulated as an *equivalent* steady flow problem, and thus known steady flow solutions for a given planform can yield low frequency data for unsteady flow. Miles also confirms the results (Eq. 14-1 and 14-2) by means of a Fourier integral formulation and indicates in [*142*] that uniqueness of the solution follows from classical wave-equation procedures. (See also, in this regard, W. P. Jones [*143*].)

Rectangular wing. The side edges of the rectangular wing involve regions like S_2 with the requirement that the pressure difference be zero at the edge. Miles [*144,145*] gives a formulation of the problem as a Wiener-Hopf integral equation and obtains its solution for harmonic conditions by concepts of analytic continuation. Although the motions prescribed are arbitrary, ready application is made only for rigid body motions. The transient flow case for rectangular planforms has also been treated by Miles and others [*142,146*].

Watkins [*147*] has found it convenient in treating the rectangular wing to expand the velocity potential in terms of a parameter containing the frequency. By also expanding the doublet solution in powers of the frequency, he formulated the boundary problem as a sequence of integral equations, solutions to which were found on the basis of starting with the steady flow solutions for the planform, and successively treating each higher order frequency term as an equivalent steady flow problem. In this way it is possible by iteration to include successively higher frequency terms, although the procedure becomes laborious. Aerodynamic coefficients for rigid body oscillations are given in [*146*]. Watkins also showed the significant changes with aspect ratio of the moment-damping term for oscillations in pitch [*148*], a problem indicated for the plane flow case in Fig. F,11a and F,11b. Nelson, Rainey, and Watkins in [*149*] improve the application of the frequency expansion method and give spanwise section aerodynamic data to the seventh power of the frequency for a rectangular wing undergoing rigid body motions. They also apply the data by strip analysis methods to the problem of the flutter of a cantilever wing.

The quarter infinite wing, which corresponds to the edge of a rectangular wing, has been treated for harmonic motion by Goodman [*150*] who used a method of Gardner, and by Stewartson [*151*] who employed Laplace transform methods. Analytic discussion of the rectangular wing of small aspect ratio ($Æ < 1$), involving the interaction of both side edges, has been given by Miles [*152*] and compared with the vanishing aspect ratio case.

An illustration. Fig. F,14b presents the spanwise distribution of the components of moment coefficients for a rectangular wing of aspect ratio *4* oscillating in pitch about midchord. The curve for R_{22} represents the component in phase with the position of the wing and thus contributes to the aerodynamic stiffness (Eq. 7-6). That for I_{22} represents the component 90° out of phase with the wing position and thus contributes to the wing aerodynamic damping. The distribution in the tip regions beyond the Mach lines, shown dashed, is significantly different from the two-dimensional distribution in the region between the Mach lines. The damping moment, for example, is negative in the two-dimensional region, whereas in the tip region it changes to positive values. Such variations

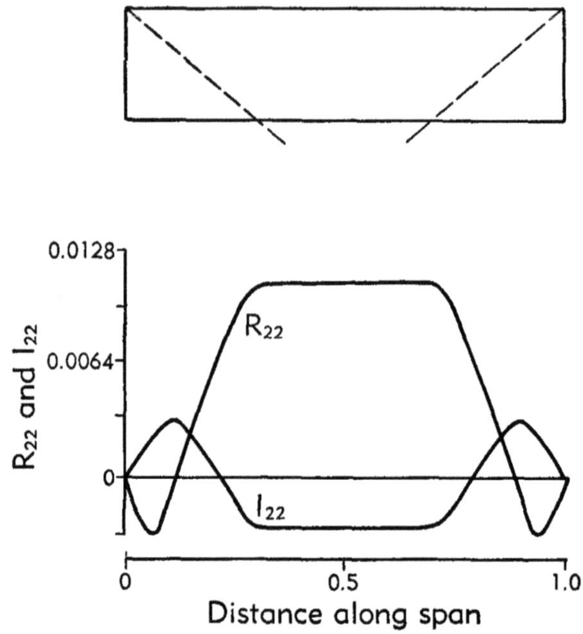

Fig. F,14b. Spanwise distribution of components of oscillating moment, $M_\infty = 1.3$; $k = 0.1$.

in the coefficients are of special significance with regard to flutter and dynamic stability.

The triangular and related wings. Results for the wide triangular planform with edges outside the Mach cone have been reported by Miles [*153*], Froehlich [*154*], and Nelson [*155*]. The more difficult problem of the narrow triangle has been studied by Stewartson [*151*] by interesting analytic methods, based on the Laplace transformation combined with Fourier integral methods. The problem has also been treated by Watkins and Berman [*156*] for rigid body motions by the method of expansion in powers of the frequency, and in a subsequent paper [*157*] the results are extended to include parabolic surface modes. Application of the methods of [*155,156*] has been made by Cunningham [*158*] to yield the lift and pitching moment on families of arrowhead wings oscillating in pitch and vertical translation.

Illustration of span effects. Two examples are given in Fig. F,14c of the span distribution of lift and phase angles for a rectangular wing and a triangular wing. The solid curve corresponds to results for an actual

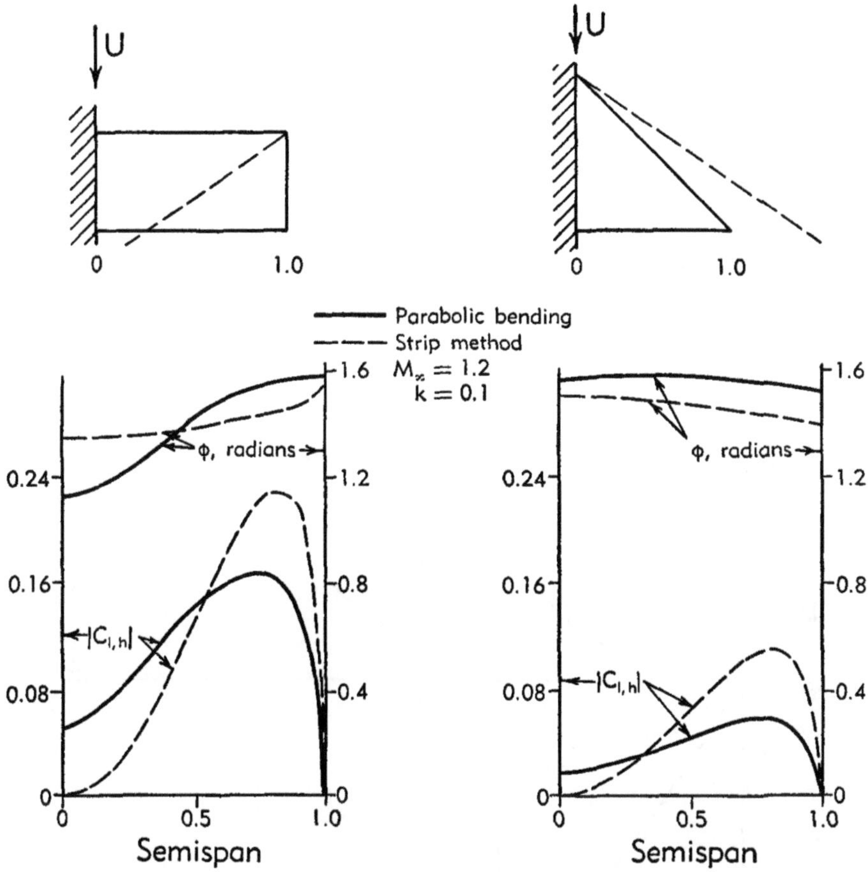

Fig. F,14c. Spanwise distribution of components of oscillating lift for a parabolic bending mode and for strip method based on uniform translation.

parabolic bending mode, and the dashed curve to the use of section data for rigid body translation weighted by a parabolic distribution, a strip theory approximation. A few examples of the effect of the planform configuration on the leading edge moment coefficient of a wing, pitching about a spanwise axis through the leading edge, are shown in Fig. F,14d.

Subsonic trailing edges. The regions which are influenced by subsonic trailing edges are the most difficult to treat, and their analysis has not been pursued very far. The Kutta condition, the floating of the shed

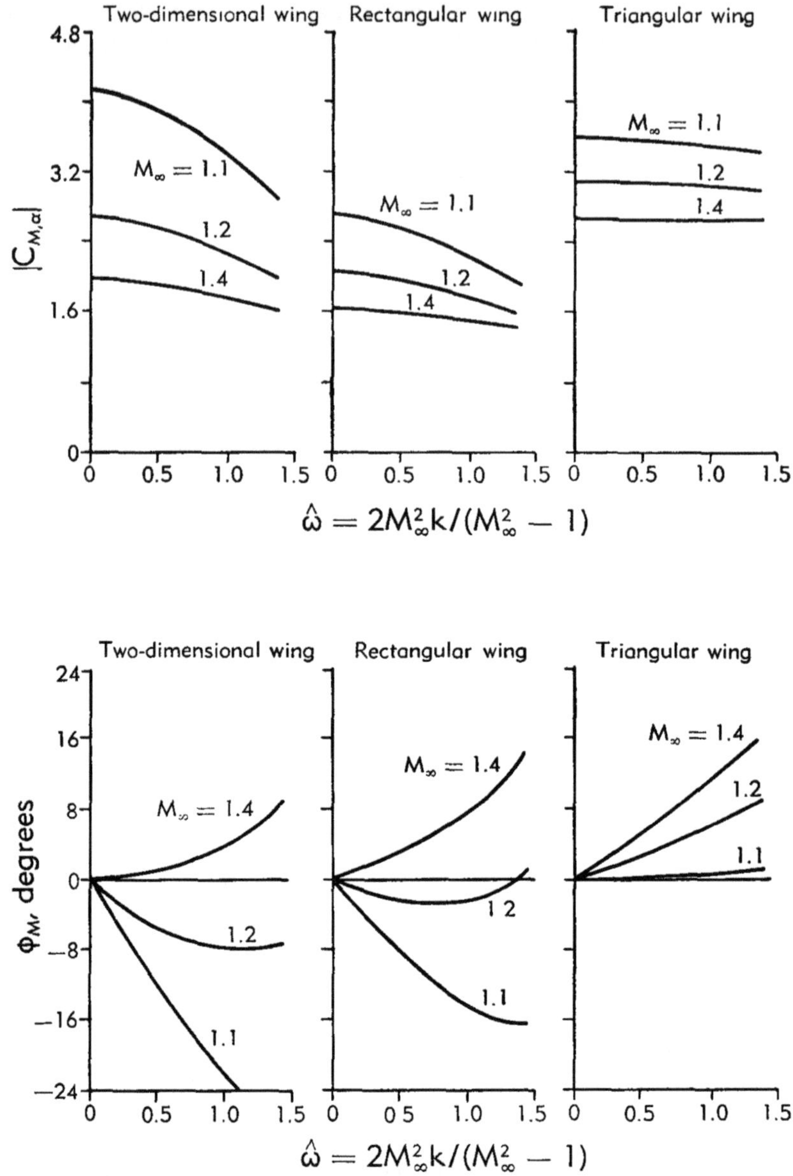

Fig. F,14d. Moment coefficient and phase angle for wing pitching about spanwise axis through leading edge: (left) two-dimensional wing section, (center) rectangular wing of aspect ratio 4, and (right) triangular wing with 45° swept leading edge.

vorticity in the wake, and edge conditions based on analogy with low speed flow are involved in the problem. The rather arbitrary nature of the edge conditions and the complex flow pattern suggest the need for further experiment to clarify the phenomenological assumptions that are implied ordinarily in linearized potential flow theory. Analytic work with subsonic trailing edges is given by Stewartson [*159*], who indicates that for regions like S_3 it is more convenient to use the acceleration potential rather than the velocity potential. It is pertinent to note that the proportions of the regions S_1, S_2, and S_3, involved for a given planform change with Mach number. For high values of M_∞, S_1 is mainly involved, whereas, for values near unity, regions S_2 and S_3 are mainly involved.

Methods proposed for elastic wing modes. It is desirable that methods be developed suitable for yielding aerodynamic data for arbitrary motions by systematic or routine numerical procedures. One such method has been termed a "box" method, proposed by Pines and others [*160,161*] for treating both purely supersonic planforms and, with some extensions, planforms with subsonic leading edges. In application it is similar to the lattice surface approach discussed for subsonic flow, but optimum or detailed methods remain to be developed. Some applications and extensions have been made by Li [*162*] and by Zartarian, Hsu, and Voss [*163,164*]. A generalized strip theory method for triangular wings is given in [*165*], and the use of reverse flow relations (see Art. 16) to alleviate the numerical labor is indicated.

Kernel function for supersonic flow. The development of a kernel function analogous to that given in Eq. 13-33 and 13-34, and involving the use of Eq. 4-29 instead of Eq. 4-24, has been made by Watkins and Berman [*166*]. The tabulation of this function for different regimes, and experience in its application, will probably lead to systematic and general procedures for elastic wing modes analogous to the procedures for subsonic flow.

F,15. Considerations on Experimental Methods.[13] The preceding articles have dealt almost exclusively with potential flows and with numerous simplifying assumptions. Very many questions arise on the degree of validity of the theoretical results and on the possible application to the many complex nonsteady aerodynamic phenomena and fields of practical interest, as mentioned in Art. 1. Experimental methods become indispensable for providing necessary data; however, there must still be a healthy interaction between experiment and theory, as one guides the other. The present article very briefly reviews some of the objectives, techniques, limitations, and accomplishments of the experimental methods with reference chiefly to the nonsteady aerodynamics.

The obtaining of complete and consistent sets of experimentally deter-

[13] Art. 15 has been prepared with the collaboration of A. A. Regier, Head of the Vibration and Flutter Branch, Langley Aeronautical Laboratory.

mined aerodynamic derivatives for various configurations and for nonsteady flight conditions has long been recognized as a desirable goal. Well-conceived experiments toward this end were devised by Frazer and Duncan [*167*] as early as 1928. Despite the time that has elapsed since investigations of this type have been started, the available results are scattered and special; consequently, the theoretically determined air forces, with some degree of empirical modification, still form the basis for most flutter studies. In the early 1930's, classical water experiments were conducted by Farren and Walker [*168*] on the validity of Wagner's theory for the growth of circulation. As it is not possible to give either the history or description of the various experimental investigations, the reader is referred for a more complete summary of available published work in this field to papers by Drescher [*169*], Victory [*170*], Williams [*33,171*], and Halfman [*172*].

Objectives and Limitations. The experimental measurements can have a threefold purpose: to determine how well the various theoretical results agree with experiment; to establish the validity of experimental techniques; and to provide useful information on configurations and flight conditions for which established theory in usable form is not available.

These objectives are not unrelated but are interwoven. For example, established theory may be used to evaluate a portion of some experimental results and the balance considered as the experimental part. Contrariwise, established experiment can in some instances circumvent laborious calculations. There are, however, a great many trials and tribulations that can plague the experimentalist dealing in nonsteady flows and that leave room for much ingenuity. For example, there are difficulties associated with the measuring of small differences between large quantities, in particular the separation of aerodynamic quantities from inertia quantities. It is necessary to establish with accuracy not only the magnitudes but also the phases of force or moment components. The small lags or phase differences are easily masked by errors of measurement or instrumentation, by friction, or by effects of unwanted flexibility leading to a coloring of the results with aeroelastic effects. The ranges of frequencies of interest for flutter are fairly high and may lead to many mechanical vibration problems, for example, to structural dynamic response corrections. Another difficulty not easily resolved is associated with the wide variety of aircraft configurations and their component parts, and the large number of modes of vibration for which information is needed.

Wind tunnels have been most often employed for conducting the required controlled experiments. These should be large enough to enable models to be sufficiently detailed and instrumented, and should be of high enough speed to yield Reynolds numbers Re and Mach numbers M_∞ of concern. Especially is it necessary to obtain data in the ranges of Re and M_∞ of interest for phenomena not amenable to theory, as for flows involving large effects of viscosity such as the presence of thick boundary layers or flow transition and separation. Phenomena involved are stall flutter, control surface oscillations, transonic flow oscillations, and buffeting. A serious problem in wind tunnel work is the matter of fluctuations or roughness of the airflow. For steady flow measurements the balances and pressure-recording systems are made to give a highly damped low frequency response and such instruments average out the fluctuations in the airflow. In the oscillating case, however, the recording system should have a high frequency response characteristic, which is then sensitive to the flow fluctuations. The threshold of noise or "hash" may lead for higher air speeds to unintelligible records. Ways to alleviate the situation involve filters, fairing, statistical analyses, and "watt meter" harmonic analyzers (see Bratt and Wight [*173*]). Modern data-reduction methods and the use of power spectral methods of generalized harmonic analysis also furnish tools for separating wanted and unwanted responses.

Air Forces on an Oscillating Wing in a Wind Tunnel. When experimental measurements of forces on oscillating wings are made in a wind tunnel, it is necessary to account for the influence of the wind tunnel walls in order to arrive at the free flight air forces required in studies of flutter, buffeting, and dynamic stability. Theoretical work on this problem has been based on the familiar method of images in which the effect of top and bottom wind tunnel walls, spaced distance H apart, is equivalent to an infinite row of alternating mirror images of airfoils, spaced at intervals equal to the tunnel height. The incompressible flow case was studied by W. P. Jones [*174*] and by Reissner [*175*] who found an appreciable correction for low reduced frequencies. Recent work of Timman [*176*] introduces elliptic functions and presents a full analytic study for incompressible two-dimensional flow. The analysis of Timman has been made use of by De Jager [*177*] in the preparation of corrections for the forces and moments on an oscillating airfoil with control surfaces, for various tunnel height-to-wing chord ratios over a range of k from 0 to 1. The treatment of a finite wing in a circular tunnel for incompressible flow has been given by Goodman [*178*] and Acum and Garner [*179*].

The compressible flow case is of considerable additional interest as it brings into focus a resonance-like phenomenon which may lead to very large corrections under some conditions, or may even preclude testing near resonance. The analysis of the two-dimensional subsonic case was made by Runyan and Watkins [*180*] and represents in effect an application of the integral equation considered in Art. 9, however with a different kernel. Thus the integral equation for the oscillating wing at the center of the tunnel takes on a form

$$w(x) = \int_{-b}^{b} l(\xi) \left[K(x - \xi) + \sum K_n(x - \xi, nH) \right] d\xi$$

The kernel consists of two parts: K, which is associated with the wing alone and which is equivalent to the two-dimensional kernel of Possio's integral equation, and the infinite series, which corresponds to the effect of the infinite row of airfoil images.

In applying the procedures described in Art. 9 to this problem, indications of possible large effects occur, associated with resonance frequency conditions defined by

$$\omega_r = \frac{\pi\sqrt{1 - M_\infty^2}}{H}(2n - 1)a \qquad (n = 1, 2, 3, \cdots)$$

These frequencies have a simple physical basis similar to organ-pipe resonance and correspond to an odd number of transverse half waves; for, significant resonance-like reactions occur for time intervals equal to, or odd multiples of, the time required for transverse pressure waves to travel to the tunnel walls and back to the model. In this regard, it is clear that the effects could be greatly modified by off-center model positions, by wave absorbing walls, or by free or open boundaries.

The integral equation has been solved for several specific wind tunnel problems, and incidentally, for some equivalent problems of oscillating airfoils in cascade [*181*]. Runyan, Woolston, and Rainey [*182*] have presented some comparisons between calculated and experimental results for the two-dimensional wind tunnel problem. Some of these results are given in Fig. F,15a which applies for a pitching wing oscillating in a wind tunnel stream of $M_\infty = 0.7$. The experimental data were obtained from measurements of oscillating pressure distributions by means of pressure cells at 24 stations on the wing. The ordinate L_α/L'_α is the ratio of lift, calculated or measured for the wind tunnel case, to the calculated lift under free space conditions. The ordinate ϕ_α is the phase angle between lift and position. The abscissa ω/ω_r is the ratio of the frequency of oscillation of the wing to the calculated resonance frequency for $n = 1$. The differences between the wind tunnel and free space theoretical results are rather large in this case, but particularly noteworthy is the agreement between theory and experiment for the wind tunnel case, because accurate measurement of oscillating air forces is indeed a difficult undertaking.

Techniques. These vary greatly in their degree of directness. Indirect and integrated comparisons with theory are afforded by experiments on flutter, in which field a great deal of effort has been expended. Since the aeroelastic phenomenon of flutter is a function of many structural and aerodynamic variables, the flutter experiment cannot readily isolate particular components of the air forces. However, controlled experiments can be made both in the wind tunnel and in flight on the theoretical predictions for flutter and for the approach to flutter. Thus one can strive to measure not only critical flutter speeds and frequencies of the aircraft components, but also changes in the damping and in the frequency spectrum with increasing air speed, in response to either forced vibration or transient decay.

More direct than flutter, and useful for isolating the components of the air forces, is the study of the response of a rigid airfoil system or component part restrained to a single degree of freedom. There are two main variations: (1) the free oscillation method, and (2) the forced oscillation method.

Free oscillation method. In the free oscillation method the model is deflected from its neutral position and its motion is recorded as the

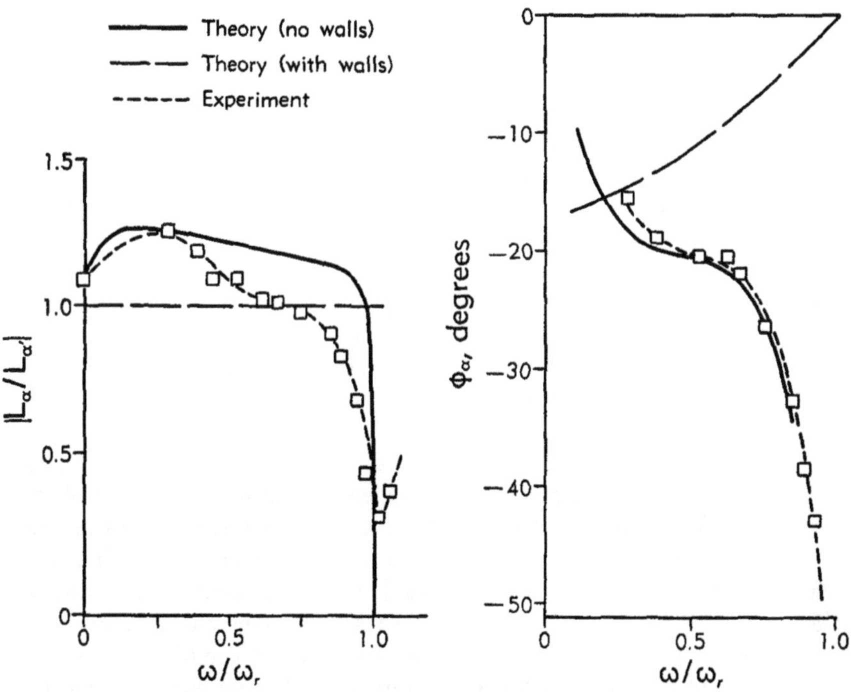

Fig. F,15a. Measured and calculated oscillating lift in a wind tunnel, $M_\infty = 0.7$.

oscillation damps out. The vibration equation for the system is of the form

$$(I + I_a)\ddot{\alpha} + (R + R_a)\dot{\alpha} + (K + K_a)\alpha = 0$$

where additional inertia, damping, and stiffness terms due to air forces are designated by the subscript.

The various air forces or moments are obtained from the differences in frequencies and in the rates of decay of the damped oscillations for air-stream-on and air-stream-off conditions, the latter preferably under low density or vacuum conditions. This method has been used by various investigators, for example, Bratt and Davis [*183*] and Clevenson and Widmayer [*184*]. If the damping is small, the air force in phase with the displacement or the aerodynamic stiffness is determined primarily from the change in the frequencies, and the component 90° out of phase or the aerodynamic damping primarily from the differences in the rates of decay.

Because the wake of the model is finite and of varying amplitude, there are some questions on the degree of validity of the transient method. However, when the damping is low, the free oscillation methods have checked well with other methods.

Forced oscillation method. In the forced oscillation method the system is put into a constant amplitude oscillation by an applied oscillating force of known magnitude. The magnitude and phase of the displacement of the system relative to the force are measured. The air forces and moments are obtained from the differences that exist in the response of the system when oscillated in the air stream, as compared with the response in air at rest. Variations of this method were used by Cicala [*185*] and other investigators. A good description of this method, and of its application to separate the various components of the air forces and moments for oscillations in both pitching and translation, has been given by van de Vooren [*186*]. For some of the measurements described in [*186*], five different axes of oscillation were chosen in order to separate the various lift and moment components and to serve as a check. Moreover, it is shown that in order to determine some of the cross aerodynamic terms, as the moment due to translation, it is necessary also to oscillate the axis of pitching.

Reactions at the supports. A still more direct procedure is a variation of the forced vibration method, which consists in oscillating the airfoil in a prescribed manner but determining the forces or moments from the reactions taking place at the supports of the model. Such a procedure has been used by Clevenson and Widmayer [*184*]. Earlier Silverstein and Joyner [*187*] oscillated an airfoil in pitch about the quarter chord point by a direct drive mechanism and measured the phase of the lift with respect to the angle of attack for a range of frequencies. It may be noted that the lift due to pitching or the moment due to translation may be measured without the necessity of taking into account inertia forces of the body, if it is properly mass-balanced. To measure moments due to pitching or lift due to translation it is necessary to compensate for the inertia forces, which may be large compared with the air forces being measured. Halfman [*172*] performed such experiments by balancing out inertia effects electrically by means of accelerometers. Bratt and Scruton [*188*] made use of a system of two wings oscillated together, one mounted in the wind tunnel, the other in a test chamber out of the air stream. The wings were driven through a mechanical balance system which measured only the difference between the forces required to oscillate the wings, that is, its output measured the effect of the air flow.

Pressure measurements. Direct measurements of the unsteady pressure distributions have also been made by means of pressure gauges, having high frequency response characteristics. Among the various types of gauges that have been used are: the flush-mounted type with the under side of the diaphragm instrumented with straingauges [*189*]; the flush-mounted crystal type (piezoelectric) gauge [*190*]; and the inductance type gauge mounted inside the wing and usually connected to top and bottom surfaces by means of short tubes, so as to record the surface pressure difference [*182,191*]. All the gauges can be made small, of the order of $\frac{1}{4}$ to $\frac{1}{2}$ inch in diameter, and about the same in thickness. The frequency response is usually flat to well over 1000 cycles per second.

Despite the small size of the gauges it is difficult to define properly the pressure distribution very near to leading edges and tips, for example, of thin tapered wing models. Fairing errors may then affect in particular the aerodynamic moments. Extraneous response of the gauges to acceleration or to wing stresses may be troublesome, and sometimes dummy gauges are used to balance acceleration effects and flexible tubes used to lead the pressures to gauges outside of the test section. This latter method may, however, introduce phase lags, attenuation, or resonance in the tubes.

A considerable amount of information on nonsteady air forces is nonetheless being obtained by means of pressure cells. Recent work by Laidlaw and Halfman [*190,192*], for example, presents the spanwise and chordwise pressure distributions on a series of low aspect ratio wings of various planforms. Pressure cells provide information on details of the flow and serve to delineate regions of the wing and test conditions where notable differences with theory exist. On the other hand, reducing the pressure data to total air force and moment coefficients requires much effort. This labor may be greatly reduced by the use of electrical integrating circuits [*193*].

Interferometric method. Another method of obtaining the pressures on an oscillating wing, as well as some knowledge of the flow field, is the optical interferometric method. It has been used primarily for two-dimensional flow, for example, in the experiments of Ritter [*194*] for determining oscillating air forces for small angles of attack. Accuracy achieved by this method particularly for the aerodynamic moments is, however, not very good, in large measure due to unsteadiness of the air flow. The method has also been used to advantage in the study of flow in cascades and on stalled airfoils (see Schnittger [*195*]).

Some experimental results. The results of well-devised experiments have in many instances substantiated the trends indicated by potential flow theory, in particular for those problems where separated flow and viscous boundary layer effects are unimportant. For example, the correlation between classical low angle of attack flutter experiments and theory has been good for wings of moderate to high aspect ratios without control surfaces. For control surfaces, comparisons have been inconsistent. Some of the available experimental results are indicated according to various categories in the next few paragraphs.

Two-Dimensional Wing. Wing of High Aspect Ratio. A substantial amount of information on the direct measurement of oscillating air forces has been obtained by various methods and on many wing sections and configurations [*33,167,169,171,172,185,186,194*]. These results usually show the same trends as obtained with the theory, but the magnitudes are usually of the order of 10 per cent lower. Some data indicate that for higher values of the reduced frequency the experimental values may reach or exceed the theoretical values of linear theory. Where large discrepancies occur, these have usually been traced to wind tunnel corrections, leakage around the ends of the airfoil, or various boundary layer interactions such as transitions between laminar and turbulent flows. In some tests made at low Reynolds number [*196*] it was found that by attaching wires near the leading edge, thereby fixing transition, many discrepancies were eliminated.

Control Surfaces. Control surface derivatives have exhibited wider deviations from theory than have the wing derivatives. This is not unexpected as the controls usually operate in the boundary layer of the main wing and introduce problems of flow leakage and sharp variations in the pressure distributions near leading edges of the control. Some data for controls are given in [*186,189,197,198*]. As indicated in Art. 7, it has sometimes been the practice to reduce theoretical derivatives by factors based on experimental considerations. For example, control stiffness derivatives may be reduced by factors based on hinge moments in steady flow and control damping moments by factors based on results of oscillatory tests. Such practice, however, must be viewed with caution since many of the tests are made with relatively small models, for which boundary layer effects are likely to be exaggerated. In some rather large scale tests made by Andreopoulos, Cheilek, and Donovan [*197*] the deviations from theory were relatively small.

Wings of Small Aspect Ratio. In many configurations of current interest with aspect ratios ranging from 3 or 4 downwards, minor modifications or extensions of two-dimensional theory for infinite aspect ratio are not warranted. Some experiments on low aspect ratio configurations are listed in [*109,183,190,192,199*]. One of the most complete published studies in which pressure measurements were made on various low aspect ratio planforms is given by Laidlaw [*192*]. These results (see also Art. 13) show fair agreement with Reissner (high aspect ratio) theory and better agreement with Lawrence and Gerber's analysis. The application of zero aspect ratio theory to aspect ratios of the order of one was not very satisfactory. For bodies of revolution in which a considerable portion of the forces arises from the flow separation on yawed bodies, the agreement with potential flow theories is poor [*109,200*].

Experiments with Separated Flow. When an oscillating airfoil has a thick boundary layer or performs oscillations within the boundary layer there may be, as already indicated, substantial differences between theory and experiment. Moreover, separation of the flow can alter the conditions to the extent that any relation between potential flow theory and experiment is hardly discernible. A few practical problems involving such phenomena are: stall flutter of propeller blades [*201,202*], the high oscillatory stresses in axial flow turbine blades near stall [*195,203,204,205*], and control surface "buzz" [*189,206*]. Of numerous other problems of this type, mention may be made only of the flutter of smoke stacks and of bridges. The spectacular failure of the Tacoma Narrows bridge in 1940 has led to a large literature [*207*].

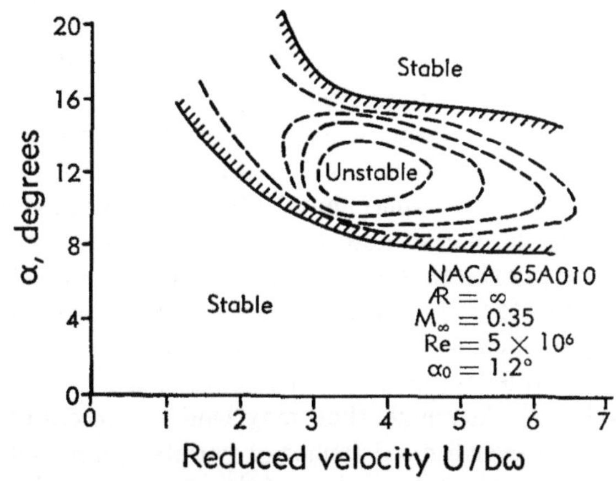

Fig. F,15b. Negative damping of a pitching airfoil at high angles of attack. (Amplitude of oscillation 1.2 degrees.)

Instabilities caused by flow separation on an airfoil have been studied on models in wind tunnels by various investigators, for example, Studer [*208*], Chuan and Magnus [*209*], and Bollay and Brown [*210*]. The instability encountered is usually a single-degree torsional instability, which occurs at angles of attack near stall and at air speeds ordinarily much lower than the classical flutter speed. Victory [*170*] has shown that, by altering the theoretically obtained damping moment due to pitching, in accordance with measurements made at high angles of attack, and in the proper Reynolds number range, the experimental flutter speed of a stalled airfoil could be predicted. A qualitative indication of how the damping in pitch varies with reduced frequency k, or rather reduced velocity $1/k$, is given in Fig. F,15b, where the dashed lines indicate contours of equal negative damping $(c_{m,\alpha})_i$. The figure shows that for the higher angles of attack the unstable range occurs over a limited velocity range, whereas for lower angles of attack it occurs for a very wide velocity range. Satisfactory theoretical explanation of this behavior of the aerodynamic damping has not been given. Related similar results were obtained in flutter experiments at high angles of attack which were studied by Chuan and Magnus [*209*] in relation to vortex shedding. It was shown that, at the very high angles of attack, the flutter speed range was limited and that the flutter frequency coincided with the frequency of the Kármán street for a rigid airfoil at the same mean angle of attack. In the intermediate angle-of-attack range, in which the flutter persists and increases with speed, the Kármán street frequency is much higher than the flutter frequency. The two phenomena are thus of different types, one resembling resonance, the other loss of dynamic stability. It is of interest that bending-type flutter has sometimes also been observed over limited speed ranges, when the bending frequency coincided with the Kármán street frequency.

The effect of mean angle of attack, as well as of the amplitude of oscillation, on the aerodynamic coefficients has been sampled experimentally in a number of studies, for example, [*173,188,205,211*]. Various semi-empirical attempts have been made to rationalize the results and to predict the dynamic behavior of airfoils in stall. Most of these methods start with the steady flow experimental lift and moment characteristics or with nonsteady potential flow characteristics, and introduce such concepts as modifying the effective camber, relaxing the Kutta condition, introduction of adjustable time lag constants, or other arbitrary assumptions. For example, quasi-steady flow conditions may apply for low values of the reduced frequency. Some of these methods are mentioned in [*195,203,204,212,213*]. Although they may lead to successful treatments in specific cases and over limited ranges of variables, none as yet may be considered sufficiently general solutions of the complex problems involved.

Supplementary Related Techniques. Brief mention may be made of additional procedures relating to time-dependent measurement of air forces. One example is the use of a different medium than air to obtain lower speeds of sound, such as the use made in the NACA Langley Flutter Research Tunnel of Freon 12 (CCl_2F_2) as a test medium in conjunction with variable density and with any mixture of air [*117,184*]. At standard conditions the speed of sound in this gas is less than half that in air. It is thus possible to extend the range of Mach number with alleviation of some of the limitations discussed earlier and to obtain trends for a wider scope of the variables.

Another method that merits mention is that of employing various electrical analogues for obtaining oscillating air forces. An example for two-dimensional incompressible flow is given in [*214*].

Some of the many methods of attack that are required in flutter research, or for dynamic stability, may also be used for nonsteady aerodynamics. For example, methods for making flutter tests in the transonic speed range are indicated in [*117,215*]. These methods involve the use of heavy bodies dropped from high altitude, or of rocket vehicles instrumented to telemeter the data to stations on the ground. The reader is referred to VIII,B and C, and to the cited references for additional material.

F,16. Miscellaneous Topics and Concluding Remarks. Attention is directed briefly in this article to a few topics in nonsteady aerodynamics, to supplement topics brought up in earlier articles, in need of further development or more conclusive application.

Application of reverse flow relations. Reciprocal relations in aerodynamics may be considered as a generalization of classical concepts of reciprocity occurring in dynamics, in structures, and in acoustics, and are related to the broad principle of conservation of energy and to the principle of virtual work. In the field of structures, one may recall the reciprocity rule that the displacement at point A due to unit load at point B is equal to the displacement at point B due to unit load at point A. In acoustics, there is a result of Helmholtz that if, in a space filled with still air, sound waves be exerted at any point A, the resulting velocity potential at a second point B is the same in magnitude and in phase as it would have been at A, had B been the source of sound.

The generalization of these reciprocal relations to aerodynamics, or for that matter to acoustics in the presence of a wind, involves the consideration of the reverse wind or the reverse flow. The most general results are due to Flax [*216*], who treated harmonically oscillating wings and to Heaslet and Spreiter [*217*], who considered, in addition, transient flows. A fairly simple discussion is given by Küssner [*218*], who derived the general results on the basis of the difference properties of the kernel function.

The theorem is of particular interest for problems of the harmonically oscillating wing and will be indicated for this case. The reverse flow theorem may be expressed in the form

$$\iint L_1 w_2 dS = \iint L_2 w_1 dS \tag{16-1}$$

where L_1 and w_1 refer to loading and normal velocity distribution for the planform in direct flow U_1, and L_2 and w_2 to those for the same planform in the reverse flow U_2, the Mach number being alike and the integrations being taken over the planform surface.

This result provides a means for obtaining integrated aerodynamic results for elastic modes from rigid body modes [*111*]. For example, let w_2 refer to uniform translation of the rigid oscillating wing in the reverse flow and let L_2 be the associated pressure distribution. Then the left-hand side of Eq. 16-1 is, except for a constant, the integration of the pressure over the surface and defines the total lift corresponding to the arbitrary normal velocity w_1, while the right-hand side gives the expression of this lift as an integration over the planform of w_1 and the relatively simple loading L_2 associated with the uniform translation. Thus total lift results for a mode of deformation may be expressed in terms of pressure results for a mode of rigid translation. With L_2 serving as appropriate influence or Green functions, it is possible to obtain innumerable solutions associated with one known distribution. Results of interest may be obtained, for example, not only for the lift but also for pitching, rolling, and aerodynamic bending moments as well as control surface aerodynamic influence or generalized forces. Other uses are: it may serve to provide numerical and analytic checks on approximate methods, and it may also be of assistance in the planning of experimental work on oscillating air forces. Application of reverse flow results to assist in determination of the generalized forces associated with elastic modes is made in [*165*]. Further applications to the whole field of aeroelasticity may be expected.

Effects of thickness. It is possible to treat effects of airfoil thickness quite generally in the case of two-dimensional incompressible potential flow; however, the methods are exact only if the wake is neglected. Interesting developments, making use of extensions of Blasius' theorem and involving complex variable methods, exist. Some of these have been made by the Russian school of Chaplygin and Sedov. Other contributions are those of Wilton, Milne-Thomson, and Morris in England, and of Couchet in France.

For supersonic flows some results are available on nonlinear effects of thickness in steady flows, such as the second order corrections of Busemann. These results have served as a guide to similar effects in the nonsteady case studied by Jones and Skan [*219*]. Carrier [*220*] has made a study of the oscillating wedge in a supersonic stream. Accelerated flow is considered in [*221*].

The work of Van Dyke [*70*] referred to in Art. 10 presents a systematic study of iteration methods for the nonlinear thickness effects in the case of two-dimensional supersonic flow. The points of special interest are the effect of thickness on damping in pitch, which effect appears to be small, and results for the "effective" center-of-pressure location, results which may have noteworthy influence on flutter characteristics.

Effects of mixed boundaries. Although wall interference effects have been studied for two-dimensional cases, it is of interest that this work should be extended to finite wings and to mixed boundaries, partly open as well as closed. As indicated in Art. 16, improved experimental techniques are in the process of being evolved.

Unsteady boundary layer and outlook. Many complex phenomena of unsteady flows, such as separated flows and unsteady wake behavior, or impulsive and accelerated motions in a fluid, involve viscosity in an essential manner, and little has been done in this field. The subject of the growth of the boundary layer in transient incompressible flows has had some attention given to it, but is in a rudimentary state. A parameter of significance in the case of periodic motion is the product of the reduced frequency and the Reynolds number. To provide a few leads to the literature, mention may be made by name and date only of the work of H. Blasius (1908), E. Boltze (1908), W. Tollmien (1924), H. Schlicting (1932), S. Goldstein and L. Rosenhead (1936), T. Sakurai (1939), K. H. Thiriot (1940 and 1942), H. Gortler (1944), A. W. Quick and K. Schroder (1947), M. Mangler (1948), W. Wuest (1949), L. Howarth (1935), J. H. Preston (1943), G. Temple (1943), and K. Stewartson (1951). The classical work of Stokes on the damping of pendulums and of Rayleigh on damping in acoustics are examples of closely related topics.

There are, of course, many unsteady flow problems associated with high speed flows which involve shocks. This large open field is mentioned here only to indicate a further class of pulsating flow problems of complex origin. While some significant initial contributions to the field have been made, much remains to be clarified. The separation of effects of random forcing functions, partly due to the flow environment, partly to the flow itself, and the effects of inherent loss of damping is a particularly difficult part of the problem. Modifications of the trailing edge condition have already been mentioned as a possible means for dealing with high angle-of-attack flow phenomena. Effects of high temperatures and high noise levels on unsteady flows are also of increasing interest.

It is clear that, although important and interesting parts of the field of nonsteady aerodynamics have been developed, much more remains undeveloped. It is hoped that the cited references and bibliography given in Art. 17, while by no means complete, may provide a guide to the rapidly growing literature.

F,17. Cited References and Bibliography.

Cited References

1. Garrick, I. E. A survey of flutter. *NACA-Univ. Conference on Aerodynamics*, June 1948. (Durand Reprinting Comm.)
2. Lin, C. C., Reissner, E., and Tsien, H. S. On two-dimensional nonsteady motion of a slender body in a compressible fluid. *J. Math. and Phys. 27*, 220–231 (1948).
3. Miles, J. W. Linearization of the equations of nonsteady flow in a compressible fluid. *J. Math. and Phys. 33*, 135–143 (1954).
4. Mollo-Christensen, E. L. An exploratory investigation of unsteady transonic flow. I. Theoretic discussion. *Mass. Inst. Technol. TACP Rept. 5*, June 1954.
5. Baker, B. B., and Copson, E. T. *The Mathematical Theory of Huygens' Principle.* Oxford Univ. Press, 1939.
6. Lamb, H. *Hydrodynamics*, 6th ed. Cambridge Univ. Press, 1932.
7. Bateman, H. *Partial Differential Equations of Mathematical Physics.* Dover, 1944.
8. Webster, A. F. *Partial Differential Equations of Mathematical Physics.* Stechert and Co., 1933.

9. Watson, G. N. *A Treatise on the Theory of Bessel Functions*, 2d ed. Macmillan, 1944.
10. Küssner, H. G. Allgemeine Tragflächentheorie. *Luftfahrtforschung 17*, 370–378 (1940). Also *NACA Tech. Mem. 979*, 1941.
11. Garrick, I. E., and Rubinow, S. I. Theoretical study of air forces on an oscillating or steady thin wing in a supersonic main stream. *NACA Tech. Rept. 872*, 1947.
12. Prandtl, L. Theorie des Flugzeugtragflügels im zusammendrückbaren Medium. *Luftfahrtforschung 13*, 313–319 (1936).
13. Garrick, I. E. Moving sources in nonsteady aerodynamics and in Kirchhoff's formula. *Proc. First U.S. Natl. Congr. Appl. Mech.*, Chicago, 1951. Also *Trans. Am. Soc. Mech. Engrs. 74*, 733–739 (1952).
14. Van der Pol, B., and Bremmer, H. *Operational Calculus*, Chap. 5. Cambridge Univ. Press, 1950.
15. Birnbaum, W. Das ebene Problem des schlagenden Flügels. *Z. angew Math. u. Mech. 4*, 277–292 (1924). Also *NACA Tech. Mem. 1364*, 1954.
16. Wagner, H. Über die Entstehung des dynamischen Auftriebs von Tragflügeln. *Z. angew. Math. u. Mech. 5*, 17–35 (1925). Also *Brit. RTP Transl. 2316.*
17. Cicala, P. Aerodynamic forces on an oscillating profile in a uniform stream. *Mem. reale. accad. Sci. Torino 68, 2*, 73–98 (1934–1935).
18. Küssner, H. G. Zusammenfassender Bericht über den instationären Auftrieb von Tragflügeln. *Luftfahrtforschung 13*, 410–424 (1936).
19. Glauert, H. The force and moment on an oscillating airfoil. *Brit. Aeronaut. Research Council Repts. and Mem. 1242*, 1929.
20. Duncan, W. J., and Collar, A. R. Resistance derivatives of flutter theory. *Brit. Aeronaut. Research Council Repts. and Mem. 1500*, 1932.
21. Durand, W. F., ed. *Aerodynamic Theory*, Vol. II, Chap. V. Springer, Berlin, 1934–1936.
22. Theodorsen, T. General theory of aerodynamic instability and the mechanism of flutter. *NACA Rept. 496*, 1935.
23. von Kármán, Th., and Sears, W. R. Airfoil theory for non-uniform motion. *J. Aeronaut. Sci. 5*, 379–390 (1938).
24. Küssner, H. G., and Schwarz, L. Der schwingende Flügel mit aerodynamisch ausgeglichenem Ruder. *Luftfahrtforschung 17*, 377–354 (1940). Also *NACA Tech. Mem. 991*, 1941.
25. Schwarz, L. Berechnung der Druckverteilung einer harmonisch sich verformenden Tragfläche in ebener Strömung. *Luftfahrtforschung 17*, 379–386 (1940). Also *Brit. RTP Transl. 1543.*
26. Söhngen, H. Bestimmung der Auftriebsverteilung für beliebige instationäre Bewegungen (ebenes Problem). *Luftfahrtforschung 17*, 401–419 (1940). Also *Brit. RTP Transl. 1446.*
27. Küssner, H. G. Das Zweidimensionale Problem der beliebig bewegten Tragfläche unter Berücksichtigung von partialbewegungen der Flussigkeit. *Luftfahrtforschung 17*, 355–361 (1940). Also *Brit. RTP Transl. 1541.*
28. Garrick, I. E. On some Fourier transforms in the theory of nonstationary flows. *Proc. Fifth Intern. Congr. Appl. Mech.*, 590–593 (1938). Also *NACA Rept. 629*, 1938.
29. Theodorsen, T., and Garrick, I. E. Non-stationary flow about a wing-aileron tab combination including aerodynamic balance. *NACA Rept. 736*, 1942.
30. Jones, W. P. Summary of formulae and notations in two-dimensional derivative theory. *Brit. Aeronaut. Research Council Repts. and Mem. 1958*, 1942.
31. Smilg, B., and Wasserman, L. S. Application of three-dimensional flutter theory to aircraft structures. *Air Materiel Command, U.S. Air Force Tech. Rept. 4798*, 1942.
32. Frazer, R. A. Flexure-torsion flutter derivatives for semi-rigid wing. *Brit. Aeronaut. Research Council Repts. and Mem. 1942*, 1941.
33. Williams, J. Aircraft flutter. *Brit. Aeronaut. Research Council. Repts. and Mem. 2492*, 1951.

34. *Lilienthal Gesellshaft für Luftfahrtforschung, Bericht 135.* Conference of March 6–8, 1941.
35. Garrick, I. E. Propulsion of a flapping and oscillating airfoil. *NACA Rept. 567*, 1936.
36. Fitzpatrick, J. L. G. Natural flight and related aeronautics. *Inst. Aeronaut. Sci. S.M.F. Fund Paper FF-7*, 1952. Also *Supplement*, 1954.
37. Küssner, H. G. Augenblicklicher Entwicklungsstand der Frage des Flügelflatterns. *Luftfahrtforschung 6*, 193–209 (1935). Also *NACA Tech. Mem. 782*, 1936.
38. Rott, N. Flügelschwingengsformen in ebener kompressibler Potentialströmung. *Z. angew. Math. u. Phys. 1*, 380–410 (1950).
39. Possio, C. Aerodynamic forces on an oscillating profile in a compressible fluid at subsonic speeds. *Aerotecnica 18*, 441–458 (1938). Also *Brit. Air Ministry Transl. 830.*
40. Dietze, F. I. The air forces of the harmonically vibrating wing in a compressible medium at subsonic velocity. II. Numerical tables and curves. *Air Materiel Command, U.S. Air Force, F-TS-506-RE and F-TS-948-RE*, 1947. Transl. from the German *ZWB, FB 1773*, 1943.
41. Schwarz, L. Numerical tables for air forces calculated for oscillating wings in two-dimensional compressible flow at subsonic velocity. *Air Materiel Command, U.S. Air Force, F-TS-599-RE*, 1946. Transl. from the German *ZWB, FB 1838*, 1943.
42. Turner, M. J., and Rabinowitz, S. Aerodynamic coefficients for an oscillating airfoil with hinged flap, with tables for a Mach number of 0.7. *NACA Tech. Note 2213*, 1950.
43. Schade, T. Beitrag zu Zahlentafeln zur Luftkraftberechnung der schwingenden Tragfläche in ebener Unterschallströmung. *ZWB, UM 3211*, 1944.
44. Eichler, M. Auflösung der Integralgleichung von Possio für den harmonisch schwingenden Tragflügel im kompressiblen Medium durch Zurüchführung auf ein linearesgleichung System. *Deutsche Luftfahrtforschung FB 1681*, 1942.
45. Fettis, H. E. An approximate method for the calculation of nonstationary air forces at subsonic speeds. *Wright Air Develop. Center Tech. Rept. 52-56*, 1952.
46. Fettis, H. E. Tables of lift and moment coefficients for an oscillating wing-aileron combination in two-dimensional subsonic flow. *U.S. Air Force Tech. Rept. 6688*, 1951. Also *Supplement 1*, 1953.
47. Reissner, E., and Sherman, S. Compressibility effects in flutter. *Curtiss-Wright Research. Lab. Rept. SB-240-S-1*, 1944.
48. Timman, R. *Beschouwingen over de luchkrachten op trillende vliegtuigvlengels. Dissertation*, Tech. Hogeschool Delft, 1946.
49. Haskind, M. D. Oscillations of a wing in a subsonic gas flow. *Prikl. Mat. i Mekh. 11*, 120–146 (1947). Also *Air Materiel Command, U.S. Air Force, Transl. A9-T-22.*
50. Reissner, E. On the application of Mathieu functions in the theory of subsonic compressible flow past oscillating airfoils. *NACA Tech. Note 2363*, 1951.
51. Billington, A. E. Harmonic oscillations of an aerofoil in subsonic flow. *Australian Aeronaut. Research. Lab. Rept. A65*, 1949.
52. Timman, R., and Van de Vooren, A. I. Theory of the oscillating wing with aerodynamically balanced control surface in a two-dimensional subsonic compressible flow. *Natl. Aeronaut. Research. Inst., Amsterdam, NLL Rept. F54*, 1949.
53. Küssner, H. G. A review of the two-dimensional problem of unsteady lifting surface theory during the last thirty years. *Univ. Maryland Lecture Series 23*, Apr. 1953.
54. Küssner, H. G. A general method for solving problems of the unsteady lifting surface theory in the subsonic range. *J. Aeronaut. Sci. 21*, 17–27 (1954).
55. McLachlan, N. W. *Theory and Application of Mathieu Functions.* Oxford Univ. Press, 1947.
56. Natl. Bur. Standards. *Tables Relating to Mathieu Functions; Characteristics, Values, Coefficients, and Joining Factors.* Columbia Univ. Press, 1951.

57. Goland, M. The quasi-steady air forces for use in low-frequency stability calculations. *J. Aeronaut. Sci. 17*, 601–608 (1950).
58. Miles, J. W. The application of unsteady flow theory to the calculation of dynamic stability derivatives. *North Amer. Aviation Rept. AL-957*, 1950.
59. Miles, J. W. Quasi-stationary airfoil theory in subsonic compressible flow. *Quart. Appl. Math. 8*, 351–358 (1951).
60. Runyan, H. L. Single-degree-of-freedom flutter calculations for a wing in subsonic potential flow and a comparison with an experiment. *NACA Rept. 1089*, 1952.
61. Timman, R., van de Vooren, A. I., and Greidanus, J. H. Aerodynamic coefficients of an oscillating airfoil in two-dimensional subsonic flow. *J. Aeronaut. Sci. 18*, 717–802 (1951). (For corrected tables see *J. Aeronaut. Sci. 21*, 499 (1954) and following reference.)
62. Tables of aerodynamic coefficients for an oscillating wing-flap system in a subsonic compressible flow. *Natl. Aeronaut. Research Inst., Amsterdam, NLL Rept. F151*, 1954.
63. Possio, C. Aerodynamic forces on an oscillating profile at supersonic speeds. *Pontif. Acad. Sci., Acta I, 11*, 93–106 (1937).
64. von Borbely, S. Aerodynamic forces on a harmonically oscillating wing at supersonic speed (two-dimensional case). *Z. angew. Math. u. Mech. 22*, 190–205 (1942).
65. Temple, G., and Jahn, H. A. Flutter at supersonic speeds. *Brit. Aeronaut. Research Council Repts. and Mem. 2140*, 1945.
66. Garrick, I. E., and Rubinow, S. I. Flutter and oscillating air force calculations for an airfoil in a two-dimensional supersonic flow. *NACA Rept. 846*, 1946.
67. Schwarz, L. Untersuchung einiger mit den Zylinderfunktionen nullter Ordnung verwandter Funktionen. *Luftfahrtforschung 20*, 341–372 (1944).
68. Huckel, V., and Durling, B. Tables of wing-aileron coefficients of oscillating air forces for two-dimensional supersonic flow. *NACA Tech. Note 2055*, 1950.
69. Lighthill, M. J. Oscillating airfoils at high Mach number. *J. Aeronaut. Sci. 20*, 402–406 (1953).
70. Van Dyke, M. D. Supersonic flow past oscillating airfoils including nonlinear thickness effects. *NACA Tech. Note 2982*, 1953.
71. Heaslet, M. A., Lomax, H., and Spreiter, J. R. Linearized compressible-flow theory for sonic flight speeds. *NACA Rept. 956*, 1950. See also *NACA Tech. Note 1767*, 1949.
72. Rott, N. Oscillating airfoils at Mach number 1. *J. Aeronaut. Sci. 16*, 380 (1949).
73. Nelson, H. C., and Berman, J. H. Calculations on the forces and moments for an oscillating wing-aileron combination in two-dimensional potential flow at sonic speed. *NACA Rept. 1128*, 1953.
74. Watkins, C. E. On transient two-dimensional flows at supersonic speeds. *J. Aeronaut. Sci. 16*, 569–570 (1949).
75. Biot, M. A. Loads on supersonic wing striking sharp-edged gust. *J. Aeronaut. Sci. 16*, 296 (1949).
76. Heaslet, M. A., and Lomax, H. Two-dimensional unsteady lift problems in supersonic flight. *NACA Rept. 945*, 1949.
77. Miles, J. W. Transient loading of airfoils at supersonic speeds. *J. Aeronaut. Sci. 15*, 592–598 (1948).
78. Mazelsky, B. Numerical determination of indicial lift. *NACA Tech. Note 2562*, 1951; *2613*, 1952.
79. Mazelsky, B., and Drischler, J. A. Numerical determination of indicial lift and moment functions for a two-dimensional sinking and pitching airfoil. *NACA Tech. Note 2739*, 1952.
80. Possio, C. Nonstationary motion of a compressible fluid. *Proc. accad. nazl. Lincei, Rome, 29, Series 62, 9*, 1939.
81. Lomax, H., Heaslet, M. A., and Sluder, L. The indicial lift and moment for a sinking or pitching two-dimensional wing flying at subsonic or supersonic speeds. *NACA Tech. Note 2403*, 1951.

82. Evvard, J. C. Use of source distributions for evaluating theoretical aerodynamics of thin finite wings at supersonic speeds. *NACA Rept. 951*, 1950.
83. Tobak, M. On the use of the indicial function concept in the analysis of unsteady motions of wings and wing-tail combinations. *NACA Rept. 1188*, 1954.
84. Biot, M. A. Transonic drag of an accelerated body. *Quart. Appl. Math. 7*, 101–105 (1949)
85. Gardner, C. S., and Ludloff, H. F. Influence of acceleration on the aerodynamic characteristics of thin wings in supersonic and transonic flight. *J. Aeronaut. Sci. 17*, 47–59 (1950).
86. Cicala, P. Sul moto non stazionario di un'ala di allungamento finito. *Rendiconti accad. nazl. Lincei, Rome, 25, Series 6a, 3-4*, 97–102 (1937).
87. Cicala, P. Comparison of theory with experiment in phenomena of wing flutter. *NACA Tech. Mem. 887*, 1939.
88. Jones, W. P., and Skan, S. W. Calculation of derivatives for rectangular wings of finite span by Cicala's method. *Brit. Aeronaut. Research Council Repts. and Mem. 1920*, 1940.
89. Jones, W. P. Aerodynamic forces on wings in simple harmonic motion. *Brit. Aeronaut. Research Council Repts. and Mem. 2026*, 1945.
90. Harvard Computation Laboratory. Tables of the complete Cicala function. *Problem Rept. 58* (*Air Force Problem 77*), 1952.
91. Dingel, M., and Küssner, H. G. Beitrage zur instationaeren Tragflaechentheorie. Pt. VIII. Die schwingende Tragflaeche grosser Streckung. *ZWB, FB 1774*, 1943. Also *Air Materiel Command, U.S. Air Force, Transl. F-TS-935-RE*, 1947.
92. Reissner, E. On the general theory of thin airfoils for non-uniform motion. *NACA Tech. Note 946*, 1944.
93. Reissner, E. Effect of finite span on airload distributions for oscillating wings, I. *NACA Tech. Note 1194*, 1947.
94. Reissner, E. Boundary value problems in aerodynamics of lifting surfaces in non-uniform motion. *Bull. Am. Math. Soc. 55*, 825–850 (1949).
95. Reissner, E., and Stevens, J. E. Effect of finite span on airload distributions for oscillating wings, II. *NACA Tech. Note 1195*, 1947.
96. Zartarian G., Fotieo, G., and Ashley, H. *Theoretical and Experimental Methods of Flutter Analysis*, Vol. VII and VIII. Mass. Inst. Technol. Press, 1949.
97. Jones, R. T. The unsteady lift of a finite wing. *NACA Tech. Note 682*, 1939.
98. Jones, R. T. The unsteady lift of a wing of finite aspect ratio. *NACA Rept. 681*, 1940.
99. Sears, W. R. A contribution to the airfoil theory for non-uniform motion. *Proc. Fifth Intern. Congr. Appl. Mech.*, 483–487 (1938).
100. Reissner, E. A problem of the theory of oscillating airfoils. *Proc. First U.S. Natl. Congr. Appl. Mech.*, 923–925 (1951).
101. Biot, M. A., and Boehnlein, C. T. Aerodynamic theory of oscillating wing of finite span. *Guggenheim Aeronaut. Lab. Calif. Inst. Technol. Rept. 5 to U.S. Army Air Force*, 1942.
102. Biot, M. A. Aerodynamic theory applied to flutter analysis. *Guggenheim Aeronaut. Lab. Calif. Inst. Technol. Rept. 6 to U.S. Army Air Force*, 1942.
103. Jones, W. P. The calculation of aerodynamic derivative coefficients for wings of any plan form in non-uniform motion. *Brit. Aeronaut. Research Council Repts. and Mem. 2470*, 1952.
104. Falkner, V. M. The calculation of aerodynamic loading on surfaces of any shape. *Brit. Aeronaut. Research Council Repts. and Mem. 1910*, 1943.
105. Jones, W. P. Aerodynamic forces on wings in non-uniform motion. *Brit. Aeronaut. Research Council Repts. and Mem. 2117*, 1945.
106. Katterbach, K. Contributions to the non-stationary wing theory. IX. Number comparisons of various approximation theories on the oscillating wing of large aspect ratio. *Air Materiel Command, U.S. Air Force F-TS-943-RE*, 1947. Transl. from the German *ZWB, FB 1789*, 1943.
107. Wasserman, L. S. Aspect ratio corrections in flutter calculations. *Air Materiel Command, U.S. Air Force MR MCREXA5-4595-8-5*, 1948.

108. Chiarulli, P. Oscillating wings of finite aspect ratio. *Air Materiel Command, U.S. Air Force Tech. Rept. 102-AC49/9-100,* 1949.
109. Ashley, H., Zartarian, G., and Neilson, D. O. Investigation of certain unsteady aerodynamic effects in longitudinal dynamic stability. *Wright Air Develop. Center, U.S. Air Force Tech. Rept. 5986,* 1951.
110. Eckhaus, W. On the theory of oscillating airfoils of finite span in subsonic flow. *Natl. Aeronaut. Research. Inst., Amsterdam, NLL Rept. F153,* 1954.
111. Garrick, I. E., and Watkins, C. E. Some recent developments in the aerodynamic theory of oscillating wings. *NACA-Univ. Conference on Aerodynamics, Construction, and Propulsion,* Oct. 1954.
112. Miles, J. W. On solving subsonic unsteady flow lifting surface problems by separating variables. *J. Aeronaut. Sci. 6,* 427–428 (1954).
113. Schade, Th., and Krienes, K. The oscillating circular airfoil on the basis of potential theory. *NACA Tech. Mem. 1098,* 1947. Also *Luftfahrtforschung 17,* 387–400 (1940); *19,* 282–291 (1943).
114. Kochin, N. E. Steady vibrations of wing of circular plan form and theory of wing of circular plan form. *NACA Tech. Mem. 1324,* 1953. Transl. from the Russian *Prikl. Mat. Mekh. 4,* 3–32 (1940); *6,* 287–316 (1942).
115. Dörr, J. Influence de l'allongement dans les calculs de flutter. Calculs comparatifs sur un empennage circulaire. *Recherche Aéronaut. 11,* 57–63 (1949).
116. Voss, H. M., and Hassig, H. J. Introductory study of flutter of low-aspect-ratio wings at subsonic speeds. *Mass. Inst. Technol. Aeroelastic and Structures Research Rept. for Bur. of Aeronautics, Navy Dept.,* 1952.
117. Garrick, I. E. Some research on high-speed flutter. *Proc. Third Anglo-American Aeronaut. Conference,* 419–446 (1951).
118. Jones, R. T. Properties of low-aspect-ratio pointed wings at speeds below and above the speed of sound. *NACA Rept. 835,* 1946.
119. Miles, J. W. On nonsteady motion of slender bodies. *Aeronaut. Quart. 2,* 183–194 (1950).
120. Voss, H. M. *Flutter of Low-Aspect-Ratio Lifting Surfaces. M.S. Thesis,* Mass. Inst. Technol., 1951.
121. Ashley, H. *Some Unsteady Aerodynamic Problems Affecting the Dynamic Stability of Aircraft. Sc.D. Thesis,* Mass. Inst. Technol., 1951.
122. Merbt, H., and Landahl, M. Aerodynamic forces on oscillating low aspect ratio wings in compressible flow. *Roy. Inst. Technol., Stockholm, KTH Aero Tech. Note 30,* 1953.
123. Merbt, H., and Landahl, M. The oscillating wing of low aspect ratio—results and tables of auxiliary functions. *Roy. Inst. Technol., Stockholm, KTH Aero Tech. Note 31,* 1954.
124. Landahl, M. T. The flow around oscillating low aspect ratio wings at transonic speeds. *Roy. Inst. Technol., Stockholm, KTH Aero Tech. Note 40,* 1954.
125. Barmby, J. G., Cunningham, H. J., and Garrick, I. E. Study of effects of sweep on the flutter of cantilever wings. *NACA Rept. 1014,* 1951.
126. Miles, J. W. Harmonic and transient motion of a swept wing in supersonic flow. *J. Aeronaut. Sci. 15,* 343–346 (1948).
127. Spielberg, I., Fettis, H. E., and Toney, H. S. Methods for calculating the flutter and vibration characteristics of swept wings. *Air Materiel Command, U.S. Air Force MR MCREXA5-4595-8-4,* 1948.
128. Turner, M. J. Aerodynamic theory of oscillating sweptback wings. *J. Math. and Phys. 28,* 280–293 (1950).
129. Lawrence, H. R., and Gerber, E. H. The aerodynamic forces on low aspect ratio wings oscillating in an incompressible flow. *J. Aeronaut. Sci. 19,* 769–781 (1952).
130. Goland, M., Luke, Y. L., and Dengler, M. A. Theoretical studies of the effects of aspect ratio on flutter. *Wright Air Develop. Center Tech. Rept. 54-29,* 1954.
131. Reissner, E. Investigation of oscillating lifting surfaces of finite span in incompressible flow. *Tech. Rept. for Air Research and Develop. Command,* Oct. 1951.
132. Miles, J. W. On the compressibility correction for subsonic unsteady flow. *J. Aeronaut. Sci. 17,* 181, 182 (1950).

133. Reissner, E. On the theory of oscillating airfoils of finite span in subsonic compressible flow. *NACA Rept. 1002*, 1950.
134. Reissner, E. Extension of the theory of oscillating airfoils of finite span in subsonic compressible flow. *NACA Tech. Note 2274*, 1951.
135. Watkins, C. E., Runyan, H. L., and Woolston, D. S. On the kernel function of the integral equation relating the lift and downwash distributions of oscillating finite wings in subsonic flow. *NACA Tech. Note 3131*, 1954.
136. Lance, G. N. The kernel function of the integral equation relating the lift and downwash distributions of oscillating finite wings in subsonic flow. *J. Aeronaut. Sci. 21*, 635, 636 (1954).
137. Temple, G. *Modern Developments in Fluid Dynamics. High Speed Flow*, Vol. I, Chap. IX. Edited by L. Howarth. Oxford Univ. Press, 1953.
138. Chang, C. C. The aerodynamic behavior of a harmonically oscillating finite sweptback wing in supersonic flow. *NACA Tech. Note 2467*, 1951.
139. Li, T. Y., and Stewart, H. J. On an integral equation in the supersonic oscillating wing theory. *J. Aeronaut. Sci. 20*, 724–726 (1953). (See also references cited therein.)
140. Krasilshchikova, E. A. Supersonic flows over slender bodies. *NACA Tech. Mem. 1383*, 1956. Transl. from *Scientific Record of Moscow State University, V.154, Mechanics*, 181–239 (1951).
141. Miles, J. W. Linearized theory of nonstationary flow at supersonic speeds. *U.S. Nav. Ord. Test Station Rept. 1930*, 1951.
142. Miles, J. W. Transient loading of thin wings. *U.S. Nav. Ord. Test Station Rept. 2077*, 1953.
143. Jones, W. P. Supersonic theory for oscillating wings of any plan form. *Brit. Aeronaut. Research Council Repts. and Mem. 2655*, 1953.
144. Miles, J. W. The oscillating rectangular airfoil at supersonic speeds. *Quart. Appl. Math. 9*, 47–65 (1951). See also *J. Aeronaut. Sci. 16*, 381 (1949) and *U.S. Nav. Ord. Rept. 1170, NOTS 226*, 1949.
145. Miles, J. W. A general solution for the rectangular airfoil in supersonic flow. *Quart. Appl. Math. 1*, 1–8 (1953).
146. Lomax, H., Fuller, F. B., and Sluder, L. Generalized indicial forces on deforming rectangular wings in supersonic flight. *NACA Tech. Note 3286*, 1954.
147. Watkins, C. E. Effect of aspect ratio on the air forces and moments of harmonically oscillating thin rectangular wings in supersonic potential flow. *NACA Tech. Note 2064*, 1950.
148. Watkins, C. E. Effect of aspect ratio on undamped torsional oscillation of a thin rectangular wing in supersonic flow. *NACA Tech. Note 1895*, 1949.
149. Nelson, H. C., Rainey, R. A., and Watkins, C. E. Lift and moment coefficients expanded to the seventh power of frequency for oscillating rectangular wings in supersonic flow and applied to a specific flutter problem. *NACA Tech. Note 3076*, 1954.
150. Goodman, T. R. The quarter-infinite wing oscillating at supersonic speeds. *Cornell Aeronaut. Lab. Rept. 36*, 1951.
151. Stewartson, K. On the linearized potential theory of unsteady supersonic motion. *Quart. J. Mech. and Appl. Math. 3*, 182–199 (1950).
152. Miles, J. W. On the low aspect ratio oscillating rectangular wing in supersonic flow. *Aeronaut. Quart. 4*, 231–244 (1953).
153. Miles, J. W. On harmonic motion of wide delta airfoils at supersonic speeds. *U.S. Nav. Ord. Test Station RRB-36*, 1950.
154. Froehlich, J. E. Non-stationary motion of purely supersonic wings. *J. Aeronaut. Sci. 18*, 298–310 (1951).
155. Nelson, H. C. Lift and moment on oscillating triangular and related wings with supersonic edges. *NACA Tech. Note 2494*, 1951.
156. Watkins, C. E., and Berman, J. H. Air forces and moments on triangular and related wings with subsonic leading edges oscillating in supersonic potential flow. *NACA Rept. 1099*, 1952.
157. Watkins, C. E., and Berman, J. H. Velocity potential and air forces associated

with a triangular wing in supersonic flow, with subsonic leading edges, and deforming harmonically according to a general quadratic equation. *NACA Tech. Note 3009*, 1953.

158. Cunningham, H. J. Total lift and pitching moment on thin arrowhead wings oscillating in supersonic potential flow. *NACA Tech. Note 3433*, 1955.
159. Stewartson, K. On the linearized potential theory of unsteady supersonic motion, II. *Quart. J. Mech. and Appl. Math. 5*, 137–154 (1952).
160. Pines, S., and Dugundji, J. Application of aerodynamic flutter derivatives to flexible wings with supersonic and subsonic edges. *Republic Aviation Corp. Rept. E-SAF-2*, 1954.
161. Liban, E., Neuringer, J., and Rabinowitz, S. Flutter analysis of an elastic wing with supersonic edges. *Republic Aviation Corp. Rept. E-SAF-1*, 1953.
162. Li, Ta. Aerodynamic influence coefficients for an oscillating finite thin wing in supersonic flow. *Inst. Aeronaut. Sci. Preprint 500*, 1955.
163. Zartarian, G., Hsu, P. T., and Voss, H. M. Comparative flutter calculations on low-aspect-ratio wings in incompressible and supersonic flows. *Mass. Inst. Technol., Aeroelastic and Structures Research Lab., Tech. Rept. 52-2*, 1954.
164. Zartarian, G., Hsu, P. T., and Voss, H. M. Application of numerical integration techniques to the low-aspect-ratio flutter problem in subsonic and supersonic flows. *Mass. Inst. Technol., Aeroelastic and Structures Research Lab., Tech. Rept. 52-3*, 1954.
165. Walsh, J., Zartarian, G., and Voss, H. M. Generalized aerodynamic forces on the delta wing with supersonic leading edges. *J. Aeronaut. Sci. 21*, 739–748 (1954).
166. Watkins, C. E., and Berman, J. H. On the kernel functions of integral equations relating lift and downwash distributions on oscillating wings in supersonic flow. *NACA Tech. Note 3438*, 1955.
167. Frazer, R. A., and Duncan, W. J. The flutter and airplane wings. *Brit. Aeronaut. Research Council Repts. and Mem. 1155*, 1928; *1255*, 1931.
168. Walker, P. B. Experiments on the growth of circulation about a wing with a description of the apparatus for measuring the fluid motion. *Brit. Aeronaut. Research Council Repts. and Mem. 1402*, 1931.
169. Drescher, H. Unsteady processes. G2—Experimental determination of unsteady lift. *AVA Monographs, Rept. and Transl. 1011*, 1948.
170. Victory, M. Flutter at high incidence. *Brit. Aeronaut. Research Council Repts. and Mem. 2048*, 1943.
171. Williams, J. An examination of experimental data relating to flexural-torsional wing derivatives. *Brit. Aeronaut. Research Council Repts. and Mem. 1944*, 1943.
172. Halfman, R. L. Experimental aerodynamic derivatives of a sinusoidally oscillating airfoil in two-dimensional flow. *NACA Report 1108*, 1952.
173. Bratt, J. B., and Wight, K. C. The effect of mean incidence, amplitude of oscillation, profile and aspect ratio on pitching moment derivatives. *Brit. Aeronaut. Research Council Repts. and Mem. 2064*, 1945.
174. Jones, W. P. Wind-tunnel interference effect on the values of experimentally determined derivative coefficients for oscillating airfoils. *Brit. Aeronaut. Research Council Repts. and Mem. 1912*, 1943.
175. Reissner, E. Wind tunnel correction for the two-dimensional theory of oscillating airfoils. *Cornell Aeronaut. Lab. Rept. SB-318-S-3*, 1947.
176. Timman, R. The aerodynamic forces on an oscillating aerofoil between two parallel walls. *Appl. Sci. Research A3, 1*, 31–57 (1951).
177. De Jager, E. M. The aerodynamic forces and moments on an oscillating aerofoil with control-surface between two parallel walls. *Natl. Aeronaut. Research Inst., Amsterdam, NLL Rept. F140*, 1953.
178. Goodman, T. R. The upwash corrections for an oscillating wing in a wind tunnel. *J. Aeronaut. Sci. 20*, 383–406 (1953).
179. Acum, W. E. A., and Garner, H. C. Approximate wall corrections for an oscillating swept wing in a wind tunnel of closed circular section. *Brit. Ministry of Supply C.P. 184*, 1955.

180. Runyan, H. L., and Watkins, C. E. Consideration on the effect of wind-tunnel walls on oscillating air forces for two-dimensional subsonic compressible flow. *NACA Tech. Note 2552*, 1951.
181. Woolston, D. S., and Runyan, H. L. Some considerations on the air forces on a wing oscillating between two tunnel walls for subsonic compressible flows. *J. Aeronaut. Sci. 22*, 41–50 (1955).
182. Runyan, H. L., Woolston, D. S., and Rainey, A. G. A theoretical and experimental investigation of the effect of tunnel walls on the forces on an oscillating airfoil in two-dimensional subsonic compressible flow. *NACA Tech. Note 3416*, 1955.
183. Bratt, J. B., and Davis, C. T. The influence of aspect ratio and taper on the fundamental damping derivative coefficients for flexural motion. *Brit. Aeronaut. Research Council Repts. and Mem. 2032*, 1945.
184. Clevenson, S. A., and Widmayer, E., Jr. Preliminary experiments on forces and moments of an oscillating wing at high-subsonic speeds. *NACA Research Mem. L9K28a*, 1950.
185. Cicala, P. Present state of research on the non-steady motion of a lifting wing. *Brit. RTP Translation 1867*. Transl. from the Italian, *Aerotecnica 19*, 1941.
186. van de Vooren, A. I. Measurement of aerodynamic forces on oscillating aerofoils. *Natl. Aeronaut. Research. Inst., Amsterdam, NLL Rept. M.P. 100*, 1954. Also *Agard Rept. AG15/P6*, 7–13 (1954).
187. Silverstein, A., and Joyner, U. T. Experimental verification of the theory of oscillating airfoils. *NACA Rept. 673*, 1939.
188. Bratt, J. B., and Scruton, C. Measurements of pitching moment derivatives for an aerofoil oscillating about the half-chord axis. *Brit. Aeronaut. Research Council Repts. and Mem. 1921*, 1938.
189. Erickson, A. L., and Robinson, R. C. Some preliminary results in the determination of aerodynamic derivatives of control surfaces in the transonic speed range by means of a flush-type electrical pressure cell. *NACA Research Mem. A8H03*, 1948.
190. Laidlaw, W. R., and Halfman, R. L. Experimental pressure distributions on oscillating low aspect ratio wings. *Inst. Aeronaut. Sci. Preprint 499*, 1955.
191. Patterson, J. L. A miniature electrical pressure gage utilizing a stretched flat diaphragm. *NACA Tech. Note 2659*, 1952.
192. Laidlaw, W. R. Theoretical and experimental pressure distributions on low aspect ratio wings oscillating in an incompressible flow. *Mass. Inst. Technol., Aeroelastic and Structures Research Lab., Tech. Rept. 51-2*, 1954.
193. Helfer, A. P. Electrical pressure integrator. *NACA Tech. Note 2607*, 1952.
194. Ritter, A. Determination of the unsteady lift-coefficients in wind tunnel tests by means of light interference methods. *Air Materiel Command, U.S. Air Force F-TR-1158-ND*, 1948.
195. Schnittger, J. R. Single degree of freedom flutter of compressor blades in separated flow. *J. Aeronaut. Sci. 21*, 27–36 (1954).
196. van de Vooren, A. I., and Bergh, H. Spontaneous oscillations of an aerofoil due to instability of the laminar boundary layer. *Natl. Aeronaut. Research. Inst., Amsterdam, NLL Rept. F96*, 1951.
197. Andreopoulos, T. C., Cheilek, H. A., and Donovan, A. F. Measurements of the aerodynamic hinge moments of an oscillating flap and tab. *Air Materiel Command, U.S. Air Force Tech. Rept. 5784*, 1949.
198. Drescher, H. Untersuchungen an einem symmetrischen Tragflügel mit spaltlos angeschlossenem Ruder bei raschen Aenderungen des Ruder-ausschlags (ebene Strömung). *Max-Planck-Institut für Strömungsforschung, Göttingen, Mitt. 6*, 1952.
199. Buchan, A. L., Harris, K. D., and Somervail, P. M. The measurement of the derivative z_w for an oscillating aerofoil. *Brit. Aeronaut. Research Council Repts. and Mem. 40*, 1950.
200. Clevenson, S. A., Widmayer, E., Jr., and Diederich, F. W. An exploratory investigation of some types of aeroelastic instability of open and closed bodies of revolution mounted on slender struts. *NACA Tech. Note 3308*, 1954.

201. Theodorsen, T., and Regier, A. A. Effect of the lift coefficient on propeller flutter. *NACA Wartime Rept. ACR L5F30*, 1945.
202. Baker, J. E. The effects of various parameters, including Mach number, on propeller blade flutter with emphasis on stall flutter. *NACA Tech. Note 3357*, 1955.
203. Sisto, F. Stall flutter in cascades. *J. Aeronaut. Sci. 20*, 598–605 (1953).
204. Mendelson, A. Aerodynamic hysteresis as a factor in critical flutter speed of compressor blades at stalling conditions. *J. Aeronaut. Sci. 16*, 645–652 (1949).
205. Johnson, D. F., and Mendelson, A. Preliminary investigation of the effects of cascading on the oscillating lift force of an airfoil vibrating in bending. *NACA Research Mem. E53F29*, 1953.
206. Smilg, B. The prevention of aileron oscillation at transonic air speeds. *Air Materiel Command, U.S. Air Force Tech. Rept. 5530*, 1946.
207. Bleich, F., McCullough, C. B., Rosecrans, R., and Vincent, G. S. *The Mathematical Theory of Vibration in Suspension Bridges*. U.S. Govt. Printing Office, 1950.
208. Studer, H. L. Experimentalle Untersuchungen über Flügelschwingen. *Inst. Aerol. Tech. H.S., Mitt. 4*, 1936.
209. Chuan, R. L., and Magnus, R. J. Study of vortex shedding as related to self-excited torsional oscillations of an airfoil. *NACA Tech. Note 2429*, 1951.
210. Bollay, W., and Brown, C. D. Some experimental results on wing flutter. *J. Aeronaut. Sci. 8*, 313–318 (1941).
211. Halfman, R. L., Johnson, H. C., and Haley, S. M. Evaluation of high angle of attack aerodynamic derivative data and stall flutter prediction techniques. *NACA Tech. Note 2533*, 1951.
212. Jones, W. P. Airfoil oscillations at high mean incidences. *Brit. Aeronaut. Research Council Repts. and Mem. 2654*, 1953.
213. George, M. B. T. A theoretical approach to the problem of stall flutter. *Cornell Univ. Grad. School of Aeronaut. Eng., Contract AF 33(038)-21406*, 1953.
214. Landahl, M. T., and Stark, V. J. E. An electrical analogy for solving the oscillating-surface problem for incompressible nonviscid flow. *Roy. Inst. Technol., Stockholm, KTH Aeronaut. Tech. Note 34*, 1953.
215. Thompson, F. L. Flight research at transonic and supersonic speeds with free-falling and rocket-propelled models. *Second Intern. Aeronaut. Conference, Inst. Aeronaut. Sci., New York*, 1949.
216. Flax, A. H. The reverse-flow theorem for nonstationary flows. *J. Aeronaut. Sci. 19*, 352–353 (1952); *20*, 120–126 (1953).
217. Heaslet, M. A., and Spreiter, J. R. Reciprocity relations in aerodynamics. *NACA Tech. Note 2700*, 1952.
218. Küssner, H. G. The difference property of the kernel of the unsteady lifting surface theory. *J. Aeronaut. Sci. 22*, 227–230 (1955).
219. Jones, W. P. The influence of thickness chord ratio on supersonic derivatives for oscillating aerofoils. *Brit. Aeronaut. Research Council Repts. and Mem. 2679*, 1954. (*ARC 10,871*, 1947.)
220. Carrier, G. F. The oscillating wedge in a supersonic stream. *J. Aeronaut. Sci. 16*, 150–152 (1949).
221. Martin, J. C., and Gerber, N. The effect of thickness on airfoils with constant vertical acceleration at supersonic speeds. *J. Aeronaut. Sci. 22*, 179–188 (1955).
222. Jahnke, E., and Emde, F. *Tables of Functions with Formulae and Curves*, 4th ed. Dover, 1945.

Bibliography

Baird, E. F., and Kelley, H. J. Formulation of the flutter problem for solution on an electronic analog computer. *J. Aeronaut. Sci. 17*, 189–190 (1950).

Barton, M. V. Stability of an oscillating airfoil in supersonic airflow. *J. Aeronaut. Sci. 15*, 371–376 (1948).

Behrbohm, H. The flat triangular wing with subsonic leading edges in steady pitch. *Svenska Aeroplan Aktiebolaget, Linkoping, Tech. Note 9*, 1952.

Bergh, H. Experimental determination of the aerodynamic coefficients of an oscillating wing in incompressible, two-dimensional flow. Part III: Experiments at zero airspeed. *Natl. Aeronaut. Research. Inst., Amsterdam, NLL Rept. F103*, 1952.

Bergh, H., and van de Vooren, A. I. Experimental determination of the aerodynamic coefficients of an oscillating wing in incompressible two-dimensional flow. Part IV: Calculation of the coefficients. *Natl. Aeronaut. Research. Inst., Amsterdam, NLL Rept. F104*, 1952.

Berndt, S. B. On the theory of slowly oscillating delta wings at supersonic speeds. *Medd. Flytekniska Forsoksanstalten, Stockholm, 43*, 1952.

Billing, H., Katterbach, K., and Küssner, H. G. Untersuchung der Bewegung einer Platte bein Eintritt in eine Strahlgrenze. *Zent. Wissensch. Berichtsw. über Luftfahrtforsch., Berlin, FB1640*, 1942.

Biot, M. A., and Arnold, L. Investigation of aileron compressibility flutter. *Air Force Tech. Rept. 6341*, 1950.

Biot, M. A., and Arnold, L. Unsteady supersonic generalized aerodynamic forces on a delta wing with arbitrary deflection mode shape. *Biot and Arnold, Contract 113583-SD, Rept. 13*, 1953.

Bisplinghoff, R. L., Ashley, H., and Halfman, R. L. *Aeroelasticity*. Addison-Wesley, 1955.

Carrier, G. F., and Ring, R. M. The lift and moment on an oscillating wedge. *Office of Air Research Rept. AF 33(038)13279, Brown Univ.*, 1951.

Chang, C. C. The transient reaction of an airfoil due to change in angle of attack at supersonic speeds. *J. Aeronaut. Sci. 15*, 635–655 (1948).

Chang, C. C. Transient aerodynamic behavior of an airfoil due to different arbitrary modes of nonstationary motions in a supersonic flow. *NACA Tech. Note 2333*, 1951.

Collar, A. R. Resistance derivatives of flutter theory, Part II. *Brit. Aeronaut. Research Council Repts. and Mem. 2139*, 1944.

Couchet, G. Les mouvements plans non stationnaires à circulation constante et les mouvements infiniment voisins. *Office natl. Etudes et Recherches aéronaut. 31*, 1949; *56*, 1952.

de Veubeke, B. F. Aérodynamique instationnaire des profils mince déformables. *Bulletin du Service Tech. Aeronaut., Brussels, Office natl. Etudes et Recherches aéronaut. 25*, 1953.

Dörr, J. Les forces aérodynamiques sur une aile vibrant harmoniquement dans un écolement supersonique. *Office natl. Etudes et Recherches aéronaut. 37*, 1949.

Duncan, W. J. *The Principles of the Control and Stability of Aircraft*. Cambridge Univ. Press, 1952.

Ellenberger, G. Luftkräfte bei beliebiger instationären Bewegung eines Tragflügels mit Querruder und bei Vorhandensein von Boen. *Z. angew. Math. u. Mech. 18*, 173–176 (1938).

Epstein, M. The effect of an axially symmetric fuselage on the non-stationary supersonic airloads acting on a wing with supersonic edges. *Wright Air Develop. Center Tech. Rept. 54-196*, 1954.

Epstein, M. The two-dimensional supersonic aerodynamic coefficients for oscillatory parabolic bending of an airfoil mean chord line. *Wright Air Develop. Center Tech. Note WCLS-53-55*, 1953.

Farren, W. S. The reaction on a wing whose angle of incidence is changing rapidly. Wind tunnel experiments with a short period recorded balance. *Brit. Aeronaut. Research Council Repts. and Mem. 1648*, 1935.

Fisher, L. R., and Fletcher, H. S. Effect of lag of sidewash on the vertical-tail contribution to oscillatory damping in yaw of airplane models. *NACA Tech. Note 3356*, 1955.

Fisher, L. R., and Wolhart, W. D. Some effects of amplitude and frequency on the aerodynamic damping of a model oscillating continuously in yaw. *NACA Tech. Note 2766*, 1952.

Frankl, F. I. Effect of the acceleration of elongated bodies of revolution upon the resistance in a compressible flow. *NACA Tech. Mem. 1230*, 1949.

Frazer, R. A. Flexure-torsion flutter derivatives for semi-rigid wings. *Brit. Aeronaut Research Council Repts. and Mem. 1942*, 1941.

Fung, Y. C. *Introduction to the Theory of Aeroelasticity*. Wiley, 1955.

Garrick, I. E. A review of unsteady aerodynamics of potential flows. *Appl. Mech. Revs. 3*, 89–91 (1952).

Germain, P., and Bader, R. Quelques remarques sur les mouvements vibratoires d'une aile en régime supersonique. *Recherche aéronaut. 11*, 1949.

Goodman, T. R. Aerodynamics of a supersonic rectangular wing striking a sharp edged gust. *J. Aeronaut. Sci. 18*, 519–526 (1951).

Greenberg, J. M. Airfoil in sinusoidal motion in a pulsating stream. *NACA Tech. Note 1326*, 1947.

Greenberg, J. M. Some considerations on an airfoil in an oscillating stream. *NACA Tech. Note 1372*, 1947.

Greidanus, J. H. Mathematical methods of flutter analysis. *Natl. Aero. Research. Inst., Amsterdam, NLL Rept. V1384*, 1947.

Greidanus, J. H., and van Heemert, A. Theory of the oscillating aerofoil in two-dimensional incompressible flow, Parts I, II. *Natl. Aero. Research. Inst., Amsterdam, NLL Report F41*, 1948.

Greidanus, J. H., van de Vooren, A. I., and Bergh, H. Experimental determination of the aerodynamic coefficients of an oscillating wing in incompressible, two-dimensional flow. Part I: Wing with fixed axis of rotation. *Natl. Aero. Research. Inst., Amsterdam, NLL Rept. F101*, 1952.

Hamel, D. Détermination théorique des coéfficients aérodynamiques d'une structure quelconque en mouvement vibratoire harmonique. *Office natl. Etudes et Recherches aéronaut. Tech. Note 6*, 1951.

Handbook of Supersonic Aerodynamics. Aeroelastic Phenomenon. Nav. Ord. Test Station Rept. 1488, Vol. 4, Sec. 12, 1952.

Hofsommer, D. J. Systematic representation of aerodynamic coefficients of an oscillating aerofoil in two-dimensional incompressible flow. *Natl. Aero. Research. Inst., Amsterdam, NLL Rept. F61*, 1950.

Hönl, H. On the sound field of a point-shaped sound source in uniform translatory motion. *NACA Tech. Mem. 1362*, 1954.

Hönl, H. Two-dimensional wing theory in the supersonic range. *NACA Tech. Mem. 1238*, 1949.

Houbolt, J. C. A recurrence matrix solution for the dynamic response of aircraft in gusts. *NACA Tech. Rept. 1010*, 1951.

Houbolt, J. C., and Kordes, E. E. Gust-response analysis of an airplane including wing bending flexibility. *NACA Tech. Note 2763*, 1952.

Huckel, V. Tabulation of the f_λ functions which occur in unsteady aerodynamic theory for supersonic flow. *NACA Tech. Note 3606*, 1955.

Isaacs, R. Airfoil theory for flows of variable velocity. *J. Aeronaut. Sci. 12*, 113–117 (1945).

Isaacs, R. Airfoil theory for rotary wing aircraft. *J. Aeronaut. Sci. 13*, 218–220 (1946).

Jaeckel, K. Forces on an oscillating airfoil in accelerated flow. *Ing.-Arch. 9*, 371–395 (1938).

Jaeckel, K. Über die Bestimmung der Zirkulationsverteilung für den zwei-dimensionalen Tragflügel bei beliebigen periodischen Bewegungen. *Luftfahrtforschung 16*, 135–138 (1939).

Jones, W. P. Theoretical air-load and derivative coefficients for rectangular wings. *Brit. Aeronaut. Research Council Repts. and Mem. 2142*, 1943.

Jones, W. P., and Skan, S. W. Aerodynamic forces on biconvex aerofoils oscillating in a supersonic airstream. *Brit. Aeronaut. Research Council Repts. and Mem. 2749*, 1953.

Jordan, P. Semi-experimental determination of aerodynamic derivatives for an oscillating wing-aileron system. *Roy. Aircraft Establishment, Tech. Note Structures 28*, 1948.

Jordan, P. Unsteady aerodynamic coefficients for supersonic flow. Part I: Analytical.

Brit. Aeronaut. Research Council 10,052, 1946; Part II: Numerical tables. *11,400*, 1946.

Karp, S. N., Shu, S. S., and Weil, H. Monograph III: Aerodynamics of the oscillating airfoil in compressible flow. *Air Materiel Command, Air Force Tech. Rept. F-TR-1167-ND*, 1947.

Karp, S. N., and Weil, H. The oscillating airfoil in compressible flow. A review of graphical and numerical data, Monograph III, Part II. *Air Materiel Command, Air Force Tech. Rept. F-TR-1195-ND*, 1948.

Küssner, H. G. Aeroelastische Probleme des Flugzeugbaus. *Z. Flügwiss. 3*, 1–18 (1955).

Küssner, H. G. The oscillating lifting surface in the range of subsonic flow. *Mitteil. Max-Planck-Inst. Stromungsforsch., Göttingen, 53/J/2*, 1953.

Küssner, H. G. Unsteady processes. *AVA Monographs*, Sec. G, (A. Betz, Ed.). *Gt. Brit. Ministry of Aircraft Production Repts. and Trans. MAP (Volkenrode) 1009–1014*, 1945–1946.

Küssner, H. G., and Billing, H. Unsteady flow. *Hydro- and Aerodynamics*, Chap. VI, (A. Betz, Ed.) *Central Air Document Office ATI 72854*, 1950.

Landahl, M. T. Unsteady flow around thin wings at high Mach numbers. *Office Sci. Research Tech. Note 55–245* (*Mass. Inst. Technol. Group Rept. 55–3*), 1955.

Lapin, E., Crookshanks, R., and Hunter, H. F. Downwash behind a two-dimensional wing oscillating in plunging motion. *J. Aeronaut. Sci. 19*, 447–450 (1952).

Lessing, H. C., Fryer, T. B., and Mead, M. H. A system for measuring the dynamic lateral stability derivatives in high-speed wind tunnels. *NACA Tech. Note 3348*, 1954.

Lomax, H. Lift developed on unrestrained rectangular wings entering gusts at subsonic and supersonic speeds. *NACA Rept. 1162*, 1954.

Lomax, H., Fuller, F. B., and Sluder, L. Generalized indicial forces on deforming rectangular wings in supersonic flight. *NACA Tech. Note 3286*, 1954.

Lomax, H., Heaslet, M. A., and Fuller, F. B. Two- and three-dimensional unsteady lift problems in high-speed flight. *NACA Rept. 1077*, 1952.

Lorell, J. The aerodynamic forces on an oscillating airfoil in supersonic flow: A comparison between two- and three-dimensional theories. *Calif. Inst. Technol. Jet Propul. Lab. Rept. 20-64*, 1952.

Luke, Y. L. Tables of coefficients for compressible flutter calculations. *Air Materiel Command, Air Force Tech. Rept. 6200*, 1950.

Lyon, H. M. Aerodynamical derivatives of flexural-torsional flutter of a wing of finite span. *Brit. Aeronaut. Research Council Repts. and Mem. 1900*, 1939.

Marini, M. Aerodynamic effects on swept wings in oscillatory motion. *Aerotecnica 4*, 275–287 (1953).

Mazelsky, B. On the noncirculatory flow about a two-dimensional airfoil at subsonic speeds. *J. Aeronaut. Sci. 19*, 848–849 (1952).

Mazet, R., and Dörr, J. Détermination par le calcul et par essais en soufflerie des vitesses critiques de vibrations d'une maquette rigide de profil constant, montée sur ressorts. *Office natl. Etudes et Recherches aéronaut. 39*, 1950.

Meller, M. G. Étude préliminaire sur le flutter des aubes de compresseur. *Office natl. Etudes et Recherches aéronaut Tech. Note 18*, 1953.

Migotsky, E. Some considerations on two-dimensional thin airfoils deforming in supersonic flow. *NACA Tech. Note 3386*, 1955.

Miles, J. W. A note on subsonic edges in unsteady supersonic flow. *Quart. Appl. Math. 3*, 363–367 (1953).

Miles, J. W. On the nonlinear thickness effect for a supersonic oscillating airfoil. *J. Aeronaut. Sci. 21*, 714–715 (1954).

Miles, J. W. Unsteady supersonic flow. *Air Research and Develop. Command, Contract AF18(600)-432*, 1955.

Miles, J. W., Naiman, I., and Schroedter, G. M. Aerodynamic derivatives for an oscillating rectangular airfoil at supersonic speeds. *Nav. Ord. Test Station, NAVORD Rept. 1292*, 1951.

Milne-Thomson, L. M. *Theoretical Hydrodynamics*. Macmillan, 1949.

Minhinnick, I. T. Subsonic aerodynamic flutter derivatives for wings and control surfaces. *Brit. Aeronaut. Research Council 14,228*, 1950.

Minhinnick, I. T. Tables of functions for a valuation of wing and control surface flutter derivatives for incompressible flow. *Brit. Aeronaut. Research Council 13,730, Rept. Structures 86*, 1950.

Molyneux, W. G. Techniques for the measurement of the aerodynamic forces on oscillating aerofoils. *Roy. Aircraft Establishment Tech. Note Structures 161*, 1955.

Morris, R. M. The two-dimensional hydrodynamical theory of moving aerofoils. *Proc. Roy. Soc. London A188*, 439–463 (1947).

Moskowitz, B., and Moeckel, W. E. First order theory for unsteady motion of thin wings at supersonic speeds. *NACA Tech. Note 2034*, 1950.

Nekrasov, A. I. Wing theory for nonstationary flow. *Doklady Akad. Nauk. SSSR*, 1947.

Nelson, H. C., and Cunningham, H. J. Theoretical investigation of flutter of two-dimensional flat panels with one surface exposed to supersonic potential flow. *NACA Tech. Note 3465*, 1955.

Neumark, S. Pressure distribution on an airfoil in nonuniform motion. *J. Aeronaut. Sci. 19*, 214–215 (1952).

Possio, G. Aerodynamic action on a wing in oscillating motion. *Atti accad. nazl. Lincei, Rome, Rendiconti (VI) 28*, 194–200 (1938).

Postel, E. E., and Leppert, E. L. Theoretical pressure distributions for a thin airfoil oscillating in incompressible flow. *J. Aeronaut. Sci. 15*, 486–492 (1948).

Radok, J. R. M. An approximate theory of the oscillating wing in a compressible subsonic flow for low frequencies. *Natl. Aeronaut. Research. Inst., Amsterdam, NLL Rept. F97*, 1951.

Radok, J. R. M. Gust loads on two-dimensional aerofoils in supersonic flow. *Australian Aeronaut. Research Lab. Rept. A66 and SM142*, 1949.

Radok, J. R. M. The theory of aerofoils in unsteady motion. *Aeronaut. Quart. 4*, 297–320 (1952).

Rainey, A. G. Preliminary study of some factors which affect the stall-flutter characteristics of thin wings. *NACA Tech. Note 3622*, 1955.

Robinson, A. On some problems of unsteady supersonic aerofoil theory. *College of Aeronautics, Cranfield, Rept. 16*, 1948.

Rott, N. On the unsteady motion of a thin rectangular wing in supersonic flow. *J. Aeronaut. Sci. 18*, 775–776 (1951).

Rott, N., and George, M. B. T. An approach to the flutter problem in real fluids. *Inst. Aeronaut. Sci. Preprint 509*, 1955.

Roumieu, C. Étude des régimes transitoires en aérodynamique supersonique à deux dimensions, aperçu théorique sur le domaine transsonique. *Recherche Aéronaut. 9*, 1949.

Runyan, H. L., Cunningham, H. J., and Watkins, C. E. Theoretical investigation of several types of single-degree-of-freedom flutter. *J. Aeronaut. Sci. 19*, 101–110 (1952).

Sauer, R. Elementare Theorie des langsam schwingenden Überschallflügels. *Z. angew. Math. u. Phys. 1*, 248–253 (1950).

Scanlan, R. H. Correction for aspect ratio of the indicial lift functions of Wagner and Küssner. *J. Aeronaut. Sci. 19*, 357–358 (1952).

Scanlan, R. H., and Rosenbaum, R. *Introduction to Study of Aircraft Vibration and Flutter*. Macmillan, 1951.

Scherer, M. Mésure en soufflerie de la résultante aérodynamique sur un profil de courant plan à incidence variable en régime harmonique. *Compt. rend. 230*, 2071–2073 (1950).

Schmeidler, W. Thrust and drag. *Z. angew. Math. u. Mech. 19*, 65–86 (1939).

Schwarz, L. A semi-experimental method for determining unsteady pressure distributions. *Brit. Aeronaut. Research Council 10, 387*, 1947.

Sears, W. R. Operational methods in the theory of airfoils in non-uniform motion *J. Franklin Inst. 130*, 95–111 (1940).

Sears, W. R. Some aspects of non-stationary airfoil theory and its practical application. *J. Aeronaut. Sci. 8*, 104–108 (1941).
Sears, W. R., and Kuethe, A. M. The growth of the circulation of an airfoil flying through a gust. *J. Aeronaut. Sci. 6*, 376–378 (1939).
Sears, W. R., and Sparks, B. O. On the reaction of an elastic wing to vertical gusts. *J. Aeronaut. Sci. 8*, 64–67 (1941).
Sedov, L. I. On the theory of the unsteady motion of an airfoil. *NACA Tech. Mem. 1156*, 1947.
Sedov, L. I. On the theory of unsteady planning and the motion of a wing with vortex separation. *NACA Tech. Mem. 942*, 1940.
Sewell, G. L. A theory of uniform supersonic flow past a thin oscillating aerofoil at appreciable incidence to the main stream. *Aeronaut. Quart. 5*, 185–194 (1954).
Sewell, G. L. Theory of an oscillating supersonic aerofoil. *Aeronaut. Quart. 2*, 34–38 (1950).
Shen, S. F. Effect of structural flexibility on aircraft loading. Part X: A new lifting-line theory for the unsteady lift of a swept or unswept wing in an incompressible fluid. *Wright Air Develop. Center Tech. Rept. 6358*, 1953.
Shen, S. F. Flutter of a two-dimensional simply-supported uniform panel in a supersonic stream. *Office of Nav. Research, Mass. Inst. Technol. Dept. Aeronaut. Eng., Contract N5ori-07833*, 1952.
Sisto, F. Unsteady aerodynamic reactions on airfoils in cascade. *J. Aeronaut. Sci. 22*, 297–302 (1955).
Spence, D. A. The characteristic frequency of small oscillations in the flow past bluff bodies. *Roy. Aircraft Establishment Rept. Aero. 2532*, 1955.
Spielberg, I. N. The two-dimensional incompressible aerodynamic coefficients for oscillatory changes in airfoil camber. *J. Aeronaut. Sci. 20*, 432–434 (1953).
Statler, I. C. Dynamic stability at high speeds from unsteady flow theory. *J. Aeronaut. Sci. 17*, 232–242 (1950).
Statler, I. C., and Easterbrook, M. Handbook for computing non-stationary flow effects on subsonic dynamic longitudinal response characteristics of an airplane. *Cornell Aeronaut. Lab. Rept. TB-495-F-12*, 1950.
Strang, W. J. A physical theory of supersonic aerofoils in unsteady flow. *Proc. Roy. Soc. London A195*, 245–264 (1948).
Strang, W. J. Gust loading of rectangular supersonic wings. *Council for Sci. and Indust. Research., Div. of Aero., A57*, 1949.
Strang, W. J. Transient lift of three-dimensional purely supersonic wings. *Proc. Roy. Soc. London A202*, 54–80 (1950).
Strang, W. J. Transient source, doublet, and vortex solutions of linearized equations of supersonic flow. *Dept. of Supply and Develop., Div. of Aero. A60*, 1949.
Temple, G. Unsteady compressible flow in two dimensions. *Brit. Aeronaut. Research Council F. M. 1089*, 1947.
Templeton, H. *Mass Balancing of Aircraft Control Surfaces.* Chapman and Hall, 1954.
Timman, R. Approximate theory of the oscillating wing in compressible subsonic flow for high frequencies. *Natl. Aeronaut. Research. Inst., Amsterdam, NLL Rept. F99*, 1951.
Timman, R. Linearized theory of the oscillating airfoil in compressible subsonic flow. *J. Aeronaut. Sci. 21*, 230–236 (1954).
Timman, R. Report on methods and results of nonstationary airfoil theory. *AGARD Rept. AG5/P6*, 50–55 (1954).
Todd (Taussky), O. On some boundary value problems on the theory of the non-uniform supersonic motion of an aerofoil. *Brit. Aeronaut. Research Council Repts. and Mem. 2141*, 1945.
Valensi, J. A review of techniques of measuring oscillatory aerodynamic forces and moments on models oscillating in wind tunnels in use on the Continent. *AGARD Paper AG15/P6*, 14–49 (1954).
van Spiegel, E., and van de Vooren, A. I. On the theory of the oscillating wing in two-dimensional subsonic flow. *Natl. Aeronaut. Research. Inst., Amsterdam, NLL Rept. F142*, 1953.

van de Vooren. A. I. Generalization of the Theodorsen function to stable oscillations. *J. Aeronaut. Sci. 19*, 209–213 (1952).

van de Vooren, A. I., and Bergh, H. Experimental determination of the aerodynamic coefficients of an oscillating wing in incompressible, two-dimensional flow. Part II: Wing with oscillating axis of rotation. *Natl. Aeronaut. Research. Inst., Amsterdam, NLL Rept. F102*, 1952.

van de Vooren, A. I., and Hofsommer, D. J. Binary aileron-tab flutter. *Natl. Aeronaut. Research Inst., Amsterdam, NLL Rept. F52*, 1949.

van de Vooren, A. I., and Yff, J. Aerodynamic coefficients of an oscillating aerofoil with a pressure-seal balanced aileron. *Natl. Aeronaut. Research. Inst., Amsterdam, NLL Rept. F115*, 1952.

Voigt, H. Further experiments on wing vibration. *RTP Transl. 874. Transl. from Jahrbuch deut. Luftfahrtforschung 1*, 249–258 (1938).

von Borbely, S. Über einen Grenzfall der instationären raumlichen Tragflügelstrommung. *Z. angew. Math. u. Mech. 18*, 319 (1938). Also *RTP Transl. 931; Brit. Aeronaut. Research Council 4090*, 1939.

Voss, H. M., Zartarian, G., and Hsu, P. T. Theoretical and experimental investigation of the aeroelastic behavior of low-aspect-ratio wings. *Mass. Inst. Technol., Aeroelastic and Structures Research, Contract NOa(s) 51-109-c*, 1953.

Wang, C. T., Vaccaro, R. J., and DeSanto, D. F. A practical approach to the problem of stall flutter. *Office of Sci. Research Tech. Note 55-98 (New York Univ.)*, 1955.

Weber, R. Tables des coéfficients aérodynamiques instationnaires régime plan supersonique. Première partie. Exposé sommaire de la méthode de L. Schwartz et introduction aux tables. *Office natl. Etudes et Recherches aéronaut. 41*, 1950.

Williams, D. E. On the integral equations of two-dimensional subsonic flutter derivative theory. *Roy. Aircraft Establishment Tech. Note Structures 181*, 1955.

Woods, L. C. The aerodynamic forces on an oscillating aerofoil in a free jet. *Proc. Roy. Soc. London A229*, 235–250 (1955).

Woods, L. C. The lift and moment acting on a thick aerofoil in unsteady motion. *Phil. Trans. Roy. Soc. London A247, 925*, 131–162 (1954).

Woods, L. C. Unsteady plane flow past curves obstacles with infinite wakes. *Proc. Roy. Soc. London A229*, 152–180 (1955).

SECTION G

THE EXPERIMENTAL AERODYNAMICS OF WINGS AT TRANSONIC AND SUPERSONIC SPEEDS

C. W. FRICK

G,1. Introduction.

Scope. Aeronautical research has as one of its objectives the formulation of the laws of fluid dynamics and the application of these laws to airplane design. Experimental observations of the nature of flows about bodies are a fundamental means by which this end is achieved. The experimental findings, however, are of real value only when they serve to provide the formal statement of the laws known as theory, either by verifying proposed theoretical relationships by comparison or by providing insight into the physical model used to formulate theory. A major research task is therefore concerned with comparison of theory and experiment.

The present section is concerned with this task. For brevity, the comparison will be confined to a study of wings at transonic and supersonic speed. The comparisons are based, in the main, on data that are available to the reader through the cited references. The sources of other data, which were released from classification by the NACA as needed to illustrate certain points, unfortunately are currently unavailable.

Assumptions of theory. Air is a complex mixture of gases. It is elastic, that is, it may be compressed or expanded. It possesses viscosity so that it can transmit shear. Its properties vary with pressure and temperature. All these items contribute to the difficulty of expressing in mathematical terms the laws governing the flow of air about an airfoil. So complex are the exact equations of motion, that only a few problems can be solved with rigor. Still, there has been evolved over a period of years a great body of theory through which many of the problems of flight have been solved.

The answer to this paradox lies in the utilization of observations of the physical nature of the flow which suggests the use of an approximate theory under conditions where the error associated with its inexactness is small. Prandtl, for example, recognized from observations that for many aerodynamic problems the action of the viscosity of the air was confined to a thin lamina of air near the airfoil surface. He concluded, therefore, that the flow outside this region closely obeys laws for the flow of inviscid fluids. Solutions could be obtained, therefore, for numerous airfoil problems through the use of the equations for inviscid flow. For other conditions, e.g. near the stall where viscous effects predominate, such theory is invalid. In another instance, Munk noted that airfoils of thin sections and at small angles of attack promoted disturbances so small that the equations of fluid motions could be linearized. The solution of many airfoil problems was thus made possible. In this way much aerodynamic theory has been accumulated. However, it is evident that extensive experiments are necessary to demonstrate the adequacy of the theory and to extend the range of variables which can be treated.

Accuracy of experiments. Another facet of the problem of measuring the accuracy of theory with experimental results concerns the question of whether or not the experimental technique reveals the true nature of the flow. In recent years, the accuracy of experimental aerodynamic research has improved greatly, partly through a great increase in size of transonic and supersonic wind tunnels so that the test Reynolds number approaches that of flight and the model contours may be more closely controlled. Also, there has been a great improvement in instrumentation in airfoil research, typified by the development of internal strain gauge balances which, by reducing the mass of the balance and by eliminating large tare effects, have improved experimental accuracy.

In addition, the investigation of model airfoils in supersonic wind tunnels has an advantage in accuracy over subsonic tests since the usual correction to the data for the presence of the wind tunnel walls is not required. On the other hand, most supersonic wind tunnels have undesirable pressure gradients which imply stream inclination and curvature, as well as model buoyant forces. The effects of stream turbulence on the extent of laminar flow are also unknown for supersonic wind tunnel air streams so that the accuracy of the skin friction drag is not known. This condition, however, is no worse than for subsonic wind tunnels.

Another factor that has influenced the development of experimental techniques in aerodynamic research and which has increased the feeling of the general reliability of wind tunnel results is associated with the fact that there has been a great deal of emphasis on free flight experimental data. The remarkable achievement in free flight instrumentation and the competitive comparison of wind tunnel and flight results has served to raise the whole level of accuracy. In many cases the comparison between wind tunnel results and both rocket model and full scale airplane results leaves nothing to be desired in the way of accuracy. In general, therefore, it is concluded that the experimental data for airfoils are of sufficient accuracy to permit an assessment of the accuracy of theory.

G,2. Lift of Wings. In general terms, there are two sources of error in the theoretical treatment of the lift of airfoils. First of all, as has been noted, there are errors implicit in the simplification necessary to achieve a solution in a mathematical sense. Secondly, there may be errors in the hypothetical flow model due to incomplete or inexact experimental flow observations. In the present section, we are concerned with the influence of these errors on the over-all accuracy of wing theory. Errors attributable to the first category serve primarily to limit the range of usefulness of theory, e.g. the angle of attack, etc., while those in the second are of a fundamental nature and are of greater concern. Both are treated in the following discussion within the limitations of the experimental evidence available.

The inaccuracies possible in the formulation of the hypothetical flow noted above are of interest in an historical sense. In an early attempt at deriving an analytical result for the lift of wings, use was made of "impact" theory (Newtonian flow) which relates the lift to the momentum change due to deflection of the airstream downward at the angle of attack. Such a calculation gives a lift proportional to the square of the angle of attack and a drag equal to the lift times the angle of attack. On this basis, the possibility of efficient flight seems remote. The basic flow model assumed was, of course, in error for the low speed flows of interest. Subsequently, the concept of potential flow was introduced, but an adequate approximation of the physical aspects of the flow was not realized until Kutta recognized the significance of tangential flow at the trailing edge of an airfoil; only then did the theory begin to give accurate estimates of the lift of an airfoil.

Lift curve slope at zero lift. Subsequent to the development of the Kutta-Joukowski flow which supplied a method for calculating the properties of airfoil sections, it was recognized that the finite extent of the wing span would influence the accuracy of estimates based on section data. Prandtl proposed a solution, based on the representation of the three dimensional effects by a model, in which the vortices bound within the wing surface turn backward at the wing tip and trail in the wake. The properties of an airfoil of finite aspect ratio are then estimated by correcting two-dimensional airfoil properties through the adjustment of the angle of attack by the flow angle induced by the vortex wake at one chordwise point. The development and verification of this theoretical model was so complete that it remains the basis for low speed airfoil theory. It has been expanded by the work of Faulkner and Weissinger to cover a wide range of geometric airfoil parameters.

An examination of the theory leads to the expectation that it applies with less and less exactness as the aspect ratio of the airfoil is reduced and that, in particular, it will not describe the load distribution with the required accuracy. Calculations necessary to improve this condition are very tedious [1] since mathematical expressions for the point by point loading cannot be derived in closed form. For very low aspect ratios, however, an important simplification can be made. If the wing is of vanishingly small aspect ratio, the influence of the vortex wake again becomes negligible as is the case for very high aspect ratios, and a much more exact calculation can be made. The flow about the wing may be considered to be two dimensional in both cases; for very high aspect ratios, the section is taken streamwise, and for very low aspect ratios, in the cross-stream direction.

The idea of the latter method stems from early work by Munk and has led to what is termed low aspect ratio wing theory. This theory was first proposed by R. T. Jones in [2] and since has become one of the most fruitful fields of aeronautical research. The exploitation of this idea has led to the important finding by Heaslet and Lomax [3] that the solutions of low aspect ratio theory are exact at a Mach number of one for triangular wings for a wide range of aspect ratio, i.e. that the limitation of low aspect ratio does not apply. This result appears reasonable in view of the fact that, as for very low aspect ratio airfoils, the vortex wake cannot influence the wing itself since its disturbance field cannot be propagated upstream in a sonic flow. Further, the fact that no upwash exists ahead of the wing in low aspect ratio theory is also consistent with sonic flow.

A subsequent finding of importance is concerned with the fact that the low aspect ratio theory applies equally to rectangular as well as to triangular wings. The theory shows that the parameter of greatest consequence is the reduced aspect ratio, $\beta Æ$, where β is $\sqrt{M_\infty^2 - 1}$. It is therefore evident that, for sonic speed, the reduced aspect ratio is zero for all wings and the theory should apply to all at this Mach number.

The first experimental data of sufficient scope to bear on the accuracy of the theoretical results are those of Nelson and McDevitt [4] for a family of affinely related rectangular wings. These data have been correlated according to transonic similarity laws [5]. In this way a single curve is obtained which permits the vast amount of data to be handled simply. For the comparison of Fig. G,2a the reduced aspect ratio of all wings is zero. It is observed that for these rectangular wings the theory is quite accurate up to a transonic aspect ratio similarity parameter of unity which corresponds, roughly, to an airfoil of aspect ratio 3 with sections 4 per cent thick. Above this value, the lift curve slope is independent of aspect ratio and depends only on the section characteristics.

For triangular wings the theory has been found to be inexact at much lower values of the aspect ratio similarity parameter than for rectangular wings, in spite of the fact that the theory was developed originally for triangular wings and would be expected to apply over a wider range of aspect ratios than for the rectangular wings. The reasons for this paradox can be discovered in a study of the correlation of the minimum drag for these airfoils with transonic similarity parameters shown in Fig. G,2b. It will be noted that the data for the rectangular wings show a linear variation with the aspect ratio similarity parameter up to the same value for which the lift curve slope agrees with theory.

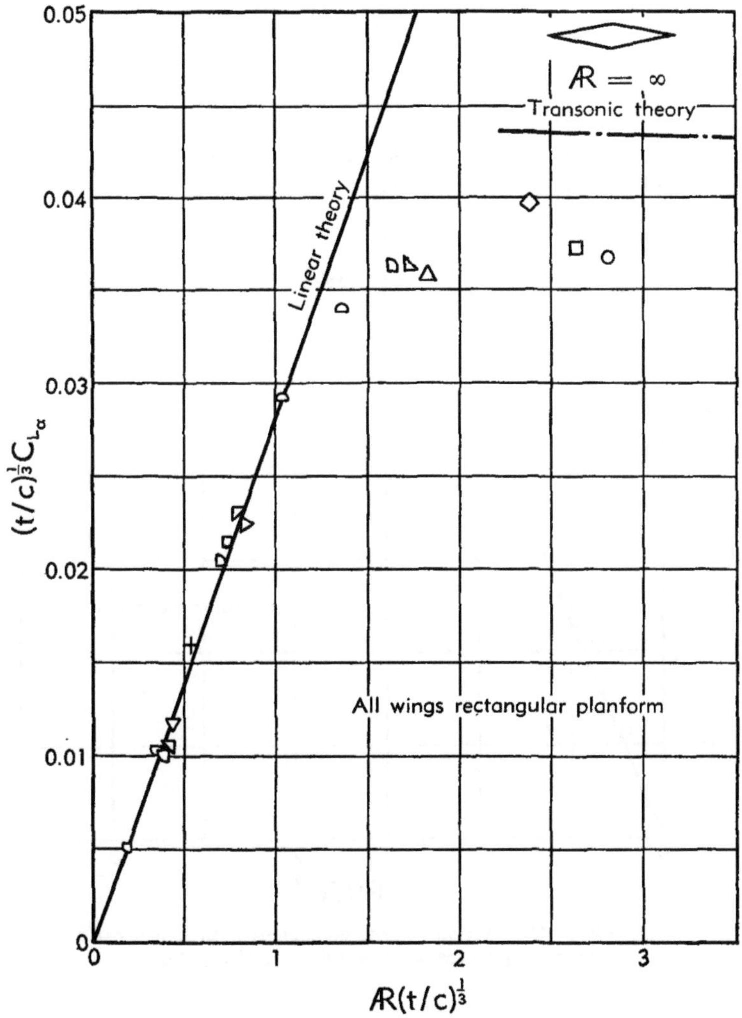

Fig. G,2a. Correlation of lift curve slope with aspect ratio according to transonic similarity. $M_\infty = 1.0$.

From a study of these data, there appears to be a relationship between the correlation of both lift and drag with similarity parameters. It should be noted first that the process of correlation with similarity parameters assures that the lift and the drag parameters will be found to be functions only of the aspect ratio similarity parameter. The correlation does not specify in what manner the two are related to the similarity parameter. It is therefore of more than passing interest that the drag parameter is found to be a *linear* function of the aspect ratio parameter for the rectangular wings over a significant range. A simple additional correlation shows the reason. If both ordinate and abscissa of Fig. G,2a and G,2b are multiplied by the thickness ratio to the two-thirds power, $(t/c)^{\frac{2}{3}}$, a

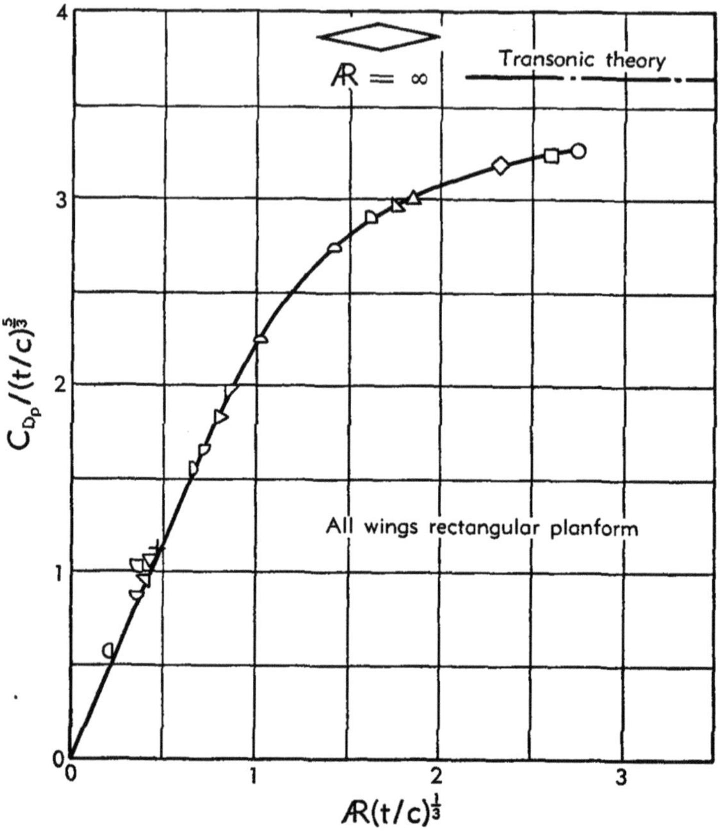

Fig. G,2b. Correlation of drag coefficient with aspect ratio according to transonic similarity. $M_\infty = 1.0$.

relationship is obtained between the drag coefficient based on the maximum frontal area of these wings and the square of an effective fineness ratio. The drag coefficients are then observed to be linear functions of the square of an effective fineness ratio. Beyond a certain value the drag coefficient no longer follows a linear variation with the fineness ratio, obviously because the type of flow changes rather abruptly at this point. It is to be expected that the change in flow which influences the drag will also influence the lift and that, beyond the aspect ratio parameter of unity, a nonlinear theory considering the interaction of thickness and lifting pressures is required.

For triangular wings, the distribution of the cross-sectional area in the streamwise direction is such as to give a steep gradient near the rear of the airfoil. It would be expected, therefore, that the flow would obey the rules for flow about slender bodies only for much lower values of the aspect ratio parameter.

Hence it is concluded that, at transonic speeds, low aspect ratio theory is adequate for estimating lift curve slope for a range of the transonic aspect ratio similarity parameters. The upper limit of this range will be determined by the value for which the flow departs from that for slender

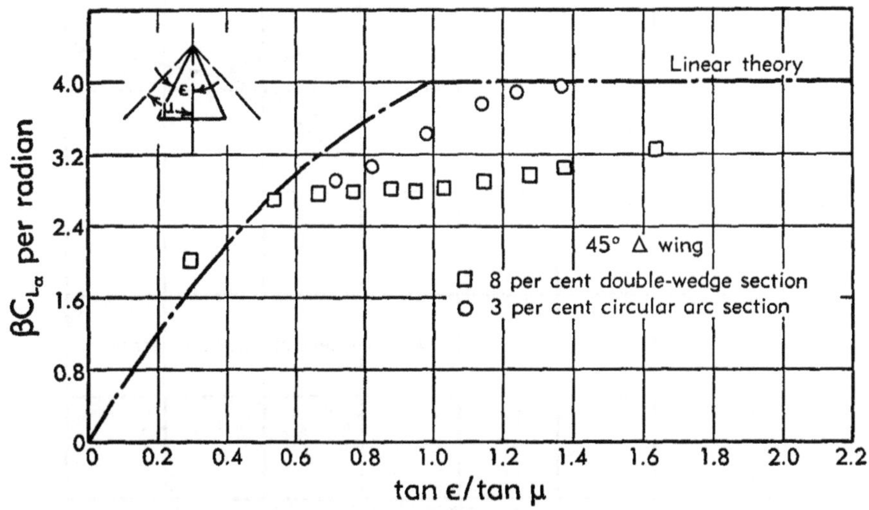

Fig. G,2c. Comparison of theoretical and experimental lift curve slope for triangular wings.

bodies. This value will depend on the effective fineness ratio and the distribution of the cross-sectional area in the stream direction.

At supersonic speed, there is reason to believe that the limitations of lifting surface theory in predicting lift curve slope are less restrictive than noted above, since in general the flow is everywhere supersonic. The regions in which the theory may fail are those for which this condition is violated and an effective transonic flow is manifest, as for flows with detached shock waves or for conditions in which elements of the airfoil are effectively at transonic speed because the airfoil generators are swept behind Mach lines.

An example of the failure of the theory under these conditions is shown in Fig. G,2c where experimental results taken from [6] and from unpublished data are compared with the theory. In this figure, the data for several triangular wings are presented in the form of the reduced lift curve slope βC_{L_α} as a function of the ratio of the tangent of the half apex angle ϵ to that of the Mach angle μ. In general the agreement is good, except for values of $\tan \epsilon/\tan \mu \cong 1$ for which the velocity component of the flow perpendicular to the leading edge is nearly sonic. For this condition, the thickness of the airfoil becomes a factor in determining the adequacy of the theory of lift. The thickness of the airfoil promotes a detached shock wave which influences the flow in such a manner as to cause a reduction in lift curve slope. The data of Fig. G,2c show that a reduction in thickness ratio from 8 per cent to 3 per cent materially improves the agreement between theory and experiment. It is expected

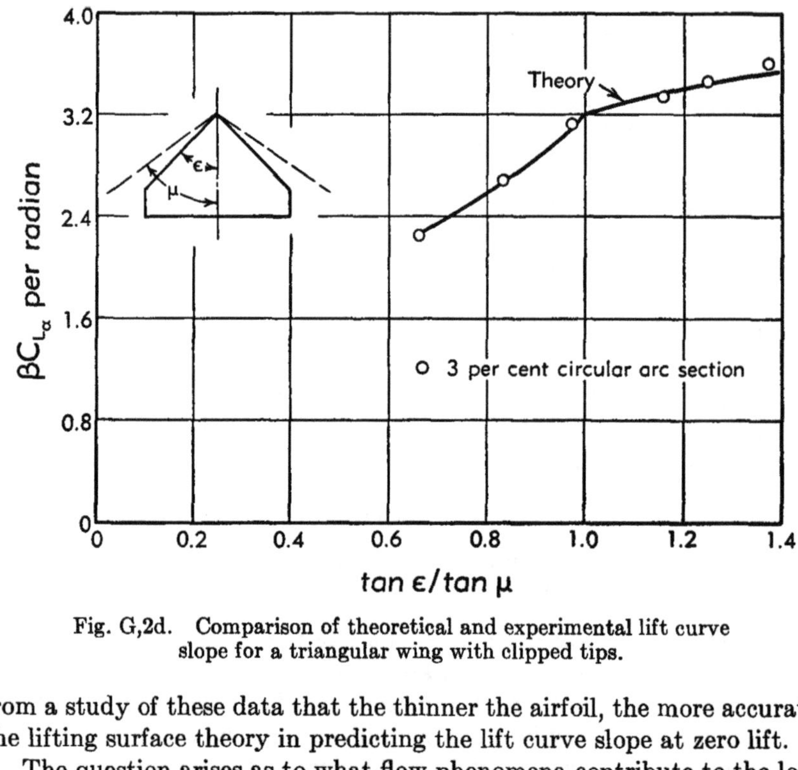

Fig. G,2d. Comparison of theoretical and experimental lift curve slope for a triangular wing with clipped tips.

from a study of these data that the thinner the airfoil, the more accurate the lifting surface theory in predicting the lift curve slope at zero lift.

The question arises as to what flow phenomena contribute to the loss in lift curve slope for airfoils of moderate thickness ($t/c = .03$) shown by Fig. G,2c. The data of Fig. G,2d, which were obtained for the 45° triangular wing of Fig. G,2c with the tips removed to give an airfoil of aspect ratio 2, show the interesting result that the theory and experiment agree for this wing even for values of $\tan \epsilon/\tan \mu$ for which the flow is sonic perpendicular to the leading edge, where formerly the theory was invalid.

These results force either of two conclusions: (1) The loss in lift curve slope found for triangular wings when the leading edge is sonic (Fig.

G,2c) occurs over the portion of the tip removed to obtain the planform of Fig. G,2d, or (2) the flow about the streamwise tip of the airfoil of Fig. G,2d is such as to provide an amount of lift enough greater than estimated by theory to supply that lost along the leading edge.

The latter is not considered a reasonable hypothesis. It is concluded that, when the velocity component perpendicular to the leading edge is nearly sonic, the thickness of a triangular airfoil promotes a bow wave of sufficient strength to cause a subsonic flow in the vicinity of the leading edge. Consequently, the portion of the tip lying within this subsonic region becomes ineffective in producing lift. As subsonic airfoils, the tip

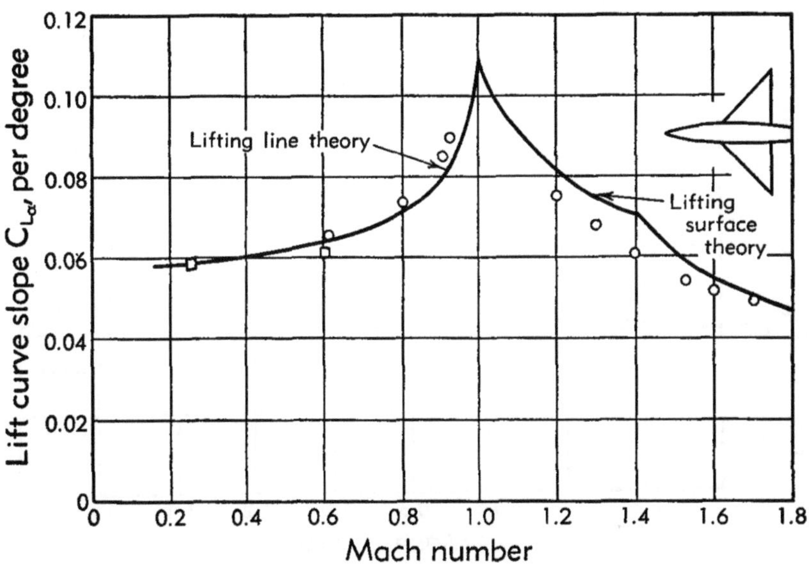

Fig. G,2e. Variation of lift curve slope with Mach number for an aspect ratio 4 triangular wing.

elements cannot sustain the large lift required by the supersonic theory and the tip may as well be removed since it serves no useful function.

The previous discussion has been concerned with a rather limited range of speeds, since the theory is generally valid except in these regions. It is of interest, however, to examine the validity of theory in general in estimating the lift curve slope of airfoils over a wide range of Mach numbers. The data of Fig. G,2e show a comparison of theory and experiment for a 45° triangular wing over a range of Mach numbers from 0.2 to 1.7. The theoretical methods used in estimating the lift curve slope of this airfoil are those of [*4,7,8,9*]. In general, the agreement is found to be satisfactory except at high subsonic speed and at supersonic speeds for which the leading edge is sonic. The latter condition has been previously discussed. The reasons for the differences between theory and experiment at high subsonic speed are believed to be associated again with the effects of airfoil thickness. The incremental velocities over the surface of the airfoil are sufficient to raise the effective Mach number to a value greater than that of the stream, so that the wing has a lift curve slope given by the theory for a slightly higher Mach number. This effect is fairly well known in airfoil theory. A reasonable, approximate theory for low supersonic speeds allowing for these higher order effects could be evolved for triangular airfoils of the shape suggested by Squire [*10*], since for these airfoils the velocity increment due to thickness is constant over the entire surface.

Another comparison for a wide range of Mach numbers is shown in Fig. G,2f for a swept wing of aspect ratio 2 obtained, as noted before, by

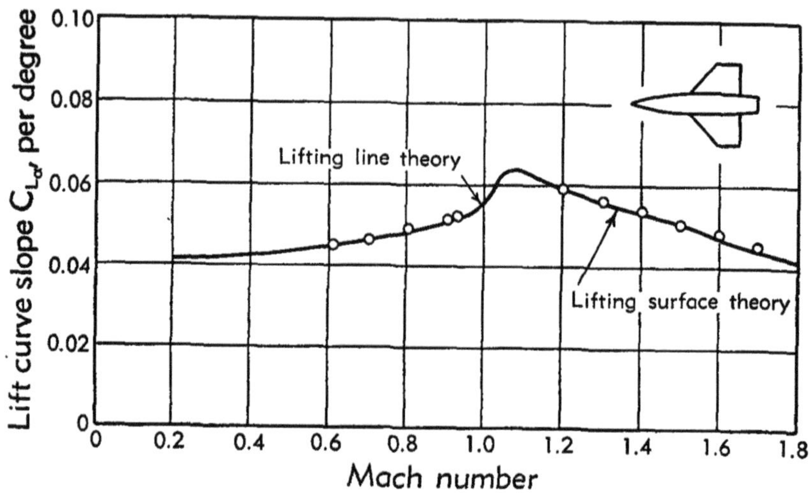

Fig. G,2f. Variation of lift curve slope with Mach number for an aspect ratio 2 clipped tip triangular wing.

clipping off the tips of the 45° triangular wing. For this airfoil the agreement between theory [*7,8,9,11*] and experiment is good at all Mach numbers. The reduction in thickness effects, resulting from removal of the tips, is sufficient to improve the accuracy of the theory at transonic speeds and is considered to be another manifestation of the thickness effects discussed in the first portion of this article.

One additional comparison of interest is shown in Fig. G,2g for an unswept tapered airfoil composed of biconvex sections of 3 per cent thickness ratio. This airfoil has an aspect ratio of 3 and a taper ratio of 0.4. For such an airfoil, lifting line theory [*7,8*] accurately predicts the lift curve slope at subsonic speed and lifting surface theory gives accurate results at supersonic speed.

High lift. In the previous discussion, it is shown that the theory of thin lifting airfoils with all its limitations is reasonably accurate in predicting the lift curve slope at zero lift. Most airplanes, however, operate over a wide range of lift coefficients, and it is necessary to discuss the adequacy of the theory for those conditions wherein the underlying assumption of small disturbances is violated, i.e. at high lift.

Two phenomena are of interest in the study of wings at high lift. The first is concerned with a condition that obtains for thin wings of moderate aspect ratios, say 2, which was originally discovered during tests of airfoil sections [*12*]. The second involves the vortex flows generated by thin low aspect ratio wings, which influence the magnitude of the load and its distribution over the airfoil surface.

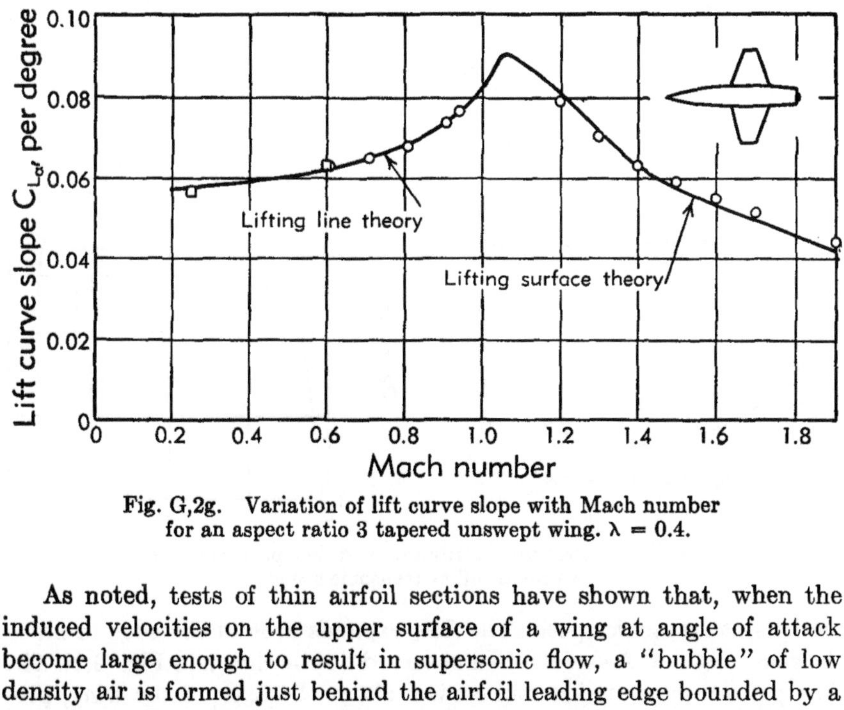

Fig. G,2g. Variation of lift curve slope with Mach number for an aspect ratio 3 tapered unswept wing. $\lambda = 0.4$.

As noted, tests of thin airfoil sections have shown that, when the induced velocities on the upper surface of a wing at angle of attack become large enough to result in supersonic flow, a "bubble" of low density air is formed just behind the airfoil leading edge bounded by a free stream line. The curvature of this streamline, and therefore the extent of this bubble, are determined by the position of the sonic line in the flow and the reflection from the sonic line of the expansion generated as the flow tends to turn the corner. A sketch of this flow together with a graph of the pressure distribution are shown in Fig. G,2h [*12*].

This phenomenon is at present one of the important unsolved problems of high speed aerodynamics. It is of importance in that it contributes a nonlinear variation of lift with angle of attack, resulting in greater lift than indicated by linear theory at transonic speed. It also contributes a marked variation of pitching moment with angle of attack which gives a large negative slope to the pitching moment curves at high angle of attack. Whether or not a flow of this character involves a buffeting of an airplane so as to make it uncomfortable or dangerous to fly is not known. Observations of the flow indicate that it is quite steady when started, though for a certain range of angles of attack and Mach number the flow may alternate periodically between the well-known viscous stall and the present transonic flow.

One interesting fact has come to light from recent research on thin triangular wings in connection with this phenomenon. Studies of pressure distributions for these wings with leading edges swept behind but close to Mach lines have shown a flow very similar to that for thin airfoil

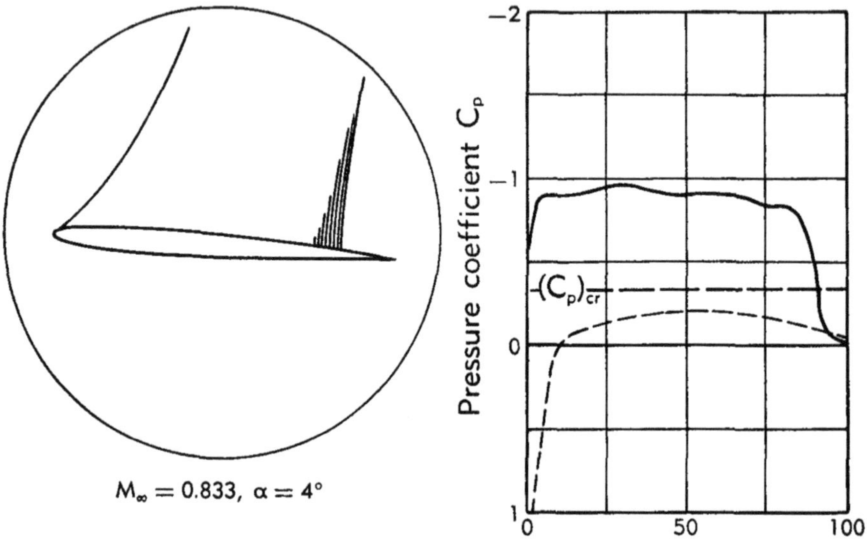

Fig. G,2h. Pressure distribution and flow pattern around a thin airfoil at transonic speed.

sections at transonic speed. The flow expands around the edge in much the same manner. A sketch of an observed flow is shown in Fig. G,2i. The flow pattern generated is still conical about the apex of the wing, provided that the thickness effects are small and effects in the vicinity of the wing tips are excluded. The lift curve is nearly linear, though the slope is now of smaller magnitude than that given by linear theory. This flow characteristic is important in that it influences the load distribution over the airfoil surface, tending to make the loading triangular, particularly at high angles of attack.

A second flow phenomenon of interest in the study of the aerodynamics of high speed wings concerns the flows generated about wings of very low aspect ratio. These flows are of fundamental interest, since they show that the mathematical model utilized previously in analytical studies of low aspect ratio wings is in error at large angles of attack. Plate G,2 shows a series of vapor screen pictures taken of the flow about a low aspect ratio triangular ($Æ = \frac{3}{4}$) wing at $M_\infty = 1.2$. (See [*13*] for details on this technique of flow visualization.) In these pictures, the model wing of aspect ratio $\frac{3}{4}$ supported on a cylindrical boom is viewed from the rear. These photographs reveal that the flow above the upper surface of the wing contains a vortex flow not considered in the usual slender wing analysis. The difference between the mathematical model previously used and the one predicated on the observed flow is shown in Fig. G,2j. Since the edges of the wing are sharp, it appears necessary to postulate tangential flow at these edges. This boundary condition can

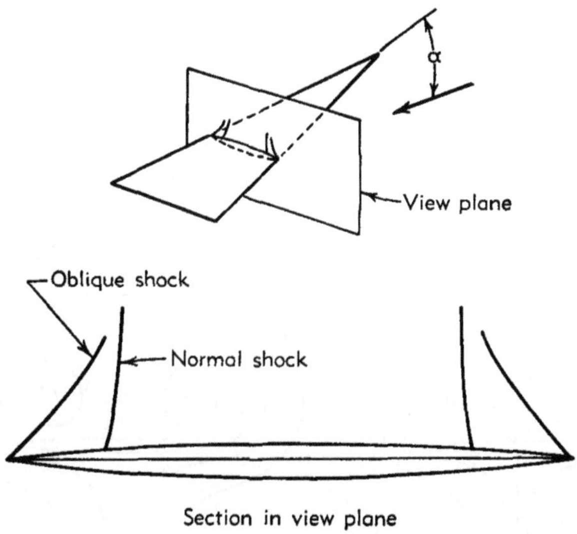

Fig. G,2i. Sketch of transonic flow in cross plane of a triangular wing.

be met either by assuming two discrete vortices in the flow above the wing or by specifying a more exact flow model, which recognizes that there is a continuous discharge of vorticity from these edges which forms the vortex pattern observed above the wing. Brown [*14*] has solved the potential flow problem with essentially the latter flow model for the case of very small reduced aspect ratios. A comparison between the lift curve given by the theory and by experiment taken from [*14*] is shown in Fig. G,2k. It will be noted that the experimental lift increases nonlinearly with angle of attack. The nonlinear theory overestimates the lift at this Mach number. The lift at high angles of attack is much greater than that given by linear theory for this triangular wing of low aspect ratio.

It should be noted that the flow discussed here will occur, when the angle of attack exceeds a certain small value, for triangular wings of low aspect ratio at all speeds for which $\beta Æ$ is small. As the Mach number becomes large and the Mach lines approach the leading edge, the flow will change gradually from the vortex flow to that involving shock waves discussed previously. The physical nature of the flow in this intermediate region needs further observation.

It has been noted that vortex flows will occur for triangular wings if $\beta Æ$ is small. The same phenomenon is to be expected for rectangular wings

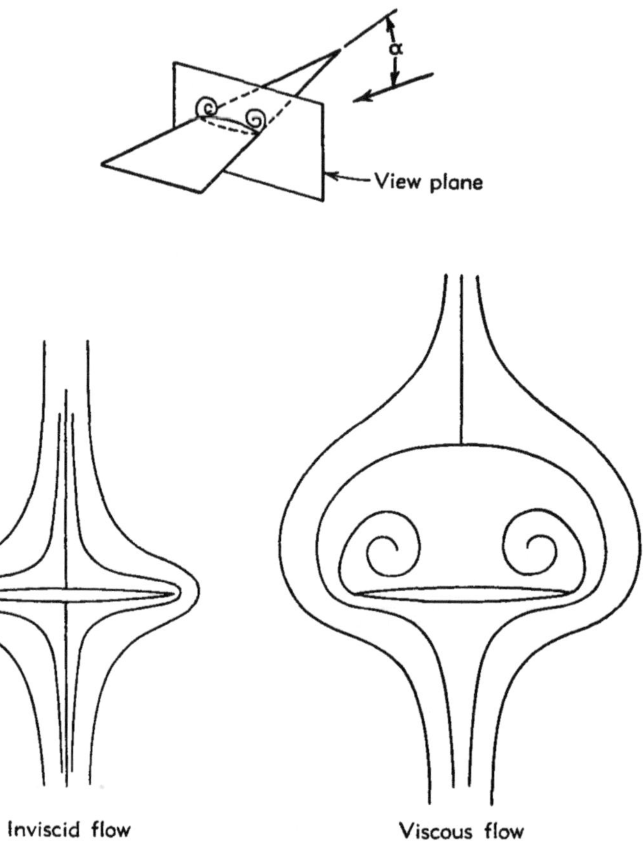

Fig. G,2j. Sketch of inviscid and viscous flow in cross plane for slender triangular wing.

of very small aspect ratio. The flows about these wings have not been investigated sufficiently to define a mathematical model of the required accuracy. Bollay has supplied a calculation [*15*] which appears to satisfy conditions at high angles of attack, but which fails to depict the aerodynamic phenomena over a range of small lift coefficients. Further, because of the difficult problem of assessing the effects of the trailing edge at subsonic speed, it is necessary to resort to the concept of the cross drag developed by the cross flow αU. While this stratagem gives adequate results for bodies of revolution [*16*], its use is considered to be too approximate for wings under the usual circumstances.

In summary, theory does not now adequately treat certain phenomena which influence the load distribution and loading at high angles of attack. Progress in the analysis of these complex flows requires further careful experimentation to afford some insight into the physical processes. Once

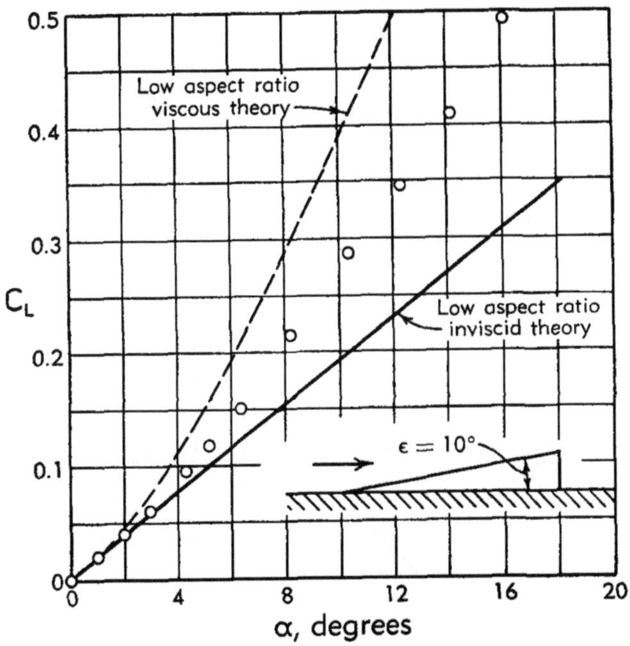

Fig. G,2k. Comparison of experiment with viscous and inviscid flow theory for a slender triangular wing. $M_\infty = 1.9$.

these are known, the problem can be treated successfully as was demonstrated by the analysis of Brown, applying to thin triangular wings of very low aspect ratio.

G,3. Drag of Wings at Zero Lift. An examination of the research at transonic and supersonic speed will show that by far the greatest concentration of effort has been directed to the investigation of the drag of airfoils at zero lift, appropriately so since the attainment of efficient supersonic flight depends on the development of airfoils of low supersonic drag. This emphasis on zero lift continues, and it is in this field of work that many of the new advances in theory have taken place.

Unswept airfoils. The earliest research on drag of supersonic airfoils at zero lift centered in the study of airfoil sections, first because this pattern of research grew naturally from the research on subsonic airfoils for which section properties are most meaningful and, second, because calculations for the two-dimensional flow were found to be relatively simple. The theory of Ackeret was utilized in the first comparisons between theory and experiment but was later supplanted by shock-expansion theory, which accounts for large thickness effect more exactly.

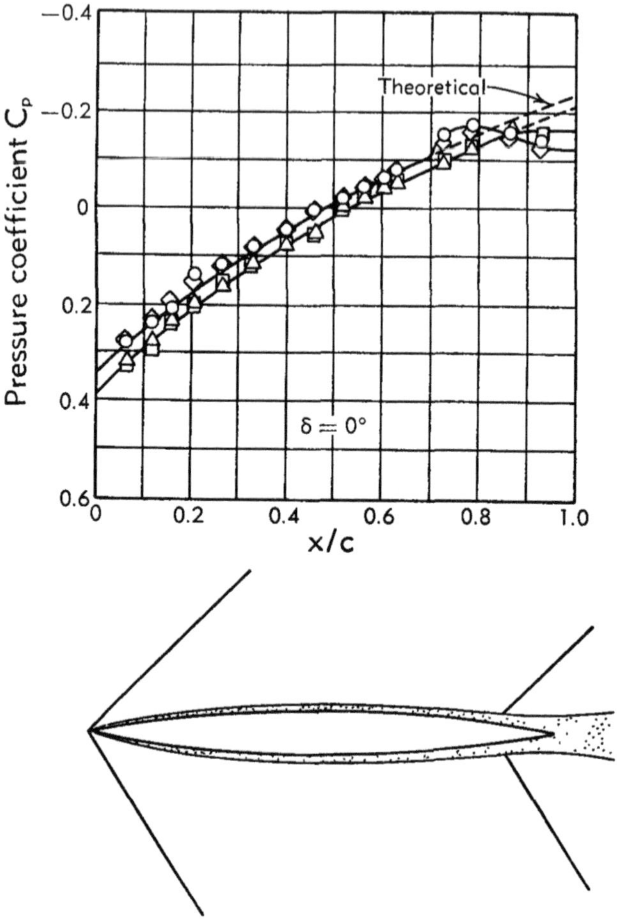

Fig. G,3a. Pressure distribution over a 9 per cent thick circular arc airfoil section. $\alpha = 0.35°$, $M_\infty = 1.62$, and $Re = 1.07 \times 10^6$.

The adequacy of shock-expansion theory for estimating the two-dimensional flow is shown by the pressure distribution data of Fig. G,3a for a 9 per cent symmetrical circular arc airfoil [*17*]. A study of these data indicates that the shock-expansion method for Mach number of 1.62 is quite satisfactory in its accuracy, except for a small region near the trailing edge of the airfoil where perhaps the boundary layer separates because of the pressure wedge acting ahead of the trailing edge shock, as indicated in the sketch. This effect of the viscous layer may be exaggerated by the low test Reynolds number of 1.07 million. The flow separation tends to reduce the drag to a value less than that given by theory.

The adequacy of the shock-expansion theory is even more strikingly illustrated by the data of Fig. G,3b, taken also from [*17*]. In this case, the

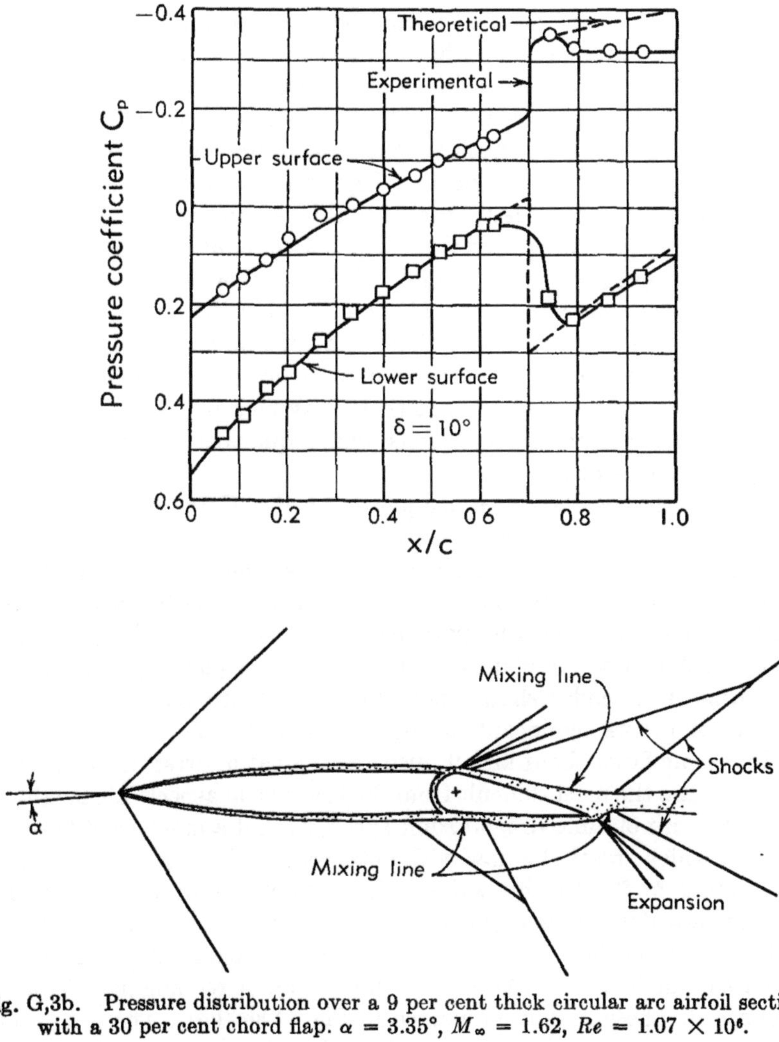

Fig. G,3b. Pressure distribution over a 9 per cent thick circular arc airfoil section with a 30 per cent chord flap. α = 3.35°, M_∞ = 1.62, Re = 1.07 × 10⁶.

30 per cent chord trailing edge flap is deflected downward 10° while the airfoil is positioned at 3.35° angle of attack. The agreement between theory and experiment is truly remarkable. Viscous effects are of importance in the vicinity of the hinge line of the flap on the lower surface, where again a pressure wedge acts to separate the boundary layer ahead

of the shock wave at the hinge. Further viscous effects cause a discrepancy between theory and experiment on the upper surface of the flap, because the presence of the trailing edge shock is felt forward through the boundary layer, as shown by the sketch. These results tend to emphasize the conclusion that has been drawn from other research, that the effects of compression waves acting on the viscous layer cause the greatest deviations of the real flow from that visualized by the inviscid theory.

One of the problems to which supersonic section theory has been applied concerns the optimization of airfoil properties from the structural point of view, that is, the solution of the problem of the greatest structural bending and torsional strength for the least drag. This problem is acute for supersonic aircraft since efficient flight requires very thin airfoils.

The most interesting development in this research was contributed by Chapman [*18*], who recognized that airfoils with blunt bases have superior structural properties, while at high Mach numbers the drag need not suffer because of the blunt base. If one notes that the maximum possible base drag for the blunt airfoil is that for a vacuum at the base, and that the base drag coefficient at this pressure becomes small at high supersonic velocities, it becomes evident that there may be a net reduction in drag from the use of blunt-based airfoils while equal bending and torsional strengths are retained.

Some of the results of [*18*] are plotted in Fig. G,3c for two airfoils of equal section modulus. It will be noted that there is a sizeable reduction in drag for the blunt base airfoil. Part of this drag reduction is due to a decrease in wave drag due to reduced section thickness (from 10 per cent to 9.1 per cent), and the other part to the decrease in wave drag due to the change in thickness distribution. Both of these are partially offset by the base drag. The reader should note that the improvement shown here may be even more significant for higher Mach numbers.

Transonic flows about airfoils also are of great interest. In many ways such flows are the most difficult from the theoretical aspect. Linear theory does not admit of finite drag at sonic speed, so that a more exact theory is required which treats the flow with detached shock waves. The research of [*19,20,21,22,23*] on this subject is summarized in Fig. G,3d. The experimental data show excellent agreement with the transonic theory.

It will be recalled that in the article on lift curve slope, data were presented which showed the experimental drag of rectangular airfoils at sonic speed. These data are repeated in Fig. G,3e (left). The drag coefficient, however, is now based on the frontal area and is plotted as a function of the square of an effective fineness ratio, a result obtained by multiplying the ordinate and abscissa of Fig. G,2b by $(t/c)^{\frac{1}{3}}$. It is of interest to note that the drag coefficient varies as the square of the effective fineness ratio over a significant range of the abscissa. These wings, therefore, exhibit a property of axially symmetric bodies at this Mach number. This result is consistent, of course, with the important findings of [23], which state the dependency of drag at sonic speed on the axial distribution of cross-sectional area. This point is illustrated in Fig. G,3e (right), where data from [23] show that a winged body and an axially symmetric body give very nearly the same drag experimentally at sonic speed (allowing

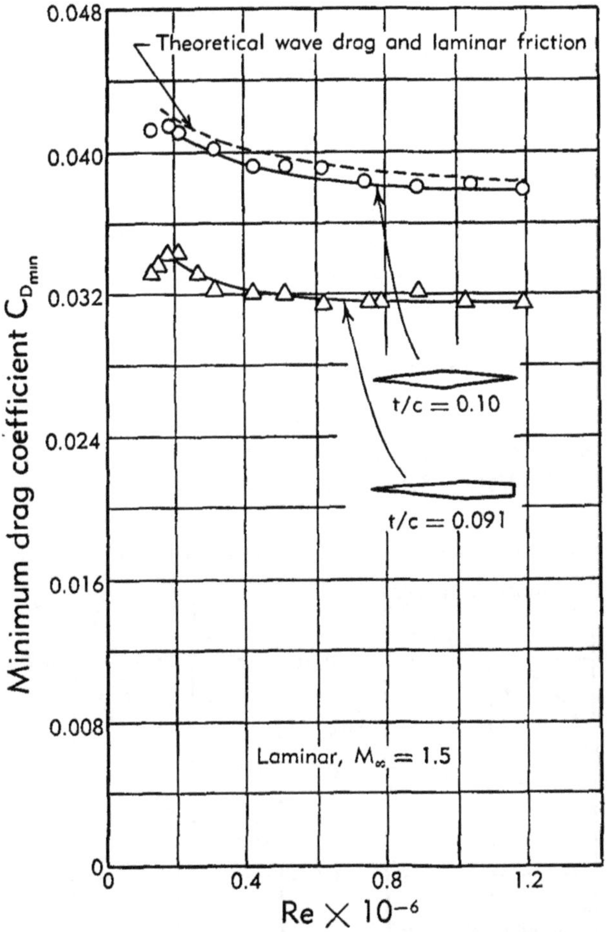

Fig. G,3c. Comparison of the drag coefficients of double-wedge and blunt-base airfoils having the same section modulus.

for the difference in skin friction drag). It is noteworthy that this result was discovered analytically, but was not considered to be characteristic of a real flow until noted experimentally as reported in [23]. This finding, of course, ranks in importance with sweepback in the list of recent aerodynamic advances.

Swept airfoils. The first interest in swept airfoils was concerned with the delay of the onset of sonic flow on the wing with its attendant drag

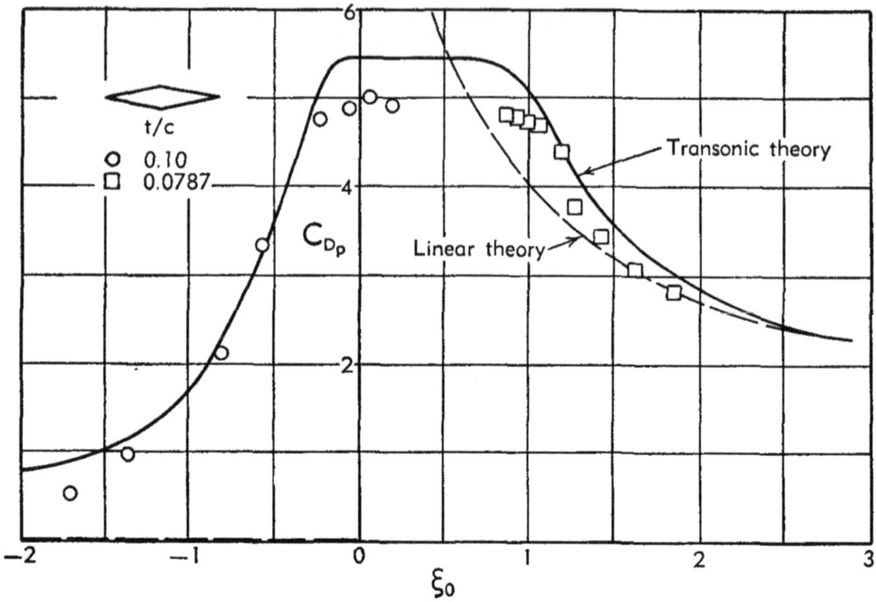

Fig. G,3d. Pressure drag of double-wedge sections at transonic speed. $\overline{C_{D_p}} = C_{D_p}[M_\infty^2(\gamma + 1)]^{1/3}/(t/c)^{2/3}$, $\xi_0 = (M_\infty^2 - 1)/[M_\infty^2(\gamma + 1)/(t/c)]^{2/3}$.

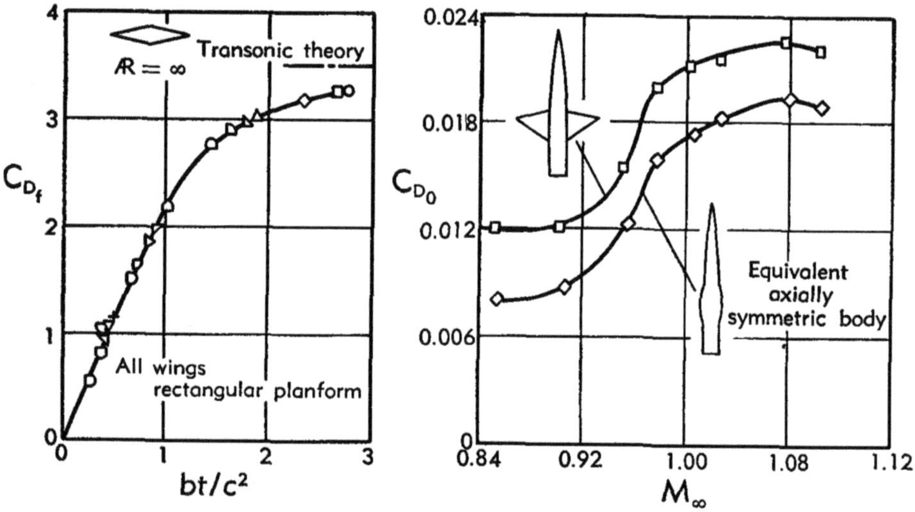

Fig. G,3e. Zero-lift drag of wings and bodies at transonic speed. Left, wave drag coefficient of rectangular wing based on frontal area; right, drag coefficient for winged body and equivalent body based on wing area.

increase. Subsequently, analytical studies of swept airfoils showed promise of reducing wave drag at supersonic speed, especially if the airfoil generators were swept behind the Mach lines. The result encouraged the experimental study of the drag of these airfoils to verify the theoretically indicated benefits.

One of the early supersonic experiments on swept airfoils is discussed in [*24*]. In this research the distribution of pressure over the airfoil surface was studied to investigate the validity of the theory of [*25*]. The results of the investigation are shown in Fig. G,3f. A study of these data shows that, ahead of the pressure minimum, the theory predicts the pressure with reasonable accuracy. Behind the minimum pressure point, the agreement between theory and experiment is poor. Three factors contribute to the inadequacy of the theory in this region. The first of these is concerned with an effect produced by the higher order terms neglected in the theory. Small pressure disturbances are propagated along a characteristic line, dependent on the local Mach number on the airfoil surface. Because of the induced velocities, the local Mach number is generally greater than that of the stream. The position of the pressure minimum is determined by the influence of the airfoil trailing edge which is propagated in the theory at the stream Mach angle. In the real flow, however, since the local Mach numbers are greater than that of the stream, the influence of the airfoil trailing edge and, therefore, the pressure minimum, lies rearward of the theoretical position. A calculation of this effect is easily made and is shown in Fig. G,3f as the dotted line.

Fig. G,3g, which defines the region of influence of the trailing edge and tip in the theoretical and real flows, gives a clearer picture of this effect. The position of the experimental minimum pressure position is also given. It will be noted that the result of the calculation considering the local Mach number agrees closely with experiment.

Another factor involved in the lack of agreement shown by the theory and experiment of Fig. G,3f is concerned with the fact that in a real flow containing compressions of finite magnitude the distributed compressions given by theory tend to form into a shock wave. This effect may be present in these results, although sufficient pressure data are not available to distinguish it with certainty. A third factor involved is the effect of the viscous boundary layer which thickens suddenly behind the pressure minimum. Studies of the flow near the surface given in [*24*] indicate a large lateral cross flow in the boundary layer in this region.

One final result of the investigation of this airfoil is given in Fig. G,3h where the drag coefficient for streamwise sections of the wing is plotted as a function of spanwise position. A comparison of theory and experiment in this case shows that the section drag departs to a greater and greater extent from the theory as the tip is approached. A major factor contributing thereto is the more rearward position of the pressure

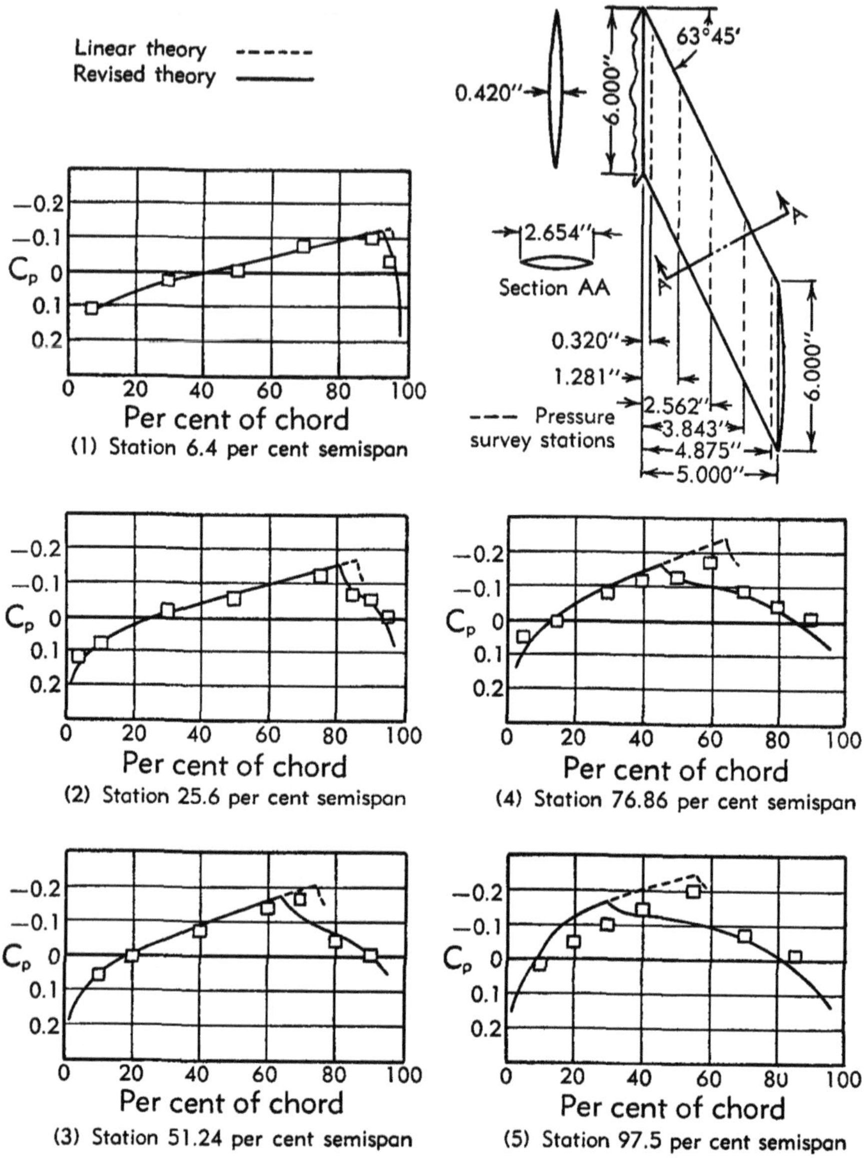

Fig. G,3f. Comparison of theoretical and experimental pressures on a constant chord swept airfoil at $M_\infty = 1.53$. Reynolds number 3.24×10^6.

minimum which occurs in the real flow. One of the interesting conclusions that may be drawn from this graph concerns the effect of aspect ratio on the agreement between theory and experiment insofar as drag is concerned. Lower aspect ratio wings of the same chord and sweepback should show better agreement between theoretical and experimental drag, and higher aspect ratio, poorer agreement.

Further experimental evidence of the inadequacy of theory in estimating the drag of wings with swept leading edges is given in Fig. G,3i which shows data at a Mach number of 1.92 for triangular wings of 8 per cent thickness ratio [6]. The drag data are compared with theory in which wave drag was computed from [26] and skin friction drag from [27] for a

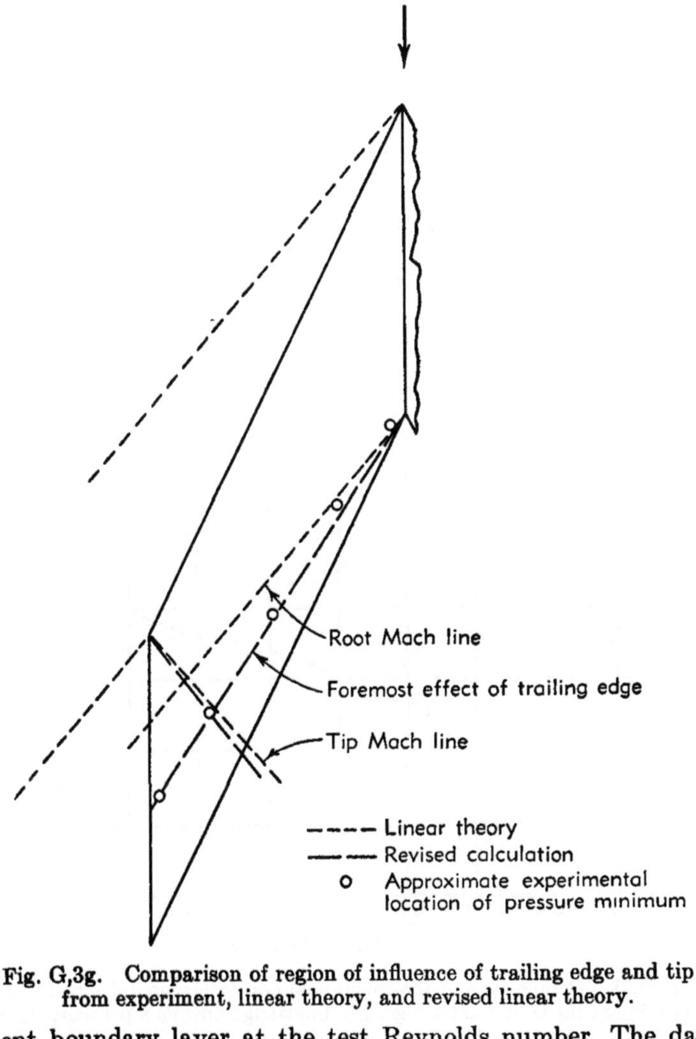

Fig. G,3g. Comparison of region of influence of trailing edge and tip from experiment, linear theory, and revised linear theory.

turbulent boundary layer at the test Reynolds number. The data are plotted as functions of tan ϵ/tan μ, which is the ratio of the tangent of the half apex angle to the tangent of the Mach angle. Values of this parameter of unity correspond to the condition in which the Mach lines lie along the leading edge.

There is a striking lack of agreement between theory and experiment near (tan ϵ/tan μ) = 1. However, in view of what is now known, it is

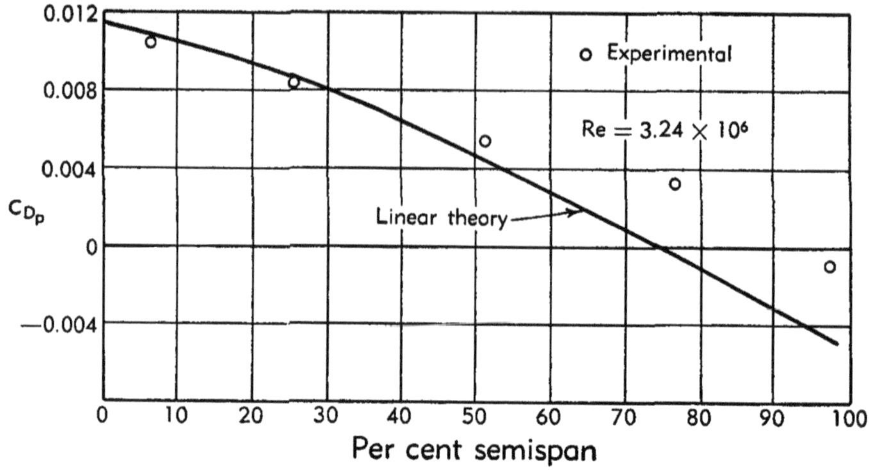

Fig. G,3h. Spanwise variation of section pressure drag coefficient for the sweptback airfoil.

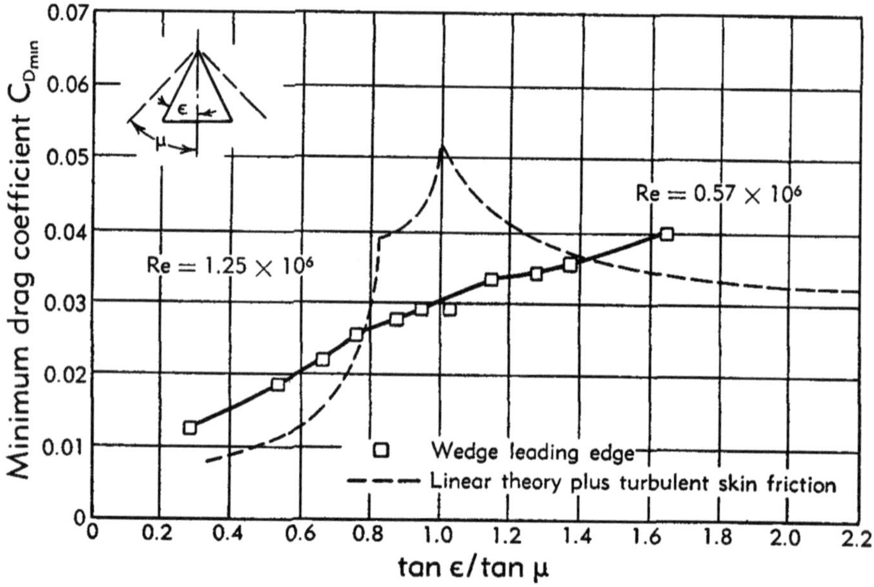

Fig. G,3i. Comparison of the theoretical and experimental zero-lift drag coefficient for triangular wings. Double-wedge airfoils 8 per cent thick.

unreasonable to expect the theory to give accurate results under conditions corresponding to transonic flow perpendicular to the wing leading edge. It was observed previously that nonlinear theory was required for accurate estimates of the drag of unswept airfoils at transonic speed. Such a theory is also required for triangular wings under these conditions. It will be observed that the agreement between theory and experiment is much better at low and high values of tan ϵ/tan μ. Again, as was observed for the case of the lift curve slope of triangular wings, better agreement between theory and experiment should be expected for thinner airfoil sections.

G,4. Drag Due to Lift. Another important division of aerodynamic research concerns the drag associated with the lift carried by the airfoil. The drag due to lift and the drag at zero lift are equally important in determining the maximum efficiency of an airplane, that is, the maximum value of the ratio of lift to drag.

Definition of aerodynamic factors. This particular field of research was treated for low speed flight by semiempirical methods for many years. Prandtl's lifting line theory shows that the drag due to lift of high aspect ratio airfoils is a function of only one variable, the aspect ratio of the wing. More refined calculations show that if the vortex sheet of the wing lies in the plane of wing the minimum drag due to lift occurs for an elliptical spanwise distribution of the lift. Departures from the idealized elliptical loading associated with rectangular planforms or other planforms of linear taper were also calculated. The result of the linearized analysis of elliptically loaded wings shows that the drag due to lift is associated with the lift by the formula

$$C_{D_i} = \frac{C_L^2}{\pi A\!\!R} \tag{4-1}$$

After experimental results in many instances were found not to agree closely with the analytical result, the formula was modified to contain an empirical constant e, or

$$C_{D_i} = \frac{C_L^2}{e\pi A\!\!R} \tag{4-2}$$

where e was called the span efficiency factor. Experimental investigations of wings were used to determine the value of the span efficiency factor, and thereby the factor included all the errors possible in wind tunnel tests, the failure of the theory to describe correctly the flow about sections of the airfoil, and effects of planform in promoting loadings different from elliptical.

This method of investigating drag due to lift was found to be unsatisfactory for high speed wings, in both theoretical and experimental work. In theoretical work, solutions for the drag of lifting wings exclusive of skin friction were obtained by calculating the drag component of the normal force on the planar airfoil, which is equal to the product of the normal force and the angle of attack, or which for small angles of attack is equal to $C_L\alpha$, and subtracting the suction force which may exist on the

20 per cent chord

40 per cent chord

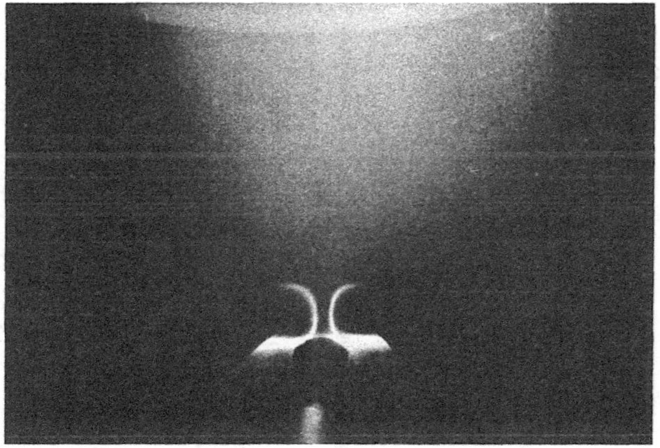

60 per cent chord

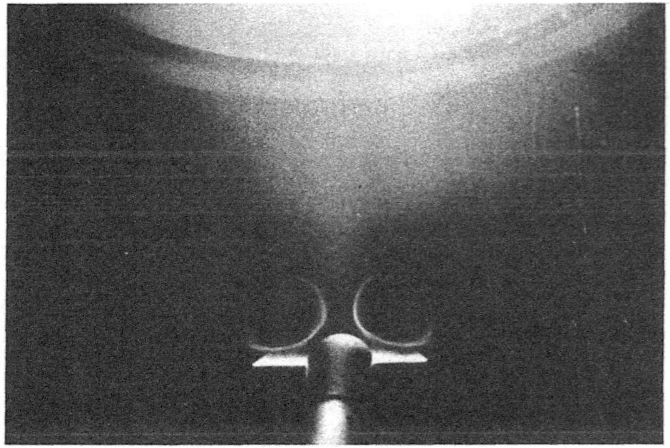

80 per cent chord

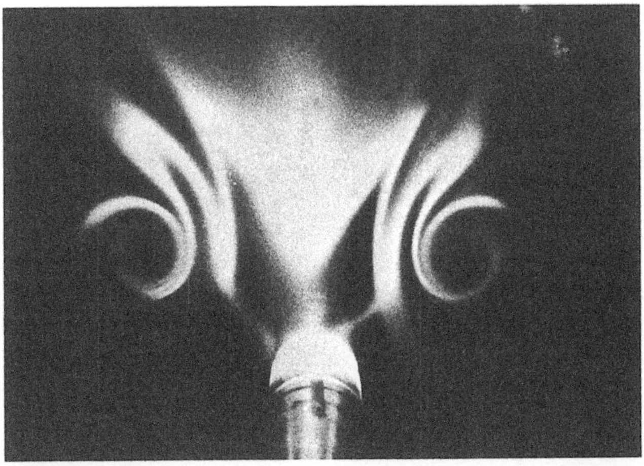

$\alpha = 15°, \phi = 0°$

$\alpha = 15°, \phi = 22.5°$

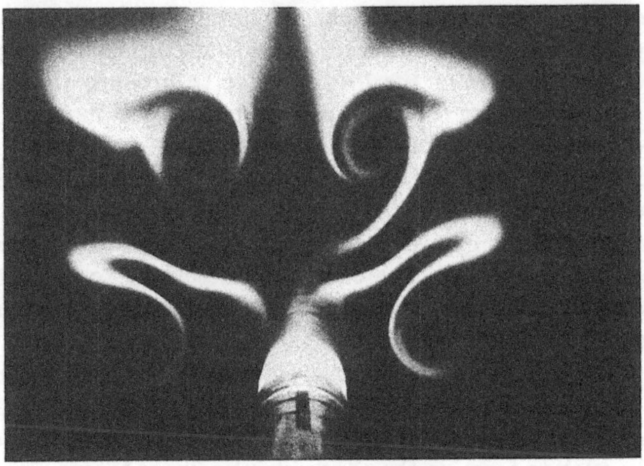

$\alpha = 15°, \phi = 45°$

Plate. G,6. Vortex flow behind cruciform wing.

leading edge, evaluated by a line integral around the singularity in the theoretical pressure at the leading edge.

For very slender triangular airfoils at all speeds, the drag due to lift so calculated is

$$C_D = \frac{C_L^2}{\pi A\!\!R} \tag{4-3}$$

which is equivalent to the drag for elliptically loaded wings at subsonic speed. If the suction force on the leading edge is not obtained because of the action of viscosity or compressibility, the drag due to lift is given by

$$C_{D_i} = \frac{C_L^2}{C_{L_\alpha}} = \frac{2C_c^2}{\pi A\!\!R} \tag{4-4}$$

that is, the drag due to lift is a function only of the lift and the lift curve slope, and is twice as large as when the suction force is present.

These calculations define the extreme limits of the drag due to lift exclusive of skin friction effects, of course. At low speeds, it is reasonable to expect that at high Reynolds numbers the full suction force on the wing leading edge will be realized for a range of lift coefficient up to and including that for a maximum lift-drag ratio. This fact is borne out by experiment for all triangular wings, except for those with very thin sections ($t/c = .03$) or those with sharp leading edges. At transonic and supersonic speeds, however, experiments show that in most cases the drag due to lift tends to be simply the product of the normal force and the angle of attack, or, in other words, the possible suction force is not realized. This result is shown by the data of Fig. G,4a for swept and unswept airfoils. The data of this figure were obtained on a transonic bump at low Reynolds number. A part of the loss in leading edge suction force exhibited at low Mach numbers is probably attributable to the low test Reynolds number, but at transonic speed compressibility accounts for most of the loss of the suction force. At transonic speed, the pressures required in the vicinity of the leading edge to produce the theoretical suction force at the lift coefficients corresponding to maximum lift-to-drag ratio, are often less than zero pressure, and are, therefore, impossible. It is evident, therefore, that the suction force cannot be obtained. This argument applies equally to swept and unswept wings though, obviously, sweepback delays the deleterious effects of compressibility to a higher Mach number. The conditions necessary to the attainment of the potential aerodynamic efficiency have thus not been realized in most cases.

Nonlinear effects at transonic speed. Since, as has been observed, the effects of compressibility are large in the vicinity of the leading edge of a lifting wing at transonic speed, so large in fact as to preclude the attainment of the suction force on the edge given by linear theory, it is reasonable to assume that a more exact nonlinear theory might lead to a method whereby the equivalent of the low drag due to lift given by linear theory may be achieved. Such a nonlinear theory is not yet available.

Experimental studies give promise of the attainment of a low drag due to lift, however, at transonic speeds. The experimental data of [*12*] are of particular interest since they illustrate that the flow requires a nonlinear theory to be properly described at the highest Mach numbers investigated, $M_\infty = 0.80$. Some of the data of [*12*] are plotted in Fig. G,4b. At these Mach numbers, it is noteworthy that the drag due to lift

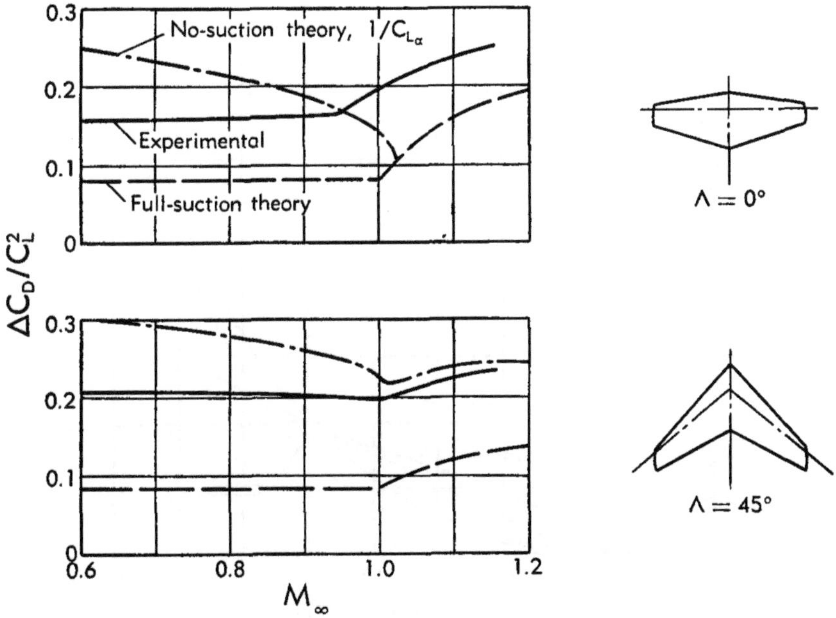

Fig. G,4a. **Drag due to lift as a function of Mach number for wings of aspect ratio 4.**

varies greatly with the thickness distribution of the airfoil sections investigated, all of which were 6 per cent thick.

There are two effects which contribute to the low drag due to lift of the airfoil sections. The first of these is due to the high velocity flow and, therefore, very low pressure generated over the upper surface of the wing at transonic speed because of the thinness of the section. This condition results in a high lift curve slope which contributes a low value of that component of the drag due to lift which is equal to the product of force normal to the wing surface and the angle of attack. The second effect is concerned with the negative chordwise force resulting from the low upper surface pressures acting on the surface slopes associated with the thickness distribution. This effect is equivalent to the force due to leading edge suction in the linear theory. The problem to be solved by the nonlinear theory is the calculation of the airfoil thickness and thickness distribution which is required to give the pressures which, when integrated over the surface, give the desired low drag due to lift.

Vortex drag and wave drag. Before discussing the drag due to lift at supersonic speed, there are two concepts which need introduction. These concepts have to do with the division of drag between vortex drag and wave drag due to lift. Vortex drag is the drag arising from the production of lift and may be obtained by integrating the kinetic energy generated by the vortex wake contained in any plane of unit width downstream

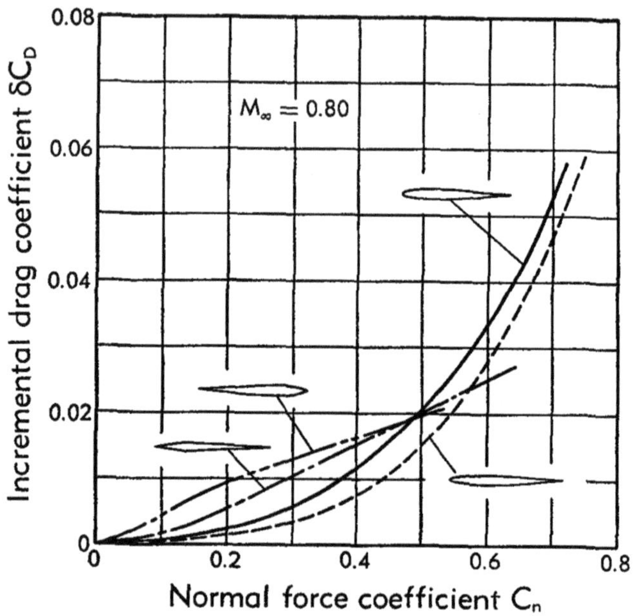

Fig. G,4b. Variation of incremental drag with normal force coefficient for airfoils of varying thickness distribution.

of the wing, taken perpendicular to the line of flight, and dividing by the flight velocity. The lift is proportional to the vertical momentum in this plane. If lift is produced, vortex drag is the result. Its magnitude is determined by the manner in which the vorticity shed by the wing, which represents this lift, is distributed in the plane. If the vorticity lies in the projected plane of the wing and is distributed so as to produce constant downflow over this plane, the vortex drag is a minimum and corresponds to elliptical span loading.

Now the reader will recall that, in a previous article, the increase in drag due to the loss in leading edge suction force was noted. This increase in drag at subsonic speed must be represented in the wake as vortex drag, and it is apparent that the discharge of vorticity from the subsonic edges of airfoils with swept leading edges contributes a vortex drag (Plate G,2).

The other concept is the wave drag due to lift of airfoils at supersonic speed. This component of drag is related to the chordwise distribution of lift and for slender airfoils has a minimum value for a planar wing of given span and chord, when the lift is distributed elliptically in both chordwise and spanwise directions. R. T. Jones has made a calculation which sets forth the minimum values obtainable for lifting bodies at supersonic speeds [*28*].

One of the important facts concerned with the wave drag due to lift, which has been pointed out by Jones, concerns the result that this miminum value of drag represents the total drag (exclusive of skin friction) of certain arrangements of bodies producing useful volumes. This fact points out a way of achieving the maximum flight efficiency at supersonic speed. There needs to be a careful analytical and experimental investigation of this subject.

Unswept airfoils. The status of research on drag due to lift of unswept airfoils may be summarized as follows:

1. Transonic speeds. At transonic speeds the drag due to lift may be estimated as simply the product of the normal force and the angle of attack. One factor contributing to a lower drag due to lift than that given by linear theory is the increase in lift attributable to the transonic flow over the upper surface. It is believed that an appropriate nonlinear transonic theory, when developed, will show how to achieve the equivalent of the full leading edge suction force indicated by linear theory.
2. Supersonic speeds. The drag due to lift here also may be estimated as the product of the normal force and the angle of attack.

Swept airfoils. Airfoils with leading edges swept ahead of the Mach lines follow the same calculation as for unswept airfoils at supersonic speed. The drag due to lift is given by the product of the normal force and the angle of attack. If the leading edge is swept behind the Mach lines, a nonlinear theory may point the way to achieving a low drag at lift. Under such conditions the flow in the vicinity of the swept edge resembles the flow about airfoil sections at transonic speed discussed previously. If the edge is swept well behind the Mach lines, it may be possible to improve the drag at lift with camber, so that the attainment of effective leading edge suction is not dependent on the high velocity flow around the leading edge.

G,5. Wing Pitching Moments. A comparison of theoretical and experimental wing pitching moments is of interest, since they are a major factor in determining the longitudinal stability of an airplane. Past research has demonstrated the general adequacy of theory which predicts that for most wing planforms the lift of the wing acts at a point on or near the quarter point of the mean aerodynamic chord at subsonic speed and at the 50 per cent point, or centroid of area, at supersonic speed. An important exception concerns wings of large sweepback for which the center of pressure at subsonic speed may be located far behind the 25 per cent chord point, as is shown later. In the transonic speed range, it is expected that the wing planform will have a large effect on the center of pressure position. The following discussion is concerned primarily with this important point.

Change with Mach number. Calculations of the position of the wing center of pressure as a function of Mach number are possible by the

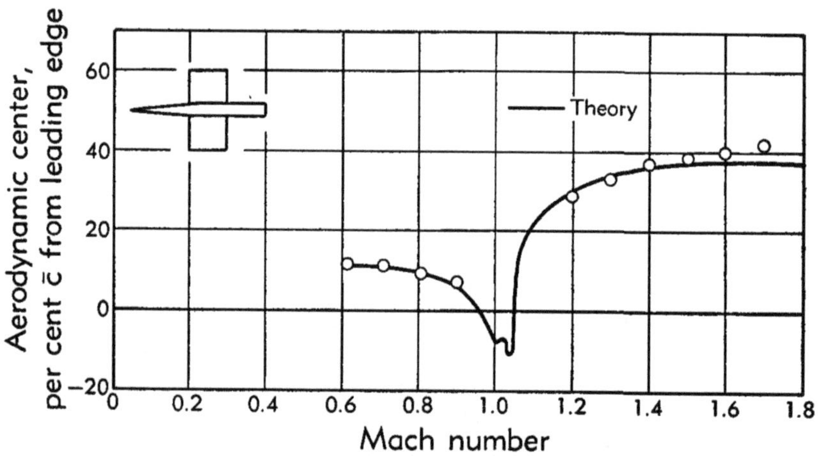

Fig. G,5a. Comparison of theoretical and experimental aerodynamic center location for a rectangular wing.

methods of [*8,29*]. However, these analytical methods are based on crude assumptions as to the spanwise distribution of lift, namely that the lift is distributed elliptically spanwise in order to solve for the chordwise distribution of lift. An experimental verification of the theory is given in Fig. G,5a which shows the position of the aerodynamic center, based on the shape of the pitching moment curve at zero lift, for the case of a rectangular wing of aspect ratio 2. This wing is composed of 3 per cent thick symmetrical biconvex sections. For this investigation, a small body of fineness ratio 12 was located on the wing to enclose the internal strain gauge balance. The pitching moment contribution of the body was estimated as equal to a change in center of pressure of 8 per cent of the mean aerodynamic chord, which was subtracted from the position of the center of pressure estimated for the wing by [*29*]. The agreement between theory and experiment is considered remarkable.

One of the interesting points involved here concerns the fact that the center of pressure of the wing is in its most forward position near the leading edge at sonic speed. At this speed, the downwash on the stabilizing tail plane is equal to the wing angle of attack, so that the tail adds very little stability. This result means that for a coplanar tandem arrangement of an unswept wing and tail, the longitudinal stability will be a minimum near sonic speed.

A comparison similar to the foregoing is made for a wing with sweptback leading edges in Fig. G,5b. This wing also has an aspect ratio of 2 but is of triangular planform. It is composed of NACA 0003 sections. In this case there is also very close agreement between theory and experiment, although there is no correction for the effect of the body. The

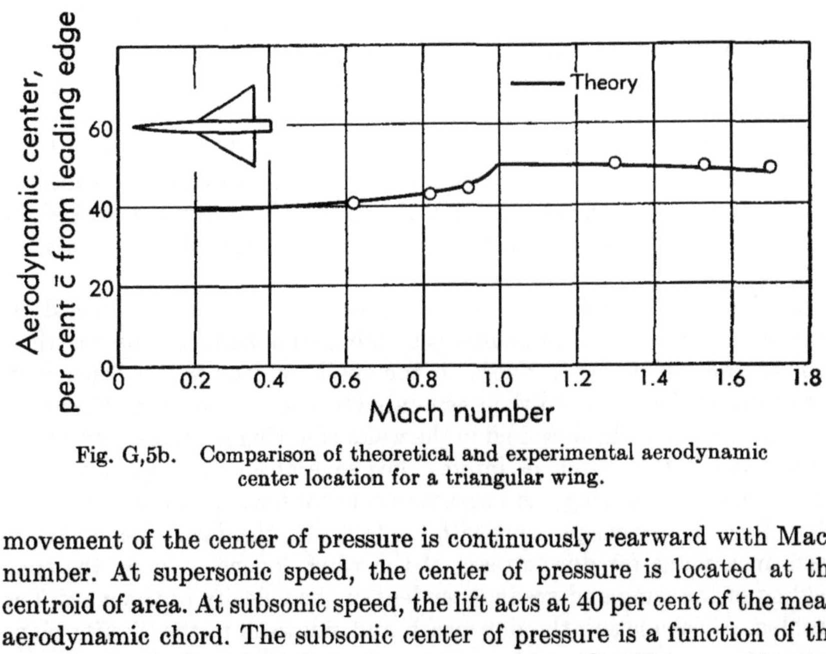

Fig. G,5b. Comparison of theoretical and experimental aerodynamic center location for a triangular wing.

movement of the center of pressure is continuously rearward with Mach number. At supersonic speed, the center of pressure is located at the centroid of area. At subsonic speed, the lift acts at 40 per cent of the mean aerodynamic chord. The subsonic center of pressure is a function of the sweepback of the triangular wing, approaching the 25 per cent mean aerodynamic chord point as the sweep decreases and the 50 per cent mean aerodynamic chord point as the sweep increases. It should be noted that the reduction in center of pressure travel with Mach number, possible with increasing sweepback, is not as great as indicated by measuring the position of the center of lift along the mean aerodynamic chord. It should also be noted that if the wing is selected to produce a certain lift at a given angle of attack, the mean aerodynamic chord length increases with sweepback. The center of pressure travel in feet is the important consideration and this factor is influenced by the absolute length of the mean aerodynamic chord.

Effects of angle of attack. In the foregoing discussion, the center of pressure position was based on the measured slope at zero lift of the curve of pitching moment coefficient as a function of lift coefficient. The variation of pitching moment with angle of attack is also of interest. For triangular wings, experiment has shown that the pitching moments are nearly linear at all Mach numbers, though there is some effect of the change in loading at subsonic speed produced when the flow separates from the leading edge and the vortex flow treated by Brown and Michael [*14*] is established. For rectangular or tapered unswept wings, it is to be expected that the center of pressure will be profoundly affected by the angle of attack at transonic speed. It has been noted before that in studying drag due to lift thin airfoils at angle of attack develop supersonic velocities over the upper surface. The flow so generated causes a very low pressure (high lift) to be distributed progressively along the chord as the angle of attack increases. This point is well illustrated by the results of [*12*]. It is, therefore, obvious that a great center of pressure variation can be expected for unswept airfoils at transonic speed. It has already been noted that the center of pressure acts near the airfoil leading edge at low lift coefficients at near sonic speed while [*12*] indicates that it is located at near the 50 per cent chord point at high lift. This problem makes the stability and control of airplanes utilizing rectangular wings difficult.

G,6. Downwash and Sidewash. Most airplanes are stabilized by horizontal surfaces placed behind the main lifting surface. The contribution of the stabilizing surfaces to the pitching moment is dependent, therefore, on the flow conditions generated by the wing at the tail location. The estimation of the flow field in the wake of a wing is a relatively simple calculation if the region of interest lies several chord lengths or more downstream of the wing. An important concept here is that of the Trefftz plane flow. This concept deals with the fact that the downwash and sidewash in a plane far downstream of the wing depend only on the span loading of the wing and are independent of the velocity of the wing. The problem of calculating the downwash and sidewash in the Trefftz plane requires:

1. The determination of the vorticity in the plane, generated by the wing. This calculation follows directly from a knowledge of the span load distribution.
2. The solution of the boundary-value problem which relates the flow velocities throughout the plane to the strength and distribution of vorticity.

It should be noted that the problem faced is one involving an incompressible two-dimensional flow.

The simplest calculation possible for the estimation of downwash and sidewash consists in replacing the wing with a horseshoe vortex. The circulation constant Γ is determined as the circulation of the maximum bound vorticity of the wing. The span of the vortex is calculated from this value of circulation, the known lift of the wing, the air density, and the velocity of the wing. Results so obtained have been found to be of reasonable accuracy for certain classes of wings if the plane in which the lateral velocities are required is far downstream. On the other hand, the determination of the lateral velocities near the wing is difficult, particularly at supersonic speed. However, methods for estimating the flow direction in this region are given in [*30*].

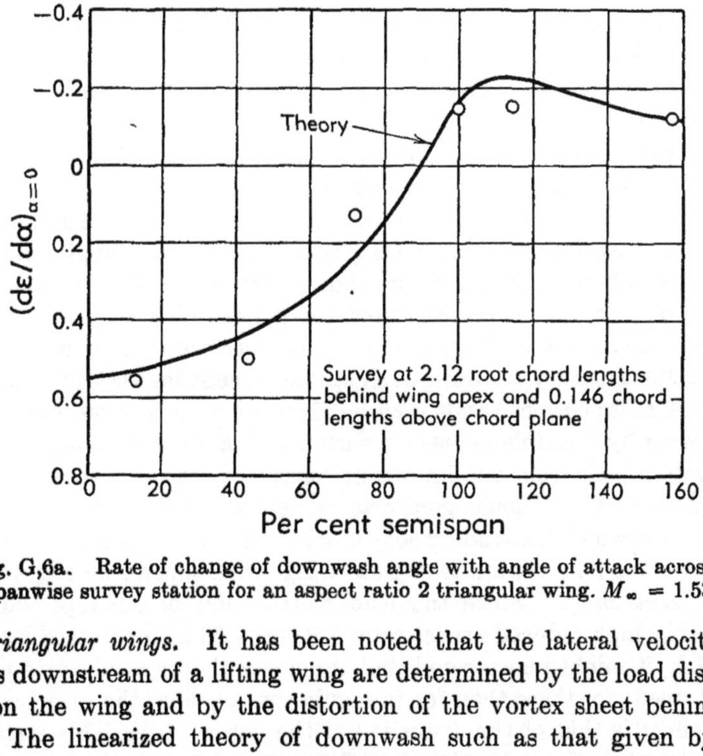

Fig. G,6a. Rate of change of downwash angle with angle of attack across a spanwise survey station for an aspect ratio 2 triangular wing. $M_\infty = 1.53$.

Triangular wings. It has been noted that the lateral velocities in planes downstream of a lifting wing are determined by the load distribution on the wing and by the distortion of the vortex sheet behind the wing. The linearized theory of downwash such as that given by [*30*] assumes: (1) that the load distribution is that for very small angles of attack, and (2) that the vortex sheet shed by the wing is not distorted as it passes downstream. It is evident then that the theory can be expected to give reasonable results only for small angles of attack. Fig. G,6a shows a comparison between theory and experiment for a triangular wing of aspect ratio 2 taken from [*31*]. This figure presents a comparison between the experimental and theoretical rate of change of downwash angle with angle of attack at zero angle of attack for this airfoil, at a Mach number of 1.53 for a position about 2 chord lengths behind and one seventh of a chord length above the extended plane of the wing. It will be noted that the agreement between theory and experiment is very good.

This agreement does not extend to higher angles of attack as can be seen by examining the data of Fig. G,6b which are the basic information from which Fig. G,6a was derived. It is evident from these data that the agreement between theory and experiment is good for all negative angles of attack. At positive angles of attack, however, the data indicate the presence of a strong vortex, between the 71 per cent span station and the tip of the wing which the linearized theory does not treat. This fact can be deduced from the very large downwash at angle of attack of 4° at the 71 per cent span station and the large upwash angles at the 100 per cent span station. It is of interest to note that a simple calculation based on replacing the wing vortex field with a horseshoe vortex shows that the centers of the downstream element of the horseshoe vortex should lie at about 80 per cent semispan and above the wing chord plane about as shown by the experimental data.

A further examination of these data shows that as the angle of attack increases to 10° the center of the vortex moves inboard of the 71 per cent span station. This fact indicates that the span load distribution of the triangular wing departs from the elliptical loading at low angles of attack and tends to become triangular at high angles of attack. This conclusion is borne out by what is known about the loading of triangular wings.

From the foregoing, it appears that the linearized theory is not a tool of sufficient accuracy for downwash studies except for very low angles of attack. A more correct result may be given by replacing the vortex field of the wing by a simple horseshoe vortex with a circulation strength and span determined from the known or measured load distribution of the wing. This method should give reasonable results at least for all Mach numbers for which the leading edge is subsonic. Further work is necessary to explore the flow behind triangular wings with supersonic edges.

Unswept wings. Since the load distribution of unswept wings of reasonably high reduced aspect ratio $\beta Æ$ tends to be unaffected by angle of attack, it might be expected that the linear theory would give a more correct result for these than for triangular wings, since the principal nonlinear effect is that of the distortion of the vortex sheet. This statement will be less true for velocities near the speed of sound, though even for those conditions a small refinement of the theory to utilize correct lift as given by experiment or nonlinear theory may be expected to give good accuracy.

These deductions are borne out by the data of [*32,33*], some of which are presented in Fig. G,6c as the rate of change of downwash angle with angle of attack at zero angle of attack. The comparison shows excellent agreement between theory and experiment for the aspect ratio 3.5 rectangular wing investigated. The agreement is good even for angles of

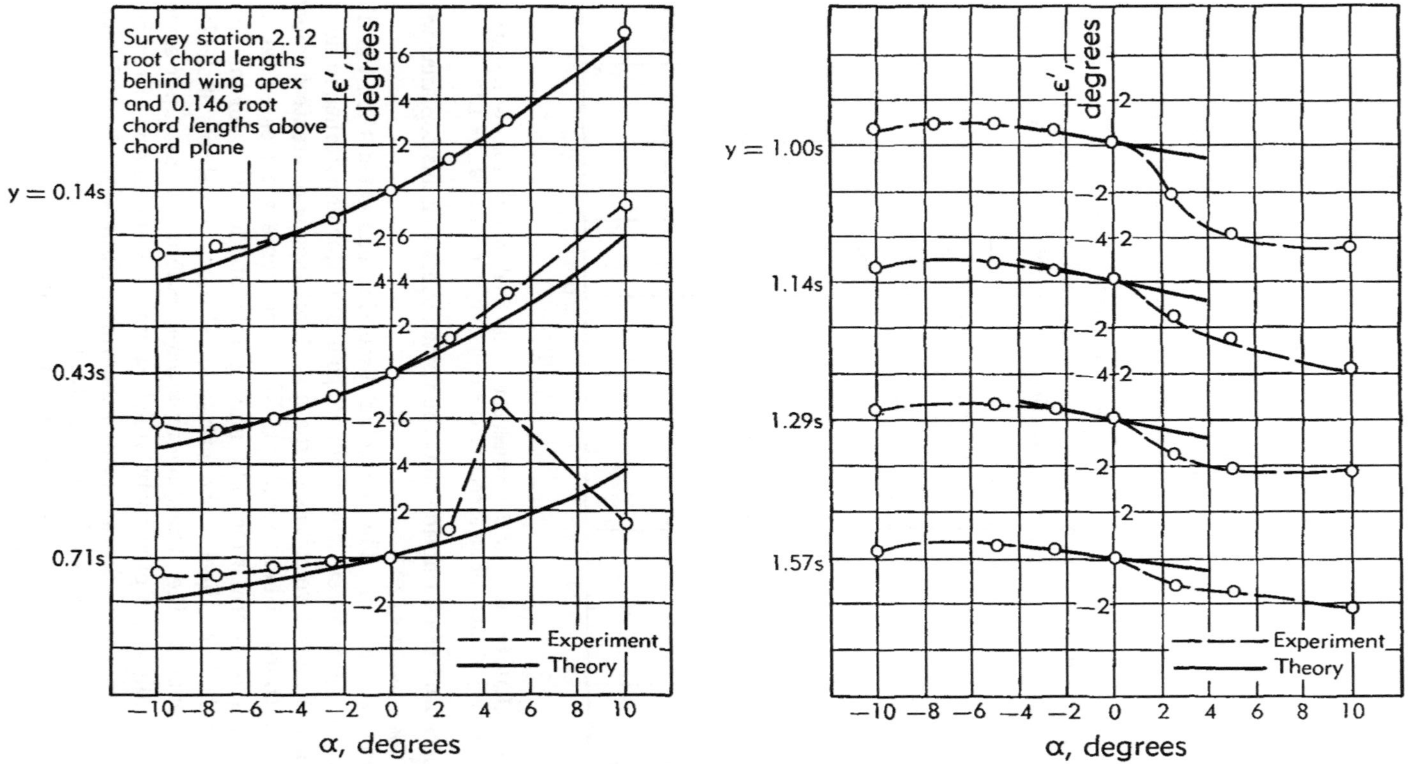

Fig. G,6b. Variation of downwash angle with angle of attack for an aspect ratio 2 triangular wing. $M_\infty = 1.53$.

attack as shown in Fig. G,6d up to 10°, except near the tip where the formation of a discrete vortex is shown outboard of the 90 per cent span station.

The flow in the region behind the tip of a rectangular wing, presented in [33], has been replotted in Fig. G,6e to show how the real flow is more closely represented at angle of attack of 7° by the flow field of a discrete vortex containing all the vortex elements shed by the wing. These data show the necessity of considering the nonlinear terms which contribute the distortion of the vortex sheet in the real flow.

It is concluded that, in general, simple considerations such as the substitution of a horseshoe vortex for the vortex field of the wing give a closer representation of the lateral velocities behind wings. Linearized theory

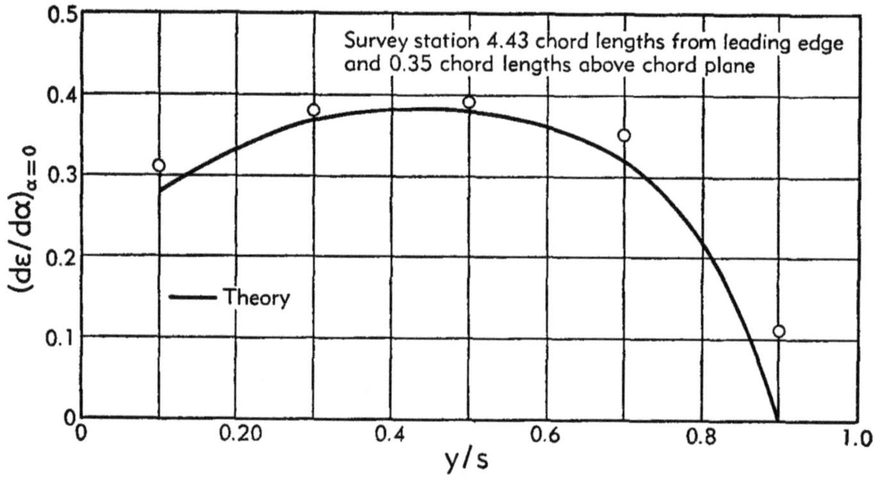

Fig. G,6c. Rate of change of downwash angle with angle of attack across a spanwise survey station for an aspect ratio 3.5 rectangular wing. $M_\infty = 1.53$.

fails because the assumption that the vortex sheet remains coplanar with the wing is not realized.

Cruciform wings. The foregoing discussion has noted that the distortion of vortex sheets shed from wings is of major importance in determining the lateral velocities in the flow. This fact is of greatest importance, of course, when nonplanar lifting elements are treated as, for instance, a cruciform wing. The photographs of Plate G,6, for instance, illustrate the large effect of the interaction of the vortex sheets of the four panels of such a wing. These pictures were made using the vapor screen technique, discussed previously, which permits a visualization of the flow several chord lengths downstream of the wing trailing edge. Several angles of bank were investigated for a constant angle of attack of the central body on which the four wing panels were mounted.

At zero angle of bank, the usual flow behind triangular wings is shown. It is observed that the vortex sheet has rolled up into two discrete vortices, the centers of which are located by the dark regions lying within the flamelike spirals. The air circulates about these centers in a direction opposite to that apparently indicated by the spirals. Experimental surveys of the flow angles indicate that vortices are positioned as shown.

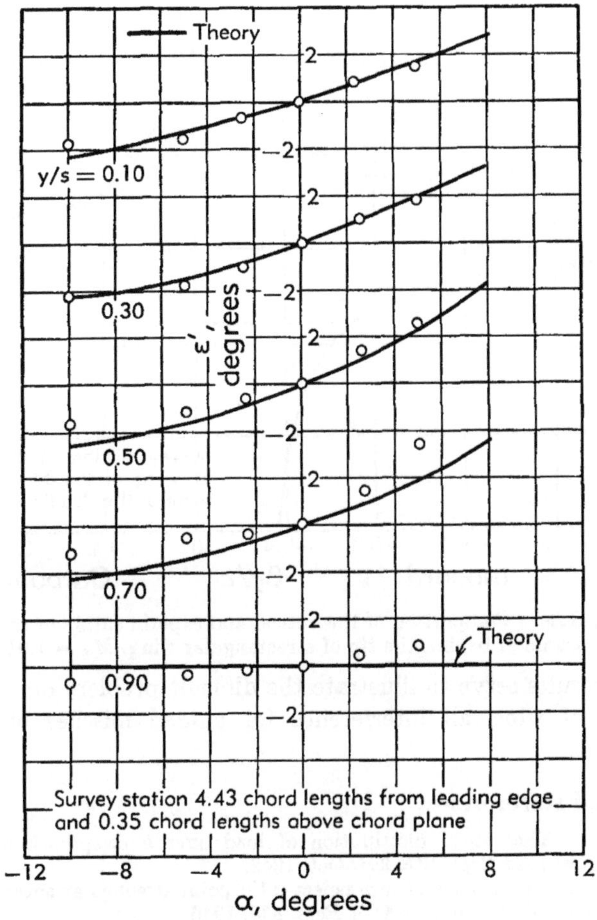

Fig. G,6d. Variation of downwash angle with angle of attack for an aspect ratio 3.5 rectangular wing. $M_\infty = 1.53$.

The study at 22.5° angle of bank shows four vortices in the flow, one of which is almost out of the picture below and to the right of the central body. All four panels develop normal forces in this condition and each discharges vorticity into the field. The interaction between the vortex fields leads to an arrangement that continually changes with downstream position.

At 45° angle of bank, the flow again has symmetry. The large effect of the mutual interaction of the vortices is evident, since the two upper vortices are found to be much closer together and the two lower much farther apart than indicated by simple calculations. As cross planes farther downstream are investigated, the two upper vortices are found to move down and in between the lower.

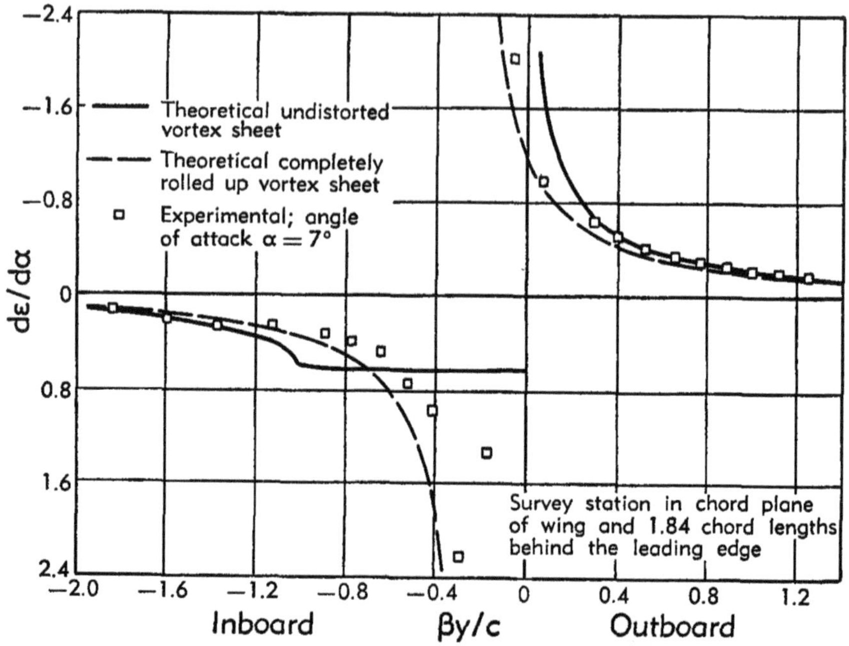

Fig. G,6e. Comparison of theoretical and experimental downwash parameter behind the tip of a rectangular wing. $M_\infty = 1.91$.

These results serve to illustrate the difficult problem involved in the estimation of wing-tail interference for guided missiles of cruciform arrangement.

G,7. Cited References.

1. Cohen, D. Theoretical distribution of load over a sweptback wing. *NACA Advance Restricted Rept. WR-221*, Oct. 1942.
2. Jones, R. T. Properties of low-aspect-ratio pointed wings at speeds below and above the speed of sound. *NACA Rept. 835*, 1946.
3. Heaslet, M. A., Lomax, H., and Spreiter, J. R. Linearized compressible-flow theory for sonic flight speeds. *NACA Tech. Note 1824*, 1949.
4. Nelson, W. H., and McDevitt, J. B. The transonic characteristics of 17 rectangular, symmetrical wing models of varying aspect ratio and thickness. *NACA Research Mem. A51A12*, 1951.
5. Spreiter, J. R. On the application of transonic rules to wings of finite span. *NACA Rept. 1153*, 1953.
6. Love, E. S. Investigations at supersonic speeds of 22 triangular wings representing two airfoil sections for each of 11 apex angles. *NACA Research Mem. L9D07*, 1949.
7. Weissinger, J. The lift distribution of sweptback wings. *NACA Tech. Mem. 1120*, 1947.
8. Lawrence, H. R. The lift distribution of low aspect ratio wings at subsonic speed. *J. Aeronaut. Sci. 18*, 683–695 (1951).

9. Brown, C. E. Theoretical lift and drag of thin triangular wings at supersonic speed. *NACA Rept. 839*, 1946.
10. Squire, H. B. An example in wing theory at supersonic speed. *Brit. Aeronaut. Research Council Repts. and Mem. 2549*, 1951.
11. Cohen, D. Formulas and charts for the supersonic lift and drag of flat sweptback wings with interacting leading and trailing edges. *NACA Tech. Note 2093*, 1950.
12. Lindsey, W. F., Daley, B. N., and Humphreys, M. D. The flow and force characteristics of supersonic airfoils at high subsonic speeds. *NACA Tech. Note 1211*, 1947.
13. Allen, H. J., and Perkins, E. W. A study of the effects of viscosity on flow over slender inclined bodies of revolution. *NACA Rept. 1048*, 1951.
14. Brown, C. E., and Michael, W. H., Jr. Effect of leading edge separation on the lift of a delta wing. *J. Aeronaut. Sci. 21*, 690–695 (1954).
15. Bollay, W. A nonlinear wing theory and its application to rectangular wings of small aspect ratio. *Jahrbuch deut. Luftfahrtforschung P. I. 139*, 1939.
16. Allen, H. Estimation of the forces and moments acting on inclined borders of revolution. *NACA Research Mem. A9I26*, 1949.
17. Czarnecki, K. R., and Muller, J. Investigation at Mach number 1.62 of the pressure distribution over a rectangular wing with symmetrical circular-arc section and 30 percent trailing edge flap. *NACA Research Mem. L9J05*, 1949.
18. Chapman, D. R. Reduction of profile drag at supersonic velocities by the use of airfoil sections having a blunt trailing edge. *NACA Research Mem. A9H11*, 1949.
19. Guderley, G., and Yoshihara, H. The flow over a wedge profile at Mach number 1. *J. Aeronaut. Sci. 17*, 723–735 (1950).
20. Vincenti, W. C., and Wagoner, C. B. Transonic flow past a wedge profile with detached bow wave. *NACA Rept. 1095*, 1952.
21. Vincenti, W. C., Dugan, D. W., and Phelps, E. R. An experimental study of the lift and pressure distribution on a double-wedge profile at Mach numbers near shock attachment. *NACA Tech. Note 3225*, 1954.
22. Humphreys, M. D. Investigation of a lifting 10-percent thick symmetrical double-wedge airfoil at Mach numbers up to 1. *NACA Tech. Note 3306*, 1954.
23. Whitcomb, R. T. A study of the zero-lift drag-rise characteristics of wing-body combinations near the speed of sound. *NACA Research Mem. L52H08*, 1952.
24. Frick, C. W., and Boyd, J. W. Investigation at supersonic speed ($M = 1.53$) of the pressure distribution over a 63° swept airfoil of biconvex section at zero lift. *NACA Research Mem. A8C22*, 1948.
25. Jones, R. T. Thin oblique airfoils at supersonic speed. *NACA Tech. Note 1107*, 1946.
26. Puckett, A. E. Supersonic wave drag of thin airfoils. *J. Aeronaut. Sci. 13*, 475–484 (1946).
27. Chapman, D. R., and Kester, R. H. Measurements of turbulent skin friction on cylinders in axial flow at subsonic and supersonic velocities. *J. Aeronaut. Sci. 20*, 441–448 (1953).
28. Jones, R. T. The minimum drag of thin wings in frictionless flow. *J. Aeronaut. Sci. 18*, 75–82 (1951).
29. Lomax, H., and Sluder, L. Chordwise and compressibility corrections to slender-wing theory. *NACA Rept. 1105*, 1952.
30. Lagerstrom, P. A., and Graham, M. E. Downwash and sidewash induced by three dimensional lifting wings in supersonic flow. *Douglas Aircraft Co. Rept. SM-13007*, Apr. 1947.
31. Perkins, E. W., and Canning, T. N. Investigation of downwash and wake characteristics at a Mach number of 1.53. II: Triangular wing. *NACA Research Mem. A9D20*, 1949.
32. Perkins, E. W., and Canning, T. N. Investigation of downwash and wake characteristics at a Mach number of 1.53. I: Rectangular wing. *NACA Research Mem. A8L16*, 1949.
33. Cummings, J. L., and Haefeli, R. C. Downwash in vortex region behind rectangular half wing at Mach number 1.91. *NACA Research Mem. E50H10*, 1950.

INDEX

GPSR Authorized Representative: Easy Access System Europe - Mustamäe tee 50, 10621 Tallinn, Estonia, gpsr.requests@easproject.com

www.ingramcontent.com/pod-product-compliance
Ingram Content Group UK Ltd.
Pitfield, Milton Keynes, MK11 3LW, UK
UKHW020100230626
472537UK00003B/27
* 9 7 8 0 6 9 1 6 2 6 7 5 8 *